## THE CHEMICAL ELEMENTS

| | | | |
|---|---|---|---|
| actinium | Ac | 89 | (227)[1] |
| aluminum | Al | 13 | 26.9815 |
| americium | Am | 95 | (243) |
| antimony | Sb | 51 | 121.760 |
| argon | Ar | 18 | 39.948 |
| arsenic | As | 33 | 74.9216 |
| astatine | At | 85 | (210) |
| barium | Ba | 56 | 137.327 |
| berkelium | Bk | 97 | (247) |
| beryllium | Be | 4 | 9.0122 |
| bismuth | Bi | 83 | 208.9804 |
| bohrium | Bh | 107 | (262) |
| boron | B | 5 | 10.811 |
| bromine | Br | 35 | 79.904 |
| cadmium | Cd | 48 | 112.411 |
| calcium | Ca | 20 | 40.078 |
| californium | Cf | 98 | (251) |
| carbon | C | 6 | 12.011 |
| cerium | Ce | 58 | 140.115 |
| cesium | Cs | 55 | 132.9054 |
| chlorine | Cl | 17 | 35.4527 |
| chromium | Cr | 24 | 51.9961 |
| cobalt | Co | 27 | 58.9332 |
| copper | Cu | 29 | 63.546 |
| curium | Cm | 96 | (247) |
| dubnium | Db | 105 | (262) |
| dysprosium | Dy | 66 | 162.50 |
| einsteinium | Es | 99 | (252) |
| erbium | Er | 68 | 167.26 |
| europium | Eu | 63 | 151.965 |
| fermium | Fm | 100 | (257) |
| fluorine | F | 9 | 18.9984 |
| francium | Fr | 87 | (223) |
| gadolinium | Gd | 64 | 157.25 |
| gallium | Ga | 31 | 69.723 |
| germanium | Ge | 32 | 72.61 |
| gold | Au | 79 | 196.9665 |
| hafnium | Hf | 72 | 178.49 |
| hassium | Hs | 108 | (265) |
| helium | He | 2 | 4.0026 |
| holmium | Ho | 67 | 164.9303 |
| hydrogen | H | 1 | 1.00794 |
| indium | In | 49 | 114.818 |
| iodine | I | 53 | 126.9045 |
| iridium | Ir | 77 | 192.22 |
| iron | Fe | 26 | 55.845 |
| krypton | Kr | 36 | 83.80 |
| lanthanum | La | 57 | 138.9055 |
| lawrencium | Lr | 103 | (262) |
| lead | Pb | 82 | 207.2 |
| lithium | Li | 3 | 6.941 |
| lutetium | Lu | 71 | 174.967 |
| magnesium | Mg | 12 | 24.3050 |
| manganese | Mn | 25 | 54.9380 |
| meitnerium | Mt | 109 | (266) |
| mendelevium | Md | [illegible] | [illegible] |
| mercury | Hg | [illegible] | [illegible] |
| molybdenum | Mo | [illegible] | 95.95 |
| neodymium | Nd | 60 | 144.24 |
| neon | Ne | 10 | 20.1797 |
| neptunium | Np | 93 | (237) |
| nickel | Ni | 28 | 58.6934 |
| niobium | Nb | 41 | 92.9064 |
| nitrogen | N | 7 | 14.0067 |
| nobelium | No | 102 | (259) |
| osmium | Os | 76 | 190.23 |
| oxygen | O | 8 | 15.9994 |
| palladium | Pd | 46 | 106.42 |
| phosphorus | P | 15 | 30.9738 |
| platinum | Pt | 78 | 195.08 |
| plutonium | Pu | 94 | (244) |
| polonium | Po | 84 | (209) |
| potassium | K | 19 | 39.0983 |
| praseodymium | Pr | 59 | 140.9076 |
| promethium | Pm | 61 | (145) |
| protactinium | Pa | 91 | 231.0359 |
| radium | Ra | 88 | (226) |
| radon | Rn | 86 | (222) |
| rhenium | Re | 75 | 186.207 |
| rhodium | Rh | 45 | 102.9055 |
| rubidium | Rb | 37 | 85.4678 |
| ruthenium | Ru | 44 | 101.07 |
| rutherfordium | Rf | 104 | (261) |
| samarium | Sm | 62 | 150.36 |
| scandium | Sc | 21 | 44.9559 |
| seaborgium | Sg | 106 | (263) |
| selenium | Se | 34 | 78.96 |
| silicon | Si | 14 | 28.0855 |
| silver | Ag | 47 | 107.868 |
| sodium | Na | 11 | 22.9898 |
| strontium | Sr | 38 | 87.62 |
| sulfur | S | 16 | 32.066 |
| tantalum | Ta | 73 | 180.948 |
| technetium | Tc | 43 | (98) |
| tellurium | Te | 52 | 127.60 |
| terbium | Tb | 65 | 158.9253 |
| thallium | Tl | 81 | 204.38 |
| thorium | Th | 90 | 232.0359 |
| thulium | Tm | 69 | 168.9342 |
| tin | Sn | 50 | 118.710 |
| titanium | Ti | 22 | 47.867 |
| tungsten | W | 74 | 183.84 |
| uranium | U | 92 | 238.0289 |
| vanadium | V | 23 | 50.9415 |
| xenon | Xe | 54 | 131.29 |
| ytterbium | Yb | 70 | 173.04 |
| yttrium | Y | 39 | 88.9058 |
| zinc | Zn | 30 | 65.39 |
| zirconium | Zr | 40 | 91.224 |

[1]The numbers in parentheses are the mass numbers for the most important isotopes for the elements with no naturally occurring isotopes.

**Your Login Name**

**Your Password**

# An Introduction to CHEMISTRY

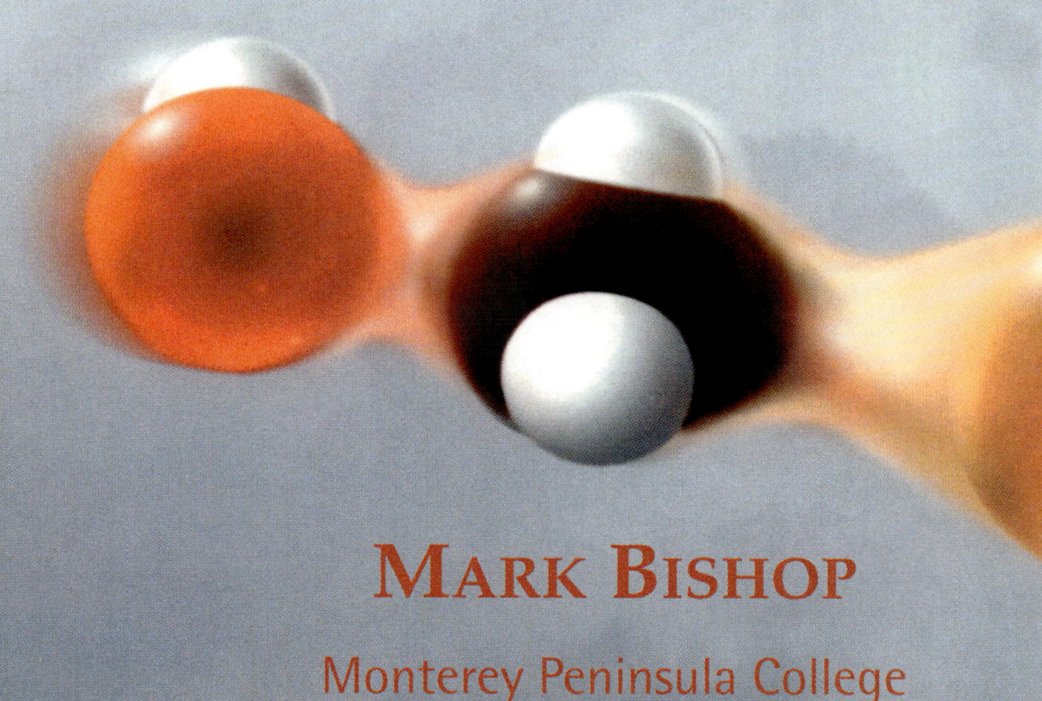


MARK BISHOP

Monterey Peninsula College

San Francisco Boston New York
Capetown Hong Kong London Madrid Mexico City
Montreal Munich Paris Singapore Sydney Tokyo Toronto

Editorial Director: Frank Ruggirello
Executive Editor: Ben Roberts
Assistant Editor: Lisa Leung
Design Manager: Blakeley Kim
Managing Editor: Joan Marsh
Senior Development Editor: Margot Otway
Development Editor: Moira Lerner Nelson
Media Producer: Claire Masson
Market Development Manager: Chalon Bridges
Marketing Manager: Christy Lawrence
Text Designer: Thompson Steele, Inc.
Cover Design: Blakeley Kim, Emiko-Rose Koike
Art Studio: Thompson Steele, Inc.
Photo Researcher: Thompson Steele, Inc.
Manufacturing Coordinator: Vivian McDougal
Project Coordination and Electronic Page Makeup: Thompson Steele, Inc.

Library of Congress Cataloging-in-Publication Data
Bishop, Mark A.
An introduction to chemistry / Mark Bishop.-- 1st ed.
p. cm.
Includes index.
ISBN 0-8053-2177-2 (student ed. : alk. paper)
1. Chemistry. I. Title.
QD33.2 .B57 2002
540--dc21

2001047343

For permission to use copyrighted material, grateful acknowledgment is made to the copyright holders on p. C-1, which is hereby made part of this copyright page.

---

1 2 3 4 5 6 7 8 9 10—QWV—04 03 02 01

www.aw.com/bc

# Brief Contents

# Contents

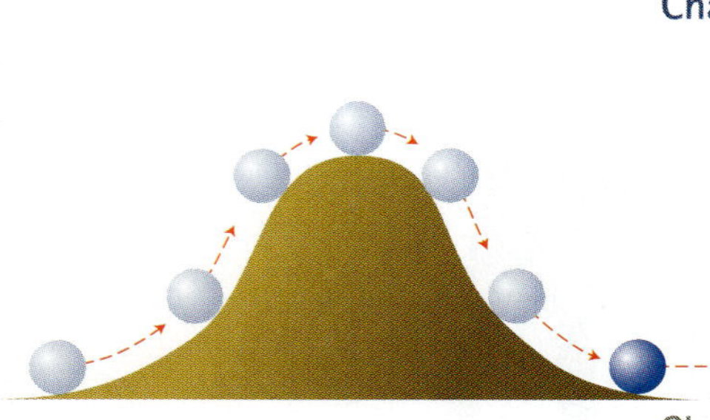



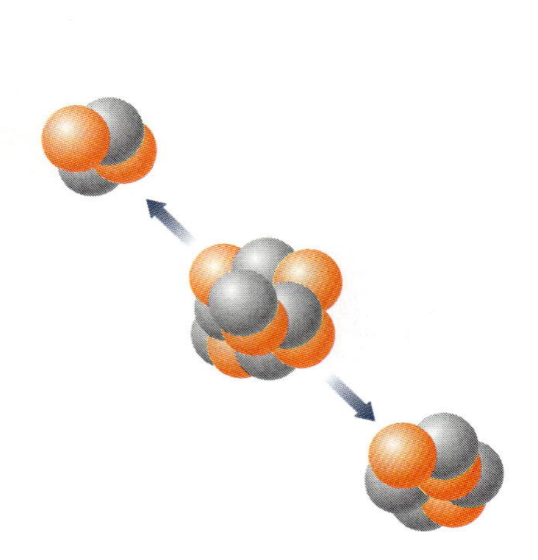

# Preface

## To the Instructor

AN INTRODUCTION TO CHEMISTRY is intended for use in beginning chemistry courses that have no chemistry prerequisite. It was written for students who want to prepare themselves for general college chemistry, for students seeking to satisfy a science requirement, and for students in health-related or other programs that require a one-semester introduction to general chemistry. No matter what your students' goals, this book will help them to learn the basics of chemistry.

I have taught introductory chemistry for over 25 years, and for much of that time I have considered writing my own text. One reason was that the existing textbooks struck me as disjointed. They read more like a list of skills to master than like a coherent story of the nature of chemistry. I thought it should be possible to organize the fundamentals of chemistry so that each would flow smoothly into the next, but it wasn't until I made some changes in my course that I began to take the prospect of writing a new textbook seriously.

The first change I tried was to move the description of unit conversions from the beginning of the course to the middle. I decided that one reason why the course felt disjointed was that I kept jumping back and forth between the description of the basic concepts of chemistry and the explanation of unit conversions. Postponing the mathematics enabled me to focus on chemistry in the first part of the course. Moreover, as a result of this change, my students develop far stronger computational skills than I was able to give them before. (The reasons for this are described below.) While I recommend this change, I know it is not an option in all courses. Therefore, I designed this text so that unit conversions can be either introduced at the start of the course or postponed (as I prefer) until later.

The second change I made was to put more emphasis on developing my students' ability to visualize the particle nature of matter. Students too often view chemistry as a set of rules for manipulating numbers, symbols, and abbreviations, never really connecting these rules to a physical reality. They can balance equations and do chemical calculations, but they cannot answer questions about what is happening on the particle level when an acid reacts with a base. Thus, whenever appropriate, I enhance the standard topics covered in introductory chemistry with corresponding descriptions of events from the particles' "point of view."

The final factor that led to the creation of this text and its supplements is that I learned to create computer-based tools myself, and it occurred to me that a package whose text and computer-based ancillaries were all produced by the same person would offer real benefits. The Web-based tools that accompany this text include animations, glossary quizzes for each chapter, tutorials to consolidate and enhance important skills, and Web pages that provide extra information. Because I have created both the tools and the text, I think you will find that they fit together seamlessly.

Read on for a more detailed discussion of how these changes have been incorporated into *An Introduction to Chemistry* and its supplements. Each innovation has been developed with the ultimate goal of making it easier for you to give your students a coherent understanding of chemistry, a positive attitude toward chemistry (and toward you and your course), and a solid foundation on which to build, should they decide to continue their chemistry studies.

## Flexible Order of Math-Related Topics

*Do you spend a lot of time in the first week or two of your course teaching unit conversions and significant figures? If so, do many students lose interest or even drop the course because they find the math-related topics boring and perhaps intimidating?*

The single most beneficial change I have made in my prep-chem course has been to shift the coverage of unit conversions from the beginning to the middle of the semester. As detailed below, this book can be used to support either that approach or a more traditional one. Delaying the coverage of unit conversions enables me to describe elements, compounds, and chemical reactions earlier than usual and, I believe, to give my students a much better understanding of what chemistry really is. Students emerge from the first lectures with a better attitude toward the course and with more confidence in their abilities—which, in my experience, has translated into significantly lower drop rates. One of the most important by-products of this change, in my assessment, is that my students end up *better* equipped with math-related skills than would otherwise be the case (see "More Emphasis on Math-Related Topics," below). Immediately after I teach them the technique of dimensional analysis, they begin using it in mole calculations. Thus, instead of learning the technique at the start of the course, and then largely forgetting it, and then trying to relearn it in haste, the students learn it well and then immediately consolidate their knowledge.

Because not everyone will choose to restructure their course in this way, I organized this book to allow several approaches to teaching unit conversions. The optional Inter-Chapter 1A, located between Chapters 1 and 2, gives a brief introduction to dimensional analysis and metric–metric conversions. Chapter 8, which covers dimensional analysis comprehensively, can be used either in its current position or early in the course.

- An instructor who wishes to introduce unit conversions briefly at an early point in the course (perhaps to prepare the students for labs), while postponing a comprehensive treatment of the topic, can use the text in its current order.

- An instructor who wishes to teach unit conversions in detail early in the course can skip Inter-Chapter 1A and instead cover Chapter 8 immediately after Chapter 1. Chapter 8 is written so that students can read it without confusion before reading Chapters 2 through 7.

- An instructor who, like myself, wishes to delay the discussion of unit conversions until the middle of the course can skip Inter-Chapter 1A and cover the remaining chapters in their current order. Chapter 8 is located so that it teaches unit analysis immediately before the students need the technique for mole calculations.

## Early Introduction to Chemical Reactions

*Are you ever frustrated that it takes so long to get to describing interesting chemical changes?*

Most prep-chem texts don't describe chemical reactions until midway through the text or even later, thereby reinforcing students' expectations that chemistry will be boring and irrelevant. In this text, chemical reactions are described in Chapters 4 through 6.

## More Emphasis on Math-Related Topics

*Do you ever wish that you could cover unit conversions in more detail but resist doing so because it would further postpone the introduction of the description of elements, compounds, and chemical changes?*

Although I postpone the math-related topics in my prep-chem courses, I think they are extremely important. Therefore, I have devoted three full chapters to them. Chapter 8 teaches unit conversions using dimensional analysis, Chapter 9 describes chemical calculations and chemical formulas, and Chapter 10 covers chemical calculations and chemical equations.

## More Logical Sequence of Topics

*In many texts, Chapter 1 or 2 asks the reader to classify substances as elements, compounds, or mixtures and to classify changes as chemical or physical. Do you find it difficult to describe compounds before your students have a clear understanding of atoms and elements? Do you find it hard to describe chemical changes before your students know about chemical bonds and chemical compounds?*

In the first week of class, I used to ask my students to classify substances as elements, compounds, or mixtures. That required me to introduce the concept of an element long before any significant discussion of atoms and to describe compounds without first presenting a clear depiction of elements. I was equally uncomfortable asking students to classify changes as chemical or physical before they had any clear definition of chemical bonds. Now I move smoothly from the kinetic molecular theory to a description of atoms and elements (Chapter 2). This flows into a description of chemical bonds and chemical compounds (Chapter 3), which in turn forms the basis for an understanding of the nature of solutions and the processes of chemical changes (Chapters 4, 5, and 6). The introductory discussions that felt so disjointed to me in the past now seem to follow a logical progression—a story, really—that flows from simple to more complex.

## Emphasis on the Development of Visualization Skills

*Do you ever worry that your students can write balanced chemical equations but do not have a clear mental image of the events that occur during a chemical reaction?*

I think it is extremely important for students to develop the ability to visualize the models that chemists use for describing the structure and behavior of matter. I want them to be able to connect a chemical equation with a visual image of what is happening in the reaction. Throughout the text, I emphasize the development of a mental image of the structure of matter and the changes it undergoes. I start with a more comprehensive description of the kinetic molecular theory than is found in most books, and I build on that description in the sections on elements, compounds, and chemical changes. To help the student visualize structures and processes, I provide the colorful and detailed illustrations that are a prominent feature of the book. Moreover, the book's Web site provides animations based on key illustrations.

## Identification of Skills to Review

*When your students have trouble with a task, is it ever because they have not completely mastered some of the lessons presented in earlier chapters?*

The Review Skills section at the start of each chapter lists skills from earlier chapters that will be needed in the present chapter. The students can test their mastery of each skill by working the problems in the Review Questions section at the end of each chapter.

Instructors who wish to teach chapters in a different order than the one in the book can use these sections to identify topics that may require supplementation. The Instructor's Manual contains a list of various possible chapter orders, with suggested detours to ensure that the students always have the skills they need.

## Sample Study Sheets

*Are the best-organized students in your class often the most successful? Do you ever wish that the text you were using helped students get more organized?*

In an introductory chemistry course, it really pays to be organized. This text helps students get organized by providing Sample Study Sheets for many of the tasks they will be expected to do on exams. Each study sheet describes how to recognize a specific kind of task ("Tip-off") and then breaks the task down into general steps. Each study sheet is accompanied by at least one worked example.

## Extensive Lists of Learning Objectives

*Do your students ever complain that they do not know what they are supposed to be able to do after studying a chapter in the text?*

The learning objectives listed at the end of each chapter are more comprehensive than the objectives in other texts. They list all the key skills taught in the chapter, thus helping students to focus on the most critical material. Objective references in the margins of the chapter denote the paragraphs that pertain to each objective, so that a student who has trouble with a particular objective can easily find the relevant text discussion. Many of the end-of-chapter problems are similarly referenced, so that students can see how each objective might be covered on an exam.

## Chapter Glossaries and Glossary Quizzes

*Do you wish your text did more to help students learn the language of chemistry?*

Learning the language of science is an important goal of the courses for which this text is designed. Most books have a glossary at the back, but I suspect that students rarely refer to it. In addition to a glossary at the back of the book, this text also has a list of new terms at the end of each chapter, where it can serve as a chapter review. Glossary quizzes for each chapter can be found on the book's Web site.

## More Real-World Examples

*Do your students feel that what they read in their textbook is far removed from the real world?*

This text is full of real-world examples, both in the chapter narrative and in the problems. For instance, after introducing the idea of limiting reactants, Section 10.2 explains why chemists design procedures for chemical reactions in such a way that some substances are limiting and others are in excess. Chapter 9 problems mention vitamins, cold medicines, throat lozenges, antacids, gemstones, asphalt roofing, fireworks, stain and rust removers, dental polishing agents, metal extraction from natural ores, explosives, mouthwashes, Alar on apples, nicotine, pesticides, heart drugs, Agent Orange, thalidomide, and more. The chemical reactions used in problems often represent actual industrial processes. Several of the Special Topics scattered throughout the book describe the achievements of "green chemistry."

## Key Ideas Questions

*Have you ever wondered whether the chapter reviews in many textbooks are useful to students?*

After the Review Questions section at the end of each chapter is a section titled Key Ideas. Students are given a list of numbers, words, and phrases that they use to fill in the blanks in a series of statements that follows the list. The statements summarize the most important ideas from the chapter—that is, they add up to a chapter review. Because this review is a game of sorts, the students get more actively involved and are more interested in recalling key ideas than they do when reading a chapter summary.

## Acknowledgments

Writing a textbook is a much bigger project than I ever imagined it would be, and to bring such a project this far requires many people, all of whom deserve my heartfelt thanks. The biggest thank you goes to my family. My loving and beautiful wife, Elizabeth, has not only done much more than her share of the tasks necessary to keep our home running smoothly, she has also kept our home a happy one. Her patience and generosity have allowed me to "disappear" to work on the project with a minimum of guilt. My kids (Meagan, Benjamin, and Claire) have had to do without their dad all too often, but they have always been understanding. I want to give a special thanks to my adult daughter, Meagan, to my brother, Bruce, and to my mother, C. Joan Ninneman. They each provided a sympathetic ear when the project got me down, and they were constant sources of good advice. I'm a truly lucky man to have been blessed with such a family.

Next on my list of those to thank are my saintly developmental editors, Sue Ewing and Moira Lerner Nelson. Sue was there at the beginning, not only helping to convert the original book that existed only in my head into a realistic text, but also giving me support and advice at every step of the way. Moira took over at the midpoint of the process, and her suggestions have led to extensive improvements in the organization and language of the text. Moira, like Sue before her, has been a caring friend as well as a constant source of good ideas. It has been a great pleasure to see the book get better and better in response to the advice of these two professionals.

I want to thank the people of Benjamin Cummings who have been essential in guiding the project to completion. First, I want to thank Anne Scanlon Rohrer, the acquisitions editor at Benjamin Cummings who had the courage to sign an unknown author to a book contract. Although she moved on to other things soon after the signing, I still appreciate her confidence in me and her belief in the value of the project. I also want to acknowledge the contributions of others at Benjamin Cummings: Linda Davis, president; Ben Roberts, executive editor; Maureen Kennedy, former acquisitions editor; Joan Marsh, managing editor; Margot Otway, senior advisor; Chalon Bridges, market development manager; Christy Lawrence, marketing manager; Frank Ruggirello, vice president and editorial director; Stacy Treco, director of marketing; Lisa Leung, assistant editor; Claire Masson, Tony Asaro, George Ellis, Nancy Gee, Claudia Herman, Blakeley Kim, and Emiko-Rose Koike.

I am grateful to the people at Thompson Steele Production Services who found the photos, improved my images, copyedited and proofread the text, and did the composition and page layout. I really enjoyed working with Andrea Fincke, my project editor. She has the toughness required to keep things moving, combined with a charming personality and a quick wit. I also want to thank Sally Thompson Steele, designer and consultant; Cia Boynton, art editor; Abby Reip, photo researcher; Connie Day, copy editor; Jeff Coolidge, photographer; Jim Atherton, illustrator; and Suzanne Kelly, page layout artist. I would also like to thank the principal and the science department at the Bromfield School for their assistance.

Another person who contributed significantly, though indirectly, to the writing of this text was Rodney Oka, my colleague and friend at Monterey Peninsula College. Rod shouldered many of the chemistry department tasks that I was just too busy to do, and he allowed me to continue to teach the same introductory courses long after he would have preferred to switch with me. (I'm sure he would rather get cash, but I thought a strong thank you in print would be more lasting.)

Next, I want to thank Ron Rinehart, another of my colleagues, and Adam Carroll for checking the solutions for all of the problems in the text. Their attention to detail was much appreciated. Last, but certainly not least, I want to thank the many people who have reviewed the text at every stage in the process. They have been my main contact with the community of chemistry instructors and, in that capacity, have given me both invaluable advice on many aspects of the work and encouragement to see the project through to the end. I want to give special thanks to Phil Reedy of Delta College and Walter Dean of Lawrence Technological University. They have been reviewing the manuscript from the beginning, and I hope they will see something of themselves between these covers. I also want to thank Donald Wink of the University of Illinois at Chicago, who, until he decided to write a competing text of his own, did his best to keep me honest.

I want all of the following reviewers to know that I greatly appreciate their contributions:

Elaine Alfonsetti, Broome Community College; Nicholas Alteri, Community College of Rhode Island; Joe Asire, Cuesta College; Caroline Ayers, East Carolina Univeristy; M.R. Barranger-Mathys, Mercyhurst College; Cheryl Baxa, Pine Manor College; Bill Bornhorst, Grossmont College; Tom Carey, Berkshire Community College; Marcus Cicerone, Brigham Young University; Juan Pablo Claude, University of Alabama at Birmingham; Denisha Dawson, Diablo Valley College; Walter Dean, Lawrence Technological University; Patrick Desrochers, University of Central Arkansas; Howard Dewald, Ohio University; Jim Diamond, Linfield College; David Dollimore, University of Toledo; Tim Donnelly, University of California, Davis; Jimmie G. Edwards, University of Toledo; Amina El-Ashmawy, Collin County Community College; Naomi Eliezer, Oakland University; Roger Frampton, Tidewater Community College; Donna Friedman, St. Louis Community College; Galen George, Santa Rosa Junior College; Kevin Gratton, Johnson Community College; Ann Gull, St. Joseph's College; Greg Guzewich; Midge Hall, Clark State Community College; James Hardcastle, Texas Women's University; Blaine Harrison, West Valley College; David Henderson, Trinity College; Jeffrey Hurlbut, Metropolitan State College; Jo Ann Jansing, Indiana University Southeast; Craig Johnson, Carlow College; James Johnson, Sinclair Community College; Sharon Kapica, County College of Morris; Roy Kennedy, Massachusetts Bay Community College; Gary Kinsel, University of Texas, Arlington; Leslie Kinsland, University of Southern Louisiana; Deborah Koeck, Southwest Texas State University; Kurtis Koll, Cameron University; Christopher Landry, University of Vermont; Joseph Lechner, Mount Vernon Nazarene College; Robley Light, Florida State University; John Long, Henderson State University; Jerome Maas, Oakton Community College; Art Maret, University of Central Florida; Jeffrey Mathys; Ken Miller, Milwaukee Area Technical College; Barbara Mowery, Thomas Nelson Community College; Kathy Nabona, Austin Community College; Ann Nalley, Cameron University; Andrea Nolan, Miami University Middletown; Rod Oka, Monterey Peninsula College (class tester); Joyce Overly, Gaston College; Maria Pacheco, Buffalo State College; Brenda Peirson; Amy Phelps, University of Northern Iowa; Morgan Ponder, Samford University; Matiur Rahman, Austin Community College; Pat Rogers, University of California, Irvine; Phil Reedy, Delta College; Ruth Russo, Whitman College; Lowell Shank, Western Kentucky University; Ike Shibley, Pennsylvania State Berks; Trudie Jo Slapar Wagner, Vincennes University; Dennis Stevens, University of Nevada–Las Vegas; Jim Swartz, Thomas More College; Sue Thornton, Montgomery College; Philip Verhalen, Panola College; Gabriela Weaver, University of Colorado, Denver; William Wilk, California State University, Dominguez Hills; Linda Wilson, Middle Tennessee State University; Donald Wink, University of Chicago; James Wood, University of Nebraska at Omaha; Jesse Yeh, South Plains College; Linda Zarzana, American River College; David Zellmer, California State University, Fresno.

If you have any questions about the text that you would like to ask me, I'd be happy to have the opportunity to answer them. Your Benjamin Cummings sales representative can provide you with my email address. I hope that your teaching experience using this book (or any other text) will be a satisfying and pleasurable experience.

*Mark Bishop*
*Monterey, California*

# Features of This Book

CHAPTER 4

## An Introduction to Chemical Reactions

NOW THAT YOU UNDERSTAND THE BASIC STRUCTURAL DIFFERENCES between different kinds of substances, you are ready to begin learning about the chemical changes that take place as one substance is converted into another. Chemical changes are chemists' primary concern. They want to know what, if anything, happens when one substance encounters another. Do the substances change? How and why? Can the conditions be altered so as to speed the changes up, slow them down, or perhaps reverse them? Once chemists understand the nature of one chemical change, they begin to explore the possibilities that arise from

an old house *as is*, with the water turned off. it will go, and all you get is a slow drip, drip, to ruin ity that ining, you our trou- change ventually 5, you ve that on your tooth- that can help fight cavities.

Chemical changes, like the ones mentioned above, are described chapter begins with a discussion of how to interpret and write chemica



*A chemical reaction causes solids to form in hot water pipes.*

Many of the chapters begin with a short introduction that describes how the topics in the chapter relate to the reader's daily life.

### Review Skills

The presentation of information in this chapter assumes that you can a listed below. You can test your readiness to proceed by answering the R the chapter. This might also be a good time to read the Chapter Objecti Questions.

- Write the formulas for the diatomic elements. (Section 2.5)
- Predict whether a bond between 2 atoms of different elements would be covalent or ionic. (Section 3.2)
- Desc
- Conv binary covalent compounds, and ionic compounds. (Sections 3.3–3.5)

The Review Skills section instructs the reader to review specific skills from earlier chapters that are necessary for success in the present chapter.

*Salt towers rise from the water in Mono Lake, California.*

Photographs give the reader visual reminders that chemistry is important in their world.

174 Chapter 4 | An Introduction to Chemical Reactions

### Review Questions

1. Write the formulas for all of the diatomic elements.
2. Predict whether atoms of each of the following pairs of elements would be expected to form ionic or covalent bonds.
   a. Mg and F  c. Fe and O
   b. O and H  d. N and Cl
3. Describe the structure of liquid water, including a description of water molecules and the attractions between them.
4. Write formulas that correspond to the following names.
   a. ammonia  c. propane
   b. methane  d. water
5. Write formulas that correspond to the following names.
   a. nitrogen dioxide  c. dibromine monoxide
   b. carbon tetrabromide  d. nitrogen monoxide
6. Write formulas that correspond to the following names.
   a. lithium fluoride  d. sodium carbonate
   b. lead(II) hydroxide  e. chromium(III) chloride
   c. potassium oxide  f. sodium hydrogen phosphate

Students can test themselves on the review skills by working the problems in the Review Questions section found at the end of each chapter.

### Key Ideas

**Complete the following statements by writing one of these words or phrases in each blank.**

| | |
|---|---|
| above | minor |
| charge | negative |
| chemical bonds | none |
| coefficients | organized, repeating |
| complete formula | partial charges |
| continuous | positive |
| converted into | precipitate |
| created | precipitates |
| delta, Δ | precipitation |
| destroyed | same proportions |
| equal to | separate ions |
| gas | shorthand description |
| homogeneous mixture | solute |
| left out | solvent |
| liquid | subscripts |
| major | very low |

7. A chemical change or chemical reaction is a process in which one or more pure substances are ________ one or more different pure

The Key Ideas section gives the reader an opportunity to review the most important ideas from the chapter.

The comprehensive list of **Chapter Objectives** at the end of each chapter identifies the key skills taught within the chapter (in the same logical order as in the chapter) so that students can concentrate on learning the most important material. The number of each objective appears in the margin next to the text where that objective is described. Many of the end-of-chapter problems are also accompanied by references to the corresponding objectives.

222 Chapter 5 | Acids, Bases, and Acid-Base Reactions

**Chapter Objectives**

The goal of this chapter is to teach you to do the following.

1. Define all of the terms in the Chapter Glossary.

**Section 5.1 Acids**

2. Identify ...
3. Describe ... added to water.
4. Identify ... and ... acetic ... id, is

... duce fewer $H_3O^+$ ions in water than strong ... umbers of acid molecules are added to equal

... ng monoprotic acids: HCl and $HNO_3$.

... protic strong acid.

... sulfuric acid is added to water.

... any acid, identify it as a strong or a weak acid.

... and formulas for binary acids and oxyacids.

**... Nomenclature**

... rmula, tell whether it represents a binary ionic ... nd with polyatomic ion(s), a binary covalent ... an oxyacid.

... and chemical formulas for binary ionic ... ds with polyatomic ion(s), binary covalent

In-chapter **Examples** provide the reader with models for how to complete important problems.

Chapter 5 | Acids, Bases, and Acid-Base Reactions

**EXAMPLE 5.8 Brønsted–Lowry Acids and Bases**

OBJECTIVE 31

Identify the Brønsted–Lowry acid and base for the forward reaction in each of the following equations.

a. $HClO_2(aq) + NaIO(aq) \rightarrow HIO(aq) + NaClO_2(aq)$

b. $HS^-(aq) + HF(aq) \rightarrow H_2S(aq) + F^-(aq)$

c. $HS^-(aq) + OH^-(aq) \rightarrow S^{2-}(aq) + H_2O(l)$

d. $H_3AsO_4(aq) + 3NaOH(aq) \rightarrow Na_3AsO_4(aq) + 3H_2O(l)$

***Solution***

a. The $HClO_2$ loses an $H^+$ ion, so it is the Brønsted–Lowry acid. The $IO^-$ in the NaIO gains the $H^+$ ion, so the NaIO is the Brønsted–Lowry base.

b. The HF loses an $H^+$ ion, so it is the Brønsted–Lowry acid. The $HS^-$ gains the $H^+$ ion, so it is the Brønsted–Lowry base.

c. The $HS^-$ loses an $H^+$ ion, so it is the Brønsted–Lowry acid. The $OH^-$ gains the $H^+$ ion, so it is the Brønsted–Lowry base.

d. The $H_3AsO_4$ loses three $H^+$ ions, so it is the Brønsted–Lowry acid. Each $OH^-$ in NaOH gains an $H^+$ ion, so the NaOH is the Brønsted–Lowry base.

**EXERCISE 5.10 Brønsted–Lowry Acids and Bases**

OBJECTIVE 31

Identify the Brønsted–Lowry acid and base in each of the following equations.

a. $HNO_2(aq) + NaBrO(aq) \rightarrow HBrO(aq) + NaNO_2(aq)$

b. $H_2AsO_4^-(aq) + HNO_2(aq) \rightleftharpoons H_3AsO_4(aq) + NO_2^-(aq)$

c. $H_2AsO_4^-(aq) + 2OH^-(aq) \rightarrow AsO_4^{3-}(aq) + 2H_2O(l)$

The Examples are followed by similar **Exercises** that provide the reader with an opportunity to test their skills. The answers for these Exercises are in the back of the book, and the complete solutions are in the Student Study Guide.

**Chapter Glossary**

**Hydronium ion** $H_3O^+$.

**Arrhenius acid** According to the Arrhenius theory, any substance that generates hydronium ions, $H_3O^+$, when added to water.

**Acidic solution** A solution with a significant concentration of hydronium ions, $H_3O^+$.

A **Chapter Glossary** of new terms at the end of each chapter makes it easy for the reader to learn these terms and provides a brief review of the chapter information.

622 Chapter 14 | Liquids: Condensation, Evaporation, and Dynamic Equilibrium

**Chapter Problems**

**Section 14.1 Change from Gas to Liquid and from Liquid to Gas—An Introduction to Dynamic Equilibrium**

OBJECTIVE 2

39. A batch of corn whiskey is being made in a backwoods still. The ingredients are mixed and heated, and because the ethyl alcohol, $C_2H_5OH$, evaporates more rapidly than the other ... mixture is enriched in $C_2H_5$... the liquid it forms has a hig... mixture. Describe the subm... $C_2H_5OH$ is converted into a... from a high-temperature ga...

40. Why is dew more likely to ... the changes that take place ...

41. Acetone, $CH_3COCH_3$, is a l... nail polish remover.

OBJECTIVE 3

a. Describe the submicros... liquid acetone when it ...

OBJECTIVE 4

b. Do all of the acetone m... liquid escape? If not, w... molecule to escape from... phase?

OBJECTIVE 5

c. If you spill some nail p... feel cold. Why?

OBJECTIVE 7

d. If you spill some aceton... rapidly than the same a...

OBJECTIVE 9

e. If you spill acetone on a ... more quickly than the same amount of acetone spilled on the cooler lab bench. Why?

42. Consider two test tubes, each containing the same amount of liquid acetone. A student leaves one of the test tubes open overnight and covers the other one with a balloon so that gas cannot escape. When the student returns to the lab the next day, all of the acetone is gone from the open test tube, but most of it remains in the covered tube.

a. Explain why the acetone is gone from one test tube and not from the other.

b. Was the initial rate at which liquid changed to gas (the rate of evaporation) greater in one test tube than in the other? Explain your answer.

c. Consider the system after 30 minutes, with liquid remaining in both test tubes. Is condensation (vapor to liquid) taking place in both test tubes? Is the rate of condensation the same in both test tubes? Explain your answer.

OBJECTIVE 10

d. Describe the submicroscopic changes in the covered test tube that lead to a constant amount of liquid and vapor.

OBJECTIVE 12

e. The balloon expands slightly after it is placed over the test tube, suggesting an initial increase in pressure in the space above the liquid. Why? After this initial expansion, the balloon stays inflated by the same amount. Why doesn't the pressure inside the balloon change after the

This text is full of real-world examples, both in the chapter narrative and in the problems. The **Chapter Problems** not only test readers on specific skills, but they also teach them how chemicals are made, how substances are used, and some of the issues that relate to chemicals.

564 Chapter 13 | Gases

## SPECIAL TOPIC 13.1 A Greener Way to Spray Paint

United States industries use an estimated 1.5 billion liters of paints and other coatings per year, and much of this is applied by spraying. Each liter sprayed from a canister releases an average of 550 grams of volatile organic compounds (VOCs), including hydrocarbons, alcohols, esters, and ketones. Some of these VOCs are hazardous air pollutants.

The mixture that comes out of the spray can has two kinds of components: (1) the solids being deposited on the surface as a coating and (2) a solvent blend that allows the solids to be sprayed and to spread evenly. The solvent blend must dissolve the coatings into a mixture that is thin enough in consistency to be easily sprayed. But a mixture that is thin enough to be easily sprayed will be too runny to remain in place when deposited on a surface. Therefore, the solvent blend contains additional components so volatile that they will evaporate from the spray droplets between the time the spray leaves the spray nozzle and the time the spray hits the surface. Still other, slower-evaporating components do not evaporate until after the spray hits the surface. They remain in the coating mixture long enough to cause it to spread out evenly. Because the more volatile solvents have escaped, the mixture that hits the surface is thick enough not to run or sag.

The Clean Air Act has set strict limitations on the emission of certain VOCs, so safer solvents are needed to replace them. One new spray system has been developed that yields a high-quality coating while emitting as much as 80% fewer VOCs of all types and none of the VOCs that are considered hazardous air pollutants. This system is called the *supercritical fluid spray process.* The solvent mixture for this process still contains some of the slowly evaporating solvents that allow the coating to spread evenly, but it replaces the rapidly-evaporating solvents with high-pressure $CO_2$.

Some gases can be converted into liquids at room temperature by being compressed into a smaller volume, but for each gas, there is a temperature above which ...

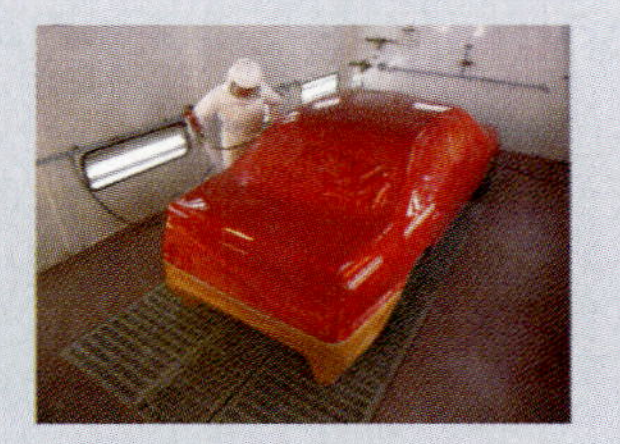

By taking advantage of the properties of gases at high temperatures and pressures, scientists have invented a new, environmentally friendly spray-painting process.

The critical temperature of $CO_2$ is 31 °C. Ab... temperature, carbon dioxide can be compress... very high-pressure and relatively high-densit... critical fluid. Like a liquid, the supercritical ca... ide will mix with or dissolve the blend of coat... low-volatility solvent to form a product that is... enough to be sprayed easily, in very small dro... supercritical $CO_2$ has a very high volatility, so... rates from the droplets almost immediately af... are emitted from the spray nozzle, leaving a m... that is thick enough not to run or sag when it h... surface. The mixture is sprayed at temperatur... 50 °C and a pressure of 100 atm (about 100 tim... room pressure).

Because carbon dioxide is much less toxic t... VOCs it replaces and because it is nonflamma... relatively inert, it is much safer to use in the w... It is also far less expensive. Moreover, the $CO_2$... obtained from the production of other chemic... new process does not lead to an increase in ca... ide in the atmosphere. In fact, because the VO...

**Special Topics** in each chapter describe important issues relating to chemistry, including environmental issues and health-related applications. Several of these Special Topics describe the achievements of "green chemistry" (or environmentally benign chemistry), which has as its goal the development of new ways to produce, process, and use chemicals so as to reduce the risks to humans and to the environment.

210 Chapter 5 | Acids, Bases, and Acid–Base Reactions

**Figure 5.14**
**Hydronium and Hydroxide Ions**

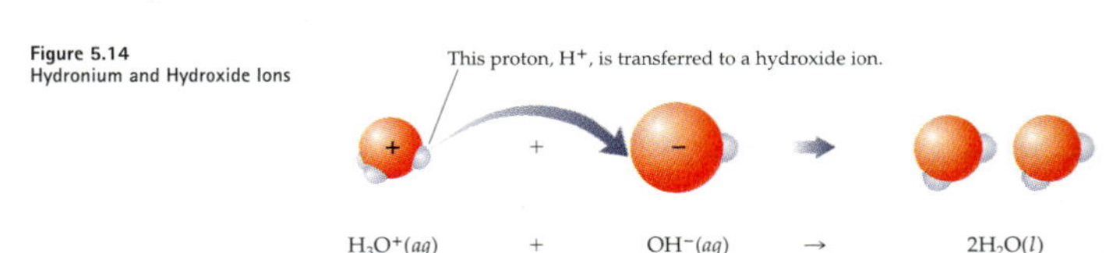


$$H_3O^+(aq) + OH^-(aq) \rightarrow 2H_2O(l)$$

**Figure 5.15**
**After Reaction of Nitric Acid and Sodium Hydroxide**

OBJECTIVE 23a

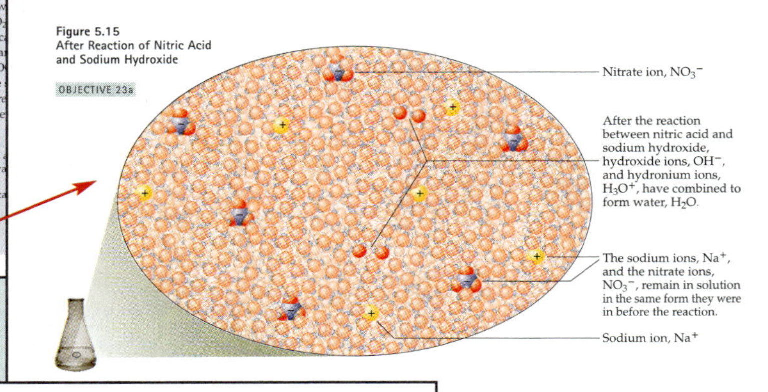

Nitrate ion, NO3−
After the reaction between nitric acid and sodium hydroxide, hydroxide ions, OH−, and hydronium ions, H3O+, have combined to form water, H2O.
The sodium ions, Na+, and the nitrate ions, NO3−, remain in solution in the same form they were in before the reaction.
Sodium ion, Na+


...d not actively participate in the reaction. In other ...ns, so they are left out of the net ionic chemical ...ation for the reaction is therefore

$\rightarrow 2H_2O(l)$

...bit of describing reactions such as this in terms ...en though hydrogen ions do not exist in a water ...e that sodium ions do. When an acid loses a ... proton immediately forms a covalent bond to ... it forms a covalent bond to a water molecule to ...on. Although $H_3O^+$ is a better description of

Many features of the book help students develop the ability to visualize the models that chemists use for describing the structure and behavior of matter. The reader is continually encouraged, with the aid of **colorful and detailed illustrations,** to visualize particle movements and particle interactions that accompany chemical changes.

13.1 Gases and Their Properties 537

**Figure 13.4**
**Relationship Between Temperature and Pressure**
Increased temperature leads to increased pressure if the moles of gas and the volume are constant.

OBJECTIVE 10b

Piston locked in position — Heat added — A constant amount of gas . . . — Constant volume — Increased temperature — Increased pressure

**Increased temperature** → Increased average velocity of the gas particles → Increased number of collisions with the walls / Increased force per collision → Increased total force of collisions → Increased $\frac{\text{force due to collisions}}{\text{area of wall}}$ → **Increased gas pressure**

### The Relationship Between Volume and Temperature

Consider the system shown in Figure 13.5 on the next page. To demonstrate the relationship between temperature and volume of gas, we must keep the moles of gas and the gas pressure constant. If our valve is closed and our system has no leaks, the moles of gas are constant. We keep the gas pressure constant by allowing the piston to move freely throughout our experiment, because then it will adjust to keep the pressure pushing on it from the inside equal to the atmospheric pressure pushing on it from the outside. The atmospheric pressure is the pressure in the air outside the container, which acts on the top of the piston because of the force of collisions between particles in the air and the surface of the piston. We can assume that it is constant throughout our experiment.

OBJECTIVE 10c

If we increase the temperature, the piston in our apparatus moves up, increasing the volume occupied by the gas. A decrease in temperature leads to a decrease in volume.

Increased temperature → Increased volume

Decreased temperature → Decreased volume

OBJECTIVE 10c

The increase in temperature of the gas leads to an increase in the average velocity of the gas particles, which leads in turn to more collisions with the walls of the container and a greater force per collision. This greater force acting on the walls of the container leads to an *initial* increase in the gas pressure. Thus, the increased temperature of our gas creates an internal pressure, acting on the bottom of the piston, that is greater than the external pressure acting on the top of the piston. The greater internal pressure causes the piston to move up, increasing the volume of the chamber. The increased volume leads to a decrease in gas pressure in the container, until

OBJECTIVE 10c

**Figures** throughout the text illustrate important ideas and act as constant reminders of the particle nature of matter. Key concepts are often summarized with logic sequences that show how one component of an explanation leads logically to the next.

A **variety of art,** combining photos and illustrations, highlight the key concepts.

like the situation depicted in Figure 16.6, where a rolling ball rolls back down the same side of a hill it started up.

If a rolling ball does not have enough energy to get to the top of a hill, it stops and rolls back down.

Figure 16.6
Not Enough Kinetic Energy to Get Over the Hill

OBJECTIVE 6

The activation energy for the oxygen–ozone reaction is 17 kJ/mole $O_3$. If the collision between reactants yields a net kinetic energy equal to or greater than the activation energy, the reaction can proceed to products (Figure 16.7). This is like a ball rolling up a hill with enough kinetic energy to reach the top of the hill, from which it can roll down the other side (Figure 16.8).

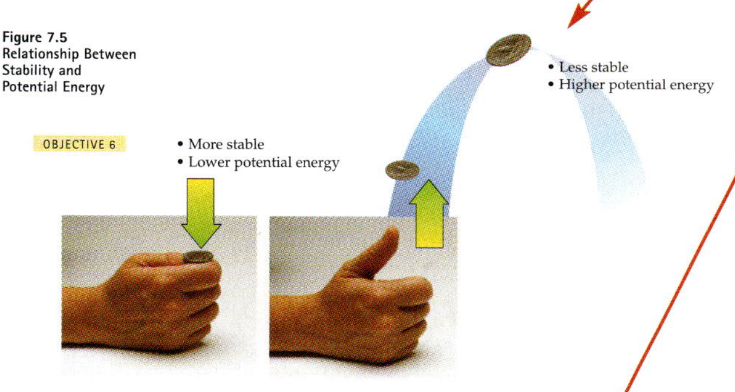
Particles that collide with a net kinetic energy greater than the activation energy can react.
Particles that collide with a net kinetic energy less than the activation energy cannot react.

Figure 16.7
Collision Energy and Activation Energy

OBJECTIVE 6

If a ball reaches the top of a hill before its energy is depleted, it will continue down the other side.

Figure 16.8
Ball with Enough Kinetic Energy to Get Over the Hill

The bonds between oxygen atoms in $O_2$ molecules are stronger and more stable than the bonds between atoms in the ozone molecules, so more energy is released in the formation of the new bonds than is

Figure 7.5
Relationship Between Stability and Potential Energy

OBJECTIVE 6

- More stable
- Lower potential energy

- Less stable
- Higher potential energy

the more stable atoms in the bond. For example, the first step in the formation of ozone in the earth's atmosphere is the breaking of the oxygen–oxygen covalent bonds in more stable oxygen molecules, $O_2$, to form less stable separate oxygen atoms. This change could not occur without an input of considerable energy—in this case, radiant energy from the sun. We call changes that absorb energy **endergonic** (or endogonic) changes (Figure 7.6).

OBJECTIVE 7

Figure 7.6
Endergonic Change

OBJECTIVE 5

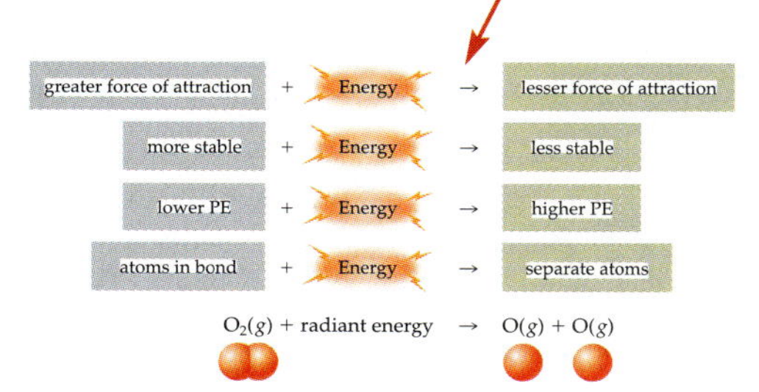
greater force of attraction + Energy → lesser force of attraction
more stable + Energy → less stable
lower PE + Energy → higher PE
atoms in bond + Energy → separate atoms
O2(g) + radiant energy → O(g) + O(g)

The attraction between the separated atoms makes it possible that they will change from their less stable separated state to the more stable bonded state. As they move together, they may bump into and move something (such as another atom), so the separated atoms have a greater capacity to

represent the combustion reactions for methane, the primary component of natural gas, and hexane, which is found in gasoline.

$$CH_4(g) + 2O_2(g) \rightarrow CO_2(g) + 2H_2O(l)$$

$$2C_6H_{14}(l) + 19O_2(g) \rightarrow 12CO_2(g) + 14H_2O(l)$$

The complete combustion of a substance, such as ethanol, $C_2H_5OH$, that contains carbon, hydrogen, and oxygen also yields carbon dioxide and water.

$$C_2H_5OH(l) + 3O_2(g) \rightarrow 2CO_2(g) + 3H_2O(l)$$

*When any substance that contains sulfur burns completely, the sulfur forms sulfur dioxide.* For example, when methanethiol, $CH_3SH$, burns completely, it forms carbon dioxide, water, and sulfur dioxide. Small amounts of this strong-smelling substance are added to natural gas to give the otherwise odorless gas a smell that can be detected if a leak occurs (Figure 6.3).

$$CH_3SH(g) + 3O_2(g) \rightarrow CO_2(g) + 2H_2O(l) + SO_2(g)$$

Combustion reactions include oxygen as a reactant and are accompanied by heat and usually light.

OBJECTIVE 8

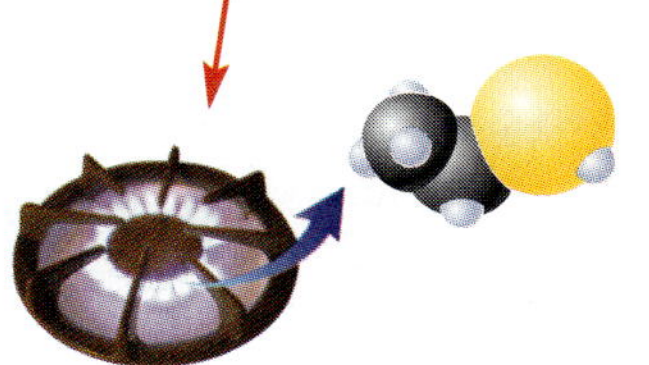

Figure 6.3
Odor to Natural Gas
The methanethiol added to natural gas warns us when there is a leak.

The following sample study sheet lists the steps for writing equations for combustion reactions.

**Sample Study Sheet 6.2**
**Writing Equations for Combustion Reactions**

OBJECTIVE 8

**Tip-off** You are asked to write an equation for the complete combustion of a substance composed of one or more of the elements carbon, hydrogen, oxygen, and sulfur.

**General Steps**

**Step 1** Write the formula for the substance combusted.
**Step 2** Write $O_2(g)$ for the second reactant.

**Sample Study Sheet 9.1**

**Converting Between Mass of Element and Mass of Compound Containing the Element**

OBJECTIVE 15

**Tip-off** When you analyze the type of unit you have and the type of unit you want, you recognize that you are converting between a unit associated with an element and a unit associated with a compound containing that element.

**General Steps** The following general procedure is summarized in Figure 9.3.

- **Convert the given unit into moles of the first substance.**

  This step often requires converting the given unit into grams, after which the grams can be converted into moles using the molar mass of the substance.

- **Convert moles of the first substance into moles of the second substance using the mole ratio derived from the formula for the compound.**

  You convert either from moles of element into moles of compound or from moles of compound into moles of element.

- **Convert moles of the second substance into the desired units of the second substance.**

  This step requires converting moles of the second substance into grams of the second substance using the molar mass of the second substance, after which the grams can be converted into the specific units that you want.

**Example** See Example 9.6.

Any unit of an element
↕ Using conversion factors from Chapter 8
Grams of element
↕ Using molar mass derived from atomic mass $\left(\frac{1 \text{ mol element}}{(\text{atomic mass}) \text{ g element}}\right)$
Moles of element
↕ Using the mole ratio from the compound's formula $\left(\frac{1 \text{ mol compound}}{(\text{number of atoms in formula}) \text{ mol element}}\right)$
Moles of compound containing the element
↕ Using molar mass derived from formula mass $\left(\frac{(\text{formula mass}) \text{ g compound}}{1 \text{ mol compound}}\right)$
Grams of compound
↕ Using conversion factors from Chapter 8
Any unit of a compound

**Figure 9.3 General Steps for Converting Between the Mass of an Element and the Mass of a Compound Containing the Element**
The calculation can be set up to convert from the mass of an element to the mass of a compound (top to bottom) or from the mass of a compound to the mass of an element (bottom to top).

OBJECTIVE 15

This text helps students get organized by providing Sample Study Sheets for many of the tasks they will be expected to do on exams. Each study sheet describes how to recognize a specific kind of task ("Tip-off") and breaks the task down into general steps.

INTER-CHAPTER 1A

# A Brief Introduction to Unit Conversions



...8 both describe the procedures for doing a ... useful in chemistry, but the inter-chapter does so ... aps you are wondering why the inter-chapter is here ... y ways to teach introductory chemistry. This text has ... rse begins the coverage of chemical principles right ... topics until mid-semester. Although many chemistry ... nstructors believe in covering the math-related topics ... for each approach. The flexible organization of this text ... e course topics in the way that they believe is best. ... ble approaches to learning chemistry's math-related ... uss unit conversions until the middle of the course, ... ncepts first. These teachers will tell you not to read ... a brief introduction to unit conversions early in the ... unit conversions in early laboratory sessions, while ... t of the subject until later. If your instructor falls into ... text, including this inter-chapter, in the order in ... ts a broader coverage of the math-related topics early ... ere to Chapter 8, which describes unit conversions in ... Chapter 8 was written in such a way that you can ... ters 2 through 7. (4) Some instructors will ask their ... he or two of the sections in Chapter 8 before proceed- ... structor t...

CHAPTER 8

# Unit Conversions



*[M]athematics . . . is the easiest of sciences, a fact which is obvious in that no one's brain rejects it . . .*

Roger Bacon (c. 1214–c. 1294)
English philosopher and scientist

*Stand firm in your refusal to remain conscious during algebra. In real life, I assure you, there is no such thing as algebra.*

Fran Lebowitz (b. 1951)
American journalist

You may agree with Roger Bacon that mathematics is the easiest of sciences, but many beginning chemistry students would not. Because they have found mathematics challenging, they wish it were not so important for learning chemistry—or for answering so many of the questions that arise in everyday life. They can better relate to Fran Lebowitz's advice in the second quotation. If you are one of the latter group, it will please you to know that even though there *is* some algebra in chemistry, this chapter teaches a technique for doing chemical calculations (and many other calculations) without it. The technique is called dimensional analysis. You will be using it throughout the rest of this book, in future chemistry and science courses, and any time you want to calculate the number of nails you need to build a fence or the number of rolls of paper necessary to cover the kitchen shelves.

*Dimensional analysis, the technique for doing unit conversions that is described in this chapter, can be used for a lot more than chemical calculations.*

### Review Skills

The presentation of information in this chapter assumes that you can already perform the tasks listed below. You can test your readiness to proceed by answering the Review Questions at the end of the chapter. This might also be a good time to read the Chapter Objectives, which precede the Review Questions.

- List the metric base units and the corresponding abbreviations for length, mass, volume, energy, and gas pressure. (Section 1.4)
- State the numbers or fractions represented by the following metric prefixes, and write their abbreviations: giga, mega, kilo, centi, milli, micro, nano, and pico. (Section 1.4)
- Given a metric unit, write its abbreviation; given an abbreviation, write the full name of the unit. (Section 1.4)
- Describe the relationships between the metric units that do not have prefixes (such as meter, gram, and liter) and units derived from them by the addition of prefixes—for example, 1 km = $10^3$ m. (Section 1.4)
- Define temperature and describe the Celsius, Fahrenheit, and Kelvin scales used to report its values. (Section 1.4)
- Given a value derived from a measurement, identify the range of possible values it represents, on the basis of the assumption that its uncertainty is ±1 in the last position reported. (For example, 8.0 mL says the value could be from 7.9 mL to 8.1 mL.) (Section 1.5)

319

This text takes a flexible approach to the coverage of chemical calculations. Chapter 8 provides a comprehensive treatment of unit conversions immediately before these calculations are used in the chemical calculations described in Chapters 9 and 10. Inter-Chapter 1A provides a briefer introduction to dimensional analysis and unit conversions. See the section in the Preface titled Flexible Order of Math-Related Topics for a description of the options that these two chapters provide.

# Supplements to This Book

## Instructor's Supplements

**Instructor's Manual and Complete Solutions (0-8053-3215-4)**
The Instructor's Manual has complete solutions to all of the in-chapter exercises and all of the end-of-chapter problems. It also contains suggestions for using the book most efficiently, possible variations in the order of coverage, lists of topics that can be skipped without causing problems for the coverage of later topics, and lists of the available computer tools for each chapter.

**Printed Test Bank (0-8053-3213-8)**
This printed test bank includes over 1500 questions that correspond to the major topics in the text.

**Computerized Test Bank (0-8053-3214-6)**
This dual-platform CD-ROM includes over 1500 questions that correspond to the major topics in the text.

**Benjamin Cummings Science Digital Library (0-8053-3209-X)**
The CD-ROM provides instructors a wealth of illustrations for incorporation into lecture presentations, student materials, and tests.

**Transparency Acetates (0-8053-3212-X)**
Includes 125 full-color acetate transparencies.

**Instructor's Manual for Lab Manual (0-8053-3217-0)**

## Student's Supplements

**Study Guide and Selected Solutions (0-8053-3211-1)**
Each chapter in the Study Guide contains introductions to every section in the corresponding text chapter, a checklist to help students study efficiently, lists of important skills to master, a concept map to help students visualize the connections among the chapter topics, solutions to the in-chapter exercises, and solutions to selected end-of-chapter problems.

**Laboratory Manual (0-8053-3217-0)**
This introductory chemistry Laboratory Manual, written to accompany the Bishop text, contains 25 labs. It is designed to help students develop data acquisition, organization, and analysis skills while teaching basic techniques. Students learn to construct their own data tables, answer conceptual questions, and make predictions before performing experiments. They also have the opportunity to visualize and describe molecular level activity and explain the results.

**Special Edition of The Chemistry Place**
www.chemplace.com/college
This special edition of The Chemistry Place engages students in interactive exploration of chemistry concepts and provides a wealth of tutorial support. Tailored to the Bishop textbook, the site includes detailed objectives for each chapter of the text, interactive activities featuring simulations, animations, and 3D visualization tools, multiple-choice and glossary quizzes, and an extensive set of Web links. For instructors, a Syllabus Manager makes it easy to create an online syllabus complete with weekly assignments, projects, and test dates that students may access on the ChemPlace site.

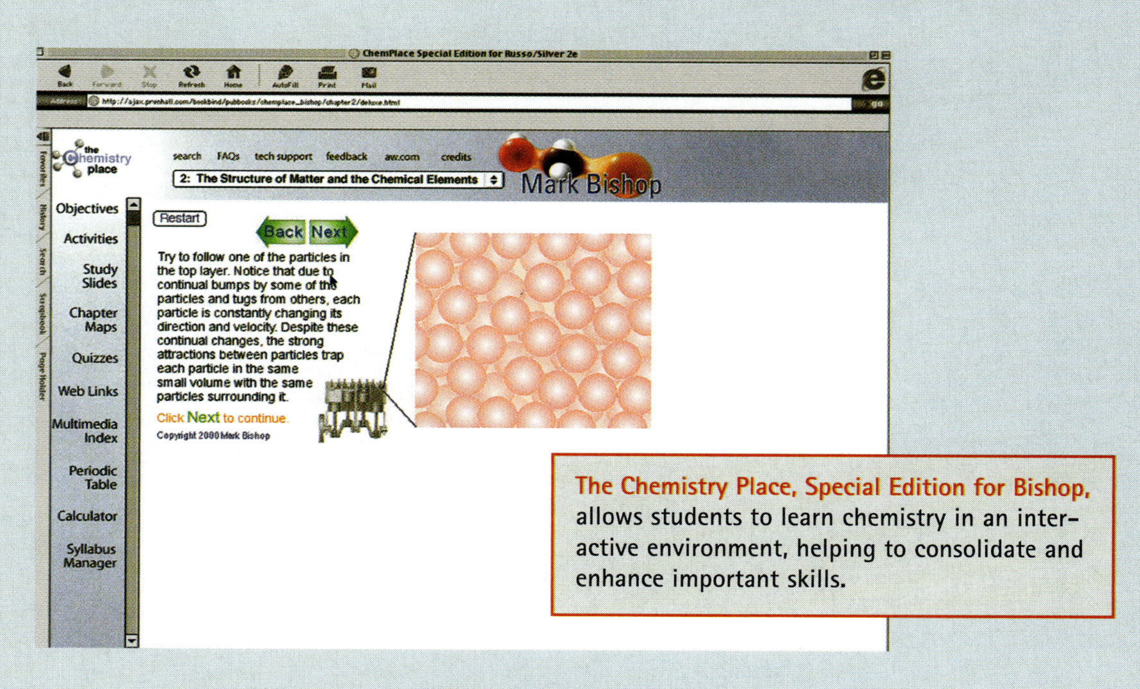


**The Chemistry Place, Special Edition for Bishop,** allows students to learn chemistry in an interactive environment, helping to consolidate and enhance important skills.

*A world of chemistry awaits you.*

# An Introduction to Chemistry

CHAPTER 1



*I would watch the buds swell in spring, the mica glint in the granite, my own hands, and I would say to myself: "I will understand this, too. I will understand everything."*

PRIMO LEVI (1919–1987)

ITALIAN CHEMIST AND AUTHOR, *THE PERIODIC TABLE*

IT IS HUMAN NATURE TO WONDER—about the origins of the universe and of life, about their workings, and even about their meanings. We look for answers in physics, biology, and the other sciences, as well as in philosophy, poetry, and religion. Primo Levi could have searched for understanding in many ways, but his wondering led him to study chemistry . . . and so has yours, although you may not yet know why.

## 1.1 What Is Chemistry, and What Can Chemistry Do for You?

One thing is certain: Once you start studying chemistry, all kinds of new questions begin to occur to you. Let's consider a typical day.

Your alarm rings early, and you are groggy from sleep but eager to begin working on a chemistry assignment that's coming due. Chemistry has taught you that there are interesting answers to questions you might once have considered silly and childish. Preparing tea, for example, now makes you wonder why the boiling water bubbles and produces steam, whereas the teakettle retains its original shape. How do the tea leaves change the color of the water while the teabag remains as full and plump as ever? Why does sugar make your tea sweet, and why is the tea itself bitter?

*An understanding of chemistry opens up a window to the world around you.*

You settle down with tea and newspaper, and the wondering continues. An article about methyl bromide, a widely used pesticide, says some scientists think it damages the ozone layer. What *are* methyl bromide and ozone? How does one destroy the other, and why should we care? How can we know whether the ozone really is being depleted?

Later, as you drive to the library to get some books you need to complete that chemistry assignment, you wonder why gasoline burns and propels your car down the road. How does gasoline pollute the air we breathe, and what does the catalytic converter do to minimize this pollution? At the library, you wonder why some books that are hundreds of years old are still in good shape, whereas the pages of other books only 50 years old are brown, brittle, and crumbling. Can the books with damaged pages be saved?

Chemists can answer all these questions and others like them. They are scientists who study the structure of material substances—collectively called matter—and the changes that they undergo. Matter can be solid like sugar, liquid like water, or gaseous like the exhaust from your car's tail pipe. **Chemistry** is often defined as the study of the structure and behavior of matter.

Chemists do a lot more than just answer questions. Industrial chemists are producing new materials to be used to build lighter and stronger airplanes, more environmentally friendly disposable cups, and more efficient antipollution devices for your car. Pharmaceutical chemists are developing new drugs to fight cancer, control allergies, and even grow hair on bald heads.

In the past, the chemists' creations have received mixed reviews. The chlorofluorocarbons (CFCs) used as propellants in aerosol cans are now known to threaten the earth's protective ozone layer. The durable plastics that chemists created have proved *too* durable; so when they are discarded, they remain in the environment for a long, long time. One of the messages you will find in this book is that, despite occasional mistakes and failures, most chemists have a strong social conscience. Not only are they actively developing new chemicals to make our lives easier, safer, and more productive, but they are also working to clean up our environment and minimize the release of chemicals that might be harmful to our surroundings (see Special Topic 1.1: *Green Chemistry*).

As you read on in this book, you will find, perhaps to your disappointment, that only limited portions of each chapter provide direct answers to real-life questions. An introductory chemistry text, like this one, must focus instead on teaching basic principles and skills. Some of the things you need to learn in order to understand chemistry may seem less than fascinating, and it will not always be easy to see why they are useful. Try to remember that the fundamental concepts and skills will soon lead you to a deeper understanding of the physical world.

Before you could run, you needed to learn how to walk. Before you could read a book or write a paper, you needed to learn the alphabet, build your vocabulary, and understand the basic rules of grammar. Chemistry has its own "alphabet" and vocabulary, as well as many standards and conventions, that enable chemists to communicate and to do efficient, safe, and meaningful work. Learning the symbols for the common chemical elements, the rules for describing measurements, or the conventions for describing chemical changes might not be as interesting as finding out how certain chemicals in our brains affect our mood, but they are necessary steps in learning chemistry.

This chapter presents some suggestions for making your learning process easier and introduces some of the methods of scientific measurement and reporting. You will then be ready for Chapter 2, which gives you a first look at some of chemistry's basic underlying concepts.

## SPECIAL TOPIC 1.1 Green Chemistry

*With knowledge comes the burden of responsibility.*

PAUL T. ANASTAS AND TRACY C. WILLIAMSON, *GREEN CHEMISTRY*

The chemical industry has made many positive contributions to modern life, but these improvements have come at a price. The chemical products themselves, the chemicals used to produce them, and the byproducts of their production have sometimes been harmful to our health and the environment. Concern about these dangers has led to a movement in the chemical industry that is sometimes called green chemistry or environmentally benign chemistry. Its goal is to produce, process, and use chemicals in new ways that pose fewer risks to humans and their environment.

In support of this goal, in March 1995 President Clinton inaugurated the Green Chemistry Program, to be coordinated by the Environmental Protection Agency. As part of this program, the Green Chemistry Challenge Awards were set up to "recognize and promote fundamental and innovative chemical methodologies that accomplish pollution prevention and that have broad application in industry."

Among the first award recipients was the Rohm and Haas Company, which received the Green Chemistry Challenge's 1996 Designing Safer Chemicals Award for the development of 4,5-dichloro-2-n-octyl-4-isothiazolin-3-one, otherwise known as Sea-Nine® antifoulant. As soon as a new ship goes into the water, various plants and animals begin to grow on its bottom. This growth slows the ship's progress and causes other problems as well, ultimately costing the shipping industry an estimated $3 billion per year. Substances called antifoulants are painted on the bottoms of ships to prevent the unwanted growth, but most of them can be harmful to the marine environment. Sea-Nine is a more environmentally friendly chemical than the alternatives because it breaks down in the ocean more quickly and does not accumulate in marine organisms.

This is only one example of the progress that green chemistry is making. Others will be presented in later chapters.

Source: P. T. Anastas and T. C. Williams, Eds. *Green Chemistry* (Washington, DC: American Chemical Society, 1996).

Vinyl/Rosin Paint

$Cu_2O$ + 3% Sea-nine 211 biocide

$Cu_2O$

New, environmentally friendly marine antifoulants protect ships without polluting the water.

## 1.2 Suggestions for Studying Chemistry

*The will to succeed is important, but what's more important is the will to prepare.*

BOBBY KNIGHT, BASKETBALL COACH

Let's face it. Chemistry has a reputation for being a difficult subject. One reason is that it includes so many different topics. Individually, they are not too difficult to understand, but collectively, they are a lot to master. Another reason is that these topics must be learned in a cumulative fashion. Topic A leads to topic B, which is important for understanding topic C, and so on. If you have a bad week and do not study topic B very carefully, topic C will not make much sense to you. Because chemistry is time-consuming and cumulative, you need to be very organized and diligent in studying it.

There is no *correct* way to study chemistry. Your study techniques will be determined by your current level of chemical knowledge, the time you have available, your strengths as a student, and your attitude toward the subject. The following list of suggestions may help you take chemistry's special challenges in stride.

*Use the Review Skills sections in this textbook.* Starting with Chapter 2, each chapter begins with a section called **Review Skills** that is complemented, at the end of the chapter, by a set of review questions that test those skills. The Review Skills section and the review questions identify the specific skills from earlier chapters that are necessary for success in the new chapter. If you have trouble with the tasks on the Review Skills list, you will have trouble with the chapter, so promise yourself that you will always review the topics listed in the Review Skills section before beginning a new chapter.

*Read each chapter in the textbook before it is covered in lecture.* There are several good reasons why a relaxed prelecture reading of each chapter is important. It provides you with a skeleton of knowledge that you can flesh out mentally during the lecture and, at the same time, guarantees that there will be fewer new ideas to absorb as you listen to your teacher talk. If you already know a few things, you will be better prepared to participate in class.

*Attend class meetings, take notes, and participate in class discussions.* You will get much more out of a lecture, discussion section, or laboratory—and will enjoy the course more—if you are actively involved. Don't hesitate to ask questions or make comments when the lecture confuses you. If you have read the chapter before coming to class and paid attention to the lecture and you still have a question, you can be fairly certain that other students have that question too and will appreciate your asking for clarification.

*Reread the chapter, marking important sections and working the practice exercises.* In the second reading, mark key segments of the chapter that you think you should reread before the exam. Do not try to do this in the first reading, because you will not know what parts are most important until you have an overview of the material the chapter contains. It is a good idea to stop after reading each section of text and ask yourself, "What have I just read?" This will keep you focused and help you to remember longer.

You will find **Examples** in each chapter that show how to do many of the tasks you will be asked to do on exams. The examples are followed by **Exercises** that you should work to test yourself on these tasks. The answers to these exercises appear at the end of the book. If you have trouble with a practice exercise, look more closely at the example that precedes it, but don't get bogged down on any one topic. When a new concept or an exercise gives you trouble, apply the *15-minute rule:* If you have spent more than 15 minutes on an idea or problem and you still do not understand it, write down what you do not understand and ask your instructor or another student to explain it.

*Use the chapter objectives as a focus of study.* At the end of each chapter is a list of **Chapter Objectives** that provides a specific description of what you should be able to do after studying the chapter. Notations in the text margins show you where to find the information needed to meet each objective. Your instructor may wish to add to the list or remove objectives from it.

Many of the objectives begin with "Explain . . . " or "Describe . . . ." You might be tested on these objectives in short-essay form, so it is a good idea to write your responses down. This will force you to organize your thoughts and develop a concise explanation that will not take too much time to write on the exam. If you do it conscientiously, perhaps using color to highlight key phrases, you will be able to visualize your study sheet while taking the exam.

Many of the objectives refer to stepwise procedures. This book will suggest one procedure, and sometimes your instructor will suggest another. Write out the steps that you think will work best for you. The act of writing will help you remember longer, and you will find it much easier to picture a study sheet written out in your own handwriting than one printed in this book.

*Use the computer-based tools that accompany the course.* This textbook is accompanied by an Internet site that supplements the text. If you have access to it, you will find it useful at this stage in your studying.

*Work some of the* ***Problems*** *at the end of the chapter.* Do not try the end-of-chapter problems too soon. They are best used as a test of what you have have not mastered in the chapter after all the previous steps have been completed. The problems with colored numbers have answers in the back of the book and complete solutions in the Student Study Guide. Many of the problems are followed by the number of the learning objective to which they correspond. Use the same *15-minute rule* for the problems as for the chapter concepts and exercises: If you have spent more than 15 minutes on any one problem, it is time to seek help.

*I hear and I forget.*
*I see and I remember.*
*I do and I understand.*

CHINESE PROVERB

*Ask for help when you need it.* Don't be shy. Sometimes 5 minutes in the instructor's office can save you an hour or more of searching for answers by other means. You might also consider starting a study group with fellow students. It can benefit those of you who are able to give help as well as those who need it. There is no better way to organize one's thoughts than to try to communicate them to someone else. There may also be other ways to get help. Ask your instructor what is available.

*He who neglects to drink of the spring of experience is likely to die of thirst in the desert of ignorance.*

LING PO

*Review for the exam.* Read the list of objectives, asking yourself whether you can meet each one. Every time the answer is "No," spend some time making it "Yes." This might mean meeting with your instructor or study group, rereading this text, or reviewing your notes. Ask your instructor which objectives are being emphasized on the exam and how you are going to be tested on them. Your lecture notes can also provide clues to this. Finally, work some more of the end-of-chapter problems. This will sharpen your skills and improve your speed on the exam.

## 1.3 The Scientific Method

*It's as simple as seeing a bug that intrigues you. You want to know where it goes at night; what it eats.*

DAVID CRONENBERG, CANADIAN FILMMAKER

Before beginning our quest for an understanding of chemistry, let's look at how science in general is done. There is no one correct way to do science. Different scientific disciplines have developed different procedures, and different scientists approach their pursuit of knowledge in different ways. Nevertheless, most scientific work has certain characteristics in common. We can see them in the story of how scientists discovered the first treatment for Parkinson's disease, a neurological condition that progressively affects muscle control. The principal steps in the process are summarized in Figure 1.1.

OBJECTIVE 2

Like most scientific work, the development of a treatment for Parkinson's disease began with *observation and the collection of data.* In the 1960s, scientists observed that South American manganese miners were developing symptoms similar to the muscle tremors and rigidity seen in Parkinson's disease. Next, the scientists made *an initial hypothesis on the basis of their observations.* Perhaps the symptoms of the manganese miners and of Parkinson's sufferers had a common cause. The initial hypothesis led to a more purposeful collection of information in the form of *systematic research or experimentation:* Systematic study of the manganese miners' brain chemistry showed that manganese interferes with the work of a brain chemical called dopamine. Because dopamine is important in the brain's control of muscle function, anyone who absorbed abnormally high levels of manganese would be expected to have troubles with movement.

*The hypothesis was refined* on the basis of the new information, and *research was designed to test the hypothesis.* Specifically, the researchers hypothesized that the brains of Parkinson's sufferers had low levels of dopamine. Brain studies showed this to be the case. *The results were published so that other scientists might repeat the research and confirm or refute the conclusions.* Because other scientists confirmed the results of the dopamine research, the *hypothesis became accepted in the scientific community.*

The discovery of the dopamine connection started a search for a drug that would elevate the levels of dopamine in the brain. This provides an example of what is very often the next step of the scientific method, *a search for useful applications of the new ideas.* Dopamine itself could not be used as a drug because it is unable to pass from the bloodstream into the brain tissue. Instead, the researchers looked for a compound that could penetrate into the brain and then be converted into dopamine. Levodopa, or L-dopa, met these requirements. (It should be noted that significant scientific research does not always lead directly to applications, or even to a published paper. One of the main driving forces of science is just the desire to understand more about ourselves and the world around us. Any research that increases this understanding is important.)

*The development of applications often leads to another round of hypothesizing and testing in order to refine the applications.* There was some initial success with L-dopa. It caused remission of Parkinson's disease in about one-third of the patients treated and brought about improvements in one-third of the others, but there were also problematic side effects, including nausea, gastrointestinal distress, reduced blood pressure, delusions, and mental disturbance. The drug's effects on blood pressure seem to be caused by the conversion of L-dopa to dopamine outside the brain. For this reason, L-dopa is now given with levocarbidopa, which inhibits that process.

And the cycle of hypothesis, experimentation, and discovery of new applications, which leads to further refinement of the hypotheses, continues . . .

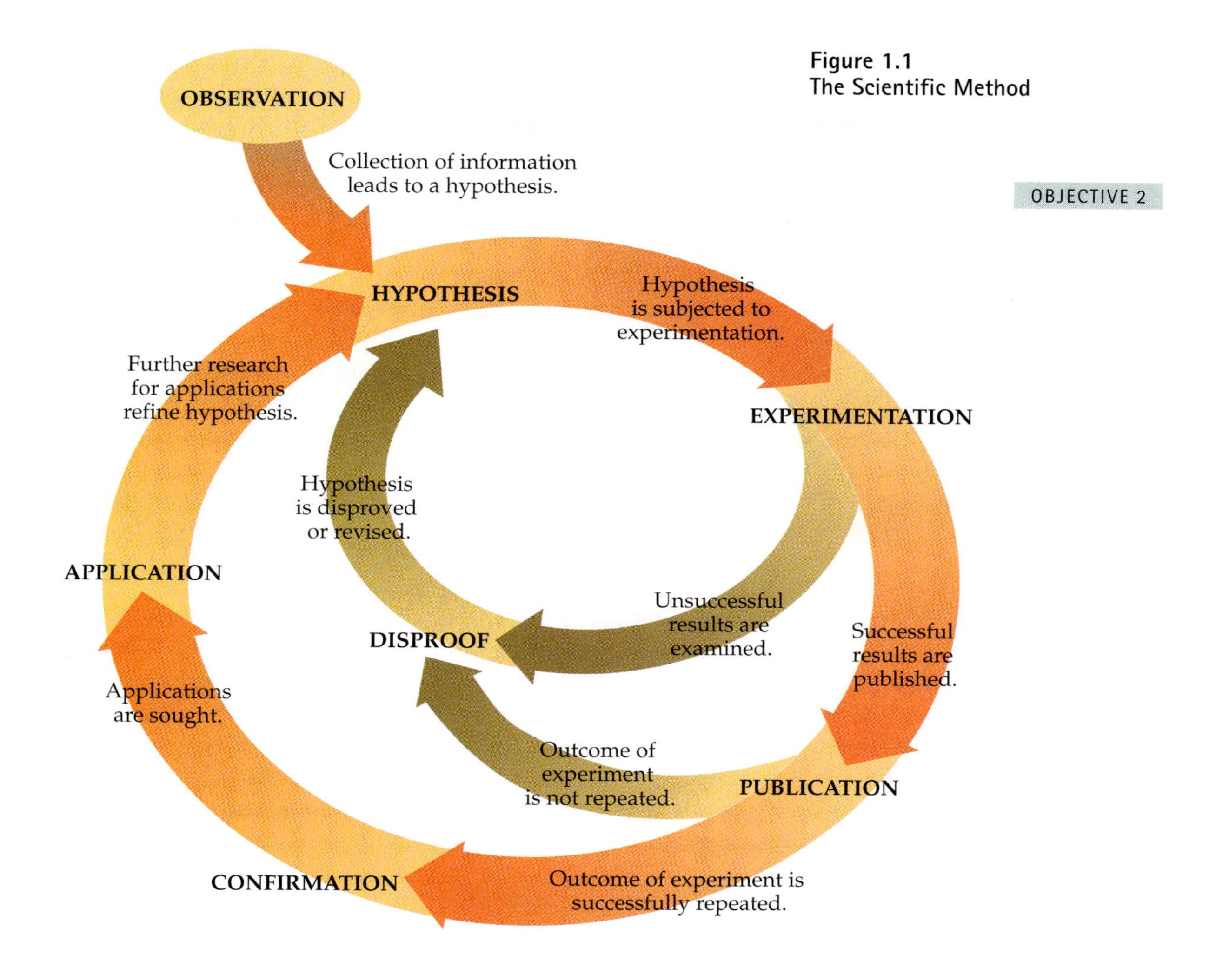
OBSERVATION
Collection of information leads to a hypothesis.
HYPOTHESIS
Hypothesis is subjected to experimentation.
EXPERIMENTATION
Successful results are published.
PUBLICATION
Outcome of experiment is successfully repeated.
CONFIRMATION
Applications are sought.
APPLICATION
Further research for applications refine hypothesis.
Unsuccessful results are examined.
DISPROOF
Outcome of experiment is not repeated.
Hypothesis is disproved or revised.

**Figure 1.1**
The Scientific Method

OBJECTIVE 2

## 1.4 Measurement and Units

*Measure what is measurable and make measurable what is not so.*

Galileo Galilei

The practice of chemistry demands both accuracy and clarity. The properties of matter must be measured correctly and reported without ambiguity. For example, a chemist has a lot of measuring to do while testing to determine whether the pesticide methyl bromide might be destroying the protective layer of ozone gas in the earth's upper atmosphere (the ozone in the upper atmosphere filters out harmful radiation from the sun). She or he might add a carefully *measured* amount of methyl bromide to a reaction vessel that already contains ozone and other chemicals found in our atmosphere. In such an experiment, the temperature, too, would be carefully *measured* and adjusted to duplicate the average temperature in the ozone layer, and then the substances in the vessel might be subjected to the same *measured* amounts and kinds of radiation to which ozone is exposed

A long-jump can be won by a centimeter.

in the atmosphere. Chemical changes would take place, after which the amounts of various substances found in the resultant mixture would be *measured* and reported at various times. All of these measurements of amounts, temperatures, and times must be carefully reported in a way that enables other scientists to judge the value of the experiment.

A measurement is always reported as a **value,** a quantitative description that includes both a number and a unit. For example, before a 100-meter race is run, the distance must be measured as precisely as possible. Its value is 100 meters. In this value, the unit is *meters* (defined below) and the number of units is *100.*

**Units** are quantities defined by standards that people have agreed to use to compare one event or object to another. For example, at one time the units for length were based on parts of the body. An inch (the unit) was the width of the thumb (the standard), and a foot was the length of a typical adult foot. Centuries ago you might have described the length of a cart you wanted made as 8 feet, meaning eight times the length of your foot. Although the length of a foot varies for different people, this description would be accurate enough to allow the carpenter to make a cart that would fit your purpose.

As measuring techniques became more precise and the demand for accuracy increased, the standards on which people based their units were improved. In the 18th century, the French invented the metric system, which was based on a more consistent, systematic, and carefully defined set of standards than had ever been used before. For example, the meter (or metre, from the Greek *metron,* "a measure") became the standard for length. The first definition for the standard meter was one ten-millionth (1/10,000,000) of the distance from the north pole to the equator. This became outdated as the precision of scientists' measuring instruments improved. Today, a meter is defined as the distance light travels in a vacuum in 1/299,792,458 second. Technical instruments for measuring length are calibrated in accordance with this very accurate definition.

## The International System of Measurement

The International System of Measurement (SI for *Système International d'Unités*), a modern elaboration of the original metric system, was set up in 1960. It was developed to provide a very organized, precise, and practical system of measurement that everyone in the world could use. The SI system is constructed using seven **base units,** from which all other units are derived (Table 1.1). The chemist is not usually interested in electric currents or luminous intensity, so only the first five of the base units on Table 1.1 will appear in this text. The meaning of *mole*, the base unit for amount of substance, is explained in Chapter 9. Until then, we will use the first four base units: meter (m), kilogram (kg), second (s), and kelvin (K).

Table 1.1
Base Units for the International System of Measurement

OBJECTIVE 3

| Type of Measurement | Base Unit | Abbreviation |
|---|---|---|
| **length** | meter | m |
| **mass** | kilogram | kg |
| **time** | second | s |
| **temperature** | kelvin | K |
| **amount of substance** | mole | mol |
| **electric current** | ampere | A |
| **luminous intensity** | candela | cd |

## SPECIAL TOPIC 1.2 Wanted: A New Kilogram

One by one, all of the SI base units except the kilogram have been defined in terms of constants of nature. For example, the speed of light in a vacuum is used to define the meter. The kilogram is the last of the standards still to be based on a manufactured object. A kilogram is defined as the mass of a cylinder, made of platinum–iridium alloy, 39 mm (1.5 in.) tall and 39 mm in diameter. This cylinder has been kept in an airtight vault in Sèvres, France, since 1883. In order to avoid using (and possibly damaging) the standard itself, 40 replicas were made in 1884 and distributed around the world to serve as a basis of comparison for mass. (The United States has kilogram number 20.) The scientists at the *Bureau International des Poids et Mesures* (BIPM) are so careful with the standard kilogram that it was used only three times in the twentieth century.

There are problems associated with defining a standard in terms of an object, such as the standard kilogram. If the object loses or gains mass as a result of contact or contamination, measurements related to the object change. For years, the standard kilogram was maintained by a French technician named Georges Girard, who kept the cylinder clean using a chamois cloth and a mixture of ethanol and ether. Girard seemed to have had just the right touch for removing contaminants without removing any metal. Now that he has retired, no one seems to be able to duplicate his procedure with the correct amount of pressure.

As the ability to measure mass becomes more exact, the need for an accurate and easily reproduced standard becomes more urgent. Already, modern instruments have shown that the copies of the standard kilogram vary slightly in mass from the original cylinder in France. Many scientists feel that the time has come to find a natural replacement for the kilogram. The International Bureau of Weights and Measures may be choosing one soon.

The copy of the standard kilogram in the United States is kilogram 20, produced in 1884.

## SI Units Derived from Standards

Many properties cannot be described directly with one of the seven SI base units. For example, chemists often need to measure volume (the amount of space that something occupies), and volume is not on the list of SI base units (Table 1.1). Rather than create a new definition for volume, we derive its units from the base unit for length, the meter. Volume can be defined as length cubed, so cubic meters, $m^3$, can be used as a volume unit. Various other units are derived in similar ways.

OBJECTIVE 3

If you have long arms, a meter is approximately the distance from the tip of your nose to the end of your fingers when you are looking forward and extending your arm fully (Figure 1.2). A cubic meter is therefore a fairly large volume. In fact, it's inconveniently large for many uses. You'd find it awkward to buy your milk by the $m^3$. Scientists, too, need smaller volume units. Chemists prefer to use the liter as the base unit for volume. A liter (L) is 1/1000 (or $10^{-3}$) of a cubic meter, so there are 1000 (or $10^3$) liters per cubic meter.[1]

OBJECTIVE 3

OBJECTIVE 4

$$1\ \text{L} = 10^{-3}\ \text{m}^3 \qquad \text{or} \qquad 10^3\ \text{L} = 1\ \text{m}^3$$

**Figure 1.2**
Liters and Cubic Meters

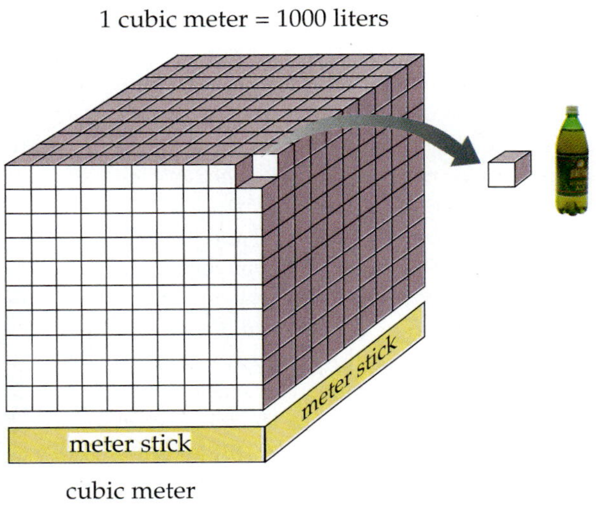

1 cubic meter = 1000 liters

meter stick

meter stick

cubic meter

## SI Units Derived from Metric Prefixes

Because the SI base units and derived units like the liter are not always a convenient size for making measurements of interest to scientists, a way of deriving new units that are larger and smaller has been developed. For example, the meter is too small to be convenient for describing the distance to the moon, and even the liter is too large for measuring the volume of a teardrop. Scientists therefore attach prefixes to the base units, which have the effect of multiplying or dividing the base unit by a power

[1]If you are unfamiliar with numbers like $10^{-3}$ and $10^3$ that are expressed using scientific notation, you should read Appendix 2 at the end of this text.

of 10. Table 1.2 lists some of the most common metric prefixes and their abbreviations.

The unit *kilometer*, for example, is composed of the prefix *kilo* and *meter*, the base unit for length. *Kilo* means $10^3$ (or 1000), so a kilometer is $10^3$ (or 1000) meters. A kilometer is therefore more appropriate for describing the average distance to the moon: 384,403 kilometers rather than 384,403,000 meters. Units derived from metric prefixes are abbreviated by combining the abbreviation for the prefix (Table 1.2) with the abbreviation for the unit to which the prefix is attached. The abbreviation for *kilo* is k and for *meter* is m, so the abbreviation for *kilometer* is km:

OBJECTIVE 6

OBJECTIVE 7

$$1 \text{ kilometer} = 10^3 \text{ meter} \quad \text{or} \quad 1 \text{ km} = 10^3 \text{ m}$$

Likewise, *micro* means $10^{-6}$ (or 0.000001 or 1/1,000,000), so a micrometer is $10^{-6}$ meter. The abbreviation for micrometer is μm. (The symbol μ is the Greek letter mu.) The micrometer can be used to describe the size of very small objects, such as the diameter of a typical human hair, which is 3 μm. This value is easier to report than 0.000003 m.

OBJECTIVE 6 OBJECTIVE 7

$$1 \text{ micrometer} = 10^{-6} \text{ meter} \quad \text{or} \quad 1\ \mu\text{m} = 10^{-6} \text{ m}$$

OBJECTIVE 5

Table 1.2
Some Common Metric Prefixes

| Prefixes for units larger than the base unit | | | Prefixes for units smaller than the base unit | | |
|---|---|---|---|---|---|
| **Prefix** | **Abbreviation** | **Definition** | **Prefix** | **Abbreviation** | **Definition** |
| giga | G | 1,000,000,000 or $10^9$ | centi | c | 0.01 or $10^{-2}$ |
| mega | M | 1,000,000 or $10^6$ | milli | m | 0.001 or $10^{-3}$ |
| kilo | k | 1000 or $10^3$ | micro | μ | 0.000001 or $10^{-6}$ |
| | | | nano | n | 0.000000001 or $10^{-9}$ |
| | | | pico | p | 0.000000000001 or $10^{-12}$ |

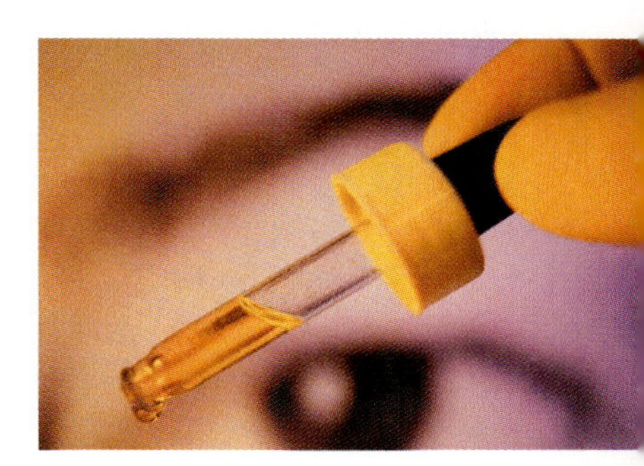

How many liters?

## EXAMPLE 1.1 Units Derived from Metric Prefixes

OBJECTIVE 6 OBJECTIVE 7

Complete the following relationships. Rewrite the relationships using abbreviations for the units.

a. 1 gigagram = ? gram

b. 1 centimeter = ? meter

c. 1 nanometer = ? meter

***Solution***
Refer to Table 1.2 until you have the prefixes memorized.

a. The prefix giga (G) means $10^9$, so

$$1 \text{ gigagram} = 10^9 \text{ gram} \quad \text{or} \quad 1 \text{ Gg} = 10^9 \text{ g}$$

b. The prefix centi (c) means $10^{-2}$, so

$$1 \text{ centimeter} = 10^{-2} \text{ meter} \quad \text{or} \quad 1 \text{ cm} = 10^{-2} \text{ m}$$

c. The prefix nano (n) means $10^{-9}$, so

$$1 \text{ nanometer} = 10^{-9} \text{ meter} \quad \text{or} \quad 1 \text{ nm} = 10^{-9} \text{ m}$$

## Exercise 1.1 Units Derived from Metric Prefixes

OBJECTIVE 6 OBJECTIVE 7

Complete the following relationships. Rewrite the relationships using abbreviations for the units.

a. 1 megagram = ? gram

b. 1 milliliter = ? liter

## More About Length Units

Although scientists rarely use the centuries-old English system of measurement, it is still commonly used in the United States to describe quantities in everyday life. Figure 1.3 may help you to learn the relationships between metric and English units.[2] A kilometer is a little more than

**Figure 1.3**
**English and Metric Length Units**

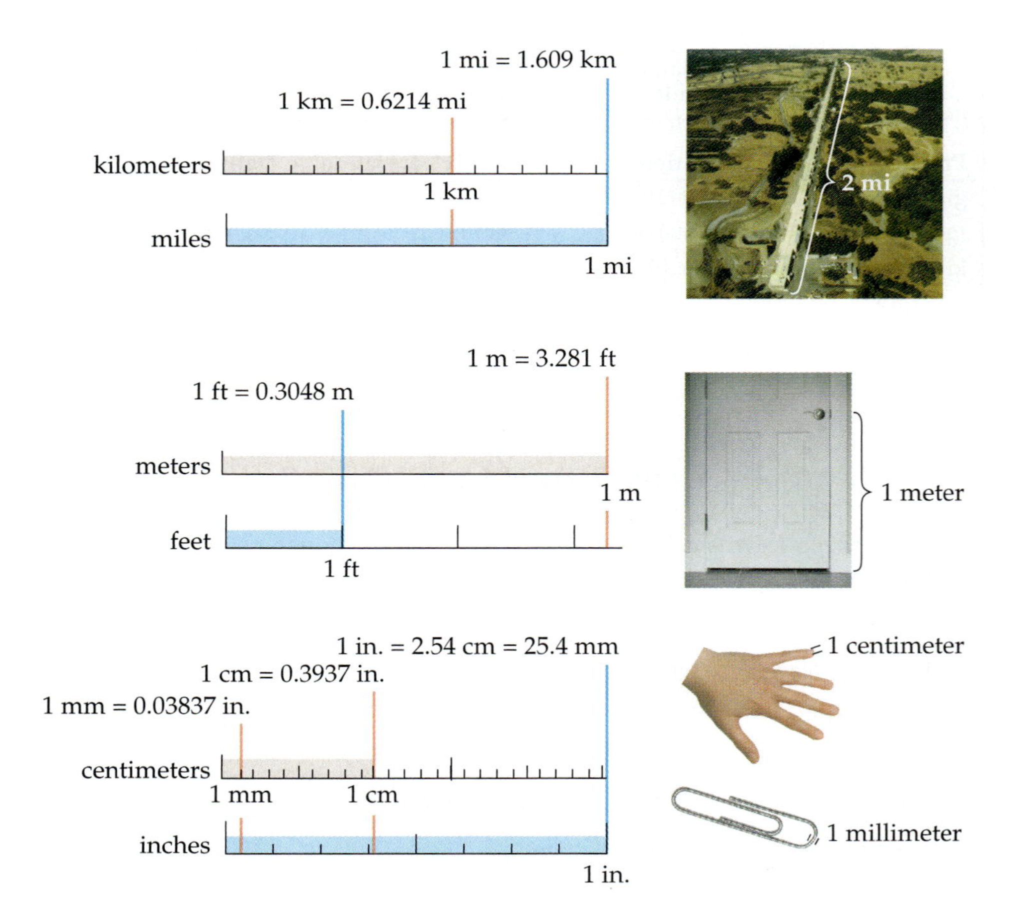


[2]The relationships between these units and many others are found in Appendix 1.

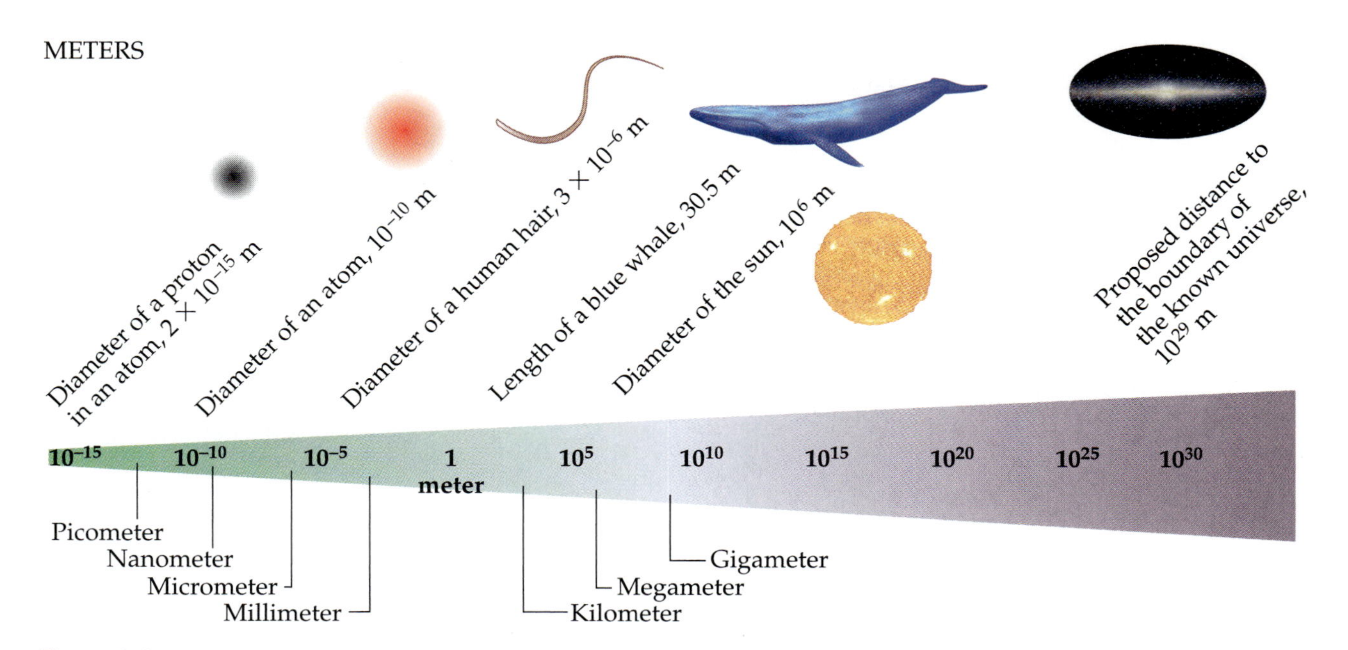


**Figure 1.4**
**The Range of Lengths**
(The relative sizes of these measurements cannot be shown on such a small page. The widening wedge is to remind you that each numbered measurement on the scale represents 100,000 times the magnitude of the preceding numbered measurement.)

½ mile. The distance between the floor and a typical doorknob is about 1 meter; the width of the fingernail on your little finger is probably about 1 centimeter; and the diameter of the wire used to make a typical paper clip is about 1 millimeter. Figure 1.4 shows the range of lengths.

OBJECTIVE 8

## More About Volume Units

The relationships shown in Figure 1.5 may help you to develop a sense of the sizes of typical metric units of volume. Another common volume unit is the cubic centimeter, $cm^3$, which is equivalent to a milliliter.

OBJECTIVE 10

$$1 \text{ cm}^3 = 1 \text{ mL}$$

**Figure 1.5**
**English and Metric Volume Units**

1 fl oz = 29.57 mL
1 mL = 0.03381 fl oz
1 gal = 3.785 L
1 qt = 0.9464 L
1 L = 1.057 qt = 0.2642 gal

1 fluid ounce (fl oz)
1 milliliter (mL)
1 gallon (gal) or 4 quarts (qt)
1 qt or 32 fl oz
1 liter (L) or 1000 mL

OBJECTIVE 9

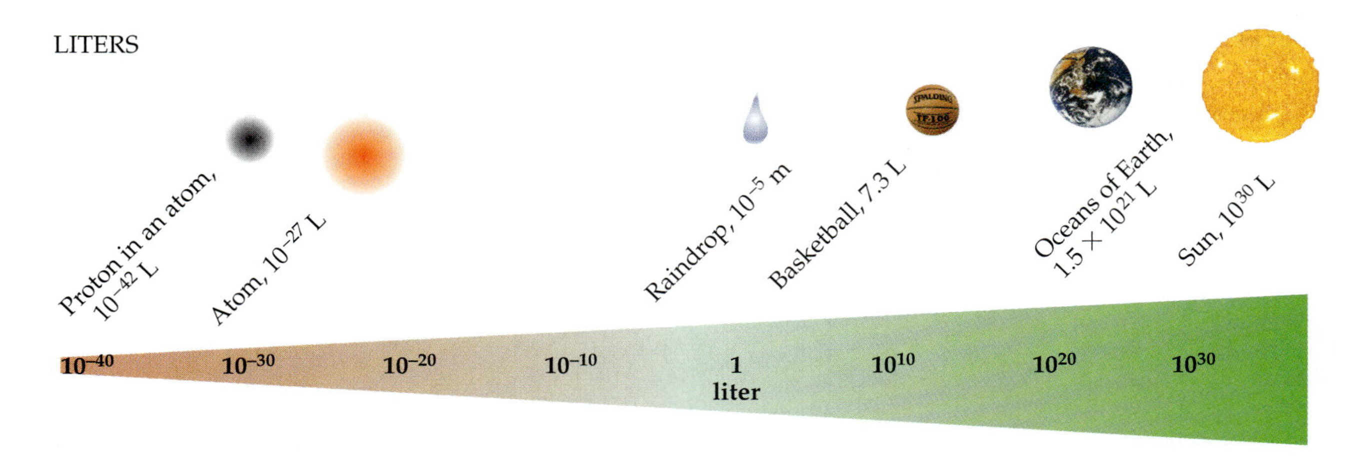


**Figure 1.6**
**The Range of Volumes**
(The relative sizes of these measurements cannot be shown on such a small page. The widening wedge is to remind you that each numbered measurement on the scale represents 10,000,000,000 times the magnitude of the preceding numbered measurement.)

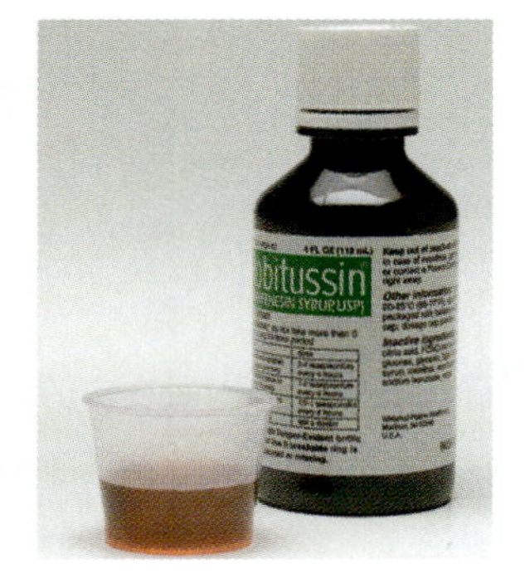

Children's cough syrup

A liter is slightly larger than a quart. There are 4.93 milliliters (or 4.93 cubic centimeters) in a teaspoon, so when the label on the bottle of a typical cough medicine suggests a dosage of one teaspoon, the volume given will be about 5 milliliters. Figure 1.6 shows the range of volumes.

## Mass and Weight

OBJECTIVE 11

OBJECTIVE 12

The terms *mass* and *weight* are often used as though they meant the same thing. The two properties are indeed related, but they are not identical. **Mass** is usually defined as a measure of the amount of matter in an object (**matter** is defined as anything that has mass and takes up space). In contrast, the **weight** of an object is a measure of the force of gravitational attraction between it and a significantly large body, such as the earth or the moon. An object's weight on the surface of the earth depends on its mass and on the distance between it and the center of the earth. In fact, *mass* can be defined as the property of matter that leads to gravitational attractions between objects and therefore gives rise to weight. As an object's mass increases, its weight increases correspondingly. However, as the distance between an object and the earth increases, the object's weight decreases whereas the amount of matter it contains, and therefore its mass, stays the same.

In the SI system, units such as gram, kilogram, and milligram are used to describe mass. People tend to use the terms *mass* and *weight* interchangeably and to describe weight with mass units too. However, because weight is actually a measure of the *force* of gravitational attraction for a body, it really should be described with force units. The accepted SI force unit is the newton, N. If your mass is 65 kg, your weight on the surface of the earth—the force with which you and the earth attract each other—is

637 N. The chemist is not generally concerned with the weight of objects, so neither weight nor its unit will be mentioned in the remaining chapters of this book.

Figure 1.7 may help you to develop a sense of the sizes of typical metric units of mass and Figure 1.8 shows the range of masses.

**Figure 1.7**
**English and Metric Mass Units**

OBJECTIVE 13 OBJECTIVE 14

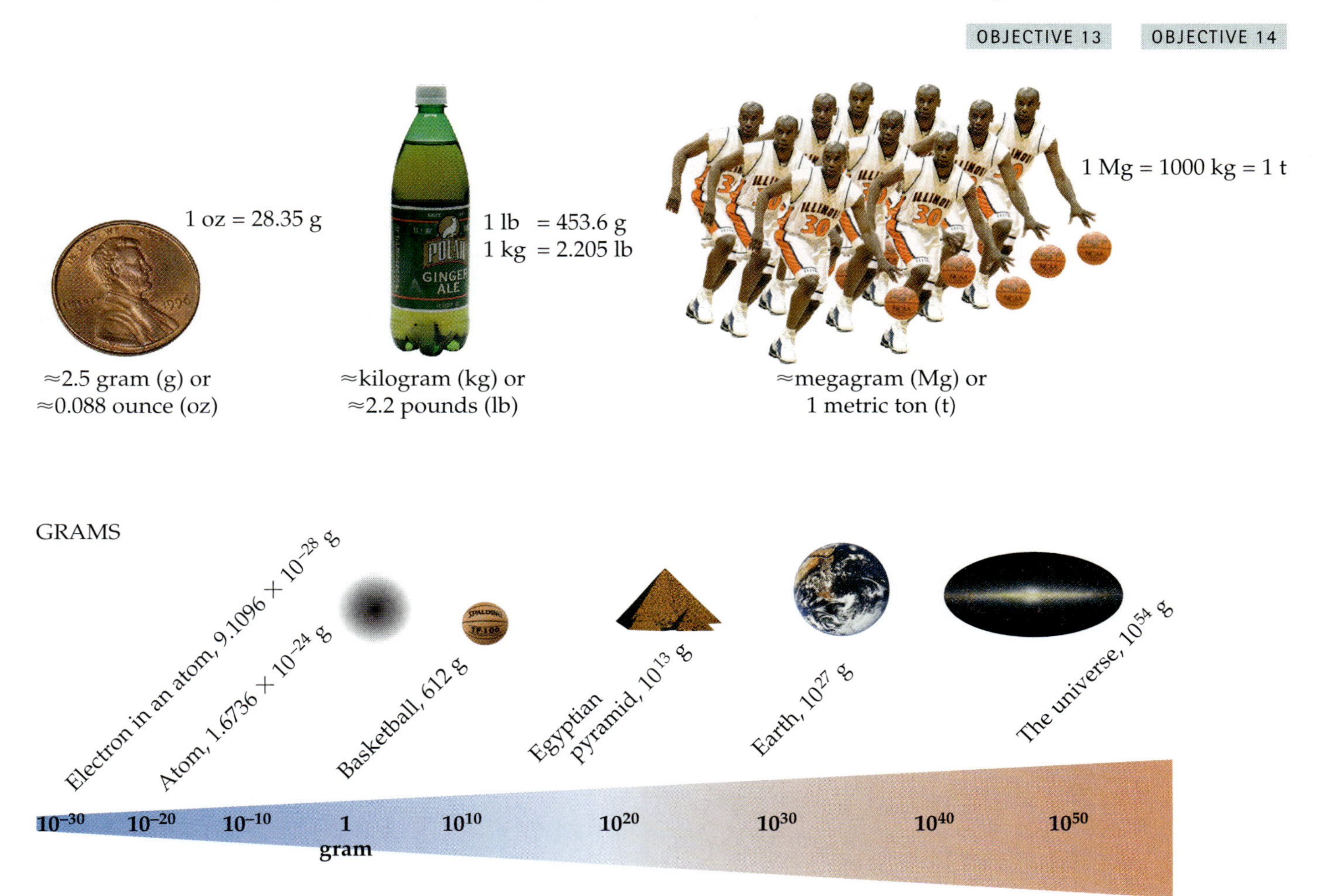


**Figure 1.8**
**The Range of Masses**
(The relative sizes of these measurements cannot be shown on such a small page. The widening wedge is to remind you that each numbered measurement on the scale represents 10,000,000,000 times the magnitude of the preceding numbered measurement.)

## Temperature

It may surprise you to learn that temperature is actually a measure of the average motion of the particles in a system. Instead of saying that the temperature is "higher" today than yesterday, you could say that the particles in the air are moving faster. By the time we return to a description of temperature measurements and temperature calculations, in Chapters 7 and 8, you will be much better prepared to understand the physical changes that take place when the temperature of an object changes, as well as how we measure these changes. For now, it is enough to have some understanding of the relationships between the three common temperature scales: Celsius, Fahrenheit, and Kelvin.

OBJECTIVE 15

For the Celsius scale, the temperature at which water freezes is defined as 0 °C, and the temperature at which water boils is defined as 100 °C. Thus a degree Celsius, °C, is 1/100 of the temperature difference between freezing and boiling water (Figure 1.9). For the Fahrenheit scale, which is still commonly used in the United States, the temperature at which water freezes is defined as 32 °F, and the temperature at which water boils is defined as 212 °F. There are 180 °F between freezing and boiling water (212 − 32 = 180), so a degree Fahrenheit, °F, is 1/180 of the temperature difference between freezing and boiling water (Figure 1.9). Note that there are 100 °C between freezing and boiling water but that there are 180 °F for the same temperature difference. Thus a degree Fahrenheit is smaller than a degree Celsius. There are 180 °F per 100 °C, or 1.8 °F per 1 °C.

OBJECTIVE 16

**Figure 1.9**
**Celsius and Fahrenheit Thermometers**

OBJECTIVE 15

OBJECTIVE 16

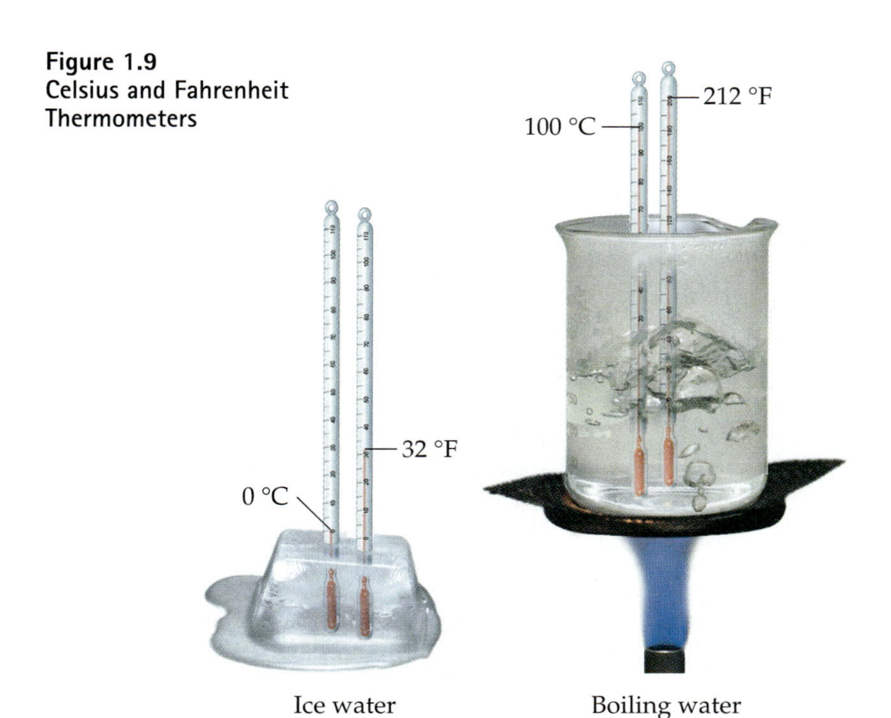


The thermometers that scientists use to measure temperature generally provide readings in degrees Celsius, but scientists usually convert Celsius values into Kelvin values to do calculations. The unit of measurement in the Kelvin scale is called the kelvin, K (not "degree kelvin," just "kelvin").

OBJECTIVE 15

The value 0 K is defined as **absolute zero**, the lowest possible temperature. Because the temperature of an object is a measure of the average motion of its particles, as the motion of the particles decreases, the temperature of the object decreases. Absolute zero is the point beyond which the motion of the particles, and therefore the temperature, cannot be decreased. Absolute zero is 0 K, −273.15 °C, and −459.67 °F. Because the zero point for the Kelvin scale is absolute zero, all Kelvin temperatures are positive. The kelvin is defined so that it is equal in size to a degree Celsius. Figure 1.10 summarizes the relationships among the three common temperature scales. In Chapter 8 we will use these relationships to convert measurements from one temperature scale to another.

OBJECTIVE 16

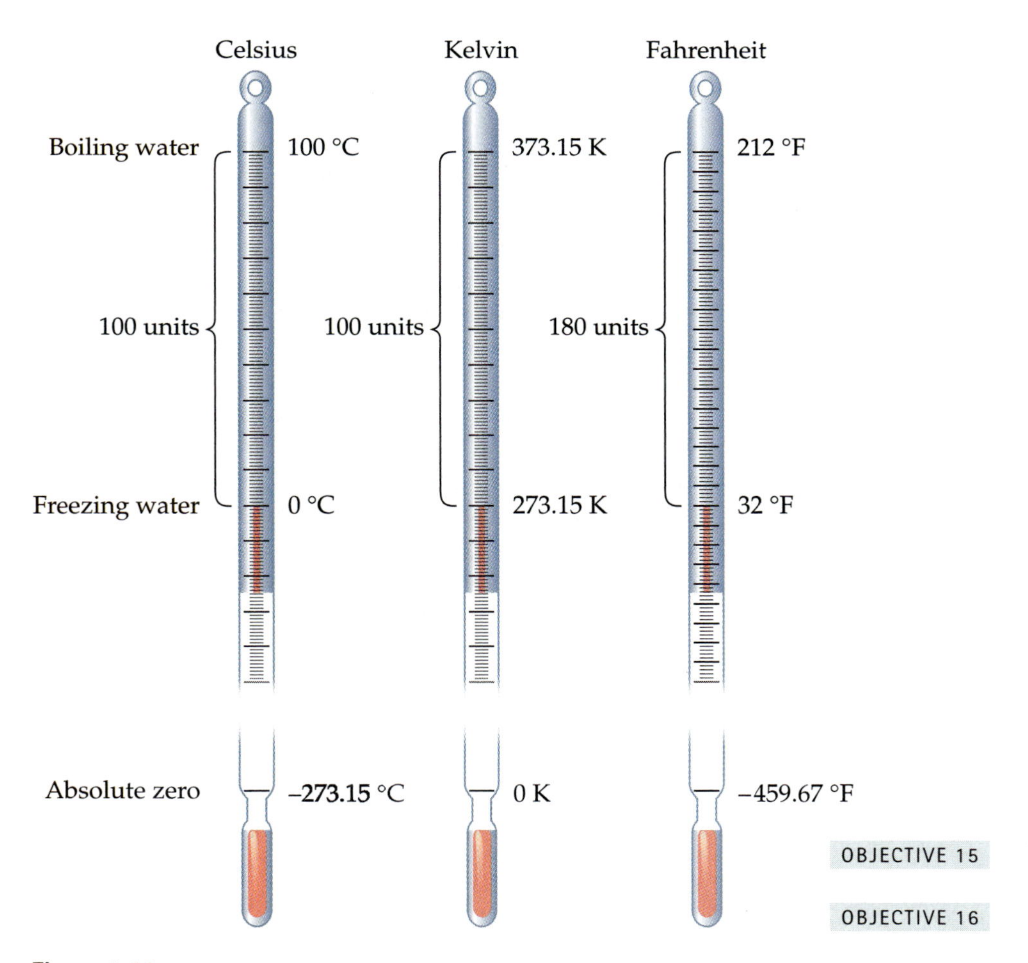


OBJECTIVE 15

OBJECTIVE 16

**Figure 1.10**
**Comparing Temperature Scales**

The highest temperatures in the universe are thought to be inside some stars, where theory predicts temperatures of about $10^9$ K (a billion kelvins). Probably the hottest thing in your home is the tungsten filament in an incandescent light bulb, with a temperature of about 2800 K. Molten lava is about 2000 K. Normal body temperature is 310.2 K (98.6 °F or 37.0 °C). Normal room temperature is about 20 °C (68 °F or 293 K). The lowest temperature achieved in the laboratory is about $2 \times 10^{-8}$ K (0.00000002 K). Figure 1.11 shows the range of temperatures.

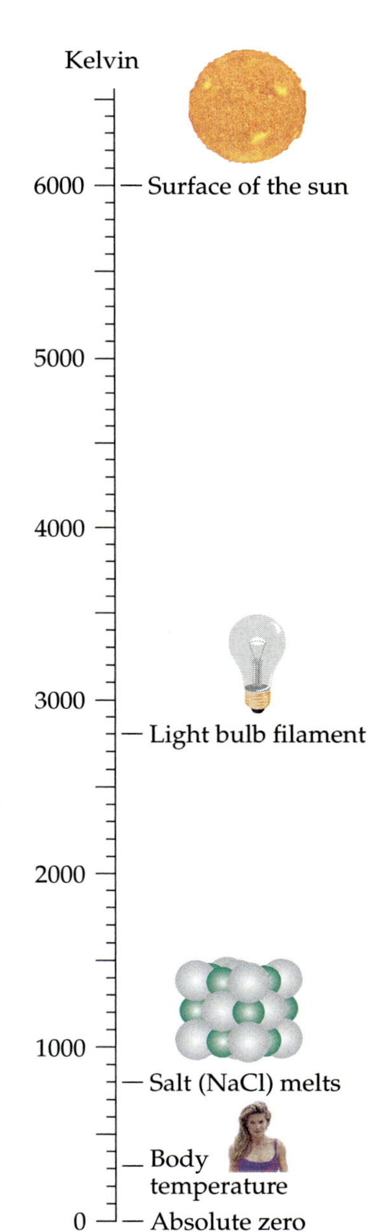


**Figure 1.11**
**The Range of Temperatures**

# 1.5 Reporting Values from Measurements

All measurements are uncertain to some degree. Scientists are very careful to report the values of measurements in a way that not only shows the measurement's magnitude but also reflects its degree of uncertainty. Because the uncertainty of a measurement is related to both the precision and the accuracy of the measuring instrument, we need to begin by explaining these terms.

## Accuracy and Precision

**Precision** describes how closely a series of measurements of the same object resemble each other. The closer the measurements are to each other, the more precise they are. The precision of a measurement is not necessarily equal to its accuracy. **Accuracy** describes how closely a measured value approaches the true value of the property.

To get a better understanding of these terms, let's imagine a penny that has a true mass of 2.525 g. If the penny is weighed by five different students, each using the same balance, the masses they report are likely to be slightly different. For example, the reported masses might be 2.680 g, 2.681 g, 2.680 g, 2.679 g, and 2.680 g. Because the range of values is ±0.001 g of 2.680 g, the precision of the balance used to weigh them is said to be ±0.001 g, but notice that none of the measured values is very accurate. Although the precision of our measurement is ±0.001 g, our measurement is inaccurate by 0.155 g (2.680 − 2.525 = 0.155). Scientists recognize that even very precise measurements are not necessarily accurate. Figure 1.12 provides another example of the difference between accuracy and precision.

OBJECTIVE 17

**Figure 1.12**
Precision and Accuracy

This archer is precise but not accurate.

This archer is precise and accurate.

This archer is imprecise and inaccurate.

## Describing Measurements

Certain standard practices and conventions make the taking and reporting of measurements more consistent. One of the conventions that scientists use for reporting measurements is to report all of the certain digits and one estimated (and thus uncertain) digit.

Consider, for example, how you might measure the volume of water shown in the graduated cylinder in Figure 1.13. (Liquids often climb a short distance up the walls of a glass container, so the surface of a liquid in a graduated cylinder is usually slightly curved. When you look from the side of the cylinder, this concave surface looks like a bubble. The surface is called a **meniscus.** Scientists follow the convention of using the bottom of the meniscus for their reading.) The graduated cylinder in Figure 1.13 has rings corresponding to milliliter values and smaller divisions corresponding to increments of 0.1 mL. When we use these marks to read the volume of the liquid shown in Figure 1.13, we are certain that the volume is between 8.7 mL and 8.8 mL. By imagining that the smallest divisions are divided into 10 equal parts, we can estimate the hundredths position. Because the bottom of the meniscus seems to be about four-tenths of the distance between 8.7 mL and 8.8 mL, we report the value as 8.74 mL. Because we are somewhat uncertain about our estimation of the hundredths position, our value of 8.74 mL represents all of the certain digits and one uncertain digit.

OBJECTIVE 18

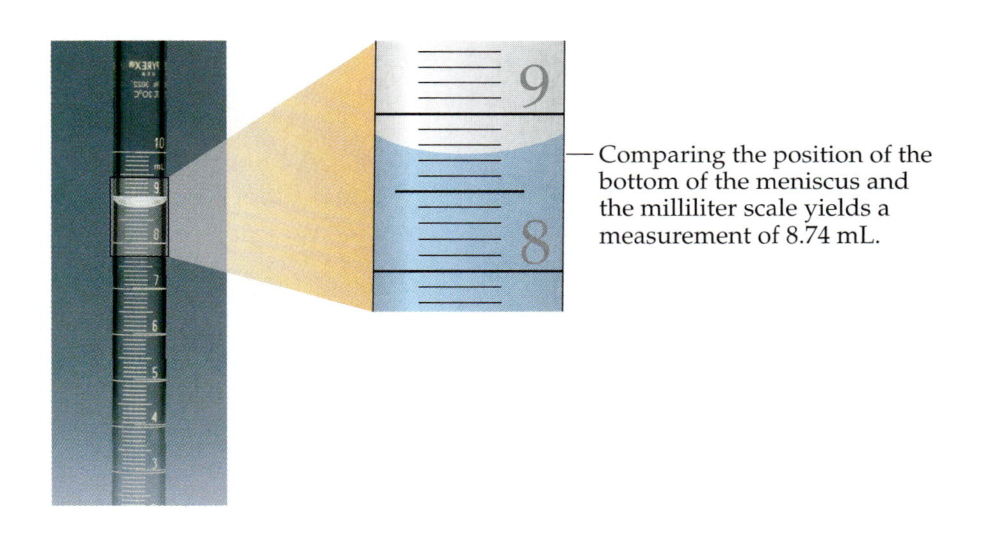

9
8
Comparing the position of the bottom of the meniscus and the milliliter scale yields a measurement of 8.74 mL.


**Figure 1.13**
**Measuring Volume with a Graduated Cylinder**

Scientists agree to assume that the number in the last reported decimal place has an uncertainty of $\pm 1$ unless stated otherwise. Example 1.2 shows how this assumption is applied.

OBJECTIVE 19

### EXAMPLE 1.2 Uncertainty

If you are given the following values that are derived from measurements, what will you assume is the range of possible values that each represents?

OBJECTIVE 19

a. 5.4 mL

b. 64 cm

c. $2.34 \times 10^3$ kg

***Solution***

We assume an uncertainty of ±1 in the last decimal place reported.

a. 5.4 mL means 5.4 ± 0.1 mL, or 5.3 to 5.5 mL.

b. 64 cm means 64 ± 1 cm, or 63 to 65 cm.

c. $2.34 \times 10^3$ kg means $(2.34 \pm 0.01) \times 10^3$ kg, or $2.33 \times 10^3$ kg to $2.35 \times 10^3$ kg.

## Exercise 1.2 Uncertainty

OBJECTIVE 19

If you are given the following values that are derived from measurements, what will you assume is the range of possible values that each represents?

a. 72 mL

b. 8.23 m

c. $4.55 \times 10^{-5}$ g

Sometimes it is necessary to use trailing zeros to show the uncertainty of a measurement. If the top of the liquid in the graduated cylinder shown in Figure 1.14 is right at the 8-mL mark, you report the measurement as 8.00 mL to indicate that the uncertainty is in the second decimal place. Someone reading 8.00 mL would recognize that the measured amount was between 7.99 mL and 8.01 mL. Reporting 8 mL would suggest an uncertainty of ±1 mL, in which case the amount would be assumed to lie anywhere between 7 mL and 9 mL.

9
8
We report 8.00 mL to show an uncertainty of ±0.1 mL.

**Figure 1.14**
**Reporting Values with Trailing Zeros**

OBJECTIVE 18

There are other ways to decide how to report values from measurements. For example, the manufacturer of the graduated cylinder in Figure 1.14 might inform you that the lines have been drawn with an accuracy of ±0.1 mL. Therefore, your uncertainty when measuring with this graduated cylinder will always be at least ±0.1 mL, and the values you report should be to the tenth position. In this case, the volumes in Figures 1.13 and 1.14 would be reported as 8.7 mL and 8.0 mL.

In general, though, unless you are told to do otherwise, the conventional practice in using instruments such as a graduated cylinder is to report all of your certain digits and one estimated digit. For example, if you were asked to measure the volumes shown in Figures 1.13 and 1.14, you would report 8.74 mL and 8.00 mL, respectively, unless you were told to report your answer to the tenth place.

OBJECTIVE 18

### EXAMPLE 1.3 Uncertainty in Measurement

OBJECTIVE 18

Consider a laboratory situation in which five students are asked to measure with a meter stick the length of a piece of tape on a lab bench. The values reported are 61.94 cm, 62.01 cm, 62.12 cm, 61.98 cm, and 62.10 cm. The average of these values is 62.03 cm. How would you report the average measurement so as to communicate the uncertainty of the value?

***Solution***

If we compare the original measured values, we see that they vary in both the tenths and the hundredths decimal places. Because we report only one uncertain decimal place, we report our answer as **62.0 cm.** The final zero must be reported to show that we are uncertain by $\pm 0.1$ cm. (Considering the uncertainties that arise from the difficulty in aligning the meter stick with the end of the tape and the difficulty estimating between the lines for the very tiny 0.1-cm divisions, it is reasonable to assume that our uncertainty is no better than $\pm 0.1$ cm.)

### EXERCISE 1.3 Uncertainty in Measurement

OBJECTIVE 18

Let's assume that four members of your class are asked to measure the mass of a dime. The reported values are 2.302 g, 2.294 g, 2.312 g, and 2.296 g. The average of these values is 2.301 g. Considering the values reported and the level of care you expect beginning chemistry students to take with their measurements, how would you report the mass so as to communicate the uncertainty of the measurement?

**Figure 1.15**
The Display of a Typical Electronic Balance

## Digital Readouts

The electronic balances found in most scientific laboratories have a digital readout that reports the mass of objects to many decimal positions. For example, Figure 1.15 shows the display of a typical electronic balance that reports values to a ten-thousandth of a gram, or a tenth of a milligram. As you become more experienced, you will realize that you are not always justified in reporting a measurement to the number of positions shown on a digital readout. For the purposes of this course, however, if you are asked to make a measurement with an instrument that has a digital readout, you should report all of the digits on the display unless instructed to do otherwise.

OBJECTIVE 18

## Chapter Glossary

**Chemistry** The study of the structure and behavior of matter.

**Value** A number and unit that together represent the result of a measurement or calculation. The distance of a particular race, for example, may be reported as a value of 100 meters.

**Unit** A defined quantity based on a standard. For example, in the value 100 meters, *meter* is the unit.

**Base units** The seven units from which all other units in the SI system of measurement are derived.

**Mass** The amount of matter in an object. *Mass* can also be defined as the property of matter that leads to gravitational attractions between objects and therefore gives rise to weight.

**Matter** Anything that has mass and occupies space.

**Weight** A measure of the force of gravitational attraction between an object and a significantly large body, such as the earth or the moon.

**Absolute zero** Zero kelvins (0 K), the lowest possible temperature, equivalent to −273.15 °C. It is the point beyond which motion can no longer be decreased.

**Precision** The closeness in value of a series of measurements of the same entity. The closer the values of the measurements, the more precise they are.

**Accuracy** How closely a measured value approaches the true value of a property.

You can test yourself on the glossary terms at the following Web site:
**www.chemplace.com/college/**

## Chapter Objectives

Notations in the text margins show where each objective is addressed in the chapter. If you do not understand an objective, or if you do not know how to meet it, find the objective number in the chapter body and reread the corresponding segment of text.

**The goal of this chapter is to teach you to do the following.**

1. Define all of the terms in the Chapter Glossary.

**Section 1.3 The Scientific Method**

2. Describe how science in general is done.

**Section 1.4 Measurement and Units**

3. Use the International System of Measurements (SI) base units and their abbreviations to describe length, mass, time, temperature, and volume.
4. Describe the relationship between liters and cubic meters.
5. State the numbers or fractions represented by the following metric prefixes, and write their abbreviations: giga, mega, kilo, centi, milli, micro, nano, and pico. (For example, kilo is $10^3$ and is represented by k.)
6. Describe the relationships between the metric units that do not have prefixes (such as meter, gram, and liter) and units derived from them by the addition of prefixes. (For example, 1 km $= 10^3$ m.)
7. Given a metric unit, write its abbreviation; given an abbreviation, write the full name of the unit. (For example, the abbreviation for *milligram* is mg.)
8. Use everyday examples to describe the approximate size of a millimeter, a centimeter, a meter, and a kilometer.
9. Use everyday examples to describe the approximate size of a milliliter, a liter, and a cubic meter.
10. Describe the relationship between cubic centimeters and milliliters.
11. Explain the relationship between mass and weight.
12. Name the two factors that cause the weight of an object to change.
13. Use everyday examples to describe the approximate size of a gram, a kilogram, and a megagram.
14. Describe the relationships between metric tons, kilograms, and megagrams.
15. Describe the Celsius, Fahrenheit, and Kelvin temperature scales.
16. Describe how a degree Celsius, a degree Fahrenheit, and a kelvin are related.

**Section 1.5 Reporting Values from Measurements**

17. Given values for a series of measurements, state the precision of the measurements.
18. Report measured values so as to show their degree of uncertainty.
19. Given a value derived from a measurement, identify the range of possible values that it represents, assuming that its uncertainty is $\pm1$ in the last position reported. (For example, 8.0 mL says that the value could be from 7.9 mL to 8.1 mL.)

## Key Ideas

**Complete the following statements by writing one of these words or phrases in each blank.**

| | |
|---|---|
| 0 °C | kg |
| 100 °C | magnitude |
| 32 °F | m |
| 212 °F | mass |
| accuracy | meter |
| applications | observation |
| base units | precision |
| Celsius | prefixes |
| certain | published |
| data | research |
| derive | research or experimentation |
| distance | second |
| estimated | standard |
| hypothesis | temperature |
| hypothesizing and testing | time |
| K | uncertain |
| kelvin | uncertainty |
| kilogram | |

1. Complete this brief description of common steps in the development of scientific ideas: The process begins with ________________ and the collection of ________________. Next, scientists make an initial ________________. This leads to a more purposeful collection of information in the form of systematic ________________. The hypothesis is refined on the basis of the new information, and ________________ is designed to test the hypothesis. The results are ________________ so that other scientists might repeat the research and confirm or refute the conclusions. If other scientists confirm the results, the hypothesis becomes accepted in the scientific community. The next step of this scientific method is a search for useful ________________ of the new ideas. This often leads to another round of ________________ in order to refine the applications.
2. In the past, as measuring techniques became more precise and the demand for accuracy increased, the ________________ on which people based their units were improved.
3. The ________________, which has an abbreviation of ________________, is the accepted SI base unit for length.
4. The ________________, which has an abbreviation of ________________, is the accepted SI base unit for mass.
5. The ________________, which has an abbreviation of s, is the accepted SI base unit for ________________.
6. The kelvin, which has an abbreviation of ________________, is the accepted SI base unit for ________________.
7. Many properties cannot be described directly with one of the seven SI ________________. Rather than create new definitions for new units, we ________________ units from the units of meter, kilogram, second, kelvin, mole, ampere, and candela.
8. Because the SI base units (such as the meter) and derived units (such as the liter) are not always a convenient size for making measurements of interest to scientists, a way of deriving new units that are larger and smaller has been developed. Scientists attach ________________ to the base units, which have the effect of multiplying or dividing the base unit by a power of 10.

9. An object's weight on the surface of the earth depends on its ____________ and on the ____________ between it and the center of the earth.

10. For the Celsius scale, the temperature at which water freezes is defined as ____________, and the temperature at which water boils is defined as ____________. For the Fahrenheit scale, which is still commonly used in the United States, the temperature at which water freezes is defined as ____________, and the temperature at which water boils is defined as ____________.

11. The thermometers that scientists use to measure temperature generally provide readings in degrees ____________, but scientists usually convert these values into ____________ values to do calculations.

12. All measurements are ____________ to some degree. Scientists are very careful to report the values of measurements in a way that not only shows the measurement's ____________ but also reflects its degree of ____________. The uncertainty of a measurement is related to both the ____________ and the ____________ of the measuring instrument.

13. One of the conventions that scientists use for reporting measurements is to report all of the ____________ digits and one ____________ (and thus uncertain) digit.

## Chapter Problems

### Section 1.3 The Scientific Method

OBJECTIVE 2

14. Describe how science in general is done.

### Section 1.4 Measurement and Units

OBJECTIVE 3

OBJECTIVE 7

15. Complete the following table by writing the property being measured (mass, length, volume, or temperature) and either the name of the unit or its abbreviation.

| Unit | Type of measurement | Abbreviation | Unit | Type of measurement | Abbreviation |
|---|---|---|---|---|---|
| megagram | | | nanometer | | |
| | | mL | | | K |

OBJECTIVE 3

OBJECTIVE 7

16. Complete the following table by writing the property being measured (mass, length, volume, or temperature), and either the name of the unit or its abbreviation.

| Unit | Type of measurement | Abbreviation | Unit | Type of measurement | Abbreviation |
|---|---|---|---|---|---|
| | | GL | kilogram | | |
| micrometer | | | | | °C |

17. Convert the following ordinary numbers to scientific notation. (See Appendix B at the end of this text if you need help with this.)
    a. 1,000
    b. 1,000,000,000
    c. 0.001
    d. 0.000000001
18. Convert the following ordinary numbers to scientific notation. (See Appendix B at the end of this text if you need help with this.)
    a. 10,000
    b. 100
    c. 0.0001
    d. 0.01
19. Convert the following numbers expressed in scientific notation to ordinary numbers. (See Appendix B at the end of this text if you need help with this.)
    a. $10^7$
    b. $10^{12}$
    c. $10^{-7}$
    d. $10^{-12}$
20. Convert the following numbers expressed in scientific notation to ordinary numbers. (See Appendix B at the end of this text if you need help with this.)
    a. $10^5$
    b. $10^6$
    c. $10^{-5}$
    d. $10^{-6}$

OBJECTIVE 6
OBJECTIVE 10
OBJECTIVE 14

21. Complete the following relationships between units.
    a. _____ m = 1 km
    b. _____ L = 1 mL
    c. _____ g = 1 Mg
    d. _____ $cm^3$ = 1 mL
    e. _____ kg = 1 t (t = metric ton)

OBJECTIVE 4
OBJECTIVE 6
OBJECTIVE 14

22. Complete the following relationships between units.
    a. _____ g = 1 Gg
    b. _____ L = 1 μL
    c. _____ m = 1 nm
    d. _____ L = $m^3$
    e. _____ Mg = 1 t (t = metric ton)

OBJECTIVE 8

23. Would each of the following distances be closest to a millimeter, a centimeter, a meter, or a kilometer?
    a. the width of a bookcase
    b. the length of an ant
    c. the width of the letter "t" in this phrase
    d. the length of the Golden Gate Bridge in San Francisco
24. Which is larger, a kilometer or a mile?
25. Which is larger, a centimeter or an inch?
26. Which is larger, a meter or a yard?

OBJECTIVE 9

27. Would the volume of each of the following be closest to a milliliter, a liter, or a cubic meter?
    a. a vitamin tablet
    b. a kitchen stove and oven
    c. this book
28. Which is larger, a liter or a quart?
29. Which is larger, a milliliter or a fluid ounce?

30. Would the mass of each of the following be closest to a gram, a kilogram, or a metric ton? OBJECTIVE 13
    a. a Volkswagen Beetle
    b. a Texas-style steak dinner
    c. a pinto bean
31. Explain the difference between mass and weight. OBJECTIVE 11
32. Which is larger, a gram or an ounce?
33. Which is larger, a kilogram or a pound?
34. Which is larger, a metric ton or an English short ton? (There are 2000 pounds per English short ton.)
35. On July 4, 1997, after a 7-month trip, the *Pathfinder* spacecraft landed on Mars and released a small robot rover called *Sojourner.* The weight of an object on Mars is about 38% of the weight of the same object on Earth. OBJECTIVE 12
    a. Explain why the weight of an object is less on Mars than on Earth.
    b. Describe the changes (if any) in the mass *and* in the weight of the rover *Sojourner* as the *Pathfinder* spacecraft moved from Earth to the surface of Mars. Explain your answer.
36. Which is larger, a degree Celsius or a degree Fahrenheit? OBJECTIVE 16
37. How does a degree Celsius compare to a kelvin? OBJECTIVE 16
38. Which is the smallest increase in temperature: 10 °C (such as from 100 °C to 110 °C), 10 K (such as from 100 K to 110 K), or 10 °F (such as from 100 °F to 110 °F)? OBJECTIVE 16
39. Is the temperature around you now closer to 100 K, 200 K, or 300 K?

### Section 1.5 Reporting Values from Measurements

40. You find an old bathroom scale at a garage sale on your way home from getting a physical exam from your doctor. You step on the scale, and it reads 135 lb. You step off and step back on, and it reads 134 lb. You do this three more times and get readings of 135 lb, 136 lb, and 135 lb.
    a. What is the precision of this old bathroom scale? Would you consider this adequate precision for the type of measurement you are making? OBJECTIVE 17
    b. The much more carefully constructed and better-maintained scale at the doctor's office reads 126 lb. Assuming that you are wearing the same clothes that you wore when the doctor weighed you, do you think the accuracy of the old bathroom scale is high or low?
41. Given the following values that are derived from measurements, what do you assume is the range of possible values that each represents? OBJECTIVE 19
    a. 30.5 m (the length of a whale)
    b. 612 g (the mass of a basketball)
    c. 1.98 m (Michael Jordan's height)
    d. $9.1096 \times 10^{-28}$ g (the mass of an electron)
    e. $1.5 \times 10^{18}$ $m^3$ (the volume of the ocean)

OBJECTIVE 19

42. If you are given the following values that come directly or indirectly from measurements, what will you assume is the range of possible values that they represent?

a. 7.3 L (the volume of a basketball)

b. 3 μm (the diameter of a hair)

c. $2.0 \times 10^{-6}$ nm (the diameter of a proton)

d. $1.98 \times 10^{5}$ kg (the mass of a whale)

OBJECTIVE 18

43. The accompanying drawings show portions of metric rulers on which the numbers correspond to centimeters. The dark bars represent the ends of objects being measured.

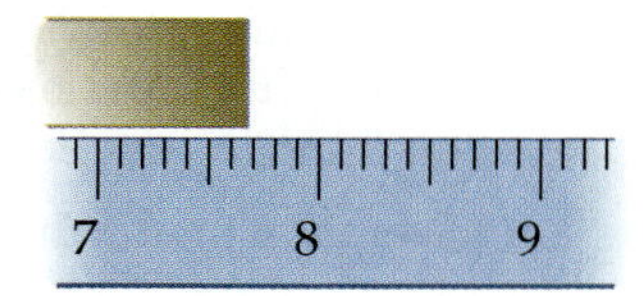


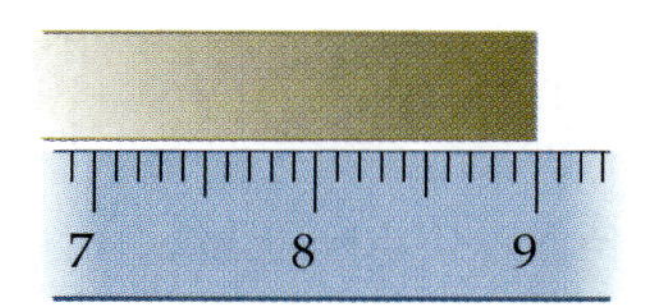


a. If you were not given any specific instructions for reporting your values, what length would you record for each of these measurements?

b. If you were told that the lines on the ruler are drawn accurately to ±0.1 cm, how would you report these two lengths?

OBJECTIVE 18

44. These images show a typical mercury-filled thermometer at two different temperatures. The units are degrees Celsius.

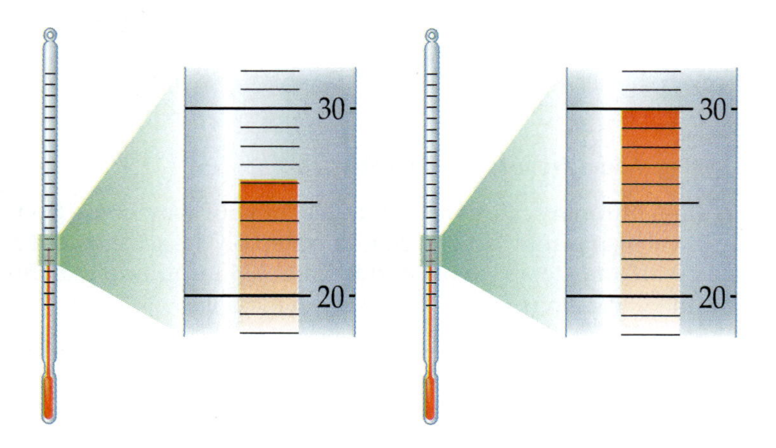


a. If you were not given any specific instructions for reporting your values, what temperature would you record for each of these readings?

b. If you were told that the lines on the thermometer are drawn accurately to ±1 °C, how would you report these two temperatures?

OBJECTIVE 18

45. At a track meet, three different timers report the times for the winner of a 100-m sprint as 10.51 s, 10.32 s, and 10.43 s. The average is 10.42 s. How would you report the time of the sprinter in a way that reflects the uncertainty of the measurements?

OBJECTIVE 18

46. Suppose that five students read a thermometer and reported temperatures of 86.6 °C, 86.8 °C, 86.6 °C, 86.8 °C, and 87.0 °C. The average of these values is 86.8 °C. How would you report this average to reflect its uncertainty?

47. The image below represents the digital display on a typical electronic balance.

OBJECTIVE 19

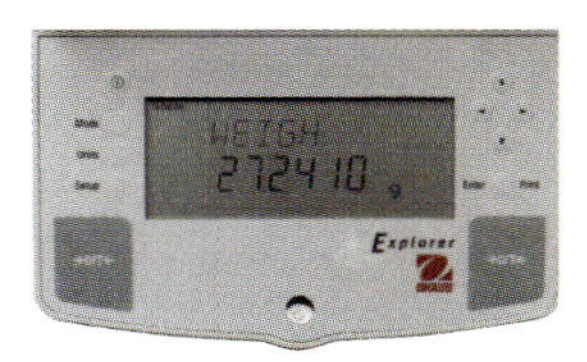


a. If the reading represents the mass of a solid object that you carefully cleaned and dried and then handled without contaminating it, how would you report this mass?

b. Now assume that the reading is for a more casually handled sample of a liquid and its container. Let's assume not only that you were less careful with your procedure this time but also that the liquid is evaporating rapidly enough for the reading to be continually decreasing. In the amount of time that the container of liquid has been sitting on the pan of the balance, the mass reading has decreased by about 0.001 g. How would you report the mass?

### Discussion Topic

48. Develop your own system of measurement for length, mass, and volume based on the objects in your immediate surroundings. Your system should have clearly defined units and abbreviations.

a. What unit could you use to measure the length of a desk? the distance from here to the moon? the diameter of an atom? What abbreviations could you use for each of these units?

b. What unit could you use to measure the mass of this book? the mass of Godzilla? the mass of a cat's whisker? What abbreviations could you use for each of these units?

c. What unit could you use to measure the volume of the water that a bathtub holds? the volume of the planet Jupiter? the volume of a pencil? What abbreviations could you use for each of these units?

INTER-CHAPTER 1A

# A Brief Introduction to Unit Conversions



THIS "INTER-CHAPTER" AND CHAPTER 8 BOTH DESCRIBE THE PROCEDURES for doing a certain kind of calculation that is very useful in chemistry, but the inter-chapter does so more briefly. That being the case, perhaps you are wondering why the inter-chapter is here at all. It is here because there are many ways to teach introductory chemistry. This text has been written by an author who in his own course begins the coverage of chemical principles right away and postpones most of the math-related topics until mid-semester. Although many chemistry instructors share this preference, many other instructors believe in covering the math-related topics early in their courses. There are good reasons for each approach. The flexible organization of this text makes it possible for all instructors to cover the course topics in the way that they believe is best.

This inter-chapter supports several possible approaches to learning chemistry's math-related topics. (1) Some instructors choose not to discuss unit conversions until the middle of the course, preferring to present various basic chemical concepts first. These teachers will tell you not to read this inter-chapter. (2) Some instructors provide a brief introduction to unit conversions early in the course, perhaps to prepare you to make simple unit conversions in early laboratory sessions, while postponing the more comprehensive treatment of the subject until later. If your instructor falls into this category, you will read the chapters of this text, including this inter-chapter, in the order in which you find them. (3) If your instructor wants a broader coverage of the math-related topics early in the course, you will be asked to skip from here to Chapter 8, which describes unit conversions in much more detail than this inter-chapter does. Chapter 8 was written in such a way that you can read it without confusion before reading Chapters 2 through 7. (4) Some instructors will ask their students to read this inter-chapter and only one or two of the sections in Chapter 8 before proceeding to Chapter 2. Be sure to check with your instructor to find out which of these options is appropriate for your course.

## Review Skills

The presentation of information in this chapter assumes that you can already perform the tasks listed below. You can test your readiness to proceed by answering the Review Questions at the end of the chapter. This might also be a good time to read the Inter-Chapter 1A Objectives.

- List the metric base units and the corresponding abbreviations for length, mass, volume, energy, and gas pressure. (Section 1.4)
- State the numbers or fractions represented by the following metric prefixes, and write their abbreviations: giga, mega, kilo, centi, milli, micro, nano, and pico. (Section 1.4)
- Given a metric unit, write its abbreviation; given an abbreviation, write the full name of the unit. (Section 1.4)
- Describe the relationships between the metric units that do not have prefixes (such as meter, gram, and liter) and units derived from them by the addition of prefixes—for example, 1 km $= 10^3$ m. (Section 1.4)
- Given a value derived from a measurement, identify the range of possible values it represents, on the basis of the assumption that its uncertainty is $\pm 1$ in the last position reported. For example, 8.0 mL says that the value could be from 7.9 mL to 8.1 mL. (Section 1.5)

## 1A.1 Common Unit Conversions

Chemists and chemistry students often need to convert values from one unit to another. For example, an American industrial chemist doing business with a European company might need to convert the mass of product that his company plans to sell each year from the English mass unit of tons into the metric unit of megagrams. And closer to home, as part of a laboratory experiment, you might be asked to convert a volume in liters into the same volume expressed in milliliters.

### Metric–Metric Unit Conversions

OBJECTIVE 3

The purpose of this section is to describe the basic steps in a unit conversion technique called **dimensional analysis.** It is a simple, reliable technique that can be applied to many different types of conversions, but we will begin by using it to convert values described with one metric unit into equivalent values described with a different metric unit. As you become familiar with dimensional analysis, you will come to appreciate the clarity with which it records the cancellation of "old" units and the generation of "new" ones. More important, you will find that it provides a stepwise thought process for solving a great many of the numerical problems that arise in chemistry. You may feel that you can make conversions between metric units without using dimensional analysis; indeed, there are other simple ways of making metric–metric conversions. However, learning to do metric–metric conversions using dimensional analysis is an important first step toward a far more fundamental goal: learning how to set up any unit conversion problem using the dimensional analysis format and learning how to do the reasoning that supports this powerful technique. In our first example, we will convert 1.275 kg into the equivalent mass in grams.

OBJECTIVE 3

The first step in the dimensional analysis procedure is to identify the unit for the value you want to calculate. In this case, you want to calculate a value in grams. You write this unit on the left side of an equals sign.

Desired unit

$$? \text{ g} =$$

OBJECTIVE 3

Next, you identify the value that you want to convert into the desired value—in this case, 1.275 kg—and you write it on the righthand side of the equals sign. (Remember that a value constitutes both a number and a unit.)

Desired unit    Given value

$$? \text{ g} = 1.275 \text{ kg}$$

Then you multiply the given value by one or more conversion factors, which, as you will see below, enables you to cancel the unwanted units and generate the desired units. A **conversion factor** is a ratio that describes a relationship between two units. One relationship between grams and kilograms comes from the definition of *kilo*, which is 1000, or $10^3$:

$$1\text{ kg} = 10^3\text{ g}$$

This relationship can be used to produce two ratios, or conversion factors:

OBJECTIVE 2

$$\frac{10^3\text{ g}}{1\text{ kg}} \quad \text{or} \quad \frac{1\text{ kg}}{10^3\text{ g}}$$

A simple organizational tool will help you decide which of these conversion factors to use. Units in a dimensional analysis setup cancel just like variables in an algebraic equation, so to cancel kg in this setup, you need a conversion factor that has kg on the bottom of a ratio. Here is the skeleton of this conversion factor:

OBJECTIVE 3

$$?\text{ g} = 1.275\text{ kg}\left(\frac{\quad}{\text{kg}}\right)$$

Skeleton of conversion factor

This skeleton reveals that you want to use the first of the two possible conversion factors displayed above to convert 1.275 kg into its equivalent mass in grams.

$$?\text{ g} = 1.275\text{ kg}\left(\frac{10^3\text{ g}}{1\text{ kg}}\right)$$

Cancels kilograms and introduces grams

If you have used correct conversion factors, and if your units cancel to yield the desired unit, then you can be confident that you will arrive at the correct answer.

OBJECTIVE 3

Desired unit — Given value — Final answer with desired unit

$$?\text{ g} = 1.275\text{ kg}\left(\frac{10^3\text{ g}}{1\text{ kg}}\right) = \mathbf{1275\text{ g}}$$

OBJECTIVE 3

Converts kilograms into grams

Example 1A.1 shows how conversion factors can be generated for metric units that have prefixes associated with negative powers of 10.

## Example 1A.1 Conversion Factors

OBJECTIVE 2

Write four conversion factors that relate centimeters and meters.

***Solution***
*Centi-* means $10^{-2}$, so *centimeter* means $10^{-2}$ meter.

$$1 \text{ cm} = 10^{-2} \text{ m}$$

There are two forms of the conversion factor that come from this relationship:

$$\frac{\mathbf{1\ cm}}{\mathbf{10^{-2}\ m}} \quad \text{and} \quad \frac{\mathbf{10^{-2}\ m}}{\mathbf{1\ cm}}$$

If a centimeter is $1/100$ ($10^{-2}$) of a meter, there must be 100 ($10^2$) centimeters per meter.

$$10^2 \text{ cm} = 1 \text{ m}$$

This equivalence leads to two more conversion factors:

$$\frac{\mathbf{10^{2}\ cm}}{\mathbf{1\ m}} \quad \text{and} \quad \frac{\mathbf{1\ m}}{\mathbf{10^{2}\ cm}}$$

## Exercise 1A.1 Conversion Factors

OBJECTIVE 2

Write four conversion factors that relate millimeters and meters.

Many calculations require the use of more than one conversion factor. For example, in converting from one metric unit with a prefix to a related metric unit with a different prefix (from millimeters to centimeters, for example), it is often a good strategy to convert from the given unit to the base unit with one conversion factor and then to convert from the base unit to the desired unit with a second conversion factor. Let's take a look at how this strategy, coupled with our dimensional analysis procedure, can be applied to the conversion of 947 mm to centimeters. The meter is the metric base unit for length, so we will convert from millimeters to the base unit of meters and then to centimeters.

$$\text{mm} \rightarrow \text{m} \rightarrow \text{cm}$$

We start our equation by setting the unit that we want equal to the given value. Next, we write the skeleton of the first conversion factor.

Desired unit Given value

$$? \text{ cm} = 947 \cancel{\text{mm}} \left( \frac{\quad}{\cancel{\text{mm}}} \right)$$

Skeleton of first conversion factor

We want this first conversion factor to convert from millimeters to meters (the base unit for length). *Milli-* means $10^{-3}$, so 1 millimeter is equal to $10^{-3}$ m, and this supplies us with the following conversion factor:

1 mm = $10^{-3}$ m

$$? \text{ cm} = 947 \text{ mm} \left(\frac{10^{-3} \text{ m}}{1 \text{ mm}}\right)$$

Converts millimeters into meters

We now write the skeleton of the next conversion factor, whose purpose is to convert meters into centimeters.

$$? \text{ cm} = 947 \text{ mm} \left(\frac{10^{-3} \text{ m}}{1 \text{ mm}}\right)\left(\frac{\quad}{\text{m}}\right)$$

Skeleton of second conversion factor

*Centi-* means $10^{-2}$, so 1 centimeter is equal to $10^{-2}$ m:

1 cm = $10^{-2}$ m

$$? \text{ cm} = 947 \text{ mm} \left(\frac{10^{-3} \text{ m}}{1 \text{ mm}}\right)\left(\frac{1 \text{ cm}}{10^{-2} \text{ m}}\right)$$

Converts meters into centimeters

We have used conversion factors that we know are written correctly, and we can see that our units have canceled to yield our desired unit. Therefore, we can be confident that we will arrive at the correct answer.

OBJECTIVE 3

Desired unit — Given value — Final answer with desired unit

$$? \text{ cm} = 947 \text{ mm} \left(\frac{10^{-3} \text{ m}}{1 \text{ mm}}\right)\left(\frac{1 \text{ cm}}{10^{-2} \text{ m}}\right) = \mathbf{94.7 \text{ cm}}$$

Converts given metric unit into metric base unit — Converts metric base unit into desired unit

OBJECTIVE 3

We get the same answer if we use the forms of the conversion factors that contain positive powers of 10.

m = $10^3$ mm
m = $10^2$ cm

$$? \text{ cm} = 947 \text{ mm} \left(\frac{1 \text{ m}}{10^3 \text{ mm}}\right)\left(\frac{10^2 \text{ cm}}{1 \text{ m}}\right) = \mathbf{94.7 \text{ cm}}$$

Appendix C shows how this calculation can be done using common calculators.

## Exercise 1A.2 Unit Conversions

OBJECTIVE 3

A basketball has a mass of 612 g. Use the dimensional analysis technique to determine its mass in kilograms.

### Exercise 1A.3 Unit Conversions

OBJECTIVE 3

A particular raindrop has a volume of 0.010 mL. Use the dimensional analysis technique to determine its approximate volume in microliters.

## Other Types of Conversions

OBJECTIVE 5

15 students = 1 class
1 student = 3 brushes

Any relationship that can be stated as "something per something" can be changed into a conversion factor. For example, consider an art instructor who wants to calculate how many brushes he needs to order for his four painting classes. He has *15 students per class*, and he needs *3 brushes per student*. Although math was never his strongest subject, the dimensional analysis technique he learned in his college chemistry class makes him confident that he can do this calculation correctly.

$$? \text{ brushes} = 4 \text{ classes}\left(\frac{15 \text{ students}}{1 \text{ class}}\right)\left(\frac{3 \text{ brushes}}{1 \text{ student}}\right) = \mathbf{180\ brushes}$$

We can use the dimensional analysis technique to convert between values with English units of measurement and values with metric units of measurement. As an example, consider a large sack of potatoes that has been weighed on a scale that is precise to ±0.1 lb and found to have a mass of 30.2 lb. Suppose we want to convert this mass into kilograms. There are *2.205 lb per kilogram*, so the following dimensional analysis setup can be used to make our conversion.

1 kg = 2.205 lb

$$? \text{ kg} = 30.2 \text{ lb}\left(\frac{1 \text{ kg}}{2.205 \text{ lb}}\right)$$

Converts pounds to kilograms

OBJECTIVE 4

Note that when a common scientific calculator is used to divide 30.2 by 2.205, it shows 13.6961451247 on the display. If we reported our answer as 13.6961451247 kg, we would be implying that the uncertainty of this value is ±0.0000000001 kg. (Remember that we assume that values are ±1 in the last position reported.) Our scale was precise only to ±0.1 lb (which is about ±0.05 kg), so we should round our answer off to yield a value that better reflects the actual precision of our measurement. Guidelines that chemists have developed for doing this are described briefly in the next section, and a more comprehensive explanation appears in Section 8.2. As you will see in the description of the rounding procedures that follow, because 30.2 has only three digits (called significant figures), we report three digits in our answer. Therefore, we round 13.6961451247 to 13.7.

$$? \text{ kg} = 30.2 \text{ lb}\left(\frac{1 \text{ kg}}{2.205 \text{ lb}}\right) = \mathbf{13.7\ kg}$$

OBJECTIVE 4

If your instructor expects you to convert between other English and metric units at this stage in your course, you should read the portion of Section 8.1 that gives additional examples of English–metric conversions. Table 1A.1 shows common English–metric conversion factors.

Table 1A.1
English-Metric Conversion Factors

OBJECTIVE 4

| Type of measurement | Probably most useful to know | Also useful to know | | |
|---|---|---|---|---|
| length | $\frac{2.54\text{ cm}}{1\text{ in.}}$ (exact) | $\frac{1.609\text{ km}}{1\text{ mi}}$ | $\frac{39.37\text{ in.}}{1\text{ m}}$ | $\frac{1.094\text{ yd}}{1\text{ m}}$ |
| mass | $\frac{453.6\text{ g}}{1\text{ lb}}$ | $\frac{2.205\text{ lb}}{1\text{ kg}}$ | | |
| volume | $\frac{3.785\text{ L}}{1\text{ gal}}$ | $\frac{1.057\text{ qt}}{1\text{ L}}$ | | |

## 1A.2 Rounding Off and Significant Figures

In Section 1.4, you discovered that all measurements are uncertain to some degree, and you learned how to report measured values to reflect this uncertainty. When measured values are used in calculations, their uncertainty leads to a corresponding degree of uncertainty in the answers to the calculations. Often, your calculator will provide more digits in an answer than you are justified, on the basis of the uncertainty of the measurement, in reporting. This section provides a brief introduction to the procedures for rounding off an answer to reflect the approximate range of certainty warranted by the numbers used in the calculation. Remember that Section 8.2 offers a more comprehensive description of these procedures.

### Rounding Off Answers Derived from Multiplication and Division

The following provides a guide to rounding off numbers calculated using multiplication and division. The procedure includes counting the number of *significant figures* in each number used in a calculation (we learn how in step 2, below) and rounding your answer off to the same number of significant figures as the value that contains the least number of significant figures.

The number of significant figures in a value reflects the value's degree of uncertainty; a *smaller* number of significant figures indicates a *greater* uncertainty. The uncertainty of an answer to a calculation is primarily determined by the most uncertain number used in the calculation, which is the number with the fewest significant figures. By restricting the number of significant figures in a reported answer to the least number of significant figures found in any of the values used in the calculation, we reflect the general uncertainty.

**Step 1** Determine whether each value is exact, and ignore exact values.

OBJECTIVE 8

- Numbers that come from definitions are exact.

  Numbers in metric–metric unit conversion factors that are derived from the metric prefixes are exact. An example is

OBJECTIVE 6

$$\frac{10^3\text{ g}}{1\text{ kg}}$$

OBJECTIVE 6

Numbers in English–English unit conversion factors with the same type of unit (for example, both length units) on the top and the bottom are exact. An example is

$$\frac{12\text{ in.}}{1\text{ ft}}$$

OBJECTIVE 6

The number 2.54 in the following conversion factor is exact.

$$\frac{2.54\text{ cm}}{1\text{ in.}}$$

OBJECTIVE 6

- Numbers derived from counting are exact. For example, there are exactly five toes on the normal foot.

$$\frac{5\text{ toes}}{1\text{ foot}}$$

OBJECTIVE 6

- Values that come from measurements are never exact.

OBJECTIVE 6

- We will assume that values derived from calculations are not exact unless otherwise indicated. (With one exception, the numbers relating English to metric units that you will see in this text have been calculated and rounded, so they are not exact. The exception is 2.54 cm/1 in. The 2.54 comes from a definition.)

**STEP 2** Determine the number of significant figures in each value that is not exact.

OBJECTIVE 8

- All nonzero digits are significant.

11.275—Five significant figures (digits numbered 1, 2, 3, 4, 5)

OBJECTIVE 7

- Zeros between nonzero digits are significant.

A zero between nonzero digits

10.275—Five significant figures (digits numbered 1, 2, 3, 4, 5)

- Zeros to the left of nonzero digits are not significant.

Not significant figures

0.000102, which can be described as $1.02 \times 10^{-4}$ (digits numbered 1, 2, 3)

Both have three significant figures

- Zeros to the right of nonzero digits in numbers that include decimal points are significant.

Five significant figures — 10.200 g (digits numbered 1 2 3 4 5)

20.0 mL (digits numbered 1 2 3) — Three significant figures

Unnecessary for reporting size of value, but do reflect degree of uncertainty

- Zeros to the right of nonzero digits in numbers without decimal points are ambiguous for significant figures.

220 kg (digits numbered 1 2 ?) — Precise to ±1 kg or ±10 kg? two or three significant figures?

Important for reporting size of value, but degree of uncertainty is unclear

$2.2 \times 10^2$ kg (digits numbered 1 2)

$2.20 \times 10^2$ kg (digits numbered 1 2 3)

Use scientific notation to remove ambiguity.

**Step 3** When multiplying and dividing, round your answer off to the same number of *significant figures* as the value that contains the fewest significant figures.

OBJECTIVE 8

- If the digit to the right of the final digit you want to retain is less than 5, round down (the last digit remains the same).

26.221 rounded to three significant figures is 26.2

First digit dropped is less than 5.

- If the digit to the right of the final digit you want to retain is 5 or greater, round up (the last significant digit increases by 1).

26.272 rounded to three significant figures is 26.3

First digit dropped is greater than 5.

26.2529 rounded to three significant figures is 26.3

First digit dropped is equal to 5.

26.15 rounded to three significant figures is 26.2

First digit dropped is equal to 5.

### EXAMPLE 1A.2 Rounding Off Answers Derived from Multiplication and Division

OBJECTIVE 8

Although the accepted SI unit of energy is the joule (J), the energy content of food is commonly described in terms of dietary calories (Cal). A dietary calorie is defined as 4184 J. The fat in our diets provides 37 kilojoules (kJ) of energy per gram of fat. If a chef weighs the fat cut off the meat she plans to serve and finds it to have a mass of 1.0 lb, how many dietary calories have been eliminated from the meal? The dimensional analysis setup is shown below. Determine the number of significant figures in each inexact value, calculate the answer, and report it to the correct number of significant figures.

$$? \text{ Cal} = 1.0 \text{ lb fat}\left(\frac{453.6 \text{ g}}{1 \text{ lb}}\right)\left(\frac{37 \text{ kJ}}{1 \text{ g fat}}\right)\left(\frac{10^3 \text{ J}}{1 \text{ kJ}}\right)\left(\frac{1 \text{ Cal}}{4184 \text{ J}}\right)$$

*Solution*

**STEP 1** The 1.0 comes from a measurement, so it is not exact. The number 453.6 is in an English–metric conversion factor. Except for 2.54 in 2.54 cm/in., the numbers relating English to metric units that you will see in this text have been calculated and rounded, so they are not exact. Thus 453.6 is not exact. The number 37 did not come from a definition, and an energy value cannot be counted. It must have come from experiments that involved measurements, so it is not exact. Both $10^3$ and 4184 come from definitions, so they are exact.

**STEP 2** Zeros to the right of nonzero digits in numbers that include decimal points are significant, so 1.0 has two significant figures. Nonzero digits are significant. Thus 453.6 has four significant figures, and 37 contains two significant figures.

**STEP 3** When multiplying and dividing, we round our answer off to the same number of significant figures as the value that contains the fewest significant figures, so we report two significant figures in this answer. The calculator reads 4011.2811, which we round off to $\mathbf{4.0 \times 10^3}$ **Cal.** We need scientific notation to report two significant figures unambiguously.

### EXERCISE 1A.4 Rounding Off Answers Derived from Multiplication and Division

OBJECTIVE 8

Plants produce carbohydrates through a process known as photosynthesis. In fact, photosynthesizing plants produce an estimated $1.7 \times 10^{11}$ Mg of carbohydrates per year, which is about $2.8 \times 10^4$ kg per person. The dimensional analysis setup that follows shows how to calculate the number of people on earth from these data. Determine the number of significant figures in each inexact value, calculate the answer, and report it to the correct number of significant figures. (You may want to consult Appendix B, "Scientific Notation," and Appendix C, "Using a Scientific Calculator," for help with this.)

$$? \text{ people} = 17 \times 10^{11} \text{ Mg carbohydrates}\left(\frac{10^6 \text{ g}}{1 \text{ Mg}}\right)\left(\frac{1 \text{ kg}}{10^3 \text{ g}}\right)\left(\frac{1 \text{ person}}{2.8 \times 10^4 \text{ kg}}\right)$$

## Rounding Off Answers Derived from Addition and Subtraction

The following steps describe how to round off numbers calculated using addition and subtraction. A more comprehensive description of these steps and an explanation of why we use this process can be found in Section 8.2.

**Step 1** Determine whether each value is exact, and ignore exact values (see Step 1 in the procedure for rounding off answers derived from multiplication and division).

OBJECTIVE 9

**Step 2** Determine the number of decimal places for each value that is not exact.

**Step 3** Round your answer to the same number of decimal places as the inexact value with the fewest decimal places.

### Example 1A.3 Rounding Off Answers Derived from Addition and Subtraction

OBJECTIVE 9

The lowest temperature ever recorded on the surface of the earth was −89.2 °C, in Antarctica. Although temperatures are often measured in degrees Celsius, the accepted SI unit for temperature is the kelvin. What is –89.2 °C on the Kelvin scale? (To convert from degrees Celsius into kelvins, add 273.15 to the value in °C.)

***Solution***

$$-89.2\ °\text{C} + 273.15 = \mathbf{184.0\ K}$$

**Step 1** The number −89.2 comes from a measurement, so it is not exact. The number 273.15 comes from a definition, so it is exact.

**Step 2** The number −89.2 is precise to the tenths position.

**Step 3** We round our answer to the tenths position.

### Exercise 1A.5 Rounding Off Answers Derived from Addition and Subtraction

OBJECTIVE 9

One way to generate electricity is to heat liquid water to steam, which can be used to turn a steam turbine generator. The hot steam is then cooled back to the liquid form in a process that makes use of water from natural systems, such as rivers or the ocean. When the water used for cooling is returned to its source, its temperature is 5 K to 10 K higher than before. Thus water removed at 287.6 K might be returned at 295.7 K, which in some cases is high enough to kill native fish and plants. Convert 295.7 K into degrees Celsius. (To convert from kelvins to degrees Celsius, subtract 273.15 from the value in K.)

## Density, Density Calculations, and Rounding

Density is a property of matter that is defined as mass divided by volume. For example, the density of water at 25 °C is *0.997 grams per milliliter,* or 0.997 g/mL. Section 8.3 describes density and density calculations in the context of a chemist's understanding of matter, but a quick introduction to density calculations here provides us with more practice in doing conversions and rounding off answers to calculations.

Densities can be used as conversion factors. Thus, if we have a beaker containing 905 mL of water (at 25 °C), we can use water's density as a conversion factor to discover the mass, in grams, of the water in the beaker.

OBJECTIVE 10

$$? \text{ g } H_2O = 905.0 \text{ mL } H_2O \left( \frac{0.997 \text{ g } H_2O)}{1 \text{ mL } H_2O} \right) = \mathbf{902 \text{ g } H_2O}$$

When you multiply 905 by 0.997 using your calculator, its display reads 902.285. Before you report your final answer, you must decide how many of these digits you are allowed to report.

**Step 1** Following the steps described above, you first determine whether 905.0 and 0.997 are exact. We can assume that the 905.0 mL came from a measurement, so it is not exact. As you will see in Example 1A.4, densities are determined by a combination of measurements and calculation, so the 0.997 g is not exact either.

**Step 2** We now count the number of significant figures in 905.0 and 0.997. Remember that zeros between nonzero digits are significant and that zeros to the right of nonzero digits in numbers with a decimal point are also significant. Thus 905.0 has four significant figures. Zeros to the left of nonzero digits are not significant, so 0.997 has three significant figures.

**Step 3** Because we are multiplying, we round our answer off to the same number of significant figures as the value that contains the fewest significant figures. Therefore, we round 902.285 to 902. Because the digit to the right of the final digit that we want to retain is less than 5 (2), we round down (keeping the last digit the same).

### Example 1A.4 Density and Rounding Off Answers Derived from Multiplication and Division

OBJECTIVE 8 OBJECTIVE 10

The density of pure diamond crystals is 3.52 g/mL, but because of impurities, the densities of gemstone diamonds vary from 3.15 g/mL to 3.53 g/mL. Assuming that the Star of Africa diamond, which weighs 530.2 carats, is pure, find its volume in milliliters. (A carat is defined so that there are exactly 5 carats per gram.)

***Solution***

$$? \text{ mL} = 530.2 \text{ carats} \left( \frac{1 \text{ g}}{5 \text{ carats}} \right) \left( \frac{1 \text{ mL}}{3.52 \text{ g}} \right) = \mathbf{30.1 \text{ mL}}$$

**Step 1** The number 530.2 comes from a measurement, so it is not exact. The number 5 comes from a definition, so it is exact. Densities are derived from a combination of measurement and calculation, so the number 3.52 is not exact.

**Step 2** Zeros between nonzero digits are significant, so 530.2 has four significant figures. Nonzero digits are significant, so 3.52 has three significant figures.

**Step 3** When multiplying and dividing, we round our answer off to the same number of significant figures as the value that contains the fewest significant figures, so we report three significant figures in this answer. The calculator reads 30.125, which we round off to **30.1 mL.**

## Exercise 1A.6 Density and Rounding Off Answers Derived from Multiplication and Division

OBJECTIVE 8 OBJECTIVE 10

High-density polyethylene (HDPE) is a strong, stiff, opaque polymer that is used to make many things, including grocery bags, pipes, and packaging. What is the mass, in kilograms, of 834.6 mL of HDPE that has a density of 0.960 g/mL?

The density of a substance can be calculated from its measured mass and volume. The examples that follow show how this can be done.

## Example 1A.5 Density Calculations

OBJECTIVE 11

Emeralds belong to the beryl family of minerals. A sample of beryl is weighed and found to have a mass of 11.7077 g. The volume of this same sample is determined by measuring the volume of a certain amount of water, placing the stone into the water, measuring the total volume of the stone and the water, and subtracting the initial volume of the water from the total to determine the volume of the stone. This procedure is called "volume by displacement." For our stone, the initial volume was 46.2 mL, and the final volume was 50.8 mL. What is the density of this sample of beryl?

***Solution***

$$? \text{ g/mL} = \frac{11.7077 \text{ g}}{(50.8 - 46.2) \text{ mL}} = \frac{11.7077 \text{ g}}{4.6 \text{ mL}} = \mathbf{2.5 \text{ g/mL}}$$

This calculation contains both division and subtraction, and the rules for rounding are different for each, so we need to round our answer off in stages. Because 50.8 and 46.2 are both precise to the tenths position, we round our answer for the subtraction to the tenths position. This yields 4.6, which has two significant figures and limits our final answer to two significant figures.

## Exercise 1A.7 Density Calculations

OBJECTIVE 11

The following data are collected in a procedure designed to determine the density of a sample of gasoline. First, a graduated cylinder is weighed and found to have a mass of 47.356 g. Next, gasoline is added to the cylinder, and the total mass is found to be 83.836 g. The difference between these two values corresponds to the mass of the gasoline. The volume reading on the graduated cylinder is 45.6 mL. What is the gasoline's density in grams per milliliter?

## A Summary of Dimensional Analysis Steps

The following steps summarize the dimensional analysis procedure for converting a value with one unit into an equivalent value with a different unit.

**STEP 1** State your question in an expression that sets the unknown unit equal to the value given.

**STEP 2** Multiply the expression to the right of the equals sign by one or more conversion factors that cancel the unwanted units and generate the desired unit.

**STEP 3** Check to be sure that you used correct conversion factors and that your units cancel to yield the desired unit.

**STEP 4** Do the calculation, rounding your answer to the correct number of significant figures and combining it with the correct unit.

Chapter 8 describes the process in more detail and introduces more of its many uses.

## Inter-Chapter Glossary

**Dimensional analysis** A general technique for doing unit conversions.

**Conversion factor** A ratio that describes the relationship between two units.

## Inter-Chapter Objectives

**The goal of this chapter is to teach you to do the following.**

1. Define all of the terms in the Inter-Chapter 1A Glossary.

### 1A.1 An Introduction to Unit Conversions

2. Write conversion factors that relate the metric base units to units derived from the metric prefixes. For example,

$$\frac{10^3 \text{ m}}{1 \text{ km}}$$

3. Use dimensional analysis to make conversions between metric units.
4. Use dimensional analysis to make conversions between English and metric units.
5. Use dimensional analysis making use of conversion factors derived from relationships that can be described as "something per something."

### 1A.2 Rounding Off and Significant Figures

6. Identify each value in a calculation as exact or not exact.

7. Write or identify the number of significant figures in any value that is not exact.
8. Round off answers derived from multiplication and division to the correct number of significant figures.
9. Round off answers derived from calculations involving addition or subtraction to the correct number of decimal positions.
10. Use density as a conversion factor to convert between mass and volume.
11. Calculate the density of a substance from its mass and volume.

## Key Ideas

**Complete the following statements by writing one of these words or phrases in each blank.**

| | |
|---|---|
| base unit | left |
| confident | never |
| conversion factors | numerical problems |
| decimal places | most uncertain |
| degree of uncertainty | relationship between two units |
| exact | right |
| fewest | same type of unit |
| fewest decimal places | "something per something" |
| fewest significant figures | unit |
| greater | unit conversion |

1. The ________________ technique called dimensional analysis is a simple, reliable technique that can be applied to many different types of conversions. It provides a stepwise thought process for solving a great many of the ________________ that arise in chemistry.
2. The first step in the dimensional analysis procedure is to identify the ____________ for the value you want to calculate. You write it on the ________________ side of an equals sign. Next, you identify the value that you want to convert into the desired value, and you write it on the ________________ side of the equals sign. Then, you multiply the given value by one or more ________________, which enables you to cancel the unwanted units and generate the desired unit.
3. A conversion factor is a ratio that describes a(n) ________________.
4. If you have used correct conversion factors in a dimensional analysis setup and if your units cancel to yield the desired unit, you can be ________________ that you will arrive at the correct answer.
5. In converting from one metric unit with a prefix to a related metric unit with a different prefix (from millimeters to centimeters, for example), it is often a good strategy to convert from the given unit to the ________________ with one conversion factor and then to convert from the base unit to the desired unit with a second conversion factor.
6. Any relationship that can be stated as ________________ can be changed into a conversion factor.

7. The number of significant figures in a value reflects the value's ____________________ ; a *smaller* number of significant figures indicates a ____________________ uncertainty.
8. In multiplication or division, the uncertainty of an answer is determined primarily by the ____________________ number used in the calculation, which is the number with the ____________________ significant figures.
9. Numbers in metric–metric unit conversion factors that are derived from the metric prefixes are ____________________ .
10. Numbers in English–English unit conversion factors with the ____________________ (for example, both length units) on the top and the bottom are exact.
11. Values that come from measurements are ____________________ exact.
12. When multiplying and dividing, round your answer off to the same number of significant figures as the value that contains the ____________________ .
13. When adding or subtracting, round your answer to the same number of ____________________ as the inexact value with the ____________________ .

## Inter-Chapter Problems

If you have not already done so, before working these problems, you might want to read Appendix B, which describes scientific notation, and Appendix C, which describes how you can use typical calculators to calculate answers to problems like those that follow.

### 1A.1 An Introduction to Unit Conversions

14. Convert the following ordinary decimal numbers into scientific notation. (See Appendix B at the end of this text if you need help.)
    a. 4,239
    b. 5,723,845
    c. 0.000415
    d. 0.000000001623
15. Convert the following ordinary decimal numbers into scientific notation. (See Appendix B at the end of this text if you need help.)
    a. 17,451.6
    b. 387,999
    c. 0.000000176
    d. 0.0871
16. Convert the following numbers expressed in scientific notation into ordinary decimal numbers. (See Appendix B at the end of this text if you need help.)
    a. $2.76423 \times 10^5$
    b. $6.342 \times 10^2$
    c. $8.8 \times 10^{-7}$
    d. $5.667 \times 10^{-4}$
17. Convert the following numbers expressed in scientific notation into ordinary decimal numbers. (See Appendix B at the end of this text if you need help.)
    a. $3.6651 \times 10^3$
    b. $1.6666667 \times 10^6$
    c. $4.11 \times 10^{-2}$
    d. $9.1 \times 10^{-5}$

18. Use your calculator to complete the following calculations. (See your calculator's instruction manual or Appendix C at the end of this text if you need help.)

    a. $27.50 \times 7.000$
    b. $7221 \div 14.5$
    c. $27 \times 0.006 \div 8.1$
    d. $(64.240 - 15.568) \div 15.6$

19. Use your calculator to complete the following calculations. (See your calculator's instruction manual or Appendix C at the end of this text if you need help.)

    a. $2.1 \div 0.025$
    b. $18 \times 3.5$
    c. $405 \div 9.00 \times 0.00080$
    d. $15. \times (55.672 - 25.672)$

20. Use your calculator to complete the following calculations. (See your calculator's instruction manual or Appendix C at the end of this text if you need help.)

    a. $10^{12} \times 10^{3}$
    b. $10^{6} \div 10^{2}$
    c. $10^{9} \times 10^{3} \div 10^{6}$
    d. $10^{12} \times 10^{-16}$
    e. $10^{9} \div 10^{-2}$
    f. $10^{-6} \times 10^{3} \div 10^{-2}$

21. Use your calculator to complete the following calculations. (See your calculator's instruction manual or Appendix C at the end of this text if you need help.)

    a. $10^{9} \div 10^{3}$
    b. $10^{2} \times 10^{12}$
    c. $10^{12} \div 10^{6} \times 10^{9}$
    d. $10^{-3} \div 10^{6}$
    e. $10^{-6} \times 10^{3}$
    f. $10^{5} \div 10^{-6} \times 10^{9}$

22. Use your calculator to complete the following calculations. (See your calculator's instruction manual or Appendix C at the end of this text if you need help.)

    a. $1.4 \times 10^{7} \cdot 6.0 \times 10^{4}$
    b. $9.39 \times 10^{23} \div 3.00 \times 10^{8}$
    c. $6.25 \times 10^{12} \times 1.6 \times 10^{5} \div 4.0 \times 10^{11}$
    d. $4.2 \times 10^{-5} \cdot 1.5 \times 10^{6}$
    e. $3.279 \times 10^{14} \div 10^{-9}$
    f. $(8.98 \times 10^{-4} - 4.82 \times 10^{-4}) \div 2.00 \times 10^{-15}$

23. Use your calculator to complete the following calculations. (See your calculator's instruction manual or Appendix C at the end of this text if you need help with this.)

    a. $4.48 \times 10^{17} \div 4.00 \times 10^{4}$
    b. $8.8 \times 10^{21} \cdot 5.0 \times 10^{11}$
    c. $4.50 \times 10^{9} \cdot 7.00 \times 10^{6} \div 3.00 \times 10^{10}$
    d. $2.0 \times 10^{13} \cdot 8.5 \times 10^{-6}$
    e. $9.27 \times 10^{-5} \div 9.00 \times 10^{-6}$
    f. $(8.634 \times 10^{3} - 1.245 \times 10^{3}) \div 10^{-15}$

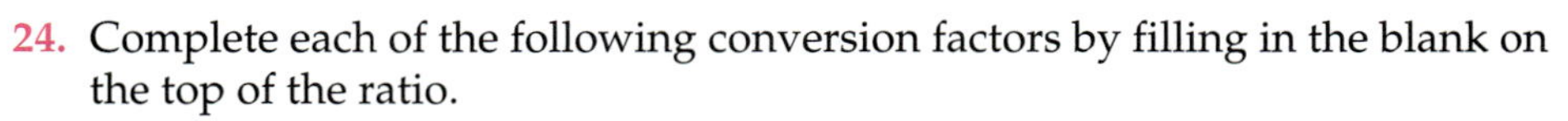

24. Complete each of the following conversion factors by filling in the blank on the top of the ratio.

a. $\left(\frac{\_\_\_ \text{ g}}{1 \text{ kg}}\right)$
b. $\left(\frac{\_\_\_ \text{ g}}{1 \text{ lb}}\right)$
c. $\left(\frac{\_\_\_ \ \mu\text{m}}{1 \text{ m}}\right)$
d. $\left(\frac{\_\_\_ \text{ pm}}{1 \text{ m}}\right)$
e. $\left(\frac{\_\_\_ \text{ min}}{1 \text{ h}}\right)$
f. $\left(\frac{\_\_\_ \text{ cm}}{1 \text{ in.}}\right)$
g. $\left(\frac{\_\_\_ \text{ in.}}{1 \text{ ft}}\right)$
h. $\left(\frac{\_\_\_ \text{ L}}{1 \text{ gal}}\right)$

25. Complete each of the following conversion factors by filling in the blank on the top of the ratio.

a. $\left(\frac{\_\_\_ \text{ g}}{1 \text{ Gg}}\right)$
b. $\left(\frac{\_\_\_ \text{ km}}{1 \text{ mi}}\right)$
c. $\left(\frac{\_\_\_ \text{ nm}}{1 \text{ m}}\right)$
d. $\left(\frac{\_\_\_ \text{ g}}{1 \text{ Mg}}\right)$
e. $\left(\frac{\_\_\_ \text{ ft}}{1 \text{ mi}}\right)$
f. $\left(\frac{\_\_\_ \text{ lb}}{1 \text{ kg}}\right)$
g. $\left(\frac{\_\_\_ \text{ h}}{1 \text{ day}}\right)$
h. $\left(\frac{\_\_\_ \text{ mL}}{1 \text{ L}}\right)$

OBJECTIVE 2 OBJECTIVE 3

26. Use the dimensional analysis technique to make the following conversions.

a. 772 g to kg
b. 0.45 m to cm
c. 1.255 kL to L
d. 84 mm to m

OBJECTIVE 2 OBJECTIVE 3

27. Use the dimensional analysis technique to make the following conversions.

a. 75 m to km
b. 0.089 L to mL
c. 0.000422 Mg to g
d. 2.2 μg to g

OBJECTIVE 2 OBJECTIVE 3

28. Use the dimensional analysis technique to make the following conversions.

a. 456.5 kg to Mg
b. 1.549 mm to μm

OBJECTIVE 2 OBJECTIVE 3

29. Use the dimensional analysis technique to make the following conversions.

a. 0.0045 GL to ML
b. 11.8 cm to mm

OBJECTIVE 2 OBJECTIVE 3

30. An Egyptian pyramid has a mass of about $10^7$ Mg. Use the dimensional analysis technique to determine its approximate mass in gigagrams.

OBJECTIVE 2 OBJECTIVE 3

31. A typical human hair has a diameter of $3 \times 10^{-3}$ mm. Use the dimensional analysis technique to determine its diameter in nanometers.

OBJECTIVE 2 OBJECTIVE 3

32. The accepted SI unit of energy is the joule (J). A 12-inch cheese pizza supplies you with 4940 kJ (kilojoules) of energy. What is this energy value expressed in megajoules?

OBJECTIVE 2 OBJECTIVE 3

33. The diameter of an atom is about 0.1 nm. Use the dimensional analysis technique to determine its approximate diameter in picometers.

### 1A.2 Rounding Off and Significant Figures

34. None of the following numbers is exact. How many significant figures does each number have? OBJECTIVE 7
    a. 327.89
    b. 0.214
    c. 1065
    d. 14.002
    e. 782.000

35. None of the following numbers is exact. How many significant figures does each number have? OBJECTIVE 7
    a. 176
    b. 1205.04
    c. 2.000
    d. 0.0040
    e. 0.0013

36. Decide whether each of the numbers shown in boldface type below is exact. If it is not exact, write the number of significant figures in it. OBJECTIVE 6 OBJECTIVE 7
    a. The estimated mass of the sun, **1.989 × 10^30^** kg → $\mathbf{1.989 \times 10^{30}}$ kg
    b. A count of **36** natural gas storage tanks at a power plant
    c. A measured volume of one of these storage tanks, $\mathbf{5.070 \times 10^{4}}$ $m^3$
    d. $\dfrac{\mathbf{3}\text{ ft}}{1\text{ yd}}$
    e. $\dfrac{\mathbf{60}\text{ s}}{1\text{ min}}$
    f. $\dfrac{\mathbf{10^3}\text{ mL}}{1\text{ L}}$
    g. $\dfrac{\mathbf{1.609}\text{ km}}{1\text{ mi}}$
    h. $\dfrac{\mathbf{2.54}\text{ cm}}{1\text{ in.}}$
    i. The density of magnesite, which is a substance used to make furnace linings, can be calculated from the measured mass of a sample and its measured volume. A portion of magnesite is found to have a mass of **34.100** g and a volume of **11.0** mL. This yields a density of **3.10** g/mL.

OBJECTIVE 6 OBJECTIVE 7

37. Decide whether each of the numbers shown in boldface type below is exact. If it is not exact, write the number of significant figures in it.

a. The volume of a moonstone, **0.00019** L

b. $\frac{\mathbf{10^6}\ \mu\text{m}}{1\ \text{m}}$

c. $\frac{\mathbf{453.6}\ \text{g}}{1\ \text{lb}}$

d. $\frac{\mathbf{16}\ \text{oz}}{1\ \text{lb}}$

e. $\frac{\mathbf{24}\ \text{h}}{1\ \text{day}}$

f. The measured mass of an anaconda, **237.0** lb

g. The volume of a sample of dolomite, which is used to make Epsom salts, can be calculated from the measured mass of the sample and its density. Using the density of **2.85** g/mL for dolomite, a **103.550**-g portion of dolomite is found to have a volume of **36.3** mL.

h. The count of **124** test tubes in a drawer of a chemistry stockroom

OBJECTIVE 9

38. Report the answers to the following calculations to the correct number of decimal places. Assume that each number is ±1 in the last decimal position reported.

a. 37 − 35.4 = b. 0.9905 + 999.06 =

OBJECTIVE 9

39. Report the answers to the following calculations to the correct number of decimal places. Assume that each number is 61 in the last decimal position reported.

a. 92.1391 + 9.08 = b. 378.4 − 352 =

***Because the ability to make unit conversions using the dimensional analysis format is an extremely important skill, be sure to set up each of the following calculations using the dimensional analysis format, even if you see another way to work the problem, and even if another technique seems easier.***

OBJECTIVE 3 OBJECTIVE 8

40. Amoebas are single-celled animals that move by extending their cellular material outward to form false feet. A typical amoeba is 2.3 mm long. What is this length in meters? in centimeters?

OBJECTIVE 3 OBJECTIVE 8

41. The chemical in hemlock that killed Socrates is called coniine. As little as 0.2 g of coniine can be deadly. What is this mass in milligrams?

OBJECTIVE 8

42. In 1866 a child found a "pebble" in the Orange River in South Africa that turned out to be a 21-carat diamond. This discovery led to the development of the greatest diamond fields in the world. Convert 21 carats to grams. (There are exactly 5 carats per gram.)

OBJECTIVE 3 OBJECTIVE 8

43. One can see details through a light microscope that are as small as 0.2 mm. What is this length in meters? in millimeters?

OBJECTIVE 3 OBJECTIVE 8

44. A chlorophyll molecule called P680 absorbs best light that has a wavelength of 680 nanometers. What is this length in meters? in picometers?

45. The energy that the sun releases comes from the fusion of hydrogen atoms to form helium atoms. In the process, some of the mass of the hydrogen atoms is converted into energy. About $5 \times 10^6$ megagrams (Mg) of matter are converted into energy each second. What is this mass in grams? in gigagrams?

OBJECTIVE 3 OBJECTIVE 8

46. The estimated mass of the sun is $1.989 \times 10^{30}$ kilograms. What is this mass in grams? in gigagrams? in pounds?

OBJECTIVE 3 OBJECTIVE 4 OBJECTIVE 8

47. The Great Pyramid in Egypt was built during the reign of King Khufu (2551 B.C. to 2528 B.C.). Its base is very nearly a perfect square, the difference between the longer and shorter sides being 19 cm. What is this length in inches?

OBJECTIVE 4 OBJECTIVE 8

48. The rainfall in tropical rain forests is over 200 cm/yr. Convert 214 cm into meters, into inches, and into feet.

OBJECTIVE 3 OBJECTIVE 4 OBJECTIVE 8

49. Anacondas, which are the largest and most powerful snakes, can weigh up to 250 lb. What is the mass of a 237.0 lb anaconda in kilograms? in megagrams?

OBJECTIVE 3 OBJECTIVE 8

50. About $2.4 \times 10^7$ Mg of asphalt are produced every year in the United States. What is this mass in grams? in kilograms? in pounds? in English short tons? (There are 2000 pounds per English short ton.)

OBJECTIVE 3 OBJECTIVE 4 OBJECTIVE 8

51. Sulfur dioxide is a pollutant that contributes to the creation of harmful "acid rain." The 1990 amendments to the Clean Air Act decreased the allowed annual release of sulfur dioxide by U.S. industry from $2.35 \times 10^7$ tons (English short tons) to $1.6 \times 10^7$ tons by 2010. Convert $2.35 \times 10^7$ tons into pounds. Convert $1.6 \times 10^7$ tons into megagrams. (There are 2000 pounds per English short ton.)

OBJECTIVE 3 OBJECTIVE 4 OBJECTIVE 8

52. The largest diamond ever found, called the Cullinan diamond, was discovered in the Premier Mine in South Africa in 1905. It weighed 3106 carats before cutting and was cut to form 128 gems, the largest being the Star of Africa. At 530.2 carats, this is the largest cut diamond in the world. What is the mass of the Star of Africa in grams? in kilograms? in pounds? (There are exactly 5 carats per gram.)

OBJECTIVE 3 OBJECTIVE 4 OBJECTIVE 8

53. The Alaska Highway is 1422 miles long. What is this length in kilometers?

OBJECTIVE 3 OBJECTIVE 4 OBJECTIVE 8

54. The longest recorded flight for a tagged monarch butterfly is $2.9 \times 10^3$ km. What is this length in meters? in miles?

OBJECTIVE 8 OBJECTIVE 10

55. Low-density polyethylene (LDPE) is an inexpensive, flexible, strong, and chemical resistant plastic used to make bottles, toys, and many other products. It has a density between 0.91 g/mL and 0.93 g/mL. What volume will 458.25 g of LDPE with a density of 0.9194 g/mL occupy?

OBJECTIVE 3 OBJECTIVE 8 OBJECTIVE 10

56. Because of their low density, plastics have replaced steel in many common objects. To get a sense of why, calculate the mass, in kilograms, of 2.2055 L of steel with a density of 7.852 g/mL and the mass, in kilograms, of 2.2055 L of a plastic with a density of 2.010 g/mL.

OBJECTIVE 3 OBJECTIVE 8 OBJECTIVE 10

57. A pillar for a bridge has a volume of 22.68 $m^3$. If it is made with concrete that has a density of 2.4 g/mL, what is its total mass in gigagrams? (There are $10^3$ L per cubic meter.)

OBJECTIVE 3 OBJECTIVE 8 OBJECTIVE 10

58. Because of the large concentration of salts dissolved in the water of the Dead Sea, its density is significantly higher than that of pure water. Pure water has a density of 1.0 g/mL, and water in the Dead Sea has a density of 1.8 g/mL. What is the volume, in liters, of $5.000 \times 10^4$ Mg of Dead Sea water?

OBJECTIVE 3 OBJECTIVE 8 OBJECTIVE 11

59. The mass of the sun is estimated to be $1.989 \times 10^{30}$ kg, and its volume is about $1.412 \times 10^{30}$ L. What is its average density in grams per milliliter?

OBJECTIVE 8 OBJECTIVE 11

60. Albite, also called soda spar, is a mineral used in the ceramics industry. Moonstone is an opalescent variety of this mineral. A 2.456-carat moonstone is found to have a volume of 0.19 mL. What is its density in grams per milliliter? (There are exactly 5 carats per gram.)

OBJECTIVE 3 OBJECTIVE 8 OBJECTIVE 11

61. Acrylonitrile is a liquid used to make acrylic fibers. If you were a purchasing agent for a chemical plant that uses 406 Mg of acrylonitrile per day, how many liters of it would you order to supply the plant for 31 days? (The density of acrylonitrile is 0.8004 g/mL.)

OBJECTIVE 5 OBJECTIVE 8

62. To avoid potentially serious problems, scuba divers are taught to ascend at a rate of about 60 feet per minute. How long will it take a diver to rise 245 feet to the surface at 55 feet per minute?

OBJECTIVE 5 OBJECTIVE 8

63. Natural gas is often stored in tanks that have a volume of $5.0 \times 10^4$ $m^3$. How many tanks would it take to store $8.74 \times 10^5$ $m^3$ of natural gas?

OBJECTIVE 5 OBJECTIVE 8

64. Albacore can swim for long distances at over 25 kilometers per hour and can exhibit short bursts of almost 90 km/h. How far will an albacore travel in 15 minutes at a speed of 27 mi/h?

OBJECTIVE 5 OBJECTIVE 8

65. Amounts of natural gas are often described in terms of the energy that it can provide. The unit commonly used is the BTU (British thermal unit), which is equivalent to 1.055 kilojoules (kJ). Each year Russia produces about 21 BTU of natural gas and consumes 14 BTU. In contrast, the United States produces about 19 BTU of natural gas per year and consumes about 22 BTU. What is the energy, in kilojoules, derived from the annual consumption of natural gas in the United States? If Russia were to continue to produce 21 BTU per year, how many BTU would be produced in 18 months?

OBJECTIVE 5 OBJECTIVE 8

66. There are about $7 \times 10^6$ Mg of gold in the ocean. If it were possible to extract the entire amount (it is not), what would its value be at $15 per gram?

OBJECTIVE 5 OBJECTIVE 8

67. The sun rotates on its axis like the earth. A point on the sun's equator takes 25.0 earth days to make a complete circle. How many rotations will this spot make in 1 year?

OBJECTIVE 5 OBJECTIVE 8

68. Deepwater currents move from 1 to 2 meters per day. If you were floating in a deepwater current moving at 1.237 m/day, how many years would it take you to travel 625 km?

OBJECTIVE 5 OBJECTIVE 8

69. Typical seawater contains 35 grams of salts per kilogram of water, but inland bodies of water can have much higher concentrations. For example, the water in the Dead Sea contains $3.0 \times 10^2$ grams of salts per kilogram of water. How many megagrams of salts are there in 625 Mg of water in the Dead Sea?

OBJECTIVE 5 OBJECTIVE 8

70. It is estimated that $7 \times 10^8$ Mg of hydrogen is converted into helium per second in the sun. How many gigagrams of hydrogen are converted into helium in 1 hour?

OBJECTIVE 5 OBJECTIVE 8

71. Although there are natural asphalt deposits, nearly all of the asphalt used in the United States comes from petroleum. About 75% of the $2.4 \times 10^7$ Mg of asphalt produced each year is used for roads. (This means that for every 100 Mg of asphalt produced, 75 Mg are used for roads.) How many megagrams of asphalt are used for roads each year in the United States?

72. In 1999 it was estimated that 40% of the $7.9 \times 10^7$ cars in the United States had air bags. (This means that for every 100 vehicles, 40 have air bags.) Use these data to calculate approximately how many cars in the United States have air bags.

OBJECTIVE 5 OBJECTIVE 8

73. Plastics make up about 9% of the $1.9 \times 10^8$ Mg of municipal waste generated each year. (This means that for every 100 Mg of waste, there are 9 Mg of plastic.) How many kilograms of plastic do these wastes contain per year?

OBJECTIVE 5 OBJECTIVE 8

74. In 1998 the average car was made of 12% plastic (for every 100 Mg of car, there are 12 Mg of plastic), for an average mass of plastic of 136 kg. What was the average mass of a car in 1998 in megagrams?

OBJECTIVE 5 OBJECTIVE 8

75. Meteorites contain small, opaque diamonds that form from graphite when an impact creates a pressure that reaches $1.022 \times 10^6$ atm. Convert $1.022 \times 10^6$ atm to kilopascals (kPa). (Although the pascal, Pa, is the accepted SI unit of pressure, pressures are often described in atmospheres, atm. There are 101.325 kPa per atmosphere.)

OBJECTIVE 5 OBJECTIVE 8

76. Atmospheric pressure decreases with increasing distance from the center of the earth. The average pressure at sea level is about 101 kPa; the average pressure in Denver, Colorado (which is at an elevation of about 1 mile), is 83 kPa; and the average pressure at 10,700 m (the altitude for normal jet flights) is 24 kPa. Convert these pressures into millimeters of mercury (mm Hg). (There are 101.325 kPa per atmosphere, and there are 760 mm Hg per atmosphere.)

OBJECTIVE 5 OBJECTIVE 8

77. Artificial diamonds are made at pressures of $7 \times 10^4$ atm and temperatures of about $2.77 \times 10^3$ °C. What is this pressure in pounds per square inch (psi) if there are 14.696 psi per atmosphere? Convert $2.77 \times 10^3$ °C into the equivalent temperature in kelvins. (To convert from degrees Celsius into kelvins, add 273.15 to the value in °C.) This problem is one of the few problems in this section that is not done using the dimensional analysis format.

OBJECTIVE 8

78. The portion of the sun where visible light is created is called the photosphere. The approximate temperature of the photosphere is $5.51 \times 10^3$ °C. Convert $5.51 \times 10^3$ °C into kelvins. (To convert from degrees Celsius into kelvins, add 273.15 to the value in °C.) You do not need to use dimensional analysis for this problem.

OBJECTIVE 8

79. The average temperature on Mercury (the hottest planet) is 477 °C. Although the accepted SI unit of temperature is the kelvin, temperatures in the laboratory are often measured in degrees Celsius. What is 477 °C in kelvins? (To convert from degrees Celsius into kelvins, add 273.15 to the value in °C.) You do not need to use dimensional analysis for this problem.

OBJECTIVE 8

### Additional Problems

80. Fireflies, which are actually beetles, produce a chemical called luciferin. When this chemical reacts with oxygen, light is produced. What is the length, in centimeters, of a firefly that is 0.55 in. long?

81. The largest aquamarine crystal ever found was discovered in Brazil in 1910. It weighed 110.3 kg. What is this mass in grams? in pounds?

82. When the human digestive tract is stretched out, it has a length of 20 to 30 ft. What is the length of a 24.6-foot digestive tract in centimeters? in meters?

83. A typical tyrannosaurus is thought to have weighed about 5 metric tons. What is this mass in grams? in kilograms? in pounds? (A metric ton is equivalent to a megagram.)

84. An electron microscope can show details as small as 0.2 nm. What is this length in meters? in micrometers?

85. It has been estimated that the annual global demand for plastics will soon exceed $2 \times 10^8$ Mg. What is this mass in grams? in kilograms? in English short tons? (There are 2000 pounds per English short ton.)

86. Albatross, which are birds that have been nicknamed "gooneys," have a wingspan of up to 3.4 meters. What is this length in centimeters? in inches? in feet?

87. Frogs' eggs have a diameter of about 0.2 cm. What is this length in meters? in millimeters? in inches?

88. The largest of the anaconda are about 10 meters long. What is the length of an anaconda that is 9.6 meters long in centimeters? in inches? in feet?

89. Conati's Comet was visible from the earth on June 2, 1858. This huge comet was estimated to be $4.5 \times 10^7$ miles long. What is this length in kilometers?

90. The planet Pluto orbits the sun at an average distance of 39.44 astronomical units (AU). An astronomical unit is a unit of length equivalent to the average distance between the earth and the sun, or $1.496 \times 10^8$ km. Convert 39.44 AU into the equivalent lengths in kilometers and miles.

91. The larvae of moths of the looper family are called measuring worms and are about 2.5 cm long. They have another common name that you might be able to guess after converting 2.5 cm into inches.

92. The shortest distance between the earth and the sun is $1.471 \times 10^8$ km. What is this length in miles?

93. The accepted SI unit of electric current is the ampere (A). For each ampere of current, $6.25 \times 10^{18}$ electrons move past any point in the current's path per second. How many electrons pass through each point per second in the electric wires of an appliance that uses 12 amperes of current?

94. The Great Pyramid in Egypt is made from about $2.3 \times 10^6$ blocks that have an average mass of 2.5 Mg. What is the total mass of the Great Pyramid?

95. Your large intestine absorbs about 6 liters of water per day. What volume of water will it absorb in 1 week?

96. The air traffic control system in the United States is responsible for $7.3 \times 10^7$ flight operations (takeoffs and landings) per year. What is the average number of flight operations in 1 day?

97. Down syndrome, which is a chromosomal disorder, occurs in about 1 in $8.0 \times 10^2$ births. In an area that has an annual birthrate of 6500, how many babies with Down syndrome are expected to be born in 1 year?

98. We hear because our ears can detect vibrations of the air. We can hear sound waves that vibrate at a rate as low as 25 vibrations per second. How many times does the air vibrate for such a sound that lasts for 2.5 minutes?

99. The sun emits about $2 \times 10^{38}$ neutrinos per second. Because they move at the speed of light, have no charge, and have little or no mass, they are very difficult to stop. Most of the neutrinos that strike the earth pass right through everything in their path and continue on unaffected. How many neutrinos does the sun emit in a month with 31 days? If you hold your hand out, about $7.0 \times 10^{10}$ solar neutrinos pass through each square centimeter of it per second. How long will it take for $1.0 \times 10^{6}$ (a million) neutrinos to pass through a square-centimeter area of your hand?

100. The sun moves around the center of our galaxy at $2.2 \times 10^{2}$ kilometers per second. It takes $2.5 \times 10^{8}$ years to complete 1 orbit. How far will it have traveled in this time? The sun is thought to be $4.6 \times 10^{9}$ years old. How many orbits of the galaxy has it made?

101. About 42% of the $2.26 \times 10^{9}$ acres of U.S. land is used for crops and livestock. (This means that for every 100 acres of land, there are 42 acres used for crops and livestock.) How many acres of land are used for this purpose?

*If you are interested in learning how to protect the world's natural ecological systems, you need to know something about the basic language and concepts of chemistry.*

CHAPTER 2

# The Structure of Matter and the Chemical Elements



*One doesn't discover new lands without consenting to lose sight of the shore for a very long time.*

André Gide
French novelist and essayist

In this chapter, we begin the journey that will lead you to an understanding of chemistry. Perhaps your ultimate educational goal is to know how the human body functions or to learn how the many parts of a shoreline ecosystem work together. You won't get very far in these studies without a basic knowledge of the chemical principles underlying them. Even before talking about basic chemical principles, though, you must learn some of the language of chemistry and develop an image of the physical world that will help you to think like a chemist.

Many important tasks in life require that you learn a new language and new skills. When you are learning to drive a car, for example, your driving instructor may tell you that when two cars reach a four-way stop at the same time, the driver on the left must yield the right of way. This statement won't mean anything to you unless you already know what a "four-way stop" is and what is meant by "yield" and "right of way." To drive safely, you need to learn which of the symbols you see on road signs means "lane merges ahead" and which means "steep grade." You need to learn procedures that will help you make lane changes and parallel park.

Chemistry, like driving a car, uses a language and skills of its own. Without a firm foundation in these fundamentals, you cannot achieve a true understanding of chemistry. This chapter begins to construct that foundation by introducing some key aspects of the chemists' view of matter.

JOHANNES KEPLER'S UPHILL BATTLE
...SO, YOU SEE THE ORBIT OF A PLANET IS ELLIPTICAL.
WHAT'S AN ORBIT?
WHAT'S A PLANET?
WHAT'S ELLIPTICAL?
S. harris

## Review Skills

The presentation of information in this chapter assumes that you can already perform the tasks listed below. You can test your readiness to proceed by answering the Review Questions at the end of the chapter. This might also be a good time to read the Chapter Objectives, which precede the Review Questions.

- Define the term *matter.* (Chapter 1 Glossary)
- Write the SI base units for mass and length and their abbreviations. (Section 1.4)
- Using everyday examples, describe the general size of a meter and a gram. (Section 1.4)

## 2.1 Solids, Liquids, and Gases

There are many situations in which a simple model helps us learn about a reality that is more complex.

As we saw in Chapter 1, a chemist's primary interest is the behavior of matter, but to understand the behavior of matter, we must first understand its internal structure. What are the internal differences between the granite of Half Dome in Yosemite, the olive oil added to your pasta sauce, and the helium in a child's balloon? A simple *model* of the structure of matter will help us begin to answer this question.

A **model** is a simplified approximation of reality. For example, architects often build a model of a construction project before actual construction begins. The architect's model is not an exact description of the project, but it is still very useful as a representation of what the structure will be like. Scientific models are like architects' models; they are simplified but useful representations of something real. In science, however, the models are not always physical entities. Sometimes they are sets of ideas instead.

In the last hundred years, there has been a tremendous increase in our understanding of the physical world, but much of that understanding is based on extremely complicated ideas and mathematics. Applying the most sophisticated forms of these modern ideas is difficult—and not very useful to those of us who are not well trained in modern physics and high-level mathematics. Therefore, scientists have developed simplified models for visualizing, explaining, and predicting physical phenomena. For example, we are about to examine a model that will help you visualize the tiny particles of the solid metal in a car's engine block, the liquid gasoline in the car's tank, and the gaseous exhaust fumes that escape from its tail pipe. The model will help you understand why **solids** have constant shape and volume at a constant temperature, why **liquids** have a constant volume but can change their shape, and why **gases** can easily change both their shape and their volume. Our model of the structure of solids, liquids, and gases says that

OBJECTIVE 2

- All matter is composed of tiny particles. (We will start by picturing these as tiny spheres.)
- These particles are in constant motion.

OBJECTIVE 3

- The amount of motion is proportional to temperature. Increased temperature means increased motion.
- Solids, gases, and liquids differ in the freedom of motion of their particles and in how strongly the particles attract each other.

### Solids

Why does the metal in a car's engine block retain its shape as you drive down the road, whereas the fuel in the car's gas tank conforms to the shape of the tank? What's happening on the submicroscopic level when solid metal is melted to a liquid, and why can molten metal take the shape of a mold used to form an engine block? Our model will help us to answer these questions.

According to our model, the particles of a solid can be pictured as spheres held closely together by strong mutual attractions (Figure 2.1). All the particles are in motion, bumping and tugging one another. Because they're so crowded and exert such strong mutual attractions, however, they only jostle in place. Picture yourself riding on particle 1 in Figure 2.1. An instant before the time captured in the figure, your particle was bumped by particle 3 and sent toward particle 5. (The curved lines in the figure represent the momentary direction of each particle's motion and its relative velocity.) This motion continues until the combination of a bump from particle 5 and tugging from particles 2, 3, and 4 quickly brings you back toward your original position. Perhaps your particle will now head toward particle 2 at a greater velocity than it had before, but again, a combination of bumps and tugs will send you back into the small space between the same particles. A ride on any of the particles in a solid would be a wild one, with constant changes in direction and velocity, but each particle continues to occupy the same small space and to have the same neighbors.

OBJECTIVE 2

OBJECTIVE 4

When a solid is heated, the average speed of the moving particles increases. Faster-moving particles collide more violently, causing each particle to push its neighbors farther away. Therefore, an increase in temperature usually causes a solid to expand somewhat (Figure 2.1).

OBJECTIVE 5

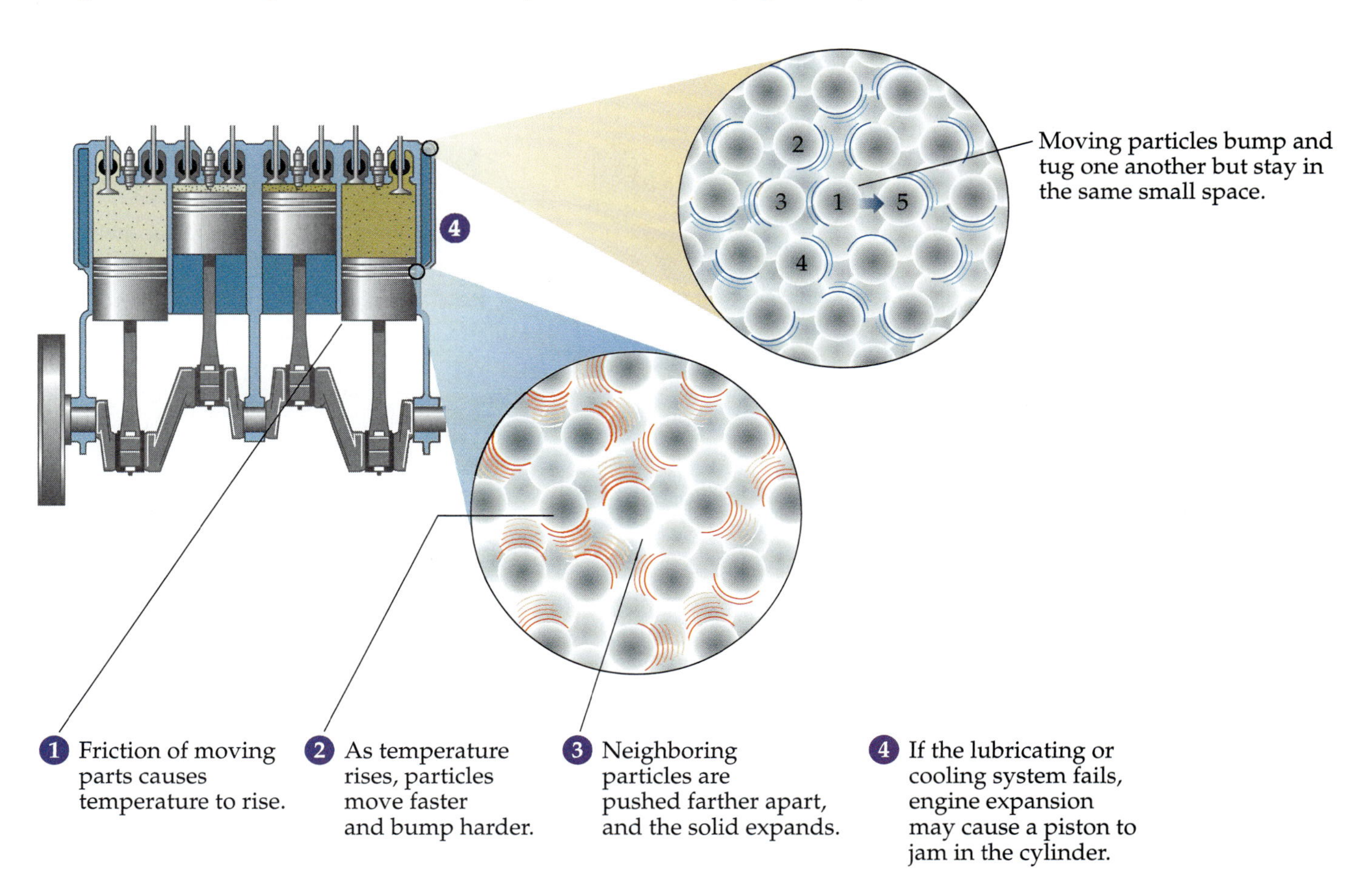

2
3
1
5
4
Moving particles bump and tug one another but stay in the same small space.
1 Friction of moving parts causes temperature to rise.
2 As temperature rises, particles move faster and bump harder.
3 Neighboring particles are pushed farther apart, and the solid expands.
4 If the lubricating or cooling system fails, engine expansion may cause a piston to jam in the cylinder.

**Figure 2.1**
Particles of a Solid

OBJECTIVE 4 OBJECTIVE 5

## Liquids

OBJECTIVE 2

When any solid is heated enough, the movements of the particles become powerful enough to push the other particles around them completely out of position. Look again at Figure 2.1. If your particle is moving fast enough, it can push adjacent particles entirely out of the way and move to a new position. For those adjacent particles to make way for yours, however, they must push the other particles around them aside. In other words, for one particle to move out of its place in a solid, all of the particles must be able to move. The organized structure collapses, and the solid becomes a liquid.

OBJECTIVE 6

Particles in a liquid are still close together, but there is generally more empty space between them than in a solid. Thus, when a solid substance melts to form a liquid, it usually expands to fill a slightly larger volume. Even so, attractions between the particles keep them a certain average distance apart, so the volume of the liquid stays constant at a constant temperature. On the other hand, because the particles in a liquid are moving faster and there is more empty space between them, the attractions are easily broken and reformed, and the particles change location freely. Eventually, each particle gets a complete tour of the container. This freedom of movement allows liquids to flow, taking on the shape of their container. It is this freedom of movement that makes it possible to pour liquid metal into a mold where it takes the shape of an engine block (Figure 2.2).

OBJECTIVE 7

OBJECTIVE 2

OBJECTIVE 8

**Figure 2.2**
**Particles of a Liquid**

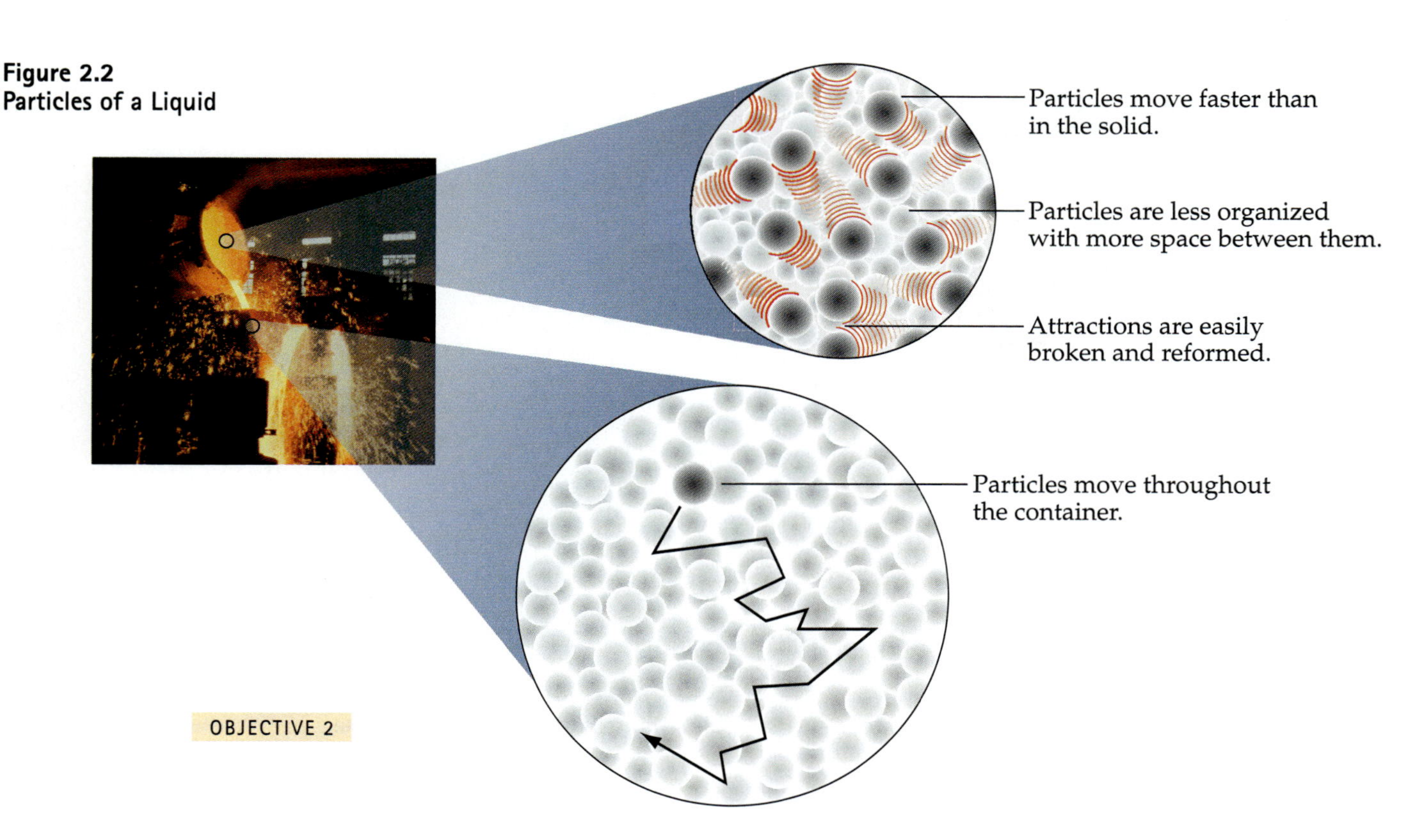
Particles move faster than in the solid.
Particles are less organized with more space between them.
Attractions are easily broken and reformed.
Particles move throughout the container.

OBJECTIVE 2

## Gases

If you've ever spilled gasoline while filling your car, you know how quickly the smell finds your nose. Our model can help you understand why.

Once again, picture yourself riding on a particle in liquid gasoline. Because the particle is moving throughout the liquid, it will eventually come to the liquid's surface. Its direction of movement may carry it beyond the surface into the space above the liquid, but the attraction of the particles behind it is likely to draw it back again. On the other hand, if your particle is moving fast enough, it can move far enough away from the other particles to break the attractions pulling it back. This is the process by which liquid is converted to gas. The conversion of liquid to gas is called **vaporization** or **evaporation** (Figure 2.3).

OBJECTIVE 9

You might also have noticed, while pumping gasoline, that the fumes smell stronger on a hot day than on a cold day. When the gasoline's temperature is higher, its particles are moving faster and are therefore more likely to escape from the liquid, so more of them reach your nose.

The particles of a gas are much farther apart than those of a solid or liquid. In the air around us, for example, the average distance between particles is about ten times the diameter of each particle. This leads to the gas particles themselves taking up only about 0.1% of the total volume. The other 99.9% of the total volume is empty space. In contrast, the particles of a liquid fill about 70% of the liquid's total volume. According to the model, each particle in a gas moves freely in a straight-line path until it collides with another gas particle or with the particles of a liquid or solid. The particles are usually moving fast enough to break any attraction that might form between them, so after two particles collide, they bounce off each other and continue on their way alone.

OBJECTIVE 2

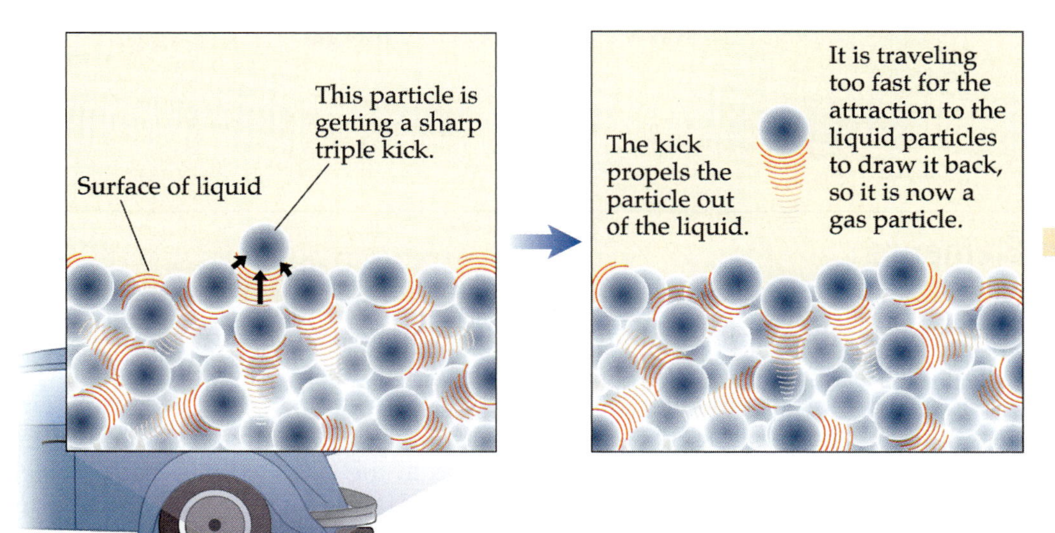

This particle is getting a sharp triple kick.
Surface of liquid
Gasoline
The kick propels the particle out of the liquid.
It is traveling too fast for the attraction to the liquid particles to draw it back, so it is now a gas particle.

**Figure 2.3**
The Process of Evaporation

OBJECTIVE 9

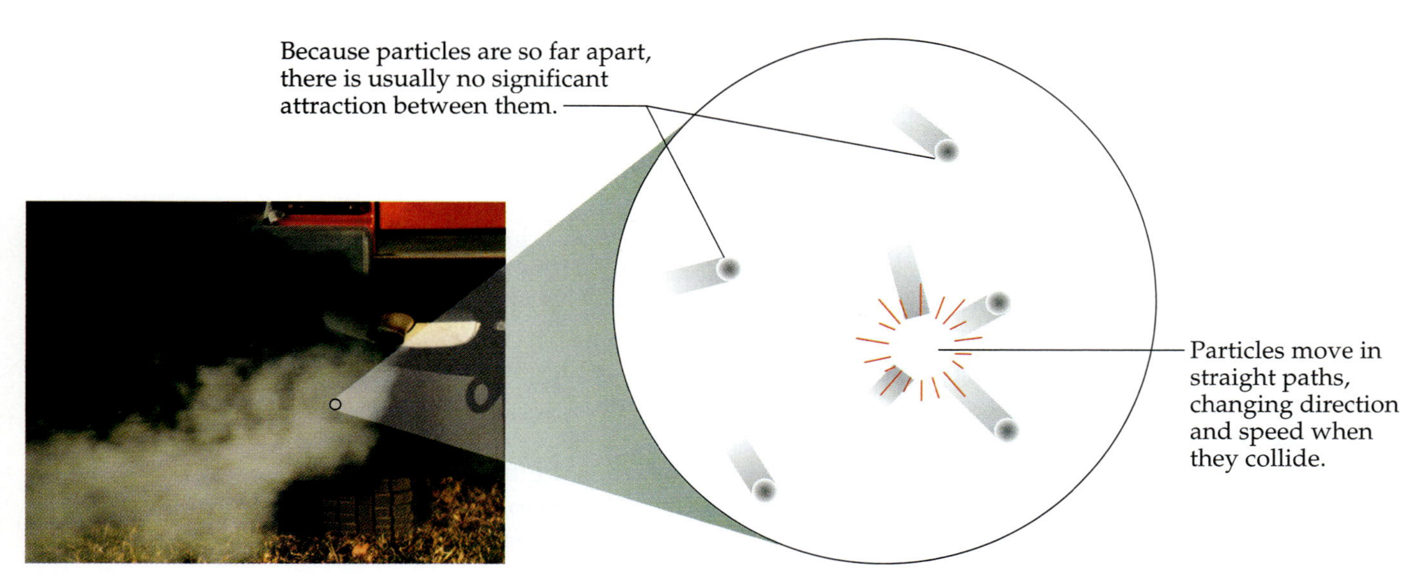

Because particles are so far apart, there is usually no significant attraction between them.
Particles move in straight paths, changing direction and speed when they collide.

**Figure 2.4**
**Particles of a Gas**

Picture yourself riding on a gas particle at the instant captured in Figure 2.4. You are so far away from any other particles that you think you are alone in the container. An instant later, you collide with a particle that seems to have come out of nowhere. The collision changes your direction and velocity. In the next instant, you are again moving freely, as though your particle were the only one in the universe.

OBJECTIVE 10

Unlike the liquid, which has a constant volume, the rapid, ever-changing, and unrestricted movement of the gas particles allows gases to expand to fill a container of any shape or volume. This movement also allows our cars' exhaust gases to move freely out of the cars and into the air we breathe.

You can review the information in this section and see particles of solids, liquids, and gases in motion at the following Web site:
**www.chemplace.com/college/**

## 2.2 The Chemical Elements

Chemists, like curious children, learn about the world around them by taking things apart. Instead of dissecting music boxes and battery-operated rabbits, however, they attempt to dismantle matter, because their goal is to understand the substances from which things are made. The model of the structure of matter presented in the last section describes the behavior of the particles in a solid, in a liquid, and in a gas. But what about the nature of the particles themselves? Are all the particles in a solid, liquid, or gas identical? And what are the particles made of? We begin our search for the answers to these questions by analyzing a simple glass of water with table salt dissolved in it.

We can separate this salt water into simpler components in three steps. First, heating can separate the salt and the water; the water will evaporate, leaving the salt behind. If we do the heating in what chemists call a distillation apparatus, the water vapor can be cooled back to its liquid form and collected in a separate container (Figure 2.5). Next, the water can be broken down into two even simpler substances—hydrogen gas and oxygen gas—by running an electric current through it. Also, we can melt the dry salt and then run an electric current through it, which causes it to break down into sodium metal and chlorine gas.

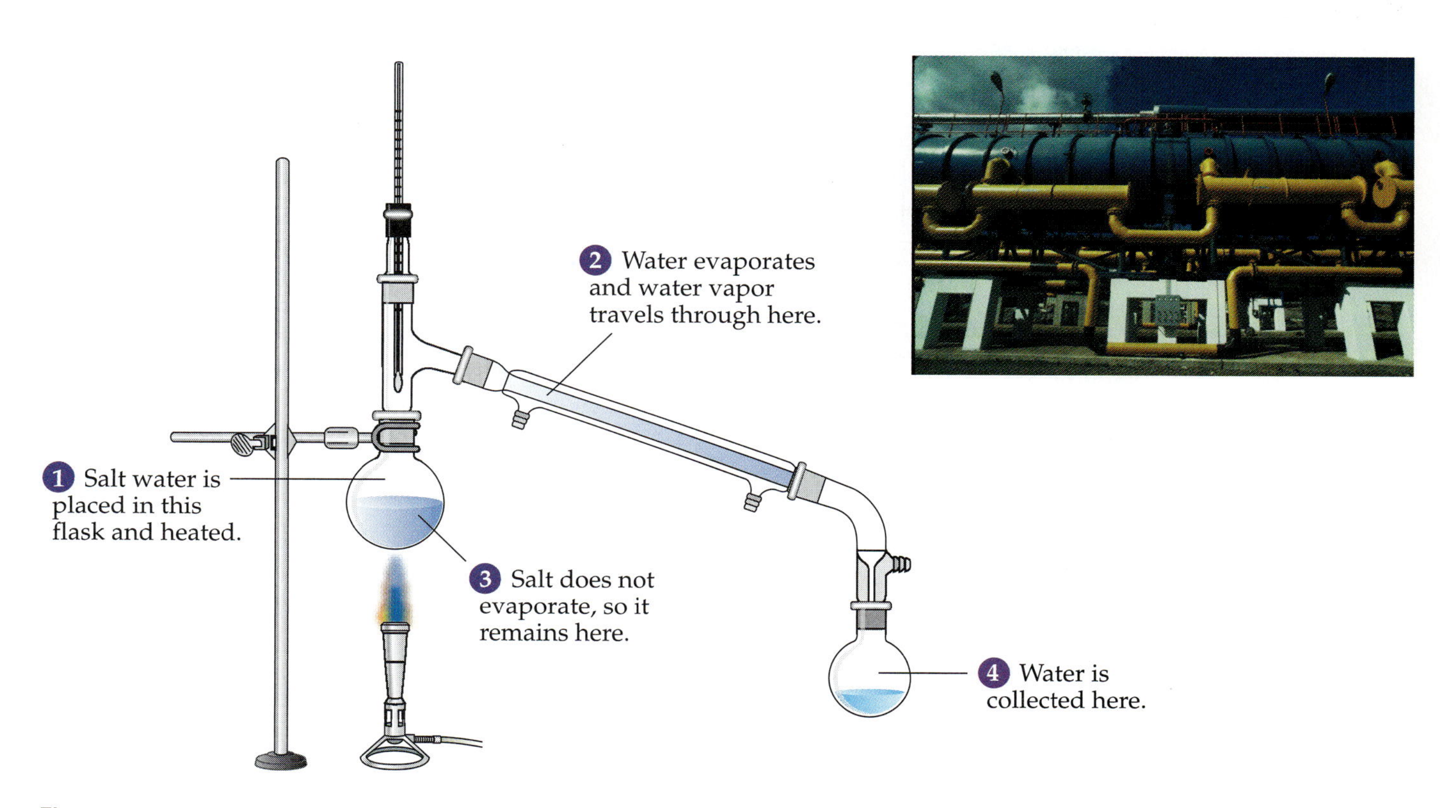

1 Salt water is placed in this flask and heated.
2 Water evaporates and water vapor travels through here.
3 Salt does not evaporate, so it remains here.
4 Water is collected here.

**Figure 2.5**
**Distillation**
Water can be separated from salt water on a small scale via the laboratory distillation apparatus on the left or on a large scale in desalinization plants like the one on the right above.

Thus the salt water can be converted into four simple substances: hydrogen, oxygen, sodium, and chlorine (Figure 2.6). Chemists are unable to convert these four substances into simpler ones. They are four of the building blocks of matter that we call **elements,** substances that cannot be chemically converted into simpler ones. (We will formulate a more precise definition of the term *elements* after we have explored the structure of these substances in more detail.)

The rest of this chapter is devoted to describing some common elements. Water, which consists of the elements hydrogen and oxygen, and salt, which consists of the elements sodium and chlorine, are examples of

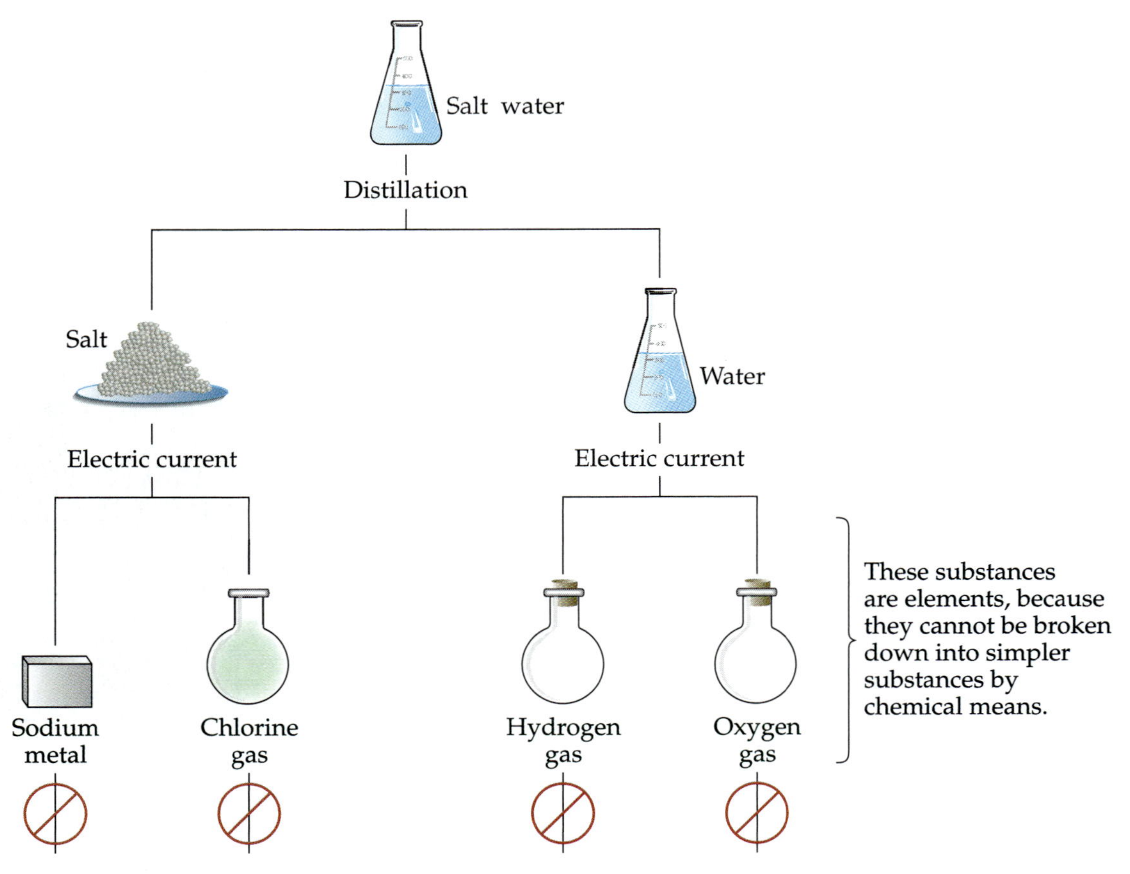


**Figure 2.6**
**Separation of Salt Water into Four Substances**

chemical compounds, which are described in Chapter 3. The mixture of salt and water is an example of a solution. Mixtures and solutions are described in Chapter 4.

Millions of simple and complex substances are found in nature or produced in the chemical laboratory, but the total number of elements that combine to form these substances is much, much smaller. By the year 2001, 115 elements had been discovered, but 25 of these elements are not found naturally on the earth, and chemists do not generally work with them. Of these 25 elements, 2 or 3 may occur in stars, but the rest are not thought to exist outside the physicist's laboratory. (See Special Topic 2.1: *Why Create New Elements?*) Some of the elements found in nature are unstable; that is, they exist for a limited time and then turn into other elements in a process called radioactive decay. Of the 83 stable elements found in nature, many are rare and will not be mentioned in this text. The most important elements for our purposes are listed in Table 2.1.

Each of the elements is known by a name and a symbol, and these can be found on the inside front cover of this book. The names were assigned in several ways. Some of the elements, such as francium and californium, were named to honor the places where they were discovered. Some have been named to honor important scientists. An element discovered in 1982 has been named meitnerium to honor Lise Meitner (1878–1968), the Austrian–Swedish physicist and mathematician who discovered the element protactinium and made major contributions to the understanding of nuclear fission. Some names reflect the source from which scientists first isolated the element. The name hydrogen came from the combination of the Greek words for "water" (*hydro*) and "forming" (*genes*). Some elements, such as the purple element iodine, are named for their appearance (iodos means "violet" in Greek).

Lise Meitner

The symbols for the elements were chosen in equally varied ways. Some are simply the first letter of the element's name. For example, C represents carbon. Other symbols are formed from the first letter and a later letter in the name. When two letters are used, the first is capitalized and the second remains lowercase. Cl is used for chlorine and Co for cobalt. Some of the symbols come from earlier, Latin names for elements. For example, Na for sodium comes from *natrium*, which is the Latin name for salt, and Au for gold comes from the Latin *aurum*, which means "shining dawn." The most recently discovered elements have not been officially named yet. They are given temporary names and three-letter symbols.

OBJECTIVE 11

In your study of chemistry, it will be useful to learn the names and symbols for as many of the elements in Table 2.1 as you can. Ask your instructor which of the element names and symbols you will be expected to know for your exams.

OBJECTIVE 11

**Table 2.1**
Common Elements

| Element | Symbol | Element | Symbol | Element | Symbol |
|---|---|---|---|---|---|
| aluminum | Al | gold | Au | oxygen | O |
| argon | Ar | helium | He | phosphorus | P |
| barium | Ba | hydrogen | H | platinum | Pt |
| beryllium | Be | iodine | I | potassium | K |
| boron | B | iron | Fe | silicon | Si |
| bromine | Br | lead | Pb | silver | Ag |
| cadmium | Cd | lithium | Li | sodium | Na |
| calcium | Ca | magnesium | Mg | strontium | Sr |
| carbon | C | manganese | Mn | sulfur | S |
| chlorine | Cl | mercury | Hg | tin | Sn |
| chromium | Cr | neon | Ne | uranium | U |
| copper | Cu | nickel | Ni | xenon | Xe |
| fluorine | F | nitrogen | N | zinc | Zn |

Jewelry can be made from the elements gold, silver, copper, and carbon in the diamond form.

You can practice converting between element names and symbols at the following Web site:
www.chemplace.com/college/

## 2.3 The Periodic Table of the Elements

Hanging on the wall of every chemistry laboratory, and emblazoned on many a chemist's favorite mug or T-shirt, is one of chemistry's most important basic tools, the periodic table of the elements (Figure 2.7). This table is like the map of the world on the wall of every geography classroom. When a geography instructor points to a country on the map, its location alone reveals what the climate is like and perhaps some of the characteristics of the culture. Likewise, you may not be familiar with the element potassium, but the position of its symbol, K, on the periodic table reveals that this element is very similar to sodium and that it will react with the element chlorine to form a substance very similar to table salt.

The elements are organized on the periodic table in a way that makes it easy to find important information about them. You will quickly come to appreciate how useful the table is when you know just a few of the details of its arrangement.

**Figure 2.7**
**Periodic Table of the Elements**

| | 1 1A | 2 2A | 3 3B | 4 4B | 5 5B | 6 6B | 7 7B | 8 8B | 9 8B | 10 8B | 11 1B | 12 2B | 13 3A | 14 4A | 15 5A | 16 6A | 17 7A | 18 8A |
|---|---|---|---|---|---|---|---|---|---|---|---|---|---|---|---|---|---|---|
| 1 | | | | | | | | | | | 1 H | | | | | | | 2 He |
| 2 | 3 Li | 4 Be | | | | | | | | | | | 5 B | 6 C | 7 N | 8 O | 9 F | 10 Ne |
| 3 | 11 Na | 12 Mg | | | | | | | | | | | 13 Al | 14 Si | 15 P | 16 S | 17 Cl | 18 Ar |
| 4 | 19 K | 20 Ca | 21 Sc | 22 Ti | 23 V | 24 Cr | 25 Mn | 26 Fe | 27 Co | 28 Ni | 29 Cu | 30 Zn | 31 Ga | 32 Ge | 33 As | 34 Se | 35 Br | 36 Kr |
| 5 | 37 Rb | 38 Sr | 39 Y | 40 Zr | 41 Nb | 42 Mo | 43 Tc | 44 Ru | 45 Rh | 46 Pd | 47 Ag | 48 Cd | 49 In | 50 Sn | 51 Sb | 52 Te | 53 I | 54 Xe |
| 6 | 55 Cs | 56 Ba | 71 Lu | 72 Hf | 73 Ta | 74 W | 75 Re | 76 Os | 77 Ir | 78 Pt | 79 Au | 80 Hg | 81 Tl | 82 Pb | 83 Bi | 84 Po | 85 At | 86 Rn |
| 7 | 87 Fr | 88 Ra | 103 Lr | 104 Rf | 105 Db | 106 Sg | 107 Bh | 108 Hs | 109 Mt | 110 Uun | 111 Uuu | 112 Uub | | 114 Uuq | | 116 Uuh | | 118 Uuo |

| | | | | | | | | | | | | | | |
|---|---|---|---|---|---|---|---|---|---|---|---|---|---|---|
| 6 | 57 La | 58 Ce | 59 Pr | 60 Nd | 61 Pm | 62 Sm | 63 Eu | 64 Gd | 65 Tb | 66 Dy | 67 Ho | 68 Er | 69 Tm | 70 Yb |
| 7 | 89 Ac | 90 Th | 91 Pa | 92 U | 93 Np | 94 Pu | 95 Am | 96 Cm | 97 Bk | 98 Cf | 99 Es | 100 Fm | 101 Md | 102 No |

Metallic elements

Metalloids

Nonmetallic elements

Black symbols: solids

Blue symbols: liquids (Br and Hg)

Red symbols: gases

OBJECTIVE 12

OBJECTIVE 15

OBJECTIVE 17

OBJECTIVE 18

The periodic table is arranged in such a way that elements in the same vertical column have similar characteristics. Therefore, it is often useful to refer to all the elements in a given column as a **group** or **family.** Each group has a number, and some have a group name. For example, the last column on the right is group 18, and the elements in this column are called noble gases.

OBJECTIVE 12

In the United States, there are two common conventions for numbering the columns (Figure 2.7). Check with your instructor to find out which numbering system you are expected to know.

- **Groups 1 to 18:** The vertical columns can be numbered from 1 to 18. This is the numbering convention we will use most often in this text.
- **Groups A and B:** Some of the groups are also commonly described with a number and the letter A or B. For example, the group headed by N is sometimes called group 15 and sometimes group 5A. The group headed by Zn can be called 12 or 2B. Because this convention is useful and is common, you will see it in this text also. Some chemists use Roman numerals with the A- and B-group convention.

OBJECTIVE 13

In short, the group headed by N can be 15, 5A, or VA. The group headed by Zn can be 12, 2B, or IIB.

The groups in the first two and the last two columns are the groups that have names as well as numbers. You should learn these names; they are used often in chemistry.

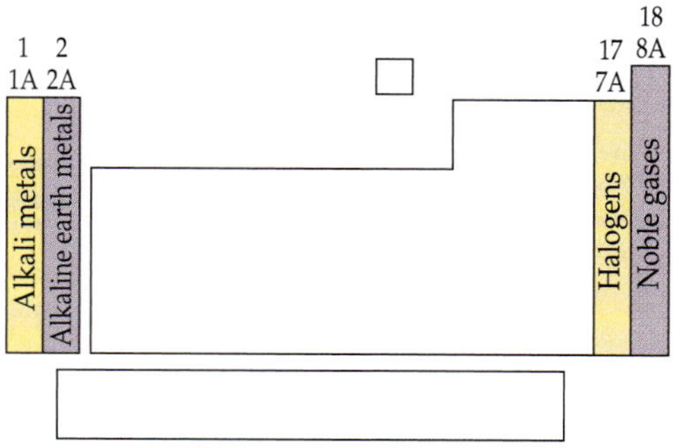

1 2
1A 2A
Alkali metals
Alkaline earth metals
17 7A
18 8A
Halogens
Noble gases

OBJECTIVE 14

Most of the elements are classified as **metals,** which means they have the following characteristics.

- Metals have a shiny, metallic luster.
- Metals conduct heat well and, in the solid form, conduct electric currents.
- Metals are **malleable,** which means they are capable of being extended or shaped by the blows of a hammer. (For example, gold, Au, can be hammered into very thin sheets without breaking.)
- At 20 °C, all metals are solids except mercury, which is a liquid.

There is more variation in the characteristics of the **nonmetal** elements. Some of them are gases at room temperature and pressure, some are solids, and one is a liquid. They have different colors and different textures. The definitive quality shared by all nonmetals is that they do not have the characteristics mentioned above for metals. For example, sulfur is a dull yellow solid that does not conduct heat or electric currents well and is not malleable. It shatters into pieces when hit with a hammer.

Sulfur is brittle, not malleable. When solid sulfur is hammered (below), it shatters into many pieces.

Gold, like other metals, is malleable; it can be hammered into thin sheets (as shown above).

A few of the elements have some but not all of the characteristics of metals. These elements are classified as **metalloids** or **semimetals.** Authorities disagree to some extent about which elements belong in this category, but the elements in yellow boxes in the image on the left are commonly classified as metalloids.

OBJECTIVE 15

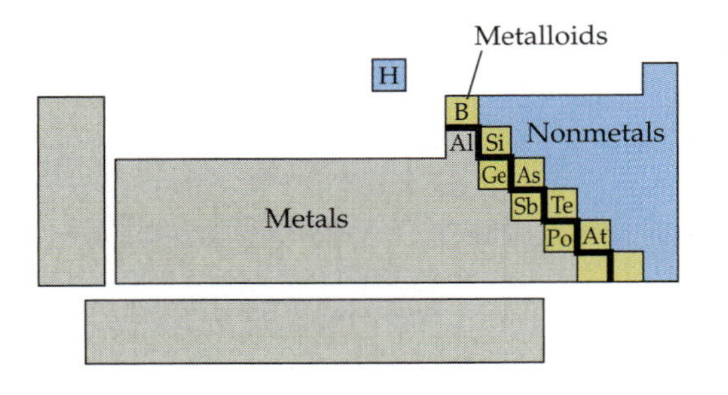


The portion of the periodic table that contains the metallic elements is shown here in gray, and the portion that contains the nonmetallic elements is shown in light blue. The stair-step line that starts between B and Al on the periodic table and descends between Al and Si, Si and Ge, and so on separates the metal elements from the nonmetal elements. The metals are below and to the left of this line, and the nonmetals are above and to the right of it. Most of the elements that have two sides of their box forming part of the stair-step line are metalloids. Aluminum is usually considered a metal.

OBJECTIVE 16

Main-group or representative elements

Transition metals

Inner transition metals

It is often useful to refer to whole blocks of elements on the periodic table. The elements in groups 1, 2, and 13 through 18 (the "A" groups) are sometimes called the **representative elements.** They are also called the **main-group elements.** The elements in groups 3 through 12 (the "B" groups) are often called the **transition metals.** The 28 elements at the bottom of the table are called the **inner transition metals.**

The horizontal rows on the periodic table are called **periods.** There are seven periods in all. The first period contains only two elements: hydrogen,[1] H, and helium, He. The second period contains eight elements: lithium, Li, through neon, Ne. The fourth period consists of eighteen elements: potassium, K, through krypton, Kr.

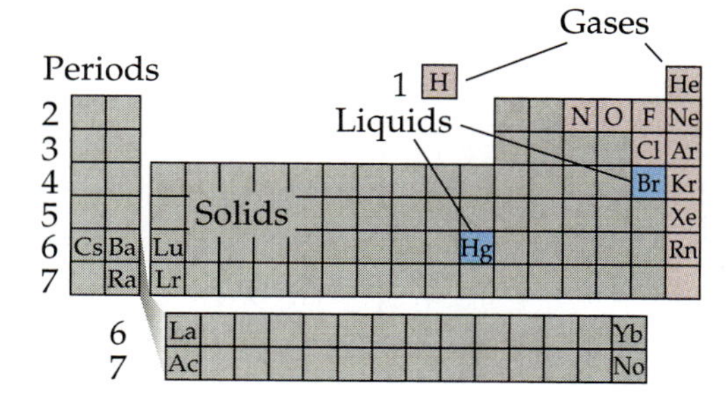


OBJECTIVE 17

OBJECTIVE 18

Note that the sixth period begins with cesium, Cs, which is element 55, and barium, Ba, which is 56, and then there is a gap that is followed by lutetium, Lu, element 71. The gap represents the proper location of the first row of the inner transition metals—that is, lanthanum, La, which is element 57, through ytterbium, Yb, which is element 70. These elements belong in the sixth period. Similarly, the second row of inner transition metals, the elements actinium, Ac, through nobelium, No, belong in the seventh period between radium, Ra, and lawrencium, Lr.

At room temperature (20 °C) and normal pressures, most of the elements are solid, two of them are liquid (Hg and Br), and eleven are gas (H, N, O, F, Cl, and the noble gases).

OBJECTIVE 18

Iodine is a solid; bromine is a liquid; chlorine is a gas. Most elements are solid at room temperature.

[1] The symbol for hydrogen is placed in different positions on different periodic tables. On some it is placed in group 1, and on others it is found at the top of group 17. Although there are reasons for placing it in these positions, there are also reasons why it does not belong in *either* position. Therefore, on our periodic table it is separate from both groups.

## Exercise 2.1 Elements and the Periodic Table

OBJECTIVE 12 OBJECTIVE 15 OBJECTIVE 16 OBJECTIVE 17 OBJECTIVE 18

Complete the following table.

| Name | Symbol | Group number | Metal, nonmetal, or metalloid? | Representative element, transition metal, or inner transition metal? | Number for period | Solid, liquid, or gas?[a] |
|---|---|---|---|---|---|---|
| | Al | | | | | |
| silicon | | | | | | |
| | Ni | | | | | |
| sulfur | | | | | | |
| | F | | | | | |
| potassium | | | | | | |
| | Hg | | | | | |
| uranium | | (no group number) | | | | |
| | Mn | | | | | |
| calcium | | | | | | |
| | | 17 | | | 4 | |
| | | 1B | | | 5 | |
| | | 14 | nonmetal | | | |

[a]At room temperature and normal pressures.

## Exercise 2.2 Group Names and the Periodic Table

OBJECTIVE 13

Write the name of the group or family on the periodic table to which each of the following elements belongs.

a. helium

b. Cl

c. magnesium

d. Na

## 2.4 The Structure of the Elements

A scanning tunneling microscope image of gold atoms in a thin gold film.

What makes one element different from another? To understand the answer to this question, you need to know about their internal structure. If you were to cut a piece of pure gold in half, and then divide one of those halves in half again and divide one of those halves in half, and continue to do that over and over, eventually the portion remaining could not be further divided and still be gold. This portion is a gold atom. The element gold consists of gold atoms, the element carbon consists of carbon atoms, and so on. To understand what makes one element different from another, we need to look inside the atom.

### The Atom

The **atom** is the smallest part of the element that retains the chemical characteristics of the element itself. (You will be better prepared to understand descriptions of the elements' chemical characteristics after reading more of this book. For now, it is enough to know that the chemical characteristics of an element include how it combines with other elements to form more complex substances.) For our purposes, we can think of the atom as a sphere with a diameter of about $10^{-10}$ meters. This is about a million times smaller than the diameter of the period at the end of this sentence. If the atoms in your body were an inch in diameter, you would have to worry about bumping your head on the moon.

A half-carat diamond contains 5,000,000,000,000,000,000,000 carbon atoms. Lined up they would stretch to the sun.

Because atoms are so small, there are a tremendous number of them in even a small sample of an element. A half-carat diamond contains about $5 \times 10^{21}$ atoms of carbon. If these atoms, tiny as they are, were arranged in a straight line with each one touching its neighbors, the line would stretch from here to the sun.

If we could look inside the gold atom, we would find that it is composed of three types of particles: protons, neutrons, and electrons.[2] Every gold atom in nature, for example, has 79 protons, 118 neutrons, and 79 electrons. Gold is different from phosphorus, because natural phosphorus atoms have 15 protons, 16 neutrons, and 15 electrons.

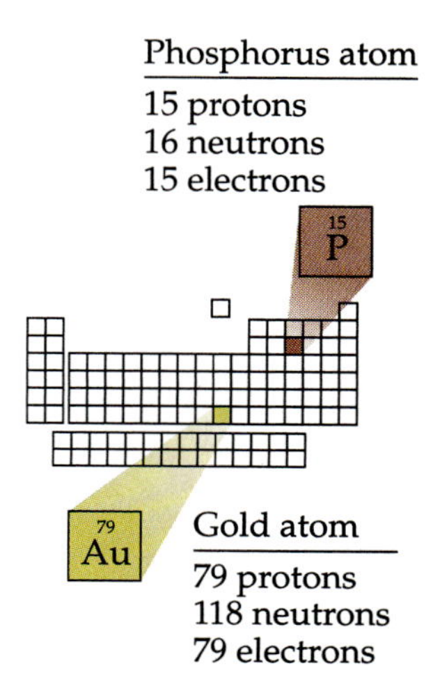


The particles within the atom are *extremely* tiny. A penny weighs about 2.5 grams, and a neutron, which is the most massive of the particles in the atom, weighs only $1.6750 \times 10^{-24}$ gram. The protons have about the same mass as the neutrons, but the electrons have about $\frac{1}{2000}$ times as much mass. Because the masses of the particles are so small, a more convenient unit of measurement has been devised for them. An **atomic mass unit** (also called the unified mass unit) is $\frac{1}{12}$ the mass of a carbon atom that has 6 protons, 6 neutrons, and 6 electrons. The modern abbreviation for atomic mass unit is u, but amu is commonly used.

**Protons** have a positive charge, **electrons** have a negative charge, and **neutrons** have no charge. Charge, a fundamental property of matter, is difficult to describe. Most definitions focus less on what it *is* than on what it *does*. For example, we know that objects of opposite charge attract each other and that objects of the same charge repel each other. An electron has a charge that is opposite but equal in magnitude to the charge of a proton. We arbitrarily assign the electron a charge of $-1$, so the charge of a proton is considered to be $+1$.

## The Nucleus

OBJECTIVE 20

Modern atomic theory tells us that even though the protons and neutrons represent most of the *mass* of the atom, they actually occupy a very small part of its *volume*. These particles cling together to form the incredibly small core of the atom, which is called the **nucleus.** Compared to the typical atom's diameter, which we described earlier as being about $10^{-10}$ meter, the diameter of a typical nucleus is about $10^{-15}$ meter. Thus nearly all the mass of the atom and all of its positive charge are found in a nucleus of about $\frac{1}{100,000}$ the diameter of the atom itself. If an atom were the size of the earth, the diameter of the nucleus would be just a little longer than a football field. If the nuclei of the atoms in your body were about an inch in diameter, you'd have to stand on the dark side of the earth to avoid burning your hair in the sun.

[2]The physicists will tell you that the proton and neutron are themselves composed of simpler particles. Because it is not useful to the chemist to describe atoms in terms of these more fundamental particles, they will not be discussed here.

## The Electron

*If I seem unusually clear to you, you must have misunderstood what I said.*

Alan Greenspan,
Chairman of the Federal Reserve Board

OBJECTIVE 20

*It is probably as meaningless to discuss how much room an electron takes up as to discuss how much room a fear, an anxiety, or an uncertainty takes up.*

Sir James Hopwood Jeans,
English mathematician, physicist, and astronomer (1877–1946)

Describing the modern view of the electron may not be as difficult as explaining the U.S. Federal Reserve Board's monetary policy, but it is still a significant challenge. We do *not* think that electrons are spherical particles orbiting the nucleus like planets around the sun. Scientists agree that electrons are outside the nucleus, but describing what they are doing out there or even explaining what they *are* turns out to be a difficult task. One way to deal with this difficulty is to disregard the question of what electrons are and how they move and focus only on the negative charge they generate. Chemists do this with the help of a model in which each electron is visualized as generating a cloud of negative charge that surrounds the nucleus. In Figure 2.8 we use the element carbon as an example.

Most of the carbon atoms in a diamond in a necklace have 6 protons, 6 neutrons, and 6 electrons. The protons and neutrons are in the nucleus, which is surrounded by a cloud of negative charge created by the 6 electrons. You will learn more about the shapes and sizes of different atoms' electron clouds in Chapter 11. For now, we will continue to picture the electron-charge clouds of all the atoms as spherical (Figure 2.8).

**Figure 2.8**
**The Carbon Atom**

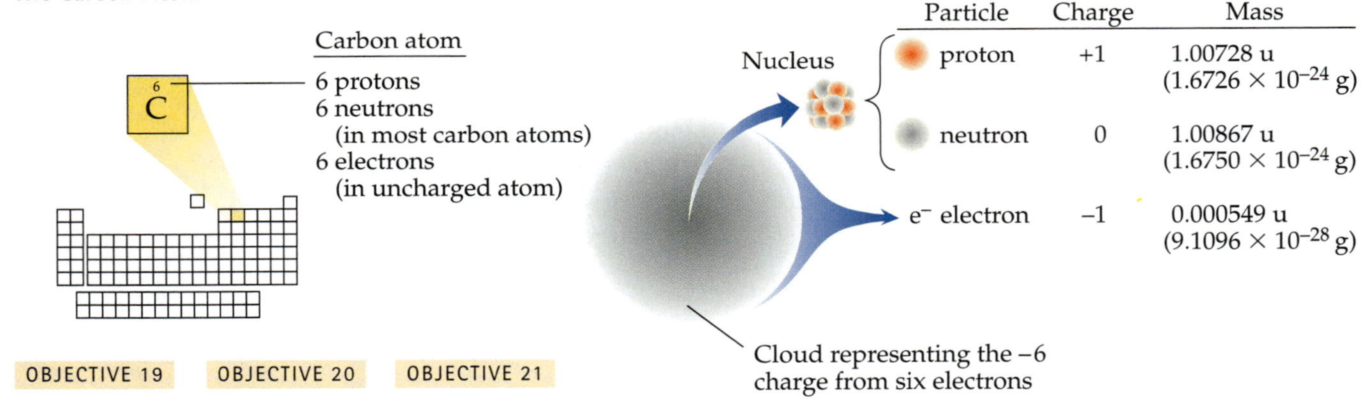


OBJECTIVE 19 OBJECTIVE 20 OBJECTIVE 21

## Ions

OBJECTIVE 22

Sometimes, when the elements form more complex substances, their atoms lose or gain electrons. Before this change, the atoms have an equal number of protons and electrons, and because protons and electrons have an equal but opposite charge, these atoms are initially uncharged overall. When an uncharged atom gains or loses one or more electrons, it forms a charged particle, called an **ion**. For example, when an atom loses one or more electrons, it will have more protons than electrons and more plus charge than minus charge. Thus it will have an overall positive charge. An atom that becomes a positively charged ion is called a **cation.** For example, uncharged sodium atoms have 11 protons and 11 electrons. They commonly lose 1 of these electrons to form +1 cations. A sodium cation's overall charge is +1 because its 11 protons have a charge of +11, and its remaining 10 electrons have a charge of −10. The sum of +11 and −10 is +1 (Figure 2.9). The symbol for a specific cation is written with the charge as a superscript on the right side of the element symbol. If the charge is +1, the convention is to write + (without the 1), so the symbol for the +1 sodium cation is $Na^+$. Aluminum atoms commonly lose 3 of their electrons to form +3 cations. The cations are +3 because each aluminum cation has a charge of +13 from its 13 protons and a charge of −10 from its 10 remaining electrons. The sum is +3. The symbol for this cation is $Al^{3+}$. (Note that the 3 comes before the +.)

Some atoms can gain electrons. When an atom gains one or more electrons, it then has more electrons than protons and more minus charge than plus charge. An atom that becomes negatively charged because of an excess of electrons is called an **anion,** a negatively charged ion. For example, uncharged chlorine atoms have 17 protons and 17 electrons. They commonly gain 1 electron to form −1 anions. The anions are −1 because their 17 protons have a charge of +17, and their 18 electrons have a charge of −18, giving a sum of −1. The anion's symbol is $Cl^-$, again without the 1. As illustrated in Figure 2.9, oxygen atoms commonly form anions with a −2 charge, $O^{2-}$, by gaining 2 electrons and therefore changing from 8 protons and 8 electrons to 8 protons (+8) and 10 electrons (−10).

OBJECTIVE 22

**Figure 2.9**
**Sodium and Oxygen Ions**

OBJECTIVE 22

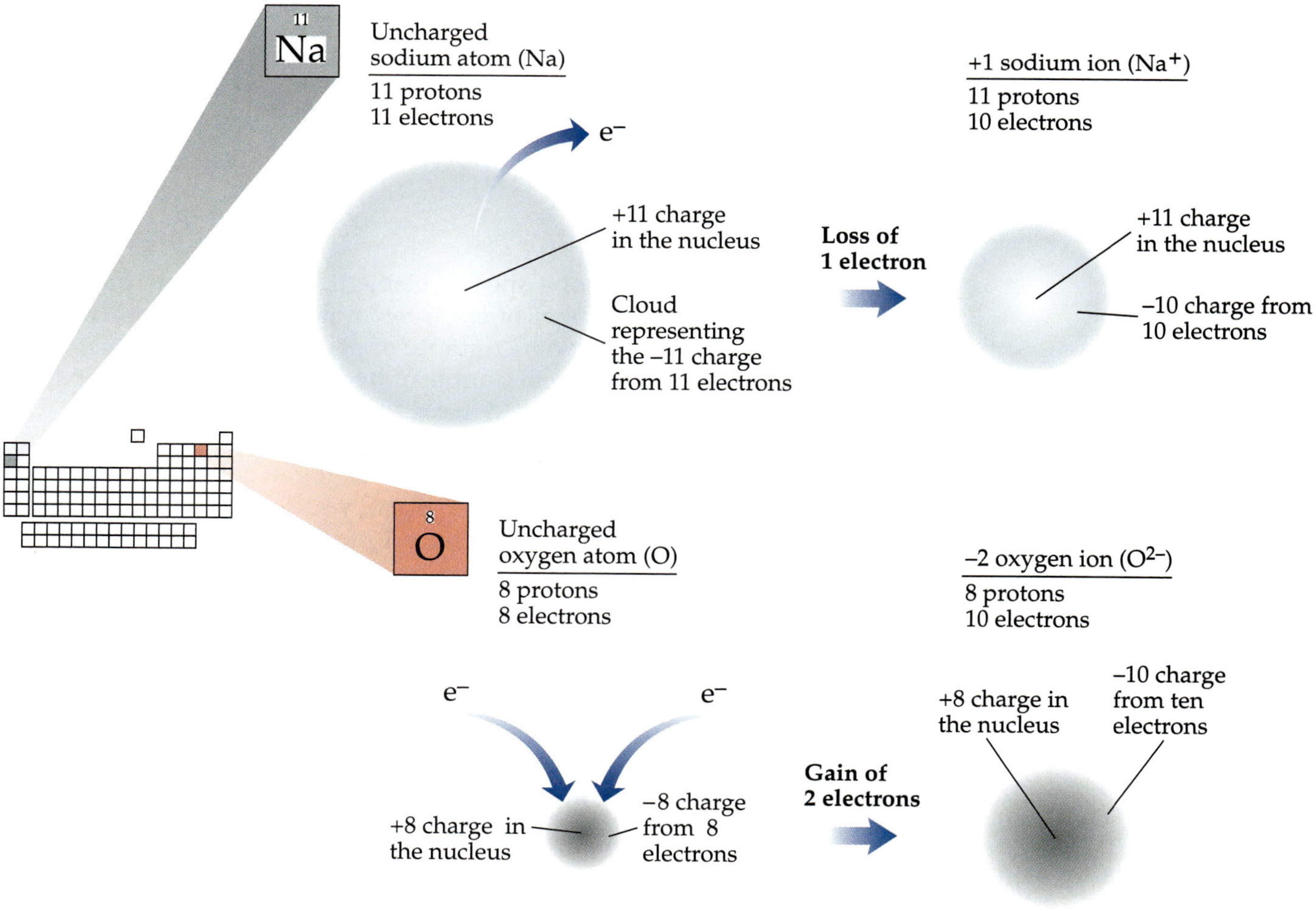

### Example 2.1 Cations and Anions

OBJECTIVE 22

Identify each of the following as a cation or an anion, and determine the charge on each.

a. a nitrogen atom with 7 protons and 10 electrons

b. a gold atom with 79 protons and 78 electrons

***Solution***

a. Seven protons have a +7 charge, and 10 electrons have a −10 total charge for a sum of **−3.** Therefore, a nitrogen atom with 7 protons and 10 electrons is an **anion.**

b. 79 protons have a charge of +79, and 78 electrons have a −78 charge, for a sum of **+1.** Therefore, a gold atom with 79 protons and 78 electrons is a **cation.**

### Exercise 2.3 Cations and Anions

OBJECTIVE 22

Identify each of the following as a cation or an anion, and determine the charge on each.

a. a magnesium atom with 12 protons and 10 electrons

b. a fluorine atom with 9 protons and 10 electrons

## Isotopes

Although all of the atoms of a specific element have the same number of protons (and the same number of electrons in uncharged atoms), they do not necessarily all have the same number of neutrons. For example, when the hydrogen atoms in a normal sample of hydrogen gas are analyzed, we find that of every 5000 atoms, 4999 have 1 proton and 1 electron, but 1 in 5000 of these atoms has 1 proton, 1 neutron, and 1 electron. This form of hydrogen is often called deuterium. Moreover, if you collected water from the cooling pond of a nuclear power plant, you would find that a very small fraction of its hydrogen atoms have 1 proton, 2 neutrons, and 1 electron (Figure 2.10). This last form of hydrogen, often called tritium, is unstable and therefore radioactive.

All of these atoms are hydrogen atoms because they have the chemical characteristics of hydrogen. For example, they all combine with oxygen atoms to form water. The chemical characteristics of an atom are determined by its number of protons (which is equal to the number of electrons if the atom is uncharged), not by its number of neutrons. Because atoms are assigned to elements on the basis of their chemical characteristics, an

**element** can be defined as a substance whose atoms have the same number of protons. When an element has two or more species of atoms, each with the same number of protons but a different number of neutrons, the different species are called **isotopes.**

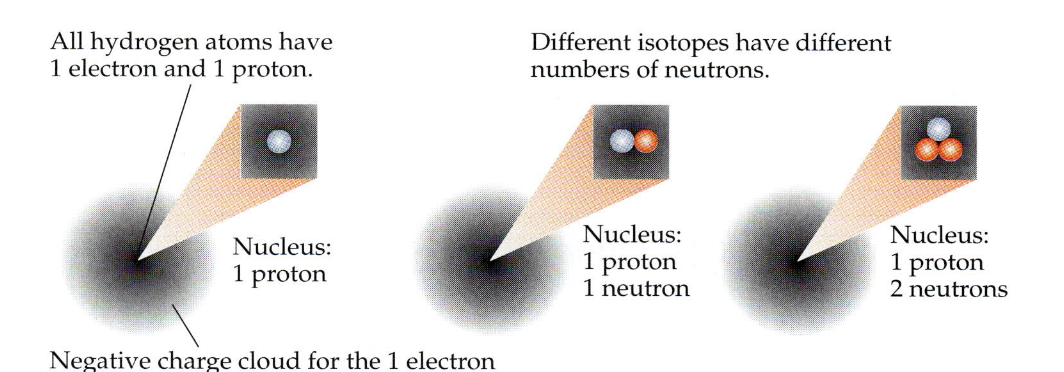


**Figure 2.10**
**Isotopes of Hydrogen**

## Atomic Number and Mass Number

OBJECTIVE 23

The number of protons in an atom—which is also the number of electrons in an uncharged atom—is known as the element's **atomic number.** Each element's atomic number can be found above its symbols on the periodic table. Because it displays the atomic numbers, the periodic table can be used to determine the number of protons and electrons in an uncharged atom of any element. For example, the atomic number of phosphorus is 15, so we know that there are 15 protons and 15 electrons in each uncharged atom of phosphorus.

The atomic number tells the number of protons in an atom of the element.

15
P

The sum of the numbers of protons and neutrons in the nucleus of an atom is called the atom's **mass number;** thus **isotopes** have the same atomic number but different mass numbers. To distinguish one isotope from another, the symbol for the element is often followed by the mass number of the isotope. For example, the mass number of the most common isotope of hydrogen, with 1 proton and no neutrons, is 1, so its symbol is H-1. The other natural isotope of hydrogen, with 1 proton and 1 neutron, has a mass number of 2 and the symbol H-2. Tritium, H-3, the radioactive form of hydrogen, has a mass number of 3. All of these isotopes of hydrogen have an atomic number of 1.

Nineteen of the elements found in nature have only one naturally occurring form. For example, all the aluminum atoms found in nature have 13 protons and 14 neutrons. Their mass number is 27.

$$\text{Number of protons} + \text{Number of Neutrons} = \text{Mass Number}$$
$$13 \quad + \quad 14 \quad = \quad 27$$

The other naturally occurring elements are composed of more than one isotope. For example, in a sample of the element tin, Sn (Figure 2.11), all

## Special Topic 2.1 Why Create New Elements?

At the Gesellschaft fur Schwerionenforschung (GSI), or Society for Heavy-Ion Research, in Germany, scientists create new elements by bombarding one kind of atom with another kind in the expectation that some of them will fuse. For example, for two weeks in 1994, the scientists bombarded a lead-208 target with a beam of nickel-62 atoms, producing four atoms of element 110, with a mass number of 269. Likewise, during 18 days in December of that year, they bombarded a bismuth-209 target with nickel-64 atoms, creating three atoms of element 111, with a mass number of 272.

Some of the best minds in physics are working on such projects and spending large amounts of money on the necessary equipment. Yet the newly created atoms are so unstable that they decay into other elements in less than a second. Do these results justify all of the time, money, and brainpower being poured into them? Would scientists' efforts be better spent elsewhere? Why do they do it?

One of the reasons these scientists devote themselves to the creation of new elements is to test their theories about matter. For example, the current model being used to describe the nucleus of the atom suggests that there are "*magic*" *numbers* of protons and neutrons that lead to relatively stable isotopes. The numbers 82, 126, and 208 are all magic numbers, exemplified by the extreme stability of lead-208, which has 82 protons and 126 neutrons. Other magic numbers suggest that an atom with 114 protons and 184 neutrons would also be especially stable. Researchers at the Flerov Laboratory of Nuclear Reactions in Dubna, Russia, were able to make two isotopes of element 114 (with 173 and 175 neutrons) by bombarding plutonium targets with calcium atoms. Both isotopes, particularly the heavier one, were significantly more stable than other isotopes of comparable size. Scientists are now attempting to make an isotope with 114 protons and 184 neutrons, in hopes that it will last even longer.

Society for Heavy-Ion Research facility in Germany

The technology developed to create these new elements is also being used for medical purposes. In a joint project with the Heidelberg Radiology Clinic and the German Cancer Research Center, GSI has constructed a heavy-ion therapy unit for the treatment of inoperable cancers. Here, the same equipment used to accelerate beams of heavy atoms toward a target in order to make new elements is put to work shooting beams of carbon atoms at tumors. When used on deep-seated, irradiation-resistant tumors, the carbon particle beam is thought to be superior to the traditional radiation therapy. Because the heavier carbon atoms are less likely to scatter, and because they release most of their energy at the end of their path, they are easier to focus on the cancerous tumor.

the atoms have 50 protons, but tin atoms can have 62, 64, 65, 66, 67, 68, 69, 70, 72, or 74 neutrons. Thus tin has ten natural isotopes with mass numbers of 112, 114, 115, 116, 117, 118, 119, 120, 122, and 124.

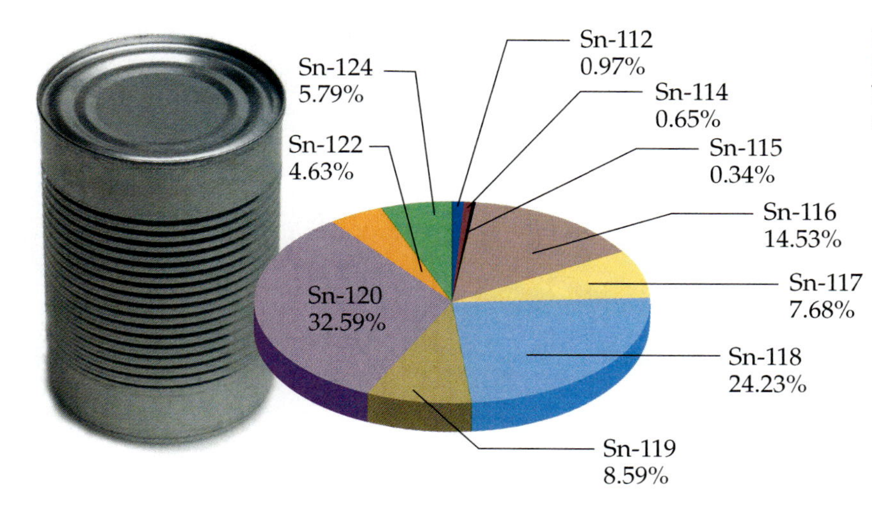


**Figure 2.11**
**Isotopes of Tin**
The tin in this tin can is composed of ten different isotopes.

To learn about a special notation used to describe isotopes, visit the following Web site:
**www.chemplace.com/college/**

## 2.5 Common Elements

Most people look at a gold nugget and see a shiny metallic substance that can be melted down and made into jewelry. A chemist looks at a substance like gold and visualizes the internal structure responsible for those external characteristics. Now that we have discussed some of the general features of atoms and elements, we can return to the model of solid, liquid, and gas structures presented in Section 2.1 and continue in our quest to visualize the particle nature of matter.

### Gas, Liquid, and Solid Elements

In Section 2.1, we pictured gases as independent spherical particles moving in straight-line paths in a container that is largely empty space. This image is most accurate for the noble gases (He, Ne, Ar, Kr, Xe, and Rn). Each noble gas particle consists of a single atom. When we picture the helium gas in a helium-filled balloon, each of the particles in our image is a helium atom containing 2 protons and 2 neutrons in a tiny nucleus surrounded by a cloud of negative charge generated by 2 electrons (Figure 2.12).

OBJECTIVE 24

The particles in hydrogen gas are quite different. Instead of the single atoms found in helium gas, the particles in hydrogen gas are pairs of hydrogen atoms. Each hydrogen atom has only 1 electron, and single, or "unpaired," electrons are less stable than electrons that are present as

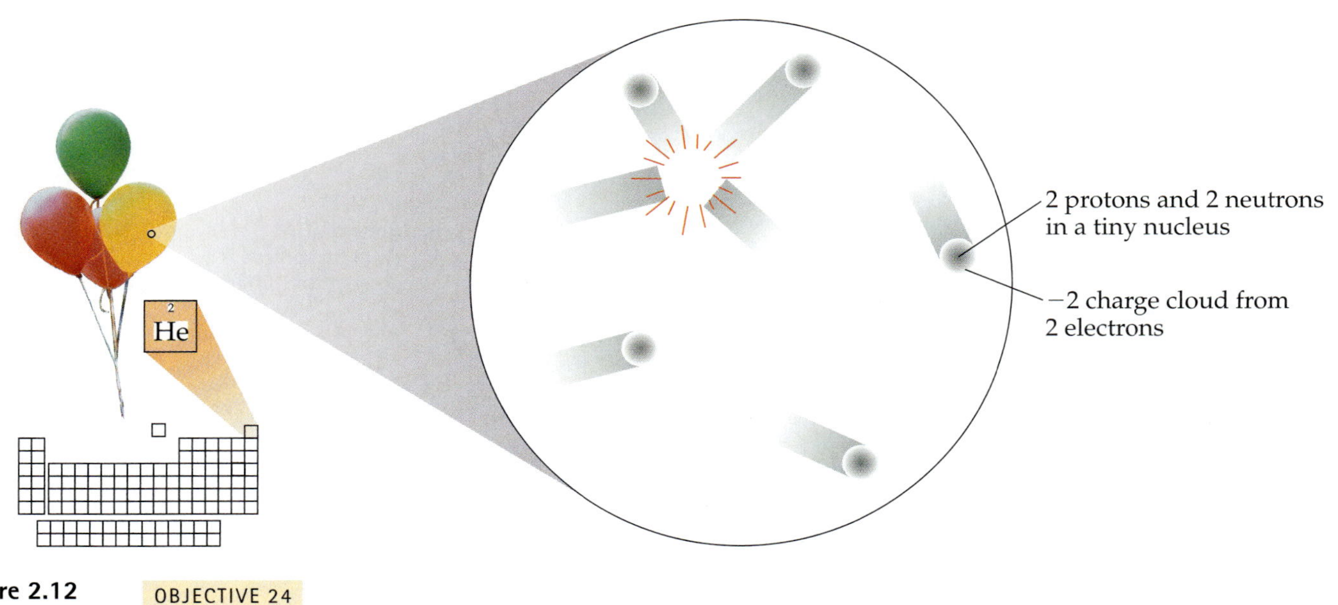


**Figure 2.12**
**Helium Gas**

OBJECTIVE 24

OBJECTIVE 25

pairs. (*Stability* is a relative term that describes resistance to change. A stable system is less likely to change than an unstable system.) To gain the greater stability conferred by pairing, the single electron of one hydrogen atom can pair up with a single electron of another hydrogen atom. The 2 electrons are then shared between the two hydrogen atoms and create a bond that holds the atoms together. Thus hydrogen gas is described as $H_2$. We call this bond between atoms that results from the sharing of 2 electrons a **covalent bond.** The pair of hydrogen atoms is a **molecule,** which is an uncharged collection of atoms held together with covalent bonds. Two hydrogen atoms combine to form one hydrogen molecule.

**Figure 2.13**
**Hydrogen Electron Cloud**

OBJECTIVE 25

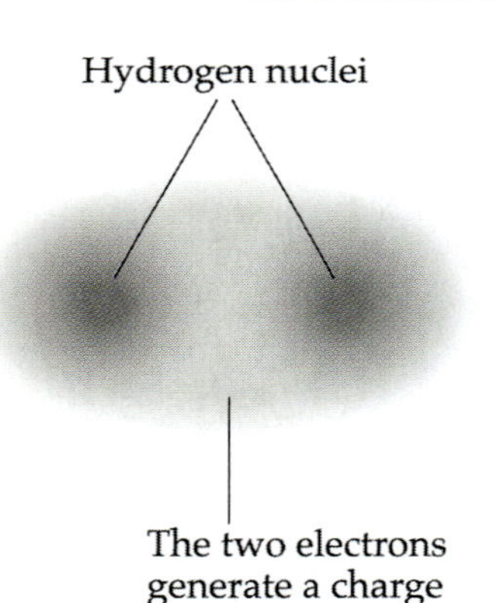

Hydrogen nuclei
The two electrons generate a charge cloud surrounding both nuclei.

The negative-charge cloud created by the electrons in the covalent bond between hydrogen atoms surrounds both of the hydrogen nuclei (Figure 2.13). Even though the shape depicted in Figure 2.13 is a better description of the $H_2$ molecule's electron cloud, there are two other common ways of illustrating the $H_2$ molecule. The first image below shows a **space-filling model.** This type of model emphasizes individual atoms in the molecule more than the image in Figure 2.13 does but still conveys a somewhat realistic idea of the electron-charge clouds that surround the atoms. The second image below is a **ball-and-stick model,** in which balls represent atoms and sticks represent covalent bonds. This model gives greater emphasis to the bond that holds the hydrogen atoms together (Figure 2.14).

**Figure 2.14**
**Molecule Models**

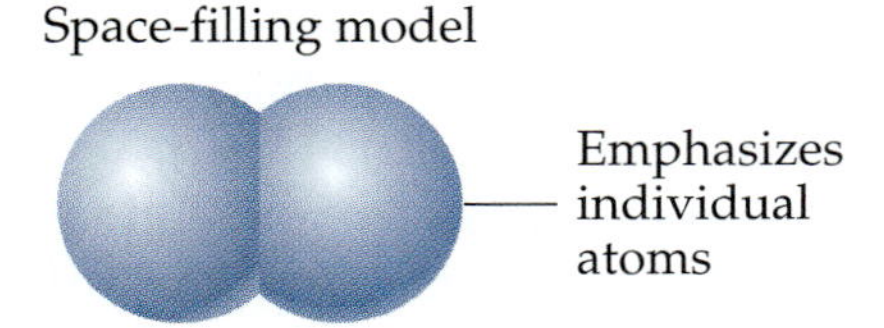

Combining space-filling molecular models with our gas model, Figure 2.15 depicts hydrogen gas as being very similar to helium gas (see Figure 2.12), except that each of the particles is a hydrogen molecule.

OBJECTIVE 24

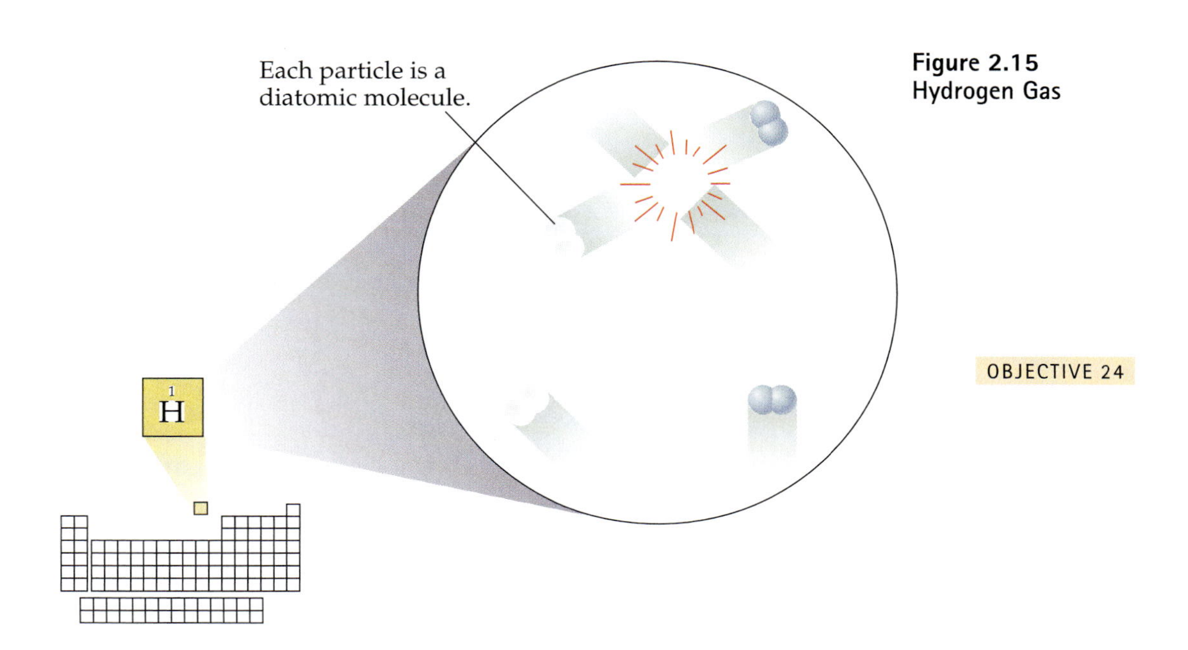
Each particle is a diatomic molecule.
1
H

**Figure 2.15** Hydrogen Gas

OBJECTIVE 24

Because hydrogen molecules are composed of two atoms, they are called **diatomic.** The elements nitrogen, oxygen, fluorine, chlorine, bromine, and iodine are also composed of diatomic molecules, so they are described as $N_2$, $O_2$, $F_2$, $Cl_2$, $Br_2$, and $I_2$. Like the hydrogen atoms in $H_2$ molecules, the two atoms in each of these molecules are held together by a covalent bond that is due to the sharing of two electrons. Nitrogen, oxygen, fluorine, and chlorine are gases at room temperature and pressure, so a depiction of gaseous $N_2$, $O_2$, $F_2$, and $Cl_2$ would be very similar to the image of $H_2$ in Figure 2.15.

OBJECTIVE 25

Bromine, which is a liquid at room temperature, is pictured like the liquid shown in Figure 2.2, except that each of the particles is a diatomic molecule (Figure 2.16, on the next page).

Solid iodine consists of a very ordered arrangement of $I_2$ molecules. To give a clearer idea of this arrangement, the first image in Figure 2.17 (next page) shows each $I_2$ as a ball-and-stick model. The second image shows the close packing of these molecules in the iodine solid. Remember that the particles of any substance, including solid iodine, are in constant motion. The solid structure presented in Figure 2.1 applies to iodine, except that we must think of each particle in it as being an $I_2$ molecule.

**Figure 2.16**
**Liquid Bromine, $Br_2$**

OBJECTIVE 25

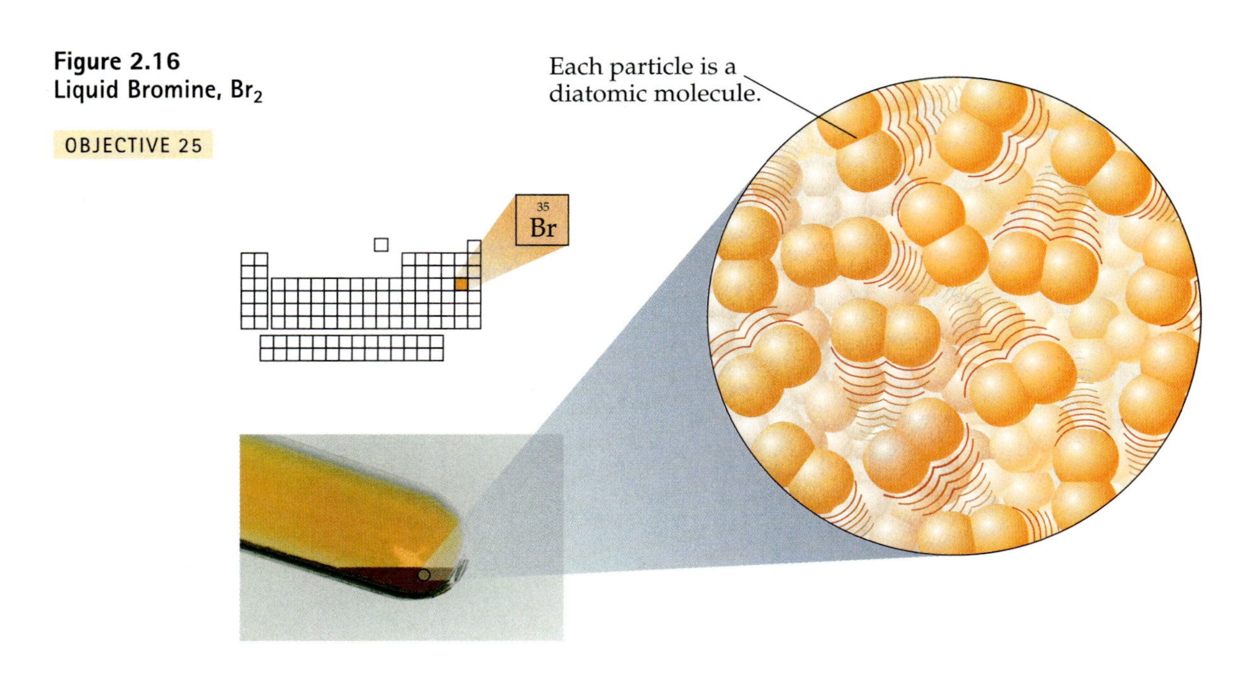


**Figure 2.17**
**Solid Iodine, $I_2$**

OBJECTIVE 25

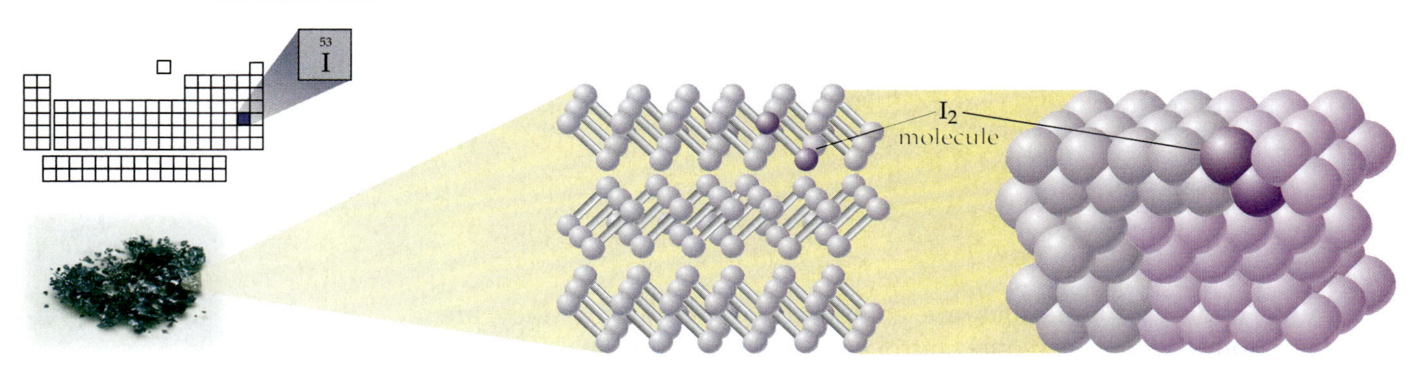


You can review the information in this section and see particles of neon, oxygen, bromine, and iodine in motion at the following Web site:

**www.chemplace.com/college/**

## Metallic Elements

The metallic elements are used for a lot more than building bridges and making jewelry. Platinum is used in a car's catalytic converter to help decrease air pollution. Titanium is mixed with other metals to construct orthopedic appliances, such as artificial hip joints. Zinc is used to make dry cell batteries. Some of the properties that give metallic elements such wide applications can be explained by an *expanded* version of the model of solids presented in Section 2.1. (One of the characteristics of a useful model is that it can be expanded to describe, explain, and predict a greater variety of phenomena.)

According to the expanded model, each atom in a metallic solid has released one or more electrons, and these electrons move freely throughout the solid. When the atoms lose the electrons, they become cations. The cations form the structure we associate with solids, and the released electrons flow between them like water flows between islands in the ocean. This model, which is often called the sea-of-electrons model, can be used to explain some of the definitive characteristics of metals. For example, the freely moving electrons make metallic elements good conductors of electric currents.

OBJECTIVE 27

Figure 2.18 shows a typical arrangement of atoms in a metallic solid and also shows how you might visualize one plane of this structure. Try to picture a cloud of negative charge, produced by mobile electrons, surrounding the cations in the solid.

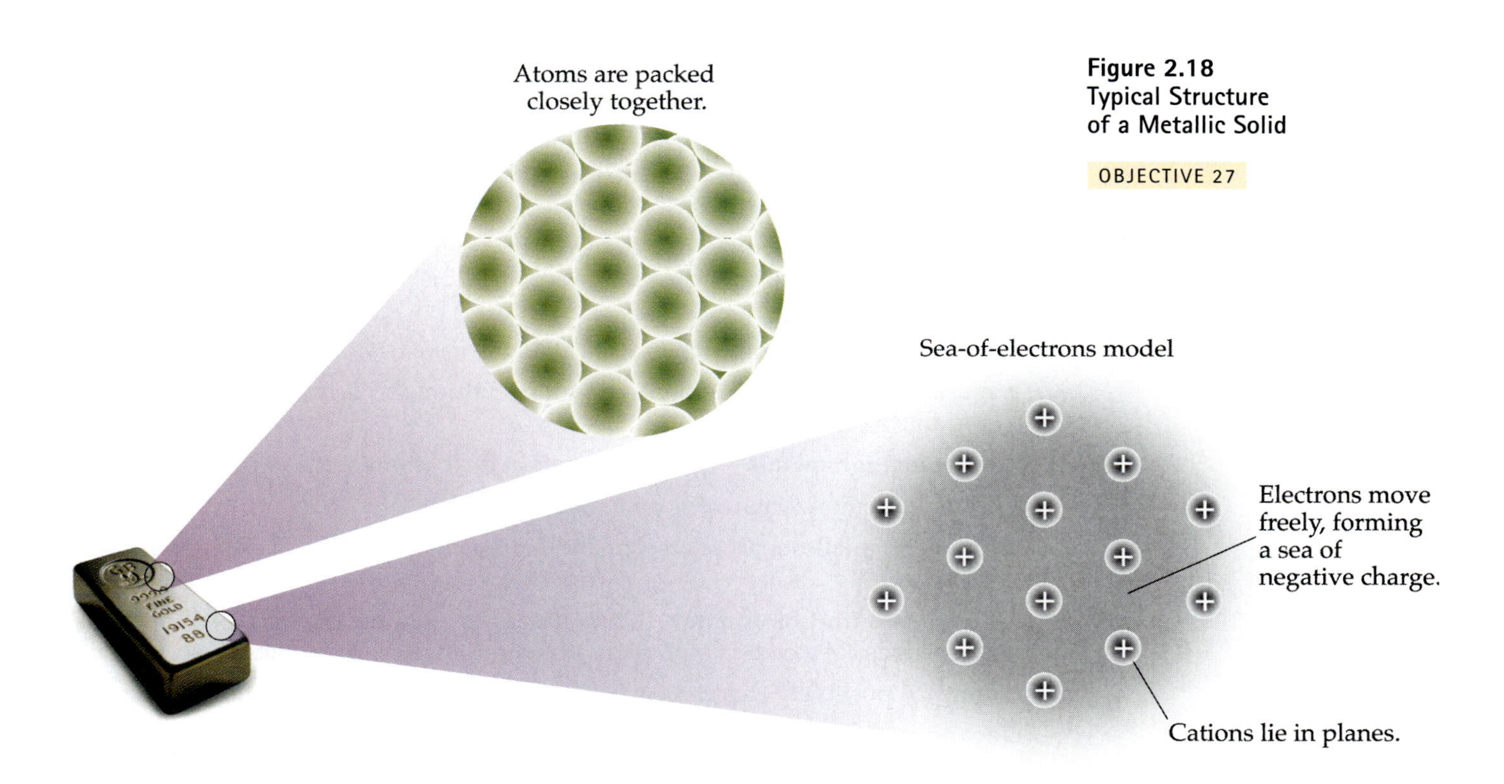
Atoms are packed closely together.
Sea-of-electrons model
Electrons move freely, forming a sea of negative charge.
Cations lie in planes.
FINE GOLD
19154
88

**Figure 2.18**
**Typical Structure of a Metallic Solid**

OBJECTIVE 27

You have just completed your first big step on the road to understanding chemistry. The new knowledge of the elements that you have gained from this chapter will help you with Chapter 3, where you will learn how elements combine to form more complex substances. An understanding of elements and the substances they form will prepare you to learn about the chemical changes that these substances undergo in yourself and the world around you.

## Chapter Glossary

**Model** A simplified approximation of reality.

**Solid** The state in which a substance has a definite shape and volume at a constant temperature.

**Liquid** The state in which a substance has a constant volume at a constant temperature but can change its shape.

**Gas** The state in which a substance can easily change shape and volume.

**Evaporation or vaporization** The conversion of a liquid to a gas.

**Element** A substance that cannot be chemically converted into simpler substances; a substance in which all of the atoms have the same number of protons and therefore the same chemical characteristics.

**Group** All the elements in a given column on the periodic table; also called a *family*.

**Family** All the elements in a given column on the periodic table; also called a *group*.

**Metals** The elements that (1) have a metallic luster, (2) conduct heat and electric currents well, (3) are malleable, and (4) are solids at 20 °C (with the exception of the metal mercury, which remains a liquid at that temperature).

**Malleable** Capable of being extended or shaped by the blows of a hammer.

**Nonmetals** The elements that do not have the characteristics of metals. Some of the nonmetals are gases at room temperature and pressure, some are solids, and one is a liquid. Various colors and textures occur among the nonmetals.

**Metalloids or semimetals** The elements that have some but not all of the characteristics of metals.

**Representative elements** The elements in groups 1, 2, and 13 through 18 (the "A" groups) on the periodic table; also called the *main-group elements*.

**Main-group elements** The elements in groups 1, 2, and 13 through 18 (the "A" groups) on the periodic table, also called the *representative elements*.

**Transition metals** The elements in groups 3 through 12 (the "B" groups) on the periodic table.

**Inner transition metals** The 28 elements at the bottom of the periodic table.

**Periods** The horizontal rows on the periodic table.

**Atom** The smallest part of an element that retains the chemical characteristics of the element.

**Atomic mass unit (u or amu)** Unit of measurement for the masses of particles; $1/12$ the mass of a carbon atom that has 6 protons, 6 neutrons, and 6 electrons.

**Proton** A positively charged particle found in the nucleus of an atom.

**Electron** A negatively charged particle found outside the nucleus of an atom.

**Neutron** An uncharged particle found in the nucleus of an atom.

**Nucleus** The extremely small, positively charged core of the atom.

**Ion** Any charged particle, whether positively or negatively charged.

**Cation** An ion formed from an atom that has lost one or more electrons and thus has become positively charged.

**Anion** An ion formed from an atom that has gained one or more electrons and thus has become negatively charged.

**Isotopes** Atoms that have the same number of protons but different numbers of neutrons. They have the same atomic number but different mass numbers.

**Atomic number** The number of protons in an atom's nucleus. It establishes the element's identity.

**Mass number** The sum of the number of protons and neutrons in an atom's nucleus.

**Covalent bond** A link between atoms that results from their sharing two electrons.

**Diatomic** Composed of paired atoms. The diatomic elements are $H_2$, $N_2$, $O_2$, $F_2$, $Cl_2$, $Br_2$, and $I_2$.

**Molecule** An uncharged collection of atoms held together with covalent bonds.

**Space-filling model** A way of representing a molecule to show a somewhat realistic image of the electron-charge clouds that surround the molecule's atoms.

**Ball-and-stick model** A representation of a molecule that uses balls for atoms and sticks for covalent bonds.

You can test yourself on the glossary terms at the following Web site:

www.chemplace.com/college/

## Chapter Objectives

**The goal of this chapter is to teach you to do the following.**

1. Define all the terms in the Chapter Glossary.

### Section 2.1 Solids, Liquids, and Gases

2. Describe solids, liquids, and gases in terms of the particle nature of matter, the degree of motion of the particles, and the degree of attraction between the particles.
3. Describe the relationship between temperature and particle motion.
4. Explain why solids have a definite shape and volume at a constant temperature.
5. Explain why solids usually expand when heated.
6. Describe the structural changes that occur when a solid is converted into a liquid by heating.
7. Explain why most substances expand when they change from a solid to a liquid.
8. Explain why liquids adjust to take the shape of their container and why they have a constant volume at a constant temperature.
9. Describe the structural changes that occur in the conversion of a liquid to a gas.
10. Explain why gases expand to take the shape and volume of their container.

### Section 2.2 The Chemical Elements

11. Give the names and symbols for the common elements. (Check with your instructor to find out which names and symbols you need to know.)

**Section 2.3 The Periodic Table of the Elements**

12. Given a periodic table, identify the number of the group to which each element belongs. (Check with your instructor to find out which numbering system you are expected to know.)
13. Given a periodic table, identify the alkali metals, alkaline earth metals, halogens, and noble gases.
14. List the characteristics of metals.
15. Given a periodic table, classify each element as a metal, a nonmetal, or a metalloid (semimetal).
16. Given a periodic table, classify each element as a representative element (or main-group element), a transition metal, or an inner transition metal.
17. Given a periodic table, write or identify the number of the period on the table to which each element belongs.
18. Classify each element as a solid, a liquid, or a gas at room temperature.

**Section 2.4 The Structure of the Elements**

19. Give the abbreviations, charges, and *relative* sizes of protons, neutrons, and electrons.
20. Describe the nuclear model of the atom, including the general location of the protons, neutrons, and electrons; the relative size of the nucleus compared to the size of the atom; and the modern description of the electron.
21. Describe the carbon atom, including a rough sketch that shows the negative charge created by its electrons.
22. Given the number of protons and electrons in a cation or anion, determine its charge.
23. Given the atomic number, state the number of protons in an atom of an element, and vice versa.

**Section 2.5 Common Elements**

24. Describe the following substances in terms of the nature of the particles that form their structure: the noble gases, hydrogen gas, nitrogen gas, oxygen gas, fluorine gas, chlorine gas, bromine liquid, and iodine solid.
25. Describe the hydrogen molecule, including a rough sketch of the charge cloud created by its electrons.
26. List the diatomic elements ($H_2$, $N_2$, $O_2$, $F_2$, $Cl_2$, $Br_2$, and $I_2$).
27. Describe the "sea-of-electrons" model for metallic structure.

## Review Questions

1. Define the term *matter*.
2. Look around you. What do you see that has a length of about a meter? What do you see that has a mass of about a gram?

## Key Ideas

**Complete the following statements by writing one these words or phrases in each blank.**

| | |
|---|---|
| $10^{-10}$ | gas |
| $10^{-15}$ | liquid |
| 1/100,000 | loses |
| 25 | molecule |
| 115 | motion |
| 0.1% | neutrons |
| 70% | particles |
| 99.9% | protons |
| atomic numbers | rapid, ever-changing, and unrestricted |
| attract | repel |
| chemical | simplified but useful |
| cloud | simpler |
| empty space | single atom |
| escape | solid |
| expand | straight-line path |
| expands | sun |
| extended or shaped | temperature |
| flow | ten times |
| gains | vertical column |

3. Scientific models are like architects' models; they are ________________ representations of something real.
4. According to the model presented in this chapter, all matter is composed of tiny ________________.
5. According to the model presented in this chapter, particles of matter are in constant ________________.
6. According to the model presented in this chapter, the amount of motion of particles is proportional to ________________.
7. Solids, gases, and liquids differ in the freedom of motion of their particles and in how strongly the particles ________________ each other.
8. An increase in temperature usually causes a solid to ________________ somewhat.
9. Particles in a liquid are still close together, but there is generally more ________________ between them than in a solid. Thus, when a solid substance melts to form a liquid, it usually ________________ to fill a slightly larger volume.
10. The freedom of movement of particles in a liquid allows liquids to ________________, taking on the shape of their container.
11. When a liquid's temperature is higher, its particles are moving faster and are therefore more likely to ________________ from the liquid.
12. The average distance between particles is about ________________ the diameter of each particle. This leads to the gas particles themselves taking up only about ________________ of the total volume. The other ________________ of the total volume is empty space. In contrast, the particles of a liquid fill about ________________ of the liquid's total volume.
13. According to our model, each particle in a gas moves freely in a(n) ________________ until it collides with another gas particle or with the particles of a liquid or solid.

14. A liquid has a constant volume, but the ________________ movement of the gas particles allows gases to expand to fill a container of any shape or volume.
15. Elements are substances that cannot be chemically converted into ________________ ones.
16. By the year 2001, ________________ elements had been discovered, but ________________ of these elements are not found naturally on the earth, and chemists do not generally work with them.
17. The periodic table is arranged in such a way that elements in the same ____________ have similar characteristics.
18. Metals are malleable, which means they are capable of being ____________ by the blows of a hammer.
19. At room temperature (20 °C) and normal pressures, most of the elements are ________________, two of them are ________________ (Hg and Br), and eleven are ________________ (H, N, O, F, Cl, and the noble gases).
20. For our purposes, we can think of the atom as a sphere with a diameter of about ________________ meter.
21. A ½-carat diamond contains about $5 \times 10^{21}$ atoms of carbon. If these atoms, tiny as they are, were arranged in a straight line with each one touching its neighbors, the line would stretch from here to the ____________.
22. We know that objects of opposite charge attract each other and that objects of the same charge ________________ each other.
23. The diameter of a typical nucleus is about ________________ meter.
24. Nearly all the mass of the atom and all of its positive charge are found in a nucleus of about ________________ the diameter of the atom itself.
25. Chemists use a model for electrons in which each electron is visualized as generating a(n) ________________ of negative charge that surrounds the nucleus.
26. When an atom ________________ one or more electrons, it will have more protons than electrons and more plus charge than minus charge. Thus it becomes a cation, which is an ion with a positive charge.
27. When an atom ________________ one or more electrons, it then has more electrons than protons and more minus charge than plus charge. Thus it becomes an anion, which is an ion with a negative charge.
28. Although all of the atoms of a specific element have the same number of ________________ (and the same number of electrons in uncharged atoms), they do not necessarily all have the same number of ____________.
29. Atoms are assigned to elements on the basis of their ________________ characteristics.
30. Because it displays the ________________, the periodic table can be used to determine the number of protons and electrons in an uncharged atom of any element.
31. Each noble gas particle consists of a(n) ________________.
32. Hydrogen gas is very similar to helium gas, except that each of the particles is a hydrogen ________________.

## Chapter Problems

### Section 2.1 Solids, Liquids, and Gases

For each of the questions in this section, illustrate your written answers with simple drawings of the particles that form the structures of the substances mentioned. You do not need to be specific about the nature of the particles. Think of them as simple spheres, and draw them as circles.

33. If you heat white sugar very carefully, it will melt.

OBJECTIVE 2 OBJECTIVE 3 OBJECTIVE 4 OBJECTIVE 6

    a. Before you begin to heat the sugar, the sugar granules maintain a constant shape and volume. Why?
    b. As you begin to heat the solid sugar, what changes are taking place in its structure?
    c. What happens to the sugar's structure when sugar melts?

34. If the pistons and cylinders in your car engine get too hot, the pistons can get stuck in the cylinders, causing major damage to the engine. Why does this happen?

OBJECTIVE 5

35. Ethylene glycol, an automobile coolant and antifreeze, is commonly mixed with water and added to car radiators. Because it freezes at a lower temperature than water and boils at a higher temperature than water, it helps to keep the liquid in your radiator from freezing or boiling.

OBJECTIVE 2 OBJECTIVE 3 OBJECTIVE 6 OBJECTIVE 8

    a. At a constant temperature, liquid ethylene glycol maintains a constant volume but takes on the shape its container. Why?
    b. The ethylene glycol–water mixture in your car's radiator heats up as you drive. What is happening to the particles in the liquid?
    c. If you spill some engine coolant on your driveway, it evaporates without leaving any residue. Describe the process of evaporation of liquid ethylene glycol, and explain what happens to the ethylene glycol particles that you spilled.

36. When a small container of liquid ammonia is opened in a classroom, in a short time everyone in the room can smell it.

OBJECTIVE 2 OBJECTIVE 9 OBJECTIVE 10

    a. Describe the changes that take place when liquid ammonia vaporizes to form a gas.
    b. Why does the gaseous ammonia expand to fill the whole room?
    c. Why does the gaseous ammonia occupy a much greater volume than the liquid ammonia?

37. As the summer sun heats up the air at the beach, what is changing for the air particles?

38. A drop of food coloring is added to water. With time, it spreads evenly through the water so that the mixture is all the same color.
    a. Describe what is happening to the food coloring and water particles as the coloring spreads into the water.
    b. When a drop of food coloring is added to two bowls of water, one at 20 °C and the other at 30 °C, the coloring spreads more quickly in the bowl at the higher temperature. Why?

39. A gaseous mixture of air and gasoline enters the cylinders of a car engine and is compressed into a smaller volume before being ignited. Explain why gases can be compressed.

### Section 2.2 The Chemical Elements and Section 2.3 The Periodic Table

OBJECTIVE 11

40. Write the chemical symbols that represent the following elements.
    a. chlorine
    b. zinc
    c. phosphorus
    d. uranium

OBJECTIVE 11

41. Write the chemical symbols that represent the following elements.
    a. hydrogen
    b. calcium
    c. mercury
    d. xenon

OBJECTIVE 11

42. Write the chemical symbols that represent the following elements.
    a. iodine
    b. platinum
    c. boron
    d. gold

OBJECTIVE 11

43. Write the element names that correspond to the following symbols.
    a. C b. Cu c. Ne d. K

OBJECTIVE 11

44. Write the element names that correspond to the following symbols.
    a. O b. Br c. N d. Si

OBJECTIVE 11

45. Write the element names that correspond to the following symbols.
    a. Ba b. F c. Sr d. Cr

OBJECTIVE 11 OBJECTIVE 12 OBJECTIVE 15 OBJECTIVE 16 OBJECTIVE 17

46. Complete the following table.

| Element name | Element symbol | Group number on periodic table | Metal, nonmetal, or metalloid? | Representative element, transition metal, or inner transition metal? | Number of period |
|---|---|---|---|---|---|
| | Na | | | | |
| tin | | | | | |
| | He | | | | |
| nickel | | | | | |
| | Ag | | | | |
| aluminum | | | | | |
| | Si | | | | |
| | | 16 | | | 3 |
| | | 2B | | | 6 |

OBJECTIVE 11 OBJECTIVE 12 OBJECTIVE 15 OBJECTIVE 16 OBJECTIVE 17

47. Complete the following table.

| Element name | Element symbol | Group number on periodic table | Metal, nonmetal, or metalloid? | Representative element, transition metal, or inner transition metal? | Number of period |
|---|---|---|---|---|---|
| | Mg | | | | |
| lead | | | | | |
| argon | | | | | |
| | Cd | | | | |
| chromium | | | | | |
| | Fe | | | | |
| | | 7B | | | 4 |
| | | 1A | | | 2 |
| | | 18 | | | 5 |

48. Write the name of the group or family to which each of the following belongs. OBJECTIVE 13
    a. bromine
    b. neon
    c. potassium
    d. beryllium
49. Write the name of the group or family to which each of the following belongs. OBJECTIVE 13
    a. strontium
    b. lithium
    c. iodine
    d. xenon
50. Identify each of the following elements as a solid, a liquid, or a gas at room temperature and pressure. OBJECTIVE 18
    a. krypton, Kr
    b. Br
    c. antimony, Sb
    d. F
    e. germanium, Ge
    f. S
51. Identify each of the following elements as a solid, a liquid, or a gas at room temperature and pressure. OBJECTIVE 18
    a. chlorine
    b. Se
    c. mercury
    d. W
    e. xenon
    f. As
52. Which two of the following elements would you expect to be most similar: lithium, aluminum, iodine, oxygen, and potassium?
53. Which two of the following elements would you expect to be most similar: nitrogen, chlorine, barium, fluorine, and sulfur?
54. Write the name and symbol for the elements that fit the following descriptions.
    a. the halogen in the third period
    b. the alkali metal in the fourth period
    c. the metalloid in the third period
55. Write the name and symbol for the elements that fit the following descriptions.
    a. the noble gas in the fifth period
    b. the alkaline earth metal in the sixth period
    c. the representative element in group 2A and the third period
56. Which element would you expect to be malleable, manganese or phosphorus? Why?
57. Which element would you expect to conduct electric currents well, aluminum or iodine? Why?

### Section 2.4 The Structure of the Elements

58. Describe the nuclear model of the atom, including the general location of the protons, neutrons, and electrons; the relative size of the nucleus compared to the size of the atom; and the modern description of the electron. OBJECTIVE 20
59. Describe the carbon atom, and make a rough sketch. OBJECTIVE 21
60. Identify each of the following as a cation or an anion, and determine the charge on each. OBJECTIVE 22
    a. a lithium ion with 3 protons and 2 electrons
    b. a sulfur ion with 16 protons and 18 electrons

OBJECTIVE 22

61. Identify each of the following as a cation or an anion, and determine the charge on each.
    a. an iodine ion with 53 protons and 54 electrons
    b. an iron ion with 26 protons and 23 electrons
62. Write definitions of the terms *atomic number* and *mass number*. Which of these can vary without changing the element? Why? Which of these cannot vary without changing the element? Why?
63. Write the atomic number for each of the following elements.
    a. oxygen
    b. Mg
    c. uranium
    d. Li
    e. lead
    f. Mn
64. Write the atomic number for each of the following elements.
    a. sodium
    b. As
    c. strontium
    d. Pu
    e. iron
    f. Se
65. Explain how two atoms of oxygen can be different.
66. Write the name and symbol for the elements that fit the following descriptions.
    a. 27 protons in the nucleus of each atom
    b. 50 electrons in each uncharged atom
    c. 18 electrons in each +2 cation
    d. 10 electrons in each −1 anion
67. Write the name and symbol for the elements that fit the following descriptions.
    a. 78 protons in the nucleus of each atom
    b. 42 electrons in each uncharged atom
    c. 80 electrons in each +3 cation
    d. 18 electrons in each −2 anion

### Section 2.5 Common Elements

OBJECTIVE 25

68. Describe the hydrogen molecule, including a rough sketch of the electron-charge cloud created by its electrons.
69. Write definitions for the terms *atom* and *molecule,* and use them to explain the difference between hydrogen atoms and hydrogen molecules.

OBJECTIVE 24

70. Describe the structure of each of the following substances, including a description of the nature of the particles that form each structure.
    a. neon gas
    b. bromine liquid
    c. nitrogen gas

OBJECTIVE 24

71. Describe the structure of each the following substances, including a description of the nature of the particles that form each structure.
    a. chlorine gas
    b. iodine solid
    c. argon gas

OBJECTIVE 27

72. Describe the "sea-of-electrons" model for metallic solids.

**Discussion Question**

73. When you heat solid iodine, it goes directly from solid to gas. Describe the process by which iodine particles escape from the solid to the gas form. What characteristics must iodine particles have to be able to escape? Draw a picture to illustrate your answer.

*Some of the substances described in this chapter provide the colors we see in paint.*

CHAPTER 3

# Chemical Compounds



LOOK AROUND YOU. Do you think you see anything composed of just one element . . . any objects consisting only of carbon, or of gold, or of hydrogen? The correct answer is almost certainly no. If you are lucky enough to have a diamond ring, you have a piece of carbon that is almost pure (although a gemologist would tell you that diamonds contain slight impurities that give each stone its unique character). If you have a "gold" ring, you have a *mixture* of gold with other metals, that were added to give the ring greater strength.

Even though a few elements, such as carbon and gold, are sometimes found in elemental form in nature, most of the substances we see around us consist of two or more elements that have combined chemically to form more complex substances called compounds. For example, in nature, the element hydrogen is combined with other elements, such as oxygen and carbon, in compounds like the water and sugar used to make a soft drink. (Perhaps you are sipping one while you read.) In this chapter, you will learn to (1) define the terms *mixture* and *compound* more precisely, (2) distinguish among elements, compounds, and mixtures, (3) describe how elements combine to form compounds, (4) construct systematic names for some chemical compounds, and (5) describe the characteristics of certain kinds of chemical compounds. The chapter will also expand your ability to visualize the basic structures of matter.

*The flecks of gold in this pan are the only pure elements visible in this scene.*

## Review Skills

The presentation of information in this chapter assumes that you can already perform the tasks listed below. You can test your readiness to proceed by answering the Review Questions at the end of the chapter. This might also be a good time to read the Chapter Objectives, which precede the Review Questions.

- Describe the particle nature of solids, liquids, and gases. (Section 2.1)
- Convert between the names and symbols for the common elements. (Table 2.1)
- Given a periodic table, write the number of the group to which each element belongs. (Figure 2.7)
- Given a periodic table, identify the alkali metals, alkaline earth metals, halogens, and noble gases. (Section 2.3)
- Using a periodic table, classify elements as metals, nonmetals, or metalloids. (Section 2.7)
- Describe the nuclear model of the atom. (Section 2.4)
- Define the terms *ion*, *cation*, and *anion*. (Section 2.4)
- Define the terms *covalent bond*, *molecule*, and *diatomic*. (Section 2.5)
- Describe the covalent bond in a hydrogen molecule, $H_2$. (Section 2.5)

## 3.1 Classification of Matter

Before getting started on your chemistry homework, you go into the kitchen to make some pasta for your six-year-old nephew. You run water into a pan, adding a few shakes of salt, and while you're waiting for it to boil, you pour a cup of coffee. When the water begins to boil, you pour in the pasta. Then you add some sugar to your coffee.

Pure water, the sucrose in white sugar, and the sodium chloride in table salt are all examples of chemical compounds. A **compound** is a substance that contains two or more elements, the atoms of those elements always combining in the same whole-number ratio (Figure 3.1). There are relatively few chemical elements, but there are millions of chemical compounds. The compounds in our food fuel our bodies, and the compounds in gasoline fuel our cars. They can alter our moods and cure our diseases.

Water is composed of molecules that contain 2 atoms of hydrogen and 1 atom of oxygen. We describe the composition of water with the chemical formula $H_2O$. White sugar is a highly purified form of sucrose, whose chemical formula is $C_{12}H_{22}O_{11}$. Its molecules are composed of 12 carbon atoms, 22 hydrogen atoms, and 11 oxygen atoms. Sodium and chlorine atoms combine in a 1:1 ratio to form sodium chloride, NaCl, which is the primary ingredient in table salt.

**Figure 3.1**
**Elements Versus Compounds**

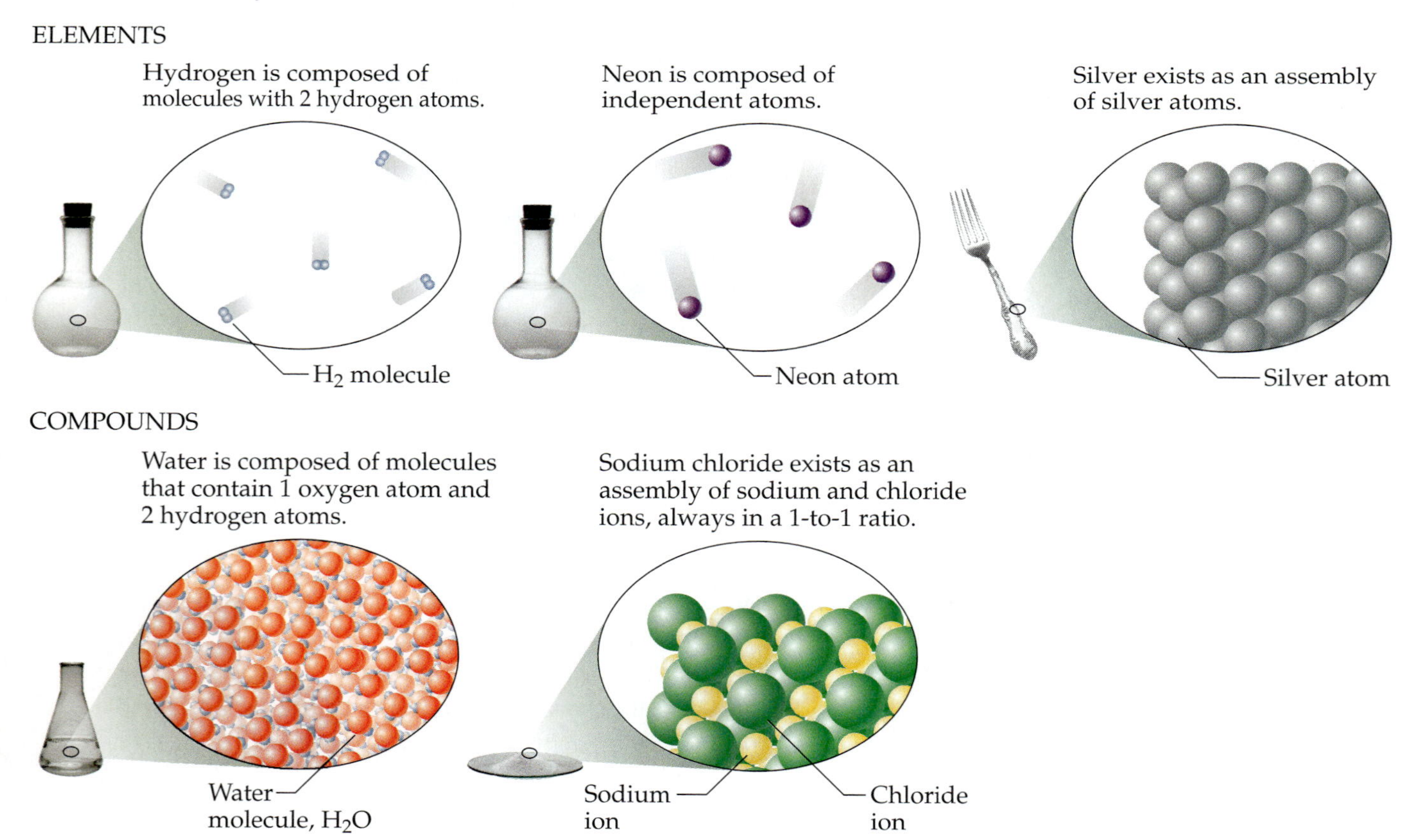
ELEMENTS
Hydrogen is composed of molecules with 2 hydrogen atoms.
H2 molecule
Neon is composed of independent atoms.
Neon atom
Silver exists as an assembly of silver atoms.
Silver atom
COMPOUNDS
Water is composed of molecules that contain 1 oxygen atom and 2 hydrogen atoms.
Water molecule, H2O
Sodium chloride exists as an assembly of sodium and chloride ions, always in a 1-to-1 ratio.
Sodium ion
Chloride ion

OBJECTIVE 2

Note that a **chemical formula** is a concise written description of the components of a chemical compound. It identifies the elements in the compound by their symbols and indicates the relative number of atoms of each element with subscripts. If an element symbol in a formula is not accompanied by a subscript, the relative number of atoms of that element is assumed to be 1.

Pure water, sodium chloride, and sucrose always have the composition described in their chemical formulas. In other words, their composition is constant. Elements, too, have a constant composition described by a chemical formula. (We have seen that the formula for hydrogen is $H_2$.) When a substance has a constant composition—when it can be described by a chemical formula—it must by definition be either an element or a compound, and it is considered a **pure substance.** For example, the symbol Na refers to pure sodium. The formula $Na_2CO_3$ refers to pure sodium carbonate and tells us that this compound is always composed of sodium, carbon, and oxygen in a constant atom ratio of 2:1:3.

OBJECTIVE 3

OBJECTIVE 4

**Mixtures** are samples of matter that contain two or more pure substances and have variable composition (Figure 3.2). For example, when salt, NaCl, and water, $H_2O$, are combined, we know that the resulting combination is a mixture because we can vary the percentage of these two pure substances. You can add 1, 2, or 10 teaspoons of salt to a pan of water, and the result will still be salt water.

OBJECTIVE 3

The following sample study sheet and Figure 3.3 show the questions you can ask to discover whether a sample of matter is an element, a compound, or a mixture.

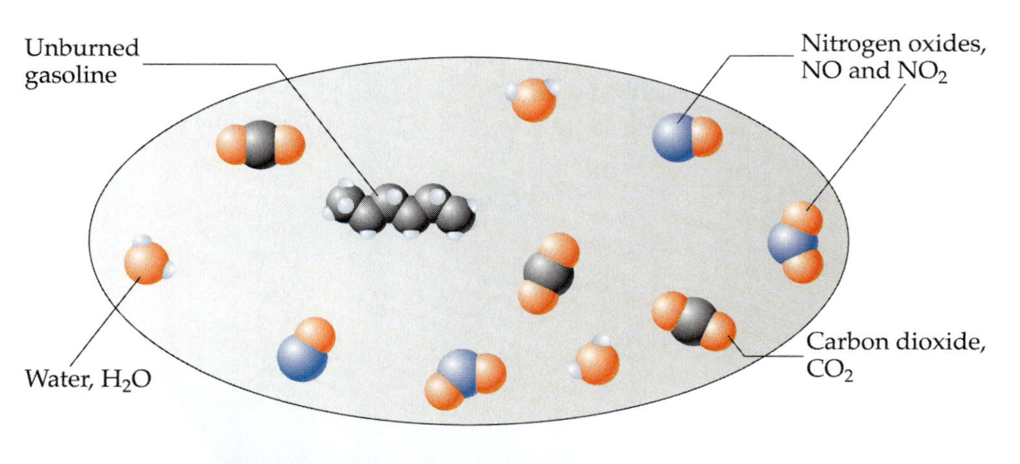


**Figure 3.2**
**Automobile Exhaust—a Mixture**
The components and amounts vary.

OBJECTIVE 3

## Sample Study Sheet 3.1

### Classification of Matter

OBJECTIVE 3

OBJECTIVE 4

**Tip-off** You are asked to classify a sample of matter as a pure substance or a mixture; or you are asked to classify a pure substance as an element or a compound.

**General Steps** The following general procedure is summarized in Figure 3.3.

- To classify a sample of matter as a pure substance or a mixture, ask one or both of the following questions:

  *Does it have a constant composition?* If it does, it is a pure substance. If it has variable composition, it is a mixture.

  *Can the sample as a whole be described with a chemical formula?* If it can, it is a pure substance. If it cannot, it is a mixture.

- To classify a pure substance as an element or a compound, ask the following question:

  *Can it be described with a single symbol?* If it can, it is an element. If its chemical formula contains two or more different element symbols, it is a compound.

**Example** See Example 3.1.

Figure 3.3
Classification of Matter

OBJECTIVE 3

OBJECTIVE 4

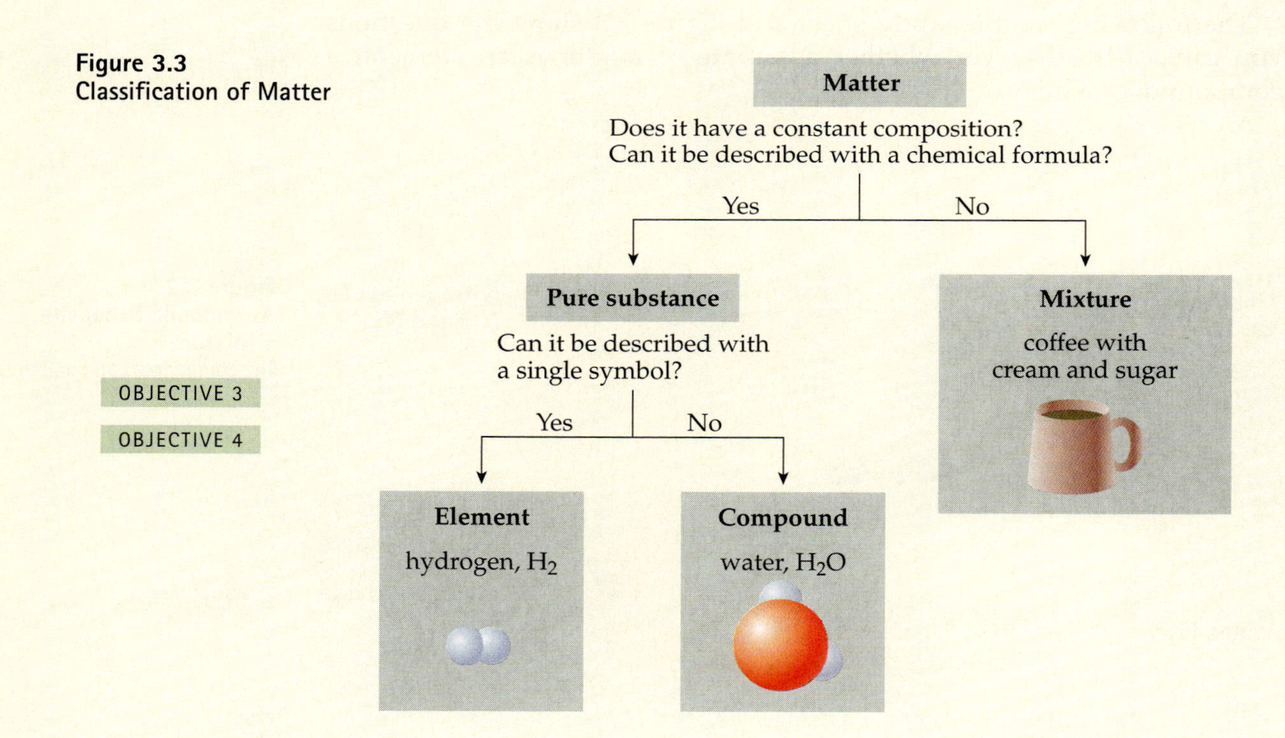
Matter
Does it have a constant composition?
Can it be described with a chemical formula?
Yes
No
Pure substance
Can it be described with
a single symbol?
Yes
No
Mixture
coffee with
cream and sugar
Element
hydrogen, $H_2$
Compound
water, $H_2O$

## EXAMPLE 3.1 Classification of Matter

OBJECTIVE 3 OBJECTIVE 4

Many of us have a bottle in our medicine cabinet containing a mild disinfectant that consists of hydrogen peroxide and water. The liquid is *about* 3% hydrogen peroxide, $H_2O_2$, and about 97% water. This percentage varies somewhat from bottle to bottle. Classify each of the following as a pure substance or a mixture. If it is a pure substance, is it an element or a compound?

a. the liquid disinfectant

b. the hydrogen peroxide, $H_2O_2$, used to make the disinfectant

c. the hydrogen used to make hydrogen peroxide

***Solution***

a. We know that the liquid disinfectant is a **mixture** for two reasons. It is composed of two pure substances ($H_2O_2$ and $H_2O$), and it has variable composition.

b. Because hydrogen peroxide can be described with a formula, $H_2O_2$, it must be a **pure substance.** Because the formula contains symbols for two elements, it represents a **compound.**

c. Hydrogen can be described with a single symbol, H or $H_2$, so it is a **pure substance** and an **element.**

## EXERCISE 3.1 Classification of Matter

OBJECTIVE 3 OBJECTIVE 4

The label on a container of double-acting baking powder tells us that it contains cornstarch, bicarbonate of soda (also called sodium hydrogen carbonate, $NaHCO_3$), sodium aluminum sulfate, and acid phosphate of calcium (which chemists call calcium dihydrogen phosphate, $Ca(H_2PO_4)_2$). Classify each of the following as a pure substance or a mixture. If it is a pure substance, is it an element or a compound?

a. calcium

b. calcium dihydrogen phosphate

c. double-acting baking powder

# 3.2 Compounds and Chemical Bonds

The percentage of $H_2O_2$ in the mixture of hydrogen peroxide and water that is used as a disinfectant can vary, but the percentage of hydrogen in the compound water is always the same. Why? One of the key reasons why the components of a given compound are always the same, and are present in the same proportions, is that the atoms in a compound are joined together by special kinds of attractions called **chemical bonds.** Because of the nature of these attractions, the atoms combine in specific ratios that give compounds their constant composition. This section will introduce the different types of chemical bonds and give you the skills necessary to predict the types of chemical bonds between atoms of different elements.

## Equal and Unequal Sharing of Electrons

Let's first consider the compound hydrogen chloride, HCl. When HCl is dissolved in water, the resulting mixture is called hydrochloric acid. Not only is this mixture a very common laboratory agent, but it is also used in food processing and to treat the water in swimming pools.

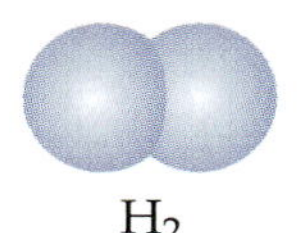


In Section 2.5, we learned about the bond between hydrogen atoms in $H_2$ molecules. We saw that the 2 electrons in the $H_2$ molecule are shared equally between the atoms and can be viewed as an electron-charge cloud surrounding the hydrogen nuclei. This sharing creates a covalent bond that holds the atoms together. There is also a covalent bond between the hydrogen atom and the chlorine atom in each molecule of HCl. It is very similar to the covalent bond in hydrogen molecules, with one important exception.

OBJECTIVE 5

The difference between the H–Cl bond and the H–H bond is that the hydrogen and chlorine atoms in HCl do not share the electrons in the bond equally. In the hydrogen–chlorine bond, the 2 electrons are attracted more strongly to the chlorine atom than to the hydrogen atom. The negatively charged electrons in the bond shift toward the chlorine atom, giving it a partial negative charge, $\delta-$, and giving the hydrogen atom a partial positive charge, $\delta+$ (Figure 3.4). The lower-case Greek delta, $\delta$, is a symbol that represents *partial* or *fractional*.

OBJECTIVE 7

When the electrons of a covalent bond are shared unequally, the bond is called a **polar covalent bond.** As a result of the unequal sharing of the electrons in the bond, a polar covalent bond has 1 atom with a partial positive charge, $\delta+$, and 1 atom with a partial negative charge, $\delta-$.

**Figure 3.4**
**Hydrogen Chloride Molecule**

OBJECTIVE 5

Electrons shift toward the chlorine atom, forming partial plus and minus charges.

$\delta+$ $\delta-$

Hydrogen attracts electrons less. H Cl Chlorine attracts electrons more.

If the electron-attracting ability of one atom in a bond is much greater than that of the others, there is a large shift in the electron cloud, and the partial charges are large. If the electron-attracting ability of one atom in a covalent bond is only slightly greater than that of the others, there is not much of a shift in the electron cloud, and the partial charges are small. When the difference in electron-attracting ability is negligible (or zero), the atoms in the bond will have no significant partial charges. We call this type of bond a **nonpolar covalent bond.** The covalent bond between hydrogen atoms in $H_2$ is an example of a nonpolar covalent bond.

OBJECTIVE 7

## Transfer of Electrons

Sometimes one atom in a bond attracts electrons so much more strongly than the other that 1 or more electrons are fully transferred from one atom to another. This commonly happens when metallic atoms combine with nonmetallic atoms. A nonmetallic atom usually attracts electrons so much more strongly than a metallic atom that 1 or more electrons shift from the metallic atom to the nonmetallic atom. For example, when the element sodium combines with the element chlorine to form sodium chloride, NaCl, the chlorine atoms attract electrons so much more strongly than the sodium atoms that 1 electron is transferred from each sodium atom to a chlorine atom.

When an electron is transferred completely from one uncharged atom to another, the atom that loses the electron is left with 1 more proton than electron and acquires a +1 charge overall. It therefore becomes a cation (Section 2.4). For example, when an uncharged sodium atom with 11 protons and 11 electrons loses an electron, it is left with 11 protons (a charge of +11) and 10 electrons (a charge of −10), yielding an overall +1 charge.

OBJECTIVE 6

$$\underset{\substack{11p/11e^- \\ +11 + (-11) = 0}}{Na} \rightarrow \underset{\substack{11p/10e^- \\ +11 + (-10) = +1}}{Na^+} + e^-$$

In contrast, an uncharged atom that gains an electron will have 1 more electron than proton, so it forms an anion with a −1 charge. When a chlorine atom gains an electron from a sodium atom, the chlorine atom changes from an uncharged atom with 17 protons and 17 electrons to an anion with 17 protons and 18 electrons and an overall −1 charge.

OBJECTIVE 6

$$\underset{\substack{17p/17e^- \\ +17 + (-17) = 0}}{Cl} + e^- \rightarrow \underset{\substack{17p/18e^- \\ +17 + (-18) = -1}}{Cl^-}$$

Atoms can transfer 1, 2, or 3 electrons. Thus cations can have a +1, +2, or +3 charge, and anions can have a −1, −2, or −3 charge.

Because particles with opposite charges attract each other, there is an attraction between cations and anions. This attraction is called an **ionic bond.** For example, when an electron is transferred from a sodium atom to a chlorine atom, the attraction between the +1 sodium cation and the −1 chlorine anion is an ionic bond (Figure 3.5).

You will see as you read more of this book that substances that have ionic bonds are very different from those that have all covalent bonds. For example, compounds that have ionic bonds, such as the sodium chloride in table salt, are solids at room temperature and pressure, but compounds with all covalent bonds, such as hydrogen chloride and water, can be gases and liquids as well as solids.

Salt, sodium chloride, is an ionic compound; water is molecular.

**Figure 3.5**
**Ionic Bond Formation**

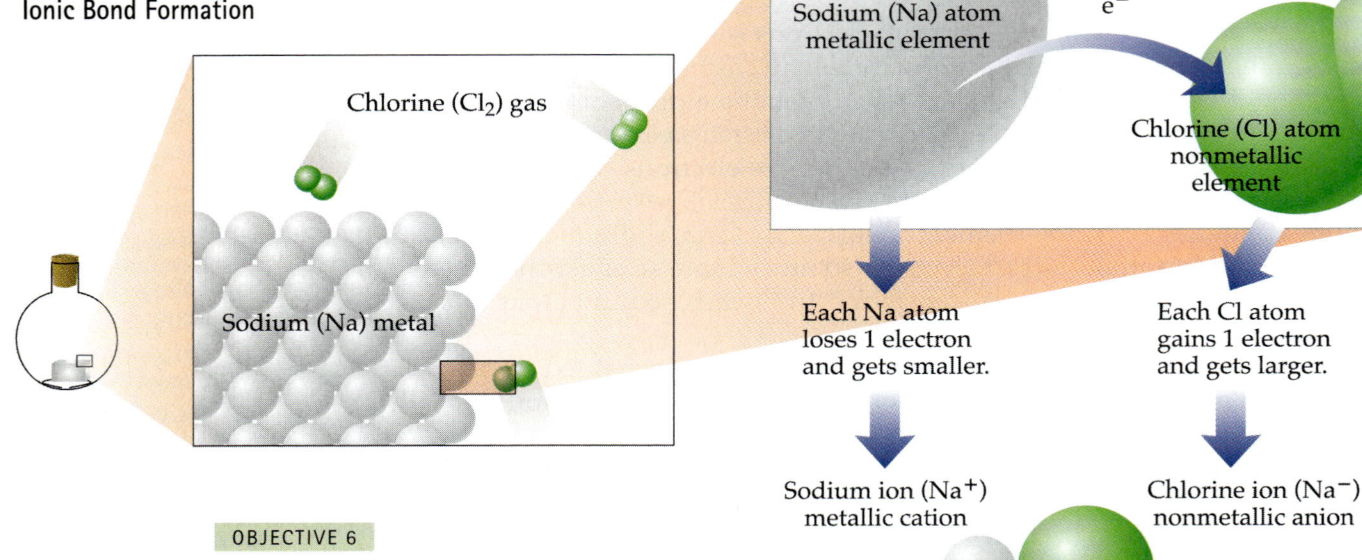


OBJECTIVE 6

## Summary of Covalent and Ionic Bond Formation

- When atoms of different elements form chemical bonds, the electrons in the bonds can shift from one bonding atom to another.
- The atom that attracts electrons more strongly acquires a negative charge, and the other atom acquires a positive charge.
- The more the atoms differ in their electron-attracting ability, the more the electron cloud shifts from one atom toward another.
- If there is a large enough difference in electron-attracting ability, 1, 2, or 3 electrons can be viewed as shifting completely from one atom to another. The atoms become positive and negative ions, and the attraction between them is called an ionic bond.
- If the electron transfer is significant but not enough to form ions, the atoms acquire partial positive and partial negative charges. The bond in this situation is called a polar covalent bond.
- If there is no shift of electrons or if the shift is negligible, no significant charges form, and the bond is a nonpolar covalent bond.

It might help, when thinking about these different kinds of bonds, to compare them to a game of tug-of-war between two people. The people are like the atoms with a chemical bond between them, and the rope is like the electrons in the bond. If the two people tugging have the same (or about the same) strength, then the rope will not move (or will not move much). This leads to a situation that is like the nonpolar covalent bond. If, on the other hand, one person is stronger than the other person, the rope will shift toward that person, the way the electrons in a polar covalent bond shift toward the atom that attracts them more. If one person can pull a lot harder than the other person can, then the stronger person pulls the rope right out of the hands of the weaker one. This is similar to the formation of ions and ionic bonds, when a nonmetallic atom pulls one or more electrons away from a metallic atom.

Figure 3.6 summarizes the general differences among nonpolar covalent bonds, polar covalent bonds, and ionic bonds.

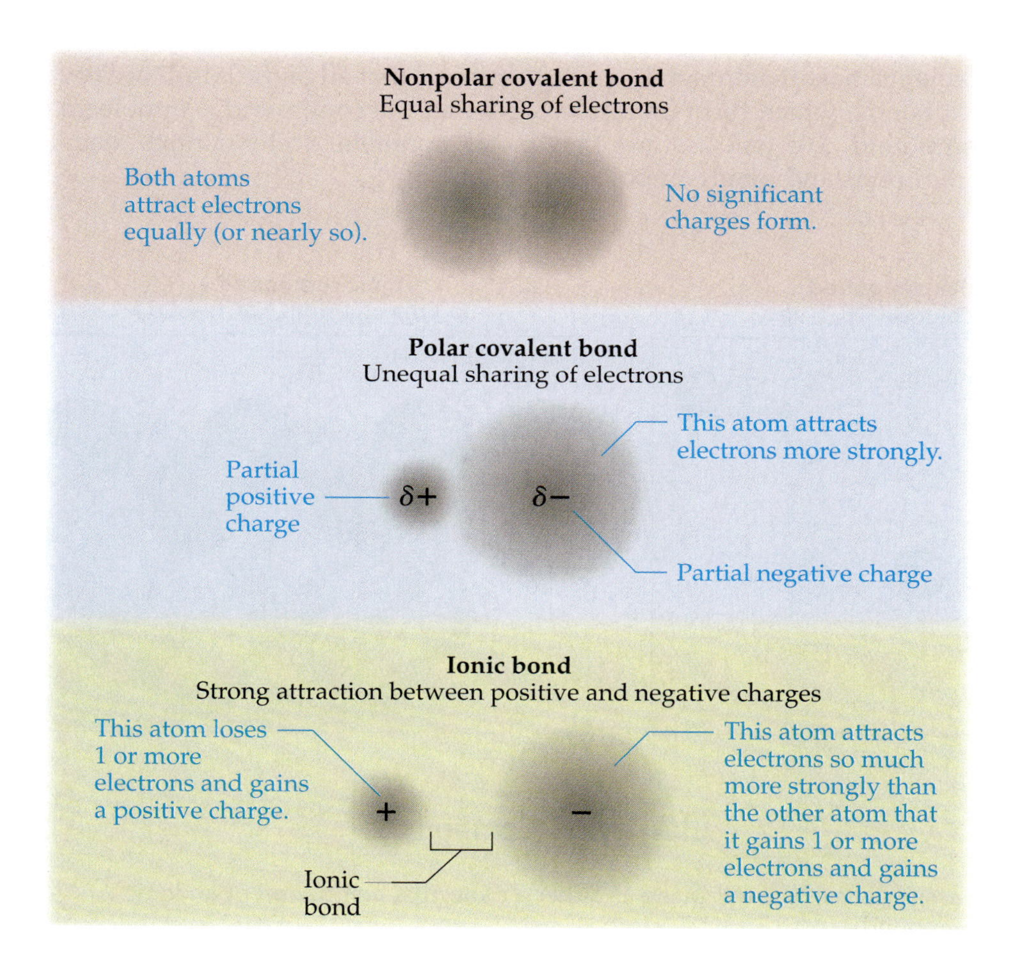
Nonpolar covalent bond
Equal sharing of electrons
Both atoms attract electrons equally (or nearly so).
No significant charges form.
Polar covalent bond
Unequal sharing of electrons
This atom attracts electrons more strongly.
Partial positive charge
δ+
δ−
Partial negative charge
Ionic bond
Strong attraction between positive and negative charges
This atom loses 1 or more electrons and gains a positive charge.
+
−
This atom attracts electrons so much more strongly than the other atom that it gains 1 or more electrons and gains a negative charge.
Ionic bond

**Figure 3.6**
**Covalent and Ionic Bonds**

OBJECTIVE 7

## Predicting Bond Type

The simplest way to predict whether a bond will be ionic or covalent is to apply the following rules.

OBJECTIVE 8

- When a nonmetallic atom bonds to another nonmetallic atom, the bond is covalent.
- When a metallic atom bonds to a nonmetallic atom, the bond is *usually* ionic.

Some bonds between a metallic atom and a nonmetallic atom are better described as covalent. For now, however, we will keep our guidelines simple. All nonmetal–nonmetal combinations lead to covalent bonds, and unless you are told otherwise, you can assume that all bonds between metallic atoms and nonmetallic atoms are ionic bonds.

## Classifying Compounds

OBJECTIVE 9

Compounds can be classified as molecular or ionic. **Molecular compounds** are composed of molecules, which are collections of atoms held together by all covalent bonds. **Ionic compounds** contain cations and anions held together by ionic bonds (Figure 3.7). You will see some exceptions in Chapter 4, but for now, if a formula for a compound indicates that all the elements in it are nonmetals, you can assume that all of the bonds are covalent bonds, which form molecules, and that the compound is a molecular compound. We will assume that metal–nonmetal combinations lead to ionic bonds and ionic compounds.

**Figure 3.7**
Classifying Compounds

OBJECTIVE 9

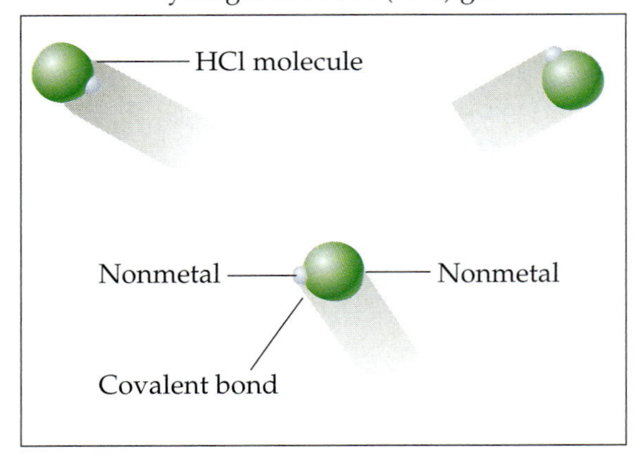


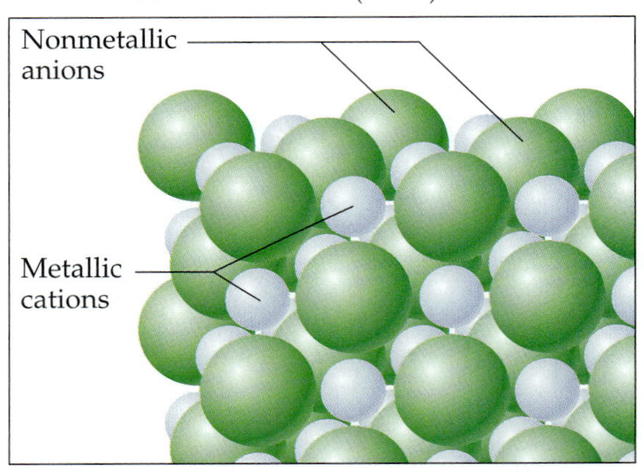


## EXAMPLE 3.2 Classifying Compounds

OBJECTIVE 9

Classify each of the following as either a molecular compound or an ionic compound.

a. calcium chloride, $CaCl_2$ (used for de-icing roads)

b. ethanethiol, $C_2H_5SH$ (a foul-smelling substance used to odorize natural gas)

***Solution***

a. Calcium, Ca, is a metal, and chlorine, Cl, is a nonmetal. We expect the bonds between them to be ionic, so calcium chloride is an **ionic compound.**

b. Carbon, hydrogen, and sulfur are all nonmetallic elements, so we expect the bonds between them to be covalent bonds. The formula, $C_2H_5SH$, tells us that ethanethiol is composed of molecules of which each contains 2 carbon atoms, 6 hydrogen atoms, and 1 sulfur atom. Ethanethiol is a **molecular compound.**

### EXERCISE 3.2 Classifying Compounds

OBJECTIVE 9

Classify each of the following substances as either a molecular compound or an ionic compound.

a. formaldehyde, $CH_2O$ (used in embalming fluids)

b. magnesium chloride, $MgCl_2$ (used in fireproofing wood and in paper manufacturing)

## 3.3 Molecular Compounds

Have you ever wondered why salt dissolves so quickly in water but oil does not? . . . why bubbles form when you open a soft drink can? . . . why a glass of water fizzes when an Alka-Seltzer tablet is plopped into it? What's going on at the submicroscopic level that makes these things happen? To answer these questions, you need to know more about the structure of water, including the spatial arrangement of atoms in water molecules. The purpose of this section is to begin to describe the three-dimensional structure of molecular compounds such as water.

Earlier we saw that when some elements form ionic and covalent bonds, their atoms gain, lose, or share electrons. This suggests an important role for electrons in chemistry. However, chemists have also found that for most elements, some electrons are more influential in the formation of chemical bonds than others are. Of chlorine's 17 electrons, for example, only 7 are important in predicting how chlorine will bond. Of sulfur's 16 electrons, only 6 are important; of phosphorus's 15 electrons, only 5 are important. Chemists noticed that the important electrons, called **valence electrons,** are equal in number to the element's "A-group" number. For example, the nonmetallic elements in group 7A (F, Cl, Br, and I) have 7 valence electrons, those in group 6A (O, S, and Se) have 6 valence electrons, those in group 5A (N and P) have 5, and carbon (C) in group 4A has 4.

OBJECTIVE 10

A more precise definition of valence electrons, and an explanation of why chlorine has 7, sulfur 6, and so on, will have to wait until we discuss atomic theory further in Chapter 11. For now, it is enough to know the numbers of valence electrons for each nonmetallic atom and how they are used to explain the bonding patterns of nonmetallic atoms.

The valence electrons for an element can be depicted visually in an **electron-dot symbol.** (Electron-dot symbols are also called electron-dot structures, electron-dot diagrams, and Lewis electron-dot symbols.) An electron-dot symbol that shows chlorine's 7 valence electrons is

$$:\ddot{\underset{\cdot\cdot}{Cl}}\cdot$$

Electron-dot symbols are derived by placing valence electrons (represented by dots) to the right, left, top, and bottom of the element's symbol. Starting on any of these four sides, we place one dot at a time until there are up to 4 unpaired electrons around the symbol. If there are more than 4 valence electrons for an atom, the remaining electrons are added one by one to the unpaired electrons to form up to 4 pairs.

OBJECTIVE 11

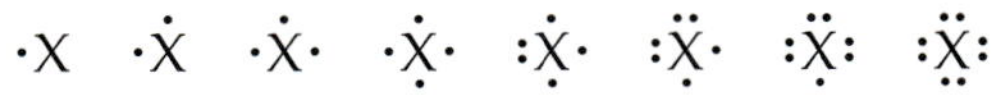

·X ·X· ·X· ·X· :X· :X· :X: :X:

There is no set convention for the placement of the paired and unpaired electrons around the symbol. For example, the electron-dot symbol for chlorine atoms could be

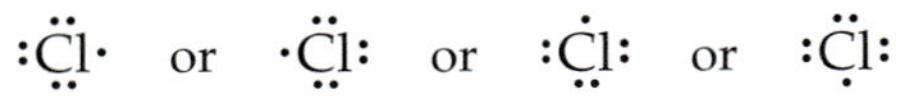

:Cl· or ·Cl: or :Cl: or :Cl:

There seems to be something special about having 8 valence electrons (often called an *octet* of electrons). For example, the noble gases (group 8A) have an octet of electrons (except for helium, which has only 2 electrons total), and they are so stable that they rarely form chemical bonds with other atoms.

:Ne: :Ar: :Kr: :Xe:

OBJECTIVE 11

When atoms other than the noble-gas atoms form bonds, they often have 8 electrons around them in total. For example, the unpaired electron of a chlorine atom often pairs with an unpaired electron of another atom to form 1 covalent bond. This gives the chlorine atom an octet of 8 electrons around it: two from the 2-electron covalent bond and six from its 3 lone pairs. This helps us explain why chlorine gas is composed of $Cl_2$ molecules.

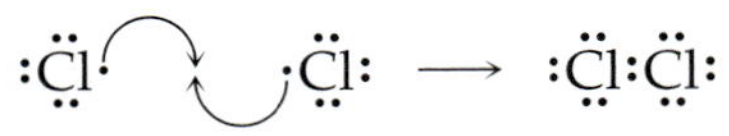

:Cl· ·Cl: ⟶ :Cl:Cl:

Note that each chlorine atom in $Cl_2$ has an octet of electrons. Apparently, the formation of an octet of electrons leads to stability.

:Cl:Cl:

OBJECTIVE 12

This way of depicting a molecule—using the elements' symbols to represent atoms and using dots to represent valence electrons—is called a **Lewis structure.** Covalent bonds are usually represented by lines in Lewis structures, so the Lewis structure of a $Cl_2$ molecule can have either of 2 forms:

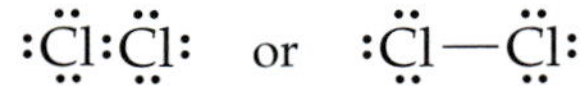

:Cl:Cl: or :Cl—Cl:

OBJECTIVE 12

The nonbonding pairs of electrons are called **lone pairs.** Each atom of chlorine in a $Cl_2$ molecule has one covalent bond and 3 lone pairs.

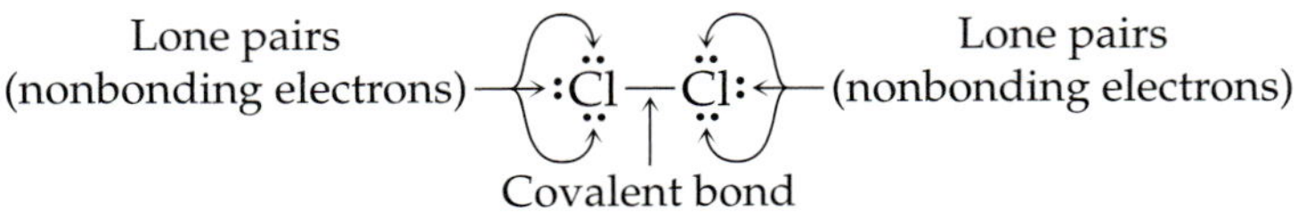

Lone pairs (nonbonding electrons) ⟶ :Cl—Cl: ⟵ Lone pairs (nonbonding electrons)

Covalent bond

OBJECTIVE 11

Some atoms do not form octets of electrons when they bond. For example, hydrogen atoms form 1 bond, achieving a total of 2 electrons around them. The reason is similar to the reason why chlorine atoms form 1 covalent bond and have 3 lone pairs. Atoms of helium, which is one of the very stable noble gases, have 2 electrons. When hydrogen atoms form

1 covalent bond, they get 2 electrons around them, like helium atoms. Knowing that hydrogen atoms form 1 covalent bond and that chlorine atoms form 1 covalent bond and have 3 lone pairs helps us to build the Lewis structure for a hydrogen chloride molecule, HCl:

$$\mathrm{H}\cdot + \cdot\underset{\cdot\cdot}{\ddot{\mathrm{Cl}}}: \longrightarrow \mathrm{H}:\underset{\cdot\cdot}{\ddot{\mathrm{Cl}}}: \quad \text{or} \quad \mathrm{H}-\underset{\cdot\cdot}{\ddot{\mathrm{Cl}}}:$$

OBJECTIVE 11

Like chlorine, the other elements in group 7A have 7 valence electrons, so their electron-dot symbols are similar to that of chlorine. The unpaired dot can be placed on any of the 4 sides of each symbol.

$$\cdot\underset{\cdot\cdot}{\ddot{\mathrm{F}}}: \qquad \cdot\underset{\cdot\cdot}{\ddot{\mathrm{Br}}}: \qquad \cdot\underset{\cdot\cdot}{\ddot{\mathrm{I}}}:$$

In order to obtain octets of electrons, these atoms tend to form compounds in which they have 1 bond and 3 lone pairs. Note how the Lewis structures of hydrogen fluoride, HF (used in the refining of uranium), hydrogen bromide, HBr (a pharmaceutical intermediate), and hydrogen iodide, HI (used to make iodine salts), resemble the structure of hydrogen chloride.

$$\mathrm{H}-\underset{\cdot\cdot}{\ddot{\mathrm{F}}}: \qquad \mathrm{H}-\underset{\cdot\cdot}{\ddot{\mathrm{Br}}}: \qquad \mathrm{H}-\underset{\cdot\cdot}{\ddot{\mathrm{I}}}:$$

Hydrogen fluoride   Hydrogen bromide   Hydrogen iodide

OBJECTIVE 11

The nonmetallic elements in group 6A (oxygen, sulfur, and selenium) have atoms with 6 valence electrons:

$$\cdot\underset{\cdot\cdot}{\ddot{\mathrm{O}}}\cdot \qquad \cdot\underset{\cdot\cdot}{\ddot{\mathrm{S}}}\cdot \qquad \cdot\underset{\cdot\cdot}{\ddot{\mathrm{Se}}}\cdot$$

(The unpaired dots can be placed on any 2 of the 4 sides of each symbol.) These elements usually gain an octet by forming 2 covalent bonds and 2 lone pairs, as in water, $H_2O$, and hydrogen sulfide, $H_2S$.

$$\mathrm{H}-\underset{\cdot\cdot}{\ddot{\mathrm{O}}}-\mathrm{H} \qquad \mathrm{H}-\underset{\cdot\cdot}{\ddot{\mathrm{S}}}-\mathrm{H}$$

Water   Hydrogen sulfide

OBJECTIVE 11

Nitrogen and phosphorus, which are in group 5A, have atoms with 5 valence electrons:

$$\cdot\underset{\cdot}{\ddot{\mathrm{N}}}\cdot \qquad \cdot\underset{\cdot}{\ddot{\mathrm{P}}}\cdot$$

They form 3 covalent bonds to pair their 3 unpaired electrons and achieve an octet of electrons around each atom. Ammonia, $NH_3$, and phosphorus trichloride, $PCl_3$, molecules are examples.

$$\begin{array}{c}\mathrm{H}-\ddot{\mathrm{N}}-\mathrm{H}\\ |\\ \mathrm{H}\end{array} \qquad \begin{array}{c}:\underset{\cdot\cdot}{\ddot{\mathrm{Cl}}}-\ddot{\mathrm{P}}-\underset{\cdot\cdot}{\ddot{\mathrm{Cl}}}:\\ |\\ :\underset{\cdot\cdot}{\mathrm{Cl}}:\end{array}$$

Ammonia   Phosphorus trichloride

$PCl_3$ is used to make pesticides and gasoline additives.

OBJECTIVE 11

Carbon, in group 4A, has 4 unpaired electrons in its electron-dot symbol.

$$\cdot\dot{\underset{\cdot}{\mathrm{C}}}\cdot$$

Predictably, carbon atoms are capable of forming 4 covalent bonds (with no lone pairs). Examples include methane, $CH_4$, the primary component of natural gas, and ethane, $C_2H_6$, and propane, $C_3H_8$, which are also found in natural gas, but in smaller quantities.

$$\begin{array}{ccccc} & \mathrm{H} & & \\ & | & & \\ \mathrm{H}- & \mathrm{C} & -\mathrm{H} \\ & | & & \\ & \mathrm{H} & & \end{array} \qquad \begin{array}{cccc} & \mathrm{H} & \mathrm{H} & \\ & | & | & \\ \mathrm{H}- & \mathrm{C}- & \mathrm{C} & -\mathrm{H} \\ & | & | & \\ & \mathrm{H} & \mathrm{H} & \end{array} \qquad \begin{array}{ccccc} & \mathrm{H} & \mathrm{H} & \mathrm{H} & \\ & | & | & | & \\ \mathrm{H}- & \mathrm{C}- & \mathrm{C}- & \mathrm{C} & -\mathrm{H} \\ & | & | & | & \\ & \mathrm{H} & \mathrm{H} & \mathrm{H} & \end{array}$$

Methane Ethane Propane

Methane, ethane, and propane are **hydrocarbons**—compounds that contain only carbon and hydrogen. The fossil fuels that we burn to heat our homes, cook our food, and power our cars are primarily hydrocarbons (Figure 3.8). For example, natural gas is a mixture of hydrocarbons with from 1 to 4 carbons, and gasoline contains hydrocarbon molecules with from 6 to 12 carbons. Like the hydrocarbons described above, many of the important compounds in nature contain a backbone of carbon–carbon bonds. These compounds are called organic compounds, and the study of carbon-based compounds is **organic chemistry.**

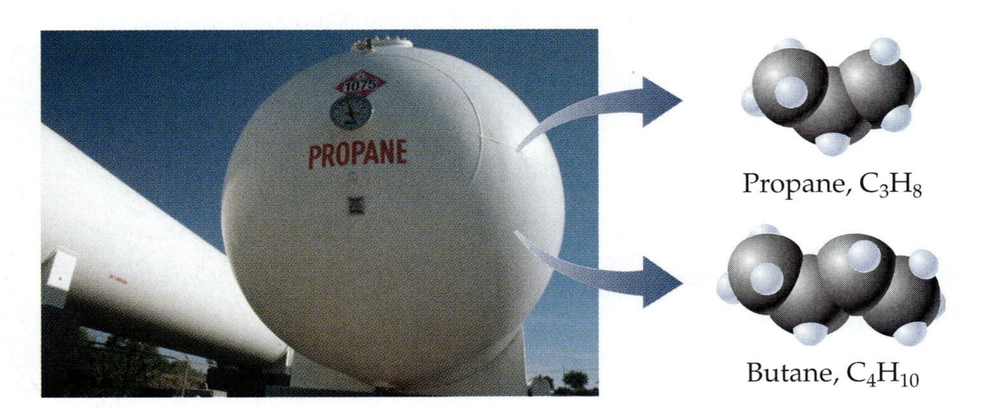


**Figure 3.8**
**Household Hydrocarbon**
Liquid petroleum gas is a mixture of the hydrocarbons propane and butane.

Table 3.1 shows electron-dot symbols for the nonmetallic atoms and lists their most common bonding patterns. Note that the sum of the numbers of bonds and lone pairs is always four for the elements in this table.

**Table 3.1**
Electron-Dot Symbols and Usual Numbers of Bonds and Lone Pairs for Nonmetallic Elements

OBJECTIVE 13

| Group 4A | | Group 5A | | Group 6A | | Group 7A | |
|---|---|---|---|---|---|---|---|
| 4 valence electrons $\cdot\dot{\underset{\cdot}{\mathrm{X}}}\cdot$ | | 5 valence electrons $\cdot\ddot{\underset{\cdot}{\mathrm{X}}}\cdot$ | | 6 valence electrons $\cdot\ddot{\underset{\cdot\cdot}{\mathrm{X}}}\cdot$ | | 7 valence electrons $:\ddot{\underset{\cdot\cdot}{\mathrm{X}}}\cdot$ | |
| 4 bonds | No lone pairs | 3 bonds | 1 lone pair | 2 bonds | 2 lone pairs | 1 bonds | 3 lone pairs |
| carbon–C $-\overset{\vert}{\underset{\vert}{\mathrm{C}}}-$ | | nitrogen–N $-\ddot{\underset{\vert}{\mathrm{N}}}-$ | | oxygen–O $-\ddot{\underset{\cdot\cdot}{\mathrm{O}}}-$ | | fluorine–F $-\ddot{\underset{\cdot\cdot}{\mathrm{F}}}:$ | |
| | | phosphorus–P $-\ddot{\underset{\vert}{\mathrm{P}}}-$ | | sulfur–S $-\ddot{\underset{\cdot\cdot}{\mathrm{S}}}-$ | | chlorine–Cl $-\ddot{\underset{\cdot\cdot}{\mathrm{Cl}}}:$ | |
| | | | | selenium–Se $-\ddot{\underset{\cdot\cdot}{\mathrm{Se}}}-$ | | bromine–Br $-\ddot{\underset{\cdot\cdot}{\mathrm{Br}}}:$ | |
| | | | | | | iodine–I $-\ddot{\underset{\cdot\cdot}{\mathrm{I}}}:$ | |

Atoms can form **double bonds,** in which 4 electrons are shared between atoms. Double bonds are represented by double lines in a Lewis structure. For example, in a carbon dioxide molecule, $CO_2$, each oxygen atom has 2 bonds to the central carbon atom.

$$:\ddot{\mathrm{O}}=\mathrm{C}=\ddot{\mathrm{O}}:$$

Note that each atom in $CO_2$ still has its most common bonding pattern. The carbon atom has 4 bonds and no lone pairs, and the oxygen atoms have 2 bonds and 2 lone pairs.

**Triple bonds,** in which 6 electrons are shared between 2 atoms, are less common than double bonds, but they do exist in such molecules as the diatomic nitrogen molecule, $N_2$. This triple bond gives each nitrogen atom its most common bonding pattern of 3 bonds and 1 lone pair.

$$:\mathrm{N}\equiv\mathrm{N}:$$

This text describes two ways to construct Lewis structures from chemical formulas. In this chapter, you will find that Lewis structures for many common substances can be drawn by giving each type of atom its most common number of covalent bonds and lone pairs. You will learn a more widely applicable procedure for drawing Lewis structures in Chapter 12.

To illustrate how Lewis structures can be drawn using the information on Table 3.1, let's figure out the Lewis structure of methanol, $CH_3OH$, which is often called methyl alcohol or wood alcohol. Methanol is a poisonous liquid used as a solvent. When drawing its Lewis structure, we assume that the carbon atom will have 4 bonds (represented by four lines), the oxygen atom will have 2 bonds and 2 lone pairs, and each hydrogen atom will have 1 bond. The Lewis structure below meets these criteria.

```
      H
      |
  H — C — Ö — H
      |   ··
      H
```

Methanol, $CH_3OH$
(methyl alcohol)

Methanol is an alcohol, which is a category of organic compounds, not just the intoxicating compound in certain drinks. **Alcohols** are organic compounds that possess one or more –OH groups attached to a hydrocarbon group (a group that contains only carbon and hydrogen). Ethanol, $C_2H_5OH$, is the alcohol in alcoholic beverages (see Special Topic 3.1: *Molecular Shapes, Intoxicating Liquids, and the Brain*), whereas the alcohol in rubbing alcohol is usually 2-propanol (Figure 3.9). These alcohols are also called methyl alcohol, $CH_3OH$; ethyl alcohol, $C_2H_5OH$; and isopropyl alcohol, $C_3H_7OH$.

OBJECTIVE 14 OBJECTIVE 15

```
                                      H
                                      |
                                  H — C — H
      H   H                       H   |
      |   |                       |   |
  H — C — C — Ö — H           H — C — C — Ö — H
      |   |   ··                  |   |   ··
      H   H                       H   H
```

Ethanol, $C_2H_5OH$
(ethyl alcohol)

2-propanol, $C_3H_7OH$
(isopropyl alcohol)

Check to see that each of these compounds follows our guidelines for drawing Lewis structures.

**Figure 3.9**
**Familiar Properties of Alcohols**

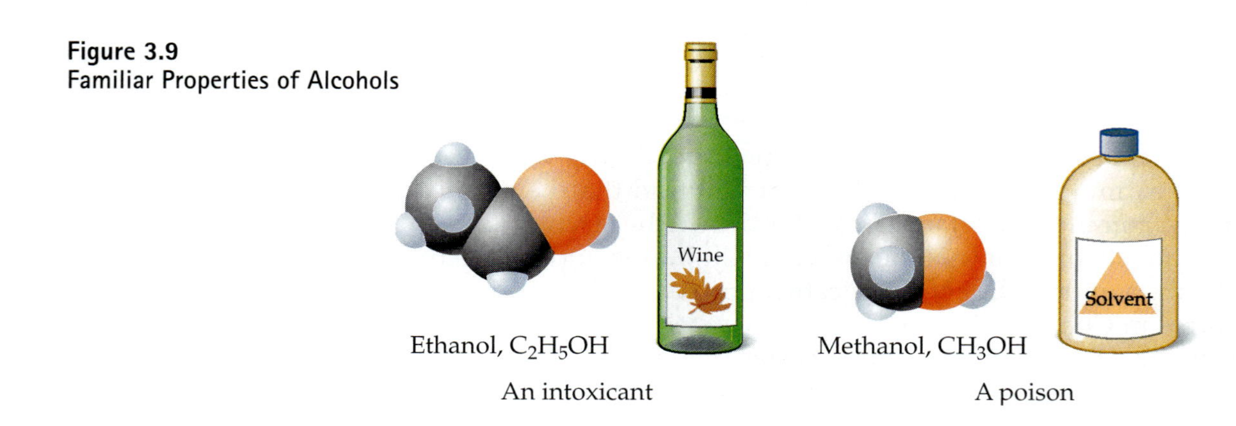

## Example 3.3 Drawing Lewis Structures from Formulas

OBJECTIVE 16

Draw a Lewis structure for each of the following formulas:

a. phosphine, $PH_3$ (used to make semiconductors)

b. hypochlorous acid, HOCl (used to bleach textiles)

c. CFC-11, $CCl_3F$ (used as a refrigerant)

d. acetylene, $C_2H_2$ (burned in oxyacetylene torches)

***Solution***

a. Phosphorus atoms usually have 3 covalent bonds and 1 lone pair, and hydrogen atoms have 1 covalent bond and no lone pairs. The following Lewis structure for $PH_3$ gives each of these atoms its most common bonding pattern.

```
    ..
H — P — H
    |
    H
```

b. Hydrogen atoms have 1 covalent bond and no lone pairs, oxygen atoms usually have 2 covalent bonds and 2 lone pairs, and chlorine atoms usually have 1 covalent bond and 3 lone pairs.

```
    ..   ..
H — O — Cl:
    ..   ..
```

c. Carbon atoms usually have 4 covalent bonds and no lone pairs. Fluorine and chlorine atoms usually have 1 covalent bond and 3 lone pairs. The fluorine atom can be put in any of the four positions around the carbon atom.

```
        ..
       :F:
 ..     |     ..
:Cl  —  C  —  Cl:
 ..     |     ..
       :Cl:
        ..
```

d. Carbon atoms form 4 bonds with no lone pairs, and hydrogen atoms form 1 bond with no lone pairs. For these bonding patterns to occur, there must be a triple bond between the carbon atoms.

H—C≡C—H

## Exercise 3.3 Drawing Lewis Structures from Formulas

OBJECTIVE 16

Draw a Lewis structure for each of the following formulas:

a. nitrogen triiodide, $NI_3$ (explodes at the slightest touch)

b. hexachloroethane, $C_2Cl_6$ (used to make explosives)

c. hydrogen peroxide, $H_2O_2$ (a common antiseptic)

d. ethylene (or ethene), $C_2H_4$ (used to make polyethylene)

## Molecular Shape

Lewis structures are useful for showing how the atoms in a molecule are connected by covalent bonds, but they do not always give a clear description of how the atoms are arranged in space. For example, the Lewis structure for methane, $CH_4$, shows the four covalent bonds connecting the central carbon atom to the hydrogen atoms:

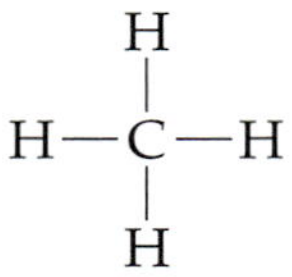

```
     H
     |
  H—C—H
     |
     H
```

However, this Lewis structure seems to indicate that the 5 atoms are all located in the same plane and that the angles between the atoms are all 90° or 180°. This is not true. The actual shape of a molecule can be more accurately predicted by recognizing that the negatively charged electrons that form covalent bonds and lone pairs repel each other. Therefore, the most stable arrangement of the electron groups is the molecular shape that keeps the groups as far away from each other as possible.

OBJECTIVE 17

OBJECTIVE 18

The best way to keep the negative charges for the 4 covalent bonds in a methane molecule as far apart as possible is to place them in a three-dimensional molecular shape called **tetrahedral,** with angles of 109.5° between the bonds.

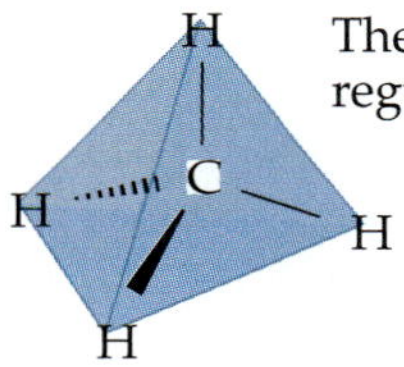

H

C

H H

H

The blue shape is a regular tetrahedron.

The angle formed by straight lines (representing bonds) connecting the nuclei of three adjacent atoms is called a **bond angle.**

OBJECTIVE 19

Three ways to represent the methane molecule are shown in Figure 3.10. The first image, a *space-filling model,* provides the most accurate representation of the electron-charge clouds for the atoms in $CH_4$. A *ball-and-stick model,* the second image, emphasizes the molecule's correct molecular shape and shows the covalent bonds more clearly. The third image, a *geometric sketch,* shows a simple technique for describing three-dimensional tetrahedral structures with a two-dimensional drawing. Two hydrogen atoms are connected to the central carbon atom with solid lines. Picture these as being in the same plane as the carbon atom. A third hydrogen atom, connected to the central carbon with a solid wedge, comes out of the plane toward you. The fourth hydrogen atom, connected to the carbon atom by a dashed wedge, is situated back behind the plane of the page.

Space-filling model

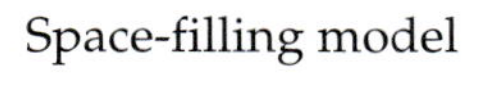

Ball-and-stick model

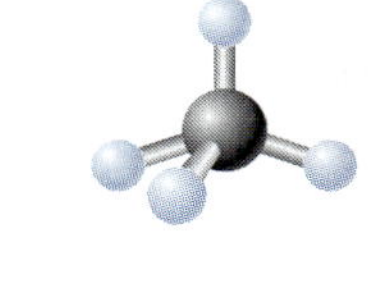

Geometric sketch

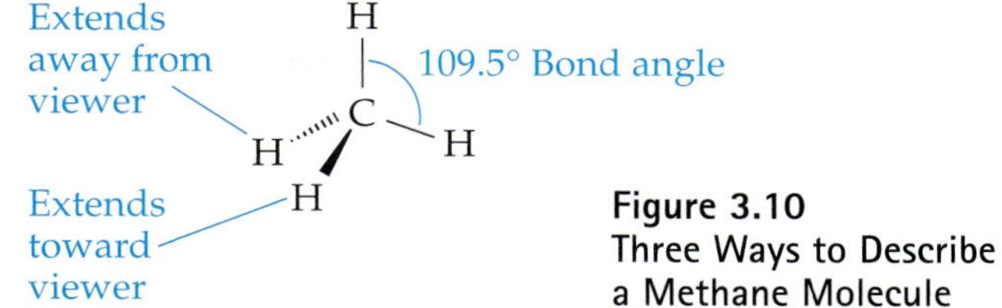


**Figure 3.10**
**Three Ways to Describe a Methane Molecule**

OBJECTIVE 19 OBJECTIVE 20

The nitrogen atom in an ammonia molecule, $NH_3$, forms 3 covalent bonds and in addition has a lone pair of electrons. A lone pair on a central atom must be considered in predicting a molecule's shape.

$$\mathrm{H{-}\ddot{N}{-}H} \quad (\text{with a third H bonded below N})$$

Like the carbon atom in a methane molecule, the nitrogen atom has 4 electron groups around it, so the ammonia molecule has a shape that is very similar to the shape of a $CH_4$ molecule. However, the lone pair on the nitrogen atom repels neighboring electron groups more strongly than the bond pairs do, so the lone pair in the ammonia molecule pushes the bond pairs closer together than the bond pairs for methane. The bond angle is about 107° instead of 109.5°. Figure 3.11 shows three ways to represent the ammonia molecule.

Space-filling model

Ball-and-stick model

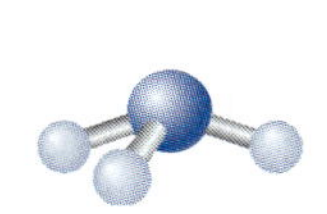

Geometric sketch

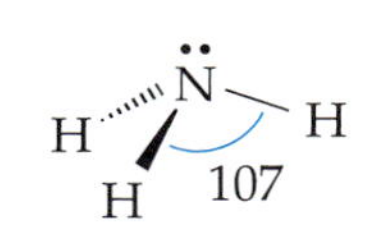


**Figure 3.11**
**Three Ways to Describe an Ammonia Molecule**

OBJECTIVE 20

## Liquid Water

A chemist's-eye view of the structure of liquid water starts with the prediction of the molecular shape of each water molecule. The Lewis structure of water shows that the oxygen atom has 4 electron-groups around it: 2 covalent bonds and 2 lone pairs.

$$\mathrm{H{-}\ddot{\underset{\cdot\cdot}{O}}{-}H}$$

We predict that the four groups will be distributed in a tetrahedral arrangement to keep their negative charges as far apart as possible. Because the lone pairs are more repulsive than the bond pairs, the angle between the bond pairs is less than 109.5°. In fact, it is about 105° (Figure 3.12).

**Figure 3.12**
**Three Ways to Describe a Water Molecule**

OBJECTIVE 20

Space-filling model

Ball-and-stick model

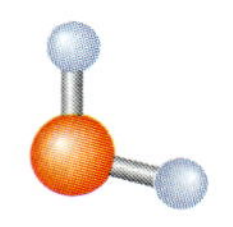

Geometric sketch

```
 H
 |  105°
:O̤—H
```

Because oxygen atoms attract electrons much more strongly than do hydrogen atoms, the O–H covalent bond is very polar, leading to a relatively large partial negative charge on the oxygen atom (represented by a $\delta-$) and a relatively large partial positive charge on the hydrogen atom (represented by a $\delta+$).

## Special Topic 3.1 Molecular Shapes, Intoxicating Liquids, and the Brain

Even if we do not all have firsthand experience with alcoholic beverages, we all know that their consumption slows brain activity. Ethanol, or ethyl alcohol, is the chemical in alcoholic beverages that causes this change.

```
    H   H
    |   |   ..
H — C — C — O — H
    |   |   ..
    H   H
```

Ethanol, $C_2H_5OH$

Information is transferred through our nervous system when one nerve cell causes the next nerve cell to fire. This firing is regulated by the attachment of molecules with *specific shapes* to large protein molecules that form part of the nerve cell's membrane. When certain small molecules of the correct shape attach to the protein structures in the cell membranes, the cell is induced to fire. When certain other molecules attach, the firing of the cell is inhibited. For example, when a molecule called gamma-aminobutanoic acid, or GABA, attaches to a protein molecule in certain nerve cells, it changes the shape of the protein and inhibits the firing of the cell.

```
         H   H   H  :O:
    ..   |   |   |   ‖   ..
H — N — C — C — C — C — O — H
    |   |   |   |       ..
    H   H   H   H
```

GABA

The GABA molecules are constantly attaching to the protein and then leaving again. When the GABA molecules are attached, the information transfer between nerve cells is inhibited. Anything that makes it easier for the GABA molecules to attach to the protein leads to a slowing of the transfer of information between nerve cells.

Ethanol molecules can attach to the same protein as the GABA molecules, but they attach to a different site on the protein molecule. They change the shape of the protein in such a way that it becomes easier for the GABA molecules to find their position on the protein. Thus, when ethanol is present, the GABA molecules become attached to the protein more often, inhibiting the firing of the cell. In this way, ethanol helps slow the transfer of nerve information.

This slowing of the transfer of nerve information to the brain may not be a big problem to someone having a glass of wine with dinner at home, but when a person has a few drinks and drives a car, the consequences can be serious. If a deer, say, runs in front of the car, we want the "put on the brake" signal sent from eyes to brain and then from brain to foot as quickly as possible.

$$\begin{array}{l} \quad H^{\delta+} \\ \quad | \\ \delta-\,:\ddot{O}-H^{\delta+} \end{array}$$

The attraction between the region of partial positive charge on 1 water molecule and the region of partial negative charge on another water molecule tends to hold water molecules close together (Figure 3.13). Remember that opposite charges attract each other and like charges repel each other.

OBJECTIVE 21

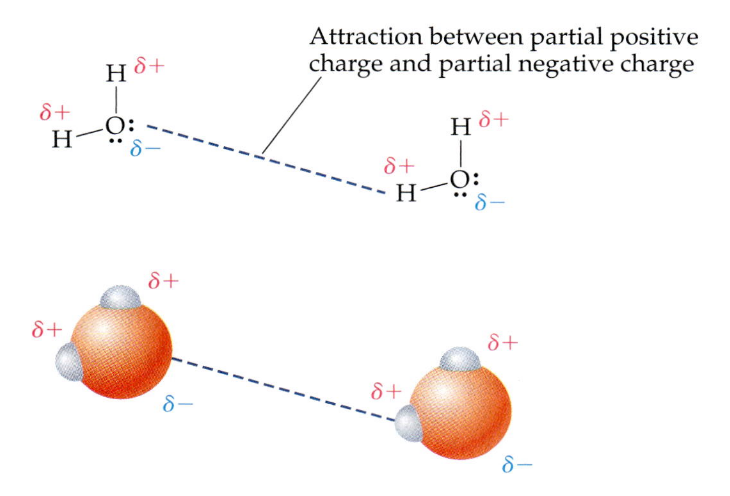


**Figure 3.13**
**Attractions between Water Molecules**

OBJECTIVE 21

As in other liquids, the attractions between water molecules are strong enough to keep them the same average distance apart but weak enough to allow each molecule to be constantly breaking the attractions that momentarily connect it to some molecules and forming new attractions to other molecules (Figure 3.14). In other chapters, this image of the structure of water will be useful as you learn what is happening when salt dissolves in your pasta water and when bubbles form in a soft drink or in a glass of Alka-Seltzer and water.

OBJECTIVE 22

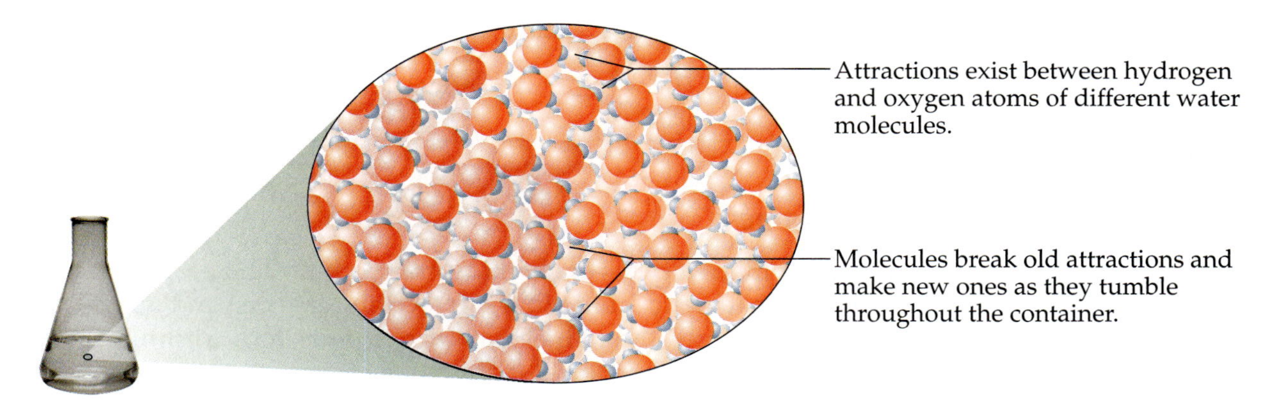

Attractions exist between hydrogen and oxygen atoms of different water molecules.
Molecules break old attractions and make new ones as they tumble throughout the container.

**Figure 3.14**
**Liquid Water**

OBJECTIVE 22

The Shockwave animation at the following Web site will help you to visualize the structure of water:
www.chemplace.com/college/

## 3.4 Naming Binary Covalent Compounds

*Please excuse Lisa for being absent; she was sick and I had her shot.*

A note to the teacher to explain why a child was absent

We can probably assume that the parent who wrote this note meant the child was taken to the doctor for an injection, but who knows? Like everyone else, chemists need to be careful about how they use language, and the names and formulas for chemical compounds are the core of the language of chemistry. The purpose of this section is to describe the guidelines for constructing the names for **binary covalent compounds,** which are pure substances that consist of two nonmetallic elements. The water, $H_2O$, that you boil to cook your potatoes and the methane, $CH_4$, in natural gas that can be burned to heat the water are examples of binary covalent compounds.

### Memorized Names

OBJECTIVE 23

Some binary covalent compounds, such as water, $H_2O$, and ammonia, $NH_3$, are known by common names that chemists have used for years. There is no systematic set of rules underlying these names, so each must simply be memorized. Organic compounds, such as methane, $CH_4$; ethane, $C_2H_6$; and propane, $C_3H_8$, are named by a systematic procedure that you may learn later in your chemical education, but for now, it will be useful to memorize some of their names and formulas also (Table 3.2).

OBJECTIVE 23

Table 3.2
Names and Formulas of Some Binary Covalent Compounds

| Name | Formula | Name | Formula |
|---|---|---|---|
| water | $H_2O$ | ammonia | $NH_3$ |
| methane | $CH_4$ | ethane | $C_2H_6$ |
| propane | $C_3H_8$ | | |

### Systematic Names

There are many different types of chemical compounds, and each type has its own set of systematic guidelines for writing names and chemical formulas. Thus, to write the name that corresponds to a formula for a compound, you need to develop the ability to recognize the formula as representing a specific type of compound. Likewise, to write a formula from a name, you need to recognize the type of compound the name represents. You also need to learn the guidelines for converting between names and formulas for that type of compound. You can recognize binary covalent compounds from their formulas, which contain symbols for only two nonmetallic elements. The general pattern of such formulas is $A_aB_b$, where "A" and "B" represent symbols for nonmetals, and "a" and "b" represent subscripts. (Remember that if one of the subscripts is absent, it is understood to be 1.) For example, because nitrogen and oxygen are nonmetallic elements, the formula $N_2O_3$ represents a binary covalent compound.

OBJECTIVE 24

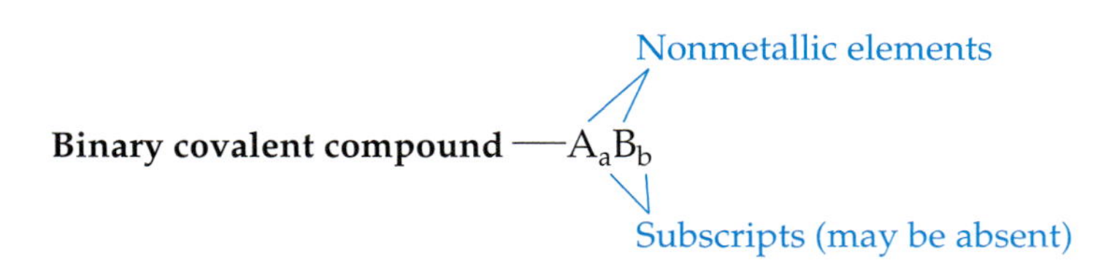


Follow these steps to write the names for binary covalent compounds:

OBJECTIVE 28

- If the subscript for the first element is greater than 1, indicate the identity of the subscript using prefixes from Table 3.3. We do not write *mono* at the beginning of a compound's name (see Example 3.4).

  *Example:* We start the name for $N_2O_3$ with *di-*.

Table 3.3
Prefixes Used in the Names of Binary Covalent Compounds

OBJECTIVE 25

| Number of atoms | Prefix | Number of atoms | Prefix |
|---|---|---|---|
| 1 | mon(o) | 6 | hex(a) |
| 2 | di | 7 | hept(a) |
| 3 | tri | 8 | oct(a) |
| 4 | tetr(a) | 9 | non(a) |
| 5 | pent(a) | 10 | dec(a) |

- Attach the selected prefix to the name of the first element in the formula. If no prefix is to be used, begin with the name of the first element.

  *Example:* We indicate the $N_2$ portion of $N_2O_3$ with *dinitrogen*.

- Select a prefix to identify the subscript for the second element (even if its subscript is understood to be 1). Leave the *a* off the end of the prefixes that end in *a* and the *o* off *mono* if they are placed in front of an element whose name begins with a vowel (oxygen or iodine).

  *Example:* The name of $N_2O_3$ grows to *dinitrogen tri-*.

- Write the root of the name of the second element in the formula as shown in Table 3.4.

  *Example:* The name of $N_2O_3$ becomes *dinitrogen triox-*.

OBJECTIVE 26

Table 3.4
Roots for the Names of the Nonmetallic Elements

| Element | Root | Element | Root | Element | Root | Element | Root |
|---|---|---|---|---|---|---|---|
| C | carb | N | nitr | O | ox | F | fluor |
| | | P | phosph | S | sulf | Cl | chlor |
| | | As | arsen | Se | selen | Br | brom |
| | | | | | | I | iod |

- Add ide to the end of the name.

  *Example:* The name of $N_2O_3$ is *dinitrogen trioxide.*

## Example 3.4 Naming Binary Covalent Compounds

OBJECTIVE 28

Write the names that correspond to the formulas (a) $N_2O_5$, (b) $NO_2$, and (c) NO.

***Solution***

These formulas are in the form of $A_aB_b$, where A and B represent symbols for nonmetallic elements, so they are binary covalent compounds.

a. The first subscript in $N_2O_5$ is 2, so the first prefix is *di.* The first symbol, N, represents nitrogen, so the name for $N_2O_5$ begins with *dinitrogen.* The second subscript is 5, and the second symbol, O, represents oxygen. Therefore, the prefix *pent* combines with the root of oxygen, *ox,* and the usual ending, *ide,* to give *pentoxide* for the second part of the name. $N_2O_5$ is **dinitrogen pentoxide.** Notice that the *a* is left off *penta,* because the root *ox* begins with a vowel.

b. $NO_2$ is **nitrogen dioxide.** We leave *mono* off the first part of the name.

c. NO is **nitrogen monoxide.** We leave *mono* off the first part of the name, but we start the second part of the name with *mon.* The *o* in *mono* is left off before the root *ox.*

OBJECTIVE 27

Hydrogen atoms always form 1 covalent bond, and halogen atoms (group 17, or 7A) usually form one bond. Thus hydrogen reacts with halogens to form compounds with the general formula of HX, with the X represents the halogen. Because this is common knowledge among scientists and science students, these compounds are often named without prefixes. For example, HF can be named hydrogen fluoride or hydrogen monofluoride. Likewise, HCl can be named hydrogen chloride or hydrogen monochloride, HBr can be named hydrogen bromide or hydrogen monobromide, and HI can be named hydrogen iodide or hydrogen moniodide. For similar reasons, $H_2S$ can be named hydrogen sulfide or dihydrogen monosulfide.

## Exercise 3.4 Naming Binary Covalent Compounds

OBJECTIVE 28

Write names that correspond to the following formulas: (a) $P_2O_5$, (b) $PCl_3$, (c) CO, (d) $H_2S$, and (e) $NH_3$.

## Converting Names of Binary Covalent Compounds to Formulas

Now let's go the other way and convert from systematic names to chemical formulas. The first step in writing formulas when given the systematic name of a binary covalent compound is to recognize the name as representing a binary covalent compound. It will have one of the following general forms.

OBJECTIVE 24

**prefix(name of nonmetal) prefix(root of name of nonmetal)** *ide*
(for example, dinitrogen pentoxide)

**(name of nonmetal) prefix(root of name of nonmetal)** *ide*
(for example, carbon dioxide)

**(name of nonmetal) (root of nonmetal)** *ide*
(for example, hydrogen fluoride)

Follow these steps for writing formulas for binary covalent compounds when you are given a systematic name. Note that they are the reverse of the steps for writing names from chemical formulas.

OBJECTIVE 28

- Write the symbols for the elements in the order mentioned in the name.
- Write the subscripts indicated by the prefixes. If the first part of the name has no prefix, assume it is *mono.*

Remember that HF, HCl, HBr, HI, and $H_2S$ are often named without prefixes. You will also be expected to write formulas for the compounds whose nonsystematic names are listed in Table 3.2.

## EXAMPLE 3.5 Writing Formulas for Binary Covalent Compounds

OBJECTIVE 28

Write the formulas that correspond to the following names: (a) dinitrogen tetroxide, (b) phosphorus tribromide, (c) hydrogen iodide, and (d) methane.

***Solution***

a. Because the name dinitrogen tetroxide has the following form, it must be binary covalent.

   prefix(name of nonmetal) prefix(root of name of nonmetal)ide

   The *di* tells us there are two nitrogen atoms, and the *tetr* tells us there are four oxygen atoms. Dinitrogen tetroxide is $\mathbf{N_2O_4}$.

b. Because the name phosphorus tribromide has the following form, it must be binary covalent.

   (name of nonmetal) prefix(root of name of nonmetal)ide

   There is no prefix attached to phosphorus, so we assume that there is 1 phosphorus atom. Phosphorus tribromide is $\mathbf{PBr_3}$.

c. Because the name hydrogen iodide has the following form, it must be binary covalent.

(name of nonmetal) (root of name of nonmetal)ide

This is one of the binary covalent compounds that does not require prefixes. Iodine usually forms 1 bond, and hydrogen always forms 1 bond, so hydrogen iodide is **HI.**

d. Methane is on our list of binary covalent compounds with names you should memorize. Methane is **$CH_4$.**

### Exercise 3.5 Writing Formulas for Binary Covalent Compounds

OBJECTIVE 28

Write formulas that correspond to the following names: (a) disulfur decafluoride, (b) nitrogen trifluoride, (c) propane, and (d) hydrogen chloride.

## 3.5 Ionic Compounds

**Ionic compounds** are substances composed of ions attracted to each other by ionic bonds. Let's consider how they play a part in a "typical" family's Fourth of July.

Before the family leaves the house to go to the holiday picnic, the kids are sent off to brush their teeth and change into clean clothes. Their toothpaste contains sodium fluoride, a common cavity-fighting ionic compound. The white shirts in their red, white, and blue outfits were bleached with the ionic compound sodium hypochlorite, the stains on their red pants were removed by potassium oxalate, and dyes were fixed to their blue hats by aluminum nitrate.

Ionic compounds play an important role in creating the excitement of a Fourth of July celebration.

While the kids are getting ready, dad and mom get the picnic dinner together. The hot dogs they are packing are cured with the ionic compound sodium nitrite, and the buns contain calcium acetate as a mold inhibitor and calcium iodate as a dough conditioner. The soft drinks have potassium hydrogen carbonate to help trap the bubbles, the mineral water contains magnesium sulfate, and the glass for the bottles was made with a variety of ionic compounds. Because it will be dark before they get home, Mom packs a flashlight as well. Its rechargeable batteries contain the ionic compounds cadmium hydroxide and nickel hydroxide.

When our family gets to the park, they find themselves a place on the lawn, which was fertilized with a mixture of ionic compounds, including iron sulfate. They eat their dinner and play in the park until it's time for the fireworks. The safety matches used to light the rockets contain barium chromate, and ionic compounds in the fireworks provide the colors: red from strontium chlorate, white from magnesium nitrate, and blue from copper(II) chloride.

## Cations and Anions

Remember the sodium fluoride in the kids' toothpaste? It could be made from the reaction of sodium metal with the nonmetallic atoms in fluorine gas. As you discovered in Section 3.2, metallic atoms hold some of their electrons relatively loosely, and as a result, they tend to lose electrons and form cations. In contrast, nonmetallic atoms attract electrons more strongly than metallic atoms, so nonmetals tend to gain electrons and form anions. Thus, when a metallic element and a nonmetallic element combine, the nonmetallic atoms often pull one or more electrons far enough away from the metallic atoms to form ions. The positive cations and the negative anions then attract each other to form ionic bonds. In the formation of sodium fluoride from sodium metal and fluoride gas, each sodium atom donates one electron to a fluorine atom to form a $Na^+$ cation and a $F^-$ anion. The $F^-$ anions in toothpaste bind to the surface of your teeth, making them better able to resist tooth decay. This section provides you with more information about other cations and anions, including how to predict their charges and how to convert between their names and formulas.

OBJECTIVE 29

Sodium fluoride is added to toothpaste to help fight tooth decay.

## Predicting Ion Charges

It is useful to be able to predict the charges that the atoms of each element are most likely to attain when they form ions. Because the periodic table can be used to predict ionic charges, it is a good idea to have one in front of you when you study this section.

We discovered in Chapter 2 that the atoms of the noble gases found in nature are uncombined with other atoms. The fact that the noble-gas atoms do not gain, lose, or share their electrons suggests that there must be something especially stable about having 2 (helium, He), 10 (neon, Ne), 18 (argon, Ar), 36 (krypton, Kr), 54 (xenon, Xe), or 86 (radon, Rn) electrons. This stability is reflected in the fact that nonmetallic atoms form anions to get the same number of electrons as the nearest noble gas.

All of the halogens (group 17) have 1 less electron than the nearest noble gas. When halogen atoms combine with metallic atoms, they tend to gain 1 electron each and form $-1$ ions (Figure 3.15). For example, uncharged chlorine atoms have 17 protons and 17 electrons. If a chlorine atom gains 1 electron, it will have 18 electrons like an uncharged argon atom. With a $-18$ charge from the electrons and a $+17$ charge from the protons, the resulting chlorine ion has a $-1$ charge. The symbol for this anion is $Cl^-$. Note that the negative charge is indicated with a $-$ near the top of the elemental symbol, not $-1$ or $1-$.

OBJECTIVE 30

$$\underset{17p/17e^-}{Cl} \quad + \quad 1e^- \quad \rightarrow \quad \underset{17p/18e^-}{Cl^-}$$

The nonmetallic atoms in group 16 (oxygen, O, sulfur, S, and selenium, Se) have 2 fewer electrons than the nearest noble gas. When atoms of these elements combine with metallic atoms, they tend to gain 2 electrons and form −2 ions (Figure 3.15). For example, oxygen, in group 16, has atoms with 8 protons and 8 electrons. Each oxygen atom can gain 2 electrons to achieve 10, the same number as its nearest noble gas, neon. The symbol for this anion is $O^{2-}$. Note that the charge is indicated with 2− not −2.

OBJECTIVE 30

$$\underset{8p/8e^-}{O} \quad + \quad 2e^- \quad \rightarrow \quad \underset{8p/10e^-}{O^{2-}}$$

Nitrogen, N, and phosphorus, P, have 3 fewer electrons than the nearest noble gas. When atoms of these elements combine with metallic atoms, they tend to gain 3 electrons and form −3 ions (Figure 3.15). For example, nitrogen atoms have 7 protons and 7 electrons. Each nitrogen atom can gain 3 electrons to achieve 10, like neon, forming a $N^{3-}$ anion.

OBJECTIVE 30

$$\underset{7p/7e^-}{N} \quad + \quad 3e^- \quad \rightarrow \quad \underset{7p/10e^-}{N^{3-}}$$

Hydrogen has one less electron than helium, so when it combines with metallic atoms, it forms a −1 ion, H−. Anions like H−, Cl−, $O^{2-}$, and $N^{3-}$, which contain single atoms with a negative charge, are called **monatomic anions.**

**Figure 3.15**
**The Making of an Anion**

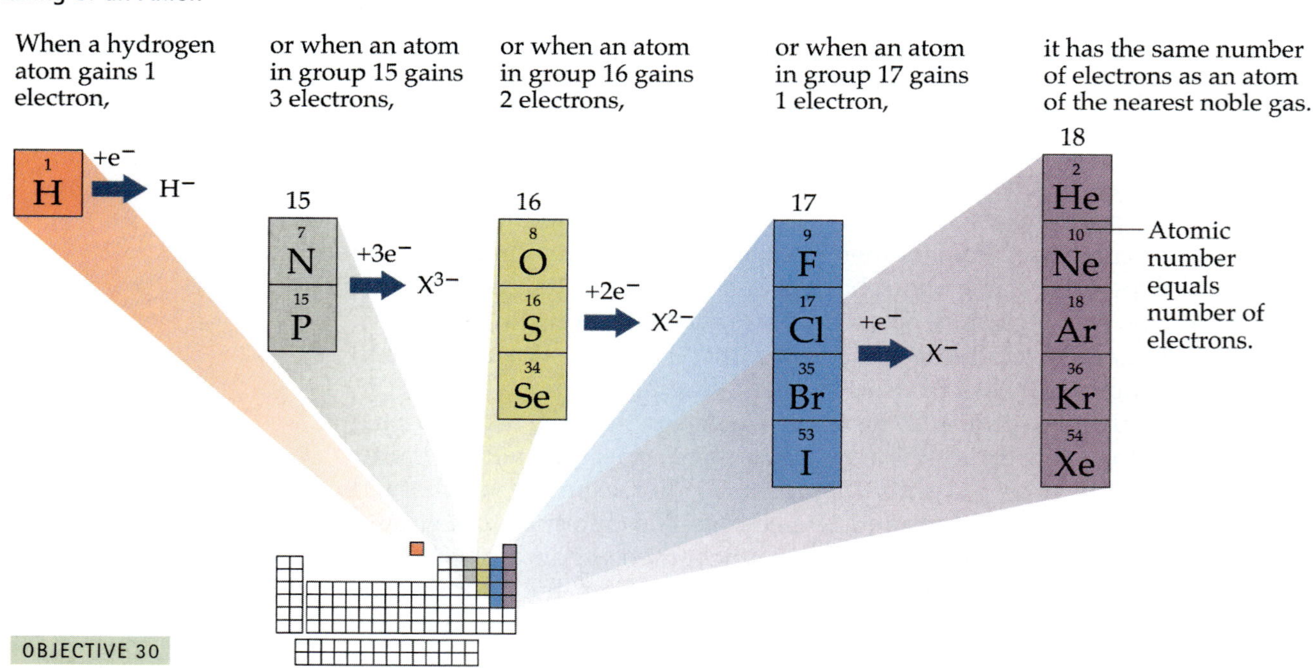
When a hydrogen atom gains 1 electron,
or when an atom in group 15 gains 3 electrons,
or when an atom in group 16 gains 2 electrons,
or when an atom in group 17 gains 1 electron,
it has the same number of electrons as an atom of the nearest noble gas.
1 H +e⁻ H⁻
15
7 N
15 P
+3e⁻ X³⁻
16
8 O
16 S
34 Se
+2e⁻ X²⁻
17
9 F
17 Cl
35 Br
53 I
+e⁻ X⁻
18
2 He
10 Ne
18 Ar
36 Kr
54 Xe
Atomic number equals number of electrons.

OBJECTIVE 30

Some metallic atoms lose enough electrons to create a cation that has the same number of electrons as the nearest smaller noble gas. For example, the alkali metals in group 1 all have 1 more electron than the nearest noble gas. When they react with nonmetallic atoms, they lose 1 electron and form +1 ions (Figure 3.16). For example, sodium has atoms with 11 protons and 11 electrons. If an atom of sodium loses 1 electron, it will have 10 electrons like uncharged neon. With a −10 charge from the electrons and a +11 charge from the protons, the sodium ions have a +1 overall charge. The symbol for this cation is $Na^+$. Note that the charge is indicated with a + instead of +1 or 1+.

OBJECTIVE 30

$$\underset{11p/11e^-}{Na} \longrightarrow \underset{11p/10e^-}{Na^+} + 1e^-$$

The alkaline earth metals in group 2 all have 2 more electrons than the nearest noble gas. When they react with nonmetallic atoms, they tend to lose 2 electrons and form +2 ions (Figure 3.16). For example, calcium has atoms with 20 protons and 20 electrons. Each calcium atom can lose 2 electrons to achieve 18, the same number as its nearest noble gas, argon. The symbol for this cation is $Ca^{2+}$. Note that the charge is indicated with 2+, not +2.

OBJECTIVE 30

$$\underset{20p/20e^-}{Ca} \longrightarrow \underset{20p/18e^-}{Ca^{2+}} + 2e^-$$

**Figure 3.16**
**The Making of a Cation**

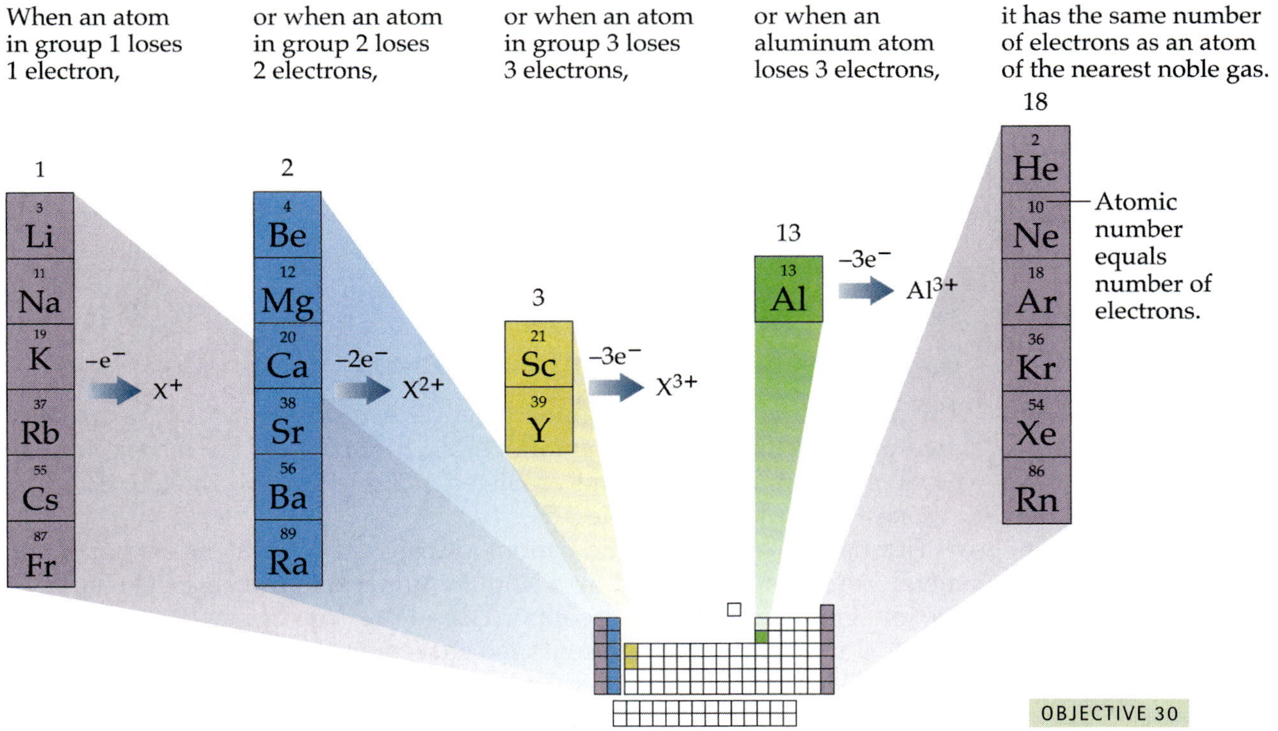

OBJECTIVE 30

OBJECTIVE 30

Aluminum atoms and the atoms of the group 3 metals have 3 more electrons than the nearest noble gas. When they react with nonmetallic atoms, they tend to lose 3 electrons and form +3 ions (Figure 3.16). For example, uncharged aluminum atoms have 13 protons and 13 electrons. Each aluminum atom can lose 3 electrons to achieve 10, like neon, forming an $Al^{3+}$ cation.

$$\underset{13p/13e^-}{Al} \rightarrow \underset{13p/10e^-}{Al^{3+}} + 3e^-$$

Cations such as $Na^+$, $Ca^{2+}$, and $Al^{3+}$, which are single atoms with a positive charge, are called **monatomic cations.**

OBJECTIVE 30

The metallic elements in groups other than 1, 2, or 3 also lose electrons to form cations, but they do so in less easily predicted ways. It will be useful to memorize some of the charges for these metals. Ask your instructor which ones you will be expected to know. To answer the questions in this text, you will need to know that iron atoms form both $Fe^{2+}$ and $Fe^{3+}$, copper atoms form $Cu^+$ and $Cu^{2+}$, zinc atoms form $Zn^{2+}$, cadmium atoms form $Cd^{2+}$, and silver atoms form $Ag^+$. Figure 3.17 summarizes the charges of the ions that you should know at this stage.

**Figure 3.17**
Common Monatomic Ions

OBJECTIVE 30

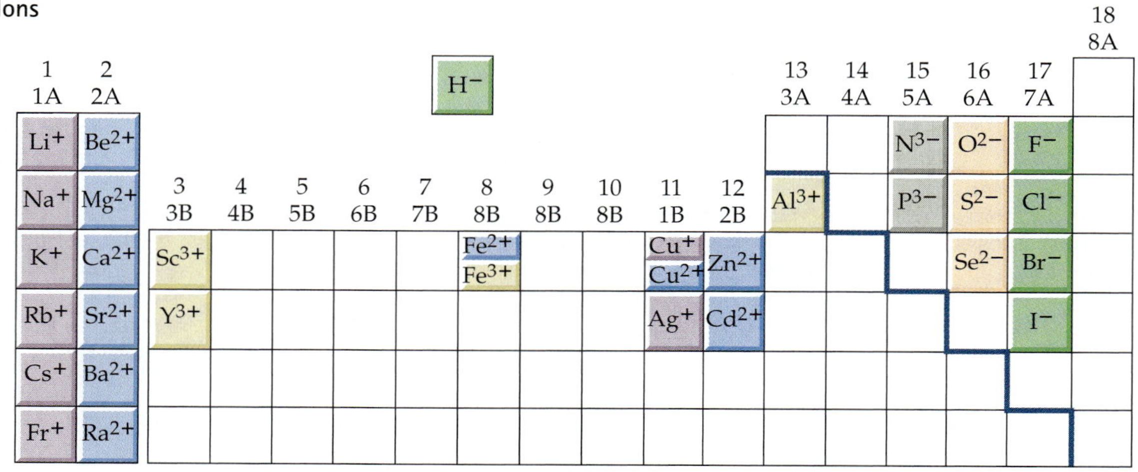

$H^-$

| 1<br>1A | 2<br>2A | 3<br>3B | 4<br>4B | 5<br>5B | 6<br>6B | 7<br>7B | 8<br>8B | 9<br>8B | 10<br>8B | 11<br>1B | 12<br>2B | 13<br>3A | 14<br>4A | 15<br>5A | 16<br>6A | 17<br>7A | 18<br>8A |
|---|---|---|---|---|---|---|---|---|---|---|---|---|---|---|---|---|---|
| $Li^+$ | $Be^{2+}$ | | | | | | | | | | | | | $N^{3-}$ | $O^{2-}$ | $F^-$ | |
| $Na^+$ | $Mg^{2+}$ | | | | | | | | | | | $Al^{3+}$ | | $P^{3-}$ | $S^{2-}$ | $Cl^-$ | |
| $K^+$ | $Ca^{2+}$ | $Sc^{3+}$ | | | | | $Fe^{2+}$<br>$Fe^{3+}$ | | | $Cu^+$<br>$Cu^{2+}$ | $Zn^{2+}$ | | | | $Se^{2-}$ | $Br^-$ | |
| $Rb^+$ | $Sr^{2+}$ | $Y^{3+}$ | | | | | | | | $Ag^+$ | $Cd^{2+}$ | | | | | $I^-$ | |
| $Cs^+$ | $Ba^{2+}$ | | | | | | | | | | | | | | | | |
| $Fr^+$ | $Ra^{2+}$ | | | | | | | | | | | | | | | | |

## Naming Monatomic Anions and Cations

OBJECTIVE 31

The monatomic anions are named by adding *ide* to the root of the name of the nonmetal that forms the anion. For example, $N^{3-}$ is the *nitride ion.* The roots of the nonmetallic atoms are listed in Table 3.4, and the names of the anions are displayed in Table 3.5.

OBJECTIVE 31

The names of monatomic cations always start with the name of the metal, sometimes followed by a Roman numeral to indicate the charge of the ion. For example, $Cu^+$ is copper(I), and $Cu^{2+}$ is copper(II). The Roman numeral in each name represents the charge on the ion and enables us to distinguish between more than one possible charge. Note that there is no space between the end of the name of the metal and the parentheses with the Roman numeral.

Table 3.5
Names of the Monatomic Anions

| Anion | Name | Anion | Name | Anion | Name |
|---|---|---|---|---|---|
| $N^{3-}$ | nitride | $O^{2-}$ | oxide | $H^-$ | hydride |
| $P^{3-}$ | phosphide | $S^{2-}$ | sulfide | $F^-$ | fluoride |
| | | $Se^{2-}$ | selenide | $Cl^-$ | chloride |
| | | | | $Br^-$ | bromide |
| | | | | $I^-$ | iodide |

OBJECTIVE 31

If the atoms of an element always have the same charge, the Roman numeral is unnecessary (and considered to be incorrect). For example, all cations formed from sodium atoms have a +1 charge, so $Na^+$ is named *sodium ion,* without the Roman numeral for the charge. The following elements have only one possible charge, so it would be incorrect to put a Roman numeral after their names.

OBJECTIVE 31

- The alkali metals in group 1 are always +1 when they form cations.
- The alkaline earth metals in group 2 are always +2 when they form cations.
- Aluminum and the elements in group 3 are always +3 when they form cations.
- Zinc and cadmium always form +2 cations.

Although silver can form both +1 and +2 cations, the +2 is so rare that we usually name $Ag^+$ *silver ion*, not *silver(I) ion.* $Ag^{2+}$ is named *silver(II) ion.*

We will assume that all of the metallic elements other than those mentioned above can have more than one charge, so their cation names will include a Roman numeral. For example, $Mn^{2+}$ is named *manganese(II).* We know that we must put the Roman numeral in the name because manganese is not on our list of metals with only one charge.

## EXAMPLE 3.6 Naming Monatomic Ions

OBJECTIVE 31

Write names that correspond to the following formulas for monatomic ions: (a) $Ba^{2+}$, (b) $S^{2-}$, and (c) $Cr^{3+}$.

***Solution***

a. Because barium is in group 2, the only possible charge is +2. When there is only one possible charge, metallic ions are named with the name of the metal. Therefore, $Ba^{2+}$ is **barium ion.**

b. Monatomic anions are named with the root of the nonmetal and *ide,* so $S^{2-}$ is **sulfide ion.**

c. Because chromium is not on our list of metals with only one possible charge, we need to indicate the charge with a Roman numeral. Therefore, $Cr^{3+}$ is **chromium(III) ion.**

### EXERCISE 3.6 Naming Monatomic Ions

OBJECTIVE 31

Write names that correspond to the following formulas for monatomic ions: (a) $Mg^{2+}$, (b) $F^-$, and (c) $Sn^{2+}$.

### EXAMPLE 3.7 Formulas for Monatomic Ions

OBJECTIVE 31

Write formulas that correspond to the following names for monatomic ions: (a) phosphide ion, (b) lithium ion, and (c) cobalt(II) ion.

***Solution***

a. We know that this is a monatomic anion because it has the form *(nonmetal root)ide*. Phosphorus atoms gain 3 electrons to get 18 electrons like the noble gas argon, Ar. Phosphide ion is $\mathbf{P^{3-}}$.

b. Lithium atoms lose 1 electron to get 2 electrons, like the noble gas helium. Lithium ion is $\mathbf{Li^+}$.

c. The Roman numeral indicates that the cobalt ion has a +2 charge. Note that we would not have been able to determine this from cobalt's position on the periodic table. Cobalt(II) is $\mathbf{Co^{2+}}$.

### EXERCISE 3.7 Formulas for Monatomic Ions

OBJECTIVE 31

Write formulas that correspond to the following names for monatomic ions: (a) bromide ion, (b) aluminum ion, and (c) gold(I) ion.

Several of the monatomic cations play important roles in our bodies. For example, we need calcium ions in our diet for making bones and teeth. Iron(II) ions are found in hemoglobin molecules in the red blood cells that carry oxygen from our lungs to the tissues of our bodies. Potassium, sodium, and chloride ions play a crucial role in the transfer of information between nerve cells. Enzymes (chemicals in the body that increase the speed of chemical reactions) often contain metallic cations, such as manganese(II) ions, iron(III) ions, copper(II) ions, and zinc ions. For example, $Zn^{2+}$ ions are in the center of alcohol dehydrogenase, which is the enzyme in our livers that accelerates the breakdown of the ethanol consumed in alcoholic beverages.

## Structure of Ionic Compounds

OBJECTIVE 32

Figure 3.18 shows the solid structure of the ionic compound sodium chloride, NaCl. We have already seen that the particles that form the structure of ionic compounds are cations and anions and that the attractions that hold them together are ionic bonds. When atoms gain electrons and form anions, they get larger. When atoms lose electrons and form cations, they get significantly smaller. Thus the chloride ions are larger than the sodium ions. The ions take the arrangement that provides the greatest cation–anion attraction while minimizing the anion–anion and cation–cation repulsions. Each sodium ion is surrounded by 6 chloride ions, and each chloride ion is surrounded by 6 sodium ions.

Any ionic compound that has the same arrangement of cations and anions as NaCl is said to have the sodium chloride crystal structure. The ionic compounds in this category include AgF, AgCl, AgBr, and the oxides and sulfides of the alkaline earth metals, such as MgO and CaS. The sodium chloride crystal structure is just one of many different possible arrangements of ions in solid ionic compounds.

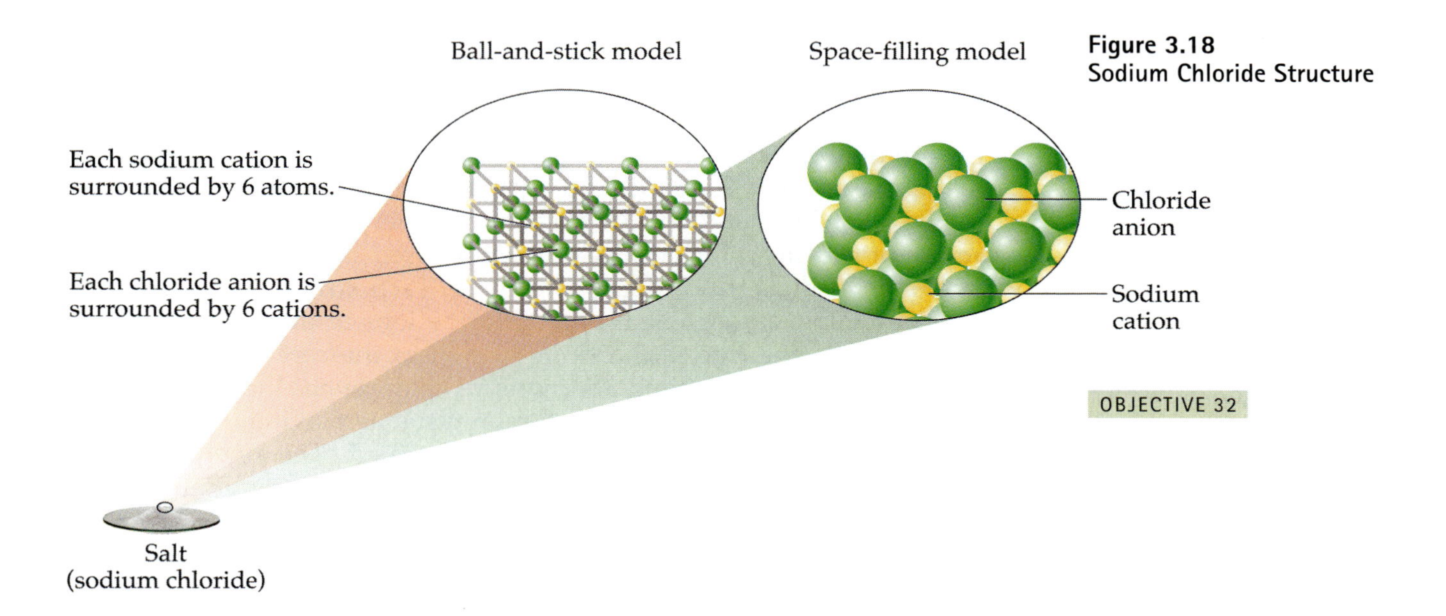
Ball-and-stick model
Space-filling model
Each sodium cation is surrounded by 6 atoms.
Each chloride anion is surrounded by 6 cations.
Chloride anion
Sodium cation
Salt (sodium chloride)

**Figure 3.18**
**Sodium Chloride Structure**

OBJECTIVE 32

## Polyatomic Ions

When an electric current is run through a purified saltwater solution (brine), hydrogen gas, chlorine gas, and an ionic compound called sodium hydroxide, NaOH, form. The sodium hydroxide, commonly called caustic soda or lye, is a very important compound that is used in paper production, vegetable oil refining, and the manufacture of many different compounds, such as soap and rayon. Like sodium chloride, NaCl, sodium hydroxide, NaOH, contains a cation and an anion, but unlike the monatomic $Cl^-$ anion in NaCl, the hydroxide ion, $OH^-$, in NaOH is a **polyatomic ion**—a charged *collection* of atoms held together by covalent bonds. To show the charge, Lewis structures of polyatomic ions are often enclosed in brackets, with the charge indicated at the top right. The Lewis structure for hydroxide is

$$\left[:\ddot{\underset{..}{O}}—H\right]^-$$

Hydroxide

Note in this Lewis structure that the oxygen atom does not have its most common bonding pattern, two bonds and two lone pairs. The gain or loss of electrons in the formation of polyatomic ions leads to one or more atoms in the ions having a different number of bonds and lone pairs than is predicted in Table 3.1.

The Lewis structure of the ammonium ion, $NH_4^+$, the only common polyatomic cation, is

$$\left[\begin{array}{ccc} & H & \\ & | & \\ H- & N & -H \\ & | & \\ & H & \end{array}\right]^+$$

Ammonium ion

The ammonium ion can take the place of a monatomic cation in an ionic crystal structure. For example, the crystal structure of ammonium chloride, $NH_4Cl$, which is found in fertilizers, is very similar to the crystal structure of cesium chloride, CsCl, which is used in brewing, in mineral waters, and to make fluorescent screens. In each structure, the chloride ions form a cubic arrangement with chloride ions at the corners of each cube. In cesium chloride, the cesium ions sit in the center of each cube, surrounded by 8 chloride ions. Ammonium chloride has the same general structure as cesium chloride, with ammonium ions playing the same role in the $NH_4Cl$ structure as cesium ions play in CsCl. The key idea is that because of its overall positive charge, the polyatomic ammonium ion acts like the monatomic cesium ion, $Cs^+$ (Figure 3.19).

OBJECTIVE 33

There are many polyatomic anions that can take the place of monatomic anions. For example, zinc hydroxide, which is used as an absorbent in surgical dressings, has a structure similar to that of zinc chloride, which is used in embalming and taxidermists' fluids. The hydroxide ion, $OH^-$, plays the same role in the structure of $Zn(OH)_2$ that the chloride ion, $Cl^-$, plays in $ZnCl_2$. (To show that there are 2 hydroxide ions for each zinc ion, the OH is in parentheses, with the subscript 2.)

It is very useful to be able to convert between the names and formulas of the common polyatomic ions listed in Table 3.6. Check with your instructor to find out which of these you will be expected to know and whether there are others you should know as well.

OBJECTIVE 34

OBJECTIVE 34

**Table 3.6**
Common Polyatomic Ions

| Ion | Name | Ion | Name | Ion | Name |
|---|---|---|---|---|---|
| $NH_4^+$ | ammonium | $PO_4^{3-}$ | phosphate | $SO_4^{2-}$ | sulfate |
| $OH^-$ | hydroxide | $NO_3^-$ | nitrate | $C_2H_3O_2^-$ | acetate |
| $CO_3^{2-}$ | carbonate | | | | |

Some polyatomic anions are formed by the attachment of 1 or more hydrogen atoms. In fact, it is common for hydrogen atoms to be transferred from 1 ion or molecule to another ion or molecule. When this happens, the hydrogen atom is usually transferred without its electron, as

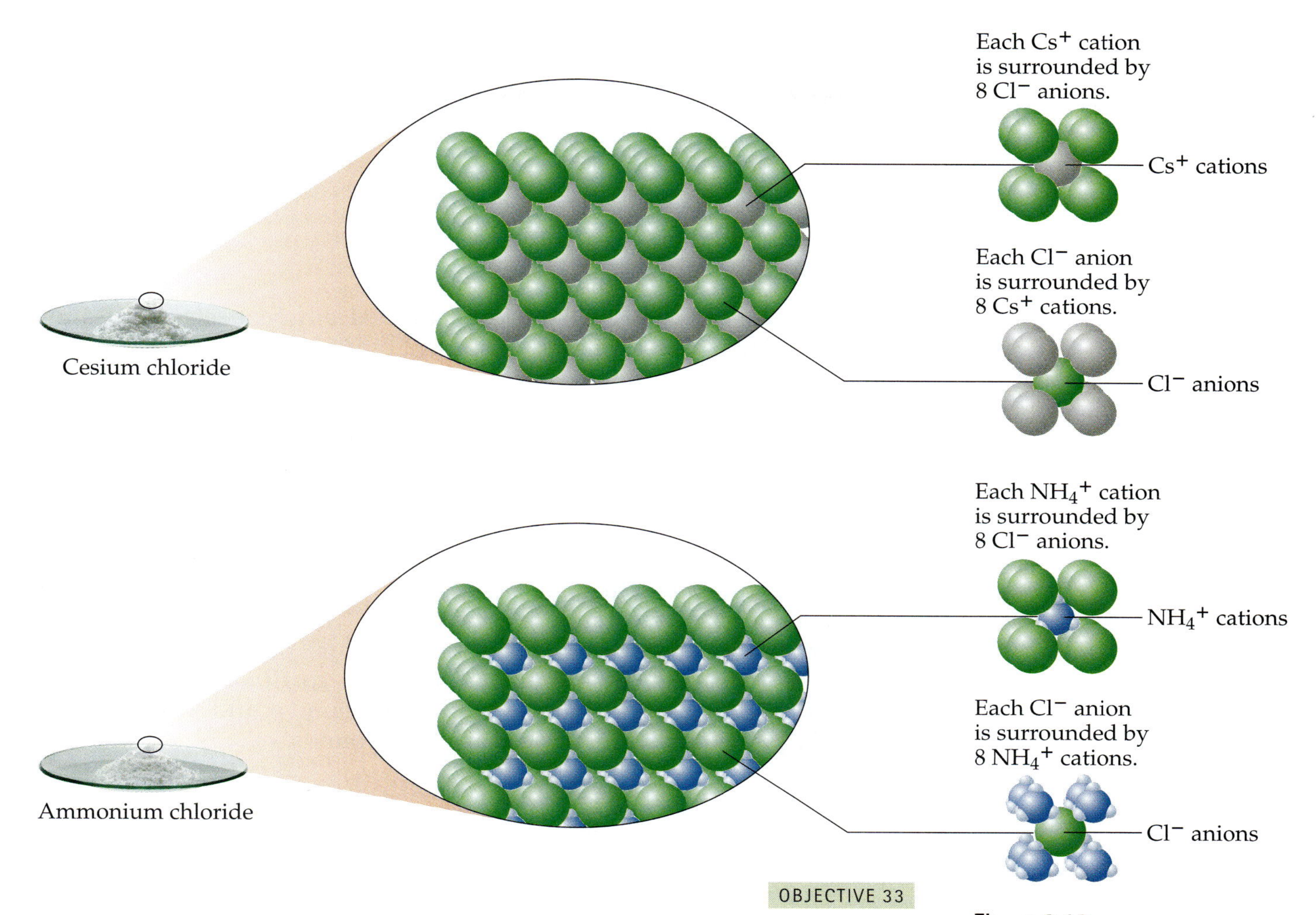
Each $Cs^+$ cation is surrounded by 8 $Cl^-$ anions.
$Cs^+$ cations
Each $Cl^-$ anion is surrounded by 8 $Cs^+$ cations.
$Cl^-$ anions
Cesium chloride
Each $NH_4^+$ cation is surrounded by 8 $Cl^-$ anions.
$NH_4^+$ cations
Each $Cl^-$ anion is surrounded by 8 $NH_4^+$ cations.
$Cl^-$ anions
Ammonium chloride

**Figure 3.19**
**Cesium Chloride Crystal Structure**

OBJECTIVE 33

$H^+$. If an anion has a charge of $-2$ or $-3$, it can gain 1 or 2 $H^+$ ions and still retain a negative charge. For example, carbonate, $CO_3^{2-}$, can gain an $H^+$ ion to form $HCO_3^-$, which is found in baking soda. The sulfide ion, $S^{2-}$, can gain 1 $H^+$ ion to form $HS^-$. Phosphate, $PO_4^{3-}$, can gain 1 $H^+$ ion to form $HPO_4^{2-}$, or it can gain 2 $H^+$ ions to form $H_2PO_4^-$. Both $HPO_4^{2-}$ and $H_2PO_4^-$ are found in flame retardants. These polyatomic ions are named with the word *hydrogen* in front of the name of the anion if there is 1 $H^+$ ion attached and with *dihydrogen* in front of the name of the anion if 2 $H^+$ ions are attached.

$HCO_3^-$ is hydrogen carbonate ion.

$HS^-$ is hydrogen sulfide ion.

OBJECTIVE 35

$HPO_4^{2-}$ is hydrogen phosphate ion.

$H_2PO_4^-$ is dihydrogen phosphate ion.

OBJECTIVE 36

Some polyatomic ions also have nonsystematic names that are often used. For example, $HCO_3^-$ is often called bicarbonate instead of hydrogen carbonate. You should avoid using this less widely accepted name, but because many people still use it, you should know it.

You can find a more comprehensive description of polyatomic ions, including a longer list of their names and formulas, at the following Web site:
**www.chemplace.com/college/**

## Converting Formulas to Names

OBJECTIVE 37

As we noted earlier, chemists have established different sets of rules for writing the names and formulas of different types of chemical compounds, so the first step in writing a name from a chemical formula is to decide what type of compound the formula represents. A chemical formula for an ionic compound will have one of the following forms.

- **Metal–nonmetal:** Ionic compounds whose formula contains one symbol for a metal and one symbol for a nonmetal are called **binary ionic compounds.** Their general formula is $M_aA_b$, where "M" represents the symbol of a metallic element, "A" represents the symbol for a nonmetallic element, and lower-case "a" and "b" represent the subscripts in the formula (unless one or more of the subscripts are assumed to be 1). For example, NaF (used to fluoridate municipal waters), $MgCl_2$ (used in floor sweeping compounds), and $Al_2O_3$ (used in ceramic glazes) represent binary ionic compounds.

Metallic element — Nonmetallic element

**Binary ionic compound** — $M_aB_b$

Subscripts (may be absent)

- **Metal–polyatomic ion:** Polyatomic ions can take the place of monatomic anions, so formulas that contain a symbol for a metallic element and the formula for a polyatomic ion represent ionic compounds. For example, $NaNO_3$ (in solid rocket propellants) and $Al_2(SO_4)_3$ (a foaming agent in fire foams) represent ionic compounds.

- **Ammonium–nonmetal or ammonium–polyatomic ion:** Ammonium ions, $NH_4^+$, can take the place of metallic cations in an ionic compound, so chemical formulas that contain the formula for ammonium with either a symbol for a nonmetallic element or a formula for a polyatomic ion represent ionic compounds. For example, $NH_4Cl$ (in dry cell batteries), $(NH_4)_2S$ (used to color brass), and $(NH_4)_2SO_4$ (in fertilizers) represent ionic compounds.

OBJECTIVE 37

The names of ionic compounds consist of the name for the cation followed by the name for the anion. Tables 3.7 and 3.8 summarize the ways in which cations and anions are named.

Table 3.7
Summary of the Ways Cations Are Named

OBJECTIVE 37

| Type of cation | General Name | Example |
|---|---|---|
| metal with one possible charge (groups 1, 2, 3—Al, Zn, and Cd) | name of metal | $Mg^{2+}$ is magnesium. |
| metal with more than one possible charge (the rest of the metals) | name of metal (Roman numeral) | $Cu^{2+}$ is copper(II). |
| polyatomic cations (For us, ammonium is the only member of this category.) | name of polyatomic cation | $NH_4^+$ is ammonium. |

Table 3.8
Summary of the Ways Anions Are Named

OBJECTIVE 37

| Type of anion | General Name | Example |
|---|---|---|
| monatomic anions | (root of nonmetal)ide | $O^{2-}$ is oxide. |
| polyatomic anions | name of polyatomic anion | $NO_3^-$ is nitrate. |

As an example of the thought process for naming ionic compounds, let's write the name for MnO (used as a food additive and dietary supplement). Our first step is to identify what type of compound MnO represents. Because it is composed of a metallic element and a nonmetallic element, we recognize it as ionic. Thus we must write the name of the cation followed by the name of the anion. Manganese is not on our list of metallic elements with only one possible charge, so its name is the name of the element followed by a Roman numeral that represents the charge. Therefore, our next step is to determine the charge on each manganese cation in MnO.

When the cation in an ionic formula is created from a metallic element whose atoms can have more than one charge, we can discover the cation's charge by identifying the charge on the anion and then figuring out what charge the cation must have to yield a formula that is uncharged overall. To discover the charge on the manganese ions in MnO, we first determine the charge on the anions. A glance at the periodic table shows oxygen to be in group 16, or 6A, whose nonmetallic members always form −2 ions. If we let $x$ represent the charge on the manganese ion, the charge on the manganese cation can be figured out as follows:

$$\text{Total cation charge} + \text{total anion charge} = 0$$

$$x + (-2) = 0$$

$$x = +2$$

Each manganese cation must therefore be +2 to balance the −2 of the oxide to yield an uncharged ionic formula. The systematic name for $Mn^{2+}$ is manganese(II). Monatomic anions are named with the root of the nonmetal followed by *ide,* so $O^{2-}$ is oxide. MnO is named manganese(II) oxide. Example 3.8 provides other demonstrations.

## EXAMPLE 3.8 Naming Ionic Compounds

OBJECTIVE 37

Write the names that correspond to the following formulas: (a) $MgO$ (used to make aircraft windshields), (b) $CoCl_2$ (used to manufacture vitamin B-12), (c) $NH_4NO_3$ (used to make explosives), and (d) $Fe_2O_3$ (in paint pigments).

***Solution***

a. The compound $MgO$ includes the cation $Mg^{2+}$ and the anion $O^{2-}$. Magnesium cations are always +2, so the name of the cation is the same as the name of the metallic element. The anion $O^{2-}$ is monatomic, so it is named by combining the root of the name of the nonmetal with *ide.* Therefore, $MgO$ is **magnesium oxide.**

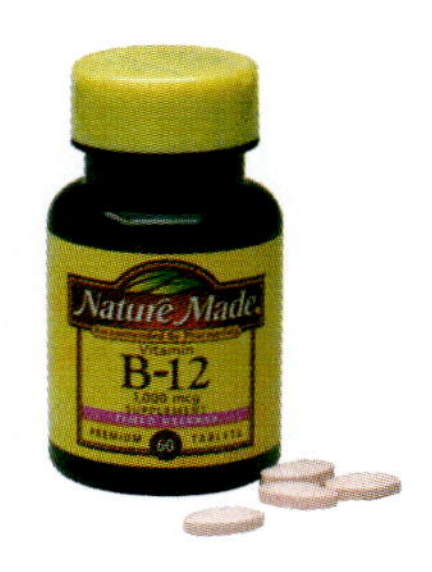

$CoCl_2$ is used to make vitamin B-12.

b. Cobalt is not on our list of metallic elements that form ions with only one charge, so we assume that it can form ions with more than one possible charge. Therefore, we need to show the charge on the cobalt ion with a Roman numeral in parentheses after the name *cobalt.* The cobalt ion must be +2 to balance the −2 from the two −1 chloride ions.

$$\text{Total cation charge} + \text{total anion charge} = 0$$
$$x + 2(-1) = 0$$
$$x = +2$$

The anion $Cl^-$ is monatomic, so its name includes the root of the name of the nonmetal and *ide.* Therefore, $CoCl_2$ is **cobalt(II) chloride.**

c. The compound $NH_4NO_3$ includes the cation $NH_4^+$ and the anion $NO_3^-$. Both of these ions are polyatomic ions with names you should memorize. $NH_4NO_3$ is **ammonium nitrate.**

d. Iron is not on our list of metallic elements that form ions with only one charge, so we need to show the charge on the iron ion with a Roman numeral. Because oxygen atoms have 2 fewer electrons than the nearest noble gas, neon, they form −2 ions. In the following equation, $x$ represents the charge on each iron ion. Because there are 2 iron ions, $2x$ represents the total cation charge. Likewise, because there are 3 oxygen ions, the total anion charge is three times the charge on each oxygen ion.

$$\text{Total cation charge} + \text{total anion charge} = 0$$
$$2x + 3(-2) = 0$$
$$x = +3$$

The iron ions must be +3 in order for them to balance the −6 from three −2 oxide ions, so the cation name is iron(III). Because $O^{2-}$ is a monatomic anion, its name includes the root of the name of the nonmetal and *ide.* Therefore, $Fe_2O_3$ is **iron(III) oxide.**

## EXERCISE 3.8 Naming Ionic Compounds

OBJECTIVE 37

Write the names that correspond to the following formulas: (a) $LiCl$ (used in soft drinks to help reduce the escape of bubbles), (b) $Cr_2(SO_4)_3$ (used in chrome plating), and (c) $NH_4HCO_3$ (used as a leavening agent for cookies, crackers, and cream puffs).

## Converting Names of Ionic Compounds to Formulas

Before you can write a chemical formula from the name of a compound, you need to recognize what type of compound the name represents. For binary ionic compounds, the first part of the name is the name of a metallic cation. This may include a Roman numeral in parentheses. The anion name starts with the root of the name of a nonmetal and ends with *ide*.

OBJECTIVE 37

(name of metal)(maybe Roman numeral) (root of nonmetal) *ide*

For example, aluminum fluoride (used in the production of aluminum) and tin(II) chloride (used in galvanizing tin) are binary ionic compounds.

You can identify other names as representing ionic compounds by recognizing that they contain the names of common polyatomic ions. For example, ammonium chloride and iron(III) hydroxide are both ionic compounds. Many of the polyatomic ions that you will be expected to recognize end in *ate*, so this ending tells you that the name represents an ionic compound. Copper(II) sulfate is an ionic compound.

Follow these steps to write formulas for ionic compounds.

OBJECTIVE 37

**Step 1** Write the formula, including the charge, for the cation. (See Figure 3.17 to review the charges on monatomic cations.)

**Step 2** Write the formula, including the charge, for the anion. (See Figure 3.17 to review the charges on monatomic anions. See Table 3.6 to review the formulas for several common polyatomic ions.)

**Step 3** Write subscripts for each formula in such a way as to yield an uncharged compound. Table 3.9 shows examples.

- Use the lowest whole-number ratio for the subscripts.
- If the subscript for a polyatomic ion is higher than 1, place the formula for the polyatomic ion in parentheses and put the subscript outside the parentheses.

Table 3.9
Possible Cation-Anion Ratios. X represents the cation, and Y represents the anion.

OBJECTIVE 37

| Ionic charges | General formula | Example ions | Example formula |
|---|---|---|---|
| $X^+$ and $Y^-$ | XY | $Na^+$ and $Cl^-$ | NaCl |
| $X^+$ and $Y^{2-}$ | $X_2Y$ | $NH_4^+$ and $SO_4^{2-}$ | $(NH_4)_2SO_4$ |
| $X^+$ and $Y^{3-}$ | $X_3Y$ | $Li^+$ and $PO_4^{3-}$ | $Li_3PO_4$ |
| $X^{2+}$ and $Y^-$ | $XY_2$ | $Mg^{2+}$ and $NO_3^-$ | $Mg(NO_3)_2$ |
| $X^{2+}$ and $Y^{2-}$ | XY | $Ca^{2+}$ and $CO_3^{2-}$ | $CaCO_3$ |
| $X^{2+}$ and $Y^{3-}$ | $X_3Y_2$ | $Ba^{2+}$ and $N^{3-}$ | $Ba_3N_2$ |
| $X^{3+}$ and $Y^-$ | $XY_3$ | $Al^{3+}$ and $F^-$ | $AlF_3$ |
| $X^{3+}$ and $Y^{2-}$ | $X_2Y_3$ | $Sc^{3+}$ and $S^{2-}$ | $Sc_2S_3$ |
| $X^{3+}$ and $Y^{3-}$ | XY | $Fe^{3+}$ and $PO_4^{3-}$ | $FePO_4$ |

## EXAMPLE 3.9 Formulas for Ionic Compounds

OBJECTIVE 37

Write the chemical formulas that correspond to the following names: (a) aluminum chloride (used in cosmetics), (b) chromium(III) oxide (a pigment for coloring pottery glazes), (c) calcium nitrate (provides a red-orange color in fireworks), and (d) ammonium sulfide (used to make synthetic flavors).

***Solution***

a. Aluminum chloride has the form *(name of metal) (root of nonmetal)ide*, so we recognize it as a binary ionic compound. Because aluminum atoms have 3 more electrons than the nearest noble gas, neon, they lose three electrons and form +3 ions. Because chlorine atoms have 1 fewer electron than the nearest noble gas, argon, they gain 1 electron to form −1 ions. The formulas for the individual ions are $Al^{3+}$ and $Cl^-$. It takes three chlorides to neutralize the +3 aluminum ion, so the formula for the compound is **$AlCl_3$.**

b. Chromium(III) oxide has the form *(name of metal)(Roman numeral) (root of nonmetal)ide*, so it represents a binary ionic compound. The (III) tells you that the chromium ions have a +3 charge. Because oxygen atoms have 2 fewer electrons than the nearest noble gas, neon, they gain 2 electrons to form −2 ions. The formulas for the ions are $Cr^{3+}$ and $O^{2-}$. When the ionic charges are +3 and −2 (or +2 and −3), a simple procedure will help you to determine the subscripts in the formula. Disregarding the signs of the charges, we use the superscript on the anion as the subscript on the cation, and the superscript on the cation as the subscript on the anion.

$$Cr^{3+} \; O^{2-}$$

Chromium(III) oxide is **$Cr_2O_3$.**

c. Calcium nitrate has the form *(name of metal) (name of a polyatomic ion)*, so it represents an ionic compound. (The *ate* at the end of nitrate tells us that it is a polyatomic ion.) Calcium is in group 2 on the periodic table. Because all metals in group 2 have 2 more electrons than the nearest noble gas, they all lose 2 electrons and form +2 ions. Nitrate is $NO_3^-$, so 2 nitrate ions are needed to neutralize the charge on the $Ca^{2+}$. Calcium nitrate is **$Ca(NO_3)_2$.** Note that in order to show that there are 2 nitrate ions, we placed the formula for nitrate in parentheses.

d. Ammonium sulfide has the form *ammonium (root of a nonmetal)ide*, so it represents an ionic compound. You should memorize the formula for ammonium, $NH_4^+$. Sulfur has 2 fewer electrons than the noble gas, argon, so it gains 2 electrons and forms a −2 anion. Two ammonium ions are necessary to neutralize the −2 sulfide. Ammonium sulfide is **$(NH_4)_2S$.**

## EXERCISE 3.9 Formulas for Ionic Compounds

OBJECTIVE 37

Write the formulas that correspond to the following names: (a) aluminum oxide (used to waterproof fabrics), (b) cobalt(III) fluoride (used to add fluorine atoms to compounds), (c) iron(II) sulfate (in enriched flour), (d) ammonium hydrogen phosphate (coats vegetation to retard forest fires), and (e) potassium bicarbonate (in fire extinguishers).

## Chapter Glossary

**Compound** A substance that contains two or more elements, the atoms of these elements always combining in the same whole-number ratio.

**Chemical formula** A concise written description of the components of a chemical compound. It identifies the elements in the compound by their symbols and indicates the relative number of atoms of each element with subscripts.

**Pure substance** A sample of matter that has constant composition. There are two types of pure substances: elements and compounds.

**Mixture** A sample of matter that contains two or more pure substances and has variable composition.

**Chemical bond** An attraction between atoms or ions in chemical compounds. Covalent bonds and ionic bonds are examples.

**Polar covalent bond** A covalent bond in which electrons are shared unequally, leading to a partial negative charge on the atom that attracts the electrons more and to a partial positive charge on the other atom.

**Nonpolar covalent bond** A covalent bond in which the difference in electron-attracting ability of two atoms in a bond is negligible (or zero), so the atoms in the bond have no significant charges.

**Ionic bond** The attraction between a cation and an anion.

**Molecular compound** A compound composed of molecules. In such compounds, all of the bonds between atoms are covalent bonds.

**Ionic compound** A compound that consists of ions held together by ionic bonds.

**Valence electrons** The electrons that are most important in the formation of chemical bonds. The number of valence electrons for the atoms of an element is equal to the element's A-group number on the periodic table. (A more comprehensive definition of valence electrons appears in Chapter 12.)

**Electron-dot symbol** A representation of an atom that consists of its elemental symbol surrounded by dots representing its valence electrons.

**Lewis structure** A representation of a molecule that consists of the elemental symbol for each atom in the molecule, lines to show covalent bonds, and pairs of dots to indicate lone pairs.

**Lone pair** Two electrons that are not involved in the covalent bonds between atoms but are important for explaining the arrangement of atoms in molecules. They are represented by pairs of dots in Lewis structures.

**Hydrocarbons** Compounds that contain only carbon and hydrogen.

**Organic chemistry** The branch of chemistry that involves the study of carbon-based compounds.

**Double bond** A link between atoms that results from the sharing of four electrons. It can be viewed as two 2-electron covalent bonds.

**Triple bond** A link between atoms that results from the sharing of six electrons. It can be viewed as three 2-electron covalent bonds.

**Alcohols** Compounds that contain a hydrocarbon group with one or more –OH groups attached.

**Tetrahedral** The molecular shape that keeps the negative charge of four electron groups as far apart as possible. This shape has angles of 109.5° between the atoms.

**Bond angle** The angle formed by straight lines (representing bonds) connecting the nuclei of three adjacent atoms.

**Binary covalent compound** A compound that consists of two nonmetallic elements.

**Monatomic anions** Negatively charged particles, such as $Cl^-$, $O^{2-}$, and $N^{3-}$, that contain single atoms with a negative charge.

**Monatomic cations** Positively charged particles, such as $Na^+$, $Ca^{2+}$, and $Al^{3+}$, that contain single atoms with a positive charge.

**Polyatomic ion** A charged collection of atoms held together by covalent bonds.

**Binary ionic compound** An ionic compound whose formula contains one symbol for a metal and one symbol for a nonmetal.

You can test yourself on the glossary terms at the following Web site:
**www.chemplace.com/college/**

## Chapter Objectives

**The goal of this chapter is to teach you to do the following.**

1. Define all of the terms in the Chapter Glossary.

### Section 3.1 Classification of Matter

2. Convert between a description of the number of atoms of each element found in a compound and its chemical formula.
3. Given a description of a form of matter, classify it as a pure substance or a mixture.
4. Given a description of a pure substance, classify it as an element or a compound.

### Section 3.2 Chemical Compounds and Chemical Bonds

5. Describe the polar covalent bond between two nonmetallic atoms, one of which attracts electrons more than the other one does. Your description should include a rough sketch of the electron cloud that represents the electrons involved in the bond.
6. Describe the process that leads to the formation of ionic bonds between metallic and nonmetallic atoms.
7. Explain how a nonpolar covalent bond, a polar covalent bond, and an ionic bond differ. Your description should include rough sketches of the electron clouds that represent the electrons involved in the formation of each bond.
8. Given the names or formulas for two elements, identify the bond that would form between them as covalent or ionic.
9. Given a formula for a compound, classify it as either a molecular compound or an ionic compound.

### Section 3.3 Molecular Compounds

10. Determine the number of valence electrons for the atoms of each of the nonmetallic elements.
11. Draw electron-dot symbols for the nonmetallic elements and use them to explain why these elements form the bonding patterns listed in Table 3.1.
12. Give a general description of the information provided in a Lewis structure.
13. Identify the most common number of covalent bonds and lone pairs for the atoms of each of the following elements: hydrogen, the halogens (group 17), oxygen, sulfur, selenium, nitrogen, phosphorus, and carbon.
14. Convert between the following systematic names and their chemical formulas: methanol, ethanol, and 2-propanol.
15. Given one of the following names for alcohols, write its chemical formula: methyl alcohol, ethyl alcohol, and isopropyl alcohol.
16. Given a chemical formula, draw a Lewis structure for it that has the most common number of covalent bonds and lone pairs for each atom.
17. Describe the tetrahedral molecular shape.
18. Explain why the atoms in the $CH_4$ molecule have a tetrahedral molecular shape.
19. Describe the information given by a space-filling model, by a ball-and-stick model, and by a geometric sketch.
20. Draw geometric sketches, including bond angles, for $CH_4$, $NH_3$, and $H_2O$.
21. Describe the attractions between $H_2O$ molecules.
22. Describe the structure of liquid water.

### Section 3.4 Naming Binary Covalent Compounds

23. Convert between the names and chemical formulas for water, ammonia, methane, ethane, and propane.
24. Given a formula or name for a compound, identify whether it represents a binary covalent compound.
25. Write or identify prefixes for the numbers 1–10. (For example, *mono* represents 1, *di* represents 2, etc.)
26. Write or identify the roots for the names of the nonmetallic elements. (For example, the root for oxygen is *ox*).
27. Convert between the complete name, the common name, and the chemical formula for HF, HCl, HBr, HI, and $H_2S$.
28. Convert between the systematic names and chemical formulas for binary covalent compounds.

### Section 3.5 Ionic Compounds

29. Explain why metals usually combine with nonmetals to form ionic bonds.
30. Write the ionic charges acquired by the following elements:
    a. group 17, halogens
    b. oxygen, sulfur, and selenium

c. nitrogen and phosphorus

d. hydrogen

e. group 1, alkali metals

f. group 2, alkaline earth metals

g. group 3 elements

h. aluminum

i. iron, silver, copper, and zinc

31. Convert between the names and chemical formulas for the monatomic ions.
32. Describe the crystal structure of sodium chloride, NaCl.
33. Describe the similarities and differences between the ionic structure of cesium chloride and that of ammonium chloride.
34. Convert between the names and chemical formulas for such common polyatomic ions as hydroxide, ammonium, acetate, sulfate, nitrate, phosphate, and carbonate. *Be sure to check with your instructor to determine which polyatomic ions you will be expected to know for your exams.*
35. Convert between the names and chemical formulas for the polyatomic anions that are derived from the addition of $H^+$ ions to anions with $-2$ or $-3$ charges. For example, $H_2PO_4^-$ is dihydrogen phosphate.
36. Write the chemical formula that corresponds to the common name *bicarbonate*.
37. Convert between the names and chemical formulas for ionic compounds.

## Review Questions

**Write in each blank the word or words that best complete each sentence.**

1. An atom or group of atoms that has lost or gained one or more electrons to create a charged particle is called a(n) ____________.
2. An atom or collection of atoms with an overall positive charge is a(n) ____________.
3. An atom or collection of atoms with an overall negative charge is a(n) ____________.
4. A(n) ____________ bond is a link between atoms that results from the sharing of two electrons.
5. A(n) ____________ is an uncharged collection of atoms held together with covalent bonds.
6. A molecule like $H_2$, which is composed of two atoms, is called ____________.
7. Describe the particle nature of solids, liquids, and gases. Your description should include the motion of the particles and the attractions between the particles.
8. Describe the nuclear model of the atom.
9. Describe the hydrogen molecule, $H_2$. Your description should include the nature of the link between the hydrogen atoms and a sketch that shows the two electrons in the molecule.

10. Complete the following table.

| Element name | Element symbol | Group number on periodic table | Metal, nonmetal, or metalloid? |
|---|---|---|---|
| | Li | | |
| carbon | | | |
| | Cl | | |
| oxygen | | | |
| | Cu | | |
| calcium | | | |
| | Sc | | |

11. Write the name of the group to which each of the following belongs.
    a. chlorine
    b. xenon
    c. sodium
    d. magnesium

## Key Ideas

**Complete the following statements by writing one these words or phrases in each blank.**

$\delta+$
$\delta-$
anion-anion and cation-cation repulsions
anions
arranged in space
binary
cation-anion attraction
cations
compounds
constant composition
constantly breaking
covalent
double
eight valence
electron-charge clouds
forming new attractions
four
four electrons
gain
gases
H+
liquids
larger
lose
molecular
molecules
monatomic cations
negative
negligible (or zero)
noble gas
nonmetallic
octet
partial positive
partial negative
positive
repel
same average distance apart
smaller
solids
specific type of compound
strongly
subscripts
symbols
usually ionic
variable
whole-number ratio

12. A compound is a substance that contains two or more elements, the atoms of those elements always combining in the same ________________.

13. There are relatively few chemical elements, but there are millions of chemical ________________.

14. A chemical formula is a concise written description of the components of a chemical compound. It identifies the elements in the compound by their ________________ and indicates the relative number of atoms of each element with ________________.

15. When a substance has a(n) ____________________, it must by definition be either an element or a compound, and it is considered a pure substance.
16. Mixtures are samples of matter that contain two or more pure substances and have ____________________ composition.
17. When the difference in electron-attracting ability between atoms in a chemical bond is ____________________, the atoms in the bond will have no significant partial charges. We call this type of bond a nonpolar covalent bond.
18. Because particles with opposite charges attract each other, there is an attraction between ____________________ and ____________________. This attraction is called an ionic bond.
19. Compounds that have ionic bonds, such as the sodium chloride in table salt, are ____________________ at room temperature and pressure, but compounds with all covalent bonds, such as hydrogen chloride and water, can be ____________________ and ____________________ as well as solids.
20. The atom in a chemical bond that attracts electrons more strongly acquires a(n) ____________________ charge, and the other atom acquires a(n) ____________________ charge. If the electron transfer is significant but not enough to form ions, the atoms acquire ____________________ and ____________________ charges. The bond in this situation is called a polar covalent bond.
21. When a nonmetallic atom bonds to another nonmetallic atom, the bond is ____________________.
22. When a metallic atom bonds to a nonmetallic atom, the bond is ____________________.
23. ____________________ compounds are composed of ____________________, which are collections of atoms held together by all covalent bonds.
24. The noble gases (group 8A) have a(n) ____________________ of electrons (except for helium, which has only two electrons total), and they are so stable that they rarely form chemical bonds with other atoms.
25. When atoms other than the noble-gas atoms form bonds, they often have ____________________ electrons around them in total.
26. The sum of the numbers of covalent bonds and lone pairs for the most common bonding patterns of the atoms of nitrogen, phosphorus, oxygen, sulfur, selenium, and the halogens is ____________________.
27. Atoms can form double bonds, in which ____________________ are shared between atoms. Double bonds are represented by ____________________ lines in a Lewis structure.
28. Lewis structures are useful for showing how the atoms in a molecule are connected by covalent bonds, but they do not always give a clear description of how the atoms are ____________________.
29. The actual shape of a molecule can be more accurately predicted by recognizing that the negatively charged electrons that form covalent bonds and lone pairs ____________________ each other. Therefore, the most stable arrangement of the electron groups is the molecular shape that keeps the groups as far away from each other as possible.
30. A space-filling model provides the most accurate representation of the ____________________ for the atoms in $CH_4$.

31. Because oxygen atoms attract electrons much more ________________ than do hydrogen atoms, the O–H covalent bond is very polar, leading to a relatively large partial negative charge on the oxygen atom (represented by a(n) ________________) and a relatively large partial positive charge on the hydrogen atom (represented by a(n) ________________).
32. As in other liquids, the attractions between water molecules are strong enough to keep them the ________________ but weak enough to allow each molecule to be ________________ the attractions that momentarily connect it to some molecules and ________________ to other molecules.
33. To write the name that corresponds to a formula for a compound, you need to develop the ability to recognize the formula as representing a(n) ________________.
34. You can recognize binary covalent compounds from their formulas, which contain symbols for only two ________________ elements.
35. Metallic atoms hold some of their electrons relatively loosely, and as a result, they tend to ________________ electrons and form cations. In contrast, nonmetallic atoms attract electrons more strongly than metallic atoms, so nonmetals tend to ________________ electrons and form anions.
36. Nonmetallic atoms form anions to get the same number of electrons as the nearest ________________.
37. The names of ________________ always start with the name of the metal, sometimes followed by a Roman numeral to indicate the charge of the ion.
38. When atoms gain electrons and form anions, they get ________________. When atoms lose electrons and form cations, they get significantly ________________.
39. The ions in ionic solids take the arrangement that provides the greatest ________________ while minimizing the ________________.
40. It is common for hydrogen atoms to be transferred from one ion or molecule to another ion or molecule. When this happens, the hydrogen atom is usually transferred without its electron, as ________________.
41. Ionic compounds whose formula contains one symbol for a metal and one symbol for a nonmetal are called ________________ ionic compounds.

## Chapter Problems

### Section 3.1 Classification of Matter

OBJECTIVE 3

OBJECTIVE 4

42. Classify each of the following as a pure substance or a mixture. If it is a pure substance, is it an element or a compound? Explain your answer.
    a. apple juice
    b. potassium (A serving of one brand of apple juice provides 6% of the recommended daily allowance of potassium.)
    c. ascorbic acid (vitamin C), $C_6H_8O_6$, in apple juice

OBJECTIVE 3

OBJECTIVE 4

43. Classify each of the following as a pure substance or a mixture. If it is a pure substance, is it an element or a compound? Explain your answer.
    a. fluorine (used to make fluorides, such as those used in toothpaste)
    b. toothpaste
    c. calcium fluoride, $CaF_2$, from the naturally occurring ore fluorite (It is used to make sodium monofluorophosphate, which is added to some toothpastes.)

44. Write the chemical formula for each of the following compounds. List the symbols for the elements in the order that the elements are mentioned in the description.
    a. a compound with molecules that consist of 2 nitrogen atoms and 3 oxygen atoms.
    b. a compound with molecules that consist of 1 sulfur atom and 4 fluorine atoms.
    c. a compound that contains 1 aluminum atom for every 3 chlorine atoms.
    d. a compound that contains 2 lithium atoms and 1 carbon atom for every 3 oxygen atoms.
45. Write the chemical formula for each of the following compounds. List the symbols for the elements in the order that the elements are mentioned in the description.
    a. a compound with molecules that consist of 2 phosphorus atoms and 5 oxygen atoms.
    b. a compound with molecules that consist of 2 hydrogen atoms and 1 sulfur atom.
    c. a compound that contains 3 calcium atoms for every 2 nitrogen atoms.
    d. a compound with molecules that consist of 12 carbon atoms, 22 hydrogen atoms, and 11 oxygen atoms.

### Section 3.2 Chemical Compounds and Chemical Bonds

OBJECTIVE 5

46. Hydrogen bromide, HBr, is used to make pharmaceuticals that require bromine in their structure. Each hydrogen bromide molecule has one hydrogen atom bonded to one bromine atom by a polar covalent bond. The bromine atom attracts electrons more than does the hydrogen atom. Draw a rough sketch of the electron cloud that represents the electrons involved in the bond.

OBJECTIVE 5

47. Iodine monochloride, ICl, is a compound used to make carbon-based (organic) compounds that contain iodine and chlorine. It consists of diatomic molecules with one iodine atom bonded to one chlorine atom by a polar covalent bond. The chlorine atom attracts electrons more than does the iodine atom. Draw a rough sketch of the electron cloud that represents the electrons involved in the bond.

OBJECTIVE 6

48. Atoms of potassium and fluorine form ions and ionic bonds in a very similar way to atoms of sodium and chlorine. Each atom of one of these elements loses one electron, and each atom of the other element gains one electron. Describe the process that leads to the formation of the ionic bond between potassium and fluorine atoms in potassium fluoride. Your answer should include mention of the charges that form on the atoms.

OBJECTIVE 6

49. Atoms of magnesium and oxygen form ions and ionic bonds in a similar way to atoms of sodium and chlorine. The difference is that instead of having each atom gain or lose one electron, each atom of one of these elements loses two electrons, and each atom of the other element gains two electrons. Describe the process that leads to the formation of the ionic bond between magnesium and oxygen atoms in magnesium oxide. Your answer should include mention of the charges that form on the atoms.

OBJECTIVE 7

50. Explain how a nonpolar covalent bond, a polar covalent bond, and an ionic bond differ. Your description should include rough sketches of the electron clouds that represent the electrons involved in the formation of each bond.

51. Write a chemical formula that represents both a molecule and a compound. Write a formula that represents a compound but not a molecule.

52. Would you expect the bonds between the following atoms to be ionic or covalent bonds? OBJECTIVE 8

    a. N–O  b. Al–Cl

53. Would you expect the bonds between the following atoms to be ionic or covalent bonds? OBJECTIVE 8

    a. Li–F  b. C–N

54. Classify each of the following as either a molecular compound or an ionic compound. OBJECTIVE 9

    a. acetone, $CH_3COCH_3$ (a common paint solvent)
    b. sodium sulfide, $Na_2S$ (used in sheep dips)

55. Classify each of the following as either a molecular compound or an ionic compound. OBJECTIVE 9

    a. cadmium fluoride, $CdF_2$ (a starting material for lasers)
    b. sulfur dioxide, $SO_2$ (a food additive that inhibits browning and bacterial growth)

## Section 3.3 Molecular Compounds

56. How many valence electrons does each atom of the following elements have? OBJECTIVE 10

    a. Cl
    b. C

57. How many valence electrons does each atom of the following elements have? OBJECTIVE 10

    a. N
    b. S

58. Draw electron-dot symbols for each of the following elements and use them to explain why each element has the bonding pattern listed in Table 3.1. OBJECTIVE 11

    a. oxygen  c. carbon
    b. fluorine  d. phosphorus

59. Draw electron-dot symbols for each of the following elements and use them to explain why each element has the bonding pattern listed in Table 3.1. OBJECTIVE 11

    a. iodine
    b. nitrogen
    c. sulfur

60. The following Lewis structure is for CFC-12, which is one of the ozone-depleting chemicals that has been used as an aerosol can propellant and as a refrigerant. Describe the information given in this Lewis structure. OBJECTIVE 12

```
        ..
       :F:
        |
  ..    |    ..
 :Cl—C—F:
  ..    |    ..
        |
       :Cl:
        ..
```

OBJECTIVE 12

61. Describe the information given in the following Lewis structure for methylamine, a compound used to make insecticides and rocket propellants.

```
     H
     |    ..
H — C — N — H
     |    |
     H    H
```

OBJECTIVE 13

62. Write the most common number of covalent bonds and lone pairs for atoms of each of the following nonmetallic elements.

a. H

b. iodine

c. sulfur

d. N

OBJECTIVE 13

63. Write the most common number of covalent bonds and lone pairs for atoms of each of the following nonmetallic elements.

a. C

b. phosphorus

c. oxygen

d. Br

OBJECTIVE 16

64. Draw a Lewis structure for each of the following formulas.

a. oxygen difluoride, $OF_2$ (an unstable, colorless gas)

b. bromoform, $CHBr_3$ (used as a sedative)

c. phosphorus triiodide, $PI_3$ (used to make organic compounds)

OBJECTIVE 16

65. Draw a Lewis structure for each of the following formulas.

a. nitrogen trifluoride, $NF_3$ (used in high-energy fuels)

b. chloroethane, $C_2H_5Cl$ (used to make the gasoline additive tetraethyl lead)

c. hypobromous acid, HOBr (used as a wastewater disinfectant)

OBJECTIVE 16

66. Draw Lewis structures for the following compounds by adding any necessary lines and dots to the skeletons given.

a. hydrogen cyanide, HCN (used to manufacture dyes and pesticides)

H — C — N

b. dichloroethene, $C_2Cl_4$ (used to make perfumes)

```
Cl — C — C — Cl
     |    |
     Cl   Cl
```

OBJECTIVE 16

67. Draw Lewis structures for the following compounds by adding any necessary lines and dots to the skeletons given.

a. formaldehyde, $H_2CO$ (used in embalming fluids)

```
     O
     |
H — C — H
```

b. 1-butyne, $C_4H_6$ (a specialty fuel)

```
            H   H
            |   |
H—C—C—C—C—H
            |   |
            H   H
```

**68.** Write two different names for each of the following alcohols.

OBJECTIVE 14

OBJECTIVE 15

a.
```
    H
    |   ..
H—C—O—H
    |   ..
    H
```

b.
```
    H   H
    |   |   ..
H—C—C—O—H
    |   |   ..
    H   H
```

c.
```
       ..
      :O—H
    H   |   H
    |   |   |
H—C—C—C—H
    |   |   |
    H   H   H
```

69. Explain why the atoms in the $CH_4$ molecule are arranged with a tetrahedral molecular shape.

OBJECTIVE 18

**70.** Compare and contrast the information given in the Lewis structure, the space-filling model, the ball-and-stick model, and the geometric sketch of a methane molecule, $CH_4$.

OBJECTIVE 19

OBJECTIVE 20

71. Compare and contrast the information given in the Lewis structure, the space-filling model, the ball-and-stick model, and the geometric sketch of an ammonia molecule, $NH_3$.

OBJECTIVE 19

OBJECTIVE 20

**72.** Compare and contrast the information given in the Lewis structure, the space-filling model, the ball-and-stick model, and the geometric sketch of a water molecule, $H_2O$.

OBJECTIVE 19

OBJECTIVE 20

73. What are the particles that form the basic structure of water? Describe the attraction that holds these particles together. Draw a rough sketch that shows the attraction between two water molecules.

OBJECTIVE 21

**74.** Describe the structure of liquid water.

OBJECTIVE 22

### Section 3.4 Naming Binary Covalent Compounds

75. What is wrong with using the name *nitrogen oxide* for NO? Why can't you be sure of the formula that corresponds to the name *phosphorus chloride?*

OBJECTIVE 28

**76.** The compound represented by the ball-and-stick model at right is used in the processing of nuclear fuels. Although bromine atoms most commonly form one covalent bond, they can form five bonds, as in the molecule shown here, in which the central sphere represents a bromine atom. The other atoms are fluorine atoms. Write this compound's chemical formula and name. List the bromine atom first in the chemical formula.

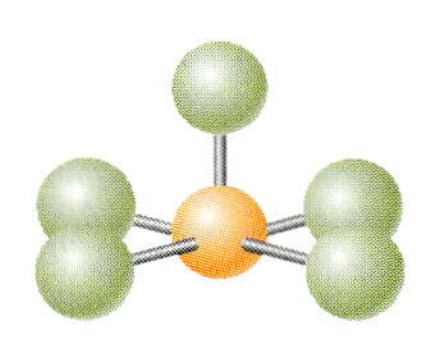

OBJECTIVE 28

77. The compound represented by the ball-and-stick model that follows is used to add chlorine atoms to other molecules. Write its chemical formula and name. The central ball represents an oxygen atom, and the other atoms are chlorine atoms. List the chlorine atom first in the chemical formula.

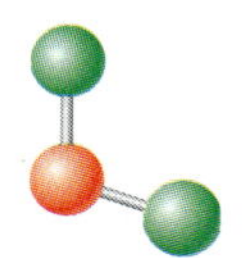

OBJECTIVE 28

78. The compound represented by the space-filling model that follows is used to vulcanize rubber and harden softwoods. Write its chemical formula and name. The central ball represents a sulfur atom, and the other atoms are chlorine atoms. List the sulfur atom first in the chemical formula.

OBJECTIVE 28

79. The compound represented by the space-filling model that follows is used in processing nuclear fuels. The central sphere represents a chlorine atom, which in most cases would form one covalent bond but is sometimes able to form three bonds. The other atoms are fluorine atoms. Write this compound's chemical formula and name. List the chlorine atom first in the chemical formula.

OBJECTIVE 28

80. Write the name for each of the following chemical formulas.
   a. $I_2O_5$ ( an oxidizing agent)
   b. $BrF_3$ (adds fluorine atoms to other compounds)
   c. IBr (used in organic synthesis)
   d. $CH_4$ (a primary component of natural gas)
   e. HBr (used to make pharmaceuticals)

OBJECTIVE 28

81. Write the name for each of the following chemical formulas.
   a. $ClO_2$ (a commercial bleaching agent)
   b. $C_2H_6$ (in natural gas)
   c. HI (when dissolved in water, used to make pharmaceuticals)
   d. $P_3N_5$ (for doping semiconductors)
   e. BrCl (an industrial disinfectant)

OBJECTIVE 28

82. Write the chemical formula for each of the following names.
   a. propane (a fuel in heating torches)
   b. chlorine monofluoride (a fluorinating agent)
   c. tetraphosphorus heptoxide (a dangerous fire risk)
   d. carbon tetrabromide (used to make organic compounds)
   e. hydrogen fluoride (an additive to liquid rocket propellants)

83. Write the chemical formula for each of the following names. OBJECTIVE 28
   a. ammonia (a household cleaner when dissolved in water)
   b. tetraphosphorus hexasulfide (used in organic chemical reactions)
   c. iodine monochloride (used for organic synthesis)
   d. hydrogen chloride (used to make hydrochloric acid)

### Section 3.5 Ionic Compounds

84. Explain why metals usually combine with nonmetals to form ionic bonds. OBJECTIVE 29
85. How may protons and electrons do each of the following ions have?
   a. $Be^{2+}$
   b. $S^{2-}$
86. How may protons and electrons do each of the following ions have?
   a. $N^{3-}$
   b. $Ba^{2+}$
87. Write the name for each of these monatomic ions. OBJECTIVE 30
   a. $Ca^{2+}$
   b. $Li^{+}$
   c. $Cr^{2+}$
   d. $F^{-}$
   e. $Ag^{+}$
   f. $Sc^{3+}$
   g. $P^{3-}$
   h. $Pb^{2+}$
88. Write the name for each of these monatomic ions. OBJECTIVE 30
   a. $Na^{+}$
   b. $Br^{-}$
   c. $Al^{3+}$
   d. $Mn^{2+}$
   e. $Se^{2-}$
   f. $Zn^{2+}$
   g. $Cr^{3+}$
89. Write the formula for each of these monatomic ions. OBJECTIVE 30 OBJECTIVE 31
   a. magnesium ion
   b. sodium ion
   c. sulfide ion
   d. iron(III) ion
   e. scandium ion
   f. nitride ion
   g. manganese(III) ion
   h. zinc ion
90. Write the formula for each of these monatomic ions. OBJECTIVE 30 OBJECTIVE 31
   a. strontium ion
   b. aluminum ion
   c. silver ion
   d. nickel(II) ion
   e. potassium ion
   f. oxide ion
   g. chloride ion
   h. copper(I) ion
   i. mercury(II) ion
91. Silver bromide, AgBr, is the compound on black and white film that causes the color change when the film is exposed to light. It has a structure similar to that of sodium chloride. What are the particles that form the basic structure of silver bromide? What type of attraction holds these particles together? Draw a rough sketch of the structure of solid silver bromide. OBJECTIVE 32
92. Describe the crystal structures of cesium chloride and ammonium chloride. How are they similar, and how are they different? OBJECTIVE 33

OBJECTIVE 34 OBJECTIVE 35

**93.** Write the name for each of these polyatomic ions.

a. $NH_4^+$ c. $HSO_4^-$

b. $C_2H_3O_2^-$

OBJECTIVE 34 OBJECTIVE 35

94. Write the name for each of these polyatomic ions.

a. $OH^-$ c. $HCO_3^-$

b. $CO_3^{2-}$

OBJECTIVE 34 OBJECTIVE 35 OBJECTIVE 36

**95.** Write the formula for each of these polyatomic ions.

a. ammonium c. hydrogen sulfate ion

b. bicarbonate ion

OBJECTIVE 37

**96.** Write the name for each of these chemical formulas.

a. $Na_2O$ (a dehydrating agent)

b. $Ni_2O_3$ (in storage batteries)

c. $Pb(NO_3)_2$ (in matches and explosives)

d. $Ba(OH)_2$ (an analytical reagent)

e. $KHCO_3$ (in baking powder and fire-extinguishing agents)

OBJECTIVE 37

97. Write the name for each of these chemical formulas.

a. $CdI_2$ (a nematocide—that is, it kills certain parasitic worms)

b. $Ca_3P_2$ (in signal flares)

c. $Au(OH)_3$ (used in gold plating)

d. $FeCl_2$ (in pharmaceutical preparations)

e. $NH_4HSO_4$ (in hair wave formulations)

OBJECTIVE 37

**98.** Write the chemical formula for each of the following names.

a. potassium sulfide (a depilatory)

b. zinc phosphide (a rodenticide)

c. nickel(II) chloride (used in nickel electroplating)

d. magnesium dihydrogen phosphate (used in fireproofing wood)

e. lithium bicarbonate (in mineral waters)

OBJECTIVE 37

99. Write the chemical formula for each of the following names.

a. barium chloride (used in the manufacture of white leather)

b. cobalt(III) oxide (used in coloring enamels)

c. manganese(II) chloride (used in pharmaceutical preparations)

d. iron(III) acetate (a medicine)

e. chromium(III) phosphate (in paint pigments)

f. magnesium hydrogen phosphate (a laxative)

OBJECTIVE 37

**100.** The ionic compounds $CuF_2$, $NH_4Cl$, CdO, and $HgSO_4$ are all used to make batteries. Write the name for each of these compounds.

OBJECTIVE 37

101. The ionic compounds $MgF_2$, $NH_4OH$, $Ba(NO_3)_2$, $Na_2HPO_4$, and $Cu_2O$ are all used to make ceramics. Write the name for each of these compounds.

OBJECTIVE 37

**102.** The ionic compounds copper(II) chloride, lithium nitrate, and cadmium sulfide are all used to make fireworks. Write the chemical formulas for these compounds.

103. The ionic compounds barium bromide, silver phosphate, and ammonium iodide are all used in photography. Write the chemical formulas for these compounds.

OBJECTIVE 37

### Discussion Topic

104. It has been suggested that there is really only one type of chemical bond—that ionic and covalent bonds are not really fundamentally different. What arguments can be made for and against this position?

*Salt towers rise from the water in Mono Lake, California.*

CHAPTER 4

# An Introduction to Chemical Reactions



NOW THAT YOU UNDERSTAND THE BASIC STRUCTURAL DIFFERENCES between different kinds of substances, you are ready to begin learning about the chemical changes that take place as one substance is converted into another. Chemical changes are chemists' primary concern. They want to know what, if anything, happens when one substance encounters another. Do the substances change? How and why? Can the conditions be altered so as to speed the changes up, slow them down, or perhaps reverse them? Once chemists understand the nature of one chemical change, they begin to explore the possibilities that arise from causing other similar changes.

For example, let's pretend that you just bought an old house *as is,* with the water turned off. On moving day, you twist the hot water tap as far as it will go, and all you get is a slow drip, drip, drip. As though the lack of hot water weren't enough to ruin your day, you also have a toothache because of a cavity that you haven't had time to get filled. As a chemist in training, you want to know what chemical changes have caused your troubles. In this chapter, you will read about the chemical change that causes a solid to form in your hot water pipes, eventually blocking the flow of water through them. In Chapter 5, you will find out about a chemical change that will dissolve that solid and a similar change that dissolves the enamel on your teeth. Chapter 5 will also reveal how fluoride in your toothpaste makes a minor chemical change in your mouth that can help fight cavities.

*A chemical reaction causes solids to form in hot water pipes.*

Chemical changes, like the ones mentioned above, are described with chemical equations. This chapter begins with a discussion of how to interpret and write chemical equations.

## Review Skills

The presentation of information in this chapter assumes that you can already perform the tasks listed below. You can test your readiness to proceed by answering the Review Questions at the end of the chapter. This might also be a good time to read the Chapter Objectives, which precede the Review Questions.

- Write the formulas for the diatomic elements.. (Section 2.5)
- Predict whether a bond between 2 atoms of different elements would be covalent or ionic. (Section 3.2)
- Describe attractions between $H_2O$ molecules. (Section 3.3)
- Describe the structure of liquid water. (Section 3.3)
- Convert between the names and formulas for alcohols, binary covalent compounds, and ionic compounds. (Sections 3.3–3.5)

## 4.1 Chemical Reactions and Chemical Equations

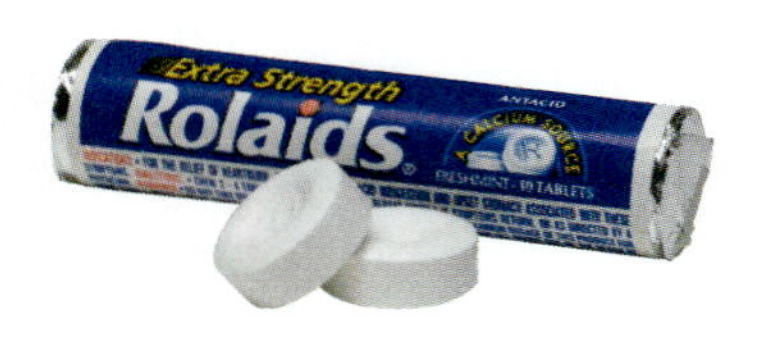


Chemical changes lead to the formation of substances that help grow our food, make our lives more productive, and cure our heartburn.

A **chemical change** or **chemical reaction** is a process by which one or more pure substances are converted into one or more different pure substances. Chemical changes lead to the formation of substances that help grow our food, make our lives more productive, cure our heartburn, and much, much more. For example, nitric acid, $HNO_3$, which is used to make fertilizers and explosives, is formed in the chemical reaction of the gases ammonia, $NH_3$, and oxygen, $O_2$. Silicon dioxide, $SiO_2$, reacts with carbon, C, at high temperature to yield silicon, Si—which can be used to make computers—and carbon monoxide, CO. An antacid tablet might contain calcium carbonate, $CaCO_3$, which combines with the hydrochloric acid in the stomach to yield calcium chloride, $CaCl_2$, water, and carbon dioxide. The chemical equations for these three chemical reactions are

$$NH_3(g) + 2O_2(g) \rightarrow HNO_3(aq) + H_2O(l)$$

$$SiO_2(s) + 2C(s) \xrightarrow{2000\ ^\circ C} Si(l) + 2CO(g)$$

$$CaCO_3(s) + 2HCl(aq) \rightarrow CaCl_2(aq) + H_2O(l) + CO_2(g)$$

Once you know how to read these chemical equations, they will tell you many details about the reactions that take place.

### Interpreting a Chemical Equation

In chemical reactions, atoms are rearranged and regrouped through the breaking and making of chemical bonds. For example, when hydrogen gas, $H_2(g)$, is burned in the presence of gaseous oxygen, $O_2(g)$, a new substance, liquid water, $H_2O(l)$, forms. The covalent bonds within the $H_2$ molecules and $O_2$ molecules break, and new covalent bonds form between oxygen atoms and hydrogen atoms (Figure 4.1).

**Figure 4.1**
**The Formation of Water from Hydrogen and Oxygen**

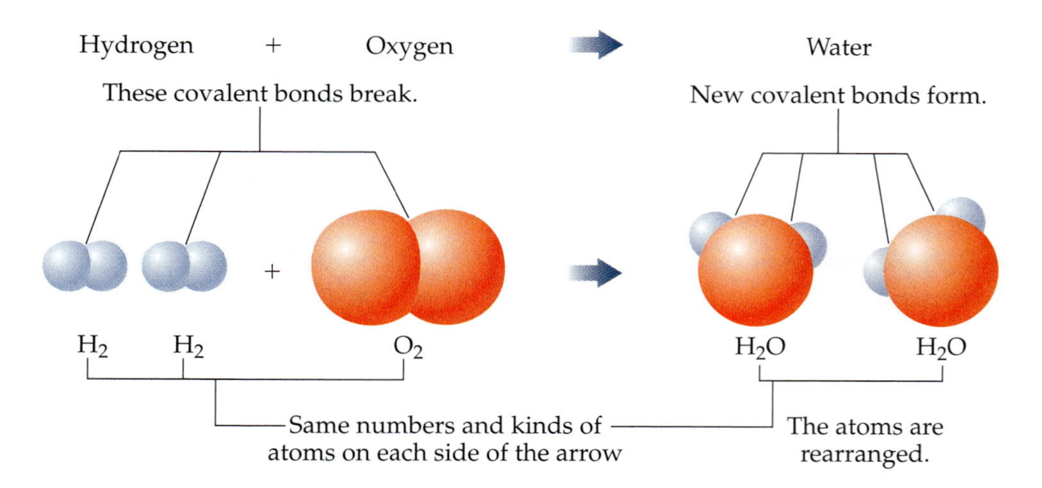


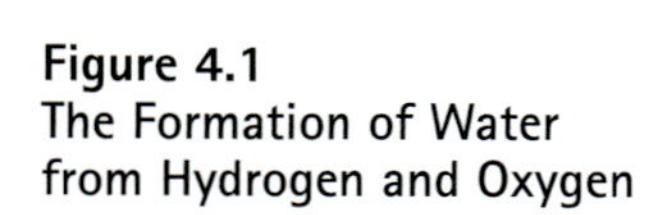

A chemical equation is a shorthand description of a chemical reaction. The following equation describes the burning of hydrogen gas to form liquid water.

$$2H_2(g) + O_2(g) \rightarrow 2H_2O(l)$$

OBJECTIVE 2

Chemical equations give the following information about chemical reactions.

- Chemical equations show the formulas for the substances that take part in the reaction. The formulas on the left side of the arrow represent the reactants, the substances that change in the reaction. The formulas on the right side of the arrow represent the **products,** the substances that are formed in the reaction. If there are more than one reactant or more than one product, they are separated by plus signs. The arrow separating the reactants from the products can be read as "goes to" or "yields" or "produces."
- The physical states of the reactants and products are provided in the equation. A *(g)* following a formula tells us the substance is a gas. Solids are described with *(s)*. Liquids are described with *(l)*. When a substance is dissolved in water, it is described with *(aq)*; this stands for *aqueous*, which means "mixed with water."

OBJECTIVE 3

- The relative numbers of particles of each reactant and product are indicated by numbers placed in front of the formulas. These numbers are called **coefficients**. An equation containing correct coefficients is called a balanced equation. For example, the 2's in front of $H_2$ and $H_2O$ in the equation we saw above are coefficients. If a formula in a balanced equation has no stated coefficient, its coefficient is understood to be 1, as is the case for oxygen in the equation above (Figure 4.2).
- If special conditions are necessary for a reaction to take place, they are often specified above the arrow. Some examples of special conditions are electric current, high temperature, high pressure, and light.

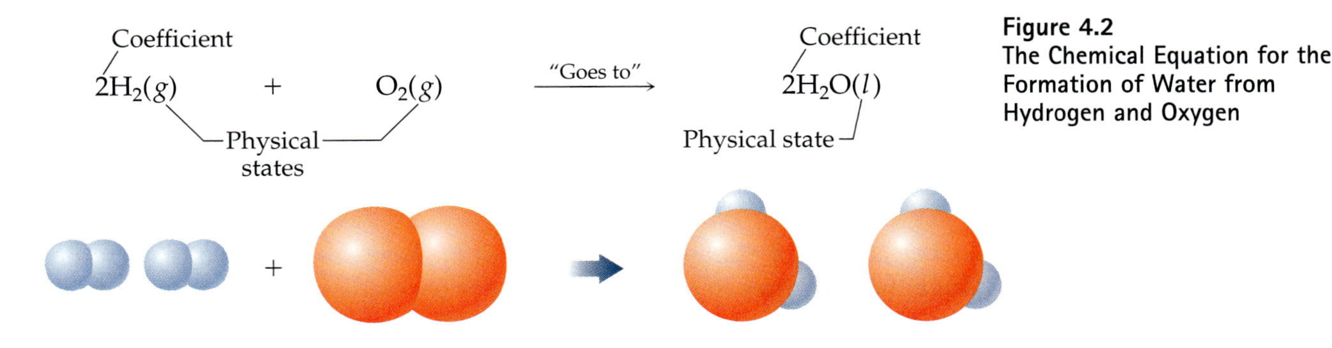


**Figure 4.2**
**The Chemical Equation for the Formation of Water from Hydrogen and Oxygen**

The burning of hydrogen gas must be started with a small flame or a spark, but that is not considered a special condition. There is no need to indicate it above the arrow in the equation for the creation of water from hydrogen and oxygen. However, the conversion of water back to hydrogen and oxygen *does* require a special condition—specifically, exposure to an electric current:

A special condition

$$2H_2O(l) \xrightarrow{\text{Electric current}} 2H_2(g) + O_2(g)$$

To indicate that a chemical reaction requires the *continuous* addition of heat in order to proceed, we place an upper case Greek delta, Δ, above the arrow in the equation. For example, the conversion of potassium chlorate (a fertilizer and food additive) to potassium chloride and oxygen requires the continuous addition of heat:

This indicates the continuous addition of heat.

$$2KClO_3(s) \xrightarrow{\Delta} 2KCl(s) + 3O_2(g)$$

## Balancing Chemical Equations

In chemical reactions, atoms are neither created nor destroyed; they merely change partners. Thus the number of atoms of an element in the reaction's products is equal to the number of atoms of that element in the original reactants. The coefficients we often place in front of one or more of the formulas in a chemical equation reflect this fact. They are used whenever necessary to balance the number of atoms of a particular element on either side of the arrow.

For an example, let's return to the reaction of hydrogen gas and oxygen gas to form liquid water. The equation for the reaction between $H_2(g)$ and $O_2(g)$ to form $H_2O(l)$ shows that there are 2 atoms of oxygen in the diatomic $O_2$ molecule to the left of the arrow, so there should also be 2 atoms of oxygen in the product to the right of the arrow. Because each water molecule, $H_2O$, contains only 1 oxygen atom, 2 water molecules must form for each oxygen molecule that reacts. The coefficient 2 in front of the $H_2O(l)$ makes this clear. But 2 water molecules contain 4 hydrogen atoms, which means that 2 hydrogen molecules must be present on the reactant side of the equation for the numbers of H atoms to balance (Figure 4.2).

$$2H_2(g) + O_2(g) \rightarrow 2H_2O(l)$$

*Note that we do not change the subscripts in the formulas,* because that would change the identities of the substances. For example, changing the formula on the right side of the arrow in the equation above to $H_2O_2$ would balance the atoms without using coefficients, but the resulting equation would be incorrect.

$$H_2(g) + O_2(g) \rightarrow \cancel{H_2O_2(l)}$$

Water is $H_2O$, whereas $H_2O_2$ is hydrogen peroxide, a very different substance from water. (You add water to your hair to clean it; you add hydrogen peroxide to your hair to bleach it.)

The following sample study sheet shows a procedure that you can use to balance chemical equations. It is an approach that chemists often call balancing equations "by inspection." Examples 4.1 through 4.5, which follow the study sheet, illustrate the process.

## Sample Study Sheet 4.1

### Balancing Chemical Equations

OBJECTIVE 4

**Tip-off** You are asked to balance a chemical equation.

**General Steps**

- Consider the first element listed in the first formula in the equation.

  If this element is mentioned in two or more formulas on the same side of the arrow, skip it until after the other elements are balanced. *(See Example 4.2.)*

  If this element is mentioned in one formula on each side of the arrow, balance it by placing coefficients in front of one or both of these formulas.
- Moving from left to right, repeat the process for each element.
- When you place a number in front of a formula that contains an element you tried to balance previously, recheck that element and put its atoms back in balance. *(See Examples 4.2 and 4.3.)*
- Continue this process until the number of atoms of each element is balanced.

The following strategies can be helpful for balancing certain equations.

**Strategy 1** Often, an element can be balanced by using the subscript for this element on the left side of the arrow as the coefficient in front of the formula that contains this element on the right side of the arrow, and vice versa (using the subscript of this element on the right side of the arrow as the coefficient in front of the formula that contains this element on the left side). *(See Example 4.3.)*

**Strategy 2** It is sometimes easiest, as a temporary measure, to balance the pure nonmetallic elements ($H_2$, $O_2$, $N_2$, $F_2$, $Cl_2$, $Br_2$, $I_2$, $S_8$, $Se_8$, and $P_4$) with a fractional coefficient ($1/2$, $3/2$, $5/2$, etc.). If you do use a fraction during the balancing process, you can eliminate it later by multiplying each coefficient in the equation by the fraction's denominator (which is usually the number 2). *(See Example 4.4.)*

**Strategy 3** If polyatomic ions do not change in the reaction, and therefore appear in the same form on both sides of the chemical equation, they can be balanced as though they were single atoms. *(See Example 4.5.)*

**Strategy 4** If you find an element difficult to balance, leave it for later.

**Example** See Examples 4.1–4.5.

## EXAMPLE 4.1 Balancing Equations

OBJECTIVE 4

Balance the following equation so that it correctly describes the reaction for the formation of dinitrogen oxide (commonly called nitrous oxide), an anesthetic used in dentistry and surgery.

$$NH_3(g) + O_2(g) \rightarrow N_2O(g) + H_2O(l)$$

***Solution***

The following table shows that the atoms are not balanced yet.

| Element | Left | Right |
|---|---|---|
| N | 1 | 2 |
| H | 3 | 2 |
| O | 2 | 2 |

Nitrogen is the first element in the first formula. It is found in one formula on each side of the arrow, so we can try to balance it now. There are 2 nitrogen atoms on the right side of the equation and only one on the left; we bring them into balance by placing a 2 in front of $NH_3$.

$$2NH_3(g) + O_2(g) \rightarrow N_2O(g) + H_2O(l)$$

There are now six hydrogen atoms on the left side of the arrow (in the two $NH_3$ molecules) and only two H's on the right, so we balance the hydrogen atoms by placing a 3 in front of the $H_2O$. This gives six atoms of hydrogen on each side.

$$2NH_3(g) + O_2(g) \rightarrow N_2O(g) + 3H_2O(l)$$

There are now two oxygen atoms on the left and four on the right (in one $N_2O$ and three $H_2O$'s), so we balance the oxygen atoms by placing a 2 in front of the $O_2$.

$$2NH_3(g) + 2O_2(g) \rightarrow N_2O(g) + 3H_2O(l)$$

The space-filling models that follow show how you might visualize the relative number of particles participating in this reaction. You can see that the atoms regroup but are neither created nor destroyed.

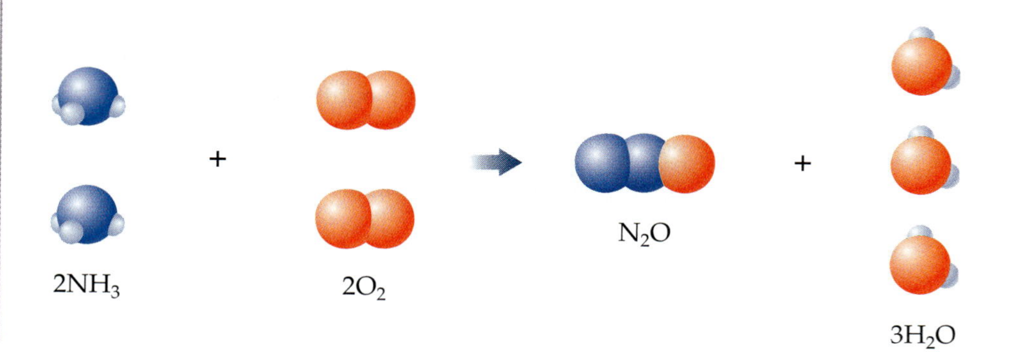

The following table shows that the atoms are now balanced.

| Element | Left | Right |
|---|---|---|
| N | 2 | 2 |
| H | 6 | 6 |
| O | 4 | 4 |

## EXAMPLE 4.2 Balancing Equations

OBJECTIVE 4

Balance the following equation.

$$N_2H_4(l) + N_2O_4(l) \rightarrow N_2(g) + H_2O(l)$$

***Solution***

Nitrogen is the first element in the equation. However, because nitrogen is found in two formulas on the left side of the arrow, we will leave the balancing of the nitrogen atoms until later.

Balance the hydrogen atoms by placing a 2 in front of $H_2O$.

$$N_2H_4(l) + N_2O_4(l) \rightarrow N_2(g) + 2H_2O(l)$$

Balance the oxygen atoms by changing the 2 in front of $H_2O$ to a 4.

$$N_2H_4(l) + N_2O_4(l) \rightarrow N_2(g) + 4H_2O(l)$$

Because we un-balanced the hydrogen atoms in the process of balancing the oxygen atoms, we need to go back and re-balance the hydrogen atoms by placing a 2 in front of the $N_2H_4$.

$$2N_2H_4(l) + N_2O_4(l) \rightarrow N_2(g) + 4H_2O(l)$$

Finally, we balance the nitrogen atoms by placing a 3 in front of $N_2$.

$$2N_2H_4(l) + N_2O_4(l) \rightarrow 3N_2(g) + 4H_2O(l)$$

The following table shows that the atoms are now balanced.

| Element | Left | Right |
|---|---|---|
| N | 6 | 6 |
| H | 8 | 8 |
| O | 4 | 4 |

### Example 4.3 Balancing Equations

OBJECTIVE 4

Balance the following equation so that it correctly describes the reaction for the formation of tetraphosphorus trisulfide (used in the manufacture of matches).

$$P_4(s) + S_8(s) \rightarrow P_4S_3(s)$$

Tetraphosphorus trisulfide is used to make matches.

***Solution***

The phosphorus atoms appear to be balanced at this stage.

We can balance the sulfur atoms by using the subscript for the sulfur on the right (3) as the coefficient for $S_8$ on the left and using the subscript for the sulfur on the left (8) as the coefficient for the sulfur compound on the right. (Strategy 1)

$$P_4(s) + S_8(s) \rightarrow P_4S_3(s)$$

$$P_4(s) + 3S_8(s) \rightarrow 8P_4S_3(s)$$

We restore the balance of the phosphorus atoms by placing an 8 in front of $P_4$.

$$8P_4(s) + 3S_8(s) \rightarrow 8P_4S_3(s)$$

### Example 4.4 Balancing Equations

OBJECTIVE 4

Balance the following equation so that it correctly describes the reaction for the formation of aluminum oxide (used to manufacture glass).

$$Al(s) + O_2(g) \rightarrow Al_2O_3(s)$$

***Solution***

Balance the aluminum atoms by placing a 2 in front of the Al.

$$2Al(s) + O_2(g) \rightarrow Al_2O_3(s)$$

There are three oxygen atoms on the right and two on the left. We can bring them into balance by placing $3/2$ in front of the $O_2$. Alternatively, we could place a 3 in front of the $O_2$ and a 2 in front of the $Al_2O_3$, but that would un-balance the aluminum atoms. By inserting only one coefficient, in front of the $O_2$, to balance the oxygen atoms, we ensure that the aluminum atoms remain balanced. We arrive at $3/2$ by asking what number times the subscript 2 of the $O_2$ would give us three atoms of oxygen on the left side: $3/2$ times 2 is 3. (Strategy 2)

$$2Al(s) + \tfrac{3}{2}O_2(g) \rightarrow Al_2O_3(s)$$

It is a good habit to eliminate the fraction by multiplying all the coefficients by the denominator of the fraction, in this case 2. (Some instructors consider fractional coefficients to be incorrect, so check with your instructor to find out whether you will be allowed to leave them in your final answer.)

$$4Al(s) + 3O_2(g) \rightarrow 2Al_2O_3(s)$$

## EXAMPLE 4.5 Balancing Equations

Balance the following equation for the chemical reaction that forms zinc phosphate (used in dental cements and in making galvanized nails).

$$Zn(NO_3)_2(aq) + Na_3PO_4(aq) \rightarrow Zn_3(PO_4)_2(s) + NaNO_3(aq)$$

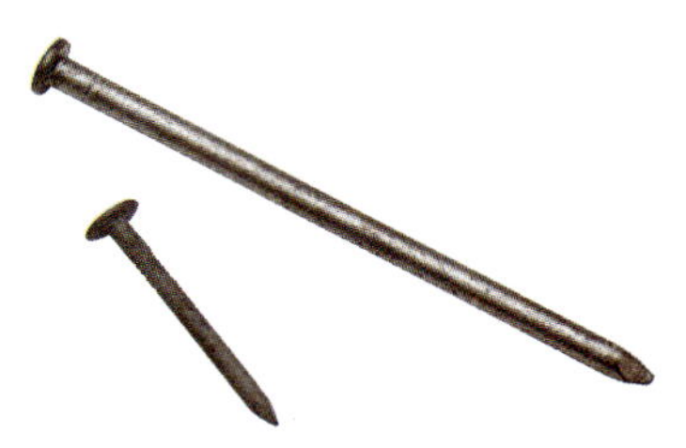

Zinc phosphate is used to galvanize nails.

***Solution***

Balance the zinc atoms by placing a 3 in front of $Zn(NO_3)_2$.

$$3Zn(NO_3)_2(aq) + Na_3PO_4(aq) \rightarrow Zn_3(PO_4)_2(s) + NaNO_3(aq)$$

The nitrate ions, $NO_3^-$, emerge unchanged from the reaction, so we can balance them as though they were single atoms. There are six $NO_3^-$ ions in three $Zn(NO_3)_2$. We therefore place a 6 in front of the $NaNO_3$ to balance the nitrates. (Strategy 3)

$$3Zn(NO_3)_2(aq) + Na_3PO_4(aq) \rightarrow Zn_3(PO_4)_2(s) + 6NaNO_3(aq)$$

Balance the sodium atoms by placing a 2 in front of the $Na_3PO_4$.

$$3Zn(NO_3)_2(aq) + 2Na_3PO_4(aq) \rightarrow Zn_3(PO_4)_2(s) + 6NaNO_3(aq)$$

The phosphate ions, $PO_4^{3-}$, do not change in the reaction, so we can balance them as though they were single atoms. There are two on each side, so the phosphate ions are balanced. (Strategy 3)

$$3Zn(NO_3)_2(aq) + 2Na_3PO_4(aq) \rightarrow Zn_3(PO_4)_2(s) + 6NaNO_3(aq)$$

## EXERCISE 4.1 Balancing Equations

OBJECTIVE 4

Balance the following chemical equations.

a. $P_4(s) + Cl_2(g) \rightarrow PCl_3(l)$
   Phosphorus trichloride, $PCl_3$, is an intermediate in the production of pesticides and gasoline additives.

b. $PbO(s) + NH_3(g) \rightarrow Pb(s) + N_2(g) + H_2O(l)$
   Lead, Pb, is used in storage batteries and as radiation shielding.

c. $P_4O_{10}(s) + H_2O(l) \rightarrow H_3PO_4(aq)$
   Phosphoric acid, $H_3PO_4$, is used to make fertilizers and detergents.

d. $Mn(s) + CrCl_3(aq) \rightarrow MnCl_2(aq) + Cr(s)$
   Manganese(II) chloride, $MnCl_2$, is used in pharmaceutical preparations.

e. $C_2H_2(g) + O_2(g) \rightarrow CO_2(g) + H_2O(l)$
   Acetylene, $C_2H_2$, is used in welding torches.

f. $Co(NO_3)_2(aq) + Na_3PO_4(aq) \rightarrow Co_3(PO_4)_2(s) + NaNO_3(aq)$
   Cobalt phosphate, $Co_3(PO_4)_2$, is used to color glass and as an additive to animal feed.

g. $CH_3NH_2(g) + O_2(g) \rightarrow CO_2(g) + H_2O(l) + N_2(g)$
   Methylamine, $CH_3NH_2$, is a fuel additive.

h. $FeS(s) + O_2(g) + H_2O(l) \rightarrow Fe_2O_3(s) + H_2SO_4(aq)$
   Iron(III) oxide, $Fe_2O_3$, is a paint pigment.

## 4.2 Solubility of Ionic Compounds and Precipitation Reactions

The reaction that forms the scale in hot water pipes, eventually leading to major plumbing bills, belongs to a category of reactions called precipitation reactions. So does the reaction of calcium and magnesium ions with soap to create a solid scum in your bathtub and washing machine. Cadmium hydroxide, which is used in rechargeable batteries, is made from the precipitation reaction between water solutions of cadmium acetate and sodium hydroxide. To understand the changes that occur in precipitation reactions, to predict when they will take place, and to describe them correctly using chemical equations, we first need a way of visualizing the behavior of ionic compounds in water.

An aqueous solution forms when dye is added to water (top). Soon the mixture will be homogeneous (bottom).

### Water Solutions of Ionic Compounds

A **solution,** also called a homogeneous mixture, is a mixture whose particles are so evenly distributed that the relative concentrations of the components are the same throughout. When salt dissolves in water, for example, the sodium and chloride ions from the salt spread out evenly throughout the water. Eventually, every part of the solution has the same proportions of water molecules and ions as every other part (Figure 4.3). Ionic compounds are often able to mix with water in this way, in which case we say they are "soluble in water." Many of the chemical reactions we will study take place in water solutions—that is, solutions in which the substances are dissolved in water. Chemists refer to these as **aqueous solutions.**

Remember how a model made it easier for us to picture the structures of solids, liquids, and gases in Chapter 2? In order to develop a mental image of the changes that take place on the microscopic level as an ionic compound dissolves in water, chemists have found it helpful to extend that model. We will use table salt (sodium chloride, NaCl) as an example in our description of the process through which an ionic compound dissolves in water. Before you read about this process, you might want to return to the end of Section 3.3 and review the description of the structure of liquid water.

OBJECTIVE 5

When solid NaCl is first added to water, it settles to the bottom of the container. Like the particles of any solid, the sodium and chloride ions at the solid's surface are constantly moving, although their mutual attractions are holding them more or less in place. For example, if you were riding on a sodium ion at the surface of the solid, you might move out and away from the surface and into the water at one instant, and back toward the surface at the next (pulled by the attractions between your sodium ion and the chloride ions near it). When you moved back toward the surface, collisions there might push you away from the surface once again. In this way, all of the ions at the solid's surface—both sodium ions and chloride ions—can be viewed as repeatedly moving out into the water and returning to the solid's surface (Figure 4.4).

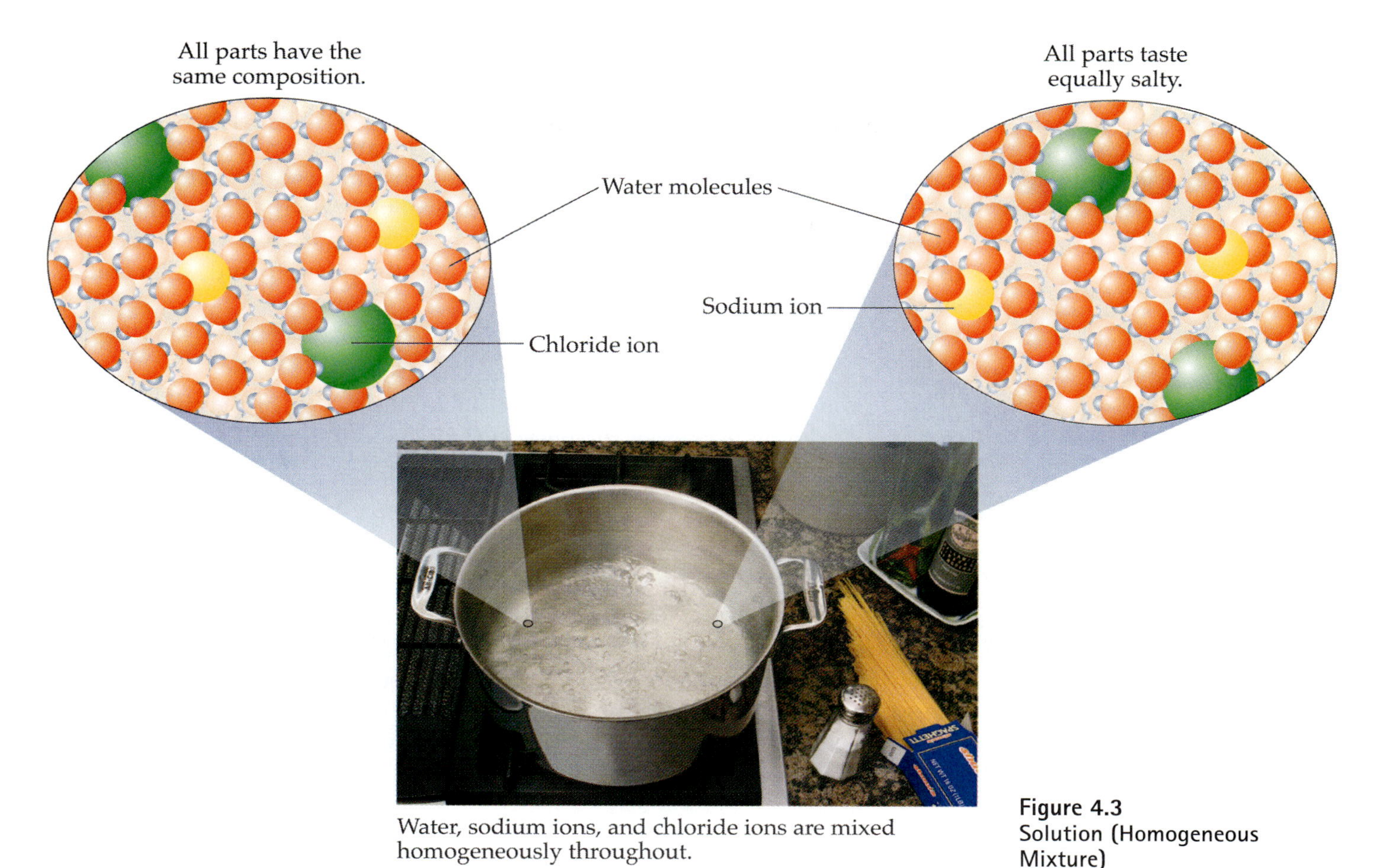
All parts have the same composition.
All parts taste equally salty.
Water molecules
Sodium ion
Chloride ion

Water, sodium ions, and chloride ions are mixed homogeneously throughout.

**Figure 4.3**
**Solution (Homogeneous Mixture)**

OBJECTIVE 5

Water molecules move to disrupt the attractions to the solid.

Attractions between the oxygen ends of water molecules and the cations

Attractions between hydrogen ends of water molecules and the anions

Cation moving out into the water

Anion moving out into the water

Collisions push the ions farther out into the water.

Collisions push the ions farther out into the water.

**Figure 4.4**
**Sodium Chloride Dissolving in Water**
This image represents the mixture immediately after sodium chloride has been added to water. Certain water molecules are highlighted to draw attention to their role in the process.

OBJECTIVE 5

Sometimes when an ion moves out into the water, a water molecule collides with it, pushing the ion out farther into the liquid and helping to break the ionic bond. Other water molecules move into the space between the ion and the solid and shield the ion from the attractions exerted by the ions at the solid's surface (Figure 4.4). The ions that escape the solid are then held in solution by attractions between their own charge and the partial charges of the polar water molecules. The negative oxygen ends of the water molecules surround the cations, and the positive hydrogen ends surround the anions (Figure 4.5).

You can find an animation that illustrates the process by which sodium chloride dissolves at the following Web site:

**www.chemplace.com/college/**

OBJECTIVE 6

The substances that mix to make a solution, such as our solution of sodium chloride and water, are called the **solute** and the **solvent.** In solutions of solids dissolved in liquids, we call the solid the **solute** and the liquid the **solvent.** Therefore, the NaCl in an aqueous sodium chloride solution is the solute, and the water is the solvent. In solutions of gases in liquids, we call the gas the **solute** and the liquid the **solvent.** Therefore, when gaseous

OBJECTIVE 5

**Figure 4.5**
**Aqueous Sodium Chloride**
This image shows a portion of the solution that forms when sodium chloride dissolves in water. Certain water molecules are highlighted to draw attention to their role in the process.

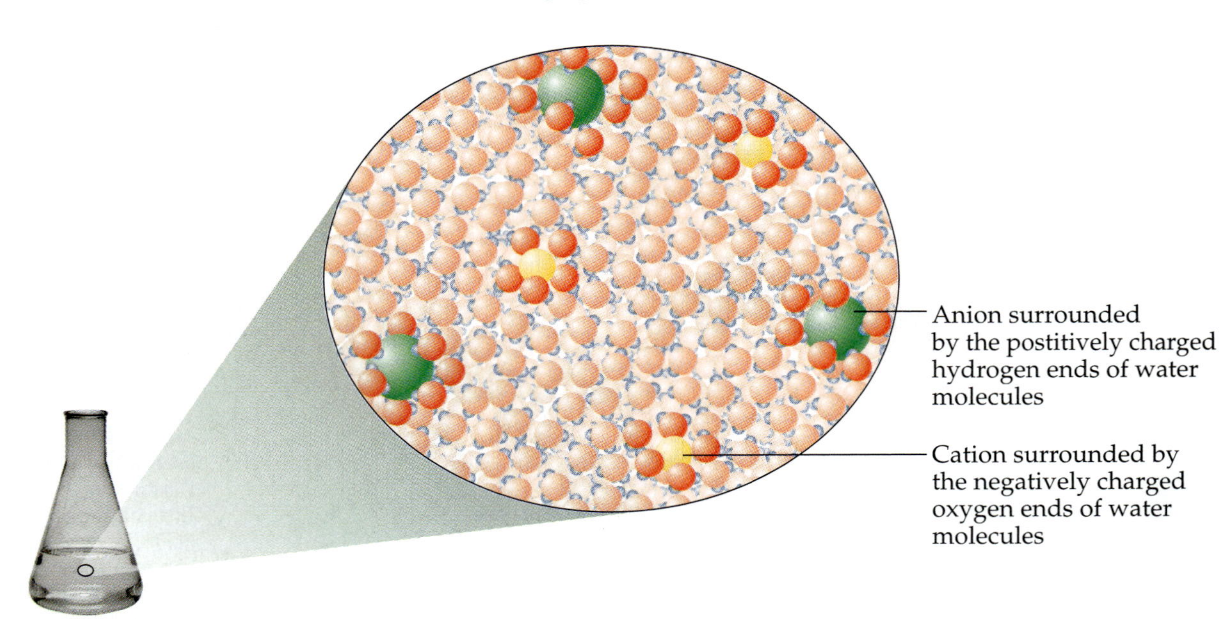

hydrogen chloride, HCl(*g*), is dissolved in liquid water to form a mixture known as hydrochloric acid, HCl(*aq*), the hydrogen chloride is the solute, and the water is the solvent. In other solutions, we call the minor component the **solute** and the major component the **solvent** (Figure 4.6). For example, in a mixture that is 5% liquid pentane, $C_5H_{12}$, and 95% liquid hexane, $C_6H_{14}$, the pentane is the solute, and the hexane is the solvent. Chapter 15 describes the formation of solutions with gaseous and liquid solutes.

**Figure 4.6**
**Liquid–Liquid Solution**
The carbon atoms in these pentane molecules are shown in green to distinguish them from the hexane molecules.

OBJECTIVE 6

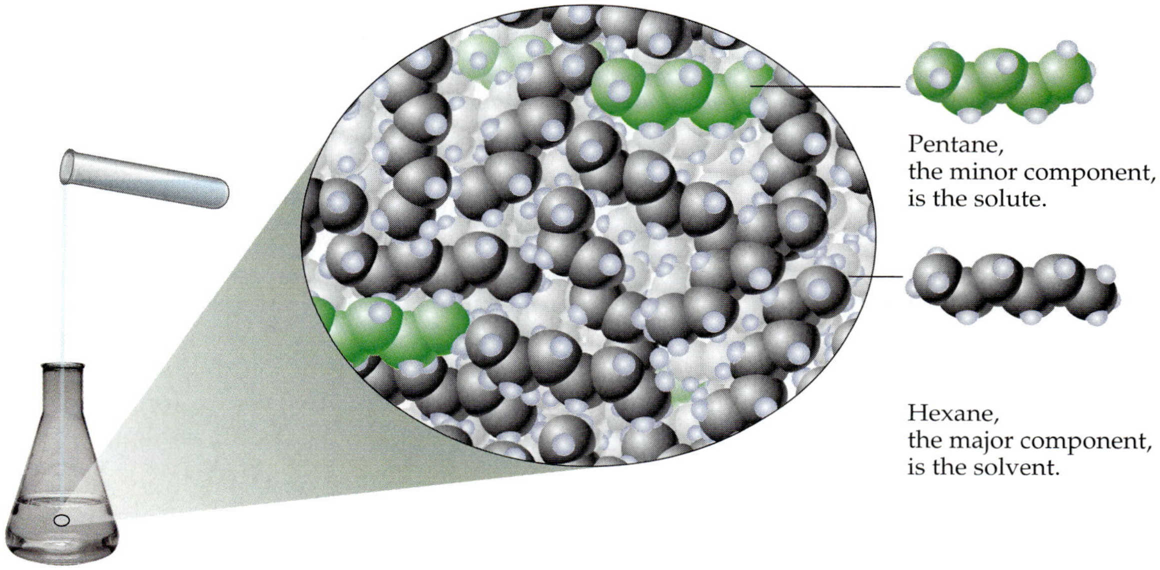

Pentane, the minor component, is the solute.

Hexane, the major component, is the solvent.

## Precipitation Reactions

Precipitation reactions, such as the ones we will see in this section, belong to a general class of reactions called **double-displacement reactions.** (Double-displacement reactions are also called **double-replacement, double-exchange,** and **metathesis reactions**.) Double-displacement reactions have the following form, signifying that the elements in two reacting compounds change partners.

$$AB + CD \longrightarrow AD + CB$$

OBJECTIVE 7

Precipitation reactions take place between ionic compounds in solution. For example, in the precipitation reactions that we will examine, A and C represent the cationic (positively charged) portions of the reactants and products, and B and D represent the anionic (negatively charged) portions of the reactants and products. The cation of the first reactant (A) combines with the anion of the second reactant (D) to form the product AD, and the cation of the second reactant (C) combines with the anion of the first reactant (B) to form the product CB.

OBJECTIVE 8

Sometimes a double-displacement reaction has one product that is insoluble in water. As that product forms, it emerges, or precipitates, from the solution as a solid. This process is called **precipitation,** such a reaction is called a **precipitation reaction**, and the solid is called the **precipitate**. For example, when water solutions of calcium nitrate and sodium carbonate are mixed, calcium carbonate precipitates from the solution while the other product, sodium nitrate, remains dissolved.

This solid precipitates from the solution. It is a precipitate.

$$Ca(NO_3)_2(aq) + Na_2CO_3(aq) \longrightarrow CaCO_3(s) + 2NaNO_3(aq)$$

One of the goals of this section is to enable you to visualize the process described by this equation. Figures 4.7, 4.8, and 4.9 will help you do this.

First, let us imagine the particles making up the $Ca(NO_3)_2$ solution. Remember that when ionic compounds dissolve, the ions separate and become surrounded by water molecules. When $Ca(NO_3)_2$ dissolves in water (Figure 4.7), the $Ca^{2+}$ ions separate from the $NO_3^-$ ions, with the oxygen ends of water molecules surrounding the calcium ions, and the hydrogen ends of water molecules surrounding the nitrate ions.

**Figure 4.7**
**Aqueous Calcium Nitrate**
Note that there are twice as many −1 nitrate ions as +2 calcium ions.

OBJECTIVE 8

When calcium nitrate, $Ca(NO_3)_2$, dissolves in water, the calcium ions, $Ca^{2+}$, become separated from the nitrate ions, $NO_3^-$.

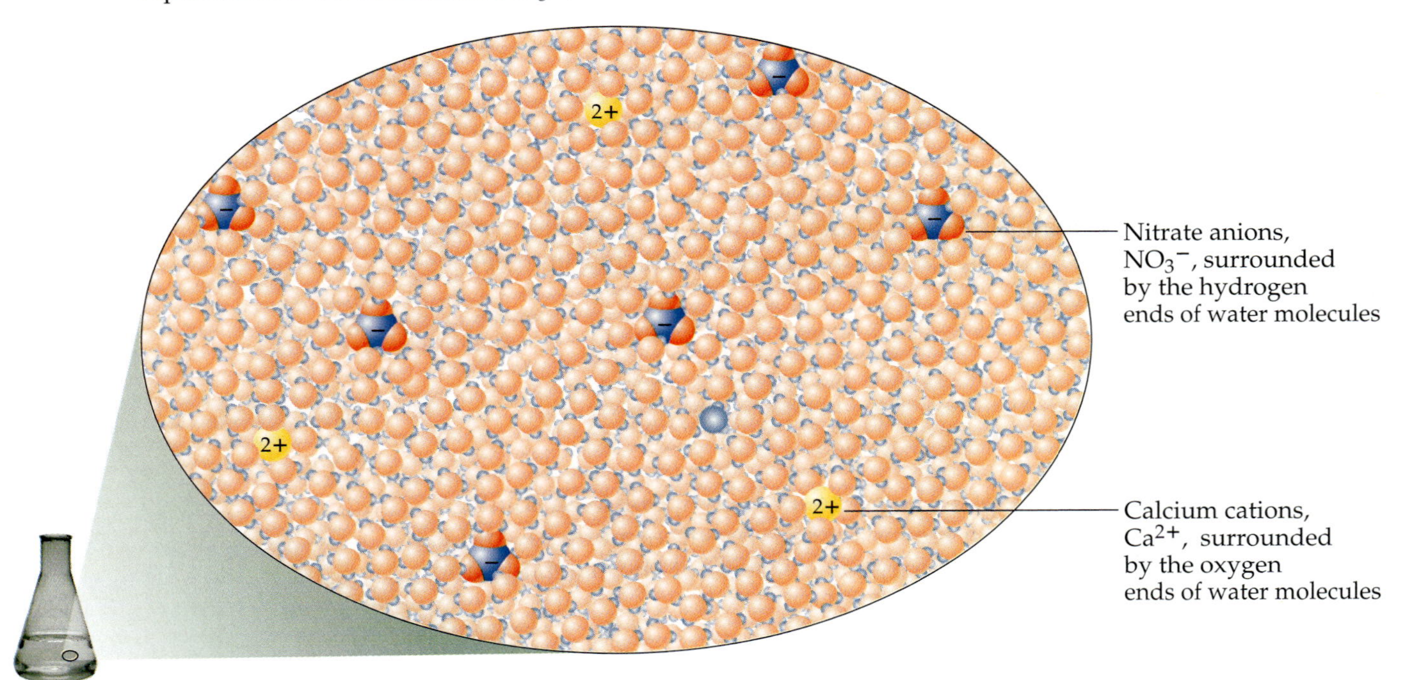

An aqueous solution of sodium carbonate also consists of ions separated and surrounded by water molecules, much like the solution of calcium nitrate. If time were to stop at the instant that the solution of sodium carbonate was added to aqueous calcium nitrate, there would be four different ions in solution surrounded by water molecules: $Ca^{2+}$, $NO_3^-$, $Na^+$, and $CO_3^{2-}$. The oxygen ends of the water molecules surround the calcium and sodium ions, and the hydrogen ends of water molecules surround the nitrate and carbonate ions. Figure 4.8 shows the system at the *hypothetical* instant just after solutions of calcium nitrate and sodium carbonate are combined and just before the precipitation reaction takes place. Because time refuses to stop, and because a chemical reaction takes place as soon as the $Ca(NO_3)_2$ and $Na_2CO_3$ solutions are combined, the four-ion system shown in this figure lasts for a very short time.

OBJECTIVE 8

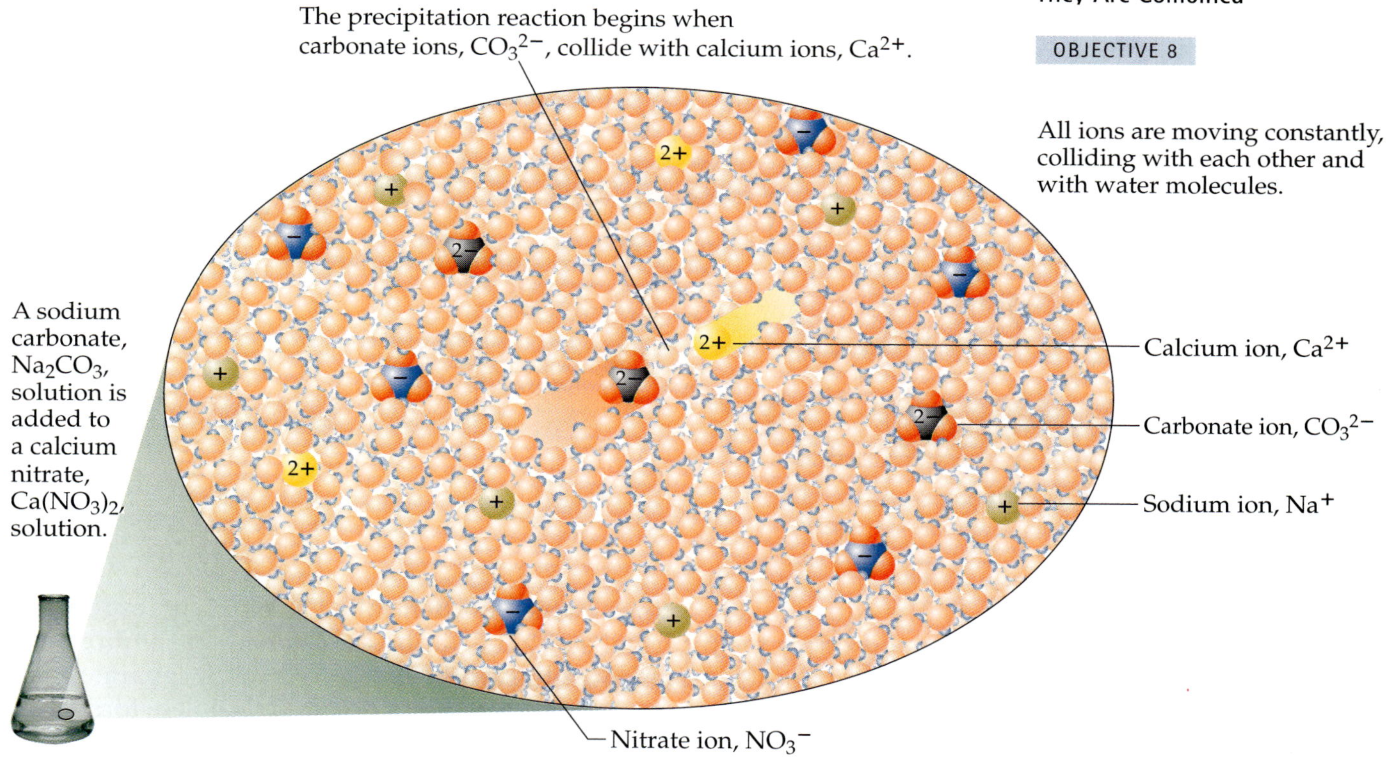
The precipitation reaction begins when carbonate ions, $CO_3^{2-}$, collide with calcium ions, $Ca^{2+}$.
A sodium carbonate, $Na_2CO_3$, solution is added to a calcium nitrate, $Ca(NO_3)_2$, solution.
Calcium ion, $Ca^{2+}$
Carbonate ion, $CO_3^{2-}$
Sodium ion, $Na^+$
Nitrate ion, $NO_3^-$
2+
2−
+
−

**Figure 4.8**
**Mixture of $Ca(NO_3)_2(aq)$ and $Na_2CO_3(aq)$ at the Instant They Are Combined**

OBJECTIVE 8

All ions are moving constantly, colliding with each other and with water molecules.

The ions in solution move in a random way, like any particle in a liquid, so they constantly collide with other ions. When two cations or two anions collide, they repel each other and move apart. When a calcium ion and a nitrate ion collide, they may stay together for a short time, but the attraction between them is too weak to keep them together when water molecules collide with them and push them apart. The same is true for the collision between sodium ions and carbonate ions. After colliding, they stay together for only an instant before water molecules break them apart again.

OBJECTIVE 8

When calcium ions and carbonate ions collide, however, they stay together longer because the attraction between them is stronger than the attractions between the other pairs of ions. They might eventually be

OBJECTIVE 8

knocked apart, but while they are together, other calcium ions and carbonate ions can collide with them. When another $Ca^{2+}$ or $CO_3^{2-}$ ion collides with a $CaCO_3$ pair, a trio forms. Other ions collide with the trio to form clusters of ions that then grow to become small **crystals**—solid particles whose component atoms, ions, or molecules are arranged in an organized, repeating pattern. Many crystals form throughout the system, so the solid $CaCO_3$ at first appears as a cloudiness in the mixture. The crystals eventually settle to the bottom of the container (Figures 4.8 and 4.9).

**Figure 4.9**
**Product Mixture for the Reaction of $Ca(NO_3)_2(aq)$ and $Na_2CO_3(aq)$**

OBJECTIVE 8

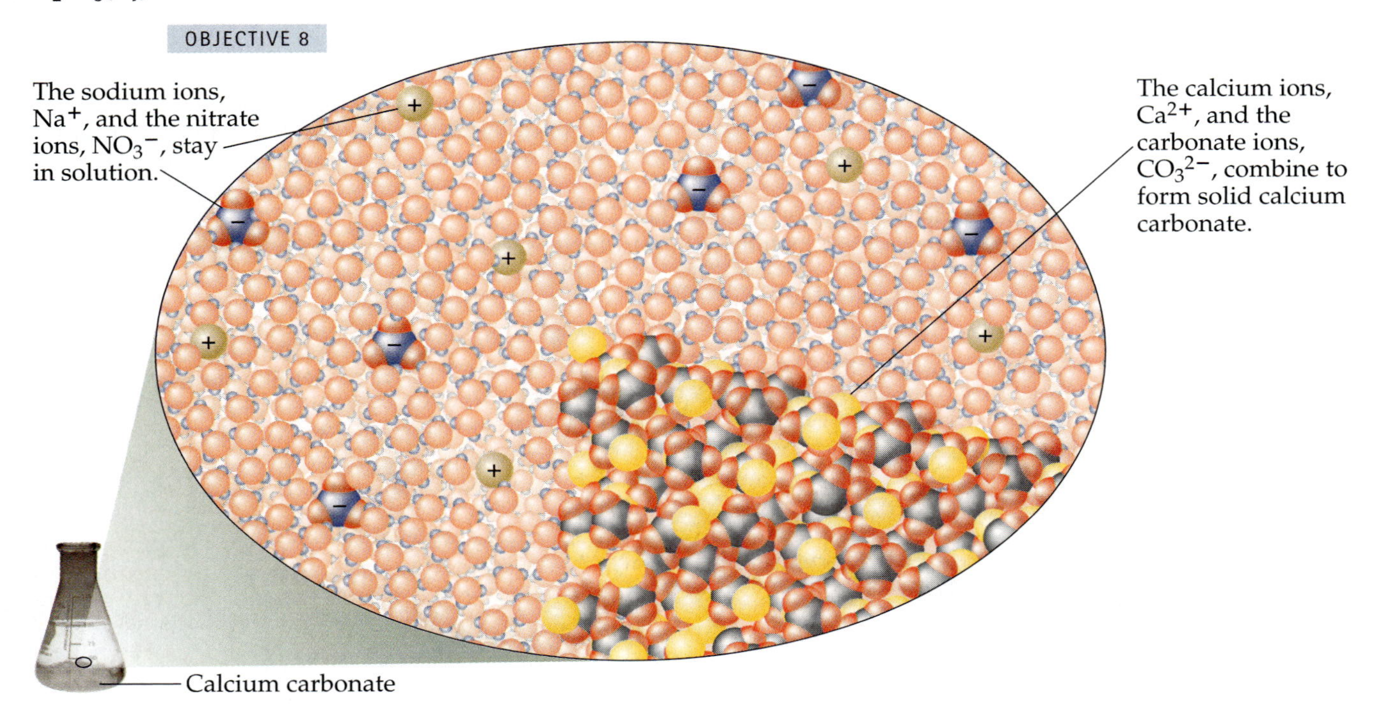

The sodium ions, Na+, and the nitrate ions, NO3−, stay in solution.
The calcium ions, Ca2+, and the carbonate ions, CO32−, combine to form solid calcium carbonate.
Calcium carbonate

The equation that follows, which is often called a **complete ionic equation,** describes the forms taken by the various substances in solution. The ionic compounds dissolved in the water are described as separate ions, and the insoluble ionic compound is described with a complete formula.

Described as separate ions — Solid precipitate — Described as separate ions

$$Ca^{2+}(aq) + 2NO_3^-(aq) + 2Na^+(aq) + CO_3^{2-}(aq) \longrightarrow CaCO_3(s) + 2Na^+(aq) + 2NO_3^-(aq)$$

The sodium and nitrate ions remain unchanged in this reaction. They were separate and surrounded by water molecules at the beginning, and they are still separate and surrounded by water molecules at the end. They were important in delivering the calcium and carbonate ions to solution (the solutions were created by dissolving solid calcium nitrate and solid sodium carbonate in water), but they did not actively participate in the reaction. When ions play this role in a reaction, we call them **spectator ions.**

Because spectator ions are not involved in the reaction, they are often left out of the chemical equation. The equation written without the spectator ions is called a **net ionic equation.**

Spectator ions are eliminated. Spectator ions

$$Ca^{2+}(aq) + 2NO_3^-(aq) + 2Na^+(aq) + CO_3^{2-}(aq) \longrightarrow CaCO_3(s) + 2Na^+(aq) + 2NO_3^-(aq)$$

Net ionic equation: $Ca^{2+}(aq) + CO_3^{2-}(aq) \longrightarrow CaCO_3(s)$

We call the equation that shows the complete formulas for all of the reactants and products the **complete equation,** or, sometimes, the **molecular equation.**

$$Ca(NO_3)_2(aq) + Na_2CO_3(aq) \rightarrow CaCO_3(s) + 2NaNO_3(aq)$$

You can find an animation that shows this precipitation reaction at the following Web site:
**www.chemplace.com/college/**

## Predicting Water Solubility

In order to predict whether a precipitation reaction will take place when two aqueous ionic compounds are mixed, you need to be able to predict whether the possible products of the double-displacement reaction are soluble or insoluble in water.

When we say that one substance is soluble in another, we mean that they can be mixed to a significant degree. More specifically, chemists describe the **solubility** of a substance as the maximum amount of it that can be dissolved in a given amount of solvent at a particular temperature. This property is often described in terms of the maximum number of grams of solute that will dissolve in 100 milliliters (or 100 grams) of solvent. For example, the water solubility of calcium nitrate is 121.2 g $Ca(NO_3)_2$ per 100 mL water at 25 °C. This means that when calcium nitrate is added steadily to 100 mL of water at 25 °C, it will dissolve until 121.2 g $Ca(NO_3)_2$ has been added. If more $Ca(NO_3)_2$ is added to the solution, it will remain in the solid form.

When we say an ionic solid is insoluble in water, we do not mean that none of the solid dissolves. There are always some ions that can escape from the surface of an ionic solid in water and go into solution. Thus, when we say that calcium carbonate is insoluble in water, what we really mean is that the solubility is very low (0.0014 g $CaCO_3$ per 100 mL $H_2O$ at 25 °C).

## SPECIAL TOPIC 4.1 Hard Water and Your Hot Water Pipes

A precipitation reaction that is a slight variation on the one depicted in Figures 4.8 and 4.9 helps explain why a solid scale forms more rapidly in your hot water pipes than in your cold water pipes.

We say water is *hard* if it contains calcium ions, magnesium ions, and (in many cases) iron ions. These ions come from rocks in the ground and dissolve into the water that passes through them. For example, limestone rock is calcium carbonate, $CaCO_3(s)$, and dolomite rock is a combination of calcium carbonate and magnesium carbonate, written as $CaCO_3 \cdot MgCO_3(s)$. Water alone will dissolve very small amounts of these minerals, but carbon dioxide dissolved in water speeds the process.

$$CaCO_3(s) + CO_2(g) + H_2O(l) \rightarrow Ca^{2+}(aq) + 2HCO_3^-(aq)$$

If the water were removed from the product mixture, calcium hydrogen carbonate, $Ca(HCO_3)_2$, would form, but this compound is much more soluble than calcium carbonate and does not precipitate from our tap water.

When hard water is heated, the reverse of this reaction occurs, and the calcium and hydrogen carbonate ions react to re-form solid calcium carbonate.

$$Ca^{2+}(aq) + 2HCO_3^-(aq) \rightarrow CaCO_3(s) + CO_2(g) + H_2O(l)$$

Thus, in your hot water pipes, solid calcium carbonate precipitates from solution and collects as scale on the inside of the pipes. After we have discussed acid–base reactions in Chapter 5, it will be possible to explain how the plumber can remove this obstruction.

Solubility is difficult to predict with confidence. The most reliable way to obtain a substance's solubility is to look it up in a table of physical properties in a reference book. When that is not possible, you can use the following guidelines for predicting whether some substances are soluble or insoluble in water. These guidelines are summarized in Table 4.1.

OBJECTIVE 9

- Ionic compounds with group 1 (or 1A) metallic cations or ammonium cations, $NH_4^+$, form soluble compounds no matter what the anion is.
- Ionic compounds with acetate ion, $C_2H_3O_2^-$, or nitrate ion, $NO_3^-$, form soluble compounds no matter what the cation is.
- Compounds containing the chloride, $Cl^-$, bromide, $Br^-$, or iodide, $I^-$, ion are water-soluble except with silver ions, $Ag^+$, and lead(II) ions, $Pb^{2+}$.
- Compounds containing the sulfate ion, $SO_4^{2-}$, are water-soluble except with barium ions, $Ba^{2+}$, and lead(II) ions, $Pb^{2+}$.
- Compounds containing carbonate, $CO_3^{2-}$, phosphate, $PO_4^{3-}$, or hydroxide, $OH^-$, ions are insoluble in water except with group 1 metallic ions and ammonium ions.

Table 4.1
Water Solubility of Ionic Compounds

OBJECTIVE 9

| Category | Ions | Except with these ions | Examples |
|---|---|---|---|
| **soluble cations** | group 1 metallic ions and ammonium, $NH_4^+$ | no exceptions | $Na_2CO_3$, LiOH, and $(NH_4)_2S$ are soluble. |
| **soluble anions** | $NO_3^-$ and $C_2H_3O_2^-$ | no exceptions | $Bi(NO_3)_3$, and $Co(C_2H_3O_2)_2$ are soluble. |
| **usually soluble anions** | $Cl^-$, $Br^-$, and $I^-$ | soluble with some exceptions, including with $Ag^+$ and $Pb^{2+}$ | $CuCl_2$ is water-soluble, but AgCl is insoluble. |
| | $SO_4^{2-}$ | soluble with some exceptions, including with $Ba^{2+}$ and $Pb^{2+}$ | $FeSO_4$ is water-soluble, but $BaSO_4$ is insoluble. |
| **usually insoluble** | $CO_3^{2-}$, $PO_4^{3-}$, and $OH^-$ | insoluble with some exceptions, including with group 1 elements and $NH_4^+$ | $CaCO_3$, $Ca_3(PO_4)_2$, and $Mn(OH)_2$ are insoluble in water, but $(NH_4)_2CO_3$, $Li_3PO_4$, and CsOH are water-soluble. |

## EXERCISE 4.2 Predicting Water Solubility

OBJECTIVE 9

Predict whether each of the following is soluble or insoluble in water.

a. $Hg(NO_3)_2$ (used to manufacture felt)

b. $BaCO_3$ (used to make radiation-resistant glass for color TV tubes)

c. $K_3PO_4$ (used to make liquid soaps)

d. $PbCl_2$ (used to make other lead salts)

e. $Cd(OH)_2$ (storage battery electrodes)

Potassium phosphate, $K_3PO_4$, contributes cleaning-enhancing phosphate ions, $PO_4^{3-}$, to detergents.

The following sample study sheet describes the steps you should take to answer questions about specific precipitation reactions.

## SAMPLE STUDY SHEET 4.2 Predicting Precipitation Reactions and Writing Precipitation Equations

OBJECTIVE 10

**TIP-OFF** You are asked to predict whether a precipitation reaction will take place between two aqueous solutions of ionic compounds and, if the answer is yes, to write the complete equation for the reaction.

**GENERAL STEPS**

**STEP 1** **Determine the formulas for the possible products using the general double-displacement equation.** (Remember to consider ion charges when writing your formulas.)

$$AB + CD \rightarrow AD + CB$$

**STEP 2** **Predict whether either of the possible products is water-insoluble.** If either possible product is insoluble, a precipitation reaction takes place, and you may continue with step 3. If neither is insoluble, write "No reaction."

**STEP 3** **Follow these steps to write the complete equation.**

- Write the formulas for the reactants separated by a + sign.

- Separate the formulas for the reactants and products with a single arrow.
- Write the formulas for the products separated by a + sign.
- Write the physical state for each formula.
  The insoluble product will be followed by (*s*).
  Water-soluble ionic compounds will be followed by (*aq*).
- Balance the equation.

EXAMPLES See Examples 4.6–4.8.

## EXAMPLE 4.6 Predicting Precipitation Reactions

OBJECTIVE 10

Predict whether a precipitate will form when water solutions of silver nitrate, $AgNO_3(aq)$, and sodium phosphate, $Na_3PO_4(aq)$, are mixed. If there is a precipitation reaction, write the complete equation that describes the reaction.

***Solution***

**STEP 1 Determine the possible products using the general double-displacement equation.**

$$AB + CD \rightarrow AD + CB$$

In $AgNO_3$, $Ag^+$ is A, and $NO_3^-$ is B. In $Na_3PO_4$, $Na^+$ is C, and $PO_4^{3-}$ is D. The possible products from the mixture of $AgNO_3(aq)$ and $Na_3PO_4(aq)$ are $Ag_3PO_4$ and $NaNO_3$. (Remember to consider charge when you determine the formulas for the possible products.)

$$AgNO_3(aq) + Na_3PO_4(aq) \quad \text{to} \quad Ag_3PO_4 \text{ and } NaNO_3$$

**STEP 2 Predict whether either of the possible products is water-insoluble.** According to our solubility guidelines, most phosphates are insoluble, and compounds with $Ag^+$ are not listed as an exception. Therefore, silver phosphate, $Ag_3PO_4$, which is used in photographic emulsions, is water-insoluble. Because compounds containing $Na^+$ and $NO_3^-$ are soluble, $NaNO_3$ is soluble.

**STEP 3 Write the complete equation.** (Don't forget to balance it.)

$$3AgNO_3(aq) + Na_3PO_4(aq) \rightarrow Ag_3PO_4(s) + 3NaNO_3(aq)$$

## EXAMPLE 4.7 Predicting Precipitation Reactions

OBJECTIVE 10

Predict whether a precipitate will form when water solutions of barium chloride, $BaCl_2(aq)$, and sodium sulfate, $Na_2SO_4(aq)$, are mixed. If there is a precipitation reaction, write the complete equation that describes the reaction.

***Solution***

**STEP 1** In $BaCl_2$, A is $Ba^{2+}$, and B is $Cl^-$. In $Na_2SO_4$, C is $Na^+$, and D is $SO_4^{2-}$. The possible products from the reaction of $BaCl_2(aq)$ and $Na_2SO_4(aq)$ are $BaSO_4$ and

NaCl. (Remember to consider charge when you determine the formulas for the possible products.)

$$BaCl_2(aq) + Na_2SO_4(aq) \quad \text{to} \quad BaSO_4 \text{ and } NaCl$$

**Step 2** According to our solubility guidelines, most sulfates are water-soluble, but $BaSO_4$ is an exception. It is insoluble and will precipitate from the mixture. Because compounds containing $Na^+$ (and most containing $Cl^-$) are soluble, NaCl is soluble.

**Step 3**

$$BaCl_2(aq) + Na_2SO_4(aq) \rightarrow BaSO_4(s) + 2NaCl(aq)$$

This is the reaction used in industry to form barium sulfate, which is used in paint preparations and in x-ray photography.

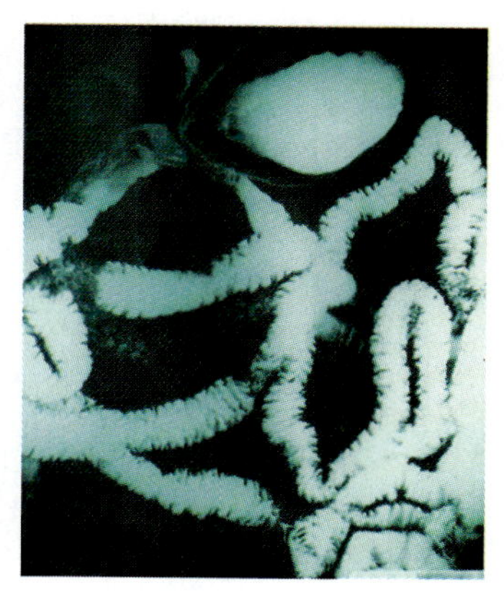

The reaction in Example 4.7 produces barium sulfate, which is used in taking X rays.

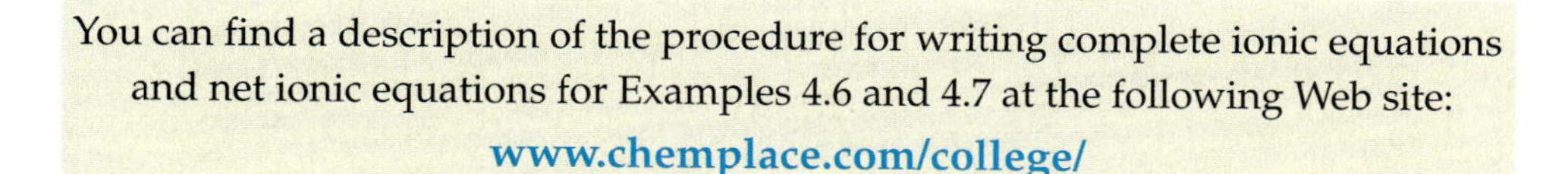

You can find a description of the procedure for writing complete ionic equations and net ionic equations for Examples 4.6 and 4.7 at the following Web site:
www.chemplace.com/college/

## Example 4.8 Predicting Precipitation Reactions

OBJECTIVE 10

Predict whether a precipitate will form when lead(II) nitrate, $Pb(NO_3)_2(aq)$, and sodium acetate, $NaC_2H_3O_2(aq)$, are mixed. If there is a precipitation reaction, write the complete equation that describes the reaction.

***Solution***

**Step 1** The possible products from the mixture of $Pb(NO_3)_2(aq)$ and $NaC_2H_3O_2(aq)$ are $Pb(C_2H_3O_2)_2$ and $NaNO_3$.

$$Pb(NO_3)_2(aq) + NaC_2H_3O_2(aq) \quad \text{to} \quad Pb(C_2H_3O_2)_2 \text{ and } NaNO_3$$

**Step 2** According to our solubility guidelines, compounds with nitrates and acetates are soluble, so both $Pb(C_2H_3O_2)_2$ and $NaNO_3$ are soluble. There is no precipitation reaction.

## Exercise 4.3 Precipitation Reactions

OBJECTIVE 10

Predict whether a precipitate will form when each of the following pairs of water solutions is mixed. If there is a precipitation reaction, write the complete equation that describes the reaction.

a. $CaCl_2(aq) + Na_3PO_4(aq)$

b. $KOH(aq) + Fe(NO_3)_3(aq)$

c. $NaC_2H_3O_2(aq) + CaSO_4(aq)$

d. $K_2SO_4(aq) + Pb(NO_3)_2(aq)$

## Having Trouble?

Are you having trouble with the topics in this chapter? People often do. To complete each of the lessons in it successfully, you need to have mastered the skills taught in previous sections. Here is a list of the things you need to know how to do in order to solve the problems at the end of this chapter. Work through these items in the order presented, and be sure you have mastered each before going on to the next.

- Convert between names and symbols for the common elements. See Table 2.1.
- Identify whether an element is a metal or a nonmetal. See Section 2.3.
- Determine the charges on many of the monatomic ions. See Figure 3.17.
- Convert between the names and formulas for polyatomic ions. See Table 3.6.
- Convert between the names and formulas for ionic compounds. See Section 3.5.
- Balance chemical equations. See Section 4.1.
- Predict the products of double-displacement reactions. See Section 4.2.
- Predict whether ionic compounds are soluble or insoluble in water. See Section 4.2.

## Chapter Glossary

**Chemical reaction or chemical change** The conversion of one or more pure substances into one or more different pure substances.

**Reactants** The substances that change in a chemical reaction. Their formulas are on the left side of the arrow in a chemical equation.

**Products** The substances that form in a chemical reaction. Their formulas are on the right side of the arrow in a chemical equation.

**Coefficients** The numbers in front of chemical formulas in a balanced chemical equation.

**Solution** A mixture whose particles are so evenly distributed that the relative concentrations of the components are the same throughout. Solutions can also be called homogeneous mixtures.

**Aqueous solution** A solution in which water is the solvent.

**Solute** The gas in a solution of a gas in a liquid. The solid in a solution of a solid in a liquid. The minor component in other solutions.

**Solvent** The liquid in a solution of a gas in a liquid. The liquid in a solution of a solid in a liquid. The major component in other solutions.

**Double-displacement reaction** A chemical reaction that has the following form: $AB + CD \rightarrow AD + CB$

**Precipitation reaction** A reaction in which one of the products is insoluble in water and comes out of solution as a solid.

**Precipitate** A solid that comes out of solution.

**Precipitation** The process of forming a solid in a solution.

**Crystals** Solid particles whose component atoms, ions, or molecules are arranged in an organized, repeating pattern.

**Complete ionic equation** A chemical equation that describes the actual form for each substance in solution. For example, ionic compounds that are dissolved in water are described as separate ions.

**Spectator ions** Ions that play a role in delivering other ions into solution to react but that do not actively participate in the reaction themselves.

**Complete equation or molecular equation** A chemical equation that includes uncharged formulas for all of the reactants and products. The formulas include the spectator ions, if any.

**Net ionic equation** A chemical equation for which the spectator ions have been eliminated, leaving only the substances actively involved in the reaction.

**Solubility** The maximum amount of solute that can be dissolved in a given amount of solvent.

You can test yourself on the glossary terms at the following Web site:
**www.chemplace.com/college/**

## Chapter Objectives

**The goal of this chapter is to teach you to do the following.**

1. Define all of the terms in the Chapter Glossary.

### Section 4.1 Chemical Reactions and Chemical Equations

2. Describe the information given by a chemical equation.
3. Write the symbols used in chemical equations to describe the physical states solid, liquid, gas, and aqueous.
4. Balance chemical equations.

### Section 4.2 Solubility of Ionic Compounds and Precipitation Reactions

5. Describe the process for dissolving an ionic compound in water. Your description should include the nature of the particles in solution and the attractions between the particles in the solution.
6. Given a description of a solution, identify the solute and the solvent.
7. Describe double-displacement reactions.
8. Describe precipitation reactions. Your description should include the nature of the particles in the system before and after the reaction, a description of the cause of the reaction, and a description of the attractions between the particles before and after the reaction.
9. Given the formula for an ionic compound, predict whether it is soluble in water.
10. Given formulas for two ionic compounds:
    a. Predict whether a precipitate will form when the water solutions of the two are mixed.
    b. If there is a reaction, predict the products of the reaction and write their formulas.
    c. If there is a reaction, write the complete equation that describes the reaction.

## Review Questions

1. Write the formulas for all of the diatomic elements.
2. Predict whether atoms of each of the following pairs of elements would be expected to form ionic or covalent bonds.
   a. Mg and F
   b. O and H
   c. Fe and O
   d. N and Cl
3. Describe the structure of liquid water, including a description of water molecules and the attractions between them.
4. Write formulas that correspond to the following names.
   a. ammonia
   b. methane
   c. propane
   d. water
5. Write formulas that correspond to the following names.
   a. nitrogen dioxide
   b. carbon tetrabromide
   c. dibromine monoxide
   d. nitrogen monoxide
6. Write formulas that correspond to the following names.
   a. lithium fluoride
   b. lead(II) hydroxide
   c. potassium oxide
   d. sodium carbonate
   e. chromium(III) chloride
   f. sodium hydrogen phosphate

## Key Ideas

**Complete the following statements by writing one of these words or phrases in each blank.**

| | |
|---|---|
| above | minor |
| charge | negative |
| chemical bonds | none |
| coefficients | organized, repeating |
| complete formula | partial charges |
| continuous | positive |
| converted into | precipitate |
| created | precipitates |
| delta, Δ | precipitation |
| destroyed | same proportions |
| equal to | separate ions |
| gas | shorthand description |
| homogeneous mixture | solute |
| left out | solvent |
| liquid | subscripts |
| major | very low |

7. A chemical change or chemical reaction is a process in which one or more pure substances are __________________ one or more different pure substances.

8. In chemical reactions, atoms are rearranged and regrouped through the breaking and making of ________________.
9. A chemical equation is a(n) ________________ of a chemical reaction.
10. If special conditions are necessary for a reaction to take place, they are often specified ________________ the arrow in the reaction's chemical equation.
11. To indicate that a chemical reaction requires the ________________ addition of heat in order to proceed, we place an upper-case Greek ________________ above the arrow in the reaction's chemical equation.
12. In chemical reactions, atoms are neither ________________ nor ________________; they merely change partners. Thus the number of atoms of an element in the reaction's products is ________________ the number of atoms of that element in the original reactants. The ________________ we often place in front of one or more of the formulas in a chemical equation reflect this fact.
13. When balancing chemical equations, we do not change the ________________ in the formulas.
14. A solution, also called a(n) ________________, is a mixture whose particles are so evenly distributed that the relative concentrations of the components are the same throughout.
15. Every part of a water solution of an ionic compound has the ________________ of water molecules and ions as every other part.
16. When an ionic compound dissolves in water, the ions that escape the solid are held in solution by attractions between their own ________________ and the ________________ of the polar water molecules. The ________________ oxygen ends of the water molecules surround the cations, and the ________________ hydrogen ends surround the anions.
17. In solutions of solids dissolved in liquids, we call the solid the ________________ and the liquid the ________________.
18. In solutions of gases in liquids, we call the ________________ the solute and the ________________ the solvent.
19. In solutions of two liquids, we call the ________________ component the solute and the ________________ component the solvent.
20. Sometimes a double-displacement reaction has one product that is insoluble in water. As that product forms, it emerges, or ________________, from the solution as a solid. This process is called ________________, and the solid is called a ________________.
21. Crystals are solid particles whose component atoms, ions, or molecules are arranged in a(n) ________________ pattern.
22. In a complete ionic equation, which describes the forms taken by the various substances in solution, the ionic compounds dissolved in the water are described as ________________, and the insoluble ionic compound is described with a(n) ________________.
23. Because spectator ions are not involved in the reaction, they are often ________________ of the chemical equation.
24. When we say an ionic solid is insoluble in water, we do not mean that ________________ of the solid dissolves. Thus, when we say that calcium carbonate is insoluble in water, what we really mean is that the solubility is ________________.

## Chapter Problems

### Section 4.1 Chemical Reactions and Chemical Equations

OBJECTIVE 2 OBJECTIVE 3

25. Describe the information given in the following chemical equation.

$$2CuHCO_3(s) \xrightarrow{\Delta} Cu_2CO_3(s) + H_2O(l) + CO_2(g)$$

OBJECTIVE 2 OBJECTIVE 3

26. Describe the information given in the following chemical equation.

$$2NaCl(l) \xrightarrow{\text{electric current}} 2Na(s) + Cl_2(g)$$

OBJECTIVE 4

27. Balance the following equations.
    a. $N_2(g) + H_2(g) \rightarrow NH_3(g)$
       $NH_3$ is used to make explosives and rocket fuel.
    b. $Cl_2(g) + CH_4(g) + O_2(g) \rightarrow HCl(g) + CO(g)$
       HCl is used to make vinyl chloride, which is then used to make polyvinyl chloride (PVC) plastic.
    c. $B_2O_3(s) + NaOH(aq) \rightarrow Na_3BO_3(aq) + H_2O(l)$
       $Na_3BO_3$ is used as an analytical reagent.
    d. $Al(s) + H_3PO_4(aq) \rightarrow AlPO_4(s) + H_2(g)$
       $AlPO_4$ is used in dental cements.
    e. $CO(g) + O_2(g) \rightarrow CO_2(g)$
       This reaction takes place in the catalytic converter of your car.
    f. $C_6H_{14}(l) + O_2(g) \rightarrow CO_2(g) + H_2O(l)$
       This is one of the chemical reactions that takes place when gasoline is burned.
    g. $Sb_2S_3(s) + O_2(g) \rightarrow Sb_2O_3(s) + SO_2(g)$
       $Sb_2O_3$ is used to flameproof cloth.
    h. $Al(s) + CuSO_4(aq) \rightarrow Al_2(SO_4)_3(aq) + Cu(s)$
       $Al_2(SO_4)_3$ has been used in paper production.
    i. $P_2H_4(l) \rightarrow PH_3(g) + P_4(s)$
       $PH_3$ is used to make semiconductors, and $P_4$ is used to manufacture phosphoric acid.

OBJECTIVE 4

28. Balance the following equations.
    a. $Fe_2O_3(s) + H_2(g) \rightarrow Fe(s) + H_2O(l)$
       Fe is the primary component in steel.
    b. $SCl_2(l) + NaF(s) \rightarrow S_2Cl_2(l) + SF_4(g) + NaCl(s)$
       $S_2Cl_2$ is used to purify sugar juices.
    c. $PCl_5(s) + H_2O(l) \rightarrow H_3PO_4(aq) + HCl(aq)$
       $H_3PO_4$ is used to make fertilizers, soaps, and detergents.
    d. $As(s) + Cl_2(g) \rightarrow AsCl_5(s)$
       $AsCl_5$ is an intermediate in the production of arsenic compounds.

e. $C_2H_5SH(l) + O_2(g) \rightarrow CO_2(g) + H_2O(l) + SO_2(g)$

$C_2H_5SH$ is added to natural gas to give it an odor. Without it or something like it, you would not know when you have a gas leak.

f. $N_2O_5(g) \rightarrow NO_2(g) + O_2(g)$

$NO_2$ is used in rocket fuels.

g. $Mg(s) + Cr(NO_3)_3(aq) \rightarrow Mg(NO_3)_2(aq) + Cr(s)$

Cr is used to make stainless steel.

h. $H_2O(g) + NO(g) \rightarrow O_2(g) + NH_3(g)$

$NH_3$ is used to make fertilizers.

i. $CCl_4(l) + SbF_3(s) \rightarrow CCl_2F_2(g) + SbCl_3(s)$

$CCl_2F_2$ is a chlorofluorocarbon called CFC-12. Although it had many uses in the past, its use has diminished greatly because of the damage it can do to our protective ozone layer.

**29.** Because of its toxicity, carbon tetrachloride is prohibited in products intended for home use, but it is used industrially for a variety of purposes, including the production of chlorofluorocarbons (CFCs). It is made in three steps. Balance their equations:

$$CS_2 + Cl_2 \rightarrow S_2Cl_2 + CCl_4$$

$$CS_2 + S_2Cl_2 \rightarrow S_8 + CCl_4$$

$$S_8 + C \rightarrow CS_2$$

30. Chlorofluorocarbons (CFCs) are compounds that contain carbon, fluorine, and chlorine. Because they destroy ozone that forms a protective shield high in earth's atmosphere, their use is being phased out, but at one time they were widely employed as aerosol propellants and as refrigerants. Balance the following equation that shows how the CFCs dichlorodifluoromethane, $CCl_2F_2$, and trichlorofluoromethane, $CCl_3F$, are produced.

$$HF + CCl_4 \rightarrow CCl_2F_2 + CCl_3F + HCl$$

**31.** Hydrochlorofluorocarbons (HCFCs), which contain hydrogen as well as carbon, fluorine, and chlorine, are less damaging to the ozone layer than the chlorofluorocarbons (CFCs) described in problem 30. HCFCs are therefore used instead of CFCs for many purposes. Balance the following equation that shows how the HCFC chlorodifluoromethane, $CHClF_2$, is made.

$$HF + CHCl_3 \rightarrow CHClF_2 + HCl$$

32. The primary use of 1,2-dichloroethane, $ClCH_2CH_2Cl$, is to make vinyl chloride, which is then converted into polyvinyl chloride (PVC) for many purposes, including plastic pipes. Balance the following equation, which describes the industrial reaction for producing 1,2-dichloroethane.

$$C_2H_4 + HCl + O_2 \rightarrow ClCH_2CH_2Cl + H_2O$$

### Section 4.2 Solubility of Ionic Compounds and Precipitation Reactions

OBJECTIVE 5

33. Describe the process for dissolving the ionic compound lithium iodide, LiI, in water, including the nature of the particles in solution and the attractions between the particles in the solution.

OBJECTIVE 5

34. Describe the process for dissolving the ionic compound potassium nitrate, $KNO_3$, in water, including the nature of the particles in solution and the attractions between the particles in the solution.

OBJECTIVE 5

35. Describe the process for dissolving the ionic compound sodium sulfate, $Na_2SO_4$, in water. Include the nature of the particles in solution and the attractions between the particles in the solution.

OBJECTIVE 5

36. Describe the process for dissolving the ionic compound calcium chloride, $CaCl_2$, in water. Include the nature of the particles in solution and the attractions between the particles in the solution.

OBJECTIVE 6

37. Solid camphor and liquid ethanol mix to form a solution. Which of these substances is the solute and which is the solvent?

OBJECTIVE 6

38. Gaseous propane and liquid diethyl ether mix to form a solution. Which of these substances is the solute and which is the solvent?

OBJECTIVE 6

39. Consider a solution of 10% liquid acetone and 90% liquid chloroform. Which of these substances is the solute and which is the solvent?

OBJECTIVE 8

40. Black-and-white photographic film has a thin layer of silver bromide deposited on it. Wherever light strikes the film, silver ions are converted to uncharged silver atoms, creating a dark image on the film. Describe the precipitation reaction that takes place between water solutions of silver nitrate, $AgNO_3(aq)$, and sodium bromide, $NaBr(aq)$, to form solid silver bromide, $AgBr(s)$, and aqueous sodium nitrate, $NaNO_3(aq)$. Include the nature of the particles in the system before and after the reaction, a description of the cause of the reaction, and a description of the attractions between the particles before and after the reaction.

OBJECTIVE 8

41. Magnesium carbonate is used as an anticaking agent in powders and as an antacid. Describe the precipitation reaction that takes place between water solutions of magnesium nitrate, $Mg(NO_3)_2(aq)$, and sodium carbonate, $Na_2CO_3(aq)$, to form solid magnesium carbonate, $MgCO_3(s)$, and aqueous sodium nitrate, $NaNO_3(aq)$. Include the nature of the particles in the system before and after the reaction, a description of the cause of the reaction, and a description of the attractions between the particles before and after the reaction.

OBJECTIVE 9

42. Predict whether each of the following substances is soluble or insoluble in water.

    a. $Na_2SO_3$ (used in water treatment)

    b. iron(III) acetate (a wood preservative)

    c. $CoCO_3$ (a red pigment)

    d. lead(II) chloride (used in the preparation of lead salts)

43. Predict whether each of the following substances is soluble or insoluble in water. OBJECTIVE 9
    a. $MgSO_4$ (used in fireproofing)
    b. barium sulfate (used in paints)
    c. $Bi(OH)_3$ (used in plutonium separation)
    d. ammonium sulfite (used in medicine and photography)
44. Predict whether each of the following substances is soluble or insoluble in water. OBJECTIVE 9
    a. zinc phosphate (used in dental cements)
    b. $Mn(C_2H_3O_2)_2$ (used in cloth dyeing)
    c. nickel(II) sulfate (used in nickel plating)
    d. AgCl (used in silver plating)
45. Predict whether each of the following substances is soluble or insoluble in water. OBJECTIVE 9
    a. copper(II) chloride (used in fireworks and as a fungicide)
    b. $PbSO_4$ (a paint pigment)
    c. potassium hydroxide (used in soap manufacture)
    d. $NH_4F$ (used as an antiseptic in brewing)
46. For each of the following pairs of formulas, predict whether the substances they represent would react in a precipitation reaction. The products formed in the reactions that take place are used in ceramics, cloud seeding, photography, electroplating, and paper coatings. If there is no reaction, write, "No Reaction." If there is a reaction, write the complete equation for the reaction. OBJECTIVE 10
    a. $Co(NO_3)_2(aq) + Na_2CO_3(aq)$
    b. $KI(aq) + Pb(C_2H_3O_2)_2(aq)$
    c. $CuSO_4(aq) + LiNO_3(aq)$
    d. $Ni(NO_3)_2(aq) + Na_3PO_4(aq)$
    e. $K_2SO_4(aq) + Ba(NO_3)_2(aq)$
47. For each of the following pairs of formulas, predict whether the substances they represent would react in a precipitation reaction. If there is no reaction, write, "No Reaction." If there is a reaction, write the complete equation for the reaction. OBJECTIVE 10
    a. $NaCl(aq) + Pb(NO_3)_2(aq)$
    b. $NH_4Cl(aq) + CaSO_3(aq)$
    c. $NaOH(aq) + Zn(NO_3)_2(aq)$
    d. $Pb(C_2H_3O_2)_2(aq) + Na_2SO_4(aq)$
48. Phosphate ions find their way into our water system from the fertilizers dissolved in the runoff from agricultural fields and from detergents that we send down our drains. Some of these phosphate ions can be removed by adding aluminum sulfate to the water and precipitating the phosphate ions as aluminum phosphate. Write the net ionic equation for the reaction that forms the aluminum phosphate.

49. The taste of drinking water can be improved by removing impurities from our municipal water by adding substances to the water that precipitate a solid (called a flocculent) that drags down impurities as it settles. One way this is done is to dissolve aluminum sulfate and sodium hydroxide in the water to precipitate aluminum hydroxide. Write the complete equation for this reaction.

50. Cadmium hydroxide is used in storage batteries. It is made from the precipitation reaction of cadmium acetate and sodium hydroxide. Write the complete equation for this reaction.

51. Chromium(III) phosphate is a paint pigment that is made in a precipitation reaction between water solutions of chromium(III) chloride and sodium phosphate. Write the complete equation for this reaction.

### Additional Problems

52. Balance the following chemical equations.
    a. $SiCl_4 + H_2O \rightarrow SiO_2 + HCl$
    b. $H_3BO_3 \rightarrow B_2O_3 + H_2O$
    c. $I_2 + Cl_2 \rightarrow ICl_3$
    d. $Al_2O_3 + C \rightarrow Al + CO_2$

53. Balance the following chemical equations.
    a. $HClO_4 + Fe(OH)_2 \rightarrow Fe(ClO_4)_2 + H_2O$
    b. $NaClO_3 \rightarrow NaCl + O_2$
    c. $Sb + Cl_2 \rightarrow SbCl_3$
    d. $CaCN_2 + H_2O \rightarrow CaCO_3 + NH_3$

54. Balance the following chemical equations.
    a. $NH_3 + Cl_2 \rightarrow N_2H_4 + NH_4Cl$
    b. $Cu + AgNO_3 \rightarrow Cu(NO_3)_2 + Ag$
    c. $Sb_2S_3 + HNO_3 \rightarrow Sb(NO_3)_3 + H_2S$
    d. $Al_2O_3 + Cl_2 + C \rightarrow AlCl_3 + CO$

55. Balance the following chemical equations.
    a. $AsH_3 \rightarrow As + H_2$
    b. $H_2S + Cl_2 \rightarrow S_8 + HCl$
    c. $Co + O_2 \rightarrow Co_2O_3$
    d. $Na_2CO_3 + C \rightarrow Na + CO$

56. Phosphoric acid, $H_3PO_4$, is an important chemical used to make fertilizers, detergents, pharmaceuticals, and many other substances. High purity phosphoric acid is made in a two-step process called the furnace process. Balance its two equations:

$$Ca_3(PO_4)_2 + SiO_2 + C \rightarrow P_4 + CO + CaSiO_3$$

$$P_4 + O_2 + H_2O \rightarrow H_3PO_4$$

57. For most applications, phosphoric acid is produced by the "wet process" (whose results are less pure than those of the furnace process described in problem 56). Balance the following equation that describes the reaction for this process.

$$Ca_3(PO_4)_2 + H_2SO_4 \rightarrow H_3PO_4 + CaSO_4$$

58. Predict whether each of the following substances is soluble or insoluble in water. OBJECTIVE 9

a. manganese(II) chloride (used as a dietary supplement)

b. $CdSO_4$ (used in pigments)

c. copper(II) carbonate (used in fireworks)

d. $Co(OH)_3$ (used as a catalyst)

59. Predict whether each of the following substances is soluble or insoluble in water. OBJECTIVE 9

a. copper(II) hydroxide (used as a pigment)

b. $BaBr_2$ (used to make photographic compounds)

c. silver carbonate (used as a laboratory reagent)

d. $Pb_3(PO_4)_2$ (used as a stabilizing agent in plastics)

60. For each of the following pairs of formulas, predict whether the substances they represent would react to yield a precipitate. (The products formed in the reactions that take place are used to coat steel, as a fire-proofing filler for plastics, in cosmetics, and as a topical antiseptic.) If there is no reaction, write, "No Reaction." If there is a reaction, write the complete equation for the reaction. OBJECTIVE 10

a. $NaCl(aq) + Al(NO_3)_3(aq)$

b. $Ni(NO_3)_2(aq) + NaOH(aq)$

c. $MnCl_2(aq) + Na_3PO_4(aq)$

d. $Zn(C_2H_3O_2)_2(aq) + Na_2CO_3(aq)$

61. For each of the following pairs of formulas, predict whether the substances they represent would react to yield a precipitate. (The products formed in the reactions that take place are used as a catalyst, as a tanning agent, as a pigment, in fertilizers, as a food additive, and on photographic film.) If there is no reaction, write, "No Reaction." If there is a reaction, write the complete equation for the reaction. OBJECTIVE 10

a. $KOH(aq) + Cr(NO_3)_3(aq)$

b. $Fe(NO_3)_3(aq) + K_3PO_4(aq)$

c. $NaBr(aq) + AgNO_3(aq)$

d. $Mg(C_2H_3O_2)_2(aq) + NaCl(aq)$

*Before working Chapter Problems 62 through 78, you might want to review the procedures for writing chemical formulas that are described in Chapter 3. Remember that some elements are described with formulas containing subscripts (as in $O_2$).*

62. Hydrochloric acid is used in the cleaning of metals (called pickling). Hydrogen chloride, used to make hydrochloric acid, is made industrially by combining hydrogen and chlorine. Write a balanced equation, without including states, for this reaction.
63. Potassium hydroxide has many uses, including the manufacture of soap. It is made by running an electric current through a water solution of potassium chloride. Both potassium chloride and water are reactants, and the products are potassium hydroxide, hydrogen, and chlorine. Write a balanced equation, without including states, for this reaction.
64. Aluminum sulfate, commonly called alum, is used to coat paper made from wood pulp. (It fills in tiny holes in the paper and thus keeps the ink from running.) Alum is made in the reaction of aluminum oxide with sulfuric acid, $H_2SO_4$, which produces aluminum sulfate and water. Write a balanced equation, without including states, for this reaction.
65. Under the right conditions, methanol reacts with oxygen to yield formaldehyde, $CH_2O$, and water. Most of the formaldehyde made in this way is used in the production of other substances, including some important plastics. Formaldehyde, itself now a suspected carcinogen, was once used in insulation foams and in plywood adhesives. Write a balanced equation, without including states, for the reaction that produces formaldehyde from methanol.
66. Hydrogen fluoride is used to make chlorofluorocarbons (CFCs) and in uranium processing. Calcium fluoride reacts with sulfuric acid, $H_2SO_4$, to form hydrogen fluoride and calcium sulfate. Write a balanced equation, without including states, for this reaction.
67. Sodium sulfate, which is used to make detergents and glass, is one product of the reaction of sodium chloride, sulfur dioxide, water, and oxygen. The other product is hydrogen chloride. Write a balanced equation, without including states, for this reaction.
68. Sodium hydroxide, which is often called caustic soda, is used to make paper, soaps, and detergents. For many years, it was made from the reaction of sodium carbonate with calcium hydroxide (also called slaked lime). The products are sodium hydroxide and calcium carbonate. Write a balanced equation, without including states, for this reaction.
69. In the modern process for making sodium hydroxide, an electric current is run through a sodium chloride solution, forming hydrogen and chlorine along with the sodium hydroxide. Both sodium chloride and water are reactants. Write a balanced equation, without including states, for this reaction.

70. For over a century, sodium carbonate (often called soda ash) was made industrially by the Solvay process. This process was designed by Ernest Solvay in 1864 and was used in the United States until extensive natural sources of sodium carbonate were found in the 1970s and 1980s. Write a balanced equation, without including states, for each step in the process:

    a. Calcium carbonate (from limestone) is heated and decomposes into calcium oxide and carbon dioxide.

    b. Calcium oxide reacts with water to form calcium hydroxide.

    c. Ammonia reacts with water to form ammonium hydroxide.

    d. Ammonium hydroxide reacts with carbon dioxide to form ammonium hydrogen carbonate.

    e. Ammonium hydrogen carbonate reacts with sodium chloride to form sodium hydrogen carbonate and ammonium chloride.

    f. Sodium hydrogen carbonate is heated and decomposes into sodium carbonate, carbon dioxide, and water.

    g. Ammonium chloride reacts with calcium hydroxide to form ammonia, calcium chloride, and water.

71. All of the equations for the Solvay process described in problem 70 can be summarized by a single equation, called a net equation, that describes the overall change for the process. This equation shows calcium carbonate reacting with sodium chloride to form sodium carbonate and calcium chloride. Write a balanced equation, without including states, for this net reaction.

72. Nitric acid, $HNO_3$, which is used to make fertilizers and explosives, is made industrially in the three steps described below. Write a balanced equation, without including states, for each of these steps.

    a. Ammonia reacts with oxygen to form nitrogen monoxide and water.

    b. Nitrogen monoxide reacts with oxygen to form nitrogen dioxide.

    c. Nitrogen dioxide reacts with water to form nitric acid and nitrogen monoxide.

73. All of the equations for the production of nitric acid described in problem 72 can be summarized in a single equation, called a net equation, that describes the overall change for the complete process. This equation shows ammonia combining with oxygen to yield nitric acid and water. Write a balanced equation, without including states, for this net reaction.

74. Ammonium sulfate, an important component in fertilizers, is made from the reaction of ammonia and sulfuric acid, $H_2SO_4$. Write a balanced equation, without including states, for the reaction that summarizes this transformation.

75. Hydrogen gas has many practical uses, including the conversion of vegetable oils into margarine. One way the gas is produced by the chemical industry is by reacting propane gas with gaseous water to form carbon dioxide gas and hydrogen gas. Write a balanced equation for this reaction, showing the states of reactants and products.

76. Sodium tripolyphosphate, $Na_5P_3O_{10}$, is called a "builder" when added to detergents. It helps to create the conditions in laundry water necessary for the detergents to work most efficiently. When phosphoric acid, $H_3PO_4$, is combined with sodium carbonate, three reactions take place in the mixture that lead to the production of sodium tripolyphosphate. Write a balanced equation, without including states, for each of the reactions:
    a. Phosphoric acid reacts with sodium carbonate to form sodium dihydrogen phosphate, water, and carbon dioxide.
    b. Phosphoric acid also reacts with sodium carbonate to form sodium hydrogen phosphate, water, and carbon dioxide.
    c. Sodium dihydrogen phosphate combines with sodium hydrogen phosphate to yield sodium tripolyphosphate and water.
77. Pig iron is iron with about 4.3% carbon in it. The carbon lowers the metal's melting point and makes it easier to shape. To produce pig iron, iron(III) oxide is combined with carbon and oxygen at high temperature. Three changes then take place to form molten iron with carbon dispersed in it. Write a balanced equation, without including states, for each of these changes:
    a. Carbon combines with oxygen to form carbon monoxide.
    b. Iron(III) oxide combines with the carbon monoxide to form iron and carbon dioxide.
    c. Carbon monoxide changes into carbon (in the molten iron) and carbon dioxide.
78. The United States chemical industry makes more sulfuric acid than any other chemical. One process uses hydrogen sulfide from "sour" natural gas wells or petroleum refineries as a raw material. Write a balanced equation, without including states, for each of the steps leading from hydrogen sulfide to sulfuric acid:
    a. Hydrogen sulfide (which could be called dihydrogen monosulfide) combines with oxygen to form sulfur dioxide and water.
    b. The sulfur dioxide reacts with more hydrogen sulfide to form sulfur and water. (Sulfur is described as both S and $S_8$ in chemical equations. Use S in this equation.)
    c. After impurities are removed, the sulfur is reacted with oxygen to form sulfur dioxide.
    d. Sulfur dioxide reacts with oxygen to yield sulfur trioxide.
    e. Sulfur trioxide and water combine to make sulfuric acid, $H_2SO_4$.
79. Assume you are given a water solution that contains either sodium ions or aluminum ions. Describe how you could determine which of these is in solution.
80. Assume you are given a water solution that contains either nitrate ions or phosphate ions. Describe how you could determine which of these is in solution.
81. Write a complete, balanced chemical equation for the reaction between water solutions of iron(III) chloride and silver nitrate.
82. Write a complete, balanced chemical equation for the reaction between water solutions of sodium phosphate and copper(II) chloride.

83. When the solid amino acid methionine, $C_5H_{11}NSO_2$, reacts with oxygen gas, the products are carbon dioxide gas, liquid water, sulfur dioxide gas, and nitrogen gas. Write a complete, balanced equation for this reaction.

84. When the explosive liquid nitroglycerin, $C_3H_5N_3O_9$, decomposes, it forms carbon dioxide gas, nitrogen gas, water vapor, and oxygen gas. Write a complete, balanced equation for this reaction.

### Discussion Problem

85. The solubility of calcium carbonate is 0.0014 g $CaCO_3$ per 100 mL of water at 25 °C, and the solubility of sodium nitrate is 92.1 $NaNO_3$ per 100 mL of water at 25 °C. We say that calcium carbonate is insoluble in water and that sodium nitrate, $NaNO_3$, is soluble.

    a. In Section 4.2, the following statement was made: *"When we say an ionic solid is insoluble in water, we do not mean that none of the solid dissolves. There are always some ions that can escape from the surface of an ionic solid in water and go into solution."* Discuss the process by which calcium ions and carbonate ions can escape from the surface of calcium carbonate solid in water and go into solution. You might want to draw a picture to illustrate this process.

    b. If the calcium and carbonate ions are constantly going into solution, why doesn't the calcium carbonate solid all dissolve?

    c. Why do you think sodium nitrate, $NaNO_3$, dissolves to a much greater degree than calcium carbonate?

    d. Why is there a limit to the solubility of even the "soluble" sodium nitrate?

*Healthy forests require the correct balance of acids and bases.*

CHAPTER 5

# Acids, Bases, and Acid–Base Reactions



IT'S TEST DAY IN CHEMISTRY CLASS—they've been learning about acids and bases—and Fran unwisely skips breakfast in order to have time for some last-minute studying. As she reads, she chews on a candy bar and sips a cup of coffee. Fran is well aware that the sugary candy sticking to her molars is providing breakfast for the bacteria in her mouth, which in turn produce an acid that will dissolve some of the enamel on her teeth. Feeling a little guilty about all that sugar from the candy, Fran drinks her coffee black, even though she doesn't like the taste. The caffeine in her coffee is a base, and like all bases, it tastes bitter.

Fran's junk-food breakfast and her worrying about the exam combine to give her an annoying case of acid indigestion, which she calms by drinking some baking soda mixed with water. The baking soda contains a base that "neutralizes" some of her excess stomach acid.

After taking the exam, Fran feels happy and confident. All those hours working problems, reviewing the learning objectives, and participating in class really paid off. Now she's ready for some lunch. Before eating, she washes her hands with soap made from the reaction of a strong base and animal fat. One of the reasons why the soap is slippery is that all bases feel slippery on the skin. To compensate for her less-than-healthful breakfast, Fran chooses a lunch of salad with a piece of lean chicken on top. Like all acids, the vinegar in her salad dressing tastes sour. Her stomach produces just enough additional acid to start the digestion of the protein from the chicken.

Read on to learn more about the acids and bases that are important in Fran's life and your own: what they are, how to construct their names and recognize their formulas, and how they react with each other.

*The vinegar in salad dressing tastes sour, as do all acids.*

## Review Skills

The presentation of information in this chapter assumes that you can already perform the tasks listed below. You can test your readiness to proceed by answering the Review Questions at the end of the chapter. This might also be a good time to read the Chapter Objectives, which precede the Review Questions.

- Describe the structure of liquid water. (Section 3.3)
- Convert between the names and the formulas for common polyatomic ions. (Table 3.5)
- Given a chemical name or formula, decide whether it represents an ionic compound. (Section 3.5)
- Convert between the names and the formulas for ionic compounds. (Section 3.5)
- Write a description of the changes that take place when an ionic compound is dissolved in water. (Section 4.2)
- Predict ionic solubility. (Section 4.2)
- Predict the products of double-displacement reactions. (Section 4.2)

## 5.1 Acids

Acids have many uses. For example, phosphoric acid is used to make gasoline additives and carbonated beverages. The textile industry uses oxalic acid (found in rhubarb and spinach) to bleach cloth, and glass is etched by the application of hydrofluoric acid. Dyes and many other chemicals are made with sulfuric acid and nitric acid; and corn syrup, which is added to a variety of foods, is processed with hydrochloric acid. The chemical reactions of acids often take place in water solutions, so after discussing what acids are, we will explore a model for visualizing the particle structure of water solutions of acids.

Acids can be found in many places in this scene, including the battery of the car, the soft drink bubbles and food sweeteners, and the napkins.

### Arrhenius Acids

You may have already noticed, in your first few weeks of studying chemistry, that the more you learn about matter, the more ways you have of grouping and classifying different substances. The most common and familiar way of classifying substances is by their noteworthy properties. For example, people long ago decided that any substance that has a sour taste is an acid. Lemons are sour because they contain citric acid, and the sour taste of old wine that has been exposed to the air is due to acetic acid. As chemists learned more about these substances, however, they developed more specific definitions that made it possible to classify them without relying on taste. A good thing, too, because many acids and bases should not be tasted—or even touched! They speed the breakdown of some of the substances that form the structure of our bodies or that help regulate chemical changes that go on in the body.

OBJECTIVE 2

Two different definitions of the term *acid* will be of use to us. For example, chemists conduct many laboratory experiments using a reagent known as "nitric acid," a substance that has been classified as an acid according to the Arrhenius definition of the term. (This classification system is named after the Swedish Nobel prize-winning chemist Svante August Arrhenius.) Arrhenius recognized that when ionic compounds dissolve, they form ions in solution. (Thus, when sodium chloride dissolves, it forms sodium ions and chloride ions.) He postulated that acids dissolve in a similar way to form $H^+$ ions and some kind of anion. For example, he predicted that when HCl is added to water, $H^+$ ions and $Cl^-$ ions form. We now know that $H^+$ ions do not persist in water; rather, they combine with water molecules to form hydronium ions, $H_3O^+$. Therefore, according to the modern form of the Arrhenius theory, an acid is a substance that produces **hydronium ions,** $H_3O^+$, when it is added to water. On the basis of this definition, an **acidic solution** is a solution with a significant concentration of $H_3O^+$. For reasons that are described in Section 5.7, chemists often find this definition too limiting, so another, broader definition of acids, called the Brønsted–Lowry definition, is commonly used instead.

OBJECTIVE 3

To get an understanding of how hydronium ions are formed when Arrhenius acids are added to water, let's consider the dissolving of gaseous hydrogen chloride, HCl(*g*), in water. When HCl molecules dissolve in water, a chemical change takes place in which water molecules pull hydrogen atoms away from HCl molecules. In each case, the hydrogen atom is transferred without its electron—that is, as an $H^+$ ion—and because most uncharged hydrogen atoms contain only one proton and one electron, most hydrogen atoms without their electrons are just protons. For this reason, the hydrogen ion, $H^+$, is often called a proton. We say that the HCl donates a proton, $H^+$, to water, forming hydronium ion, $H_3O^+$, and chloride ion, $Cl^-$.

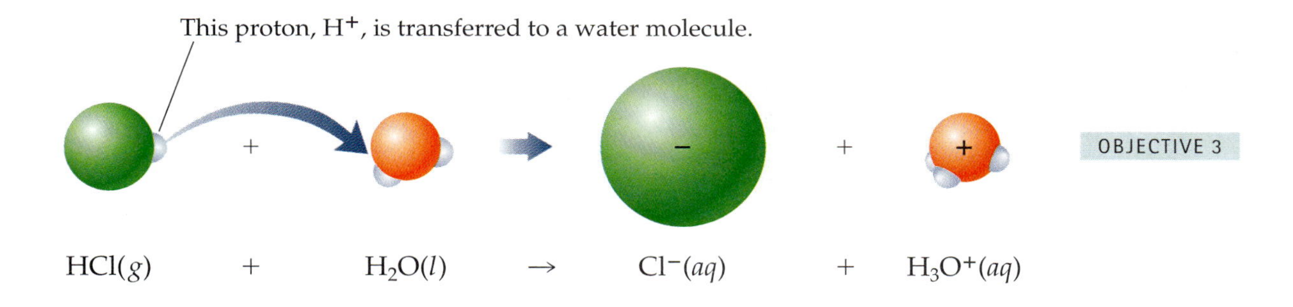
This proton, H+, is transferred to a water molecule.
+
−
+
+
HCl(g) + H2O(l) → Cl−(aq) + H3O+(aq)

OBJECTIVE 3

OBJECTIVE 3

Because HCl produces hydronium ions when added to water, it is an **Arrhenius acid.** Once the chloride ion and the hydronium ion are formed, the negatively charged oxygen atoms of the water molecules surround the hydronium ion, and the positively charged hydrogen atoms of the water molecules surround the chloride ion. Figure 5.1, on the next page, shows how you can picture this solution. Hydrochloric acid solutions are used in the chemical industry to remove impurities from metal surfaces (this is called pickling), to process food, to increase the permeability of limestone (an aid in oil drilling), and to make many important chemicals.

## Types of Arrhenius Acids

In terms of chemical structure, Arrhenius acids can be divided into several different subcategories. We will look at three of them here: binary acids, oxyacids, and organic acids. The **binary acids** are HF(*aq*), HCl(*aq*), HBr(*aq*), and HI(*aq*); all have the general formula HX(*aq*), where X is one of the first four halogens. The formulas for the binary acids will be followed by (*aq*) in this text to show that they are dissolved in water. The most common binary acid is hydrochloric acid, HCl(*aq*).

OBJECTIVE 4

**Oxyacids** (often called oxoacids) are molecular substances that have the general formula $H_aX_bO_c$. In other words, they contain hydrogen, oxygen, and one other element represented by X; a, b, and c represent numerical subscripts. The most common oxyacids in the chemical laboratory are nitric acid, $HNO_3$, and sulfuric acid, $H_2SO_4$.

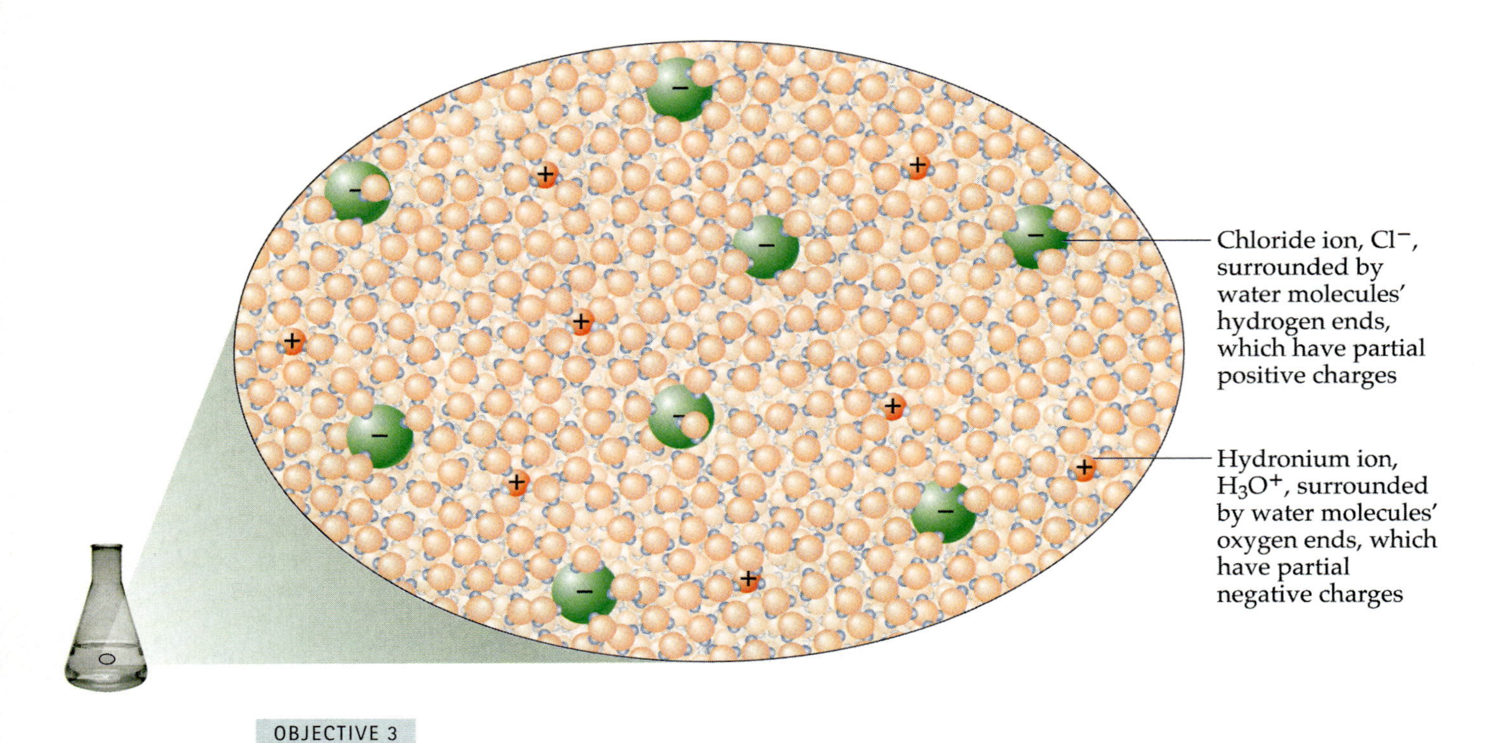


OBJECTIVE 3

**Figure 5.1**
**Hydrochloric Acid in Water**

Acetic acid, the acid responsible for the properties of vinegar, contains hydrogen, oxygen, and carbon and therefore fits the criteria for classification as an oxyacid, but it is more commonly described as an **organic** (or carbon-based) acid. It can also be called a carboxylic acid. (This type of acid is described in more detail in Section 17.1.) The formula for acetic acid can be written as $HC_2H_3O_2$, as $CH_3CO_2H$, or as $CH_3COOH$. The reason for keeping one H in these formulas separate from the others is that the hydrogen atoms in acetic acid are not all equal. Only one of them can be transferred to a water molecule. That hydrogen is known as the acidic hydrogen. We will use the formula $HC_2H_3O_2$ because it is more consistent with the formulas for other acids presented in this chapter. The Lewis structure, space-filling model, and ball-and-stick model for acetic acid (Figure 5.2) show why $CH_3CO_2H$, and $CH_3COOH$ are also commonly used. The acidic hydrogen is the one connected to an oxygen atom.

OBJECTIVE 5

**Figure 5.2**
**Acetic Acid**

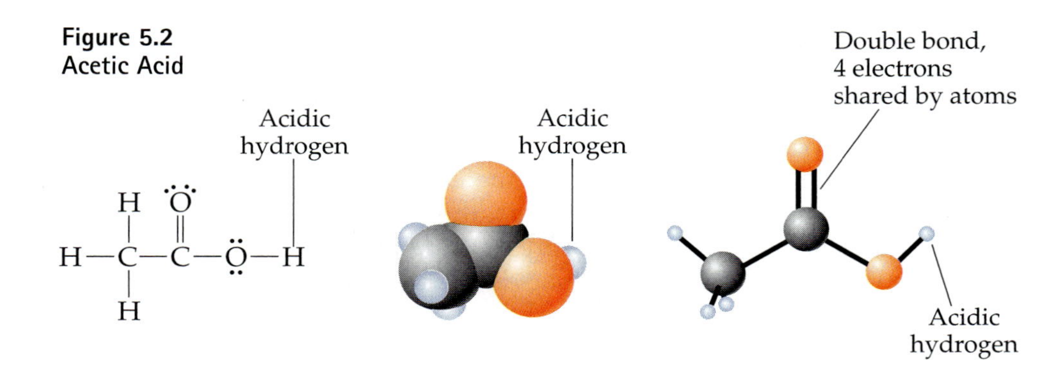

Pure acetic acid freezes at 17 °C (63 °F), so it is a liquid at normal room temperature, but if you put it outside on a cold day, it will freeze. The solid has layered crystals that look like tiny glaciers, so pure acetic acid is called glacial acetic acid. The chemical industry uses acetic acid to make several substances necessary for producing latex paints, safety-glass layers, photographic film, cigarette filters, magnetic tapes, and clothing. Acetic acid is also used to make esters, which are substances that have very pleasant odors and are added to candy and other foods.

Acids can have more than one acidic hydrogen. If each molecule of an acid can donate one hydrogen ion, the acid is called a **monoprotic acid.** If each molecule can donate two or more hydrogen ions, the acid is a **polyprotic acid.** A **diprotic acid,** such as sulfuric acid, $H_2SO_4$, has two acidic hydrogen atoms. Some acids, such as phosphoric acid, $H_3PO_4$, are **triprotic acids.** Most of the phosphoric acid produced by the chemical industry is used to make fertilizers and detergents, but it is also used to make pharmaceuticals, to refine sugar, and in water treatment. The tartness of some foods and beverages comes from acidifying them by adding phosphoric acid. The space-filled model in Figure 5.3 shows the three acidic hydrogen atoms of phosphoric acid.

**Figure 5.3**
**Phosphoric Acid, $H_3PO_4$**
The phosphate in this fertilizer was made from phosphoric acid.

## Strong and Weak Acids

Although hydrochloric acid and acetic acid are both acids according to the Arrhenius definition, the solutions created by dissolving the same numbers of HCl and $HC_2H_3O_2$ molecules in water have very different acid properties. You wouldn't hesitate to put a solution of the weak acid $HC_2H_3O_2$ (vinegar) on your salad, but putting a solution of the strong acid HCl on your salad would have a very different effect on the lettuce. With hydrochloric acid, you will get a brown, fuming mess rather than a crisp, green salad. *Strong acids* form nearly one $H_3O^+$ ion in solution for each acid molecule dissolved in water, whereas, *weak acids* yield significantly less than one $H_3O^+$ ion in solution for each acid molecule dissolved in water.

When an acetic acid molecule, $HC_2H_3O_2$, collides with an $H_2O$ molecule, an $H^+$ can be transferred to the water to form a hydronium ion, $H_3O^+$, and an acetate ion, $C_2H_3O_2^-$. The acetate ion, however, is less stable in solution than the chloride ion formed when the strong acid HCl dissolves in water. Because of this instability, the $C_2H_3O_2^-$ reacts with the hydronium ion, pulling the $H^+$ ion back to reform $HC_2H_3O_2$ and $H_2O$. A reaction in which the reactants are constantly forming products and, at the same time, the products are re-forming the reactants is called a **reversible reaction.** The chemical equations for reactions that are significantly reversible are written with double arrows as illustrated in Figure 5.4.

OBJECTIVE 6

**Figure 5.4**
**Reversible Reaction**

OBJECTIVE 6

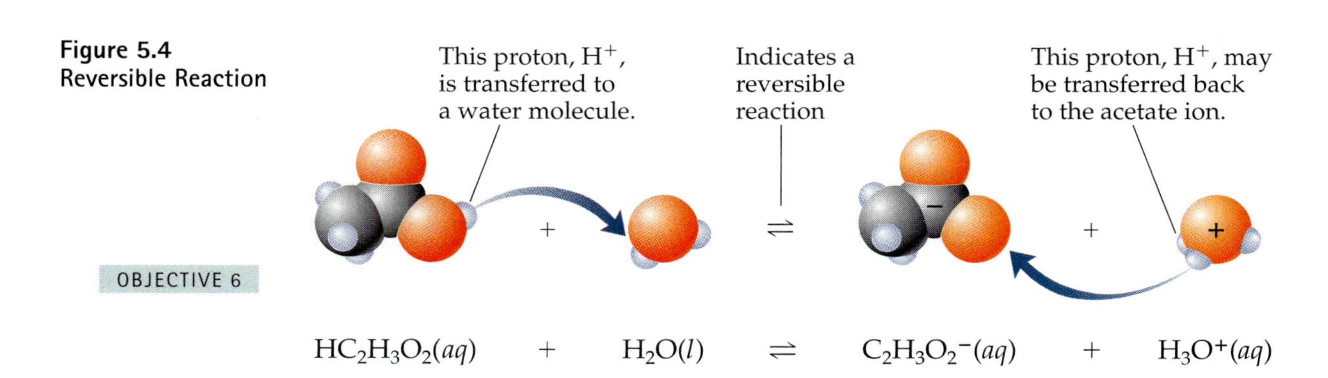
This proton, H+, is transferred to a water molecule.
Indicates a reversible reaction
This proton, H+, may be transferred back to the acetate ion.
+
⇌
+
HC2H3O2(aq) + H2O(l) ⇌ C2H3O2−(aq) + H3O+(aq)

If you were small enough to be riding on one of the carbon atoms in $HC_2H_3O_2$ or $C_2H_3O_2^-$, you would find that your atom was usually in the $HC_2H_3O_2$ form but was often in the $C_2H_3O_2^-$ form and was continually changing back and forth. The forward and reverse reactions would be taking place simultaneously all around you. When acetic acid is added to water, the relative amounts of the different products and reactants soon reach levels at which the opposing reactions proceed at equal rates. This means that the forward reaction is producing $C_2H_3O_2^-$ as quickly as the reverse reaction is producing $HC_2H_3O_2(aq)$. At this point, there is no more net change in the amounts of $HC_2H_3O_2$, $H_2O$, $C_2H_3O_2^-$, or $H_3O^+$ in the solution. For example, for each 1000 molecules of acetic acid added to water, the solution will eventually contain about 996 acetic acid molecules ($HC_2H_3O_2$), 4 hydronium ions ($H_3O^+$), and 4 acetate ions ($C_2H_3O_2^-$). Acetic acid is therefore a **weak acid**—a substance that is incompletely ionized in water because of the reversibility of its reaction with water that forms hydronium ion, $H_3O^+$. Figure 5.5 shows a simple model that will help you picture this solution.

OBJECTIVE 6

The products formed from the reaction of a strong acid and water do not recombine at a significant rate to re-form the uncharged acid molecules and water. For example, when HCl molecules react with water, the $H_3O^+$ and $Cl^-$ ions that form do not react to a significant degree to re-form HCl and $H_2O$. (Look again at Figure 5.1 to see the behavior of a strong acid in solution.) Reactions like this that are not significantly reversible are often called **completion reactions.** The chemical equations for completion reac-

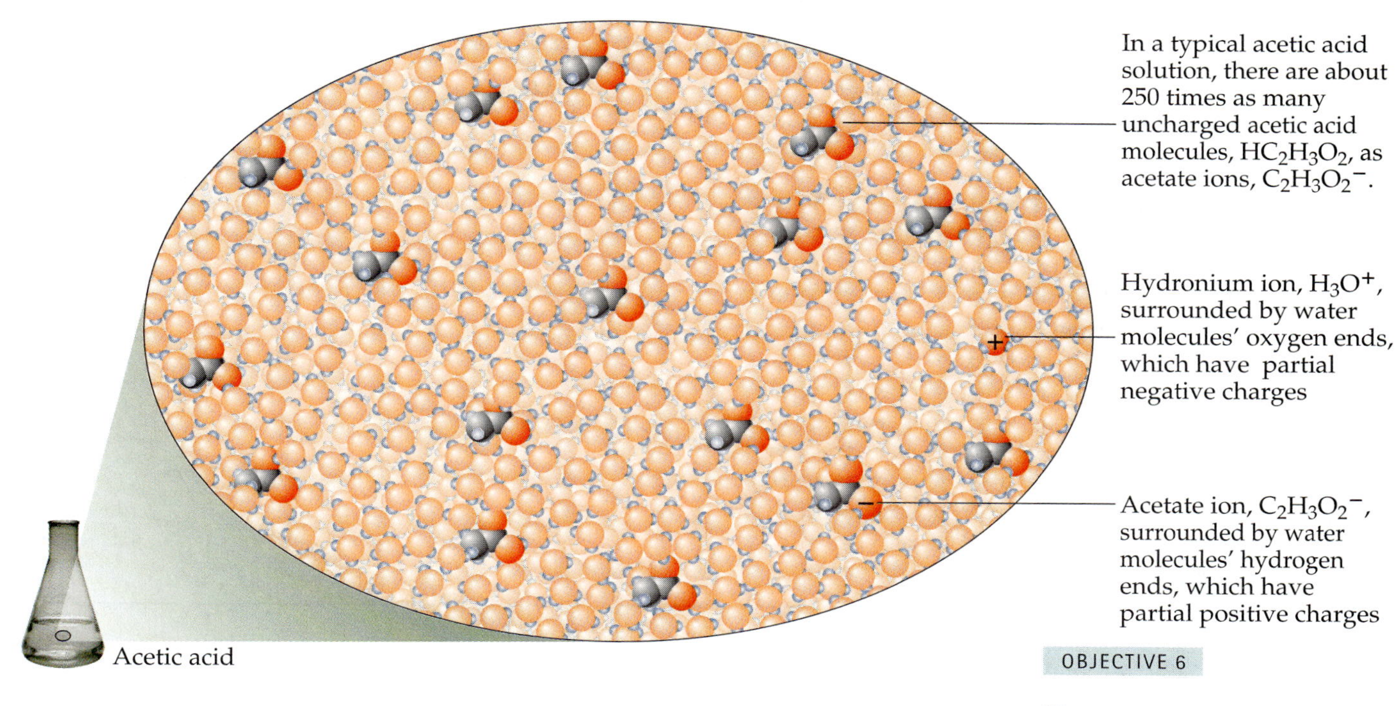

In a typical acetic acid solution, there are about 250 times as many uncharged acetic acid molecules, $HC_2H_3O_2$, as acetate ions, $C_2H_3O_2^-$.
Hydronium ion, $H_3O^+$, surrounded by water molecules' oxygen ends, which have partial negative charges
Acetate ion, $C_2H_3O_2^-$, surrounded by water molecules' hydrogen ends, which have partial positive charges
Acetic acid


OBJECTIVE 6

**Figure 5.5**
Acetic Acid in Water

tions are written with single arrows to indicate that the reaction proceeds to form almost 100% products.

Indicates a completion reaction

$$HCl(g) + H_2O(l) \rightarrow Cl^-(aq) + H_3O^+(aq)$$

OBJECTIVE 7

OBJECTIVE 8

Therefore, a **strong acid** is a substance that undergoes a completion reaction with water such that each acid particle reacts to form a hydronium ion, $H_3O^+$. The strong monoprotic acids that you will be expected to recognize are nitric acid, $HNO_3$, and hydrochloric acid, $HCl(aq)$. (There are others that you may be expected to recognize later in your chemical education.) If we were to examine equal volumes of two aqueous solutions, one made with a certain number of molecules of a strong acid and one made with the same number of molecules of a weak acid, we would find fewer hydronium ions in the solution of weak acid than in the solution of strong acid (Figure 5.6).

OBJECTIVE 9

OBJECTIVE 10

Sulfuric acid, $H_2SO_4$, is a strong diprotic acid. When it is added to water, each $H_2SO_4$ molecule loses its first hydrogen ion completely. This is why $H_2SO_4$ is classified as a strong acid. Note the single arrow to indicate a completion reaction.

$$H_2SO_4(aq) + H_2O(l) \rightarrow H_3O^+(aq) + HSO_4^-(aq)$$

The hydrogen sulfate ion, $HSO_4^-$, which is a product of this reaction, is a weak acid. It reacts with water in a reversible reaction to form a hydronium

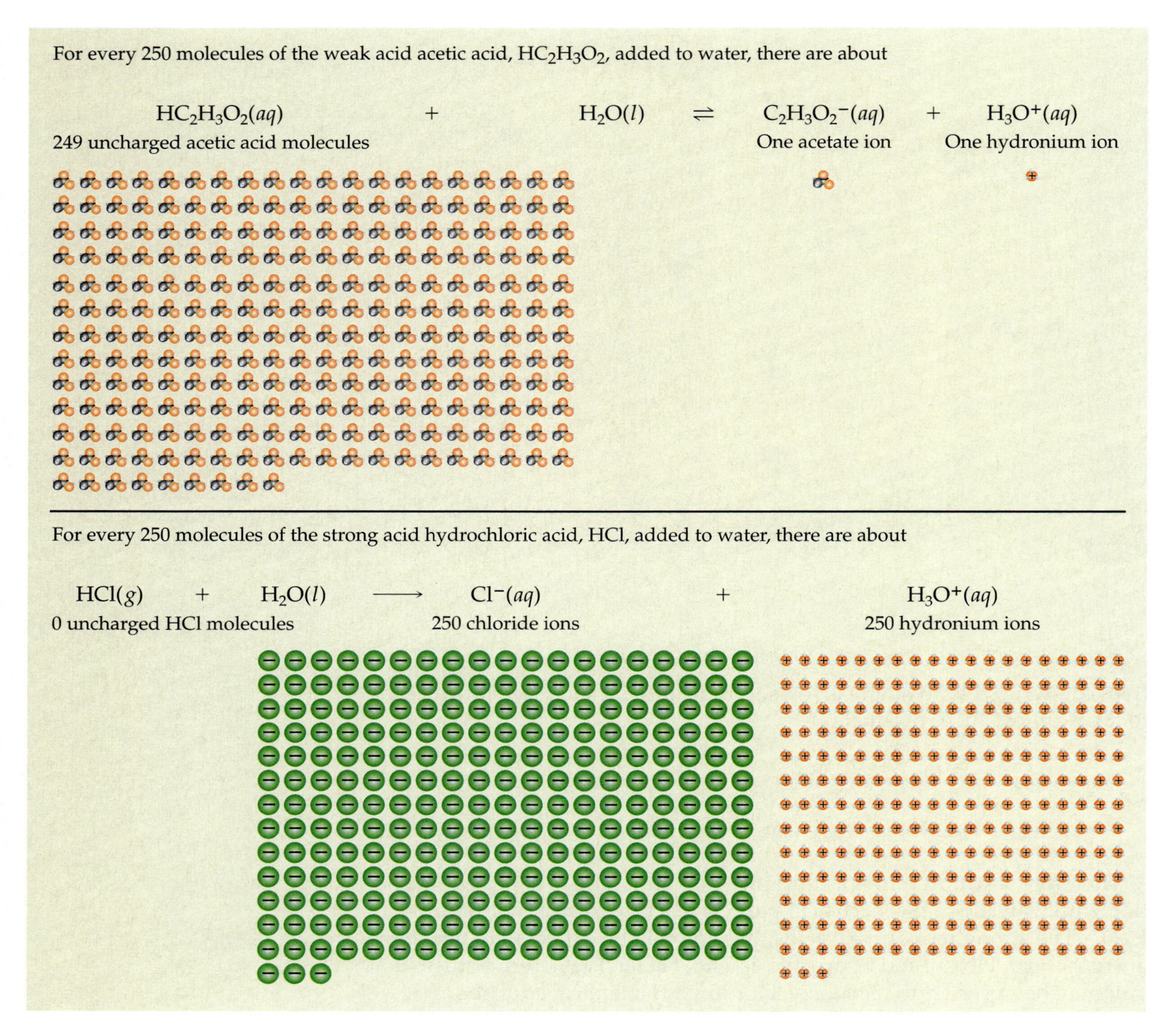
For every 250 molecules of the weak acid acetic acid, $HC_2H_3O_2$, added to water, there are about
$HC_2H_3O_2(aq)$ + $H_2O(l)$ ⇌ $C_2H_3O_2^-(aq)$ + $H_3O^+(aq)$
249 uncharged acetic acid molecules
One acetate ion
One hydronium ion
For every 250 molecules of the strong acid hydrochloric acid, HCl, added to water, there are about
$HCl(g)$ + $H_2O(l)$ ⟶ $Cl^-(aq)$ + $H_3O^+(aq)$
0 uncharged HCl molecules
250 chloride ions
250 hydronium ions

**Figure 5.6**
**Weak and Strong Acids**

ion and a sulfate ion. Note the double arrow to indicate a reversible reaction.

$$HSO_4^-(aq) + H_2O(l) \rightleftharpoons H_3O^+(aq) + SO_4^{2-}(aq)$$

For each 100 sulfuric acid molecules added to water, the solution will eventually contain about 101 hydronium ions ($H_3O^+$), 99 hydrogen sulfate ions ($HSO_4^-$), and 1 sulfate ion ($SO_4^{2-}$).

Sulfuric acid, $H_2SO_4$, is produced by the United States chemical industry in greater mass than any other chemical. Over 40 billion kilograms of $H_2SO_4$ are

## SPECIAL TOPIC 5.1 Acid Rain

Normal rainwater is very slightly acidic as a result of several reactions between substances dissolved in the water and the water itself. For example, carbon dioxide, nitrogen dioxide, and sulfur trioxide—all of which are natural constituents of air—react with water to form carbonic acid, nitric acid, and sulfuric acid.

Nitrogen dioxide is produced in nature in many ways, including a reaction between the oxygen and nitrogen in the air during electrical storms.

$$N_2(g) + O_2(g) \rightarrow 2NO(g)$$

$$2NO(g) + O_2(g) \rightarrow 2NO_2(g)$$

Sulfur dioxide also has natural sources, including the burning of sulfur-containing compounds in volcanic eruptions and forest fires. Sulfur dioxide is converted into sulfur trioxide, $SO_3$, by reaction with the nitrogen dioxide in the air, among other mechanisms.

$$SO_2(g) + NO_2(g) \rightarrow SO_3(g) + NO(g)$$

We humans have added considerably to the levels of $NO_2(g)$ and $SO_2(g)$ in our air, causing a steady increase in the acidity of rain. Coal, for example, contains a significant amount of sulfur; when coal is burned, the sulfur is converted into sulfur dioxide, $SO_2(g)$. The sulfur dioxide is converted into sulfur trioxide, $SO_3(g)$, in the air, and that compound dissolves in rainwater and becomes sulfuric acid, $H_2SO_4(aq)$. As individuals, we also contribute to acid rain every time we drive a car around the block. When air, which contains nitrogen and oxygen, is heated in the cylinders of the car, the two gases combine to yield nitrogen monoxide, $NO(g)$, which is then converted into nitrogen dioxide, $NO_2(g)$, in the air. The $NO_2$ reacts with water in rain to form nitric acid, $HNO_3(aq)$. There are about 25 times more $H_3O^+$ ions in the rain falling in the northeastern United States than would be expected without human contributions.

The increased acidity of the rain leads to many problems. For example, the acids in acid rain react with the calcium carbonate in marble statues and buildings, causing them to dissolve. (Marble is compressed limestone, which is composed of solid calcium carbonate, $CaCO_3$.)

$$CaCO_3(s) + 2HNO_3(aq) \rightarrow Ca(NO_3)_2(aq) + CO_2(g) + H_2O(l)$$

A similar reaction allows a plumber to remove the calcium carbonate scale in your hot water pipes. If the pipes are washed in an acidic solution, the calcium carbonate dissolves.

The Renaissance statue on the left was transported by William Randolph Hearst to his home in San Simeon, California. Because it so rarely rains there, and because San Simeon is far from any major sources of pollution, these statues are in much better condition than the similar statues found elsewhere, such as the one on the right, that have been damaged by acid rain.

produced each year to make phosphate fertilizers, plastics, and many other substances. Sulfuric acid is also used in ore processing, petroleum refining, pulp and paper making and for a variety of other purposes. Most cars are started by lead–acid storage batteries, which contain about 33.5% $H_2SO_4$.

OBJECTIVE 11

To do the Chapter Problems at the end of this chapter, you will need to identify important acids as being either strong or weak. The strong acids that you will be expected to recognize are hydrochloric acid, HCl(*aq*), nitric acid, $HNO_3$, and sulfuric acid, $H_2SO_4$. An acid is considered weak if it is not on the list of strong acids. Table 5.1 summarizes this information.

OBJECTIVE 11

Table 5.1
Arrhenius Acids

| | **Strong** | **Weak** |
|---|---|---|
| Binary acids | hydrochloric acid, HCl(*aq*) | hydrofluoric acid, HF(*aq*) |
| Oxyacids | nitric acid, $HNO_3$, sulfuric acid, $H_2SO_4$ | other acids with the general formula $H_aX_bO_c$ |
| Organic acids | none | acetic acid, $HC_2H_3O_2$, and others you will see in Section 17.1 |

Special Topic 5.1, on the previous page, tells how acids are formed in the earth's atmosphere and how these acids can be damaging to our environment.

## 5.2 Acid Nomenclature

Before exploring how different kinds of acids react with compounds other than water, you need a little more familiarity with their names and formulas. Remember that the names of Arrhenius acids usually end in *acid* (hydrochloric acid, sulfuric acid, nitric acid ) and that their formulas fit one of two general patterns:

HX(*aq*) X = F, Cl, Br, or I

$H_aX_bO_c$

For example, HCl(*aq*) (hydrochloric acid), $H_2SO_4$ (sulfuric acid), and $HNO_3$ (nitric acid) represent acids.

### Names and Formulas of Binary Acids

Binary acids are named by writing *hydro* followed by the root of the name of the halogen, then *-ic*, and finally *acid* (Table 5.2):

OBJECTIVE 12

hydro(root)ic acid

The only exception to remember is that the *o* in *hydro* is left off for HI(*aq*), so its name is hydriodic acid (an acid used to make pharmaceuticals).

Most chemists refer to pure HCl gas as hydrogen chloride, but when HCl gas is dissolved in water, HCl(*aq*), the solution is called hydrochloric acid. We will follow the same rule in this text, calling HCl or HCl(*g*) hydrogen chloride and calling HCl(*aq*) hydrochloric acid. The same pattern holds for the other binary acids as well.

You will be expected to be able to write formulas and names for the binary acids listed in Table 5.2. Remember that it is a good habit to write (*aq*) after the formula.

Table 5.2
Names and Formulas for Compounds with Hydrogen and a Halogen

OBJECTIVE 12

| Formula | Named as binary covalent compound | Acid formula | Named as binary acid |
|---|---|---|---|
| HF or HF(*g*) | hydrogen monofluoride or hydrogen fluoride | HF(*aq*) | hydrofluoric acid |
| HCl or HCl(*g*) | hydrogen monochloride or hydrogen chloride | HCl(*aq*) | hydrochloric acid |
| HBr or HBr(*g*) | hydrogen monobromide or hydrogen bromide | HBr(*aq*) | hydrobromic acid |
| HI or HI(*g*) | hydrogen moniodide or hydrogen iodide | HI(*aq*) | hydriodic acid |

## Names and Formulas of Oxyacids

OBJECTIVE 12

To name oxyacids, you must first be able to recognize them by the general formula $H_aX_bO_c$, with X representing an element other than hydrogen or oxygen (Section 5.1). It will also be useful for you to know the names of the polyatomic oxyanions (Table 3.7), because many oxyacid names can be derived from them. If enough $H^+$ ions are added to a (root)ate polyatomic ion to neutralize its charge completely, the (root)ic acid is formed (Table 5.3).

- If one $H^+$ ion is added to nitrate, $NO_3^-$, then nitric acid, $HNO_3$, is formed.
- If two $H^+$ ions are added to sulfate, $SO_4^{2-}$, then sulfuric acid, $H_2SO_4$, is formed.
- If three $H^+$ ions are added to phosphate, $PO_4^{3-}$, then phosphoric acid, $H_3PO_4$, is formed.

Note that the whole name *sulfur*, not just the root, *sulf-*, is found in the name *sulfuric acid*. Similarly, although the usual root for phosphorus is *phosph-*, the root *phosphor-* is used for phosphorus-containing oxyacids, as in the name *phosphoric acid*.

OBJECTIVE 12

Table 5.3
Relationship Between (Root)ate Polyatomic Ions and (Root)ic Acids

| Oxyanion formula | Oxyanion name | Oxyacid formula | Oxyacid name |
|---|---|---|---|
| $NO_3^-$ | nitrate | $HNO_3$ | nitric acid |
| $C_2H_3O_2^-$ | acetate | $HC_2H_3O_2$ | acetic acid |
| $SO_4^{2-}$ | sulfate | $H_2SO_4$ | sulfuric acid (Note that the whole name *sulfur* is used in the oxyacid name.) |
| $CO_3^{2-}$ | carbonate | $H_2CO_3$ | carbonic acid |
| $PO_4^{3-}$ | phosphate | $H_3PO_4$ | phosphoric acid (Note that the root of phosphorus in an oxyacid name is *phosphor-*.) |

There is a more complete description of acid nomenclature at the following Web address:
www.chemplace.com/college/

## Example 5.1 Formulas for Acids

OBJECTIVE 12

Write the chemical formulas that correspond to the following names: (a) hydrobromic acid and (b) sulfuric acid.

***Solution***

a. The name *hydrobromic acid* has the form of a binary acid, hydro(root)ic acid. Binary acids have the formula HX(*aq*), so hydrobromic acid is **HBr(*aq*).** We follow the formula with (*aq*) to distinguish hydrobromic acid from a pure sample of hydrogen bromide, HBr.

b. Sulfuric acid is **$H_2SO_4$.** Sulfuric acid is a very common acid, one whose formula, $H_2SO_4$, you ought to memorize. We recognize *sulfuric acid* as a name for an oxyacid, because it has the form (root)ic acid. You can also figure out its formula from the formula for sulfate, $SO_4^{2-}$, by adding enough $H^+$ ions to neutralize the charge. Among the many uses of $H_2SO_4$ are the manufacture of explosives and the reprocessing of spent nuclear fuel.

## Example 5.2 Naming Acids

OBJECTIVE 12

Write the names that correspond to the chemical formulas (a) $HNO_3$ and (b) HF(*aq*).

***Solution***

a. The first step in writing a name from a chemical formula is to decide which type of compound the formula represents. This formula represents an oxyacid. Remember that the (root)ate polyatomic ion leads to the (root)ic acid. The name for $NO_3^-$ is nitrate, so $HNO_3$ is **nitric acid.**

b. The first step in writing a name from a chemical formula is to determine what type of compound the formula represents. This one, HF(*aq*), has the form of a binary acid, HX(*aq*), so its name is *hydro-* followed by the root of the name of the halogen, then by *-ic* and *acid*: **hydrofluoric acid.** This acid is used to make the chlorofluorocarbons, CFCs.

### Exercise 5.1 Formulas for Acids

OBJECTIVE 12

Write the chemical formulas that correspond to the names (a) hydrofluoric acid and (b) phosphoric acid.

### Exercise 5.2 Naming Acids

OBJECTIVE 12

Write the names that correspond to the chemical formulas (a) HI(*aq*) and (b) $HC_2H_3O_2$.

## 5.3 Summary of Chemical Nomenclature

Perhaps at this point you are feeling confused by the many different conventions for naming different kinds of chemical compounds. Here is an overview of the guidelines for naming and writing formulas for all the types of compounds described in this chapter and in Chapter 3.

Some names and formulas for compounds can be constructed from general rules, but others must be memorized. Table 5.4 lists some commonly encountered names and formulas that must be memorized. Check with your instructor to see which of these you need to know. Your instructor may also want to add others to the list.

OBJECTIVE 14

Table 5.4
Compound Names and Formulas

| Name | Formula | Name | Formula |
|---|---|---|---|
| water | $H_2O$ | ammonia | $NH_3$ |
| methane | $CH_4$ | methanol (methyl alcohol) | $CH_3OH$ |
| ethane | $C_2H_6$ | ethanol (ethyl alcohol) | $C_2H_5OH$ |
| propane | $C_3H_8$ | 2-propanol (isopropyl alcohol) | $C_3H_7OH$ |

OBJECTIVE 14

The general procedure for naming other compounds consists of two steps:

**Step 1** Decide what type of compound the name or formula represents.

**Step 2** Apply the rules for writing the name or formula for that type of compound.

Table 5.5 summarizes the distinguishing features of different kinds of formulas and names (step 1) and lists the sections in this chapter and in Chapter 3 where you can find instructions for converting names to formulas and formulas to names (step 2).

**Table 5.5**
Nomenclature for Some Types of Compounds

| Type of compound | General formula | Examples | General name | Examples |
|---|---|---|---|---|
| binary covalent Section 3.4 | $A_aB_b$ | $N_2O_5$ or $CO_2$ | (prefix unless *mono*) (name of first element in formula) (prefix) (root of second element)ide | dinitrogen pentoxide or carbon dioxide |
| binary ionic Section 3.5 | $M_aA_b$ | NaCl or $FeCl_3$ | (name of metal) (root of nonmetal)ide or (name of metal)(Roman num.) (root of nonmetal)ide | sodium chloride or iron(III) chloride |
| ionic with polyatomic ion(s) Section 3.5 | $M_aY_b$ or $(NH_4)_aY_b$ Y = formula of polyatomic ion | $Li_2HPO_4$ or $CuSO_4$ or $NH_4Cl$ or $(NH_4)_2SO_4$ | (name of metal) (name of polyatomic ion) or (name of metal)(Roman num.) (name of polyatomic ion) or ammonium (root of nonmetal)ide or ammonium (name of polyatomic ion) | lithium hydrogen phosphate or copper(II) sulfate or ammonium chloride or ammonium sulfate |
| binary acid Section 5.2 | $HX(aq)$ | $HCl(aq)$ | hydro(root)ic acid | hydrochloric acid |
| oxyacid Section 5.2 | $H_aX_bO_c$ | $HNO_3$ or $H_2SO_4$ or $H_3PO_4$ | (root)ic acid | nitric acid or sulfuric acid or phosphoric acid |

M = symbol of metal
X = some element other than H or O
A and B = symbols of nonmetals
The letters a, b, and c represent numerical subscripts.

OBJECTIVE 13 OBJECTIVE 14

## EXERCISE 5.3 Formulas to Names

OBJECTIVE 14

Write the names that correspond to the following chemical formulas.

a. $AlF_3$
b. $PF_3$
c. $H_3PO_4$
d. $CaSO_4$
e. $Ca(HSO_4)_2$
f. $CuCl_2$
g. $NH_4F$
h. $HCl(aq)$
i. $(NH_4)_3PO_4$

## EXERCISE 5.4 Names to Formulas

OBJECTIVE 14

Write the chemical formulas that correspond to the following names.

a. ammonium nitrate
b. acetic acid
c. sodium hydrogen sulfate
d. potassium bromide
e. magnesium hydrogen phosphate
f. hydrofluoric acid
g. diphosphorus tetroxide
h. aluminum carbonate
i. sulfuric acid

## 5.4 Strong and Weak Arrhenius Bases

Each year, the U.S. chemical industry produces over 10 billion kilograms of the base sodium hydroxide, NaOH, which is then used for many purposes, including water treatment, vegetable oil refining, the peeling of fruits and vegetables in the food industry, and the manufacture of numerous other chemical products, including soaps and detergents. Likewise, over 15 billion kilograms of the base ammonia, $NH_3$, is produced each year. Although a water solution of ammonia is a common household cleaner, most of the $NH_3$ produced in the United States is used to make fertilizers and explosives. As you read this section, you will learn about the chemical properties of basic compounds that make them so useful to chemists and others.

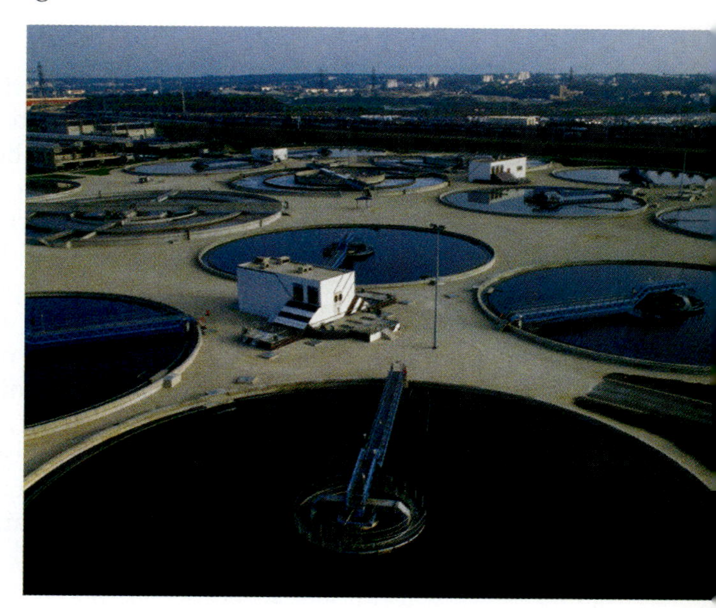

This water treatment plant uses the base sodium hydroxide, NaOH, to remove impurities from the water.

According to the modern version of the Arrhenius theory of acids and bases, a base is a substance that produces **hydroxide ions,** $OH^-$, when it is added to water. A solution that has a significant concentration of hydroxide ions is called a **basic solution.** Sodium hydroxide, NaOH, is the most common laboratory base. It is designated a **strong base** because for every NaOH unit dissolved, one hydroxide ion is formed in solution.

$$NaOH(aq) \rightarrow Na^+(aq) + OH^-(aq)$$

Compounds that contain hydroxide ions are often called **hydroxides**. All water-soluble hydroxides are strong bases. Other examples include lithium hydroxide, LiOH, which is used in storage batteries and as a carbon dioxide absorbent in space vehicles, and potassium hydroxide, KOH, which is used to make some soaps, liquid fertilizers, and paint removers.

OBJECTIVE 16

When ammonia, $NH_3$, dissolves in water, some hydrogen ions, $H^+$, are transferred from water molecules to ammonia molecules, $NH_3$, producing ammonium ions, $NH_4^+$, and hydroxide ions, $OH^-$. The reaction is reversible, so when an ammonium ion and a hydroxide ion meet in solution, the $H^+$ ion can be passed back to the $OH^-$ to re-form an $NH_3$ molecule and a water molecule (Figure 5.7).

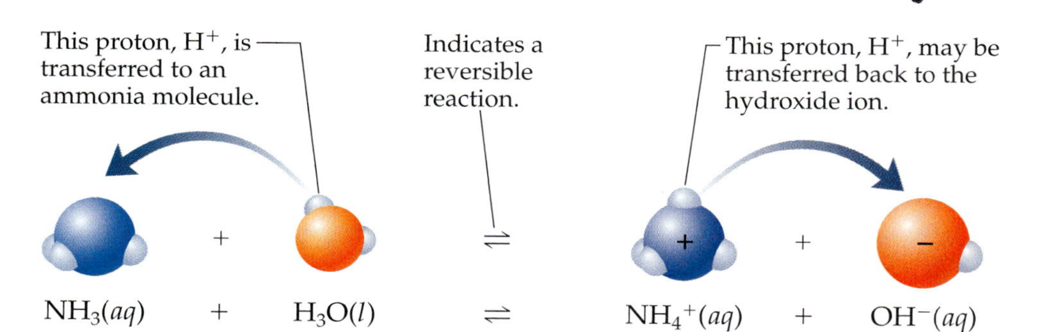

This proton, H+, is transferred to an ammonia molecule.
Indicates a reversible reaction.
This proton, H+, may be transferred back to the hydroxide ion.
+
⇌
+
NH3(aq) + H3O(l) ⇌ NH4+(aq) + OH−(aq)

**Figure 5.7**
The Reversible Reaction of Ammonia and Water

OBJECTIVE 16

Ammonia is an **Arrhenius base** because it produces $OH^-$ ions when it is added to water. Because the reaction is reversible, however, only some ammonia molecules have acquired protons (creating $OH^-$) at any given time, so an ammonia solution contains fewer hydroxide ions than would be found in a solution made using an equivalent amount of a strong base. Therefore, we classify ammonia as a **weak base,** which is a base that produces fewer hydroxide ions in water solution than there are particles of base dissolved.

OBJECTIVE 16

To visualize the reaction between ammonia and water at the molecular level, imagine that you are taking a ride on a nitrogen atom. Your nitrogen would usually be bonded with three hydrogen atoms in an $NH_3$ molecule, but occasionally it would gain an extra $H^+$ ion from a water molecule to form $NH_4^+$ for a short time. When your $NH_4^+$ ion collides with an $OH^-$ ion, an $H^+$ ion is transferred to the $OH^-$ ion to form $H_2O$ and $NH_3$. Ammonia molecules are constantly gaining and losing $H^+$ ions, but soon after the initial addition of ammonia to water, both changes proceed at an equal rate. At this point, there will be no more net change in the amounts of ammonia, water, hydroxide, and ammonium ion in the solution. When a typical solution of ammonia stops changing, it is likely to contain about 200 $NH_3$ molecules for each $NH_4^+$ ion. As you study the ammonia solution depicted in Figure 5.8, try to picture about 200 times as many $NH_3$ molecules as $NH_4^+$ or $OH^-$ ions.

There are many other weak Arrhenius bases, but the only ones you will be expected to recognize are ionic compounds that contain carbonate (for example, sodium carbonate, $Na_2CO_3$) and hydrogen carbonate (for example, sodium hydrogen carbonate, $NaHCO_3$). When sodium carbonate, which is used to make glass, soaps, and detergents, dissolves in water,

**Figure 5.8**
Ammonia in Water

OBJECTIVE 16

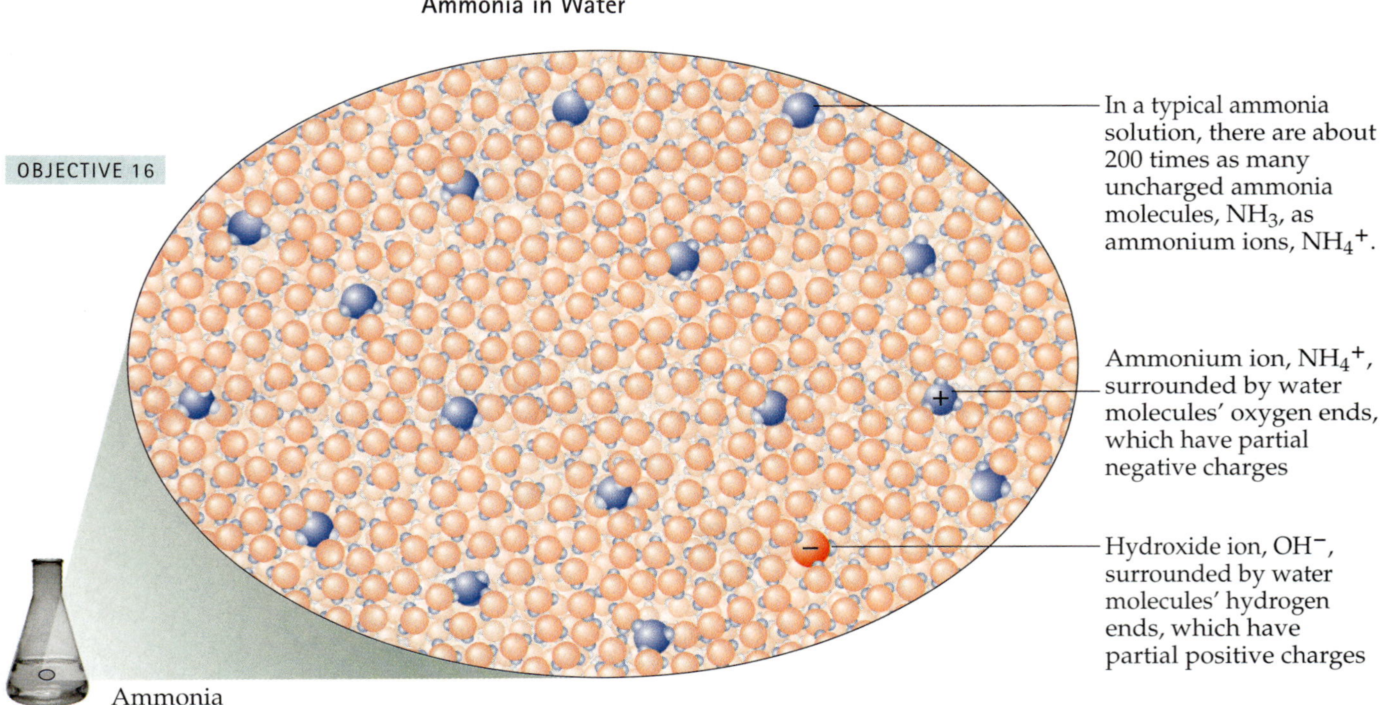
In a typical ammonia solution, there are about 200 times as many uncharged ammonia molecules, $NH_3$, as ammonium ions, $NH_4^+$.
Ammonium ion, $NH_4^+$, surrounded by water molecules' oxygen ends, which have partial negative charges
Hydroxide ion, $OH^-$, surrounded by water molecules' hydrogen ends, which have partial positive charges
Ammonia

the carbonate ions, $CO_3^{2-}$, react with water in a reversible way to yield hydroxide ions.

OBJECTIVE 17

$$Na_2CO_3(s) \rightarrow 2Na^+(aq) + CO_3^{2-}(aq)$$

$$CO_3^{2-}(aq) + H_2O(l) \rightleftharpoons HCO_3^-(aq) + OH^-(aq)$$

In a similar reaction, the hydrogen carbonate ions, $HCO_3^-$, formed when $NaHCO_3$ dissolves in water, react to yield hydroxide ions.

$$NaHCO_3(s) \rightarrow Na^+(aq) + HCO_3^-(aq)$$

$$HCO_3^-(aq) + H_2O(l) \rightleftharpoons H_2CO_3(aq) + OH^-(aq)$$

Sodium hydrogen carbonate is found in fire extinguishers, baking powders, antacids, and mouthwashes. Table 5.6 summarizes how you can recognize substances as bases and how you can classify them as strong or weak bases. (There are other Arrhenius bases that you may learn about later.)

These products all contain the weak base sodium hydrogen carbonate.

**Table 5.6**
Arrhenius Bases

| | Strong | Weak |
|---|---|---|
| Ionic compounds | metal hydroxides, such as NaOH | ionic compounds with $CO_3^{2-}$ and $HCO_3^-$, such as $Na_2CO_3$ and $NaHCO_3$ |
| Certain uncharged molecules | none | $NH_3$ |

The following sample study sheet summarizes the ways you can recognize strong and weak acids and bases.

## Sample Study Sheet 5.1
### Identification of Strong and Weak Acids and Bases

OBJECTIVE 18

**Tip-off** You are asked to identify a substance as (1) an Arrhenius strong acid, (2) an Arrhenius weak acid, (3) an Arrhenius strong base, or (4) an Arrhenius weak base.

**General Steps**

**Step 1** Identify the substance as an Arrhenius acid or base by using the following criteria.

- The names of the acids end in *acid*. Acid formulas have one of these forms: $HX(aq)$ or $H_aX_bO_c$.
- Ionic compounds that contain hydroxide, carbonate, or hydrogen carbonate anions are basic. Ammonia, $NH_3$, is also a base.

**Step 2** If the substance is an acid or a base, determine whether it is strong or weak.

- We will consider all acids except $HCl(aq)$, $HNO_3$, and $H_2SO_4$ to be weak.
- We will consider all bases except metal hydroxides to be weak.

**Example** See Example 5.3.

## SPECIAL TOPIC 5.2 Chemicals and Your Sense of Taste

*[T]hat formed of bodies round and smooth are things which touch the senses sweetly, while those which harsh and bitter do appear, are held together bound with particles more hooked, and for this cause are wont to tear their way into our senses, and on entering in to rend the body.*

LUCRETIUS
ROMAN PHILOSOPHER AND POET, ABOUT 2000 YEARS AGO

Lucretius was mistaken in certain details, but it is true that the shape of molecules is important in determining whether compounds taste sweet or bitter. Your tongue has about 3000 taste buds, each of which is an onion-shaped collection of 50 to 150 taste cells. Each taste bud is specialized for tasting sweet, sour, salt, or bitter. It has been suggested that the tongue can also perceive another taste, umami, which is a subtle taste most commonly associated with monosodium glutamate, MSG. At the tips of the bitter and sweet taste cells are receptor molecules shaped to fit parts of certain molecules in our food.

When chocolate, for example, is roasted, caffeine and other compounds are formed that stimulate the bitter taste cells. The molecules of these compounds have a shape that allows them to attach to the taste cell receptors and cause an adjacent nerve cell to fire. This event sends the bitter signal to the brain.

Sugar is added to chocolate to counteract the bitter taste. The arrangement of atoms in sugar molecules allows them to fit into the receptor sites of sweet taste cells. When a sugar molecule such as glucose or sucrose attaches to a receptor of a sweet taste cell, the sweet signal is sent to the brain.

The salt taste is thought to be perceived through different mechanisms than the sweet and bitter tastes. It is the presence of sodium ions, $Na^+$, in the sodium chloride, NaCl, of table salt that causes the taste. The interior of a salt taste cell is negatively charged. When such a cell is bathed in saliva that contains dissolved sodium ions, the $Na^+$ ions enter the cell and make its interior less negative. This change triggers the release of chemicals called neurotransmitters into the space between the taste cells and nerve cells. The neurotransmitters cause the nerve cells to fire, sending the salt signal to the brain.

**Bases taste bitter.**

Acids cause the sour taste in foods. Vinegar contains acetic acid, sour milk contains lactic acid, and lemons contain citric acid. What these acids have in common is that they can lose $H^+$ ions in water solutions such as our saliva. Different animal species have different mechanisms for sending the sour signal. In amphibians, the $H^+$ ions block the normal release of potassium ions from sour taste cells, changing the cells' charge balance and causing them to release neurotransmitters. The neurotransmitters in turn tell the sour nerve cells to fire.

It has been suggested that there are good reasons for the evolution of our sense of taste. The four main tastes either lead us to food we need or warn us away from substances that might be harmful. We need sugar for energy and salt to replace the sodium and potassium ions lost in exercise. On the other hand, spoiled foods produce bitter-tasting substances, and numerous poisons are bitter. And many a bellyache from unripe fruit has been avoided by the warning signal provided by the sour taste.

### EXAMPLE 5.3 Identification of Acids and Bases

OBJECTIVE 18

Identify (a) $H_2SO_4$, (b) oxalic acid, (c) $NaHCO_3$, (d) potassium hydroxide, (e) $HCl(aq)$, and (f) ammonia as an Arrhenius strong acid, an Arrhenius weak acid, an Arrhenius strong base, or an Arrhenius weak base.

***Solution***

a. $H_2SO_4$ is an acid because it has the form of an oxyacid, $H_aX_bO_c$. It is on the list of **strong acid**s.

b. Oxalic acid is not on the list of strong acids—$HCl(aq)$, $HNO_3$, and $H_2SO_4$—so it is a **weak acid.**

c. Ionic compounds that contain hydrogen carbonate, such as $NaHCO_3$, are **weak base**s.

d. Ionic compounds that contain hydroxide, such as potassium hydroxide, are **strong base**s.

e. We know that hydrochloric acid, $HCl(aq)$, is an acid because its name ends in acid, and its formula has the form of a binary acid. It is found on the list of **strong acid**s.

f. Ammonia, $NH_3$, is our one example of an uncharged **weak base.**

### EXERCISE 5.5 Identification of Acids and Bases

OBJECTIVE 18

Identify each of the following as an Arrhenius strong acid, an Arrhenius weak acid, an Arrhenius strong base, or an Arrhenius weak base.

a. $HNO_3$

b. lithium hydroxide

c. $K_2CO_3$

d. hydrofluoric acid

Special Topic 5.2, on the previous page, describes the role that acids and bases play in the tastes of food and drink.

## 5.5 pH and Acidic and Basic Solutions

The scientific term *pH* has crept into our everyday language. Advertisements encourage us to choose products that are "pH-balanced," and environmentalists point to the lower pH of rain in certain parts of the country as a cause of ecological damage (Figure 5.9). The term was originated by chemists to describe the acidic and basic strengths of solutions.

We know that an Arrhenius acid donates $H^+$ ions to water to create $H_3O^+$ ions. The resulting solution is called an acidic solution. We also know that when you add a certain amount of a strong acid to one sample of water—say the water's volume is a liter—and add the same amount of a weak acid to another sample of water whose volume is also a liter, the strong acid generates more $H_3O^+$ ions in solution. Because the concentration of $H_3O^+$ ions in the strong acid solution is higher (there are more

A pH-balanced cleaner.

**Figure 5.9**
**Acid Rain**
The map on the right shows the pH of rain in different parts of the United States. The scale below shows the effect on fish of decreasing pH.

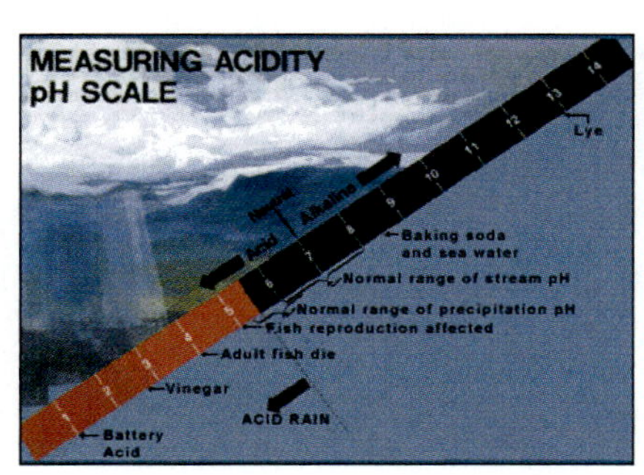


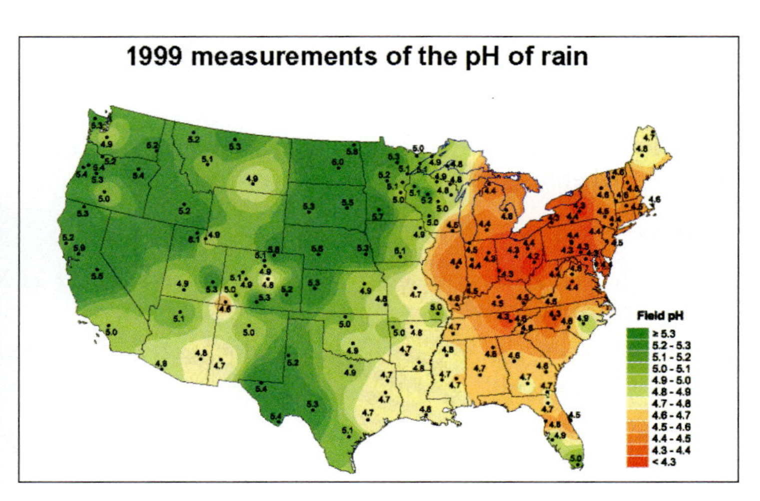


$H_3O^+$ ions per liter of solution), we say it is more acidic than the weak acid solution. A solution can also be made more acidic by the addition of more acid (while the amount of water remains the same). The pH scale can be used to describe the relative acidity of solutions.

OBJECTIVE 19

OBJECTIVE 20

If you take other chemistry courses, you will probably learn how pH is defined and how the pH values of solutions are determined. For now, all you need to remember is that acidic solutions have pH values less than 7 and that the more acidic a solution is, the lower its pH. A change of 1 pH unit reflects a 10-fold change in $H_3O^+$ ion concentration. For example, a solution with a pH of 5 has a concentration of $H_3O^+$ ions 10 times as great as does a solution with a pH of 6. The pH values of some common solutions are listed in Figure 5.10. Note that the gastric juice in our stomach has a pH of about 1.4, and that orange juice has a pH of about 2.8. Thus gastric juice is more than 10 times as concentrated in $H_3O^+$ ions as is orange juice.

The pH scale is also used to describe basic solutions, which are formed when an Arrhenius base is added to water, generating $OH^-$ ions. When you add a certain amount of a strong base to one sample of water—again, let's say a liter—and add the same amount of a weak base to another sample of water whose volume is the same, the strong base generates more $OH^-$ ions in solution. Because the concentration of $OH^-$ ions in the strong base solution is higher (there are more $OH^-$ ions per liter of solution), we say it is more basic than the weak base solution. A solution can also be made more basic by adding more base while holding the amount of water constant.

OBJECTIVE 19

OBJECTIVE 20

Basic solutions have pH values greater than 7, and the more basic the solution is, the higher its pH. A change of 1 pH unit reflects a 10-fold change in $OH^-$ ion concentration. For example, a solution with a pH of 12 has a concentration of $OH^-$ ions 10 times as great as does a solution with a pH of 11. The pH difference of about 4 between household ammonia solutions (pH about 11.9) and seawater (pH about 7.9) shows that household ammonia has about 10,000 ($10^4$) times the hydroxide ion concentration of seawater.

In nature, fresh water contains dissolved substances that make it slightly acidic, but pure water is neutral and has a pH of 7.

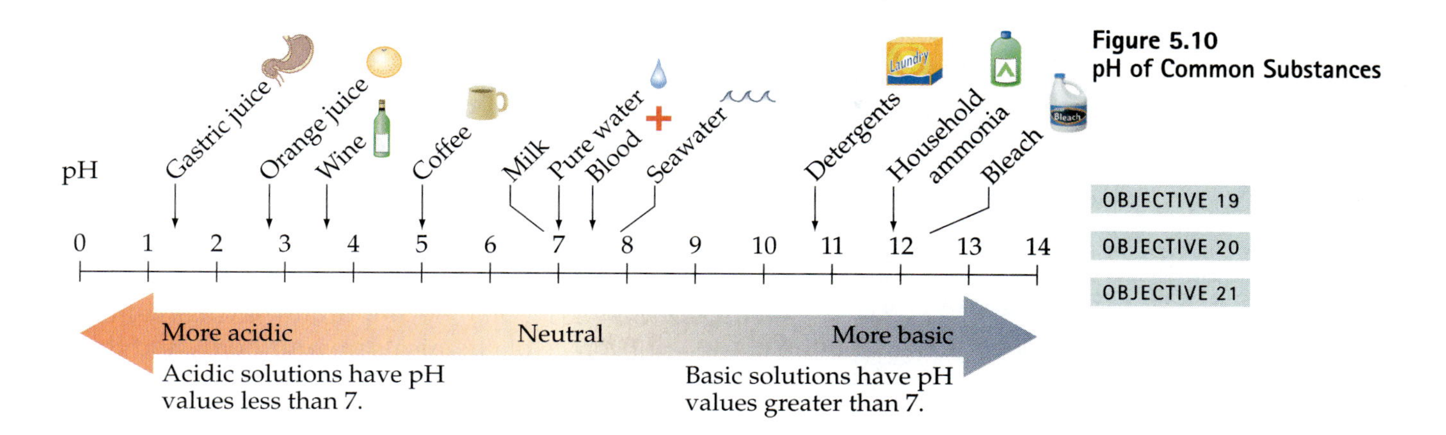


**Figure 5.10**
**pH of Common Substances**

OBJECTIVE 19

OBJECTIVE 20

OBJECTIVE 21

In the laboratory, we can detect acids and bases in solution in several ways. Perhaps the simplest test uses a substance called litmus, a natural dye derived from lichen, which turns red in acidic conditions and blue in basic conditions. Litmus paper is paper that has been coated with litmus. To test whether a liquid is acidic, we add a drop of the liquid to blue litmus paper, which is litmus paper that has been made slightly basic and therefore blue. If the paper turns red, then the liquid is acidic. To test whether a liquid is basic, we add a drop of the liquid to red litmus paper, which is litmus paper that has been made slightly acidic and therefore red. If the paper turns blue, then the liquid is basic (Figure 5.11).

OBJECTIVE 22

**Figure 5.11**
**Litmus**
Litmus, whose natural source is lichen (shown on the right), can be applied to the surface of paper that is then used to identify acidic and basic solutions. At left, acids turn blue litmus paper to red, and bases turn red litmus paper to blue.

OBJECTIVE 22

## 5.6 Arrhenius Acid–Base Reactions

When an Arrhenius acid is combined with an Arrhenius base, we say that they neutralize each other. By this we mean that the acid counteracts the properties of the base and that the base counteracts the properties of the acid. For example, a strong acid, such as nitric acid, must be handled with extreme caution, because if it gets on your skin, it can cause severe chemical burns. If you accidentally spilled nitric acid on a laboratory

This emergency crew is neutralizing an acid spill on the highway by covering it with a basic foam.

bench, however, you could quickly pour a solution of a weak base, such as sodium hydrogen carbonate, on top of the spill to neutralize the acid and make it safer to wipe. In a similar way, a solution of a weak acid, such as acetic acid, can be poured on a strong base spill to neutralize the base before cleanup. Therefore, reactions between Arrhenius acids and bases are often called **neutralization reactions.**

Neutralization reactions are important in maintaining the necessary balance of chemicals in your body, and they help preserve a similar balance in our oceans and lakes. Neutralization reactions are used in industry to make a wide range of products, including pharmaceuticals, food additives, and fertilizers. Let's look at some of the different forms of Arrhenius acid–base reactions, consider how they can be visualized, and learn how to describe them with chemical equations.

## Reactions of Aqueous Strong Arrhenius Acids and Aqueous Strong Arrhenius Bases

OBJECTIVE 23a

The reaction between the strong acid nitric acid and the strong base sodium hydroxide is our first example. Figure 5.12 shows the behavior of nitric acid in solution. Nitric acid is a strong acid, so virtually every $HNO_3$ molecule donates an $H^+$ ion to water to form a hydronium ion, $H_3O^+$, and a nitrate ion, $NO_3^-$. Because the reaction goes essentially to completion, you can picture the solution as containing $H_2O$, $NO_3^-$, and $H_3O^+$, with no $HNO_3$ remaining. The negatively charged oxygen ends of the water molecules surround the positive hydronium ions, and the positively charged hydrogen ends of the water molecules surround the nitrate ions.

OBJECTIVE 23a

Like a water solution of any ionic compound, a solution of sodium hydroxide, NaOH, consists of ions separated and surrounded by water molecules. At the instant that the solution of sodium hydroxide is added to the aqueous nitric acid, there are four different ions in solution surrounded by water molecules: $H_3O^+$, $NO_3^-$, $Na^+$, and $OH^-$ (Figure 5.13).

Neutralization reactions keep our bodies in chemical balance and also maintain the "health" of the world around us.

The ions in solution move in a random way, like any particles in a liquid, so they constantly collide with other ions. When two cations or two anions collide, they repel each other and move apart. When a hydronium ion and a nitrate ion collide, it is possible that the $H_3O^+$ ion will return an $H^+$ ion to the $NO_3^-$ ion, but nitrate ions are stable enough in water to make this unlikely. When a sodium ion collides with a hydroxide ion, they may stay together for a short time, but their attraction is too weak, and water molecules collide with them and push them apart. When hydronium ions and hydroxide ions collide, however, they react to form water (Figure 5.14), so more water molecules are shown in Figure 5.15 than in Figure 5.13.

The sodium and nitrate ions are unchanged in the reaction. They were separate and surrounded by water molecules at the beginning of the reaction, and they are still separate and surrounded by water molecules after the reaction. They were important in delivering the hydroxide and hydronium

**Figure 5.12**
**Aqueous Nitric Acid**
In a nitric acid solution, nitrate ions are separated from hydronium ions. These ions are constantly moving, colliding with each other and with water molecules.

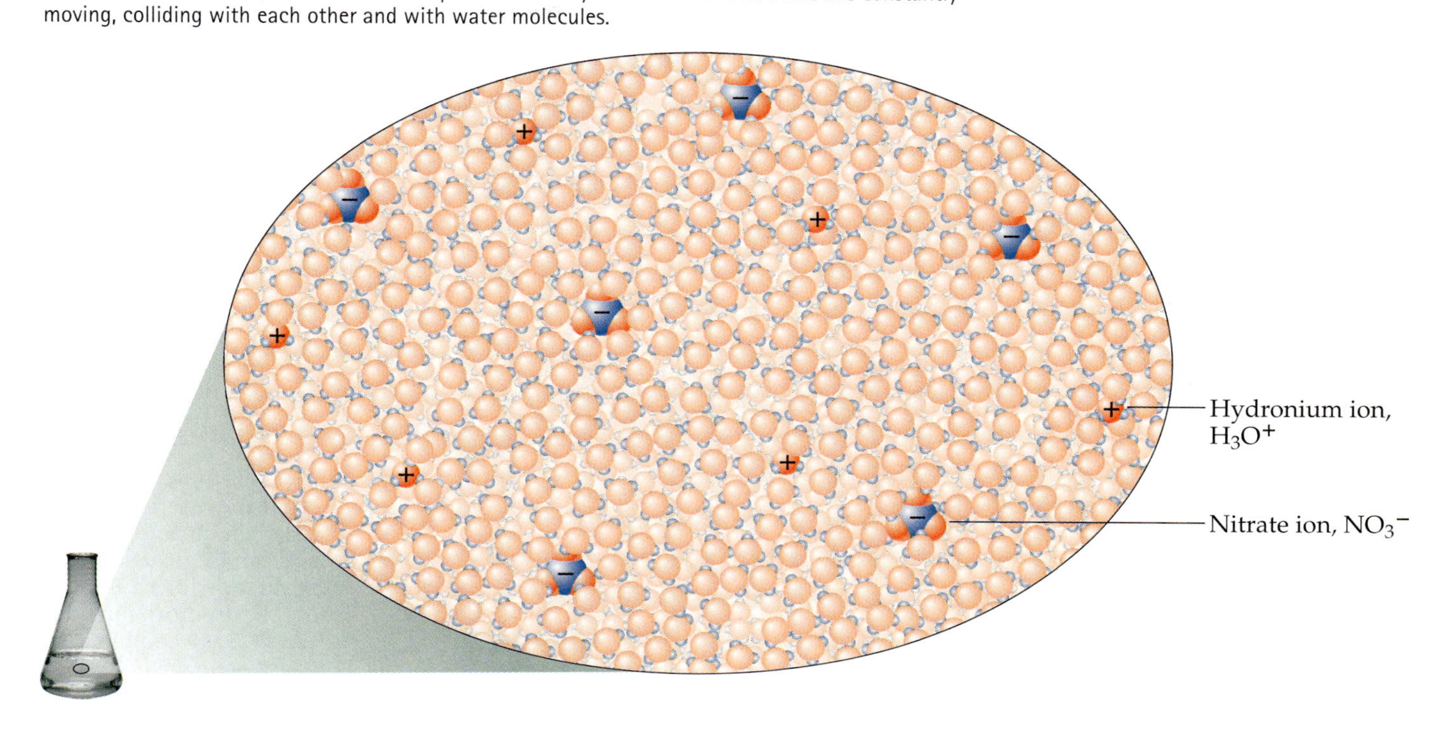
Hydronium ion, $H_3O^+$
Nitrate ion, $NO_3^-$

**Figure 5.13**
**Water Solution of Nitric Acid and Sodium Hydroxide Before Reaction**
At the instant after nitric acid and sodium hydroxide solutions are mixed and before the reaction, 4 separate ions move throughout the solution, breaking and making attractions and constantly colliding with each other.

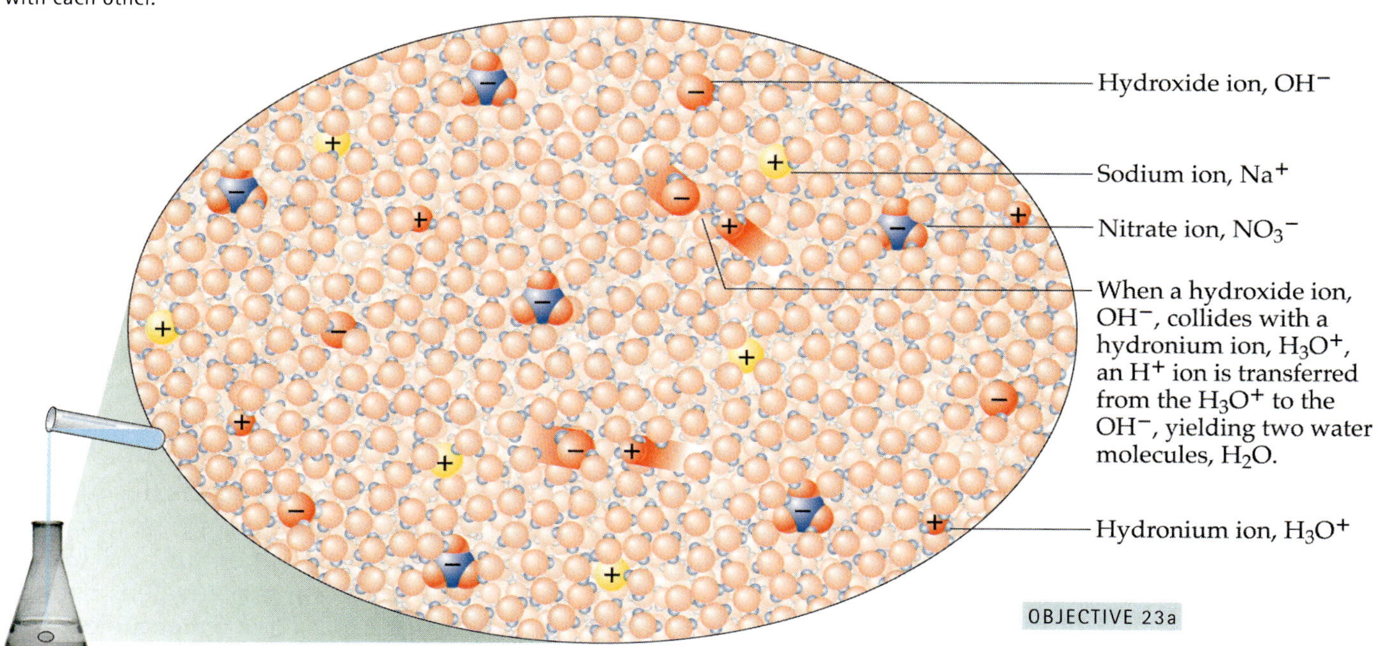
Hydroxide ion, $OH^-$
Sodium ion, $Na^+$
Nitrate ion, $NO_3^-$
When a hydroxide ion, $OH^-$, collides with a hydronium ion, $H_3O^+$, an $H^+$ ion is transferred from the $H_3O^+$ to the $OH^-$, yielding two water molecules, $H_2O$.
Hydronium ion, $H_3O^+$

OBJECTIVE 23a

**Figure 5.14**
**Hydronium and Hydroxide Ions**

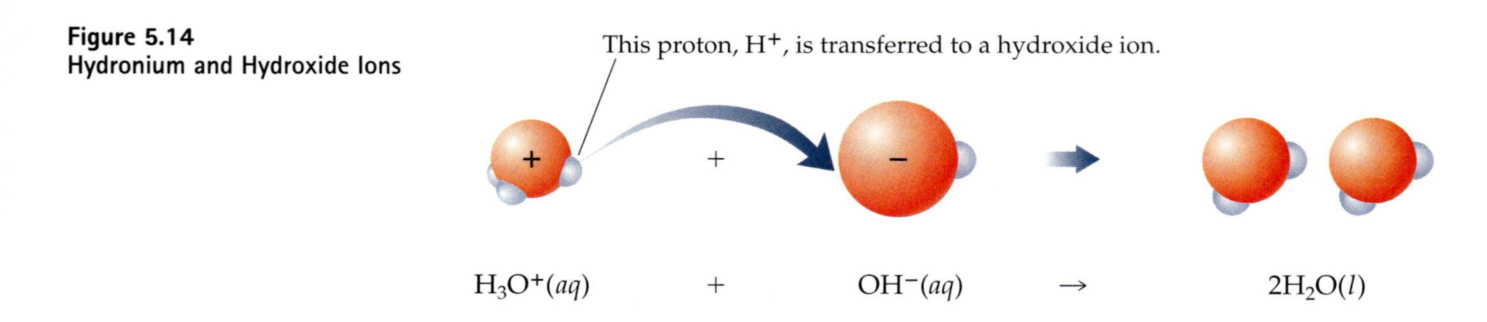


**Figure 5.15**
**After Reaction of Nitric Acid and Sodium Hydroxide**

OBJECTIVE 23a

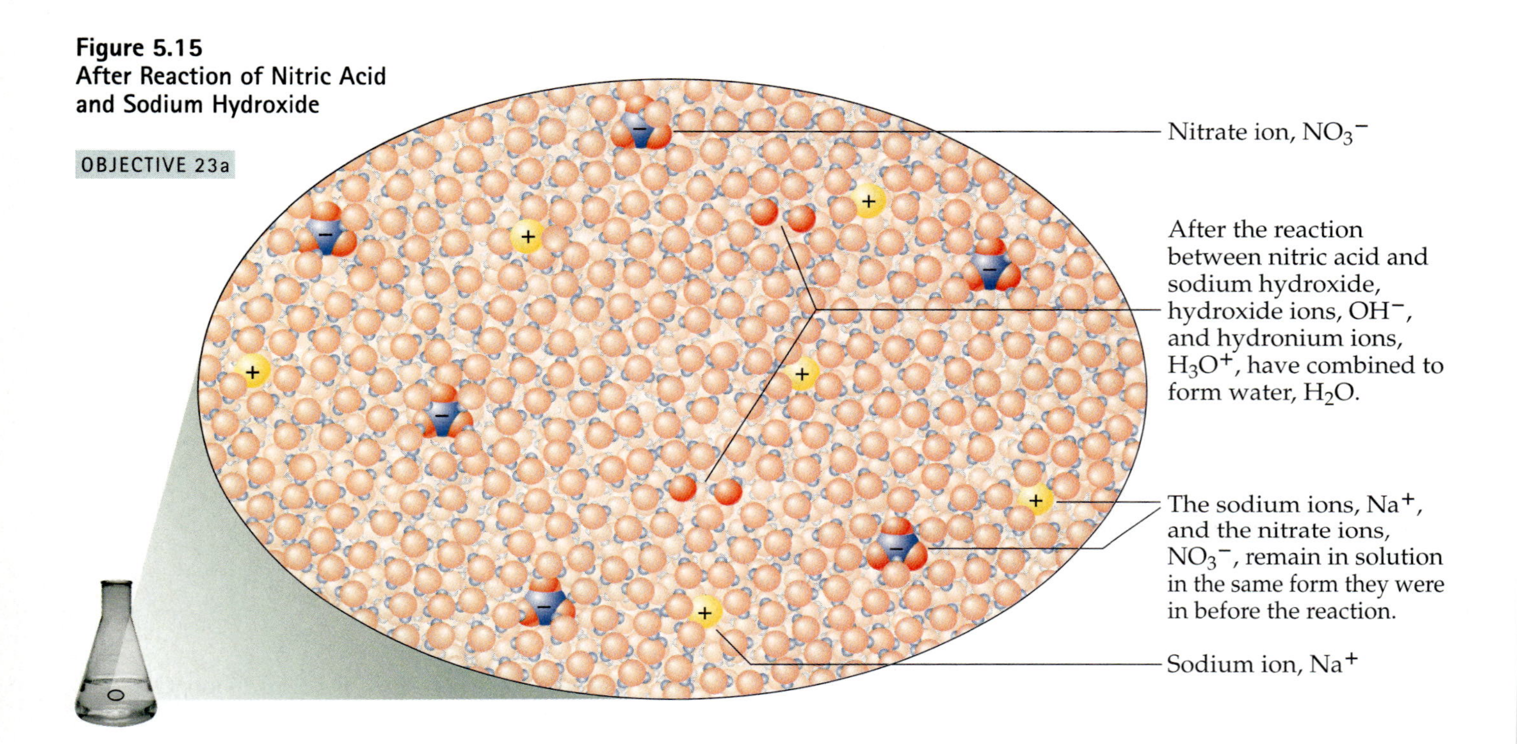

Nitrate ion, $NO_3^-$
After the reaction between nitric acid and sodium hydroxide, hydroxide ions, $OH^-$, and hydronium ions, $H_3O^+$, have combined to form water, $H_2O$.
The sodium ions, $Na^+$, and the nitrate ions, $NO_3^-$, remain in solution in the same form they were in before the reaction.
Sodium ion, $Na^+$


OBJECTIVE 23a

ions to solution, but they did not actively participate in the reaction. In other words, they are spectator ions, so they are left out of the net ionic chemical equation. The net ionic equation for the reaction is therefore

$$H_3O^+(aq) + OH^-(aq) \rightarrow 2H_2O(l)$$

Most chemists are in the habit of describing reactions such as this in terms of $H^+$ rather than $H_3O^+$, even though hydrogen ions do not exist in a water solution in the same sense that sodium ions do. When an acid loses a hydrogen atom as $H^+$, the proton immediately forms a covalent bond to some other atom. In water, it forms a covalent bond to a water molecule to produce the hydronium ion. Although $H_3O^+$ is a better description of what is found in acid solutions, it is still convenient and conventional to write $H^+$ in equations instead. You can think of $H^+$ as a shorthand notation for $H_3O^+$. Therefore, the following net ionic equation is commonly used to describe the net ionic equation above.

$$H^+(aq) + OH^-(aq) \rightarrow H_2O(l)$$

## Writing Equations for Reactions Between Acids and Bases

The procedure for writing equations for acid–base reactions is very similar to that used to write equations for precipitation reactions in Section 4.2.

The first step in writing an equation for the reaction between nitric acid, $HNO_3$, and the base sodium hydroxide, NaOH, is to predict the formulas for the products by recognizing that most Arrhenius neutralization reactions, like the reaction between nitric acid and sodium hydroxide, are double-displacement reactions.

OBJECTIVE 24

$$\text{AB} + \text{CD} \rightarrow \text{AD} + \text{CB}$$

$$HNO_3(aq) + NaOH(aq) \rightarrow H_2O(l) + NaNO_3(aq)$$

We consider the positive portion of the acid to be $H^+$, so for the reaction above, A is $H^+$, B is $NO_3^-$, C is $Na^+$, and D is $OH^-$. When $H^+$ ions combine with $OH^-$ ions, they form HOH (water, $H_2O$). The ion formulas $Na^+$ and $NO_3^-$ are combined in the complete equation as the CB formula, $NaNO_3$.

OBJECTIVE 24

In picturing reactions of a polyprotic acid with a strong base, we shall assume that enough base is added to react with all of the acidic hydrogen atoms. The following complete equations describe the reactions of the diprotic acid sulfuric acid and the triprotic acid phosphoric acid with sodium hydroxide. Each equation represents the sum of a series of reactions in which the acidic hydrogen atoms are removed one at a time.

$$H_2SO_4(aq) + 2NaOH(aq) \rightarrow 2H_2O(l) + Na_2SO_4(aq)$$

$$H_3PO_4(aq) + 3NaOH(aq) \rightarrow 3H_2O(l) + Na_3PO_4(aq)$$

The problems at the end of the chapter ask you to write complete equations for reactions like these. Note that these too are double-displacement reactions. In each of these examples, A is $H^+$, C is $Na^+$, and D is $OH^-$. In the first reaction, B is $SO_4^{2-}$, and in the second reaction, B is $PO_4^{3-}$.

OBJECTIVE 24

One of the useful properties of acids is that they will react with insoluble ionic compounds that contain basic anions. Because the products of such reactions are soluble, acids can be used to dissolve normally insoluble ionic compounds (see Special Topic 5.3: *Precipitation, Acid–Base Reactions, and Tooth Decay*). For example, water-insoluble aluminum hydroxide dissolves in a hydrochloric acid solution.

OBJECTIVE 23b

$$Al(OH)_3(s) + 3HCl(aq) \rightarrow AlCl_3(aq) + 3H_2O(l)$$

## SPECIAL TOPIC 5.3 Precipitation, Acid–Base Reactions, and Tooth Decay

Teeth have a protective coating of hard enamel that is about 2 mm thick and consists of about 98% hydroxyapatite, $Ca_5(PO_4)_3OH$. Like any ionic solid surrounded by a water solution, the hydroxyapatite is constantly dissolving and reprecipitating.

$$Ca_5(PO_4)_3OH(s) \rightleftharpoons 5Ca^{2+}(aq) + 3PO_4^{3-}(aq) + OH^-(aq)$$

Your saliva provides the calcium ions and the phosphate ions for this process, and as long as your saliva does not get too acidic, it will contain enough hydroxide to keep the rate of solution and the rate of precipitation about equal. Thus there is no net change in the amount of enamel on your teeth.

Unfortunately, certain foods can upset this balance. The bacteria in your mouth break down your food, especially food high in sugar, to form acids such as acetic acid and lactic acid. These acids neutralize the hydroxide in your saliva, slowing the precipitation of enamel. The $Ca_5(PO_4)_3OH$ continues to go into solution, however, so there is a net loss of the protective coating on the teeth.

Fluoride in our drinking water and our toothpaste can help minimize this damage. The fluoride ion takes the place of the hydroxide ion to precipitate fluorapatite, $Ca_5(PO_4)_3F$, a compound very similar to the original enamel.

$$5Ca^{2+}(aq) + 3PO_4^{3-}(aq) + F^-(aq) \rightleftharpoons Ca_5(PO_4)_3F(s)$$

Fluorapatite is 100 times less soluble than hydroxyapatite, so it is less likely to be affected by the acid formed by the bacteria.

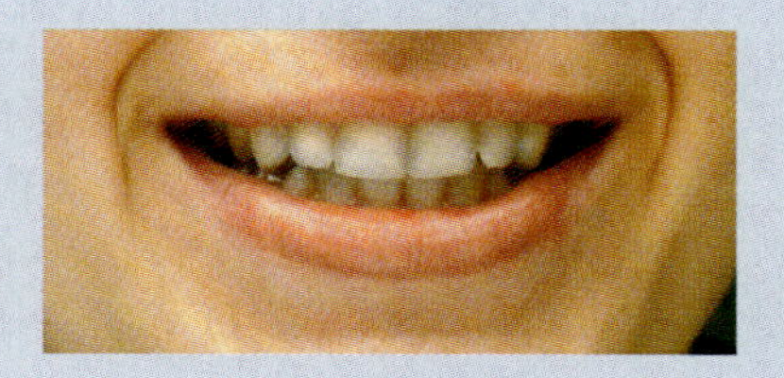

The natural enamel that coats teeth is mostly $Ca_5(PO_4)_3OH$. Fluoride in our water or toothpaste leads to the less soluble $Ca_5(PO_4)_3F$ replacing $Ca_5(PO_4)_3OH$ in tooth enamel, helping to protect our teeth from tooth decay.

### EXAMPLE 5.4 Neutralization Reactions

OBJECTIVE 24

Write the complete equations for the neutralization reactions that take place when the following are mixed. (If an acid has more than one acidic hydrogen, assume that there is enough base to remove all of them. Assume that there is enough acid to neutralize all of the basic hydroxide ions.)

a. $HCl(aq) + KOH(aq)$

b. $H_2SO_4(aq) + KOH(aq)$

c. $HNO_3(aq) + Mn(OH)_2(s)$

***Solution***

a. Neutralization reactions between strong monoprotic acids, such as $HCl(aq)$, and ionic compounds, such as KOH, are double-displacement reactions, so they have the form

$$AB + CD \rightarrow AD + CB$$

For HCl, A is $H^+$, and B is $Cl^-$. For KOH, C is $K^+$, and D is $OH^-$. Thus AD is HOH or $H_2O$, which we know is a liquid, and CB is KCl, which is a water-soluble ionic compound and thus aqueous.

$$HCl(aq) + KOH(aq) \rightarrow H_2O(l) + KCl(aq)$$

b. For $H_2SO_4$ in a double-displacement reaction, A is $H^+$, and B is $SO_4^{2-}$. (In neutralization reactions, you can assume that all of the acidic hydrogen atoms are lost to the base. Monoprotic acids lose one $H^+$ ion, diprotic acids such as $H_2SO_4$ lose two $H^+$ ions, and triprotic acids such as $H_3PO_4$ lose three $H^+$ ions.) For KOH, C is $K^+$, and D is $OH^-$. Thus AD is $H_2O$, and CB is $K_2SO_4$, a water-soluble ionic compound. The two $H^+$ ions from the diprotic acid $H_2SO_4$ react with the two $OH^-$ ions from two units of KOH to form two $H_2O$ molecules.

$$\begin{array}{ccccccc} AB & + & CD & \rightarrow & AD & + & CB \\ H_2SO_4(aq) & + & 2KOH(aq) & \rightarrow & 2H_2O(l) & + & K_2SO_4(aq) \end{array}$$

c. For $HNO_3$ in a double-displacement reaction, A is $H^+$, and B is $NO_3^-$. For $Mn(OH)_2$, C is $Mn^{2+}$, and D is $OH^-$. Thus AD is $H_2O$, and CB is $Mn(NO_3)_2$, a water-soluble ionic compound. Two $H^+$ ions from two nitric acid molecules react with the two $OH^-$ ions from the $Mn(OH)_2$ to form two $H_2O$ molecules.

$$\begin{array}{ccccccc} AB & + & CD & \rightarrow & AD & + & CB \\ 2HNO_3(aq) & + & Mn(OH)_2(s) & \rightarrow & 2H_2O(l) & + & Mn(NO_3)_2(aq) \end{array}$$

## Exercise 5.6 Neutralization Reactions

OBJECTIVE 24

Write the complete equation for the neutralization reactions that take place when the following are mixed. (If an acid has more than one acidic hydrogen, assume that there is enough base to remove all of them. Assume that there is enough acid to neutralize all of the basic hydroxide ions.)

a. $HCl(aq) + NaOH(aq)$

b. $HF(aq) + LiOH(aq)$

c. $H_3PO_4(aq) + LiOH(aq)$

d. $Fe(OH)_3(s) + HNO_3(aq)$

## Reactions of Arrhenius Acids and Ionic Compounds That Contain Carbonate or Hydrogen Carbonate

The reaction between an acid and an ionic compound containing either carbonate or hydrogen carbonate leads to carbon dioxide and water as products. The addition of $H^+$ ions to $CO_3^{2-}$ or $HCO_3^-$ forms carbonic acid, $H_2CO_3$. Carbonic acid, however, is unstable in water, so when it forms, it decomposes into carbon dioxide, $CO_2(g)$, and water, $H_2O(l)$.

OBJECTIVE 23c

$$2H^+(aq) + CO_3^{2-}(aq) \rightarrow H_2CO_3(aq) \rightarrow H_2O(l) + CO_2(g)$$

OBJECTIVE 25

$$H^+(aq) + HCO_3^-(aq) \rightarrow H_2CO_3(aq) \rightarrow H_2O(l) + CO_2(g)$$

Thus when $H_2CO_3$ would be predicted as a product for a double-displacement reaction, write "$H_2O(l) + CO_2(g)$" instead. Three examples follow.

$$2HCl(aq) + Na_2CO_3(aq) \rightarrow H_2O(l) + CO_2(g) + 2NaCl(aq)$$

$$HCl(aq) + NaHCO_3(aq) \rightarrow H_2O(l) + CO_2(g) + NaCl(aq)$$

$$2HCl(aq) + CaCO_3(s) \rightarrow H_2O(l) + CO_2(g) + CaCl_2(aq)$$

The third equation above describes a reaction that the oil industry exploits to extract more oil from a well. For oil to be pumped from deep in the earth to the surface, it must first seep through underground rock formations to the base of the oil well's pipes. Limestone, which is composed of $CaCO_3$, can be made more permeable to oil by pumping hydrochloric acid down into the limestone formations, converting the insoluble calcium carbonate to soluble calcium chloride.

**When acids are added to solutions that contain carbonate ions or hydrogen carbonate ions, carbon dioxide gas bubbles out of the solution. This reaction can be used to make limestone more permeable to oil by converting solid calcium carbonate into water-soluble calcium chloride.**

## Example 5.5 Neutralization Reactions with Compounds Containing Carbonate

OBJECTIVE 24

Write the complete equation for the reaction between $HNO_3(aq)$ and water-insoluble solid $MgCO_3$.

***Solution***

Translated into the general format of double-displacement reactions, A is $H^+$, B is $NO_3^-$, C is $Mg^{2+}$, and D is $CO_3^{2-}$. Compound AD would therefore be $H_2CO_3$, but this decomposes to form $H_2O(l)$ and $CO_2(g)$. Compound CB is $Mg(NO_3)_2$, which is a water-soluble ionic compound and thus aqueous.

$$2HNO_3(aq) + MgCO_3(s) \rightarrow H_2O(l) + CO_2(g) + Mg(NO_3)_2(aq)$$

## Exercise 5.7 Neutralization Reactions with Compounds Containing Carbonate

OBJECTIVE 24

Write the complete equation for the neutralization reaction that takes place when water solutions of sodium carbonate, $Na_2CO_3$, and hydrobromic acid, HBr, are mixed.

Special Topic 5.4 explains why some of the paper used to make books in the nineteenth and twentieth centuries was acidic, why this caused the paper to quickly become brittle, and how a neutralization reaction can stop the deterioration associated with this brittleness.

## Special Topic 5.4 Saving Valuable Books

Before the nineteenth century, paper in Europe was made from linen or old rags. Supplies of these materials dwindled as the demand for paper soared, and new manufacturing methods and raw materials were sought. Paper began to be made from wood pulp, but the first such products contained microscopic holes that caused the ink to bleed and blur. To fill these holes, the paper was saturated with "alum," which is aluminum sulfate, $Al_2(SO_4)_3$. The new process seemed to make a suitable paper, but as time passed, serious problems emerged. The aluminum ions in alum, like many metal ions, are acidic in the Arrhenius sense, reacting with moisture from the air to release $H^+$ ions.

$$Al^{3+}(aq) + H_2O(l) \rightleftharpoons AlOH^{2+}(aq) + H^+(aq)$$

The $H^+$ ions react in turn with the paper and weaken it. Many valued books are so brittle that they cannot be handled without their pages crumbling.

Several techniques are now being developed to neutralize the acid in the paper. As we have seen, most acid–base reactions take place in water, and there are obvious problems with dunking a book in an aqueous solution of base. The challenge, then, has been to develop a technique in which a gas is used to neutralize acid in the paper without causing further damage.

One such technique is called the DEZ treatment. DEZ, or diethyl zinc, $(CH_3CH_2)_2Zn$, can be made gaseous near room temperature. It reacts with either oxygen or water vapor to form zinc oxide, ZnO(s), which is deposited evenly on the paper.

$$(CH_3CH_2)_2Zn(g) + 7O_2(g) \rightarrow ZnO(s) + 4CO_2(g) + 5H_2O(l)$$

$$(CH_3CH_2)_2Zn(g) + H_2O(g) \rightarrow ZnO(s) + 2CH_3CH_3(g)$$

The zinc oxide contains the basic anion oxide, $O^{2-}$, which reacts with $H^+$ ions to neutralize the acid in the paper.

$$ZnO + 2H^+ \rightarrow Zn^{2+} + H_2O$$

Damage that has already been done cannot be reversed, so the goal is to save as many books as possible before they deteriorate so much that they cannot be handled.

The acid in the paper used to make some books damages the paper and leaves it brittle. The paper in the book on the right was made with a process that did not leave the paper acidic. The book on the left was made with a different process that left the paper acidic.

## Special Topic 5.5 Be Careful with Bleach

Common bleach, used for household cleaning and laundering, is a solution of sodium hypochlorite, NaClO(*aq*). The hypochlorite is made by reacting chlorine gas with a basic solution.

$$Cl_2(g) + 2OH^-(aq) \rightleftharpoons OCl^-(aq) + Cl^-(aq) + H_2O(l)$$

This reaction is reversible, so the chlorine atoms are constantly switching back and forth from $Cl_2$ to $OCl^-$. In a basic solution, the forward reaction is fast enough to ensure that most of the chlorine in the bottle of bleach is in the $OCl^-$ form.

If the bleach is added to an acidic solution, the hydroxide ions in the basic solution of bleach react with the acidic $H^+$ ions to form water. With fewer hydroxide ions available, the reaction between the $OH^-$ and the $Cl_2$ slows down, but the reverse reaction continues at the same pace. This creates potentially dangerous levels of chlorine gas and is the reason why the labels on bleach bottles warn against mixing bleach with other cleaning agents such as toilet bowl cleaners. Toilet bowl cleaners are usually acidic, containing such acids as phosphoric acid, $H_3PO_4$, or hydrogen sulfate, $HSO_4^-$.

Mixing bleach and toilet bowl cleaners can be dangerous.

Do you want to know why bleach bottles have a warning label that tells you not to mix the bleach with acidic cleaning agents, such as toilet bowl cleaners? The explanation is in Special Topic 5.5 above.

## 5.7 Brønsted–Lowry Acids and Bases

Although the Arrhenius definitions of *acid*, *base*, and *acid–base* reaction are very useful, an alternative set of definitions is also commonly employed. In this system, a **Brønsted–Lowry acid** is a proton ($H^+$) donor, a **Brønsted–Lowry base** is a proton acceptor, and a **Brønsted–Lowry acid–base reaction** is a proton transfer. Table 5.7 summarizes the definitions of *acid* and *base* in the Arrhenius and Brønsted–Lowry systems.

Table 5.7
Definition of *acid* and *base*

| System | definition of *acid* | definition of *base* |
|---|---|---|
| Arrhenius | generates $H_3O^+$ when added to water | generates $OH^-$ when added to water |
| Brønsted–Lowry | proton ($H^+$) donor in reaction | proton ($H^+$) acceptor in reaction |

To grasp the differences and to understand why new definitions were suggested, consider the following reactions:

$$NH_3(aq) + HC_2H_3O_2(aq) \rightarrow NH_4^+(aq) + C_2H_3O_2^-(aq)$$

$$H_2O(l) + HC_2H_3O_2(aq) \rightleftharpoons H_3O^+(aq) + C_2H_3O_2^-(aq)$$

OBJECTIVE 26

$$NH_3(aq) + H_2O(l) \rightleftharpoons NH_4^+(aq) + OH^-(aq)$$

These reactions are very similar, but only the first reaction would be considered an acid–base reaction in the Arrhenius system. In each of the reactions, an $H^+$ is transferred from one reactant to another, but only the first is a reaction between an Arrhenius acid and an Arrhenius base. In the first reaction, an $H^+$ is transferred from the Arrhenius weak acid acetic acid, $HC_2H_3O_2(aq)$, to the Arrhenius weak base ammonia, $NH_3(aq)$. In the second reaction, an $H^+$ is transferred from the Arrhenius weak acid acetic acid, $HC_2H_3O_2(aq)$, to water, which is not considered an acid or a base in the Arrhenius sense. In the third reaction, an $H^+$ is transferred from water, which is not considered an acid or base in the Arrhenius sense, to the Arrhenius weak base ammonia, $NH_3(aq)$.

The Brønsted–Lowry system allows us to describe all of these reactions as acid–base reactions. They are repeated below, with the Brønsted–Lowry acids and bases labeled. Note that in each case, the acid loses an $H^+$ ion as it reacts, and the base gains an $H^+$ ion.

$$\underset{\text{B/L base}}{NH_3(aq)} + \underset{\text{B/L acid}}{HC_2H_3O_2(aq)} \rightarrow NH_4^+(aq) + C_2H_3O_2^-(aq)$$

$$\underset{\text{B/L base}}{H_2O(l)} + \underset{\text{B/L acid}}{HC_2H_3O_2(aq)} \rightleftharpoons H_3O^+(aq) + C_2H_3O_2^-(aq)$$

OBJECTIVE 26

$$\underset{\text{B/L base}}{NH_3(aq)} + \underset{\text{B/L acid}}{H_2O(l)} \rightleftharpoons NH_4^+(aq) + OH^-(aq)$$

Acetic acid reacts with the dihydrogen phosphate polyatomic ion, $H_2PO_4^-$, in a reversible reaction. In the forward reaction, acetic acid acts as the Brønsted–Lowry acid, and dihydrogen phosphate acts as the Brønsted–Lowry base.

$$\underset{\text{B/L acid}}{HC_2H_3O_2(aq)} + \underset{\text{B/L base}}{H_2PO_4^-(aq)} \rightleftharpoons C_2H_3O_2^-(aq) + H_3PO_4(aq)$$

The reverse reaction, too, is a Brønsted–Lowry acid–base reaction. An $H^+$ ion is transferred from $H_3PO_4$ (the acid) to a $C_2H_3O_2^-$ ion (the base). The Brønsted–Lowry base for the forward reaction ($H_2PO_4^-$) gains an $H^+$ ion to form $H_3PO_4$, which then acts as a Brønsted–Lowry acid in the reverse reaction and returns the $H^+$ ion to $C_2H_3O_2^-$. Chemists say that $H_3PO_4$ is the conjugate acid of $H_2PO_4^-$. The **conjugate acid** of a molecule or ion is the molecule or ion that forms when one $H^+$ ion is added. The formulas $H_3PO_4$ and $H_2PO_4^-$ represent a **conjugate acid–base pair,** molecules or ions that differ by one $H^+$ ion.

OBJECTIVE 27

Likewise, the Brønsted–Lowry acid for the forward reaction ($HC_2H_3O_2$) loses an $H^+$ ion to form $C_2H_3O_2^-$, which acts as a Brønsted–Lowry base in the reverse reaction and regains the $H^+$ ion. Chemists say that $C_2H_3O_2^-$ is the conjugate base of $HC_2H_3O_2$. The **conjugate base** of a molecule or ion is the molecule or ion that forms when one $H^+$ ion is removed. The formulas $HC_2H_3O_2$ and $C_2H_3O_2^-$ represent a conjugate acid–base pair (Figure 5.16).

OBJECTIVE 28

**Figure 5.16**
**Conjugate Acid–Base Pairs**

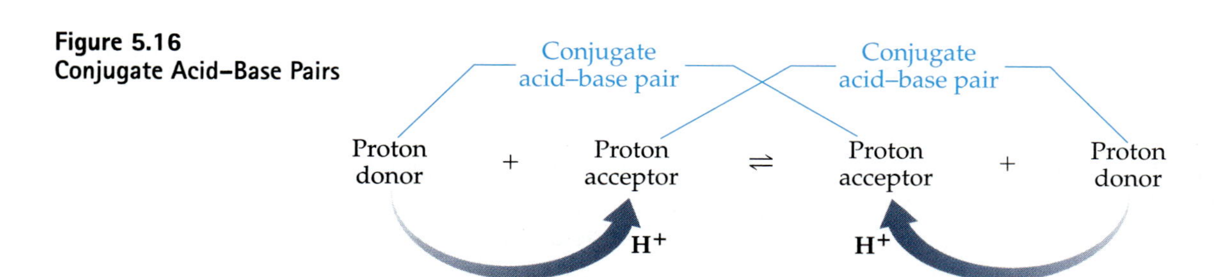


## EXAMPLE 5.6 Conjugate Acids

OBJECTIVE 27

Write the formula for the conjugate acid of (a) $F^-$, (b) $NH_3$, (c) $HSO_4^-$, and (d) $CrO_4^{2-}$.

***Solution***

In each case, the formula for the conjugate acid is derived by adding one $H^+$ ion to the formulas above.

a. $HF$   b. $NH_4^+$   c. $H_2SO_4$   d. $HCrO_4^-$

## EXERCISE 5.8 Conjugate Acids

OBJECTIVE 27

Write the formula for the conjugate acid of (a) $NO_2^-$, (b) $HCO_3^-$, (c) $H_2O$, and (d) $PO_4^{3-}$.

## EXAMPLE 5.7 Conjugate Bases

OBJECTIVE 28

Write the formula for the conjugate base of (a) $HClO_3$, (b) $H_2SO_3$, (c) $H_2O$, and (d) $HCO_3^-$.

***Solution***

In each case, the formula for the conjugate base is derived by removing one $H^+$ ion from the formulas above.

a. $ClO_3^-$   b. $HSO_3^-$   c. $OH^-$   d. $CO_3^{2-}$

## EXERCISE 5.9 Conjugate Bases

OBJECTIVE 28

Write the formula for the conjugate base of (a) $H_2C_2O_4$, (b) $HBrO_4$, (c) $NH_3$, and (d) $H_2PO_4^-$.

Some substances can act as a Brønsted–Lowry acid in one reaction and as a Brønsted–Lowry base in another. Consider the following net ionic equations for the reaction of dihydrogen phosphate ion with either the acid hydrochloric acid or the strong base hydroxide.

OBJECTIVE 29

$$\underset{\text{B/L base}}{H_2PO_4^-(aq)} + \underset{\text{B/L acid}}{HCl(aq)} \rightarrow H_3PO_4(aq) + Cl^-(aq)$$

$$\underset{\text{B/L acid}}{H_2PO_4^-(aq)} + \underset{\text{B/L base}}{2OH^-(aq)} \rightarrow PO_4^{3-}(aq) + 2H_2O(l)$$

In the first reaction, the dihydrogen phosphate acts as a Brønsted–Lowry base, and in the second reaction, it acts as a Brønsted–Lowry acid. A substance that can act as either a Brønsted–Lowry acid or a Brønsted–Lowry base, depending on the circumstances, is called an **amphoteric substance.**

The hydrogen carbonate ion is another example of an amphoteric substance. In the first reaction below, it acts as a Brønsted–Lowry base, and in the second reaction, it acts as a Brønsted-Lowry acid.

$$\underset{\text{B/L base}}{HCO_3^-(aq)} + \underset{\text{B/L acid}}{HC_2H_3O_2(aq)} \rightarrow H_2O(l) + CO_2(g) + C_2H_3O_2^-(aq)$$

$$\underset{\text{B/L acid}}{HCO_3^-(aq)} + \underset{\text{B/L base}}{OH^-(aq)} \rightarrow CO_3^{2-}(aq) + H_2O(l)$$

Because both dihydrogen phosphate and hydrogen carbonate can be Brønsted–Lowry acids or bases, neither can be described as a Brønsted–Lowry acid or base except with reference to a specific acid–base reaction. For this reason, the Arrhenius definitions of acids and bases are the ones used to categorize isolated substances on the stockroom shelf. A substance generates hydronium ions, hydroxide ions, or neither when added to water, so it is always an acid, a base, or neutral in the Arrhenius sense. Hydrogen carbonate is an Arrhenius base because it yields hydroxide ions when it is added to water. Dihydrogen phosphate is an Arrhenius acid because it generates hydronium ions when it is added to water.

OBJECTIVE 30

$$HCO_3^-(aq) + H_2O(l) \rightleftharpoons H_2CO_3(aq) + OH^-(aq)$$

$$H_2PO_4^-(aq) + H_2O(l) \rightleftharpoons HPO_4^{2-}(aq) + H_3O^+(aq)$$

Thus we have two systems for describing acids, bases, and acid–base reactions. The Brønsted–Lowry system is often used to describe specific acid–base reactions, but the Arrhenius system is used to describe whether isolated substances are acids, bases, or neither.

### EXAMPLE 5.8 Brønsted–Lowry Acids and Bases

OBJECTIVE 31

Identify the Brønsted–Lowry acid and base for the forward reaction in each of the following equations.

a. $HClO_2(aq) + NaIO(aq) \rightarrow HIO(aq) + NaClO_2(aq)$

b. $HS^-(aq) + HF(aq) \rightarrow H_2S\,(aq) + F^-(aq)$

c. $HS^-(aq) + OH^-(aq) \rightarrow S^{2-}(aq) + H_2O(l)$

d. $H_3AsO_4(aq) + 3NaOH(aq) \rightarrow Na_3AsO_4(aq) + 3H_2O(l)$

***Solution***

a. The $HClO_2$ loses an $H^+$ ion, so it is the Brønsted–Lowry acid. The $IO^-$ in the NaIO gains the $H^+$ ion, so the NaIO is the Brønsted–Lowry base.

b. The HF loses an $H^+$ ion, so it is the Brønsted–Lowry acid. The $HS^-$ gains the $H^+$ ion, so it is the Brønsted–Lowry base.

c. The $HS^-$ loses an $H^+$ ion, so it is the Brønsted–Lowry acid. The $OH^-$ gains the $H^+$ ion, so it is the Brønsted–Lowry base.

d. The $H_3AsO_4$ loses three $H^+$ ions, so it is the Brønsted–Lowry acid. Each $OH^-$ in NaOH gains an $H^+$ ion, so the NaOH is the Brønsted–Lowry base.

### EXERCISE 5.10 Brønsted–Lowry Acids and Bases

OBJECTIVE 31

Identify the Brønsted–Lowry acid and base in each of the following equations.

a. $HNO_2(aq) + NaBrO(aq) \rightarrow HBrO(aq) + NaNO_2(aq)$

b. $H_2AsO_4^-(aq) + HNO_2(aq) \rightleftharpoons H_3AsO_4(aq) + NO_2^-(aq)$

c. $H_2AsO_4^-(aq) + 2OH^-(aq) \rightarrow AsO_4^{3-}(aq) + 2H_2O(l)$

## Chapter Glossary

**Hydronium ion** $H_3O^+$.

**Arrhenius acid** According to the Arrhenius theory, any substance that generates hydronium ions, $H_3O^+$, when added to water.

**Acidic solution** A solution with a significant concentration of hydronium ions, $H_3O^+$.

**Binary acids** Substances that have the general formula of $HX(aq)$, where X is one of the first four halogens: $HF(aq)$, $HCl(aq)$, $HBr(aq)$, and $HI(aq)$.

**Oxyacids (or oxoacids)** Molecular substances that have the general formula $H_aX_bO_c$. In other words, they contain hydrogen, oxygen, and one other element represented by X; the letters a, b, and c represent subscripts.

**Organic acids** Carbon-based acids.

**Monoprotic acid** An acid that donates one hydrogen ion per molecule in a reaction.

**Polyprotic acid** An acid that can donate more than one hydrogen ion per molecule in a reaction.

**Diprotic acid** An acid that can donate two hydrogen ions per molecule in a reaction.

**Triprotic acid** An acid that can donate three hydrogen ions per molecule in a reaction.

**Reversible reaction** A chemical reaction in which the reactants are constantly forming products and, at the same time, the products are re-forming the reactants.

**Weak acid** A substance that is incompletely ionized in water due to the reversibility of the reaction that forms hydronium ions, $H_3O^+$, in water. Weak acids yield significantly less than one $H_3O^+$ ion in solution for each acid molecule dissolved in water.

**Completion reaction** A chemical reaction that is not significantly reversible.

**Strong acid** An acid that donates its $H^+$ ions to water in a reaction that goes completely to products. Such a compound produces close to one $H_3O^+$ ion in solution for each acid molecule dissolved in water.

**Arrhenius base** A substance that produces hydroxide ions, $OH^-$, when added to water.

**Basic solution** A solution with a significant concentration of hydroxide ions, $OH^-$.

**Strong base** A substance that generates at least one hydroxide ion in solution for every unit of substance added to water.

**Hydroxides** Compounds that contain hydroxide ions.

**Weak base** A substance that produces fewer hydroxide ions in water solution than particles of the substance added.

**Neutralization reactions** Chemical reactions between Arrhenius acids and bases.

**Brønsted–Lowry acid-base reaction** A chemical reaction in which a proton, $H^+$, is transferred.

**Brønsted–Lowry acid** A substance that donates protons, $H^+$, in a Brønsted–Lowry acid–base reaction.

**Brønsted–Lowry base** A substance that accepts protons, $H^+$, in a Brønsted–Lowry acid–base reaction.

**Conjugate acid** The molecule or ion that forms when one $H^+$ ion is added to a molecule or ion.

**Conjugate base** The molecule or ion that forms when one $H^+$ ion is removed from a molecule or ion.

**Conjugate acid–base pair** Two molecules or ions that differ by one $H^+$ ion.

**Amphoteric substance** A substance that can act as either a Brønsted–Lowry acid and a Brønsted–Lowry base, depending on the circumstances.

You can test yourself on the glossary terms at the following Web address:

**www.chemplace.com/college/**

## Chapter Objectives

**The goal of this chapter is to teach you to do the following.**

1. Define all of the terms in the Chapter Glossary.

### Section 5.1 Acids

2. Identify acids as substances that taste sour.
3. Describe what occurs when a strong, monoprotic acid, such as HCl, is added to water.
4. Identify the following acids as binary acids: HF(*aq*), HCl(*aq*), HBr(*aq*), and HI(*aq*).
5. Write or identify three different formulas that can be used to describe acetic acid, and explain why each is used.
6. Describe what occurs when a weak, monoprotic acid, such as acetic acid, is added to water.
7. Explain why weak acids produce fewer $H_3O^+$ ions in water than strong acids, even when the same numbers of acid molecules are added to equal volumes of water.
8. Identify the following as strong monoprotic acids: HCl and $HNO_3$.
9. Identify sulfuric acid as a diprotic strong acid.
10. Describe what occurs when sulfuric acid is added to water.
11. Given a formula or name for any acid, identify it as a strong or a weak acid.

### Section 5.2 Acid Nomenclature

12. Convert between the names and formulas for binary acids and oxyacids.

### Section 5.3 Summary of Chemical Nomenclature

13. Given a name or chemical formula, tell whether it represents a binary ionic compound, an ionic compound with polyatomic ion(s), a binary covalent compound, a binary acid, or an oxyacid.
14. Convert between the names and chemical formulas for binary ionic compounds, ionic compounds with polyatomic ion(s), binary covalent compounds, binary acids, and oxyacids.

### Section 5.4 Strong and Weak Arrhenius Bases

15. Identify ionic compounds that contain hydroxide ions as strong bases.
16. Describe the changes that take place when ammonia, $NH_3$, is dissolved in water, and use this description to explain why ammonia is a weak Arrhenius base.
17. Describe the changes that take place when an ionic compound that contains carbonate or hydrogen carbonate ions is dissolved in water, and use this description to explain why these anions are weak Arrhenius bases.
18. Given a name or formula for a substance, identify it as (1) an Arrhenius strong acid, (2) an Arrhenius weak acid, (3) an Arrhenius strong base, or (4) an Arrhenius weak base.

### Section 5.5 pH and Acidic and Basic Solutions

19. Given the pH of a solution, identify the solution as acidic, basic, or neutral.
20. Given the pH of two acidic solutions, identify which solution is more acidic.
21. Given the pH of two basic solutions, identify which solution is more basic.
22. Describe how litmus paper can be used in the laboratory to identify whether a solution is acidic or basic.

### Section 5.6 Arrhenius Acid–Base Reactions

23. Describe the process that takes place at the molecular level for the following reactions. Your description should include mention of the particles in solution before and after the reaction. It should also include a description of the process that leads to the reaction.
    a. a strong, monoprotic acid, such as $HNO_3$, and an aqueous strong base, such as NaOH
    b. a strong monoprotic acid, such as $HCl(aq)$, and an insoluble ionic compound, such as $Al(OH)_3$
    c. any monoprotic acid and a solution containing carbonate ions or hydrogen carbonate ions
24. Given the names or formulas for a monoprotic or polyprotic acid and an ionic compound containing hydroxide, carbonate, or hydrogen carbonate ions, write the complete balanced equation that describes the neutralization reaction that takes place between them.
25. Identify $H_2O(l)$ and $CO_2(g)$ as the products of the reaction of an acid with carbonate, $CO_3^{2-}$, or hydrogen carbonate, $HCO_3^-$.

### Section 5.7 Brønsted–Lowry Acids and Bases

26. Explain why the Brønsted–Lowry definitions for *acid* and *base* are often used, instead of the Arrhenius definitions, to describe acid–base reactions.
27. Given a formula for a molecule or ion, write the formula for its conjugate acid.
28. Given a formula for a molecule or ion, write the formula for its conjugate base.
29. Explain why a substance can be a Brønsted–Lowry acid in one reaction and a Brønsted–Lowry base in a different reaction. Give an example to illustrate your explanation.
30. Explain why the Arrhenius definitions for *acid* and *base,* and not the Brønsted–Lowry definitions, are used to describe whether an isolated substance is an acid or base.
31. Given a Brønsted–Lowry acid–base equation, identify the Brønsted–Lowry acid and the Brønsted–Lowry base.

## Review Questions

1. Define the following terms.
   a. aqueous
   b. spectator ion
   c. double-displacement reaction
   d. net ionic equation
2. Write the name of the polyatomic ions represented by the formulas $CO_3^{2-}$ and $HCO_3^-$.
3. Write the formulas for the polyatomic ions dihydrogen phosphate ion and acetate ion.
4. Which of the following formulas represents an ionic compound?
   a. $MgCl_2$
   b. $PCl_3$
   c. $KHSO_4$
   d. $Na_2SO_4$
   e. $H_2SO_4$
5. Write the names that correspond to the formulas KBr, $Cu(NO_3)_2$, and $(NH_4)_2HPO_4$.
6. Write the formulas that correspond to the names nickel(II) hydroxide, ammonium chloride, and calcium hydrogen carbonate.
7. Predict whether each of the following is soluble or insoluble in water.
   a. iron(III) hydroxide
   b. barium sulfate
   c. aluminum nitrate
   d. copper(II) chloride
8. Describe the process by which the ionic compound sodium hydroxide dissolves in water.
9. Write the complete equation for the precipitation reaction that takes place when water solutions of zinc chloride and sodium phosphate are mixed.

## Key Ideas

**Complete the following statements by writing one of these words or phrases in each blank.**

10-fold
acceptor
acid
acidic
added
amphoteric
Arrhenius
basic
blue
Brønsted–Lowry
carbon dioxide, $CO_2$,
diprotic
donor
double
double-displacement
fewer
hydro
hydronium ions, $H_3O^+$,
hydroxide ions, $OH^-$,
hydroxides
-ic
less than 7
lower
nearly one
neutralize
one hydrogen ion
red
re-forming
removed
(root)ate
significantly less than one
single

| | |
|---|---|
| forming | sour |
| greater than 7 | strong bases |
| $H_aX_bO_c$ | transfer |
| halogens | weak |
| higher | water |

10. Any substance that has a(n) ________________ taste is an acid.
11. According to the modern form of the Arrhenius theory, an acid is a substance that produces ________________ when it is added to water.
12. On the basis of the Arrhenius definitions, a(n) ________________ solution is a solution with a significant concentration of $H_3O^+$.
13. The binary acids have the general formula of HX(*aq*), where X is one of the first four ________________.
14. Oxyacids (often called oxoacids) are molecular substances that have the general formula ________________.
15. If each molecule of an acid can donate ________________, the acid is called a monoprotic acid. A(n) ____________ acid, such as sulfuric acid, $H_2SO_4$, has two acidic hydrogen atoms.
16. Strong acids form ________________ $H_3O^+$ ion in solution for each acid molecule dissolved in water, whereas weak acids yield ________________ $H_3O^+$ ion in solution for each acid molecule dissolved in water.
17. A reaction in which the reactants are constantly ________________ products and, at the same time, the products are ________________ the reactants is called a reversible reaction. The chemical equations for reactions that are significantly reversible are written with ________________ arrows.
18. A(n) ________________ acid is a substance that is incompletely ionized in water because of the reversibility of its reaction with water that forms hydronium ion, $H_3O^+$.
19. The chemical equations for completion reactions are written with ________________ arrows to indicate that the reaction proceeds to form almost 100% products.
20. Binary acids are named by writing ________________ followed by the root of the name of the halogen, then ________________, and finally ________________.
21. If enough $H^+$ ions are added to a(n) ________________ polyatomic ion to completely neutralize its charge, the (root)ic acid is formed.
22. According to the modern version of the Arrhenius theory of acids and bases, a base is a substance that produces ________________ when it is added to water.
23. A solution that has a significant concentration of hydroxide ions is called a ________________ solution.
24. Compounds that contain hydroxide ions are often called ________________.
25. All water-soluble hydroxides are ________________.
26. A weak base is a base that produces ________________ hydroxide ions in water solution than there are particles of base dissolved.
27. Acidic solutions have pH values ________________, and the more acidic a solution is, the ________________ its pH. A change of 1 pH unit reflects a(n) ________________ change in $H_3O^+$ ion concentration.

28. Basic solutions have pH values ________________, and the more basic the solution is, the ________________ its pH.

29. Litmus, a natural dye, is derived from lichen. It turns ________________ in acidic conditions and ________________ in basic conditions.

30. When an Arrhenius acid is combined with an Arrhenius base, we say that they ________________ each other.

31. When hydronium ions and hydroxide ions collide in solution they react to form ________________.

32. Most Arrhenius neutralization reactions, like the reaction between nitric acid and sodium hydroxide, are ________________ reactions.

33. Carbonic acid is unstable in water, so when it forms in aqueous solutions, it decomposes into ________________ and water, $H_2O(l)$.

34. A Brønsted–Lowry acid is a proton ($H^+$) ________________, a Brønsted–Lowry base is a proton ________________, and a Brønsted–Lowry acid–base reaction is a proton ________________.

35. The conjugate acid of a molecule or ion is the molecule or ion that forms when one $H^+$ ion is ________________.

36. The conjugate base of a molecule or ion is the molecule or ion that forms when one $H^+$ ion is ________________.

37. A substance that can act as either a Brønsted–Lowry acid or a Brønsted–Lowry base, depending on the circumstances, is called a(n) ________________ substance.

38. The ________________ system is often used to describe specific acid–base reactions, but the ________________ system is used to describe whether isolated substances are acids, bases, or neither.

## Chapter Problems

### Section 5.1 Acids

OBJECTIVE 3

39. Describe how the strong monoprotic acid nitric acid, $HNO_3(aq)$ (used in the reprocessing of spent nuclear fuels) acts when it is added to water, including a description of the nature of the particles in solution before and after the reaction with water. If there is a reversible reaction with water, describe the forward and the reverse reactions.

OBJECTIVE 6

40. Describe how the weak monoprotic acid hydrofluoric acid, HF (used in aluminum processing) acts when it is added to water, including a description of the nature of the particles in solution before and after the reaction with water. If there is a reversible reaction with water, describe the forward and the reverse reactions.

OBJECTIVE 10

41. Describe how the strong diprotic acid sulfuric acid, $H_2SO_4$ (used to make industrial explosives) acts when it is added to water, including a description of the nature of the particles in solution before and after the reaction with water. If there is a reversible reaction with water, describe the forward and the reverse reactions.

OBJECTIVE 5

42. Explain why acetic acid is described with three different formulas: $HC_2H_3O_2$, $CH_3COOH$, and $CH_3CO_2H$.

OBJECTIVE 7

43. Explain why weak acids produce fewer $H_3O^+$ ions in water than strong acids, even when the same number of acid molecules are added to equal volumes of water.

44. Classify each of the following acids as monoprotic, diprotic, or triprotic.
    a. $HCl(aq)$ (used in food processing)
    b. $H_2SO_4$ (used in petroleum refining)
    c. $HC_2H_3O_2$ (solvent in the production of polyesters)
    d. $H_3PO_4$ (catalyst for the production of ethanol)

45. Identify each of the following as strong or weak acids. OBJECTIVE 11
    a. sulfurous acid (for bleaching straw)
    b. $H_2SO_4$ (used to make plastics)
    c. oxalic acid (in car radiator cleaners)

46. Identify each of the following as strong or weak acids. OBJECTIVE 11
    a. $HCl(aq)$ (used to make dyes)
    b. nitrous acid (source of nitrogen monoxide, NO, used to bleach rayon)
    c. $H_2CO_3$ (formed when $CO_2$ dissolves in water)

47. Identify each of the following as strong or weak acids. OBJECTIVE 11
    a. $H_3PO_4$ (added to animal feeds)
    b. hypophosphorous acid (in electroplating baths)
    c. $HF(aq)$ (used to process uranium)

48. Identify each of the following as strong or weak acids. OBJECTIVE 11
    a. benzoic acid (used to make a common food preservative, sodium benzoate)
    b. $HNO_3$ (used to make explosives)
    c. hydrocyanic acid (used to make rodenticides and pesticides)

49. For each of the following, write the chemical equation for its reaction with water.
    a. The monoprotic weak acid nitrous acid, $HNO_2$
    b. The monoprotic strong acid hydrobromic acid, HBr

50. For each of the following, write the chemical equation for its reaction with water.
    a. The monoprotic weak acid chlorous acid, $HClO_2$
    b. The monoprotic strong acid perchloric acid, $HClO_4$

### Sections 5.2 and 5.3 Acid Nomenclature and Summary of Chemical Nomenclature

51. Write the formulas and names of the acids that are derived from adding enough $H^+$ ions to the following ions to neutralize their charge.
    a. $NO_3^-$
    b. $CO_3^{2-}$
    c. $PO_4^{3-}$

52. Write the formulas and names of the acids that are derived from adding enough $H^+$ ions to the following ions to neutralize their charge.
    a. $SO_4^{2-}$
    b. $C_2H_3O_2^-$

OBJECTIVE 12 OBJECTIVE 13 OBJECTIVE 14

53. Classify each of the following compounds as either (1) a binary ionic compound, (2) an ionic compound with polyatomic ion(s), (3) a binary covalent compound, (4) a binary acid, or (5) an oxyacid. Write the chemical formula that corresponds to each name.

a. phosphoric acid
b. ammonium bromide
c. diphosphorus tetriodide
d. lithium hydrogen sulfate
e. hydrochloric acid
f. magnesium nitride
g. acetic acid
h. lead(II) hydrogen phosphate

OBJECTIVE 12 OBJECTIVE 13 OBJECTIVE 14

54. Classify each of the following compounds as either (1) a binary ionic compound, (2) an ionic compound with polyatomic ion(s), (3) a binary covalent compound, (4) a binary acid, or (5) an oxyacid. Write the chemical formula that corresponds to each name.

a. potassium sulfide
b. sulfuric acid
c. ammonium nitrate
d. iodine pentafluoride
e. copper(I) sulfate
f. hydrofluoric acid
g. sodium hydrogen carbonate

OBJECTIVE 12 OBJECTIVE 13 OBJECTIVE 14

55. Classify each of the following formulas as either (1) a binary ionic compound, (2) an ionic compound with polyatomic ion(s), (3) a binary covalent compound, (4) a binary acid, or (5) an oxyacid. Write the name that corresponds to each formula.

a. $HBr(aq)$
b. $ClF_3$
c. $CaBr_2$
d. $Fe_2(SO_4)_3$
e. $H_2CO_3$
f. $(NH_4)_2SO_4$
g. $KHSO_4$

OBJECTIVE 12 OBJECTIVE 13 OBJECTIVE 14

56. Classify each of the following formulas as either (1) a binary ionic compound, (2) an ionic compound with polyatomic ion(s), (3) a binary covalent compound, (4) a binary acid, or (5) an oxyacid. Write the name that corresponds to each formula.

a. $HNO_3$
b. $Ca(OH)_2$
c. $(NH_4)_2HPO_4$
d. $Ni_3P_2$
e. $HI(aq)$
f. $Li_2O$
g. $Br_2O$

### Section 5.4 Arrhenius Bases

OBJECTIVE 16

57. Describe the changes that take place when ammonia, $NH_3$, is dissolved in water, and use this description to explain why ammonia is a weak Arrhenius base.

OBJECTIVE 18

58. Classify each of the substances as a weak acid, strong acid, weak base, or strong base in the Arrhenius acid–base sense.

a. $H_2CO_3$
b. cesium hydroxide
c. $HF(aq)$
d. sodium carbonate
e. $NH_3$
f. chlorous acid
g. $HCl(aq)$
h. benzoic acid

59. Classify each of the substances as a weak acid, strong acid, weak base, or strong base in the Arrhenius acid–base sense.

OBJECTIVE 18

a. $HNO_3$
b. ammonia
c. LiOH
d. phosphorous acid
e. $H_2SO_4$
f. nitrous acid
g. $NaHCO_3$

### Section 5.5 pH and Acidic and Basic Solutions

60. Classify each of the following solutions as acidic, basic, or neutral.

OBJECTIVE 19

a. Tomato juice with a pH of 4.53
b. Milk of magnesia with a pH of 10.4
c. Urine with a pH of 6.8

61. Classify each of the following solutions as acidic, basic, or neutral.

OBJECTIVE 19

a. Saliva with a pH of 7.0
b. Beer with a pH of 4.712
c. A solution of a drain cleaner with a pH of 14.0

62. Which is more acidic, carbonated water with a pH of 3.95 or milk with a pH of 6.3?

OBJECTIVE 20

63. Which is more basic, a soap solution with a pH of 10.0 or human tears with a pH of 7.4?

OBJECTIVE 21

64. Identify each of the following characteristics as associated with acids or bases.

OBJECTIVE 2 OBJECTIVE 22

a. tastes sour
b. turns litmus red
c. reacts with $HNO_3$

65. Identify each of the following properties as characteristic of acids or of bases.

OBJECTIVE 22

a. turns litmus blue
b. reacts with carbonate to form $CO_2(g)$

### Section 5.6 Arrhenius Acid-Base Reactions

66. Describe the process that takes place in the following neutralization reaction, mentioning the nature of the particles in the solution before and after the reaction: The strong acid hydrochloric acid, $HCl(aq)$, and the strong base sodium hydroxide, $NaOH(aq)$, form water and sodium chloride, $NaCl(aq)$.

OBJECTIVE 23

67. Describe the process that takes place in the following neutralization reaction, mentioning the nature of the particles in the solution before and after the reaction: The strong acid nitric acid, $HNO_3(aq)$, and the strong base potassium hydroxide, $KOH(aq)$, form water and potassium nitrate, $KNO_3(aq)$.

OBJECTIVE 23

68. Describe the process that takes place in the following neutralization reaction, mentioning the nature of the particles in the solution before and after the reaction: The strong acid nitric acid, $HNO_3(aq)$, and water insoluble nickel(II) hydroxide, $Ni(OH)_2(s)$, form nickel(II) nitrate, $Ni(NO_3)_2(aq)$, and water.

OBJECTIVE 23

OBJECTIVE 23

69. Describe the process that takes place in the following neutralization reaction, mentioning the nature of the particles in the solution before and after the reaction: The strong acid hydrochloric acid, $HCl(aq)$, and water insoluble chromium(III) hydroxide, $Cr(OH)_3(s)$, form chromium(III) chloride, $CrCl_3(aq)$, and water.

OBJECTIVE 23

70. Describe the process that takes place in the following neutralization reaction, mentioning the nature of the particles in the solution before and after the reaction: The strong acid hydrochloric acid, $HCl(aq)$, and the weak base potassium carbonate, $K_2CO_3(aq)$, form water, carbon dioxide, $CO_2(g)$, and potassium chloride, $KCl(aq)$.

OBJECTIVE 23

71. Describe the process that takes place in the following neutralization reaction, mentioning the nature of the particles in the solution before and after the reaction: The strong acid nitric acid, $HNO_3(aq)$, and the weak base lithium hydrogen carbonate, $LiHCO_3(aq)$, form water, carbon dioxide, $CO_2(g)$, and lithium nitrate, $LiNO_3(aq)$.

OBJECTIVE 24

72. Write the complete equation for the neutralization reactions that take place when the following water solutions are mixed. (If an acid has more than one acidic hydrogen, assume that there is enough base to remove all of them. Assume that there is enough acid to neutralize all of the basic hydroxide ions.)

a. $HCl(aq) + LiOH(aq)$

b. $H_2SO_4(aq) + NaOH(aq)$

c. $KOH(aq) + HF(aq)$

d. $Cd(OH)_2(s) + HCl(aq)$

OBJECTIVE 24

73. Write the complete equation for the neutralization reactions that take place when the following water solutions are mixed. (If an acid has more than one acidic hydrogen, assume that there is enough base to remove all of them. Assume that there is enough acid to neutralize all of the basic hydroxide ions.)

a. $LiOH(aq) + HNO_2(aq)$

b. $Co(OH)_2(s) + HNO_3(aq)$

c. $H_3PO_4(aq) + KOH(aq)$

OBJECTIVE 24 OBJECTIVE 25

74. Write the complete equation for the reaction between $HI(aq)$ and water-insoluble solid $CaCO_3$.

OBJECTIVE 24 OBJECTIVE 25

75. Write the complete equation for the reaction between $HCl(aq)$ and water-insoluble solid $Al_2(CO_3)_3$.

OBJECTIVE 24

76. Iron(III) sulfate is made in industry by the neutralization reaction between solid iron(III) hydroxide and aqueous sulfuric acid. The iron(III) sulfate is then added with sodium hydroxide to municipal water in water treatment plants. These compounds react to form a precipitate that settles to the bottom of the holding tank, taking impurities with it. Write the complete equations for both the neutralization reaction that forms iron(III) sulfate and the precipitation reaction between water solutions of iron(III) sulfate and sodium hydroxide.

OBJECTIVE 24

77. Industrial chemists make hydrofluoric acid (which is used in aluminum and uranium processing, to etch glass, and to make CFCs) from the reactions of aqueous calcium fluoride and aqueous sulfuric acid. Write the complete equation for this reaction.

78. Complete the following equations by writing the formulas for the acid and base that could form the given products.

a. ____ + ____ $\rightarrow$ $H_2O(l) + NaCl(aq)$

b. ____ + ____ $\rightarrow$ $2H_2O(l) + Li_2SO_4(aq)$

c. ____ + ____ $\rightarrow$ $H_2O(l) + CO_2(g) + 2KCl(aq)$

79. Complete the following equations by writing the formulas for the acid and base that could form the given products.

   a. ____ + ____ $\rightarrow$ $H_2O(l) + NaC_2H_3O_2(aq)$

   b. ____ + ____ $\rightarrow$ $H_2O(l) + CO_2(g) + LiNO_3(aq)$

   c. ____ + ____ $\rightarrow$ $H_2O(l) + KNO_3(aq)$

### Section 5.7 Brønsted–Lowry Acids and Bases

80. Explain why the Brønsted–Lowry definitions for acid and base are often used instead of the Arrhenius definitions to describe acid–base reactions. OBJECTIVE 36

81. Write the formula for the conjugate acid of each of the following. OBJECTIVE 27

   a. $IO_3^-$ b. $HSO_3^-$ c. $PO_3^{3-}$ d. $H^-$

82. Write the formula for the conjugate acid of each of the following. OBJECTIVE 27

   a. $HC_2O_4^-$ b. $SO_3^{2-}$ c. $BrO^-$ d. $NH_2^-$

83. Write the formula for the conjugate base of each of the following. OBJECTIVE 28

   a. $HClO_4$ b. $HSO_3^-$ c. $H_3O^+$ d. $H_3PO_2$

84. Write the formula for the conjugate base of each of the following. OBJECTIVE 28

   a. $NH_4^+$ b. $H_2S$ c. $HNO_2$ d. $HC_2O_4^-$

85. Explain why a substance can be a Brønsted–Lowry acid in one reaction and a Brønsted–Lowry base in a different reaction. Give an example to illustrate your explanation. OBJECTIVE 29

86. Explain why the Arrhenius definitions for acid and base and not the Brønsted-Lowry definitions are used to describe whether an isolated substance is an acid or base. OBJECTIVE 30

87. For each of the following equations, identify the Brønsted–Lowry acid and base for the forward reaction. OBJECTIVE 31

   a. $NaCN(aq) + HC_2H_3O_2(aq) \rightarrow NaC_2H_3O_2(aq) + HCN(aq)$

   b. $H_2PO_3^-(aq) + HF(aq) \rightleftharpoons H_3PO_3(aq) + F^-(aq)$

   c. $H_2PO_3^-(aq) + 2OH^-(aq) \rightarrow PO_3^{3-}(aq) + 2H_2O(l)$

   d. $3NaOH(aq) + H_3PO_3(aq) \rightarrow 3H_2O(l) + Na_3PO_3(aq)$

88. For each of the following equations, identify the Brønsted–Lowry acid and base for the forward reaction. OBJECTIVE 31

   a. $3NaOH(aq) + H_3PO_4(aq) \rightarrow 3H_2O(l) + Na_3PO_4(aq)$

   b. $HS^-(aq) + HIO_3(aq) \rightarrow H_2S(aq) + IO_3^-(aq)$

   c. $HS^-(aq) + OH^-(aq) \rightarrow S^{2-}(aq) + H_2O(l)$

89. Butanoic acid, $CH_3CH_2CH_2CO_2H$, is a monoprotic weak acid that is responsible for the smell of rancid butter. Write the formula for the conjugate base of this acid. Write the equation for the reaction between this acid and water, and indicate the Brønsted–Lowry acid and base for the forward reaction. (The acidic hydrogen atom is on the right side of the formula.)

90. One of the substances that give wet goats and dirty gym socks their characteristic odors is hexanoic acid, $CH_3CH_2CH_2CH_2CH_2CO_2H$, which is a monoprotic weak acid. Write the formula for the conjugate base of this acid. Write the equation for the reaction between this acid and water, and indicate the Brønsted–Lowry acid and base for the forward reaction. (The acidic hydrogen atom is on the right side of the formula.)

91. Identify the amphoteric substance in each of the following equations.
    a. $HCl(aq) + HS^-(aq) \rightarrow Cl^-(aq) + H_2S(aq)$
    b. $HS^-(aq) + OH^-(aq) \rightarrow S^{2-}(aq) + H_2O(l)$
92. Identify the amphoteric substance in each of the following equations.
    a. $HSO_3^-(aq) + HF(aq) \rightarrow H_2SO_3(aq) + F^-(aq)$
    b. $NH_3(aq) + HSO_3^-(aq) \rightarrow NH_4^+(aq) + SO_3^{2-}(aq)$

### Additional Problems

93. For each of the following pairs of compounds, write the complete equation for the neutralization reaction that takes place when the substances are mixed. (You can assume that there is enough base to remove all of the acidic hydrogen atoms, that there is enough acid to neutralize all of the basic hydroxide ions, and that each reaction goes to completion.)
    a. $HBr(aq) + NaOH(aq)$
    b. $H_2SO_3(aq) + LiOH(aq)$
    c. $KHCO_3(aq) + HF(aq)$
    d. $Al(OH)_3(s) + HNO_3(aq)$
94. For each of the following pairs of compounds, write the complete equation for the neutralization reaction that takes place when the substances are mixed. (You can assume that there is enough base to remove all of the acidic hydrogen atoms, that there is enough acid to neutralize all of the basic hydroxide ions, and that each reaction goes to completion.)
    a. $Ni(OH)_2(s) + HBr(aq)$
    b. $K_2CO_3(aq) + HC_2H_3O_2(aq)$
    c. $HOCl(aq) + NaOH(aq)$
    d. $H_3PO_3\ (aq) + KOH(aq)$
95. Classify each of the following substances as acidic, basic, or neutral.
    a. An apple with a pH of 2.9
    b. Milk of magnesia with a pH of 10.4
    c. Fresh egg white with a pH of 7.6
96. Classify each of the following substances as acidic, basic, or neutral.
    a. A liquid detergent with a pH of 10.1
    b. Maple syrup with a pH of 7.0
    c. Wine with a pH of 3.2
97. The pH of processed cheese is kept at about 5.7 to prevent it from spoiling. Is this acidic, basic, or neutral?
98. Is it possible for a weak acid solution to have a lower pH than a strong acid solution? If so, how?
99. The walls of limestone caverns are composed of solid calcium carbonate. The ground water that makes its way down from the surface into these caverns is often acidic. The calcium carbonate and the $H^+$ ions from the acidic water react to dissolve the limestone. If this happens to the ceiling of the cavern, the ceiling can collapse, leading to what is called a sinkhole. Write the net ionic equation for the reaction between the solid calcium carbonate and the aqueous $H^+$ ions.

**100.** Magnesium sulfate, a substance used for fireproofing and paper sizing, is made in industry from the reaction of aqueous sulfuric acid and solid magnesium hydroxide. Write the complete equation for this reaction.

101. Manganese(II) phosphate is used to coat steel, aluminum, and other metals to prevent corrosion. It is produced in the reaction between solid manganese(II) hydroxide and aqueous phosphoric acid. Write the complete equation for this reaction.

**102.** The smell of Swiss cheese is, in part, due to the monoprotic weak acid propanoic acid, $CH_3CH_2CO_2H$. Write the equation for the complete reaction between this acid and sodium hydroxide. (The acidic hydrogen atom is on the right.)

103. Lactic acid, $CH_3CH(OH)CO_2H$, is used in cosmetic lotions, some of which claim to remove wrinkles. The lactic acid is thought to speed the removal of dead skin cells. Write the equation for the complete reaction between this acid and potassium hydroxide. (The acidic hydrogen atom is on the right.)

**104.** Malic acid, $HO_2CCH_2CH(OH)CO_2H$, is a diprotic weak acid found in apples and watermelon. Write the equation for the complete reaction between this acid and sodium hydroxide. (The acidic hydrogen atoms are on each end of the formula.)

105. One of the substances used to make nylon is hexanedioic acid, $HO_2CCH_2CH_2CH_2CH_2CO_2H$. This diprotic weak acid is also called adipic acid. Write the equation for the complete reaction between this acid and sodium hydroxide. (The acidic hydrogen atoms are on each end of the formula.)

**106.** For the following equation, identify the Brønsted–Lowry acid and base for the forward reaction, and write the formulas for the conjugate acid–base pairs.

$$NaHS(aq) + NaHSO_4(aq) \rightarrow H_2S(g) + Na_2SO_4(aq)$$

107. For the following equation, identify the Brønsted–Lowry acid and base for the forward reaction, and write the formulas for the conjugate acid–base pairs.

$$HF(aq) + NaHSO_3(aq) \rightarrow NaF(aq) + H_2SO_3(aq)$$

### Discussion Problems

108. Assume you are given a water solution and told that it contains either hydrochloric acid or sodium chloride. Describe how you could determine which of these is present.

109. Assume that you are given a water solution that contains either sodium hydroxide or sodium chloride. Describe how you could determine which is in solution.

110. Assume that you are given a water solution that contains either sodium carbonate or sodium hydroxide. Describe how you could determine which is in solution.

*The distinctive color of the Statue of Liberty comes from oxidation–reduction reactions.*

CHAPTER 6

# Oxidation–Reduction Reactions



In many important chemical reactions, electrons are transferred from atom to atom. We are surrounded, inside and out, by these reactions, which are commonly called oxidation–reduction (or redox) reactions. Let's consider a typical "new millennium" family, sitting around the dining room table after the dishes have been cleared.

Mom, a computer programmer, is typing away on her portable computer. She's very eager to see whether the idea she got while driving home will fix a glitch in the accounting program at work. Chris, the seven-year-old, is fighting the bad guys on her Game Boy®. The electric currents from the batteries that power the computer and the game are generated by oxidation–reduction reactions. Buddy, who is fifteen, has recently become interested in studying Eastern philosophy. Just now, he's gazing meditatively out the window, but redox reactions are powering his activity as well; they are crucial to the storage and release of energy in all our bodies. Dad's an engineer in charge of blasting a tunnel under the bay for the city's new rapid-transit project. Each of the explosions that he triggers is created by oxidation–reduction reactions. The silverware he has just cleared from the table is tarnishing via redox reactions, and the combustion of natural gas in the heater warming the room is a redox reaction as well. . . .

*Oxidation–reduction reactions power the batteries that run this Game Boy.*

## Review Skills

The presentation of information in this chapter assumes that you can already perform the tasks listed below. You can test your readiness to proceed by answering the Review Questions at the end of the chapter. This might also be a good time to read the Chapter Objectives, which precede the Review Questions.

- Determine the charge on a monatomic ion in an ionic formula. (Section 3.5)
- Determine the formulas, including the charges, for common polyatomic ions. (Section 3.5)
- Identify a chemical formula as representing an element, a binary ionic compound, an ionic compound with one or two polyatomic ions, or a molecular compound. (Section 5.3)

## 6.1 An Introduction to Oxidation–Reduction Reactions

Zinc oxide is a white substance used as a pigment in rubber, sun-blocking ointments, and paint. It is added to plastics to make them less likely to be damaged by ultraviolet radiation and is also used as a dietary supplement. It can be made from the reaction of pure zinc and oxygen:

$$2Zn(s) + O_2(g) \rightarrow 2ZnO(s)$$

Zinc oxide ointment protects noses from the harmful rays of the sun.

In a similar reaction that occurs every time you drive your car around the block, nitrogen monoxide is formed from some of the nitrogen and oxygen that are drawn into your car's engine:

$$N_2(g) + O_2(g) \rightarrow 2NO(g)$$

This nitrogen monoxide in turn produces other substances that lead to acid rain and help create the brown haze above our cities.

When an element, such as zinc or nitrogen, combines with oxygen, chemists say it is oxidized (or undergoes oxidation). They also use the term **oxidation** to describe many similar reactions that do not have oxygen as a reactant. This section explains the meaning of oxidation and shows why oxidation is always coupled with a corresponding chemical change called reduction.

### Oxidation, Reduction, and the Formation of Binary Ionic Compounds

Zinc oxide is an ionic compound made up of zinc cations, $Zn^{2+}$, and oxide anions, $O^{2-}$. When uncharged zinc and oxygen atoms react to form zinc oxide, electrons are transferred from the zinc atoms to the oxygen atoms to form these ions. Each zinc atom loses two electrons, and each oxygen atom gains two electrons.

OBJECTIVE 2

Overall reaction: $2Zn(s) + O_2(g) \rightarrow 2ZnO(s)$

What happens to Zn: $Zn \rightarrow Zn^{2+} + 2e^-$ or $2Zn \rightarrow 2Zn^{2+} + 4e^-$

What happens to O: $O + 2e^- \rightarrow O^{2-}$ or $O_2 + 4e^- \rightarrow 2O^{2-}$

As we saw in Chapter 3, this transfer of electrons from metal atoms to nonmetal atoms is the general process for the formation of any binary ionic compound from its elements. For example, when sodium chloride is formed from the reaction of metallic sodium with gaseous chlorine, each sodium atom loses an electron, and each chlorine atom gains one.

Overall reaction: $2Na(s) + Cl_2(g) \rightarrow 2NaCl(s)$

What happens to Na: $Na \rightarrow Na^+ + e^-$ or $2Na \rightarrow 2Na^+ + 2e^-$

What happens to Cl: $Cl + e^- \rightarrow Cl^-$ or $Cl_2 + 2e^- \rightarrow 2Cl^-$

The reactions that form sodium chloride and zinc oxide from their elements are so similar that chemists find it useful to describe them using the same terms. Zinc atoms that lose electrons in the reaction with oxygen are said to be oxidized; therefore, when sodium atoms undergo a similar change in their reaction with chlorine, chemists say that they too are *oxidized*, even though no oxygen is present. According to the modern convention, any chemical change in which an element loses electrons is called an **oxidation** (Figure 6.1).

OBJECTIVE 2

**Figure 6.1**
**Oxidation and the Formation of Binary Ionic Compounds**

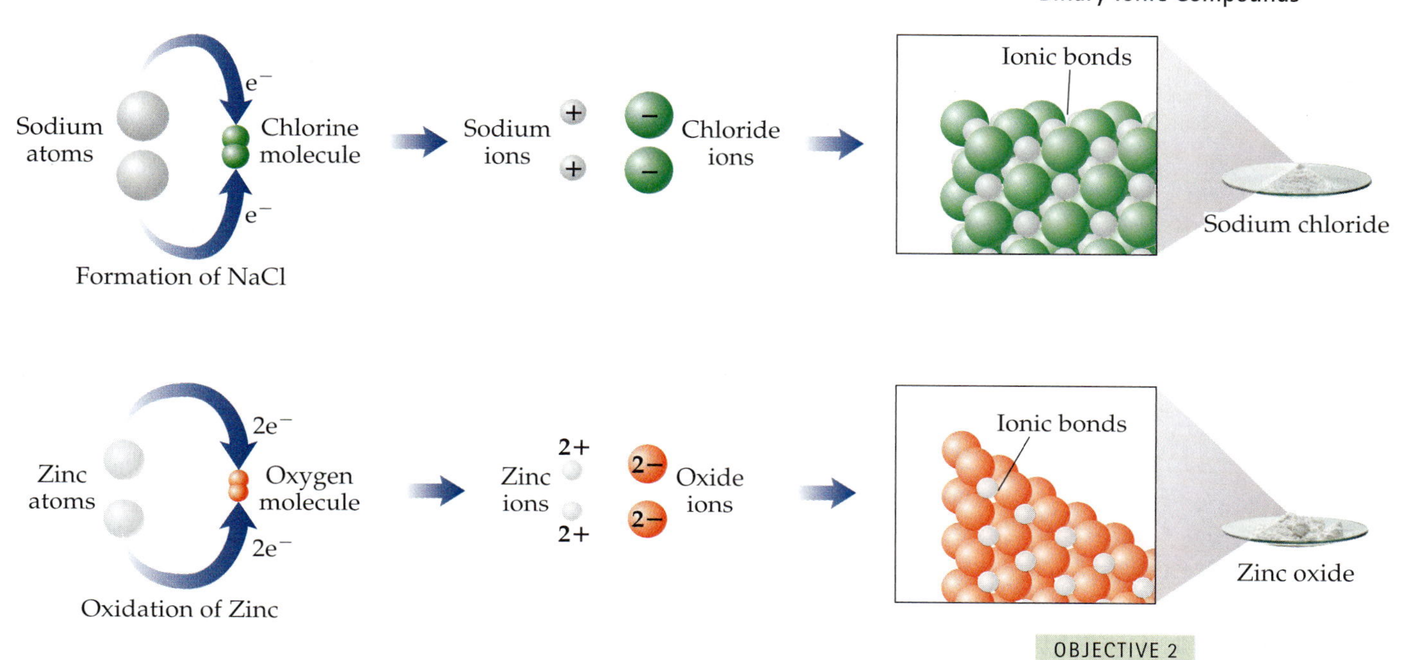


OBJECTIVE 2

The concept of reduction has undergone a similar evolution. At high temperature, zinc oxide, ZnO, reacts with carbon, C, to form molten zinc and carbon monoxide gas.

$$ZnO(s) + C(s) \xrightarrow{\Delta} Zn(l) + CO(g)$$

Bonds between zinc atoms and oxygen atoms are lost in this reaction, so chemists say that the zinc has been *reduced*. Like the term *oxidation*, the term *reduction* has been expanded to include similar reactions, even when oxygen is not a participant. The zinc ions in zinc oxide have a 2+ charge, and the atoms in metallic zinc are uncharged. Thus, in the conversion of zinc oxide to metallic zinc, each zinc ion must gain two electrons. According to the modern definition, any chemical change in which an element gains electrons is called a **reduction**. (Yes, *reduction* means a *gain* of electrons.) Because this can be confusing, some people use a memory aid to remember what oxidation and reduction mean in terms of the electron transfer. One device is the phrase **oil rig** which stands for **o**xidation **i**s **l**oss (of electrons)" and "**r**eduction **i**s **g**ain (of electrons)."

OBJECTIVE 2

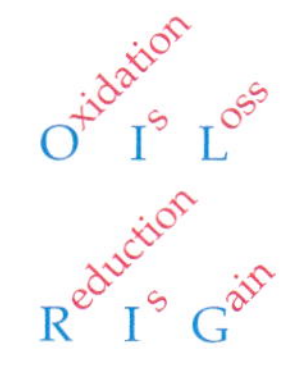

When an electric current passes through molten sodium chloride, the sodium ions, $Na^+$, are converted into uncharged sodium atoms, and the chloride ions, $Cl^-$, are converted into uncharged chlorine molecules, $Cl_2$. Because sodium ions gain one electron each, we say they are reduced. Chloride ions lose one electron each, so they are oxidized.

$$2NaCl(l) \xrightarrow{\text{electric current}} 2Na(l) + Cl_2(g)$$

Oxidation: $2Cl^- \rightarrow Cl_2 + 2e^-$

Reduction: $2Na^+ + 2e^- \rightarrow 2Na$

Electrons are rarely found unattached to atoms. Thus, for one element or compound to lose electrons and be oxidized, another element or compound must be there to gain the electrons and be reduced. In other words, oxidation (loss of electrons) must be accompanied by reduction (gain of electrons). In the reaction that forms ZnO from Zn and $O_2$, the uncharged zinc atoms cannot easily lose electrons and be oxidized unless something like oxygen is there to gain the electrons and be reduced. In the reaction that converts NaCl to Na and $Cl_2$, the chloride ions can lose electrons and be oxidized because the sodium ions are available to gain the electrons and be reduced.

By similar reasoning, we can say that reduction requires oxidation. Because electrons are not likely to be found separated from an element or compound, a substance cannot gain electrons and be reduced unless there is another substance that is able to transfer the electrons and be oxidized. Oxidation and reduction take place together.

Reactions in which electrons are transferred, resulting in oxidation and reduction, are called **oxidation–reduction reactions.** Because the term *oxidation–reduction* is a bit cumbersome, we usually call these reactions **redox reactions.**

Even though the oxidation and reduction of a redox reaction take place simultaneously, each making the other possible, chemists often have reason to describe the reactions separately. The separate oxidation and reduction equations are called **half-reactions.** For example, in the reaction

$$2Zn(s) + O_2(g) \rightarrow 2ZnO(s)$$

the oxidation half-reaction is

$$2Zn \rightarrow 2Zn^{2+} + 4e^-$$

and the reduction half-reaction is

$$O_2 + 4e^- \rightarrow 2O^{2-}$$

Because the zinc atoms lose the electrons that make it possible for the oxygen atoms to gain electrons and be reduced, the zinc is called the reducing agent. A **reducing agent** is a substance that loses electrons, making it possible for another substance to gain electrons and be reduced. *The oxidized substance is always the reducing agent.*

Because the oxygen atoms gain electrons and make it possible for the zinc atoms to lose electrons and be oxidized, the oxygen is called the oxidizing agent. An **oxidizing agent** is a substance that gains electrons, making it possible for another substance to lose electrons and be oxidized. *The reduced substance is always the oxidizing agent.*

In the reaction that forms sodium chloride from the elements sodium and chlorine, sodium is oxidized, and chlorine is reduced. Because sodium makes it possible for chlorine to be reduced, sodium is the reducing agent in this reaction. Because chlorine makes it possible for sodium to be oxidized, chlorine is the oxidizing agent.

$$2Na(s) + Cl_2(g) \rightarrow 2NaCl(s)$$

reducing agent (2Na), oxidizing agent ($Cl_2$)

Oxidation half-reaction: $2Na \rightarrow 2Na^+ + 2e^-$

Reduction half-reaction: $Cl_2 + 2e^- \rightarrow 2Cl^-$

## Oxidation–Reduction and Molecular Compounds

The oxidation of nitrogen to form nitrogen monoxide is very similar to the oxidation of zinc to form zinc oxide.

$$N_2(g) + O_2(g) \rightarrow 2NO(g)$$

$$2Zn(s) + O_2(g) \rightarrow 2ZnO(s)$$

The main difference between these reactions is that as the nitrogen monoxide forms, electrons are not transferred completely, as occurs in the formation of zinc oxide, and no ions are formed. Nitrogen monoxide is a molecular compound, and the bonds between the nitrogen and the oxygen are covalent bonds, in which electrons are shared. Because the oxygen atoms attract electrons more strongly than do the nitrogen atoms, there is a partial transfer of electrons from the nitrogen atoms to the oxygen atoms in the formation of NO molecules, leading to polar bonds with a partial negative charge on each oxygen atom and a partial positive charge on each nitrogen atom.

OBJECTIVE 2

$$^{\delta+}N \rightarrow O^{\delta-}$$

OBJECTIVE 2

Because the reactions are otherwise so much alike, chemists have expanded the definition of oxidation–reduction reactions to include *partial* as well as *complete* transfer of electrons. Thus **oxidation** is defined as the complete or partial loss of electrons, **reduction** as the complete or partial gain of electrons. The nitrogen in the reaction that forms NO from nitrogen and oxygen is oxidized, and the oxygen is reduced. Because the nitrogen makes it possible for the oxygen to be reduced, the nitrogen is the reducing agent. The oxygen is the oxidizing agent (Figure 6.2).

$$\underset{\text{the reducing agent}}{\underset{\text{Oxidized;}}{N_2(g)}} + \underset{\text{the oxidizing agent}}{\underset{\text{Reduced;}}{O_2(g)}} \rightarrow 2NO(g)$$

Special Topic 6.1 describes how oxidizing agents might play a role in aging and how a good healthy diet might slow the aging process.

## SPECIAL TOPIC 6.1 Oxidizing Agents and Aging

In some of the normal chemical reactions that take place in the human body, strong oxidizing agents, such as hydrogen peroxide, are formed. These highly reactive substances cause chemical changes in cell DNA that can be damaging unless the changes are reversed. Fortunately, in healthy cells, normal repair reactions occur that convert the altered DNA back to its normal form.

$$\text{Normal DNA} \underset{\text{normal repair}}{\overset{\text{oxidizing agents (such as } H_2O_2)}{\rightleftharpoons}} \text{Altered DNA}$$

The repair mechanisms are thought to slow down with age. Some medical researchers believe that this slowing down of DNA repair is connected to certain diseases associated with aging, such as cancer, heart disease, cataracts, and brain dysfunction.

$$\text{Normal DNA} \underset{\text{normal repair slowed with aging}}{\overset{\text{oxidizing agents (such as } H_2O_2)}{\rightleftharpoons}} \text{Altered DNA}$$

Substances called *antioxidants* that are found in food react with oxidizing agents (such as hydrogen peroxide) and thus remove them from our system. This is believed to slow the alteration of DNA, so the slower rate of normal repair can balance it.

Five portions of fruits and vegetables per day have enough of the antioxidants vitamin C and vitamin E to mitigate some of the problems of aging.

$$\text{Normal DNA} \underset{\text{normal repair slowed with aging}}{\overset{\text{fewer oxidizing agents}}{\rightleftharpoons}} \text{Altered DNA}$$

Vitamins C and E are antioxidants, and foods that contain relatively high amounts of them are considered important in slowing the onset of some of the medical problems associated with aging. Five servings of fruits and vegetables per day are thought to supply enough antioxidants to provide reasonable protection from the damage done by oxidizing agents.

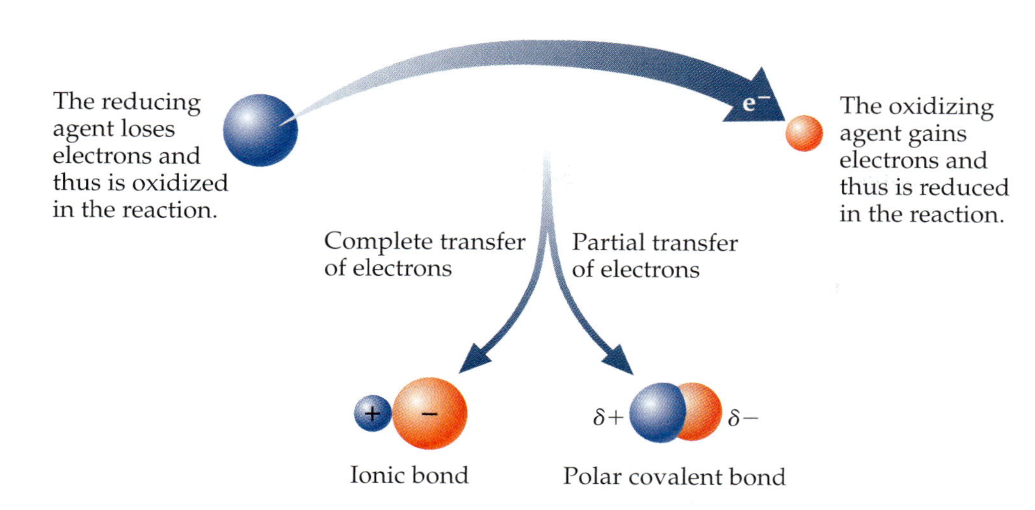
The reducing agent loses electrons and thus is oxidized in the reaction.
e⁻
The oxidizing agent gains electrons and thus is reduced in the reaction.
Complete transfer of electrons
Partial transfer of electrons
+
−
δ+
δ−
Ionic bond
Polar covalent bond

**Figure 6.2**
**Redox Reactions**
In oxidation–reduction (redox) reactions, electrons are completely or partially transferred.

OBJECTIVE 2

## 6.2 Oxidation Numbers

Phosphates, such as ammonium phosphate, are important components of fertilizers used to stimulate the growth of agricultural crops and to make our gardens green. Their commercial synthesis requires elemental phosphorus, which can be acquired by heating phosphate rock (containing calcium phosphate) with sand (containing silicon dioxide) and coke (a carbon-rich mixture produced by heating coal). This method for isolating phosphorus, which is called the *furnace process*, is summarized in the first equation below. The other equations show how phosphorus can be converted into ammonium phosphate.

$$2Ca_3(PO_4)_2 + 6SiO_2 + 10C \rightarrow P_4 + 10CO + 6CaSiO_3$$

$$P_4 + 5O_2 + 6H_2O \rightarrow 4H_3PO_4$$

$$H_3PO_4 + 3NH_3 \rightarrow (NH_4)_3PO_4$$

This plant houses electric phosphate smelting furnaces used to make elemental phosphorus.

Are these reactions oxidation–reduction reactions? Are electrons transferred? Simply reading a chemical equation does not always tell us whether oxidation and reduction have occurred, so chemists have developed a numerical system to help identify a reaction as redox. For redox reactions, this system also shows us which element is oxidized, which is reduced, what the oxidizing agent is, and what the reducing agent is.

The first step in this system is to assign an **oxidation number** to each atom in the reaction equation. As you become better acquainted with the procedure, you will gain a better understanding of what the numbers signify, but for now, just think of them as tools for keeping track of the flow of electrons in redox reactions. Oxidation numbers are also called **oxidation states.**

OBJECTIVE 4 OBJECTIVE 5 OBJECTIVE 6

If any element undergoes a change of oxidation number in the course of a reaction, the reaction is a redox reaction. If an element's oxidation number increases in a reaction, that element is oxidized. If an element's oxidation number decreases in a reaction, that element is reduced. The reactant containing the element that is oxidized is the reducing agent. The reactant containing the element that is reduced is the oxidizing agent (Table 6.1).

OBJECTIVE 4 OBJECTIVE 5 OBJECTIVE 6

Table 6.1
Questions Answered by the Determination of Oxidation Numbers

| Question | Answer |
|---|---|
| **Is the reaction redox?** | If any atoms change their oxidation number, the reaction is redox. |
| **Which element is oxidized?** | The element that increases its oxidation number is oxidized. |
| **Which element is reduced?** | The element that decreases its oxidation number is reduced. |
| **What's the reducing agent?** | The reactant that contains the element that is oxidized is the reducing agent. |
| **What's the oxidizing agent?** | The reactant that contains the element that is reduced is the oxidizing agent. |

The following sample study sheet describes how you can assign oxidation numbers to individual atoms.

## Sample Study Sheet 6.1
### Assignment of Oxidation Numbers

**Tip-off** You are asked to determine the oxidation number of an atom, or you need to assign oxidation numbers to atoms to determine whether a reaction is a redox reaction and, if it is, to identify which element is oxidized, which is reduced, what the oxidizing agent is, and what the reducing agent is.

**General Steps**

OBJECTIVE 3

Use the following guidelines to assign oxidation numbers to as many atoms as you can. (Table 6.2 provides a summary of these guidelines, with examples.)

- The oxidation number for each atom in a pure element is zero.
- The oxidation number of a monatomic ion is equal to its charge.
- When fluorine atoms are combined with atoms of other elements, their oxidation number is $-1$.
- When oxygen atoms are combined with atoms of other elements, their oxidation number is $-2$, except in peroxides, such as hydrogen peroxide, $H_2O_2$, where their oxidation number is $-1$. (*There are other exceptions that you will not see in this text.*)
- The oxidation number for each hydrogen atom in a molecular compound or a polyatomic ion is $+1$.

If a compound's formula contains one element for which you cannot assign an oxidation number using the guidelines listed above, calculate the oxidation number according to the following rules:

- The sum of the oxidation numbers for the atoms in an uncharged formula is equal to zero.
- The sum of the oxidation numbers for the atoms in a polyatomic ion is equal to the overall charge on the ion.

**EXAMPLE** See Example 6.1.

OBJECTIVE 3

**Table 6.2**
Guidelines for Assigning Oxidation Numbers

| | Oxidation number | Examples | Exceptions |
|---|---|---|---|
| **Pure element** | 0 | The oxidation number for each atom in Zn, $H_2$, and $S_8$ is zero. | none |
| **Monatomic ions** | charge on ion | Cd in $CdCl_2$ is +2. Cl in $CdCl_2$ is −1. H in LiH is −1. | none |
| **Fluorine in the combined form** | −1 | F in $AlF_3$ is −1. F in $CF_4$ is −1. | none |
| **Oxygen in the combined form** | −2 | O in ZnO is −2. O in $H_2O$ is −2. | O is −1 in peroxides, such as $H_2O_2$. |
| **Hydrogen in the combined form** | +1 | H in $H_2O$ is +1. | H is −1 when combined with a metal. |

Example 6.1 shows how we can use our new tools.

## EXAMPLE 6.1 Oxidation Numbers and Redox Reactions

OBJECTIVE 3 OBJECTIVE 4 OBJECTIVE 5 OBJECTIVE 6

The following equations represent the reactions that lead to the formation of ammonium phosphate for fertilizers. Determine the oxidation number for each atom in the formulas. Decide whether each reaction is a redox reaction, and if it is, identify what element is oxidized, what is reduced, what the oxidizing agent is, and what the reducing agent is.

a. $2Ca_3(PO_4)_2 + 6SiO_2 + 10C \rightarrow P_4 + 10CO + 6CaSiO_3$

b. $P_4 + 5O_2 + 6H_2O \rightarrow 4H_3PO_4$

c. $H_3PO_4 + 3NH_3 \rightarrow (NH_4)_3PO_4$

***Solution***

a. The first step is to determine the oxidation number for as many atoms as possible using Table 6.2.

Combined oxygen, oxidation number −2 | Pure elements, oxidation number 0 | Monatomic ion, oxidation number equal to its charge (2+)

$$2Ca_3(PO_4)_2 + 6SiO_2 + 10C \rightarrow P_4 + 10CO + 6CaSiO_3$$

Monatomic ion, oxidation number equal to its charge (2+) | Combined oxygen, oxidation number −2

Because the sum of the oxidation numbers for the atoms in an uncharged molecule is zero, the oxidation number of the carbon atom in CO is +2:

$$(\text{ox \# C}) + (\text{ox \# O}) = 0$$

$$(\text{ox \# C}) + -2 = 0$$

$$(\text{ox \# C}) = +2$$

Using a similar process, we can assign a +4 oxidation number to the silicon atom in $SiO_2$:

$$(\text{ox \# Si}) + 2(\text{ox \# O}) = 0$$

$$(\text{ox \# Si}) + 2(-2) = 0$$

$$(\text{ox \# Si}) = +4$$

Calcium phosphate, $Ca_3(PO_4)_2$, is an ionic compound that contains monatomic calcium ions, $Ca^{2+}$, and polyatomic phosphate ions, $PO_4^{3-}$. The oxidation number of each phosphorus atom can be determined in two ways. The following shows how it can be done by considering the whole formula.

$$3(\text{ox \# Ca}) + 2(\text{ox \# P}) + 8(\text{ox \# O}) = 0$$

$$3(+2) + 2(\text{ox \# P}) + 8(-2) = 0$$

$$(\text{ox \# P}) = +5$$

The oxidation number for the phosphorus atom in $PO_4^{3-}$ is always the same, no matter what cation balances its charge. Thus we could also have determined the oxidation number of each phosphorus atom by considering the phosphate ion separately from the calcium ion.

$$(\text{ox \# P}) + 4(\text{ox \# O}) = -3$$

$$(\text{ox \# P}) + 4(-2) = -3$$

$$(\text{ox \# P}) = +5$$

The silicon atoms in $CaSiO_3$ must have an oxidation number of +4.

(ox # Ca) + (ox # Si) + 3(ox # O) = 0

(+2) + (ox # Si) + 3(−2) = 0

(ox # Si) = +4

The oxidation numbers for the individual atoms in the first reaction are as follows:

Oxidation number increases, oxidized

$$\overset{+2\ +5\ -2}{2Ca_3(PO_4)_2} + \overset{+4\ -2}{6SiO_2} + \overset{0}{10C} \rightarrow \overset{0}{P_4} + \overset{+2-2}{10CO} + \overset{+2+4\ -2}{6CaSiO_3}$$

Oxidation number decreases, reduced

Phosphorus atoms and carbon atoms change their oxidation numbers, so **reaction is redox.** Each phosphorus atom changes its oxidation number from +5 to 0, so the **phosphorus atoms in $Ca_3(PO_4)_2$ are reduced, and $Ca_3(PO_4)_2$ is the oxidizing agent.** Each carbon atom changes its oxidation number from 0 to +2, so the **carbon atoms are oxidized, and carbon is the reducing agent.**

b. Now let's consider the second reaction.

Combined oxygen, oxidation number −2

$$P_4 + 5O_2 + 6H_2O \rightarrow H_3PO_4$$

Pure elements, oxidation number 0

Hydrogen in a molecular compound, oxidation number +1

The following shows how we can determine the oxidation number of the phosphorus atom in $H_3PO_4$.

3(ox # H) + (ox # P) + 4(ox # O) = 0

3(+1) + (ox # P) + 4(−2) = 0

(ox # P) = +5

The oxidation numbers for the individual atoms in the second reaction are as follows:

Oxidation number increases, oxidized

$$\overset{0}{P_4} + \overset{0}{5O_2} + \overset{+1\ -2}{6H_2O} \rightarrow \overset{+1+5\ -2}{4H_3PO_4}$$

Oxidation number decreases, reduced

Phosphorus atoms and oxygen atoms change their oxidation numbers, so **this reaction is redox.** Each phosphorus atom changes its oxidation number from 0 to +5, so the **phosphorus atoms in $P_4$ are oxidized, and $P_4$ is the reducing agent.** Each oxygen atom in $O_2$ changes its oxidation number from 0 to −2, so the **oxygen atoms in $O_2$ are reduced, and $O_2$ is the oxidizing agent.**

c. Finally, let's consider the third reaction.

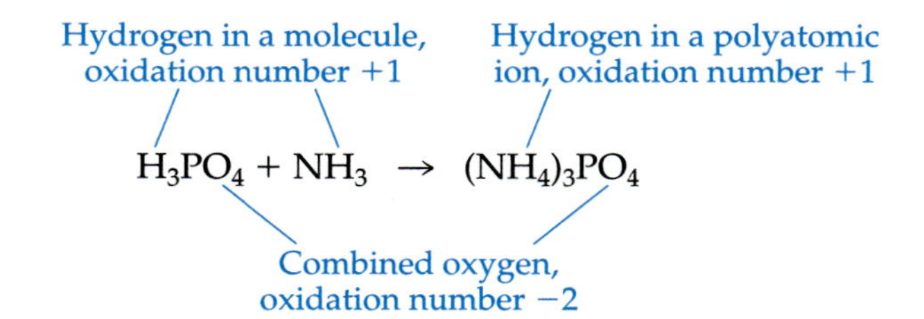


We know from part (b) that the oxidation number of the phosphorus atoms in $H_3PO_4$ is +5.
The oxidation number of the nitrogen atom in $NH_3$ is calculated as follows:

$$(\text{ox \# N}) + 3(\text{ox \# H}) = 0$$

$$(\text{ox \# N}) + 3(+1) = 0$$

$$(\text{ox \# N}) = -3$$

We can determine the oxidation number of each nitrogen atom in $(NH_4)_3PO_4$ in two ways, either from the whole formula or from the formula for the ammonium ion alone.

$$3(\text{ox \# N}) + 12(\text{ox \# H}) + (\text{ox \# P}) + 4(\text{ox \# O}) = 0$$

$$3(\text{ox \# N}) + 12(+1) + (+5) + 4(-2) = 0$$

$$(\text{ox \# N}) = -3$$

or

$$(\text{ox \# N}) + 4(\text{ox \# H}) = +1$$

$$(\text{ox \# N}) + 4(+1) = +1$$

$$(\text{ox \# N}) = -3$$

The oxidation numbers for the individual atoms in this reaction are as follows:

$$\overset{+1\,+5\,-2}{H_3PO_4} + \overset{-3\,+1}{3NH_3} \rightarrow \overset{-3\,+1\;\;+5\,-2}{(NH_4)_3PO_4}$$

None of the atoms change their oxidation number, so **this reaction is not redox.**

## Exercise 6.1 Oxidation Numbers

OBJECTIVE 3

In one part of the steel manufacturing process, carbon is combined with iron to form pig iron. Pig iron is easier to work with than pure iron because it has a lower melting point (about 1130 °C, compared to 1539 °C for pure iron) and is more pliable. The following equations describe its formation. Determine the oxidation number for each atom in the formulas. Decide whether each reaction is a redox reaction, and if it is, identify what is oxidized, what is reduced, what the oxidizing agent is, and what the reducing agent is.

$$2C(s) + O_2(g) \rightarrow 2CO(g)$$

$$Fe_2O_3(s) + CO(g) \rightarrow 2Fe(l) + 3CO_2(g)$$

$$2CO(g) \rightarrow C(\text{in iron}) + CO_2(g)$$

Equations for redox reactions can be difficult to balance, but your ability to determine oxidation numbers can help. You can find a description of the process for balancing redox equations at the following Web site:

**www.chemplace.com/college/**

# 6.3 Types of Chemical Reactions

Chemists often group reactions into general categories, rather than treating each chemical change as unique. For example, you saw in Chapter 4 that many chemical changes can be assigned to the category of precipitation reactions. Understanding the general characteristics of this type of reaction helped you to learn how to predict products and write equations for specific precipitation reactions. You developed similar skills in Chapter 5 for neutralization reactions. Because several types of chemical reactions can also be redox reactions, we continue the discussion of types of chemical reactions here.

## Combination Reactions

In **combination reactions,** two or more elements or compounds combine to form one compound. Combination reactions are also called **synthesis** reactions. The following are examples of combination reactions.

OBJECTIVE 7

$$2Na(s) + Cl_2(g) \rightarrow 2NaCl(s)$$

$$C(s) + O_2(g) \rightarrow CO_2(g)$$

$$MgO(s) + H_2O(l) \rightarrow Mg(OH)_2(s)$$

Two of these combination reactions are also redox reactions. Can you tell which one is not?

## Decomposition Reactions

OBJECTIVE 7

In **decomposition reactions,** one compound is converted into two or more simpler substances. The products can be either elements or compounds. For example, when an electric current is passed through liquid water or molten sodium chloride, these compounds decompose to form their elements.

The decomposition of nitroglycerin.

$$2H_2O(l) \xrightarrow{\text{electric current}} 2H_2(g) + O_2(g)$$

$$2NaCl(l) \xrightarrow{\text{electric current}} 2Na(l) + Cl_2(g)$$

Another example is the decomposition of nitroglycerin. When this compound decomposes, it produces large amounts of gas and heat; consequently, nitroglycerin is a dangerous explosive.

$$4C_3H_5N_3O_9(l) \rightarrow 12CO_2(g) + 6N_2(g) + 10H_2O(g) + O_2(g)$$

As is true of combination reactions, not all decomposition reactions are redox reactions. The following equation represents a decomposition reaction that is not a redox reaction.

$$CaCO_3(s) \xrightarrow{\Delta} CaO(s) + CO_2(g)$$

## Combustion Reactions

OBJECTIVE 7

A log burns in the fireplace as a result of a **combustion reaction,** a redox reaction in which oxidation is very rapid and is accompanied by heat and usually light. The combustion reactions that you will be expected to recognize have oxygen, $O_2$, as one of the reactants. For example, the elements carbon, hydrogen, and sulfur react with oxygen in combustion reactions.

$$C(s) + O_2(g) \rightarrow CO_2(g)$$

$$2H_2(g) + O_2(g) \rightarrow 2H_2O(l)$$

$$S_8(s) + 8O_2(g) \rightarrow 8SO_2(g)$$

OBJECTIVE 8

*When any substance that contains carbon is combusted (or burned) completely, the carbon forms carbon dioxide. When a substance that contains hydrogen is burned completely, the hydrogen forms water.* Therefore, when hydrocarbons found in natural gas, gasoline, and other petroleum products burn completely, the only products are $CO_2$ and $H_2O$. The following equations represent the combustion reactions for methane, the primary component of natural gas, and hexane, which is found in gasoline.

$$CH_4(g) + 2O_2(g) \rightarrow CO_2(g) + 2H_2O(l)$$

$$2C_6H_{14}(l) + 19O_2(g) \rightarrow 12CO_2(g) + 14H_2O(l)$$

The complete combustion of a substance, such as ethanol, $C_2H_5OH$, that contains carbon, hydrogen, and oxygen also yields carbon dioxide and water.

$$C_2H_5OH(l) + 3O_2(g) \rightarrow 2CO_2(g) + 3H_2O(l)$$

*When any substance that contains sulfur burns completely, the sulfur forms sulfur dioxide.* For example, when ethanethiol, $C_2H_5SH$, burns completely, it forms carbon dioxide, water, and sulfur dioxide. Small amounts of this strong-smelling substance are added to natural gas to give the otherwise odorless gas a smell that can be detected if a leak occurs (Figure 6.3).

$$2C_2H_5SH(g) + 9O_2(g) \rightarrow 4CO_2(g) + 6H_2O(l) + 2SO_2(g)$$

**Combustion reactions include oxygen as a reactant and are accompanied by heat and usually light.**

OBJECTIVE 8

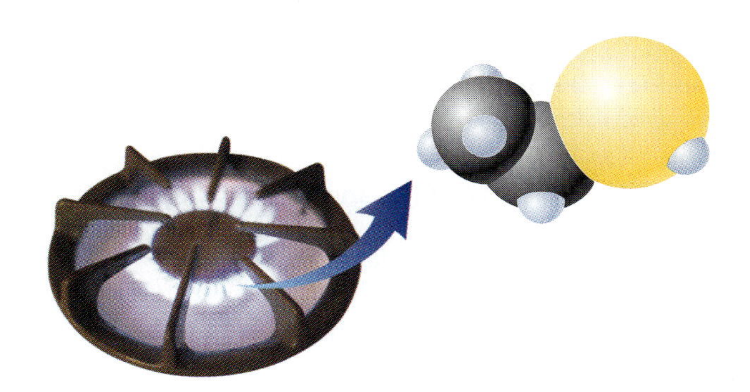

**Figure 6.3**
**Odor to Natural Gas**
The ethanethiol added to natural gas warns us when there is a leak.

The following sample study sheet lists the steps for writing equations for combustion reactions.

## Sample Study Sheet 6.2

### Writing Equations for Combustion Reactions

OBJECTIVE 8

**Tip-off** You are asked to write an equation for the complete combustion of a substance composed of one or more of the elements carbon, hydrogen, oxygen, and sulfur.

**General Steps**

**Step 1** Write the formula for the substance combusted.

**Step 2** Write $O_2(g)$ for the second reactant.

**STEP 3** Predict the products using the following guidelines:

a. If the compound contains carbon, one product will be $CO_2(g)$.

b. If the compound contains hydrogen, one product will be $H_2O(l)$.

(Even though water may be gaseous when it is first formed in a combustion reaction, we usually describe it as a liquid in the equation. By convention, we describe the state of each reactant and product as its state at room temperature and pressure. When water returns to room temperature, it is a liquid.)

c. If the compound contains sulfur, one product will be $SO_2(g)$.

d. Any oxygen in the combusted substance will be distributed between the products already mentioned.

**STEP 4** Balance the equation.

**EXAMPLE** See Example 6.2.

## EXAMPLE 6.2 Combustion Reactions

OBJECTIVE 8

Write balanced equations for the complete combustion of (a) $C_8H_{18}(l)$, (b) $CH_3OH(l)$, and (c) $C_3H_7SH(l)$.

***Solution***

a. The carbon in $C_8H_{18}$ goes to $CO_2(g)$, and the hydrogen goes to $H_2O(l)$.

$$2C_8H_{18}(l) + 25O_2(g) \rightarrow 16CO_2(g) + 18H_2O(l)$$

b. The carbon in $CH_3OH$ goes to $CO_2(g)$, the hydrogen goes to $H_2O(l)$, and the oxygen is distributed between the $CO_2$ and the $H_2O$.

$$2CH_3OH(l) + 3O_2(g) \rightarrow 2CO_2(g) + 4H_2O(l)$$

c. The carbon in $C_3H_7SH$ goes to $CO_2(g)$, the hydrogen goes to $H_2O(l)$, and the sulfur goes to $SO_2(g)$.

$$C_3H_7SH(l) + 6O_2(g) \rightarrow 3CO_2(g) + 4H_2O(l) + SO_2(g)$$

## EXERCISE 6.2 Combustion Reactions

OBJECTIVE 8

Write balanced equations for the complete combustion of (a) $C_4H_{10}(g)$, (b) $C_3H_7OH(l)$, and (c) $C_4H_9SH(l)$.

If there is insufficient oxygen to burn a carbon-containing compound completely, the incomplete combustion converts some of the carbon into carbon monoxide, CO. Because the combustion of gasoline in a car's

## Special Topic 6.2 Air Pollution and Catalytic Converters

When gasoline, which is a mixture of hydrocarbons, burns in the cylinders of a car engine, the primary products are carbon dioxide and water. Unfortunately, the hydrocarbons are not burned completely, so the exhaust leaving the cylinders also contains some unburned hydrocarbons and some carbon monoxide, which are serious pollutants. When the unburned hydrocarbon molecules escape into the atmosphere, they combine with other substances to form eye irritants and other problematic compounds.

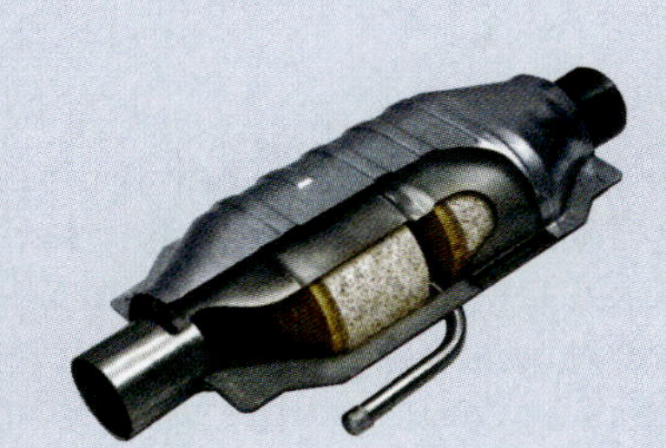
Catalytic Converter

Carbon monoxide is a colorless, odorless, and poisonous gas. When inhaled, it deprives the body of oxygen. Normally, hemoglobin molecules in the blood carry oxygen throughout the body, but when carbon monoxide is present, it combines with hemoglobin and prevents oxygen from doing so. As little as 0.001% CO in air can lead to headache, dizziness, and nausea. Concentrations of 0.1% can cause death.

Catalytic converters in automobile exhaust systems were developed to remove some of the carbon monoxide and unburned hydrocarbons from automobile exhaust. A catalyst is any substance that speeds a chemical reaction without being permanently altered itself. Some of the transition metals, such as platinum, palladium, iridium, and rhodium, and some transition metal oxides, such as $V_2O_5$, CuO, and $Cr_2O_3$, speed up oxidation–reduction reactions. An automobile's catalytic converter contains tiny beads coated with a mixture of transition metals and transition metal oxides. When the exhaust passes over these beads, the oxidation of CO into $CO_2$ and the conversion of unburned hydrocarbons to $CO_2$ and $H_2O$ are both promoted.

The conversion of CO to $CO_2$ and of unburned hydrocarbons to $CO_2$ and $H_2O$ is very slow at normal temperatures, but it takes place quite rapidly at temperatures around 600 °F. This means that the removal of the CO and hydrocarbon pollutants does not take place efficiently when you first start your engine. The pollutants released when a car is first started are called "cold-start emissions." Recent research has focused on developing a way to insulate the catalytic converter so that it retains its heat even when the car is not running. Some of the prototypes can maintain their operating temperatures for almost a day while the engine is shut off. With these new, insulated catalytic converters installed in most cars, cold-start emissions could be reduced by at least 50%. This could have a significant effect on air pollution.

engine is not complete, the exhaust that leaves the engine contains both carbon dioxide and carbon monoxide. However, before this exhaust escapes out the tail pipe, it passes through a catalytic converter, which further oxidizes much of the carbon monoxide to carbon dioxide. See Special Topic 6.2: *Air Pollution and Catalytic Converters*.

## Single-Displacement Reactions

OBJECTIVE 7

Now consider another type of chemical reaction. In **single-displacement reactions**, atoms of one element in a compound are displaced (or replaced) by atoms from a pure element. These reactions are also called single-replacement reactions. All of the following are single-displacement reactions, in which a pure element displaces an element in a compound.

$$Zn(s) + CuSO_4(aq) \rightarrow ZnSO_4(aq) + Cu(s)$$

$$Cd(s) + H_2SO_4(aq) \rightarrow CdSO_4(aq) + H_2(g)$$

$$Cl_2(g) + 2NaI(aq) \rightarrow 2NaCl(aq) + I_2(s)$$

Our model of particle behavior enables us to visualize the movements of atoms and ions participating in single-displacement reactions. For example, consider the first equation above, a reaction in which atoms of zinc replace ions of copper in a copper sulfate solution. Because copper(II) sulfate is a water-soluble ionic compound, the $CuSO_4$ solution consists of free $Cu^{2+}$ ions surrounded by the negatively charged oxygen ends of water molecules and free $SO_4^{2-}$ ions surrounded by the positively charged hydrogen ends of water molecules. These ions move throughout the solution, colliding with each other, with water molecules, and with the walls of their container. Now imagine that a lump of solid zinc is added to the solution. Copper ions begin to collide with the surface of the zinc. When the $Cu^{2+}$ ions collide with the uncharged zinc atoms, two electrons are transferred from the zinc atoms to the copper(II) ions. The resulting zinc ions move into solution, where they become surrounded by the negatively charged ends of water molecules, and the uncharged copper solid forms on the surface of the zinc (Figure 6.4).

OBJECTIVE 9

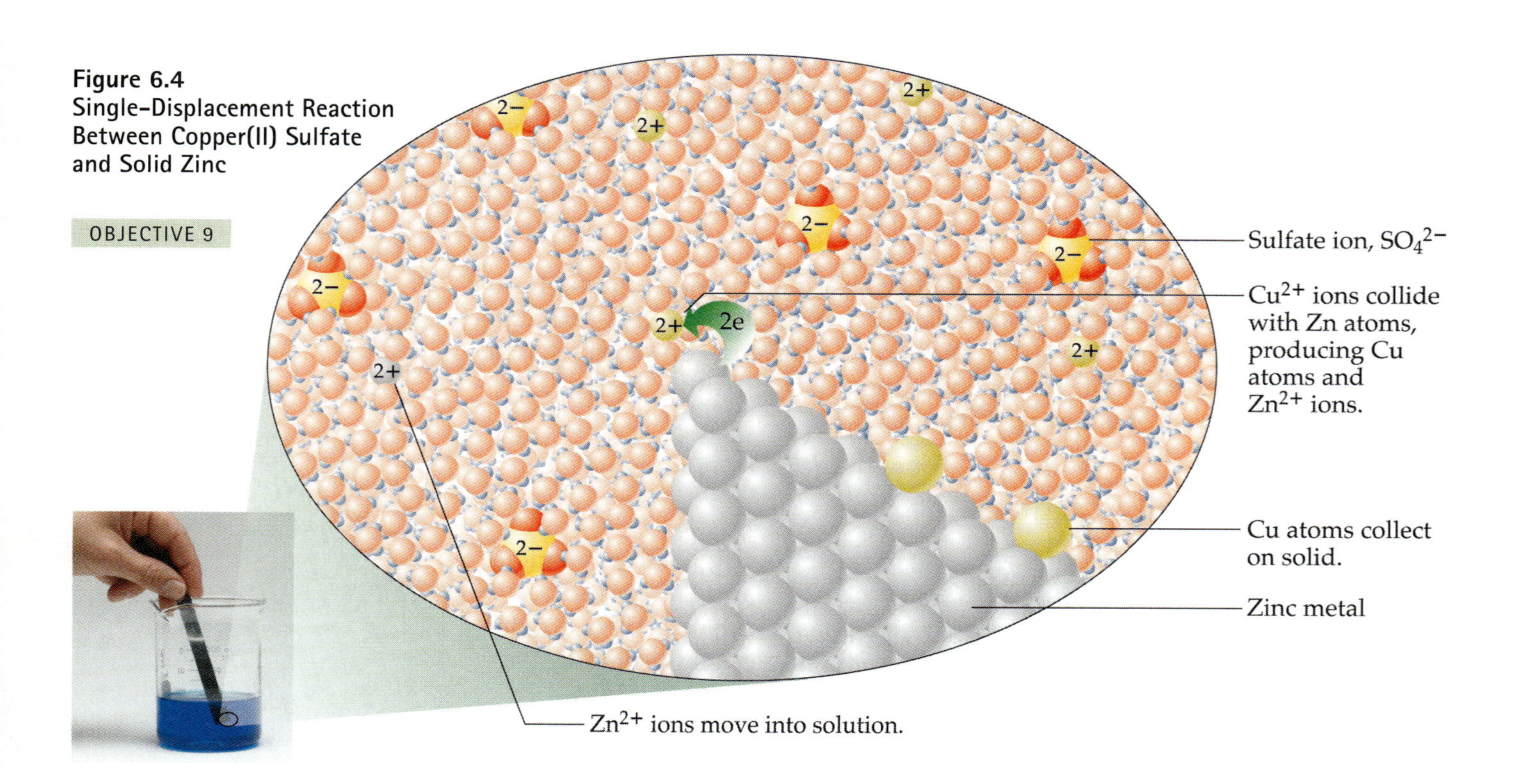
2−
2+
2+
2−
2−
2−
2+
2e
2+
2+
2−
Sulfate ion, $SO_4^{2-}$
$Cu^{2+}$ ions collide with Zn atoms, producing Cu atoms and $Zn^{2+}$ ions.
Cu atoms collect on solid.
Zinc metal
$Zn^{2+}$ ions move into solution.

**Figure 6.4**
**Single-Displacement Reaction Between Copper(II) Sulfate and Solid Zinc**

OBJECTIVE 9

Because the zinc atoms lose electrons in this reaction and change their oxidation number from 0 to +2, they are oxidized, and zinc is the reducing agent. The $Cu^{2+}$ ions gain electrons and decrease their oxidation number from +2 to 0, so they are reduced and act as the oxidizing agent. The half-reaction equations and the net ionic equation for this reaction are

Oxidation: $Zn(s) \rightarrow Zn^{2+}(aq) + 2e^-$

Reduction: $Cu^{2+}(aq) + 2e^- \rightarrow Cu(s)$

Net ionic equation: $Zn(s) + Cu^{2+}(aq) \rightarrow Zn^{2+}(aq) + Cu(s)$

Example 6.3 shows how chemical changes can be classified with respect to the categories described in this section.

## EXAMPLE 6.3 Classification of Chemical Reactions

OBJECTIVE 7

Classify each of these reactions as a combination reaction, a decomposition reaction, a combustion reaction, or a single-displacement reaction.

a. $2Fe_2O_3(s) + 3C(s) \xrightarrow{\Delta} 4Fe(l) + 3CO_2(g)$

b. $Cl_2(g) + F_2(g) \rightarrow 2ClF(g)$

c. $2Pb(NO_3)_2(s) \xrightarrow{\Delta} 4NO_2(g) + 2PbO(s) + O_2(g)$

d. $C_5H_{11}SH(l) + 9O_2(g) \rightarrow 5CO_2(g) + 6H_2O(l) + SO_2(g)$

***Solution***

a. $2Fe_2O_3(s) + 3C(s) \rightarrow 4Fe(l) + 3CO_2(g)$

Because the atoms of the pure element carbon are displacing (or replacing) the atoms of iron in iron(III) oxide, this is a **single-displacement reaction.** This particular reaction is used to isolate metallic iron from iron ore.

b. $Cl_2(g) + F_2(g) \rightarrow 2ClF(g)$

Because two substances are combining to form one substance, this is a **combination reaction.**

c. $2Pb(NO_3)_2(s) \xrightarrow{\Delta} 4NO_2(g) + 2PbO(s) + O_2(g)$

Because one substance is being converted into more than one substance, this is a **decomposition reaction.**

d. $C_5H_{11}SH(l) + 9O_2(g) \rightarrow 5CO_2(g) + 6H_2O(l) + SO_2(g)$

Because a substance combines with oxygen, and because we see carbon in that substance going to $CO_2$, hydrogen going to $H_2O$, and sulfur going to $SO_2$, we classify this reaction as a **combustion reaction.** The compound $C_5H_{11}SH$ is 3-methyl-1-butanethiol, a component of the spray produced by skunks.

### Exercise 6.3 Classification of Chemical Reactions

OBJECTIVE 7

Classify each of these reactions as a combination reaction, a decomposition reaction, a combustion reaction, or a single-displacement reaction.

a. $2HgO(s) \xrightarrow{\Delta} 2Hg(l) + O_2(g)$

b. $C_{12}H_{22}O_{11}(s) + 12O_2(g) \rightarrow 12CO_2(g) + 11H_2O(l)$

c. $B_2O_3(s) + 3Mg(s) \xrightarrow{\Delta} 2B(s) + 3MgO(s)$

d. $C_2H_4(g) + H_2(g) \rightarrow C_2H_6(g)$

## 6.4 Voltaic Cells

We're people on the go—with laptop computers, portable drills, and electronic games that kids can play in the car. To keep all of these tools and toys working, we need batteries, and because the newer electronic devices require more power in smaller packages, scientists are constantly searching for stronger and more efficient batteries. The goal of this section is to help you understand how batteries work and to describe some that may be familiar to you and some that may be new.

A **battery** is a device that converts chemical energy into electrical energy using redox reactions. To discover what this means and how batteries work, let's examine a simple system that generates an electric current via the reaction between zinc metal and copper(II) ions described above (see Figure 6.4). In this redox reaction, uncharged zinc atoms are oxidized to zinc ions, and copper(II) ions are reduced to uncharged copper atoms.

$$Zn(s) + Cu^{2+}(aq) \rightarrow Zn^{2+}(aq) + Cu(s)$$

Oxidation: $Zn(s) \rightarrow Zn^{2+}(aq) + 2e^-$

Reduction: $Cu^{2+}(aq) + 2e^- \rightarrow Cu(s)$

In the last section, we saw how this reaction takes place when zinc metal is added to a solution of copper(II) sulfate. When a $Cu^{2+}$ ion collides with the zinc metal, two electrons are transferred from a zinc atom directly to the copper(II) ion.

OBJECTIVE 10

A clever arrangement of the reaction components enables us to harness this reaction to produce electrical energy. The setup, shown in Figure 6.5, keeps the two half-reactions separated, causing the electrons released in the oxidation half of the reaction to pass through a wire connecting the two halves of the apparatus. The proper name for a setup of this type is a **voltaic cell.** Strictly speaking, a **battery** is a series of voltaic cells joined in such a way that they work together.

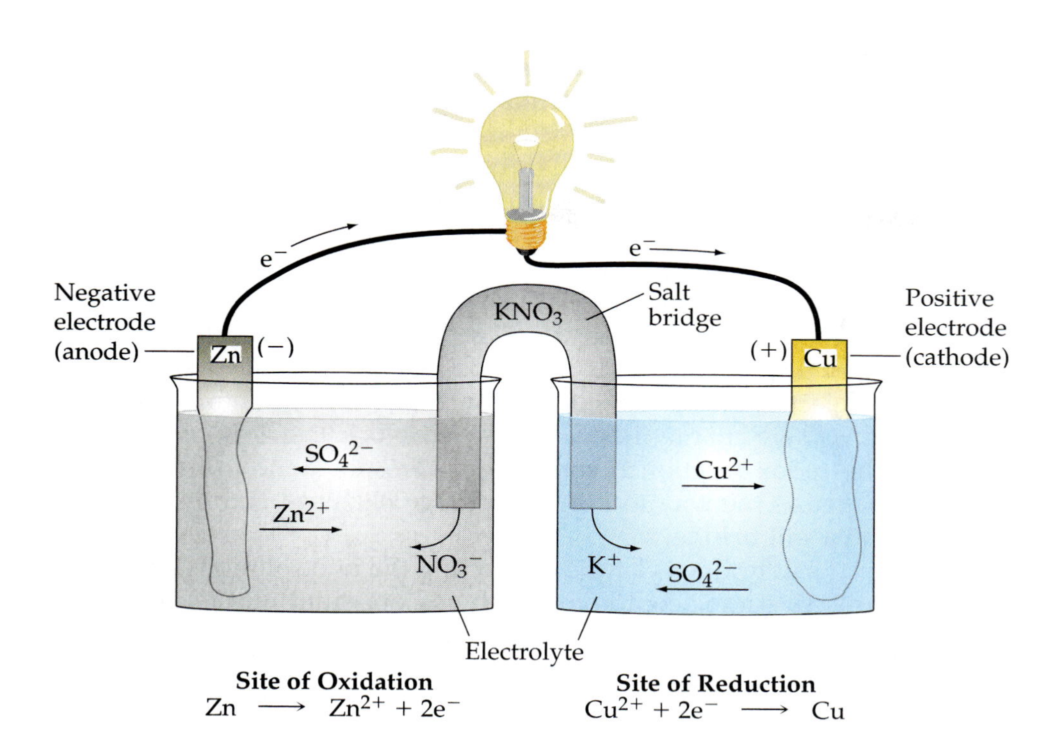


**Figure 6.5**
**A Voltaic Cell**
This diagram shows the key components of a voltaic cell composed of the $Zn/Zn^{2+}$ and $Cu/Cu^{2+}$ half-cells.

OBJECTIVE 10

Messages were sent through telegraph wires by voltaic cells composed of $Zn/Zn^{2+}$ and $Cu/Cu^{2+}$ half-cells.

A voltaic cell, then, is composed of two separate half-cells. In the case of the zinc and copper(II) redox reaction, the first half-cell consists of a strip of zinc metal in a solution of zinc sulfate. The second half-cell consists of a strip of copper metal in a solution of copper(II) sulfate. In the $Zn/Zn^{2+}$ half-cell, zinc atoms lose two electrons and are converted into zinc ions. The two electrons pass through the wire to the $Cu/Cu^{2+}$ half-cell, where $Cu^{2+}$ ions gain the two electrons to form uncharged copper atoms. The zinc metal and copper metal strips are called **electrodes,** the general name for electrical conductors placed in half-cells of voltaic cells.

OBJECTIVE 10

The electrode at which oxidation occurs in a voltaic cell is called the **anode**. Because electrons are lost in oxidation, the anode is the source of electrons. For this reason, the anode of a voltaic cell is designated the negative electrode. Because electrons are lost, forming more positive (or less negative) species at the anode, the anode surroundings tend to become more positive. Thus anions are attracted to the anode. In our voltaic cell, Zn is oxidized to $Zn^{2+}$ at the zinc electrode, so this electrode is the anode. The solution around the zinc metal becomes positive as a result of there being an excess of $Zn^{2+}$ ions, so anions are attracted to the zinc electrode.

OBJECTIVE 11

OBJECTIVE 10

The **cathode** is the electrode in a voltaic cell at which reduction occurs. By convention, the cathode is designated the positive electrode. Because electrons flow along the wire to the cathode, and because substances gain those electrons to become more negative (or less positive), the cathode surroundings tend to become more negative. Thus cations are attracted to the cathode. In our voltaic cell, $Cu^{2+}$ ions are reduced to uncharged copper atoms at the copper strip, so metallic copper is the cathode. Because

OBJECTIVE 10

cations are removed from the solution there and anions are not, the solution around the copper cathode tends to become negative and attracts cations.

The component of the voltaic cell through which ions are able to flow is called the **electrolyte.** For our voltaic cell, the zinc sulfate solution is the electrolyte in the anode half-cell, and the copper(II) sulfate solution is the electrolyte in the cathode half-cell.

As described above, the $Zn/Zn^{2+}$ half-cell tends to become positive because of the loss of electrons as uncharged zinc atoms are converted into $Zn^{2+}$ ions, and the $Cu/Cu^{2+}$ half-cell tends to become negative because of the gain of electrons and the conversion of $Cu^{2+}$ ions into uncharged copper atoms. This charge imbalance would block the flow of electrons and stop the redox reaction if something were not done to balance the growing charges. One way to keep the charge balanced is to introduce a device called a **salt bridge.**

OBJECTIVE 10

One type of salt bridge is a tube connecting the two solutions and filled with an unreactive ionic compound such as potassium nitrate in a semisolid support such as gelatin or agar. For each negative charge lost at the anode because of the loss of an electron, an anion like $NO_3^-$ moves from the salt bridge into the solution to replace it. For example, when 1 zinc atom oxidizes at the anode and loses 2 electrons to form $Zn^{2+}$, 2 nitrates enter the solution to keep the solution electrically neutral overall. For each negative charge gained at the cathode as a result of the gain of an electron, a cation, such as $K^+$ ion, moves into the solution to keep the solution uncharged. For example, each time a copper ion gains 2 electrons and forms an uncharged copper atom, 2 potassium ions enter the solution to replace the $Cu^{2+}$ lost.

## Dry Cells

Although a voltaic cell of the kind just described, containing liquid solutions of zinc and copper ions, was used in early telegraph systems, there are problems with this sort of cell. The greatest problem is that the cell cannot be moved easily because the electrolyte solutions are likely to spill. The Leclanché cell, or dry cell, was developed in the 1860s to solve this problem. It contained a paste, or semisolid, electrolyte. The reactions in the dry cell can be thought of as consisting of the following half-reactions (although they are a bit more complicated than described here).

Anode oxidation: $Zn(s) \rightarrow Zn^{2+}(aq) + 2e^-$

Cathode reduction:

$$2MnO_2(s) + 2NH_4^+(aq) + 2e^- \rightarrow Mn_2O_3(s) + 2NH_3(aq) + H_2O(l)$$

Overall reaction:

$$Zn(s) + 2MnO_2(s) + 2NH_4^+(aq) \rightarrow Zn^{2+}(aq) + Mn_2O_3(s) + 2NH_3(aq) + H_2O(l)$$

These inexpensive and reliable cells have served as the typical "flashlight battery" for many years. Their outer wrap surrounds a zinc metal cylinder

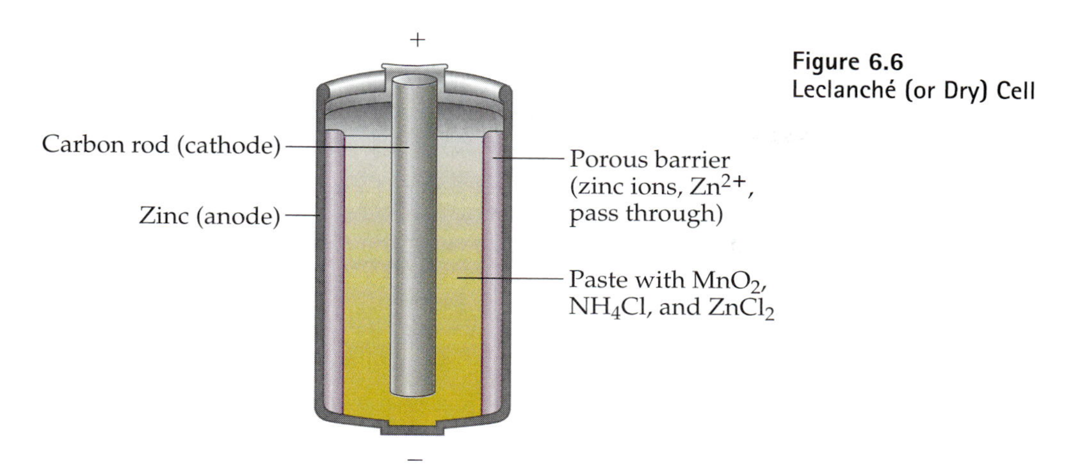


**Figure 6.6**
Leclanché (or Dry) Cell

that acts as the anode (Figure 6.6). Inside the zinc cylinder is a porous barrier that separates the zinc metal from a paste containing $NH_4Cl$, $ZnCl_2$, and $MnO_2$. The porous barrier allows $Zn^{2+}$ ions to pass through, but it keeps the $MnO_2$ from coming into direct contact with the zinc metal. In the center of the cell, a carbon rod that can conduct an electric current acts as the cathode.

## Electrolysis

Voltage, a measure of the strength of an electric current, represents the force that moves electrons from the anode to the cathode in a voltaic cell. When a greater force (voltage) is applied in the opposite direction, electrons can be pushed from what would normally be the cathode toward the voltaic cell's anode. This process is called electrolysis. In a broader sense, **electrolysis** is the process by which a redox reaction is made to occur in the nonspontaneous direction. For example, sodium metal reacts readily with chlorine gas to form sodium chloride, but we do not expect sodium chloride, as it sits in our saltshakers, to decompose into sodium metal and chlorine gas. We say that the forward reaction below is spontaneous and the reverse reaction is nonspontaneous.

$$2Na(s) + Cl_2(g) \rightarrow 2NaCl(s)$$

Sodium metal and chlorine gas can, however, be formed from the electrolysis of salt, in which an electric current is passed through the molten sodium chloride.

$$2NaCl(l) \xrightarrow{\text{electrolysis}} 2Na(l) + Cl_2(g)$$

Electrolysis is used in industry to purify metals, such as copper and aluminum, and in electroplating—the process used, for example, to deposit the chrome on the bumper of a 1955 Chevy.

As we shall soon see, a similar process is used to refresh rechargeable batteries.

## Nickel–Cadmium Batteries

Leclanché cells are not rechargeable; once the reactants are depleted, the battery must be discarded and replaced. Batteries that are not rechargeable are called **primary batteries.** A rechargeable battery is called a **secondary battery** or a **storage battery.** The nickel–cadmium (NiCd) battery is a popular rechargeable battery that uses the following redox reaction:

Anode reaction: $Cd(s) + 2OH^-(aq) \rightarrow Cd(OH)_2(s) + 2e^-$

Cathode reaction:
$NiO(OH)(s) + H_2O(l) + e^- \rightarrow Ni(OH)_2(s) + OH^-(aq)$

Net reaction:
$Cd(s) + 2NiO(OH)(s) + 2H_2O(l) \rightarrow Cd(OH)_2(s) + 2Ni(OH)_2(s)$

To recharge a secondary battery, an opposing external voltage is applied that is greater than the voltage of the cell, pushing the electrons in the opposite direction from the way they move in the normal operation of the cell. In this process, the original chemical reaction is reversed. Because the $Cd(OH)_2$ and $Ni(OH)_2$ produced during the normal operation of the nickel–cadmium battery are solids, they stay at the electrodes where they are formed and are available to be converted back to the original reactants.

$$Cd(s) + 2NiO(OH)(s) + 2H_2O(l) \underset{\text{when being charged}}{\overset{\text{normal operation}}{\rightleftharpoons}} Cd(OH)_2(s) + 2Ni(OH)_2(s)$$

NiCd batteries' long life and low price make them the rechargeable battery of choice for flashlights, portable radios, power tools, and many other common devices.

Leclanché cells and nickel–cadmium batteries both have their drawbacks. One problem with the Leclanché cell is that over time, the zinc metal and the ammonium ions react to corrode the anode, so dry cell batteries have a shorter "shelf life" than other batteries do. Another problem is that Leclanché cells are not rechargeable. Nickel–cadmium batteries are rechargeable, but they are too bulky to be used for many purposes. The manufacturers of electronic devices are constantly trying to make their products smaller and more powerful. Thus they want batteries that are smaller, lighter, and more powerful than the dry cell and nickel–cadmium batteries available today. They want batteries with the highest possible energy-to-mass ratio and energy-to-volume ratio, and dry cells and NiCd batteries have very low ratios. To solve these and other problems, many additional types of batteries have been developed, including the lithium battery (Table 6.3).

Extensive research is being done to develop new types of batteries to power electric cars and trucks. See Special Topic 6.3 for an example.

**Many flashlights use NiCd, rechargeable, batteries.**

Table 6.3
Common Voltaic Cells (Batteries)

| Voltaic cell | Anode reaction | Cathode reaction | Pros | Cons |
|---|---|---|---|---|
| **dry cell** | $Zn(s) \rightarrow Zn^{2+}(aq) + 2e^-$ | $2MnO_2(s) + 2NH_4^+(aq) + 2e^- \rightarrow Mn_2O_3(s) + 2NH_3(aq) + H_2O(l)$ | Inexpensive; reliable | short shelf-life; large and heavy; not rechargeable |
| **alkaline battery** | $Zn(s) + 2OH^-(aq) \rightarrow ZnO(s) + H_2O(l) + 2e^-$ | $2MnO_2(s) + H_2O(l) + 2e^- \rightarrow Mn_2O_3(s) + 2OH^-(aq)$ | longer-lasting and slightly higher energy density than dry cells | more expensive than dry cells; not rechargeable |
| **nickel–cadmium battery** | $Cd(s) + 2OH^-(aq) \rightarrow Cd(OH)_2(s) + 2e^-$ | $NiO(OH)(s) + H_2O(l) + e^- \rightarrow Ni(OH)_2(s) + OH^-(aq)$ | rechargeable; lower price than other recharge-able batteries; long life | relatively low energy density; less efficient with repeated use |
| **lead–acid battery** | $Pb(s) + HSO_4^-(aq) + H_2O(l) \rightarrow PbSO_4(s) + H_3O^+(aq) + 2e^-$ | $PbO_2(s) + HSO_4^-(aq) + 3H_3O^+(aq) + 2e^- \rightarrow PbSO_4(s) + 5H_2O(l)$ | rechargeable; very reliable; inexpensive; durable | very heavy, yielding a very low energy density |
| **lithium battery** | $Li \rightarrow Li^+ + 1e^-$ | $MnO_2 + 1e^- \rightarrow MnO_2^-$ or $FeS + 2e^- \rightarrow Fe + S^{2-}$ or $\frac{1}{2}I_2 + 1e^- \rightarrow I^-$ | rechargeable; reliable; very high energy density | expensive |

## Special Topic 6.3 Zinc–Air Batteries

One type of battery that is being considered for use in electric cars is the zinc–air battery, in which zinc acts as the anode and air provides the oxygen for the cathode reaction. This battery is low in mass because of its use of oxygen as the cathode reactant. The overall reaction is

$$2Zn(s) + O_2(g) \rightarrow 2ZnO(s)$$

Most of the batteries being considered for electric cars take several hours to recharge—a major drawback in comparison to the ease of refilling a gas tank. It is possible that the use of zinc–air batteries will avoid this problem. The reactions are not directly reversible in the zinc–air batteries, so the materials would need to be removed to be recharged. You could stop at your local zinc station, have the spent electrolyte and zinc oxide particles quickly suctioned from the cells and replaced with a slurry of fresh particles and electrolyte, and be ready to go another 200–300 miles. The zinc oxide would later be taken to a central plant where it would be converted back to zinc metal and returned to the stations.

Vans powered by zinc–air batteries are already being used by the European postal services. In 1997, one of them made a 152-mile trip from Chambria, France, over the Alps to Turin, Italy. It not only succeeded in crossing the 6874-ft peak but also averaged 68 mph on the highway.

## Chapter Glossary

**Oxidation** Any chemical change in which at least one element loses electrons, either completely or partially.

**Reduction** Any chemical change in which at least one element gains electrons, either completely or partially.

**Oxidation–reduction reactions** The chemical reactions in which there is a complete or partial transfer of electrons, resulting in oxidation and reduction. These reactions are also called ***redox reactions.***

**Half-reactions** Separate oxidation and reduction reaction equations in which electrons are shown as a reactant or product.

**Reducing agent** A substance that loses electrons, making it possible for another substance to gain electrons and be reduced.

**Oxidizing agent** A substance that gains electrons, making it possible for another substance to lose electrons and be oxidized.

**Oxidation number** A tool for keeping track of the flow of electrons in redox reactions (also called ***oxidation state***).

**Combination or synthesis reaction** The joining of two or more elements or compounds into one product.

**Decomposition reaction** The conversion of one compound into two or more simpler substances.

**Combustion reaction** Rapid oxidation accompanied by heat and (usually) light.

**Single-displacement reaction** Chemical change in which atoms of one element displace (or replace) atoms of another element in a compound. These reactions are also called ***single-replacement reactions.***

**Voltaic cell** A system in which two half-reactions for a redox reaction are separated, allowing the electrons transferred in the reaction to be passed between them through a wire.

**Battery** A device that has two or more voltaic cells connected together. The term is also used to describe any device that converts chemical energy into electrical energy via redox reactions.

**Electrodes** The electrical conductors placed in the half-cells of a voltaic cell.

**Anode** The electrode at which oxidation occurs in a voltaic cell. It is the source of electrons and is the negative electrode.

**Cathode** The electrode at which reduction occurs in a voltaic cell. It is the positive electrode.

**Salt bridge** A device used to keep the charges in a voltaic cell balanced by permitting ions to flow from one compartment to another.

**Electrolyte** The portion of a voltaic cell that allows ions to flow.

**Primary battery** A battery that is not rechargeable.

**Secondary battery or storage battery** A rechargeable battery.

**Electrolysis** The process by which a redox reaction is pushed in the nonspontaneous direction or the process of applying an external voltage to a voltaic cell, causing electrons to move from what would normally be the cell's cathode toward its anode.

You can test yourself on the glossary terms at the following Web site:

**www.chemplace.com/college/**

## Chapter Objectives

**The goal of this chapter is to teach you to do the following.**

1. Define all of the terms in the Chapter Glossary.

### Section 6.1 An Introduction to Oxidation–Reduction Reactions

2. Describe the difference between the redox reactions that form binary ionic compounds, such as zinc oxide, from their elements and the similar redox reactions that form molecular compounds, such as nitrogen monoxide, from their elements. (Your description should include the degree to which the electrons are transferred in the reactions.)

### Section 6.2 Oxidation Numbers

3. Given a formula for a substance, determine the oxidation number (or oxidation state) for each atom in the formula.
4. Given an equation for a chemical reaction, identify whether the equation represents a redox reaction.
5. Given an equation for a redox reaction, identify the substance that is oxidized and the substance that is reduced.
6. Given an equation for a redox reaction, identify the substance that is the reducing agent and the substance that is the oxidizing agent.

### Section 6.3 Types of Chemical Reactions

7. Given an equation for a reaction, indicate whether it represents a combination reaction, a decomposition reaction, a combustion reaction, or a single-displacement reaction.
8. Given the formula for a substance that contains one or more of the elements carbon, hydrogen, oxygen, and sulfur, write the balanced equation for the complete combustion of the substance.
9. Describe the reaction that takes place between an uncharged metal, such as zinc, and a cation, such as the copper(II) ion, in a water solution. Your description should include the nature of the particles in the system before the reaction takes place, the nature of the reaction itself, and the nature of the particles in the system after the reaction. Your description should also include the equations for the half-reactions and the net ionic equation for the overall reaction.

### Section 6.4 Voltaic Cells

10. Describe how a voltaic cell can be made using the redox reaction between a metal, such as zinc, and an ionic substance containing a metallic ion, such as the copper(II) ion. Your description should include a sketch that shows the key components of the two half-cells, labels to indicate which electrode is the cathode and which is the anode, signs to indicate which electrode is negative and which is positive, arrows to show the direction of movement of the electrons in the wire between the electrodes, and arrows to show the direction of movement of the ions in the system. (See Figure 6.5.)
11. Given a description of a voltaic cell, including the nature of the half-reactions, indicate which electrode is the cathode and which is the anode.

## Review Questions

1. For each of the following ionic formulas, write the formula for the cation and the formula for the anion.
   a. $FeBr_3$
   b. $Co_3(PO_4)_2$
   c. $AgCl$
   d. $(NH_4)_2SO_4$
2. Classify each of the following formulas as representing a binary ionic compound, an ionic compound with polyatomic ions, or a molecular compound.
   a. $CF_4$
   b. $Pb(C_2H_3O_2)_2$
   c. $CoCl_2$
   d. $C_2H_5OH$
   e. $H_2S$
   f. $ClF$
   g. $Cr(OH)_3$
   h. $H_3PO_4$
3. Balance the following equations. ($C_8H_{18}$ is a component of gasoline, and $P_2S_5$ is used to make the insecticides parathion and malathion.)
   a. $C_8H_{18}(l) + O_2(g) \rightarrow CO_2(g) + H_2O(l)$
   b. $P_4(s) + S_8(s) \rightarrow P_2S_5(s)$

## Key Ideas

**Complete the following statements by writing one of these words or phrases in each blank.**

| | |
|---|---|
| carbon dioxide | oxidizing |
| change | oxidizing agent |
| decreases | partial |
| flow of electrons | pure element |
| gains | rarely |
| half-reactions | reduced |
| heat | reducing |
| increases | reducing agent |
| light | reduction |
| loses | sulfur dioxide |
| one | transferred |
| oxidation | two or more |
| oxidized | water |

4. According to the modern convention, any chemical change in which an element ________________ electrons is called an oxidation.
5. According to the modern definition, any chemical change in which an element ________________ electrons is called a reduction.

6. Electrons are ________________ found unattached to atoms. Thus, for one element or compound to lose electrons and be ________________, another element or compound must be there to gain the electrons and be ________________. In other words, ________________ (loss of electrons) must be accompanied by ________________ (gain of electrons).
7. Reactions in which electrons are ________________, resulting in oxidation and reduction, are called oxidation–reduction reactions.
8. The separate oxidation and reduction equations are called ________________.
9. A(n) ________________ is a substance that loses electrons, making it possible for another substance to gain electrons and be reduced.
10. A(n) ________________ is a substance that gains electrons, making it possible for another substance to lose electrons and be oxidized.
11. Oxidation is defined as the complete or ________________ loss of electrons, reduction as the complete or partial gain of electrons.
12. Just think of oxidation numbers as tools for keeping track of the ________________ in redox reactions.
13. If any element undergoes a(n) ________________ of oxidation number in the course of a reaction, the reaction is a redox reaction. If an element's oxidation number ________________ in a reaction, that element is oxidized. If an element's oxidation number ________________ in a reaction, that element is reduced. The reactant containing the element that is oxidized is the ________________ agent. The reactant containing the element that is reduced is the ________________ agent.
14. In combination reactions, ________________ elements or compounds combine to form one compound.
15. In decomposition reactions, ________________ compound is converted into two or more simpler substances.
16. In a combustion reaction, oxidation is very rapid and is accompanied by ________________ and usually ________________.
17. When any substance that contains carbon is combusted (or burned) completely, the carbon forms ________________.
18. When a substance that contains hydrogen is burned completely, the hydrogen forms ________________.
19. When any substance that contains sulfur burns completely, the sulfur forms ________________.
20. In single-displacement reactions, atoms of one element in a compound are displaced (or replaced) by atoms from a(n) ________________.

**Complete the following statements by writing one of these words or phrases in each blank.**

| | |
|---|---|
| chemical energy | not rechargeable |
| electrical conductors | oxidation |
| electrical energy | positive |
| electrolysis | positive electrode |
| flow | rechargeable |
| force | reduction |
| half-reactions | voltaic cells |

21. Strictly speaking, a battery is a series of ________________ joined in such a way that they work together. A battery can also be described as a device that converts ________________ into ________________ using redox reactions.
22. A voltaic cell keeps two oxidation–reduction ________________ separated, causing the electrons released in the oxidation half of the reaction to pass through a wire connecting the two halves of the apparatus.
23. Metal strips in voltaic cells are called electrodes, which is the general name for ________________ placed in half-cells of voltaic cells.
24. The electrode at which ________________ occurs in a voltaic cell is called the anode. Because electrons are lost, forming more positive (or less negative) species at the anode, the anode surroundings tend to become more ________________.
25. The cathode is the electrode in a voltaic cell at which ________________ occurs. By convention, the cathode is designated the ________________. Because electrons flow along the wire to the cathode, and because substances gain those electrons to become more negative (or less positive), the cathode surroundings tend to become more negative. Thus, cations are attracted to the cathode.
26. The component of the voltaic cell through which ions are able to ________________ is called the electrolyte.
27. Voltage, a measure of the strength of an electric current, represents the ________________ that moves electrons from the anode to the cathode in a voltaic cell. When a greater voltage is applied in the opposite direction, electrons can be pushed from what would normally be the cathode toward the voltaic cell's anode. This process is called ________________.
28. Batteries that are ________________ are called primary batteries. A(n) ________________ battery is called a secondary battery or a storage battery.

## Chapter Problems

OBJECTIVE 2

### Section 6.1 An Introduction to Oxidation–Reduction Reactions

29. Describe the difference between the redox reactions that form binary ionic compounds, like zinc oxide, from their elements and the similar redox reactions that form molecular compounds, like nitrogen monoxide, from their elements.

30. Are the electrons in the following redox reactions transferred completely from the atoms of one element to the atoms of another, or are they only partially transferred?
    a. $4Al(s) + 3O_2(g) \rightarrow 2Al_2O_3(s)$
    b. $C(s) + O_2(g) \rightarrow CO_2(g)$

31. Are the electrons in the following redox reactions transferred completely from the atoms of one element to the atoms of another or are they only partially transferred?
    a. $2K(s) + F_2(g) \rightarrow 2KF(s)$
    b. $2H_2(g) + O_2(g) \rightarrow 2H_2O(l)$

32. Are the electrons in the following redox reactions transferred completely from the atoms of one element to the atoms of another, or are they only partially transferred?
    a. $S_8(s) + 8O_2(g) \rightarrow 8SO_2(g)$
    b. $P_4(s) + 6Cl_2(g) \rightarrow 4PCl_3(l)$

33. Are the electrons in the following redox reactions transferred completely from the atoms of one element to the atoms of another or are they only partially transferred?
    a. $Ca(s) + Cl_2(g) \rightarrow CaCl_2(s)$
    b. $4Cu(s) + O_2(g) \rightarrow 2Cu_2O(s)$

34. Aluminum bromide, $AlBr_3$, which is used to add bromine atoms to organic compounds, can be made by passing gaseous bromine over hot aluminum. Which of the following half-reactions for this oxidation–reduction reaction describes the oxidation, and which one describes the reduction?

$$2Al \rightarrow 2Al^{3+} + 6\,e^-$$

$$3Br_2 + 6e^- \rightarrow 6Br^-$$

35. Iodine, $I_2$, has many uses, including the production of dyes, antiseptics, photographic film, pharmaceuticals, and medicinal soaps. It forms when chlorine, $Cl_2$, reacts with iodide ions in a sodium iodide solution. Which of the following half-reactions for this oxidation–reduction reaction describes the oxidation, and which one describes the reduction?

$$Cl_2 + 2e^- \rightarrow 2Cl^-$$

$$2I^- \rightarrow I_2 + 2e^-$$

### Section 6.2 Oxidation Numbers

OBJECTIVE 3

36. Determine the oxidation number for the atoms of each element in the following formulas.
    a. $S_8$
    b. $S^{2-}$
    c. $Na_2S$
    d. $FeS$

OBJECTIVE 3

37. Determine the oxidation number for the atoms of each element in the following formulas.
    a. $P_4$
    b. $PF_3$
    c. $PH_3$
    d. $P_2O_3$
    e. $H_3PO_4$

OBJECTIVE 3

38. Determine the oxidation number for the atoms of each element in the following formulas.
    a. $Sc_2O_3$
    b. $RbH$
    c. $N_2$
    d. $NH_3$

OBJECTIVE 3

39. Determine the oxidation number for the atoms of each element in the following formulas.
    a. $N^{3-}$
    b. $N_2O_5$
    c. $K_3N$

OBJECTIVE 3

40. Determine the oxidation number for the atoms of each element in the following formulas.
    a. $CHF_3$
    b. $H_2O_2$
    c. $H_2SO_4$

OBJECTIVE 3

41. Determine the oxidation number for the atoms of each element in the following formulas.
    a. $Co_3N_2$
    b. $Na$
    c. $NaH$

OBJECTIVE 3

42. Determine the oxidation number for the atoms of each element in the following formulas.
    a. $HPO_4{}^{2-}$
    b. $NiSO_4$
    c. $N_2O_4{}^{2-}$
    d. $Mn_3(PO_2)_2$

OBJECTIVE 3

43. Determine the oxidation number for the atoms of each element in the following formulas.
    a. $HSO_3{}^-$
    b. $Cu(NO_3)_2$
    c. $Cu_3(PO_4)_2$

44. The following partial reactions represent various means by which bacteria obtain energy. Determine the oxidation number for each atom other than oxygen and hydrogen atoms, and decide whether the half-reaction represents oxidation or reduction. None of the oxygen or hydrogen atoms are oxidized or reduced.

a. $2Fe^{2+} \rightarrow Fe_2O_3 + \text{energy}$

b. $2NH_3 \rightarrow 2NO_2^- + 3H_2O + \text{energy}$

c. $NO_2^- \rightarrow NO_3^- + \text{energy}$

d. $8H_2S \rightarrow S_8 + 8H_2O + \text{energy}$

e. $S_8 + 8H_2O \rightarrow SO_4^{2-} + 16H^+ + \text{energy}$

45. About 47% of the hydrochloric acid produced in the United States is used for cleaning metallic surfaces. Hydrogen chloride, HCl, which is dissolved in water to make the acid, is formed in the reaction of chlorine gas and hydrogen gas, displayed below. Determine the oxidation number for each atom in the equation, and decide whether the reaction is a redox reaction. If it is redox, indicate which substance is oxidized, which substance is reduced, the oxidizing agent, and the reducing agent.

OBJECTIVE 3 OBJECTIVE 4 OBJECTIVE 5 OBJECTIVE 6

$$Cl_2(g) + H_2(g) \rightarrow 2HCl(g)$$

46. The following equation describes the reaction that produces hydrofluoric acid, which is used to make chlorofluorocarbons (CFCs). Determine the oxidation number for each atom in the equation, and decide whether the reaction is a redox reaction. If it is redox, indicate which substance is oxidized, which substance is reduced, the oxidizing agent, and the reducing agent.

OBJECTIVE 3 OBJECTIVE 4 OBJECTIVE 5 OBJECTIVE 6

$$CaF_2 + H_2SO_4 \rightarrow 2HF + CaSO_4$$

47. Water and carbon dioxide fire extinguishers should not be used on magnesium fires, because both substances react with magnesium and generate enough heat to intensify the fire. Determine the oxidation number for each atom in the equations that describe these reactions (displayed below), and decide whether each reaction is a redox reaction. If it is redox, indicate which substance is oxidized, which substance is reduced, the oxidizing agent, and the reducing agent.

OBJECTIVE 3 OBJECTIVE 4 OBJECTIVE 5 OBJECTIVE 6

$$Mg(s) + 2H_2O(l) \rightarrow Mg(OH)_2(aq) + H_2(g) + \text{heat}$$

$$2Mg(s) + CO_2(g) \rightarrow 2MgO(s) + C(s) + \text{heat}$$

48. Potassium nitrate is used in the production of fireworks, explosives, and matches. It is also used in curing foods and to modify the burning properties of tobacco. The reaction for the industrial production of $KNO_3$ is summarized below. Determine the oxidation number for each atom, and decide whether the reaction is a redox reaction. If it is redox, indicate which substance is oxidized, which substance is reduced, the oxidizing agent, and the reducing agent.

OBJECTIVE 3 OBJECTIVE 4 OBJECTIVE 5 OBJECTIVE 6

$$4KCl + 4HNO_3 + O_2 \rightarrow 4KNO_3 + 2Cl_2 + 2H_2O$$

OBJECTIVE 3 OBJECTIVE 4 OBJECTIVE 5 OBJECTIVE 6

**49.** Formaldehyde, $CH_2O$, which is used in embalming fluids, is made from methanol in the reaction described below. Determine the oxidation number for each atom in this equation, and decide whether the reaction is a redox reaction. If it is redox, indicate which substance is oxidized, which substance is reduced, the oxidizing agent, and the reducing agent.

$$2CH_3OH + O_2 \rightarrow 2CH_2O + 2H_2O$$

OBJECTIVE 3 OBJECTIVE 4 OBJECTIVE 5 OBJECTIVE 6

50. The weak acid hydrofluoric acid, HF(*aq*), is used to frost light bulbs. It reacts with the silicon dioxide in glass on the inside of light bulbs to form a white substance, $H_2SiF_6$, that deposits on the glass and reduces the glare from the bulb. The same reaction is run on a larger scale to produce $H_2SiF_6$ used for fluoridating drinking water. Determine the oxidation number for each atom in the reaction, and decide whether the reaction is a redox reaction. If it is redox, indicate which substance is oxidized, which substance is reduced, the oxidizing agent, and the reducing agent.

$$6HF + SiO_2 \rightarrow H_2SiF_6 + 2H_2O$$

OBJECTIVE 3 OBJECTIVE 4 OBJECTIVE 5 OBJECTIVE 6

**51.** For each of the following equations, determine the oxidation number for each atom in the equation, and indicate whether the reaction is a redox reaction. If the reaction is redox, identify what is oxidized, what is reduced, the oxidizing agent, and the reducing agent.

a. $Co(s) + 2AgNO_3(aq) \rightarrow Co(NO_3)_2(aq) + 2Ag(s)$

b. $V_2O_5(s) + 5Ca(l) \xrightarrow{\Delta} 2V(l) + 5CaO(s)$

c. $CaCO_3(aq) + SiO_2(s) \rightarrow CaSiO_3(s) + CO_2(g)$

d. $2NaH(s) \xrightarrow{\Delta} 2Na(s) + H_2(g)$

e. $5As_4O_6(s) + 8KMnO_4(aq) + 18H_2O(l) + 52KCl(aq)$
$\rightarrow 20K_3AsO_4(aq) + 8MnCl_2(aq) + 36HCl(aq)$

OBJECTIVE 3 OBJECTIVE 4 OBJECTIVE 5 OBJECTIVE 6

52. For each of the following equations, determine the oxidation number for each atom in the equation, and indicate whether the reaction is a redox reaction. If the reaction is redox, identify what is oxidized, what is reduced, the oxidizing agent, and the reducing agent.

a. $2Na(s) + 2H_2O(l) \rightarrow 2NaOH(aq) + H_2(g)$

b. $HCl(aq) + NH_3(aq) \rightarrow NH_4Cl(aq)$

c. $2Cr(s) + 3CuSO_4(aq) \rightarrow Cr_2(SO_4)_3(aq) + 3Cu(s)$

d. $3H_2SO_3(aq) + 2HNO_3(aq) \rightarrow 2NO(g) + H_2O(l) + 3H_2SO_4(aq)$

e. $CaO(s) + H_2O(l) \rightarrow Ca^{2+}(aq) + 2OH^-(aq)$

OBJECTIVE 3 OBJECTIVE 4 OBJECTIVE 5 OBJECTIVE 6

**53.** The following equations summarize the steps in the process used to make about 99% of the sulfuric acid produced in the United States. Determine the oxidation number for each atom in each of the following equations, and indicate whether each reaction is a redox reaction. For the redox reactions, identify what is oxidized, what is reduced, the oxidizing agent, and the reducing agent.

$$\tfrac{1}{8}S_8(s) + O_2(g) \rightarrow SO_2(g)$$

$$SO_2 + \tfrac{1}{2}O_2 \rightarrow SO_3$$

$$SO_3 + H_2O \rightarrow H_2SO_4$$

54. Because of its superior hiding ability, titanium dioxide has been the best-selling white pigment since 1939. In 1990 there were 2.16 billion pounds of it sold in the United States for a variety of purposes, including surface coatings (paint), plastics, and paper. The following equations show how impure $TiO_2$ is purified. Determine the oxidation number for each atom in them, and indicate whether each reaction is a redox reaction. For the redox reactions, identify what is oxidized, what is reduced, the oxidizing agent, and the reducing agent.

OBJECTIVE 3 OBJECTIVE 4 OBJECTIVE 5 OBJECTIVE 6

$$3TiO_2 + 4C + 6Cl_2 \rightarrow 3TiCl_4 + 2CO + 2CO_2$$

$$TiCl_4 + O_2 \rightarrow TiO_2 + 2Cl_2$$

## Section 6.3 Types of Chemical Reactions

55. Classify each of these reactions as a combination reaction, a decomposition reaction, a combustion reaction, or a single-displacement reaction.

OBJECTIVE 7

a. $2NaH(s) \rightarrow 2Na(s) + H_2(g)$

b. $2KI(aq) + Cl_2(g) \rightarrow 2KCl(aq) + I_2(s)$

c. $2C_2H_5SH(l) + 9O_2(g) \rightarrow 4CO_2(g) + 6H_2O(l) + 2SO_2(g)$

d. $H_2(g) + CuO(s) \xrightarrow{\Delta} Cu(s) + H_2O(l)$

e. $P_4(s) + 5O_2(g) \rightarrow P_4O_{10}(s)$

56. Classify each of these reactions with respect to the following categories: combination reaction, decomposition reaction, combustion reaction, and single-displacement reaction.

OBJECTIVE 7

a. $Fe_2(CO_3)_3(s) \xrightarrow{\Delta} Fe_2O_3(s) + 3CO_2(g)$

b. $2C_6H_{11}OH(l) + 17O_2(g) \rightarrow 12CO_2(g) + 12H_2O(l)$

c. $P_4O_{10}(s) + 6H_2O(l) \rightarrow 4H_3PO_4(aq)$

d. $2C(s) + MnO_2(s) \xrightarrow{\Delta} Mn(s) + 2CO(g)$

e. $2NaClO_3(s) \xrightarrow{\Delta} 2NaCl(s) + 3O_2(g)$

57. Classify each of these reactions as a combination reaction, a decomposition reaction, a combustion reaction, or a single-displacement reaction.

OBJECTIVE 7

a. $4B(s) + 3O_2(g) \rightarrow 2B_2O_3(s)$

b. $(C_2H_5)_2O(l) + 6O_2(g) \rightarrow 4CO_2(g) + 5H_2O(l)$

c. $2Cr_2O_3(s) + 3Si(s) \xrightarrow{\Delta} 4Cr(s) + 3SiO_2(s)$

d. $C_6H_{11}SH(l) + 10O_2(g) \rightarrow 6CO_2(g) + 6H_2O(l) + SO_2(g)$

e. $2NaHCO_3(s) \xrightarrow{\Delta} Na_2CO_3(s) + H_2O(l) + CO_2(g)$

58. Classify each of these reactions with respect to the following categories: combination reaction, decomposition reaction, combustion reaction, and single-displacement reaction.

OBJECTIVE 7

a. $3H_2(g) + WO_3(s) \xrightarrow{\Delta} W(s) + 3H_2O(l)$

b. $2I_4O_9(s) \rightarrow 2I_2O_6(s) + 2I_2(s) + 3O_2(g)$

c. $2NaNO_3(s) \xrightarrow{\Delta} 2NaNO_2(s) + O_2(g)$

d. $Cl_2(g) + 2KBr(aq) \rightarrow 2KCl(aq) + Br_2(l)$

OBJECTIVE 8

59. Write balanced equations for the complete combustion of each of the following substances.

a. $C_3H_8(g)$
b. $C_4H_9OH(l)$
c. $CH_3COSH(l)$

OBJECTIVE 8

60. Write balanced equations for the complete combustion of each of the following substances.

a. $C_{13}H_{28}(l)$
b. $C_{12}H_{22}O_{11}(s)$
c. $C_2H_5SO_3H(l)$

OBJECTIVE 9

61. The following pairs react in single-displacement reactions that are similar to the reaction between uncharged zinc metal and the copper(II) ions in a copper(II) sulfate solution. Describe the changes in these reactions, including the nature of the particles in the system before the reaction takes place, the nature of the reaction itself, and the nature of the particles in the system after the reaction. Your description should also include the equations for the half-reactions and the net ionic equation for the overall reaction.

a. magnesium metal and nickel(II) nitrate, $Ni(NO_3)_2(aq)$
b. calcium metal and cobalt(II) chloride, $CoCl_2(aq)$

### Section 6.4 Voltaic Cells

62. We know that the following reaction can by used to generate an electric current in a voltaic cell.

$$Zn(s) + CuSO_4(aq) \rightarrow Cu(s) + ZnSO_4(aq)$$

OBJECTIVE 10 OBJECTIVE 11

Sketch similar voltaic cells made from each of the reactions presented below, showing the key components of the two half-cells and indicating the cathode electrode and the anode electrode, the negative and positive electrodes, the direction of movement of the electrons in the wire between the electrodes, and the direction of movement of the ions in the system. Show a salt bridge in each sketch, and show the movement of ions out of the salt bridge.

a. $Mn(s) + Pb(NO_3)_2(aq) \rightarrow Pb(s) + Mn(NO_3)_2\ (aq)$
b. $Mg(s) + 2AgNO_3(aq) \rightarrow 2Ag(s) + Mg(NO_3)_2(aq)$

63. The following equation summarizes the chemical changes that take place in a typical dry cell.

$$Zn(s) + 2MnO_2(s) + 2NH_4^+(aq) \rightarrow Zn^{2+}(aq) + Mn_2O_3(s) + 2NH_3(aq) + H_2O(l)$$

OBJECTIVE 3 OBJECTIVE 4 OBJECTIVE 5 OBJECTIVE 6

Determine the oxidation number for each atom in this equation, and identify what is oxidized, what is reduced, the oxidizing agent, and the reducing agent.

OBJECTIVE 3 OBJECTIVE 4 OBJECTIVE 5 OBJECTIVE 6

64. The following equation summarizes the chemical changes that take place in a nickel–cadmium battery. Determine the oxidation number for each atom in the equation, and identify what is oxidized, what is reduced, the oxidizing agent, and the reducing agent.

$$Cd(s) + 2NiO(OH)(s) + 2H_2O(l) \rightarrow Cd(OH)_2(s) + 2Ni(OH)_2(s)$$

OBJECTIVE 3 OBJECTIVE 4 OBJECTIVE 5 OBJECTIVE 6

65. The following equation summarizes the chemical changes that take place in a lead–acid battery. Determine the oxidation number for each atom in the equation, and identify what is oxidized, what is reduced, the oxidizing agent, and the reducing agent.

$$Pb(s) + PbO_2(s) + 2HSO_4^-(aq) + 2H_3O^+(aq) \rightarrow 2PbSO_4(s) + 4H_2O(l)$$

### Additional Problems

66. Nitric acid, which is used to produce fertilizers and explosives, is made in the following three steps. Determine the oxidation number for each atom in the three equations, and indicate whether each reaction is a redox reaction. If a reaction is redox, identify what is oxidized and what is reduced.

$$4NH_3 + 5O_2 \rightarrow 4NO + 6H_2O$$

$$2NO + O_2 \rightarrow 2NO_2$$

$$3NO_2 + H_2O \rightarrow 2HNO_3 + NO$$

67. Sodium hydrogen carbonate, $NaHCO_3$, best known as the active ingredient in baking soda, is used in several ways in food preparation. It is also added to animal feeds and used to make soaps and detergents. Baking soda can be used to put out small fires on your stovetop. The heat of the flames causes the $NaHCO_3$ to decompose to form carbon dioxide, which displaces the air above the flames, depriving the fire of the oxygen necessary for combustion. The equation for the reaction is

$$2NaHCO_3(s) \xrightarrow{\Delta} CO_2(g) + Na_2CO_3(s) + H_2O(g)$$

68. Swimming pools can be chlorinated by adding either calcium hypochlorite, $Ca(OCl)_2$, or sodium hypochlorite, NaOCl. The active component is the hypochlorite ion, $OCl^-$. Because the rate of the following reaction is increased by ultraviolet radiation in sunlight, it is best to chlorinate pools in the evening to avoid the hypochlorite ion's decomposition.

$$OCl^-(aq) \rightarrow Cl^-(aq) + \frac{1}{2}O_2(g)$$

Determine the oxidation number for each atom in this equation, and indicate whether the reaction is a redox reaction. If the reaction is redox, identify what is oxidized and what is reduced.

69. Not too long ago mercury batteries were commonly used to power electronic watches and small appliances. The overall reaction for this type of battery is

$$HgO(s) + Zn(s) \rightarrow ZnO(s) + Hg(l)$$

Determine the oxidation number for each atom in the equation and decide whether the reaction is a redox reaction or not. If it is redox, identify which substance is oxidized, which substance is reduced, the oxidizing agent, and the reducing agent.

70. Silver batteries have been used to run heart pacemakers and hearing aids. The overall reaction for this type of battery is

$$Ag_2O(s) + Zn(s) \rightarrow ZnO(s) + 2Ag(s)$$

Determine the oxidation number for each atom in the equation and decide whether the reaction is a redox reaction or not. If it is redox, identify which substance is oxidized, which substance is reduced, the oxidizing agent, and the reducing agent.

71. $Mn_3(PO_4)_2$, which is used to make corrosion resistant coatings on steel, aluminum, and other metals, is made from the reaction of $Mn(OH)_2$ with $H_3PO_4$.

$$3Mn(OH)_2(s) + 2H_3PO_4(aq) \rightarrow Mn_3(PO_4)_2(s) + 6H_2O(l)$$

Determine the oxidation number for each atom in the equation and identify whether the reaction is a redox reaction or not. If it is redox, identify what is oxidized, what is reduced, the oxidizing agent, and the reducing agent.

72. One of the ways that plants generate oxygen is represented by the following reaction.

$$2Mn^{4+} + 2H_2O \rightarrow 2Mn^{2+} + 4H^+ + O_2$$

Determine the oxidation number for each atom in the equation and decide whether the reaction is a redox reaction or not. If it is redox, identify which substance is oxidized, which substance is reduced, the oxidizing agent, and the reducing agent.

73. The *noble* gases in group 18 on the periodic table used to be called the *inert* gases because they were thought to be incapable of forming compounds. Their name has been changed to noble gases because although they resist combining with the more common elements to their left on the periodic table, they do mingle with them on rare occasions. The following equations describe reactions that form xenon compounds. Determine the oxidation number for each atom in the reactions, and identify each reaction as redox or not. If it is redox, identify which substance is oxidized, which substance is reduced, the oxidizing agent, and the reducing agent.

$$Xe + 3F_2 \rightarrow XeF_6$$

$$XeF_6 + H_2O \rightarrow XeOF_4 + 2HF$$

$$XeF_6 + OPF_3 \rightarrow XeOF_4 + PF_5$$

74. Sometimes one of the elements in a reactant appears in more than one product of a reaction and has in one product, a higher oxidation number than before the reaction and in the other product, a lower oxidation number than before the reaction. In this way, the same element is both oxidized and reduced, and the same compound is both the oxidizing agent and the reducing agent. This process is called disproportionation. For example iodine monofluoride, IF, disproportionates into iodine and iodine pentafluoride in the following reaction:

$$5IF \rightarrow 2I_2 + IF_5$$

Determine the oxidation number for each atom in this equation and show that iodine is both oxidized and reduced and that iodine monofluoride is both the oxidizing agent and the reducing agent.

75. Sodium perbromate is an oxidizing agent that can be made in the two ways represented by the equations below. The first equation shows the way it was made in the past, and the second equation represents the technique used today.

$$NaBrO_3 + XeF_2 + H_2O \rightarrow NaBrO_4 + 2HF + Xe$$

$$NaBrO_3 + F_2 + 2NaOH \rightarrow NaBrO_4 + 2NaF + H_2O$$

Determine the oxidation number for each atom in each of these equations, and decide whether each reaction is a redox reaction or not. If a reaction is a redox reaction, identify which substance is oxidized, which substance is reduced, the oxidizing agent, and the reducing agent.

76. Calcium hydrogen sulfite, $Ca(HSO_3)_2$, which is used as a paper pulp preservative and as a disinfectant, is made by reacting sulfur dioxide with calcium hydroxide.

$$2SO_2 + Ca(OH)_2 \rightarrow Ca(HSO_3)_2$$

Determine the oxidation number for each atom in the equation and decide whether the reaction is a redox reaction or not. If the reaction is a redox reaction, identify which substance is oxidized, which substance is reduced, the oxidizing agent, and the reducing agent.

77. The following equations represent reactions that involve only halogen atoms. Iodine pentafluoride, $IF_5$, is used to add fluorine atoms to other compounds, bromine pentafluoride, $BrF_5$, is an oxidizing agent in liquid rocket propellants, and chlorine trifluoride, $ClF_3$, is used to reprocess nuclear reactor fuels.

$$IF(g) + 2F_2(g) \rightarrow IF_5$$

$$BrF(g) + 2F_2(g) \rightarrow BrF_5(g)$$

$$Cl_2(g) + 3F_2(g) \rightarrow 2ClF_3(g)$$

Determine the oxidation number for each atom in these equations, and decide whether each reaction is a redox reaction or not. If a reaction is a redox reaction, identify which substance is oxidized, which substance is reduced, the oxidizing agent, and the reducing agent.

78. The water solutions of hydrogen peroxide, $H_2O_2$, used as an antiseptic (3%) and as a bleach (6%) are stored in dark plastic bottles because the following reaction is accelerated by the metal ions found in glass and by light.

$$2H_2O_2(aq) \rightarrow 2H_2O(l) + O_2(g)$$

Determine the oxidation number for each atom in this equation, and indicate whether the reaction is a redox reaction. If the reaction is redox, identify what is oxidized and what is reduced.

79. Sodium sulfate, which is used to make detergents and glass, is formed in the following reaction.

$$4NaCl + 2SO_2 + 2H_2O + O_2 \rightarrow 2Na_2SO_4 + 4HCl$$

Determine the oxidation number for each atom in the equation and decide whether the reaction is a redox reaction or not. If a reaction is a redox reaction, identify which substance is oxidized, which substance is reduced, the oxidizing agent, and the reducing agent.

80. Hydrogen gas can be made in two steps:

$$CH_4(g) + H_2O(g) \rightarrow CO(g) + 3H_2(g)$$

$$CO(g) + H_2O(g) \rightarrow CO_2(g) + H_2(g)$$

Determine the oxidation number for each atom in these equations and decide whether each reaction is a redox reaction or not. If a reaction is redox, identify which substance is oxidized and which substance is reduced. (You do not need to identify the oxidizing agent and the reducing agent.)

81. Elemental sulfur is produced by the chemical industry from naturally occurring hydrogen sulfide in the following steps.

$$2H_2S + 3O_2 \rightarrow 2SO_2 + 2H_2O$$

$$SO_2 + 2H_2S \rightarrow 3S + 2H_2O$$

Determine the oxidation number for each atom in these equations and decide whether each reaction is a redox reaction or not. If a reaction is a redox reaction, identify which substance is oxidized, which substance is reduced, the oxidizing agent, and the reducing agent.

82. Sodium chlorate, $NaClO_3$, which is used to bleach paper, is made in the following reactions. Determine the oxidation number for each atom in the equations, and indicate whether each reaction is a redox reaction. If a reaction is redox, identify what is oxidized and what is reduced.

$$Cl_2 + 2NaOH \rightarrow NaOCl + NaCl + H_2O$$

$$3NaOCl \rightarrow NaClO_3 + 2NaCl$$

83. Leaded gasoline, originally developed to decrease pollution, is now banned because the lead(II) bromide, $PbBr_2$, emitted when it burns decomposes in the atmosphere into two serious pollutants, lead and bromine. The equation for this reaction follows. Determine the oxidation number for each atom in the equation, and indicate whether the reaction is a redox reaction. If the reaction is redox, identify what is oxidized and what is reduced.

$$PbBr_2 \xrightarrow{\text{sunlight}} Pb + Br_2$$

84. When leaded gasoline was banned, there was a rush to find safer ways to reduce emissions of unburned hydrocarbons from gasoline engines. One alternative is to add methyl t-butyl ether (MTBE). In 1990 about 25% of the methanol, $CH_3OH$, produced by the U.S. chemical industry was used to make methyl t-butyl ether. The equations that follow show the steps used to make methanol. Determine the oxidation number for each atom in the equations, and indicate whether the reactions are redox reactions. For any redox reaction, identify what is oxidized and what is reduced.

$$3CH_4 + 2H_2O + CO_2 \rightarrow 4CO + 8H_2$$

$$CO + 2H_2 \rightarrow CH_3OH$$

85. When the calcium carbonate, $CaCO_3$, in limestone is heated to a high temperature, it decomposes into calcium oxide (called lime or quicklime) and carbon dioxide. Lime was used by the early Romans, Greeks, and Egyptians to make cement, and it is used today to make over 150 different chemicals. In another reaction, calcium oxide and water form calcium hydroxide, $Ca(OH)_2$ (called slaked lime), which is used to remove the sulfur dioxide from smokestacks above power plants that burn high-sulfur coal. The equations for all these reactions follow. Determine the oxidation number for each atom in the equations, and indicate whether the reactions are redox reactions. For each redox reaction, identify what is oxidized and what is reduced.

$$CaCO_3 \xrightarrow{\Delta} CaCO + CO_2$$

$$CaO + H_2O \rightarrow Ca(OH)_2$$

$$SO_2 + H_2O \rightarrow H_2SO_3$$

$$Ca(OH)_2 + H_2SO_3 \rightarrow CaSO_3 + 2H_2O$$

86. Potassium hydroxide, which is used to make fertilizers and soaps, is produced by running an electric current through a potassium chloride solution. The equation for this reaction follows. Is this a redox reaction? What is oxidized and what is reduced?

$$2KCl(aq) + 2H_2O(l) \rightarrow 2KOH(aq) + H_2(g) + Cl_2(g)$$

87. The space shuttle's solid rocket boosters get their thrust from the reaction of aluminum metal with ammonium perchlorate, $NH_4ClO_4$, which generates a lot of gas and heat. The billowy white smoke is due to the formation of very finely divided solid aluminum oxide. One of the reactions that takes place is

$$10Al(s) + 6NH_4ClO_4(s) \rightarrow 5Al_2O_3(s) + 6HCl(g) + 3N_2(g) + 9H_2O(g)$$

Is this a redox reaction? What is oxidized, and what is reduced?

88. For each of the following equations, determine the oxidation number for each atom in the equation, and indicate whether the reaction is a redox reaction. If the reaction is redox, identify what is oxidized what is reduced, the oxidizing agent, and the reducing agent.

    a. $2HNO_3(aq) + 3H_2S(aq) \rightarrow 2NO(g) + 3S(s) + 4H_2O(l)$

    b. $3CuSO_4(aq) + 2Na_3PO_4(aq) \rightarrow Cu_3(PO_4)_2(s) + 3Na_2SO_4(aq)$

89. Determine the oxidation number for each atom in the following equations and decide whether each reaction is a redox reaction. If a reaction is redox, indicate which substance is oxidized, which is reduced, the oxidizing agent, and the reducing agent.

    a. $K_2Cr_2O_7(aq) + 14HCl(aq) \rightarrow 2KCl(aq) + 2CrCl_3(aq) + 7H_2O(l) + 3Cl_2(g)$

    b. $Ca(s) + 2H_2O(l) \rightarrow Ca(OH)_2(s) + H_2(g)$

90. Determine the oxidation number for each atom in the following equations and decide whether each reaction is a redox reaction. If the reaction is redox, indicate which substance is oxidized, which is reduced, the oxidizing agent, and the reducing agent.

    a. $2Ag_2CrO_4(s) + 4HNO_3(aq) \rightarrow 4AgNO_3(aq) + H_2Cr_2O_7(aq) + H_2O(l)$

    b. $2MnO_4^-(aq) + 3IO_3^-(aq) + H_2O(l) \rightarrow 2MnO_2(s) + 3IO_4^-(aq) + 2OH^-(aq)$

91. For each of the following equations, determine the oxidation number for each atom in the equation and identify whether the reaction is a redox reaction or not. If the reaction is redox, identify what is oxidized, what is reduced, the oxidizing agent, and the reducing agent.

    a. $Ca(s) + F_2(g) \rightarrow CaF_2(s)$

    b. $2Al(s) + 3H_2O(g) \rightarrow Al_2O_3(s) + 3H_2(g)$

92. Determine the oxidation number for each atom in the following equations and decide whether each reaction is a redox reaction. If a reaction is redox, indicate which substance is oxidized, which is reduced, the oxidizing agent, and the reducing agent.

    a. $Cr_2O_7^{2-}(aq) + 6Cl^-(aq) + 14H^+(aq) \rightarrow 2Cr^{3+}(aq) + 3Cl_2(g) + 7H_2O(l)$

    b. $5H_2C_2O_4(aq) + 2KMnO_4(aq) + 3H_2SO_4(aq)$
    $\rightarrow 10CO_2(g) + 2MnSO_4(aq) + 8H_2O(l) + K_2SO_4(aq)$

93. The following equations represent reactions used by the U.S. chemical industry. Classify as a combination reaction, a decomposition reaction, a combustion reaction, or a single-displacement reaction.

    a. $P_4 + 5O_2 + 6H_2O \rightarrow 4H_3PO_4$

    b. $TiCl_4 + O_2 \rightarrow TiO_2 + 2Cl_2$

    c. $CH_3CH_3 \xrightarrow{\Delta} CH_2CH_2 + H_2$

94. `The following equations represent reactions used by the U.S. chemical industry. Classify each as a combination reaction, a decomposition reaction, a combustion reaction, or a single-displacement reaction.

    a. $2HF + SiF_4 \rightarrow H_2SiF_6$

    b. $2H_2S + 3O_2 \rightarrow 2SO_2 + 2H_2O$

    c. $CH_3OH \xrightarrow{\Delta} CH_2O + H_2$

95. Write a balanced equation for the redox reaction of carbon dioxide gas and hydrogen gas to form carbon solid and water vapor.

96. Phosphorus pentachloride, which is used to add chlorine atoms to other substances, can be made from the reaction of phosphorus trichloride and chlorine. The phosphorus pentachloride is the only product. Write a balance equation, without including states, for this redox reaction.
97. Titanium metal is used to make metal alloys for aircraft, missiles, and artificial hip joints. It is formed in the reaction of titanium(IV) chloride with magnesium metal. The other product is magnesium chloride. Write a balanced equation, without including states, for this redox reaction.
98. Dichlorine monoxide, which is used to add chlorine atoms to other substances, is made from mercury(II) oxide and chlorine. The products are dichlorine monoxide and mercury. Write a balanced equation, without including states, for this redox reaction.
99. Write a balanced equation for the redox reaction of solid potassium with liquid water to form aqueous potassium hydroxide and hydrogen gas.
100. Write a balanced equation for the redox reaction of aqueous chlorine with aqueous potassium iodide to form aqueous potassium chloride and solid iodine.
101. Write a balanced equation for the redox reaction at room temperature of calcium metal and bromine liquid to form solid calcium bromide.
102. Write a balanced equation for the redox reaction of solid copper(II) sulfide with oxygen gas to form solid copper(II) oxide and sulfur dioxide gas.
103. Magnesium chloride is used to make disinfectants, fire extinguishers, paper, and floor sweeping compounds. It is made from the redox reaction of hydrochloric acid with solid magnesium hydroxide. Write a balanced equation for this reaction, which yields aqueous magnesium chloride and liquid water.
104. Write a balanced equation for the redox reaction at room temperature of chromium metal with hydrochloric acid to form aqueous chromium(III) chloride and hydrogen gas.

### Discussion Question

105. What makes one battery better than another? Find a reference book that tells you about the properties of the elements. Why do you think lithium batteries are superior to batteries that use lead? What other elements might be considered for new batteries?

*Energy changes accompany chemical changes.*

CHAPTER 7

# Energy and Chemical Reactions



ENERGY . . . IT MAKES THINGS HAPPEN. To get an idea of the role energy plays in our lives, let's spend some time with Joan, a college student in one of the coastal towns in California. She wakes up in the morning to a beautiful sunny day and decides to take her chemistry book to the beach. Before leaving, she fries up some scrambled eggs, burns some toast, and pops a cup of day-old coffee into the microwave oven. After finishing her breakfast, she shoves her chemistry textbook into her backpack and jumps on her bike for the short ride to the seashore. Once at the beach, she reads two pages of her chemistry assignment and then, despite the fascinating topic, gets drowsy and drops off to sleep. When she wakes up an hour later, she's sorry that she forgot to put on her sunscreen. Her painful sunburn drives her off the beach and back to her apartment to spend the rest of the day inside.

All of Joan's actions required energy. It took energy for her to get out of bed, make breakfast, pedal to the beach, and (as you well know) read her chemistry book; Joan gets that energy from the chemical changes that her body induces in the food she eats. It also took heat energy to cook her eggs and burn her toast. The radiant energy from microwaves raised the temperature of her coffee, and the radiant energy from the sun caused her sunburn.

What is energy, and what different forms does it take? Why do some chemical changes release energy whereas others absorb it? This chapter attempts to answer such questions and then to apply our understanding of energy to some of the important environmental issues that people face today.

*Radiant energy from the sun causes sunburn.*

## Review Skills

The presentation of information in this chapter assumes that you can already perform the tasks listed below. You can test your readiness to proceed by answering the Review Questions at the end of the chapter. This might also be a good time to read the Chapter Objectives, which precede the Review Questions.

- Describe the similarities and differences among solids, liquids, and gases with reference to the particle nature of matter, the degree of motion of the particles, and the degree of attraction between the particles. (Section 2.1)
- Describe the relationship between temperature and motion. (Section 2.1)

## 7.1 Energy

All chemical changes are accompanied by energy changes. Some reactions, such as the combustion of methane (a component of natural gas) release energy. This is why natural gas can be used to heat our homes:

$$CH_4(g) + 2O_2(g) \rightarrow CO_2(g) + 2H_2O(l) + \text{Energy}$$

Other reactions absorb energy. For example, when energy from the sun strikes oxygen molecules, $O_2$, in the earth's atmosphere, some of the energy is absorbed by the molecules, causing them to break apart into separate atoms (Figure 7.1).

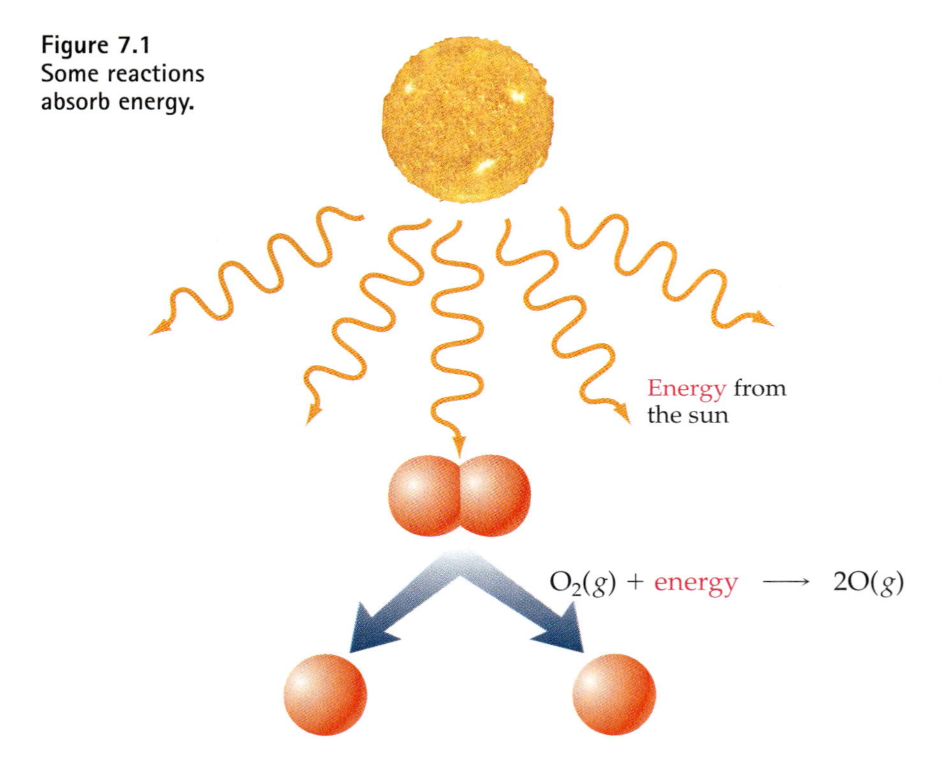


**Figure 7.1** Some reactions absorb energy.

Before we can begin to explain the role that energy plays in these and other chemical reactions, we need to get a better understanding of what energy is and the different forms it can take.

You probably have a general sense of what energy is. When you get up in the morning after a good night's sleep, you feel that you have plenty of energy to get your day's work done. After a long day of studying chemistry, you might feel you hardly have enough energy to drag yourself to bed. The main goal of this section is to give you a more specific, scientific understanding of energy.

The simplest definition of **energy** is that it is the capacity to do work. Work, in this context, may be defined as what is done to move an object against some sort of resistance. For example, when you push this book across a table, the work you do overcomes the resistance caused by the contact between the book and the table. Likewise, when you lift this book, you do work to overcome the gravitational attraction that causes the book and the earth to resist being separated. When two oxygen atoms are linked together in a covalent bond, work must be done to separate them. Anything that has the capacity to do such work must, by definition, have energy (Figure 7.2).

**Figure 7.2**
**Energy:**
**The Capacity to Do Work**

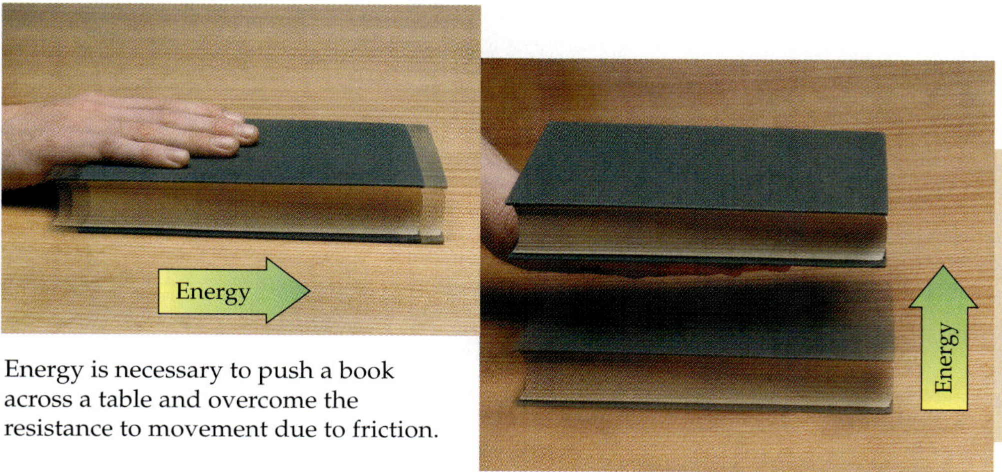


Energy is necessary to push a book across a table and overcome the resistance to movement due to friction.

Energy is necessary to lift a book and overcome the resistance to movement due to gravity.

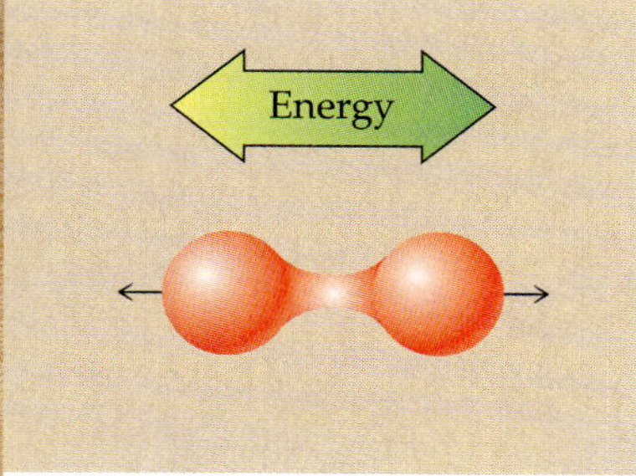


Energy is necessary to separate two atoms in a molecule and overcome the resistance to movement due to the chemical bond between them.

## Kinetic Energy

It takes work to move a brick wall. A bulldozer moving at 20 miles per hour has the capacity to do this work, but when the same bulldozer is sitting still, it's not going to get the work done. The movement of the bulldozer gives it the capacity to do work, so this movement must be a form of energy. Any object that is in motion can collide with another object and move it, so any object in motion has the capacity to do work. This capacity to do work resulting from the motion of an object is called **kinetic energy (KE).**

OBJECTIVE 2

The amount of an object's kinetic energy is related to its mass and to its velocity. If two objects are moving at the same velocity, the one with the greater mass will have a greater capacity to do work and thus a greater kinetic energy. For example, a bulldozer moving at 20 miles per hour can do more work than a scooter moving at the same velocity. If these two objects were to collide with a brick wall, the bulldozer would do more of the work of moving the wall than the scooter.

OBJECTIVE 3

If two objects have equal mass but different velocities, the one with the greater velocity has the greater kinetic energy. A bulldozer moving at 20 miles per hour can do more work than an identical bulldozer moving at 5 miles per hour (Figure 7.3).

**Figure 7.3**
**Factors That Affect Kinetic Energy**

A stationary bulldozer does not have the capacity to do the work of moving a wall.

The more rapidly moving bulldozer does more of the work of moving the wall. The greater an object's velocity, the more work it can do, and the more kinetic energy it has.

A scooter moving at the same velocity as a bulldozer will do less work and therefore has less energy.

OBJECTIVE 2 OBJECTIVE 3

## Potential Energy

Energy can be transferred from one object to another. Picture the coin toss that precedes a football game. A coin starts out resting in the referee's hand. After he flips it, sending it moving up into the air, it has some kinetic energy that it did not have before it was flipped. Where did the coin get this energy? From the referee's moving thumb. When scientists analyze such energy transfers, they find that all of the energy still exists. The **Law of Conservation of Energy** states that energy can be neither created nor destroyed, but it can be transferred from one system to another and changed from one form to another.

OBJECTIVE 4

As the coin rises, it slows down and eventually stops. At this point, the kinetic energy it got from the referee's moving thumb is gone, but the Law of Conservation of Energy says that energy cannot be destroyed. Where did the kinetic energy go? Although some of it has been transferred to the air particles the coin bumps into on its flight, most of the energy is still there in the coin in a form called **potential energy (PE),** which is the retrievable, stored form of energy an object possesses by virtue of its position or state. We get evidence of this transformation when the coin falls back down toward the grass on the field. The potential energy it had at the peak of its flight is converted into the kinetic energy of its downward movement, and this kinetic energy does the work of flattening a few blades of grass when the coin hits the field (Figure 7.4).

There are many kinds of potential energy. An alkaline battery contains potential energy that can be used to move a toy car. A plate of pasta provides potential energy to allow your body to move. Knowing the relationships between potential energy and stability can help you to recognize changes in potential energy and to decide whether the potential energy has increased or decreased as a result of each change.

Let's look at the relationship between potential energy and *stability*. A system's stability is a measure of its tendency to change. A more stable

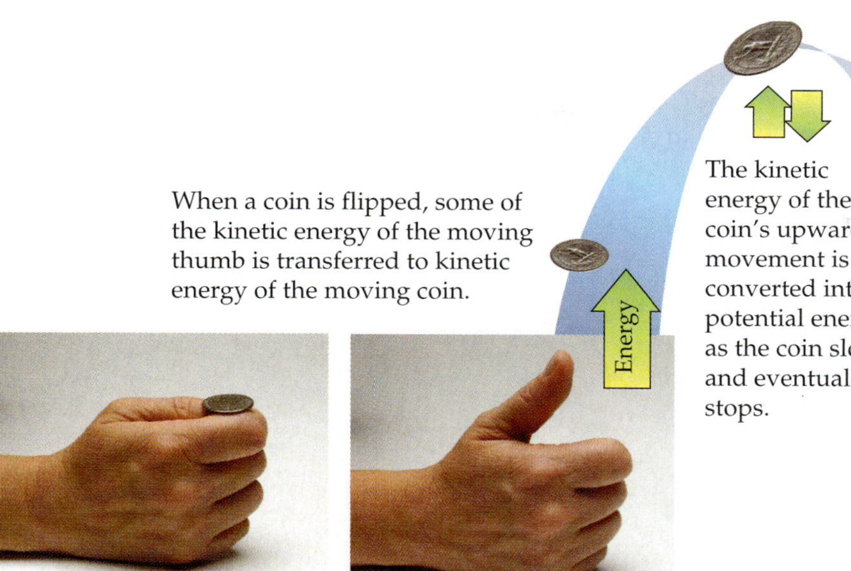

When a coin is flipped, some of the kinetic energy of the moving thumb is transferred to kinetic energy of the moving coin.
Energy
The kinetic energy of the coin's upward movement is converted into potential energy as the coin slows and eventually stops.


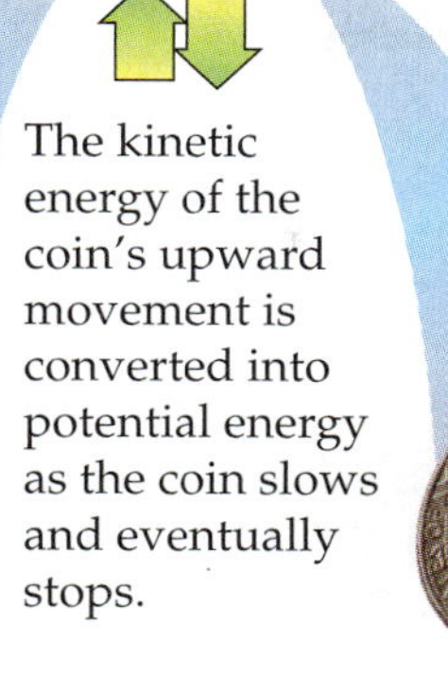

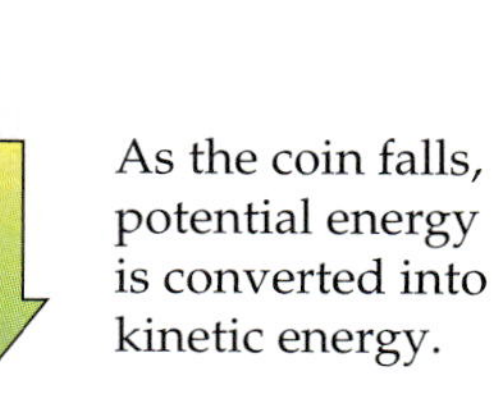

As the coin falls, potential energy is converted into kinetic energy.


OBJECTIVE 4

**Figure 7.4**
Law of Conservation of Energy

system is less likely to change than a less stable system. As an object moves from a less stable state to a more stable state, it can do work. Thus, as an object becomes less stable, it gains a greater capacity to do work and, therefore, a greater potential energy. For example, a coin in your hand is less likely to move than a flipped coin at the peak of its flight, so we say that the coin in the hand is more stable than the coin in the air. As the coin moves from its less stable state in the air to a more stable state on the ground, it collides with and moves particles in the air and blades of grass. Therefore, the coin at the peak of its flight has a greater capacity to do the work of moving the objects—and, therefore, a greater potential energy than the more stable coin in the hand (Figure 7.5, on the next page). *Any time a system shifts from a more stable state to a less stable state, the potential energy of the system increases.* We have already seen that kinetic energy is converted into potential energy as the coin is moved from the more stable position in the hand to the less stable position in the air.

OBJECTIVE 5 OBJECTIVE 6

OBJECTIVE 5

more stable + energy → less stable system

lesser capacity to do work + energy → greater capacity to do work

lower PE + energy → higher PE

coin in hand + energy → coin in air above hand

Just as energy is needed to propel a coin into the air and increase its potential energy, energy is also necessary to separate two atoms being held together by mutual attraction in a chemical bond. The energy supplied increases the potential energy of the less stable separate atoms, compared to

**Figure 7.5**
**Relationship Between Stability and Potential Energy**

OBJECTIVE 6

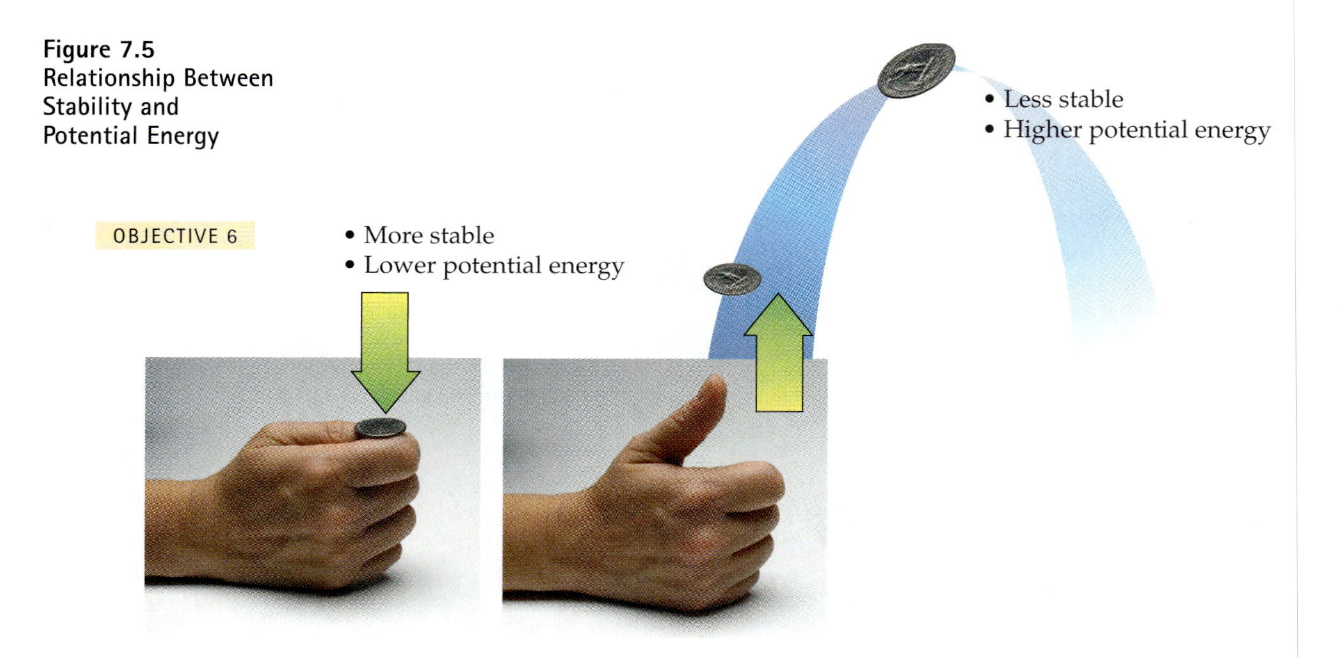


the more stable atoms in the bond. For example, the first step in the formation of ozone in the earth's atmosphere is the breaking of the oxygen–oxygen covalent bonds in more stable oxygen molecules, $O_2$, to form less stable separate oxygen atoms. This change could not occur without an input of considerable energy—in this case, radiant energy from the sun. We call changes that absorb energy **endergonic** (or endogonic) changes (Figure 7.6).

OBJECTIVE 7

**Figure 7.6**
**Endergonic Change**

OBJECTIVE 5

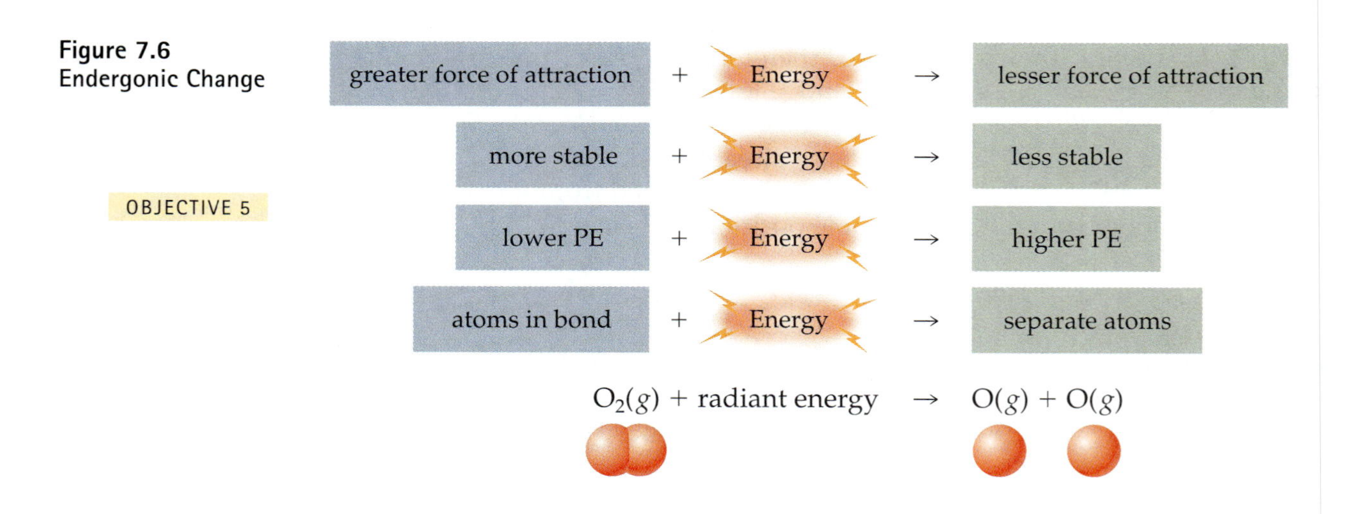


The attraction between the separated atoms makes it possible that they will change from their less stable separated state to the more stable bonded state. As they move together, they may bump into and move something (such as another atom), so the separated atoms have a greater capacity to do work and a greater potential energy than the atoms in the bond. This is why energy must be supplied to break chemical bonds.

*When objects shift from less stable states to more stable states, energy is released.* For example, when a coin moves from the less stable peak of its flight to the more stable position on the ground, potential energy is released as kinetic energy. Likewise, energy is released when separate atoms come together to form a chemical bond. Because the less stable separate atoms have higher potential energy than the more stable atoms that participate in a bond, the change from separate atoms to atoms in a bond corresponds to a decrease in potential energy. Ozone, $O_3$, forms in the stratosphere when an oxygen atom, O, and an oxygen molecule, $O_2$, collide. The energy released in this change comes from the formation of a new O–O bond in ozone, $O_3$. We call changes that release energy **exergonic** (or exogonic) changes (Figure 7.7).

OBJECTIVE 5

OBJECTIVE 8

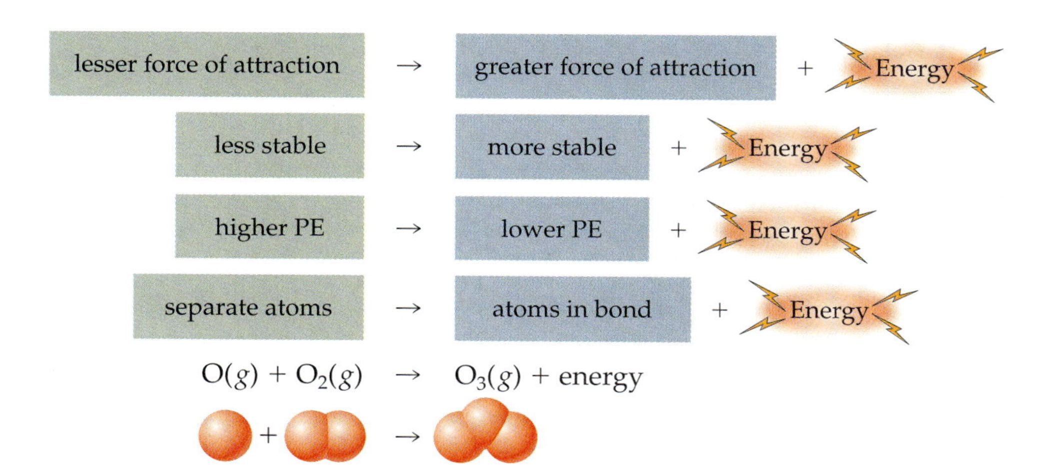


**Figure 7.7**
Exergonic Change

Some bonds are more stable than others. The products of the chemical reactions that take place in an alkaline battery, and those produced in our bodies when the chemicals in pasta are converted into other substances, have more stable chemical bonds between their atoms than the reactants do. Therefore, in each case, the potential energy of the products is lower than that of the reactants, and the lost potential energy supplies the energy to move a toy car across the carpet *and* propel a four-year-old along behind it.

## Example 7.1 Energy

OBJECTIVE 2 OBJECTIVE 3

OBJECTIVE 5

For each of the following situations, you are asked which of two objects or substances has greater energy. Explain your answer with reference to the capacity of each to do work, and indicate whether the energy that distinguishes them is kinetic energy or potential energy.

a. Incandescent light bulbs burn out because the tungsten filament gradually evaporates, weakening until it breaks. Argon gas is added to these bulbs to reduce the rate of evaporation. Which has greater energy: an argon atom, Ar, with a velocity of 428 m/s or the same atom moving with a velocity of 456 m/s? (These are the average velocities of argon atoms at 20 °C and 60 °C.)

b. Krypton, Kr, gas does a better job than argon of reducing the rate of evaporation of the tungsten filament in an incandescent light bulb. Because of its higher cost, however, krypton is used only when longer life is worth the extra cost. Which has greater energy: an argon atom with a velocity of 428 m/s or a krypton atom moving at the same velocity?

c. According to our model for ionic solids, the ions at the surface of the crystal are constantly moving out and away from the other ions and then being attracted back to the surface. Which has greater energy, a stationary sodium ion separated farther from the chloride ions at the surface of a sodium chloride crystal or a stationary sodium ion located closer to the chloride ions on the surface of the crystal?

d. The chemical reactions that lead to the formation of polyvinyl chloride (PVC), which is used to make rigid plastic pipes, are initiated by the decomposition of peroxides. The general reaction is shown below. The simplest peroxide is hydrogen peroxide, $H_2O_2$ or HOOH. Which has greater energy, a hydrogen peroxide molecule or two separate HO molecules that form when the relatively weak O–O bond in an HOOH molecule is broken?

$$\text{HOOH} \rightarrow 2\text{HO}$$

e. Hydrogen atoms react with oxygen molecules in the earth's upper atmosphere to form $HO_2$ molecules. Which has greater energy, a separate H atom and $O_2$ molecule or an $HO_2$ molecule?

$$\text{H}(g) + \text{O}_2(g) \rightarrow \text{HO}_2(g)$$

f. Dry ice—solid carbon dioxide—sublimes, which means that it changes directly from solid to gas. Assuming that the temperature of the system remains constant, which has greater energy, the dry ice or the gaseous carbon dioxide?

*Solution*

a. Any object in motion can collide with another object and move it, so any object in motion has the capacity to do work. This capacity to do work resulting from the motion of an object is called kinetic energy (KE). The particle with the higher velocity will move another object (such as another atom) farther, so it can do more work. It must therefore have greater energy. Thus, an **argon atom with a velocity of 456 m/s has greater kinetic energy** than the same atom with a velocity of 428 m/s.

b. The moving particle with the higher mass can move another object (such as another molecule) farther, so it can do more work and must therefore have energy. Thus the **more massive krypton atoms moving at 428 m/s have greater kinetic energy** than the less massive argon atoms with the same velocity.

c. Separated ions are less stable than atoms in an ionic bond, so the **separated sodium and chloride ions have greater potential energy** than the ions that are closer together. The attraction between the separated sodium cation and the chloride anion pulls them together; as they approach each other, they could conceivably bump into another object, move it, and do work.

d. Separated atoms are less stable and have greater potential energy than atoms in a chemical bond, so energy is required to break a chemical bond. Thus energy is required to separate the two oxygen atoms of HOOH being held together by mutual attraction in a chemical bond. The energy supplied is represented in the greater potential energy of separate HO molecules, compared to the HOOH molecule. If the bond were reformed, the potential energy would be converted into a form of energy that could be used to do work. In short, **two HO molecules have greater potential energy** than an HOOH molecule.

e. Atoms in a chemical bond are more stable and have lower potential energy than separated atoms, so energy is released when chemical bonds form. When H and $O_2$ are converted into an $HO_2$ molecule, a new bond is formed, and some of the potential energy of the separate particles is released. The energy could be used to do some work.

$$H(g) + O_2(g) \rightarrow HO_2(g)$$

Therefore, **separated hydrogen atoms and oxygen molecules have greater potential energy** than the $HO_2$ molecules that they form.

f. When carbon dioxide sublimes, the attractions that link the $CO_2$ molecules together are broken. The energy that the dry ice must absorb to break these attractions goes to increase the potential energy of the $CO_2$ gas. If the $CO_2$ returns to the solid form, attractions are re-formed, and the potential energy is converted into a form of energy that could be used to do work. Therefore, **gaseous $CO_2$ has greater potential energy** than solid $CO_2$.

## Exercise 7.1 Energy

OBJECTIVE 2 OBJECTIVE 3 OBJECTIVE 5

For each of the following situations, you are asked which of two objects or substances has greater energy. Explain your answer with reference to the capacity of each to do work, and indicate whether the energy that distinguishes them is kinetic energy or potential energy.

a. Nitric acid molecules, $HNO_3$, in the upper atmosphere decompose to form HO molecules and $NO_2$ molecules by the breaking of a bond between the nitrogen atom and one of the oxygen atoms. Which has greater energy, a nitric acid molecule or the HO molecule and $NO_2$ molecule that come from its decomposition?

$$HNO_3(g) \rightarrow HO(g) + NO_2(g)$$

b. Nitrogen oxides, $NO(g)$ and $NO_2(g)$, are released into the atmosphere in the exhaust of our cars. Which has greater energy, a $NO_2$ molecule moving at 439 m/s or the same $NO_2$ molecule moving at 399 m/s? (These are the average velocities of $NO_2$ molecules at 80 °C and 20 °C, respectively.)

c. Which has greater energy, a nitrogen monoxide molecule, NO, emitted from your car's tail pipe at 450 m/s or a nitrogen dioxide molecule, $NO_2$, moving at the same velocity?

d. Liquid nitrogen is used for a number of purposes, including the removal (by freezing) of warts. Assume that the temperature remains constant. Which has greater energy, liquid nitrogen or gaseous nitrogen?

e. Halons, such as halon-1301 ($CF_3Br$) and halon-1211 ($CF_2ClBr$), which have been used as fire-extinguishing agents, are a potential threat to the earth's protective ozone layer, partly because they lead to the production of $BrONO_2$, which is created from the combination of BrO and $NO_2$. Which has greater energy, separate BrO and $NO_2$ molecules or the $BrONO_2$ that they form?

f. The so-called alpha particles released by large radioactive elements such as uranium are helium nuclei consisting of two protons and two neutrons. Which has greater energy, an uncharged helium atom or an alpha particle and two separate electrons?

## Units of Energy

OBJECTIVE 9

The accepted SI unit for energy is the **joule (J),** but another common unit is the **calorie (cal).** The calorie has been defined in several different ways. One early definition described it as the energy necessary to increase the temperature of 1 gram of water from 14.5 °C to 15.5 °C. There are 4.186 J/cal according to this definition. Today, however, the U.S. National Institute of Standards and Technology defines the calorie as 4.184 joules:

$$4.184 \text{ J} = 1 \text{ cal}$$

OBJECTIVE 10

or $$4.184 \text{ kJ} = 1 \text{ kcal}$$

The "calories" spoken of in the context of dietary energy—the energy supplied by food—are actually kilocalories, kcal, equivalent to 4184 J or 4.184 kJ. This dietary calorie is often written **Calorie** (using an upper-case C) and abbreviated **Cal.**

$$4184 \text{ J} = 1 \text{ Cal}$$

or $$4.184 \text{ kJ} = 1 \text{ Cal}$$

Thus a meal of about 1000 dietary calories (Calories) provides about 4184 kJ of energy. Table 7.1 shows the energy provided by various foods.

We will use joules and kilojoules to describe energy in this text. Figure 7.8 shows some approximate values, in joules, for the energy represented by various events.

**Table 7.1**
Approximate Energy Provided by Various Foods

| Food | Dietary calories (kcal) | Kilojoules (kJ) | Food | Dietary calories (kcal) | Kilojoules (kJ) |
|---|---|---|---|---|---|
| cheese pizza (12-inch diameter) | 1180 | 4940 | unsweetened apple juice (1 cup) | 120 | 500 |
| roasted cashew nuts (1 cup) | 780 | 3260 | butter (1 tablespoon) | 100 | 420 |
| white granular sugar (1 cup) | 770 | 3220 | raw apple (medium size) | 100 | 420 |
| raw black beans (1 cup) | 680 | 2850 | beer (8 fl oz) | 100 | 420 |
| dry rice (1 cup) | 680 | 2850 | chicken's egg (extra large) | 90 | 380 |
| wheat flour (1 cup) | 400 | 1670 | cheddar cheese (1-inch cube) | 70 | 290 |
| ice cream, 10% fat (1 cup) | 260 | 1090 | whole wheat bread (1 slice) | 60 | 250 |
| raw broccoli (1 pound) | 140 | 590 | black coffee (6 fl oz) | 2 | 8 |

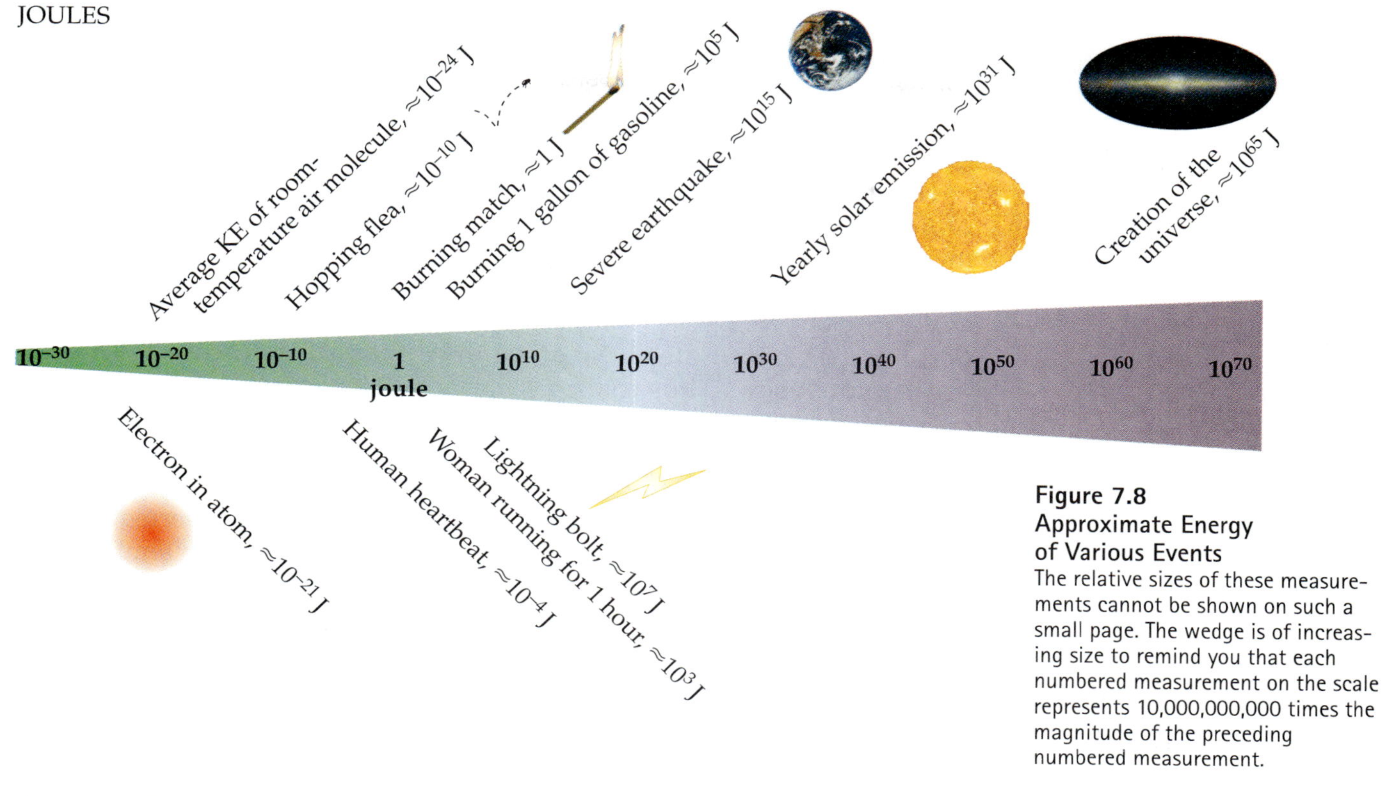


**Figure 7.8**
**Approximate Energy of Various Events**
The relative sizes of these measurements cannot be shown on such a small page. The wedge is of increasing size to remind you that each numbered measurement on the scale represents 10,000,000,000 times the magnitude of the preceding numbered measurement.

## Thermal Energy and Heat

OBJECTIVE 11

An object's kinetic energy can be classified as internal or external. For example, a falling coin has a certain external kinetic energy that is related to its overall mass and to its velocity as it falls. The coin is also composed of particles that, like all particles, are moving in a random way, independent of the overall motion (or position) of the coin. The particles in the coin are constantly moving, colliding, changing direction, and changing velocity. The energy associated with this internal motion can be called either internal kinetic energy or thermal energy. **Thermal energy** is the energy associated with the random motion of particles (Figure 7.9).

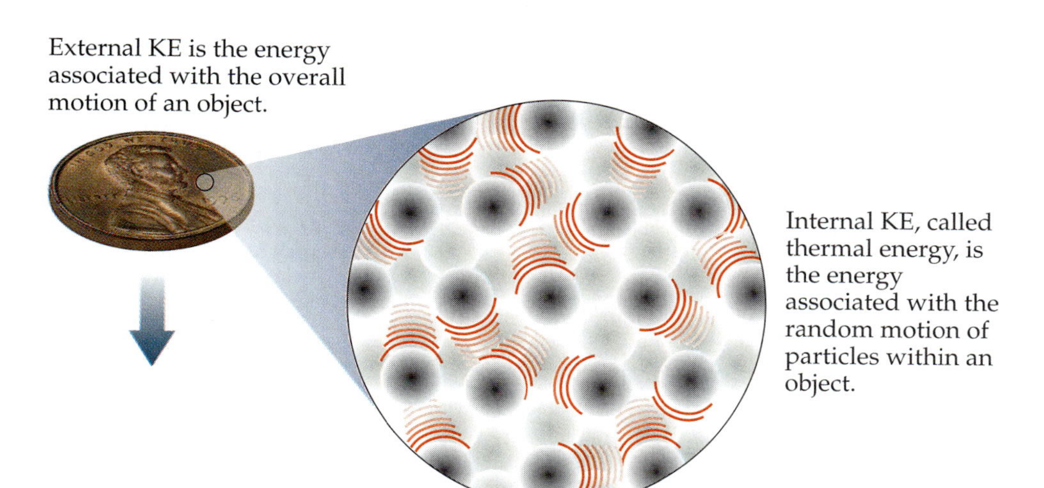


**Figure 7.9**
**External Kinetic Energy and Internal Kinetic Energy (Thermal Energy)**

OBJECTIVE 12

The amount of thermal energy in an object can be increased in three general ways. The first way is to rub, compress, or distort the object. For example, after a good snowball fight, you can warm your hands by rubbing them together. Likewise, if you beat on metal with a hammer, it gets hot.

OBJECTIVE 12

OBJECTIVE 13

OBJECTIVE 14

The second way to increase the thermal energy of an object is to put it in contact with another object at a higher temperature. **Temperature** is a measure of the average internal kinetic energy of an object, so higher temperature means a greater average internal energy for the particles within the object. The particles in a higher-temperature object collide with other particles with greater average force than the particles of a lower-temperature object. Thus collisions between the particles of two objects at different temperatures cause the particles of the lower-temperature object to speed up, increasing the object's internal kinetic energy (or thermal energy), and cause the particles of the higher-temperature object to slow down, decreasing this object's thermal energy. In this way, thermal energy is transferred from the higher-temperature object to the lower-temperature object. We call thermal energy that is transferred in this way heat. The thermal energy that is transferred through an object, such as from the bottom of a cooking pan to its handle, is also called heat. **Heat** is the thermal energy that is transferred from a region of higher temperature to a region of lower temperature as a consequence of the collisions of particles (Figure 7.10).

**Figure 7.10**
Heat Transfer

OBJECTIVE 14

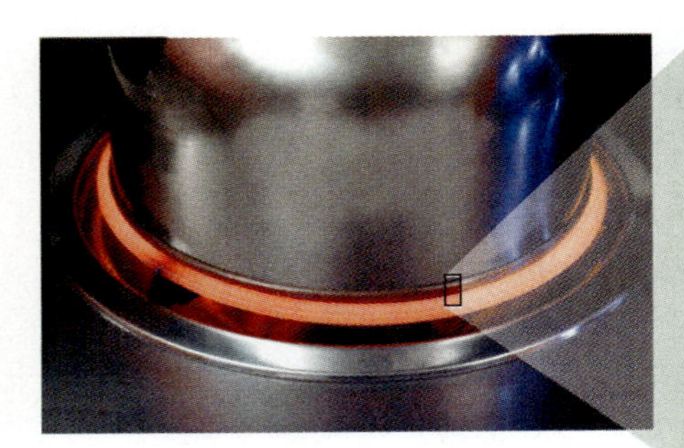

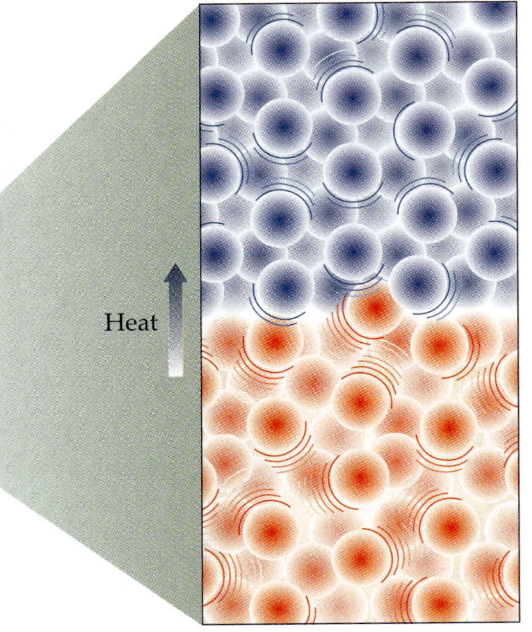

The third way in which an object's thermal energy is increased is by exposure to radiant energy, such as the energy coming from the sun. The radiant energy is converted to kinetic energy of the particles in the object. This is why we get hot in the sun.

OBJECTIVE 12

## Radiant Energy

Gamma rays, X rays, ultraviolet radiation, visible light, infrared radiation, microwaves, and radio and TV waves are all examples of radiant energy. Although we know a great deal about radiant energy, we still have trouble describing what it is. For example, it seems to have a dual nature, with both particle and wave characteristics. It is difficult to visualize both of these aspects of radiant energy at the same time, so sometimes we focus on its particle nature and sometimes on its wave character, depending on which is more suitable in a given context. Accordingly, we can describe the light that comes from a typical flashlight either as a flow of about $10^{17}$ particles of energy leaving the bulb per second or as waves of a certain length.

In the particle view, radiant energy is a stream of tiny, massless packets of energy called **photons.** The light from the flashlight contains photons of many different energies, so you might try to picture the beam as a stream of photons of many different sizes. (It is difficult to picture a particle without mass, but that is just one of the problems we have in describing what light is!)

OBJECTIVE 15

The wave view says that as radiant energy moves away from its source, it has an effect on the space around it that can be described as a wave consisting of an oscillating electric field perpendicular to an oscillating magnetic field (Figure 7.11).

OBJECTIVE 16

Because radiant energy seems to have both wave and particle characteristics, some experts have suggested that it is probably neither a wave nor a stream of particles. Perhaps the simplest model that includes both aspects of radiant energy says that as the photons travel, they somehow affect the space around them in such a way as to create the electric and magnetic fields.

**Radiant energy,** then, is energy that can be described in terms of oscillating electric and magnetic fields or in terms of photons. It is often called **electromagnetic radiation.** Because all forms of radiant energy have these qualities, we can distinguish one form of radiant energy from another either by the energy of its photons or by the characteristics of its waves. The energies of the photons of radiant energy range from about $10^{-8}$ J per photon for the very high-energy gamma rays released in radioactive decay to about $10^{-31}$ J per photon, or even smaller, for low-energy radio waves. The different forms of radiant energy are listed in Figure 7.12, at the end of this section.

One distinguishing difference in the waves of radiant energy is wavelength, λ, the distance between two peaks on the wave of electromagnetic radiation. A more specific definition of **wavelength** is the distance in space over which a wave completes one cycle of its repeated form. Between two successive peaks, the wave has gone through all of its possible combinations of magnitude and direction and has begun to repeat the cycle again (Figure 7.11).

**Figure 7.11**
**A Light Wave's Electric and Magnetic Fields**

OBJECTIVE 16

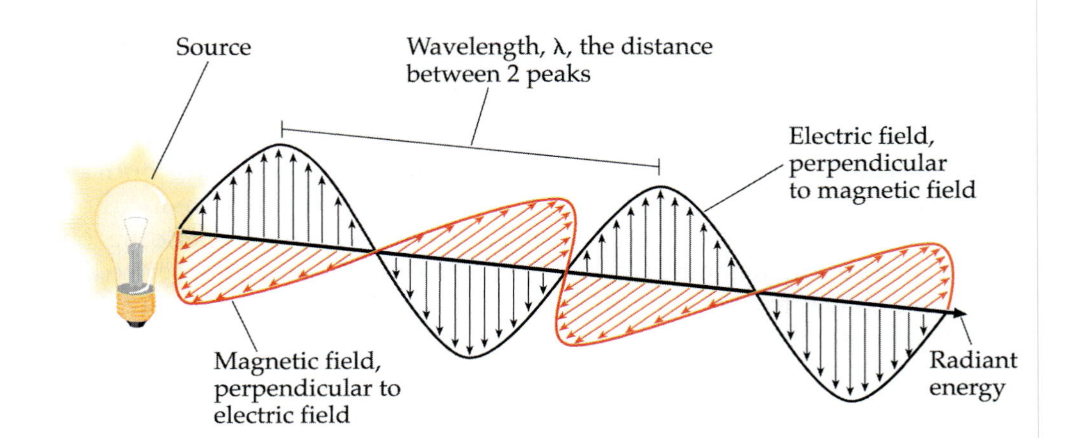


Gamma rays, with very high-energy photons, have very short wavelengths (Figure 7.12), on the order of $10^{-14}$ meter ($10^{-5}$ nm). Short wavelengths are often described with nanometers, nm, which are $10^{-9}$ m. In contrast, the radio waves on the low-energy end of the AM radio spectrum have wavelengths of about 500 m (about one-third of a mile). If you look at the energy and wavelength scales in Figure 7.12, you will see that longer wavelength corresponds to lower-energy photons. The shorter the wavelength of a wave of electromagnetic radiation, the greater the energy of its photons. In other words, the energy, ϵ, of a photon is inversely proportional to the radiation's wavelength, λ. (The symbol ϵ is a lower-case Greek epsilon, and the λ is a lower-case Greek lambda.)

OBJECTIVE 17

$$\epsilon \propto \frac{1}{\lambda}$$

As Figure 7.12 illustrates, all forms of radiant energy are part of a continuum, and there are no precise dividing lines between one form and the next. In fact, there is some overlap between categories. Note that visible light is only a small portion of the radiant-energy spectrum. The different colors of visible light are due to different photon energies and associated wavelengths.

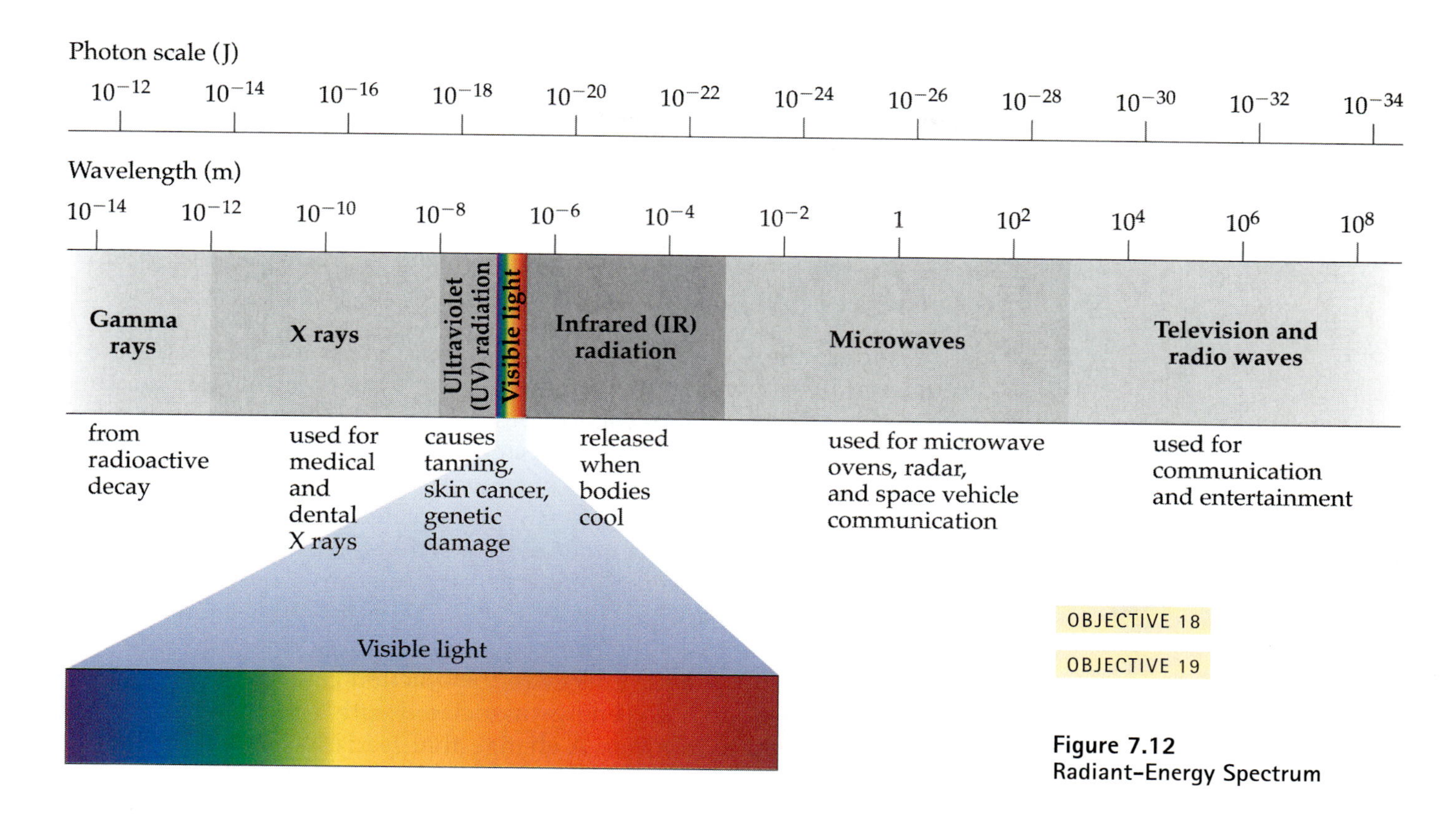


OBJECTIVE 18

OBJECTIVE 19

**Figure 7.12**
Radiant-Energy Spectrum

## 7.2 Chemical Changes and Energy

OBJECTIVE 20

In most chemical reactions, bonds are broken and new ones formed. In Section 7.1 we saw that the breaking of bonds requires energy and that the formation of bonds releases it. If, in a chemical reaction, more energy is released in the formation of new bonds than was necessary to break old bonds, energy is evolved overall, and the reaction is exergonic. The burning of hydrogen gas is an exergonic process.

$$2H_2(g) \; + \; O_2(g) \; \rightarrow \; 2H_2O(g) \; + \; \text{energy}$$

On the one hand, energy is required to break the bonds between the hydrogen atoms in the hydrogen molecules, $H_2$, and between the oxygen atoms in the oxygen molecules, $O_2$. On the other hand, energy is released in the formation of the H–O bonds in water. Because the bonds in the product are more stable, they are stronger than the bonds in the reactants, and the product has lower potential energy than the reactants. Energy is released overall.

weaker bonds → stronger bonds + energy

OBJECTIVE 20

higher PE → lower PE + energy

$2H_2(g) + O_2(g) \rightarrow 2H_2O(g)$ + energy

To visualize these energy changes at the molecular level, let's picture a container of hydrogen and oxygen that is initially at the same temperature as its surroundings. As the reaction proceeds and forms water molecules, some of the potential energy of the system is converted into kinetic energy. (This is similar to the conversion of potential energy into kinetic energy when a coin falls toward the ground.) The greater average kinetic energy of the particles in the product mixture means that the product mixture is at a higher temperature than the initial reactant mixture—and therefore at a higher temperature than the surroundings. Thus the higher-temperature products transfer heat to the surroundings. Because the conversion of potential energy into kinetic energy in the reaction can lead to heat being released from the system, the energy associated with a chemical reaction is often called the *heat of reaction*. A change that leads to heat energy being released from the system to the surroundings is called **exothermic.** (*Exergonic* means that *any form of energy* is released; *exothermic* means *heat energy* is released.)

OBJECTIVE 20

If less energy is released in the formation of the new bonds than is necessary to break the old bonds, energy must be absorbed from the surroundings for the reaction to proceed. The reaction that forms calcium oxide, often called quicklime, from calcium carbonate is an example of this sort of change. (Quicklime has been used as a building material since 1500 B.C. It is used today to remove impurities from iron ores. Calcium oxide is also used in air pollution control and water treatment.) In the industrial production of quicklime, this reaction is run at over 2000 °C to provide enough energy to convert the calcium carbonate into calcium oxide and carbon dioxide.

OBJECTIVE 21

stronger bonds + energy → weaker bonds

OBJECTIVE 21

lower PE + energy → higher PE

$CaCO_3(s)$ + energy → $CaO(s) + CO_2(g)$

Collectively, the bonds formed in the products are weaker and, therefore, less stable than those of the reactant, so the products have greater potential energy than the reactants. The high temperature is necessary to provide energy for the change to the greater-potential-energy products. Because energy is absorbed in the reaction, it is endergonic.

OBJECTIVE 21

Endergonic changes can lead to a transfer of heat *from* the surroundings. Cold packs used to cool a sprained ankle quickly are an example of this kind of change. One kind of cold pack contains a small pouch of ammonium nitrate, $NH_4NO_3$, inside a larger pouch of water. When the cold pack is twisted, ammonium nitrate is released into the water, and as it dissolves, the water cools.

$$NH_4NO_3(s) + \text{energy} \rightarrow NH_4^+(aq) + NO_3^-(aq)$$

The attractions between the particles in the final mixture are less stable and have higher potential energy than the attractions in the separate ammonium nitrate solid and liquid water. The energy necessary to increase the potential energy of the system comes from some of the kinetic energy of the moving particles, so the particles in the final mixture have a lower average kinetic energy and a lower temperature than the original substances. Heat is transferred from the higher-temperature sprained ankle to the lower-temperature cold pack. A change such as the one in the cold pack, that leads a system to absorb heat energy from the surroundings, is called an **endothermic** change. (Endergonic means that *any form of energy* is absorbed, and *endothermic* means *heat energy* is absorbed.)

OBJECTIVE 21

Figure 7.13 shows the logical sequence that summarizes why chemical reactions either release or absorb energy. It also shows why, in many cases, the change leads to a transfer of heat either to or from the surroundings, with a corresponding increase or decrease in the temperature of the surroundings.

OBJECTIVE 22

**Figure 7.13**
**Energy and Chemical Reactions**

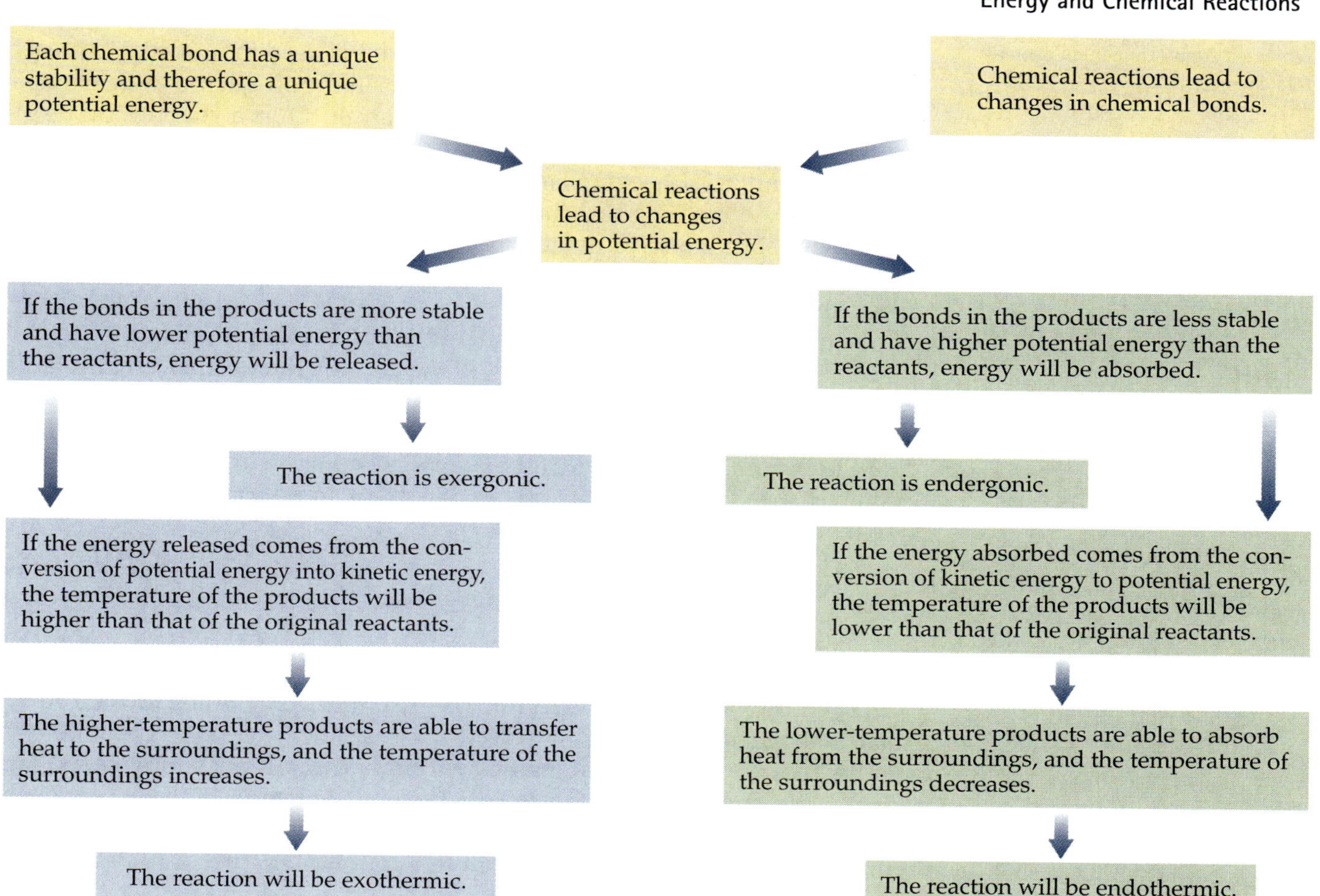

Each chemical bond has a unique stability and therefore a unique potential energy.
Chemical reactions lead to changes in chemical bonds.
Chemical reactions lead to changes in potential energy.
If the bonds in the products are more stable and have lower potential energy than the reactants, energy will be released.
If the bonds in the products are less stable and have higher potential energy than the reactants, energy will be absorbed.
The reaction is exergonic.
The reaction is endergonic.
If the energy released comes from the conversion of potential energy into kinetic energy, the temperature of the products will be higher than that of the original reactants.
If the energy absorbed comes from the conversion of kinetic energy to potential energy, the temperature of the products will be lower than that of the original reactants.
The higher-temperature products are able to transfer heat to the surroundings, and the temperature of the surroundings increases.
The lower-temperature products are able to absorb heat from the surroundings, and the temperature of the surroundings decreases.
The reaction will be exothermic.
The reaction will be endothermic.

## 7.3 Ozone: Pollutant and Protector

These grape plants were damaged by ground level ozone. The plant on the right has leaves damaged by ground-level ozone; the other healthy plant, on the left, was protected from ozone in an ozone-free chamber.

The discoveries that have caused scientists to worry about ozone levels in the earth's atmosphere offer some excellent examples of the relationship between energy and chemical changes. Perhaps the news media's handling of these issues has caused you some confusion. One day you might see a newspaper headline about an ozone alert in Los Angeles, triggered by the concentration of ozone in the air rising to *too high* a level. Schoolteachers are warned to keep students off the playgrounds to prevent damage to their lungs. Turning the page in the same newspaper, you discover another article that describes problems resulting from *decreasing* amounts of ozone in the "ozone layer" of the upper atmosphere. *So which is it, too much or too little?* Is ozone a pollutant or a protector? What is this substance, and why are we so worried about it?

Two forms of the element oxygen are found in nature: the life-sustaining diatomic oxygen, $O_2$, and ozone, $O_3$, which is a pale blue gas with a strong odor. The concentrations of ozone in the air around us are usually too low for the color and the odor to be apparent, but sometimes when we stand next to an electric motor, we notice ozone's characteristic smell. This is because an electric spark passing through oxygen gas creates ozone.

OBJECTIVE 23

Ozone is a very powerful oxidizing agent. Sometimes this property can be used to our benefit, and sometimes it is a problem. Ozone mixed with oxygen can be used to sanitize hot tubs, and it is used in industry to bleach waxes, oils, and textiles, but when the levels in the air get too high, ozone's high reactivity becomes a problem. For example, $O_3$ is a very strong respiratory irritant that can lead to shortness of breath, to chest pain when inhaling, and to wheezing and coughing. Anyone who has lived in a smoggy city will recognize these symptoms. Not only can ozone oxidize lung tissue, but it also damages rubber and plastics, leading to premature deterioration of products made with these materials. Furthermore, according to the Agricultural Research Service of North Carolina State University, ozone damages plants more than all other pollutants combined.

OBJECTIVE 24

OBJECTIVE 25

The highest concentrations of $O_3$ in the air we breathe are found in large industrial cities with lots of cars and lots of sun. The explanation why this is true begins with a description of the source of nitrogen oxides. Any time air (which contains nitrogen and oxygen) is heated to high temperature (as occurs in the cylinders of our cars and in many industrial processes), nitrogen oxides are formed (NO and $NO_2$).

$$N_2(g) + O_2(g) \rightarrow 2NO(g)$$

$$2NO(g) + O_2(g) \rightarrow 2NO_2(g)$$

Nitrogen dioxide, $NO_2$, is a red-brown gas that contributes to the brown haze associated with smog.

OBJECTIVE 24

OBJECTIVE 25

The radiant energy that passes through the air on sunny days can supply the energy necessary to break covalent bonds between nitrogen atoms and oxygen atoms in $NO_2$ molecules, converting $NO_2$ molecules into NO molecules and oxygen atoms. Remember that the shorter the

OBJECTIVE 26

wavelength of light, the higher the energy. Radiant energy of wavelengths less than 400 nm has enough energy to break N–O bonds in $NO_2$ molecules, but radiant energy with wavelengths longer than 400 nm does not supply enough energy to separate the atoms.

$$NO_2(g) \xrightarrow{\lambda < 400\text{ nm}} NO(g) + O(g)$$

The oxygen atoms react with oxygen molecules to form ozone molecules.

$$O(g) + O_2(g) \rightarrow O_3(g)$$

OBJECTIVE 24

Because the process of ozone formation is initiated by light photons, the pollutant mixture is called photochemical smog. Figure 7.14 shows typical ozone levels in different areas of the United States. Note that the concentrations are highest in southern California. Cities such as Los Angeles, with its sunny weather and hundreds of thousands of cars, have ideal conditions for the production of photochemical smog. This smog is worst from May to September, when the days are long and the sunlight intense.

OBJECTIVE 25

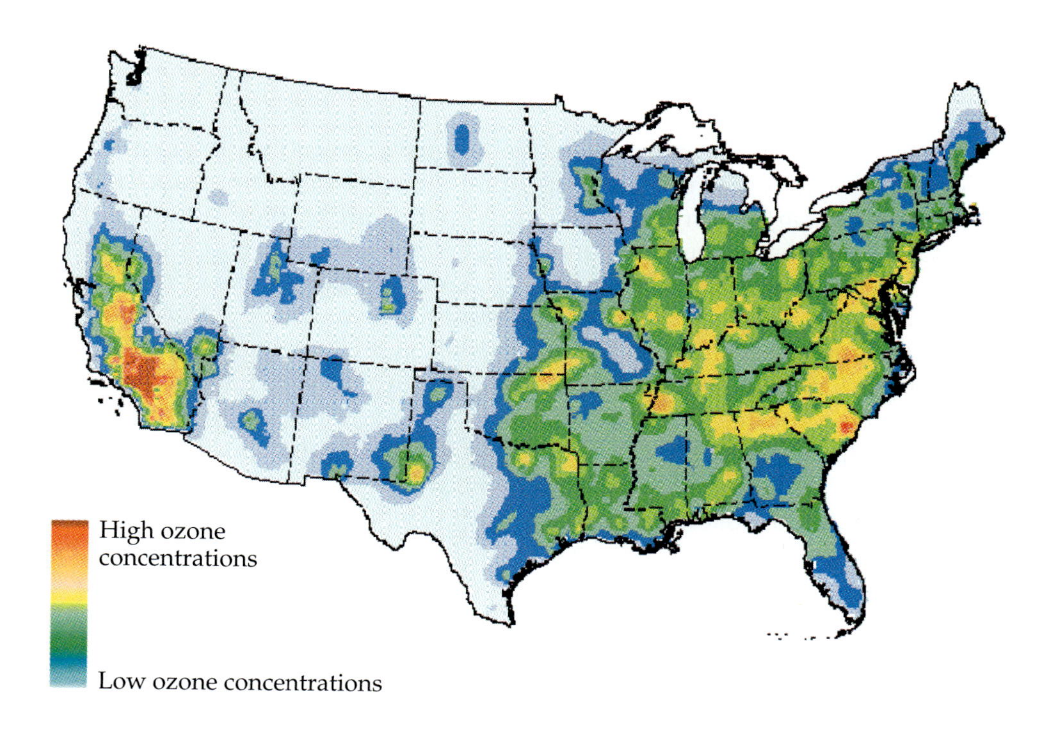


**Figure 7.14**
Typical Ozone Concentrations in the United States

Now that we have seen what conditions lead to *too much* $O_3$ in the air we breathe and understand why that is a problem, we need to know why depleting the ozone in the upper atmosphere, creating *too little* ozone there, can also be a problem. Let's start with a little information about our atmosphere. Atmospheric scientists view the atmosphere as consisting of layers, each with its own characteristics. The lowest layer, which extends from the surface of the earth to about 10 km (about 6 miles) above sea level, is called the **troposphere.** For our discussion of ozone, we are more interested in the next lowest layer, the **stratosphere,** which extends from about 10 km to about 50 km above sea level (Figure 7.15).

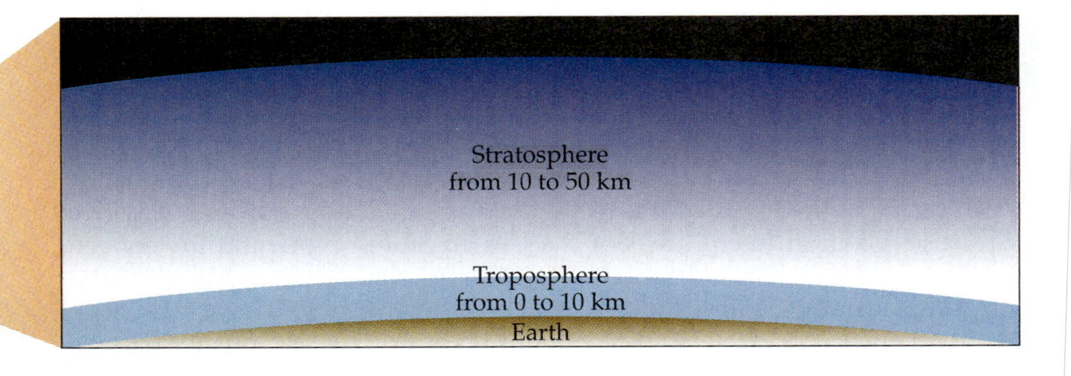


**Figure 7.15**
The Earth's Atmosphere

The stratosphere contains a mixture of gases, including oxygen molecules, $O_2$, and ozone molecules, $O_3$, that play a very important role in protecting the earth from the sun's high-energy ultraviolet radiation. The ultraviolet portion of the sun's energy spectrum can be divided into three parts: UV-A, UV-B, and UV-C. Not all UV radiation is harmful. **UV-A,** which includes radiant energy of wavelengths from about 320 to 400 nm, passes through the stratosphere and reaches us on the surface of the earth. We are glad it does, because UV-A radiation provides energy that our bodies use to produce vitamin D.

OBJECTIVE 27

OBJECTIVE 28

The shorter-wavelength **UV-B** radiation (from about 290 to 320 nm) has greater energy than the UV-A radiation. Some UV-B radiation is removed by the gases in the stratosphere, but some of it reaches the surface of the earth. Radiation in this portion of the spectrum has energy great enough that excessive exposure can cause sunburn, premature skin aging, and skin cancer.

OBJECTIVE 27

OBJECTIVE 29

The greatest-energy ultraviolet radiation is **UV-C,** with wavelengths from about 40 to 290 nm. We are very fortunate that this radiant energy is almost completely removed by the gases in the atmosphere, because UV-C is energetic enough to cause serious damage not only to us but to all life on earth. One reason why it is so dangerous is that DNA, the substance that carries genetic information in living cells, absorbs UV radiation of about 260 nm. Likewise, proteins, which are vital structural and functional components of living systems, absorb radiation with wavelengths of about 280 nm. If these wavelengths were to reach the earth in significant quantity, the changes they would cause by altering DNA and protein molecules would lead to massive crop damage and general ecological disaster.

OBJECTIVE 27

OBJECTIVE 30

## Removal of UV Radiation by Oxygen and Ozone Molecules

Some of the dangerous radiation removed in the stratosphere is absorbed by the $O_2$ molecules there. Radiant-energy wavelengths must be shorter than 242 nm to have enough energy to break the O–O bond, and UV-C radiation has wavelengths in the proper range.

$$O_2(g) \xrightarrow{\text{UV with } \lambda < 242 \text{ nm}} 2O(g)$$

OBJECTIVE 31

UV radiation can also provide the energy to break a bond between oxygen atoms in ozone molecules. Because less energy is needed to break a bond in the $O_3$ molecule than to break the bond in $O_2$, the UV photons that break the bond in $O_3$ are associated with longer wavelengths. The $O_3$ molecules will absorb UV radiation of wavelengths from 240 nm to 320 nm.

OBJECTIVE 31

$$O_3(g) \xrightarrow{\text{UV with } \lambda \text{ from 240 to 320 nm}} O(g) + O_2(g)$$

OBJECTIVE 31

Thus oxygen molecules, $O_2$, and ozone molecules, $O_3$, work together to absorb high-energy UV radiation. Oxygen molecules absorb UV radiation with wavelengths less than 242 nm, and ozone molecules absorb radiant energy with wavelengths from 240 nm to 320 nm (Figure 7.16). We have seen that wavelengths in the range of 240 to 320 nm can cause problems that include skin aging, skin cancer, and crop failure. Because $O_2$ does not remove this radiation from the atmosphere, it is extremely important that the ozone layer be preserved.

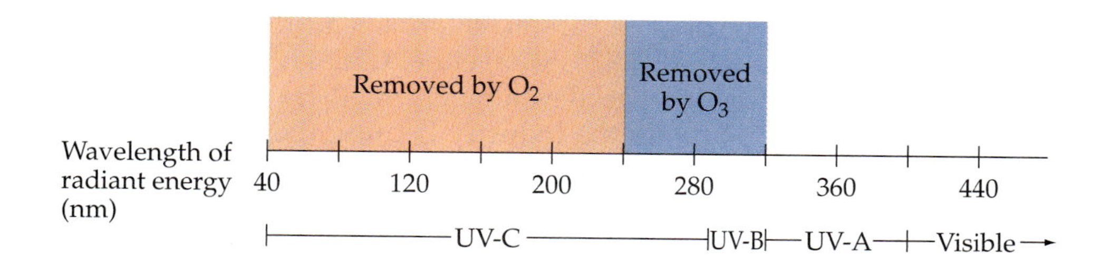


**Figure 7.16**
Removal of Ultraviolet Radiation in the Stratosphere

OBJECTIVE 31

## The Natural Destruction of Ozone

Ozone is constantly being generated and destroyed in the stratosphere as part of a natural cycle. The reactions for the synthesis of ozone are

$$O_2(g) \xrightarrow{\text{UV radiation}} 2O(g)$$

$$O(g) + O_2(g) \rightarrow O_3(g)$$

Several natural processes destroy ozone in the stratosphere. Perhaps the most important are

OBJECTIVE 32

$$NO(g) + O_3(g) \rightarrow NO_2(g) + O_2(g)$$

and $$NO_2(g) + O(g) \rightarrow NO(g) + O_2(g)$$

The first reaction destroys one ozone molecule directly. The second reaction destroys an oxygen atom that might have become part of an ozone molecule. (Because oxygen atoms can collide with oxygen molecules to form ozone molecules, the ozone concentration is depleted indirectly through the removal of oxygen atoms.) The main reason why this pair of reactions is so efficient at destroying ozone molecules, however, is that the $NO(g)$ that is destroyed in the first reaction is regenerated in the second reaction. Therefore, in the overall reaction, an ozone molecule and an oxygen atom are converted into two oxygen molecules with no change in the number of NO molecules. This makes $NO(g)$ a catalyst for the reaction. A **catalyst** is a substance that speeds a chemical reaction without being permanently altered itself. The equation for the net reaction is

OBJECTIVE 32

$$\text{Net reaction:} \quad O_3(g) + O(g) \xrightarrow{\text{NO catalyst}} 2O_2(g)$$

## 7.4 Chlorofluorocarbons: A Chemical Success Story Gone Wrong

In 1972 the chemical industry was producing about 700,000 metric tons (about 1.5 billion pounds) of **chlorofluorocarbons** (CFCs) per year—compounds composed entirely of carbon, chlorine, and fluorine. Most of the CFCs produced in the early 1970s were either CFC-11, which is $CFCl_3$, or CFC-12, which is $CF_2Cl_2$. The development of these chemicals was considered a major triumph for the industry because they seemed to be perfect for use as aerosol propellants, solvents, expansion gases in foams, heat-exchanging fluids in air conditioners, and temperature-reducing fluids in refrigerators.

One of the reasons why CFCs were so successful is that they are extremely stable compounds; very few substances react with them. As a result, they are nontoxic and nonflammable. Another important characteristic is that they are gases at normal room temperatures and pressures but become liquids at pressures slightly above normal. These were precisely the characteristics needed for the applications listed above.

*The announcement today, suggesting a worse situation than we thought, affirms the warning I issued last Spring: Upper atmosphere ozone depletion remains one of the world's most pressing environmental threats.*

William K. Reilly, Administrator of the U.S. Environmental Protection Agency, 1991

### CFCs and the Ozone Layer

Generally, gases are removed from the lower atmosphere in two ways. Either they dissolve in the clouds and are rained out, or they react chemically to be converted into other substances. Neither of these mechanisms is important for CFCs. Chlorofluorocarbons are insoluble in water, and they are so stable that they can persist in the lower atmosphere for years. For example, CFC-11 molecules have an average life of 50 years in the atmosphere, and CFC-12 molecules have an average life of about 102 years. During this time, the CFC molecules wander about, moving wherever the air currents take them. They can eventually make their way into the stratosphere, where they encounter radiation with enough energy to break them down. For example, radiant energy of wavelengths less than 215 nm will break the covalent bond between a chlorine atom and the carbon atom in $CF_2Cl_2$.

OBJECTIVE 33

OBJECTIVE 34

$$CF_2Cl_2(g) \xrightarrow{\lambda < 215\text{ nm}} CF_2Cl(g) + Cl(g)$$

The chlorine atoms released in this sort of reaction can destroy ozone molecules in a process similar to the catalytic reactions between NO, $O_3$, and O described in the last section.

$$Cl(g) + O_3(g) \rightarrow ClO(g) + O_2(g)$$

$$ClO(g) + O(g) \rightarrow Cl(g) + O_2(g)$$

OBJECTIVE 35

Each chlorine atom destroys one ozone molecule directly. In addition, the resulting ClO goes on to react with an oxygen atom that might otherwise have collided with an oxygen molecule to form an ozone molecule, so a second ozone molecule is prevented from forming. Because the chlorine atom is regenerated in the second reaction, its role is that of a catalyst for the reaction. The equation for the net reaction is

$$\text{Net reaction:} \quad O_3(g) + O(g) \xrightarrow{\text{Cl catalyst}} 2O_2(g)$$

Each chlorine atom is thought to destroy an average of 1000 ozone molecules before being temporarily incorporated into a compound such as HCl or $ClONO_2$.

$$CH_4(g) + Cl(g) \rightarrow CH_3(g) + HCl(g)$$

and $ClO(g) + NO_2(g) \rightarrow ClONO_2(g)$

In 1985 scientists discovered a large decrease in the atmospheric concentration of ozone over Antarctica. This "ozone hole" could not be explained with the models used to describe atmospheric chemistry at that time, but it has since been explained in terms of an unexpectedly rapid re-formation of chlorine atoms from chlorine compounds such as HCl and $ClONO_2$. The new model suggests that reactions like the following take place on the surface of ice crystals that form in the cold air of the stratosphere over Antarctica.

$$ClONO_2(g) + HCl(s) \rightarrow Cl_2(g) + HNO_3(s)$$

$$ClONO_2(g) + H_2O(s) \rightarrow HOCl(g) + HNO_3(s)$$

$$HOCl(g) + HCl(s) \rightarrow Cl_2(g) + H_2O(s)$$

$$HOCl(g) \xrightarrow{\text{radiant energy}} Cl(g) + OH(g)$$

$$Cl_2(g) \xrightarrow{\text{radiant energy}} 2Cl(g)$$

The chlorine atoms freed in these reactions can once again react with ozone molecules and oxygen atoms. Many scientists fear that each chlorine atom that reaches the stratosphere may destroy tens of thousands of ozone molecules before escaping from the stratosphere. One way in which chlorine atoms finally escape is by migrating back into the lower atmosphere in HCl molecules that dissolve in the clouds and return to the earth in rain.

Special Topic 7.1 describes substitutes for chlorofluorocarbons, and Special Topic 7.2 describes other ozone-depleting chemicals.

## SPECIAL TOPIC 7.1 Green Chemistry—Substitutes for Chlorofluorocarbons

Any television, computer, or other fragile item you have purchased in recent years has probably come packaged in polystyrene foam (Styrofoam®) for protection, the same foam likely to be serving as insulation in the cooler you take along on a picnic. This stiff, low-density, non-heat-conducting solid is produced by blowing gas into polystyrene liquid as it solidifies. Over 700 million pounds were manufactured in 1995.

Chlorofluorocarbons have been used as the blowing agents in the production of polystyrene foam, but because of the damage CFCs may be causing to the ozone layer, chemists are actively seeking other ways of doing the job. In 1996 the Dow Chemical Company received a Presidential Green Chemistry Challenge Award—specifically, the Alternative Solvents/Reaction Conditions Award—for developing a process in which polystyrene foam is made using 100% carbon dioxide, $CO_2$, as the blowing agent. (See Special Topic 1.1: *Green Chemistry*.) Carbon dioxide is nonflammable and nontoxic and does not deplete the ozone layer. The process does not even increase the level of $CO_2$ in the atmosphere, because the carbon dioxide it uses comes from other commercial or natural sources, such as ammonia plants and natural-gas wells. This new technology reduces the use of CFCs by 3.5 million pounds per year.

## Special Topic 7.2 Other Ozone-Depleting Chemicals

Although CFCs have received the most attention as a threat to the ozone layer, other chemicals are also thought to pose a danger to it. Halons, which are similar to CFCs but contain at least one bromine atom, are one such group of compounds. Halon-1301 ($CF_3Br$) and halon-1211 ($CF_2ClBr$) have been used as fire-extinguishing agents, but the bromine atoms they release have been shown to be even more efficient at destroying ozone than chlorine atoms are.

Governments are in general agreement that halons should be banned, but the status of another bromine-containing compound is more ambiguous. This compound is methyl bromide, $CH_3Br$. The controversy revolves around whether methyl bromide has a significant effect on the ozone layer and whether the benefits of using it outweigh the potential hazards. This ozone-depleting compound is different from CFCs and halons because it is produced not by humans only but also by many prolific natural sources. For example, the ocean is thought to both release and absorb significant amounts of $CH_3Br$, and wildfires generate methyl bromide as well. The actual contribution of the various sources of methyl bromide to the atmosphere is still uncertain, but many scientists agree that significant amounts do come from human activities (so-called *anthropogenic sources*) such as the burning of rain forests; the use of insecticides, herbicides, and fungicides; and the use of leaded gasoline that contains ethylene dibromide. One of the reasons why methyl bromide has been considered less threatening to the ozone layer than CFCs or halons is that it has a much shorter lifetime. The best estimates predict its average lifetime to be 1 to 2 years, compared to over 50 years for the shortest-lived common CFC and over 20 years for the most common halons. Although research on the possible effects of methyl bromide continues, experts now consider it damaging enough that steps have been taken to begin phasing it out.

Since the discovery of the damaging effects of CFCs, alternatives have been developed that offer many of the desirable characteristics of CFCs but are less stable in the lower atmosphere and less likely to reach the stratosphere. These chemicals, called hydrochlorofluorocarbons (HCFCs), are similar in structure to CFCs but contain at least one hydrogen atom. For example, HCFC-22 is $CF_2HCl$, and HCFC-123 is $CF_3CHCl_2$. Although these chemicals are thought to be less damaging to the ozone layer, they too can reach the stratosphere and lead to some depletion of $O_3$. Thus they are viewed as transitional compounds to be used until better substitutes have been found. HCFCs, too, are being phased out over time.

## Chapter Glossary

**Energy** The capacity to do work.

**Kinetic energy (KE)** The capacity to do work as a consequence of the motion of an object.

**Law of Conservation of Energy** Energy can be neither created nor destroyed, but it can be transferred from one system to another and changed from one form to another.

**Potential energy (PE)** A retrievable, stored form of energy that an object possesses by virtue of its position or state.

**Endergonic (endogonic) change** Change that absorbs energy.

**Exergonic (exogonic) change** Change that releases energy.

**Joule (J)** The accepted SI unit for energy.

**calorie (with a lower-case c)** A common energy unit. There are 4.184 joules per calorie (abbreviated *cal*).

**Calorie (with an upper-case C)** The dietary calorie (abbreviated **Cal**). In fact, a Calorie is a kilocalorie, the equivalent of 4184 joules.

**Thermal energy** The energy associated with the random motion of particles.

**Temperature** A measure of the average internal kinetic energy of an object.

**Heat** The thermal energy transferred from a region of higher temperature to a region of lower temperature as a result of collisions between particles.

**Radiant energy or electromagnetic radiation** Energy that can be described either in terms of oscillating electric and magnetic fields or in terms of a stream of tiny packets of energy with no mass.

**Photons** A tiny packet or particle of radiant energy.

**Wavelength ($\lambda$)** The distance in space over which a wave completes one cycle of its repeated form.

**Exothermic change** Change that leads to heat energy being released from the system to the surroundings.

**Endothermic change** Change that leads the system to absorb heat energy from the surroundings.

**Catalyst** A substance that speeds a chemical reaction without being permanently altered itself.

**Troposphere** The lowest layer of the earth's atmosphere. It extends from the surface of the earth to about 10 km above the earth.

**Stratosphere** The second layer of the earth's atmosphere, which extends from about 10 km to about 50 km above sea level.

**UV-A** Ultraviolet radiation in the range of about 320–400 nm wavelengths. This is the part of the ultraviolet spectrum that reaches the earth and provides energy for the production of vitamin D.

**UV-B** Ultraviolet radiation in the range of about 290–320 nm wavelengths. Most of this radiation is filtered out by the earth's atmosphere, but some reaches the surface of the earth. Excessive exposure can cause sunburn, premature skin aging, and skin cancer.

**UV-C** Ultraviolet radiation in the range of about 40–290 nm wavelengths. Almost all UV-C is filtered out by our atmosphere. Because DNA and proteins absorb radiation in this range, UV-C could cause crop damage and general ecological disaster if it were to reach the earth's surface in significant amounts.

**Chlorofluorocarbon (CFC)** A compound composed of just carbon, chlorine, and fluorine.

You can test yourself on the glossary terms at the following Web site:
**www.chemplace.com/college/**

## Chapter Objectives

**The goal of this chapter is to teach you to do the following.**

1. Define all of the terms in the Chapter Glossary.

### Section 7.1 Energy

2. Explain why a more massive object, such as a bulldozer, has greater energy than a less massive object, such as a scooter, moving at the same velocity.
3. Explain why an object, such as a bulldozer, has greater energy when it is moving at a higher velocity than the same object moving at a lower velocity.
4. State the Law of Conservation of Energy.
5. Describe the relationship among stability, capacity to do work, and potential energy.
6. Explain why an object, such as a coin, gains potential energy as it moves farther from the earth.
7. Explain why energy must be absorbed to break a chemical bond.
8. Explain why energy is released when a chemical bond is formed.
9. Convert between the names and abbreviations for joule (J), calorie (cal), and dietary calorie (Calorie or Cal).
10. Explain the relative sizes of the joule, calorie, and dietary calorie (Calorie).
11. Explain the difference between external kinetic energy and internal kinetic energy.
12. List the three ways in which an object's thermal energy can be increased.
13. Explain what we imply about the average internal kinetic energy of two objects when we say that one of them has a higher temperature than the other.
14. Describe the changes that take place during heat transfer between objects at different temperatures.
15. Write a brief description of radiant energy in terms of its particle nature.

16. Write a brief description of radiant energy in terms of its wave nature.
17. Identify the relationship between the wavelength of radiant energy and the energy of its photons.
18. Cite the relative energies and wavelengths of the following forms of radiant energy: gamma rays, X rays, ultraviolet (UV) radiation, visible light, infrared (IR) radiation, microwaves, and radio waves.
19. Indicate the relative energies and wavelengths of the following colors of visible light: violet, blue, green, yellow, orange, and red.

### Section 7.2 Chemical Changes and Energy

20. Explain why some chemical reactions release heat to their surroundings.
21. Explain why some chemical reactions absorb heat from their surroundings.
22. Explain why chemical reactions either absorb or release energy.

### Section 7.3 Ozone: Pollutant and Protector

23. Explain why the same characteristic that makes ozone useful in industry also leads to health problems.
24. Describe how ozone is produced in the air we breathe.
25. Explain why the highest concentrations of ozone in the air we breathe are found in large industrial cities with lots of cars and lots of sun.
26. Explain why UV radiation less than 400 nm in wavelength is able to break N–O bonds in $NO_2$ molecules, and explain why radiant energy greater than 400 nm in wavelength cannot break these bonds.
27. Describe the three types of ultraviolet radiation (UV-A, UV-B, and UV-C) including their relative wavelengths and energies.
28. Explain why it is beneficial to humans for UV-A radiation to reach the surface of the earth.
29. Explain why UV-B radiation can be damaging to us and our environment if it reaches the earth in higher quantities than it does now.
30. Explain why we are fortunate that UV-C radiation is almost completely filtered out by the gases in the atmosphere.
31. Explain how oxygen molecules, $O_2$, and ozone molecules, $O_3$, work together to protect us from high-energy ultraviolet radiation.
32. Explain how nitrogen monoxide, NO, is able to catalyze the conversion of ozone molecules, $O_3$, and oxygen atoms, O, to oxygen molecules, $O_2$.

### Section 7.4 Chlorofluorocarbons: A Chemical Success Story Gone Wrong

33. Explain why CFCs eventually make their way into the stratosphere even though most chemicals released into the atmosphere do not.
34. Explain why the radiant energy found in the troposphere is unable to liberate chlorine atoms from CFC molecules, but the radiant energy in the stratosphere is able to do this.
35. Explain why the chlorine atoms liberated from CFCs are thought to be a serious threat to the ozone layer.

## Review Questions

For questions 1 and 2, illustrate your answers with simple drawings of the particles that form the structures of the substances mentioned. You do not need to be specific about the nature of the particles. Think of them as simple spheres, and draw them as circles. Provide a general description of the arrangement of the particles, the degree of interaction between them, and their freedom of movement.

1. A pressurized can of a commercial product used to blow the dust off computer components contains tetrafluoroethane, $C_2H_2F_4$. At room temperature, this substance is a liquid at pressures slightly above normal pressure and a gas at normal pressures. Although most of the tetrafluoroethane in the can is in the liquid form, $C_2H_2F_4$ evaporates rapidly, resulting in a significant amount of vapor above the liquid. When the valve on the top of the can is pushed, the tetrafluoroethane gas rushes out, blowing dust off the computer. When the valve closes, more of the liquid $C_2H_2F_4$ evaporates to replace the vapor released. If the can is heated, the liquid evaporates more quickly, and the increase in gas causes the pressure to build up to possibly dangerous levels.
   a. Describe the general structure of liquid tetrafluoroethane.
   b. Describe the general structure of gaseous tetrafluoroethane.
   c. Describe the process by which particles move from the liquid form to the gaseous form.
   d. Describe the changes that take place in the liquid when it is heated, and explain why these changes lead to a greater rate of evaporation of the liquid.
2. Sodium metal can be made by running an electric current through molten sodium chloride.
   a. Describe the general structure of solid sodium chloride.
   b. Describe the changes that take place when the temperature of NaCl solid increases.
   c. Describe the changes that take place when sodium chloride melts.

## Key Ideas

**Complete the following statements by writing one of these words or phrases in each blank.**

| | |
|---|---|
| 4.184 | joule |
| absorbed | less |
| average | magnetic |
| calorie | mass |
| change | massless |
| changed | more |
| collisions | motion |
| created | particle |
| cycle | peaks |
| decrease | position or state |
| destroyed | potential |
| electric | released |
| energy | resistance |
| external | retrievable, stored |
| from | shorter |
| greater | thermal |
| heat | transferred |
| increases | wave |

3. The simplest definition of ________________ is that it is the capacity to do work. Work, in this context, may be defined as what is done to move an object against some sort of ________________.
4. The capacity to do work resulting from the ________________ of an object is called kinetic energy, KE.
5. If two objects are moving at the same velocity, the one with the greater ________________ will have a greater capacity to do work and thus a greater kinetic energy.
6. If two objects have equal mass but different velocities, the one with the greater velocity has the ________________ kinetic energy.
7. The Law of Conservation of Energy states that energy can be neither ________________ nor ________________, but it can be ________________ from one system to another and ________________ from one form to another.
8. Potential energy (PE) is a(n) ________________ form of energy an object possesses by virtue of its ________________.
9. A system's stability is a measure of its tendency to ________________.
10. As an object moves from a(n) ________________ stable state to a(n) ________________ stable state, it can do work.
11. Any time a system shifts from a more stable state to a less stable state, the potential energy of the system ________________.

12. Energy is necessary to separate two atoms being held together by mutual attraction in a chemical bond. The energy supplied increases the ________________ energy of the less stable separate atoms compared to the more stable atoms in the bond.
13. Because less stable separate atoms have higher potential energy than the more stable atoms that participate in a bond, the change from separate atoms to atoms in a bond corresponds to a(n) ________________ in potential energy.
14. The accepted SI unit for energy is the ________________, but another common unit is the ________________.
15. The U.S. National Institute of Standards and Technology defines the calorie as ________________ joules.
16. A falling coin has a certain ________________ kinetic energy that is related to its overall mass and to its velocity as it falls.
17. The energy associated with internal motion of particles that compose an object can be called either internal kinetic energy or ________________ energy.
18. Temperature is a measure of the ________________ internal kinetic energy of an object.
19. Heat is the thermal energy that is transferred from a region of higher temperature to a region of lower temperature as a consequence of the ________________ of particles.
20. Radiant energy seems to have a dual nature, with both ________________ and ________________ characteristics.
21. Radiant energy can be viewed as a stream of tiny, ________________ packets of energy called photons.
22. The wave view says that as radiant energy moves away from its source, it has an effect on the space around it that can be described as a wave consisting of an oscillating ________________ field perpendicular to an oscillating ________________ field.
23. One distinguishing characteristic of the waves of radiant energy is wavelength, $\lambda$, the distance between two ________________ on the wave of electromagnetic radiation. A more specific definition of wavelength is the distance in space over which a wave completes one ________________ of its repeated form.
24. The ________________ the wavelength of a wave of electromagnetic radiation, the greater the energy of its photons.
25. If in a chemical reaction, more energy is released in the formation of new bonds than was necessary to break old bonds, energy is ________________ overall, and the reaction is exergonic.
26. A change that leads to ________________ energy being released from the system to the surroundings is called exothermic.
27. If less energy is released in the formation of the new bonds than is necessary to break the old bonds, energy must be ________________ from the surroundings for the reaction to proceed.
28. Endergonic changes can lead to a transfer of heat ________________ the surroundings.

**Complete the following statements by writing one these words or phrases in each blank.**

| | |
|---|---|
| 10 | permanently |
| 242 nm | premature aging |
| 240 nm to 320 nm | proteins |
| 50 | shorter |
| 1000 | skin cancer |
| DNA | speeds |
| cars | stable |
| light photons | sun |
| longer | sunburn |
| nitrogen oxides | UV-A |
| oxidizing | UV-B |
| oxygen, $O_2$ | UV-C |
| ozone, $O_3$ | |

29. Two forms of the element oxygen are found in nature: the life-sustaining diatomic ________ and ________ which is a pale blue gas with a strong odor.
30. Ozone is a very powerful ________ agent.
31. The highest concentrations of $O_3$ in the air we breathe are found in large industrial cities with lots of ________ and lots of ________.
32. Any time air (which contains nitrogen and oxygen) is heated to high temperature (as occurs in the cylinders of our cars and in many industrial processes), ________ are formed.
33. Radiant energy of wavelengths ________ than 400 nm has enough energy to break N–O bonds in $NO_2$ molecules, but radiant energy with wavelengths ________ than 400 nm does not supply enough energy to separate the atoms.
34. Because the process of ozone formation is initiated by ________, the pollutant mixture is called photochemical smog.
35. The stratosphere extends from about ________ km to about ________ km above sea level.
36. ________ radiation, which includes radiant energy of wavelengths from about 320 to 400 nm, passes through the stratosphere and reaches us on the surface of the earth. We are glad it does, because it provides energy that our bodies use to produce vitamin D.
37. ________ radiation has wavelengths from about 290 to 320 nm. Some of it is removed by the gases in the stratosphere, but some of it reaches the surface of the earth. Radiation in this portion of the spectrum has energy great enough that excessive exposure can cause ________, ________, and ________.

38. The highest-energy ultraviolet radiation is ______________, with wavelengths from about 40 to 290 nm. It is energetic enough to cause serious damage not only to us but to all life on earth. One reason it is so dangerous is that ______________, the substance that carries genetic information in living cells, absorbs UV radiation of about 260 nm. Likewise, ______________, vital structural and functional components of living systems, absorb(s) radiation with wavelengths of about 280 nm. If these wavelengths were to reach the earth in significant quantity, the changes they would cause would lead to massive crop damage and general ecological disaster.

39. Oxygen molecules, $O_2$, and ozone molecules, $O_3$, work together to absorb high-energy UV radiation. Oxygen molecules absorb UV radiation with wavelengths less than ______________, and ozone molecules absorb radiant energy with wavelengths from ______________.

40. A catalyst is a substance that ______________ a chemical reaction without being ______________ altered itself.

41. One of the reasons why CFCs were so successful is that they are extremely ______________ compounds; very few substances react with them.

42. Each chlorine atom is thought to destroy an average of ______________ ozone molecules before being temporarily incorporated into a compound such as HCl or $ClONO_2$.

## Chapter Problems

### Section 7.1 Energy

OBJECTIVE 2 OBJECTIVE 3 OBJECTIVE 5

43. For each of the following situations, you are asked which of two objects or substances has the higher energy. Explain your answer with reference to the capacity of each to do work and say whether the energy that distinguishes them is kinetic energy or potential energy.

    a. An ozone molecule, $O_3$, with a velocity of 393 m/s or the same molecule moving with a velocity of 410 m/s. (These are the average velocities of ozone molecules at 25 °C and 50 °C.)

    b. An ozone molecule, $O_3$, moving at 300 m/s or an oxygen molecule, $O_2$, moving at the same velocity.

    c. A proton and an electron close together or a proton and an electron farther apart.

    d. An HOCl molecule or an OH molecule and a chlorine atom formed from breaking the chlorine-oxygen bond in the HOCl molecule. (The conversion of HOCl into Cl and OH takes place in the stratosphere.)

    e. Two separate chlorine atoms in the stratosphere or a chlorine molecule, $Cl_2$, that can form when they collide.

    f. Water in the liquid form or water in the gaseous form. (Assume that the two systems are at the same temperature.)

OBJECTIVE 2 OBJECTIVE 3 OBJECTIVE 5

44. For each of the following situations, you are asked which of two objects or substances has the higher energy. Explain your answer with reference to the capacity of each to do work and say whether the energy that distinguishes them is kinetic energy or potential energy.
    a. A methane molecule, $CH_4$, in the stratosphere or a $CH_3$ molecule and a hydrogen atom formed from breaking one of the carbon-hydrogen bonds in a $CH_4$ molecule.
    b. A water molecule moving at $1.63 \times 10^3$ mi/h or the same water molecule moving at $1.81 \times 10^3$ mi/h. (These are the average velocities of water molecules at 110 °C and 200 °C.)
    c. Iodine solid or iodine gas. (Assume that the two systems are at the same temperature.)
    d. A nitrogen monoxide, NO, molecule and an oxygen atom in the stratosphere or the $NO_2$ molecule that can form when they collide.
    e. Two bar magnets pushed together with the north pole of one magnet almost touching the south pole of the other magnet or the same magnets farther apart.
    f. A water molecule moving at $1.63 \times 10^3$ mi/h or a uranium hexafluoride, $UF_6$, molecule moving at the same velocity.
45. At 20 °C, ozone molecules, $O_3$, have an average velocity of 390 m/s, and oxygen molecules, $O_2$, have an average velocity of 478 m/s. If these gases are at the same temperature, they have the same average kinetic energy. Explain in qualitative terms how these gases could have the same average kinetic energy but different average velocities.
46. Energy is the capacity to do work. With reference to this definition, describe how you would demonstrate that each of the following has potential energy. (There is no one correct answer in these cases. There are many ways to demonstrate that a system has potential energy.)
    a. A brick on the top of a tall building
    b. A stretched rubber band
    c. Alcohol molecules added to gasoline
47. Energy is the capacity to do work. With reference to this definition, describe how you would demonstrate that each of the following has potential energy. (There is no one correct answer in these cases. There are many ways to demonstrate that a system has potential energy.)
    a. A paper clip an inch away from a magnet
    b. A candy bar
    c. A baseball popped up to the catcher at the peak of its flight.
48. For each of the following changes, describe whether (1) kinetic energy is being converted into potential energy, (2) potential energy is being converted into kinetic energy, (3) kinetic energy is transferred from one object to another. (More than one of these changes may be occurring.)
    a. An archer pulls back a bow with the arrow in place.
    b. The archer releases the arrow, and it speeds toward the target.

49. For each of the following changes, describe whether (1) kinetic energy is being converted into potential energy, (2) potential energy is being converted into kinetic energy, (3) kinetic energy is transferred from one object to another. (More than one of these changes may be occurring.)
    a. A car in an old wooden roller coaster is slowly dragged up a steep incline to the top of the first big drop.
    b. After the car passes the peak of the first hill, it falls down the backside at high speed.
    c. As it goes done the hill, the car makes the whole wooden structure shake.
    d. By the time the car reaches the bottom of the first drop, it is moving fast enough to go up to the top of the next smaller hill on its own.
50. When a child swings on a swing, energy is constantly being converted back and forth between kinetic energy and potential energy. At what point (or points) in the child's motion is the potential energy at a maximum? At what point (or points) is the kinetic energy at a maximum? If a parent stops pushing a swinging child, why does the child eventually stop? Where has the energy of the swinging child gone?
51. Methyl bromide is an agricultural soil fumigant that can make its way into the stratosphere, where bromine atoms are stripped away by radiant energy from the sun. The bromine atoms react with ozone molecules (thus diminishing the earth's protective ozone layer) to produce BrO, which in turn reacts with nitrogen dioxide , $NO_2$, to form $BrONO_2$. For each of these reactions, identify whether energy would be absorbed or released. Explain why. Describe how energy is conserved in each reaction.

    OBJECTIVE 7 OBJECTIVE 8

    a. $CH_3Br(g) \rightarrow CH_3(g) + Br(g)$
    b. $BrO(g) + NO_2(g) \rightarrow BrONO_2(g)$
52. The following chemical changes take place in the air over sunny industrial cities, like Los Angeles. Identify whether energy would be absorbed or released in each reaction. Explain why. Describe how energy is conserved in each reaction.

    OBJECTIVE 7 OBJECTIVE 8

    a. $2O(g) \rightarrow O_2(g)$
    b. $NO_2(g) \rightarrow NO(g) + O(g)$
53. A silver bullet speeding toward a vampire's heart has both external kinetic energy and internal kinetic energy. Explain the difference between the two.

    OBJECTIVE 11
54. Describe three different ways to increase the thermal energy of your skin.

    OBJECTIVE 12
55. When a room-temperature thermometer is placed in a beaker of boiling water, heat is transferred from the hot water to the glass of the thermometer and then to the liquid mercury inside the thermometer. With reference to the motion of the particles in the water, glass, and mercury, describe the changes that are taking place during this heat transfer. What changes in thermal energy and average internal kinetic energy are happening for each substance? Why do you think the mercury moves up the thermometer?

    OBJECTIVE 14
56. With reference to both their particle and their wave nature, describe the similarities and differences between visible light and ultraviolet radiation.

    OBJECTIVE 15 OBJECTIVE 16 OBJECTIVE 17 OBJECTIVE 18

OBJECTIVE 15 OBJECTIVE 16 OBJECTIVE 17 OBJECTIVE 19

57. With reference to both their particle and their wave nature, describe the similarities and differences between red light and blue light.

OBJECTIVE 18

58. Consider the following forms of radiant energy: microwaves, infrared, ultraviolet, X rays, visible light, radio waves, and gamma rays.
    a. List them in order of increasing energy.
    b. List them in order of increasing wavelength.

OBJECTIVE 19

59. Consider the following colors of visible light: green, yellow, violet, red, blue, and orange.
    a. List these in order of increasing energy.
    b. List these in order of increasing wavelength.

### Section 7.2 Chemical Changes and Energy

OBJECTIVE 21

60. Consider the following endergonic reaction. In general terms, explain why energy is absorbed in the process of this reaction.

$$N_2(g) + O_2(g) \rightarrow 2NO(g)$$

OBJECTIVE 20

61. The combustion of propane is an exergonic reaction. In general terms, explain why energy is released in the process of this reaction.

$$C_3H_8(g) + 5O_2(g) \rightarrow 3CO_2(g) + 4H_2O(l)$$

OBJECTIVE 14 OBJECTIVE 20

62. Hydrazine, $N_2H_4$, is used as rocket fuel. Consider a system in which a sample of hydrazine is burned in a closed container, followed by heat transfer from the container to the surroundings.

$$N_2H_4(g) + O_2(g) \rightarrow N_2(g) + 2H_2O(g)$$

    a. In general terms, explain why energy is released in the reaction.
    b. Before heat energy is transferred to the surroundings, describe the average internal kinetic energy of the product particles compared to the reactant particles. If the product's average internal kinetic energy is higher than for the reactants, from where did this energy come? If the average internal kinetic energy is lower than for the reactants, to where did this energy go?
    c. Describe the changes in particle motion as heat is transferred from the products to the surroundings.

63. Cinnabar is a natural mercury(II) sulfide, HgS, found near volcanic rocks and hot springs. It is the only important source of mercury, which has many uses, including dental amalgams, thermometers, and mercury vapor lamps. Mercury is formed in the following endothermic reaction when mercury(II) sulfide is heated. Consider a system in which a sample of HgS in a closed container is heated with a Bunsen burner flame.

OBJECTIVE 14 OBJECTIVE 21

$$8HgS(s) \rightarrow 8Hg(l) + S_8(s)$$

a. Describe the changes in particle motion as heat is transferred from the hot gases of the Bunsen burner flame to the container to the HgS(*s*).

b. In general terms, explain why energy is absorbed in the reaction.

c. Into what form of energy is the heat energy converted for this reaction?

64. Even on a hot day, you get cold when you step out of a swimming pool. Suggest a reason why your skin cools as water evaporates from it. Is evaporation an exothermic or endothermic process?

OBJECTIVE 21

65. Classify each of the following changes as exothermic or endothermic.

a. Leaves decaying in a compost heap.

b. Dry ice (solid carbon dioxide) changing to carbon dioxide gas.

c. Dew forming on a lawn at night.

66. Classify each of the following changes as exothermic or endothermic.

a. The nuclear reaction that takes place in a nuclear electrical generating plant.

b. Cooking an egg in boiling water

c. The breakdown of plastic in the hot sun.

67. Explain why chemical reactions either absorb or evolve energy.

OBJECTIVE 22

### Section 7.3 Ozone: Pollutant and Protector

68. What characteristic of ozone makes it useful for some purposes and a problem in other situations? What do people use it for in industry? What health problems does it cause?

OBJECTIVE 23

69. Having aced your finals, you decide to spend your summer vacation visiting a friend in Southern California. Unfortunately, during your first day at an amusement park there, you begin to have some chest pain, slight wheezing, and shortness of breath. This reminds you of what you read about ozone in Chapter 7 of your chemistry textbook. When you tell your friend that ozone is the likely cause of your problems, she asks you to explain how the ozone you are breathing is created and why the ozone levels are higher in Los Angeles than where you live, in rural Minnesota. What do you tell her? How did you contribute to the increase in ozone as you drove your rental car from your hotel to the park?

OBJECTIVE 24 OBJECTIVE 25

70. Explain why UV radiation of wavelength less than 400 nm is able to break N–O bonds in $NO_2$ molecules, and explain why radiant energy of wavelength longer than 400 nm cannot break these bonds.

OBJECTIVE 26

71. Explain why it is beneficial to us to have UV-A radiation reach the surface of the earth.

OBJECTIVE 28

OBJECTIVE 29

72. Explain why UV-B radiation can be damaging to us and our environment if it reaches the earth in higher quantities than it does now.

OBJECTIVE 30

73. Explain why we are fortunate that UV-C radiation is almost completely filtered out by the gases in the atmosphere.

OBJECTIVE 31

74. Explain how oxygen molecules, $O_2$, and ozone molecules, $O_3$, work together to protect us from high-energy ultraviolet radiation.

OBJECTIVE 32

75. Explain how nitrogen monoxide, NO, is able to catalyze the conversion of ozone molecules, $O_3$, and oxygen atoms, O, to oxygen molecules, $O_2$.

### Section 7.4 Chlorofluorocarbons: A Chemical Success Story Gone Wrong

OBJECTIVE 33

76. Explain why CFCs eventually make their way into the stratosphere even though most chemicals released into the atmosphere do not.

OBJECTIVE 34

77. Explain why the radiant energy found in the troposphere is unable to liberate chlorine atoms from CFC molecules but the radiant energy in the stratosphere is able to do this.

OBJECTIVE 35

78. Explain why the chlorine atoms liberated from CFCs are thought to be a serious threat to the ozone layer.

### Additional Problems

79. Energy is the capacity to do work. With reference to this definition, describe how you would demonstrate that each of the following objects or substances has potential energy. (There is no one correct answer in these cases. There are many ways to demonstrate that a system has potential energy.)
    a. Natural gas used to fuel a city bus
    b. A compressed spring
    c. A pinecone at the top of a tall tree
80. Energy is the capacity to do work. With reference to this definition, describe how you would demonstrate that each of the following objects or substances has potential energy. (There is no one correct answer in these cases. There are many ways to demonstrate that a system has potential energy.)
    a. Two bar magnets with the north pole of one magnet about an inch away from the north pole of the second magnet (like poles of two magnets repel each other)
    b. Uranium atoms in the reactor fuel of a nuclear power plant
    c. A suitcase on the top shelf of a garage
81. For each of the following changes, describe whether (1) kinetic energy is being converted into potential energy, (2) potential energy is being converted into kinetic energy, (3) kinetic energy is transferred from one object to another. (More than one of these changes may be occurring.)
    a. Using an elaborate system of ropes and pulleys, a piano mover hoists a piano up from the sidewalk to outside a large third floor window of a city apartment building.
    b. His hands slip and the piano drops 6 feet before he is able to stop the rope from unwinding.

82. For each of the following changes, describe whether (1) kinetic energy is being converted into potential energy, (2) potential energy is being converted into kinetic energy, (3) kinetic energy is transferred from one object to another. (More than one of these changes may be occurring.)
    a. A pinball player pulls back the knob, compressing a spring, in preparation for releasing the ball into the machine.
    b. The player releases the knob, and the ball shoots into the machine.
83. The following changes combine to help move a car down the street. For each change, describe whether (1) kinetic energy is being converted into potential energy, (2) potential energy is being converted into kinetic energy, (3) kinetic energy is transferred from one object to another. (More than one of these changes may be occurring.)
    a. The combustion of gasoline in the cylinder releases heat energy, increasing the temperature of the gaseous products of the reaction.
    b. The hot gaseous products push the piston down in the cylinder.
    c. The moving piston turns the crankshaft, which ultimately turns the wheels.
84. For each of the following changes, describe whether (1) kinetic energy is being converted into potential energy, (2) potential energy is being converted into kinetic energy, (3) kinetic energy is transferred from one object to another. (More than one of these changes may be occurring.)
    a. Wind turns the arms of a windmill.
    b. The windmill pumps water from below the ground up into a storage tank at the top of a hill.
85. Classify each of the following changes as exothermic or endothermic.
    a. Fuel burning in a camp stove
    b. The melting of ice in a camp stove to provide water on a snow-camping trip
86. Classify each of the following changes as exothermic or endothermic.
    a. Fireworks exploding
    b. Water evaporating on a rain-drenched street
87. Why does your body temperature rise when you exercise?

### Discussion Questions

88. You know that the Law of Conservation of Energy states that energy is conserved. When you put on the brakes of a car to stop it, where does the energy associated with the moving car go?
89. Only a fraction of the energy released in the combustion of gasoline is converted into kinetic energy of the moving car. Where does the rest go?

*The technique described in this chapter for making unit conversions is useful to both chemists and carpenters.*

CHAPTER 8

# Unit Conversions



*[M]athematics . . . is the easiest of sciences, a fact which is obvious in that no one's brain rejects it . . .*

ROGER BACON (C. 1214–C. 1294)
ENGLISH PHILOSOPHER AND SCIENTIST

*Stand firm in your refusal to remain conscious during algebra. In real life, I assure you, there is no such thing as algebra.*

FRAN LEBOWITZ (B. 1951)
AMERICAN JOURNALIST

YOU MAY AGREE WITH ROGER BACON THAT MATHEMATICS IS THE EASIEST OF SCIENCES, but many beginning chemistry students would not. Because they have found mathematics challenging, they wish it were not so important for learning chemistry—or for answering so many of the questions that arise in everyday life. They can better relate to Fran Lebowitz's advice in the second quotation. If you are one of the latter group, it will please you to know that even though there *is* some algebra in chemistry, this chapter teaches a technique for doing chemical calculations (and many other calculations) without it. The technique is called dimensional analysis. You will be using it throughout the rest of this book, in future chemistry and science courses, and any time you want to calculate the number of nails you need to build a fence or the number of rolls of paper necessary to cover the kitchen shelves.

*Dimensional analysis, the technique for doing unit conversions that is described in this chapter, can be used for a lot more than chemical calculations.*

## Review Skills

The presentation of information in this chapter assumes that you can already perform the tasks listed below. You can test your readiness to proceed by answering the Review Questions at the end of the chapter. This might also be a good time to read the Chapter Objectives, which precede the Review Questions.

- List the metric base units and the corresponding abbreviations for length, mass, volume, energy, and gas pressure. (Section 1.4)
- State the numbers or fractions represented by the following metric prefixes, and write their abbreviations: giga, mega, kilo, centi, milli, micro, nano, and pico. (Section 1.4)
- Given a metric unit, write its abbreviation; given an abbreviation, write the full name of the unit. (Section 1.4)
- Describe the relationships between the metric units that do not have prefixes (such as meter, gram, and liter) and units derived from them by the addition of prefixes—for example, 1 km = $10^3$ m. (Section 1.4)
- Define temperature and describe the Celsius, Fahrenheit, and Kelvin scales used to report its values. (Section 1.4)
- Given a value derived from a measurement, identify the range of possible values it represents, on the basis of the assumption that its uncertainty is ±1 in the last position reported. (For example, 8.0 mL says the value could be from 7.9 mL to 8.1 mL.) (Section 1.5)

## 8.1 Dimensional Analysis

Many of the questions asked in chemistry and in everyday life can be answered by converting one unit of measure into another. For example, suppose you are taking care of your nephew for the weekend, and he breaks his arm. The doctor sets the arm, puts it in a cast, and prescribes an analgesic to help control the pain. Back at home, after filling the prescription, you realize that the label calls for 2 teaspoons of medicine every 6 hours, but the measuring device that the pharmacy gave you is calibrated in milliliters. You can't find a measuring teaspoon in your kitchen, so you've got to figure out how many milliliters are equivalent to 2 tsp. It's been a rough day, there's a crying boy on the couch, and you're really tired. Is there a technique for doing the necessary calculation that is simple and reliable? Dimensional analysis to the rescue!

The main purpose of this chapter is to show you how to make many different types of unit conversions, such as the one required above. You will find that the stepwise thought process associated with the procedure called **dimensional analysis** not only guides you in figuring out how to set up unit conversion problems but also gives you confidence that your answers are correct. If you studied Inter-Chapter 1A, which introduces the procedure, some of this chapter will be review, although the description presented here is much more comprehensive than the one provided in the inter-chapter. If you were instructed to skip Inter-Chapter 1A at the beginning of your course, you do not need to go back and read it now. This chapter discusses dimensional analysis without assuming you have done it before.

*In every affair, consider what precedes and follows, and then undertake it.*

EPICTETUS (C. 55–C. 135)
GREEK PHILOSOPHER

### An Overview of the General Procedure

You will see many different types of unit conversions in this chapter, but they can all be worked using the same basic procedure. To illustrate the process, we will convert 2 teaspoons to milliliters and solve the problem of how much medicine to give the little boy who broke his arm.

The first step in the procedure is to identify the unit for the value we want to calculate. We write this on the left side of an equals sign. Next, we identify the value that we will convert into the desired value, and we write it on the right side of the equals sign. (Remember that a value constitutes both a number and a unit.) We want to know how many milliliters are equivalent to 2 tsp. We express this question as

$$? \text{ mL} = 2 \text{ tsp}$$

Desired unit (mL) — Given unit (tsp)

Next, we multiply by one or more conversion factors that enable us to cancel the unwanted units and generate the desired units. A **conversion**

**factor** is a ratio that describes the relationship between two units. To create a conversion factor for converting teaspoons to milliliters, we can look in any modern cookbook (check its index under "metric conversions") and discover that the relationship between teaspoons and milliliters is

$$1 \text{ tsp} = 5 \text{ mL}$$

This relationship can be used to produce two ratios, or conversion factors:

$$\frac{5 \text{ mL}}{1 \text{ tsp}} \quad \text{or} \quad \frac{1 \text{ tsp}}{5 \text{ mL}}$$

The first of these can be used to convert teaspoons to milliliters, and the second can be used to convert milliliters to teaspoons.

The final step in the procedure is to multiply the known unit (2 tsp) by the proper conversion factor, the one that converts teaspoons to milliliters.

Desired unit — Conversion factor

$$? \text{ mL} = 2 \text{ tsp} \left(\frac{5 \text{ mL}}{1 \text{ tsp}}\right) = 10 \text{ mL}$$

Given unit — Answer in desired unit

Because 1 teaspoon is equivalent to 5 milliliters, multiplying by 5 mL/1 tsp is the same as multiplying by 1. The volume associated with 2 tsp does not change when we multiply by the conversion factor, but the value (number and unit) does. Because 1 milliliter is one-fifth the volume of 1 teaspoon, there are five times as many milliliters for a given volume. Therefore, 2 tsp and 10 mL represent the same volume.

Note that the units in a dimensional analysis setup cancel just like variables in an algebraic equation. Therefore, when we want to convert tsp to mL, we choose the ratio that has tsp on the bottom to cancel the tsp unit in our original value and leave us with the desired unit, mL. If you have used correct conversion factors, and if your units cancel to yield the desired unit or units, you can be confident that you will arrive at the correct answer.

## Metric–Metric Unit Conversions

As you saw in Chapter 1, one of the convenient features of the metric system is that the relationships between metric units can be derived from the metric prefixes. These relationships can easily be translated into conversion factors. For example, *milli* means $10^{-3}$ (or 0.001 or 1/1000), so a milliliter (mL) is $10^{-3}$ liters (L). Thus there are 1000, or $10^3$, milliliters in a liter. (A complete list of the prefixes that you need to know to solve the problems in this text appears in Table 1.2.) Two possible sets of conversion factors for relating milliliters to liters can be obtained from these relationships.

OBJECTIVE 2

$$10^3 \text{ mL} = 1 \text{ L} \quad \text{leads to} \quad \frac{10^3 \text{ mL}}{1 \text{ L}} \quad \text{or} \quad \frac{1 \text{ L}}{10^3 \text{ mL}}$$

$$1 \text{ mL} = 10^{-3} \text{ L} \quad \text{leads to} \quad \frac{1 \text{ mL}}{10^{-3} \text{ L}} \quad \text{or} \quad \frac{10^{-3} \text{ L}}{1 \text{ mL}}$$

In the remainder of this text, metric–metric conversion factors will have positive exponents like those found in the first set of conversion factors above.

### EXAMPLE 8.1 Conversion Factors

OBJECTIVE 2

Write two conversion factors that relate nanometers and meters. Use positive exponents in each.

***Solution***

*Nano* means $10^{-9}$, so a nanometer is $10^{-9}$ meter.

$$1 \text{ nm} = 10^{-9} \text{ m} \quad \text{and} \quad 10^9 \text{ nm} = 1 \text{ m}$$

Because we want our conversion factors to have positive exponents, we will build our ratios from the equation on the right ($10^9$ nm = 1 m):

$$\frac{\mathbf{10^9 \text{ nm}}}{\mathbf{1 \text{m}}} \quad \text{and} \quad \frac{\mathbf{1 \text{ m}}}{\mathbf{10^9 \text{ nm}}}$$

### EXERCISE 8.1 Conversion Factors

OBJECTIVE 2

Write two conversion factors that relate the following pairs of metric units. Use positive exponents for each.

a. joule and kilojoule

b. meter and centimeter

c. liter and gigaliter

d. gram and microgram

e. gram and megagram

### EXAMPLE 8.2 Unit Conversions

OBJECTIVE 3

Convert 365 nanometers to kilometers.

***Solution***

We want the answer in kilometers (km), and the units we are given are nanometers (nm), so we are converting from one metric length unit to another metric length unit. We begin by writing

$$? \text{ km} = 365 \text{ m}$$

We continue constructing the dimensional analysis setup by writing the "skeleton" of a conversion factor: the parentheses, the line dividing the numerator and the denominator, and the unit that we know we want to cancel. This step helps us organize our thoughts by showing us that our first conversion factor must have the nm unit on the bottom to cancel the nm unit associated with 365 nm.

Desired unit — Skeleton

$$? \text{ km} = 365 \text{ nm} \left( \frac{\quad}{\text{nm}} \right)$$

Given value — Unit to be cancelled

Note that in this problem, both the desired metric unit and the known metric unit have prefixes. One way to make this type of conversion is to change the given unit to the corresponding metric base unit and then change that metric base unit to the desired unit. We write two conversion factors, one for each of these changes. Sometimes it is useful to write a simple description of this plan first. Our plan here is

$$\text{nm} \rightarrow \text{m} \rightarrow \text{km}$$

The dimensional analysis setup is therefore

Converts given metric unit to metric base unit — Converts metric base unit to desired metric unit

$$? \text{ km} = 365 \text{ nm} \left( \frac{1 \text{ m}}{10^9 \text{ nm}} \right) \left( \frac{1 \text{ km}}{10^3 \text{ m}} \right) = \mathbf{3.65 \times 10^{-10} \text{ km}}$$

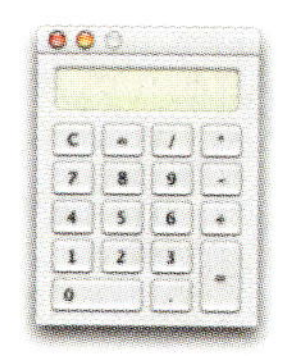

**See Appendix C for a description of how to do this calculation using common scientific calculators.**

The nm and m units cancel, leaving the answer in km.

## Exercise 8.2 Unit Conversions

OBJECTIVE 3

Convert 4.352 micrograms to megagrams.

## English–Metric Unit Conversions

English units[1] are still common in some countries, whereas people in other countries (and the scientific community everywhere), use metric units almost exclusively. Dimensional analysis provides a convenient method for converting between English and metric units. Several of the most commonly needed English–metric conversion factors are listed in

[1]Table A.5 in Appendix A shows some useful English–English unit conversion factors.

Table 8.1. Because the English inch is defined as 2.54 cm, the number 2.54 in this value is exact. The numbers in the other conversion factors in Table 8.1 are not exact.

OBJECTIVE 4

**Table 8.1**
English–Metric Unit Conversion Factors

| Type of measurement | Probably most useful to know | Also useful to know | | |
|---|---|---|---|---|
| length | $\frac{2.54 \text{ cm}}{1 \text{ in.}}$ (exact) | $\frac{1.609 \text{ km}}{1 \text{ mi}}$ | $\frac{39.37 \text{ in.}}{1 \text{ m}}$ | $\frac{1.094 \text{ yd}}{1 \text{ m}}$ |
| mass | $\frac{453.6 \text{ g}}{1 \text{ lb}}$ | $\frac{2.205 \text{ lb}}{1 \text{ kg}}$ | | |
| volume | $\frac{3.785 \text{ L}}{1 \text{ gal}}$ | $\frac{1.057 \text{ qt}}{1 \text{ L}}$ | | |

Sometimes the British are obstinate about the change from English to metric units. Greengrocer Steve Thoburn went to jail for refusing to switch from pounds to kilograms.

## Example 8.3 Unit Conversions

OBJECTIVE 5

The mass of a hydrogen atom is $1.67 \times 10^{-18}$ micrograms. Convert this mass into pounds.

***Solution***
We start with

$$? \text{ lb} = 1.67 \times 10^{-18} \text{ μg}$$

We then add the "skeleton" of the first conversion factor.

$$? \text{ lb} = 1.67 \times 10^{-18} \text{ μg} \left( \frac{\quad}{\text{μg}} \right)$$

We are converting from metric mass to English mass. Table 8.1 contains a conversion factor that is convenient for most English–metric mass conversions, the one relating grams to pounds. Our given unit is micrograms. If we convert the micrograms into grams, we can then convert the gram unit into pounds.

$$\text{μg} \rightarrow \text{g} \rightarrow \text{lb}$$

Converts metric mass unit into English mass unit

$$? \text{ lb} = 1.67 \times 10^{-18} \text{ μg} \left( \frac{1 \text{ g}}{10^6 \text{ μg}} \right) \left( \frac{1 \text{ lb}}{453.6 \text{ g}} \right) = \mathbf{3.68 \times 10^{-27} \text{ lb}}$$

See Appendix C for a description of how to do this calculation using common scientific calculators.

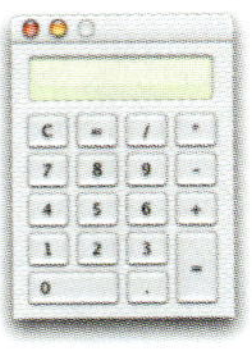

## EXERCISE 8.3 Unit Conversions

OBJECTIVE 5

The volume of the earth's oceans is estimated to be $1.5 \times 10^{18}$ kiloliters. What is this volume in gallons?

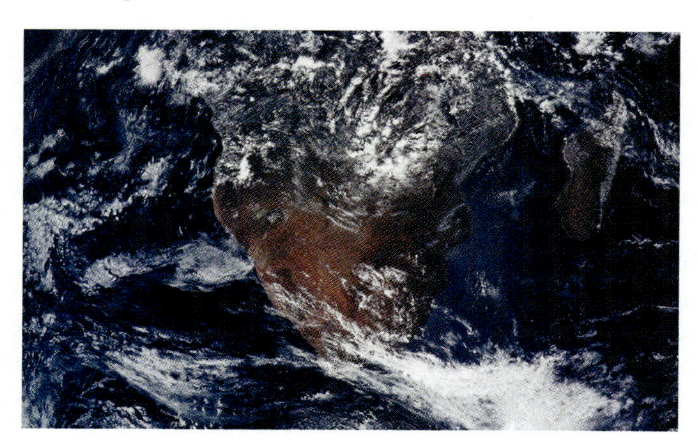

## 8.2 Rounding Off and Significant Figures

Most of your calculations in chemistry are likely to be done with a calculator, and calculators often provide more digits in the answer than you would be justified in reporting as scientific data. This section shows you how to round off an answer to reflect the approximate range of certainty warranted by the data.

### Measurements, Calculations, and Uncertainty

In Section 1.5, you read about the issue of uncertainty in measurement and learned to report measured values to reflect this uncertainty. For example, an inexpensive letter scale might show you that the mass of a nickel is 5 grams, but this is not an exact measurement. It is reasonable to assume that the letter scale measures mass with a precision of $\pm 1$ g and that the nickel therefore has a mass between 4 grams and 6 grams. You could use a more sophisticated instrument with a precision of $\pm 0.1$ g and report the mass of the nickel as 5.0 g. The purpose of the zero in this value is to show that this measurement of the nickel's mass has an uncertainty of $\pm 0.1$ g. With this instrument, you can assume that the mass of the nickel is between 4.9 g and 5.1 g. *Unless we are told otherwise, we assume that values from measurements have an uncertainty of plus or minus one in the last decimal place reported.* Using a far more precise balance found in a chemistry laboratory, you could determine the mass to be 4.9800 g, but this measurement still has an uncertainty of $\pm 0.0001$ g. *Measurements never give exact values.*

If a calculation is performed using all exact values and if the answer is not rounded off, the answer is exact, but this is a rare occurrence. The values used in calculations are usually *not* exact, and therefore the answers must be expressed with the proper degree of uncertainty. Consider the conversion of the mass of our nickel from grams into pounds. (There are 453.6 g per pound.)

$$? \text{ lb} = 4.9800 \text{ g} \left( \frac{1 \text{ lb}}{453.6 \text{ g}} \right) = \mathbf{0.01098 \text{ lb (or } 1.098 \times 10^{-2} \text{ lb)}}$$

The number 4.9800 is somewhat uncertain because it derives from a measurement. The number 453.6 was derived from a calculation, and the answer to that calculation was rounded off to four digits. Therefore, the number 453.6 is also uncertain. Thus any answer we obtain using these numbers is inevitably going to be uncertain as well.

Different calculators report different numbers of decimal places in their answers. For example, perhaps your calculator reports the answer to 4.9800 divided by 453.6 as 0.010978835978884. If we were to report this result as the mass of our nickel, we would be suggesting that we were certain of the mass to a precision of $\pm 0.00000000000001$, which is not the case. Instead, we report 0.01098 lb (or $1.098 \times 10^{-2}$ lb), which is a better reflection of the uncertainty in the numbers we used to calculate our answer. Figure 8.1 compares the precision of three different scales.

**Figure 8.1**
**Measurement Precision.**
Even highly precise measurements have some uncertainty.

mass = 5 g
meaning 4 g to 6 g

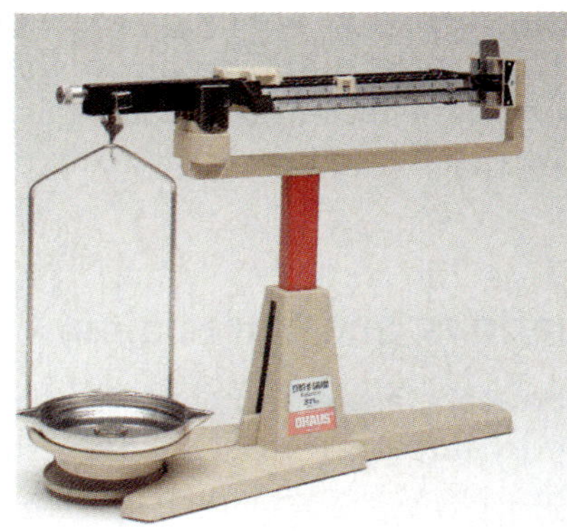


mass = 5.0 g
meaning 4.9 g to 5.1 g

mass = 4.9800 g
meaning 4.9799 g to 4.9801 g

## Rounding Off Answers Derived from Multiplication and Division

There are three general steps to rounding off answers so that they reflect the precision of the values used in a calculation. Consider the following example, which shows how the mass of a hydrogen atom in micrograms can be converted into the equivalent mass in pounds.

$$? \text{ lb} = 1.67 \times 10^{-18} \,\mu\text{g} \left( \frac{1 \text{ g}}{10^6 \,\mu\text{g}} \right) \left( \frac{1 \text{ lb}}{453.6 \text{ g}} \right)$$

The first step in rounding off is to decide which of the numbers in the calculation affect the uncertainty of the answer. We can assume that $1.67 \times 10^{-18}$ μg comes from a measurement, and all measurements are uncertain to some extent. Thus $1.67 \times 10^{-18}$ affects the uncertainty of our answer. The $10^6$ number comes from the definition of the metric prefix *micro*, so it is exact. Because it has no effect on the uncertainty of our answer, we will not consider it when we are deciding how to round off our answer. The 453.6 comes from a calculation that was rounded off (as we saw above), so it is not exact. It affects the uncertainty of our answer and must be considered when we round our answer.

The second step in rounding off is to consider the degree of uncertainty in each of our inexact values. We can determine their relative uncertainties by counting the numbers of significant figures: three in $1.67 \times 10^{-18}$ and four in 453.6. The number of **significant figures,** which is equal to the number of meaningful digits in a value, reflects the degree of uncertainty in the value (this is discussed more specifically in Study Sheet 8.1, below). A larger number of significant figures indicates a smaller uncertainty.

The final step is to round off our answer to reflect the most uncertain value used in our calculation. *When an answer is calculated by multiplying or dividing, we round it off to the same number of significant figures as the inexact value with the fewest significant figures.* For our example, that value is $1.67 \times 10^{-18}$ μg, with three significant figures, so we round off the calculated result, $3.681657848325 \times 10^{-27}$, to $3.68 \times 10^{-27}$.

The following sample study sheet provides a detailed guide to rounding off numbers calculated via multiplication and division. (Addition and subtraction will be covered in the subsequent discussion.) Examples 8.4 and 8.5 and Figure 8.2 demonstrate these steps.

## Sample Study Sheet 8.1

### Rounding Off Numbers Calculated Using Multiplication and Division

OBJECTIVE 8

OBJECTIVE 6

TIP-OFF After calculating a number using multiplication and division, you need to round it off to the correct number of significant figures.

GENERAL STEPS

STEP 1 **Determine whether each value is exact, and ignore exact values.**

- Numbers that come from definitions are exact.

  Numbers in metric–metric unit conversion factors that are derived from the metric prefixes are exact. An example is

$$\frac{10^3\text{ g}}{1\text{ kg}}$$

  Numbers in English–English unit conversion factors with the same type of unit (for example, both length units) on the top and the bottom are exact. An example is

$$\frac{12\text{ in.}}{1\text{ ft}}$$

The number 2.54 in the following conversion factor is exact.

$$\frac{2.54 \text{ cm}}{1 \text{ in.}}$$

- Numbers derived from counting are exact. For example, there are exactly five toes on the normal foot.

$$\frac{5 \text{ toes}}{1 \text{ foot}}$$

- Values that come from measurements are never exact.
- We will assume that values derived from calculations are not exact unless otherwise indicated. (With one exception, the numbers relating English to metric units that you will see in this text have been calculated and rounded, so they are not exact. The exception is 2.54 cm/1 in. The 2.54 comes from a definition.)

OBJECTIVE 8

OBJECTIVE 7

**STEP 2 Determine the number of significant figures in each value that is not exact.**

- *All nonzero digits are significant.*

11.275 — Five significant figures

- *Zeros between nonzero digits are significant.*

A zero between nonzero digits

10.275 — Five significant figures

- *Zeros to the left of nonzero digits are not significant.*

Not significant figures

0.000102, which can be described as $1.02 \times 10^{-4}$

Both have three significant figures.

- *Zeros to the right of nonzero digits in numbers that include decimal points are significant.*

Five significant figures — 10.200 g    20.0 mL — Three significant figures

Unnecessary for reporting size of value, but do reflect degree of uncertainty

- *Zeros to the right of nonzero digits in numbers without decimal points are ambiguous for significant figures.*

1 ?
220 kg — Precise to ±1 kg or ±10 kg? Two or three significant figures?
2
Important for reporting size of value, but unclear about degree of uncertainty

1
$2.2 \times 10^2$ kg
2
1 3
$2.20 \times 10^2$ kg
2
Use scientific notation to remove ambiguity.

OBJECTIVE 8

**STEP 3 When multiplying and dividing, round your answer off to the same number of *significant figures* as the value that contains the fewest significant figures.**

- If the digit to the right of the final digit you want to retain is less than 5, round down (the last digit remains the same).

26.221 rounded to three significant figures is 26.2
First digit dropped is less than 5.

- If the digit to the right of the final digit you want to retain is 5 or greater, round up (the last significant digit increases by 1).

26.272 rounded to three significant figures is 26.3
First digit dropped is greater than 5.

26.2529 rounded to three significant figures is 26.3
First digit dropped is equal to 5.

26.15 rounded to three significant figures is 26.2
First digit dropped is equal to 5.

**EXAMPLE** See Examples 8.4 and 8.5.

## EXAMPLE 8.4 Rounding Off Answers Derived from Multiplication and Division

OBJECTIVE 8

The average human body contains 5.2 L of blood. What is this volume in quarts? The dimensional analysis setup for this conversion follows. Identify whether each value in the setup is exact. Then determine the number of significant figures in each inexact value, calculate the answer, and report it to the correct number of significant figures.

$$? \text{ qt} = 5.2 \text{ L}\left(\frac{1 \text{ gal}}{3.785 \text{ L}}\right)\left(\frac{4 \text{ qt}}{1 \text{ gal}}\right)$$

***Solution***

A typical calculator shows the answer to this calculation to be 5.4953765, a number with far too many decimal places, considering the uncertainty of the values used in the calculation. It needs to be rounded to the correct number of significant figures.

**STEP 1** The 5.2 L is based on measurement, so it is not exact. The 3.785 L is part of an English–metric unit conversion factor, and we assume those factors are not exact, except for 2.54 cm/in. On the other hand, 4 qt/gal is an English–English conversion factor based on the definition of quart and gallon; thus the 4 is exact.

**STEP 2** Because 5.2 contains two nonzero digits, it has two significant figures. The number 3.785 contains four nonzero digits, so it has four significant figures.

**STEP 3** Because the value with the fewest significant figures has two significant figures, we report two significant figures in our answer, rounding 5.4953765 to 5.5.

$$? \text{ qt} = 5.2 \text{ L}\left(\frac{1 \text{ gal}}{3.785 \text{ L}}\right)\left(\frac{4 \text{ qt}}{1 \text{ gal}}\right) = \mathbf{5.5 \text{ qt}}$$

## EXAMPLE 8.5 Rounding Off Answers Derived from Multiplication and Division

OBJECTIVE 8

How many minutes does it take for an ant walking at a rate of 0.01 m/s to travel 6.0 feet across a picnic table? The dimensional analysis setup for this conversion follows. Identify whether each value in the setup is exact. Then determine the number of significant figures in each inexact value, calculate the answer, and report it to the correct number of significant figures.

$$? \text{ min} = 6.0 \text{ ft}\left(\frac{12 \text{ in.}}{1 \text{ ft}}\right)\left(\frac{2.54 \text{ cm}}{1 \text{ in.}}\right)\left(\frac{1 \text{ m}}{10^2 \text{ cm}}\right)\left(\frac{1 \text{ s}}{0.01 \text{ m}}\right)\left(\frac{1 \text{ min}}{60 \text{ s}}\right)$$

***Solution***

**STEP 1** The table's length and the ant's velocity come from measurements, so 6.0 and 0.01 are not exact. The other numbers are exact because they are derived from definitions. Thus only 6.0 and 0.01 can limit our significant figures.

**STEP 2** Zeros to the right of nonzero digits in numbers that have decimal points are significant, so 6.0 contains two significant figures. Zeros to the left of nonzero digits are not significant, so 0.01 contains one significant figure.

**STEP 3** A typical calculator shows 3.048 for the answer. Because the value with the fewest significant figures has one significant figure, we report one significant figure in our answer. Our final answer of 3 minutes signifies that it could take 2 to 4 minutes for the ant to cross the table.

See Appendix C for a description of how to do this calculation using common scientific calculators.

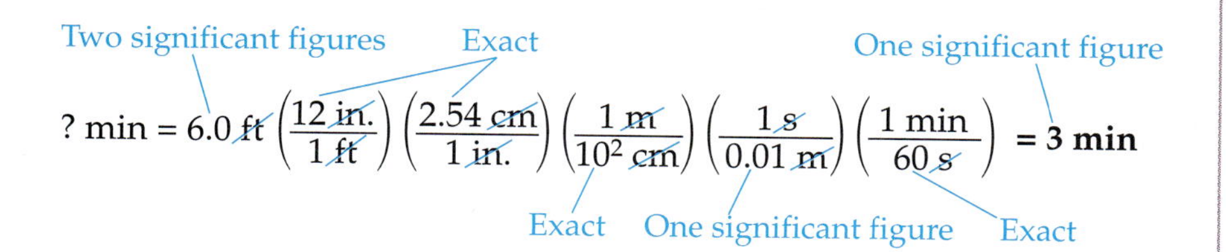


$$? \text{ min} = 6.0 \text{ ft}\left(\frac{12 \text{ in.}}{1 \text{ ft}}\right)\left(\frac{2.54 \text{ cm}}{1 \text{ in.}}\right)\left(\frac{1 \text{ m}}{10^2 \text{ cm}}\right)\left(\frac{1 \text{ s}}{0.01 \text{ m}}\right)\left(\frac{1 \text{ min}}{60 \text{ s}}\right) = \mathbf{3 \text{ min}}$$

With dimensional analysis, we don't have to wait for the ant to cross the entire table to know how long it will take.

## EXERCISE 8.4 Rounding Off Answers Derived from Multiplication and Division

OBJECTIVE 8

A first-class stamp allows you to send letters weighing up to 1 oz. (There are 16 ounces per pound.) You weigh a letter and find that it has a mass of 10.5 g. Can you mail this letter with one stamp? The dimensional analysis setup for converting 10.5 g to ounces follows. Identify whether each value in the setup is exact. Then determine the number of significant figures in each inexact value, calculate the answer, and report it to the correct number of significant figures.

$$? \text{ oz} = 10.5 \text{ g}\left(\frac{1 \text{ lb}}{453.6 \text{ g}}\right)\left(\frac{16 \text{ oz}}{1 \text{ lb}}\right)$$

### EXERCISE 8.5 Rounding Off Answers Derived from Multiplication and Division

OBJECTIVE 8

The re-entry speed of the Apollo 10 space capsule was 11.0 km/s. How many hours would it have taken for the capsule to fall through 25.0 miles of the stratosphere? The dimensional analysis setup for this calculation follows. Identify whether each value in the setup is exact. Then determine the number of significant figures in each inexact value, calculate the answer, and report it to the correct number of significant figures.

$$? \text{ hr} = 25.0 \text{ mi} \left(\frac{5280 \text{ ft}}{1 \text{ mi}}\right)\left(\frac{12 \text{ in.}}{1 \text{ ft}}\right)\left(\frac{2.54 \text{ cm}}{1 \text{ in.}}\right)\left(\frac{1 \text{ m}}{10^2 \text{ cm}}\right)\left(\frac{1 \text{ km}}{10^3 \text{ m}}\right)\left(\frac{1 \text{ s}}{11.0 \text{ km}}\right)\left(\frac{1 \text{ min}}{60 \text{ s}}\right)\left(\frac{1 \text{ hr}}{60 \text{ min}}\right)$$

## Rounding Off Answers Derived from Addition and Subtraction

The following sample study sheet provides a guide to rounding off numbers calculated using addition and subtraction.

### SAMPLE STUDY SHEET 8.2 Rounding Off Numbers Calculated Using Addition and Subtraction

OBJECTIVE 9

**TIP-OFF** After calculating a number using addition and subtraction, you need to round it off to the correct number of decimal places.

**GENERAL STEPS**

**STEP 1 Determine whether each value is exact, and ignore exact values (see Study Sheet 8.1).**

**STEP 2 Determine the number of decimal places for each value that is not exact.**

**STEP 3 Round your answer to the same number of *decimal places* as the inexact value with the fewest decimal places.**

**EXAMPLE** See Example 8.6.

### EXAMPLE 8.6 Rounding Off Answers Derived from Addition and Subtraction

OBJECTIVE 9

A laboratory procedure calls for you to determine the mass of an unknown liquid. Let's suppose that you weigh a 100-mL beaker on a new electronic balance and record its mass as 52.3812 g. You then add 10 mL of the unknown liquid to the beaker and discover that the electronic balance has stopped working. You find a 30-year-old balance in a cupboard, use it to weigh the beaker of liquid, and record that mass as 60.2 g. What is the mass of the unknown liquid?

*Solution*

You can calculate the mass of the liquid by subtracting the mass of the beaker from the mass of the beaker and the liquid.

$$60.2 \text{ g beaker with liquid} - 52.3812 \text{ g beaker} = 7.8188 \text{ g liquid}$$

We can use the steps outlined in Sample Study Sheet 8.2 to decide how to round off our answer.

STEP 1 The numbers 60.2 and 52.3812 come from measurements, so they are not exact.

STEP 2 We assume that values given to us have uncertainties of $\pm1$ in the last decimal place reported, so 60.2 has an uncertainty of $\pm0.1$ g, and 52.3812 has an uncertainty of $\pm0.0001$ g. The first value is precise to the tenths place, and the second value is precise to four places to the right of the decimal point.

STEP 3 We round answers derived from addition and subtraction to the same number of decimal places as the inexact value with the *fewest* decimal places. Therefore, we report our answer to the tenths place—rounding it off if necessary—to reflect this uncertainty. The answer is **7.8 g.**

Be sure to remember that the guidelines for rounding answers derived from addition or subtraction are different from the guidelines for rounding answers from multiplication or division. Note that when we are adding or subtracting, we are concerned with *decimal places* in the numbers used rather than with the number of significant figures. Let's take a closer look at why. In Example 8.6, we subtracted the mass of a beaker (52.3812 g) from the mass of the beaker and an unknown liquid (60.2 g) to get the mass of the liquid. If the reading of 60.2 has an uncertainty of $\pm0.1$ g, the actual value could be anywhere between 60.1 g and 60.3 g. The range of possible values for the mass of the beaker is 52.3811 g to 52.3813 g. This leads to a range of possible values for our answer of from 7.7187 g to 7.9189 g.

$$\begin{array}{r} 60.1 \text{ g} \\ -52.3813 \text{ g} \\ \hline 7.7187 \text{ g} \end{array} \qquad \begin{array}{r} 60.3 \text{ g} \\ -52.3811 \text{ g} \\ \hline 7.9189 \text{ g} \end{array}$$

Our possible values vary from about 7.7 g to about 7.9 g, or $\pm0.1$ of our reported answer of 7.8 g. Because our least precise value (60.2 g) has an uncertainty of $\pm0.1$ g, our answer can be no more precise than $\pm0.1$ g.

Use the same reasoning to prove that the following addition and subtraction problems are rounded to the correct number of decimal positions.

OBJECTIVE 9

$$97.40 + 31 = 128$$

$$1035.67 - 989.2 = 46.5$$

Note that although the numbers in the addition problem have four and two significant figures, the answer is reported with three significant figures. This answer is limited to the ones place by the number 31, which we assume has an uncertainty of $\pm1$. Note also that although the numbers in the subtraction problem have six and four significant figures, the answer has only three. The answer is limited to the tenths place by 989.2, which we assume has an uncertainty of ±0.1.

### EXERCISE 8.6 Rounding Off Answers Derived from Addition and Subtraction

OBJECTIVE 9

Report the answers to the following calculations to the correct number of decimal positions. Assume that each number is ±1 in the last decimal position reported.

a. 684 − 595.325 = b. 92.771 + 9.3 =

## 8.3 Density and Density Calculations

A truckload of logs is much heavier (has a greater mass) than a pea-sized amount of lead, but the density of lead is much greater than that of wood.

When people say that lead is heavier than wood, they do not mean that a pea-sized piece of lead weighs more than a truckload of pine logs. What they mean is that a sample of lead will have a greater mass than an equal volume of wood. A more concise way of putting this is to say that lead is more dense than wood. This type of density, formally known as **mass density,** is defined as mass divided by volume. It is what people usually mean by the term **density.**

$$\text{Density} = \frac{\text{mass}}{\text{volume}}$$

The density of lead is 11.34 g/mL, and the density of pinewood is about 0.5 g/mL. In other words, a milliliter of lead contains 11.34 g of matter, whereas a milliliter of pine contains only about half a gram of matter. See Table 8.2 for the densities of other common substances.

Although there are exceptions, the densities of liquids and solids generally decrease with increasing temperature.[2] Thus, when chemists report densities, they generally state the temperature at which the density was measured. For example, the density of ethanol is 0.806 g/mL at 0 °C but 0.789 g/mL at 20 °C.[3]

OBJECTIVE 10

The densities of liquids and solids are usually described in grams per milliliter or grams per cubic centimeter. (Remember that a milliliter and a cubic centimeter are the same volume.) The particles of a gas are much farther apart than the particles of a liquid or solid, so gases are much less dense than solids and liquids. Thus it is more convenient to describe the densities of gases as grams per liter. The density of air at sea level and 20 °C is about 1.2 g/L.

Because the density of a substance depends on the substance's identity and its temperature, it is possible to identify an unknown substance by

[2]The density of liquid water actually increases as its temperature rises from 0 °C to 4 °C. Such exceptions are very rare.

[3]The temperature effect on the density of gases is more complicated, but it, too, changes with changes in temperature. This effect will be described in Chapter 13.

comparing its density at a particular temperature to the densities of known substances at the same temperature (Table 8.2). For example, we can determine whether an object is pure gold by measuring its density at 20 °C and comparing that to the known density of gold at 20 °C, which is 19.31 g/mL.

**Table 8.2**
Mass Densities of Some Common Substances (at 20 °C unless otherwise stated)

| Substance | Density, g/mL |
|---|---|
| air at sea level | 0.0012 (or 1.2 g/L) |
| Styrofoam | 0.03 |
| pinewood | 0.4–0.6 |
| gasoline | 0.70 |
| ethanol | 0.7893 |
| ice | 0.92 |
| water, $H_2O$, at 20 °C | 0.9982 |
| water, $H_2O$, at 0 °C | 0.9998 |
| water, $H_2O$, at 3.98 °C | 1.0000 |
| seawater | 1.025 |
| whole blood | 1.05 |
| bone | 1.5–2.0 |
| glass | 2.4–2.8 |
| aluminum, Al | 2.702 |
| the planet Earth (average) | 5.25 |
| iron, Fe | 7.86 |
| silver, Ag | 10.5 |
| gold, Au | 19.31 |
| lead, Pb | 11.34 |
| platinum, Pt | 21.45 |
| atomic nucleus | $\approx 10^{14}$ |
| a black hole (not 20 °C) | $\approx 10^{16}$ |

Figure 8.2 shows densities of some common substances.

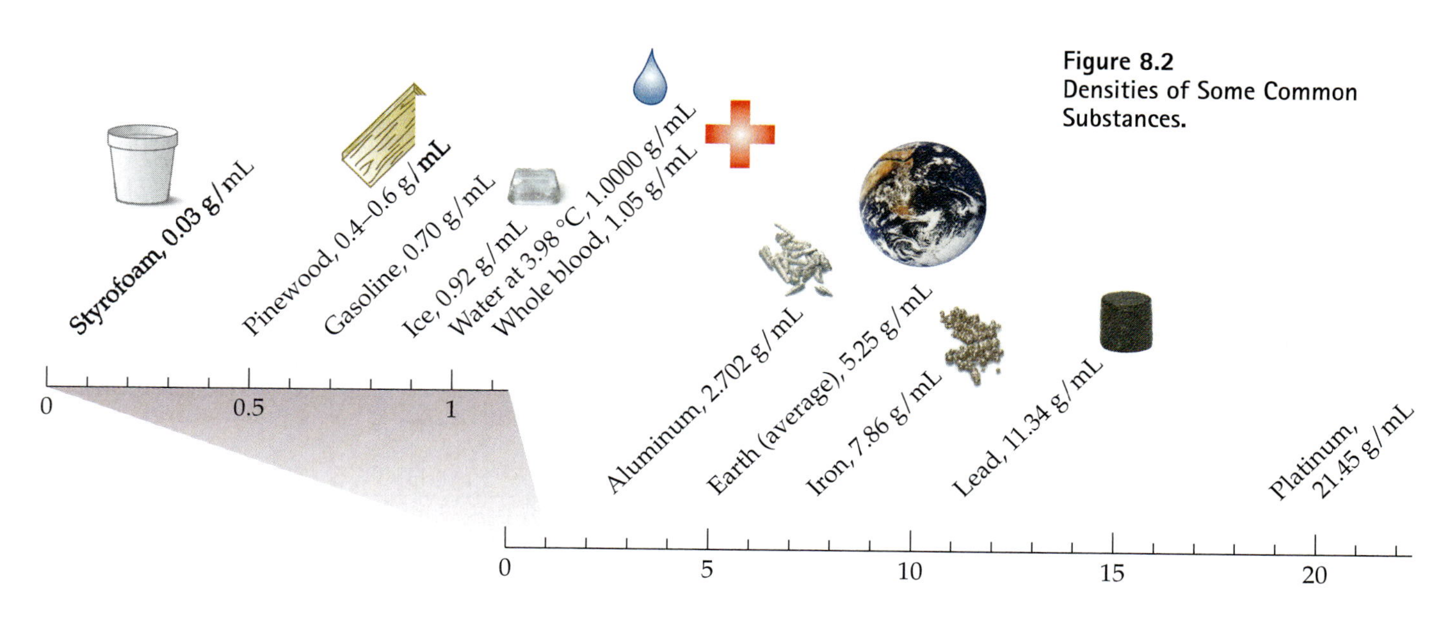


**Figure 8.2**
Densities of Some Common Substances.

## Using Density as a Conversion Factor

OBJECTIVE 11

Because density is reported as a ratio that describes a relationship between two units, the density of a substance can be used in dimensional analysis to convert between the substance's mass and its volume. For example, the density of water at 20 °C can be used to convert between the mass in grams of a given sample of water and the sample's volume in milliliters.

### Example 8.7 Density Conversions

OBJECTIVE 11

What is the mass in grams of 75.0 mL of water at 20 °C?

***Solution***

The dimensional analysis setup for this problem begins

$$? \text{ g } H_2O = 75.0 \text{ mL } H_2O \left( \frac{\qquad}{\text{mL } H_2O} \right)$$

Being asked to convert from volume into mass is the tip-off that we can use the density of water as a conversion factor in solving this problem. We can find water's density on a table of densities, such as Table 8.2.

The density of water at 20 °C

$$? \text{ g } H_2O = 75.0 \cancel{\text{mL } H_2O} \left( \frac{0.9982 \text{ g } H_2O}{1 \cancel{\text{mL } H_2O}} \right) = \mathbf{74.9 \text{ g } H_2O}$$

### Example 8.8 Density Conversions

OBJECTIVE 11

What is the volume of 25.00 kilograms of water at 20 °C?

***Solution***

Like any other conversion factor, density can be used in the inverted form to make conversions:

$$\frac{0.9982 \text{ g } H_2O}{1 \text{ mL } H_2O} \quad \text{or} \quad \frac{1 \text{ mL } H_2O}{0.9982 \text{ g } H_2O}$$

The conversion factor on the right enables us to convert from mass in grams to volume in milliliters. First, however, we need to convert the given unit, kilograms, into grams. Then, after using the density of water to convert grams into milliliters, we need to convert milliliters into liters.

$$? \text{ L } H_2O = 25.00 \cancel{\text{kg } H_2O} \left( \frac{10^3 \cancel{\text{g}}}{1 \cancel{\text{kg}}} \right) \left( \frac{1 \cancel{\text{mL}} \text{ } H_2O}{0.9982 \cancel{\text{g } H_2O}} \right) \left( \frac{1 \text{ L}}{10^3 \cancel{\text{mL}}} \right)$$

$$= \mathbf{25.05 \text{ L } H_2O}$$

## EXERCISE 8.7 Density Conversions

OBJECTIVE 11

a. What is the mass in kilograms of 15.6 gallons of gasoline?

b. A shipment of iron to a steel-making plant has a mass of 242.6 metric tons. What is the volume in liters of this iron?

## Determination of Mass Density

The density of a substance can be calculated from its measured mass and volume. The examples that follow show how this can be done and demonstrate more of the dimensional analysis thought process.

## EXAMPLE 8.9 Density Calculations

OBJECTIVE 12

An empty 2-L graduated cylinder is found to have a mass of 1124.2 g. Liquid methanol, $CH_3OH$, is added to the cylinder, and its volume is measured as 1.20 L. The total mass of the methanol and the cylinder is 2073.9 g, and the temperature of the methanol is 20 °C. What is the density of methanol at this temperature?

***Solution***

We are not told the specific units desired for our answer, but we will follow the usual convention of describing the density of liquids in grams per milliliter. It is a good idea to write these units in a way that reminds us we want *g* to be on the top of our ratio, and *mL* on the bottom, when we are finished.

$$\frac{?\ \text{g}}{\text{mL}} =$$

Because we want our answer to contain a ratio of two units, we start the right side of our setup with a ratio of two units. Because we want mass to be on the top and volume on the bottom when we are finished, we put our mass unit on the top and our volume unit on the bottom. The mass of the methanol is found by subtracting the mass of the cylinder from the total mass of the cylinder and the methanol.

$$\frac{?\ \text{g}}{\text{mL}} = \frac{(2073.9 - 1124.2)\ \text{g}}{1.20\ \text{L}}$$

Now that we have placed units for the desired properties (mass and volume) in the correct positions, we convert the units we have been given to the specific units we want. For our problem, this means converting liters to milliliters. Because we want to cancel L and it is on the bottom of the first ratio, the skeleton of the next conversion factor has the L on top.

$$\frac{?\ \text{g}}{\text{mL}} = \frac{(2073.9 - 1124.2)\ \text{g}}{1.20\ \text{L}}\left(\frac{\quad \text{L}}{\quad}\right)$$

The completed setup, and the answer, are

$$\frac{?\ \text{g}}{\text{mL}} = \frac{(2073.9 - 1124.2)\ \text{g}}{1.20\ \cancel{\text{L}}}\left(\frac{1\ \cancel{\text{L}}}{10^3\ \text{mL}}\right) = \frac{949.7\ \text{g}}{1.20\ \cancel{\text{L}}}\left(\frac{1\ \cancel{\text{L}}}{10^3\ \text{mL}}\right)$$

$$= \mathbf{0.791\ g/mL}$$

See Appendix C for a description of how to do this calculation using common scientific calculators.

## Example 8.10 Density Calculations

OBJECTIVE 12

You could find out whether a bracelet is made of silver or platinum by determining its density and comparing that density to the densities of silver (10.5 g/mL) and platinum (21.45 g/mL) (Figure 8.3). A bracelet is placed directly on the pan of a balance, and its mass is found to be 50.901 g. Water is added to a graduated cylinder, and the volume of the water is recorded as 18.2 mL. The bracelet is added to the water, and the volume rises to 23.0 mL. What is the density of the bracelet? Is it more likely to be silver or platinum?

***Solution***

We can find the volume of the bracelet by subtracting the volume of the water from the total volume of water and bracelet. We can then calculate the density using the following setup.

$$\frac{?\ \text{g bracelet}}{\text{mL bracelet}} = \frac{50.901\ \text{g bracelet}}{(23.0 - 18.2)\ \text{mL bracelet}}$$

Rounding off an answer that is derived from a mixture of subtraction (or addition) and division (or multiplication) is more complex than when these calculations are done separately. We need to recognize the different components of the problem and follow the proper rules for each.

Because 23.0 and 18.2 are both uncertain in the tenths place, the answer from the subtraction is reported to the tenths place only. This answer is 4.8.

When we divide 50.901, which has five significant figures, by 4.8, which has two significant figures, we report two significant figures in our answer.

$$\frac{?\ \text{g bracelet}}{\text{mL bracelet}} = \frac{50.901\ \text{g bracelet}}{(23.0 - 18.2)\ \text{mL bracelet}} = \frac{50.901\ \text{g bracelet}}{4.8\ \text{mL bracelet}}$$

$$= \mathbf{11\ g/mL\ bracelet}$$

The density is closer to that of silver, so the bracelet is more likely to be **silver** than platinum.

**Figure 8.3**
**Silver or Platinum?**

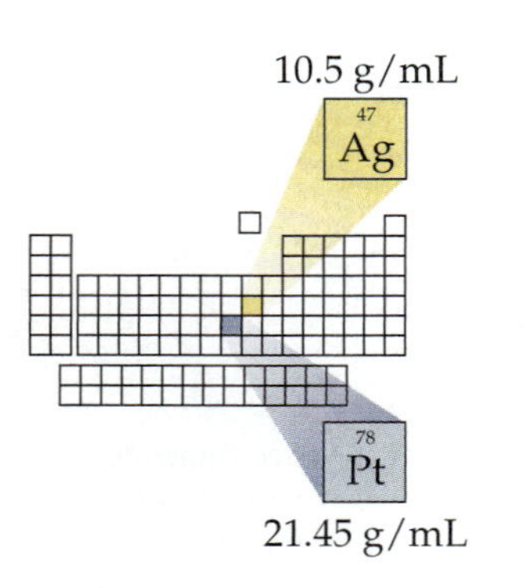


Is this bracelet silver or platinum?

❶ Determine its mass.

❷ Determine its volume.

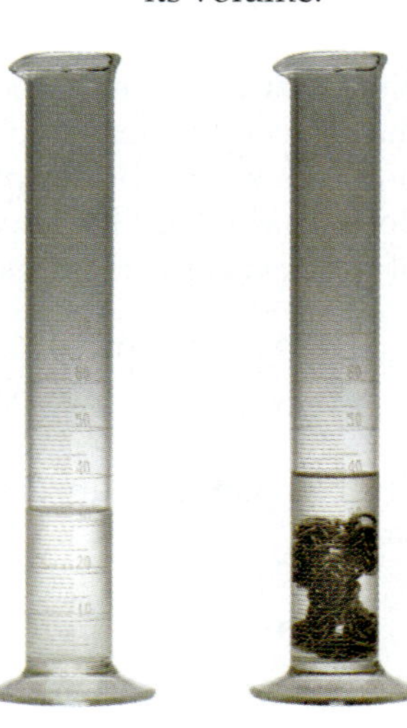

## Exercise 8.8 Density Calculations

OBJECTIVE 12

a. A graduated cylinder is weighed and found to have a mass of 48.737 g. A sample of hexane, $C_6H_{14}$, is added to the graduated cylinder, and the total mass is measured as 57.452 g. The volume of the hexane is 13.2 mL. What is the density of hexane?

b. A tree trunk is found to have a mass of $1.2 \times 10^4$ kg and a volume of $2.4 \times 10^4$ L. What is the density of the tree trunk in g/mL?

## 8.4 Percentage and Percentage Calculations

What does it mean to say that your body is about 8.0% blood and that when you work hard, between 3% and 4% by volume of your blood goes to your brain? Once you understand their meaning, percentage values such as these will provide you with ratios that can be used as conversion factors.

*Percentage by mass,* the most common form of percentage used in chemical descriptions, is a value that tells us the number of mass units of the part for each 100 mass units of the whole. You may assume that any percentage in this book is a percentage by mass unless you are specifically told otherwise. Thus, when we are told that our bodies are 8.0% blood, we assume that means 8.0% by mass, which is to say that for every 100 grams of body, there are 8.0 grams of blood and that for every 100 kilograms of body, there are 8.0 kilograms of blood. We can use any mass units we want in the ratio as long as the units are the same for the part and for the whole. Consequently, a percentage by mass can be translated into any number of conversion factors.

$$\frac{8.0 \text{ kg blood}}{100 \text{ kg body}} \quad \text{or} \quad \frac{8.0 \text{ g blood}}{100 \text{ g body}} \quad \text{or} \quad \frac{8.0 \text{ lb blood}}{100 \text{ lb body}} \quad \text{or} \quad \ldots$$

The general form for conversion factors derived from mass percentages is

OBJECTIVE 13

$$\text{For } X\% \text{ by mass} \quad \frac{X \text{ (any mass unit) part}}{100 \text{ (same mass unit) whole}}$$

Another frequently encountered form of percentage is *percentage by volume* (or % by volume). Because we assume that all percentages in chemistry are mass percentages unless told otherwise, volume percentages should always be designated as such. The general form for conversion factors derived from volume percentages is

OBJECTIVE 13

$$\text{For } X\% \text{ by volume} \quad \frac{X \text{ (any volume unit) part}}{100 \text{ (same volume unit) whole}}$$

For example, the statement that 3.2% by volume of your blood goes to your brain provides you with conversion factors to convert between volume of blood to the brain and total volume of blood.

$$\frac{3.2 \text{ L blood to brain}}{100 \text{ L blood total}} \quad \text{or} \quad \frac{3.2 \text{ mL blood to brain}}{100 \text{ mL blood total}} \quad \text{or}$$

$$\frac{3.2 \text{ qt blood to brain}}{100 \text{ qt blood total}} \quad \text{or} \quad \ldots$$

## Example 8.11 Unit Conversions

OBJECTIVE 14

Your body is about 8.0% blood. If you weigh 145 pounds, what is the mass of your blood in kilograms?

***Solution***

Our setup begins with

$$? \text{ kg blood} = 145 \text{ lb body} \left( \frac{\phantom{x}}{\text{lb body}} \right)$$

Because we are not told what type of percentage "8.0% blood" represents, we assume that it is a mass percentage. Both pounds and kilograms are mentioned in the problem, so we could use either one of the following conversion factors.

$$\frac{8.0 \text{ lb blood}}{100 \text{ lb body}} \quad \text{or} \quad \frac{8.0 \text{ kg blood}}{100 \text{ kg body}}$$

Both of the following setups lead to the correct answer.

$$? \text{ kg blood} = 145 \cancel{\text{lb body}} \left( \frac{8.0 \cancel{\text{lb}} \text{ blood}}{100 \cancel{\text{lb body}}} \right) \left( \frac{453.6 \cancel{\text{g}}}{1 \cancel{\text{lb}}} \right) \left( \frac{1 \text{ kg}}{10^3 \cancel{\text{g}}} \right) = \mathbf{5.3 \text{ kg blood}}$$

or

$$? \text{ kg blood} = 145 \cancel{\text{lb body}} \left( \frac{453.6 \cancel{\text{g}}}{1 \cancel{\text{lb}}} \right) \left( \frac{1 \cancel{\text{kg}}}{10^3 \cancel{\text{g}}} \right) \left( \frac{8.0 \cancel{\text{kg}} \text{ blood}}{100 \cancel{\text{kg body}}} \right) = \mathbf{5.3 \text{ kg blood}}$$

8.0% limits our significant figures to two.

**Figure 8.4**
**Unit Conversions.**
When doing heavy work, your muscles get about 75 to 80% by volume of your blood.

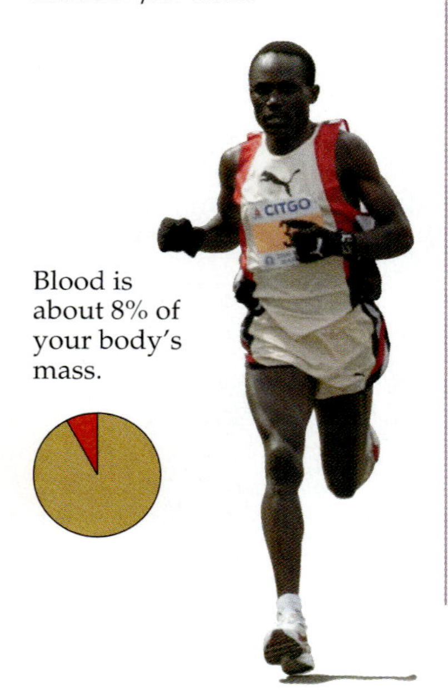

CITGO
Blood is about 8% of your body's mass.

## Exercise 8.9 Unit Conversions

OBJECTIVE 14

a. The mass of the ocean is about $1.8 \times 10^{21}$ kg. If the ocean contains 0.014% by mass hydrogen carbonate ions, $HCO_3^-$, how many pounds of $HCO_3^-$ are in the ocean?

b. When you are doing heavy work, your muscles get about 75 to 80% by volume of your blood (Figure 8.4). If your body contains 5.2 liters of blood, how many liters of blood are in your muscles when you are working hard enough to send them 78% by volume of your blood?

## 8.5 A Summary of the Dimensional Analysis Process

You have seen some of the many uses of dimensional analysis and looked at various kinds of information that provide useful conversion factors for chemical calculations. Now it is time for you to practice a general procedure for navigating your way through unit conversion problems so that you will be able to do them efficiently on your own. Sample Study Sheet 8.3 describes a stepwise thought process that can help you to decide what conversion factors to use and how to assemble them into a dimensional analysis format.

*The winds and waves are always on the side of the ablest navigators.*
EDWARD GIBBON
ENGLISH HISTORIAN

### SAMPLE STUDY SHEET 8.3
**Calculations Using Dimensional Analysis**

OBJECTIVE 14

**TIP-OFF** You wish to express a given value in terms of a different unit or units.

**GENERAL STEPS**

**STEP 1** State your question in an expression that sets the unknown unit(s) equal to one or more of the values given.

- To the left of the equals sign, show the unit(s) you want in your answer.
- To the right of the equals sign, start with an expression composed of the given unit(s) that parallels in kind and placement the units you want in your answer.

  If you want a single unit in your answer, start with a value that has a single unit.

  If you want a ratio of two units in your answer, start with a value that has a ratio of two units, or start with a ratio of two values, each of which has one unit. Put each type of unit in the position you want it to have in the answer.

**STEP 2** Multiply the expression to the right of the equals sign by conversion factors that cancel unwanted units and generate the desired units.

If you are not certain which conversion factor to use, ask yourself, "What is the fundamental conversion the problem requires, and what conversion factor do I need to make that type of conversion?" Figure 8.5, on the next page, provides a guide to useful conversion factors.

**STEP 3** Do a quick check to be sure that you used correct conversion factors and that your units cancel to yield the desired unit(s).

**STEP 4** Do the calculation, rounding your answer to the correct number of significant figures and combining it with the correct unit.

**EXAMPLE** See Examples 8.12 to 8.16.

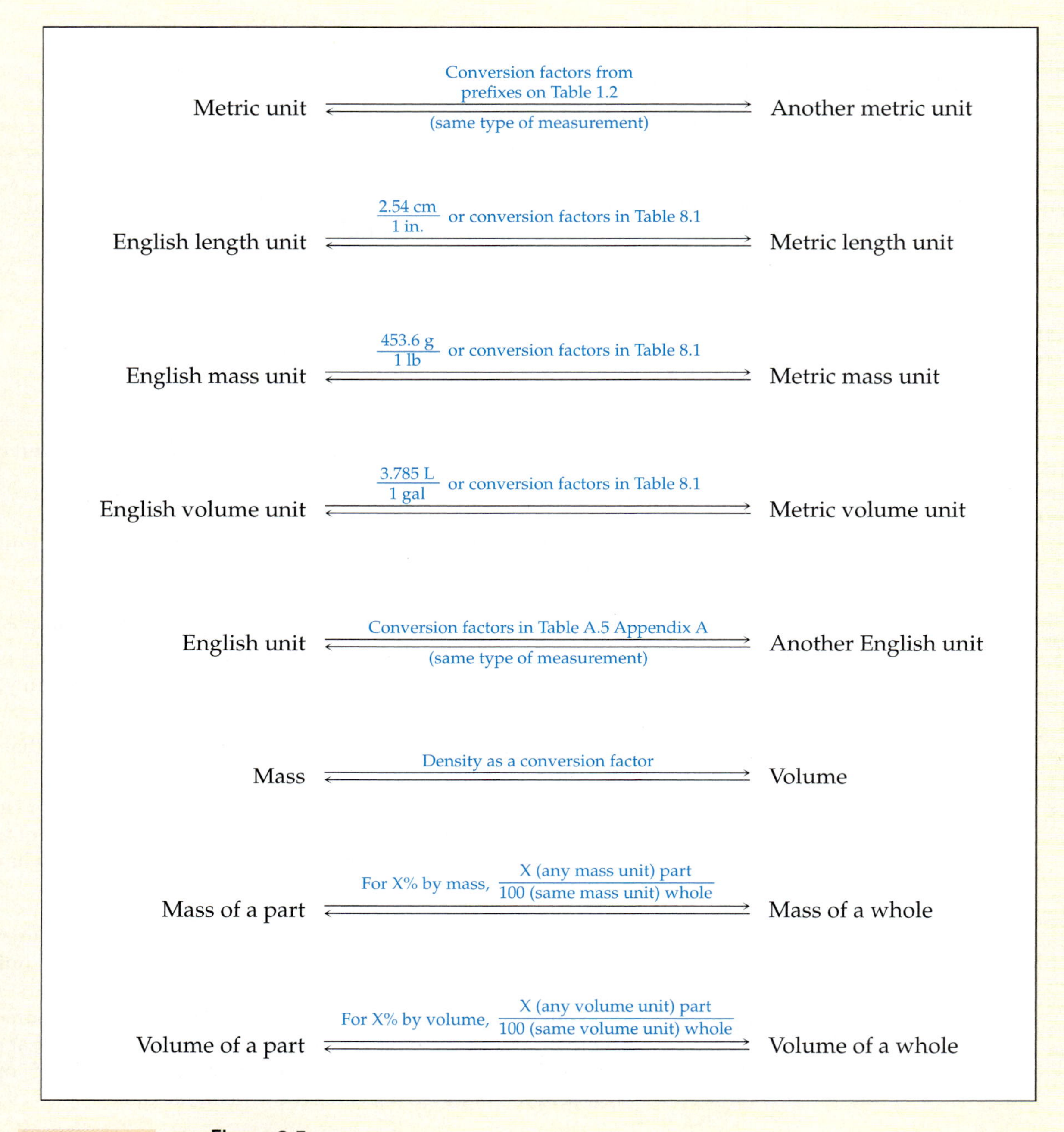


OBJECTIVE 15

**Figure 8.5**
**Types of Unit Conversions.**
Here is a summary of some of the basic types of conversions that are common in chemistry and the types of conversion factors used to make them.

Here are more examples of the most useful types of dimensional analysis conversions.

OBJECTIVE 15

## Example 8.12 Metric-Metric Unit Conversions

Convert 4567.36 micrograms to kilograms.

***Solution***

When converting from one metric unit to another, convert from the given unit to the base unit and then from the base unit to the unit that you want.

Desired unit — Given value

$$? \text{ kg} = 4567.36 \, \mu\text{g} \left(\frac{1 \text{ g}}{10^6 \, \mu\text{g}}\right)\left(\frac{1 \text{ kg}}{10^3 \text{ g}}\right) = \mathbf{4.56736 \times 10^{-6} \ kg}$$

Converts given metric unit to metric base unit — Converts metric base unit to desired metric unit

## Example 8.13 English–Metric Unit Conversions

Convert 475 miles to kilometers.

***Solution***

The conversion factor 2.54 cm/in. can be used to convert from an English to a metric unit of length.

English–metric conversion factor

$$? \text{ km} = 475 \text{ mi} \left(\frac{5280 \text{ ft}}{1 \text{ mi}}\right)\left(\frac{12 \text{ in.}}{1 \text{ ft}}\right)\left(\frac{2.54 \text{ cm}}{1 \text{ in.}}\right)\left(\frac{1 \text{ m}}{10^2 \text{ cm}}\right)\left(\frac{1 \text{ km}}{10^3 \text{ m}}\right) = \mathbf{764 \ km}$$

English–English unit conversions — Metric–metric unit conversions

Memorizing other English–metric unit conversion factors will save you time and effort. For example, if you know that 1.609 km = 1 mi, the problem becomes much easier.

$$? \text{ km} = 475 \text{ mi} \left(\frac{1.609 \text{ km}}{1 \text{ mi}}\right) = \mathbf{764 \ km}$$

### Example 8.14 Unit Conversions Using Density

What is the volume in liters of 64.567 pounds of ethanol at 20 °C?

***Solution***

Pound is a mass unit, and we want volume. Density provides a conversion factor that converts between mass and volume. You can find the density of ethanol in a table like Table 8.2. It is 0.7893 g/mL at 20 °C.

$$? \text{ L ethanol} = 64.567 \text{ lb ethanol}\left(\frac{453.6 \text{ g}}{1 \text{ lb}}\right)\left(\frac{1 \text{ mL}}{0.7893 \text{ g}}\right)\left(\frac{1 \text{ L}}{10^3 \text{ mL}}\right)$$

$$= \mathbf{37.11 \text{ L ethanol}}$$

OBJECTIVE 15

Before we go on to the next example, let's look at one more way to generate dimensional analysis conversion factors. Anything that can be read as "something per something" can be used as a dimensional analysis conversion factor. For example, if a car is moving at 55 miles *per* hour, we can use the following conversion factor to convert from distance traveled in miles to time in hours or from time in hours to distance traveled in miles.

$$\frac{55 \text{ mi}}{1 \text{ hr}}$$

If you are building a fence and plan to use four nails *per* board, the following conversion factor enables you to calculate the number of nails necessary to nail up 94 fence boards.

$$\frac{4 \text{ nails}}{1 \text{ fence board}}$$

### Example 8.15 Unit Conversions Using Percentage

The label on a can of cat food tells you there are 0.94 lb of cat food per can and that this brand of cat food is 0.15% calcium. If there are 3 servings per can, how many grams of calcium are in each serving?

***Solution***

Note that two phrases in this question can be read as "something per something" and therefore can be used as a dimensional analysis conversion factors. The phrase *3 servings per can* leads to the first conversion factor used on the facing page, and *0.94 lb of cat food per can* leads to the second.

Percentages also provide ratios that can be used as dimensional analysis conversion factors. Because percentages are assumed to be mass percentages unless otherwise indicated, they tell us the number of mass units of the part for each 100 mass units of the whole. The ratio can be constructed using any unit of mass as long as the same unit is written in both the numerator and the denominator. This leads to the third conversion factor in our setup. The fourth conversion factor changes pounds to grams.

$$? \text{ g} = \text{One serving}\left(\frac{1 \text{ can}}{3 \text{ servings}}\right)\left(\frac{0.94 \text{ lb food}}{1 \text{ can}}\right)\left(\frac{0.15 \text{ lb Ca}}{100 \text{ lb food}}\right)\left(\frac{453.6 \text{ g}}{1 \text{ lb}}\right)$$

$$= \mathbf{0.21 \text{ g Ca}}$$

## EXAMPLE 8.16 Converting a Ratio of Two Units

When 2.3942 kg of the sugar glucose is burned (combusted), 37.230 kJ of heat is evolved. What is the heat of combustion of glucose in J/g? (Heat evolved is described with a negative sign.)

***Solution***

When the answer you want is a ratio of two units, start your dimensional analysis setup with a ratio of two units.

Put the correct type of unit in the correct position in the ratio. For this problem, we put the heat unit on the top and the mass unit on the bottom.

$$\frac{? \text{ J}}{\text{g glucose}} = \frac{-37.230 \text{ kJ}}{2.3942 \text{ kg glucose}}\left(\frac{10^3 \text{ J}}{1 \text{ kJ}}\right)\left(\frac{1 \text{ kg}}{10^3 \text{ g}}\right) = \mathbf{-15.550 \text{ J/g}}$$

## EXERCISE 8.10 Unit Conversions

OBJECTIVE 15

a. The diameter of a proton is $2 \times 10^{-15}$ meter. What is this diameter in nanometers?

b. The mass of an electron is $9.1093897 \times 10^{-31}$ kg. What is this mass in nanograms?

c. There are $4.070 \times 10^6$ lb of sulfuric acid used to make Jell-O each year. Convert this to kilograms.

d. A piece of Styrofoam has a mass of 88.978 g and a volume of 2.9659 L. What is its density in g/mL?

e. The density of blood plasma is 1.03 g/mL. A typical adult has about 2.5 L of blood plasma. What is the mass in kilograms of this amount of blood plasma?

f. Pain signals are transferred through the nervous system at a speed between 12 and 30 meters per second. If a student drops a textbook on her toe, how long will it take for the signal, traveling at a velocity of 18 meters per second, to reach her brain 6.0 feet away?

g. An electron takes $6.2 \times 10^{-9}$ second to travel across a TV set that is 22 inches wide. What is the velocity of the electron in km/hr?

h. The mass of the ocean is about $1.8 \times 10^{21}$ kg. If the ocean contains 0.041% by mass calcium ions, $Ca^{2+}$, how many tons of $Ca^{2+}$ are in the ocean? (There are 2000 pounds per ton.)

i. When you are at rest, your heart pumps about 5.0 liters of blood per minute. Your brain gets about 15% by volume of your blood. What volume of blood, in liters, is pumped through your brain in 1.0 hour of rest?

## 8.6 Temperature Conversions

Section 1.4 presented the three most frequently used scales for describing temperature: Celsius, Fahrenheit, and Kelvin. In this section, we will use the following equations to convert a temperature reported in one of these systems to the equivalent temperature in another. Note that the numbers 1.8, 32, and 273.15 in these equations all come from definitions, so they are all exact.

$$? \text{ °F} = \text{number of °C}\left(\frac{1.8 \text{ °F}}{1 \text{ °C}}\right) + 32 \text{ °F}$$

OBJECTIVE 16

$$? \text{ °C} = (\text{number of °F} - 32 \text{ °F})\left(\frac{1 \text{ °C}}{1.8 \text{ °F}}\right)$$

$$? \text{ K} = \text{number of °C} + 273.15$$

$$? \text{ °C} = \text{number of K} - 273.15$$

### Example 8.17 Temperature Conversions

OBJECTIVE 16

"Heavy" water contains the heavy form of hydrogen called deuterium, each atom of which has one proton, one neutron, and one electron. Heavy water freezes at 38.9 °F. What is this temperature in °C?

***Solution***

We use the equation for converting Fahrenheit temperatures to Celsius.

$$? \text{ °C} = (38.9 \text{ °}\cancel{\text{F}} - 32 \text{ °}\cancel{\text{F}})\left(\frac{1 \text{ °C}}{1.8 \text{ °}\cancel{\text{F}}}\right)$$

Rounding off the answer can be tricky here. When you subtract 32 from 38.9, you get 6.9. The 32 is exact, so it is ignored when considering how to round off the answer. The 38.9 is precise to the first number after the decimal point, so the answer to the subtraction is reported to the tenths place. There are two significant figures in 6.9, so when we divide by the exact value of 1.8 °F, we round our answer to two significant figures.

$$? \text{ °C} = (38.9 \text{ °}\cancel{\text{F}} - 32 \text{ °}\cancel{\text{F}})\left(\frac{1 \text{ °C}}{1.8 \text{ °}\cancel{\text{F}}}\right) = (6.9 \text{ °}\cancel{\text{F}})\left(\frac{1 \text{ °C}}{1.8 \text{ °}\cancel{\text{F}}}\right) = \mathbf{3.8 \text{ °C}}$$

## EXAMPLE 8.18 Temperature Conversions

OBJECTIVE 16

The compound 1-chloropropane, $CH_3CH_2CH_2Cl$, boils at 46.6 °C. What is this temperature in °F?

***Solution***

The equation for converting Celsius temperatures to Fahrenheit is

$$? \text{ °F} = 46.6 \text{ °C}\left(\frac{1.8 \text{ °F}}{1 \text{ °C}}\right) + 32 \text{ °F}$$

Because the calculation involves multiplication as well as addition, we need to apply two different rules for rounding off the answer. When we multiply 46.6, which has three significant figures, by the exact value of 1.8 °F, our answer should have three significant figures. The answer on the display of the calculator, 83.88, would therefore be rounded off to 83.9. We then add the exact value of 32 °F and round off that answer to the tenths place.

$$? \text{ °F} = 46.6 \text{ °C}\left(\frac{1.8 \text{ °F}}{1 \text{ °C}}\right) + 32 \text{ °F} = 83.9 \text{ °F} + 32 \text{ °F} = \mathbf{115.9 \text{ °F}}$$

## EXAMPLE 8.19 Temperature Conversions

OBJECTIVE 16

Silver melts at 961 °C. What is this temperature in K?

***Solution***

$$? \text{ K} = 961 \text{ °C} + 273.15 = \mathbf{1234 \text{ K}}$$

For rounding off our answer, we assumed that 961 °C came from a measurement and so is not exact. It is precise to the ones place. On the other hand, 273.15 is exact and has no effect on the uncertainty of our answer. We therefore report the answer for our addition to the ones place, rounding off 1234.15 to 1234.

## EXAMPLE 8.20 Temperature Conversions

OBJECTIVE 16

Tin(II) sulfide, SnS, melts at 1155 K. What is this temperature in °C?

***Solution***

$$? \text{ °C} = 1155 \text{ K} - 273.15 = \mathbf{882 \text{ °C}}$$

Because 1155 is precise to the ones place and 273.15 is exact, we report the answer for our subtraction to the ones place.

## Exercise 8.11 Temperature Conversions

OBJECTIVE 16

a. N,N-dimethylaniline, $C_6H_5N(CH_3)_2$, melts at 2.5 °C. What is N,N-dimethylaniline's melting point in °F and K?

b. Benzenethiol, $C_6H_5SH$, melts at 5.4 °F. What is benzenethiol's melting point in °C and K?

c. The hottest part of the flame on a Bunsen burner is found to be $2.15 \times 10^3$ K. What is this temperature in °C and °F?

## Chapter Glossary

**Dimensional analysis** A general technique for doing unit conversions.

**Conversion factor** A ratio that describes the relationship between two units.

**Significant figures** The number of meaningful digits in a value. The number of significant figures in a value reflects the value's degree of uncertainty. A larger number of significant figures indicates a smaller degree of uncertainty.

**Mass density** Mass divided by volume (usually called ***density***).

You can test yourself on the glossary terms at the following Web site:
**www.chemplace.com/college/**

## Chapter Objectives

**The goal of this chapter is to teach you to do the following.**

1. Define all of the terms in the Chapter Glossary.

### Section 8.1 Dimensional Analysis

2. Write conversion factors that relate the metric base units to units derived from the metric prefixes—for example,

$$\frac{10^3 \text{ m}}{1 \text{ km}}$$

3. Use dimensional analysis to make conversions from one metric unit to another.
4. Write the English–metric unit conversion factors listed in Table 8.1.
5. Use dimensional analysis to make conversions between English mass, volume, or length units and metric mass, volume, or length units.

### Section 8.2 Rounding Off and Significant Figures

6. Identify each value in a calculation as exact or not exact.
7. Write or identify the number of significant figures in any value that is not exact.
8. Round off answers derived from multiplication and division to the correct number of significant figures.
9. Round off answers derived from calculations involving addition or subtraction to the correct number of decimal positions.

### Section 8.3 Density and Density Calculations

10. Provide or recognize the units commonly used to describe the density of solids, liquids, and gases.
11. Use density as a conversion factor to convert between mass and volume.
12. Calculate the density of a substance from its mass and volume.

### Section 8.4 Percentage and Percentage Calculations

13. Given a percentage by mass or a percentage by volume, write a conversion factor based on it.
14. Use conversion factors derived from percentages to convert between units of a part and units of the whole.

### Section 8.5 A Summary of the Dimensional Analysis Process

15. Use dimensional analysis to make unit conversions using conversion factors derived from any relationship that can be described as "something per something."

### Section 8.6 Temperature Conversions

16. Convert a temperature reported in the Celsius, Fahrenheit, or Kelvin scale to both of the other two scales.

## Review Questions

1. Write the metric base units and their abbreviations for length, mass, volume, energy, and gas pressure. (See Section 1.4.)
2. Complete the following table by writing the type of measurement that the unit represents (mass, length, volume, or temperature) and either the name or the abbreviation for the unit. (See Section 1.4.)

| Unit | Type of measurement | Abbreviation | Unit | Type of measurement | Abbreviation |
|---|---|---|---|---|---|
| milliliter | | | kilojoule | | |
| | | μg | | | K |

3. Complete the following relationships between units.

   a. _____ m = 1 μm
   b. _____ g = 1 Mg
   c. _____ L = 1 mL
   d. _____ m = 1 nm
   e. _____ $cm^3$ = 1 mL
   f. _____ L = $m^3$
   g. _____ kg = 1 t (t = metric ton)
   h. _____ Mg = 1 t (t = metric ton)

4. An empty 2-L graduated cylinder is weighed on a balance and found to have a mass of 1124.2 g. Liquid methanol, $CH_3OH$, is added to the cylinder, and its volume is measured as 1.20 L. The total mass of the methanol and the cylinder is measured as 2073.9 g. On the basis of the way these data are reported, what do you assume is the range of possible values that each represents?

## Key Ideas

**Complete the following statements by writing one of these words or phrases in each blank.**

| | |
|---|---|
| cancel | inexact |
| correct | known |
| counting | left |
| decimal places | less dense |
| decrease | mass |
| defined | never exact |
| definitions | one |
| desired | part |
| fewest decimal places | "something per something" |
| fewest | uncertainty |
| exact | unit conversion |
| given value | unwanted |
| grams per cubic centimeter | variables |
| grams per liter | volume |
| grams per milliliter | whole |
| identity | |

5. You will find that the stepwise thought process associated with the procedure called dimensional analysis not only guides you in figuring out how to set up ____________ problems but also gives you confidence that your answers are ____________.
6. The first step in the dimensional analysis procedure is to identify the unit for the value we want to calculate. We write this on the ____________ side of an equals sign. Next, we identify the ____________ that we will convert into the desired value, and we write it on the other side of the equals sign.
7. Next, we multiply by one or more conversion factors that enable us to cancel the ____________units and generate the ____________ units.
8. Note that the units in a dimensional analysis setup cancel just like ____________ in an algebraic equation.
9. If you have used correct conversion factors in a dimensional analysis setup, and if your units ____________ to yield the desired unit or units, you can be confident that you will arrive at the correct answer.
10. Because the English inch is ____________ as 2.54 cm, the number 2.54 in this value is exact.
11. Unless we are told otherwise, we assume that values from measurements have an uncertainty of plus or minus ____________ in the last decimal place reported.
12. If a calculation is performed using all exact values and if the answer is not rounded off, the answer is ____________.
13. When an answer is calculated by multiplying or dividing, we round it off to the same number of significant figures as the ____________ value with the ____________ significant figures.
14. The number of significant figures, which is equal to the number of meaningful digits in a value, reflects the degree of ____________ in the value.

15. Numbers that come from definitions and from ____________ are exact.

16. Values that come from measurements are ____________.

17. When adding or subtracting, round your answer to the same number of ____________ as the inexact value with the ____________.

18. Although there are exceptions, the densities of liquids and solids generally ____________ with increasing temperature.

19. The densities of liquids and solids are usually described in ____________ or ____________.

20. The particles of a gas are much farther apart than the particles of a liquid or solid, so gases are much ____________ than solids and liquids. Thus, it is more convenient to describe the densities of gases as ____________.

21. Because the density of a substance depends on the substance's ____________ and its temperature, it is possible to identify an unknown substance by comparing its density at a particular temperature to the ____________ densities of substances at the same temperature.

22. Because density is reported as a ratio that describes a relationship between two units, the density of a substance can be used in dimensional analysis to convert between the substance's ____________ and its ____________.

23. Percentage by mass, the most common form of percentage used in chemical descriptions, is a value that tells us the number of mass units of the ____________ for each 100 mass units of the ____________.

24. Anything that can be read as ____________ can be used as a dimensional analysis conversion factor.

25. The numbers 1.8, 32, and 273.15 in the equations used for temperature conversions all come from ____________, so they are all exact.

## Chapter Problems

Because this chapter and Inter-Chapter 1A cover similar topics, the problems they contain are also similar. If you have not already worked the problems in the inter-chapter, you can obtain additional practice by working them now.

### Problems Relating to Appendices B and C

If you have not yet read Appendix B, which describes scientific notation, and Appendix C, which contains pointers for using calculators, you might want to read them before working the problems that follow.

26. Convert the following ordinary decimal numbers to scientific notation. (See Appendix B at the end of this book if you need help with scientific notation.)
    a. 67,294
    b. 438,763,102
    c. 0.000073
    d. 0.0000000435

27. Convert the following ordinary decimal numbers to scientific notation. (See Appendix B at the end of this book if you need help with scientific notation.)
    a. 1,346.41
    b. 429,209
    c. 0.000002056
    d. 0.00488

28. Convert the following numbers expressed in scientific notation to ordinary decimal numbers. (See Appendix B at the end of this book if you need help with scientific notation.)

a. $4.097 \times 10^3$
b. $1.55412 \times 10^4$
c. $2.34 \times 10^{-5}$
d. $1.2 \times 10^{-8}$

29. Convert the following numbers expressed in scientific notation to ordinary decimal numbers. (See Appendix B at the end of this book if you need help with scientific notation.)

a. $6.99723 \times 10^5$
b. $2.333 \times 10^2$
c. $3.775 \times 10^{-3}$
d. $5.1012 \times 10^{-6}$

30. Use your calculator to complete the following calculations. (See your calculator's instruction manual or Appendix C at the end of this book if you need help using a calculator.)

a. $34.25 \times 84.00$
b. $2607 \div 8.25$
c. $425 \div 17 \times 0.22$
d. $(27.001 - 12.866) \div 5.000$

31. Use your calculator to complete the following calculations. (See your calculator's instruction manual or Appendix C at the end of this book if you need help using a calculator.)

a. $36.6 \div 0.0750$
b. $848.8 \times 0.6250$
c. $575.0 \div 5.00 \times 0.20$
d. $2.50 \times (33.141 + 5.099)$

32. Use your calculator to complete the following calculations. (See your calculator's instruction manual or Appendix C at the end of this book if you need help using a calculator.)

a. $10^9 \times 10^3$
b. $10^{12} \div 10^3$
c. $10^3 \times 10^6 \div 10^2$
d. $10^9 \times 10^{-4}$
e. $10^{23} \div 10^{-6}$
f. $10^{-4} \times 10^2 \div 10^{-5}$

33. Use your calculator to complete the following calculations. (See your calculator's instruction manual or Appendix C at the end of this book if you need help using a calculator.)

a. $10^{12} \div 10^9$
b. $10^{14} \times 10^5$
c. $110^{17} \div 10^3 \times 10^6$
d. $10^{-8} \div 10^5$
e. $10^{-11} \times 10^8$
f. $10^{17} \div 10^{-9} \times 10^3$

34. Use your calculator to complete the following calculations. (See your calculator's instruction manual or Appendix C at the end of this book if you need help using a calculator.)

a. $9.5 \times 10^5 \cdot 8.0 \times 10^9$
b. $6.12 \times 10^{19} \div 6.00 \times 10^3$
c. $2.75 \times 10^4 \times 6.00 \times 10^7 \div 5.0 \times 10^6$
d. $8.50 \times 10^{-7} \cdot 2.20 \times 10^3$
e. $8.203 \times 10^9 \div 10^{-4}$
f. $(7.679 \times 10^{-4} - 3.457 \times 10^{-4}) \div 2.000 \times 10^{-8}$

35. Use your calculator to complete the following calculations. (See your calculator's instruction manual or Appendix C at the end of this book if you need help using a calculator.)
    a. $1.206 \times 10^{13} \div 6.00 \times 10^{6}$
    b. $5.00 \times 10^{23} \cdot 4.4 \times 10^{17}$
    c. $7.500 \times 10^{3} \cdot 3.500 \times 10^{9} \div 2.50 \times 10^{15}$
    d. $1.85 \times 10^{4} \cdot 2.0 \times 10^{-12}$
    e. $1.809 \times 10^{-9} \div 9.00 \times 10^{-12}$
    f. $(7.131 \times 10^{6} - 4.006 \times 10^{6}) \div 10^{-12}$

### Section 8.1 Dimensional Analysis

36. Complete each of the following conversion factors by filling in the blank on the top of the ratio. OBJECTIVE 2 OBJECTIVE 4
    a. $\left(\frac{\_\_\_\_\ \text{m}}{1\ \text{km}}\right)$
    b. $\left(\frac{\_\_\_\_\ \text{cm}}{1\ \text{m}}\right)$
    c. $\left(\frac{\_\_\_\_\ \text{mm}}{1\ \text{m}}\right)$
    d. $\left(\frac{\_\_\_\_\ \text{cm}^3}{1\ \text{mL}}\right)$
    e. $\left(\frac{\_\_\_\_\ \text{cm}}{1\ \text{in.}}\right)$
    f. $\left(\frac{\_\_\_\_\ \text{g}}{1\ \text{lb}}\right)$

37. Complete each of the following conversion factors by filling in the blank on the top of the ratio. OBJECTIVE 2 OBJECTIVE 4
    a. $\left(\frac{\_\_\_\_\ \mu\text{m}}{1\ \text{m}}\right)$
    b. $\left(\frac{\_\_\_\_\ \text{nm}}{1\ \text{m}}\right)$
    c. $\left(\frac{\_\_\_\_\ \text{kg}}{1\ \text{metric ton}}\right)$
    d. $\left(\frac{\_\_\_\_\ \text{L}}{1\ \text{gal}}\right)$
    e. $\left(\frac{\_\_\_\_\ \text{km}}{1\ \text{mile}}\right)$
    f. $\left(\frac{\_\_\_\_\ \text{in.}}{1\ \text{m}}\right)$

38. Complete each of the following conversion factors by filling in the blank on the top of the ratio. OBJECTIVE 2 OBJECTIVE 4
    a. $\left(\frac{\_\_\_\_\ \text{g}}{1\ \text{kg}}\right)$
    b. $\left(\frac{\_\_\_\_\ \text{mg}}{1\ \text{g}}\right)$
    c. $\left(\frac{\_\_\_\_\ \text{yd}}{1\ \text{m}}\right)$
    d. $\left(\frac{\_\_\_\_\ \text{lb}}{1\ \text{kg}}\right)$

39. Complete each of the following conversion factors by filling in the blank on the top of the ratio. OBJECTIVE 2 OBJECTIVE 4
    a. $\left(\frac{\_\_\_\_\ \mu\text{g}}{1\ \text{g}}\right)$
    b. $\left(\frac{\_\_\_\_\ \text{mL}}{1\ \text{L}}\right)$
    c. $\left(\frac{\_\_\_\_\ \mu\text{L}}{1\ \text{L}}\right)$
    d. $\left(\frac{\_\_\_\_\ \text{g}}{1\ \text{oz}}\right)$
    e. $\left(\frac{\_\_\_\_\ \text{qt}}{1\ \text{L}}\right)$

40. The mass of an electron is $9.1093897 \times 10^{-31}$ kg. What is this mass in grams? OBJECTIVE 4

OBJECTIVE 3

41. The diameter of a human hair is 2.5 micrometers. What is this diameter in meters?

OBJECTIVE 3

42. The diameter of typical bacteria cells is 0.00032 centimeters. What is this diameter in micrometers?

OBJECTIVE 3

43. The mass of a proton is $1.6726231 \times 10^{-27}$ kg. What is this mass in micrograms?

OBJECTIVE 5

44. The thyroid gland is the largest of the endocrine glands, with a mass between 20 and 25 grams. What is the mass in pounds of a thyroid gland measuring 22.456 grams?

OBJECTIVE 5

45. The average human body contains 5.2 liters of blood. What is this volume in gallons?

OBJECTIVE 5

46. The mass of a neutron is $1.674929 \times 10^{-27}$ kg. Convert this to ounces. (There are 16 oz/lb.)

OBJECTIVE 5

47. The earth weighs about $1 \times 10^{21}$ tons. Convert this to gigagrams. (There are 2000 lb/ton.)

OBJECTIVE 5

48. A red blood cell is $8.7 \times 10^{-5}$ inches thick. What is this thickness in micrometers?

OBJECTIVE 5

49. The gallbladder has a capacity of between 1.2 and 1.7 fluid ounces. What is the capacity in milliliters of a gallbladder that can hold 1.42 fluid ounces? (There are 32 fl oz/qt.)

### Section 8.2 Rounding Off and Significant Figures

OBJECTIVE 6 OBJECTIVE 7

50. Decide whether each of the numbers shown in bold type below is exact or not. If it is not exact, write the number of significant figures in it.
   a. The approximate volume of the ocean, $\mathbf{1.5 \times 10^{21}}$ **L.**
   b. A count of **24** instructors in the physical science division of a state college.
   c. The **54**% of the instructors in the physical science division who are women (determined by counting 13 women in the total of 24 instructors and then calculating the percentage).
   d. The **25**% of the instructors in the physical science division who are left handed (determined by counting 6 left-handed instructors in the total of 24 and then calculating the percentage).
   e. $\dfrac{\mathbf{16}\text{ oz}}{1\text{ lb}}$
   f. $\dfrac{\mathbf{10^6}\text{ mm}}{1\text{ m}}$
   g. $\dfrac{\mathbf{1.057}\text{ qt}}{1\text{ L}}$
   h. A measurement of **107.200** g water
   i. A mass of **0.2363** lb water (calculated from Part h, using $\dfrac{453.6\text{ g}}{1\text{ lb}}$ as a conversion factor)
   j. A mass of $\mathbf{1.182 \times 10^{-4}}$ tons (calculated from the 0.2363 lb of the water described in Part i.)

OBJECTIVE 2 OBJECTIVE 4

51. Decide whether each of the numbers shown in bold type below is exact or not. If it is not exact, write the number of significant figures in it.

a. 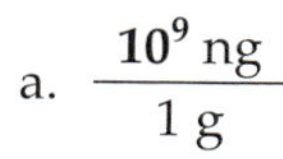 $\frac{\mathbf{10^9}\text{ ng}}{1\text{ g}}$

b. $\frac{\mathbf{32}\text{ fl oz}}{1\text{ qt}}$

c. $\frac{\mathbf{1.094}\text{ yd}}{1\text{ m}}$

d. The diameter of the moon, $\mathbf{3.480 \times 10^3}$ km

e. A measured volume of **8.0** mL water

f. A volume **0.0080** L water (calculated from the volume in Part e, using $\frac{1\text{ L}}{10^3\text{ mL}}$)

g. A volume of **0.0085** qt water (calculated from the volume in Part f, using $\frac{1.057\text{ qt}}{1\text{ L}}$)

h. The count of **112** symbols for elements on a periodic table

i. The **40%** of halogens that are gases at normal room temperature and pressure (determined by counting 2 gaseous halogens out of the total of 5 halogens and then calculating the percentage)

j. The **9.8%** of the known elements that are gases at normal room temperature and pressure (determined by counting 11 gaseous elements out of the 112 elements total and then calculating the percentage)

OBJECTIVE 7

52. Assuming that the following numbers are not exact, how many significant figures does each number have?

a. 13.811
b. 0.0445
c. 505
d. 9.5004
e. 81.00

OBJECTIVE 7

53. Assuming that the following numbers are not exact, how many significant figures does each number have?

a. 9,875
b. 102.405
c. 10.000
d. 0.00012
e. 0.411

OBJECTIVE 7

54. Assuming that the following numbers are not exact, how many significant figures does each number have?

a. $4.75 \times 10^{23}$
b. $3.009 \times 10^{-3}$
c. $4.000 \times 10^{13}$

55. Assuming that the following numbers are not exact, how many significant figures does each number have?

a. $2.00 \times 10^{8}$
b. $1.998 \times 10^{-7}$
c. $2.0045 \times 10^{-5}$

**56.** Convert each of the following numbers to a number having 3 significant figures.

a. 34.579
b. 193.405
c. 23.995
d. 0.003882
e. 0.023
f. 2,846.5
g. $7.8354 \times 10^4$

57. Convert each of the following numbers to a number having 4 significant figures.

a. 4.30398
b. 0.000421
c. $4.44802 \times 10^{-19}$
d. 99.9975
e. 11,687.42
f. 874.992

OBJECTIVE 8

**58.** Complete the following calculations and report your answers to the correct number of significant figures. The exponential factors, such as $10^3$, are exact, and the 2.54 in part (c) is exact. All the other numbers are not exact.

a. $\dfrac{2.45 \times 10^{-5}(10^{12})}{(10^3)237.00} =$

b. $\dfrac{16.050(10^3)}{(24.8 - 19.4)(1.057)(453.6)} =$

c. $\dfrac{4.77 \times 10^{11}(2.54)^3(73.00)}{(10^3)} =$

OBJECTIVE 8

59. Complete the following calculations and report your answers to the correct number of significant figures. The exponential factors, such as $10^3$, are exact, and the 5280 in part (c) is exact. All the other numbers are not exact.

a. $\dfrac{8.9932 \times 10^{-2}\,(10^3)0.0048}{(10^{-6})7.140} =$

b. $\dfrac{(44.945 - 23.775)(10^3)3.785412}{(15.200)(453.59237)} =$

c. $\dfrac{456.8(5280)^2}{(10^3)^2(1.609)^2} =$

OBJECTIVE 9

**60.** Report the answers to the following calculations to the correct number of decimal positions. Assume that each number is precise to $\pm$ 1 in the last decimal position reported.

a. $0.8995 + 99.24 =$
b. $88 - 87.3 =$

OBJECTIVE 9

61. Report the answers to the following calculations to the correct number of decimal positions. Assume that each number is precise to $\pm$ 1 in the last decimal position reported.

a. $23.40 - 18.2 =$
b. $948.75 + 62.45 =$

### Section 8.3 Density and Density Calculations

*Because the ability to make unit conversions using the dimensional analysis format is an extremely important skill, be sure to set up each of the following calculations using the dimensional analysis format, even if you see another way to work the problem, and even if another technique seems easier.*

OBJECTIVE 12

**62.** A piece of balsa wood has a mass of 15.196 g and a volume of 0.1266 L. What is its density in g/mL?

OBJECTIVE 12

63. A ball of clay has a mass of 2.65 lb and a volume of 0.5025 qt. What is its density in g/mL?

OBJECTIVE 11

**64.** The density of water at 0 °C is 0.99987 g/mL. What is the mass in kilograms of 185.0 mL of water?

65. The density of water at 3.98 °C is 1.00000 g/mL. What is the mass in pounds of 16.785 L of water?

OBJECTIVE 11

66. The density of a piece of ebony wood is 1.174 g/mL. What is the volume in quarts of a 2.1549-lb piece of this ebony wood?

OBJECTIVE 11

67. The density of whole blood is 1.05 g/mL. A typical adult has about 5.5 L of whole blood. What is the mass in pounds of this amount of whole blood?

OBJECTIVE 11

### Section 8.4 Percentage and Percentage Calculations

68. The mass of the ocean is about $1.8 \times 10^{21}$ kg. If the ocean contains 1.076% by mass sodium ions, $Na^+$, what is the mass in kilograms of $Na^+$ in the ocean?

OBJECTIVE 14

69. While you are at rest, your brain gets about 15% by volume of your blood. If your body contains 5.2 L of blood, how many liters of blood are in your brain at rest? . . . how many quarts?

OBJECTIVE 14

70. While you are doing heavy work, your heart pumps up to 25.0 L of blood per minute. Your brain gets about 3–4% by volume of your blood under these conditions. What volume of blood in liters is pumped through your brain in 125 minutes of work that causes your heart to pump 22.0 L per minute, 3.43% of which goes to your brain?

OBJECTIVE 14

71. While you are doing heavy work, your heart pumps up to 25.0 L of blood per minute. Your muscles get about 80% by volume of your blood under these conditions. What volume of blood in quarts is pumped through your muscles in 105 minutes of work that causes your heart to pump 21.0 L per minute, 79.25% by volume of which goes to your muscles?

OBJECTIVE 14

72. In chemical reactions that release energy, from $10^{-8}$% to $10^{-7}$% of the mass of the reacting substances is converted to energy. Consider a chemical reaction for which $1.8 \times 10^{-8}$% of the mass is converted into energy. What mass in milligrams is converted into energy when $1.0 \times 10^3$ kilograms of substance reacts?

OBJECTIVE 14

73. In nuclear fusion, about 0.60% of the mass of the fusing substances is converted to energy. What mass in grams is converted into energy when 22 kilograms of substance undergoes fusion?

OBJECTIVE 14

### Section 8.5 A Summary of the Dimensional Analysis Process

74. If an elevator moves 1340 ft to the 103rd floor of the Sears Tower in Chicago in 45 seconds, what is the velocity (distance traveled divided by time) of the elevator in kilometers per hour?

OBJECTIVE 15

75. The moon orbits the sun with a velocity of $2.2 \times 10^4$ miles per hour. What is this velocity in meters per second?

OBJECTIVE 15

76. Sound travels at a velocity of 333 m/s. How long does it take for sound to travel the length of a 100-yard football field?

OBJECTIVE 15

77. How many miles can a commercial jetliner flying at 253 meters per second travel in 6.0 hours?

OBJECTIVE 15

78. A peanut butter sandwich provides about $1.4 \times 10^3$ kJ of energy. A typical adult uses about 95 kcal/hr of energy while sitting. If all of the energy in one peanut butter sandwich were to be burned off by sitting, how many hours would it be before this energy was used? (A kcal is a dietary calorie. There are 4.184 J/cal.)

OBJECTIVE 15

OBJECTIVE 15

79. One-third cup of vanilla ice cream provides about 145 kcal of energy. A typical adult uses about 195 kcal/hr of energy while walking. If all of the energy in one-third of a cup of vanilla ice cream were to be burned off by walking, how many minutes would it take for this energy to be used? (A kcal is a dietary calorie.)

OBJECTIVE 15

80. When one gram of hydrogen gas, $H_2(g)$, is burned, 141.8 kJ of heat are released. How much heat is released when 2.3456 kg of hydrogen gas are burned?

OBJECTIVE 15

81. When one gram of liquid ethanol, $C_2H_5OH(l)$, is burned, 29.7 kJ of heat are released. How much heat is released when 4.274 pounds of liquid ethanol are burned?

OBJECTIVE 15

82. When one gram of carbon in the graphite form is burned, 32.8 kJ of heat are released. How many kilograms of graphite must be burned to release $1.456 \times 10^4$ kJ of heat?

OBJECTIVE 15

83. When one gram of methane gas, $CH_4(g)$, is burned, 55.5 kJ of heat are released. How many pounds of methane gas must be burned to release $2.578 \times 10^3$ kJ of heat?

OBJECTIVE 15

84. The average adult male needs about 58 g of protein in the diet each day. A can of vegetarian refried beans has 6.0 g of protein per serving. Each serving is 128 g of beans. If your only dietary source of protein were vegetarian refried beans, how many pounds of beans would you need to eat each day?

OBJECTIVE 15

85. The average adult needs at least $1.50 \times 10^2$ g of carbohydrates in the diet each day. A can of vegetarian refried beans has 19 g of carbohydrate per serving. Each serving is 128 g of beans. If your only dietary source of carbohydrate were vegetarian refried beans, how many pounds of beans would you need to eat each day?

OBJECTIVE 15

86. About $6.0 \times 10^5$ tons of 30% by mass hydrochloric acid, $HCl(aq)$, are used to remove metal oxides from metals to prepare them for painting or for the addition of a chrome covering. How many kilograms of pure HCl would be used to make this hydrochloric acid? (Assume that 30% has two significant figures. There are 2000 lb/ton.)

OBJECTIVE 15

87. Normal glucose levels in the blood are from 70 to 110 mg glucose per 100 mL of blood. If the level falls too low, there can be brain damage. If a person has a glucose level of 108 mg/100 mL, what is the total mass of glucose in grams in 5.10 L of blood?

OBJECTIVE 15

88. A typical non-obese male has about 11 kg of fat. Each gram of fat can provide the body with about 38 kJ of energy. If this person requires $8.0 \times 10^3$ kJ of energy per day to survive, how many days could he survive on his fat alone?

OBJECTIVE 15

89. The kidneys of a normal adult male filter 125 mL of blood per minute. How many gallons of blood are filtered in one day?

OBJECTIVE 15

90. During quiet breathing, a person breathes in about 6 L of air per minute. If a person breathes in an average of 6.814 L of air per minute, what volume of air in liters does this person breathe in 1 day?

OBJECTIVE 15

91. During exercise, a person breathes in between 100 and 200 L of air per minute. If a person is exercising enough to breathe in an average of 125.6 L of air per minute, what total volume of air in liters is breathed in exactly one hour of exercise?

OBJECTIVE 15

92. The kidneys of a normal adult female filter 115 mL of blood per minute. If this person has 5.345 quarts of blood, how many minutes will it take to filter all of her blood once?

93. A normal hemoglobin concentration in the blood is 15 g/100 mL of blood. How many kilograms of hemoglobin are there in a person who has 5.5 L of blood? OBJECTIVE 15

**94.** We lose between 0.2 and 1 liter of water from our skin and sweat glands each day. For a person who loses an average of 0.89 L $H_2O$/day in this manner, how many quarts of water are lost from the skin and sweat glands in 30 days? OBJECTIVE 15

95. Normal blood contains from 3.3 to 5.1 mg of amino acids per 100 mL of blood. If a person has 5.33 L of blood and 4.784 mg of amino acids per 100 mL of blood, how many grams of amino acids does the blood contain? OBJECTIVE 15

**96.** The average heart rate is 75 beats/min. How many times does the average person's heart beat in a week? OBJECTIVE 15

97. The average heart rate is 75 beats/min. Each beat pumps about 75 mL of blood. How many liters of blood does the average person's heart pump in a week? OBJECTIVE 15

**98.** In optimum conditions, one molecule of the enzyme carbonic anhydrase can convert $3.6 \times 10^5$ molecules per minute of carbonic acid, $H_2CO_3$, to carbon dioxide, $CO_2$, and water, $H_2O$. How many molecules could be converted by one of these enzyme molecules in one week? OBJECTIVE 15

99. In optimum conditions, one molecule of the enzyme fumarase can convert $8 \times 10^2$ molecules per minute of fumarate to malate. How many molecules could be converted by one of these enzyme molecules in 30 days? OBJECTIVE 15

**100.** In optimum conditions, one molecule of the enzyme amylase can convert $1.0 \times 10^5$ molecules per minute of starch to the sugar maltose. How many days would it take one of these enzyme molecules to convert a billion ($1.0 \times 10^9$) starch molecules? OBJECTIVE 15

101. There are about $1 \times 10^5$ chemical reactions per second in the 10 billion nerve cells in the brain. How many chemical reactions take place in a day in a single nerve cell? OBJECTIVE 15

**102.** When you sneeze, you close your eyes for about 1.00 s. If you are driving 65 miles per hour on the freeway and you sneeze, how many feet do you travel with your eyes closed? OBJECTIVE 15

### Section 8.6 Temperature Conversions

**103.** Butter melts at 31 °C. What is this temperature in °F? . . . in K? OBJECTIVE 16

104. Dry ice freezes at –79 °C. What is this temperature in °F? . . . in K? OBJECTIVE 16

**105.** A saturated salt solution boils at 226 °F. What is this temperature in °C? . . . in K? OBJECTIVE 16

106. Table salt, sodium chloride, melts at 801 °C. What is this temperature in °F? . . . in K? OBJECTIVE 16

**107.** Iron boils at 3023 K. What is this temperature in °C? . . . in °F? OBJECTIVE 16

108. Absolute zero, the lowest possible temperature, is exactly 0 K. What is this temperature in °C? . . . in °F? OBJECTIVE 16

**109.** The surface of the sun is $1.0 \times 10^4$ °F. What is this temperature in °C? . . . in K? OBJECTIVE 16

*Chemists often ask questions that begin with, "How much . . . ?"*

CHAPTER 9

# Chemical Calculations and Chemical Formulas



IN ORDER TO EXPLORE AND MAKE USE OF THE SEEMINGLY LIMITLESS CHANGES that matter can undergo, chemists and chemistry students often need to answer questions that begin with "*How much . . .* ?" The research chemist who is developing a new cancer drug wants to know, "*How much* boron-10 do I need to make 5 g of the drug?" At a plant where the fat substitute Olestra is manufactured from white sugar and vegetable oil, a business manager asks a chemist, "*How much* sucrose and cottonseed oil should I order if we need to produce 500 Mg of Olestra per day?" In an experiment for a chemistry course you are taking, you might be asked, "How much magnesium oxide can be formed from the reaction of your measured mass of magnesium with the oxygen in the air?" This chapter and the chapters that follow provide you with the tools necessary to answer these questions and many others like them.

## Review Skills

The presentation of information in this chapter assumes that you can already perform the tasks listed below. You can test your readiness to proceed by answering the Review Questions at the end of the chapter. This might also be a good time to read the Chapter Objectives, which precede the Review Questions.

- Write or recognize the definition of *isotope*. (Section 2.4)
- Describe the general structure of molecular and ionic compounds. (Sections 3.3 and 3.5)
- Convert between the names of compounds and their chemical formulas. (Section 5.3)
- Report the answers to calculations to the correct number of significant figures. (Section 8.2)
- Use percentages as conversion factors. (Section 8.4)
- Make unit conversions. (Section 8.5)

*All of these people ask chemical questions that begin, "How much . . . ?"*

## 9.1 A Typical Problem

Imagine you are a chemist at a company that makes phosphoric acid, $H_3PO_4$, for use in the production of fertilizers, detergents, and pharmaceuticals. The goal for your department is to produce 84.0% $H_3PO_4$ (and 16.0% water) in the last stage of a three-step process known as the furnace method. (We will be using the furnace method as a source of practical examples throughout this chapter and in Chapter 10.)

The first step in the furnace method is the extraction of phosphorus from phosphate rock by heating the rock with sand and coke to 2000 °C. The phosphate rock contains calcium phosphate, $Ca_3(PO_4)_2$; the sand contains silicon dioxide, $SiO_2$; and coke is a carbon-rich substance that can be produced by heating coal to a high temperature. At 2000 °C, these three substances react as follows:[1]

$$2Ca_3(PO_4)_2 + 6SiO_2 + 10C \rightarrow 4P + 10CO + 6CaSiO_3$$

In the second step, the step on which we will focus in this chapter, the phosphorus, P, is reacted with oxygen in air to form tetraphosphorus decoxide, $P_4O_{10}$:

$$4P(s) + 5O_2(g) \rightarrow P_4O_{10}(s)$$

The third and final step is the reaction of $P_4O_{10}$ with water to form phosphoric acid:

$$P_4O_{10}(s) + 6H_2O(l) \rightarrow 4H_3PO_4(aq)$$

Your colleagues who are responsible for the first step, the production of pure phosphorus, estimate that they can supply you with $1.09 \times 10^4$ kg of phosphorus per day. The production manager asks you to figure out the maximum mass of $P_4O_{10}$ that can be made from this amount of phosphorus. The tools described in this chapter, in combination with dimensional analysis, enable you to satisfy this request.

Let's begin by thinking about how dimensional analysis might be used to solve your problem. Your production manager is asking you to convert from amount of phosphorus, P, to amount of tetraphosphorus decoxide, $P_4O_{10}$. To do this, you need a conversion factor relating amount of P to amount of $P_4O_{10}$. The chemical formula, $P_4O_{10}$, provides such a conversion factor. It shows that there are four atoms of phosphorus for each molecule of $P_4O_{10}$.

$$\frac{4 \text{ atoms P}}{1 \text{ molecule } P_4O_{10}} \quad \text{or} \quad \frac{1 \text{ molecule } P_4O_{10}}{4 \text{ atoms P}}$$

[1]To avoid potential confusion, the states—(*s*), (*l*), (*g*), and (*aq*)—are not mentioned in equations describing industrial reactions in Chapters 9 and 10. Many industrial reactions are run at temperatures and pressures under which the states of substances differ from what would be expected at room temperatures and pressures. For example, the $Ca_3(PO_4)_2$ in the first equation on this page is a solid under normal conditions but a liquid at 2000 °C; $SiO_2$ is also solid under normal conditions but a glass (or semisolid) at 2000 °C.

But how many atoms of phosphorus are in our $1.09 \times 10^4$ kg P? Until we know that, how can we tell how much $P_4O_{10}$ we'll be able to produce? Unfortunately, atoms and molecules are so small and numerous that they cannot be counted directly. *We need a conversion factor that converts back and forth between any mass of the element and the number of atoms contained in that mass.*

If we can determine the number of phosphorus atoms in $1.09 \times 10^4$ kg P, we can use the second conversion factor displayed above to determine the number of molecules of $P_4O_{10}$. But your production manager doesn't want to know the number of $P_4O_{10}$ *molecules* that you can make. She wants to know the *mass* of $P_4O_{10}$ that can be made. That means *we also need a conversion factor that converts back and forth between any mass of the compound and the number of molecules contained in that mass*. The main goal of the first half of this chapter is to develop conversion factors that convert between mass and number of particles.

An industrial chemist, working in a chemical plant such as this, must be able to calculate the amounts of reactants necessary to run an efficient reaction and predict the amount of product that will form.

## 9.2 Relating Mass to Number of Particles

Before we develop conversion factors to convert back and forth between any mass of an element and the number of atoms contained in that mass, let's consider a similar task that may be easier to visualize. We will calculate the number of carpenter's nails in a hardware store bin.

Imagine that you have decided to make a little money for schoolbooks by taking a temporary job doing inventory at the local hardware store. Your first task is to determine how many nails there are in a bin filled with nails. There are a lot of nails in the bin, and you do not want to take the time to count them one by one. It would be much easier to weigh the nails and calculate the number of nails from the mass. To do this you need a conversion factor that enables you to convert from the mass of the nails in the bin to the number of nails in the bin.

Let's assume that you individually weigh 100 of the nails from the bin and find that 82 of them have a mass of 3.80 g each, 14 of them have a mass of 3.70 g each, and the last 4 have a mass of 3.60 g each. From this information, you can calculate the average mass of the nails in this sample, taking into consideration that 82% of the nails have a mass of 3.80 g, 14% have a mass of 3.70 g, and 4% have a mass of 3.60 g. Such an average is known as a **weighted average mass.** You can calculate the weighted average of the nails' masses by multiplying the decimal fraction of each subgroup of nails times the mass of one of its members and adding the results of these multiplications.

$$0.82(3.80\ \text{g}) + 0.14(3.70\ \text{g}) + 0.04(3.60\ \text{g}) = 3.78\ \text{g}$$

Thus the weighted average mass of the nails in this sample is 3.78 g. It is possible that *none* of the nails in our bin has exactly this mass, but this is a good description of what we can expect the average mass of each nail in a large number of nails to be. It can be used as a conversion factor to convert between mass of nails and number of nails.

$$\left(\frac{3.78 \text{ g nails}}{1 \text{ nail}}\right)$$

We can measure the total mass of the nails in the bin and then use our conversion factor to determine their number. For example, if the nails in the bin are found to weigh 218 pounds, the number of nails in the bin is found as follows

$$? \text{ nails} = 218 \cancel{\text{lb nails}} \left(\frac{453.6 \cancel{\text{g}}}{1 \cancel{\text{lb}}}\right)\left(\frac{1 \text{ nail}}{3.78 \cancel{\text{g nails}}}\right) = 2.62 \times 10^4 \text{ nails}$$

Instead of counting every nail individually, we can count by weighing. We measure their total mass and convert it into a number by using the weighted average mass.

There is some uncertainty in this result because we relied on measuring rather than counting, but this procedure is a lot faster than the alternative of counting over 26,000 nails. The key point to remember is that this procedure enables us to determine the number of objects in a sample of a large number of those objects without actually counting them. Our procedure enables us to count by weighing.

It is often convenient to describe numbers of objects in terms of a collective unit such as a dozen (12) or a gross (144). The number $2.62 \times 10^4$ is large and inconvenient to use. We might therefore prefer to describe the number of nails in another way. For example, we could describe them in terms of dozens of nails: 218 pounds of our nails is $2.18 \times 10^3$ dozen nails, which is still an awkward number. We could use gross instead of dozen. A gross of objects is 144 objects. The following calculation shows how to create a conversion factor that converts between mass of nails and gross of nails:

$$\frac{? \text{ g nails}}{1 \text{ gross nails}} = \left(\frac{3.78 \text{ g nails}}{1 \cancel{\text{nail}}}\right)\left(\frac{144 \cancel{\text{nails}}}{1 \text{ gross nails}}\right) = \frac{544 \text{ g nails}}{1 \text{ gross nails}}$$

We can use this conversion factor to determine the number of gross of nails in 218 pounds of nails.

$$? \text{ gross nails} = 218 \cancel{\text{lb nails}} \left(\frac{453.6 \cancel{\text{g}}}{1 \cancel{\text{lb}}}\right)\left(\frac{1 \text{ gross nails}}{544 \cancel{\text{g nails}}}\right) = 182 \text{ gross of nails}$$

## Atomic Mass and Counting Atoms by Weighing

Now let's take similar steps to "count" atoms of the element carbon. Because of the size and number of carbon atoms in any normal sample of carbon, it is impossible to count the atoms directly. Therefore, we want to develop a way of converting from mass of carbon, which we can measure, to the number of carbon atoms. To do this, we will follow steps that are similar to those we followed to "count" nails by weighing.

First, we need to know the masses of individual atoms of carbon. To describe the mass of something as small as an atom of carbon, we need a unit whose magnitude (or lack of it) is correspondingly small. The unit most often used to describe the mass of atoms is the atomic mass unit, whose symbol is **u** or **amu.** An **atomic mass unit** is defined as exactly one-twelfth the mass of an atom of carbon-12. Carbon-12 is the isotope of carbon that contains six protons, six neutrons, and six electrons. (You might want to review Section 2.4, which describes isotopes.) One atomic mass unit is equivalent to $1.660540 \times 10^{-24}$ grams.

$$1 \text{ atomic mass unit (u)} = \frac{1}{12} \text{ mass of one carbon-12 atom} = 1.660540 \times 10^{-24} \text{ g}$$

To generate a relationship between mass of carbon and number of carbon atoms, we need to know the weighted average mass of the carbon atoms found in nature. Experiments show that 98.90% of the carbon atoms in natural carbon are carbon-12 and that 1.10% are carbon-13, with six protons, seven neutrons, and six electrons. Related experiments show that each carbon-13 atom has a mass of 13.003355 u. From the definition of the atomic mass unit, we know that the mass of each carbon-12 atom is 12 u. The following setup shows how the weighted average mass of carbon atoms is calculated.

$$0.9890(12 \text{ u}) + 0.0110(13.003355 \text{ u}) = 12.011 \text{ u}$$

This value is carbon's atomic mass. Because an element's atomic mass is often described without units, carbon's atomic mass[2] is usually described as 12.011 instead of 12.011 u. The **atomic mass** of any element is the weighted average of the masses of the naturally occurring isotopes of the element. (This property is commonly called *atomic weight*, but because it describes the masses of the atoms, not their weights, this text will use the term *atomic mass*.) Scientists have calculated the atomic masses of all elements that have stable isotopes, and they can be found on any standard periodic table, including the table inside the front cover of this book.

6
C
12.011
Atomic mass

Note that no carbon atom has a mass of 12.011 u. This value is the weighted average mass of the carbon atoms found in nature. It leads to the following conversion factor for natural carbon.

$$\left(\frac{12.011 \text{ u C}}{1 \text{ C atom}}\right)$$

[2]The atomic mass without units is perhaps more accurately called the *relative atomic mass* (relative to the mass of carbon-12 as 12 u). We will take the common approach of calling both 12.011 u and 12.011 the atomic mass of carbon.

Although we can use the conversion factor shown above to convert between mass of carbon in atomic mass units and number of carbon atoms, let's wait to do this type of calculation until we take the next step: describing the number of atoms with a convenient collective unit, analogous to a dozen or a gross.

Just 1 gram of carbon has over $10^{22}$ carbon atoms. A dozen and a gross are both too small to be useful for conveniently describing this number of atoms. Thus chemists have created a special collective unit, called the mole, which is similar to but much greater than a dozen or a gross. A **mole** (which is abbreviated mol) is an amount of substance that contains the same number of particles as there are atoms in 12 g of carbon-12. To four significant figures, there are $6.022 \times 10^{23}$ atoms in 12 g of carbon-12. Thus a mole of natural carbon is the amount of carbon that contains $6.022 \times 10^{23}$ carbon atoms. The number $6.022 \times 10^{23}$ is often called **Avogadro's number.**

OBJECTIVE 2

The mole is used in very much the same way as we use the collective units of trio and dozen. There are 3 items in 1 trio, as in 3 musicians in a jazz trio.

$$\left(\frac{3 \text{ musicians}}{1 \text{ jazz trio}}\right) \quad \text{or} \quad \left(\frac{3 \text{ anything}}{1 \text{ trio of anything}}\right)$$

There are 12 items in 1 dozen, as in 12 eggs in a dozen eggs.

$$\left(\frac{12 \text{ eggs}}{1 \text{ dozen eggs}}\right) \quad \text{or} \quad \left(\frac{12 \text{ anything}}{1 \text{ dozen anything}}\right)$$

There are $6.022 \times 10^{23}$ items in 1 mole, as in $6.022 \times 10^{23}$ carbon-12 atoms in a mole of carbon-12.

$$\left(\frac{6.022 \times 10^{23}\ {}^{12}_{6}\text{C atoms}}{1 \text{ mol } {}^{12}_{6}\text{C}}\right) \quad \text{or} \quad \left(\frac{6.022 \times 10^{23} \text{ anything}}{1 \text{ mol anything}}\right)$$

Avogadro's number is unimaginably huge. For example, even though a carbon atom is extremely small, if you were to arrange the atoms contained in 12 grams (1 mole or $6.022 \times 10^{23}$ atoms) of carbon in a straight line, the string of atoms would stretch over 300 times the average distance from the earth to the sun (Figure 9.1).

According to the definition of the mole, 1 mol of carbon-12 has a mass of 12 g, so the following conversion factor could be used to convert between mass of carbon-12 and moles of carbon-12.

$$\left(\frac{12 \text{ g C-12}}{1 \text{ mol C-12}}\right)$$

Unfortunately, natural carbon always contains carbon-13 atoms as well as carbon-12 atoms, so the conversion factor we just found is not very useful. We need a conversion factor that relates mass and moles of natural carbon instead. Because the average mass of the atoms in natural carbon (12.011 u) is slightly greater than the mass of each carbon-12 atom (12 u), a

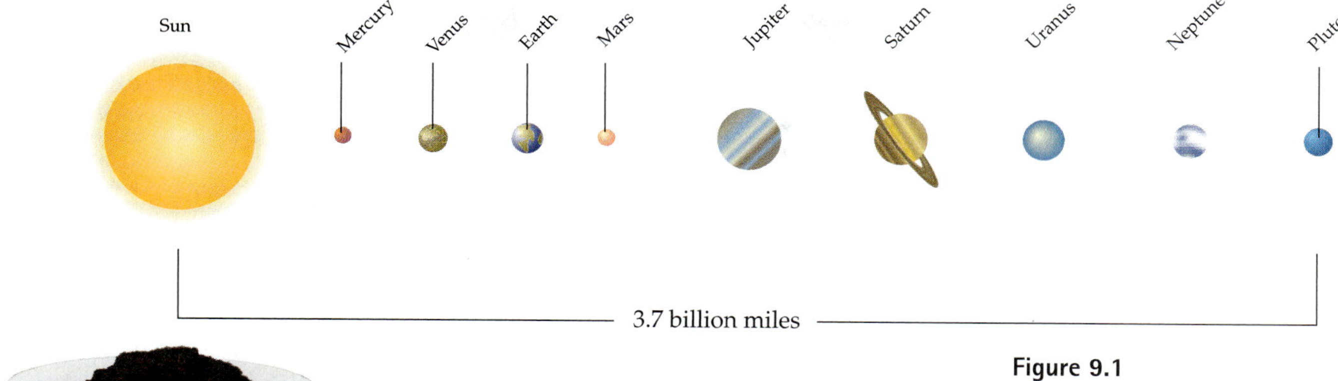


**Figure 9.1**
**Avogadro's Number**
If the extremely tiny atoms in just 12 grams of carbon were arranged in a line, the line would extend over 300 times the distance between the earth and the sun.

mole of natural carbon atoms has a mass slightly greater than the mass of a mole of carbon-12 atoms (12.011 g compared to 12 g). (Which weighs more, a dozen medium eggs or a dozen jumbo eggs?) The following conversion factor can be used to convert between mass of natural carbon and number of moles of carbon atoms.

$$\left(\frac{12.011\text{ g C}}{1\text{ mol C}}\right)$$

## Example 9.1 Converting to Moles

The masses of diamonds and other gemstones are measured in carats. There are exactly 5 carats per gram. How many moles of carbon atoms are in a 0.55-carat diamond? (Assume that the diamond is pure carbon.)

***Solution***

$$?\text{ mol C} = 0.55\ \cancel{\text{carat}}\ \cancel{\text{C}}\left(\frac{1\ \cancel{\text{g}}}{5\ \cancel{\text{carat}}}\right)\left(\frac{1\text{ mol C}}{12.011\ \cancel{\text{g}}\ \cancel{\text{C}}}\right) = 9.2 \times 10^{-3}\text{ mol C}$$

A 0.55-carat diamond contains over $10^{21}$ carbon atoms.

Note that just as we counted the nails by weighing them, we have also developed a method of counting carbon atoms by weighing. For the nails, the technique was merely convenient. (If we counted one nail per second, it would take over 7 hours to count 26,000 nails, but we could do it if we wanted to take the time.) For the carbon atoms, however, we have accomplished what would otherwise have been an impossible task. Even if we had the manual dexterity to pick up one carbon atom at a time, it would take us about $10^{14}$ centuries to count the atoms in the diamond described in Example 9.1.

## Molar Mass

The mass in grams of 1 mole of substance is called **molar mass.** Each element has its own unique molar mass. For example, carbon's molar mass is 12.011 g/mol, and magnesium's molar mass is 24.3050 g/mol. To see why these elements have different molar masses, we need only remember that the atoms of different elements contain different numbers of protons, neutrons, and electrons. The atomic masses given in the periodic table inside the front cover of this book represent the different weighted average masses of the naturally occurring atoms of each element. Different atomic masses lead to different molar masses.

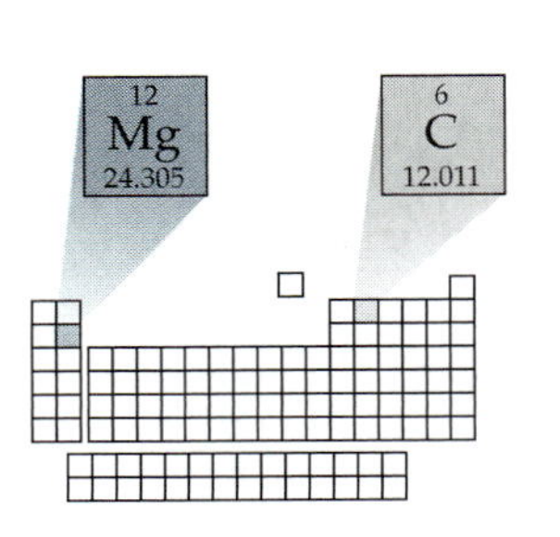


For example, the atomic mass of magnesium (24.305) shows us that the average mass of magnesium atoms is about twice the average mass of carbon atoms (12.011), so the mass of $6.022 \times 10^{23}$ magnesium atoms (the number of atoms in 1 mole of magnesium) is about twice the mass of $6.022 \times 10^{23}$ carbon atoms (the number of atoms in 1 mole of carbon). Thus the molar mass of magnesium is 24.305 g/mol, compared to carbon's molar mass of 12.011 g/mol.

The number of grams in the molar mass of an element is the same as the atomic mass. Translating atomic masses into molar masses, you can construct conversion factors that convert between the mass of an element and the number of moles of the element.

OBJECTIVE 3

$$\textbf{Molar mass of an element} = \left(\frac{\text{(atomic mass from periodic table) g element}}{\text{1 mol element}}\right)$$

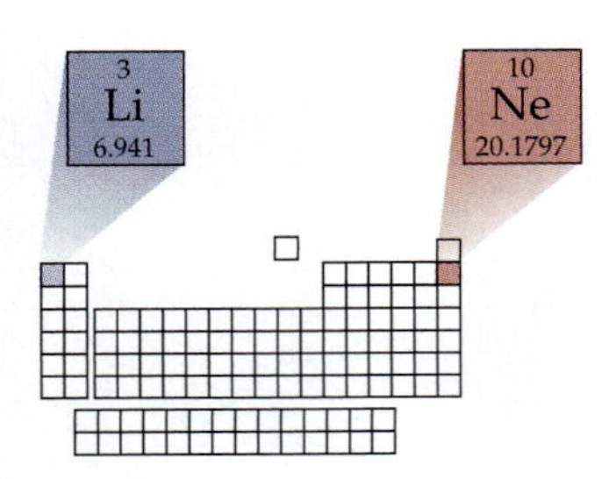


For example, the atomic mass of the element neon listed in the periodic table inside the front cover of this book is 20.1797, so neon has a molar mass of 20.1797 g/mol. This measurement provides the following conversion factors for converting between grams and moles of neon.

$$\frac{20.1797 \text{ g Ne}}{1 \text{ mol Ne}} \quad \text{or} \quad \frac{1 \text{ mol Ne}}{20.1797 \text{ g Ne}}$$

Lithium's atomic mass is 6.941, so the conversion factors for converting between mass and moles of lithium are

$$\frac{6.941 \text{ g Li}}{1 \text{ mol Li}} \quad \text{or} \quad \frac{1 \text{ mol Li}}{6.941 \text{ g Li}}$$

Example 9.2 shows how an atomic mass translates into a molar mass that enables us to convert between mass of an element and number of moles of that element.

## EXAMPLE 9.2 Atomic Mass Calculations

OBJECTIVE 3

OBJECTIVE 4

The element boron is used as a neutron absorber in nuclear reactors. It is also used to make semiconductors and rocket propellants.

a. Write the molar mass of boron as a conversion factor that can be used to convert between grams of boron and moles of boron.

b. Calculate the mass in kilograms of 219.9 moles of boron.

c. Calculate how many moles of boron are in 0.1532 lb B.

***Solution***

a. The molar mass of an element comes from its atomic mass. The atomic mass of boron can be found in the periodic table inside the front cover of this text. It is 10.811. The atomic mass of any element tells you the number of grams of that element per mole.

$$\left(\frac{\mathbf{10.811\ g\ B}}{\mathbf{1\ mol\ B}}\right)$$

c. $? \text{ kg B} = 219.9 \cancel{\text{mol B}} \left(\frac{10.811 \cancel{\text{g}} \text{ B}}{1 \cancel{\text{mol B}}}\right)\left(\frac{1 \text{ kg}}{10^3 \cancel{\text{g}}}\right) = \mathbf{2.377\ kg\ B}$

d. $? \text{ mol B} = 0.1532 \cancel{\text{lb B}} \left(\frac{453.6 \cancel{\text{g}}}{1 \cancel{\text{lb}}}\right)\left(\frac{1 \text{ mol B}}{10.811 \cancel{\text{g B}}}\right) = \mathbf{6.428\ mol\ B}$

## EXERCISE 9.1 Atomic Mass Calculations

OBJECTIVE 3

OBJECTIVE 4

Gold is often sold in units of troy ounces. There are 31.10 grams per troy ounce.

a. What is the atomic mass of gold?

b. What is the mass in grams of $6.022 \times 10^{23}$ gold atoms?

c. Write the molar mass of gold as a conversion factor that can be used to convert between grams of gold and moles of gold.

d. What is the mass in grams of 0.20443 mole of gold?

e. What is the mass in milligrams of $7.046 \times 10^{-3}$ mole of gold?

f. How many moles of gold are in 1.00 troy ounce of pure gold?

Exercise 9.1 asks you to calculate the number of moles of gold, assuming that this 1-troy-ounce sample is pure gold.

## 9.3 Molar Mass and Chemical Compounds

In Section 9.2, you learned how to calculate the number of atoms—expressed in moles of atoms—in a sample of an element. Molecular substances are composed of molecules, and in this section you will learn how to calculate the number of molecules, expressed in moles of molecules, in a sample of a molecular substance. Remember that, like *dozen*, the collective unit *mole* can be used to describe the number of anything. There are $6.022 \times 10^{23}$ atoms in a mole of carbon atoms, there are $6.022 \times 10^{23}$ electrons in a mole of electrons, and there are $6.022 \times 10^{23}$ $H_2O$ molecules in a mole of water.

## Molecular Mass and Molar Mass of Molecular Compounds

Because counting individual molecules is as impossible as counting individual atoms, we need to develop a way of converting back and forth between the number of moles of molecules in a sample of a molecular compound and the mass of the sample. To develop the tools necessary for the conversion between numbers of atoms and mass for elements, we had to determine the atomic mass of each element, which is the weighted average mass of the element's naturally occurring *atoms*. Likewise, for molecular compounds, our first step is to determine the **molecular mass** of the compound, which is the weighted average mass of the compound's naturally occurring *molecules*. This is found by adding the atomic masses of the atoms in each molecule.

OBJECTIVE 5

**Molecular mass** = the sum of the atomic masses of each atom in the molecule

Therefore, the molecular mass of water, $H_2O$, is equal to the sum of the atomic masses of two hydrogen atoms and one oxygen atom, which can be found in the periodic table.

$$\text{Molecular mass } H_2O = 2(1.00794) + 15.9994 = 18.0153$$

Note that the atomic mass of each element is multiplied by the number of atoms of that element in a molecule of the compound.

The number of grams in the molar mass (grams per mole) of a molecular compound is the same as its molecular mass.

OBJECTIVE 6

$$\textbf{Molar mass of a molecular compound} = \left(\frac{\text{(molecular mass) g compound}}{\text{1 mol compound}}\right)$$

$$\text{or, for water, } \left(\frac{18.0153 \text{ g } H_2O}{1 \text{ mol } H_2O}\right)$$

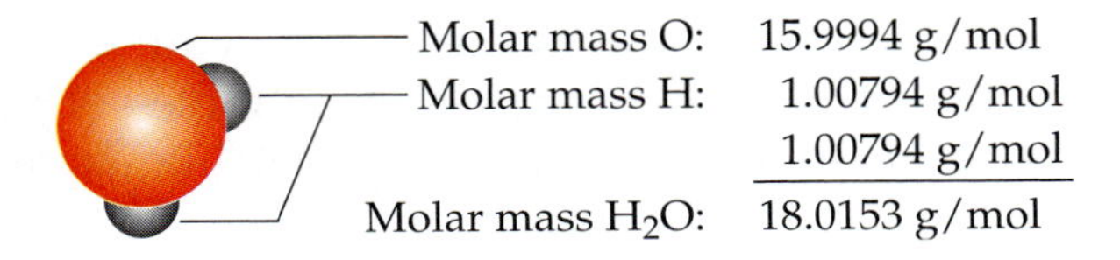

Example 9.3 shows how a molecular mass translates into a molar mass that enables us to convert between mass of a molecular compound and number of moles of that compound.

## EXAMPLE 9.3 Molecular Mass Calculations

OBJECTIVE 5 OBJECTIVE 6 OBJECTIVE 7

We have seen that the molecular compound tetraphosphorus decoxide, $P_4O_{10}$, is one of the substances needed for the production of phosphoric acid.

a. Write a conversion factor to convert between grams of $P_4O_{10}$ and moles of $P_4O_{10}$ molecules.

b. What is the mass in kilograms of $8.80 \times 10^4$ moles of tetraphosphorus decoxide, $P_4O_{10}$?

***Solution***

a. The molar mass of a molecular compound, such as $P_4O_{10}$, provides a conversion factor that converts back and forth between grams and moles of compound. This molar mass comes from the compound's molecular mass, which is the sum of the atomic masses of the atoms in a molecule. The atomic masses of phosphorus and oxygen are found in the periodic table. The atomic mass of each element is multiplied by the number of atoms of that element in a molecule of the compound.

$$\begin{aligned}\text{Molecular mass of } P_4O_{10} &= 4(\text{atomic mass P}) + 10(\text{atomic mass O})\\ &= 4(30.9738) + 10(15.9994)\\ &= 123.895 + 159.994 = 283.889 \text{ or } 283.889 \text{ u}\end{aligned}$$

Thus there are 283.889 g $P_4O_{10}$ in 1 mole of $P_4O_{10}$.

$$\left(\frac{\mathbf{283.889\ g\ P_4O_{10}}}{\mathbf{1\ mol\ P_4O_{10}}}\right)$$

b. $$? \text{ kg } P_4O_{10} = 8.80 \times 10^4 \cancel{\text{mol } P_4O_{10}} \left(\frac{283.889 \cancel{\text{g}}\ P_4O_{10}}{1 \cancel{\text{mol } P_4O_{10}}}\right)\left(\frac{1 \text{ kg}}{10^3 \cancel{\text{g}}}\right) = \mathbf{2.50 \times 10^4\ kg\ P_4O_{10}}$$

## EXERCISE 9.2 Molecular Mass Calculations

OBJECTIVE 5 OBJECTIVE 6 OBJECTIVE 7

A typical glass of wine contains about 16 g of ethanol, $C_2H_5OH$.

a. What is the molecular mass of $C_2H_5OH$?

b. What is the mass of 1 mole of $C_2H_5OH$?

c. Write a conversion factor that will convert between mass and moles of $C_2H_5OH$.

d. How many moles of ethanol are in 16 grams of $C_2H_5OH$?

e. What is the volume in milliliters of 1.0 mole of pure $C_2H_5OH$? (The density of ethanol is 0.7893 g/mL.)

## Ionic Compounds, Formula Units, and Formula Mass

The chemist also needs to be able to convert between mass and moles for ionic compounds. The calculations are the same as for molecular compounds, but some of the terminology is different. Remember that solid ionic compounds and molecular compounds differ in the way their particles are organized and held together (Figure 9.2). Water, a molecular substance, is composed of discrete $H_2O$ molecules, each of which contains 2 hydrogen atoms and 1 oxygen atom. The ionic compound sodium chloride, NaCl, does not contain separate molecules. Its particles form a continuous lattice, with each sodium ion surrounded by 6 chloride ions, and each chloride ion surrounded by 6 sodium ions. There are no sodium–chlorine atom pairs that belong together, separate from the other parts of the crystal. For this reason, we avoid using the term *molecule* when referring to an ionic compound. (You might want to review Sections 3.3 and 3.5, which describe the structure of molecular and ionic compounds.)

OBJECTIVE 8

It is still useful to have a term analogous to *molecule* to use when describing the composition of ionic compounds. In this text, the term *formula unit* will be used to describe ionic compounds in situations where *molecule* is used to describe molecular substances. A **formula unit** of a substance is the group represented by the substance's chemical formula—that is, a group containing the kinds and numbers of atoms or ions listed in the chemical formula. *Formula unit* is a general term that can be used in reference to elements, molecular compounds, or ionic compounds. One formula unit of the noble gas neon, Ne, contains one neon atom. In this case, the formula unit is an atom. One formula unit of water, $H_2O$, contains two hydrogen atoms and one oxygen atom. In this case, the formula unit is a molecule. One formula unit of ammonium chloride, $NH_4Cl$, contains one ammonium ion and one chloride ion—or one nitrogen atom, four hydrogen atoms, and one chloride ion (Figure 9.2).

OBJECTIVE 9

This photograph shows 1 mole of carbon in the graphite form at right, 1 mole of water, and 1 mole of sodium chloride, NaCl, at left. What is the mass of each of these samples?

A sample of an element is often described in terms of the number of atoms it contains, a sample of a molecular substance can be described in terms of the number of molecules it contains, and a sample of an ionic compound is often described in terms of the number of formula units it contains. Thus 1 mole of carbon contains $6.022 \times 10^{23}$ carbon atoms, 1 mole of water contains $6.022 \times 10^{23}$ water molecules, and 1 mole of sodium chloride contains $6.022 \times 10^{23}$ NaCl formula units.

Just as we saw with mass and moles of elements and molecular compounds, it is important to be able to convert between mass and moles of ionic substances. The development of the tools for this conversion starts with the determination of the **formula mass,** which is the weighted average of the masses of the naturally occurring formula units of the substance. (It is analogous to the atomic mass for an element and to the molecular mass for a molecular substance.)

OBJECTIVE 10

**Formula mass** = the sum of the atomic masses of all the atoms in a formula unit

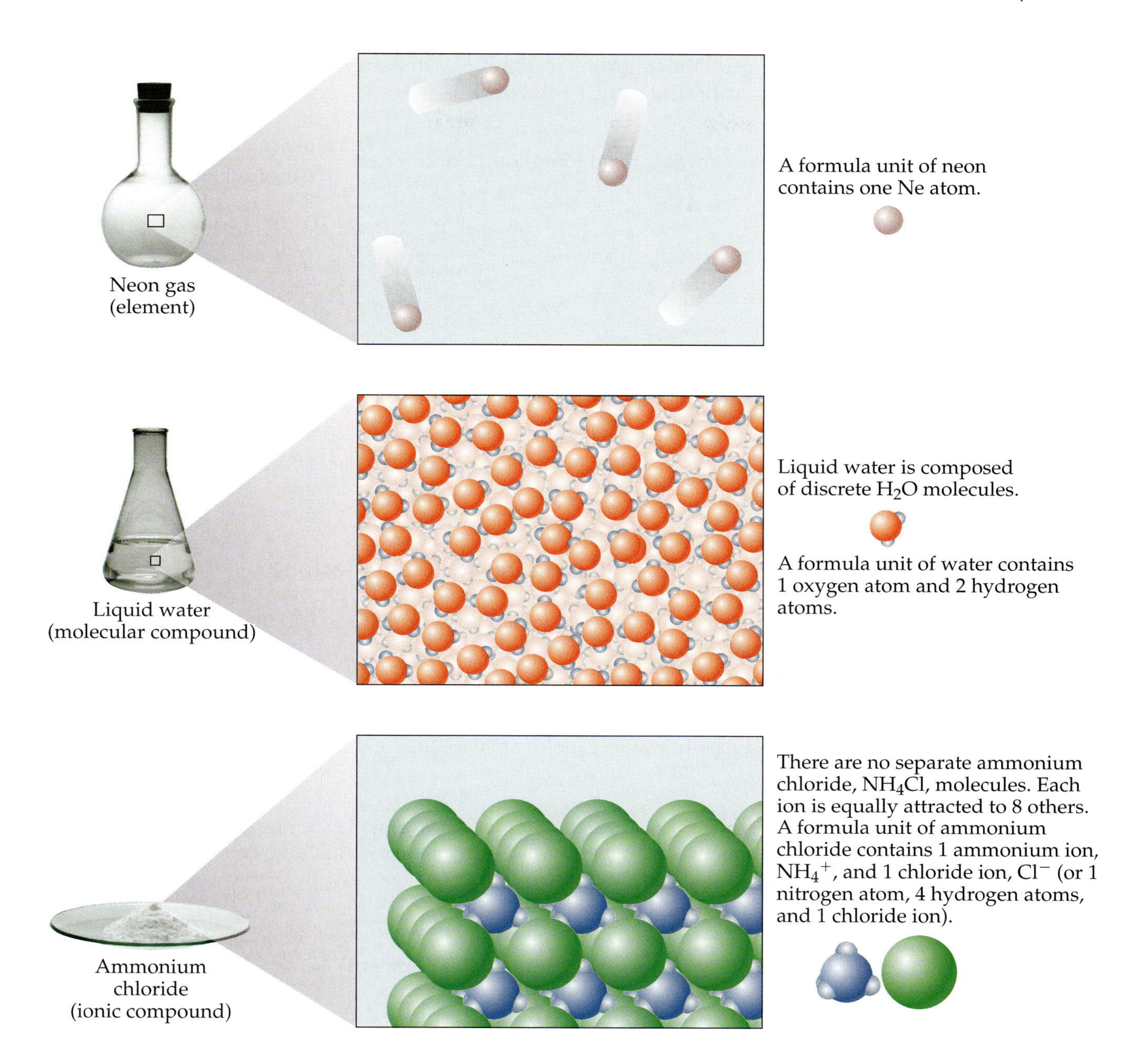

Neon gas
(element)
A formula unit of neon contains one Ne atom.
Liquid water
(molecular compound)
Liquid water is composed of discrete $H_2O$ molecules.
A formula unit of water contains 1 oxygen atom and 2 hydrogen atoms.
Ammonium chloride
(ionic compound)
There are no separate ammonium chloride, $NH_4Cl$, molecules. Each ion is equally attracted to 8 others. A formula unit of ammonium chloride contains 1 ammonium ion, $NH_4^+$, and 1 chloride ion, $Cl^-$ (or 1 nitrogen atom, 4 hydrogen atoms, and 1 chloride ion).


**Figure 9.2**
**Atoms, Molecules, and Formula Units**

OBJECTIVE 9

The formula mass of sodium chloride is equal to the sum of the atomic masses of sodium and chlorine, which can be found in the periodic table.

$$\textbf{Formula mass NaCl} = 22.9898 + 35.4527 = 58.4425$$

OBJECTIVE 11

*Formula mass*, like *formula unit*, is a general term. The atomic mass of carbon, C, could also be called its formula mass. The molecular mass of water, $H_2O$, could also be called water's formula mass.

The number of grams in the molar mass (grams per mole) of any ionic compound is the same as its formula mass.

OBJECTIVE 11

$$\text{Molar mass of an ionic compound} = \left(\frac{\text{(formula mass) g substance}}{\text{1 mol substance}}\right)$$

$$\text{or, for sodium chloride, } \left(\frac{\text{58.4425 g NaCl}}{\text{1 mol NaCl}}\right)$$

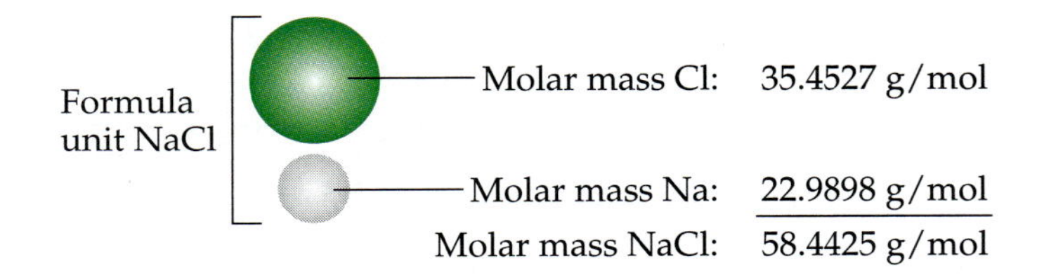


## EXAMPLE 9.4 Formula Mass Calculations

OBJECTIVE 10 OBJECTIVE 11 OBJECTIVE 12

Water from coal mines is contaminated with sulfuric acid that forms when water reacts with iron(III) sulfate, $Fe_2(SO_4)_3$.

a. Write a conversion factor to convert between grams of the ionic compound iron(III) sulfate and moles of $Fe_2(SO_4)_3$ formula units.

b. Calculate how many moles of $Fe_2(SO_4)_3$ are in 2.672 lb of $Fe_2(SO_4)_3$.

***Solution***

a. The molar mass, which provides a conversion factor for converting back and forth between grams and moles of an ionic compound, comes from the compound's formula mass.

Formula mass of $Fe_2(SO_4)_3$

$$= 2(\text{atomic mass Fe}) + 3(\text{atomic mass S}) + 12(\text{atomic mass O})$$

$$= 2(55.845) + 3(32.066) + 12(15.9994)$$

$$= 111.69 + 96.198 + 191.993 = 399.88 \text{ or } 399.88 \text{ u}$$

Thus the conversion factor that converts back and forth between grams and moles of iron(III) sulfate is

$$\left(\frac{\mathbf{399.88\ g\ Fe_2(SO_4)_3}}{\mathbf{1\ mol\ Fe_2(SO_4)_3}}\right)$$

b. $$? \text{ mol } Fe_2(SO_4)_3 = 2.672 \cancel{\text{lb } Fe_2(SO_4)_3}\left(\frac{453.6 \cancel{\text{g}}}{1 \cancel{\text{lb}}}\right)\left(\frac{1 \text{ mol } Fe_2(SO_4)_3}{399.88 \cancel{\text{g } Fe_2(SO_4)_3}}\right) = \mathbf{3.031\ mol\ Fe_2(SO_4)_3}$$

## Exercise 9.3 Formula Mass Calculations

OBJECTIVE 10 OBJECTIVE 11 OBJECTIVE 12

A quarter teaspoon of a typical baking powder contains about 0.4 g of sodium hydrogen carbonate, $NaHCO_3$.

a. Calculate the formula mass of sodium hydrogen carbonate.

b. What is the mass in grams of 1 mole of $NaHCO_3$?

c. Write a conversion factor to convert between mass and moles of $NaHCO_3$.

d. How many moles of $NaHCO_3$ are in 0.4 g of $NaHCO_3$?

## 9.4 Relationships Between Masses of Elements and Compounds

Many conversions you will be doing in chemistry require you to convert amount of one substance (substance 1) into amount of another substance (substance 2). Such conversions can be done in three basic steps:

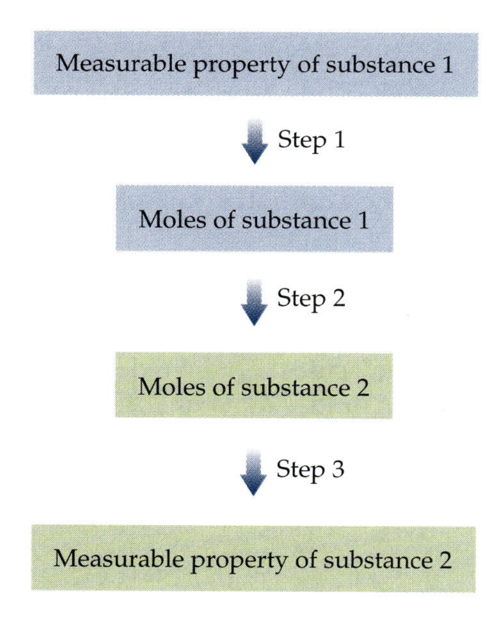


Mass is often the most easily measured property, and we now know how to convert between mass in grams and moles of a substance using the substance's molar mass. The solutions to many problems will therefore follow these steps:

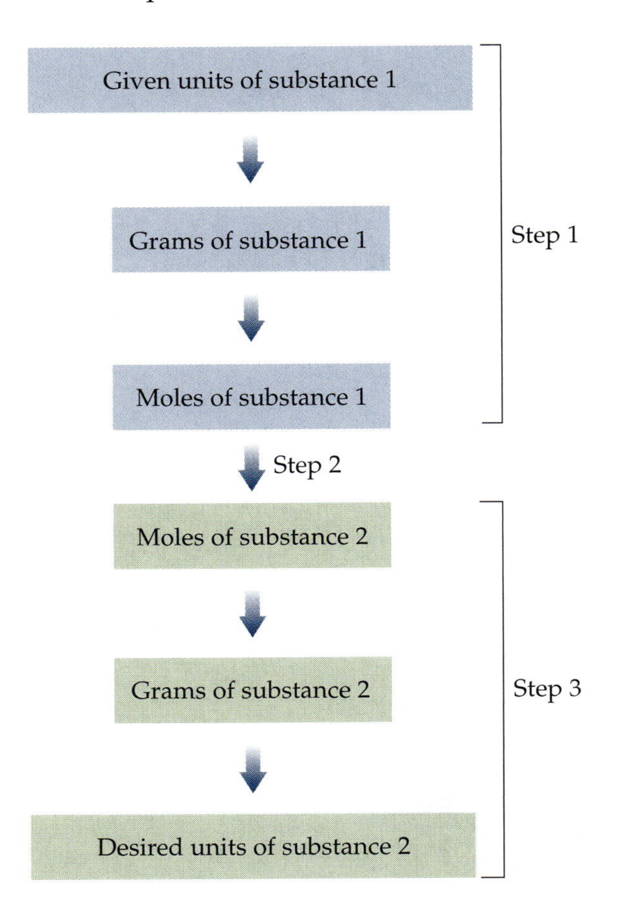


To complete these steps, we need one additional kind of conversion factor that converts between moles of an element and moles of a compound containing that element. We obtain this conversion factor from the compound's chemical formula. For example, the formula for hexane, $C_6H_{14}$, tells us that each hexane molecule contains 6 atoms of carbon and 14 atoms of hydrogen. A dozen $C_6H_{14}$ molecules contain 6 dozen atoms of carbon and 14 dozen atoms of hydrogen, and 1 mole of $C_6H_{14}$ contains 6 moles of carbon atoms and 14 moles of hydrogen atoms. These relationships lead to the following conversion factors:

OBJECTIVE 13

$$\frac{6 \text{ mol C}}{1 \text{ mol } C_6H_{14}} \quad \text{and} \quad \frac{14 \text{ mol H}}{1 \text{ mol } C_6H_{14}}$$

One mole of the oxygen found in air, $O_2$, contains $6.022 \times 10^{23}$ molecules and $1.204 \times 10^{24}$ (2 times $6.022 \times 10^{23}$) oxygen atoms. There are 2 moles of oxygen atoms in 1 mole of oxygen molecules.

$$\frac{2 \text{ mol O}}{1 \text{ mol } O_2}$$

Similarly, we can use ionic formulas to generate conversion factors that convert between moles of atoms of each element in an ionic compound and moles of compound. For example, the formula for calcium nitrate, $Ca(NO_3)_2$, yields the following conversion factors:

$$\frac{1 \text{ mol Ca}}{1 \text{ mol } Ca(NO_3)_2} \quad \text{and} \quad \frac{2 \text{ mol N}}{1 \text{ mol } Ca(NO_3)_2} \quad \text{and} \quad \frac{6 \text{ mol O}}{1 \text{ mol } Ca(NO_3)_2}$$

The collective unit mole can also be used to describe ions. Thus the following conversion factors also come from the formula of calcium nitrate, $Ca(NO_3)_2$.

$$\frac{1 \text{ mol } Ca^{2+} \text{ ions}}{1 \text{ mol } Ca(NO_3)_2} \quad \text{and} \quad \frac{2 \text{ mol } NO_3^{-} \text{ ions}}{1 \text{ mol } Ca(NO_3)_2}$$

## EXAMPLE 9.5 Molar Ratios of Element to Compound

OBJECTIVE 13

OBJECTIVE 14

Consider the molecular compound tetraphosphorus decoxide, $P_4O_{10}$, and the ionic compound iron(III) sulfate, $Fe_2(SO_4)_3$.

a. Write a conversion factor that converts between moles of phosphorus and moles of $P_4O_{10}$.

b. Write a conversion factor that converts between moles of iron and moles of iron(III) sulfate.

c. How many moles of sulfate are in 1 mole of iron(III) sulfate?

***Solution***

a. The formula $P_4O_{10}$ shows that there are 4 atoms of phosphorus per molecule of $P_4O_{10}$ and 4 moles of phosphorus per mole of $P_4O_{10}$.

$$\frac{\mathbf{4 \text{ mol P}}}{\mathbf{1 \text{ mol } P_4O_{10}}}$$

b. $$\frac{\mathbf{2 \text{ mol Fe}}}{\mathbf{1 \text{ mol } Fe_2(SO_4)_3}}$$

c. There are **3 moles of $SO_4^{2-}$** in 1 mole of $Fe_2(SO_4)_3$.

## EXERCISE 9.4 Molar Ratios of Element to Compound

OBJECTIVE 13

OBJECTIVE 14

Write the requested conversion factors.

a. Write a conversion factor that converts between moles of hydrogen and moles of $C_2H_5OH$.

b. Write a conversion factor that converts between moles of oxygen and moles of $NaHCO_3$.

c. How many moles of hydrogen carbonate ions, $HCO_3^-$, are in 1 mole of $NaHCO_3$?

We are now ready to work the "typical problem" presented in Section 9.1.

### EXAMPLE 9.6 Using Molar Mass in a Conversion

OBJECTIVE 15

What is the maximum mass of tetraphosphorus decoxide, $P_4O_{10}$, that could be produced from $1.09 \times 10^4$ kg of phosphorus, P?

***Solution***

The steps for this conversion are

$$1.09 \times 10^4 \text{ kg P} \rightarrow \text{g P} \rightarrow \text{mol P} \rightarrow \text{mol } P_4O_{10} \rightarrow \text{g } P_4O_{10} \rightarrow \text{kg } P_4O_{10}$$

Notice that these follow the general steps we have been discussing.

To find a conversion factor that converts from moles of phosphorus to moles of $P_4O_{10}$, we look at the formula for tetraphosphorus decoxide, $P_4O_{10}$. It shows that each molecule of tetraphosphorus decoxide contains 4 atoms of phosphorus. By extension, one dozen $P_4O_{10}$ molecules contains 4 dozen P atoms; and 1 mole of $P_4O_{10}$ ($6.022 \times 10^{23}$ $P_4O_{10}$ molecules) contains 4 moles of phosphorus (4 times $6.022 \times 10^{23}$ P atoms). Thus the formula $P_4O_{10}$ provides us with the following conversion factor:

$$\left(\frac{1 \text{ mol } P_4O_{10}}{4 \text{ mol P}}\right)$$

The molar mass of phosphorus can be used to convert grams of phosphorus to moles of phosphorus, and the molar mass of $P_4O_{10}$ can be used to convert moles of $P_4O_{10}$ to grams. Conversion of kilograms to grams and of conversion grams to kilograms complete our setup.

Converts given mass unit into grams

Converts moles of element into moles of compound

Converts grams into desired mass unit

$$? \text{ kg } P_4O_{10} = 1.09 \times 10^4 \text{ kg P} \left(\frac{10^3 \text{ g}}{1 \text{ kg}}\right)\left(\frac{1 \text{ mol P}}{30.9738 \text{ g P}}\right)\left(\frac{1 \text{ mol } P_4O_{10}}{4 \text{ mol P}}\right)\left(\frac{283.889 \text{ g } P_4O_{10}}{1 \text{ mol } P_4O_{10}}\right)\left(\frac{1 \text{ kg}}{10^3 \text{ g}}\right)$$

Converts grams of element into moles

Converts moles of compound into grams

$$= \mathbf{2.50 \times 10^4 \text{ kg } P_4O_{10}}$$

The following sample study sheet describes the general procedure for this approach to converting between the mass of an element and the mass of a compound containing the element. Remember that a given problem can often be solved in several different ways. This text may suggest one general approach to a certain type of problem, and your instructor may recommend another. You will need to decide which technique best fits your style. Thus *your* study sheet might be different from the sample provided here.

## Sample Study Sheet 9.1

### Converting Between Mass of Element and Mass of Compound Containing the Element

OBJECTIVE 15

**Tip-off** When you analyze the type of unit you have and the type of unit you want, you recognize that you are converting between a unit associated with an element and a unit associated with a compound containing that element.

**General Steps** The following general procedure is summarized in Figure 9.3.

- **Convert the given unit into moles of the first substance.**

  This step often requires converting the given unit into grams, after which the grams can be converted into moles using the molar mass of the substance.

- **Convert moles of the first substance into moles of the second substance using the molar ratio derived from the formula for the compound.**

  You convert either from moles of element into moles of compound or from moles of compound into moles of element.

- **Convert moles of the second substance into the desired units of the second substance.**

  This step requires converting moles of the second substance into grams of the second substance using the molar mass of the second substance, after which the grams can be converted into the specific units that you want.

**Example** See Example 9.6.

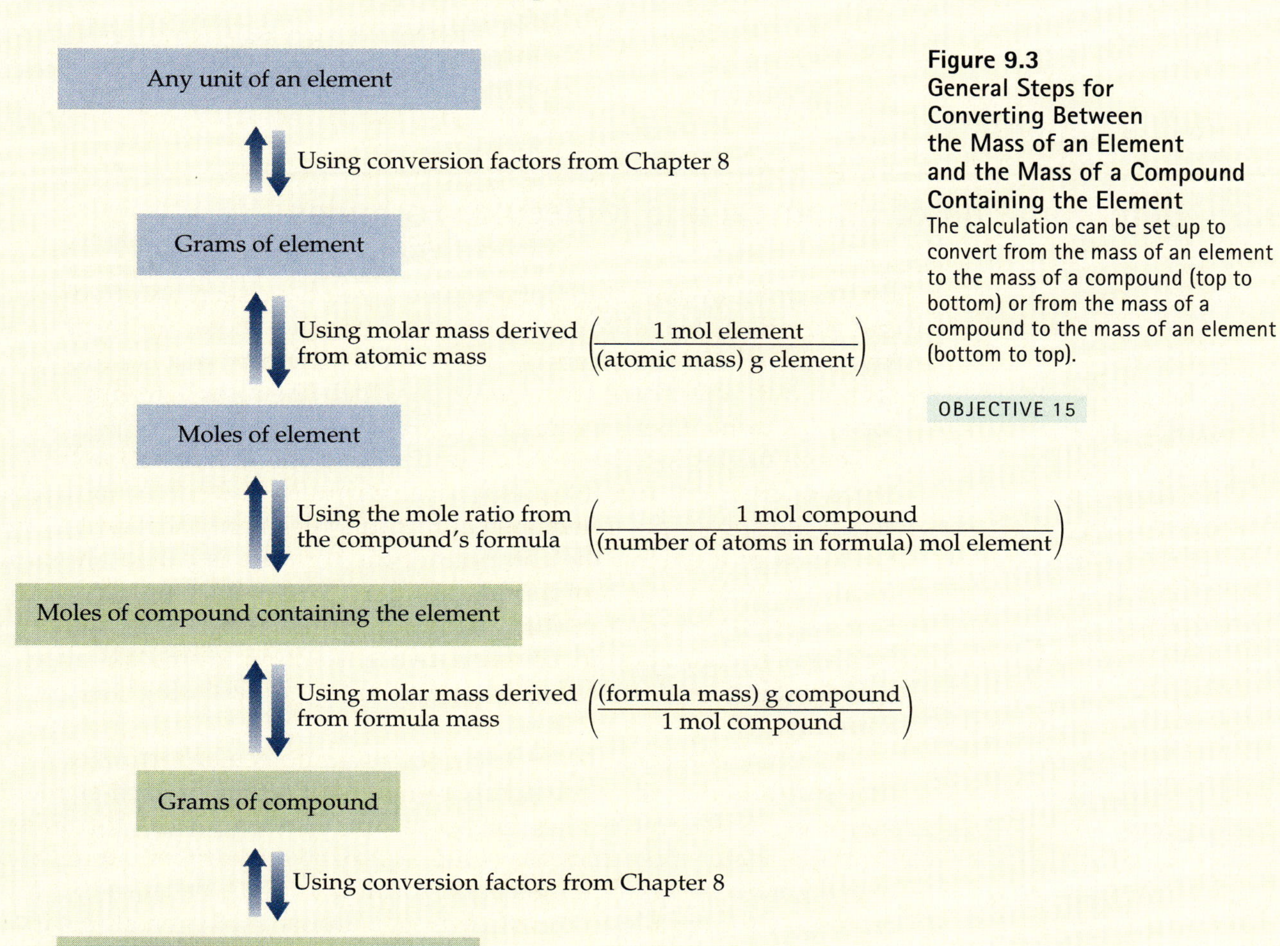


**Figure 9.3**
**General Steps for Converting Between the Mass of an Element and the Mass of a Compound Containing the Element**
The calculation can be set up to convert from the mass of an element to the mass of a compound (top to bottom) or from the mass of a compound to the mass of an element (bottom to top).

OBJECTIVE 15

### Exercise 9.5 Molar Mass Calculations

OBJECTIVE 15

Disulfur dichloride, $S_2Cl_2$, is used in vulcanizing rubber and in hardening soft woods. It can be made from the reaction of pure sulfur with chlorine gas. What is the mass of $S_2Cl_2$ that contains 123.8 g S?

### Exercise 9.6 Molar Mass Calculations

OBJECTIVE 15

Vanadium metal, which is used as a component of steel and to catalyze various industrial reactions, is produced from the reaction of vanadium(V) oxide, $V_2O_5$, and calcium metal. What is the mass (in kilograms of vanadium) in 2.3 kilograms of $V_2O_5$?

The skills developed in this chapter can be used to calculate the percentage of each element in a compound. To see how this is done, visit the following Web site: www.chemplace.com/college/

## 9.5 Determination of Empirical and Molecular Formulas

By this point in your study of chemistry, you have seen hundreds of chemical formulas. Have you wondered where they come from or how we know the relative numbers of atoms of each element in a compound? This section describes some of the ways in which chemists determine chemical formulas from experimental data.

Before beginning, we need to understand the distinction between two types of chemical formulas: empirical formulas and molecular formulas. When the subscripts in a chemical formula represent the simplest ratio of the kinds of atoms in the compound, the formula is called an **empirical formula.** Most ionic compounds are described with empirical formulas. For example, chromium(III) oxide's formula, $Cr_2O_3$, is an empirical formula. The compound contains two chromium atoms for every three oxide atoms, and there is no lower ratio representing these relative amounts.

Molecular compounds are described with molecular formulas. A **molecular formula** describes the actual numbers of atoms of each element in a molecule. Some molecular formulas are also empirical formulas. For example, water molecules are composed of two hydrogen atoms and one oxygen atom, so water's molecular formula is $H_2O$. Because this formula represents the simplest ratio of hydrogen atoms to oxygen atoms in water, it is also an empirical formula.

On the other hand, many molecular formulas are not empirical formulas. Hydrogen peroxide molecules contain two hydrogen atoms and two oxygen atoms, so hydrogen perioxide's molecular formula is $H_2O_2$. Its empirical formula, however, is HO. The molecular formula for glucose is $C_6H_{12}O_6$, and its empirical formula is $CH_2O$.

Hydrogen peroxide
molecular formula, $H_2O_2$,
empirical formula, HO

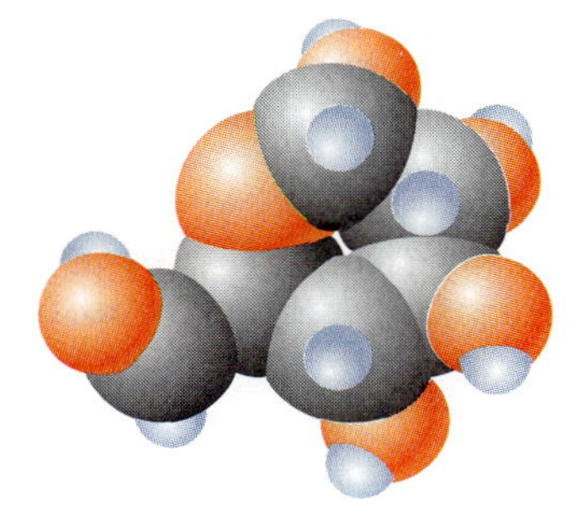

Glucose
molecular formula, $C_6H_{12}O_6$,
empirical formula, $CH_2O$

## Determining Empirical Formulas

If we know the relative mass or the mass percentage of each element in a compound, we can determine the compound's empirical formula. The general procedure is summarized in Study Sheet 9.2, but first let's reason it out using a substance that is sometimes called photophor, an ingredient in signal fires, torpedoes, fireworks, and rodent poison.

The subscripts in an empirical formula are positive integers that represent the simplest ratio of the atoms of each element in the formula. For now, we can describe the empirical formula for photophor in the following way:

$Ca_aP_b$ where *a* and *b* represent positive integers

(Remember that in binary compounds containing a metal and a nonmetal, the symbol for the metal is listed first.) The *a* and *b* in this formula represent the subscripts in the empirical formula. For every *a* atoms of calcium, there are *b* atoms of phosphorus, and for every *a* moles of Ca, there are *b* moles of P. Thus the ratio of *a* to *b* describes the molar ratio of these elements in the compound.

If we can determine the number of moles of each element in any amount of a substance, we can calculate the molar ratio of these elements in the compound, which can then be simplified to the positive integers that represent the simplest molar ratio. We found in Sections 9.2 and 9.3 that we can calculate moles of a substance from grams. Thus one path to determining the empirical formula for a compound is

OBJECTIVE 16

Grams of each element in a specific amount of compound

↓

Moles of each element in that amount of compound

↓

Molar ratio of the elements

↓

Simplest molar ratio of the elements

Imagine that a sample of photophor has been analyzed and found to contain 12.368 grams of calcium and 6.358 grams of phosphorus. These masses can be converted into moles using the molar masses of the elements.

$$? \text{ mol Ca} = 12.368 \text{ g Ca}\left(\frac{1 \text{ mol Ca}}{40.078 \text{ g Ca}}\right) = 0.30860 \text{ mol Ca}$$

$$? \text{ mol P} = 6.358 \text{ g P}\left(\frac{1 \text{ mol P}}{30.9738 \text{ g P}}\right) = 0.2053 \text{ mol P}$$

The molar ratio of Ca to P is therefore 0.30860 mole of calcium to 0.2053 mole of phosphorus. We next simplify it to the integers that represent the simplest molar ratio by performing the following steps.

- Divide each mole value by the smallest mole value and round the answer to the nearest positive integer or common mixed fraction. (A mixed fraction contains an integer and a fraction. For example, 2½ is a mixed fraction.)
- If one of the values is a fraction, then multiply all the values by the denominator of the fraction.

Calcium phosphide, often called photophor, is used to make signal flares.

In this case, we divide the mole values for Ca and P by 0.2053. This step always leads to 1 mole for the smallest value. We then round the other mole values to positive integers or common mixed fractions.

$$\frac{0.30860 \text{ mol Ca}}{0.2053} = 1.503 \text{ mol Ca} \approx 1\tfrac{1}{2} \text{ mol Ca}$$

$$\frac{0.2053 \text{ mol P}}{0.2053} = 1 \text{ mol P}$$

(The mole value for calcium has been rounded from 1.503 to 1½.)

We will restrict the examples in this text to compounds with relatively simple formulas. Thus, when you work the end-of-chapter problems, the values you will obtain at this stage will always be within 0.02 of a positive integer or a common mixed fraction, and the denominators for your fractions will be 4 or smaller.

To complete the determination of the empirical formula of photophor, we multiply each mole value by 2 to get rid of the fraction:

$$1\tfrac{1}{2} \text{ mol Ca} \times 2 = 3 \text{ mol Ca}$$

$$1 \text{ mol P} \times 2 = 2 \text{ mol P}$$

Our empirical formula is $Ca_3P_2$, which represents calcium phosphide. Sample Study Sheet 9.2 summarizes these steps.

## Sample Study Sheet 9.2

### Calculating Empirical Formulas

OBJECTIVE 16

OBJECTIVE 17

**Tip-off** You wish to calculate an empirical formula.

**General Procedure** The following procedure is summarized in Figure 9.4.

1. If you are not given the mass of each element in grams, convert the data you *are* given into the mass of each element in grams.
   - In some cases, this can be done with simple unit conversions. For example, you may be given pounds or milligrams, which can be converted to grams via dimensional analysis.
   - Sometimes you are given the percentage of each element in the compound. If so, assume that you have 100 g of compound, and change the percentages to grams (see Example 9.7).
2. Convert grams of each element into moles by dividing by the atomic mass of the element.
3. Divide each mole value by the smallest mole value, and round your answers to positive integers or common mixed fractions.
4. If you have a fraction after the last step, multiply all the mole values by the denominator of the fraction.
5. The resulting mole values represent the subscripts in the empirical formula.

**Example** See Example 9.7.

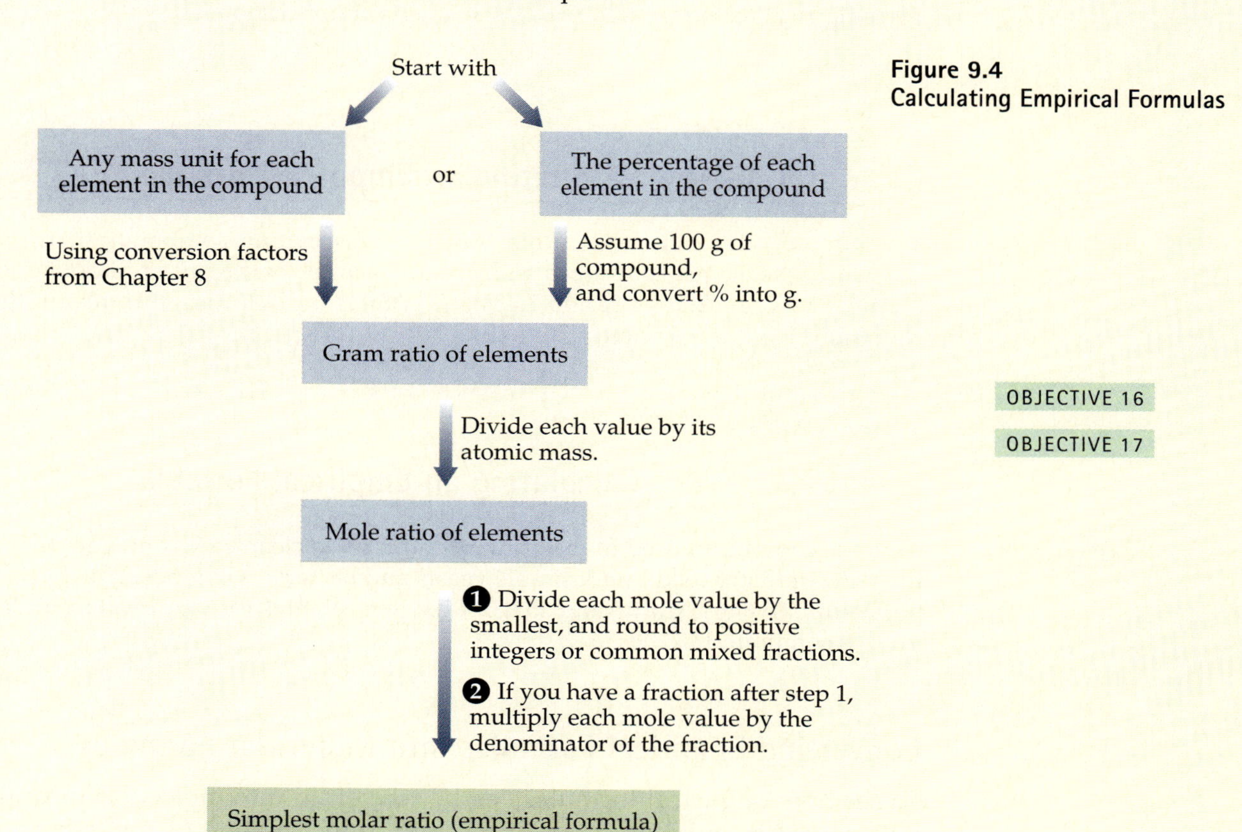
Start with
Any mass unit for each element in the compound
or
The percentage of each element in the compound
Using conversion factors from Chapter 8
Assume 100 g of compound, and convert % into g.
Gram ratio of elements
Divide each value by its atomic mass.
Mole ratio of elements
❶ Divide each mole value by the smallest, and round to positive integers or common mixed fractions.
❷ If you have a fraction after step 1, multiply each mole value by the denominator of the fraction.
Simplest molar ratio (empirical formula)

**Figure 9.4**
Calculating Empirical Formulas

OBJECTIVE 16

OBJECTIVE 17

### Example 9.7 Calculating an Empirical Formula from the Percentage of Each Element

OBJECTIVE 17

An ionic compound sometimes called pearl ash is used to make special glass for color TV tubes. A sample of this compound is analyzed and found to be 56.50% potassium, 8.75% carbon, and 34.75% oxygen. What is the empirical formula for this compound? What is its chemical name?

***Solution***
If we assume that we have a 100-g sample, the conversion from percentages to a gram ratio becomes very simple. We just change each "%" to "g". Thus 100 g of pearl ash would contain 56.50 g K, 8.75 g C, and 34.75 g O.

Now we can proceed with the steps described in Sample Study Sheet 9.2 to convert from grams of each element to the empirical formula.

$$? \text{ mol K} = 56.50 \cancel{\text{g K}} \left( \frac{1 \text{ mol K}}{39.098 \cancel{\text{g K}}} \right) = 1.445 \text{ mol K} \div 0.728 = 1.985 \text{ mol K} \cong 2 \text{ mol K}$$

$$? \text{ mol C} = 8.75 \cancel{\text{g C}} \left( \frac{1 \text{ mol C}}{12.011 \cancel{\text{g C}}} \right) = 0.728 \text{ mol C} \div 0.728 = 1 \text{ mol C}$$

$$? \text{ mol O} = 34.75 \cancel{\text{g O}} \left( \frac{1 \text{ mol O}}{15.9994 \cancel{\text{g O}}} \right) = 2.172 \text{ mol O} \div 0.728 \cong 2.984 \text{ mol O} \cong 3 \text{ mol O}$$

The empirical formula is $\mathbf{K_2CO_3}$, which is **potassium carbonate.**

### Exercise 9.7 Calculating an Empirical Formula

OBJECTIVE 16

Bismuth ore, often called bismuth glance, contains an ionic compound that consists of the elements bismuth and sulfur. A sample of the pure compound is found to contain 32.516 g Bi and 7.484 g S. What is the empirical formula for this compound? What is its name?

### Exercise 9.8 Calculating an Empirical Formula

OBJECTIVE 17

An ionic compound used in the brewing industry to clean casks and vats and in the wine industry to kill undesirable yeasts and bacteria is composed of 35.172% potassium, 28.846% sulfur, and 35.982% oxygen. What is the empirical formula for this compound?

## Converting Empirical Formulas Into Molecular Formulas

To see how empirical formulas can be converted into molecular formulas, let's consider an analysis of adipic acid, which is one of the substances used to make Nylon-66. (Nylon-66, originally used as a replacement for silk in women's stockings, was invented in 1935 and was produced commercially starting in 1940.) Experiment shows that adipic acid is 49.31% carbon, 6.90% hydrogen, and 43.79% oxygen, which converts into an empirical formula of $C_3H_5O_2$. From a separate experiment, the molecular mass of adipic acid is found to be 146.144. We can determine the molecular formula for adipic acid from these data.

The subscripts in a molecular formula are always whole-number multiples of the subscripts in the empirical formula. The molecular formula for adipic acid can therefore be described in the following way:

$$C_{3n}H_{5n}O_{2n} \qquad n = \text{some positive integer, such as } 1, 2, 3, \ldots$$

The $n$ can be calculated from adipic acid's molecular mass and its empirical formula mass. A substance's empirical formula mass can be calculated from the subscripts in its empirical formula and the atomic masses of the elements. The empirical formula mass of adipic acid is

$$\text{Empirical formula mass} = 3(12.011) + 5(1.00794) + 2(15.9994)$$

$$= 73.072$$

Because the subscripts in the molecular formula are always a positive-integer multiple of the subscripts in the empirical formula, the molecular mass is always equal to a positive-integer multiple of the empirical formula mass.

$$\text{Molecular mass} = n\ (\text{empirical formula mass})$$

$$n = \text{some positive integer, such as } 1, 2, 3, \ldots$$

Therefore, we can calculate $n$ by dividing the molecular mass by the empirical formula mass.

$$n = \frac{\text{molecular mass}}{\text{empirical formula mass}}$$

For our example, the given molecular mass of $C_3H_5O_2$ divided by the empirical formula mass is

$$n = \frac{\text{molecular mass}}{\text{empirical formula mass}} = \frac{146.144}{73.072} = 2$$

Once we have calculated $n$, we can determine the molecular formula by multiplying each of the subscripts in the empirical formula by $n$. This gives a molecular formula for adipic acid of $C_{3(2)}H_{5(2)}O_{2(2)}$, or $C_6H_{10}O_4$ (see Special Topic 9.1).

Can nylon be made from sugar? See Special Topic 9.1.

## SPECIAL TOPIC 9.1 Green Chemistry— Making Chemicals from Safer Reactants

Because benzene, $C_6H_6$, is a readily available and inexpensive substance that can be easily converted into many different substances, it is a common industrial starting material for making a wide range of important chemicals. A problem with using benzene, however, is that it is known to cause cancer and to be toxic in other ways as well. Thus there is strong incentive to find alternative methods for making chemicals that have traditionally been made from benzene.

You discovered in this section that adipic acid is used to make nylon, but it is also used to make paint, synthetic lubricants, and plasticizers (substances that make plastics more flexible). Adipic acid is one of the important industrial chemicals conventionally produced from benzene, but recently a new process has been developed that forms adipic acid from the much safer, simple sugar glucose instead. If you were a worker in a chemical plant making adipic acid, you would certainly prefer working with sugar to working with benzene!

## SAMPLE STUDY SHEET 9.3 Calculating Molecular Formulas

OBJECTIVE 18

**TIP-OFF** You want to calculate a molecular formula and have been given the molecular mass of the substance and either the empirical formula or enough data to calculate the empirical formula.

**GENERAL PROCEDURE** The following procedure is summarized in Figure 9.5.

- If necessary, calculate the empirical formula of the compound from the data given. (See Sample Study Sheet 9.2.)
- Divide the molecular mass by the empirical formula mass.

$$n = \frac{\text{molecular mass}}{\text{empirical formula mass}}$$

- Multiply each of the subscripts in the empirical formula by $n$ to get the molecular formula.

**EXAMPLE** See Example 9.8.

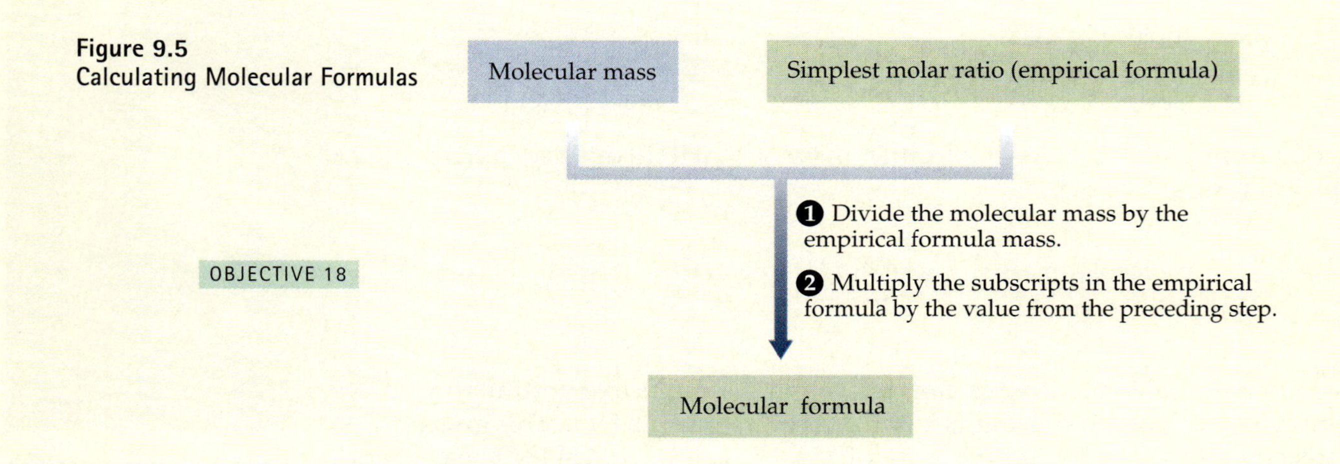
Molecular mass
Simplest molar ratio (empirical formula)
1 Divide the molecular mass by the empirical formula mass.
2 Multiply the subscripts in the empirical formula by the value from the preceding step.
Molecular formula

**Figure 9.5**
Calculating Molecular Formulas

OBJECTIVE 18

## EXAMPLE 9.8 Calculating a Molecular Formula Using the Percentage of Each Element in a Compound

OBJECTIVE 18

A chemical called BD (or sometimes BDO), which is used in the synthesis of Spandex, has controversial uses as well. In 1999 it was added to various products that their manufacturers claimed would stimulate the body's immune system, reduce tension, heighten sexual experience, repair muscle tissue, and cure insomnia (see Special Topic 9.2: *Safe and Effective?*). The FDA seized these products because of suspicions that BD caused at least three deaths and many severe adverse reactions. BD is 53.31% carbon, 11.18% hydrogen, and 35.51% oxygen and has a molecular mass of 90.122. What is its molecular formula?

***Solution***

$$? \text{ mol C} = 53.31 \cancel{\text{g C}} \left( \frac{1 \text{ mol C}}{12.011 \cancel{\text{g C}}} \right) = 4.438 \text{ mol C} \div 2.219 = 2 \text{ mol C}$$

$$? \text{ mol H} = 11.18 \cancel{\text{g H}} \left( \frac{1 \text{ mol H}}{1.00794 \cancel{\text{g H}}} \right) = 11.09 \text{ mol H} \div 2.219 \cong 5 \text{ mol H}$$

$$? \text{ mol O} = 35.51 \cancel{\text{g O}} \left( \frac{1 \text{ mol O}}{15.9994 \cancel{\text{g O}}} \right) = 2.219 \text{ mol O} \div 2.219 = 1 \text{ mol O}$$

$$\text{Empirical formula: } C_2H_5O \quad n = \frac{\text{molecular mass}}{\text{empirical formula mass}} = \frac{90.122}{45.061} = 2$$

Molecular formula: $\mathbf{C_4H_{10}O_2}$

## EXERCISE 9.9 Calculating a Molecular Formula Using the Percentage of Each Element in a Compound

OBJECTIVE 18

Compounds called polychlorinated biphenyls (PCBs) have structures similar to chlorinated insecticides such as DDT. They have been used in the past for a variety of purposes, but because they have been identified as serious pollutants, their only legal use today is as insulating fluids in electrical transformers. This is a use for which no suitable substitute has been found. One PCB is 39.94% carbon, 1.12% hydrogen, and 58.94% chlorine and has a molecular mass of 360.88. What is its molecular formula?

Empirical and molecular formulas can be derived from a process called combustion analysis. To learn about this process, visit the following Web site:

www.chemplace.com/college/

## Special Topic 9.2 Safe and Effective?

The public is continually bombarded with claims for new products containing "miracle" ingredients "guaranteed" to improve our health, strength, happiness, and sex life and to give us a good night's sleep. In 1999 such claims were made on behalf of products with names like Revitalize Plus, Serenity, Enliven, and Thunder Nectar, all of which contained a substance called 1,4-butanediol (BD). They were marketed on the Internet, sold in health food stores, and advertised in muscle-building magazines.

The claims were enticing, but according to the U.S. Food and Drug Administration (FDA), they were also unfounded. Perhaps more important, the FDA decided that this unapproved new drug could cause dangerously low respiratory rates, unconsciousness, seizures, and even death. Because the FDA connected BD with at least three deaths and many severe adverse reactions, it was designated a Class I health hazard, which means that "its use could pose potentially life-threatening risks." The FDA seized products containing BD to prevent their causing further illness and death. Advocates of the drug submitted anecdotal evidence to show that many people have taken the substance without ill effects. As of this writing, no one knows which side is right.

Because humans are very complex chemical factories, the positive and negative effects of a drug can be difficult to determine. To some extent, they depend on each person's unique biochemistry, as well as on interactions with other chemicals that may be present in the body. For example, the effects of BD are thought to be enhanced by alcohol and other depressants, so an individual takes one of the "party drugs" containing BD (drugs with names like Cherry fX Bombs) and drinks too much beer, the combined depressant effect can lead to loss of consciousness, coma, and even death. To minimize the uncertainties associated with individual reactions to a drug, scientists run carefully controlled tests, first on animals and only much later on humans. Until these tests are done for BD, its true effects (both positive and negative) cannot be known with confidence.

How do you decide whether to consume a product containing a chemical like BD? Let's consider a student named Fred who is surfing the Internet for ideas about how to relax before his final exams. One site that he finds describes a product guaranteed to calm his nerves and recharge his immune system. The product seems to contain a lot of good ingredients, such as vitamins and minerals, but the most important component is tetramethylene glycol. The description of how the substance works is written in unfamiliar terminology, but Fred thinks he gets the gist, and it seems to make sense. Fortunately, he passes up the opportunity to buy this product and takes a walk in the woods to calm his nerves instead. He has never heard of the product's distributor, he knows that false and unproven advertising claims are often cloaked in pseudoscientific explanations, and he remembers reading in his chemistry book that tetramethylene glycol is another name for 1,4-butanediol (BD), a potentially dangerous substance.

The important points in this story are that it is best to stick to products from known manufacturers who have a reputation for carefully screening their ingredients; to be skeptical of claims made in advertising and on the Internet; and to keep yourself educated about substances that are suspected of being harmful. When in doubt, ask your doctor. It's part of his or her job to know about the safety and effectiveness of health products.

## Chapter Glossary

**Weighted average mass** A mass calculated by multiplying the decimal fraction of each component in a sample by its mass and adding the results of these multiplications together.

**Atomic mass unit** One-twelfth of the mass of a carbon-12 atom. It is sometimes called a ***unified mass unit.*** Its accepted abbreviation is ***u,*** but ***amu*** is sometimes used.

**Atomic mass** The weighted average of the masses of the naturally occurring isotopes of an element.

**Mole** The amount of substance that contains the same number of particles as there are atoms in 12 g of carbon-12.

**Avogadro's number** The number of atoms in 12 g of carbon-12. To four significant figures, it is $\boldsymbol{6.022 \times 10^{23}}$.

**Molar mass** The mass in grams of 1 mole of substance. (The number of grams in the molar mass of an element is the same as its atomic mass. The number of grams in the molar mass of a molecular compound is the same as its molecular mass. The number of grams in the molar mass of an ionic compound is the same as its formula mass.)

**Molecular mass** The weighted average of the masses of the naturally occurring molecules of a molecular substance. It is the sum of the atomic masses of the atoms in a molecule.

**Formula unit** A group represented by a substance's chemical formula—that is, a group containing the kinds and numbers of atoms or ions listed in the chemical formula. This general term can be used in reference to elements, molecular compounds, or ionic compounds.

**Formula mass** The weighted average of the masses of the naturally occurring formula units of the substance. It is the sum of the atomic masses of the atoms in a formula unit.

**Empirical formula** A chemical formula that includes positive integers that describe the simplest ratio of the kinds of atoms in a compound.

**Molecular formula** The chemical formula that describes the actual numbers of atoms of each element in a molecule of a compound.

You can test yourself on the glossary terms at the following Web site:

**www.chemplace.com/college/**

## Chapter Objectives

**The goal of this chapter is to teach you to do the following.**

1. Define all of the terms in the Chapter 9 Glossary.

### Section 9.2 Relating Mass to Number of Particles

2. Describe how a mole is similar to a dozen.
3. Given an atomic mass for an element, write a conversion factor that converts between mass and moles of that element.

4. Given a periodic table that shows atomic masses of the elements, convert between mass of an element and moles of that element.

#### Section 9.3 Molar Mass and Chemical Compounds

5. Given a formula for a molecular substance and a periodic table that includes atomic masses for the elements, calculate the substance's molecular mass.
6. Given enough information to calculate a molecular substance's molecular mass, write a conversion factor that converts between mass and moles of the substance.
7. Given enough information to calculate a molecular substance's molecular mass, convert between mass and moles of the substance.
8. Explain why the term *molecule* is better applied to molecular substances, such as water, than to ionic compounds, such as NaCl.
9. Given the name or chemical formula of a compound, identify whether it would be better to use the term *molecule* or the term *formula unit* to describe the group that contains the number of atoms or ions of each element that is equal to the subscript for that element in its chemical formula.
10. Given a formula for an ionic compound and a periodic table that includes atomic masses for the elements, calculate the compound's formula mass.
11. Given enough information to calculate an ionic compound's formula mass, write a conversion factor that converts between mass and moles of the compound.
12. Given enough information to calculate an ionic compound's formula mass, convert between mass and moles of the compound.

#### Section 9.4 Relationships Between Masses of Elements and Compounds

13. Given a formula for a compound, write conversion factors that convert between moles of atoms of each element in the compound and moles of compound.
14. Given the name or formula for an ionic compound, determine the number of moles of each ion in one mole of the compound.
15. Make conversions between the mass of a compound and the mass of an element in the compound.

#### Section 9.5 Determination of Empirical and Molecular Formulas

16. Given the masses of each element in a sample of a compound, calculate the compound's empirical formula.
17. Given the mass percentage of each element in a compound, calculate the compound's empirical formula.
18. Given an empirical formula for a molecular compound (or enough information to calculate the empirical formula) and given the molecular mass of the compound, determine its molecular formula.

## Review Questions

1. Complete each of the following conversion factors by filling in the blank on the top of the ratio.

   a. $\left(\dfrac{\_\_\_\_\_\_\ \text{g}}{1\ \text{kg}}\right)$

   b. $\left(\dfrac{\_\_\_\_\_\_\ \text{mg}}{1\ \text{g}}\right)$

   c. $\left(\dfrac{\_\_\_\_\_\_\ \text{kg}}{1\ \text{metric ton}}\right)$

   d. $\left(\dfrac{\_\_\_\_\_\_\ \mu\text{g}}{1\ \text{g}}\right)$

2. Convert $3.45 \times 10^4$ kg into grams.
3. Convert 184.570 g into kilograms.
4. Convert $4.5000 \times 10^6$ g into megagrams.
5. Convert 871 Mg into grams.
6. Surinam bauxite is an ore that is 54–57% aluminum oxide, $Al_2O_3$. What is the mass (in kilograms) of $Al_2O_3$ in 1256 kg of Surinam bauxite that is 55.3% $Al_2O_3$?

## Key Ideas

**Complete the following statements by writing one of these words or phrases in each blank.**

| | |
|---|---|
| atomic mass | molecules |
| atoms | mole |
| atoms of each element | one-twelfth |
| formula | naturally |
| formula unit | numbers |
| formula units | simplest |
| impossible | weighted |
| kinds | whole-number |

7. Because of the size and number of carbon atoms in any normal sample of carbon, it is ____________ to count the atoms directly.
8. The unit most often used to describe the mass of atoms is the atomic mass unit, whose symbol is u or amu. An atomic mass unit is defined as exactly ____________ the mass of an atom of carbon-12.
9. The atomic mass of any element is the ____________ average of the masses of the ____________ occurring isotopes of the element.
10. A mole (which is abbreviated mol) is an amount of substance that contains the same number of particles as there are ____________ in 12 g of carbon-12.
11. The number of grams in the molar mass of an element is the same as the element's ____________.
12. The molecular mass of the compound is the weighted average mass of the compound's naturally occurring ____________.

13. In this text, the term ____________ is used to describe ionic compounds in situations where molecule is used to describe molecular substances. It is the group represented by the substance's chemical formula, that is, a group containing the ____________ and ____________ of atoms or ions listed in the chemical formula.
14. One ____________ of carbon contains 6.022 x $10^{23}$ carbon atoms.
15. Formula mass is the weighted average of the masses of the naturally occurring ____________ of the substance.
16. We can convert between moles of element and moles of a compound containing that element by using the molar ratio derived from the ____________ for the compound.
17. When the subscripts in a chemical formula represent the ____________ ratio of the kinds of atoms in the compound, the formula is called an empirical formula.
18. A molecular formula describes the actual numbers of ____________ in a molecule.
19. The subscripts in a molecular formula are always ____________ multiples of the subscripts in the empirical formula.

## Chapter Problems

### Section 9.2 Relating Mass to Number of Particles

OBJECTIVE 2

20. Describe how a mole is similar to a dozen.
21. What is the weighted average mass in atomic mass units (u) of each atom of the following elements?
    a. sodium b. oxygen
22. What is the weighted average mass in atomic mass units (u) of each atom of the following elements?
    a. calcium b. neon
23. What is the weighted average mass in grams of 6.022 x $10^{23}$ atoms of the following elements?
    a. sulfur b. fluorine
24. What is the weighted average mass in grams of 6.022 x $10^{23}$ atoms of the following elements?
    a. bromine b. nickel
25. What is the molar mass for each of the following elements?
    a. zinc b. aluminum
26. What is the molar mass for each of the following elements?
    a. chlorine b. silver

OBJECTIVE 3

27. For each of the following elements, write a conversion factor that converts between mass in grams and moles of the substance.
    a. iron b. krypton

OBJECTIVE 3

28. For each of the following elements, write a conversion factor that converts between mass in grams and moles of the substance.
    a. manganese b. silicon

29. A vitamin supplement contains 50 micrograms of the element selenium in each tablet. How many moles of selenium does each tablet contain? OBJECTIVE 4

30. A multivitamin tablet contains 40 milligrams of potassium. How many moles of potassium does each tablet contain? OBJECTIVE 4

31. A multivitamin tablet contains $1.6 \times 10^{-4}$ mole of iron per tablet. How many milligrams of iron does each tablet contain? OBJECTIVE 4

32. A multivitamin tablet contains $1.93 \times 10^{-6}$ mole of chromium. How many micrograms of chromium does each tablet contain? OBJECTIVE 4

### Section 9.3 Molar Mass and Chemical Compounds

33. For each of the following molecular substances, calculate its molecular mass and write a conversion factor that converts between mass in grams and moles of the substance. OBJECTIVE 5 OBJECTIVE 6

    a. $H_3PO_2$  b. $C_6H_5NH_2$

34. For each of the following molecular substances, calculate its molecular mass and write a conversion factor that converts between mass in grams and moles of the substance. OBJECTIVE 5 OBJECTIVE 6

    a. $CF_3CHCl_2$  b. $SO_2Cl_2$

35. Each dose of a nighttime cold medicine contains 1000 mg of the analgesic acetaminophen. Acetaminophen, or N-acetyl-p-aminophenol, has the general formula $C_8H_9NO$. OBJECTIVE 7

    a. How many moles of acetaminophen are in each dose?
    b. What is the mass in grams of 15.0 moles of acetaminophen?

36. A throat lozenge contains 5.0 mg of menthol, which has the formula $C_{10}H_{20}O$. OBJECTIVE 7

    a. How many moles of menthol are in 5.0 mg of menthol?
    b. What is the mass in grams of 1.56 moles of menthol?

37. A group of atoms that contains one atom of nitrogen and three atoms of hydrogen is called an ammonia molecule. A group of ions that contains one potassium ion and one fluoride ion is called a potassium fluoride formula unit instead of a molecule. Why? OBJECTIVE 8

38. For each of the following examples, decide whether it would be better to use the term *molecule* or *formula unit.* OBJECTIVE 9

    a. $Cl_2O$
    b. $Na_2O$
    c. $(NH_4)_2SO_4$
    d. $HC_2H_3O_2$

39. For each of the following examples, decide whether it would be better to use the term *molecule* or *formula unit.* OBJECTIVE 9

    a. $K_2SO_3$
    b. $H_2SO_3$
    c. $CCl_4$
    d. $NH_4Cl$

OBJECTIVE 10 OBJECTIVE 11

**40.** For each of the following ionic substances, calculate its formula mass and write a conversion factor that converts between mass in grams and moles of the substance.

a. $BiBr_3$ b. $Al_2(SO_4)_3$

OBJECTIVE 10 OBJECTIVE 11

41. For each of the following ionic substances, calculate its formula mass and write a conversion factor that converts between mass in grams and moles of the substance.

a. $Co_2O_3$ b. $Fe_2(C_2O_4)_3$

OBJECTIVE 12

**42.** A common antacid tablet contains 500 mg of calcium carbonate, $CaCO_3$.

a. How many moles of $CaCO_3$ does each tablet contain?

b. What is the mass in kilograms of 100.0 mole of calcium carbonate?

OBJECTIVE 12

43. An antacid contains 200 mg of aluminum hydroxide and 200 mg of magnesium hydroxide per capsule.

a. How many moles of $Al(OH)_3$ does each capsule contain?

b. What is the mass in milligrams of 0.0457 mole of magnesium hydroxide?

OBJECTIVE 12

**44.** Rubies and other minerals in the durable corundum family are primarily composed of aluminum oxide, $Al_2O_3$, with trace impurities that lead to their different colors. For example, the red color in rubies comes from a small amount of chromium replacing some of the aluminum. If a 0.78-carat ruby were pure aluminum oxide, how many moles of $Al_2O_3$ would be in the stone? (There are exactly 5 carats per gram.)

OBJECTIVE 12

45. Many famous "rubies" are in fact spinels, which look like rubies but are far less valuable. Spinels consist primarily of $MgAl_2O_4$, whereas rubies are primarily $Al_2O_3$. If the Timur Ruby, a 361-carat spinel, were pure $MgAl_2O_4$, how many moles of $MgAl_2O_4$ would it contain? (There are exactly 5 carats per gram.)

### Section 9.4 Relationships Between Masses of Elements and Compounds

OBJECTIVE 13

**46.** Write a conversion factor that converts between moles of nitrogen in nitrogen pentoxide, $N_2O_5$, and moles of $N_2O_5$.

OBJECTIVE 13

47. Write a conversion factor that converts between moles of oxygen in phosphoric acid, $H_3PO_4$, and moles of $H_3PO_4$.

OBJECTIVE 13 OBJECTIVE 14

**48.** The green granules on older asphalt roofing are chromium(III) oxide. Write a conversion factor that converts between moles of chromium atoms in chromium(III) oxide, $Cr_2O_3$, and moles of $Cr_2O_3$.

OBJECTIVE 13 OBJECTIVE 14

49. Calcium phosphide is used to make fireworks. Write a conversion factor that converts between moles of calcium ions in calcium phosphide, $Ca_3P_2$, and moles of $Ca_3P_2$.

OBJECTIVE 14

**50.** Ammonium oxalate is used for stain and rust removal. How many moles of ammonium ions are in 1 mole of ammonium oxalate, $(NH_4)_2C_2O_4$?

OBJECTIVE 14

51. Magnesium phosphate is used as a dental polishing agent. How many moles of ions (cations and anions together) are in 1 mole of magnesium phosphate, $Mg_3(PO_4)_2$?

52. A nutritional supplement contains 0.405 g of $CaCO_3$. The recommended daily value of calcium is 1.000 g Ca. OBJECTIVE 13 OBJECTIVE 15

 a. Write a conversion factor that relates moles of calcium to moles of calcium carbonate.

 b. Calculate the mass in grams of calcium in 0.405 g of $CaCO_3$.

 c. What percentage of the daily value of calcium comes from this tablet?

53. A multivitamin tablet contains 0.479 g of $CaHPO_4$ as a source of phosphorus. The recommended daily value of phosphorus is 1.000 g of P. OBJECTIVE 13 OBJECTIVE 15

 a. Write a conversion factor that relates moles of phosphorus to moles of calcium hydrogen phosphate.

 b. Calculate the mass in grams of phosphorus in 0.479 g of $CaHPO_4$.

 c. What percentage of the daily value of phosphorus comes from this tablet?

54. A multivitamin tablet contains 10 μg of vanadium in the form of sodium metavanadate, $NaVO_3$. How many micrograms of $NaVO_3$ does each tablet contain? OBJECTIVE 15

55. A multivitamin tablet contains 5 μg of nickel in the form of nickel(II) sulfate. How many micrograms of $NiSO_4$ does each tablet contain? OBJECTIVE 15

56. There are several natural sources of the element titanium. One is the ore called rutile, which contains oxides of iron and titanium, FeO and $TiO_2$. Titanium metal can be made by first converting the $TiO_2$ in rutile to $TiCl_4$ by heating the ore to high temperature in the presence of carbon and chlorine. The titanium in $TiCl_4$ is then reduced from its +4 oxidation state to its zero oxidation state by reaction with a good reducing agent such as magnesium or sodium. What is the mass of titanium, in kilograms, in 0.401 Mg of $TiCl_4$? OBJECTIVE 15

57. Manganese metal is produced from the manganese(III) oxide, $Mn_2O_3$, which is found in manganite, a manganese ore. The manganese is reduced from its +3 oxidation state in $Mn_2O_3$ to the zero oxidation state of the uncharged metal by reacting the $Mn_2O_3$ with a reducing agent such as aluminum or carbon. How many pounds of manganese are in 1.261 tons of $Mn_2O_3$? (1 ton = 2000 pounds) OBJECTIVE 15

### Section 9.5 Determination of Empirical and Molecular Formulas

58. Explain the difference between molecular formulas and empirical formulas. Give an example of a substance whose empirical formula is different from its molecular formula. Give an example of a substance whose empirical formula and molecular formula are the same.

59. An extremely explosive ionic compound is made from the reaction of silver compounds with ammonia. A sample of this compound is found to contain 17.261 g of silver and 0.743 g of nitrogen. What is the empirical formula for this compound? What is its chemical name? OBJECTIVE 16

60. A sample of an ionic compound that is often used as a dough conditioner is analyzed and found to contain 7.591 g of potassium, 15.514 g of bromine, and 9.319 g of oxygen. What is the empirical formula for this compound? What is its chemical name? OBJECTIVE 16

OBJECTIVE 16

61. A sample of a compound used to polish dentures and as a nutrient and dietary supplement is analyzed and found to contain 9.2402 g of calcium, 7.2183 g of phosphorus, and 13.0512 g of oxygen. What is the empirical formula for this compound?

OBJECTIVE 16

62. A sample of an ionic compound that is used in the semiconductor industry is analyzed and found to contain 53.625 g of indium and 89.375 g of tellurium. What is the empirical formula for this compound?

OBJECTIVE 17

63. An ionic compound that is 38.791% nickel, 33.011% arsenic, and 28.198% oxygen is employed as a catalyst for hardening fats used to make soap. What is the empirical formula for this compound?

OBJECTIVE 17

64. An ionic compound sometimes called TKPP is used as a soap and detergent builder. It is composed of 47.343% potassium, 18.753% phosphorus, and 33.904% oxygen. What is the empirical formula for this compound?

OBJECTIVE 17

65. An ionic compound that contains 10.279% calcium, 65.099% iodine, and 24.622% oxygen is used in deodorants and in mouthwashes. What is the empirical formula for this compound?

OBJECTIVE 17

66. An ionic compound that is 62.56% lead, 8.46% nitrogen, and 28.98% oxygen is used as a mordant in the dyeing industry. A mordant helps to bind a dye to the fabric. What is the empirical formula for this compound? What do you think its name is? (Consider the possibility that this compound contains more than one polyatomic ion.)

OBJECTIVE 18

67. In 1989 a controversy arose concerning the chemical daminozide, or Alar®, which was sprayed on apple trees to yield redder, firmer, and more shapely apples. Concerns about Alar's safety stemmed from the suspicion that one of its breakdown products, unsymmetrical dimethylhydrazine (UDMH), was carcinogenic. Alar is no longer sold for food uses. UDMH has the empirical formula of $CNH_4$ and has a molecular mass of 60.099. What is the molecular formula for UDMH?

OBJECTIVE 18

68. It would be advisable for the smokers of the world to consider that nicotine was once used as an insecticide but is no longer used for that purpose because of safety concerns. Nicotine has an empirical formula of $C_5H_7N$ and a molecular mass of 162.23. What is the molecular formula for nicotine?

OBJECTIVE 18

69. Lindane is one of the chlorinated pesticides the use of which is now restricted in the United States. It is 24.78% carbon, 2.08% hydrogen, and 73.14% chlorine and has a molecular mass of 290.830. What is lindane's molecular formula?

OBJECTIVE 18

70. Hydralazine is a drug used to treat heart disease. It is 59.99% carbon, 5.03% hydrogen, and 34.98% nitrogen and has a molecular mass of 160.178. What is the molecular formula for hydralazine?

OBJECTIVE 18

71. Melamine is a compound used to make the melamine–formaldehyde resins in very hard surface materials such as Formica®. It is 28.57% carbon, 4.80% hydrogen, and 66.63% nitrogen and has a molecular mass of 126.121. What is melamine's molecular formula?

OBJECTIVE 18

72. About 40 different compounds called organophosphorus compounds are registered in the United States as insecticides. They are considered less damaging to the environment than some other insecticides because of the relative rapidity of their breakdown in the environment. The first of these organophosphorus insecticides to be produced was tetraethyl pyrophosphate, TEPP, which is 33.11% carbon, 6.95% hydrogen, 38.59% oxygen, and 21.35% phosphorus and has a molecular mass of 290.190. What is the molecular formula for TEPP?

### Additional Problems

73. Your boss at the hardware store points you to a bin of screws and asks you to find out the approximate number of screws it contains. You weigh the screws and find that their total mass is 68 pounds. You take out 100 screws and weigh them individually, and you find that 7 screws weigh 2.65 g, 4 screws weigh 2.75 g, and 89 screws weigh 2.90 g. Calculate the weighted average mass of each screw. How many screws are in the bin? How many gross of screws are in the bin?

74. Atomic masses are derived from calculations using experimental data. As the experiments that provide this data get more precise, the data get more accurate, and the atomic masses values reported on the periodic table are revised. One source of data from 1964 reports that the element potassium is 93.10% potassium-39, which has atoms with a mass of 38.963714 u (atomic mass units), 0.0118% potassium-40, which has atoms with a mass of 39.964008 u, and 6.88% potassium-41, which has atoms with a mass of 40.961835 u. Using this data, calculate the weighted average mass of potassium atoms, in atomic mass units. Report your answer to the fourth decimal position. The weighted average mass of potassium atoms is potassium's atomic mass. How does your calculated value compare to potassium's reported atomic mass on the periodic table in this text?

75. As a member of the corundum family of minerals, sapphire (the September birthstone) consists primarily of aluminum oxide, $Al_2O_3$. Small amounts of iron and titanium give it its rich dark blue color. Gem cutter Norman Maness carved a giant sapphire into the likeness of Abraham Lincoln. If this 2302-carat sapphire were pure aluminum oxide, how many moles of $Al_2O_3$ would it contain? (There are exactly 5 carats per gram.)

76. Emeralds are members of the beryl family, which are silicates of beryllium and aluminum with a general formula of $Be_3Al_2(SiO_3)_6$. The emerald's green color comes from small amounts of chromium in the crystal. The Viennese treasury has a cut emerald that weighs 2205 carats. If this stone were pure $Be_3Al_2(SiO_3)_6$, how many moles of $Be_3Al_2(SiO_3)_6$ would it contain? (There are exactly 5 carats per gram.)

77. Aquamarine (the March birthstone) is a light blue member of the beryl family, which is made up of natural silicates of beryllium and aluminum that have the general formula $Be_3Al_2(SiO_3)_6$. Aquamarine's bluish color is caused by trace amounts of iron(II) ions. A 43-pound aquamarine mined in Brazil in 1910 remains the largest gem-quality crystal ever found. If this stone were pure $Be_3Al_2(SiO_3)_6$, how many moles of beryllium would it contain?

78. In 1985 benitoite became the California "state gemstone." Found only in a tiny mine near Coalinga, California, it is a silicate of barium and titanium with trace impurities that cause a range of hues from colorless to blue to pink. Its general formula is $BaTi(SiO_3)_3$. If a 15-carat stone were pure $BaTi(SiO_3)_3$, how many moles of silicon would it contain?

79. November's birthstone is citrine, a yellow member of the quartz family. It is primarily silicon dioxide, but small amounts of iron(III) ions give it its yellow color. A high-quality citrine containing about 0.040 moles of $SiO_2$ costs around $225. If this stone were pure $SiO_2$, how many carats would it weigh? (There are exactly 5 carats per gram.)

80. The gemstone tanzanite was first discovered in Tanzania in 1967. Like other gemstones, it contains impurities that give it distinct characteristics, but it is primarily $Ca_2Al_3Si_3O_{12}(OH)$. The largest tanzanite stone ever found contains about 0.0555 mole of $Ca_2Al_3Si_3O_{12}(OH)$. What is the mass of this stone in kilograms?

81. A common throat lozenge contains 29 mg of phenol, $C_6H_5OH$.
    a. How many moles of $C_6H_5OH$ are there in 5.0 mg of phenol?
    b. What is the mass in kilograms of 0.9265 mole of phenol?

82. Some forms of hematite, a mineral composed of iron(III) oxide, can be used to make jewelry. Because of its iron content, hematite jewelry has a unique problem among stone jewelry: It shows signs of rusting. How many moles of iron are there in a necklace that contains 78.435 g of $Fe_2O_3$?

83. Beryl, $Be_3Al_2(SiO_3)_6$, is a natural source of beryllium, a known carcinogen. What is the mass in kilograms of beryllium in 1.006 Mg of $Be_3Al_2(SiO_3)_6$?

84. The element antimony is used to harden lead for use in lead–acid storage batteries. One of the principal antimony ores is stibnite, which contains antimony in the form of $Sb_2S_3$. Antimony is obtained through the reduction that occurs (from the +3 oxidation state to the zero oxidation state of pure antimony) when $Sb_2S_3$ reacts with the iron in iron scrap. What is the mass of antimony in 14.78 lb of $Sb_2S_3$?

85. Cermets (for *cer*amic plus *met*al) are synthetic substances with both ceramic and metallic components. They combine the strength and toughness of metal with the resistance to heat and oxidation that ceramics offer. One cermet containing molybdenum and silicon is used to coat molybdenum engine parts on space vehicles. A sample of this compound is analyzed and found to contain 14.212 g of molybdenum and 8.321 g of silicon. What is the empirical formula for this compound?

86. Blue vitriol is a common name for an ionic compound that has many purposes in industry, including the production of germicides, pigments, pharmaceuticals, and wood preservatives. A sample contains 20.238 g of copper, 10.213 g of sulfur, and 20.383 g of oxygen. What is its empirical formula? What is its name?

87. A compound that is sometimes called sorrel salt can be used to remove ink stains or to clean wood. It is 30.52% potassium, 0.787% hydrogen, 18.75% carbon, and 49.95% oxygen. What is the empirical formula for this compound?

88. The ionic compound sometimes called uranium yellow is used to produce colored glazes for ceramics. It is 7.252% sodium, 75.084% uranium, and 17.664% oxygen. What is the empirical formula for this compound?

89. An ionic compound that is 24.186% sodium, 33.734% sulfur, and 42.080% oxygen is used as a food preservative. What is its empirical formula?

90. A defoliant is an herbicide that removes leaves from trees and growing plants. One ionic compound used for this purpose is 12.711% magnesium, 37.083% chlorine, and 50.206% oxygen. What is the empirical formula for this compound? What do you think its chemical name is? (Consider the possibility that this compound contains more than one polyatomic ion.)

91. An ionic compound that is 22.071% manganese, 1.620% hydrogen, 24.887% phosphorus, and 51.422% oxygen is used as a food additive and dietary supplement. What is the empirical formula for this compound? What do you think its chemical name is? (Consider the possibility that this compound contains more than one polyatomic ion.)

92. Agent Orange, which was used as a defoliant in the Vietnam War, contained a mixture of two herbicides called 2,4,5-T and 2,4-D. (Agent Orange got its name from the orange barrels in which it was stored.) Some of the controversy that surrounds the use of Agent Orange is related to the discovery in 2,4,5-T of a trace impurity called TCDD. Although recent studies suggest that it is much more harmful to animals than to humans, TCDD has been called the most toxic small molecule known, and all uses of 2,4,5-T were banned in 1985. TCDD is 44.77% carbon, 1.25% hydrogen, 9.94% oxygen, and 44.04% chlorine and has a molecular mass of 321.97. What is the molecular formula for TCDD?

93. Thalidomide was used as a tranquilizer and flu medicine for pregnant women in Europe until it was found to cause birth defects. (The horrible effects of this drug played a significant role in the passage of the Kefauver–Harris Amendment to the Food and Drug Act, requiring that drugs be proved safe before they are put on the market.) Thalidomide is 60.47% carbon, 3.90% hydrogen, 24.78% oxygen, and 10.85% nitrogen and has a molecular mass of 258.23. What is the molecular formula for thalidomide?

94. Nabam is a fungicide used on potato plants. It is 17.94% sodium, 18.74% carbon, 2.36% hydrogen, 10.93% nitrogen, and 50.03% sulfur and has a formula mass of 256.35. Nabam is an ionic compound with a formula that is not an empirical formula. What is the formula for nabam?

### Challenge Problems

95. Calamine is a naturally occurring zinc silicate that contains the equivalent of 67.5% zinc oxide, ZnO. (The term *calamine* also refers to a substance used to make calamine lotion.) What is the mass, in kilograms, of zinc in $1.347 \times 10^4$ kg of natural calamine that is 67.5% ZnO?

96. Zirconium metal, which is used to coat nuclear fuel rods, can be made from the zirconium(IV) oxide, $ZrO_2$, in the zirconium ore called baddeleyite (or zirconia). What maximum mass, in kilograms, of zirconium metal can be extracted from $1.2 \times 10^3$ kg of baddeleyite that is 53% $ZrO_2$?

97. Flue dust from the smelting of copper and lead contains $As_2O_3$. (Smelting is the heating of a metal ore until it melts, so that its metallic components can be separated.) When this flue dust is collected, it contains 90% to 95% $As_2O_3$. The arsenic in $As_2O_3$ can be reduced to the element arsenic by reaction with charcoal. What is the maximum mass, in kilograms, of arsenic that can be formed from 67.3 kg of flue dust that is 93% $As_2O_3$?

98. Thortveitite is a natural ore that contains from 37% to 42% scandium oxide, $Sc_2O_3$. Scandium metal is made by first reacting the $Sc_2O_3$ with ammonium hydrogen fluoride, $NH_4HF_2$, to form scandium fluoride, $ScF_3$. The scandium in $ScF_3$ is reduced to metallic scandium in a reaction with calcium. What is the maximum mass, in kilograms, of scandium metal that can be made from 1230.2 kilograms of thortveitite that is 39% $Sc_2O_3$?

99. Magnesium metal, which is used to make die-cast auto parts, missiles, and space vehicles, is obtained by the electrolysis of magnesium chloride. Magnesium hydroxide forms magnesium chloride when it reacts with hydrochloric acid. There are two common sources of magnesium hydroxide.
    a. Magnesium ions can be precipitated from seawater as magnesium hydroxide, $Mg(OH)_2$. Each kiloliter of seawater yields about 3.0 kg of the compound. How many metric tons of magnesium metal can be made from the magnesium hydroxide derived from $1.0 \times 10^5$ kL of seawater?
    b. Brucite is a natural form of magnesium hydroxide. A typical crude ore containing brucite is 29% $Mg(OH)_2$. What minimum mass, in metric tons, of this crude ore is necessary to make 34.78 metric tons of magnesium metal?

100. Spodumene is a lithium aluminum silicate containing the equivalent of 6.5% to 7.5% lithium oxide, $Li_2O$. Crude ore mined in North Carolina contains 15% to 20% spodumene. What maximum mass, in kilograms, of lithium could be formed from 2.538 megagrams of spodumene containing the equivalent of 7.0% $Li_2O$?

101. The element fluorine can be obtained by the electrolysis of combinations of hydrofluoric acid and potassium fluoride. These compounds can be made from the calcium fluoride, $CaF_2$, found in nature as the mineral fluorite. Fluorite's commercial name is fluorspar. Crude ores containing fluorite have 15% to 90% $CaF_2$. What minimum mass, in metric tons, of crude ore is necessary to make 2.4 metric tons of fluorine if the ore is 72% $CaF_2$?

102. Chromium metal is used in metal alloys and as a surface plating on other metals to minimize corrosion. It can be obtained by reducing the chromium(III) in chromium(III) oxide, $Cr_2O_3$, to the uncharged metal with finely divided aluminum. The $Cr_2O_3$ is found in an ore called chromite. What is the maximum mass, in kilograms, of chromium that can be made from 143.0 metric tons of Cuban chromite ore that is 38% $Cr_2O_3$?

103. What mass of baking powder that is 36% $NaHCO_3$ contains 1.0 mole of sodium hydrogen carbonate?

104. Roscoelite is a vanadium-containing form of mica used to make vanadium metal. Although the fraction of vanadium in the ore is variable, roscoelite can be described as $K_2V_4Al_2Si_6O_{20}(OH)_4$. Another way to describe the vanadium content of this mineral is to say that it has the equivalent of up to 28% $V_2O_3$.

    a. What is the mass, in grams, of vanadium in 123.64 g of roscoelite?

    b. What is the mass, in kilograms, of vanadium in 6.71 metric tons of roscoelite that contains the equivalent of 28% $V_2O_3$?

105. Hafnium metal is used to make control rods in water-cooled nuclear reactors and to make filaments in light bulbs. The hafnium is found with zirconium in zircon sand, which is about 1% hafnium(IV) oxide, $HfO_2$. What minimum mass, in metric tons, of zircon sand is necessary to make 120.5 kg of hafnium metal if the sand is 1.3% $HfO_2$?

*The techniques described in this chapter allow scientists to calculate the amounts of raw materials necessary to create computer chips.*

CHAPTER 10

# Chemical Calculations and Chemical Equations



Although Chapter 9 was full of questions that began with "*How much . . . ?*" we are not finished with such questions yet. In Chapter 9, our questions focused on chemical formulas. For example, we answered such questions as, "*How much* of the element vanadium can be obtained from 2.3 metric tons of the compound $V_2O_5$?" The chemical formula $V_2O_5$ told us that there are 2 moles of vanadium, V, in each mole of $V_2O_5$. We used this molar ratio to convert from moles of $V_2O_5$ to moles of vanadium.

In this chapter, we encounter questions that focus instead on chemical reactions. These questions ask us to convert from amount of one substance in a given chemical reaction to amount of another substance participating in the same reaction. For example, a business manager, budgeting for the production of silicon-based computer chips, wants to know *how much* silicon can be produced from 16 kg of carbon and 32 kg of silica, $SiO_2$, in the reaction

$$SiO_2(s) + 2C(s) \xrightarrow{2000\ ^\circ C} Si(l) + 2CO(g)$$

A safety engineer in a uranium-processing plant wants to know *how much* water needs to be added to 25 lb of uranium hexafluoride to maximize the synthesis of $UO_2F_2$ by the reaction

$$UF_6 + 2H_2O \rightarrow UO_2F_2 + 4HF$$

A chemistry student working in the lab might be asked to calculate *how much* 1-bromo-2-methylpropane, $C_4H_9Br$, could be made from 6.034 g of 2-methyl-2-propanol, $C_4H_9OH$, in the reaction

$$3C_4H_9OH + PBr_3 \rightarrow 3C_4H_9Br + H_3PO_3$$

In these calculations, we will be generating conversion factors from the coefficients in the balanced chemical equation.

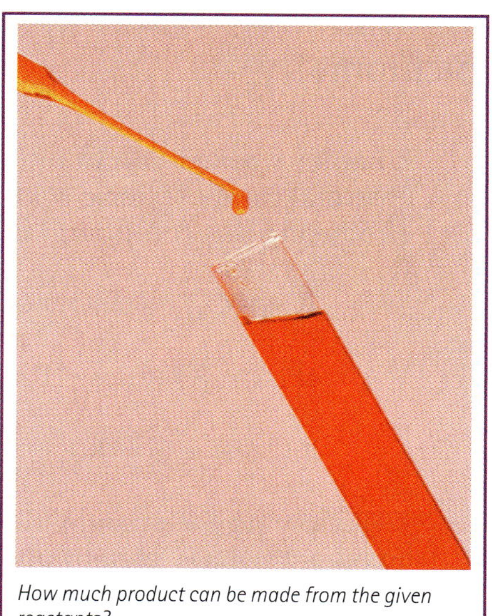

*How much product can be made from the given reactants?*

## Review Skills

The presentation of information in this chapter assumes that you can already perform the tasks listed below. You can test your readiness to proceed by answering the Review Questions at the end of the chapter. This might also be a good time to read the Chapter Objectives, which precede the Review Questions.

- Balance chemical equations. (Section 4.1)
- Write or identify the definitions of *solution*, *solute*, and *solvent*. (Section 4.2)
- Given a description of a solution, identify the solute and solvent. (Section 4.2)
- Describe water solutions in terms of the nature of the particles in solution and the attractions between them. (Sections 4.2 and 5.1)
- Given formulas for two ionic compounds, predict whether a precipitate will form when water solutions of the two are mixed, and write the complete equation that describes the reaction. (Section 4.2)
- Convert between names of chemical substances and their chemical formulas. (Section 5.3)
- Given the names or formulas for a monoprotic or polyprotic acid and an ionic compound containing hydroxide, carbonate, or hydrogen carbonate ions, write the complete balanced equation for the neutralization reaction that takes place between them. (Section 5.6)
- Make general unit conversions. (Sections 8.1 and 8.5)
- Report answers to calculations using the correct number of significant figures. (Section 8.2)
- Find the atomic mass for elements, the molecular mass for molecular compounds, and the formula mass for ionic compounds. (Sections 9.2 and 9.3)
- Use the atomic mass for elements, the molecular mass for molecular compounds, and the formula mass for ionic compounds to convert between mass and moles of substance. (Sections 9.2 and 9.3)

# 10.1 Equation Stoichiometry

Chapter 9 asked you to pretend to be an industrial chemist at a company that makes phosphoric acid, $H_3PO_4$. The three-step "furnace method" for producing this compound is summarized by these three equations:

$$2Ca_3(PO_4)_2 + 6SiO_2 + 10C \rightarrow 4P + 10CO + 6CaSiO_3$$

$$4P(s) + 5O_2(g) \rightarrow P_4O_{10}(s)$$

$$P_4O_{10}(s) + 6H_2O(l) \rightarrow 4H_3PO_4(aq)$$

Following the strategy demonstrated in Example 9.6, we calculated the maximum mass of tetraphosphorus decoxide, $P_4O_{10}$, that can be made from $1.09 \times 10^4$ kilograms of phosphorus in the second of these three reactions. The answer is $2.50 \times 10^4$ kg $P_4O_{10}$. We used the following steps:

$1.09 \times 10^4$ kg P → g P → mol P → mol $P_4O_{10}$ → g $P_4O_{10}$ → kg $P_4O_{10}$

$$? \text{ kg } P_4O_{10} = 1.09 \times 10^4 \text{ kg P} \left(\frac{10^3 \text{ g}}{1 \text{ kg}}\right)\left(\frac{1 \text{ mol P}}{30.9738 \text{ g P}}\right)\left(\frac{1 \text{ mol } P_4O_{10}}{4 \text{ mol P}}\right)\left(\frac{283.889 \text{ g } P_4O_{10}}{1 \text{ mol } P_4O_{10}}\right)\left(\frac{1 \text{ kg}}{10^3 \text{ g}}\right)$$

$$= \mathbf{2.50 \times 10^4 \text{ kg } P_4O_{10}}$$

The ratio of moles of $P_4O_{10}$ to moles of P (which came from the subscripts in the chemical formula, $P_4O_{10}$) provided the key conversion factor that enabled us to convert from units of phosphorus to units of tetraphosphorus decoxide.

Now let's assume that you have been transferred to the division responsible for the final stage of the process, the step in which tetraphosphorus decoxide is converted into phosphoric acid in the third reaction in the list displayed above. Your first assignment there is to calculate the mass of water, in kilograms, that would be necessary to react with $2.50 \times 10^4$ kg $P_4O_{10}$. The steps for this conversion are very similar to those in Example 9.6:

$2.50 \times 10^4$ kg $P_4O_{10}$ → g $P_4O_{10}$ → mol $P_4O_{10}$ → mol $H_2O$ → g $H_2O$ → kg $H_2O$

As part of our calculation, we convert from moles of one substance ($P_4O_{10}$) to moles of another ($H_2O$), so we need a conversion factor that relates the numbers of particles of these substances. The coefficients in the balanced chemical equation provide us with information that we can use to build this conversion factor. They tell us that six molecules of $H_2O$ are needed to react with one molecule of $P_4O_{10}$ in order to produce four molecules of phosphoric acid:

$$P_4O_{10}(s) + 6H_2O(l) \rightarrow 4H_3PO_4(aq)$$

Thus the ratio of amount of $H_2O$ to amount of $P_4O_{10}$ is

$$\left(\frac{6 \text{ molecules } H_2O}{1 \text{ molecule } P_4O_{10}}\right)$$

We found in Chapter 9 that it is convenient to describe numbers of molecules in terms of moles. If the reaction requires 6 molecules of water for each molecule of $P_4O_{10}$, it would require 6 dozen $H_2O$ molecules for each dozen $P_4O_{10}$ molecules, or 6 moles of $H_2O$ for each mole of $P_4O_{10}$ (Table 10.1).

$$\left(\frac{6 \text{ dozen } H_2O}{1 \text{ dozen } P_4O_{10}}\right) \quad \text{or} \quad \left(\frac{6 \text{ mol } H_2O}{1 \text{ mol } P_4O_{10}}\right)$$

OBJECTIVE 2

**Table 10.1**
Information Derived from the Coefficients in the Balanced Equation for the Reaction That Produces Phosphoric Acid

| $P_4O_{10}(s)$ | + | $6H_2O(l)$ | → | $4H_3PO_4(aq)$ |
|---|---|---|---|---|
| 1 molecule of $P_4O_{10}$ | | 6 molecules of $H_2O$ | | 4 molecules of $H_3PO_4$ |
| 1 dozen $P_4O_{10}$ molecules | | 6 dozen $H_2O$ molecules | | 4 dozen $H_3PO_4$ molecules |
| $6.022 \times 10^{23}$ molecules of $P_4O_{10}$ | | $6(6.022 \times 10^{23})$ molecules of $H_2O$ | | $4(6.022 \times 10^{23})$ molecules of $H_3PO_4$ |
| 1 mole of $P_4O_{10}$ | | 6 moles of $H_2O$ | | 4 moles of $H_3PO_4$ |

Example 10.1 shows how the coefficients in a balanced chemical equation provide a number of conversion factors that enable us to convert from moles of any reactant or product to moles of any other reactant or product.

## EXAMPLE 10.1 Equation Stoichiometry

OBJECTIVE 2

Write three different conversion factors that relate moles of one reactant or product in the reaction that follows to moles of another reactant or product in this reaction.

$$P_4O_{10}(s) + 6H_2O(l) \rightarrow 4H_3PO_4(aq)$$

***Solution***

Any combination of two coefficients from the equation leads to a conversion factor.

$$\left(\frac{1 \text{ mol } P_4O_{10}}{6 \text{ mol } H_2O}\right) \quad \left(\frac{1 \text{ mol } P_4O_{10}}{4 \text{ mol } H_3PO_4}\right) \quad \left(\frac{6 \text{ mol } H_2O}{4 \text{ mol } H_3PO_4}\right)$$

Let's return to our conversion of $2.50 \times 10^4$ kg of $P_4O_{10}$ to kilograms of water. Like so many chemistry calculations, this problem can be worked using the dimensional analysis thought process and format. We start by identifying the unit that we want to arrive at (kg $H_2O$) and a known value that can be used to start setting up the dimensional analysis ($2.50 \times 10^4$ kg $P_4O_{10}$). We have already decided that we will convert from amount of $P_4O_{10}$ to amount of $H_2O$ using the molar ratio derived from the balanced equation, but before we can convert from moles of $P_4O_{10}$ to moles of $H_2O$, we need to convert from mass of $P_4O_{10}$ to number of moles of $P_4O_{10}$.

mass $P_4O_{10}$ → mol $P_4O_{10}$ → mol $H_2O$ → mass $H_2O$

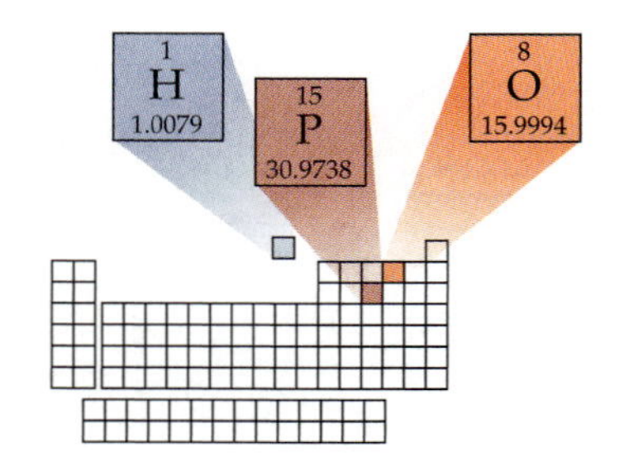


$P_4O_{10}$ is a molecular compound, and we discovered in Section 9.3 that we can convert from mass of a molecular substance to moles using its molar mass, which comes from its molecular mass. Because we are starting with a mass measured in kilograms, our equation also needs a conversion factor for converting kilograms into grams. We can convert from moles of $H_2O$ to grams of $H_2O$ using the molar mass of water (which we determine from the molecular mass of $H_2O$). We then convert from grams to kilograms to complete the calculation.

$$? \text{ kg } H_2O = 2.50 \times 10^4 \text{ kg } P_4O_{10}\left(\frac{10^3 \text{ g}}{1 \text{ kg}}\right)\left(\frac{1 \text{ mol } P_4O_{10}}{283.889 \text{ g } P_4O_{10}}\right)\left(\frac{6 \text{ mol } H_2O}{1 \text{ mol } P_4O_{10}}\right)\left(\frac{18.0153 \text{ g } H_2O}{1 \text{ mol } H_2O}\right)\left(\frac{1 \text{ kg}}{10^3 \text{ g}}\right)$$

OBJECTIVE 3

OBJECTIVE 4

$$= \mathbf{9.52 \times 10^3 \text{ kg } H_2O}$$

There is a shortcut for this calculation. We can collapse all five of the conversion factors above into one. The reaction equation tells us that there are six moles of $H_2O$ for each mole of $P_4O_{10}$. The molecular masses of these substances tell us that each mole of $H_2O$ weighs 18.0153 g, and each mole of $P_4O_{10}$ weighs 283.889 g. Thus, the mass ratio of $H_2O$ to $P_4O_{10}$ is 6 times 18.0153 g to 1 times 283.889 g.

$$\left(\frac{6 \times 18.0153 \text{ g } H_2O}{1 \times 283.889 \text{ g } P_4O_{10}}\right)$$

We can describe this mass ratio using any mass units we want.

$$\left(\frac{6 \times 18.0153 \text{ g } H_2O}{1 \times 283.889 \text{ g } P_4O_{10}}\right)$$

or

$$\left(\frac{6 \times 18.0153 \text{ kg } H_2O}{1 \times 283.889 \text{ kg } P_4O_{10}}\right)$$

or

$$\left(\frac{6 \times 18.0153 \text{ lb } H_2O}{1 \times 283.889 \text{ lb } P_4O_{10}}\right)$$

Thus our setup for this example can be simplified to

OBJECTIVE 3

OBJECTIVE 4

$$? \text{ kg } H_2O = 2.50 \times 10^4 \text{ kg } P_4O_{10}\left(\frac{6 \times 18.0153 \text{ kg } H_2O}{1 \times 283.889 \text{ kg } P_4O_{10}}\right)$$

$$= \mathbf{9.52 \times 10^3 \text{ kg } H_2O}$$

Calculations like this are called **equation stoichiometry** problems, or just stoichiometry problems. Stoichiometry, from the Greek words for "measure" and "element," refers to the quantitative relationships between substances. Calculations like those in Chapter 9, in which we convert between an amount of compound and an amount of element in the compound, are one kind of stoichiometry problem, but the term is rarely used in that context. Instead, it is usually reserved for calculations like the one above, which deal with conversion of the amount of one substance in a chemical reaction into the amount of a different substance in the reaction.

The following is a sample study sheet for equation stoichiometry problems.

## Sample Study Sheet 10.1

### Basic Equation Stoichiometry—Converting Mass of One Compound in a Reaction to Mass of Another

OBJECTIVE 3

OBJECTIVE 4

**Tip-off** The calculation calls for you to convert from an amount of one substance in a given chemical reaction to the corresponding amount of another substance that participates in the same reaction.

**General Steps** Use a dimensional analysis format. Set it up around a mole-to-mole conversion in which the coefficients from a balanced equation are used to generate a mole ratio. (See Figure 10.1 for a summary.) The general steps are

**Step 1** If you are not given it, write and balance the chemical equation for the reaction.

**Step 2** Start your dimensional analysis in the usual way.

You want to calculate amount of substance 2, so you set that unknown equal to the given amount of substance 1. (In this section, the given units will be mass of an element or compound, but as you will see later in this chapter and in Chapter 13, the given units might instead be the volume of a solution or the volume of a gas.)

**Step 3** If you are given a unit of mass other than grams for substance 1, convert from the unit that you are given to grams. This may require one or more conversion factors.

**Step 4** Convert from grams of substance 1 to moles of substance 1, using the substance's molar mass.

**Step 5** Convert from moles of substance 1 to moles of substance 2, using their coefficients from the balanced equation to create a molar ratio to use as a conversion factor.

**Step 6** Convert from moles of substance 2 to grams of substance 2, using the substance's molar mass.

**Step 7** If necessary, convert from grams of substance 2 to the desired unit for substance 2. This may require one or more conversion factors.

**Step 8** Calculate your answer and report it with the correct number of significant figures (in scientific notation, if necessary) and the correct unit.

The general form of the dimensional analysis setup follows. (Here we are using the symbol # to stand for "substance.")

$$? \text{ (unit) \#2} = \text{(given)(unit) \#1}\left(\frac{\_\_\ \text{g}}{\_\_\ \text{(unit)}}\right)\left(\frac{1 \text{ mol \#1}}{\text{(formula mass \#1) g \#1}}\right)$$

$$\left(\frac{\text{(coef. \#2) mol \#2}}{\text{(coef. \#1) mol \#1}}\right)\left(\frac{\text{(formula mass \#2) g \#2}}{1 \text{ mol \#2}}\right)\left(\frac{\_\_\ \text{(unit)}}{\_\_\ \text{g}}\right)$$

**SHORTCUT STEPS** If the mass unit desired for substance 2 is the same mass unit given for substance 1, the general steps described above can be condensed into a shortcut. (See Figure 10.2, page 410, for a summary.)

**STEP 1** If you are not given it, write and balance the chemical equation for the reaction.

**STEP 2** Start setting up your dimensional analysis in the usual way. (See step 2 above.)

**STEP 3** Convert directly from the mass units of substance 1 that you have been given to the same mass units of substance 2, using a conversion factor that has the following general form, where # stands for "substance."

$$? \text{ (unit) \#2} = \text{(given) (unit) \#1} \left( \frac{\text{coefficient \#2 (formula mass \#2)(any mass unit) substance 2}}{\text{coefficient \#1 (formula mass \#1)(same mass unit) substance 1}} \right)$$

**STEP 4** Calculate your answer and report it with the correct number of significant figures, in scientific notation if necessary, and the correct unit.

**EXAMPLE** See Example 10.2.

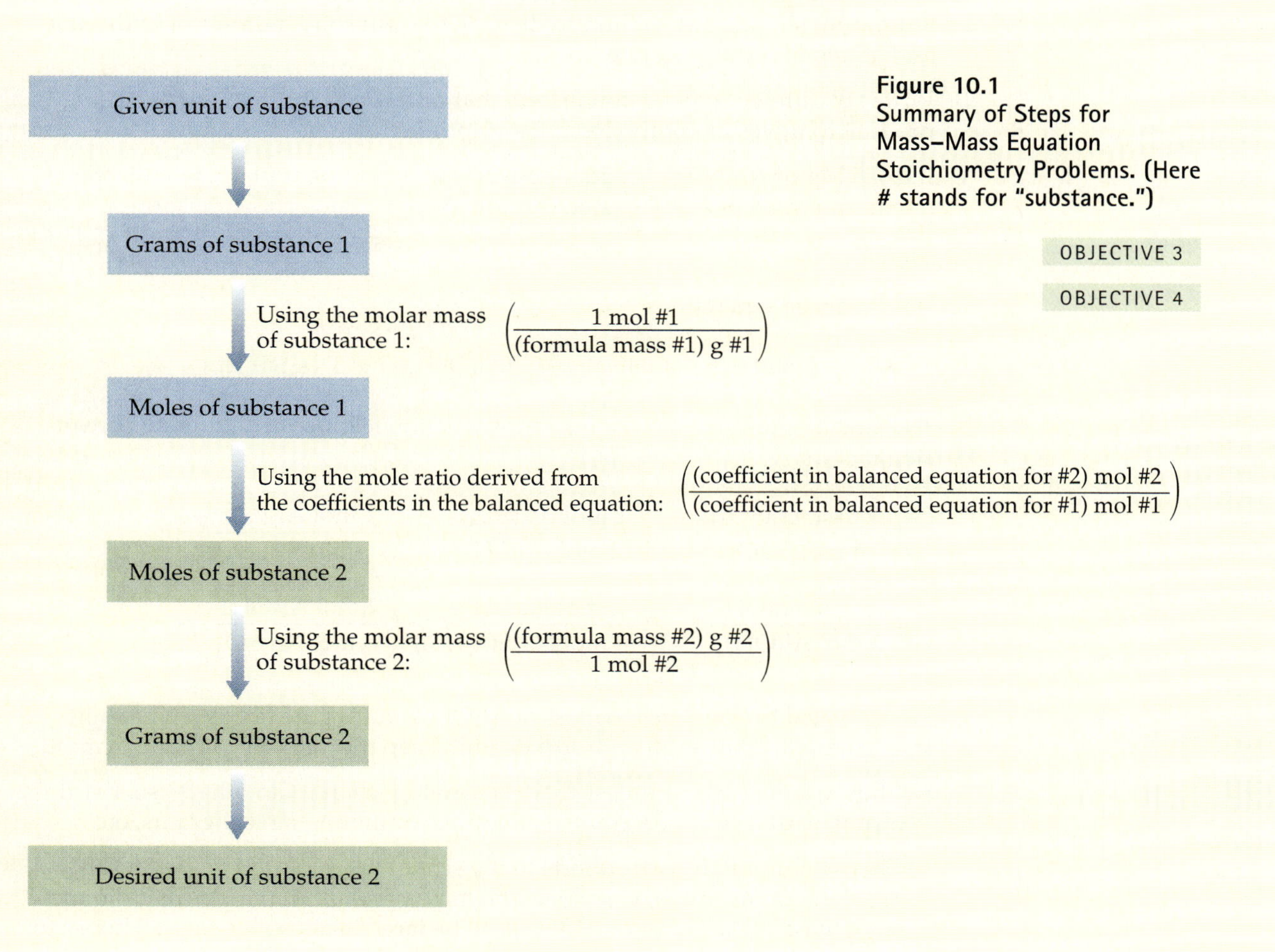


**Figure 10.1**
Summary of Steps for Mass-Mass Equation Stoichiometry Problems. (Here # stands for "substance.")

OBJECTIVE 3

OBJECTIVE 4

**Figure 10.2**
**Shortcut for Mass–Mass Equation Stoichiometry Problems.**

OBJECTIVE 3

OBJECTIVE 4

Given mass of substance 1

Using $\left(\frac{\text{coefficient \#2 (formula mass \#2)(any mass unit) substance 2)}}{\text{coefficient \#1 (formula mass \#1)(s ame mass unit) substance 1)}}\right)$

Same mass unit of substance 2

## EXAMPLE 10.2 Equation Stoichiometry

OBJECTIVE 3

OBJECTIVE 4

Aluminum sulfate is used in water purification as a coagulant that removes phosphate and bacteria and as a pH conditioner. It acts as a coagulant by reacting with hydroxide to form aluminum hydroxide, which precipitates from the solution and drags impurities down with it as it settles.

a. Write a complete, balanced equation for the reaction of water solutions of aluminum sulfate and sodium hydroxide to form solid aluminum hydroxide and aqueous sodium sulfate.

b. Write six different conversion factors that relate moles of one reactant or product to moles of another reactant or product.

c. If 0.655 Mg of $Al_2(SO_4)_3$ is added to water in a treatment plant, what is the maximum mass of $Al(OH)_3$ that can form?

***Solution***

a. The balanced equation is

$$Al_2(SO_4)_3(aq) + 6NaOH(aq) \rightarrow 2Al(OH)_3(s) + 3Na_2SO_4(aq)$$

b. The stoichiometric relationships in the reaction lead to the following conversion factors.

$$\left(\frac{1 \text{ mol } Al_2(SO_4)_3}{6 \text{ mol NaOH}}\right) \quad \left(\frac{1 \text{ mol } Al_2(SO_4)_3}{2 \text{ mol } Al(OH)_3}\right) \quad \left(\frac{1 \text{ mol } Al_2(SO_4)_3}{3 \text{ mol } Na_2SO_4}\right)$$

$$\left(\frac{6 \text{ mol NaOH}}{2 \text{ mol } Al(OH)_3}\right) \quad \left(\frac{6 \text{ mol NaOH}}{3 \text{ mol } Na_2SO_4}\right) \quad \left(\frac{2 \text{ mol } Al(OH)_3}{3 \text{ mol } Na_2SO_4}\right)$$

c. We are asked to calculate the mass of $Al(OH)_3$, but we are not told what units to calculate. To choose an appropriate unit, keep the following criteria in mind.

- Choose a metric unit unless there is a good reason to do otherwise. For this problem, that could be grams, kilograms, milligrams, megagrams, etc.
- Choose a unit that corresponds to the size of the expected value. In this problem, for example, we expect the mass of $Al(OH)_3$ that forms from the large mass of 0.655 Mg of $Al_2(SO_4)_3$ to be large itself, so we might choose to calculate kilograms or megagrams instead of grams or milligrams.

- Choose a unit that keeps the calculation as easy as possible. This usually means picking a unit that is mentioned in the problem. In this example, megagrams are mentioned, so we will calculate megagrams.

| 13 | 8 | 1 | 16 |
|---|---|---|---|
| Al | O | H | S |
| 26.9815 | 15.9994 | 1.0079 | 32.066 |

We are asked to convert from amount of one compound in a reaction to amount of another compound in the reaction: an equation stoichiometry problem.

$$? \text{ Mg Al(OH)}_3 = 0.655 \text{ Mg Al}_2(\text{SO}_4)_3 \left(\frac{10^6 \text{ g}}{1 \text{ Mg}}\right)\left(\frac{1 \text{ mol Al}_2(\text{SO}_4)_3}{342.154 \text{ g Al}_2(\text{SO}_4)_3}\right)\left(\frac{2 \text{ mol Al(OH)}_3}{1 \text{ mol Al}_2(\text{SO}_4)_3}\right)\left(\frac{78.0035 \text{ g Al(OH)}_3}{1 \text{ mol Al(OH)}_3}\right)\left(\frac{1 \text{ Mg}}{10^6 \text{ g}}\right)$$

$$= \mathbf{0.299 \text{ Mg Al(OH)}_3}$$

The setup for the shortcut is

$$? \text{ Mg Al(OH)}_3 = 0.655 \text{ Mg Al}_2(\text{SO}_4)_3 \left(\frac{2 \times 78.0035 \text{ Mg Al(OH)}_3}{1 \times 342.154 \text{ Mg Al}_2(\text{SO}_4)_3}\right)$$

$$= \mathbf{0.299 \text{ Mg Al(OH)}_3}$$

Note that this setup follows the steps described in Sample Study Sheet 10.1 and Figure 10.2.

## Exercise 10.1 Equation Stoichiometry

OBJECTIVE 3

OBJECTIVE 4

Tetrachloroethene, $C_2Cl_4$, often called perchloroethylene (perc), is a colorless liquid used in dry cleaning. It can be formed in several steps from the reaction of dichloroethane, chlorine gas, and oxygen gas. The equation for the net reaction is

$$8C_2H_4Cl_2(l) + 6Cl_2(g) + 7O_2(g) \rightarrow 4C_2HCl_3(l) + 4C_2Cl_4(l) + 14H_2O(l)$$

a. Fifteen different conversion factors for relating moles of one reactant or product to moles of another reactant or product can be derived from this equation. Write five of them.

b. How many grams of water form when 362.47 grams of tetrachloroethene, $C_2Cl_4$, are made in the reaction above?

c. What is the maximum mass of perchloroethylene, $C_2Cl_4$, that can be formed from 23.75 kilograms of dichloroethane, $C_2H_4Cl_2$?

Because substances are often found in mixtures, equation stoichiometry problems often include conversions between masses of pure substances and masses of mixtures containing the pure substances, using percentages as conversion factors. To see calculations like these, visit the following Web site:

www.chemplace.com/college/

## 10.2 Real-World Applications of Equation Stoichiometry

Let's return to the reaction of solid tetraphosphorus decoxide and water.

$$P_4O_{10}(s) + 6H_2O(l) \rightarrow 4H_3PO_4(aq)$$

Imagine that it is your job to design an industrial procedure for running this reaction. Whenever such a procedure is developed, some general questions must be answered, including the following:

- *How much of each reactant should be added to the reaction vessel?* This might be determined by the amount of product that you want to make or by the amount of one of the reactants that you have available.
- *What level of purity is desired for the final product? If the product is mixed with other substances (such as excess reactants), how will this purity be achieved?*

To understand some of the issues related to these questions, let's take a closer look at the reaction of $P_4O_{10}$ and $H_2O$, keeping in mind that we want to react $2.50 \times 10^4$ kg of $P_4O_{10}$ per day. What if you used a shovel to transfer solid $P_4O_{10}$ into a large container and then added water with a garden hose? Could you expect both of these reactants to react completely? When the reaction is finished, would the only substance in the container be phosphoric acid?

The coefficients in the balanced equation show us that to achieve the complete reaction of *both* reactants, we would need to add *exactly* six times as many water molecules as $P_4O_{10}$ molecules. With the precision expected from using a shovel and a hose to add the reactants, this seems unlikely. In fact, it is virtually impossible. The only way we could ever achieve the complete reaction of both reactants is by controlling the addition of reactants with a precision of plus or minus one molecule, and that is impossible (or at least highly improbable). No matter how careful we are to add the reactants in the correct ratio, we will always end up with at least a slight excess of one component compared to the other.

For some chemical reactions, chemists want to mix reactants in amounts that are as close as possible to the ratio that would lead to the complete reaction of each. This ratio is sometimes called the *stoichiometric ratio*. For example, in the production of phosphoric acid, the balanced equation shows that six moles of $H_2O$ react with each mole of $P_4O_{10}$, so for efficiency's sake, or to avoid leaving an excess of one of the reactants contaminating the product, we might want to add a molar ratio of $P_4O_{10}$ to $H_2O$ as close to the 1:6 stoichiometric ratio as possible.

$$P_4O_{10}(s) + 6H_2O(l) \rightarrow 4H_3PO_4(aq)$$

Sometimes the chemist deliberately uses a limited amount of one reactant and excessive amounts of others. There are many practical reasons for doing so. For example, an excess of one or more of the reactants will increase the likelihood that the other reactant or reactants will be used up.

Thus, if one reactant is more expensive than the others, adding an excess of the less expensive reactants will ensure the greatest conversion possible of the one that is costly. For our reaction that produces phosphoric acid, water is much less expensive than $P_4O_{10}$, so it makes sense to add water in excess.

OBJECTIVE 5

Sometimes one product is more important than others are, and the amounts of reactants are chosen to optimize its production. For example, the following reactions are part of the process that extracts pure silicon from silicon dioxide, $SiO_2$, for use in the semiconductor industry.

OBJECTIVE 5

$$SiO_2(s) + 2C(s) \rightarrow Si(l) + 2CO(g)$$

$$Si(s) + 3HCl(g) \rightarrow SiCl_3H(g) + H_2(g)$$

$$SiCl_3H(g) + H_2(g) \rightarrow Si(s) + 3HCl(g)$$

The ultimate goal of these reactions is to convert the silicon in $SiO_2$ to pure silicon, and the most efficient way to do this is to add an excess of carbon in the first reaction, an excess of HCl in the second reaction, and an excess of hydrogen gas in the last reaction.

Any component added in excess will remain when the reaction is complete. If one reactant is more dangerous to handle than others are, chemists would rather not have that reactant remaining at the reaction's end. For example, phosphorus, P, is highly reactive and dangerous to handle. Thus, in bringing about the reaction between P and $O_2$ to form $P_4O_{10}$, chemists would add the much safer oxygen in excess. When the reaction stops, the phosphorus is likely to be gone, leaving a product mixture that is mostly $P_4O_{10}$ and oxygen.

OBJECTIVE 5

$$P(s) + 5O_2(g) \text{ (with excess)} \rightarrow P_4O_{10}(s) + \text{excess } O_2(g)$$

Another consideration is that when a reaction ends, some of the reactant that was added in excess is likely to be mixed in with the product. Chemists would prefer that the substance in excess be a substance that is easy to separate from the primary product. For example, if we add excess carbon in the following reaction, some of it will remain after the silica has reacted.

OBJECTIVE 5

$$SiO_2(s) + 2C(s) \text{ (with excess)} \rightarrow Si(l) + 2CO(g) + \text{excess } C(s)$$

This excess carbon can be removed by converting it into carbon monoxide gas or carbon dioxide gas, which is easily separated from the silicon product.

## Limiting Reactants

The reactant that runs out first in a chemical reaction limits the amount of product that can form. This reactant is called the **limiting reactant.** For example, say we add an excess of water to the reaction that forms phosphoric acid from water and $P_4O_{10}$:

$$P_4O_{10}(s) + 6H_2O(l) \rightarrow 4H_3PO_4(aq)$$

The amount of $H_3PO_4$ that can be formed will be determined by the amount of $P_4O_{10}$. Therefore, the $P_4O_{10}$ will be the limiting reactant. When the $P_4O_{10}$ is gone, the reaction stops, leaving an excess of water with the product.

A simple analogy might help to clarify the idea of the limiting reactant. To build a bicycle, you need one frame and two wheels (and a few other components that we will ignore). The following equation describes the production of bicycles from these two components.

$$1 \text{ frame} + 2 \text{ wheels} \rightarrow 1 \text{ bicycle}$$

Let's use this formula for bicycles to do a calculation very similar to the calculations we will do for chemical reactions. Assume that you are making bicycles to earn money for college. If your storeroom contains 7 frames and 12 wheels, what is the maximum number of bicycles you can make? To decide how many bicycles you can make, you need to determine which of your components will run out first. You can do this by first determining the maximum number of bicycles that each component can make. Whichever component makes the fewer bicycles must run out first and limit the amount of product that can be made. From the formula for producing a bicycle, you obtain the necessary conversion factors:

$$\left(\frac{1 \text{ bicycle}}{1 \text{ frame}}\right) \quad \text{and} \quad \left(\frac{1 \text{ bicycle}}{2 \text{ wheels}}\right)$$

$$? \text{ bicycles} = 7 \cancel{\text{frames}} \left(\frac{1 \text{ bicycle}}{1 \cancel{\text{frame}}}\right) = 7 \text{ bicycles}$$

$$? \text{ bicycles} = 12 \cancel{\text{wheels}} \left(\frac{1 \text{ bicycle}}{2 \cancel{\text{wheels}}}\right) = 6 \text{ bicycles}$$

You have enough frames for 7 bicycles but enough wheels for only 6. The wheels will run out first. Even if you had 7000 frames, you could make only 6 bicycles with your 12 wheels. Thus, the wheels are limiting, and the frames are in excess (Figure 10.3).

Now let's apply what we have learned from the bicycle example to a calculation that deals with a chemical reaction. Electronic-grade (EG) silicon used in the electronics industry is a purified form of metallurgical-grade silicon, which is made from the reaction of silica, $SiO_2$, with carbon in the form of coke at 2000 °C. (Silica is found in nature as quartz or quartz sand.)

$$SiO_2(s) + 2C(s) \xrightarrow{2000\ °C} Si(l) + 2CO(g)$$

If 1000 moles of carbon are heated with 550 moles of silica, what is the maximum number of moles of metallurgical-grade silicon, Si, that can be formed? This example is similar to the bicycle example. We need 2 times

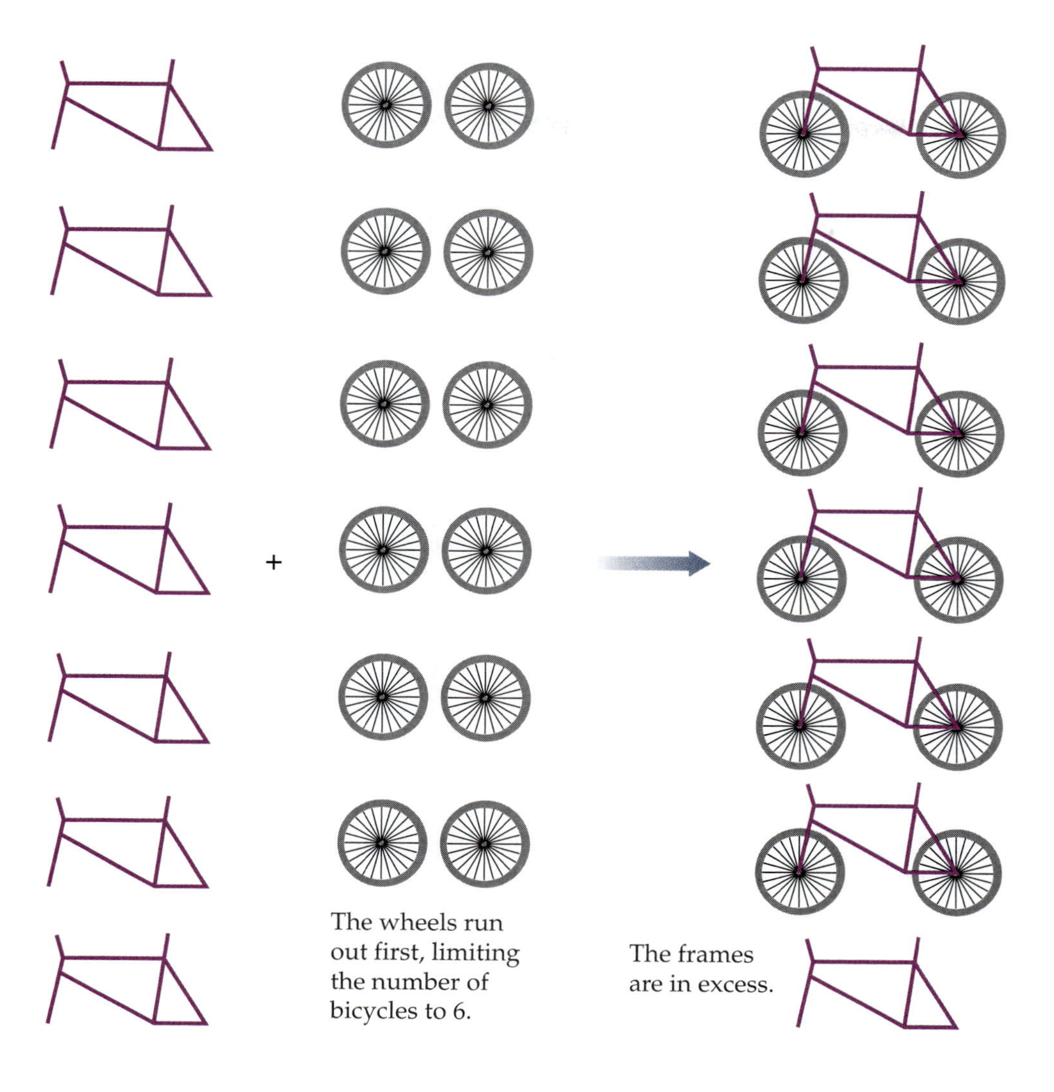
+
The wheels run out first, limiting the number of bicycles to 6.
The frames are in excess.

**Figure 10.3**
Limiting Component

as many wheels as frames to build bicycles, and to get a complete reaction of silicon dioxide and carbon, we need two atoms (or moles of atoms) of carbon for every formula unit (or mole of formula units) of silicon dioxide.

$$\text{1 frame} + \text{2 wheels} \rightarrow \text{1 bicycle}$$

$$SiO_2(s) + 2C(s) \rightarrow Si(l) + 2CO(g)$$

In the reaction between carbon and silicon dioxide, we can assume that one of the reactants is in excess and the other is limiting, but we do not yet know which is which. With the bicycle example, we discovered which component was limiting (and also the maximum number of bicycles that can be made) by calculating the maximum number of bicycles we could make from each component. The component that could make the fewer bicycles was limiting, and that number of bicycles was the maximum number of bicycles that could be made.

For our reaction between carbon and silicon dioxide, we can determine which reactant is the limiting reactant by first calculating the maximum amount of silicon that can be formed from the given amount of each reactant. The reactant that forms the least product will run out first and limit the amount of product that can form. The coefficients in the balanced equation provide us with conversion factors to convert from moles of reactants to moles of products.

$$? \text{ mol Si} = 1000 \text{ mol C}\left(\frac{1 \text{ mol Si}}{2 \text{ mol C}}\right) = \mathbf{500 \text{ mol Si}}$$

$$? \text{ mol Si} = 550 \text{ mol SiO}_2\left(\frac{1 \text{ mol Si}}{1 \text{ mol SiO}_2}\right) = 550 \text{ mol Si}$$

The carbon will be used up after 500 moles of silicon have been formed. The silicon dioxide would not be used up until 550 moles of silicon had formed. Therefore, the carbon is the limiting reactant, and the maximum amount of product that can form is 500 moles Si.

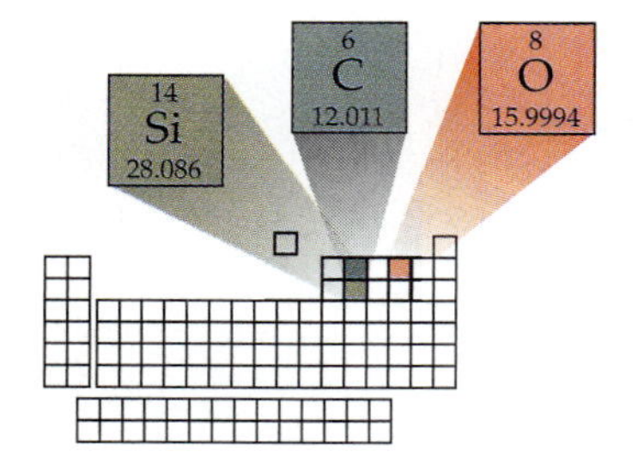


Now let's work a problem that is similar but deals with masses of reactants and products rather than moles. If 16.491 g of carbon is heated with 32.654 g of silica, what is the maximum mass of metallurgical grade silicon, Si, that can be formed?

For this calculation, we follow a procedure that is very similar to the procedure we would follow to calculate the moles of silicon that can be made from the reaction of 1000 moles of carbon and 550 moles of $SiO_2$. We calculate the amount of silicon that can be made from 16.491 g of C and also the amount that can be made from 32.654 g of $SiO_2$. Whichever forms the least product is the limiting reactant and therefore determines the maximum amount of product that can form. These two calculations are equation stoichiometry problems, so we will use the procedure described in Sample Study Sheet 10.1.

$$? \text{ g Si} = 16.491 \text{ g C}\left(\frac{1 \text{ mol C}}{12.011 \text{ g C}}\right)\left(\frac{1 \text{ mol Si}}{2 \text{ mol C}}\right)\left(\frac{28.0855 \text{ g Si}}{1 \text{ mol Si}}\right)$$

or $$? \text{ g Si} = 16.491 \text{ g C}\left(\frac{1 \times 28.0855 \text{ g Si}}{2 \times 12.011 \text{ g C}}\right) = 19.281 \text{ g Si}$$

OBJECTIVE 6

$$? \text{ g Si} = 32.654 \text{ g SiO}_2\left(\frac{1 \text{ mol SiO}_2}{60.0843 \text{ g SiO}_2}\right)\left(\frac{1 \text{ mol Si}}{1 \text{ mol SiO}_2}\right)\left(\frac{28.0855 \text{ g Si}}{1 \text{ mol Si}}\right)$$

or $$? \text{ g Si} = 32.654 \text{ g SiO}_2\left(\frac{1 \times 28.0855 \text{ g Si}}{1 \times 60.0843 \text{ g SiO}_2}\right) = \mathbf{15.264 \text{ g Si}}$$

The $SiO_2$ will be used up after 15.264 g of silicon have been formed. The carbon would not be used up until 19.281 g of silicon had formed. Because the $SiO_2$ produces the least silicon, it runs out first and limits the amount of product that can form. Therefore, the silicon dioxide is the limiting reactant, and the maximum amount of product that can form is 15.264 g Si.

The following is a sample study sheet that summarizes the procedure for working limiting reactant problems.

## Sample Study Sheet 10.2 Limiting Reactant Problems

OBJECTIVE 6

**Tip-off** Given two or more amounts of reactants in a chemical reaction, you are asked to calculate the maximum amount of product that they can form.

**General Procedure** Follow these steps.

- Do separate calculations of the maximum amount of product that can form from each reactant. (These calculations are equation stoichiometry problems, so you can use the procedure described in Sample Study Sheet 10.1 for each calculation.)
- The smallest of the values calculated in the step above is your answer. It is the maximum amount of product that can be formed from the given amounts of reactants.

**Example** See Example 10.3.

## Example 10.3 Limiting Reactant

Titanium carbide, TiC, is the hardest of the known metal carbides. It can be made by heating titanium(IV) oxide, $TiO_2$, with carbon black to 2000 to 2200 °C. (Carbon black is a powdery form of carbon that is produced when vaporized heavy oil is burned with 50% of the air required for complete combustion.)

$$TiO_2 + 3C \rightarrow TiC + 2CO$$

a. What is the maximum mass of titanium carbide, TiC, that can be formed from the reaction of 985 kg of titanium(IV) oxide, $TiO_2$, with 500 kg of carbon, C?

OBJECTIVE 6

b. Why do you think the reactant in excess was chosen to be in excess?

OBJECTIVE 7

***Solution***

a. Because we are given amounts of two reactants and asked to calculate an amount of product, we recognize this as a limiting reactant problem. Thus we first calculate the amount of product that can form from each reactant. The reactant that forms the least product is the limiting reactant and determines the maximum amount of product that can form from the given amounts of reactants.

$$? \text{ kg TiC} = 985 \cancel{\text{kg TiO}_2} \left( \frac{1 \times 59.878 \text{ kg TiC}}{1 \times 79.866 \cancel{\text{kg TiO}_2}} \right) = \mathbf{738 \text{ kg TiC}}$$

$$? \text{ kg TiC} = 500 \cancel{\text{kg C}} \left( \frac{1 \times 59.878 \text{ kg TiC}}{3 \times 12.011 \cancel{\text{kg C}}} \right) = 831 \text{ kg TiC}$$

The limiting reactant is $TiO_2$ because it results in the least amount of product.

b. We are not surprised that the carbon is provided in excess. We expect it to be less expensive than titanium dioxide, and the excess carbon can be easily separated from the solid product by burning to form gaseous CO or $CO_2$.

### Exercise 10.2 Limiting Reactant

The uranium(IV) oxide, $UO_2$, which is used as fuel in nuclear power plants, has a higher percentage of the fissionable isotope uranium-235 than is present in the $UO_2$ found in nature. To make fuel-grade $UO_2$, chemists first convert uranium oxides to uranium hexafluoride, $UF_6$, whose concentration of uranium-235 can be increased by a process called gas diffusion. The enriched $UF_6$ is then converted back to $UO_2$ in a series of reactions, beginning with

$$UF_6 + 2H_2O \rightarrow UO_2F_2 + 4HF$$

OBJECTIVE 6

a. How many megagrams of $UO_2F_2$ can be formed from the reaction of 24.543 Mg $UF_6$ with 8.0 Mg of water?

OBJECTIVE 7

b. Why do you think the reactant in excess was chosen to be in excess?

## Percent Yield

In Examples 10.2 and 10.3, we determined the maximum amount of product that could be formed from the given amounts of reactants. This is the amount of product that could be obtained if 100% of the limiting reactant were converted to product and if this product could be isolated from the other components in the product mixture without any loss. This calculated maximum yield is called the **theoretical yield.** Often, somewhat less than 100% of the limiting reactant is converted to product, and somewhat less than the total amount of product is isolated from the mixture, so the **actual yield** of the reaction, the amount of product that one actually obtains, is less than the theoretical yield. The actual yield is sometimes called the experimental yield. The efficiency of a reaction can be evaluated by calculating the **percent yield,** the ratio of the actual yield to the theoretical yield expressed as a percentage.

There are many reasons why the actual yield in an experiment is less than the maximum possible yield.

$$\text{Percent yield} = \frac{\text{Actual yield}}{\text{Theoretical yield}} \times 100\%$$

There are many reasons why the actual yield in a reaction might be less than the theoretical yield. One key reason is that *many chemical reactions are significantly reversible.* As soon as some products are formed, they begin to convert back to reactants, which then react again to re-form products.

OBJECTIVE 8

$$\text{Reactants} \rightleftharpoons \text{Products}$$

For example, we found in Chapter 5 that the reaction between a weak acid and water is reversible:

$$\underset{\text{acetic acid}}{HC_2H_3O_2(aq)} + \underset{\text{water}}{H_2O(l)} \rightleftharpoons \underset{\text{hydronium ion}}{H_3O^+(aq)} + \underset{\text{acetate}}{C_2H_3O_2^-(aq)}$$

When an $HC_2H_3O_2$ molecule collides with an $H_2O$ molecule, an $H^+$ ion is transferred to the water molecule to form an $H_3O^+$ ion and a $C_2H_3O_2^-$ ion. However, when an $H_3O^+$ ion and a $C_2H_3O_2^-$ ion collide, an $H^+$ ion can be transferred back again to re-form $HC_2H_3O_2$ and $H_2O$. Therefore, the reaction never goes to completion; there are always both reactants and products in the final mixtures.

OBJECTIVE 8

A reaction's yield is also affected by the occurrence of *side reactions*. This is particularly common in reactions of organic (carbon-based) compounds. Side reactions are reactions occurring in the reaction mixture that form products other than the desired product. For example, in one reaction, butane, $CH_3CH_2CH_2CH_3$, reacts with chlorine gas in the presence of light to form 2-chlorobutane and hydrogen chloride:

$$\begin{array}{ccccccccc} & H & & H & & H & & H & \\ & | & & | & & | & & | & \\ H- & C & - & C & - & C & - & C & -H \\ & | & & | & & | & & | & \\ & H & & H & & H & & H & \end{array} + Cl_2 \rightarrow \begin{array}{ccccccccc} & H & & H & & H & & H & \\ & | & & | & & | & & | & \\ H- & C & - & C & - & C & - & C & -H \\ & | & & | & & | & & | & \\ & H & & :\ddot{\underset{..}{Cl}}: & & H & & H & \end{array} + HCl$$

2-chlorobutane (72%)

However, this reaction does not lead to 100% yield of 2-chlorobutane because of an alternative reaction in which the butane and chlorine react to form 1-chlorobutane instead of 2-chlorobutane.

$$\begin{array}{ccccccccc} & H & & H & & H & & H & \\ & | & & | & & | & & | & \\ H- & C & - & C & - & C & - & C & -H \\ & | & & | & & | & & | & \\ & H & & H & & H & & H & \end{array} + Cl_2 \rightarrow \begin{array}{ccccccccc} & H & & H & & H & & H & \\ & | & & | & & | & & | & \\ H- & C & - & C & - & C & - & C & -H \\ & | & & | & & | & & | & \\ & :\ddot{\underset{..}{Cl}}: & & H & & H & & H & \end{array} + HCl$$

1-chlorobutane (28%)

Under normal conditions, the result of mixing butane with chlorine gas will be a mixture that is about 72% 2-chlorobutane and 28% 1-chlorobutane.

OBJECTIVE 8

Another factor that affects the actual yield is a reaction's *rate*. Sometimes a reaction is so slow that it has not reached the maximum yield by the time the product is isolated. Finally, as we saw above, even if 100% of the limiting reactant proceeds to products, usually the product still *needs to be separated* from the other components in the product mixture (excess reactants, products of side reactions, and other impurities). This separation process generally leads to some loss of product.

OBJECTIVE 8

Example 10.4 shows how percent yield calculations can be combined with equation stoichiometry problems.

### EXAMPLE 10.4 Percent Yield

OBJECTIVE 9

Phosphorus tribromide, $PBr_3$, can be used to add bromine atoms to alcohol molecules such as 2-methyl-1-propanol. In a student experiment, 5.393 g of 1-bromo-2-methylpropane forms when an excess of $PBr_3$ reacts with 6.034 g of 2-methyl-1-propanol:

$$\underset{\text{2-methyl-1-propanol}}{3CH_3CH(CH_3)CH_2OH} + PBr_3 \rightarrow \underset{\text{1-bromo-2-methylpropane}}{3CH_3CH(CH_3)CH_2Br} + H_3PO_3$$

What is the percent yield?

***Solution***

We calculate percent yield from the following general equation.

$$\text{Percent yield} = \frac{\text{Actual yield}}{\text{Theoretical yield}} \times 100\%$$

The actual yield is the amount of product that the chemist is able to isolate after running a reaction. It is given to you in problems like this one. Our actual yield is 5.393 g.

$$\text{Percent yield} = \frac{5.393 \text{ g } C_4H_9Br}{\text{Theoretical yield}} \times 100\%$$

We need to calculate the theoretical (or maximum) yield we would get if the 2-methyl-1-propanol, $C_4H_9OH$, which is the limiting reactant, were converted completely to 1-bromo-2-methylpropane, $C_4H_9Br$, and if the student were able to isolate the product with 100% efficiency. Because we are converting from amount of one substance in a chemical reaction to amount of another substance in the reaction, the calculation is an equation stoichiometry problem. We can use the shortcut for our setup.

$$? \text{ g } CH_3CH(CH_3)CH_2Br = 6.034 \cancel{\text{ g } CH_3CH(CH_3)CH_2OH} \left( \frac{3 \times 137.01 \text{ g } CH_3CH(CH_3)CH_2Br}{3 \times 74.123 \cancel{\text{ g } CH_3CH(CH_3)CH_2OH}} \right)$$

$$= 11.15 \text{ g } CH_3CH(CH_3)CH_2Br$$

The theoretical yield, 11.15 g $C_4H_9Br$, can be divided into the given actual yield, 5.393 g, to calculate the percent yield.

$$\text{Percent yield} = \frac{\text{Actual yield}}{\text{Theoretical yield}} \times 100\% = \frac{5.393 \text{ g } C_4H_9Br}{11.15 \text{ g } C_4H_9Br} \times 100\%$$

$$\mathbf{= 48.37\% \text{ yield}}$$

## Exercise 10.3 Percent Yield

OBJECTIVE 9

The raw material used as a source of chromium and chromium compounds is a chromium–iron ore called chromite. For example, sodium chromate, $Na_2CrO_4$, is made by roasting chromite with sodium carbonate, $Na_2CO_3$. (Roasting means heating in the presence of air or oxygen.) A simplified version of the net reaction is

$$4FeCr_2O_4 + 8Na_2CO_3 + 7O_2 \rightarrow 8Na_2CrO_4 + 2Fe_2O_3 + 8CO_2$$

What is the percent yield if 1.2 kg of $Na_2CrO_4$ is produced from ore that contains 1.0 kg of $FeCr_2O_4$?

## Special Topic 10.1 Big Problems Require Bold Solutions—Global Warming and Limiting Reactants

There is general agreement in the scientific community that the average temperature of the earth is increasing and that this global warming is potentially a major problem. The increase in temperature is expected to cause more frequent and intense heat waves, ecological disruptions that could lead certain types of forests to disappear and some species to become extinct, a decline in agricultural production that could result in hunger and famine, an expansion of deserts, and a rise in sea level.

The global system that regulates the earth's temperature is very complex, but many scientists believe the increase in temperature is caused by an increase of certain gases in the atmosphere that trap energy that would otherwise escape into space. These gases, which are called *greenhouse gases,* include carbon dioxide, methane, nitrous oxide, chlorofluorocarbons (CFCs), and the ozone in the lower atmosphere. You can visit the following Web site to see a description of how these gases contribute to global warming.

www.chemplace.com/college/

Because carbon dioxide plays a major role in global warming, many of the proposed solutions to the problem are aimed at reducing the levels of carbon dioxide in the atmosphere. One suggestion is to try altering the chemistry of the earth's oceans such that they will absorb more $CO_2$ from the air.

Huge amounts of carbon dioxide from the air are constantly dissolving in the ocean, and at the same time, carbon dioxide is leaving the ocean and returning to the air. The ocean takes about 100 billion tons of carbon dioxide out of the atmosphere per year and ultimately returns 98 billion tons of it. The remaining 2 billion tons end up as organic deposits on the sea floor. If the rate of escape from the ocean to the air could be slowed, the net shift of $CO_2$ from the air to the ocean would increase, and the levels of $CO_2$ in the air would fall. The goal is to find some way to increase the use of $CO_2$ in the ocean before it can escape.

Phytoplankton, the microorganisms that form the base of the food web in the ocean, take $CO_2$ from the air and convert it into more complex organic compounds. When the phytoplankton die, they fall to the sea floor, and the carbon they contain becomes trapped in the sediment. If the rate of growth and reproduction of these organisms could be increased they would use more carbon dioxide before it could escape from the ocean into the air. The late John Martin of the Moss Landing Marine Laboratories in California suggested a way of making this happen.

It has been known since the 1980s that large stretches of the world's ocean surface receive plenty of sunlight and possess an abundance of the major nutrients and yet contain fairly low levels of phytoplankton. One possible explanation for this low level of growth was that the level of some trace nutrient in the water was low. This nutrient would be acting as a *limiting reactant* in chemical changes necessary for the growth and reproduction of organisms.

Dr. Martin hypothesized that the limiting factor was iron. Iron is necessary for a number of crucial functions

*(continued)*

SPECIAL TOPIC 10.1 (CONTINUED)

of phytoplankton, including the production of chlorophyll. He suggested that an increase in the iron concentration of the ocean would stimulate phytoplankton growth and that more carbon dioxide would be drawn from the atmosphere to fuel that growth. In a 1988 seminar at Woods Hole Oceanographic Institution, Dr. Martin ventured a bold statement: *"Give me a tankerload of iron, and I'll give you an ice age."*

The first tests done in the laboratory had positive results, but the conditions in the laboratory could not fully duplicate the conditions in the ocean. Dr. Martin's next bold suggestion was to do a large-scale test in the ocean itself. Because of the difficulty of controlling conditions in the real world, ocean tests on the scale he suggested had never been done.

Sadly, Dr. Martin did not live long enough to see the results of the ocean experiments that have now been done. Two of these experiments were conducted in the Pacific Ocean about 300 miles south and 800 miles west of the Galapagos Islands. In October of 1993, about 450 kg of iron(II) sulfate was spread over about 64 square kilometers of ocean, and the chemistry and biology of the area were monitored for nine days. The rate of phytoplankton growth increased significantly, and the levels of $CO_2$ in the water decreased. These results confirmed Dr. Martin's main hypothesis, that iron was a limiting factor in the growth and reproduction of phytoplankton. Unfortunately, the $CO_2$ concentration leveled off after only one day at about 10% of the expected drop. A second test in 1995 led to 20 times the normal abundance of phytoplankton and to a significant drop in $CO_2$ levels in the ocean, but as soon as the iron was gone, the levels returned to normal.

Because the decrease in $CO_2$ in the water was less than expected and because the increase in phytoplankton growth requires a constant addition of iron, it seems unlikely now that fertilizing the oceans with iron is going to solve the problem of global warming. This is disappointing, but it should not surprise us that a problem this big cannot be reversed by a single solution. The fact that fertilizing the oceans with iron did markedly stimulate the phytoplankton growth encourages others to seek similarly bold solutions without being discouraged by the magnitude of the task.

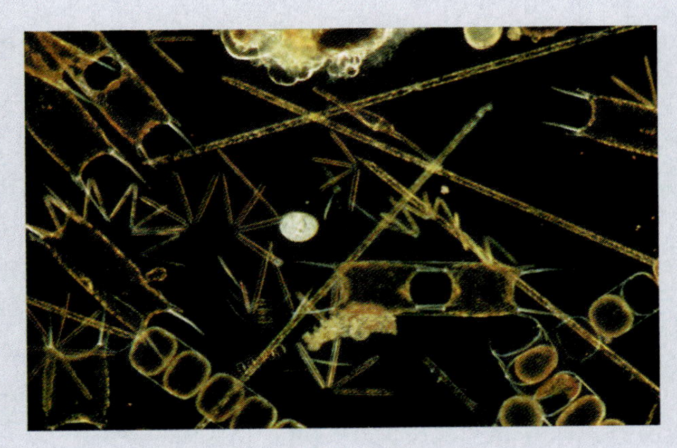

If you can find a way to increase the abundance of phytoplankton, you may be instrumental in ending global warming.

## 10.3 Molarity and Equation Stoichiometry

The general steps we have been following in setting up equation stoichiometry calculations can be summarized as follows, where # again stands for "substance."

For pure liquids and solids, as we have seen, the most convenient measurable property is mass. It is very easy to measure the mass of a pure solid or liquid, and we can convert between that and the number of particles it represents using the molar mass as a conversion factor.

Although these steps describe many important calculations, their usefulness is limited by the fact that many reactions are run in the gas phase or in solution, where the determination of a reactant's mass or of a product's mass is more difficult. Equation stoichiometry involving gases is described in Chapter 13. This section shows how we can do equation stoichiometry problems for reactions run in solution.

## Reactions in Solution and Molarity

To see why many reactions are run in solution and why we need a new component for our equation stoichiometry problems, let's assume that we want to make silver phosphate, $Ag_3PO_4$, a substance used in photographic emulsions and in the production of pharmaceuticals. It is made by reacting silver nitrate, $AgNO_3$, and sodium phosphate, $Na_3PO_4$. Both of these substances are solids at room temperature. If the solids are mixed, no reaction takes place.

$$AgNO_3(s) + Na_3PO_4(s) \quad \text{No reaction}$$

In the solid form, the $Ag^+$ ions are still tightly bound to the $NO_3^-$ ions, and the $PO_4^{3-}$ ions are still linked to the $Na^+$ ions, so the $Ag^+$ ions and the $PO_4^{3-}$ ions are not able to move together and form silver phosphate.

For the reaction between $AgNO_3$ and $Na_3PO_4$ to proceed, the two reactants must first be dissolved in water. In solution, the $Ag^+$ ions are separate from the $NO_3^-$ ions, and the $PO_4^{3-}$ ions are separate from the $Na^+$ ions, and all of these ions move freely throughout the liquid. This allows the $Ag^+$ and $PO_4^{3-}$ ions to find each other in solution and combine to precipitate from the solution as the yellow solid, $Ag_3PO_4$.

$$3AgNO_3(aq) + Na_3PO_4(aq) \rightarrow Ag_3PO_4(s) + 3NaNO_3(aq)$$

These two substances, silver nitrate and sodium phosphate, will not react in solid form (top). They must first be dissolved in water (bottom).

When two solutions are mixed to start a reaction, it is more convenient to measure their volumes than their masses. Therefore, in equation stoichiometry problems for such reactions, it is common for the chemist to want to convert back and forth between volume of solution and moles of the reacting substance in the solution. For example, you might be asked to calculate the volume of an $AgNO_3$ solution that must be added to 25.00 mL of a solution of $Na_3PO_4$ to precipitate all of the phosphate as $Ag_3PO_4$. Because we want to convert from amount of one substance in a given chemical reaction to amount of another substance participating in the same reaction, we recognize this as an equation stoichiometry problem. Thus we know that in the center of our conversion, we will convert from moles of $Na_3PO_4$ to moles of $AgNO_3$. The steps in our calculation will be

The general steps for similar calculations (# stands for "substance") will be

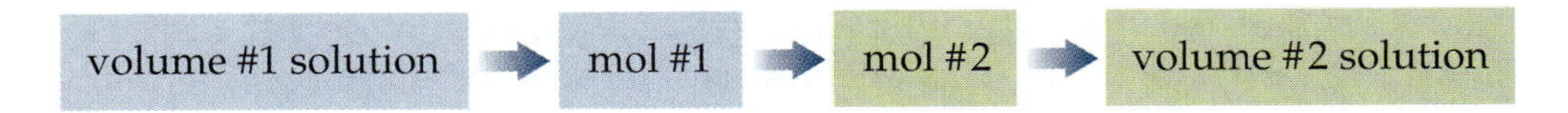


Therefore, we need conversion factors that convert back and forth between volumes of solutions and moles of solute in those solutions. **Molarity** (abbreviated M), which is defined as moles of solute per liter of solution, provides such conversion factors.

$$\text{Molarity} = \frac{\text{moles of solute}}{\text{liter of solution}}$$

For example, you might be told that a solution is 0.500 M $Na_3PO_4$. (This is read "0.500 molar $Na_3PO_4$" or "0.500 molar sodium phosphate.") This means that there are 0.500 mole of $Ag_3PO_4$ in 1 liter of this solution. Because 1 liter is $10^3$ milliliters, two useful conversion factors can be derived from the molarity of $Ag_3PO_4$:

$$0.500 \text{ M } Ag_3PO_4 \quad \text{means}$$

$$\frac{0.500 \text{ mol } Ag_3PO_4}{1 \text{ L } Ag_3PO_4 \text{ solution}} \quad \text{or} \quad \frac{0.500 \text{ mol } Ag_3PO_4}{10^3 \text{ mL } Ag_3PO_4 \text{ solution}}$$

Example 10.5 shows how to use dimensional analysis to determine the molarity of a solution.

### EXAMPLE 10.5 Calculating a Solution's Molarity

OBJECTIVE 10

A solution was made by dissolving 8.20 g of sodium phosphate in water and then diluting the mixture with water to achieve a total volume of 100.0 mL. What is the solution's molarity?

***Solution***

Because the answer we want, molarity, is a ratio of two units (moles of solute—in this case, $Na_3PO_4$—per liter of solution), we start our dimensional analysis with a ratio of two units. Because we want amount of $Na_3PO_4$ on the top when we are finished, we start with 8.20 g $Na_3PO_4$ on the top. Because we want volume of solution on the bottom when we are finished, we start with 100.0 mL of solution on the bottom. To convert mass of $Na_3PO_4$ to moles of $Na_3PO_4$, we use the molar mass of $Na_3PO_4$. We finish our conversion with a conversion factor that converts milliliters to liters.

Molarity expressed with more specific units — Given amount of solute — Converts mass to moles — Converts the given volume unit into the desired volume unit — Given amount of solution

$$\text{Molarity} = \frac{?\ \text{mol Na}_3\text{PO}_4}{1\ \text{L Na}_3\text{PO}_4\ \text{soln}} = \frac{8.20\ \text{g Na}_3\text{PO}_4}{100\ \text{mL Na}_3\text{PO}_4\ \text{soln}}\left(\frac{1\ \text{mol Na}_3\text{PO}_4}{163.9408\ \text{g Na}_3\text{PO}_4}\right)\left(\frac{10^3\ \text{mL}}{1\ \text{L}}\right)$$

$$= \frac{0.500\ \text{mol Na}_3\text{PO}_4}{1\ \text{L Na}_3\text{PO}_4\ \text{soln}} = \mathbf{0.500\ M\ Na_3PO_4}$$

## EXERCISE 10.4 Calculating a Solution's Molarity

OBJECTIVE 10

A silver perchlorate solution was made by dissolving 29.993 g of pure $AgClO_4$ in water and then diluting the mixture with additional water to achieve a total volume of 50.00 mL. What is the solution's molarity?

## Equation Stoichiometry and Reactions in Solution

Conversion factors constructed from molarities can be used in stoichiometric calculations in very much the same way as conversion factors from molar mass are used. When a substance is pure, its molar mass can be used to convert back and forth between the measurable property of mass and moles. When a substance is in solution, its molarity can be used to convert between the measurable property of volume of solution and moles of solute.

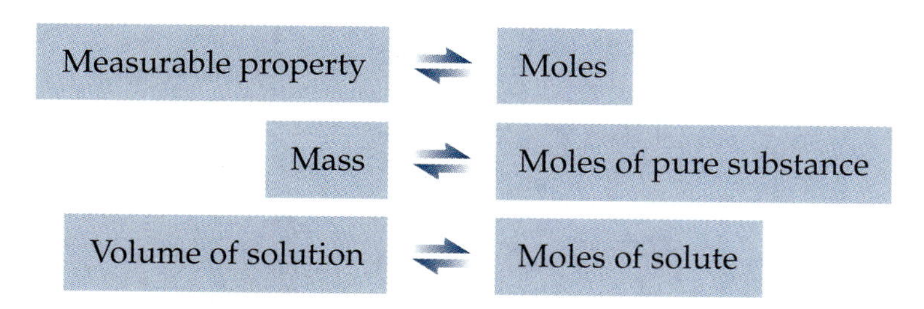


To see how molarity can be used in equation stoichiometry problems, let's take a look at the thought process for calculating the number of milliliters of 1.00 M $AgNO_3$ necessary to precipitate the phosphate from 25.00 mL of 0.500 M $Na_3PO_4$. The problem asks us to convert from amount of one substance in a chemical reaction to amount of another substance in the reaction, so we know it is an equation stoichiometry problem. The core of our setup will be the conversion factor for changing moles of sodium phosphate to moles of silver nitrate. To construct it, we need to know the molar ratio of $AgNO_3$ to $Na_3PO_4$, which comes from the balanced equation for the reaction.

$$3AgNO_3(aq) + Na_3PO_4(aq) \rightarrow Ag_3PO_4(s) + 3NaNO_3(aq)$$

The balanced equation tells us that three formula units of $AgNO_3$ are required for each formula unit of $Na_3PO_4$, or three moles of $AgNO_3$ for each mole of $Na_3PO_4$.

Our general steps will be

mL $Na_3PO_4$ soln → mol $Na_3PO_4$ → mol $AgNO_3$ → mL $AgNO_3$ soln

We start our dimensional analysis in the usual way, setting what we want to know (the milliliters of silver nitrate solution) equal to one of the values we are given. Because we want a single unit in our final answer, we set milliliters of $AgNO_3$ solution equal to the only given value that has a single unit: 25.00 mL $Na_3PO_4$ solution. (1.00 M $AgNO_3$ and 0.500 M $Na_3PO_4$ may look like they have single units, but remember that molarity is actually a ratio of two units: moles and liters)

$$? \text{ mL } AgNO_3 \text{ soln} = 25.00 \text{ mL } Na_3PO_4 \text{ soln}\left(\frac{}{\text{mL } Na_3PO_4 \text{ soln}}\right)$$

We need to convert from volume of $Na_3PO_4$ solution to moles of $Na_3PO_4$. The molarity of sodium phosphate provides two possible conversion factors for this conversion.

OBJECTIVE 11

0.500 M $Na_3PO_4$ means

$$\frac{0.500 \text{ mol } Na_3PO_4}{1 \text{ L } Na_3PO_4 \text{ solution}} \quad \text{or} \quad \frac{0.500 \text{ mol } Na_3PO_4}{10^3 \text{ mL } Na_3PO_4 \text{ solution}}$$

These two conversion factors provide two ways to convert from milliliters of $H_3PO_4$ solution to moles of $H_3PO_4$.

OBJECTIVE 12

$$? \text{ mL } AgNO_3 \text{ soln} = 25.00 \text{ mL } Na_3PO_4 \text{ soln}\left(\frac{1 \text{ L}}{10^3 \text{ mL}}\right)\left(\frac{0.500 \text{ mol } Na_3PO_4}{1 \text{ L } Na_3PO_4 \text{ soln}}\right)$$

$$? \text{ mL } AgNO_3 \text{ soln} = 25.00 \text{ mL } Na_3PO_4 \text{ soln}\left(\frac{0.500 \text{ mol } Na_3PO_4}{10^3 \text{ mL } Na_3PO_4 \text{ soln}}\right)$$

The second setup requires one less conversion factor than the first setup, so we will use it. We can now use the molar ratio to convert from moles of $Na_3PO_4$ to moles of $AgNO_3$.

$$? \text{ mL } AgNO_3 \text{ soln} = 25.00 \text{ mL } Na_3PO_4 \text{ soln}\left(\frac{0.500 \text{ mol } Na_3PO_4}{10^3 \text{ mL } Na_3PO_4 \text{ soln}}\right)\left(\frac{3 \text{ mol } AgNO_3}{1 \text{ mol } Na_3PO_4}\right)$$

We complete the problem by using the molarity of $AgNO_3$ to convert from moles of $AgNO_3$ to volume of $AgNO_3$ solution. The value 1.00 M

$AgNO_3$ provides us with two possible conversion factors. Like any conversion factors, they can be used in the form you see here or inverted.

$$\frac{1.00 \text{ mol } AgNO_3}{1 \text{ L } AgNO_3 \text{ solution}} \quad \text{or} \quad \frac{1.00 \text{ mol } AgNO_3}{10^3 \text{ mL } AgNO_3 \text{ solution}}$$

OBJECTIVE 11

The two possible setups for the problem are below.

OBJECTIVE 12 OBJECTIVE 13

Molarity as a conversion factor converts milliliters into moles. Coefficients from balanced equation convert moles of 1 substance into moles of another substance. Molarity as a conversion factor converts moles into milliliters.

$$? \text{ mL } AgNO_3 \text{ soln} = 25.00 \text{ mL } Na_3PO_4 \text{ soln}\left(\frac{0.500 \text{ mol } Na_3PO_4}{10^3 \text{ mL } Na_3PO_4 \text{ soln}}\right)\left(\frac{3 \text{ mol } AgNO_3}{1 \text{ mol } Na_3PO_4}\right)\left(\frac{10^3 \text{ mL } AgNO_3 \text{ soln}}{1.00 \text{ mol } AgNO_3}\right)$$

or

Converts given unit into liters. Coefficients from balanced equation convert moles of 1 substance into moles of another substance. Converts liters into desired unit

$$? \text{ mL } AgNO_3 \text{ soln} = 25.00 \text{ mL } Na_3PO_4 \text{ soln}\left(\frac{1 \text{ L}}{10^3 \text{ mL}}\right)\left(\frac{0.500 \text{ mol } Na_3PO_4}{1 \text{ L } Na_3PO_4 \text{ soln}}\right)\left(\frac{3 \text{ mol } AgNO_3}{1 \text{ mol } Na_3PO_4}\right)\left(\frac{1 \text{ L } AgNO_3 \text{ soln}}{1.00 \text{ mol } AgNO_3}\right)\left(\frac{10^3 \text{ mL}}{1 \text{ L}}\right)$$

Molarity as a conversion factor converts liters into moles. Molarity as a conversion factor converts moles into liters.

**= 37.5 mL $AgNO_3$ solution**

Example 10.6 shows a very similar case.

## EXAMPLE 10.6 Molarity and Equation Stoichiometry

OBJECTIVE 11 OBJECTIVE 12 OBJECTIVE 13

How many milliliters of 2.00 M sodium hydroxide are necessary to neutralize 25.00 mL of 1.50 M phosphoric acid?

***Solution***

Because we are asked to convert from the amount of one substance in a chemical reaction to the amount of another substance in the reaction, we recognize the problem as an equation stoichiometry problem. Our general steps are

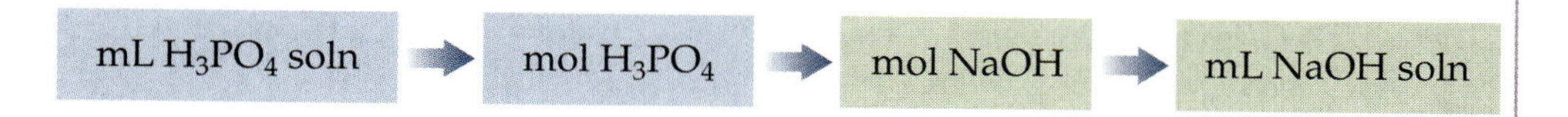


The molarity of phosphoric acid provides the conversion factor that converts from volume of $H_3PO_4$ solution to moles of $H_3PO_4$, and the molarity of sodium hydroxide provides the conversion factor that converts from moles of NaOH to volume of NaOH solution. The conversion from moles of $H_3PO_4$ to moles of

NaOH is made with the molar ratio that comes from the coefficients in the balanced equation for the reaction.

$$3NaOH(aq) + H_3PO_4(aq) \rightarrow Na_3PO_4(aq) + 3H_2O(l)$$

The two possible setups for the problem are

$$? \text{ mL NaOH soln} = 25.00 \text{ mL } H_3PO_4 \text{ soln} \left(\frac{1.50 \text{ mol } H_3PO_4}{10^3 \text{ mL } H_3PO_4 \text{ soln}}\right)\left(\frac{3 \text{ mol NaOH}}{1 \text{ mol } H_3PO_4}\right)\left(\frac{10^3 \text{ mL NaOH soln}}{2.00 \text{ mol NaOH}}\right)$$

or

$$? \text{ mL NaOH soln} = 25.00 \text{ mL } H_3PO_4 \text{ soln} \left(\frac{1 \text{ L}}{10^3 \text{ mL}}\right)\left(\frac{1.50 \text{ mol } H_3PO_4}{1 \text{ L } H_3PO_4 \text{ soln}}\right)\left(\frac{3 \text{ mol NaOH}}{1 \text{ mol } H_3PO_4}\right)\left(\frac{1 \text{ L NaOH soln}}{2.00 \text{ mol NaOH}}\right)\left(\frac{10^3 \text{ mL}}{1 \text{ L}}\right)$$

$$= \mathbf{56.3 \text{ mL NaOH solution}}$$

Molar mass conversions and molarity conversions can be combined to solve equation stoichiometry problems, as we see in Example 10.7.

## EXAMPLE 10.7 Molarity and Equation Stoichiometry

OBJECTIVE 13

What volume of 6.00 M HCl is necessary to neutralize and dissolve 31.564 g of solid aluminum hydroxide?

***Solution***

Our steps for this equation stoichiometry problem are

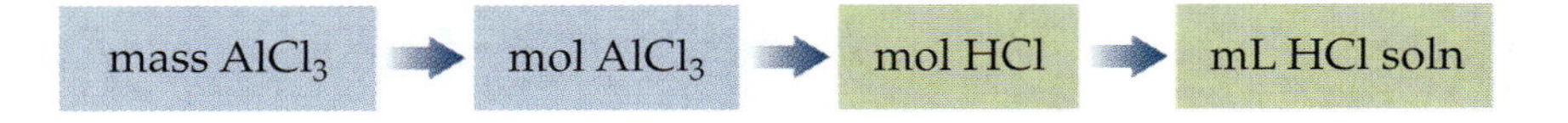


The dimensional analysis is built around a mole-to-mole conversion factor created from the coefficients in the balanced equation. The reaction is a neutralization reaction, so you can use the skills you developed in Chapter 5 to write the balanced equation.

$$3HCl(aq) + Al(OH)_3(s) \rightarrow 3H_2O(l) + AlCl_3(aq)$$

We use molar mass to convert mass of $Al(OH)_3$ to moles of $Al(OH)_3$, and we use molarity to convert moles of HCl to volume of HCl solution.

$$? \text{ mL HCl soln} = 31.564 \text{ g } Al(OH)_3 \left(\frac{1 \text{ mol } Al(OH)_3}{78.0036 \text{ g } Al(OH)_3}\right)\left(\frac{3 \text{ mol HCl}}{1 \text{ mol } Al(OH)_3}\right)\left(\frac{10^3 \text{ mL HCl soln}}{6.00 \text{ mol HCl}}\right)$$

$$= \mathbf{202 \text{ mL HCl solution}}$$

The following is a general study sheet for the types of equation stoichiometry problems that we have considered in this chapter.

## SAMPLE STUDY SHEET 10.3

### Equation Stoichiometry Problems

OBJECTIVE 13

**TIP-OFF** You are asked to convert from amount of one substance in a chemical reaction to amount of another substance in the reaction.

**GENERAL PROCEDURE** Use the dimensional analysis format to make the following conversions. (Figure 10.4, on the following page, summarizes these steps.)

1. If you are not given it, write and balance the chemical equation for the reaction.
2. Start your dimensional analysis in the usual way, setting the desired units of substance 2 equal to the given units of substance 1.
3. Convert from the units that you are given for substance 1 to moles of substance 1.
   - For pure solids and liquids, this means converting grams to moles using the molar mass of the substance. (It might be necessary to insert one or more additional conversion factors to convert from the given mass unit to grams.)
   - Molarity can be used to convert from volume of solution to moles of solute. (It might be necessary to insert one or more additional conversion factors to convert from the given volume unit to liters or milliliters.)
4. Convert from moles of substance 1 to moles of substance 2 using the coefficients from the balanced equation.
5. Convert from moles of substance 2 to the desired units for substance 2.
   - For pure solids and liquids, this means converting from moles to mass using the molar mass of substance 2.
   - Molarity can be used to convert from moles of solute to volume of solution.
6. Calculate your answer and report it with the correct number of significant figures (in scientific notation, if necessary) and with the correct unit.

**EXAMPLES** See Examples 10.6 and 10.7.

## EXERCISE 10.5 Molarity and Equation Stoichiometry

OBJECTIVE 13

How many milliliters of 6.00 M $HNO_3$ are necessary to neutralize the carbonate in 75.0 mL of 0.250 M $Na_2CO_3$?

## EXERCISE 10.6 Molarity and Equation Stoichiometry

OBJECTIVE 13

What is the maximum number of grams of silver chloride that will precipitate from a solution made by mixing 25.00 mL of 0.050 M $MgCl_2$ with an excess of $AgNO_3$ solution?

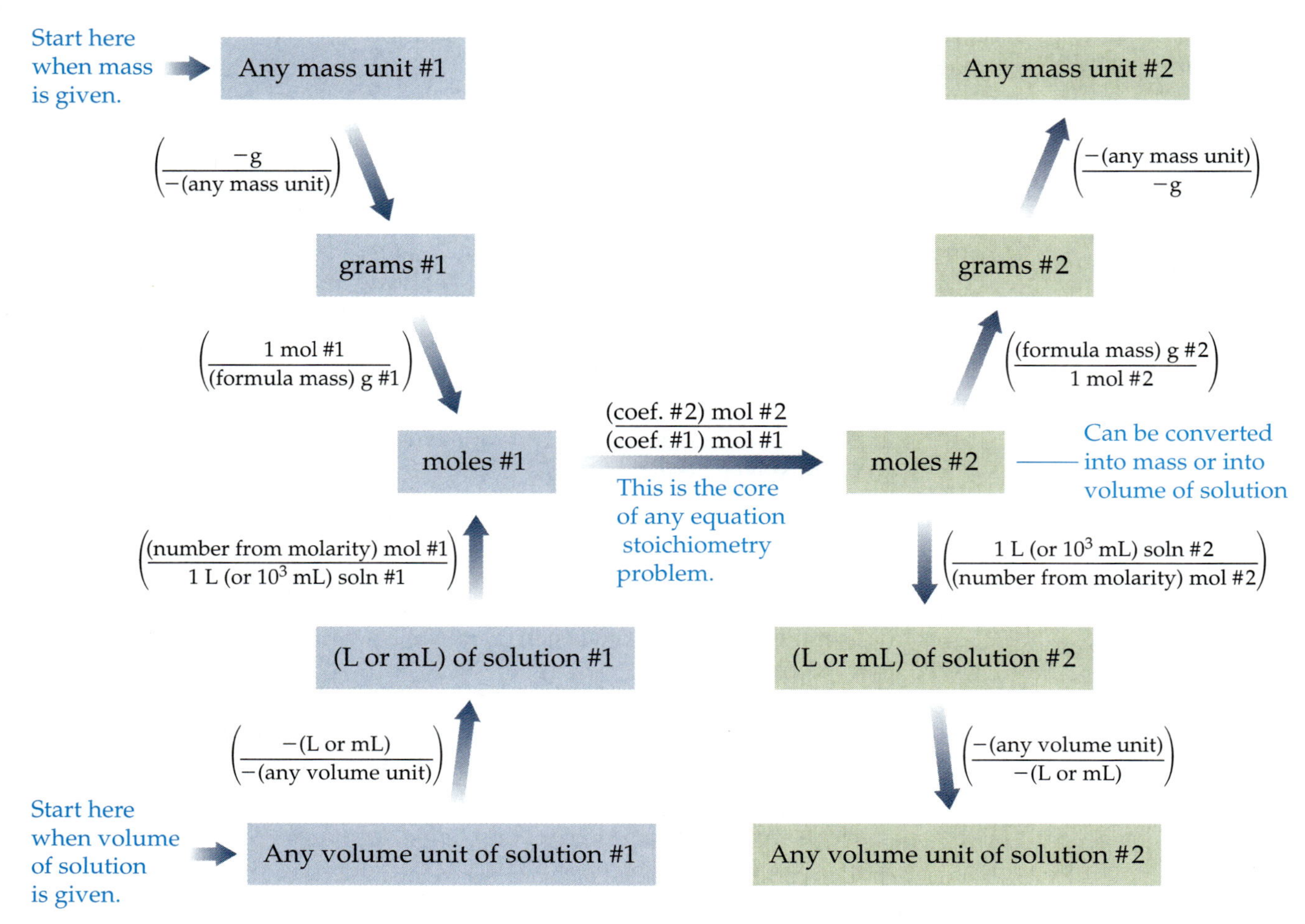


**Figure 10.4**
General Steps for Equation Stoichiometry. (Here # stands for "substance.")

Visit the following Web site to see a description of a procedure called titration that can be used to determine the molarities of acidic and basic solutions:

**www.chemplace.com/college/**

## Chapter Glossary

**Equation stoichiometry** Calculations that make use of the quantitative relationships between the substances in a chemical reaction to convert the amount of one substance in the chemical reaction to the amount of a different substance in the reaction.

**Limiting reactant** The reactant that runs out first and hence limits the amount of product that can form.

**Theoretical yield** The calculated maximum amount of product that can form in a chemical reaction.

**Actual yield** The amount of product that is actually obtained in a chemical reaction.

**Percent yield** The actual yield divided by the theoretical yield times 100.

**Molarity** (abbreviated M) Moles of solute per liter of solution.

You can test yourself on the glossary terms at the following Web site:

**www.chemplace.com/college/**

## Chapter Objectives

**The goal of this chapter is to teach you to do the following.**

1. Define all of the terms in the Chapter 10 Glossary.

### Section 10.1 Equation Stoichiometry

2. Given a balanced chemical equation (or enough information to write one), construct conversion factors that relate moles of any two reactants or products in the reaction.
3. Given a balanced chemical equation (or enough information to write one), convert from moles of one reactant or product to moles of any other reactant or product.
4. Given a balanced chemical equation (or enough information to write one), convert from mass of one reactant or product to mass of any other reactant or product.

### Section 10.2 Real-World Applications of Equation Stoichiometry

5. Describe four reasons for adding an excess of one or more of the reactants in a chemical reaction.
6. Given the masses of two or more reactants, calculate the maximum mass of product that could form from the reaction between them.
7. Given a description of a specific chemical reaction, explain why a chemist might decide to add one of the reactants (rather than the others) in excess.

8. Describe four reasons why the actual yield in a chemical reaction is less than the theoretical yield.
9. Given an actual yield and a theoretical yield (or enough information to calculate a theoretical yield) for a chemical reaction, calculate the percent yield for the reaction.

**Section 10.3 Molarity and Equation Stoichiometry**

10. Given the mass of substance dissolved in a solution and the total volume of the solution, calculate the molarity of the solution.
11. Given the molarity of a substance, write a conversion factor that relates the moles of substance to the liters of solution and a conversion factor that relates the moles of solute to the milliliters of solution.
12. Use molarity to convert between moles of solute and volume of solution.
13. Convert between amount of one substance in a chemical reaction and amount of another substance in the reaction, when the amounts of the substances are described by either of the following:
    a. Mass of pure substance
    b. Volume of a solution that contains one of the substances, along with the molarity of the solution

## Review Questions

This chapter requires many of the same skills that were listed as review skills for Chapter 9. Thus you should make sure that you can do the Review Questions at the end of Chapter 9 before continuing with the questions that follow.

1. Write balanced equations for the following reactions. You do *not* need to include the substance's states.
    a. Hydrofluoric acid reacts with silicon dioxide to form silicon tetrafluoride and water.
    b. Ammonia reacts with oxygen gas to form nitrogen monoxide and water.
    c. Water solutions of nickel(II) acetate and sodium phosphate react to form solid nickel(II) phosphate and aqueous sodium acetate.
    d. Phosphoric acid reacts with potassium hydroxide to form water and potassium phosphate.
2. Write complete equations, including states, for the precipitation reaction that takes place between the reactants in part (a) and the neutralization reaction that takes place in part (b).
    a. $Ca(NO_3)_2(aq) + Na_2CO_3(aq) \rightarrow$
    b. $HNO_3(aq) + Al(OH)_3(s) \rightarrow$
3. How many moles of phosphorous acid, $H_3PO_3$, are there in 68.785 g of phosphorous acid?
4. What is the mass in kilograms of 0.8459 mole of sodium sulfate?

## Key Ideas

**Complete the following statements by writing one of these words or phrases in each blank.**

| | |
|---|---|
| 100% | remaining |
| actual | reversible |
| amount of product | runs out first |
| balanced chemical equation | separated |
| costly | side reactions |
| easy to separate | slow |
| excess | solute |
| less | solution |
| loss | stoichiometric |
| mass | stoichiometry |
| molar mass | two or more amounts |
| optimize | volume of solution |
| remain | volumes |

5. If a calculation calls for you to convert from an amount of one substance in a given chemical reaction to the corresponding amount of another substance participating in the same reaction, it is an equation ________________ problem.
6. The coefficients in a(n) ________________ provide us with information that we can use to build conversion factors that convert from an amount of one substance in a given chemical reaction to the corresponding amount of another substance participating in the same reaction.
7. For some chemical reactions, chemists want to mix reactants in amounts that are as close as possible to the ratio that would lead to the complete reaction of each. This ratio is sometimes called the ________________ ratio.
8. A(n) ________________ of one or more of the reactants will increase the likelihood that the other reactant or reactants will be used up. Thus, if one reactant is more expensive than the others, adding an excess of the ________________ expensive reactants will ensure the greatest conversion possible of the one that is ________________.
9. Sometimes one product is more important than others are, and the amounts of reactants are chosen to ________________ its production.
10. Any component added in excess will ________________ when the reaction is complete. If one reactant is more dangerous to handle than others are, chemists would rather not have that reactant ________________ at the reaction's end.
11. Because some of the reactant that was added in excess is likely to be mixed in with the product, chemists would prefer that the substance in excess be a substance that is ________________ from the primary product.
12. The reactant that ________________ in a chemical reaction limits the amount of product that can form. This reactant is called the limiting reactant.
13. The tip-off for limiting reactant problems is that you are given ________________ of reactants in a chemical reaction, and you are asked to calculate the maximum ________________ that they can form.

14. The theoretical yield is the maximum amount of product that could be formed from the given amounts of reactants. This is the amount of product that could be obtained if ________________ of the limiting reactant were converted to product and if this product could be isolated from the other components in the product mixture without any ________________. The efficiency of a reaction can be evaluated by calculating the percent yield, the ratio of the ________________ yield to the theoretical yield expressed as a percentage.
15. There are many reasons why the actual yield in a reaction might be less than the theoretical yield. One key reason is that many chemical reactions are significantly ________________.
16. A reaction's yield is also affected by the occurrence of ________________.
17. Another factor that affects the actual yield is a reaction's rate. Sometimes a reaction is so ________________ that it has not reached the maximum yield by the time the product is isolated.
18. Even if 100% of the limiting reactant proceeds to products, usually the product still needs to be ________________ from the other components in the product mixture (excess reactants, products of side reactions, and other impurities). This process generally leads to some loss of product.
19. When two solutions are mixed to start a reaction, it is more convenient to measure their ________________ than their masses.
20. Molarity (abbreviated M) is defined as moles of ________________ per liter of ________________.
21. Conversion factors constructed from molarities can be used in stoichiometric calculations in very much the same way conversion factors from ________________ are used. When a substance is pure, its molar mass can be used to convert back and forth between the measurable property of ________________ and moles. When a substance is in solution, its molarity can be used to convert between the measurable property of ________________ and moles of solute.

## Chapter Problems

### Section 10.1 Equation Stoichiometry
### Section 10.2 Real-World Applications of Equation Stoichiometry

OBJECTIVE 2 OBJECTIVE 3

22. Because the bond between fluorine atoms in $F_2$ is relatively weak while the bonds between fluorine atoms and atoms of other elements are relatively strong, it is difficult to make diatomic fluorine, $F_2$. One way it can be made is to run an electric current through liquid hydrogen fluoride, HF. This reaction yields hydrogen gas, $H_2$, and fluorine gas, $F_2$.
    a. Write a complete balanced equation, including states, for this reaction.
    b. Draw a picture of the reaction, using rough sketches of space-filling models in place of the coefficients and formulas in the equation. Fluorine atoms have a little more than twice the diameter of hydrogen atoms.
    c. Write a conversion factor that could be used to convert between moles of HF and moles of $F_2$.
    d. How many moles of $F_2$ form when one mole of HF reacts completely?
    e. How many moles of HF react to yield 3.452 moles of $H_2$?

23. Hydrogen gas is used for many purposes, including the hydrogenation of vegetable oils to make margarine. The most common industrial process for producing hydrogen is "steam-reforming," in which methane gas, $CH_4$, from natural gas reacts with water vapor to form carbon monoxide gas and hydrogen gas.

OBJECTIVE 2 OBJECTIVE 3

   a. Write a complete balanced equation, including states, for this reaction.
   b. Draw a picture of the reaction, using rough sketches of space-filling models in place of the coefficients and formulas in the equation. Draw the carbon atoms a little larger than the oxygen atoms. Both carbon and oxygen atoms have over twice the diameter of hydrogen atoms.
   c. Write a conversion factor that could be used to convert between moles of methane and moles of hydrogen.
   d. How many moles of hydrogen form when 4 moles of methane react completely?
   e. How many moles of water vapor react to yield 174.82 moles of hydrogen?

24. The bond between nitrogen atoms in $N_2$ molecules is very strong, making $N_2$ very unreactive. Because of this, magnesium is one of the few metals that react with nitrogen gas directly. This reaction yields solid magnesium nitride.

OBJECTIVE 2 OBJECTIVE 3

   a. Write a complete balanced equation, without including states, for the reaction between magnesium and nitrogen to form magnesium nitride.
   b. Write a conversion factor that could be used to convert between moles of magnesium and moles of magnesium nitride.
   c. How many moles of magnesium nitride form when 1.0 mole of magnesium reacts completely?
   d. Write a conversion factor that could be used to convert between moles of nitrogen and moles of magnesium nitride.
   e. How many moles of nitrogen react to yield 3.452 moles of magnesium nitride?

25. Fluorine gas is an important chemical because it is used to add fluorine atoms to many different compounds. As mentioned in Problem 22, it is difficult to make, but the following two-step process produces fairly high yields of $F_2$.

OBJECTIVE 2 OBJECTIVE 3

$$2KMnO_4 + 2KF + 10HF + 3H_2O_2 \rightarrow 2K_2MnF_6 + 8H_2O + 3O_2$$

$$2K_2MnF_6 + 4SbF_5 \rightarrow 4KSbF_6 + 2MnF_3 + F_2$$

For the second of these two reactions:

   a. Write a conversion factor that could be used to convert between moles of antimony pentafluoride, $SbF_5$, and moles of fluorine, $F_2$.
   b. How many moles of $F_2$ form when 8 moles of $SbF_5$ react completely?
   c. What is the maximum number of moles of $F_2$ that could form in the combination of 2.00 moles of $K_2MnF_6$ and 5.00 moles of $SbF_5$?
   d. What is the maximum number of moles of $F_2$ that could form in the combination of 2 moles of $K_2MnF_6$ and 5,000 moles of $SbF_5$?
   e. Write a conversion factor that could be used to convert between moles of manganese(III) fluoride, $MnF_3$, and moles of $F_2$.
   f. How many moles of $F_2$ form along with 0.802 mole of $MnF_3$?

OBJECTIVE 2 OBJECTIVE 3

26. For many years, it was thought that no reaction could produce sodium perbromate, but the discovery of a reaction producing the equally elusive xenon difluoride, $XeF_2$, led to the discovery of the following reaction that yields sodium perbromate.

$$NaBrO_3 + XeF_2 + H_2O \rightarrow NaBrO_4 + 2HF + Xe$$

a. Write a conversion factor that could be used to convert between moles of xenon difluoride, $XeF_2$, and moles of hydrogen fluoride, HF.

b. How many moles of $XeF_2$ are necessary to form 16 moles of hydrogen fluoride?

c. What is the maximum number of moles of $NaBrO_4$ that could form in the combination of 2 moles of $NaBrO_3$ and 3 moles of $XeF_2$?

d. What is the maximum number of moles of $NaBrO_4$ that could form in the combination of 2 moles of $NaBrO_3$ and 3 million moles of $XeF_2$?

e. Write a conversion factor that could be used to convert between moles of sodium perbromate, $NaBrO_4$, and moles of hydrogen fluoride, HF.

f. How many moles of HF form along with 5.822 moles of sodium perbromate, $NaBrO_4$?

OBJECTIVE 2 OBJECTIVE 3

27. The thiocyanate polyatomic ion, $SCN^-$, is commonly called a pseudohalogen because it acts very much like halide ions. For example, we know that the pure halogens consist of diatomic molecules, such as $Cl_2$. Thiocyanate ions form similar molecules in the following reaction:

$$2NaSCN + 2H_2SO_4 + MnO_2 \rightarrow (SCN)_2 + 2H_2O + MnSO_4 + Na_2SO_4$$

a. Write a conversion factor that could be used to convert between moles of NaSCN and moles of $(SCN)_2$.

b. How many moles of $(SCN)_2$ form when 0.50 moles of NaSCN react completely?

c. What is the maximum number of moles of $(SCN)_2$ that could form in the combination of 4 moles of NaSCN and 3 moles of $MnO_2$?

d. Write a conversion factor that could be used to convert between moles of sulfuric acid, $H_2SO_4$, and moles of manganese(II) sulfate, $MnSO_4$.

e. What is the minimum number of moles of $H_2SO_4$ that must react to form 1.7752 moles of manganese(II) sulfate?

OBJECTIVE 2 OBJECTIVE 3

28. In chapter 3, you were told that halogen atoms generally form a single covalent bond, but there are many compounds in which halogen atoms form more than one bond. For example, bromine pentafluoride (used as an oxidizing agent in rocket propellants) has bromine atoms with five covalent bonds. Liquid bromine pentafluoride is the only product in the reaction of gaseous bromine monofluoride with fluorine gas.

a. Write a complete balanced equation, including states, for this reaction.

b. Write a conversion factor that could be used to convert between moles of fluorine and moles of bromine pentafluoride.

c. How many moles of bromine pentafluoride form when 6 moles of fluorine react completely?

d. What is the maximum number of moles of bromine pentafluoride that could form in the combination of 8 moles of bromine monofluoride with 12 moles of fluorine?

e. Write a conversion factor that could be used to convert between moles of bromine monofluoride and moles of bromine pentafluoride.

f. How many moles of bromine monofluoride must react to yield 0.78 mole of bromine pentafluoride?

29. Potassium chlorate, $KClO_3$, acts as an oxidizing agent in matches, explosives, flares, and fireworks. In the equation below, it is formed from the element chlorine and potassium hydroxide.

OBJECTIVE 2 OBJECTIVE 3

$$3Cl_2 + 6KOH \rightarrow KClO_3 + 5KCl + 3H_2O$$

a. Write a conversion factor that could be used to convert between moles of potassium hydroxide and moles of potassium chlorate.

b. How many moles of potassium chlorate form when 2 moles of potassium hydroxide react completely?

c. What is the maximum number of moles of $KClO_3$ that could form in the combination of 6.0 moles of $Cl_2$ with 9.0 moles of KOH?

d. Write a conversion factor that could be used to convert between moles of chlorine and moles of potassium chloride.

e. How many moles of chlorine react when 2.006 moles of potassium chloride form?

30. Potassium perchlorate, $KClO_4$, is a better oxidizing agent than the potassium chlorate, $KClO_3$, described in the previous problem. Potassium perchlorate, which is used in explosives, fireworks, flares, and solid rocket propellants, is made by carefully heating potassium chlorate to between 400 °C and 500 °C. The *unbalanced* equation for this reaction is

OBJECTIVE 2 OBJECTIVE 3

$$KClO_3 \rightarrow KClO_4 + KCl$$

a. Balance this equation.

b. Write a conversion factor that could be used to convert between moles of potassium chlorate and moles of potassium perchlorate.

c. How many moles of potassium perchlorate form from the complete reaction of 12 moles of potassium chlorate?

d. Write a conversion factor that could be used to convert between moles of potassium perchlorate and moles of potassium chloride.

e. How many moles of potassium chloride form along with 11.875 moles of potassium perchlorate?

OBJECTIVE 2 OBJECTIVE 3 OBJECTIVE 4 OBJECTIVE 9

31. Aniline, $C_6H_5NH_2$, is used to make many different chemicals, including dyes, photographic chemicals, antioxidants, explosives, and herbicides. It can be formed from nitrobenzene, $C_6H_5NO_2$, in the following reaction with iron(II) chloride as a catalyst.

$$4C_6H_5NO_2 + 9Fe + 4H_2O \xrightarrow{FeCl_2} 4C_6H_5NH_2 + 3Fe_3O_4$$

a. Write a conversion factor that could be used to convert between moles of iron and moles of nitrobenzene.

b. What is the minimum mass of iron that would be necessary to react completely with 810.5 g of nitrobenzene, $C_6H_5NO_2$?

c. Write a conversion factor that could be used to convert between moles of aniline and moles of nitrobenzene.

d. What is the maximum mass of aniline, $C_6H_5NH_2$, that can be formed from 810.5 g of nitrobenzene, $C_6H_5NO_2$, with excess iron and water?

e. Write a conversion factor that could be used to convert between moles of $Fe_3O_4$ and moles of aniline.

f. What is the mass of $Fe_3O_4$ formed with the amount of aniline, $C_6H_5NH_2$, calculated in part (d)?

g. If 478.2 g of aniline, $C_6H_5NH_2$, are formed from the reaction of 810.5 g of nitrobenzene, $C_6H_5NO_2$, with excess iron and water, what is the percent yield?

OBJECTIVE 2 OBJECTIVE 3 OBJECTIVE 4 OBJECTIVE 9

32. Tetraboron carbide, $B_4C$, is a very hard substance used for grinding purposes, ceramics, and armor plating. Because boron is an efficient absorber of neutrons, $B_4C$ is also used as a protective material in nuclear reactors. The reaction that forms $B_4C$ from boron(III) oxide, $B_2O_3$, is

$$2B_2O_3 + 7C \xrightarrow{2400\ °C} B_4C + 6CO$$

a. Write a conversion factor that could be used to convert between moles of carbon and moles of boron(III) oxide.

b. What is the minimum mass of carbon, in grams, necessary to react completely with 24.675 g of boron(III) oxide, $B_2O_3$?

c. Write a conversion factor that could be used to convert between moles of $B_4C$ and moles of boron(III) oxide.

d. What is the maximum mass, in grams, of $B_4C$ that can be formed from the reaction of 24.675 g of boron(III) oxide, $B_2O_3$ with an excess of carbon?

e. If 9.210 g of $B_4C$ is isolated in the reaction of 24.675 g of boron(III) oxide, $B_2O_3$, with an excess of carbon, what is the percent yield?

**33.** Because of its red-orange color, sodium dichromate, $Na_2Cr_2O_7$, is used in the manufacture of pigments. It can be made by reacting sodium chromate, $Na_2CrO_4$, with sulfuric acid. The products other than sodium dichromate are sodium sulfate and water.

OBJECTIVE 2 OBJECTIVE 3 OBJECTIVE 4

a. Write a balanced equation for this reaction. (You do not need to write the states.)

b. How many kilograms of sodium chromate, $Na_2CrO_4$, are necessary to produce 84.72 kg of sodium dichromate, $Na_2Cr_2O_7$?

c. How many kilograms of sodium sulfate are formed with 84.72 kg of $Na_2Cr_2O_7$?

34. Chromium(III) oxide, often called chromic oxide, is used as a green paint pigment, as a catalyst in organic synthesis, as a polishing powder, and to make metallic chromium. One way to make chromium(III) oxide is by reacting sodium dichromate, $Na_2Cr_2O_7$, with ammonium chloride at 800 to 1000 °C to form chromium(III) oxide, sodium chloride, nitrogen, and water.

OBJECTIVE 2 OBJECTIVE 3 OBJECTIVE 4 OBJECTIVE 9

a. Write a balanced equation for this reaction. (You do not need to write the states.)

b. What is the minimum mass, in megagrams, of ammonium chloride necessary to react completely with 275 Mg of sodium dichromate, $Na_2Cr_2O_7$?

c. What is the maximum mass, in megagrams, of chromium(III) oxide that can be made from 275 Mg of sodium dichromate, $Na_2Cr_2O_7$, and excess ammonium chloride?

d. If 147 Mg of chromium(III) oxide is formed in the reaction of 275 Mg of sodium dichromate, $Na_2Cr_2O_7$, with excess ammonium chloride, what is the percent yield?

**35.** The tanning agent, $Cr(OH)SO_4$, is formed in the reaction of sodium dichromate ($Na_2Cr_2O_7$), sulfur dioxide, and water. Tanning protects animal hides from bacterial attack, reduces swelling, and prevents the fibers from sticking together when the hides dry. This leads to a softer, more flexible leather.

OBJECTIVE 2 OBJECTIVE 3 OBJECTIVE 4

$$Na_2Cr_2O_7 + 3SO_2 + H_2O \rightarrow 2Cr(OH)SO_4 + Na_2SO_4$$

a. How many kilograms of sodium dichromate, $Na_2Cr_2O_7$, are necessary to produce 2.50 kg of $Cr(OH)SO_4$?

b. How many megagrams of sodium sulfate are formed with 2.50 Mg of $Cr(OH)SO_4$?

OBJECTIVE 2 OBJECTIVE 3 OBJECTIVE 4 OBJECTIVE 9

36. The mineral hausmannite, $Mn_3O_4$, which contains both manganese(II) and manganese(III) ions, is formed from heating manganese(IV) oxide to 890 °C.

$$3MnO_2(l) \xrightarrow{890\ °C} Mn_3O_4(s) + O_2(g)$$

a. What is the maximum mass, in megagrams, of $Mn_3O_4$ that can be formed from the decomposition of 31.85 Mg of manganese(IV) oxide, $MnO_2$?

b. If 24.28 Mg of $Mn_3O_4$ is isolated in the decomposition reaction of 31.85 Mg of manganese(IV) oxide, $MnO_2$, what is the percent yield?

OBJECTIVE 5

37. Describe four reasons for adding an excess of one or more of the reactants in a chemical reaction.

OBJECTIVE 6

38. Chromium(III) oxide can be made from the reaction of sodium dichromate and ammonium chloride. What is the maximum mass, in grams, of chromium(III) oxide that can be produced from the complete reaction of 123.5 g of sodium dichromate, $Na_2Cr_2O_7$, with 59.5 g of ammonium chloride? The other products are sodium chloride, nitrogen gas, and water.

OBJECTIVE 6

39. The equation for one process for making aluminum fluoride follows. What is the maximum mass, in grams, of aluminum fluoride, $AlF_3$, that can be produced from the complete reaction of $1.4 \times 10^3$ g of aluminum hydroxide, $Al(OH)_3$, with $1.0 \times 10^3$ g of $H_2SiF_6$?

$$2Al(OH)_3 + H_2SiF_6 \xrightarrow{100\ °C} 2AlF_3 + SiO_2 + 4H_2O$$

OBJECTIVE 6 OBJECTIVE 7

40. Tetraboron carbide, $B_4C$, which is used as a protective material in nuclear reactors, can be made from boric acid, $H_3BO_3$.

$$4H_3BO_3 + 7C \xrightarrow{2400\ °C} B_4C + 6CO + 6H_2O$$

a. What is the maximum mass, in kilograms, of $B_4C$ formed in the reaction of 30.0 kg of carbon with 54.785 kg of $H_3BO_3$?

b. Explain why one of the substances in part (a) is in excess and one is limiting.

OBJECTIVE 6 OBJECTIVE 7 OBJECTIVE 9

41. Potassium permanganate, $KMnO_4$, a common oxidizing agent, is made from various ores that contain manganese(IV) oxide, $MnO_2$. The following equation shows the net reaction for one process that forms potassium permanganate.

$$2MnO_2 + 2KOH + O_2 \rightarrow 2KMnO_4 + H_2$$

a. What is the maximum mass, in kilograms, of $KMnO_4$ that can be made from the reaction of 835.6 g of $MnO_2$ with 585 g of KOH and excess oxygen gas?

b. Explain why the oxygen gas is in excess.

c. If 1.18 kg of $KMnO_4$ are isolated from the product mixture of the reaction of 835.6 g of $MnO_2$ with 585 g of KOH and excess oxygen gas, what is the percent yield?

42. Aniline, $C_6H_5NH_2$, which is used to make antioxidants, can be formed from nitrobenzene, $C_6H_5NO_2$, in the following reaction.

OBJECTIVE 6 OBJECTIVE 7

$$4C_6H_5NO_2 + 9Fe + 4H_2O \xrightarrow{FeCl_2} 4C_6H_5NH_2 + 3Fe_3O_4$$

a. What is the maximum mass of aniline, $C_6H_5NH_2$, formed in the reaction of 810.5 g of nitrobenzene, $C_6H_5NO_2$, with 985.0 g of Fe and 250 g of $H_2O$?

b. Explain why two of these substances are in excess and one is limiting.

43. Uranium is distributed in a form called yellow cake, which is made from uranium ore. In the second step of the reactions that form yellow cake from uranium ore, uranyl sulfate, $UO_2SO_4$, is converted to $(NH_4)_2U_2O_7$.

OBJECTIVE 6 OBJECTIVE 7

$$2UO_2SO_4 + 6NH_3 + 3H_2O \rightarrow (NH_4)_2U_2O_7 + 2(NH_4)_2SO_4$$

a. What is the maximum mass, in kilograms, of $(NH_4)_2U_2O_7$ that can be formed from the reaction of 100 kg of water and 100 kg of ammonia with 481 kg of $UO_2SO_4$?

b. Explain why two of these substances are in excess and one is limiting.

44. Calcium carbide, $CaC_2$, is formed in the reaction between calcium oxide and carbon. The other product is carbon monoxide.

OBJECTIVE 6

a. Write a balanced equation for this reaction. (You do not need to write the states.)

b. If you were designing the procedure for producing calcium carbide from calcium oxide and carbon, which of the reactants would you have as the limiting reactant? Why?

c. Assuming 100% yield from the limiting reactant, what are the approximate amounts of CaO and carbon that you would combine to form 860.5 g of $CaC_2$?

45. Calcium carbide, $CaC_2$, reacts with water to form acetylene, $C_2H_2$, and calcium hydroxide.

OBJECTIVE 6

a. Write a balanced equation for this reaction. (You do not need to write the states.)

b. If you were designing the procedure for producing acetylene from calcium carbide and water, which of the reactants would you have as the limiting reactant? Why?

c. Assuming 100% yield from the limiting reactant, what are the approximate amounts of $CaC_2$ and water that you would combine to form 127 g of $C_2H_2$?

46. Give four reasons why the actual yield in a chemical reaction is less than the theoretical yield.

OBJECTIVE 8

47. When determining the *theoretical* yield for a reaction, why must we first determine which reactant is the limiting reactant?

48. Does the reactant in excess affect the *actual* yield for a reaction? If it does, explain how.

49. Can the calculated percent yield ever be above 100%? If it can, explain how.

### Section 10.3 Molarity and Equation Stoichiometry

OBJECTIVE 10

50. What is the molarity of a solution made by dissolving 37.452 g of aluminum sulfate, $Al_2(SO_4)_3$, in water and diluting with water to 250.0 mL total?

OBJECTIVE 10

51. What is the molarity of a solution made by dissolving 18.476 g of potassium carbonate, $K_2CO_3$, in water and diluting with water to 100.0 mL total?

OBJECTIVE 11 OBJECTIVE 12 OBJECTIVE 13

52. The following equation represents the first step in the conversion of $UO_3$, found in uranium ore, into the uranium compounds called "yellow cake."

$$UO_3 + H_2SO_4 \rightarrow UO_2SO_4 + H_2O$$

a. How many milliliters of 18.0 M $H_2SO_4$ are necessary to react completely with 249.6 g of $UO_3$?

b. What is the maximum mass, in grams, of $UO_2SO_4$ that forms from the complete reaction of 125 mL of 18.0 M $H_2SO_4$?

OBJECTIVE 11 OBJECTIVE 12 OBJECTIVE 13

53. Most of the sodium chlorate, $NaClO_3$, produced in the United States is converted into chlorine dioxide, which is then used for bleaching wood pulp.

$$NaClO_3(aq) + 2HCl(aq) \rightarrow ClO_2(g) + \tfrac{1}{2}Cl_2(g) + NaCl(aq) + H_2O(l)$$

a. How many milliliters of 12.1 M HCl are necessary to react completely with 35.09 g of sodium chlorate, $NaClO_3$?

b. What is the maximum mass, in grams, of $ClO_2$ that can be formed from the complete reaction of 65 mL of 12.1 M HCl?

OBJECTIVE 11 OBJECTIVE 12 OBJECTIVE 13

54. When a water solution of sodium sulfite, $Na_2SO_3$, is added to a water solution of iron(II) chloride, $FeCl_2$, iron(II) sulfite, $FeSO_3$, precipitates from the solution.

a. Write a balanced equation for this reaction.

b. What is the maximum mass of iron(II) sulfite that will precipitate from a solution prepared by adding an excess of a $Na_2SO_3$ solution to 25.00 mL of 1.009 M $FeCl_2$?

OBJECTIVE 11 OBJECTIVE 12 OBJECTIVE 13

55. Consider the precipitation reaction that takes place when a water solution of aluminum nitrate, $Al(NO_3)_3$, is added to a water solution of potassium phosphate, $K_3PO_4$.

a. Write a balanced equation for this reaction.

b. What is the maximum mass of aluminum phosphate that will precipitate from a solution prepared by adding an excess of an $Al(NO_3)_3$ solution to 50.00 mL of 1.525 M $K_3PO_4$?

OBJECTIVE 11 OBJECTIVE 12 OBJECTIVE 13

56. Consider the neutralization reaction that takes place when nitric acid reacts with aqueous potassium hydroxide.

a. Write a conversion factor that relates moles of $HNO_3$ to moles of KOH for this reaction.

b. What is the minimum volume of 1.50 M $HNO_3$ necessary to neutralize completely the hydroxide in 125.0 mL of 0.501 M KOH?

OBJECTIVE 11 OBJECTIVE 12 OBJECTIVE 13

57. Consider the neutralization reaction that takes place when hydrochloric acid reacts with aqueous sodium hydroxide.

a. Write a conversion factor that relates moles of HCl to moles of NaOH for this reaction.

b. What is the minimum volume of 6.00 M HCl necessary to neutralize completely the hydroxide in 750.0 mL of 0.107 M NaOH?

58. Consider the neutralization reaction that takes place when sulfuric acid reacts with aqueous sodium hydroxide.

OBJECTIVE 11 OBJECTIVE 12 OBJECTIVE 13

a. Write a conversion factor that relates moles of $H_2SO_4$ to moles of NaOH for this reaction.

b. What is the minimum volume of 6.02 M $H_2SO_4$ necessary to neutralize completely the hydroxide in 47.5 mL of 2.5 M NaOH?

59. Consider the neutralization reaction that takes place when phosphoric acid reacts with aqueous potassium hydroxide.

OBJECTIVE 11 OBJECTIVE 12 OBJECTIVE 13

a. Write a conversion factor that relates moles of $H_3PO_4$ to moles of KOH for this reaction.

b. What is the minimum volume of 2.02 M $H_3PO_4$ necessary to neutralize completely the hydroxide in 183 mL of 0.550 M KOH?

60. Consider the neutralization reaction that takes place when hydrochloric acid reacts with solid cobalt(II) hydroxide.

OBJECTIVE 11 OBJECTIVE 12 OBJECTIVE 13

a. Write a conversion factor that relates moles of HCl to moles of $Co(OH)_2$ for this reaction.

b. What is the minimum volume of 6.14 M HCl necessary to react completely with 2.53 kg of solid cobalt(II) hydroxide, $Co(OH)_2$?

61. Consider the neutralization reaction that takes place when hydrochloric acid reacts with solid nickel(II) carbonate.

OBJECTIVE 11 OBJECTIVE 12 OBJECTIVE 13

a. Write a conversion factor that relates moles of HCl to moles of $NiCO_3$ for this reaction.

b. What is the minimum volume of 6.0 M HCl necessary to react completely with 14.266 g of solid nickel(II) carbonate, $NiCO_3$?

62. Consider the neutralization reaction that takes place when nitric acid reacts with solid chromium(III) hydroxide.

OBJECTIVE 19 OBJECTIVE 20 OBJECTIVE 21

a. Write a conversion factor that relates moles of $HNO_3$ to moles of $Cr(OH)_3$ for this reaction.

b. What is the minimum volume of 2.005 M $HNO_3$ necessary to react completely with 0.5187 kg of solid chromium(III) hydroxide, $Cr(OH)_3$?

63. Consider the neutralization reaction that takes place when nitric acid reacts with solid iron(II) carbonate.

OBJECTIVE 11 OBJECTIVE 12 OBJECTIVE 13

a. Write a conversion factor that relates moles of $HNO_3$ to moles of $FeCO_3$ for this reaction.

b. What is the minimum volume of 2.00 M $HNO_3$ necessary to react completely with 1.06 kg of solid iron(II) carbonate, $FeCO_3$?

### Additional Problems

64. Because nitrogen and phosphorus are both nonmetallic elements in group 15 on the periodic table, we expect them to react with other elements in similar ways. To some extent, they do, but there are also distinct differences in their chemical behavior. For example, nitrogen atoms form stable triple bonds to carbon atoms in substances such as hydrogen cyanide (often called hydrocyanic acid), HCN. Phosphorus atoms also form triple bonds to carbon atoms, in substances such as HCP, but those substances are much less stable. The compound HCP can be formed in the following reaction.

$$CH_4 + PH_3 \xrightarrow{\text{electric arc}} HCP + 3H_2$$

   a. Write a conversion factor that could be used to convert between moles of HCP and moles of $H_2$.
   b. How many moles of HCP form along with 9 moles of $H_2$?
   c. Write a conversion factor that could be used to convert between moles of methane, $CH_4$, and moles of hydrogen, $H_2$.
   d. How many moles of hydrogen gas form when 1.8834 moles of $CH_4$ react with an excess of $PH_3$?

65. Because carbon and silicon are both elements in group 14 on the periodic table, we expect them to react with other elements in similar ways. To some extent, they do, but in some cases, carbon and silicon compounds that seem to have analogous structures have very different chemical characteristics. For example, carbon tetrachloride, $CCl_4$, is very stable in the presence of water, but silicon tetrachloride, $SiCl_4$, reacts quickly with water. The *unbalanced* equation for this reaction is

$$SiCl_4 + H_2O \rightarrow Si(OH)_4 + HCl$$

   a. Balance this equation.
   b. Write a conversion factor that could be used to convert between moles of $SiCl_4$ and moles of $H_2O$.
   c. How many moles of $SiCl_4$ react with 24 moles of water?
   d. Write a conversion factor that could be used to convert between moles of $Si(OH)_4$ and moles of water.
   e. How many moles of $Si(OH)_4$ form when 4.01 moles of $H_2O$ react with an excess of $SiCl_4$?

66. Iodine pentafluoride is an incendiary agent, a substance that ignites combustible materials. This compound is usually made by passing fluorine gas over solid iodine, but it also forms when iodine monofluoride changes into the element iodine and iodine pentafluoride.

   a. Write a balanced equation, without including states, for the conversion of iodine monofluoride into iodine and iodine pentafluoride.
   b. How many moles of the element iodine form when 15 moles of iodine monofluoride react completely?
   c. How many moles of iodine pentafluoride form when 7.939 moles of iodine monofluoride react completely?

67. The first laboratory experiments to produce compounds containing noble gas atoms aroused great excitement, not because the compounds might be useful but because they demonstrated that the noble gases were not completely inert. Since that time, however, important uses have been found for a number of noble gas compounds. For example, xenon difluoride, $XeF_2$, is an excellent fluorinating agent (a substance that adds fluorine atoms to other substances). One reason it is preferred over certain other fluorinating agents is that the products of its fluorinating reactions are easily separated (since one of them is gaseous xenon). The following *unbalanced* equation represents one such reaction:

$$S_3O_9 + XeF_2 \rightarrow S_2O_6F_2 + Xe$$

a. Balance this equation.

b. What is the minimum number of moles of $XeF_2$ necessary to react with 4 moles of $S_3O_9$?

c. What is the maximum number of moles of $S_2O_6F_2$ that can form from the complete reaction of 4 moles of $S_3O_9$ and 7 moles of $XeF_2$?

d. How many moles of xenon gas form from the complete reaction of 0.6765 mole of $S_3O_9$?

68. Xenon hexafluoride is a better fluorinating agent than the xenon difluoride described in the previous problem, but it must be carefully isolated from any moisture. This is because xenon hexafluoride reacts with water to form hydrogen fluoride (hydrogen monofluoride) and the dangerously explosive xenon trioxide.

a. Write a balanced equation, without including states, for the reaction of xenon hexafluoride and water to form xenon trioxide and hydrogen fluoride.

b. How many moles of hydrogen fluoride form when 0.50 mole of xenon hexafluoride reacts completely?

c. What is the maximum number of moles of xenon trioxide that can form in the combination of 7 moles of xenon hexafluoride and 18 moles of water?

69. It is fairly easy to make the fluorides of xenon by combining xenon gas with fluorine gas. Unfortunately, the products of the reaction—$XeF_2$, $XeF_4$, and $XeF_6$—are difficult to separate. The percentage of $XeF_2$ can be raised by adding a large excess of xenon and removing the product mixture soon after the reaction has begun. The percentage of $XeF_6$ can be raised by running the reaction at 700–800 °C in the presence of a nickel catalyst with a large excess of fluorine. If $XeF_4$ is desired, a different reaction can be used. For example, $XeF_4$ can be made from the reaction of xenon gas with dioxygen difluoride. The reaction also produces oxygen gas.

a. Write a balanced equation, including states, for the reaction of xenon gas and dioxygen difluoride gas to form xenon tetrafluoride gas and oxygen gas.

b. How many moles of xenon tetrafluoride form from 7.50 moles of dioxygen difluoride?

c. What is the maximum number of moles of xenon tetrafluoride gas that can form in the combination of 4.75 moles of xenon and 9.00 moles of dioxygen difluoride?

70. Hydriodic acid is produced industrially by the reaction of hydrazine, $N_2H_4$, with iodine, $I_2$. HI(*aq*) is used to make iodine salts such as AgI, which are used to seed clouds to promote rain. What is the minimum mass of iodine, $I_2$, necessary to react completely with 87.0 g of hydrazine, $N_2H_4$?

$$N_2H_4 + 2I_2 \rightarrow 4HI + N_2$$

71. Calcium dihydrogen phosphate, which is used in the production of triple superphosphate fertilizers, can be formed from the reaction of apatite, $Ca_5(PO_4)_3F$, in apatite ore with phosphoric acid. How many grams of calcium dihydrogen phosphate can be formed from 6.78 g of $Ca_5(PO_4)_3F$?

$$2Ca_5(PO_4)_3F + 14H_3PO_4 \rightarrow 10Ca(H_2PO_4)_2 + 2HF$$

72. Because plants need nitrogen compounds, potassium compounds, and phosphorus compounds to grow, these are often added to the soil as fertilizers. Potassium sulfate, which is used to make fertilizers, is made industrially by reacting potassium chloride with sulfur dioxide gas, oxygen gas, and water. Hydrochloric acid is formed with the potassium sulfate.
    a. Write a balanced equation for this reaction. (You do not need to include states.)
    b. What is the maximum mass, in kilograms, of potassium sulfate that can be formed from $2.76 \times 10^5$ kg of potassium chloride with excess sulfur dioxide, oxygen, and water?
    c. If $2.94 \times 10^5$ kg of potassium sulfate is isolated from the reaction of $2.76 \times 10^5$ kg of potassium chloride, what is the percent yield?

73. Sodium hydrogen sulfate is used as a cleaning agent and as a flux (a substance that promotes the fusing of metals and prevents the formation of oxides). One of the ways in which sodium hydrogen sulfate is manufactured is by reacting sodium dichromate, $Na_2Cr_2O_7$, with sulfuric acid. This process also forms water and chromium(VI) oxide, $CrO_3$, which is often called chromic acid.
    a. Write a balanced equation for this reaction. (You do not need to include states.)
    b. How many kilograms of sodium dichromate, $Na_2Cr_2O_7$, are necessary to produce 130.4 kg of sodium hydrogen sulfate?
    c. How many kilograms of chromium(VI) oxide are formed when 130.4 kg of sodium hydrogen sulfate is made?
    d. What is the minimum volume of 18.0 M $H_2SO_4$ solution necessary to react with 874.0 kg of sodium dichromate?
    e. What is the maximum mass of sodium hydrogen sulfate, $NaHSO_4$, that can be formed from the reaction of 874.0 kg of sodium dichromate with 400.0 L of 18.0 M $H_2SO_4$?

74. The element phosphorus can be made by reacting carbon in the form of coke with calcium phosphate, $Ca_3(PO_4)_2$, which is found in phosphate rock.

$$Ca_3(PO_4)_2 + 5C \rightarrow 3CaO + 5CO + 2P$$

a. What is the minimum mass of carbon, C, necessary to react completely with 67.45 Mg of $Ca_3(PO_4)_2$?

b. What is the maximum mass of phosphorus produced from the reaction of 67.45 Mg of $Ca_3(PO_4)_2$ with an excess of carbon?

c. What mass of calcium oxide, CaO, is formed with the mass of phosphorus calculated in part (b)?

d. If 11.13 Mg of phosphorus is formed in the reaction of 67.45 Mg of $Ca_3(PO_4)_2$ with an excess of carbon, what is the percent yield?

75. When coal is burned, the sulfur it contains is converted into sulfur dioxide. This $SO_2$ is a serious pollutant, so it needs to be removed before it escapes from the stack of a coal-fired plant. One way to remove the $SO_2$ is to add limestone, which contains calcium carbonate, $CaCO_3$, to the coal before it is burned. The heat of the burning coal converts the $CaCO_3$ to calcium oxide, CaO. The calcium oxide reacts with the sulfur dioxide in the following reaction:

$$2CaO + 2SO_2 + O_2 \rightarrow 2CaSO_4$$

The solid calcium sulfate does not escape from the stack as the gaseous sulfur dioxide would. What mass of calcium sulfate forms for each 1.00 Mg of $SO_2$ removed by this technique?

76. Thionyl chloride, $SOCl_2$, is a widely used source of chlorine in the formation of pesticides, pharmaceuticals, dyes, and pigments. It can be formed from disulfur dichloride in the following reaction.

$$2SO_2 + S_2Cl_2 + 3Cl_2 \rightarrow 4SOCl_2$$

If 1.140 kg of thionyl chloride is isolated from the reaction of 457.6 grams of disulfur dichloride, $S_2Cl_2$, with excess sulfur dioxide and chlorine gas, what is the percent yield?

77. Chromium(III) oxide, which can be converted into metallic chromium, is formed in the following reaction:

$$Na_2Cr_2O_7 + S \xrightarrow{800\text{–}1000\ °C} Cr_2O_3 + Na_2SO_4$$

a. How many grams of chromium(III) oxide, $Cr_2O_3$, are formed in the reaction of 981 g of sodium dichromate, $Na_2Cr_2O_7$, with 330 g of sulfur, S?

b. Explain why one of these substances is in excess and one is limiting.

78. Sodium dichromate, $Na_2Cr_2O_7$, is converted to chromium(III) sulfate, which is used in the tanning of animal hides. Sodium dichromate can be made by reacting sodium chromate, $Na_2CrO_4$, with water and carbon dioxide.

$$2Na_2CrO_4 + H_2O + 2CO_2 \rightleftharpoons Na_2Cr_2O_7 + 2NaHCO_3$$

a. Show that the sodium chromate is the limiting reactant when 87.625 g of $Na_2CrO_4$ reacts with 10.008 g of water and excess carbon dioxide.

b. Explain why the carbon dioxide and water are in excess and sodium chromate is limiting.

79. What is the molarity of a solution made by dissolving 37.895 g of $CoCl_2$ in water and diluting with water to 250.0 mL total?

80. What is the molarity of a solution made by dissolving 100.065 g of $SnBr_2$ in water and diluting with water to 1.00 L total?

81. Sodium dichromate, $Na_2Cr_2O_7$, can be made by reacting sodium chromate, $Na_2CrO_4$, with sulfuric acid.

$$2Na_2CrO_4 + H_2SO_4 \rightarrow Na_2Cr_2O_7 + Na_2SO_4 + H_2O$$

a. What is the minimum volume of 18.0 M $H_2SO_4$ necessary to react completely with 15.345 kg of sodium chromate, $Na_2CrO_4$?

b. What is the maximum mass, in kilograms, of sodium dichromate that can be formed from the reaction of 203 L of 18.0 M $H_2SO_4$?

82. A precipitation reaction takes place when a water solution of sodium carbonate, $Na_2CO_3$, is added to a water solution of chromium(III) nitrate, $Cr(NO_3)_3$.

a. Write a balanced equation for this reaction.

b. What is the maximum mass of chromium(III) carbonate that will precipitate from a solution prepared by adding an excess of a $Na_2CO_3$ solution to 10.00 mL of 0.100 M $Cr(NO_3)_3$?

83. A precipitation reaction takes place when a water solution of potassium phosphate, $K_3PO_4$, is added to a water solution of cobalt(II) chloride, $CoCl_2$.

a. Write a balanced equation for this reaction.

b. What is the maximum mass of cobalt(II) phosphate that will precipitate from a solution prepared by adding an excess of a $K_3PO_4$ solution to 5.0 mL of 1.0 M $CoCl_2$?

84. Consider the neutralization reaction between nitric acid and aqueous barium hydroxide.

a. Write a conversion factor that shows the ratio of moles of nitric acid to moles of barium hydroxide.

b. What volume of 1.09 M nitric acid would be necessary to neutralize the hydroxide in 25.00 mL of 0.159 M barium hydroxide?

85. Consider the neutralization reaction between sulfuric acid and aqueous lithium hydroxide.

    a. Write a conversion factor that shows the ratio of moles of sulfuric acid to moles of lithium hydroxide.

    b. What volume of 0.505 M sulfuric acid would be necessary to neutralize the hydroxide in 25.00 mL of 2.87 M lithium hydroxide?

86. Consider the neutralization reaction between hydrochloric acid and solid zinc carbonate.

    a. Write a conversion factor that shows the ratio of moles of hydrochloric acid to moles of zinc carbonate.

    b. What volume of 0.500 M hydrochloric acid would be necessary to neutralize and dissolve 562 milligrams of solid zinc carbonate?

87. Consider the neutralization reaction between nitric acid and solid cadmium hydroxide.

    a. Write a conversion factor that shows the ratio of moles of nitric acid to moles of cadmium hydroxide.

    b. What volume of 3.00 M nitric acid would be necessary to neutralize and dissolve 2.56 kg of solid cadmium hydroxide?

### Challenge Problems

*Some of problems 88–98 include conversions between masses of pure substances and masses of mixtures that contain the pure substances, using percentages as conversion factors. If you need some help with them, visit the Web site below:*

**www.chemplace.com/college/**

88. A solution is made by adding 22.609 g of a solid that is 96.3% NaOH to a beaker of water. What volume of 2.00 M $H_2SO_4$ is necessary to neutralize the NaOH in this solution?

89. Potassium hydroxide can be purchased as a solid that is 88.0% KOH. What is the minimum mass of this solid necessary to neutralize all of the HCl in 25.00 mL of 3.50 M HCl?

90. Aluminum sulfate, often called alum, is used in paper making to increase the paper's stiffness and smoothness and to help keep the ink from running. It is made from the reaction of sulfuric acid with the aluminum oxide found in bauxite ore. The products are aluminum sulfate and water. Bauxite ore is 30% to 75% aluminum oxide.

    a. Write a balanced equation for this reaction. (You do not need to write the states.)

    b. What is the maximum mass, in kilograms, of aluminum sulfate that could be formed from $2.3 \times 10^3$ kilograms of bauxite ore that is 62% aluminum oxide?

91. The element phosphorus can be made by reacting carbon in the form of coke with $Ca_3(PO_4)_2$ found in phosphate ore. When 8.0 Mg of ore that is 68% $Ca_3(PO_4)_2$ is combined with an excess of carbon in the form of coke, what is the maximum mass, in megagrams, of phosphorus that can be formed?

$$Ca_3(PO_4)_2 + 5C \rightarrow 3CaO + 5CO + 2P$$

92. Sodium tripolyphosphate, or STPP, $Na_5P_3O_{10}$, is used in detergents. It is made by combining phosphoric acid with sodium carbonate at 300 to 500 °C. What is the minimum mass, in kilograms, of sodium carbonate that would be necessary to react with excess phosphoric acid to make enough STPP to produce $1.025 \times 10^5$ kg of a detergent that is 32% $Na_5P_3O_{10}$?

$$6H_3PO_4 + 5Na_2CO_3 \rightarrow 2Na_5P_3O_{10} + 9H_2O + 5CO_2$$

93. Hydrazine, $N_2H_4$, is a liquid with many industrial purposes, including the synthesis of herbicides and pharmaceuticals. It is made from urea in the following reaction at 100 °C.

$$NH_2CONH_2 + NaOCl + 2NaOH \rightarrow N_2H_4 + NaCl + Na_2CO_3 + H_2O$$

If the percent yield for the reaction is 90.6%, how many kilograms of hydrazine, $N_2H_4$, are formed from the reaction of 243.6 kg of urea, $NH_2CONH_2$, with excess sodium hypochlorite and sodium hydroxide?

94. Urea, $NH_2CONH_2$, is a common nitrogen source used in fertilizers. When urea is made industrially, its temperature must be carefully controlled because heat turns urea into biuret, $NH_2CONHCONH_2$, a compound that is harmful to plants. Consider a pure sample of urea that has a mass of 92.6 kg. If 0.5% of the urea in this sample decomposes to form biuret, what mass, in grams, of $NH_2CONHCONH_2$ will it contain?

$$2NH_2CONH_2 \rightarrow \underset{\text{biuret}}{NH_2CONHCONH_2} + NH_3$$

95. Chilean niter deposits are mostly sodium nitrate, but they also contain 0.3% iodine in the form of calcium iodate, $Ca(IO_3)_2$. After the sodium nitrate in the niter is dissolved and recrystallized, the remaining solution contains 9 g/L sodium iodate, $NaIO_3(aq)$. The $NaIO_3$ is converted to iodine when it reacts with sulfur dioxide and water.

$$2NaIO_3 + 5SO_2 + 4H_2O \rightarrow Na_2SO_4 + 4H_2SO_4 + I_2$$

a. How many liters of sodium iodate solution that contains 9 g of $NaIO_3$ per liter would be necessary to form 127.23 kg of iodine, $I_2$?

b. What mass, in megagrams, of Chilean niter that is 0.3% I would be necessary to form the volume of sodium iodate solution you calculated in part (a)?

96. The white pigment titanium(IV) oxide (often called titanium dioxide), $TiO_2$, is made from rutile ore that is about 95% $TiO_2$. Before the $TiO_2$ can be used, it must be purified. The equation that follows represents the first step in this purification.

$$3TiO_2(s) + 4C(s) + 6Cl_2(g) \xrightarrow{900\ °C} 3TiCl_4(l) + 2CO(g) + 2CO_2(g)$$

a. How many pounds of $TiCl_4$ can be made from the reaction of $1.250 \times 10^5$ pounds of rutile ore that is 95% $TiO_2$ with $5.0 \times 10^4$ pounds of carbon?

b. Explain why two of these substances are in excess and one is limiting.

97. The tanning agent $Cr(OH)SO_4$ is formed in the reaction of sodium dichromate, $Na_2Cr_2O_7$, sulfuric acid, and the sucrose in molasses:

$$8Na_2Cr_2O_7 + 24H_2SO_4 + C_{12}H_{22}O_{11} \rightarrow 16Cr(OH)SO_4 + 8Na_2SO_4 + 12CO_2 + 22H_2O$$

What is the maximum mass of $Cr(OH)SO_4$ formed from the reaction of 431.0 kg of sodium dichromate with 292 L of 18.0 M $H_2SO_4$ and 90.0 kg of $C_{12}H_{22}O_{11}$?

98. What is the maximum mass of calcium hydrogen phosphate, $CaHPO_4$, that can form from the mixture of 12.50 kg of a solution that contains 84.0% $H_3PO_4$, 25.00 kg of $Ca(NO_3)_2$, 25.00 L of 14.8 M $NH_3$, and an excess of $CO_2$ and $H_2O$?

$$3H_3PO_4 + 5Ca(NO_3)_2 + 10NH_3 + 2CO_2 + 2H_2O \rightarrow 10NH_4NO_3 + 2CaCO_3 + 3CaHPO_4$$

*Modern imaging techniques allow scientists to create pictures of atoms and molecules.*

CHAPTER 11

# Modern Atomic Theory



*To see a World in a Grain of Sand*
*And a Heaven in a Wild Flower*
*Hold Infinity in the palm of your hand*
*And Eternity in an hour*

WILLIAM BLAKE (1757–1827)
*AUGURIES OF INNOCENCE*

SCIENTISTS' ATTEMPTS TO UNDERSTAND THE ATOM have led them into the unfamiliar world of the unimaginably small, where the rules of physics seem to be different from the rules that apply in the world we can see and touch. Scientists explore this world through the use of mathematics. Perhaps this is similar to the way a writer uses poetry to express ideas and feelings beyond the reach of everyday language: Mathematics enables the scientist to explore beyond the boundaries of the world we can experience directly. Just as scholars then try to analyze the poems and share ideas about them in everyday language, scientists try to translate the mathematical description of the atom into words that more of us can understand. Although both kinds of translations are fated to fall short of capturing the fundamental truths of human nature and the physical world, the attempt is worthwhile for the occasional glimpse of those truths that it provides.

This chapter offers a brief, qualitative introduction to the mathematical description of electrons and examines the highly utilitarian model of atomic structure that chemists have constructed from it. Because we are reaching beyond the world of our senses, we should not be surprised that the model we create is uncertain and, when described in normal language, a bit vague. In spite of these limitations, however, you will return from your journey into the strange, new world of the extremely small with a useful tool for explaining and predicting the behavior of matter.

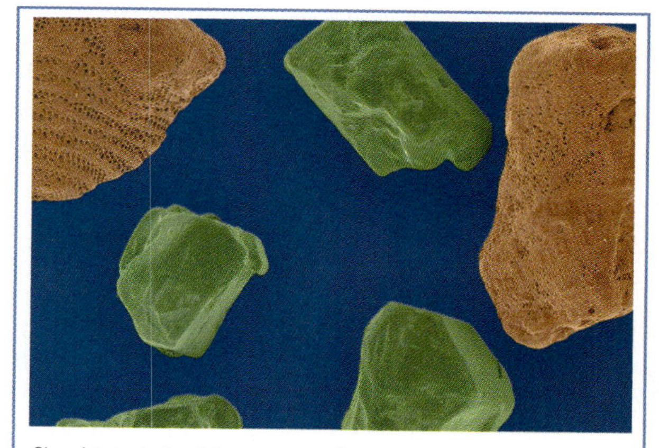

*Chemists try to "see" the structure of matter even more closely than it can be seen in this microscopic picture of a grain of sand.*

## Review Skills

The presentation of information in this chapter assumes that you can already perform the tasks listed below. You can test your readiness to proceed by answering the Review Questions at the end of the chapter. This might also be a good time to read the Chapter Objectives, which precede the Review Questions.

- Describe the nuclear model of the atom. (Section 2.4)
- Describe the relationship between stability and potential energy. (Section 7.1)

## 11.1 The Mysterious Electron

*Where there is an open mind, there will always be a frontier.*

CHARLES F. KETTERING (1876–1958)
AMERICAN ENGINEER AND INVENTOR

Scientists have known for a long time that it is incorrect to think of electrons as tiny particles orbiting the nucleus like planets around the sun. Nevertheless, nonscientists have become used to picturing them in this way. The "solar system" model of the atom may be useful in some circumstances, but you should know that the electron is much more unusual than that model suggests. The electron is extremely tiny, and modern physics tells us that strange things happen in the realm of the very, very small.

The modern description of the electron is based on complex mathematics and on the discoveries of modern physics. The mathematical complexity alone makes an accurate verbal portrayal of the electron challenging, but our difficulty in describing the electron goes beyond complexity. Modern physics tells us that it is *impossible* to know exactly where an electron is and what it is doing. As your mathematical and scientific knowledge increases, you will be able to understand more sophisticated descriptions of the electron, but the problem of describing exactly where the electron is and what it is doing never goes away. It is a problem inherent in the study of very tiny objects. Thus, complete confidence in our description of the nature of the electron is beyond our reach.

There are two ways in which scientists deal with the problems associated with the complexity and fundamental uncertainty of the modern description of the electron:

**Analogies** In order to communicate something of the nature of the electron, scientists often use analogies, comparing the electron to objects with which we are more familiar. For example, in this chapter we will look at the ways in which electrons are *like* vibrating guitar strings.

**Probabilities** In order to accommodate the uncertainty of the electron's position and motion, scientists talk about where the electron *probably is* within the atom, instead of where it *definitely is.*

Through the use of analogies and a discussion of probabilities, this chapter attempts to give you an idea of what scientists are learning about the electron's character.

### Standing Waves and Guitar Strings

Each electron seems to have a dual nature in which both particle and wave characteristics are apparent. It is difficult to describe these two aspects of an electron at the same time, so we focus sometimes on its particle nature and sometimes on its wave character, depending on which is more suitable in a given context. In the particle view, electrons are tiny, negatively charged particles with a mass of about $9.1096 \times 10^{-28}$ gram, or 0.000549 atomic mass unit. In the wave view, an electron has an effect on the space around it that can be described as a wave of negative charge varying in its intensity. To gain a better understanding of this electron-wave character, let's compare it to the wave character of guitar strings. Because the guitar string is easier to visualize than an electron, its vibrations serve as a useful analogy for the wave character of electrons.

When a guitar string is plucked, the string vibrates up and down in a wave pattern. Figure 11.1 shows one way in which it can vibrate; the seven images on the left represent the position of the string at various isolated moments, and the final image shows all those positions combined. If you squint a bit while looking at a vibrating guitar string, what you see is a blur with a shape determined by the varying intensity of the vibration along the string. This blur, which we will call the **waveform,** appears to be stationary. Although the string is constantly moving, the waveform is not, so this wave pattern is called a standing or stationary wave. Note that as your eye moves along the string, the intensity, or amount, of the string's movement varies. The points in the waveform where there is no motion are called **nodes.**

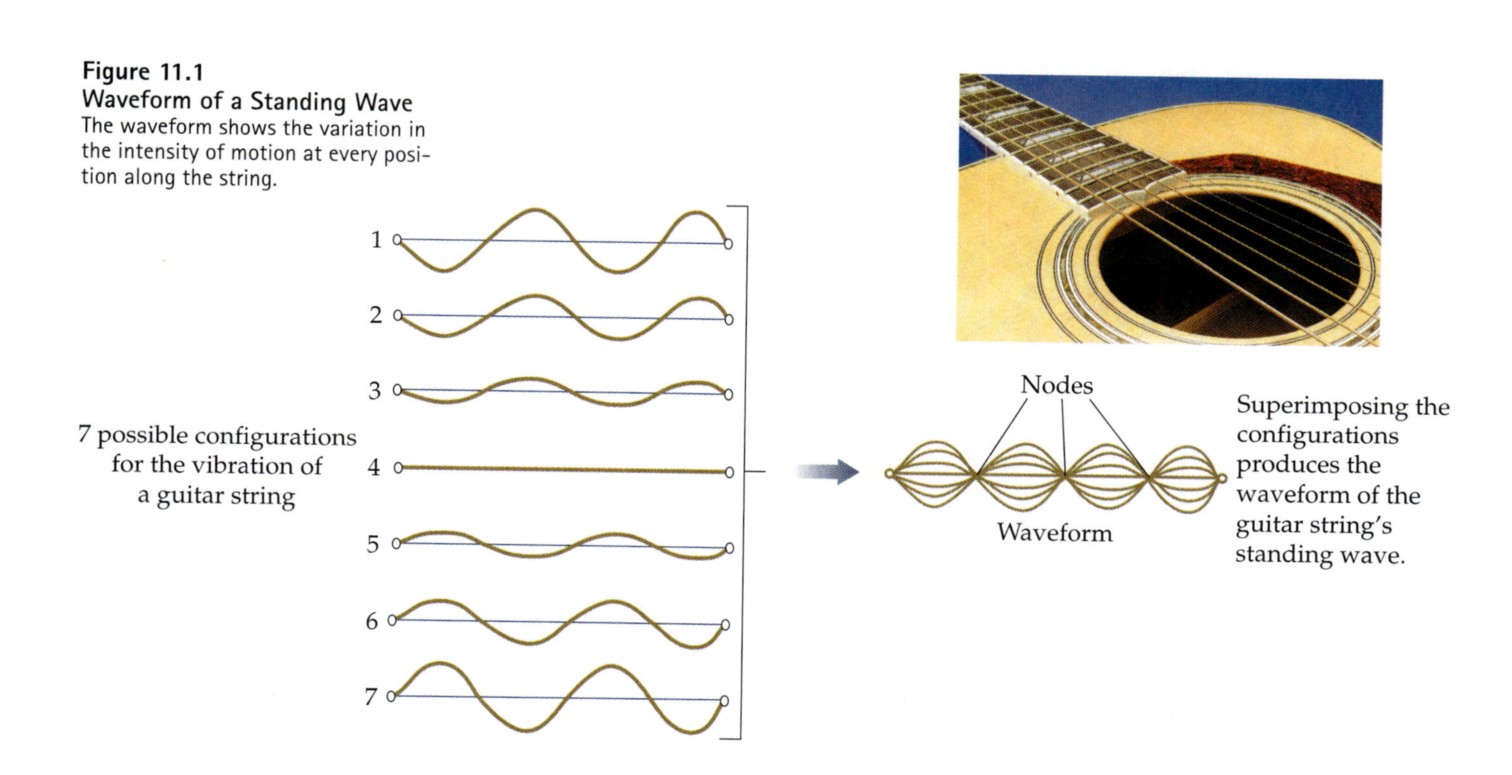
1
2
3
4
5
6
7
7 possible configurations for the vibration of a guitar string
Nodes
Waveform
Superimposing the configurations produces the waveform of the guitar string's standing wave.

**Figure 11.1**
**Waveform of a Standing Wave**
The waveform shows the variation in the intensity of motion at every position along the string.

Although many waveforms are possible, the possibilities are limited by the fact that the string is tied down and cannot move at the ends. In theory, there is an infinite number of possible waveforms, but they all allow the string to remain stationary at the ends. Figure 11.2 shows various allowed waveforms.

OBJECTIVE 2

**Figure 11.2**
**Some Possible Waveforms for a Vibrating Guitar String**

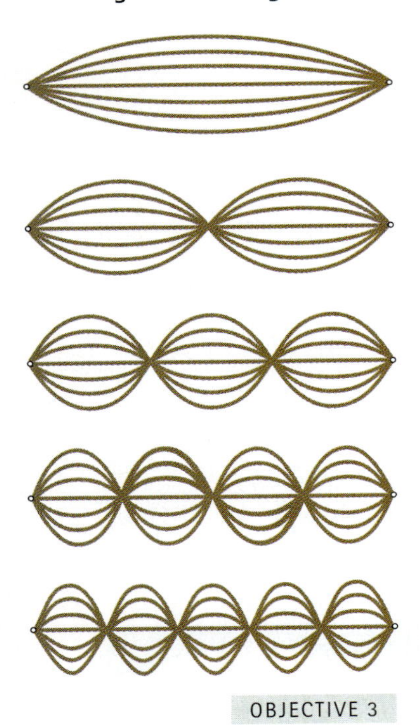

## Electrons as Standing Waves

The wave character of the guitar string is represented by the movement of the string. We can focus our attention on the blur of the waveform and forget the material the string is made of. The waveform describes the *motion* of the string over time, not the string itself.

In a similar way, the wave character of the electron is represented by the waveform of its negative charge, on which we can focus without concerning ourselves about the electron's particle nature. This frees us from asking questions about where the electrons are in the atom and how they are moving—questions that we are unable to answer. The waveforms for electrons in an atom describe the variation in intensity of negative charge within the atom, with respect to the location of the nucleus. This can be described without mentioning the positions and motion of the electron particle itself.

The following statements represent the core of the modern description of the wave character of the electron.

OBJECTIVE 3

- Just as the intensity of the movement of a guitar string can vary, so can the intensity of the negative charge of the electron vary at different positions outside the nucleus.
- The variation in the intensity of the electron charge can be described in terms of a three-dimensional standing wave *like* the standing wave of the guitar string.
- As in the case of the guitar string, only certain waveforms are possible for the electron in an atom.
- We can focus our attention on the waveform of varying charge intensity without having to think about the actual physical nature of the electron.

*Thus, the task is not so much to see what no one has yet seen, but to think what nobody has yet thought, about that which everybody sees.*
ERWIN SCHRÖDINGER (1887–1961)
AUSTRIAN PHYSICIST AND NOBEL LAUREATE

## Waveforms for Hydrogen Atoms

Most of the general descriptions of electrons found in the rest of this chapter are based on the wave mathematics for the one electron in a hydrogen atom. The comparable calculations for other elements are too difficult to lead to useful results, so, as you will see in the next section, the information calculated for the hydrogen electron is used to describe the other elements as well. Fortunately, this approximation works quite well.

The wave equation for the one electron of a hydrogen atom predicts waveforms for the electron that are similar to the allowed waveforms for a vibrating guitar string. For example, the simplest allowed waveform for the guitar string looks something like

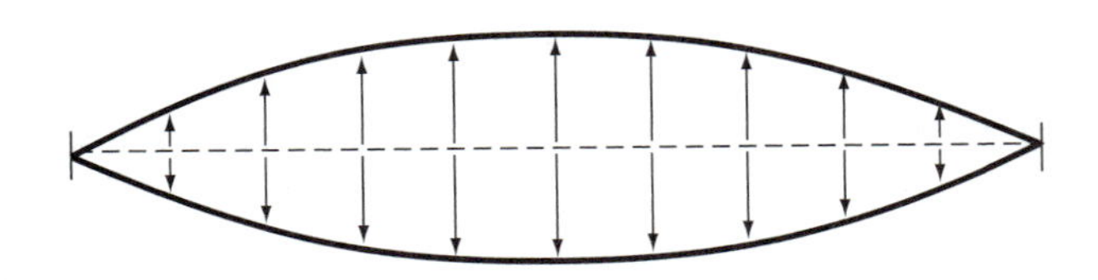

The simplest allowed waveform for an electron in a hydrogen atom looks like the image in Figure 11.3. The cloud that you see surrounds the nucleus and represents the variation in the intensity of the negative charge at different positions outside the nucleus. The negative charge is most intense at the nucleus and diminishes with increasing distance from the nucleus. The variation in charge intensity for this waveform is the same in all directions, so the waveform is a sphere. The allowed waveforms for the electron are also called **orbitals.** The orbital shown in Figure 11.3 is called the 1*s* orbital.

OBJECTIVE 4

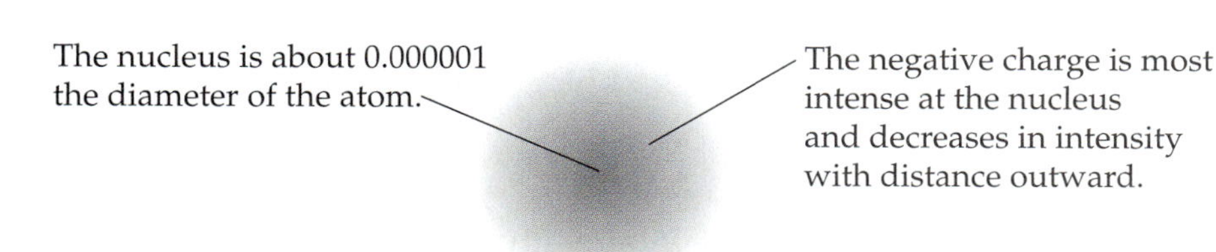
The nucleus is about 0.000001 the diameter of the atom.
The negative charge is most intense at the nucleus and decreases in intensity with distance outward.

**Figure 11.3**
**Waveform of the 1*s* Electron**

OBJECTIVE 4

Theoretically, the charge intensity depicted in Figure 11.3 decreases toward zero as the distance from the nucleus approaches infinity. This suggests the amusing possibility that some of the negative charge created by an electron in a hydrogen atom is felt an infinite distance from the atom's nucleus. The more practical approach taken by chemists, however, is to specify a volume that contains most of the electron charge and to focus their attention on that, forgetting about the small negative charge felt outside the specified volume. For example, we can focus on a sphere containing 90% of the charge of the 1*s* electron. If we wanted to include more of the electron charge, we could enlarge the sphere so that it encloses 99% (or 99.9%) of the electron charge (Figure 11.4). This leads us to another definition of **orbital** as the volume that contains a given high percentage of the electron charge.

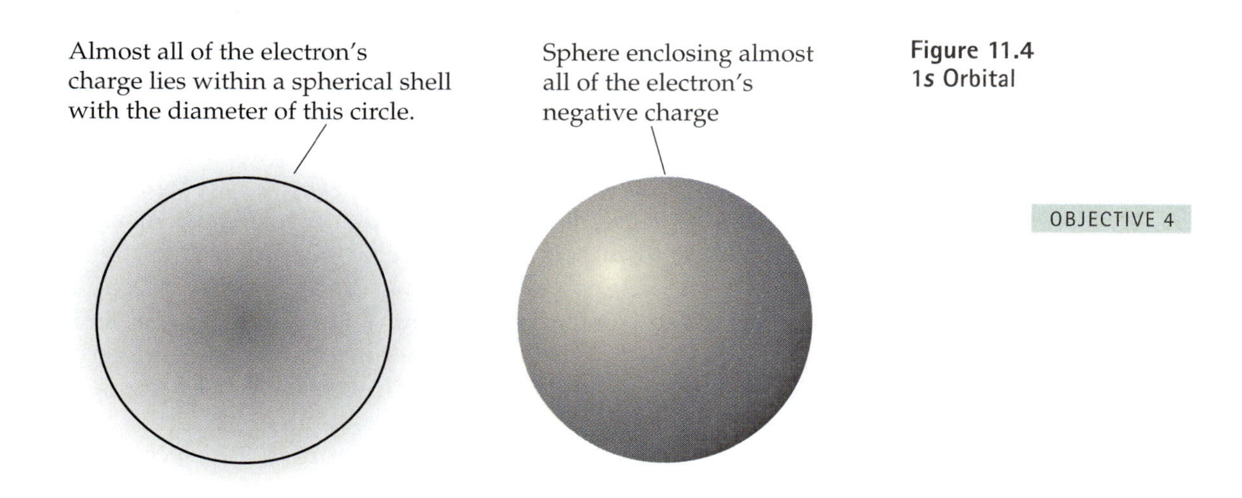
Almost all of the electron's charge lies within a spherical shell with the diameter of this circle.
Sphere enclosing almost all of the electron's negative charge

**Figure 11.4**
**1*s* Orbital**

OBJECTIVE 4

OBJECTIVE 4

Most of the pictures you will see of orbitals represent the hypothetical surfaces that surround a high percentage of the negative charge of an electron of a given waveform. The 1*s* orbital, for example, can be represented either by a fuzzy sphere depicting the varying intensity of the negative charge (Figure 11.3) or by a smooth spherical surface depicting the boundary within which most of the charge is to be found (Figure 11.4).

Is the sphere in Figure 11.3 the 1*s* electron? This is like asking whether the guitar string is the blur that you see when the string vibrates. When we describe the standing wave that represents the motion of a guitar string, we generally do not refer to the material composition of the string. The situation is very similar for the electron. We are able to describe the variation in intensity of the negative charge created by the electron without thinking too much about what the electron is and what it is doing.

*Scientific knowledge is a body of statements of varying certainty—about some of them we are mostly unsure, some are nearly certain, none are absolutely certain.*

RICHARD FEYNMAN
1965 WINNER OF THE NOBEL PRIZE IN PHYSICS

## Particle Interpretation of the Wave Character of the Electron

Where *is* the electron, then, and what *is* it doing? We do not really know, and modern physics tells us that we never will know with 100% certainty. However, with information derived from wave mathematics for an electron, we can predict where the electron *probably* is. According to the particle interpretation of the wave character of the electron, the surface that surrounds 90% of an electron's charge is the surface within which we have a 90% probability of finding the electron. We could say that the electron spends 90% of its time within the space enclosed by that surface. Thus an **orbital** can also be defined as the volume within which an electron has a high probability of being found.

OBJECTIVE 4

It is not difficult to visualize a reasonable path for an electron in a 1*s* orbital, but as you will see later in this section, the shapes of orbitals get much more complex than the simple sphere of the 1*s* orbital. For this reason, we do not attempt to visualize the path an electron follows in an orbital.

How should we picture the electron? Perhaps the best image is the one we used in Chapter 2: a cloud. Our interpretation of the nature of the electron cloud will differ, however, depending on whether we are considering the charge of the electron or its particle nature.

OBJECTIVE 5

If we wish to consider the charge of the electron, the electron cloud represents the continuous variation in the intensity of the negative charge; it is stronger near the nucleus and weaker with increasing distance away from the nucleus, like the fuzzy sphere pictured in Figure 11.3.

In the particle view, the electron cloud can be compared to a multiple-exposure photograph of the electron. (Once again, we must resort to an analogy to describe electron behavior.) If we were able to take a series of sharply focused photos of an electron over a period of time without advancing the film, our final picture would look like Figure 11.5. We would find a high density of dots near the nucleus (because most of the times when the shutter snapped, the electron would be near the nucleus) and a decrease in density with increasing distance from the nucleus (because some of the times when the shutter snapped, the electron would be farther away from the nucleus). This arrangement of dots would illustrate the wave equation's prediction of the probability of finding the electron at any given distance from the nucleus.

A multiple-exposure picture of the electron in a 1*s* orbital of a hydrogen atom might look like this.

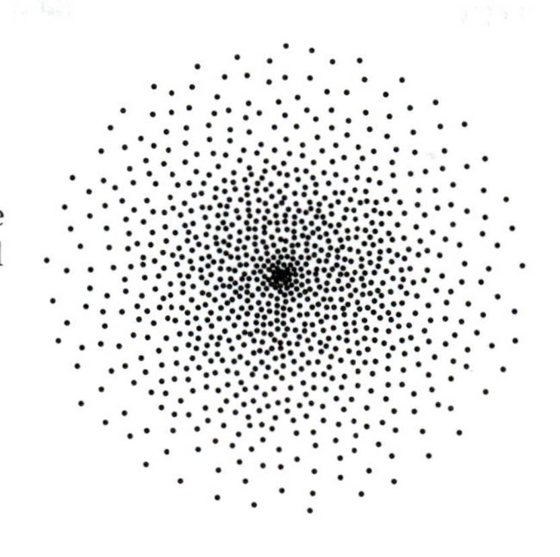

**Figure 11.5**
**Particle Interpretation of a 1*s* Orbital**

OBJECTIVE 4

If enough photos of these dancers are combined, we see a blur instead of individual dancers. This dancer blur is like an electron cloud.

## Other Important Waveforms

Just as the guitar string can have different waveforms, the one electron in a hydrogen atom can also have different waveforms, or orbitals. The shapes and sizes for these orbitals are predicted by the mathematics associated with the wave character of the hydrogen electron. Figure 11.6 shows some of them.

$3d_{z^2}$ $3d_{xz}$ $3d_{yz}$ $3d_{xy}$ $3d_{x^2-y^2}$

$3s$ $3p_x$ $3p_y$ $3p_z$

$2s$ $2p_x$ $2p_y$ $2p_z$

$1s$

**Figure 11.6**
**Some Possible Waveforms, or Orbitals, for an Electron in a Hydrogen Atom**

Before considering the second possible orbital for the electron of a hydrogen atom, let's look at another of the possible ways in which a guitar string can vibrate. The guitar-string waveform that follows has a node in the center where there is no movement of the string.

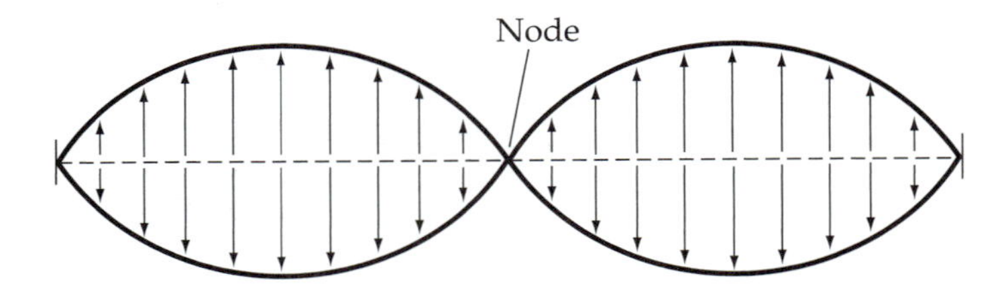


The electron-wave calculations predict that an electron in a hydrogen atom can have a waveform called the 2*s* orbital that is analogous to the guitar-string waveform above. The 2*s* orbital for an electron in a hydrogen atom is spherical like the 1*s* orbital, but it is a larger sphere. All spherical electron waveforms are called *s* orbitals. For an electron in the 2*s* orbital, the charge is most intense at the nucleus. With increasing distance from the nucleus, the charge diminishes in intensity until it reaches a minimum at a certain distance from the nucleus; it then increases again to a maximum, and finally it diminishes again. The region within the 2*s* orbital where the charge intensity decreases to zero is called a **node.** Figure 11.7 shows cutaway, quarter-section views of the 1*s* and 2*s* orbitals.

OBJECTIVE 6

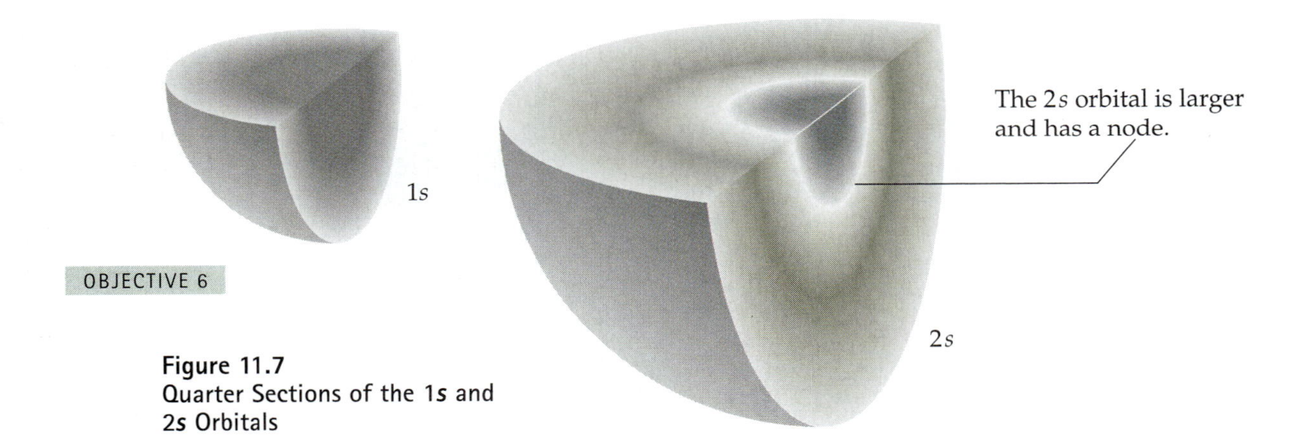

1s
2s
The 2s orbital is larger and has a node.

OBJECTIVE 6

**Figure 11.7**
Quarter Sections of the 1*s* and 2*s* Orbitals

The average distance between the positive charge of the nucleus and the negative charge of a 2*s* electron cloud is greater than the average distance between the nucleus and the charge of a 1*s* electron cloud. Because the strength of the attraction between positive and negative charges decreases with increasing distance between the charges, an electron is more strongly attracted to the nucleus—and therefore is more stable—when it has the smaller 1*s* waveform than when it has the larger 2*s* waveform. As you discovered in Section 7.1, increased stability is associated with decreased

OBJECTIVE 7

potential energy, so a 1*s* electron has lower potential energy than a 2*s* electron.[1] We describe this energy difference by saying that the 1*s* electron is in the first principal energy level and the 2*s* electron is in the second principal energy level. All of the orbitals that have the same potential energy for a hydrogen atom are said to be in the same **principal energy level.** The principal energy levels are often called **shells.** The 1 in 1*s* and the 2 in 2*s* show the principal energy levels, or shells, for these orbitals.

Chemists sometimes draw **orbital diagrams,** such as the following, wherein lines are used to represent the orbitals in an atom and arrows (which we will be adding later) are used to represent electrons:

2*s* __

1*s* __

The line representing the 2*s* orbital is higher on the page to indicate its higher potential energy.

Because electrons seek the lowest energy level possible, we expect the electron in a hydrogen atom to have the 1*s* waveform, or electron cloud. We say that the electron is *in* the 1*s* orbital. But the electron in a hydrogen atom does not need to stay in the 1*s* orbital at all times. Just as an input of energy (a little arm work on our part) can lift a book that is resting on a table and raise it to a position that has greater potential energy, so can the waveform of an electron in a hydrogen atom be changed from the 1*s* shape to the 2*s* shape by the addition of energy to the atom. We say that the electron can be *excited* from the 1*s* orbital to the 2*s* orbital. Hydrogen atoms with their electron in the 1*s* orbital are said to be in their **ground state.** A hydrogen atom with its electron in the 2*s* orbital is in an **excited state.**

The energy generated by the explosives in fireworks supplies the energy to excite electrons to higher energy levels. When the electrons return to their ground state, they give off energy in the form of light.

If you lift a book from a table to above the table and then release it, it falls back down to its lower-energy position on the table. The same is true for the electron. After the electron is excited from the 1*s* orbital to the 2*s* orbital, it spontaneously returns to its lower-energy 1*s* form.

OBJECTIVE 8

An electron in a hydrogen atom can be excited to orbitals other than the 2*s*. For example, an electron in a hydrogen atom can be excited from the 1*s* to a 2*p* orbital. There are actually three possible 2*p* orbitals. They are identical in shape and size, but each lies at a 90° angle to the other two. Because they can be viewed as lying on the *x*, *y*, and *z* axes of a three-dimensional coordinate system, they are often called the $2p_x$, $2p_y$, and $2p_z$ orbitals. An electron with a 2*p* waveform has its negative charge distributed in two lobes on opposite sides of the nucleus. Figure 11.8 shows the shape of a $2p_z$ orbital.

**Figure 11.8**
Realistic View of a $2p_z$ Orbital

OBJECTIVE 8

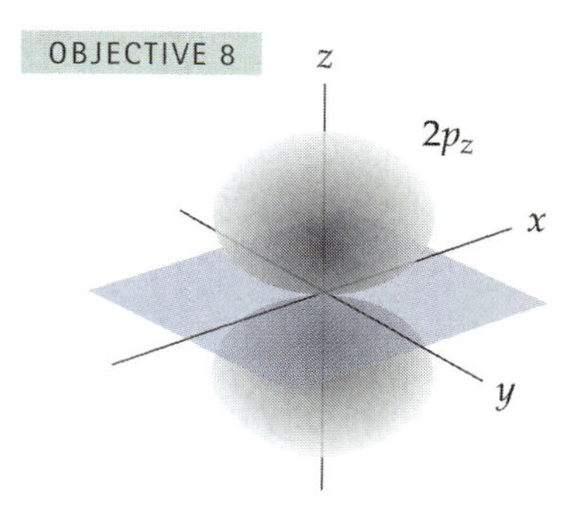


[1] If you are reading this chapter before studying Chapter 7, you may want to read the portions of Section 7.1 that describe energy, kinetic energy, and potential energy.

**Figure 11.9**
Stylized View of a $2p_z$ Orbital

OBJECTIVE 8

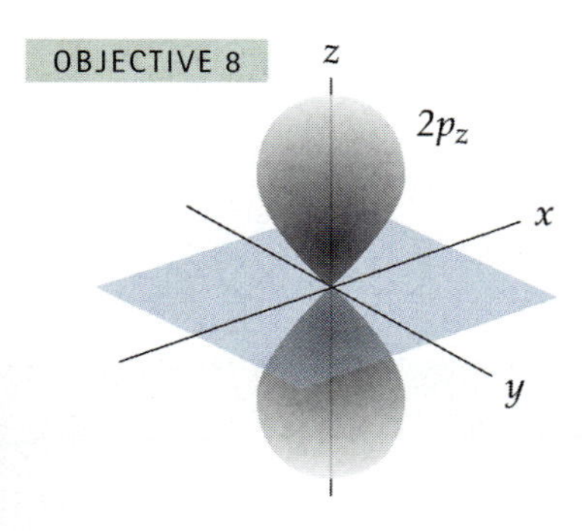


**Figure 11.10**
Stylized View of the $2p_x$, $2p_y$, and $2p_z$ Orbitals Combined

OBJECTIVE 8

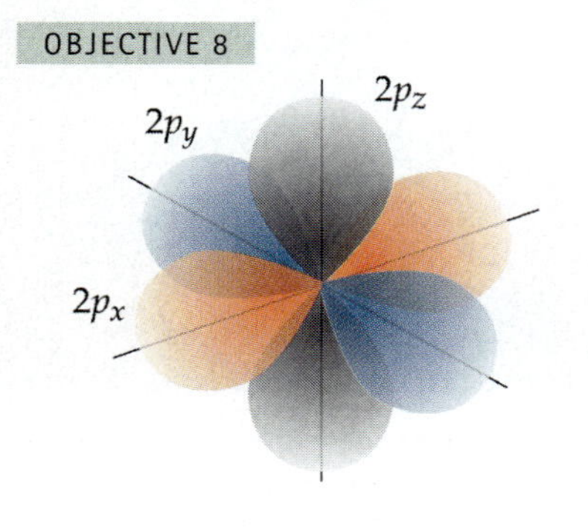


In order to show more easily how the 2*p* orbitals fit together, we often draw them in a more elongated and stylized manner (Figures 11.9 and 11.10).

In a hydrogen atom, the average distance between the negative charge of a 2*p* electron cloud and the nucleus is the same as for a 2*s* electron cloud. Therefore, an electron in a hydrogen atom has the same attraction to the nucleus and the same stability when it has a 2*p* form as when it has the 2*s* form. Therefore, a 2*s* electron has the same potential energy as a 2*p* electron.[2] These orbitals are in the same principal energy level.

Because the shapes of the 2*s* and 2*p* electron clouds are different, we distinguish between them by saying that an electron with the 2*s* waveform is in the 2*s* sublevel and that an electron with any of the three 2*p* waveforms is in the 2*p* sublevel. Orbitals that have the same potential energy, the same size, and the same shape are in the same **sublevel.** The sublevels are sometimes called **subshells.** Thus there is one orbital in the 1*s* sublevel, there is one orbital in the 2*s* sublevel, and there are three orbitals in the 2*p* sublevel. The following orbital diagram shows these orbitals and sublevels.

2*s* __ 2*p* __ __ __

1*s* __

The lines for the 2*s* and 2*p* orbitals are drawn side by side to show that they have the same potential energy for the one electron in a hydrogen atom. You will see in the next section that the 2*s* and 2*p* orbitals have different potential energies for atoms larger than hydrogen atoms.

In the third principal energy level, there are nine possible waveforms for an electron, in three different sublevels. The 3*s* sublevel has one orbital that has a spherical shape, like the 1*s* and the 2*s*, but it has a larger average radius and two nodes. Its greater average distance from the positive charge of the nucleus makes it less stable and higher in energy than the 1*s* and 2*s* orbitals. The calculations predict that the third principal energy level has a 3*p* sublevel, with three 3*p* orbitals. These have the same general shape as the 2*p* orbitals, but they are larger, which leads to less attraction to the nucleus, less stability, and higher potential energy than for a 2*p* orbital. The third principal energy level also has a 3*d* sublevel, with five 3*d* orbitals. Four of these orbitals have four lobes whose shape is similar to the lobes of a 3*p* orbital. We will call these "double dumbbells." An electron in a 3*d* orbital has its negative charge spread out in these four lobes. The fifth 3*d* orbital has a different shape, as shown in Figure 11.11.

OBJECTIVE 9

When an electron in a hydrogen atom is excited to the fourth principal energy level, it can be in any one of four sublevels: 4*s*, 4*p*, 4*d*, or 4*f*. There is one 4*s* orbital, with a spherical shape and a larger volume than the 3*s*.

[2] The next section shows how chemists use the orbitals predicted for hydrogen to describe atoms of other elements. There you will see that when an atom has more than one electron, the 2*s* orbital is lower in potential energy than the 2*p* orbitals.

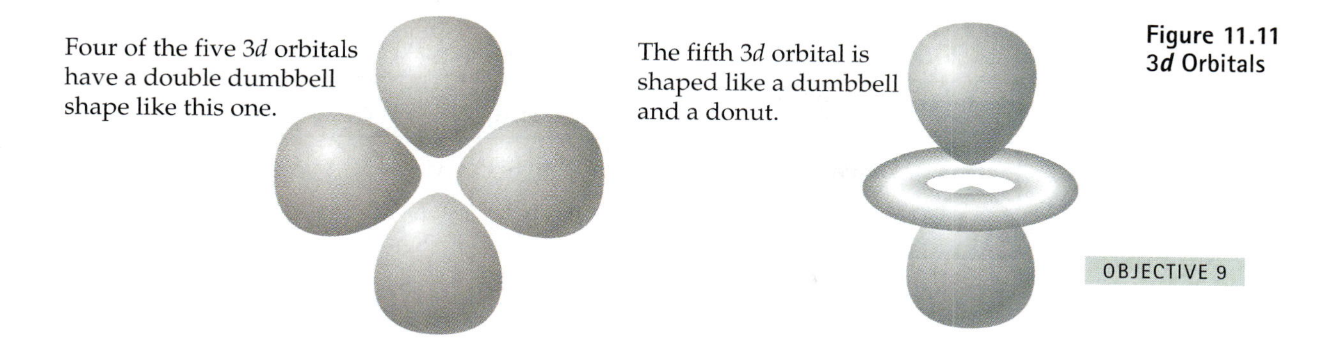

Four of the five 3*d* orbitals have a double dumbbell shape like this one.
The fifth 3*d* orbital is shaped like a dumbbell and a donut.


**Figure 11.11**
**3*d* Orbitals**

OBJECTIVE 9

There are three 4*p* orbitals, similar in shape to the 3*p* orbitals but with a larger volume than the 3*p* orbitals. There are five 4*d* orbitals, similar to but larger than the 3*d* orbitals. The 4*f* sublevel has seven possible orbitals.

## Overall Organization of Principal Energy Levels, Sublevels, and Orbitals

Table 11.1 shows all the orbitals predicted for the first seven principal energy levels. Note that the first principal energy level has one sublevel, the second has two, the third has three, and the fourth has four. If *n* is the number associated with the principal energy level, then each principal energy level has *n* sublevels. Thus there are five sublevels on the fifth principal energy level: 5*s*, 5*p*, 5*d*, 5*f*, and 5*g*. The 5*s*, 5*p*, 5*d*, and 5*f* orbitals have shapes similar to the 4*s*, 4*p*, 4*d*, and 4*f* orbitals, but they are larger and have higher potential energy.

Each *s* sublevel has one orbital, each *p* sublevel has three orbitals, each sublevel *d* has five orbitals, and each *f* sublevel has seven orbitals. Thus there are one 5*s* orbital, three 5*p* orbitals, five 5*d* orbitals, and seven 5*f* orbitals. The trend of increasing the number of orbitals by two for each succeeding sublevel continues for 5*g* and beyond. There are nine 5*g* orbitals with shapes more complex than the shapes of the 4*f* orbitals.

In the next section, where we use the orbitals predicted for hydrogen to describe atoms of other elements, you will see that none of the known elements has electrons in the 5*g* sublevel for their most stable state (ground state). Thus we are not very interested in describing the 5*g* orbitals. Likewise, although the sixth principal energy level has six sublevels and the seventh has seven, only the 6*s*, 6*p*, 6*d*, 7*s*, and 7*p* sublevels are important for describing the ground states of the known elements. The reason for this will be explained in the next section.

None of the known elements in its ground state has any electrons in a principal energy level higher than the seventh, so we are not concerned with the principal energy levels above seven. The orbital diagram in Figure 11.12 shows only the sublevels (or subshells) and orbitals that are necessary for describing the ground states of the known elements.

**Table 11.1**
Possible Sublevels and Orbitals for the First Seven Principal Energy Levels

| Principal energy level (shell) | Sublevels (subshells) | Number of orbitals |
|---|---|---|
| 1 | 1*s* | 1 |
| 2 | 2*s* | 1 |
| | 2*p* | 3 |
| 3 | 3*s* | 1 |
| | 3*p* | 3 |
| | 3*d* | 5 |
| 4 | 4*s* | 1 |
| | 4*p* | 3 |
| | 4*d* | 5 |
| | 4*f* | 7 |
| 5 | 5*s* | 1 |
| | 5*p* | 3 |
| | 5*d* | 5 |
| | 5*f* | 7 |
| | (5*g*)[a] | 9 |
| 6 | 6*s* | 1 |
| | 6*p* | 3 |
| | 6*d* | 5 |
| | (6*f*) | 7 |
| | (6*g*) | 9 |
| | (6*h*) | 11 |
| 7 | 7*s* | 1 |
| | 7*p* | 3 |
| | (7*d*) | 5 |
| | (7*f*) | 7 |
| | (7*g*) | 9 |
| | (7*h*) | 11 |
| | (7*i*) | 13 |

[a] The sublevels in parentheses are not necessary for describing any of the known elements.

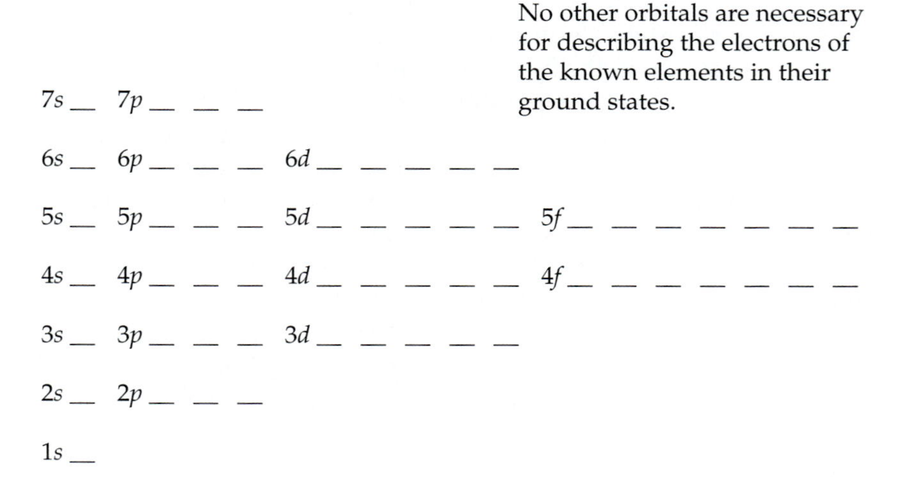

**Figure 11.12**
**Diagram of the Orbitals for an Electron in a Hydrogen Atom**

No other orbitals are necessary for describing the electrons of the known elements in their ground states.

7*s* __ 7*p* __ __ __

6*s* __ 6*p* __ __ __ 6*d* __ __ __ __ __

5*s* __ 5*p* __ __ __ 5*d* __ __ __ __ __ 5*f* __ __ __ __ __ __ __

4*s* __ 4*p* __ __ __ 4*d* __ __ __ __ __ 4*f* __ __ __ __ __ __ __

3*s* __ 3*p* __ __ __ 3*d* __ __ __ __ __

2*s* __ 2*p* __ __ __

1*s* __

## 11.2 Multi-Electron Atoms

It is possible to solve the wave equation and determine the shapes and sizes of electron orbitals for the hydrogen atom and for any one-electron ion (for example, $He^+$ or $Li^{2+}$), but the calculations for any atom with two or more electrons are extremely complex. Fortunately, scientists have found that when they take the information derived from solving the wave equation for hydrogen atoms and apply it to other atoms, the resulting descriptions are sufficiently accurate to explain many of the chemical characteristics of those elements. Thus the following assumption lies at the core of the modern description of atoms other than hydrogen.

> **All of the elements have the same set of possible principal energy levels, sublevels, and orbitals that has been calculated for hydrogen.**

In other words, we assume that all atoms can have the sublevels and orbitals listed in Table 11.1. Now let us look at how the electrons of elements other than hydrogen are distributed in those hydrogen-like orbitals.

### Helium and Electron Spin

OBJECTIVE 10

Helium, with atomic number 2 and two electrons, is the next element after hydrogen in the periodic table. Both of helium's electrons are in the 1*s* orbital. This allows them to be as close to the positive charge of the nucleus as possible. Even though the two electron charge clouds occupy the same space, the two electrons are not identical. The property in which they differ is called **electron spin.** Although the true nature of electron spin is uncertain, it is useful to think of it as though electrons can spin on their axis. We can visualize the two electrons in a helium atom as spinning in opposite directions (Figure 11.13).

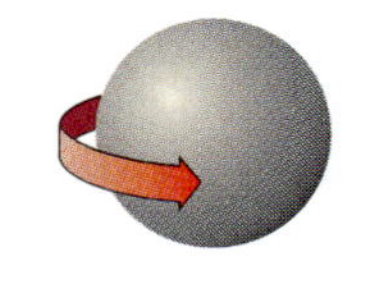
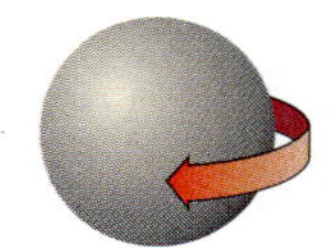

**Figure 11.13**
Electron Spin

OBJECTIVE 10

Arrows are added to an **orbital diagram** to show the distribution of electrons in the possible orbitals and the relative spin of each electron. The following is an orbital diagram for a helium atom.

$$1s\ \underline{\uparrow\downarrow}$$

The orbital distribution of electrons can also be expressed in a shorthand notation that describes the arrangement of electrons in the sublevels without reference to their spin. This shorthand for helium's configuration is written $1s^2$ and is commonly called an electron configuration. The 1 represents the first principal energy level, the *s* indicates an electron cloud with a spherical shape, and the 2 shows that there are two electrons in that 1*s* sublevel.

$$1s^2$$

Represents the principal energy level

Shows the number of electrons in the orbital

Indicates the shape of the orbital

In the broadest sense, an **electron configuration** is any description of the complete distribution of electrons in atomic orbitals. Although this can mean either an orbital diagram or the shorthand notation, this text will follow the common convention of referring to only the shorthand notation as an **electron configuration.** For example, if you are asked for an electron configuration for helium, you should write $1s^2$. If you are expected to describe the orbitals with lines and the electrons with arrows, you will be asked to draw an orbital diagram.

## The Second-Period Elements

Electrons do not fill the available sublevels in the order we might expect. Thus, to predict the electron configurations of elements larger than helium, we need a way of remembering the actual order in which sublevels fill. Probably the *least* reliable method is to memorize the following list (even though it shows the order of filling of all the orbitals necessary for describing the ground state electron configurations of all the known elements).

$$1s\ 2s\ 2p\ 3s\ 3p\ 4s\ 3d\ 4p\ 5s\ 4d\ 5p\ 6s\ 4f\ 5d\ 6p\ 7s\ 5f\ 6d\ 7p$$

Instead of relying on memorization, you can use the memory aid shown in Figure 11.14 to remind you of the correct order of filling of the sublevels. The following steps explain how to write it and use it yourself.

- Write the possible sublevels for each energy level in organized rows and columns, as in Figure 11.14. To do this, you need to remember that there is one sublevel on the first principal energy level, two on the second, three on the third, and so on. Every principal energy level has an *s* orbital. The second principal energy level and all higher energy levels have a *p* sublevel. The *d* sublevels start on the third principal energy level, the *f* sublevels start on the fourth principal energy level, and so on.
- Draw arrows like those you see in Figure 11.14.
- Starting with the top arrow, follow the arrows one by one in the direction they point, listing the sublevels as you pass through them.

The sublevels that are not needed for describing the known elements are enclosed in parentheses in Figure 11.14. Later in this section, you will learn how to determine the order of filling of the sublevels by using a periodic table.

*An atomic orbital may contain 2 electrons at most, and the electrons must have different spins.* Because each *s* sublevel has 1 orbital and each orbital contains a maximum of 2 electrons, each *s* sublevel contains a maximum of 2 electrons. Because each *p* sublevel has 3 orbitals and each orbital contains a maximum of 2 electrons, each *p* sublevel contains a maximum of 6 electrons. Using similar reasoning, we can determine that each *d* sublevel contains a maximum of 10 electrons and that each *f* sublevel contains a maximum of 14 electrons (Table 11.2).

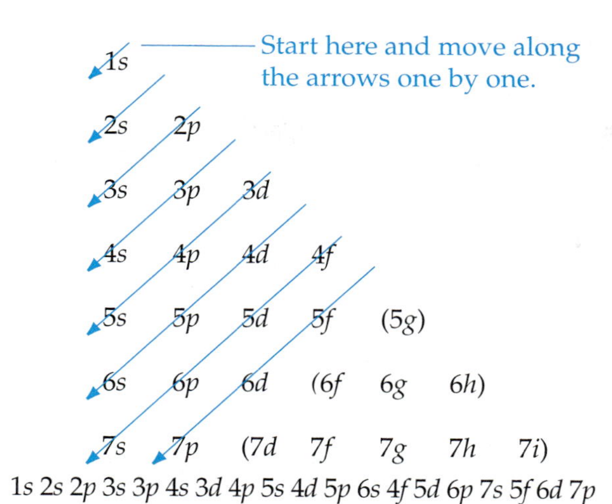


**Figure 11.14**
Aid for Remembering the Order of Sublevel Filling

**Table 11.2**
General Information About Sublevels

| Type of sublevel | Number of orbitals | Maximum number of electrons |
|---|---|---|
| *s* | 1 | 2 |
| *p* | 3 | 6 |
| *d* | 5 | 10 |
| *f* | 7 | 14 |

We can now predict the electron configurations and orbital diagrams for the ground state of lithium, which has three electrons, and beryllium, which has four electrons:

OBJECTIVE 11

OBJECTIVE 12

Lithium $1s^2\,2s^1$

2*s* ↑

1*s* ⇅

Beryllium $1s^2\,2s^2$

2*s* ⇅

1*s* ⇅

A boron atom has five electrons. The first four fill the 1*s* and 2*s* orbitals, and the fifth electron goes into a 2*p* orbital. The electron configuration and orbital diagram for the ground state of boron atoms are shown below. Even though only one of three possible 2*p* orbitals contains an electron, we show all three 2*p* orbitals in the orbital diagram. The lines for the 2*p* orbitals are drawn higher on the page than the line for the 2*s* orbital to show that electrons in the 2*p* orbitals have a higher potential energy than electrons in the 2*s* orbital of the same atom.[3]

OBJECTIVE 11

OBJECTIVE 12

Boron $1s^2\,2s^2\,2p^1$

2*s* ⇅ 2*p* ↑ __ __

1*s* ⇅

[3] The 2*s* and 2*p* orbitals available for the 1 electron of a hydrogen atom have the same potential energy. However, for reasons that are beyond the scope of this discussion, when an atom has more than 1 electron, the 2*s* orbital is lower in potential energy than the 2*p* orbitals.

*When electrons are filling orbitals of the same energy, they enter orbitals in such a way as to maximize the number of unpaired electrons, all with the same spin. In other words, they enter empty orbitals first, and all electrons in half-filled orbitals have the same spin.* The orbital diagram for the ground state of carbon atoms is

OBJECTIVE 12

2s ⇅ 2p ↑ __ __

1s ⇅

There are two ways to write the electron configuration of carbon atoms. One way emphasizes that the two electrons in the 2*p* sublevel are in different orbitals:

$1s^2\ 2s^2\ 2p_x^{\ 1}\ 2p_y^{\ 1}$

Often, however, the arrangement of the 2*p* electrons is taken for granted, and the following notation is used:

OBJECTIVE 11

$1s^2\ 2s^2\ 2p^2$

Unless you are told otherwise, you may assume that "a complete electron configuration" means a configuration written in the second form, without the subscripts *x*, *y*, and *z*.

The electron configurations and orbital diagrams for the rest of the elements of the second period show the sequential filling of the 2*p* orbitals:

OBJECTIVE 11

OBJECTIVE 12

Nitrogen $1s^2\ 2s^2\ 2p^3$

2s ⇅ 2p ↑ ↑ ↑

1s ⇅

Oxygen $1s^2\ 2s^2\ 2p^4$

2s ⇅ 2p ⇅ ↑ ↑

1s ⇅

OBJECTIVE 11

OBJECTIVE 12

Fluorine $1s^2\ 2s^2\ 2p^5$

2s ⇅ 2p ⇅ ⇅ ↑

1s ⇅

Neon $1s^2\ 2s^2\ 2p^6$

2s ⇅ 2p ⇅ ⇅ ⇅

1s ⇅

Figure 11.15 shows how the electron clouds for the electrons of the atoms of first- and second-period elements are envisioned. Note that the electron-charge waveforms for the electrons in different orbitals of each atom are all superimposed on each other.

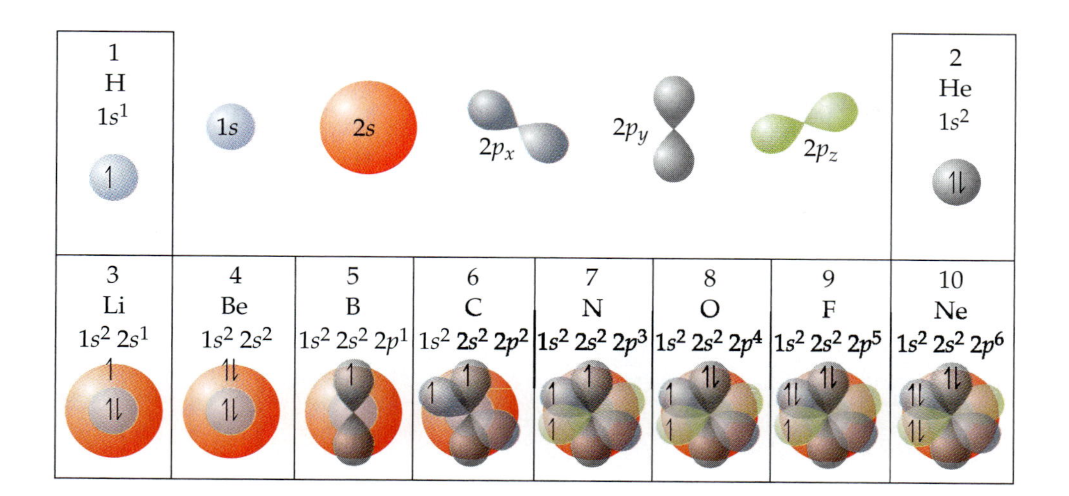


**Figure 11.15**
Electron Configurations and Orbital Models for the First Ten Elements

## The Periodic Table and the Modern Model of the Atom

The periodic table itself can be used as a guide for predicting the electron configurations of most of the elements. Conversely, the electron configurations of the elements can be used to explain the table's structure and the similarities and differences that were the basis for its creation.

The organization of the periodic table reflects the modern model of the atom. For example, the highest-energy electrons for all of the elements in groups 1 (1A) and 2 (2A) in the periodic table are in *s* orbitals. That is, the highest-energy electrons for lithium, Li, and beryllium, Be, atoms are in the 2*s* orbital, and the highest-energy electrons for sodium, Na, and magnesium, Mg, atoms are in the 3*s* orbital. This continues down to francium, Fr, and radium, Ra, which have their highest-energy electrons in the 7*s* orbital. Therefore, the first two columns in the periodic table are called the *s block.* Because hydrogen and helium have their electrons in the 1*s* orbital, they belong in the *s* block too (Figure 11.16).

All of the elements in the block with boron, B, neon, Ne, thallium, Tl, and radon, Rn, at the corners have their highest-energy electrons in *p* orbitals, so this is called the *p block* (Figure 11.16). The second principal energy level is the first to contain *p* orbitals, so atoms of elements in the first row of the *p* block have their highest-energy electrons in the 2*p* sublevel. The highest-energy electrons for elements in the second row of the *p* block are in the 3*p* sublevel. This trend continues, so we can predict that the 81st through the 86th electrons for the elements thallium, Tl, through radon, Rn, are added to the 6*p* sublevel. Moreover, we can predict that the new elements that have been made with atoms larger than element 112 have their highest-energy electrons in the 7*p* sublevel.

The last electrons to be added to an orbital diagram for the atoms of the transition metal elements go into *d* orbitals. For example, the last electrons added to atoms of scandium, Sc, through zinc, Zn, are added to 3*d* orbitals. The elements yttrium, Y, through cadmium, Cd, have their highest-energy electrons in the 4*d* sublevel. The elements directly below them in rows 6

**Figure 11.16**
The Periodic Table and the Modern Model of the Atom

| | 1 1A | 2 2A | 3 3B | 4 4B | 5 5B | 6 6B | 7 7B | 8 8B | 9 8B | 10 8B | 11 1B | 12 2B | 13 3A | 14 4A | 15 5A | 16 6A | 17 7A | 18 8A |
|---|---|---|---|---|---|---|---|---|---|---|---|---|---|---|---|---|---|---|
| 1 | 1 H | | | | | | | | | | | | | | | | | 2 He |
| 2 | 3 Li | 4 Be | | | | | | | | | | | 5 B | 6 C | 7 N | 8 O | 9 F | 10 Ne |
| 3 | 11 Na | 12 Mg | | | | | | | | | | | 13 Al | 14 Si | 15 P | 16 S | 17 Cl | 18 Ar |
| 4 | 19 K | 20 Ca | 21 Sc | 22 Ti | 23 V | 24 Cr | 25 Mn | 26 Fe | 27 Co | 28 Ni | 29 Cu | 30 Zn | 31 Ga | 32 Ge | 33 As | 34 Se | 35 Br | 36 Kr |
| 5 | 37 Rb | 38 Sr | 39 Y | 40 Zr | 41 Nb | 42 Mo | 43 Tc | 44 Ru | 45 Rh | 46 Pd | 47 Ag | 48 Cd | 49 In | 50 Sn | 51 Sb | 52 Te | 53 I | 54 Xe |
| 6 | 55 Cs | 56 Ba | 71 Lu | 72 Hf | 73 Ta | 74 W | 75 Re | 76 Os | 77 Ir | 78 Pt | 79 Au | 80 Hg | 81 Tl | 82 Pb | 83 Bi | 84 Po | 85 At | 86 Rn |
| 7 | 87 Fr | 88 Ra | 103 Lr | 104 Rf | 105 Db | 106 Sg | 107 Bh | 108 Hs | 109 Mt | 110 Uun | 111 Uuu | 112 Uub | | 114 Uuq | | 116 Uuh | | 118 Uuo |

| | | | | | | | | | | | | | | |
|---|---|---|---|---|---|---|---|---|---|---|---|---|---|---|
| 6 | 57 La | 58 Ce | 59 Pr | 60 Nd | 61 Pm | 62 Sm | 63 Eu | 64 Gd | 65 Tb | 66 Dy | 67 Ho | 68 Er | 69 Tm | 70 Yb |
| 7 | 89 Ac | 90 Th | 91 Pa | 92 U | 93 Np | 94 Pu | 95 Am | 96 Cm | 97 Bk | 98 Cf | 99 Es | 100 Fm | 101 Md | 102 No |

**s block**

| | 1 1A | 2 2A |
|---|---|---|
| 1s | 1 | |
| 2s | 3 | 4 |
| 3s | 11 | 12 |
| 4s | 19 | 20 |
| 5s | 37 | 38 |
| 6s | 55 | 56 |
| 7s | 87 | 88 |

**d block**

| | 3 3B | 4 4B | 5 5B | 6 6B | 7 7B | 8 8B | 9 8B | 10 8B | 11 1B | 12 2B |
|---|---|---|---|---|---|---|---|---|---|---|
| 3d | 21 | 22 | 23 | 24 | 25 | 26 | 27 | 28 | 29 | 30 |
| 4d | 39 | 40 | 41 | 42 | 43 | 44 | 45 | 46 | 47 | 48 |
| 5d | 71 | 72 | 73 | 74 | 75 | 76 | 77 | 78 | 79 | 80 |
| 6d | 103 | 104 | 105 | 106 | 107 | 108 | 109 | 110 | 111 | 112 |

**p block**

| | 13 3A | 14 4A | 15 5A | 16 6A | 17 7A | 18 8A |
|---|---|---|---|---|---|---|
| | | | | | | 2 |
| 2p | 5 | 6 | 7 | 8 | 9 | 10 |
| 3p | 13 | 14 | 15 | 16 | 17 | 18 |
| 4p | 31 | 32 | 33 | 34 | 35 | 36 |
| 5p | 49 | 50 | 51 | 52 | 53 | 54 |
| 6p | 81 | 82 | 83 | 84 | 85 | 86 |
| 7p | 113 | 114 | 115 | 116 | 117 | 118 |

**f block**

| | | | | | | | | | | | | | | |
|---|---|---|---|---|---|---|---|---|---|---|---|---|---|---|
| 4f | 57 | 58 | 59 | 60 | 61 | 62 | 63 | 64 | 65 | 66 | 67 | 68 | 69 | 70 |
| 5f | 89 | 90 | 91 | 92 | 93 | 94 | 95 | 96 | 97 | 98 | 99 | 100 | 101 | 102 |

and 7 add electrons to the 5*d* and 6*d* orbitals. The transition metals can be called the *d block.* (Figure 11.16).

The section of the periodic table that contains the inner transition metals is called the *f block.* Thus we can predict that the last electrons added to the orbital diagrams of elements with atomic numbers 57 through 70 go into the 4*f* sublevel. Elements 89 through 102 are in the second row of the *f* block. Because the fourth principal energy level is the first to have an *f* sublevel, we can predict that the highest-energy electrons for these elements go to the 5*f* sublevel.

We can also use the block organization of the periodic table, as shown in Figure 11.16, to remind us of the order in which sublevels are filled. To do this, we move through the elements in order of increasing atomic number, listing new sublevels as we come to them. The type of sublevel (*s*, *p*, *d*, or *f* ) is determined from the block in which the atomic number is found. The number for the principal energy level (for example, the 3 in 3*p*) is determined from the row in which the element is found and from our knowledge that the *s* sublevels start on the first principal energy level, the *p* sublevels start on the second principal energy level, the *d* sublevels start on the third principal energy level, and the *f* sublevels start on the fourth principal energy level.

- We know that the first two electrons added to an atom go to the 1*s* sublevel.
- Atomic numbers 3 and 4 are in the second row of the *s* block (look for them in the bottom half of Figure 11.16), signifying that the 3rd and 4th electrons are in the 2*s* sublevel.
- Atomic numbers 5 through 10 are in the first row of the *p* block, and the *p* sublevels start on the second energy level. Therefore, the 5th through 10th electrons go into the 2*p* sublevel.
- Atomic numbers 11 and 12 are in the third row of the *s* block, so the 11th and 12th electrons go into the 3*s* sublevel.
- Because atomic numbers 13 through 18 are in the *p* block, we know they go into a *p* sublevel. Because the *p* sublevels begin on the second principal energy level and atomic numbers 13 through 18 are in the second row of the *p* block, the 13th through 18th electrons must go into the 3*p* sublevel.
- The position of atomic numbers 19 and 20 in the fourth row of the *s* block and the position of atomic numbers 21 through 30 in the first row of the *d* block show that the 4*s* sublevel fills before the 3*d* sublevel.

Moving through the periodic table in this fashion produces the following order of sublevels up through 6*s*:

1*s* 2*s* 2*p* 3*s* 3*p* 4*s* 3*d* 4*p* 5*s* 4*d* 5*p* 6*s*

Note that atomic numbers 57 through 70 in the periodic table in Figure 11.17 are in the 4*f* portion of the table. It is a common mistake to forget that the $4f$ sublevel is filled after the $6s$ sublevel and before the $5d$ sublevel. In order to make the overall shape of the table more compact and convenient to display, scientists have adopted the convention of removing the elements with atomic numbers 57 through 70 and 89 through 102 (the latter being the $5f$ portion of the table) from their natural position between the *s* and *d* blocks and placing them at the bottom of the table. Electrons go into the $5f$ sublevel after the $7s$ sublevel and before the $6d$ sublevel. The periodic table in Figure 11.17 shows how the blocks in the periodic table would fit together if the inner transition metals—the *f* block—were left in their natural position.

**Figure 11.17**
**Periodic Table with the Inner Transition Metals in Their Natural Position**

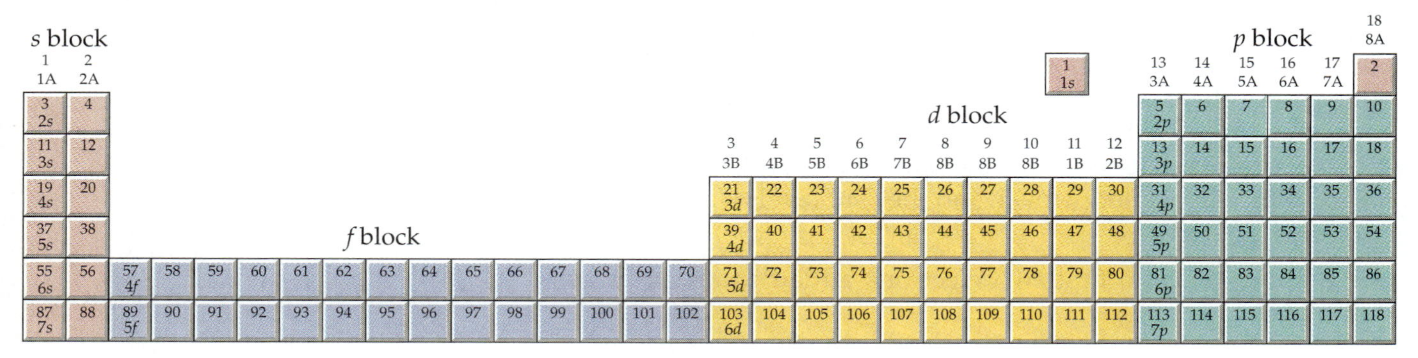


The following sample study sheet shows the general steps for writing complete electron configurations and orbital diagrams for uncharged atoms.

## Sample Study Sheet 11.1

### Writing Complete Electron Configurations and Orbital Diagrams for Uncharged Atoms

OBJECTIVE 11

OBJECTIVE 12

**Tip-off** If you are asked to write a complete electron configuration or an orbital diagram, you can use the following guidelines.

**General Steps** To write a complete electron configuration for an uncharged atom:

**Step 1** Determine the number of electrons in the atom from its atomic number.

**Step 2** Add electrons to the sublevels in the correct order of filling.

Add 2 electrons to each *s* sublevel, 6 to each *p* sublevel, 10 to each *d* sublevel, and 14 to each *f* sublevel.

**Step 3** To check your complete electron configuration, look to see whether the location of the last electron added corresponds to the element's position in the periodic table. (See Example 11.1.)

To draw an orbital diagram for an uncharged atom:

**Step 1** Write the complete electron configuration for the atom. (This step is not absolutely necessary, but it can help guide you to the correct orbital diagram.)

**Step 2** Draw a line for each orbital of each sublevel mentioned in the complete electron configuration.

Draw one line for each *s* sublevel, three lines for each *p* sublevel, five lines for each *d* sublevel, and seven lines for each *f* sublevel.

As a guide to the order of filling, draw your lines so that the orbitals that fill first are lower on the page than the orbitals that fill later.

Label each sublevel.

**Step 3** For orbitals containing two electrons, draw one arrow up and one arrow down to indicate the electrons' opposite spin.

**Step 4** For unfilled sublevels, add electrons to empty orbitals whenever possible, giving them the same spin.

The arrows for the first 3 electrons to enter a *p* sublevel should all be placed pointing up in different orbitals. The fourth, fifth, and sixth are then placed pointing down in the same sequence so as to fill these orbitals.

The first 5 electrons to enter a *d* sublevel should be drawn pointing up in different orbitals. The next 5 electrons are drawn as arrows pointing down and fill these orbitals (again, following the same sequence as for the first 5 *d* electrons).

The first 7 electrons to enter an *f* sublevel should be drawn as arrows pointing up in different orbitals. The next 7 electrons are paired with the first 7 (in the same order of filling) and are drawn as arrows pointing down.

**Example** See Example 11.1.

## Example 11.1 Electron Configurations and Orbital Diagrams

OBJECTIVE 11

OBJECTIVE 12

Write the complete electron configuration and draw an orbital diagram for iron, Fe.

***Solution***

We follow the steps described in Study Sheet 11.1 to write the complete electron configuration.

**Determine the number of electrons in the atom from its atomic number.**

The periodic table shows us that iron, Fe, has an atomic number of 26, so an uncharged atom of iron has 26 electrons.

**Add electrons to the sublevels in the correct order of filling.**

We can determine the order of filling by memorizing it, by figuring it out from the memory aid shown in Figure 11.15, or by using the periodic table and our knowledge of *s*, *p*, *d*, and *f* blocks. The order of filling is

1*s* 2*s* 2*p* 3*s* 3*p* 4*s* 3*d* 4*p* 5*s* 4*d* 5*p* 6*s* 4*f* 5*d* 6*p* 7*s* 5*f* 6*d* 7*p*

Next, we fill the orbitals according to this sequence, putting 2 electrons in each *s* sublevel, 6 in each *p* sublevel, 10 in each *d* sublevel, and 14 in each *f* sublevel until we reach the desired number of electrons. For iron, we get the following complete electron configuration:

$1s^2\ 2s^2\ 2p^6\ 3s^2\ 3p^6\ 4s^2\ 3d^6$

**To check your complete electron configuration, look to see whether the location of the last electron added corresponds to the element's position in the periodic table.**

Because it is fairly easy to forget a sublevel or to miscount the electrons added, it is a good idea to check your complete electron configuration quickly by looking to see whether the last electrons added correspond to the element's location in the periodic table. The symbol for iron, Fe, is found in the sixth column of the d block. This shows that there are 6 electrons in a *d* sublevel. Because iron is in the first row of the *d* block, and because the *d* sublevels begin on the third principal energy level, these six electrons are correctly described as $3d^6$.

To draw the orbital diagram, we draw a line for each orbital of each sublevel mentioned in the complete electron configuration above. For orbitals containing two electrons, we draw one arrow up and one arrow down to indicate the electrons' opposite spin. For unfilled sublevels, we add electrons to empty orbitals first with the same spin. The orbital diagram for iron atoms follows. Note that 4 of the 6 electrons in the 3*d* sublevel are in different orbitals and have the same spin.

4*s* ⇅ 3*d* ⇅ ↑ ↑ ↑ ↑

3*s* ⇅ 3*p* ⇅ ⇅ ⇅

2*s* ⇅ 2*p* ⇅ ⇅ ⇅

1*s* ⇅

## EXERCISE 11.1 Electron Configurations and Orbital Diagrams

OBJECTIVE 11

OBJECTIVE 12

Write the complete electron configuration and draw an orbital diagram for antimony, Sb.

## Abbreviated Electron Configurations

We learned in Chapter 2 that the noble gases rarely form chemical bonds. This can now be explained in terms of the stability of their electron configurations. The formation of chemical bonds requires atoms to gain, lose, or share electrons, and the electron configurations of the noble gases (displayed below) are so stable that their atoms do not easily undergo any of those changes.

**He** $1s^2$

**Ne** $1s^2\ 2s^2\ 2p^6$

**Ar** $1s^2\ 2s^2\ 2p^6\ 3s^2\ 3p^6$

**Kr** $1s^2\ 2s^2\ 2p^6\ 3s^2\ 3p^6\ 4s^2\ 3d^{10}\ 4p^6$

**Xe** $1s^2\ 2s^2\ 2p^6\ 3s^2\ 3p^6\ 4s^2\ 3d^{10}\ 4p^6\ 5s^2\ 4d^{10}\ 5p^6$

**Rn** $1s^2\ 2s^2\ 2p^6\ 3s^2\ 3p^6\ 4s^2\ 3d^{10}\ 4p^6\ 5s^2\ 4d^{10}\ 5p^6\ 6s^2\ 4f^{14}\ 5d^{10}\ 6p^6$

The atoms of other elements (other than hydrogen) contain noble-gas configurations as part of their own electron configurations. For example, the configuration of a sodium atom is the same as that of neon with the addition of 1 more electron to the 3*s* orbital.

**Ne** $1s^2\ 2s^2\ 2p^6$

**Na** $1s^2\ 2s^2\ 2p^6\ 3s^1$

The 3*s* electron of sodium is much more important in the description of sodium's chemical reactions than the other electrons are. There are two reasons why this is true. We know from the stability of neon atoms that the $1s^2\ 2s^2\ 2p^6$ configuration is very stable, so these electrons are not lost or shared in chemical reactions involving sodium. These electrons are called the noble-gas inner core of sodium. Another reason is that an electron in the larger 3*s* orbital is less strongly attracted to the nuclear charge than are electrons in smaller orbitals in the first and second principal energy levels. As a result, the 3*s* electron is easier to remove.

Because we are not as interested in the noble-gas inner core of electrons for sodium atoms as we are in the 3*s* electron, sodium atoms are often described with the following abbreviated electron configuration, in which [Ne] represents the electron configuration of neon:

**Na** [Ne] $3s^1$

Sometimes the symbol for the noble gas is left out of abbreviated electron configurations, but in this text, all abbreviated electron configurations will consist of the symbol, in brackets, for the noble-gas element at the end of the previous row, followed by the electron configuration of the remaining electrons (Figure 11.18).

Abbreviated electron configurations are especially useful for describing the chemistry of larger atoms. For example, atoms of cesium, Cs, have one more electron than atoms of xenon, Xe. Therefore, the 55th electron of a cesium atom is the only electron that is likely to participate in the formation of chemical bonds. Because atomic number 55 is in the sixth horizontal row of the *s* block, the 55th electron must be in the 6*s* sublevel, and this makes the abbreviated electron configuration for cesium atoms

OBJECTIVE 13

**Cs** [Xe] $6s^1$

This is much easier to determine and much less time-consuming to write than the complete electron configuration:

$$1s^2\ 2s^2\ 2p^6\ 3s^2\ 3p^6\ 4s^2\ 3d^{10}\ 4p^6\ 5s^2\ 4d^{10}\ 5p^6\ 6s^1$$

**Figure 11.18**
**Group 1 Abbreviated Electron Configurations**

The following sample study sheet shows the steps for writing abbreviated electron configurations.

## Sample Study Sheet 11.2

### Abbreviated Electron Configurations

OBJECTIVE 13

**Tip-off** If you are asked to write an abbreviated electron configuration, you can use the following steps.

**General Steps**

**Step 1** Find the symbol for the element in the periodic table.

For example, to write an abbreviated electron configuration for zinc atoms, we first find Zn in the periodic table (Figure 11.19).

**Step 2** Write the symbol in brackets for the noble gas located at the far right of the preceding horizontal row of the table.

For zinc, we move up to the third period and across to Ar (Figure 11.19). To describe the first 18 electrons of a zinc atom, we write

[Ar]

**Step 3** Move back down a row (to the row containing the element you wish to describe) and to the far left. Following the elements in the row from left to right, write the outer-electron configuration associated with each column until you reach the element you are describing.

For zinc, we need to describe the 19th through the 30th electrons. The atomic numbers 19 and 20 are in the fourth row of the *s* block, so the 19th and 20th electrons for each zinc atom enter the $4s^2$ sublevel. The atomic numbers 21 through 30 are in the first row of the *d* block, so the 21st to the 30th electrons for each zinc atom fill the $3d$ sublevel (Figure 11.19). Zinc, with atomic number 30, has the abbreviated configuration

**[Ar] $4s^2$ $3d^{10}$**

**Example** See Example 11.2.

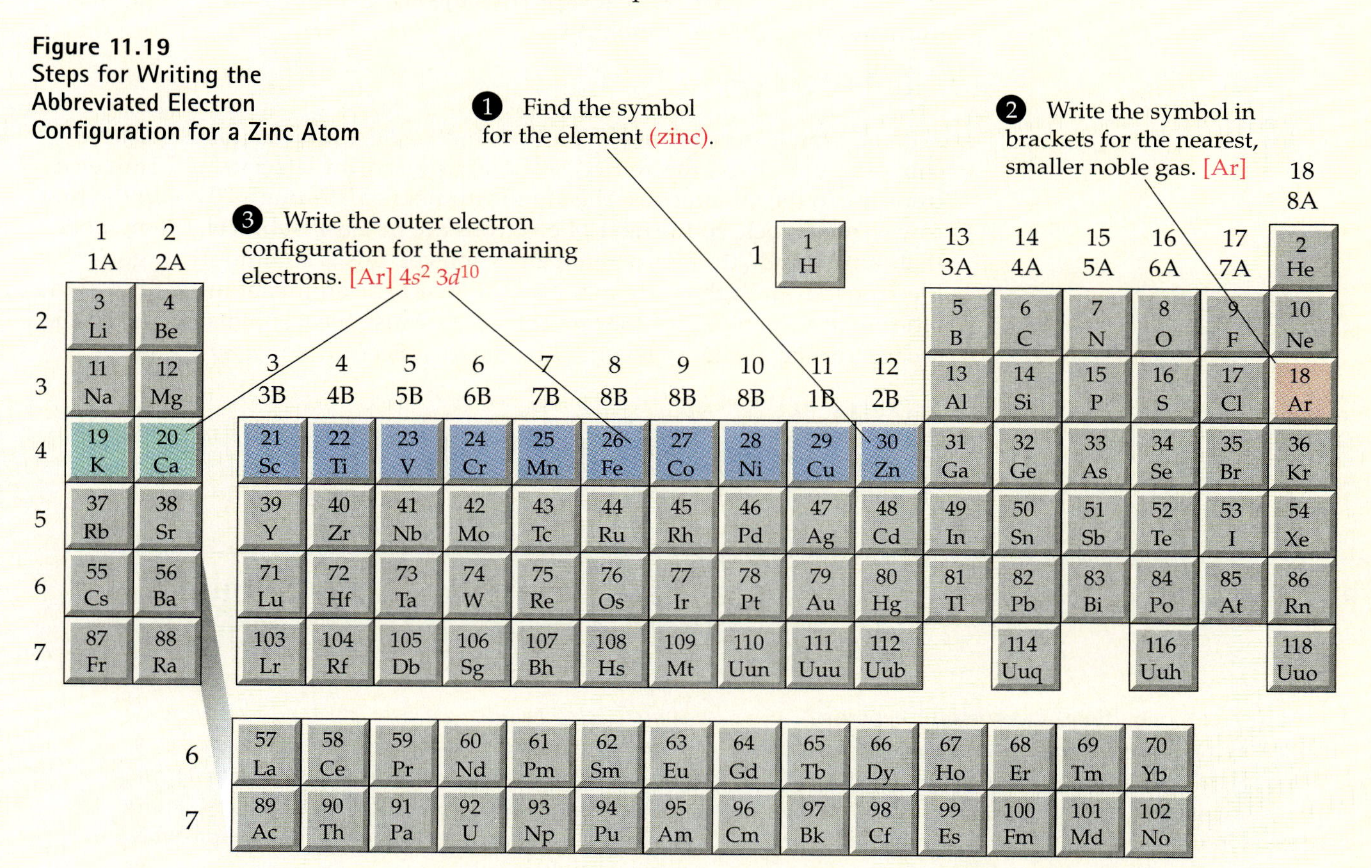


**Figure 11.19**
**Steps for Writing the Abbreviated Electron Configuration for a Zinc Atom**

## EXAMPLE 11.2 Abbreviated Electron Configurations

OBJECTIVE 13

Write abbreviated electron configurations for (a) strontium, Sr, (b) germanium, Ge, and (c) thallium, Tl.

***Solution***

a. With an atomic number of 38, strontium has 38 electrons. The noble gas at the end of the previous row is krypton, so putting Kr in brackets describes strontium's first 36 electrons. The atomic numbers 37 and 38 are in the fifth row of the *s* block, indicating that the 37th and 38th electrons enter and fill the 5*s* sublevel:

**[Kr] $5s^2$**

b. With an atomic number of 32, germanium has 32 electrons. The noble gas at the end of the previous row is argon, so putting Ar in brackets describes germanium's first 18 electrons. The atomic numbers 19 and 20 are in the fourth row of the *s* block, so the 19th and 20th electrons enter the 4*s* sublevel. The atomic numbers 21 through 30 are in the first row of the *d* block. Because the *d* sublevels start on the third principal energy level, the 21st through 30th electrons are in the 3*d* sublevel. The atomic numbers 31 and 32 are in the third row of the *p* block, or the 4*p* sublevel (because the *p* sublevels start on the second principal energy level). Thus the last two electrons for a germanium atom are in the 4*p* sublevel.

**[Ar] $4s^2\ 3d^{10}\ 4p^2$**

c. With an atomic number of 81, thallium has 81 electrons. Its noble-gas inner core of electrons has the xenon, Xe, configuration, and its 55th and 56th electrons are in the 6*s* sublevel. The atomic numbers 57 through 70 are in the first row of the *f* block, so the next 14 electrons are in the 4*f* sublevel. (Remember that the 6th and 7th rows contain an *f* sublevel; this is commonly forgotten.) Electrons 71 through 80 are in the 5*d* sublevel. The atomic number 81 is in the 5th row of the *p* block, or the 6*p* sublevel (because the *p* sublevels start on the second principal energy level). Thus the 81st electron is in a 6*p* orbital.

**[Xe] $6s^2\ 4f^{14}\ 5d^{10}\ 6p^1$**

## SPECIAL TOPIC 11.1 Why Does Matter Exist, and Why Should We Care About Answering This Question?

A great deal of money, time, and scientific expertise are spent on trying to answer such questions as, "*What is the nature of matter, and why does it exist?*" When it comes time for lawmakers to distribute tax money and for executives in industry and academia to plan their research goals, it is natural for them, in turn, to wonder, "*Shouldn't we direct our energies and money toward more practical goals? Isn't it much more important to develop procedures for early diagnosis and treatment of Alzheimer's disease than to spend our limited resources on understanding how the universe came to be?*" These sorts of questions are at the core of an on-going debate between people who believe strongly in basic research, the seeking of know-

*(continued)*

## SPECIAL TOPIC 11.1 (CONTINUED)

ledge for its own sake without a specific application in mind, and those who prefer to focus on applied research, the attempt to develop a specific procedure or product.

Often, basic research leads to unforeseen practical applications. Quantum mechanics, the wave mathematics used to predict the shapes and sizes of atomic orbitals, provides an example. In 1931 Paul Adrien Dirac, one of the pioneers of quantum mechanics, found an unexpected minus sign in the equations he was developing to describe the nature of matter. He could have ignored this problem in the hope that it would disappear with further refinement of the equations, but Dirac took a much bolder approach instead. He suggested that the minus sign indicated the existence of a particle identical to the electron except that it has a positive charge instead of a negative charge. This "anti-electron," which was detected in the laboratory by Carl Anderson in 1932, came to be called a positron, $e^+$.

Apparently, every particle has a twin antiparticle, and they are formed together from very concentrated energy. In 1955 and 1956, antiprotons and antineutrons were created and detected. The general name for antiparticles is *antimatter.* As the terminology suggests, when a particle meets its antimatter counterpart, they annihilate each other, leaving pure energy in their place. For example, when a positron collides with an electron, they both disappear, sending out two gamma (γ)-ray photons in opposite directions.

$e^+ \longrightarrow \quad \longleftarrow e^-$

positron–electron collision followed by the creation of 2 gamma ray photons

$\longleftarrow \gamma\text{-ray} \qquad \gamma\text{-ray} \longrightarrow$

Originally, there was no practical application for the discovery of antimatter, although its existence suggested possible answers to some very big questions. Perhaps the universe was originally created from the conversion of some extremely concentrated energy source into matter and antimatter. However, while some scientists continue to do basic research on the nature of matter and antimatter, others are using the knowledge and techniques developed through this basic research to seek practical applications. One use of the antimatter created

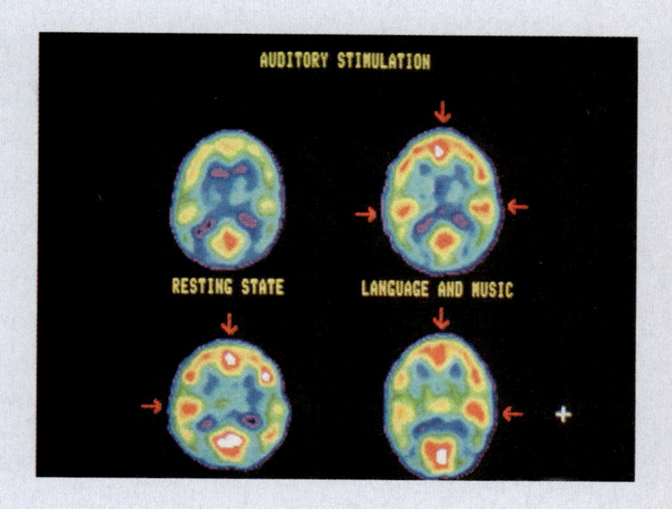


in the laboratory is called positron emission tomography (PET). With the help of a large team of scientists and computer experts, physicians use PET to scan the brain in search of biochemical abnormalities that signal the presence of Alzheimer's disease, schizophrenia, epilepsy, brain tumors, and other brain disorders.

In PET, radioactive substances that emit positrons are introduced into a patient's bloodstream. As the radioactive atoms decay, the positrons that they emit collide with electrons, producing gamma rays that escape from the body and are detected by an array of instruments surrounding the patient. Computer analysis of the amount and direction of gamma ray production, followed by comparison of the data collected for people with and without certain brain disorders, provides doctors with valuable information. For example, PET scans of the brain have been used to study the movement of the medication L-dopa in the brains of people suffering from Parkinson's disease. In these procedures, fluorine-18 atoms are attached to L-dopa molecules, which are then injected into a patient. Each fluorine-18 decays and emits a positron that generates gamma rays when it meets an electron.

Similar experiments are providing information about physiological processes such as glucose metabolism, the effects of opiate drugs, and the mechanisms of memory retrieval. Without the basic research, PET technology could never have been imagined, and it is just one of the many reasons why we should continue to ask fundamental questions such as, *"Why are we here?"* and *"How do things work?"*

### Exercise 11.2 Abbreviated Electron Configurations

OBJECTIVE 13

Write abbreviated electron configurations for (a) rubidium, Rb, (b) nickel, Ni, and (c) bismuth, Bi.

Some of the elements have electron configurations that differ slightly from what our general procedure would lead us to predict. You can read more about these at the following Web site:

**www.chemplace.com/college/**

You can also see how to predict charges on monatomic ions and how to write their electron configurations.

Now that you have learned something about the elements' electron configurations (even though you cannot say precisely what an electron *is*), Chapter 12 will show you how electron configurations can be used to explain patterns of bonding and the structure of molecules.

## Chapter Glossary

**Waveform** A representation of the shape of a wave.

**Nodes** The locations in a waveform where the intensity of the wave is always zero.

**Orbitals** The allowed waveforms for the electron. This term can also be defined as a volume that contains a high percentage of the electron charge and as a volume within which an electron has a high probability of being found.

**Principal energy level or shell** A collection of orbitals that have the same potential energy for a hydrogen atom, except for the first (lowest) principal energy level, which contains only one orbital (1*s*). For example, the 2*s* and 2*p* orbitals are in the second principal energy level.

**Orbital diagram** A drawing that uses lines or squares to show the distribution of electrons in orbitals and arrows to show the relative spin of each electron.

**Ground state** The condition of an atom whose electrons are in the orbitals that give it the lowest possible potential energy.

**Excited state** The condition of an atom that has at least one of its electrons in orbitals that do not represent the lowest possible potential energy.

**Sublevel or subshell** A given type (or shape) of orbital available at a given principal energy level. For example, the second principal energy level contains a 2*s* sublevel (with one spherical orbital) and a 2*p* sublevel (with three dumbbell-shaped orbitals).

**Electron configuration** A description of the complete distribution of an element's electrons in atomic orbitals. Although a configuration can be described either with an orbital diagram or with its shorthand notation, this text will use the term *electron configuration* to mean the shorthand notation that describes the distribution of electrons in sublevels without reference to the spin of the electrons.

You can test yourself on the glossary terms at the following Web site:

www.chemplace.com/college/

## Chapter Objectives

**The goal of this chapter is to teach you to do the following.**

1. Define all of the terms in the Chapter Glossary.

### Section 11.1 The Mysterious Electron

2. Explain why, in theory, a guitar string can vibrate with an infinite number of possible waveforms, but not all waveforms are possible.
3. Describe how electrons are like vibrating guitar strings.
4. Describe the 1*s* orbital in a hydrogen atom in terms of negative charge and in terms of the electron as a particle.
5. Explain why electrons in atoms are often described in terms of electron clouds.
6. Describe a 2*s* orbital for a hydrogen atom.
7. Explain why an electron in a hydrogen atom has lower potential energy in the 1*s* orbital than in the 2*s* orbital.
8. Describe the three 2*p* orbitals for a hydrogen atom.
9. Draw, describe, or recognize descriptions or representations of the 3*s*, 3*p*, and 3*d* orbitals.

### Section 11.2 Multi-Electron Atoms

10. Describe the difference between any two electrons in the same atomic orbital.
11. Write complete electron configurations for all of the elements in the periodic table that follow the normal order of filling of the sublevels.
12. Draw orbital diagrams for all of the elements in the periodic table that follow the normal order of filling of the sublevels.
13. Write abbreviated electron configurations for all of the elements in the periodic table that follow the normal order of filling of the sublevels.

## Review Questions

1. Describe the nuclear model of the atom.
2. Describe the relationship between stability and potential energy.

## Key Ideas

**Complete the following statements by writing one of these words or phrases in each blank.**

| | |
|---|---|
| analogies | particle |
| calculated for hydrogen | particle interpretation |
| cloud | positions |
| *d* | possible |
| decreased | principal |
| definitely | probabilities |
| diminishes | probably |
| electron configuration | *s* orbitals |
| exactly | seven |
| five | seventh |
| high percentage | spherical |
| hydrogen | spinning |
| impossible | spins |
| inner transition metals | spontaneously returns |
| intense | strange |
| intensity of the movement | strength |
| intensity of the negative charge | sublevel |
| mathematics | three |
| modern physics | three-dimensional |
| motion | two |
| *n* | unpaired |
| negative charge | volume |
| nine | wave |
| other elements | waveforms |
| *p* block | |

3. The electron is extremely tiny, and modern physics tells us that ____________ things happen in the realm of the very, very small.
4. The modern description of the electron is based on complex ____________ and on the discoveries of ____________.
5. Modern physics tells us that it is ____________ to know ____________ where an electron is and what it is doing.
6. There are two ways in which scientists deal with the problems associated with the complexity and fundamental uncertainty of the modern description of the electron, ____________ and ____________.
7. In order to accommodate the uncertainty of the electron's position and motion, scientists talk about where the electron ____________ is within the atom, instead of where it ____________ is.
8. Each electron seems to have a dual nature in which both ____________ and ____________ characteristics are apparent.

9. In the wave view, an electron has an effect on the space around it that can be described as a wave of ________________ varying in its intensity.
10. The ________________ for electrons in an atom describe the variation in intensity of negative charge within the atom, with respect to the location of the nucleus. This can be described without mentioning the ________________ and ________________ of the electron particle itself.
11. Just as the ________________ of a guitar string can vary, so can the ________________ of the electron vary at different positions outside the nucleus.
12. The variation in the intensity of the electron charge can be described in terms of a(n) ________________ standing wave *like* the standing wave of the guitar string.
13. As in the case of the guitar string, only certain waveforms are ________________ for the electron in an atom.
14. Most of the general descriptions of electrons found in this chapter are based on the wave mathematics for the one electron in a(n) ________________ atom.
15. The information calculated for the hydrogen electron is used to describe the ________________ as well.
16. For the 1*s* orbital, the negative charge is most ________________ at the nucleus and ________________ with increasing distance from the nucleus.
17. The allowed waveforms for the electron are also called orbitals. Another definition of orbital as the volume that contains a given ________________ of the electron charge. An orbital can also be defined as the ________________ within which an electron has a high probability of being found.
18. According to the ________________ of the wave character of the electron, the surface that surrounds 90% of an electron's charge is the surface within which we have a 90% probability of finding the electron.
19. In the particle view, the electron ________________ can be compared to a multiple-exposure photograph of the electron.
20. All ________________ electron waveforms are called *s* orbitals.
21. Because the ________________ of the attraction between positive and negative charges decreases with increasing distance between the charges, an electron is more strongly attracted to the nucleus and therefore is more stable when it has the smaller 1*s* waveform than when it has the larger 2*s* waveform. Increased stability is associated with ________________ potential energy, so a 1*s* electron has lower potential energy than a 2*s* electron.
22. All of the orbitals that have the same potential energy for a hydrogen atom are said to be in the same ________________ energy level.
23. After the electron is excited from the 1*s* orbital to the 2*s* orbital, it ________________ to its lower-energy 1*s* form.
24. There are ________________ possible 2*p* orbitals.
25. Orbitals that have the same potential energy, the same size, and the same shape are in the same ________________.
26. In the third principal energy level, there are ________________ possible orbitals for an electron, in three different sublevels.

27. Note that the first principal energy level has one sublevel, the second has two, the third has three, and the fourth has four. If ________________ is the number associated with the principal energy level, each principal energy level has *n* sublevels.
28. Each *s* sublevel has one orbital, each *p* sublevel has three orbitals, each *d* sublevel has ________________ orbitals, and each *f* sublevel has ________________ orbitals.
29. None of the known elements in its ground state has any electrons in a principal energy level higher than the ________________.
30. Scientist assume that all of the elements have the same set of possible principal energy levels, sublevels, and orbitals that has been ________________.
31. We can visualize the two electrons in a helium atom as ________________ in opposite directions.
32. In the broadest sense, a(n) ________________ is any description of the complete distribution of electrons in atomic orbitals.
33. An atomic orbital may contain ________________ electrons at most, and the electrons must have different ________________.
34. When electrons are filling orbitals of the same energy, they enter orbitals in such a way as to maximize the number of ________________ electrons, all with the same spin. In other words, they enter empty orbitals first, and all electrons in half-filled orbitals have the same spin.
35. The highest-energy electrons for all of the elements in groups 1 (1A) and 2 (2A) in the periodic table are in ________________.
36. All of the elements in the block with boron, B, neon, Ne, thallium, Tl, and radon, Rn, at the corners have their highest-energy electrons in *p* orbitals, so this is called the ________________.
37. The last electrons to be added to an orbital diagram for the atoms of the transition metal elements go into ________________ orbitals.
38. The section of the periodic table that contains the ________________ is called the *f* block.

## Chapter Problems

### Section 11.1 The Mysterious Electron

OBJECTIVE 2

39. Explain why, in theory, a guitar string can vibrate with an infinite number of possible waveforms but not all waveforms are possible.

OBJECTIVE 3

40. Describe how electrons are like vibrating guitar strings.

OBJECTIVE 4

41. Describe the 1*s* orbital in a hydrogen atom in terms of negative charge and in terms of the electron as a particle.

OBJECTIVE 5

42. Explain why electrons in atoms are often described in terms of electron clouds.

OBJECTIVE 6

43. Describe a 2*s* orbital for a hydrogen atom.

OBJECTIVE 7

44. Explain why an electron in a hydrogen atom has lower potential energy in the 1*s* orbital than in the 2*s* orbital.

45. Which is larger, a 2*p* orbital or a 3*p* orbital? Would the one electron in a hydrogen atom be more strongly attracted to the nucleus in a 2*p* orbital or in a 3*p* orbital? Would the electron be more stable in a 2*p* orbital or in a 3*p* orbital? Would the electron have higher potential energy when it is in a 2*p* orbital or a 3*p* orbital?

46. Which is larger, a 3*d* orbital or a 4*d* orbital? Would the one electron in a hydrogen atom be more strongly attracted to the nucleus in a 3*d* orbital or in a 4*d* orbital? Would the electron be more stable in a 3*d* orbital or in a 4*d* orbital? Would the electron have higher potential energy when it is in a 3*d* orbital or a 4*d* orbital?

47. Describe the three 2*p* orbitals for a hydrogen atom. OBJECTIVE 8

48. Write descriptions of the 3*s*, 3*p*, and 3*d* orbitals. OBJECTIVE 9

49. How many sublevels are in the fourth principal energy level for the hydrogen atom? What is the shorthand notation used to describe them? (For example, there is one sublevel in the first principal energy level, and it is described as 1*s*.)

50. How many orbitals are there in the 3*p* sublevel for the hydrogen atom?

51. How many orbitals are there in the 4*d* sublevel for the hydrogen atom?

52. How many orbitals are there in the third principal energy level for the hydrogen atom?

53. How many orbitals are there in the fourth principal energy level for the hydrogen atom?

54. Which of the following sublevels do not exist?

 a. 5*p*

 b. 2*s*

 c. 3*f*

 d. 6*d*

55. Which of the following sublevels do not exist?

 a. 1*p*

 b. 5*d*

 c. 6*f*

 d. 1*s*

### Section 11.2 Multi-Electron Atoms

56. Describe the difference between any two electrons in the same atomic orbital. OBJECTIVE 10

57. What is the maximum number of electrons that can be placed in a 3*p* orbital? . . . in a 3*d* orbital?

58. What is the maximum number of electrons that can be placed in a 5*s* orbital? . . . in a 5*f* orbital?

59. What is the maximum number of electrons that can be placed in a 3*p* sublevel? . . . in a 3*d* sublevel?

60. What is the maximum number of electrons that can be placed in a 5*s* sublevel? . . . in a 5*f* sublevel?

61. What is the maximum number of electrons that can be placed in the third principal energy level?

62. What is the maximum number of electrons that can be placed in the fourth principal energy level?

63. For each of the following pairs, identify the sublevel that is filled first.
   a. $2s$ or $3s$
   b. $3p$ or $3s$
   c. $3d$ or $4s$
   d. $4f$ or $6s$

64. For each of the following pairs, identify the sublevel that is filled first.
   a. $2p$ or $3p$
   b. $5s$ or $5d$
   c. $4d$ or $5s$
   d. $4f$ or $5d$

OBJECTIVE 11 OBJECTIVE 12

65. Write the complete electron configuration and orbital diagram for each of the following.
   a. carbon, C
   b. phosphorus, P
   c. vanadium, V
   d. iodine, I
   e. mercury, Hg

OBJECTIVE 11 OBJECTIVE 12

66. Write the complete electron configuration and orbital diagram for each of the following.
   a. oxygen, O
   b. sulfur, S
   c. manganese, Mn
   d. tellurium, Te
   e. radon, Rn

67. Which element is associated with each of the ground state electron configurations listed below?
   a. $1s^2\ 2s^2$
   b. $1s^2\ 2s^2\ 2p^6\ 3s^1$
   c. $1s^2\ 2s^2\ 2p^6\ 3s^2\ 3p^6\ 4s^2\ 3d^{10}\ 4p^5$
   d. $1s^2\ 2s^2\ 2p^6\ 3s^2\ 3p^6\ 4s^2\ 3d^{10}\ 4p^6\ 5s^2\ 4d^{10}\ 5p^6\ 6s^2\ 4f^{14}\ 5d^{10}\ 6p^2$

68. Which element is associated with each of the ground state electron configurations listed below?
   a. $1s^2\ 2s^2\ 2p^3$
   b. $1s^2\ 2s^2\ 2p^6\ 3s^2\ 3p^6$
   c. $1s^2\ 2s^2\ 2p^6\ 3s^2\ 3p^6\ 4s^2\ 3d^{10}$
   d. $1s^2\ 2s^2\ 2p^6\ 3s^2\ 3p^6\ 4s^2\ 3d^{10}\ 4p^6\ 5s^2\ 4d^{10}\ 5p^6\ 6s^2$

69. Would the following electron configurations represent ground states or excited states?

a. $1s^2\ 2s^1\ 2p^5$

b. $1s^2\ 2s^2\ 2p^4$

c. $1s^2\ 2s^2\ 2p^4\ 3s^1$

d. $1s^2\ 2s^2\ 2p^5$

70. Would the following electron configurations represent ground states or excited states?

a. $1s^2\ 2s^2\ 2p^6\ 3s^2$

b. $1s^2\ 2s^2\ 2p^6\ 3s^1\ 3p^1$

c. $1s^2\ 2s^2\ 2p^6\ 3s^2\ 3p^6\ 4s^2\ 3d^8\ 4p^1$

d. $1s^2\ 2s^2\ 2p^6\ 3s^2\ 3p^6\ 4s^2\ 3d^9$

71. Write the abbreviated electron configurations for each of the following. OBJECTIVE 13

a. fluorine, F

b. silicon, Si

c. cobalt, Co

d. indium, In

e. polonium, Po

f. palladium, Pd

72. Write the abbreviated electron configurations for each of the following. OBJECTIVE 13

a. chlorine, Cl

b. boron, B

c. scandium, Sc

d. yttrium, Y

e. astatine, At

### Additional Problems

73. Which sublevel contains:

a. the highest-energy electron for francium, Fr?

b. the 25th electron added to an orbital diagram for elements larger than chromium, Cr?

c. the 93rd electron added to an orbital diagram for elements larger than uranium, U?

d. the 82nd electron added to an orbital diagram for elements larger than lead, Pb?

74. Which sublevel contains:

a. the highest-energy electron for strontium, Sr?

b. the 63rd electron added to an orbital diagram for elements larger than samarium, Sm?

c. the 33rd electron added to an orbital diagram for elements larger than germanium, Ge?

d. the 75th electron added to an orbital diagram for elements larger than tungsten, W?

75. What is the first element on the periodic table to have
    a. an electron in the $3p$ sublevel?
    b. a filled $4s$ sublevel?
    c. a half-filled $3d$ sublevel?
76. What is the first element on the periodic table to have
    a. an electron in the $5s$ sublevel?
    b. a filled $4d$ sublevel?
    c. a half-filled $6p$ sublevel?
77. Which pair of the following ground-state, abbreviated electron configurations corresponds to elements in the same group on the periodic table? What elements are they? What is the name of the group to which they belong?
    a. [Ne] $3s^2$
    b. [Ar] $4s^2\ 3d^{10}$
    c. [Kr] $5s^2$
    d. [Xe] $6s^2\ 4f^{14}\ 5d^{10}\ 6p^1$
78. Which pair of the following ground-state, abbreviated electron configurations corresponds to elements in the same group on the periodic table? What elements are they? What is the name of the group to which they belong?
    a. [Ar] $4s^2\ 3d^{10}\ 4p^3$
    b. [Ne] $3s^2\ 3p^5$
    c. [Xe] $6s^2$
    d. [Kr] $5s^2\ 4d^{10}\ 5p^5$
79. What is the maximum number of electrons in each of the following?
    a. the $8j$ sublevel
    b. a $6h$ orbital
    c. the $n = 8$ principal energy level
80. What is the maximum number of electrons in each of the following?
    a. the $9k$ sublevel
    b. a $12n$ orbital
    c. the $n = 9$ principal energy level
81. Draw a sketch of how the orbitals for the electron clouds for all of the electrons in a phosphorus atom are superimposed on each other.
82. Write the expected abbreviated electron configuration for the as-yet-undiscovered element with an atomic number of 121. Use Uuo for the symbol of the noble gas below xenon, Xe. (Hint: See Figure 11.18.)
83. Write the expected abbreviated electron configuration for the as-yet-undiscovered element with an atomic number of 139. Use Uuo for the symbol of the noble gas below xenon, Xe. (Hint: See Figure 11.17.)
84. Draw a periodic table like the one in Figure 11.17 but showing the $5g$ sublevel in its correct position.

**Discussion Topics**

85. Do you think electrons are more like baseballs or guitar strings?
86. What do you think of the following statement? "We will never know about the true nature of the electron."
87. Do you think that there is some understanding of nature that is just beyond our ability to attain? What do you think about the following statement? "Just as all dogs have a limit to their ability to understand things, humans also have a limit."
98. With which of the following two statements do you most agree?

    "The main criterion for accepting a scientific model like our model for the electron is whether or not it is *true*."

    "The main criterion for accepting a scientific model like our model for the electron is whether or not it is *useful*."

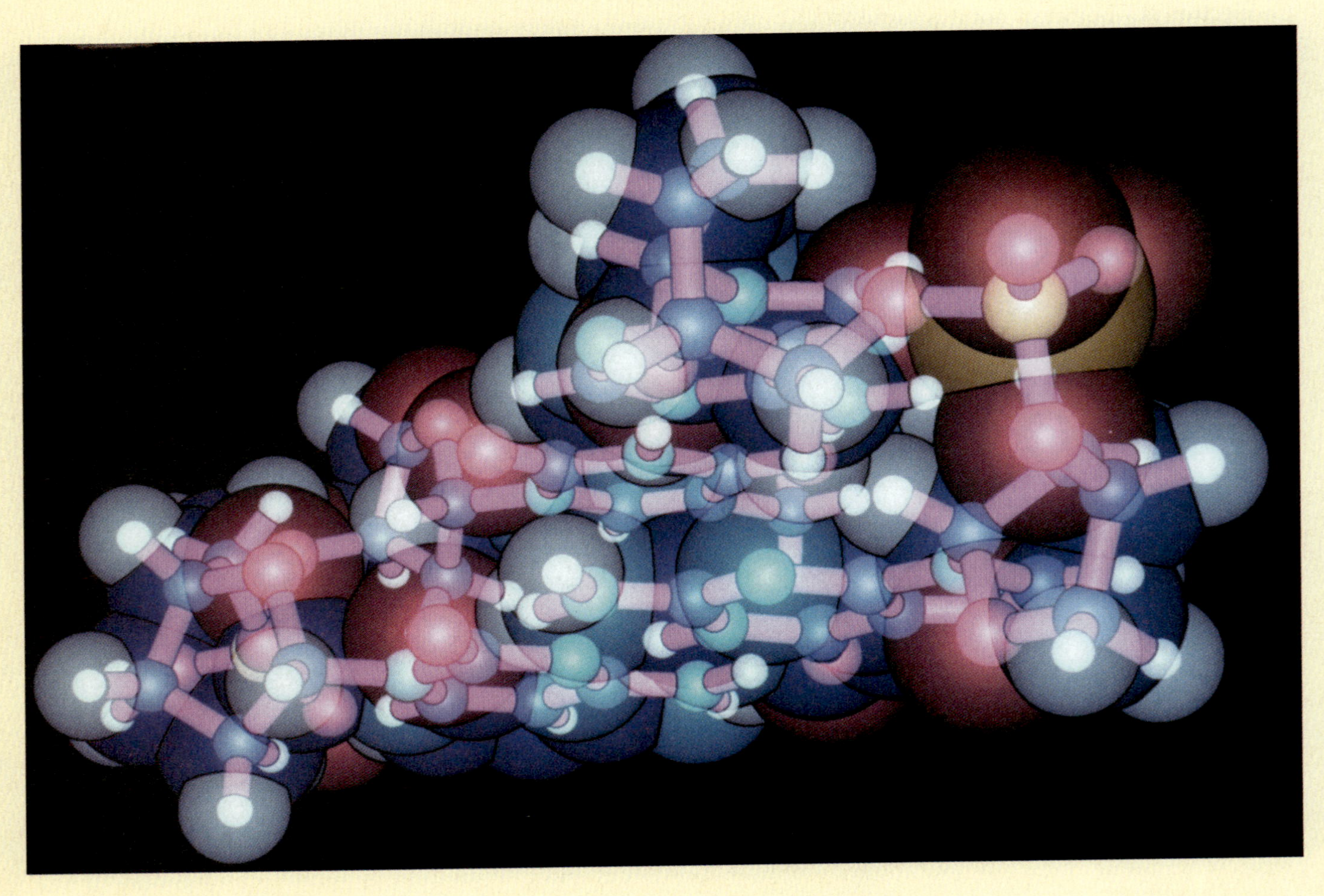

*In this chapter, you will learn more about the arrangements of atoms in molecules.*

CHAPTER 12

# Molecular Structure



IT'S MONDAY MORNING, AND YOU'D LIKE A CUP OF COFFEE, but when you try cranking up the stove to reheat yesterday's brew, nothing happens. Apparently, the city gas line has sprung a leak and has been shut down for repairs. The coffee cravings are strong, so you rummage in the garage until you find that can of Sterno left over from your last camping trip. You're saved. Both Sterno and natural gas contain compounds that burn and release heat, but the primary compounds in these substances are different. Natural gas is mostly methane, $CH_4$, whereas the primary flammable substance in Sterno is methanol, $CH_3OH$.

```
    H               H
    |               |
H—C—H         H—C—Ö—H
    |               |  ¨
    H               H
 Methane        Methanol
```

The oxygen atom in methanol molecules makes methanol's properties very different from those of methane. Methane is a colorless, odorless, and tasteless gas. Methanol, or wood alcohol, is a liquid with a distinct odor, and it is poisonous in very small quantities.

Chemists have discovered that part of the reason the small difference in structure leads to large differences in properties lies in the nature of covalent bonds and the arrangement of those bonds in space. This chapter provides a model for explaining how covalent bonds form, teaches you how to describe the resulting molecules with Lewis structures, and shows how Lewis structures can be used to predict the three-dimensional geometric arrangement of atoms in molecules.

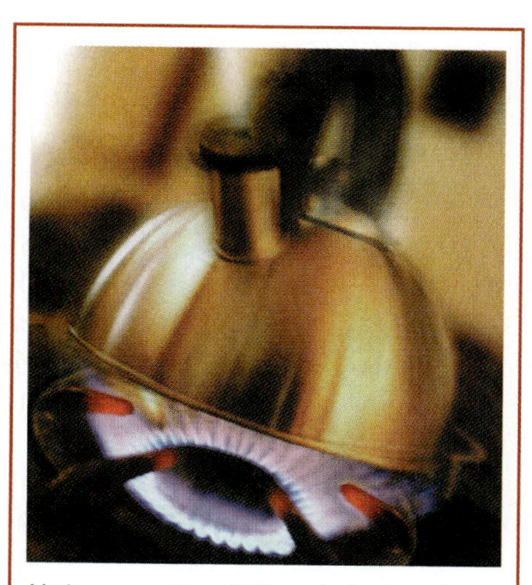

*Methane or methanol? We use both to heat food. How do their molecules—and properties—differ?*

## Review Skills

The presentation of information in this chapter assumes that you can already perform the tasks listed below. You can test your readiness to proceed by answering the Review Questions at the end of the chapter. This might also be a good time to read the Chapter 12 Objectives, which precede the Review Questions.

- Given a periodic table, identify the number of the group to which each element belongs. (Section 2.3)
- Given a chemical formula, draw for it the Lewis structure that has the most common number of covalent bonds and lone pairs for each atom. (Section 3.3)
- Write or identify the definitions of valence electrons, electron-dot symbol, lone pairs, Lewis structure, double bond, and triple bond. (Chapter 3 Glossary)
- Write or identify the definition of atomic orbital. (Section 11.1)
- Write electron configurations and orbital diagrams for the nonmetallic elements. (Section 11.2)

## 12.1 A New Look at Molecules and the Formation of Covalent Bonds

In Chapter 3, we saw that carbon atoms usually have four bonds, that oxygen atoms usually have two bonds and two lone pairs, and that hydrogen atoms form one bond. Using guidelines such as these, we can predict that there are two possible arrangements of the atoms of $C_2H_6O$.

```
    H   H                     H         H
    |   |                     |   ..    |
H — C — C — Ö — H         H — C — O — C — H
    |   |   ..                |   ..    |
    H   H                     H         H
```

Ethanol Dimethyl ether

In Chapter 3, these bonding characteristics were described without explanation, because you did not yet have the tools necessary for understanding them. Now that you know more about the electron configurations of atoms, you can begin to understand why atoms form bonds as they do. To describe the formation of covalent bonds in molecules, we use a model called the valence-bond model, but before the assumptions of this model are described, let's revisit some of the important issues related to the use of models for describing the physical world.

### The Strengths and Weaknesses of Models

When developing a model of physical reality, scientists take what they think is true and simplify it enough to make it useful. Such is the case with their description of the nature of molecules. Scientific understanding of molecular structure has advanced tremendously in the last few years, but the most sophisticated descriptions are too complex and mathematical to be understood by anyone but the most highly trained chemists and physicists. To be useful to the rest of us, these descriptions have been translated into simplified versions of what scientists consider to be true.

OBJECTIVE 2

Such models have advantages and disadvantages. They help us to visualize, explain, and predict chemical changes, but we need to remind ourselves now and then that they are only models and that as models, they have their limitations. For example, because a model is a simplified version of what we think is true, the processes it depicts are sometimes described using the phrase *as if.* When you read, "It is *as if* an electron were promoted from one orbital to another," the phrase is a reminder that we do not necessarily think this is what really happens. We merely find it *useful* to talk about the process *as if* this is the way it happens.

*Every path has its puddle.*
ENGLISH PROVERB

One characteristic of models is that they change with time. Because our models are simplifications of what we think is real, we are not surprised when they sometimes fail to explain experimental observations. When this happens, the model is altered to fit the new observations.

The valence-bond model for covalent bonds, described below, has its limitations, but it is still extremely useful. For example, we will see in Chapter 14 that it helps us to understand the attractions between molecules and to predict the relative melting points and boiling points of substances. The model is also extremely useful in describing the mechanisms of chemical changes. Therefore, even though it strays a bit from what scientists think is the most accurate description of real molecules, the valence-bond model is the most popular model for explaining covalent bonding.

## The Valence-Bond Model

The *valence-bond model,* which is commonly used to describe the formation of covalent bonds, is based on the following assumptions:

OBJECTIVE 3

- Only the highest-energy electrons participate in bonding.
- Covalent bonds usually form to pair unpaired electrons.

Fluorine is our first example. Take a look at its electron configuration and orbital diagram.

$$\text{F}\quad 1s^2\,2s^2\,2p^5 \qquad \begin{array}{l} 2s\ \underline{\uparrow\downarrow}\quad 2p\ \underline{\uparrow\downarrow}\ \underline{\uparrow\downarrow}\ \underline{\uparrow} \\ 1s\ \underline{\uparrow\downarrow} \end{array}$$

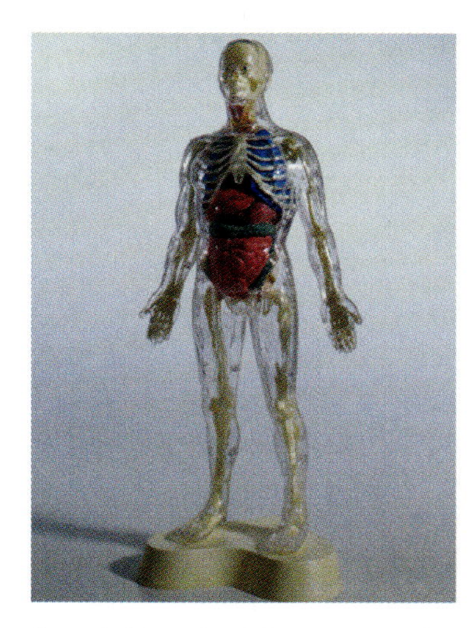

A model is a simplified version of reality that helps us to visualize, explain, and make predictions about real objects.

The first assumption of our model states that only the highest-energy electrons of fluorine atoms participate in bonding. There are two reasons why this is a reasonable assumption. First, we know from the unreactive nature of helium atoms that the $1s^2$ electron configuration is very stable, so we assume that these electrons in a fluorine atom are less important than others for the creation of bonds. The second reason is that the electrons in $2s$ and $2p$ orbitals have larger electron clouds and are therefore more readily available for interaction with other atoms.

The $2s^2$ and $2p^5$ electrons are the fluorine atom's valence electrons, the important electrons that we learned about in Section 3.3. Now we can define them more precisely. **Valence electrons** are the highest-energy *s* and *p* electrons in an atom. As we saw in Chapter 3, the following guideline can be used to determine the number of valence electrons in each atom of a representative (or main-group) element.

> When the columns in the periodic table are numbered by the A-group convention, the number of valence electrons in each atom of a representative element is equal to the element's group number in the periodic table (Figure 12.1).

OBJECTIVE 4

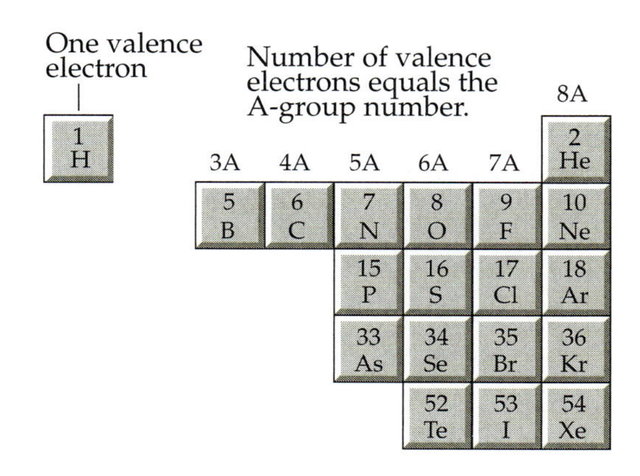


**Figure 12.1**
When the columns in the periodic table are numbered by the A-group convention, the number of valence electrons in each atom of a representative element is equal to the element's group number in the periodic table.

OBJECTIVE 4

Fluorine is in group 7A, so it has 7 valence electrons. The orbital diagram for the valence electrons of fluorine is

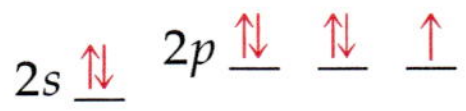

2s ⇅ 2p ⇅ ⇅ ↑

OBJECTIVE 5(a)

When atoms pair their unpaired electrons by forming chemical bonds, the atoms become more stable. We know that one way for fluorine to pair its 1 unpaired electron is to gain an electron from another atom and form a fluoride ion, $F^-$. This is possible when an atom is available that can easily lose an electron. For example, a sodium atom can transfer an electron to a fluorine atom to form a sodium ion, $Na^+$, and a fluoride ion, $F^-$.

If no atoms are available that can donate electrons to fluorine, the fluorine atoms will *share* electrons with other atoms to form electron pairs. For example, if we had a container of separate fluorine atoms, then each fluorine atom would very quickly bind to another fluorine atom, allowing each of them to pair its unpaired valence electron.

OBJECTIVE 6

To visualize this process, we can use the electron-dot symbols introduced in Chapter 3. The electron-dot symbol or electron-dot structure of an element shows the valence electrons as dots. Electrons that are paired in an orbital are shown as a pair of dots, and unpaired electrons are shown as single dots. The paired valence electrons are called lone pairs (because they do not participate in bonding). In an electron-dot symbol, the lone pairs and the single dots are arranged to the right, left, top, and bottom of the element's symbol. The electron-dot symbol for fluorine can be drawn with the single dot in any of the four positions:

OBJECTIVE 6

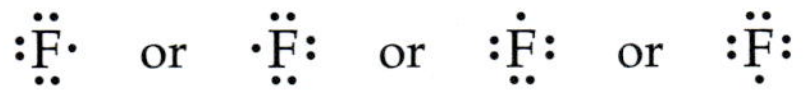

:F̤̈· or ·F̤̈: or :F̤̈: or :F̣̈:

According to the valence-bond model, two fluorine atoms bond covalently when their unpaired electrons form an electron pair that is then shared between the fluorine atoms.

OBJECTIVE 5(a)

:F̤̈· + ·F̤̈: → :F̤̈:F̤̈:

Usually, the covalent bonds in the electron-dot symbols for molecules are indicated with lines. Structures that show how the valence electrons of a molecule or polyatomic ion form covalent bonds and lone pairs are called Lewis structures. These are the same Lewis structures that we used for drawing molecular structures in Chapter 3. Although the bonds in Lewis structures can be described either with lines or with dots, in this text they will be described with lines. The Lewis structure for a fluorine molecule, $F_2$, is

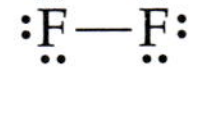

:F̤̈—F̤̈:

Each hydrogen atom in its ground state has one valence electron in a 1*s* orbital.

1s ↑

Its electron-dot symbol is therefore

H·

Because atoms become more stable when they pair their unpaired electrons, hydrogen atoms combine to form hydrogen molecules, $H_2$, which allow each atom to share two electrons.

OBJECTIVE 5(b)

H· + ·H → H:H or H—H

Hydrogen atoms can also combine with fluorine atoms to form HF molecules.

H· + ·F: → H:F: or H—F:

Carbon is in group 4A in the periodic table, so we predict that it has four valence electrons. Looking at the orbital diagram for these electrons, we might expect carbon to form 2 covalent bonds (to pair its 2 unpaired electrons) and have 1 lone pair.

2*s* ⇅ 2*p* ↑ ↑ — :C·

Carbon atoms do exhibit this bonding pattern in very rare circumstances, but in most cases, they form 4 bonds and have no lone pairs. Methane, $CH_4$, is a typical example. When forming 4 bonds to hydrogen atoms in a methane molecule, each carbon atom behaves *as if* it has four unpaired electrons. It is *as if*[1] 1 electron is promoted from the 2*s* orbital to the 2*p* orbital.

OBJECTIVE 5(c)

2*s* ⇅ 2*p* ↑ ↑ — → 2*s* ↑ 2*p* ↑ ↑ ↑

:C· → ·Ċ·

The following describes the bond formation in methane using electron-dot symbols.

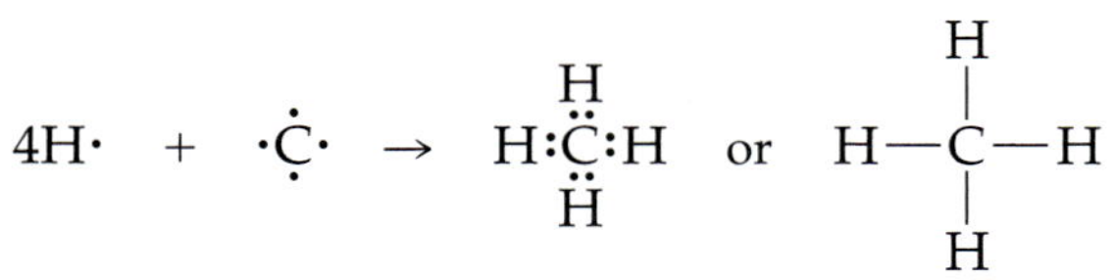


Carbon atoms also frequently form double bonds, in which they share 4 electrons with another atom, often another carbon atom. Ethene (commonly called ethylene), $C_2H_4$, is an example.

[1]Remember that the valence-bond model is a simplification of what really happens in bond formation. The use of phrases like *as if* reminds us that the real process is more complex.

$$4H\cdot + 2\cdot\dot{\underset{\cdot}{C}}\cdot \rightarrow H:\underset{\underset{H}{\cdot\cdot}}{C}::\underset{\underset{H}{\cdot\cdot}}{C}:H \quad \text{or} \quad H-\underset{\underset{H}{|}}{C}=\underset{\underset{H}{|}}{C}-H$$

Note that each carbon atom in $C_2H_4$ has a total of 4 bonds; 2 single bonds to hydrogen atoms and 2 bonds to the other carbon atom.

The bond between the carbon atoms in ethyne (commonly called acetylene), $C_2H_2$, is a triple bond, which can be viewed as the sharing of 6 electrons between 2 atoms.

$$2H\cdot + 2\cdot\dot{\underset{\cdot}{C}}\cdot \rightarrow H:C\vdots\vdots C:H \quad \text{or} \quad H-C\equiv C-H$$

Note that each carbon atom in $C_2H_2$ has a total of 4 bonds; 1 single bond to a hydrogen atom and 3 bonds to the other carbon atom.

Nitrogen is in group 5A, so it has 5 valence electrons. Its orbital diagram and electron-dot symbol are

$$2s \,\underline{\uparrow\downarrow} \quad 2p \,\underline{\uparrow}\ \underline{\uparrow}\ \underline{\uparrow} \qquad \cdot\ddot{\underset{\cdot}{N}}\cdot$$

OBJECTIVE 5(d)

Each nitrogen atom has 3 unpaired electrons, and as the model predicts, it forms 3 covalent bonds. For example, a nitrogen atom can bond to 3 hydrogen atoms to form an ammonia molecule, $NH_3$:

$$3H\cdot + \cdot\ddot{\underset{\cdot}{N}}\cdot \rightarrow H:\ddot{\underset{\underset{H}{\cdot\cdot}}{N}}:H \quad \text{or} \quad H-\ddot{\underset{\underset{H}{|}}{N}}-H$$

OBJECTIVE 5(e)

Another common bonding pattern for nitrogen atoms is 4 bonds with no lone pairs. The nitrogen atom in an ammonium polyatomic ion, $NH_4^+$, is an example. This pattern and the positive charge on the ion can be explained by the loss of 1 electron from the nitrogen atom. It *as if* an uncharged nitrogen atom loses 1 electron from the 2*s* orbital, leaving it 4 unpaired electrons and the ability to make 4 bonds.

$$2s \,\underline{\uparrow\downarrow} \quad 2p \,\underline{\uparrow}\ \underline{\uparrow}\ \underline{\uparrow} \;\; (\cdot\ddot{\underset{\cdot}{N}}\cdot) \xrightarrow{-1e^-} 2s \,\underline{\uparrow} \quad 2p \,\underline{\uparrow}\ \underline{\uparrow}\ \underline{\uparrow} \;\; (\cdot\dot{\underset{\cdot}{N}}\cdot)$$

$$4H\cdot + \cdot\dot{\underset{\cdot}{N}}\cdot \rightarrow H:\overset{\overset{H}{\cdot\cdot}}{\underset{\underset{H}{\cdot\cdot}}{N}}:H$$

Because nitrogen must lose an electron to form this bonding pattern, the overall structure of the ammonium ion has a +1 charge. The Lewis structures of polyatomic ions are usually enclosed in brackets, with the overall charge written outside the brackets on the upper right.

$$\left[\begin{array}{c} \text{H} \\ | \\ \text{H—N—H} \\ | \\ \text{H} \end{array}\right]^{+}$$

Phosphorus is in group 5A, so its atoms also have 5 valence electrons—in this case, in the $3s^23p^3$ configuration. Arsenic, also in group 5A, has atoms with a $4s^24p^3$ configuration. Because these valence configurations are similar to nitrogen's, $2s^22p^3$, the model correctly predicts that phosphorus and arsenic atoms will form bonds as nitrogen does. For example, each forms 3 bonds to hydrogen atoms and has 1 lone pair in $NH_3$, $PH_3$, and $AsH_3$.

OBJECTIVE 5(d)

$$\text{H—}\ddot{\text{N}}\text{—H} \quad \text{H—}\ddot{\text{P}}\text{—H} \quad \text{H—}\ddot{\text{As}}\text{—H}$$
(each with a third H bonded below the central atom)

On the other hand, phosphorus and arsenic atoms exhibit bonding patterns that are not possible for nitrogen atoms. For example, molecules such as $PCl_5$ and $AsF_5$ have 5 bonds and no lone pairs. If you take other chemistry courses, you are likely to see compounds with bonding patterns like this, but because they are somewhat uncommon, you will not see them again in this text.

The most common bonding pattern for oxygen atoms is 2 covalent bonds and 2 lone pairs. Our model explains this in terms of the valence electrons' $2s^22p^4$ electron configuration.

OBJECTIVE 5(f)

$$2s \ \underline{\uparrow\downarrow} \quad 2p \ \underline{\uparrow\downarrow} \ \underline{\uparrow} \ \underline{\uparrow} \qquad :\dot{\ddot{\text{O}}}\cdot$$

The 2 unpaired electrons are able to participate in 2 covalent bonds, and the 2 pairs of electrons remain 2 lone pairs. The oxygen atom in a water molecule has this bonding pattern.

$$2\text{H}\cdot \ + \ :\dot{\ddot{\text{O}}}\cdot \ \rightarrow \ :\ddot{\text{O}}:\text{H} \ (\text{with H below}) \quad \text{or} \quad :\ddot{\text{O}}\text{—H} \ (\text{with —H below})$$

In another common bonding pattern, oxygen atoms gain 1 electron and form 1 covalent bond with 3 lone pairs. The oxygen atom in the hydroxide ion has this bonding pattern.

OBJECTIVE 5(g)

$$2s \ \underline{\uparrow\downarrow} \quad 2p \ \underline{\uparrow\downarrow} \ \underline{\uparrow} \ \underline{\uparrow} \ \ :\dot{\ddot{\text{O}}}\cdot \quad \xrightarrow{+1e^-} \quad 2s \ \underline{\uparrow\downarrow} \quad 2p \ \underline{\uparrow\downarrow} \ \underline{\uparrow\downarrow} \ \underline{\uparrow} \ \ :\ddot{\underset{\cdot\cdot}{\text{O}}}\cdot$$

$$\text{H}\cdot \ + \ :\ddot{\underset{\cdot\cdot}{\text{O}}}\cdot \ \rightarrow \ :\ddot{\underset{\cdot\cdot}{\text{O}}}:\text{H} \quad \text{or} \quad \left[:\ddot{\underset{\cdot\cdot}{\text{O}}}\text{—H}\right]^-$$

In rare circumstances, carbon and oxygen atoms can form triple bonds, leaving each atom with 1 lone pair. The carbon monoxide molecule, CO, is an example.

:C≡O:

OBJECTIVE 5(h)

According to the valence-bond model, it is *as if* an electron is transferred from the oxygen atom to the carbon atom as the bonding in CO occurs. This gives each atom 3 unpaired electrons to form the triple bond, with 1 lone pair each left over.

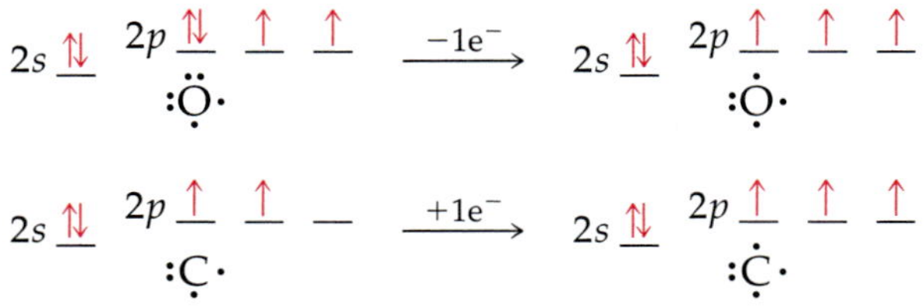

$2s$ ⇅ $2p$ ⇅ ↑ ↑ :Ö· $\xrightarrow{-1e^-}$ $2s$ ⇅ $2p$ ↑ ↑ ↑ :Ȯ·

$2s$ ⇅ $2p$ ↑ ↑ — :C· $\xrightarrow{+1e^-}$ $2s$ ⇅ $2p$ ↑ ↑ ↑ :Ċ·

OBJECTIVE 5(f)

Like oxygen, sulfur and selenium are in group 6A in the periodic table, so they too have 6 valence electrons, with $3s^23p^4$ and $4s^24p^4$ electron configurations, respectively. We therefore expect sulfur atoms and selenium atoms to have bonding patterns similar to oxygen's. For example, they all commonly form 2 bonds and have 2 lone pairs, as in molecules such as $H_2O$, $H_2S$, and $H_2Se$.

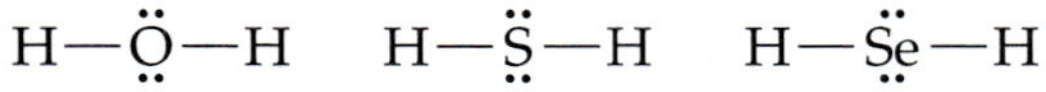

H—Ö—H  H—S̈—H  H—S̈e—H

Sulfur and selenium atoms have additional bonding patterns that are not possible for oxygen atoms. For example, they can form 6 bonds in molecules like $SF_6$ and $SeF_6$. You will not see these somewhat uncommon bonding patterns again in this text.

OBJECTIVE 5(i)

When the 3 equivalent B–F bonds form in boron trifluoride, $BF_3$, it is *as if* one of the boron atom's valence electrons were promoted from its $2s$ orbital to an empty $2p$ orbital. This leaves 3 unpaired electrons to form 3 covalent bonds.

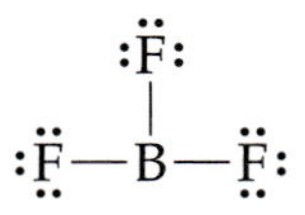

$2s$ ⇅ $2p$ ↑ — — :B· → $2s$ ↑ $2p$ ↑ ↑ — ·Ḃ·

:F̤̈:
|
:F̤̈—B—F̤̈:

OBJECTIVE 5(j)

The elements in group 7A all have the $ns^2np^5$ configuration for their valence electrons. Thus they all commonly form 1 covalent bond and have

3 lone pairs. For example, their atoms all form 1 bond to a hydrogen atom to form HF, HCl, HBr, and HI.

$ns \, \underline{\uparrow\downarrow} \quad np \, \underline{\uparrow\downarrow} \; \underline{\uparrow\downarrow} \; \underline{\uparrow} \qquad \cdot\ddot{\underset{..}{X}}:$

$H\cdot \; + \; \cdot\ddot{\underset{..}{X}}: \; \rightarrow \; H:\ddot{\underset{..}{X}}: \quad \text{or} \quad H{-}\ddot{\underset{..}{X}}: \quad X = F, Cl, Br, \text{or } I$

$H{-}\ddot{\underset{..}{F}}: \quad H{-}\ddot{\underset{..}{Cl}}: \quad H{-}\ddot{\underset{..}{Br}}: \quad H{-}\ddot{\underset{..}{I}}:$

Chlorine, bromine, and iodine atoms have additional, less common bonding patterns that you might see in other chemistry courses.

Table 12.1 summarizes the bonding patterns described in this section. The patterns listed there are not the only ones possible for these elements, but any patterns not listed are rare. In Section 12.2, you will be asked to draw Lewis structures from formulas, and knowledge of the common bonding patterns will help you to propose structures and evaluate their stability.

**Table 12.1**
Covalent Bonding Patterns

| **Element** | **Frequency of pattern** | **Number of bonds** | **Number of lone pairs** | **Example** |
|---|---|---|---|---|
| H | always | 1 | 0 | H— |
| B | most common | 3 | 0 | —B— (with one bond down) |
| C | most common | 4 | 0 | —C— (four single bonds) or —C= (with one bond down) or —C≡ |
|  | rare | 3 | 1 | ≡C: |
| N, P, and As | most common | 3 | 1 | $-\ddot{N}-$ (with one bond down) |
|  | common | 4 | 0 | —N— (four single bonds) |
| O, S, and Se | most common | 2 | 2 | $-\ddot{\underset{..}{O}}-$ or $\ddot{O}$ with double bond (=) |
|  | common | 1 | 3 | $-\ddot{\underset{..}{O}}:$ |
|  | rare | 3 | 1 | ≡O: |
| F, Cl, Br, and I | most common | 1 | 3 | $-\ddot{\underset{..}{X}}:$ |

## 12.2 Drawing Lewis Structures

After studying chemistry all morning, you go out to mow the lawn. While adding gasoline to the lawnmower's tank, you spill a bit, so you go off to get some soap and water to clean it up. By the time you get back, the gasoline has all evaporated, which starts you wondering. Why does gasoline evaporate so much faster than water? We are not yet ready to explain this, but part of the answer is found by comparing the substances' molecular structures and shapes. Lewis structures provide this information. You will see in Chapter 14 that the ability to draw Lewis structures for the chemical formulas of water, $H_2O$, and hexane, $C_6H_{14}$ (one of the major components of gasoline), will help you to explain their relative rates of evaporation. The ability to draw Lewis structures will be important for many other purposes as well, including explaining why soap would have helped clean up the spill if the gasoline had not evaporated so quickly.

### General Procedure

In Chapter 3, you learned to draw Lewis structures for many common molecules by trying to give each atom its most common bonding pattern (Table 12.2). For example, to draw a Lewis structure for methanol, $CH_3OH$, you would ask yourself how you can get 1 bond for each hydrogen atom, 4 bonds for the carbon atom, and 2 bonds and 2 lone pairs for the oxygen atom. The structure that follows shows how this can be done.

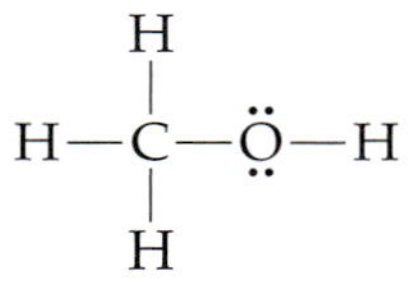

```
      H
      |    ..
  H — C — O — H
      |    ..
      H
```

**Table 12.2**
The Most Common Bonding Pattern for Each Nonmetallic Element

| Elements | Number of covalent bonds | Number of lone pairs |
|---|---|---|
| C | 4 | 0 |
| N, P, and As | 3 | 1 |
| O, S, and Se | 2 | 2 |
| F, Cl, Br, and I | 1 | 3 |

The shortcut just described works well for many simple uncharged molecules, but it does not work reliably for molecules that are more complex or for polyatomic ions. To draw Lewis structures for these, you can use the stepwise procedure described in the following sample study sheet.

## Sample Study Sheet 12.1

### Drawing Lewis Structures from Formulas

OBJECTIVE 7

**Tip-off** In this chapter, you may be given a chemical formula for a molecule or polyatomic ion and asked to draw a Lewis structure, but there are other, more subtle tip-offs that you will see in later chapters.

**General Steps** See Figure 12.2 for a summary of these steps.

**Step 1** Determine the total number of valence electrons for the molecule or polyatomic ion. (Remember that the number of valence electrons for a representative element is equal to its group number, using the A-group convention for numbering groups. For example, chlorine, Cl, is in group 7A, so it has seven valence electrons. Hydrogen has one valence electron.)

- For uncharged molecules, the total number of valence electrons is the sum of the valence electrons of each atom.
- For polyatomic cations, the total number of valence electrons is the sum of the valence electrons for each atom minus the charge.
- For polyatomic anions, the total number of valence electrons is the sum of the valence electrons for each atom plus the charge.

**Step 2** Draw a reasonable skeletal structure, using single bonds to join all the atoms. One or more of the following guidelines might help with this step. (They are clarified in the examples that follow.)

- Try to arrange the atoms to yield the most typical number of bonds for each atom. Table 12.2 lists the most common bonding patterns for the nonmetallic elements.
- Apply the following guidelines in deciding what element belongs in the center of your structure.

  Hydrogen and fluorine atoms are never in the center.

  Oxygen atoms are rarely in the center.

  The element with the fewest atoms in the formula is often in the center.

  The atom that is capable of making the most bonds is often in the center.
- Oxygen atoms rarely bond to other oxygen atoms.
- The molecular formula often reflects the molecular structure. (See Example 12.4.)
- Carbon atoms commonly bond to other carbon atoms.

**Step 3** Subtract two electrons from the total for each of the single bonds (lines) described in step 2 above. This tells us the number of electrons that still need to be distributed.

**Step 4** Try to distribute the remaining electrons as lone pairs to obtain a total of 8 electrons around each atom except hydrogen and boron. We saw in Chapter 3 that the atoms in reasonable Lewis structures are often surrounded by an octet of electrons. The following are some helpful observations pertaining to octets.

- In a reasonable Lewis structure, carbon, nitrogen, oxygen, and fluorine always have 8 electrons around them.
- Hydrogen will always have a total of 2 electrons from its 1 bond.
- Boron can have fewer than 8 electrons but never more than 8.
- The nonmetallic elements in periods beyond the second period (P, S, Cl, Se, Br, and I) usually have 8 electrons around them, but they can have more. (In this text, they will always have 8 electrons around them, but if you go on to take other chemistry courses, it will be useful to know that they can have more.)
- The bonding properties of the metalloids arsenic, As, and tellurium, Te, are similar to those of phosphorus, P, and sulfur, S, so they usually have 8 electrons around them but can have more.

**STEP 5** Do one of the following.

- If in step 4 you were able to obtain an octet of electrons around each atom other than hydrogen and boron, and if you used all of the remaining valence electrons, go to step 6.
- If you have electrons remaining after each of the atoms other than hydrogen and boron have their octet, you can put more than 8 electrons around elements in periods beyond the second period. (You will not need to use this procedure for any of the structures in this text, but if you take more advanced chemistry courses, it will be useful.)
- If you do not have enough electrons to obtain octets of electrons around each atom (other than hydrogen and boron), convert 1 lone pair into a multiple bond for each 2 electrons that you are short. (See Example 12.2.)

    If you need 2 more electrons to get octets, convert 1 lone pair in your structure into a second bond between 2 atoms.

    If you need 4 more electrons to get octets, convert 2 lone pairs into bonds. This could mean creating a second bond in 2 different places or creating a triple bond in 1 place.

    If you need 6 more electrons to get octets, convert 3 lone pairs into bonds.

    And so on.

**STEP 6** Check your structure to see whether every atom has its most common bonding pattern (Table 12.2).

- If each atom has its most common bonding pattern, then your structure is a reasonable structure. Skip step 7.
- If one or more atoms do not have their most common bonding pattern, continue to step 7.

**STEP 7** If necessary, try to rearrange your structure to give each atom its most common bonding pattern. One way to do this is to return to step 2 and try another skeleton. (This step is unnecessary if all of the atoms in your structure have their most common bonding pattern.)

**EXAMPLE** See Examples 12.1 to 12.4.

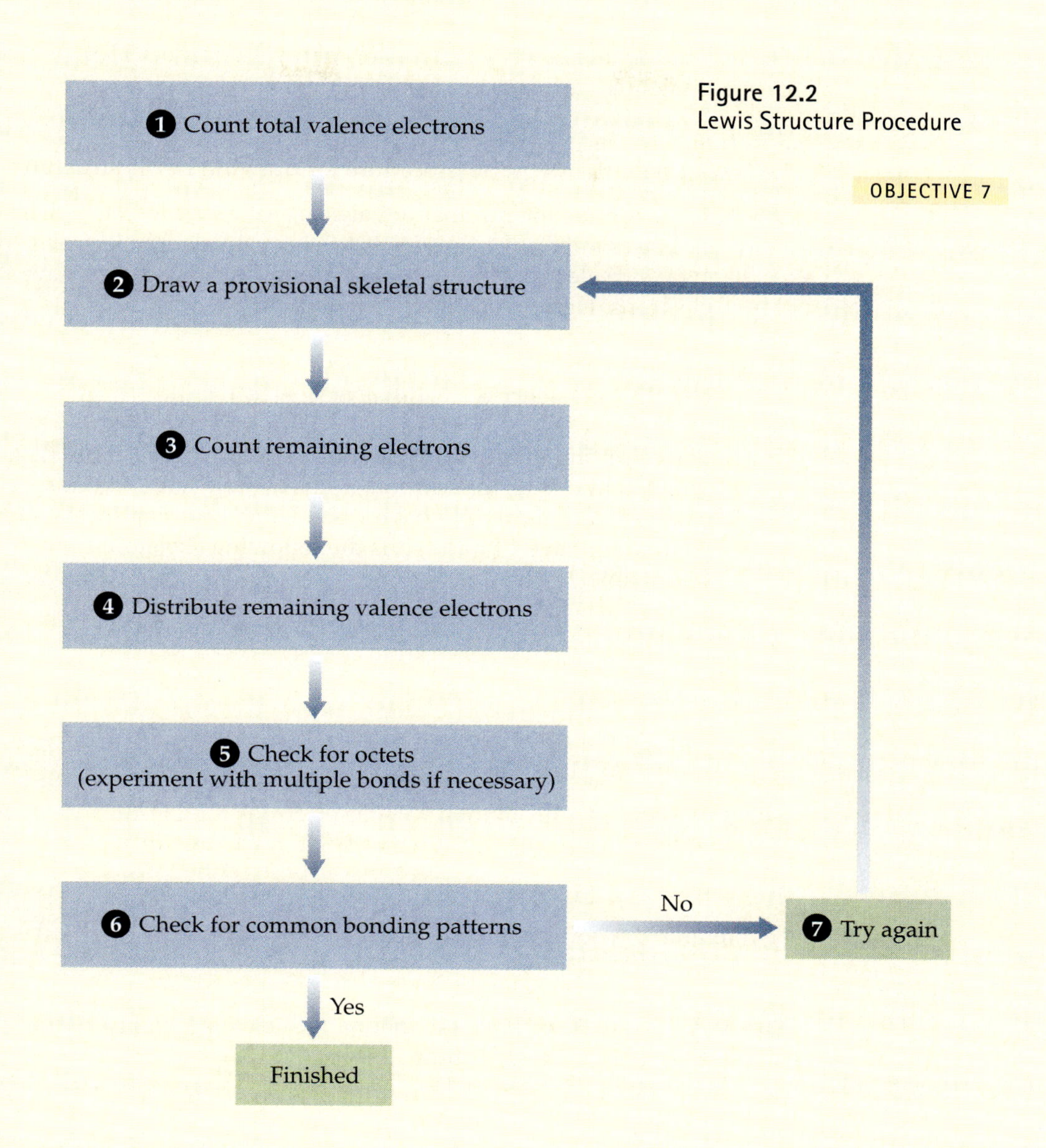


**Figure 12.2**
Lewis Structure Procedure

OBJECTIVE 7

## EXAMPLE 12.1 Drawing Lewis Structures

OBJECTIVE 7

Draw a reasonable Lewis structure for methyl bromide, $CH_3Br$, which is an ozone-depleting gas used as a fumigant.

***Solution***

Let's start with the stepwise procedure for drawing Lewis structures.

**STEP 1** To determine the number of valence electrons for $CH_3Br$, we note that carbon is in group 4A, so its atoms have 4 valence electrons; hydrogen has 1 valence electron; bromine is in group 7A, so its atoms have 7 valence electrons. Here # stands for "number of."

$$CH_3Br \quad \text{\# valence } e^- = \overset{C}{1(4)} + \overset{H}{3(1)} + \overset{Br}{1(7)} = 14$$

**STEP 2** Before setting up the skeleton for $CH_3Br$, we check Table 12.2, which reminds us that carbon atoms usually have 4 bonds, hydrogen atoms always have 1 bond, and bromine atoms most commonly have 1 bond. Thus the following skeleton is most reasonable.

```
      H
      |
  H—C—Br
      |
      H
```

**STEP 3** We started with 14 total valence electrons for $CH_3Br$, and we have used 8 of them for the 4 bonds in the skeleton we drew in step 2.

$$\text{\# e remaining} = \text{Total valence } e^- \text{ from step 1} - \text{\# bonds}\left(\frac{2\ e^-}{1\ \text{bond}}\right) = 14 - 4(2) = 6$$

**STEP 4** After step 3, we have the following skeleton for $CH_3Br$ and 6 valence electrons still to distribute.

```
      H
      |
  H—C—Br
      |
      H
```

Hydrogen atoms never have lone pairs, and the carbon atom has an octet of electrons around it from its 4 bonds. In contrast, the bromine atom needs 6 more electrons to obtain an octet, so we put the remaining 6 electrons around the bromine atom as 3 lone pairs.

```
      H
      |   ..
  H—C—Br:
      |   ..
      H
```

**STEP 5** The structure drawn in step 4 for $CH_3Br$ has an octet of electrons around the carbon and bromine atoms. Hydrogen has its 1 bond. We have also used all of the valence electrons. Thus we move on to step 6.

**STEP 6** All of the atoms in the structure drawn for $CH_3Br$ have their most common bonding pattern, so we have a reasonable Lewis structure.

```
     H
     |    ..
  H—C—Br:
     |    ..
     H
```

**STEP 7** Because the atoms in our structure for $CH_3Br$ have their most common bonding pattern, we skip this step.

**Shortcut** The shortcut to drawing Lewis structures that is described in Section 3.3 can often be used for uncharged molecules such as $CH_3Br$. Carbon atoms usually have 4 bonds and no lone pairs, hydrogen atoms always have 1 bond, and bromine atoms most commonly have 1 bond and 3 lone pairs. The only way to give these atoms their most common bonding patterns is with the following Lewis structure, which is the same Lewis structure we arrived at with the step-wise procedure.

```
     H
     |    ..
  H—C—Br:
     |    ..
     H
```

## EXAMPLE 12.2 Drawing Lewis Structures

OBJECTIVE 7

Formaldehyde, $CH_2O$, has many uses, including the preservation of biological specimens. Draw a reasonable Lewis structure for formaldehyde.

***Solution***

**STEP 1** Carbon is in group 4A, so its atoms have 4 valence electrons. Hydrogen has 1 valence electron. Oxygen is in group 6A, so its atoms have 6 valence electrons.

$$CH_2O \quad \#\text{ valence } e^- = \overset{C}{1(4)} + \overset{H}{2(1)} + \overset{O}{1(6)} = 12$$

**STEP 2** There are two possible skeletons for this structure. We will work with skeleton 1 first.

```
                          O
                          |
H—C—O—H      or      H—C—H

Skeleton 1           Skeleton 2
```

**STEP 3** (Skeleton 1) $\# e^-$ remaining $= 12 - 3(2) = 6$

**Step 4** (Skeleton 1) The most common bonding pattern for oxygen atoms is 2 bonds and 2 lone pairs. It is possible for carbon atoms to have 1 lone pair, although that is rare. Thus we might use our 6 remaining electrons as 2 lone pairs on the oxygen atom and 1 lone pair on the carbon atom.

H—C̈—Ö—H (with a second lone pair below O)

**Step 5** (Skeleton 1) The structure above leaves the carbon atom without its octet. Because we are short 2 electrons for octets, we convert 1 lone pair into another bond. Converting the lone pair on the carbon into a multiple bond would still leave the carbon with only 6 electrons around it and would put 10 electrons around the oxygen atom. Carbon and oxygen atoms always have 8 electrons around them in a reasonable Lewis structure. Thus we instead try converting 1 of the lone pairs on the oxygen atom into another C–O bond.

H—C̈—Ö—H to H—C̈=Ö—H

Structure 1

**Step 6** In structure 1, the carbon atom and the oxygen atom have rare bonding patterns. This suggests that there might be a better way to arrange the atoms. We proceed to step 7.

**Step 7** In an effort to give each atom its most common bonding pattern, we return to step 2 and try another skeleton.

**Step 2** The only alternative to skeleton 1 for $CH_2O$ is

O (single bond down to C)
H—C—H

Skeleton 2

**Step 3** (Skeleton 2) $\# e^-$ remaining $= 12 - 3(2) = 6$

**Step 4** (Skeleton 2) Oxygen atoms commonly have 1 bond and 3 lone pairs, so we might use our remaining electrons as 3 lone pairs on the oxygen atom.

:Ö: (single bond down to C)
H—C—H

**Step 5** (Skeleton 2) In step 4, we used all of the remaining electrons but left the carbon atom with only 6 electrons around it. Because we are short 2 electrons needed to obtain octets of electrons around each atom, we convert 1 lone pair into another bond.

:Ö: (single bond down to C) or Ö (double bond down to C)
H—C—H H—C—H

Structure 2

**Step 6** (Skeleton 2) All of the atoms have their most common bonding pattern, so we have a reasonable Lewis structure.

**STEP 7** (Skeleton 2) Each atom in structure 2 has its most common bonding pattern, so we skip this step.

Because structure 2 is the only arrangement that gives each atom its most common bonding pattern, we could have skipped the stepwise procedure and used the shortcut for this molecule.

## EXAMPLE 12.3 Drawing Lewis Structures

OBJECTIVE 7

The cyanide polyatomic ion, $CN^-$, is similar in structure to carbon monoxide, CO. Both are poisons that can disrupt the transfer of oxygen, $O_2$, by red blood cells. Draw a reasonable Lewis structure for the cyanide ion.

***Solution***

The shortcut does not work for polyatomic ions, so we use the stepwise procedure for $CN^-$.

**STEP 1** Carbon is in group 4A, so its atoms have 4 valence electrons. Nitrogen is in group 5A, so its atoms have 5 valence electrons. Remember to add 1 electron for the −1 charge.

$$CN^- \quad \# \text{ valence } e^- = \underset{C}{1(4)} + \underset{N}{1(5)} + \underset{[-]}{1} = 10$$

**STEP 2** There is only one way to arrange two atoms.

C–N

**STEP 3** $\# e^-$ remaining $= 10 - 1(2) = 8$

**STEP 4** Both the carbon atom and the nitrogen atom need 6 more electrons. We do not have enough electrons to provide octets by the formation of lone pairs. You will see in the next step that we will make up for this lack of electrons with multiple bonds. For now, we might put 2 lone pairs on each atom.

:C̈—N̈:

**STEP 5** Because we are short 4 electrons (or 2 pairs) in our effort to obtain octets, we convert 2 lone pairs into bonds. If we convert 1 lone pair from each atom into another bond, we get octets of electrons around each atom.

:C̈—N̈: or :C≡N:

**STEP 6** The nitrogen atom now has its most common bonding pattern: 3 bonds and 1 lone pair. The carbon atom has a rare bonding pattern, so we proceed to step 7.

**STEP 7** There is no other way to arrange the structure and still have an octet of electrons around each atom, so the Lewis structure for $CN^-$ is

$[:C{\equiv}N:]^-$

Remember to put the Lewis structures for polyatomic ions in brackets and to show the charge on the outside upper right.

## EXAMPLE 12.4 Drawing Lewis Structures

OBJECTIVE 7

Draw a reasonable Lewis structure for $CF_3CHCl_2$, the molecular formula for HCFC-123, which is one of the hydrochlorofluorocarbons used as a replacement for more damaging chlorofluorocarbons. (See Special Topic 7.2: *Other Ozone-Depleting Chemicals*.)

***Solution***

**STEP 1** Carbon is in group 4A, so its atoms have 4 valence electrons. Chlorine and fluorine are in group 7A, so their atoms have 7 valence electrons. Hydrogen has 1 valence electron.

$$CF_3CHCl_2 \quad \#\text{ valence } e^- = \overset{\text{C}}{2(4)} + \overset{\text{F}}{3(7)} + \overset{\text{H}}{1(1)} + \overset{\text{Cl}}{2(7)} = 44$$

**STEP 2** The best way to start our skeleton is to remember that carbon atoms often bond to other carbon atoms. Thus we link the 2 carbon atoms together in the center of the skeleton. We expect hydrogen, fluorine, and chlorine atoms to form 1 bond, so we attach all of them to the carbon atoms. There are many ways in which they could be arranged, but the way the formula has been written (especially the separation of the 2 carbons) gives us clues: $CF_3CHCl_2$, tells us that the 3 fluorine atoms are on 1 carbon atom and that the hydrogen and chlorine atoms are on the second carbon.

```
    F   H
    |   |
F — C — C — Cl
    |   |
    F   Cl
```

**STEP 3** $\#\, e^-$ remaining $= 44 - 7(2) = 30$

**STEP 4** We expect halogen atoms to have 3 lone pairs, so we can use the 30 remaining electrons to give us 3 lone pairs for each fluorine and chlorine atom.

```
     ..
    :F:  H
     |   |
 ..      ..
:F — C — C — Cl:
 ..      ..
     |   |
    :F: :Cl:
     ..  ..
```

Structure 1

We could also represent $CF_3CHCl_2$ with the following Lewis structures:

```
     ..  ..                     ..  ..
    :F: :Cl:                   :F: :Cl:
     |   |                      |   |
 ..                         ..          ..
:F — C — C — H      or     :F — C — C — Cl:
 ..                         ..          ..
     |   |                      |   |
    :F: :Cl:                   :F:  H
     ..  ..                     ..
```

Structure 2 Structure 3

Structures 2 and 3 might appear to represent different molecules, but they actually represent the same molecule as structure 1. To confirm that this is true, picture yourself sitting on the hydrogen atom in either the space-filling or the ball-and-stick model shown in Figure 12.3 for $CF_3CHCl_2$. Atoms connected to each other by single bonds, like the 2 carbon atoms in this molecule, are constantly rotating with respect to each other. Thus your hydrogen atom is sometimes turned toward the top of the molecule (somewhat as in structure 1), sometimes toward the bottom of the structure (somewhat like structure 3), and sometimes in one of the many possible positions in between. (Section 12.4 describes how you can predict molecular shapes. If you do not see at this point why structures 1, 2, and 3 all represent the same molecule, you might return to this example after reading that section.)

**Step 5** We have used all of the valence electrons, and we have obtained octets of electrons around each atom other than hydrogen. Thus we move to step 6.

**Step 6** All of the atoms in our structure have their most common bonding pattern, so we have a reasonable Lewis structure.

**Step 7** Because each atom in our structure has its most common bonding pattern, we skip this step.

**Figure 12.3**
Models of $CF_3CHCl_2$

## More Than One Possible Structure

It is possible to generate more than one reasonable Lewis structure for some formulas. When this happens, remember that the more common the bonding pattern (as summarized in Table 12.1), the more stable the structure. Even after you apply this criterion, you may still be left with two or more Lewis structures that are equally reasonable. For example, the following Lewis structures for $C_2H_6O$ both have the most common bonding pattern for all of their atoms.

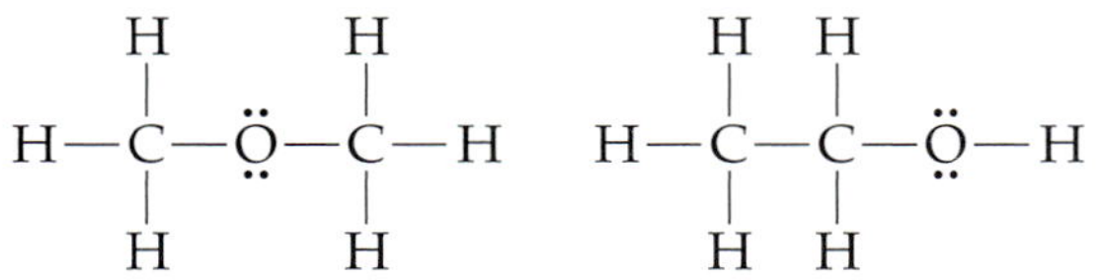

OBJECTIVE 7

```
    H       H            H   H
    |  ..   |            |   |   ..
H — C — O — C — H    H — C — C — O — H
    |  ..   |            |   |   ..
    H       H            H   H
```

In fact, both of these structures describe actual compounds. The first structure is dimethyl ether, and the second is ethanol. Substances that have the same molecular formula but different structural formulas are called **isomers.** We can write the formulas for these two isomers in such a way as to distinguish between them: $CH_3OCH_3$ represents dimethyl ether, and $CH_3CH_2OH$ represents ethanol.

## EXAMPLE 12.5 Drawing Lewis Structures

OBJECTIVE 7

Acetaldehyde can be converted into the sedative chloral hydrate (the "Mickey Finn," or knockout drops, often mentioned in detective stories). In the first step of the reaction that forms chloral hydrate, acetaldehyde, $CH_3CHO$, changes to its isomer, $CH_2CHOH$. Draw reasonable Lewis structures for each of these isomers.

***Solution***

**STEP 1** Both molecules have the molecular formula $C_2H_4O$.

$$\#\text{ valence } e^- = \overset{C}{2(4)} + \overset{H}{4(1)} + \overset{O}{1(6)} = 18$$

**STEP 2** Because we expect carbon atoms to bond to other carbon atoms, we can link the two carbon atoms together to start each skeleton, which is then completed to try to match each formula given.

```
    H  O
    |  |
 H—C—C—H      H—C—C—O—H
    |            |  |
    H            H  H
```

$CH_3CHO$
Skeleton 1

$CH_2CHOH$
Skeleton 2

**STEP 3** $\#e^-$ remaining $= 18 - 6(2) = 6$

**STEP 4** For skeleton 1, we can add 3 lone pairs to the oxygen atom to give it its octet. For skeleton 2, we can add 2 lone pairs to the oxygen atom to get its most common bonding pattern of 2 bonds and 2 lone pairs. We can place the remaining 2 electrons on one of the carbon atoms as a lone pair.

```
    H :Ö:
    |  |            ..      ..
 H—C—C—H       H—C—C—O—H
    |              |  |  ..
    H              H  H
```

**STEP 5** In each case, we have 1 carbon atom with only 6 electrons around it, so we convert 1 lone pair into another bond.

```
    H :Ö:              H  Ö
    |  |               |  ‖
 H—C—C—H    to   H—C—C—H
    |                  |
    H                  H
```

Structure 1

```
    ..      ..                   ..
 H—C—C—O—H    to   H—C=C—O—H
    |  |  ..             |  |  ..
    H  H                 H  H
```

Structure 2

**STEP 6** Every atom in both structures has its most common bonding pattern, so we have 2 reasonable Lewis structures representing isomers. Structure 1 is acetaldehyde (or ethanal), and structure 2 is ethenol.

**STEP 7** Because each atom in our structures has its most common bonding pattern, we skip this step.

### EXERCISE 12.1 Lewis Structures

OBJECTIVE 7

Draw a reasonable Lewis structure for each of the following formulas.

a. $CCl_4$

b. $Cl_2O$

c. $COF_2$

d. $C_2Cl_6$

e. $BCl_3$

f. $N_2H_4$

g. $H_2O_2$

h. $NH_2OH$

i. $NCl_3$

## 12.3 Resonance

As the valence-bond model was being developed, chemists came to recognize that it did not always describe all molecules and polyatomic ions adequately. For example, if we followed the procedures in Section 12.2 to draw a Lewis structure for the nitrate ion, $NO_3^-$, we would obtain a structure that chemists have discovered is not an accurate description of the ion's bonds:

```
       O
       ‖
[ :O—N—O: ]⁻
```

This Lewis structure shows two different types of bonds, single and double. Because more energy is required to break a double bond than to break a single bond, we say that a double bond is stronger than a single bond. Double bonds also have a shorter bond length (the distance between the nuclei of the two atoms in the bond) than single bonds do. Thus, if the above Lewis structure for nitrate were correct, the nitrate polyatomic ion would have one bond that is shorter and stronger than the other two.

This is not the case. Laboratory analyses show all three of the bonds in the nitrate ion to be the same strength and the same length. Interestingly, analysis of the bonds suggests that they are longer than double bonds and shorter than single bonds. They are also stronger than single bonds but not so strong as double bonds. In order to explain how these characteristics are possible for the nitrate ion and for molecules and polyatomic ions like it, the valence-bond model had to be expanded.

The model now enables us to view certain molecules and polyatomic ions *as if* they were able to resonate—to switch back and forth—between two or more different structures. For example, the nitrate ion can be viewed *as if* it resonates among the three different structures below. Each of these structures is called a **resonance structure.** The hypothetical switching from one resonance structure to another is called **resonance.** In chemical notation, the convention is to separate the resonance structures with double-headed arrows.

$$\left[\begin{array}{c} \ddot{O} \\ \| \\ :\ddot{O}-N-\ddot{O}: \end{array}\right]^- \leftrightarrow \left[\begin{array}{c} :\ddot{O}: \\ | \\ \ddot{O}=N-\ddot{O}: \end{array}\right]^- \leftrightarrow \left[\begin{array}{c} :\ddot{O}: \\ | \\ :\ddot{O}-N=\ddot{O} \end{array}\right]^-$$

It is important to stress that the nitrate ion is *not really changing from one resonance structure to another*, but chemists find it *useful*, as an intermediate stage in the process of developing a better description of the nitrate ion, to think of it *as if* it were doing so. In actuality, the ion behaves as if it were a blend of the three resonance structures.

OBJECTIVE 8

We can draw a Lewis-like structure that provides a better description of the actual character of the nitrate ion by blending the resonance structures into a single **resonance hybrid:**

**STEP 1** Draw the skeletal structure, using solid lines for the bonds that are found in all of the resonance structures.

**STEP 2** Where there is sometimes a bond and sometimes not, draw a dotted line.

**STEP 3** Draw only those lone pairs that are found on every one of the resonance structures. (Leave off the lone pairs that are on one or more resonance structures but are not on all of them.)

The resonance hybrid for the nitrate polyatomic ion is

$$\left[\begin{array}{c} \ddot{O} \\ \vdots| \\ \ddot{O}\cdots N \cdots \ddot{O} \end{array}\right]^-$$

A bond found in at least 1, but not all, of the resonance structures

A bond found in all the resonance structures

A lone pair found in all the resonance structures

OBJECTIVE 8

Resonance is possible whenever a Lewis structure has a multiple bond and an adjacent atom with at least 1 lone pair. The arrows in the following generalized structures show how you can think of the electrons shifting as one resonance form changes to another.

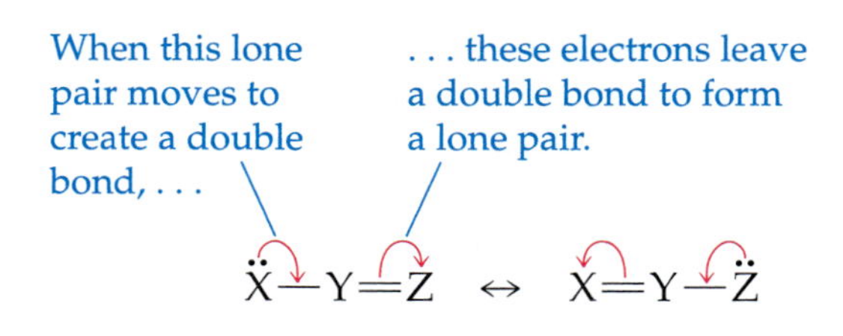

For example, the two resonance structures for the formate ion, $HCO_2^-$, are

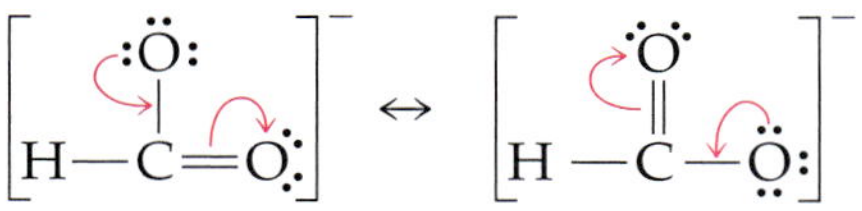

To generate the second resonance structure from the first, we imagine 1 lone pair dropping down to form another bond and pushing an adjacent bond off to form a lone pair. The arrows show this *hypothetical* shift of electrons, which leads to the resonance hybrid.

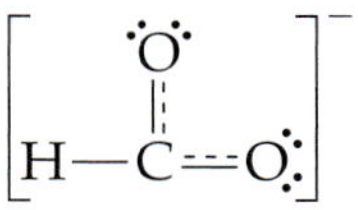

## Exercise 12.2 Resonance

OBJECTIVE 8

Draw all of the reasonable resonance structures and the resonance hybrid for the carbonate ion, $CO_3^{2-}$. A reasonable Lewis structure for the carbonate ion is

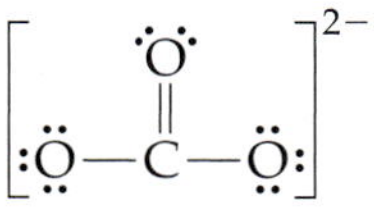

You can find a more comprehensive description of resonance at the following Web site:
**www.chemplace.com/college/**

## 12.4 Molecular Geometry from Lewis Structures

The shapes of molecules play a major role in determining their function. For example, Special Topic 3.1 describes how the shape of ethanol molecules enables them to attach to specific sites on nerve cell membranes and slow the transfer of information from one neuron to another. Special Topic 5.2 describes how the shapes of the molecules in our food determine whether they taste sweet or bitter. You will discover in Special Topic 17.2 that the fat substitute Olestra is indigestible because it does not fit into the enzyme that digests natural fat. The purpose of this section is to show you how to use Lewis structures to predict three-dimensional shapes of simple molecules and polyatomic ions. Let's start with a review of some of the information from Section 3.1, where this topic was first introduced.

The carbon atoms in methane, $CH_4$, molecules are surrounded by four electron groups, each composed of a two-electron covalent bond to a hydrogen atom.

```
     H
     |
 H — C — H
     |
     H
```

OBJECTIVE 9

Because each electron group is negatively charged and because negative charges repel each other, the most stable arrangement for the electron groups is one in which they are as far apart as possible. In Chapter 3 we saw that the geometric arrangement in which four electron groups are as far apart as possible is a *tetrahedral arrangement,* with bond angles of 109.5°. A tetrahedron is a four-sided solid for which each side is an equilateral triangle, and a **bond angle** is the angle formed by any three adjacent atoms in a molecule. The carbon in a methane molecule can be viewed as sitting in the center of a tetrahedron with a hydrogen atom at each of the four corners.

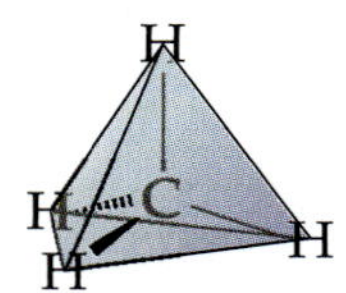


Figure 12.4 shows four ways to describe the methane molecule. The space-filling model in that figure is probably the best representation of the actual structure of the $CH_4$ molecule, but when scientists are focusing on the bonds in a molecule, they often prefer to think of the structure as represented by the ball-and-stick model, the Lewis structure, and the geometric sketch. Note that the Lewis structure for $CH_4$ does not show the correct bond angles. In order to display bond angles correctly with a Lewis-like structure, chemists use geometric sketches, such as the last image in Figure 12.4. Remember that in a geometric sketch, a solid line represents a bond whose central axis lies in the plane of the paper, a solid wedge shape represents a bond whose axis is extending out of the plane of the paper toward you, and a dashed wedge shape represents a bond whose axis extends out behind the plane of the paper, away from you.

**Figure 12.4**
Four Ways to Describe a Methane, $CH_4$, Molecule

OBJECTIVE 9

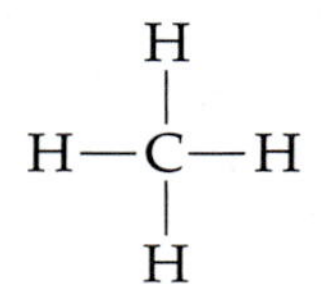


Lewis structure

Space-filling model

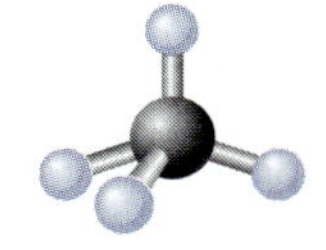

Ball-and-stick model

H
109.5°
C
H
H
H

Geometric sketch

The tetrahedral arrangement is very common and important in chemistry, so be sure that you have a clear image in your mind of the relationship between the geometric sketch in Figure 12.4 and the three-dimensional shape of a molecule such as methane. You can make a ball-and-stick model of this molecule by using an apple for the carbon atom, four toothpicks to represent the axes through each of the four bonds, and four grapes to represent the hydrogen atoms. Try to arrange the toothpicks and grapes so that each grape is an equal distance from the other three. When you think you have successfully constructed a $CH_4$-fruit molecule, turn your model over and over, so that a different grape is on top each time. If your model is really tetrahedral, it will always look the same, no matter which grape is on top (Figure 12.5).

**Figure 12.5**
**The Apple–Grape Ball-and-Stick Model of a $CH_4$ Molecule**

The nitrogen atom in each ammonia molecule, $NH_3$, has 4 electron groups around it (3 bond groups and 1 lone pair), and, as with methane molecules, the best way to get these 4 groups as far apart as possible is in a tetrahedral arrangement. The electrons in the lone pair exert a greater repulsive force than the electrons in the bonds, however, so the bond groups are pushed a little closer together. As a result, the bond angles for ammonia are 107° (rather than the 109.5° found in the methane molecule).

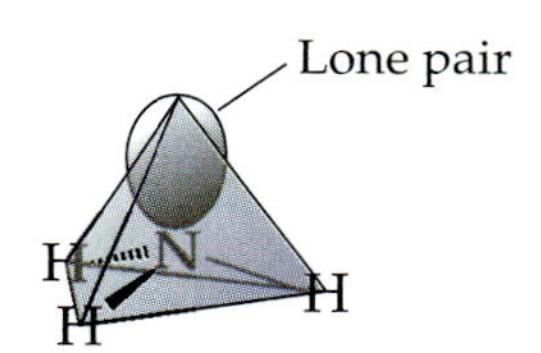


There are two ways to describe the geometry of the ammonia molecule. We can describe the arrangement of all the electron groups, including the lone pair, and call the shape tetrahedral. Or we can describe the arrangement of the atoms only, without considering the lone pair, and call ammonia's shape a **trigonal pyramid** (Figure 12.6). The three hydrogen atoms represent the corners of the pyramid's base, and the nitrogen atom forms the peak. In this book, we will call the geometry that describes all of the electron groups, including the lone pairs, the **electron group geometry.** The shape that describes the arrangement of the atoms only—treating the lone pairs as invisible—will be called the **molecular geometry.**

OBJECTIVE 9

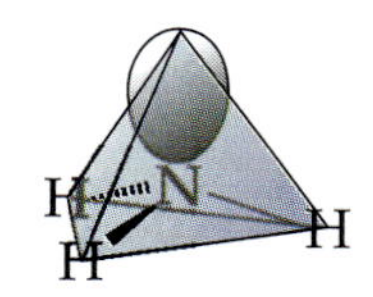


Electron group geometry
(tetrahedral)

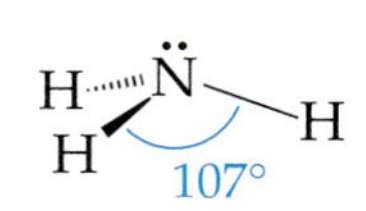


Molecular geometry
(trigonal pyramid)

**Figure 12.6**
**Geometry of Ammonia**

OBJECTIVE 9

The water molecule's electron group geometry, which is determined by the arrangement of its 4 electron groups around the oxygen atom, is tetrahedral (Figure 12.7).

**Figure 12.7**
Geometry of Water

OBJECTIVE 9

Lone pairs

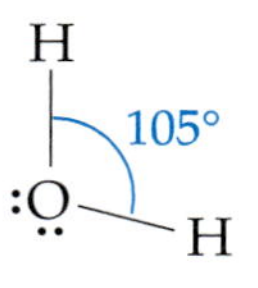


Electron group geometry (tetrahedral)

Molecular geometry (bent)

Its molecular geometry, which describes the arrangement of the atoms only, is called **bent.** The two lone pairs on the oxygen atom exert a stronger repulsion than the two bond groups and push the bond groups closer together than they would be if all of the electron groups around the oxygen atom were identical. The bond angle becomes 105° instead of the 109.5° predicted for four identical electron groups.

The boron atom in the center of a $BF_3$ molecule is surrounded by 3 electron groups, which are each composed of a 2-electron covalent bond to a fluorine atom.

```
      :F:
       |
  :F—B—F:
```

OBJECTIVE 9

The geometric arrangement that keeps 3 electron groups as far apart as possible is called **trigonal planar** (often called **triangular planar**) and leads to angles of 120° between the groups. Therefore, the B–F bonds in $BF_3$ molecules have a trigonal planar arrangement with 120° angles between any two fluorine atoms (Figure 12.8).

**Figure 12.8**
Four Ways to Describe a Boron Trifluoride, $BF_3$, Molecule

OBJECTIVE 9

```
      :F:
       |
  :F—B—F:
```

Lewis structure

Space-filling model

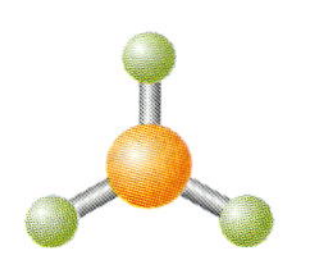

Ball-and-stick model

:F: — B (120°) — :F: :F:

Geometric sketch

Molecules quite often have more than one central atom (an atom with two or more atoms bonded to it). To describe the arrangement of atoms in such molecules, we need to consider the central atoms one at a time. Let's look at the ethene, $C_2H_4$, molecule.

```
H—C=C—H
  |  |
  H  H
```

There are 3 electron groups around each of ethene's carbon atoms (2 single bonds to hydrogen atoms and the double bond to the other carbon). Because these groups repel each other, the most stable arrangement around each carbon atom—with the electron groups as far apart as possible—is a triangle with all 3 groups in the same plane and with angles of about 120° between the groups. That is, ethene has a trigonal planar arrangement of atoms around each carbon.

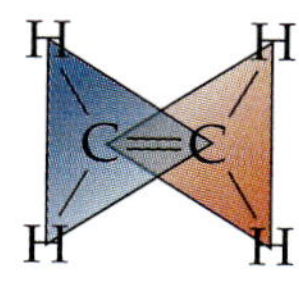


Figure 12.9 shows four ways to describe a $C_2H_4$ molecule. Because the bond groups around each carbon are not identical (2 are single bonds and 1 is a double bond), the angles are only approximately 120°.

Lewis structure

H—C═C—H (each C also bonded to an H below)

Space-filling model

Ball-and-stick model

Geometric sketch

H ≈120° H, C═C ≈120°, H H

Trigonal planar geometry around each carbon

**Figure 12.9**
Four Ways to Describe an Ethene (Ethylene), $C_2H_4$, Molecule

OBJECTIVE 9

When atoms have only 2 electron groups around them, the groups are arranged in a line, with bond angles of 180°. This arrangement is called **linear.** For example, the carbon atom in HCN has 2 electron groups around it (one single bond and one triple bond), and the best way to keep these groups apart is in a linear shape.

OBJECTIVE 9

H—C≡N: H—C≡N: (180°)

There are two central atoms in an acetylene, $C_2H_2$, molecule.

H—C≡C—H

Because each of acetylene's central atoms has 2 electron groups around it, the arrangement of atoms around each is linear. Figure 12.10 shows four ways to describe a $C_2H_2$ molecule.

OBJECTIVE 9

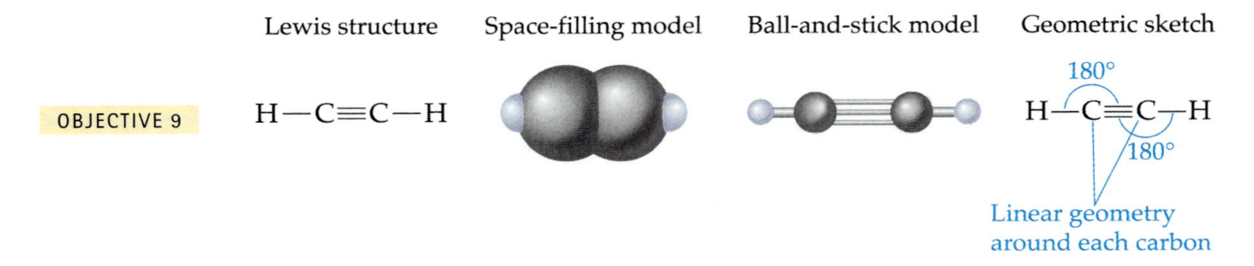


**Figure 12.10**
**Four Ways to Describe an Ethyne (Acetylene), $C_2H_2$, Molecule**

The following sample study sheet summarizes a systematic procedure for predicting electron group geometries, drawing geometric sketches, and describing molecular geometries.

## Sample Study Sheet 12.2

### Predicting Molecular Geometry

OBJECTIVE 9

**Tip-off** In this chapter, you are given a Lewis structure for a molecule or polyatomic ion (or a formula from which you can draw a Lewis structure), and asked to (1) name the electron group geometry around 1 or more atoms in the structure, (2) draw a geometric sketch of the structure, and/or (3) name the molecular geometry around 1 or more of the atoms in the structure. You will find in later chapters that there are other tip-offs for these tasks.

**General Steps**

**Step 1** To determine the name of the electron group geometry around each atom that is attached to two or more other atoms, count the number of electron groups around each "central" atom and apply the guidelines found in Table 12.3. An electron group can be a single bond, a lone pair, or a multiple bond. (Double bonds and triple bonds count as one group.)

**Step 2** Use one or more of the geometric sketches shown in Table 12.3 as models for the geometric sketch of your molecule. If the groups around the central atom are identical, the bond angle is exact. If the groups are not identical, the angles are approximate.

**Step 3** To determine the name of the molecular geometry around each atom that has two or more atoms attached to it, count the number of bond groups and lone pairs, and then apply the guidelines found in Table 12.3. Single, double, and triple bonds all count as 1 bond group. Note that if all of the electron groups attached to the atom are bond groups (no lone pairs), the name of the molecular geometry is the same as the name of the electron group geometry.

**Example** See Examples 12.6 and 12.7.

**Table 12.3**
Electron Group and Molecular Geometry

OBJECTIVE 9

| $e^-$ groups | $e^-$ group geometry | General geometric sketch | Bond angles | Bond groups | Lone pairs | Molecular geometry |
|---|---|---|---|---|---|---|
| 2 | linear | 180° | 180° | 2 | 0 | linear |
| 3 | trigonal planar | 120° | 120° | 3 | 0 | trigonal planar |
| | | | | 2 | 1 | bent |
| 4 | tetrahedral | 109.5 | 109.5° | 4 | 0 | tetrahedral |
| | | | | 3 | 1 | trigonal pyramid |
| | | | | 2 | 2 | bent |

## EXAMPLE 12.6 Predicting Molecular Geometry

OBJECTIVE 9

Nitrosyl fluoride, NOF, is used as an oxidizer in rocket fuels. Its Lewis structure is

:O=N—F:

Using its Lewis structure, (a) write the name of the electron group geometry around the nitrogen atom, (b) draw a geometric sketch of the molecule, including bond angles, and (c) write the name of the molecular geometry around the nitrogen atom.

***Solution***

a. Table 12.3 tells us that because there are 3 electron groups around the nitrogen in nitrosyl fluoride (2 bond groups and 1 lone pair), the electron group geometry around the nitrogen atom is **trigonal planar.**

b. The geometric sketch for NOF is

O
‖ ≈120°
N
F

c. The nitrogen atom in NOF has 1 single bond, 1 double bond, and 1 lone pair. According to Table 12.3, an atom with 2 bond groups and 1 lone pair has the molecular geometry called **bent.**

## EXAMPLE 12.7 Predicting Molecular Geometry

OBJECTIVE 9

Methyl cyanoacrylate is the main ingredient in Super Glue.

(a) Write the name of the electron group geometry around each atom that has 2 or more atoms attached to it, (b) draw a geometric sketch of the molecule, including bond angles, and (c) write the name of the molecular geometry around each atom that has two or more atoms attached to it. (This example was chosen because it contains several different types of electron groups and molecular shapes. The structures in the exercises and problems that accompany this chapter are much simpler.)

*Solution*

a. 2 electron groups, linear

3 electron groups, trigonal planar

4 electron groups, tetrahedral

b and c. There are many ways to sketch the molecule and still show the correct geometry. The following is just one example. (Remember that your assigned exercises will be simpler than this one.)

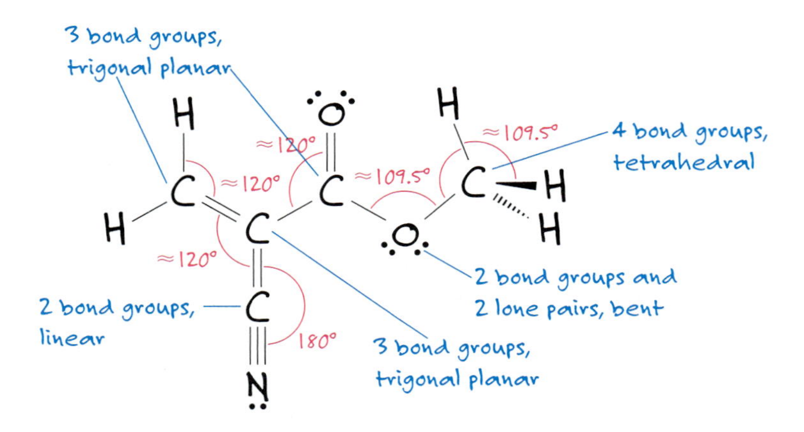

## EXERCISE 12.3 Molecular Geometry

OBJECTIVE 9

For each of the Lewis structures that follow, (a) write the name of the electron group geometry around each atom that has two or more atoms attached to it, (b) draw a geometric sketch of the molecule, including bond angles, and (c) write the name of the molecular geometry around each atom that has two or more atoms attached to it.

a. $CBr_4$ (C bonded to four Br atoms, each Br with three lone pairs)

b. $AsF_3$ (:F—As—F: with F above As; As has one lone pair)

c. $\ddot{S}=C=\ddot{S}$

d. :F—O—F: (O has two lone pairs)

e. $BI_3$ (:I—B—I: with I above B)

f. :Cl—C—Cl: with O double-bonded to C

g. H—C—C≡C—H (first C bonded to H above and H below)

h. :I—N—I: with I above N; N has one lone pair

## Chapter Glossary

**Valence electrons** The highest-energy *s* and *p* electrons for an atom.

**Isomers** Compounds that have the same molecular formula but different molecular structures.

**Resonance structures** Two or more Lewis structures for a single molecule or polyatomic ion that differ in the positions of lone pairs and multiple bonds but not in the positions of the atoms in the structure. It is *as if* the molecule or ion were able to shift from one of these structures to another by shifting pairs of electrons from one position to another.

**Resonance** The hypothetical switching from one resonance structure to another.

**Resonance hybrid** A structure that represents the average of the resonance structures for a molecule or polyatomic ion.

**Bond angle** The angle between any three adjacent atoms in a molecule.

**Trigonal pyramid** The molecular geometry formed around an atom with 3 bonds and 1 lone pair.

**Electron group geometry** A description of the arrangement of all the electron groups around a central atom in a molecule or polyatomic ion, including the lone pairs.

**Molecular geometry** A description of the arrangement of all the atoms around a central atom in a molecule or polyatomic ion. This description does not consider lone pairs.

**Trigonal planar (often called triangular planar)** The geometric arrangement that keeps three electron groups as far apart as possible. It leads to angles of 120° between the groups.

**Linear geometry** The geometric arrangement that keeps two electron groups as far apart as possible. It leads to angles of 180° between the groups.

**Bent** The molecular geometry formed around an atom with 2 bonds and 2 lone pairs.

You can test yourself on the glossary terms at the following Web site:
www.chemplace.com/college/

## Chapter Objectives

**The goal of this chapter is to teach you to do the following.**

1. Define all of the terms in the Chapter 12 Glossary.

### Section 12.1 A New Look at Molecules and the Formation of Covalent Bonds

2. Describe the advantages and disadvantages of using models to describe physical reality.
3. Describe the assumptions that lie at the core of the valence-bond model.
4. Write or identify the number of valence electrons in an atom of any representative element.
5. Use the valence-bond model to explain how (a) fluorine atoms form 1 covalent bond and have 3 lone pairs in molecules like $F_2$; (b) hydrogen atoms can form 1 covalent bond and have no lone pairs in molecules like HF; (c) carbon atoms can form 4 covalent bonds and have no lone pairs in molecules like $CH_4$; (d) nitrogen, phosphorus, and arsenic atoms can form 3 covalent bonds and have 1 lone pair in molecules like $NH_3$, $PH_3$, and $AsH_3$; (e) nitrogen atoms can form 4 covalent bonds and have no lone pairs in structures like $NH_4^+$; (f) oxygen, sulfur, and selenium atoms can form 2 covalent bonds and have 2 lone pairs in molecules like $H_2O$, $H_2S$, and $H_2Se$; (g) oxygen atoms can form 1 covalent bond and have 3 lone pairs in structures like $OH^-$; (h) carbon and oxygen atoms can form 3 covalent bonds and have 1 lone pair in molecules like CO; (i) boron atoms can form 3 covalent bonds and have no lone pairs in molecules like $BF_3$; and (j) halogen atoms can form 1 covalent bond and have 3 lone pairs in molecules like HF, HCl, HBr, and HI.
6. Write electron-dot symbols for the representative elements.

### Section 12.2 Drawing Lewis Structures

7. Given a formula for a molecule or polyatomic ion, draw a reasonable Lewis structure for it.

### Section 12.3 Resonance

8. Given a Lewis structure or enough information to draw one, predict whether it is best described in terms of resonance, and if it is, draw all of the reasonable resonance structures and the resonance hybrid for the structure.

### Section 12.4 Molecular Geometry from Lewis Structures

9. Given a Lewis structure or enough information to write one, do the following:
   a. Write the name of the electron group geometry around each atom that has 2 or more atoms attached to it.
   b. Draw a geometric sketch for the molecule, including bond angles (or approximate bond angles).
   c. Write the name of the molecular geometry around each atom that has 2 or more atoms attached to it.

## Review Questions

1. Using the A-group convention, what is the group number of the column in the periodic table that includes the element chlorine, Cl?
2. Draw Lewis structures for $CH_4$, $NH_3$, and $H_2O$.
3. Define the term *orbital.*
4. Write a complete electron configuration and an orbital diagram for each of the following.
   a. oxygen, O
   b. phosphorus, P

## Key Ideas

**Complete the following statements by writing one of these words or phrases in each blank.**

| | |
|---|---|
| 109.5° | most |
| A-group | most common |
| anions | never |
| as far apart | oxygen |
| as if | pair unpaired |
| cations | polyatomic ions |
| change | predict |
| eight | rarely |
| electron group | *s* and *p* |
| explain | structural |
| fewest | triple |
| four | true |
| function | two |
| highest-energy | uncharged |
| lone pairs | useful |
| molecular | visualize |
| more | |

5. When developing a model of physical reality, scientists take what they think is ____________ and simplify it enough to make it ____________.

6. Models have advantages and disadvantages. They help us to __________, __________, and __________ chemical changes, but we need to remind ourselves now and then that they are only models and that as models, they have their limitations.
7. One characteristic of models is that they __________ with time.
8. The valence-bond model, which is commonly used to describe the formation of covalent bonds, is based on the assumptions that only the __________ electrons participate in bonding, and covalent bonds usually form to __________ electrons.
9. Valence electrons are the highest-energy __________ electrons in an atom.
10. When the columns in the periodic table are numbered by the __________ convention, the number of valence electrons in each atom of a representative element is equal to the element's group number in the periodic table.
11. Paired valence electrons are called __________.
12. When forming 4 bonds to hydrogen atoms in a methane molecule, each carbon atom behaves __________ 1 electron is promoted from the $2s$ orbital to the $2p$ orbital.
13. Carbon atoms frequently form double bonds, in which they share __________ electrons with another atom, often another carbon atom.
14. A(n) __________ bond can be viewed as the sharing of 6 electrons between 2 atoms.
15. The shortcut for drawing Lewis structures for which we try to give each atom its __________ bonding pattern works well for many simple uncharged molecules, but it does not work reliably for molecules that are more complex or for __________.
16. For __________ molecules, the total number of valence electrons is the sum of the valence electrons of each atom.
17. For polyatomic __________, the total number of valence electrons is the sum of the valence electrons for each atom minus the charge.
18. For polyatomic __________, the total number of valence electrons is the sum of the valence electrons for each atom plus the charge.
19. Hydrogen and fluorine atoms are __________ in the center of a Lewis structure.
20. Oxygen atoms are __________ in the center of a Lewis structure.
21. The element with the __________ atoms in the formula is often in the center of a Lewis structure.
22. The atom that is capable of making the __________ bonds is often in the center of a Lewis structure.
23. Oxygen atoms rarely bond to other __________ atoms.
24. In a reasonable Lewis structure, carbon, nitrogen, oxygen, and fluorine always have __________ electrons around them.

25. In a reasonable Lewis structure, hydrogen will always have a total of ____________ electrons from its one bond.

26. Boron can have fewer than 8 electrons around it in a reasonable Lewis structure but never ____________ than 8.

27. Substances that have the same molecular formula but different ____________ formulas are called isomers.

28. The shapes of molecules play a major role in determining their ____________.

29. The most stable arrangement for electron groups is one in which they are ____________ as possible.

30. The geometric arrangement in which 4 electron groups are as far apart as possible is a tetrahedral arrangement, with bond angles of ____________.

31. In this book, we call the geometry that describes all of the electron groups, including the lone pairs, the ____________ geometry. The shape that describes the arrangement of the atoms only—treating the lone pairs as invisible—will be called the ____________ geometry.

## Chapter Problems

### Section 12.1 A New Look at Molecules and the Formation of Covalent Bonds

32. Describe the advantages and disadvantages of using models to describe physical reality. OBJECTIVE 2

33. Describe the assumptions that lie at the core of the valence-bond model. OBJECTIVE 3

34. How many valence electrons do the atoms of each of the following elements have? Write the electron configuration for these electrons. (For example, fluorine has 7 valence electrons, which can be described as $2s^2\ 2p^5$.) OBJECTIVE 4

    a. nitrogen, N
    b. sulfur, S
    c. iodine, I
    d. argon, Ar

35. How many valence electrons do the atoms of each of the following elements have? Write the electron configuration for these electrons. (For example, fluorine has 7 valence electrons, which can be described as $2s^2\ 2p^5$.) OBJECTIVE 4

    a. oxygen, O
    b. boron, B
    c. neon, Ne
    d. phosphorus, P
    e. carbon, C

36. Draw electron-dot symbols for each of the following elements. OBJECTIVE 6

    a. nitrogen, N
    b. sulfur, S
    c. iodine, I
    d. argon, Ar

37. Draw electron-dot symbols for each of the following elements. OBJECTIVE 6

    a. oxygen, O
    b. boron, B
    c. neon, Ne
    d. phosphorus, P
    e. carbon, C

**38.** To which group on the periodic table would atoms with the following electron dot symbols belong? List the group numbers using the 1–18 convention and using the A-group convention.

a. $\cdot\underset{\cdot}{\ddot{\mathrm{X}}}\cdot$

b. $:\underset{\cdot\cdot}{\ddot{\mathrm{X}}}:$

c. $\cdot\underset{\cdot}{\mathrm{X}}\cdot$

39. To which group on the periodic table would atoms with the following electron dot symbols belong? List the group numbers using the 1-18 convention and using the A-group convention.

a. $\cdot\underset{\cdot\cdot}{\ddot{\mathrm{X}}}:$

b. $\cdot\underset{\cdot\cdot}{\ddot{\mathrm{X}}}\cdot$

c. $\cdot\underset{\cdot}{\dot{\mathrm{X}}}\cdot$

**40.** For each of the following elements, sketch *all* of the ways mentioned in Section 12.1 that their atoms could look in a Lewis structure. For example, fluorine has only one bonding pattern, and it looks like $-\underset{\cdot\cdot}{\ddot{\mathrm{F}}}:$ .

a. nitrogen, N

b. boron, B

c. carbon, C

41. For each of the following elements, sketch all of the ways mentioned in Section 12.1 that their atoms could look in a Lewis structure. For example, fluorine has only one bonding pattern, and it looks like $-\underset{\cdot\cdot}{\ddot{\mathrm{F}}}:$ .

a. hydrogen, H

b. oxygen, O

c. chlorine, Cl

OBJECTIVE 5

**42.** Use the valence-bond model to explain the following observations.

a. Fluorine atoms have 1 bond and 3 lone pairs in $F_2$.

b. Carbon atoms have 4 bonds and no lone pairs in $CH_4$.

c. Nitrogen atoms have 3 bonds and 1 lone pair in $NH_3$.

d. Sulfur atoms have 2 bonds and 2 lone pairs in $H_2S$.

e. Oxygen atoms have 1 bond and 3 lone pairs in $OH^-$.

OBJECTIVE 5

43. Use the valence-bond model to explain the following observations.

a. Phosphorus atoms have 3 bonds and 1 lone pair in $PH_3$.

b. Nitrogen atoms have 4 bonds and no lone pairs in $NH_4^+$.

c. Oxygen atoms have 2 bonds and 2 lone pairs in $H_2O$.

d. Boron atoms have 3 bonds and no lone pairs in $BF_3$.

e. Chlorine atoms have 1 bond and 3 lone pairs in HCl.

44. Based on your knowledge of the most common bonding patterns for the nonmetallic elements, predict the formulas with the lowest subscripts for the compounds that would form from the following pairs of elements. (For example, hydrogen and oxygen can combine to form $H_2O$ and $H_2O_2$, but $H_2O$ has lower subscripts.)

   a. C and H

   b. S and H

   c. B and F

45. Based on your knowledge of the most common bonding patterns for the nonmetallic elements, predict the formulas with the lowest subscripts for the compounds that would form from the following pairs of elements. (For example, hydrogen and oxygen can combine to form $H_2O$ and $H_2O_2$, but $H_2O$ has lower subscripts.)

   a. P and I

   b. O and Br

   c. N and Cl

### Section 12.2 Drawing Lewis Structures

46. Copy the following Lewis structure and identify the single bonds, the double bond, and the lone pairs.

      H   H
      |   |
   H—C=C—F̤̈:

47. Copy the following Lewis structure and identify the single bonds, the triple bond, and the lone pairs.

            H
            |
   H—C≡C—C—B̤̈r:
            |
            H

48. For each of the following molecular compounds, identify the atom that is most likely to be found in the center of its Lewis structure. Explain why.

   a. $CBr_4$

   b. $SO_2$

   c. $H_2S$

   d. NOF

49. For each of the following molecular compounds, identify the atom that is most likely to be found in the center of its Lewis structure. Explain why.

   a. $BI_3$

   b. $SO_3$

   c. $AsH_3$

   d. HCN

50. Calculate the total number of valence electrons for each of the following formulas.

a. $HNO_3$

b. $CH_2CHF$

51. Calculate the total number of valence electrons for each of the following formulas.

a. $H_3PO_4$

b. $HC_2H_3O_2$

OBJECTIVE 7

52. Draw a reasonable Lewis structure for each of the following formulas.

a. $CI_4$

b. $O_2F_2$

c. $HC_2F$

d. $NH_2Cl$

e. $PH_3$

f. $S_2F_2$

g. $HNO_2$

h. $N_2F_4$

i. $CH_2CHCH_3$

OBJECTIVE 7

53. Draw a reasonable Lewis structure for each of the following formulas.

a. $H_2S$

b. $CHBr_3$

c. $NF_3$

d. $Br_2O$

e. $H_2CO_3$

f. $H_2S_2$

g. HOCl

h. $BBr_3$

i. $CH_3CH_2CHCH_2$

**Section 12.3 Resonance**

OBJECTIVE 8

54. Draw a reasonable Lewis structure for the ozone molecule, $O_3$, using the skeleton that follows. The structure is best described in terms of resonance, so draw all of its reasonable resonance structures and the resonance hybrid that summarizes these structures.

O–O–O

OBJECTIVE 8

55. Draw a reasonable Lewis structure for a nitric acid, $HNO_3$, using the skeleton that follows. The structure is best described in terms of resonance, so draw all of its reasonable resonance structures and the resonance hybrid that summarizes these structures.

```
         O
         |
H—O—N—O
```

**Section 12.4 Molecular Geometry from Lewis Structures**

56. Although both $CO_2$ molecules and $H_2O$ molecules have three atoms, $CO_2$ molecules are linear, and $H_2O$ molecules are bent. Why?

57. Although both $BF_3$ molecules and $NH_3$ molecules have four atoms, the $BF_3$ molecules are planar, and $NH_3$ molecules are pyramidal. Why?

58. Using the symbol X for the central atom and Y for the outer atoms, draw the general geometric sketch for a 3-atom molecule with linear geometry.

59. Using the symbol X for the central atom and Y for the outer atoms, draw the general geometric sketch for a molecule with trigonal planar geometry.

60. Using the symbol X for the central atom and Y for the outer atoms, draw the general geometric sketch for a molecule with tetrahedral geometry.

61. For each of the Lewis structures given below:

OBJECTIVE 9

- Write the name of the electron group geometry around each atom that has 2 or more atoms attached to it.
- Draw a geometric sketch of the molecule, including bond angles (or approximate bond angles).
- Write the name of the molecular geometry around each atom that has 2 or more atoms attached to it.

a. :F:—C—F: with :F: above and :F: below C

b. :O=N—Cl:

c. :Br—B—Br: with :I: above B

d. :Br—As—Br: with :Br: above As

e. :Br—C—Br: with O double-bonded above C

f. :O=C=S:

62. For each of the Lewis structures given below,

OBJECTIVE 9

- Write the name of the electron group geometry around each atom that has 2 or more atoms attached to it.
- Draw a geometric sketch of the molecule, including bond angles (or approximate bond angles).
- Write the name of the molecular geometry around each atom that has 2 or more atoms attached to it.

a. :I—B—I: with :Br: above B

b. :Cl—As—Cl: with :Cl: above As

c. :Cl—C—Cl: with :F: above and :Cl: below C

d. :Se=C=S:

e. H—O—O—H

*Studying this chapter will help you to explain many things that relate to gases, including why this person has more trouble breathing here than at sea level.*

# Gases

CHAPTER 13



It's Monday morning, and Lilia is walking out of the chemistry building, thinking about the introductory lecture on gases that her instructor just presented. Dr. Scanlon challenged the class to try to visualize gases in terms of the model she described, so Lilia looks at her hand and tries to picture the particles in the air bombarding each square centimeter of her skin at a rate of $10^{23}$ collisions per second. Lilia has high hopes that a week of studying gases will provide her with answers to the questions her brothers and sisters posed to her the night before at dinner.

When Ted, who is a mechanic for a Formula One racing team, learned that Lilia was going to be studying gases in her chemistry class, he asked her to find out how to calculate gas density. He knows that when the density of the air changes, he needs to adjust the car's brakes and other components to improve its safety and performance. John, who is an environmental scientist, wanted to be reminded why the balloons that carry his scientific instruments into the upper atmosphere expand as they rise. Amelia is an artist who has recently begun to add neon lights to her work. After bending the tubes into the desired shape, she fills them with gas from a high-pressure cylinder. She wanted to know how to determine the number of tubes she can fill with one cylinder.

Lilia's sister Rebecca, the oldest, is a chemical engineer who could answer Ted's, John's, and Amelia's questions, but to give Lilia an opportunity to use her new knowledge, she keeps quiet except to describe a gas-related issue of her own. Rebecca is helping to design an apparatus in which two gases will react at high temperature, and her responsibility is to equip the reaction vessel with a valve that will keep the pressure from rising to dangerous levels. She started to explain to Lilia why increased temperature leads to increased pressure, but when Lilia asked what gas pressure was and what caused it, Rebecca realized that she had better save her explanation for the next family dinner. Lilia (like yourself) will learn about gas pressures and many other gas-related topics by reading Chapter 13 of her textbook carefully and listening closely in lecture.

*The gas particles in the air around us are constantly colliding with our skin.*

## Review Skills

The presentation of information in this chapter assumes that you can already perform the tasks listed below. You can test your readiness to proceed by answering the Review Questions at the end of the chapter. This might also be a good time to read the Chapter Objectives, which precede the Review Questions.

- Describe the particle nature of gases. (Section 2.1)
- Convert between temperatures in the Celsius and Kelvin scales. (Section 8.6)
- Convert the amount of one substance in a given reaction to the amount of another substance in the same reaction, whether the amounts are described by mass of pure substance or volume of a solution containing a given molarity of one of the substances. (Sections 10.1 and 10.3)
- Given an actual yield and a theoretical yield for a chemical reaction (or enough information to calculate a theoretical yield), calculate the percent yield for the reaction. (Section 10.2)

## 13.1 Gases and Their Properties

If you want to understand how gases behave—such as why fresh air rushes into your lungs when certain chest muscles contract or how gases in a car's engine move the pistons and power the car—you need a clear mental image of the model chemists use to explain the properties of gases and the relationships between them. The model was introduced in Section 2.1, but we'll be adding some new components to it in the review presented here.

Gases consist of tiny particles widely spaced (Figure 13.1). Under typical conditions, the average distance between gas particles is about ten times their diameter. Because of these large distances, the volume occupied by the particles themselves is very small compared to the volume of the empty space around them. For a gas at room temperature and pressure, the gas particles themselves occupy about 0.1% of the total volume. The other 99.9% of the total volume is empty space (whereas in liquids, about 70% of the volume is occupied by particles). Because of the large distances between gas particles, there are very few attractions or repulsions between them.

OBJECTIVE 2

The particles in a gas are in rapid and continuous motion. For example, the average velocity of nitrogen molecules, $N_2$, at 20 °C is about 500 m/s. As the temperature of a gas rises, the particles' velocity increases. The average velocity of nitrogen molecules at 100 °C is about 575 m/s.

The particles in a gas are constantly colliding with the walls of the container and with each other. Because of these collisions, the gas particles are constantly changing their direction of motion and their velocity. In a typical situation, a gas particle moves a very short distance between collisions. For example, oxygen, $O_2$, molecules at normal temperatures and pressures move an average of $10^{-7}$ m between collisions.

OBJECTIVE 3

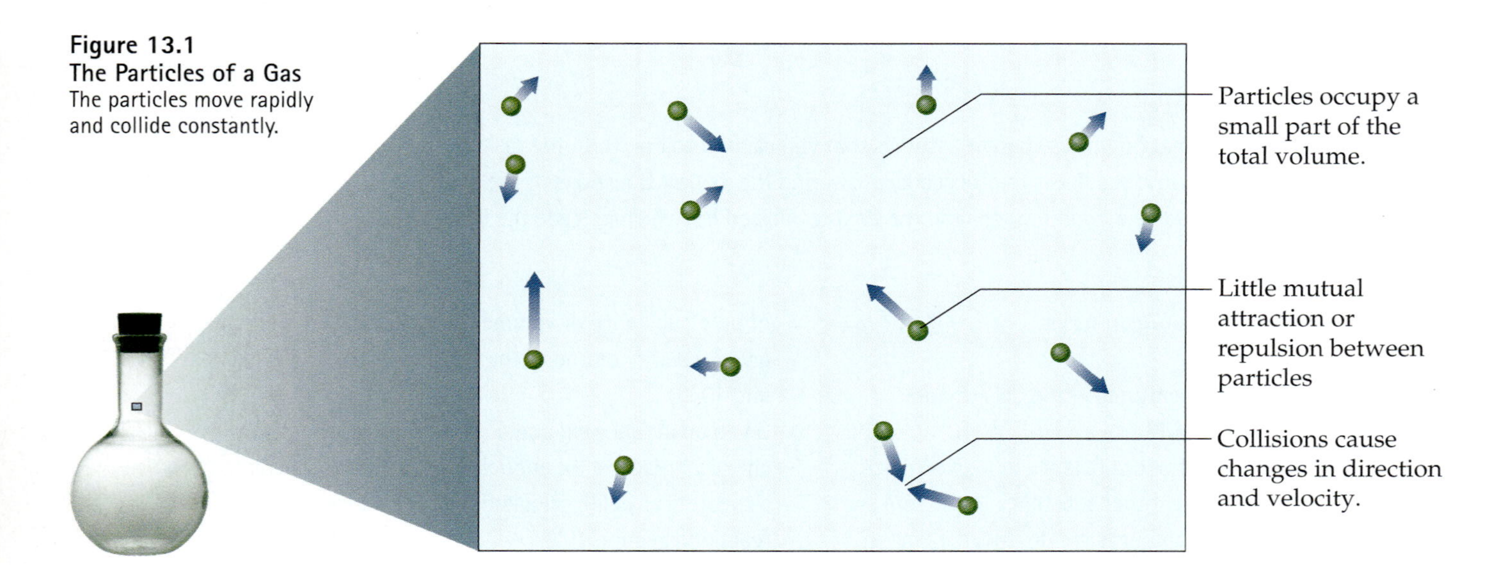
Particles occupy a small part of the total volume.
Little mutual attraction or repulsion between particles
Collisions cause changes in direction and velocity.

**Figure 13.1**
**The Particles of a Gas**
The particles move rapidly and collide constantly.

## Ideal Gases

The model described above applies to *real* gases, but chemists often simplify the model further by imagining the behavior of an *ideal* gas. An ideal gas differs from a real gas in that

OBJECTIVE 4

- The particles are assumed to be point-masses—that is, particles that have a mass but occupy no volume.
- There are no attractive or repulsive forces at all between the particles.

When we add these assumptions to our model for gases, we call it the **ideal gas model.** As the name implies, the ideal gas model describes an "ideal" of gas behavior that is only approximated by reality. Nevertheless, the model succeeds in explaining and predicting the behavior of typical gases under typical conditions. In fact, some actual gases do behave very much in accordance with the model, and scientists may call them **ideal gases.** The ideal gas assumptions make it easier for chemists to describe the relationships between the properties of gases and enable us to calculate values for these properties.

## Properties of Gases

The ideal gas model is used to predict changes in four related gas properties: volume, number of particles, temperature, and pressure. Volumes of gases are usually described in liters, L, or cubic meters, $m^3$, and numbers of particles are usually described in moles, mol. Although gas temperatures are often measured with thermometers that report temperatures in degrees Celsius, °C, scientists generally use Kelvin temperatures for calculations. Remember that you can convert between degrees Celsius, °C, and kelvins, K, by using the following equations:

$$? \text{ K} = °\text{C} + 273.15$$

$$? \text{ °C} = \text{K} - 273.15$$

To understand gas pressure, picture a typical gas in a closed container. Each time a gas particle collides with and ricochets off one of the walls of its container, it exerts a force against the wall. The sum of the forces of these ongoing collisions of gas particles against all the container's interior walls creates a continuous pressure upon those walls. **Pressure** is force divided by area.

OBJECTIVE 5

$$\text{Pressure} = \frac{\text{Force}}{\text{Area}}$$

$$\text{Gas pressure} = \frac{\text{Force due to particle collisions with the walls}}{\text{Area of the walls}}$$

The accepted SI unit for gas pressure is the pascal, Pa. A pascal is a very small amount of pressure, so the kilopascal, kPa, is more commonly used.

OBJECTIVE 6

OBJECTIVE 7

Other units used to describe gas pressure include atmospheres (atm), torr, and millimeters of mercury (mmHg). The relationships between these pressure units are

OBJECTIVE 8

$$1 \text{ atm} = 101{,}325 \text{ Pa} = 101.325 \text{ kPa} = 760 \text{ mmHg} = 760 \text{ torr}$$

The numbers in these relationships come from definitions, so they are all exact. At sea level on a typical day, the atmospheric pressure is about 101 kPa, or about 1 atm.

OBJECTIVE 9

In calculations, the variables $P$, $T$, $V$, and $n$ are commonly used to represent pressure, temperature, volume, and moles of gas, respectively.

## Discovering the Relationships Between Properties

If we want to explain why a weather balloon carrying instruments into the upper atmosphere expands as it rises, we need to consider changes in the properties of the gases (pressure, volume, temperature, and moles) inside and outside the balloon. For example, as the balloon rises, the pressure outside of it, which is called the atmospheric pressure, decreases. But there are also variations in temperature, and the balloon might have small leaks that change the moles of gas it contains.

In a real situation, pressure, temperature, and moles of gas may all be changing, and predicting the effect of such a blend of changing properties on gas volume is tricky. Therefore, before we tackle predictions for real-world situations, such as the weather balloon, we will consider simpler systems in which two of the four gas properties are held constant, a third property is varied, and the effect of this variation on the fourth property is observed. For example, it is easier to understand the relationship between volume and pressure when the moles of gas and temperature are held constant. The volume can be varied, and the effect that this change has on the pressure can be measured. An understanding of the relationships between gas properties in controlled situations will help us to explain and predict the effects of changing gas properties in more complicated, real situations.

Figure 13.2 shows a laboratory apparatus that can be used to demonstrate all the relationships we are going to be discussing. It consists of a cylinder with a movable piston, a thermometer, a pressure gauge, and a valve through which gas may be added to the cylinder's chamber or removed from it.

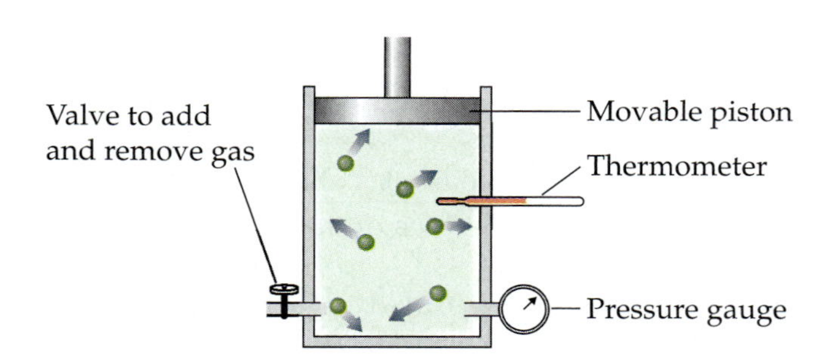


**Figure 13.2**
Apparatus Used to Demonstrate Relationships Between the Properties of Gases

## The Relationship Between Volume and Pressure

Figure 13.3 shows how our demonstration apparatus would be used to determine the relationship between gas volume and pressure. While holding moles of gas constant (by closing the valve) and holding the temperature constant (by allowing heat to transfer in or out so that the apparatus remains the same temperature as the surrounding environment), we move the piston to change the volume, and then we observe the change in pressure. When we decrease the gas volume, the pressure gauge on our system shows us that the gas pressure increases. When we increase the gas volume, the gauge shows that the pressure goes down.

OBJECTIVE 10a

Decreased volume $\rightarrow$ Increased pressure

Increased volume $\rightarrow$ Decreased pressure

OBJECTIVE 10a

For an ideal gas (in which the particles occupy no volume and experience no attractions or repulsions), gas pressure and volume are inversely proportional. This means that if the temperature and moles of gas are constant and if the volume is decreased to one-half its original value, the pressure of the gas will double. If the volume is doubled, the pressure decreases to one-half its original value. The following expression summarizes this inverse relationship:[1]

$$P \propto \frac{1}{V} \quad \text{if } n \text{ and } T \text{ are constant}$$

OBJECTIVE 10a

Real gases deviate somewhat from this mathematical relationship, but the general trend of increased pressure with decreased volume (or decreased pressure with increased volume) is true for any gas.

The observation that *the pressure of an ideal gas is inversely proportional to the volume it occupies if the moles of gas and the temperature are constant* is a statement of **Boyle's Law.** This relationship can be explained in the following way. When the volume of the chamber decreases but the moles of gas remain constant, there is an increase in the concentration (moles per liter) of the gas. This leads to an increase in the number of particles near any given area of the container walls at any time and to an increase in the number of collisions against the walls per unit area in a given time. More collisions mean an increase in the force per unit area—or pressure—of the gas. The logic sequence presented in Figure 13.3 on the next page summarizes this explanation. The arrows in the logic sequence can be read as "leads to." Take the time to read the sequence carefully to confirm that each phrase leads logically to the next.

When this boy stomps on his rocket launcher, the volume of air inside it decreases, increasing the pressure enough to shoot his rocket toward the moon.

[1]The symbol $\alpha$ represents "is proportional to." The statement $x \alpha y$ means that if $y$ doubles, then $x$ doubles or if $y$ is cut in half, $x$ becomes one-half its original value.

**Figure 13.3**
**Relationship Between Volume and Pressure**
Decreased volume leads to increased pressure if the moles of gas and the temperature are constant.

OBJECTIVE 10a

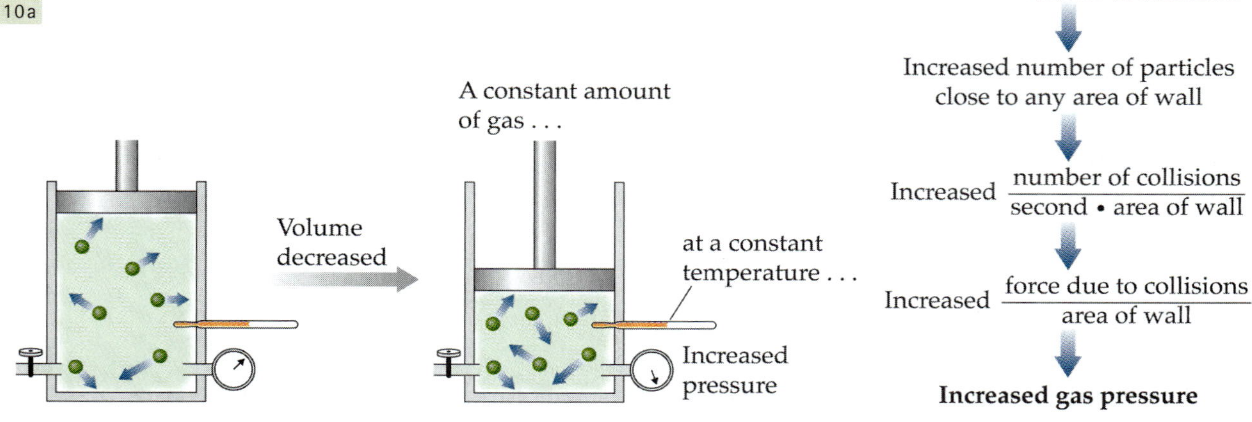


## The Relationship Between Pressure and Temperature

OBJECTIVE 10b

In order to examine the relationship between pressure and temperature, we must adjust our demonstration apparatus so that the other two properties (moles of gas and volume) are held constant. This can be done by locking the piston so that it cannot move and closing the valve tightly so that no gas leaks in or out (Figure 13.4). When the temperature of a gas trapped inside the chamber is increased, the measured pressure increases. When the temperature is decreased, the pressure decreases.

OBJECTIVE 10b

Increased temperature → Increased pressure

Decreased temperature → Decreased pressure

We can explain the relationship between temperature and pressure using our model for gas. Increased temperature means increased motion of the particles. If the particles are moving faster in the container, they will collide with the walls more often and with greater force per collision. This leads to a greater overall force pushing on the walls and to a greater force per unit area—or pressure (Figure 13.4).

If the gas is behaving like an ideal gas, a doubling of the Kelvin temperature doubles the pressure. If the temperature decreases to 50% of the original Kelvin temperature, the pressure decreases to 50% of the original pressure. This relationship can be expressed by saying that *the pressure of an ideal gas is directly proportional to the Kelvin temperature of the gas if the volume and moles of gas are constant*. This relationship is sometimes called **Gay-Lussac's Law.**

OBJECTIVE 10b

$P \propto T$ if $n$ and $V$ are constant

**Figure 13.4**
**Relationship Between Temperature and Pressure**
Increased temperature leads to increased pressure if the moles of gas and the volume are constant.

OBJECTIVE 10b

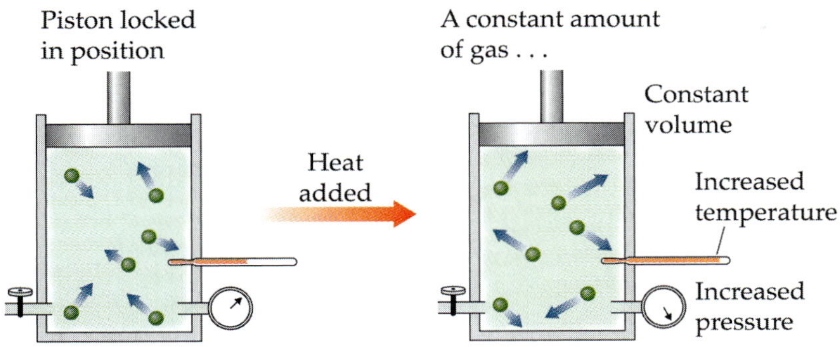

Piston locked in position

Heat added

A constant amount of gas . . .

Constant volume

Increased temperature

Increased pressure

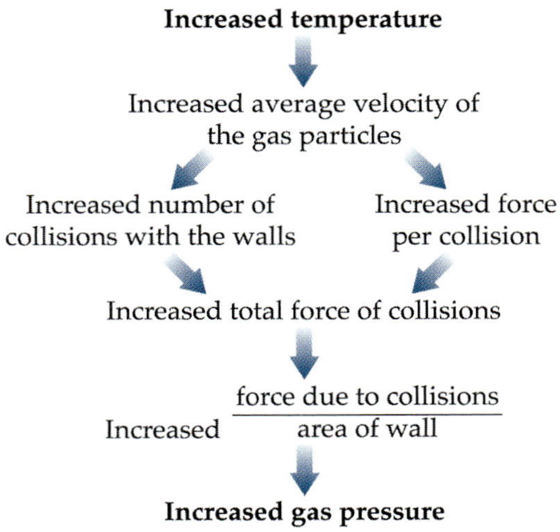

**Increased temperature**

Increased average velocity of the gas particles

Increased number of collisions with the walls

Increased force per collision

Increased total force of collisions

Increased $\frac{\text{force due to collisions}}{\text{area of wall}}$

**Increased gas pressure**

## The Relationship Between Volume and Temperature

Consider the system shown in Figure 13.5 on the next page. To demonstrate the relationship between temperature and volume of gas, we must keep the moles of gas and the gas pressure constant. If our valve is closed and our system has no leaks, the moles of gas are constant. We keep the gas pressure constant by allowing the piston to move freely throughout our experiment, because then it will adjust to keep the pressure pushing on it from the inside equal to the atmospheric pressure pushing on it from the outside. The atmospheric pressure is the pressure in the air outside the container, which acts on the top of the piston because of the force of collisions between particles in the air and the surface of the piston. We can assume that it is constant throughout our experiment.

OBJECTIVE 10c

If we increase the temperature, the piston in our apparatus moves up, increasing the volume occupied by the gas. A decrease in temperature leads to a decrease in volume.

OBJECTIVE 10c

Increased temperature $\rightarrow$ Increased volume

Decreased temperature $\rightarrow$ Decreased volume

The increase in temperature of the gas leads to an increase in the average velocity of the gas particles, which leads in turn to more collisions with the walls of the container and a greater force per collision. This greater force acting on the walls of the container leads to an *initial* increase in the gas pressure. Thus the increased temperature of our gas creates an internal pressure, acting on the bottom of the piston, that is greater than the external pressure acting on the top of the piston. The greater internal pressure causes the piston to move up, increasing the volume of the chamber. The increased volume leads to a decrease in gas pressure in the container, until the internal pressure is once again equal to the constant external pressure (Figure 13.5). Similar reasoning can be used to explain why decreased

OBJECTIVE 10c

temperature leads to decreased volume when moles of gas and pressure are held constant.

For an ideal gas, *volume and temperature described in kelvins are directly proportional if moles of gas and pressure are constant.* This is a statement of **Charles's Law.**

OBJECTIVE 10c

$$V \propto T \quad \text{if } n \text{ and } P \text{ are constant}$$

**Figure 13.5**
**Relationship Between Temperature and Volume**
Increased temperature leads to increased volume if the moles of gas and pressure are constant.

OBJECTIVE 10c

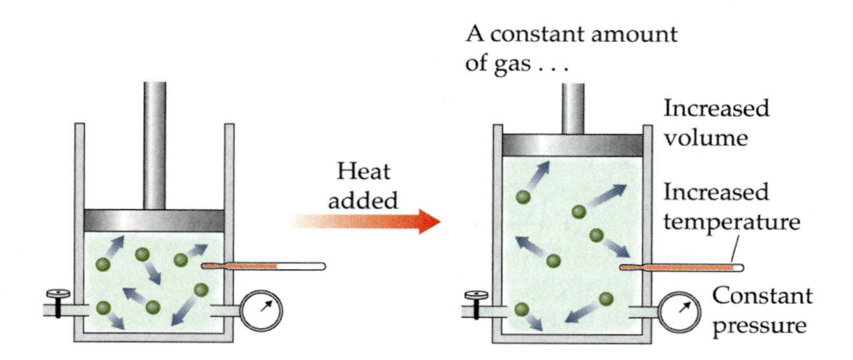


**Increased temperature**
↓
Increased average velocity of the gas particles
↙ ↘
Increased number of collisions with the walls | Increased force per collision
↘ ↙
Initial increase in force per area or pressure
↓
Inside pressure > external pressure
↓
Container expands → **Increased volume**
↓
Decreased pressure until the inside pressure equals the external pressure

Placing the balloon in liquid nitrogen lowers the temperature of the gas and causes an initial decrease in pressure. With its internal pressure now lower than the pressure of the air outside, the balloon shrinks to a much smaller volume.

## The Relationship Between Moles of Gas and Pressure

To explore the relationship between gas pressure and moles of gas, we could set up our experimental system as shown in Figure 13.6. The volume is held constant by locking the piston so that it cannot move. The temperature is kept constant by allowing heat to flow in or out of the cylinder in order to keep the temperature of the gas in the cylinder equal to the external temperature. When the moles of gas are increased by adding gas through the valve on the left of the cylinder, the pressure gauge shows an increase in pressure. When gas is allowed to escape from

OBJECTIVE 10d

the valve, the decrease in the moles of gas causes a decrease in the pressure of the gas.

Increased moles of gas → Increased pressure

OBJECTIVE 10d

Decreased moles of gas → Decreased pressure

The increase in the moles of gas in the container leads to an increase in the number of collisions with the walls per unit time. This leads to an increase in the force per unit area—that is, to an increase in gas pressure.

*If the temperature and the volume of an ideal gas are held constant, the moles of gas in a container and the gas pressure are directly proportional.*

$P \propto n$ if $T$ and $V$ are constant

OBJECTIVE 10d

**Figure 13.6**
**Relationship Between Moles of Gas and Pressure**
Increased moles of gas lead to increased pressure if the temperature and volume are constant.

OBJECTIVE 10d

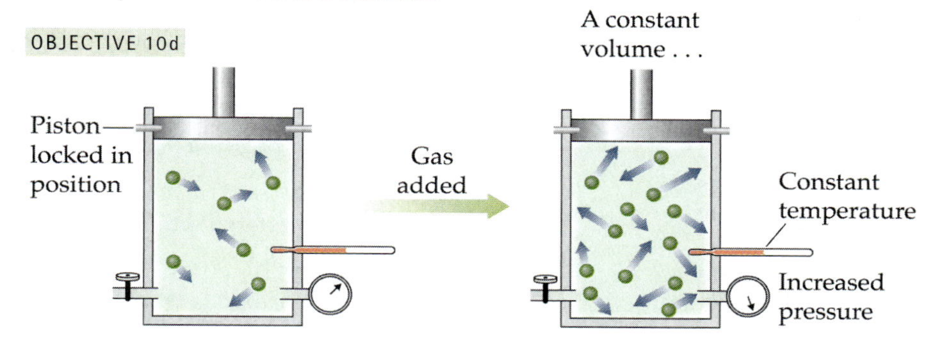


**Increased moles of gas**
↓
Increased number of collisions with the walls
↓
Increased total force of collisions
↓
**Increased gas pressure**

## The Relationship Between Moles of Gas and Volume

Figure 13.7 on the next page shows how the relationship between moles of gas and volume can be demonstrated using our apparatus. Temperature is held constant by allowing heat to move into or out of the system, thus keeping the internal temperature equal to the constant external temperature. Pressure is held constant by allowing the piston to move freely to keep the internal pressure equal to the atmospheric pressure. When we increase the moles of gas in the cylinder by adding gas through the valve on the left of the apparatus, the piston rises, increasing the volume available to the gas. If the gas is allowed to escape from the valve, the volume decreases again.

Increased moles of gas → Increased volume

OBJECTIVE 10e

Decreased moles of gas → Decreased volume

The explanation why an increase in the moles of gas increases volume starts with the recognition that the increase in the number of gas particles

When this person blows more air into the balloon, the increase in the moles of gas initially leads to an increased pressure. Because the internal pressure is now greater than the pressure of the air outside the balloon, the balloon expands to a larger volume.

results in more collisions per second against the walls of the container. The greater force due to these collisions creates an initial increase in the force per unit area—or gas pressure—acting on the walls. This will cause the piston to rise, increasing the gas volume and decreasing the pressure until the internal and external pressure are once again equal (Figure 13.7). Take a minute or two to work out a similar series of steps to explain why a decrease in the moles of gas leads to decreased volume.

OBJECTIVE 10e

The relationship between moles of an ideal gas and volume is summarized by **Avogadro's Law,** which states that *the volume and moles of gas are directly proportional if the temperature and pressure are constant.*

$$V \propto n \quad \text{if } T \text{ and } P \text{ are constant}$$

**Figure 13.7**
**Relationship Between Moles of Gas and Volume**
Increased moles of gas lead to increased volume if the temperature and pressure are constant.

OBJECTIVE 10e

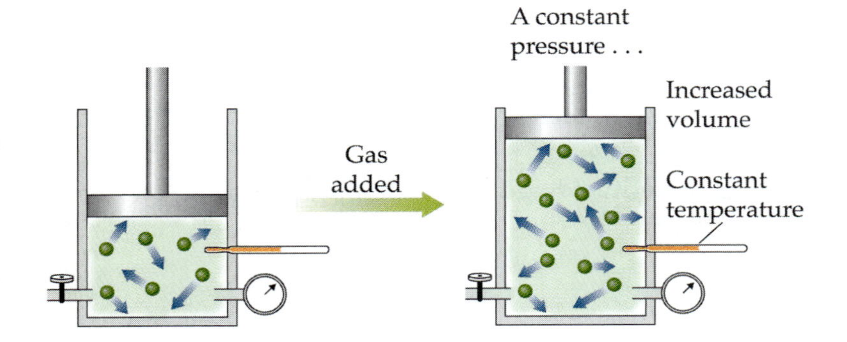


**Increased moles of gas**
↓
Increased number of collisions with the walls
↓
Increased total force of collisions
↓
Initial increase in force per area or pressure
↓
Inside pressure > external pressure
↓
Container expands → **Increased volume**
↓
Decreased pressure until the inside pressure equals the external pressure

## Gases and the Internal Combustion Engine

Now we are ready to apply our model for gases to a real situation. Because the changes that take place in a typical car engine illustrate many of the characteristics of gases, let's take a look at how the internal combustion engine works (Figure 13.8).

Liquid gasoline is a mixture of hydrocarbons, with from 5 to 12 carbon atoms in each molecule. It evaporates to form a gas, which is mixed with air and injected into the engine's cylinders (cylinder 1 in Figure 13.8). The movement of the engine's pistons turns a crankshaft that causes the piston in one of the cylinders containing the gasoline–air mixture to move up, compressing the gases and increasing the pressure of the gas mixture in the cylinder (cylinder 2 in Figure 13.8).

OBJECTIVE 11

A spark ignites the mixture of compressed gases in a cylinder, and the hydrocarbon compounds in the gasoline react with the oxygen in the air to form carbon dioxide gas and water vapor (cylinder 3 in Figure 13.8). A typical reaction is

$$2C_8H_{18}(g) + 25O_2(g) \rightarrow 16CO_2(g) + 18H_2O(g) + \text{Energy}$$

2 mol + 25 mol = 27 mol    16 mol + 18 mol = 34 mol

In this representative reaction, a total of 27 moles of gas are converted into 34 moles of gas. The increase in moles of gas leads to an increase in the number of collisions per second with the walls of the cylinder, which creates greater force acting against the walls—that is, a greater gas pressure in the cylinder. The pressure is increased even more by the increase in the temperature of the gas due to the energy released in the reaction. The increased temperature increases the average velocity of the gas particles, which leads to more frequent collisions with the walls and to a greater average force per collision.

OBJECTIVE 11

The increased pressure of the gas pushes the piston down, thus increasing the volume available to the gas in the cylinder (cylinder 4 in Figure 13.8). This movement of the pistons turns the crankshaft, which, through a series of mechanical connections, turns the wheels of the car.

OBJECTIVE 11

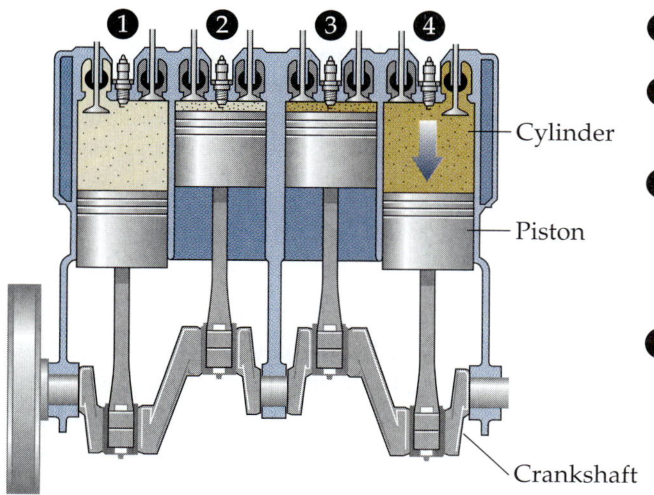


1 Gaseous gasoline mixed with air moves into the cylinder.

2 Decreased volume leads to increased gas pressure inside the cylinder.

3 The combustion of the gasoline leads to an increase in moles of gas, which also causes the gas pressure to increase. The reaction is exothermic, so the temperature of the product gases increases, contributing to the increased gas pressure.

4 The increased pressure pushes the piston down, turning the crankshaft and ultimately the car's wheels.

OBJECTIVE 11

**Figure 13.8**
Gases and the Internal Combustion Engine

## Explanations for Other Real-World Situations

The relationships between gas properties can be used to explain how we breathe in and out. When the muscles of your diaphragm contract, your ribs expand, and the volume of your lungs increases. This change leads to a decrease in the number of particles per unit volume inside the lungs, leaving fewer particles near any given area of the inner surface of the lungs. Fewer particles mean fewer collisions per second per unit area of lungs and a decrease in force per unit area, or gas pressure. During quiet, normal breathing, this increase in volume decreases the pressure in the

OBJECTIVE 12

lungs to about 0.4 kilopascal lower than the atmospheric pressure. As a result, air moves into the lungs faster than it moves out, bringing in fresh oxygen. When the muscles relax, the lungs return to their original, smaller volume, causing the pressure in the lungs to increase to about 0.4 kilopascal above atmospheric pressure. Air now moves out of the lungs faster than it moves in (Figure 13.9).

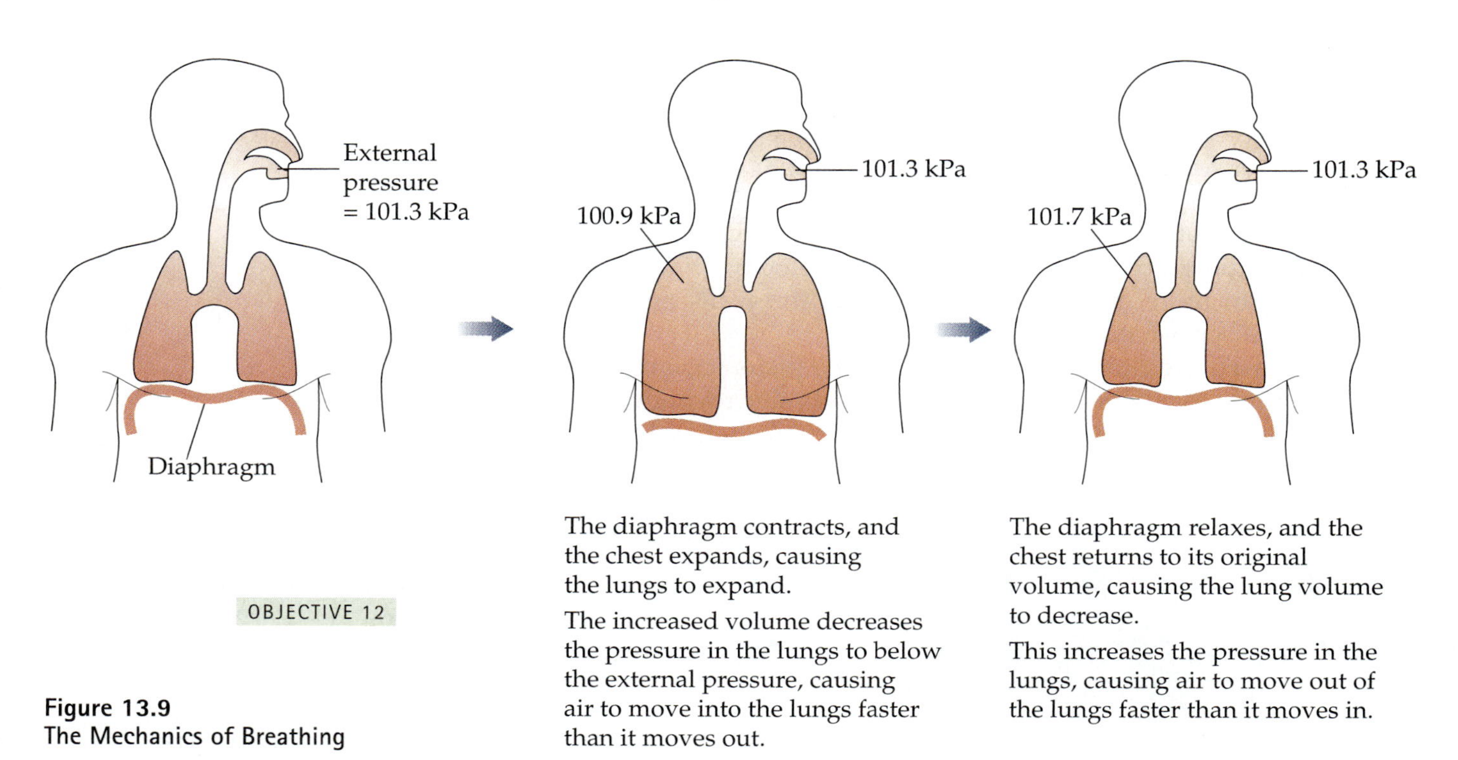

External
pressure
= 101.3 kPa
Diaphragm
100.9 kPa
101.3 kPa
101.7 kPa
101.3 kPa
The diaphragm contracts, and the chest expands, causing the lungs to expand.
The increased volume decreases the pressure in the lungs to below the external pressure, causing air to move into the lungs faster than it moves out.
The diaphragm relaxes, and the chest returns to its original volume, causing the lung volume to decrease.
This increases the pressure in the lungs, causing air to move out of the lungs faster than it moves in.


OBJECTIVE 12

**Figure 13.9**
The Mechanics of Breathing

The volume of this balloon will increase as it moves up into the atmosphere.

Let's return to the gas-related issues discussed at Lilia's family dinner. When Lilia's brother John adds helium gas to one of his instrument-carrying balloons, its volume increases. Increased moles of gas lead to increased gas pressure, making the internal pressure greater than the external pressure acting on the outside surface of the balloon. The increase of internal pressure leads to an expansion of the balloon. When the balloon is released and rises into the air, the concentration of gases *outside* the balloon decreases with increased distance from the earth, leading to a decrease in the atmospheric pressure acting on the *outside* of the balloon. The greater pressure *inside* the balloon causes the balloon to expand as it rises. This increase in volume is diminished somewhat by the slight loss of gas from tiny holes in the balloon and by the general decrease in temperature with increased distance from the earth.

When the temperature is increased for the gases in the reaction vessel that Lilia's sister Rebecca is helping design, the gas pressure rises. This is due to an increase in the average velocity of the particles and the resulting increase in (1) the rate of collisions of gas particles with the constant area of the walls and (2) the average force per collision. Rebecca's pressure release valve allows gas to escape if the pressure gets to a certain level. The decrease in moles of gas when gas escapes from the valve keeps the pressure below dangerous levels.

To answer Ted's question about how to calculate gas density and to see how Amelia can estimate the number of tubes she can fill for her neon-light sculptures, we need to continue on to the next section.

## 13.2 Ideal Gas Calculations

This section shows how to do calculations like those necessary to answer Ted's and Amelia's questions about gas density and volume, and in addition, it considers some of the gas-related issues that Lilia's sister Rebecca and her co-workers need to resolve. For one thing, the design team needs to know the amount of gas that they can safely add to their reaction vessel, and then Rebecca needs to determine the maximum temperature at which the reaction can be run without causing the pressure of that amount of gas to reach dangerous levels.

All these calculations, and others like them, can be done with the aid of two useful equations that we will now derive from the relationships described in Section 13.1.

### Calculations Using the Ideal Gas Equation

We discovered in Section 13.1 that the pressure of an ideal gas is directly proportional to the moles of gas, directly proportional to the temperature, and inversely proportional to the volume of the container.

$$P \propto n \qquad \text{if } T \text{ and } V \text{ are constant}$$

$$P \propto T \qquad \text{if } n \text{ and } V \text{ are constant}$$

$$P \propto \frac{1}{V} \qquad \text{if } T \text{ and } n \text{ are constant}$$

These three relationships can be summarized in a single equation:

$$P \propto \frac{nT}{V}$$

Another way to express the same relationship is

$$P = (\text{a constant})\frac{nT}{V}$$

The constant in this equation is the same for all ideal gases. It is called the **universal gas constant** and is represented by the symbol ***R***. The value of $R$

depends on the units of measure one wishes to use in a given calculation. The two choices that follow show $R$ for different pressure units (atmospheres, atm, and kilopascals, kPa).

OBJECTIVE 4

$$R = \frac{0.082058 \text{ L} \cdot \text{atm}}{\text{K} \cdot \text{mol}} \quad \text{or} \quad \frac{8.3145 \text{ L} \cdot \text{kPa}}{\text{K} \cdot \text{mol}}$$

Substituting $R$ for "a constant" and rearranging the equation yield the *ideal gas equation* (often called the ideal gas law) in the form that is most commonly memorized and written:

OBJECTIVE 15

$$PV = nRT$$

Because we will often be interested in the masses of gas samples, it is useful to remember an expanded form of the ideal gas equation that uses mass in grams ($g$) divided by molar mass ($M$) instead of moles ($n$).

$$n = \text{moles} = \frac{\text{grams}}{\frac{\text{grams}}{\text{mole}}} = \frac{\text{mass in grams}}{\text{molar mass}} = \frac{g}{M}$$

OBJECTIVE 16

$$PV = \frac{g}{M}RT$$

The following sample study sheet describes how calculations can be done using the two forms of the ideal gas equation.

## Study Sheet 13.1

### Using the Ideal Gas Equation

OBJECTIVE 15

OBJECTIVE 16

**Tip-off** The usual tip-off that you can use the ideal gas equation to answer a question is that you are given three properties of a sample of gas and asked to calculate the fourth. A more general tip-off is that only one gas is mentioned and there are no changing properties.

**General Steps**

**Step 1** Assign variables to the values given and to the value that is unknown. Use $P$ for pressure, $V$ for volume, $n$ for moles, $T$ for temperature, $g$ for mass, and $M$ for molar mass.

**Step 2** Write the appropriate form of the Ideal Gas Equation.

- If the number of particles is given or desired in moles, use the most common form of the ideal gas equation.

$$PV = nRT$$

$$R = \frac{0.082058 \text{ L} \cdot \text{atm}}{\text{K} \cdot \text{mol}} \quad \text{or} \quad \frac{8.3145 \text{ L} \cdot \text{kPa}}{\text{K} \cdot \text{mol}}$$

- If mass or molar mass is given or desired, use the expanded form of the ideal gas equation.

$$PV = \frac{g}{M} RT$$

**STEP 3** Rearrange the equation to isolate the unknown.

**STEP 4** Plug in the known values, including units. Be sure to use Kelvin temperatures.

**STEP 5** Make any necessary unit conversions and cancel your units.

**STEP 6** Calculate your answer and report it to the correct number of significant figures and with the correct unit.

**EXAMPLE** See Examples 13.1 and 13.2.

## EXAMPLE 13.1 Using the Ideal Gas Equation

OBJECTIVE 15

Incandescent light bulbs "burn out" because their tungsten filament evaporates, weakening the thin wire until it breaks. Argon gas is added inside the bulbs to reduce the rate of evaporation. (Argon is chosen because, as a noble gas, it will not react with the components of the bulb and because it is easy to obtain in significant quantities. It is the third most abundant element in air.) What is the pressure in atmospheres of $3.4 \times 10^{-3}$ mole of argon gas in a 75-mL incandescent light bulb at 20 °C?

***Solution***

We are given three of the four properties of ideal gases (moles, volume, and temperature), and we are asked to calculate the fourth (pressure). Therefore, we use the ideal gas equation to solve this problem.

**STEP 1** We assign variables to the values that we are given and to the unknown value. Remember to use Kelvin temperatures in gas calculations.

$$P = ? \quad n = 3.4 \times 10^{-3}\ \text{mol} \quad V = 75\ \text{mL}$$

$$T = 20\ °\text{C} + 273.15 = 293\ \text{K}$$

**Step 2** We pick the appropriate form of the ideal gas equation. Because moles of gas are mentioned in the problem, we use

$$PV = nRT$$

**Step 3** We rearrange our equation to isolate the unknown property.

$$P = \frac{nRT}{V}$$

**Step 4** We plug in the values that are given, including their units. We use the value for $R$ that contains the unit we want to calculate, atmospheres.

$$P = \frac{3.4 \times 10^{-3}\,\cancel{\text{mol}}\left(\dfrac{0.082058\ \text{L} \cdot \text{atm}}{\cancel{\text{K}} \cdot \cancel{\text{mol}}}\right)293\,\cancel{\text{K}}}{75\ \text{mL}}$$

**Steps 5 & 6** If the units do not cancel to yield the desired unit, we do the necessary unit conversions to make them cancel. In the example that follows, we need to convert 75 mL to liters so that the volume units cancel. We finish the problem by calculating the answer and reporting it with the correct number of significant figures and the correct unit.

$$P = \frac{3.4 \times 10^{-3}\,\cancel{\text{mol}}\left(\dfrac{0.082058\,\cancel{\text{L}} \cdot \text{atm}}{\cancel{\text{K}} \cdot \cancel{\text{mol}}}\right)293\,\cancel{\text{K}}}{75\,\cancel{\text{mL}}}\left(\frac{10^3\,\cancel{\text{mL}}}{\cancel{\text{L}}}\right) = \mathbf{1.1\ atm}$$

To see a practical application of calculations like Example 13.1, let's take a closer look at one of the gas-related issues that Lilia's sister Rebecca and her co-workers are dealing with. The apparatus Rebecca is helping to design will be used in the first step of the commercial process that makes nitric acid for fertilizers. In this step, the gases ammonia and oxygen are converted into gaseous nitrogen monoxide and water.

$$4NH_3(g) + 5O_2(g) \rightarrow 4NO(g) + 6H_2O(g)$$

When the team of chemists and chemical engineers meet to design the new process, they begin with some preliminary calculations. First, they want to determine how much gas they can safely add to their reaction vessel. They know that the optimum conditions for this reaction are a temperature of about 825 °C and a pressure of about 700 kPa, and they have been told by the plant architect that the maximum volume of the reaction vessel will be 2500 $m^3$. Note that the value for $R$ that contains the unit kilopascal is used.

$$PV = nRT$$

$$n = \frac{PV}{RT} = \frac{700\ \text{kPa}\ (2500\ \text{m}^3)}{\left(\frac{8.3145\ \text{L} \cdot \text{kPa}}{\text{K} \cdot \text{mol}}\right)(825 + 273.15)\ \text{K}}\left(\frac{10^3\ \text{L}}{1\ \text{m}^3}\right)$$

$$= \mathbf{1.92 \times 10^5\ mol}$$

As you practice using the ideal gas equation, you may be tempted to save time by plugging in the numbers without including their accompanying units. Be advised, however, that it is always a good idea to include the units as well. If the units cancel to yield a reasonable unit for the unknown property, you can feel confident that

- You picked the correct equation.
- You did the algebra correctly to solve for your unknown.
- You have made the necessary unit conversions.

If the units do not cancel to yield the desired unit, check to be sure that you have done the algebra correctly and that you have made the necessary unit conversions.

## Example 13.2 Using the Ideal Gas Equation

OBJECTIVE 16

Incandescent light bulbs were described in Example 13.1. At what temperature will 0.0421 g of Ar in a 23.0-mL incandescent light bulb have a pressure of 952 mmHg?

***Solution***

$T = ?$ $g = 0.0421\ \text{g}$ $V = 23.0\ \text{mL}$ $P = 952\ \text{mmHg}$

The tip-off that this is an ideal gas equation problem is that we are given three properties of a gas and are asked to calculate the fourth.

Because the gas's mass is given, we choose the expanded form of the ideal gas equation. We rearrange the equation to isolate the unknown variable, plug in given values, and cancel our units. In order to cancel the pressure and volume units, we must convert 952 mmHg into atmospheres and convert 23.0 mL into liters. When this is done, the units cancel to yield an answer in kelvins. Because the kelvin is a reasonable temperature unit, we can assume that we have picked the correct equation, have done the algebraic manipulation correctly, and have made all of the necessary unit conversions.

$$PV = \frac{g}{M}RT$$

$$T = \frac{PVM}{gR} = \frac{952\ \text{mmHg}\ (23.0\ \text{mL})\ 39.948\ \frac{\text{g}}{\text{mol}}}{0.0421\ \text{g}\left(\frac{0.082058\ \text{L} \cdot \text{atm}}{\text{K} \cdot \text{mol}}\right)}\left(\frac{1\ \text{atm}}{760\ \text{mmHg}}\right)\left(\frac{1\ \text{L}}{10^3\ \text{mL}}\right)$$

$$= \mathbf{333\ K,\ or\ 60\ °C}$$

## Example 13.3 Using the Ideal Gas Equation

OBJECTIVE 17

In the incandescent light bulb described in Example 13.1, what is the density of argon gas at 80.8 °C and 131 kPa?

***Solution***

Because only one gas is mentioned and because there are no changing properties, we recognize this as an ideal gas equation problem.

Density is mass divided by volume. Using variables from the expanded form of the ideal gas equation, it can be expressed as $g/V$.

$$\frac{g}{V} = ? \quad P = 131 \text{ kPa} \quad T = 80.8\ ^\circ\text{C} + 273.15 = 354.0 \text{ K}$$

$$PV = \frac{g}{M} RT$$

To save a little time, we can use the value for $R$ that contains the pressure unit given in the problem, kilopascal.

$$\frac{g}{V} = \frac{PM}{RT} = \frac{131 \cancel{\text{kPa}} \left(\frac{39.948 \text{ g}}{\cancel{\text{mol}}}\right)}{\left(\frac{8.3145 \text{ L} \cdot \cancel{\text{kPa}}}{\cancel{\text{K}} \cdot \cancel{\text{mol}}}\right)(354.0 \cancel{\text{K}})} = \mathbf{1.78\ g/L}$$

Alternatively, we could use the same value for $R$ that we used in Examples 13.1 and 13.2, along with a unit conversion.

$$\frac{g}{V} = \frac{PM}{RT} = \frac{131 \cancel{\text{kPa}} \left(\frac{39.948 \text{ g}}{\cancel{\text{mol}}}\right)}{\left(\frac{0.082058 \text{ L} \cdot \cancel{\text{atm}}}{\cancel{\text{K}} \cdot \cancel{\text{mol}}}\right)(354.0 \cancel{\text{K}})} \left(\frac{1 \cancel{\text{atm}}}{101.325 \cancel{\text{kPa}}}\right) = \mathbf{1.78\ g/L}$$

A disastrous series of crashes in the 1992 Indianapolis 500 race has been traced to an unfortunate combination of factors, including an unusually and unexpectedly low temperature, which led to a higher-than-expected air density that day. A change in air density causes changes in aerodynamic forces, engine power, and brake performance. If such a change is anticipated, car designers and mechanics, such as Lilia's brother Ted, can make adjustments. For example, because of its altitude, Denver has about 15% lower air density than does a city at sea level. Therefore, Indy cars that race in Denver are fitted with larger-than-normal brakes to compensate for the lower air resistance and greater difficulty stopping. Ted can do calculations similar to those we saw in Example 13.3 to determine the density of air on different days and at different racetracks. The following set of equations calculate the density of air at 762 mmHg and 32 °C and at 758 mmHg and 4 °C. (The average molar mass for the gases in air is 29 g/mol.)

$$PV = \frac{g}{M}RT$$

$$\frac{g}{V} = \frac{PM}{RT} = \frac{762\ \text{mmHg}\left(\dfrac{29\ \text{g}}{1\ \text{mol}}\right)}{\left(\dfrac{0.082058\ \text{L} \cdot \text{atm}}{\text{K} \cdot \text{mol}}\right)305\ \text{K}}\left(\frac{1\ \text{atm}}{760\ \text{mmHg}}\right) = \mathbf{1.2\ g/L}$$

OBJECTIVE 17

$$\frac{g}{V} = \frac{PM}{RT} = \frac{758\ \text{mmHg}\left(\dfrac{29\ \text{g}}{1\ \text{mol}}\right)}{\left(\dfrac{0.082058\ \text{L} \cdot \text{atm}}{\text{K} \cdot \text{mol}}\right)277\ \text{K}}\left(\frac{1\ \text{atm}}{760\ \text{mmHg}}\right) = \mathbf{1.3\ g/L}$$

### EXERCISE 13.1 Using the Ideal Gas Equation

OBJECTIVE 15 OBJECTIVE 16 OBJECTIVE 17

Krypton, Kr, gas does a better job than argon of slowing the evaporation of the tungsten filament in an incandescent light bulb. Because of its higher cost, however, krypton is used only when longer life is considered to be worth the extra expense.

a. How many moles of krypton gas must be added to a 175-mL incandescent light bulb to yield a gas pressure of 117 kPa at 21.6 °C?

b. What is the volume of an incandescent light bulb that contains 1.196 g of Kr at a pressure of 1.70 atm and a temperature of 97 °C?

c. What is the density of krypton gas at 18.2 °C and 762 mmHg?

## When Properties Change

Another useful equation, which is derived from the ideal gas equation, can be used to calculate changes in the properties of a gas. In the first step of the derivation, the ideal gas equation is rearranged to isolate $R$.

$$\frac{PV}{nT} = R$$

In this form, the equation shows that no matter how the pressure, volume, moles of gas, or temperature of an ideal gas may change, the other properties will adjust so that the ratio of $PV/nT$ always remains the same.

$$P_1V_1 = n_1RT_1 \quad \text{so} \quad \frac{P_1V_1}{n_1T_1} = R$$

$$P_2V_2 = n_2RT_2 \quad \text{so} \quad \frac{P_2V_2}{n_2T_2} = R$$

Therefore,

OBJECTIVE 18

$$\frac{P_1V_1}{n_1T_1} = \frac{P_2V_2}{n_2T_2}$$

$P_1$, $V_1$, $n_1$, and $T_1$ are the initial pressure, volume, moles of gas, and temperature.

$P_2$, $V_2$, $n_2$, and $T_2$ are the final pressure, volume, moles of gas, and temperature.

The equation above is often called the *combined gas law equation*. It can be used to do calculations such as those that will be required in Example 13.4 and Exercise 13.2, which follow the sample study sheet below.

## Sample Study Sheet 13.2

### Using the Combined Gas Law Equation

OBJECTIVE 18

**Tip-off** The problem requires calculating a value for a gas property that has changed. In other words, you are asked to calculate a new pressure, temperature, moles (or mass), or volume of gas, given sufficient information about the initial properties and the other final properties.

**General Steps**

**Step 1** Assign the variables $P$, $T$, $n$, and $V$ to the values you are given and to the unknown value. Use the subscripts 1 and 2 to show initial (1) and final (2) conditions.

**Step 2** Write out the combined gas law equation, but eliminate the variables for any constant properties. (You can assume that the properties not mentioned in the problem remain constant.) See Example 13.4.

**Step 3** Rearrange the equation to isolate the unknown property.

**Step 4** Plug in the values for the given properties.

**Step 5** Make any necessary unit conversions and cancel your units.

**Step 6** Calculate your answer and report it with the correct units and to the correct number of significant figures.

**Example** See Example 13.4.

## EXAMPLE 13.4 Using the Combined Gas Law Equation

OBJECTIVE 18

Neon gas in luminous tubes radiates red light—the original "neon light." The standard gas containers used to fill the tubes have a volume of 1.0 L and store neon gas at a pressure of 101 kPa at 22 °C. A typical luminous neon tube contains enough neon gas to exert a pressure of 1.3 kPa at 19 °C. If all the gas from a standard container is allowed to expand until it exerts a pressure of 1.3 kPa at 19 °C, what will its final volume be? If Lilia's sister Amelia is adding this gas to luminous tubes that have an average volume of 500 mL, what is the approximate number of tubes that she can fill?

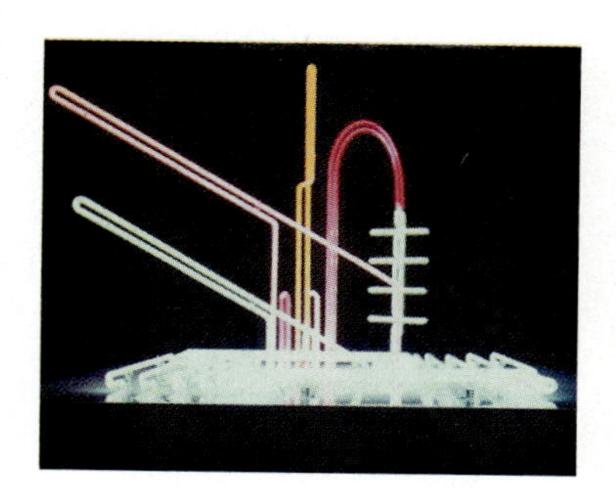

Neon tubes

***Solution***

We recognize this as a combined gas law problem because it requires calculating a value for a gas property that has changed. In this case, that property is volume.

**STEP 1** We assign variables to the given values and to the unknown.

$$V_1 = 1.0\text{ L} \qquad P_1 = 101\text{ kPa} \qquad T_1 = 22\text{ °C} + 273.15 = 295\text{ K}$$

$$V_2 = ? \qquad P_2 = 1.3\text{ kPa} \qquad T_2 = 19\text{ °C} + 273.15 = 292\text{ K}$$

**STEP 2** We write the combined gas law equation, eliminating variables for properties that are constant. Because moles of gas are not mentioned, we assume that they are constant ($n_1 = n_2$).

$$\frac{P_1V_1}{n_1T_1} = \frac{P_2V_2}{n_2T_2} \quad \text{becomes} \quad \frac{P_1V_1}{T_1} = \frac{P_2V_2}{T_2}$$

**STEPS 3 & 4** We rearrange the equation to solve for our unknown and plug in the given values.

$$V_2 = V_1\left(\frac{T_2}{T_1}\right)\left(\frac{P_1}{P_2}\right) = 1.0\text{ L}\left(\frac{292\text{ K}}{295\text{ K}}\right)\left(\frac{101\text{ kPa}}{1.3\text{ kPa}}\right)$$

**STEP 5** Our units cancel to yield liters, which is a reasonable volume unit, so we do not need to make any unit conversions.

**STEP 6** We finish the problem by calculating our answer and reporting it to the correct number of significant figures.

$$V_2 = V_1\left(\frac{T_2}{T_1}\right)\left(\frac{P_1}{P_2}\right) = 1.0\text{ L}\left(\frac{292\text{ K}}{295\text{ K}}\right)\left(\frac{101\text{ kPa}}{1.3\text{ kPa}}\right) = \mathbf{77\text{ L}}$$

We can now answer Amelia's question about approximately how many neon tubes she can expect to fill from each neon cylinder.

$$?\text{ tubes} = 77\text{ L}\left(\frac{10^3\text{ mL}}{1\text{ L}}\right)\left(\frac{1\text{ tube}}{500\text{ mL}}\right) = \mathbf{1.5 \times 10^2\text{ tubes, or about 150}}$$

Let's look at another gas-related issue that Lilia's sister Rebecca needs to consider in designing the pressure valve for the reaction vessel at her chemical plant. She knows that, for safety reasons, the overall pressure must be kept below 1000 kPa, and she knows that the most likely cause of increased pressure is increased temperature. To get an idea of how high the temperature can go safely, she could use the combined gas law equation to calculate the temperature at which the pressure of the gas will reach 1000 kPa if the initial temperature was 825 °C (1098 K) and the initial pressure was 700.0 kPa. The reaction vessel has a constant volume, and we can assume that it has no leaks. Therefore, the volume and moles of gas remain constant.

$$\frac{P_1V_1}{n_1T_1} = \frac{P_2V_2}{n_2T_2} \quad \text{or} \quad \frac{P_1}{T_1} = \frac{P_2}{T_2} \quad \text{because} \quad V_2 = V_1 \text{ and } n_1 = n_2$$

$$T_2 = T_1\left(\frac{P_2}{P_1}\right) = 1098\text{ K}\left(\frac{1000\text{ kPa}}{700.0\text{ kPa}}\right) = \mathbf{1569\ K,\ or\ 1295\ °C}$$

Although Rebecca designs her pressure release valve to allow gas to escape when the pressure reaches 1000 kPa, she hopes it will never need to be used. To avoid the release of gases, she informs the rest of her team that they should be careful to design the process to keep the temperature well below 1295 °C.

### EXERCISE 13.2 Using the Combined Gas Law Equation

OBJECTIVE 18

A helium weather balloon is filled in Monterey, California, on a day when the atmospheric pressure is 102 kPa and the temperature is 18 °C. Its volume under these conditions is $1.6 \times 10^4$ L. Upon being released, it rises to an altitude where the temperature is −8.6 °C, and its volume increases to $4.7 \times 10^4$ L. Assuming that the internal pressure of the balloon equals the atmospheric pressure, what is the pressure at this altitude?

## 13.3 Equation Stoichiometry and Ideal Gases

Chemists commonly want to convert from the amount of one substance in a chemical reaction to an amount of another substance in that reaction. For example, they might want to calculate the maximum amount of a product that can be formed from a certain amount of a reactant, or they might want to calculate the minimum amount of one reactant that would be needed to use up a known amount of another reactant. Chapter 10 showed how to do conversions like these. In Example 10.1, we calculated the minimum mass of water, in kilograms, necessary to react with $2.50 \times 10^4$ kg of tetraphosphorus decoxide, $P_4O_{10}$, in the reaction

$$P_4O_{10}(s) + 6H_2O(l) \rightarrow 4H_3PO_4(aq)$$

This kind of problem, an equation stoichiometry problem, is generally solved in the following steps, where the symbol # stands for "substance."

Measurable property of #1 → moles of #1 → moles of #2 → measurable property of #2

For substances that are pure solids and liquids, such as $P_4O_{10}$ and $H_2O$, mass is an easily measured property, so the specific steps used in solving Example 10.1 were

$$\text{Mass of } P_4O_{10} \rightarrow \text{moles of } P_4O_{10} \rightarrow \text{moles of } H_2O \rightarrow \text{mass of } H_2O$$

Amounts of gas, however, are more commonly described in terms of volume at a particular temperature and pressure. (It is far easier to measure the volume of a gas than to measure its mass.) Therefore, to solve equation stoichiometry problems involving gases, we must be able to convert between the volume of a gaseous substance and the corresponding number of moles. For example, if both substances involved in the equation stoichiometry problem are gases, the general steps would be

$$\text{Volume of gas 1} \rightarrow \text{moles of \#1} \rightarrow \text{moles of \#2} \rightarrow \text{volume of gas 2}$$

There are several ways to convert between moles of a gaseous substance and its volume. One approach is to use as a conversion factor the molar volume at STP. *STP* stands for "standard temperature and pressure." *Standard temperature* is 0 °C, or 273.15 K, and *standard pressure* is 1 atm, or 101.325 kPa, or 760 mmHg. The **molar volume,** or liters per mole of an ideal gas, at STP can be calculated from the ideal gas equation.

This box has a volume of 22.414 L, the volume of one mole of gas at standard temperature (273.15 K) and pressure (101.325 kPa). Three basketballs have approximately the same volume.

$$PV = nRT$$

$$\frac{V}{n} = \frac{RT}{P} = \frac{\left(\dfrac{8.3145 \text{ L} \cdot \cancel{\text{kPa}}}{\cancel{\text{K}} \cdot \text{mol}}\right)(273.15 \cancel{\text{K}})}{101.325 \cancel{\text{kPa}}} = \left(\frac{22.414 \text{ L}}{1 \text{ mol}}\right)_{\text{STP}}$$

Because the ideal gas equation applies to all ideal gases, the molar volume at STP applies to all gases that exhibit the characteristics of the ideal gas model. In equation stoichiometry, the molar volume at STP is used in much the way we use molar mass. Molar mass converts between moles and the measurable property of mass; molar volume at STP converts between moles and the measurable property of volume of gas. Note that although every substance has a different molar mass, all ideal gases have the same molar volume at STP. Example 13.5 provides a demonstration.

OBJECTIVE 19

## EXAMPLE 13.5 Gas Stoichiometry

OBJECTIVE 19

How many liters of carbon dioxide at STP will be formed from the complete combustion of 82.60 g of ethanol, $C_2H_5OH(l)$?

***Solution***

Because we are converting from units of one substance to units of another substance, both involved in a chemical reaction, we recognize this problem as an equation stoichiometry problem. We can therefore use dimensional analysis for our calculation.

Equation stoichiometry problems have at their core the conversion of moles of one substance into moles of another substance. The conversion factor that accomplishes this part of the calculation comes from the coefficients in the balanced equation. Although we have not been given the balanced equation for the combustion of ethanol, we can supply it ourselves by remembering (from Chapter 6) that when a hydrocarbon compound burns completely, all of its carbon forms carbon dioxide and all of its hydrogen forms water.

$$C_2H_5OH(l) + 3O_2(g) \rightarrow 2CO_2(g) + 3H_2O(l)$$

First, however, we convert from mass of $C_2H_5OH$ to moles, using the compound's molar mass. Then we set up the mole-to-mole conversion, using the molar ratio of ethanol to carbon dioxide derived from the coefficients in the balanced equation. The next step is the new one: We use the molar volume at STP to convert from moles of $CO_2$ to volume of $CO_2$ at STP. The sequence as a whole is

$$\text{Mass } C_2H_5OH \rightarrow \text{moles } C_2H_5OH \rightarrow \text{moles } CO_2 \rightarrow \text{volume of } CO_2 \text{ at STP}$$

$$?\ \text{L } CO_2 = 82.60\ \cancel{\text{g } C_2H_5OH}\left(\frac{1\ \cancel{\text{mol } C_2H_5OH}}{46.0692\ \cancel{\text{g } C_2H_5OH}}\right)\left(\frac{2\ \cancel{\text{mol } CO_2}}{1\ \cancel{\text{mol } C_2H_5OH}}\right)\left(\frac{22.414\ \text{L } CO_2}{1\ \cancel{\text{mol } CO_2}}\right)_{\text{STP}}$$

$$= \mathbf{80.37\ L\ CO_2}$$

The molar volume at STP—22.414 L—is useful for conversions as long as the gases are at 0 °C and 1 atm of pressure. When the temperature or pressure changes, the volume of a mole of gas will also change, so 22.414 L/mol cannot be used as a conversion factor for conditions other than STP. One way to convert between volume of gas and moles of gas at temperatures and pressures other than 0 °C and 1 atm is to use the ideal gas equation. We convert from volume to moles by solving the ideal gas equation for $n$ and plugging in the given values for $V$, $T$, and $P$:

$$PV = nRT \quad \text{leads to} \quad n = \frac{PV}{RT}$$

We convert from moles to volume by solving the ideal gas equation for $V$ and plugging in the given values for $n$, $T$, and $P$:

$$PV = nRT \quad \text{leads to} \quad V = \frac{nRT}{P}$$

This technique is demonstrated in Examples 13.6 and 13.7.

## EXAMPLE 13.6 Gas Stoichiometry for Conditions Other Than STP

OBJECTIVE 19

Ammonia is produced when nitrogen gas and hydrogen gas react at high pressure and temperature:

$$N_2(g) + 3H_2(g) \rightleftharpoons 2NH_3(g)$$

At intervals, the system is cooled to between −10 °C and −20 °C, causing some of the ammonia to liquefy so that it can be separated from the remaining nitrogen and hydrogen gases. The gases are then recycled to make more ammonia. An average ammonia plant might make 1000 metric tons of ammonia per day. When $4.0 \times 10^7$ L of hydrogen gas at 503 °C and 155 atm reacts with an excess of nitrogen, what is the maximum volume of gaseous ammonia that can be formed at 20.6 °C and 1.007 atm?

Approximately 70 million metric tons of ammonia are synthesized yearly in the United States at plants like this. About 80% of it is used to make fertilizers, and most of the rest is used to make explosives (such as nitroglycerine) and other chemicals.

***Solution***

This equation stoichiometry problem can be solved in three steps:

$$\text{Volume of } H_2(g) \rightarrow \text{moles } H_2 \rightarrow \text{moles } NH_3 \rightarrow \text{volume of } NH_3(g)$$

The first step is to use the ideal gas equation to convert from volume of $H_2$ to moles of $H_2$ under the initial conditions.

$$PV = nRT$$

$$n_{H_2} = \frac{PV}{RT} = \frac{155\,\cancel{\text{atm}}\,(4.0 \times 10^7\,\cancel{\text{L}})}{\dfrac{0.082058\,\cancel{\text{L}} \cdot \cancel{\text{atm}}}{\cancel{\text{K}} \cdot \text{mol}}(776\,\cancel{\text{K}})} = \mathbf{9.7 \times 10^7\ mol\ H_2}$$

The second step is the typical equation stoichiometry mole-to-mole conversion, using the coefficients from the balanced equation.

$$?\ \text{mol } NH_3 = 9.7 \times 10^7\,\cancel{\text{mol } H_2}\left(\frac{2\ \text{mol } NH_3}{3\,\cancel{\text{mol } H_2}}\right) = 6.5 \times 10^7\ \text{mol } NH_3$$

In the third step, we use the ideal gas equation to convert from moles of $NH_3$ to volume of $NH_3(g)$ under the final conditions.

$$V_{NH_3} = \frac{nRT}{P} = \frac{6.5 \times 10^7\,\cancel{\text{mol}}\left(\dfrac{0.082058\ \text{L} \cdot \cancel{\text{atm}}}{\cancel{\text{K}} \cdot \cancel{\text{mol}}}\right)293.8\,\cancel{\text{K}}}{1.007\,\cancel{\text{atm}}}$$

$$= \mathbf{1.6 \times 10^9\ L\ NH_3}$$

An alternative technique enables us to work Example 13.6 and problems like it using a single dimensional analysis setup. This technique, illustrated in Example 3.7, uses the universal gas constant, $R$, as a conversion factor.

## EXAMPLE 13.7 Gas Stoichiometry for Conditions Other Than STP (alternative technique)

OBJECTIVE 19

The question is the same as in Example 13.6: When $4.0 \times 10^7$ L of hydrogen gas at 503 °C and 155 atm reacts with an excess of nitrogen, what is the maximum volume of gaseous ammonia that can be formed at 20.6 °C and 1.007 atm?

***Solution***

We can use *R* as a conversion factor to convert from volume in liters of $H_2$ to moles of $H_2$. We use it in the inverted form, with liters (L) on the bottom, so that liters will cancel out.

$$? \text{ L NH}_3 = 4.0 \times 10^7 \cancel{\text{L}} \text{ H}_2 \left(\frac{\text{K} \cdot \text{mol}}{0.082058 \cancel{\text{L}} \cdot \text{atm}}\right)$$

The universal gas constant, *R*, is different from other conversion factors in that it contains four units rather than two. When we use it to convert from liters to moles, its presence introduces the units atmosphere, atm, and kelvin, K. We can cancel these units, however, with a ratio constructed from the temperature and pressure values that we are given for $H_2$:

$$? \text{ L NH}_3 = 4.0 \times 10^7 \cancel{\text{L}} \text{ H}_2 \left(\frac{\cancel{\text{K}} \cdot \text{mol}}{0.082058 \cancel{\text{L}} \cdot \cancel{\text{atm}}}\right)\left(\frac{155 \cancel{\text{atm}}}{776 \cancel{\text{K}}}\right)$$

We now insert a factor to convert from moles of $H_2$ to moles of $NH_3$, using the coefficients from the balanced equation.

$$? \text{ L NH}_3 = 4.0 \times 10^7 \cancel{\text{L}} \cancel{\text{H}_2} \left(\frac{\cancel{\text{K}} \cdot \cancel{\text{mol}}}{0.082058 \cancel{\text{L}} \cdot \cancel{\text{atm}}}\right)\left(\frac{155 \cancel{\text{atm}}}{776 \cancel{\text{K}}}\right)\left(\frac{2 \text{ mol NH}_3}{3 \cancel{\text{mol}} \cancel{\text{H}_2}}\right)$$

We complete the sequence by inserting *R* once again, this time to convert moles of $NH_3$ into volume of $NH_3$ in liters. We eliminate the units K and atm by using a ratio constructed from the temperature and pressure of $NH_3$.

$$? \text{ L NH}_3 = 4.0 \times 10^7 \cancel{\text{L}} \cancel{\text{H}_2} \left(\frac{\cancel{\text{K}} \cdot \cancel{\text{mol}}}{0.082058 \cancel{\text{L}} \cdot \cancel{\text{atm}}}\right)\left(\frac{155 \cancel{\text{atm}}}{776 \cancel{\text{K}}}\right)\left(\frac{2 \cancel{\text{mol}} \text{ NH}_3}{3 \cancel{\text{mol}} \cancel{\text{H}_2}}\right)\left(\frac{0.082058 \text{ L} \cdot \cancel{\text{atm}}}{\cancel{\text{K}} \cdot \cancel{\text{mol}}}\right)\left(\frac{293.8 \cancel{\text{K}}}{1.007 \cancel{\text{atm}}}\right)$$

$$= \mathbf{1.6 \times 10^9 \text{ L NH}_3}$$

The Internet link below takes you to a Web site that describes a shortcut for working equation stoichiometry problems in which volume of one gas is converted into volume of another gas at the same temperature and pressure.

**www.chemplace.com/college/**

Between this chapter and Chapter 10, we have now seen three different ways to convert between a measurable property and moles in equation stoichiometry problems. The different paths are summarized in Figure 13.10 in the sample study sheet on the next page. For pure liquids and solids, we can convert between mass and moles, using the molar mass as a conversion factor. For gases, we can convert between volume of gas and moles using the methods described above. For solutions, molarity provides a conversion factor that enables us to convert between moles of solute and volume of solution. Equation stoichiometry problems can contain any combination of two of these conversions, such as we see in Example 13.8.

## EXAMPLE 13.8 Equation Stoichiometry

OBJECTIVE 19

The gastric glands of the stomach secrete enough hydrochloric acid to give the stomach juices a hydrochloric acid concentration of over 0.01 M. Some of this stomach acid can be neutralized by the antacid Tums, which contains calcium carbonate. The $CaCO_3$ reacts with HCl in the stomach to form aqueous calcium chloride, carbon dioxide gas, and water. A regular Tums tablet will react with about 80 mL of 0.050 M HCl. How many milliliters of $CO_2$ gas at 37 °C and 1.02 atm will form from the complete reaction of 80 mL of 0.050 M HCl?

$$CaCO_3(s) + 2HCl(aq) \rightarrow CaCl_2(aq) + CO_2(g) + H_2O(l)$$

***Solution***

The general steps for this conversion are

$$\text{Volume of HCl solution} \rightarrow \text{moles of HCl} \rightarrow \text{moles of } CO_2 \rightarrow \text{volume of } CO_2 \text{ gas}$$

$$? \text{ mL } CO_2 = 80 \cancel{\text{mL HCl soln}} \left(\frac{0.050 \cancel{\text{mol HCl}}}{10^3 \cancel{\text{mL HCl soln}}}\right)\left(\frac{1 \cancel{\text{mol}} \text{ } CO_2}{2 \cancel{\text{mol HCl}}}\right)\left(\frac{0.082058 \cancel{\text{L}} \cdot \cancel{\text{atm}}}{\cancel{\text{K}} \cdot \cancel{\text{mol}}}\right)\left(\frac{310 \cancel{\text{K}}}{1.02 \cancel{\text{atm}}}\right)\left(\frac{10^3 \text{ mL}}{1 \cancel{\text{L}}}\right)$$

$$= \mathbf{50 \text{ mL } CO_2}$$

The following sample study sheet summarizes the methods we have learned for solving equation stoichiometry problems.

## Sample Study Sheet 13.3

### Equation Stoichiometry Problems

**Tip-off** You are asked to convert an amount of one substance (#1) to an amount of another substance (#2), both involved in a chemical reaction.

**General Steps** These steps are summarized in Figure 13.10.

**Step 1** Convert the amount of #1 to moles of #1.

- For a pure solid or liquid, this will be a molar mass conversion from mass to moles.

$$\_\_\_\_ \text{ g \#1} \left( \frac{\text{1 mol \#1}}{\text{(molar mass) g \#1}} \right)$$

- For solutions, molarity can be used to convert from volume of solution to moles.

$$\_\_\_\_ \text{ L \#1} \left( \frac{\text{(number from molarity) mol \#1}}{\text{1 L \#1 solution}} \right)$$

$$\text{or } \_\_\_\_ \text{ mL \#1} \left( \frac{\text{(number from molarity) mol \#1}}{10^3 \text{ mL \#1 solution}} \right)$$

- For a gas at STP, the molar volume at STP can be used as a conversion factor to convert from volume of gas at STP to moles.

OBJECTIVE 19

$$\_\_\_\_ \text{ L \#1} \left( \frac{\text{1 mol \#1}}{\text{22.414 L \#1}} \right)_{STP}$$

- For a gas under conditions other than STP, the ideal gas equation can be used to convert volume to moles, or the universal gas constant, $R$, can be used as a conversion factor.

$$n_{\#1} = \frac{P_{\#1}V_{\#1}}{RT_{\#1}}$$

OBJECTIVE 19

$$\text{or } \_\_\_\_ \text{ L \#1} \left( \frac{\text{K} \cdot \text{mol}}{0.082058 \text{ L} \cdot \text{atm}} \right) \left( \frac{\_\_\_ \text{ atm}}{\_\_\_ \text{ K}} \right)$$

$$\text{or } \_\_\_\_ \text{ L \#1} \left( \frac{\text{K} \cdot \text{mol}}{8.3145 \text{ L} \cdot \text{kPa}} \right) \left( \frac{\_\_\_ \text{ kPa}}{\_\_\_ \text{ K}} \right)$$

**Step 2** Convert moles of #1 to moles of #2, using the coefficients from the balanced equation.

**Step 3** Convert moles of #2 to amount of #2.

- For a pure solid or liquid, this will be a molar mass conversion from moles to mass.

$$\_\_\_\_ \left( \frac{\text{(coefficient \#2) mol \#2}}{\text{(coefficient \#1) mol \#1}} \right) \left( \frac{\text{(molar mass) g \#2}}{\text{1 mol \#2}} \right)$$

- For solutions, molarity can be used to convert from moles to volume of solution.

$$\_\_\_ \left(\frac{(\text{coefficient \#2}) \text{ mol \#2}}{(\text{coefficient \#1}) \text{ mol \#1}}\right)\left(\frac{1 \text{ L \#2 solution}}{(\text{number from molarity}) \text{ mol \#2}}\right)$$

$$\text{or } \_\_\_ \left(\frac{(\text{coefficient \#2}) \text{ mol \#2}}{(\text{coefficient \#1}) \text{ mol \#1}}\right)\left(\frac{10^3 \text{ mL \#2 solution}}{(\text{number from molarity}) \text{ mol \#2}}\right)$$

- For a gas at STP, the molar volume at STP can be used as a conversion factor to convert from moles to volume of gas at STP.

OBJECTIVE 19

$$\_\_\_ \left(\frac{(\text{coefficient \#2}) \text{ mol \#2}}{(\text{coefficient \#1}) \text{ mol \#1}}\right)\left(\frac{22.414 \text{ L \#2 solution}}{1 \text{ mol \#2}}\right)_{\text{STP}}$$

- For a gas under conditions other than STP, the ideal gas equation or the universal gas constant, $R$, can be used to convert from moles to volume of gas.

$$V_{\#2} = \frac{n_{\#2}RT_{\#2}}{P_{\#2}}$$

OBJECTIVE 19

$$\text{or } \_\_\_ \left(\frac{(\text{coefficient \#2}) \text{ mol \#2}}{(\text{coefficient \#1}) \text{ mol \#1}}\right)\left(\frac{0.082058 \cdot \text{L atm}}{\text{K} \cdot \text{mol}}\right)\left(\frac{\_\text{ K}}{\_\text{ atm}}\right)$$

$$\text{or } \_\_\_ \left(\frac{(\text{coefficient \#2}) \text{ mol \#2}}{(\text{coefficient \#1}) \text{ mol \#1}}\right)\left(\frac{8.3145 \text{ L} \cdot \text{kPa}}{\text{K} \cdot \text{mol}}\right)\left(\frac{\_\text{ K}}{\_\text{ kPa}}\right)$$

**EXAMPLE** See Examples 13.5, 13.6, 13.7, and 13.8.

**Figure 13.10**
Summary of Equation Stoichiometry Possibilities

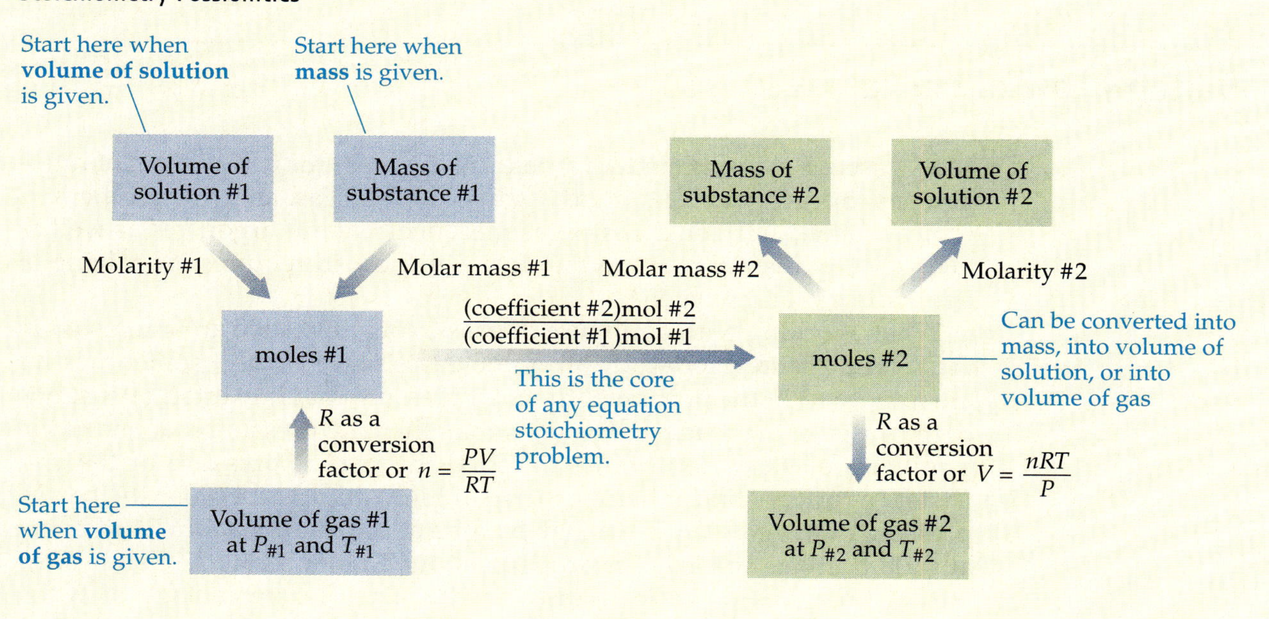

### Exercise 13.3 Equation Stoichiometry

OBJECTIVE 19

Iron is combined with carbon in a series of reactions to form pig iron, which is about 4.3% carbon.

$$2C + O_2 \rightarrow 2CO$$

$$Fe_2O_3 + 3CO \rightarrow 2Fe + 3CO_2$$

$$2CO \rightarrow C\text{ (in iron)} + CO_2$$

Pig iron is easier to shape than pure iron, and the presence of carbon lowers its melting point from the 1539 °C required to melt pure iron to 1130 °C.

a. In the first reaction, what minimum volume of oxygen at STP is necessary to convert 125 Mg of carbon to carbon monoxide?

b. In the first reaction, what is the maximum volume of carbon monoxide at 1.05 atm and 35 °C that could form from the conversion of $8.74 \times 10^5$ L of oxygen at 0.994 atm and 27 °C?

### Exercise 13.4 Equation Stoichiometry

OBJECTIVE 19

Sodium hypochlorite, NaOCl, which is found in household bleaches, can be made from a reaction using chlorine gas and aqueous sodium hydroxide:

$$Cl_2(g) + 2NaOH(aq) \rightarrow NaOCl(aq) + NaCl(aq) + H_2O(l)$$

What minimum volume of chlorine gas at 101.4 kPa and 18.0 °C must be used to react with all the sodium hydroxide in 3525 L of 12.5 M NaOH?

## 13.4 Dalton's Law of Partial Pressures

Most gaseous systems contain a mixture of gases. Air is a mixture of nitrogen gas, oxygen gas, xenon gas, carbon dioxide gas, and many others. A typical "neon" light on a Las Vegas marquee contains argon gas as well as neon. The industrial reaction that forms nitric acid requires the mixing of ammonia gas and oxygen.

When working with a mixture of gases, sometimes we are interested in the total pressure exerted by all of the gases together, and sometimes we are interested in the portion of the total pressure that is exerted by just one of the gases in the mixture. The portion of the total pressure that one gas in a mixture of gases contributes is called the **partial pressure** of the gas. Distance runners find it harder to run at high altitudes because the partial pressure of oxygen in the air they breathe at sea level is about 21 kPa, but at 6000 ft above sea level, it drops to about 17 kPa. Table 13.1 shows the partial pressures of various gases in dry air. (The percentage of water vapor in air varies from place to place and from day to day.)

Table 13.1
Gases in Dry Air at STP

| Gas | Percentage of total particles | Partial pressure (kPa) |
|---|---|---|
| nitrogen | 78.081 | 79.116 |
| oxygen | 20.948 | 21.226 |
| argon | 0.934 | 0.995 |
| carbon dioxide | 0.035 | 0.035 |
| neon | 0.002 | 0.002 |
| helium | 0.0005 | 0.0005 |
| methane | 0.0001 | 0.0001 |
| krypton | 0.0001 | 0.0001 |
| nitrous oxide | 0.00005 | 0.00005 |
| hydrogen | 0.00005 | 0.00005 |
| ozone | 0.000007 | 0.000007 |
| xenon | 0.000009 | 0.000009 |

To get an understanding of the relationship between the total pressure of a mixture of gases and the partial pressure of each gas in the mixture, let's picture a luminous tube being filled with neon and argon gases. Let's say that the neon gas is added to the tube first, and let's picture ourselves riding on one of the neon atoms. The particle is moving rapidly, colliding constantly with other particles and with the walls of the container. Each collision with a wall exerts a small force pushing out against the wall. The total pressure (force per unit area) due to the collisions of all of the neon atoms with the walls is determined by the rate of collision of the neon atoms with the walls and the average force per collision. The neon gas pressure can be changed in only two ways: by a change in the rate of collisions with the walls or by a change in the average force per collision.

OBJECTIVE 20

Now consider the effect on gas pressure of the addition of argon gas—at the same temperature—to the tube that already contains neon. Assuming that the mixture of gases acts as an ideal gas, there are no significant attractions or repulsions between the particles, and the volume occupied by the particles is very small. Except for collisions between particles, each gas particle acts independently of all of the other particles in the tube. Now that there are also a certain number of argon atoms in the tube, the neon atoms (including the one we are riding on) have more collisions with other particles and change their direction of motion more often, but they still collide with the walls at the same rate as before. If the temperature stays the same, the average velocity of the neon atoms and their average force per collision with the walls remain the same.

If the neon atoms in the mixture are colliding with the walls of the tube at the same rate and with the same average force per collision as they did when alone, they are exerting the same pressure against the walls now as they did then. Therefore, the partial pressure of neon in the argon–neon mixture is the same as the pressure that the neon exerted when it was alone. Assuming ideal gas character, the **partial pressure** of any gas in a mixture is the pressure that the gas would exert if it were alone in the container.

Because the argon atoms in the mixture are also acting independently, the partial pressure of the argon gas is also the same as the pressure it would exert if it were alone. The total pressure in the luminous tube that now contains both neon gas and argon gas is equal to the sum of their partial pressures (Figure 13.11).

$$P_{total} = P_{Ne} + P_{Ar}$$

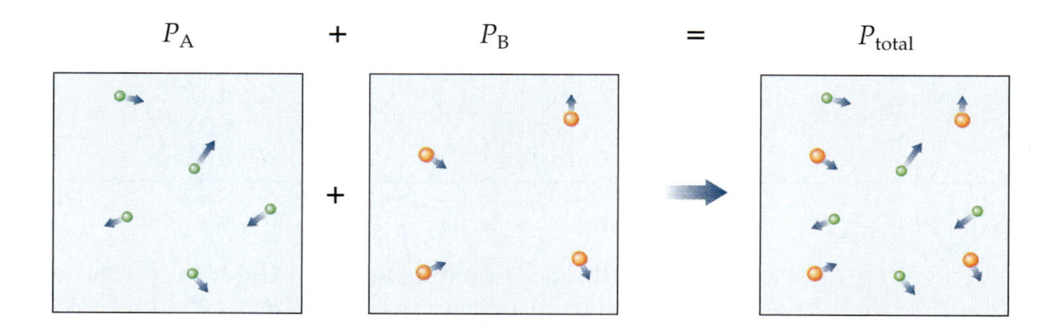


**Figure 13.11**
**Dalton's Law of Partial Pressures**
The total pressure of a mixture of ideal gases is equal to the sum of the partial pressures of the gases considered separately.

The situation is summarized in **Dalton's Law of Partial Pressures:** *the total pressure of a mixture of gases is equal to the sum of the partial pressures of all the gases.*

OBJECTIVE 21

$$P_{total} = \Sigma P_{partial}$$

Dalton's Law of Partial Pressures can be combined with the ideal gas equation to solve many kinds of problems involving mixtures of gases. One way of combining them is to begin by rearranging the ideal gas equation to express the partial pressures of gases in a mixture—for example, the partial pressures of neon gas and argon gas in a mixture of argon and neon gases.

$$PV = nRT$$

$$P = \frac{nRT}{V}$$

Therefore,

$$P_{Ne} = \frac{n_{Ne}RT_{Ne}}{V_{Ne}} \qquad P_{Ar} = \frac{n_{Ar}RT_{Ar}}{V_{Ar}}$$

$$P_{total} = P_{Ne} + P_{Ar} = \frac{n_{Ne}RT_{Ne}}{V_{Ne}} + \frac{n_{Ar}RT_{Ar}}{V_{Ar}}$$

Because both neon gas and argon gas expand to fill the whole container, the volume for the argon gas and the volume for the neon gas are equal. Similarly, because the collisions between gas particles in the same container will lead the gases to have the same temperature, the temperatures for the argon gas and the neon gas are also equal. Therefore, the common $RT/V$ can be factored out to yield a simplified form of the equation.

$$V_{Ar} = V_{Ne} = V \qquad T_{Ar} = T_{Ne} = T$$

$$P_{total} = \frac{n_{Ne}RT}{V} + \frac{n_{Ar}RT}{V} = (n_{Ne} + n_{Ar})\frac{RT}{V}$$

In other words, the total pressure of a mixture of gases is equal to the sum of the moles of all gases in the mixture times $RT/V$. The general form of this equation is

$$P_{total} = (\Sigma n_{each\ gas})\frac{RT}{V}$$

OBJECTIVE 22

The following sample study sheet describes calculations that can be done using Dalton's Law of Partial Pressures.

## Sample Study Sheet 13.4

### Using Dalton's Law of Partial Pressures

OBJECTIVE 21

OBJECTIVE 22

**Tip-off** The problem involves a mixture of gases and no chemical reaction. You are asked to calculate a value for one of the variables in the equations below, and you are given (directly or indirectly) values for the other variables.

**General Steps** The following steps can be used to work these problems.

**Step 1** Assign variables to the values that are given and to the value that is unknown.

**Step 2** From the following equations, choose the one that best fits the variables assigned in step 1.

$$P_{total} = \Sigma P_{partial}$$

$$P_{total} = (\Sigma n_{each\ gas})\frac{RT}{V}$$

**Step 3** Rearrange the equation to solve for your unknown.

**Step 4** Plug in the values for the given properties.

**Step 5** Ensure that the equation yields the correct units. Make any necessary unit conversions.

**Step 6** Calculate your answer and report it with the correct units and to the correct number of significant figures.

**Example** See Examples 13.9 and 13.10.

## SPECIAL TOPIC 13.1 A Greener Way to Spray Paint

United States industries use an estimated 1.5 billion liters of paints and other coatings per year, and much of this is applied by spraying. Each liter sprayed from a canister releases an average of 550 grams of volatile organic compounds (VOCs), including hydrocarbons, alcohols, esters, and ketones. Some of these VOCs are hazardous air pollutants.

The mixture that comes out of the spray can has two kinds of components: (1) the solids being deposited on the surface as a coating and (2) a solvent blend that allows the solids to be sprayed and to spread evenly. The solvent blend must dissolve the coatings into a mixture that is thin enough in consistency to be easily sprayed. But a mixture that is thin enough to be easily sprayed will be too runny to remain in place when deposited on a surface. Therefore, the solvent blend contains additional components so volatile that they will evaporate from the spray droplets between the time the spray leaves the spray nozzle and the time the spray hits the surface. Still other, slower-evaporating components do not evaporate until after the spray hits the surface. They remain in the coating mixture long enough to cause it to spread out evenly. Because the more volatile solvents have escaped, the mixture that hits the surface is thick enough not to run or sag.

The Clean Air Act has set strict limitations on the emission of certain VOCs, so safer solvents are needed to replace them. One new spray system has been developed that yields a high-quality coating while emitting as much as 80% fewer VOCs of all types and none of the VOCs that are considered hazardous air pollutants. This system is called the *supercritical fluid spray process.* The solvent mixture for this process still contains some of the slowly evaporating solvents that allow the coating to spread evenly, but it replaces the rapidly evaporating solvents with high-pressure $CO_2$.

Some gases can be converted into liquids at room temperature by being compressed into a smaller volume, but for each gas, there is a temperature above which the particles are moving too fast to allow a liquid to form, no matter how much they are compressed. The temperature above which a gas cannot be liquefied is called the *critical temperature.* The very high pressures that are possible for a gas above its critical temperature allow the formation of gas systems in which the density approaches the densities of liquids. At this high pressure and density, the substance takes on characteristics of both gases and liquids and is called a *supercritical fluid.*

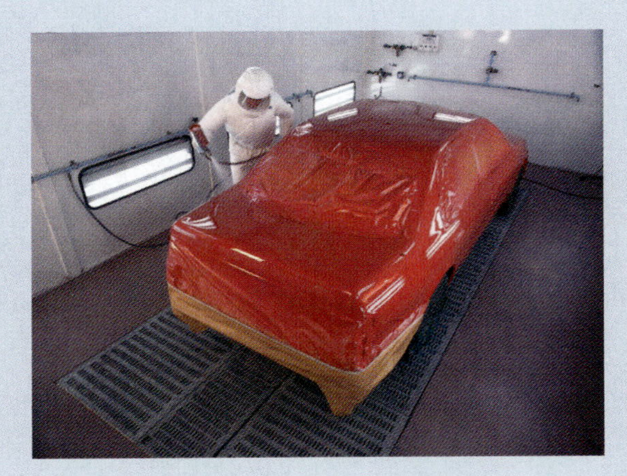

**By taking advantage of the properties of gases at high temperatures and pressures, scientists have invented a new, environmentally friendly spray-painting process.**

The critical temperature of $CO_2$ is 31 °C. Above this temperature, carbon dioxide can be compressed to a very high-pressure and relatively high-density supercritical fluid. Like a liquid, the supercritical carbon dioxide will mix with or dissolve the blend of coating and low-volatility solvent to form a product that is thin enough to be sprayed easily, in very small droplets. The supercritical $CO_2$ has a very high volatility, so it evaporates from the droplets almost immediately after they are emitted from the spray nozzle, leaving a mixture that is thick enough not to run or sag when it hits the surface. The mixture is sprayed at temperatures of about 50 °C and a pressure of 100 atm (about 100 times normal room pressure).

Because carbon dioxide is much less toxic than the VOCs it replaces and because it is nonflammable and relatively inert, it is much safer to use in the workplace. It is also far less expensive. Moreover, the $CO_2$ can be obtained from the production of other chemicals, so the new process does not lead to an increase in carbon dioxide in the atmosphere. In fact, because the VOCs that it replaces can form $CO_2$ in the atmosphere, the supercritical fluid spray process actually leads to a *decrease* in atmospheric $CO_2$ levels compared to what they would be with the traditional spray process.

**Source:** M. D. Donohue, J. L. Geiger, A. A. Kiamos, and K. A. Nielsen, "Reduction of VOC Emissions During Spray Painting." In *Green Chemistry*; ed. P. T. Anastas and T. C. Williams (Washington, DC: American Chemical Society 1996), pp. 153–167.

## EXAMPLE 13.9 Dalton's Law of Partial Pressures

OBJECTIVE 21

A typical 100-watt light bulb contains a mixture of argon gas and nitrogen gas. In a light bulb with a total gas pressure of 111 kPa, enough argon is present to yield a partial pressure of 102 kPa. What is the partial pressure of the nitrogen gas?

***Solution***

**STEP 1** Assign variables.

$$P_T = 111 \text{ kPa} \qquad P_{Ar} = 102 \text{ kPa} \qquad P_{N_2} = ?$$

**STEP 2** Our variables fit the general equation

$$P_{total} = \Sigma P_{partial} \quad \text{or} \quad P_{total} = P_{Ar} + P_{N_2}$$

**STEPS 3–6** Solve the equation for the unknown variable, plug in the given values, check the units, and calculate the answer.

$$P_{N_2} = P_{total} - P_{Ar} = 111 \text{ kPa} - 102 \text{ kPa} = \mathbf{9\ kPa}$$

## EXAMPLE 13.10 Dalton's Law of Partial Pressures

OBJECTIVE 22

If an 85-mL light bulb contains 0.140 grams of argon and 0.011 grams of nitrogen at 20 °C, what is the total pressure of the mixture of gases?

***Solution***

**STEP 1** Although there are no mass variables in our partial-pressure equations, we do know that we can convert mass to moles using molar masses.

$$n_{Ar} = ? \text{ mol Ar} = 0.140 \cancel{\text{g Ar}} \left(\frac{1 \text{ mol Ar}}{39.948 \cancel{\text{g Ar}}}\right) = 0.00350 \text{ mol Ar}$$

$$n_{N_2} = ? \text{ mol } N_2 = 0.011 \cancel{\text{g } N_2} \left(\frac{1 \text{ mol } N_2}{28.0134 \cancel{\text{g } N_2}}\right) = 0.00039 \text{ mol } N_2$$

$$V = 85 \text{ mL} \qquad T = 20 \text{ °C} + 273.15 = 293 \text{ K}$$

**STEP 2** Our variables fit the following form of Dalton's Law of Partial Pressures:

$$P_{total} = (\Sigma n_{each\ gas}) \frac{RT}{V} \qquad P_T = (n_{Ar} + n_{N_2}) \frac{RT}{V}$$

**STEPS 3–6** We do not need to rearrange the equation algebraically, but when we plug in our values with their units, we see that we need to convert 85 mL into liters to get our units to cancel correctly.

$$P_{total} = (0.00350 \cancel{\text{mol Ar}} + 0.00039 \cancel{\text{mol } N_2}) \left(\frac{\frac{0.082058 \cancel{\text{L}} \cdot \text{atm}}{\cancel{\text{K}} \cdot \cancel{\text{mol}}}(293 \cancel{\text{K}})}{85 \cancel{\text{mL}}}\right)\left(\frac{10^3 \cancel{\text{mL}}}{1 \cancel{\text{L}}}\right)$$

$$= \mathbf{1.1\ atm\ (or\ 1.1 \times 10^2\ kPa)}$$

## Exercise 13.5 Dalton's Law of Partial Pressures

OBJECTIVE 21

OBJECTIVE 22

A typical "neon light" contains neon gas mixed with argon gas.

a. If the total pressure of the mixture of gases is 1.30 kPa and the partial pressure of neon gas is 0.27 kPa, what is the partial pressure of the argon gas?

b. If 6.3 mg of Ar and 1.2 mg Ne are added to the 375-mL tube at 291 K, what is the total pressure of the gases in millimeters of mercury?

## Special Topic 13.2 Green Decaf Coffee

Caffeine, the primary stimulant found naturally in coffee, tea, and chocolate, is artificially added to many other products, such as soft drinks, over-the-counter stimulants, cold remedies, and pain relievers. It is the world's most widely used drug. Different types of coffee contain different amounts of caffeine, and different brewing techniques lead to further variations in content. Depending on the type of coffee (an arabica coffee contains about half the caffeine of a robusta) and the fineness of the grounds, the drip method of brewing leads to 110 to 150 mg of caffeine per cup. Instant coffee has 40–108 mg of caffeine per cup. Thus the average cup of coffee contains from 40 to 150 mg of caffeine.

Some people like the taste of coffee and the rituals associated with it (the customary cup of coffee after dinner) but do not like the effects of the caffeine. They may choose to drink decaffeinated coffee, which has 2–4 mg caffeine per cup. A coffee must have at least 97% of its caffeine removed to be marketed as decaffeinated coffee.

One of the commercial methods for decaffeinating coffee is the direct-contact method. The unroasted coffee beans are first softened with steam and then brought into direct contact with a decaffeinating agent, such as dichloromethane (most often called methylene chloride in this context), $CH_2Cl_2$. The caffeine dissolves in the dichloromethane, after which the dichloromethane–caffeine solution is removed from the beans.

There is some evidence that dichloromethane is carcinogenic. According to the United States Food and Drug Administration (FDA), most decaffeinated coffee has less than 0.1 part per million (ppm) of residual dichloromethane, which is 100 times less than the maximum level of 10 ppm allowed by the FDA. Despite this low residual concentration of dichloromethane, there is still some concern about a possible health hazard, so substitutes are being sought for many of its applications.

One alternative to the current principal methods of decaffeination uses supercritical carbon dioxide. The unroasted beans are softened by steam and then immersed in carbon dioxide at very high temperature and pressure. The $CO_2$ penetrates the beans more than a typical liquid would and then, like a liquid, dissolves the caffeine. When the supercritical carbon dioxide is later separated from the beans, it carries about 97% of the caffeine along with it. Any residue of carbon dioxide remaining on the beans is quickly lost as gaseous $CO_2$. The workers employed in the decaffeination process are safer because they have no contact with dichloromethane, and the coffee drinkers do not have to worry about the very slight residue of dichloromethane that remains when the coffee is decaffeinated via the direct-contact method.

All of these products contain caffeine.

## Chapter Glossary

**Ideal gas model** The model for gases that assumes that (1) the particles are point-masses (they have mass but no volume) and (2) there are no attractive or repulsive forces between the particles.

**Ideal gas** A gas for which the ideal gas model is a good description.

**Pressure** Force per unit area.

**Boyle's Law** The pressure of a gas is inversely proportional to the volume it occupies if the moles of gas and the temperature are constant.

**Gay–Lussac's Law** The pressure of an ideal gas is directly proportional to the Kelvin temperature of the gas if the volume and moles of gas are constant.

**Charles's Law** The volume of a gas and its temperature are directly proportional if the moles of gas and its pressure are constant.

**Avogadro's Law** The volume and moles of a gas are directly proportional if its temperature and pressure are constant.

**Universal gas constant, *R*** The constant in the ideal gas equation.

**Partial pressure** The portion of the total pressure that one gas in a mixture of gases contributes. Assuming ideal gas character, the partial pressure of any gas in a mixture is the pressure that the gas would yield if it were alone in the container.

**Dalton's Law of Partial Pressures** The total pressure of a mixture of gases is equal to the sum of the partial pressures of all the gases in the mixture.

You can test yourself on the glossary terms at the following Web site:

**www.chemplace.com/college/**

## Chapter Objectives

**The goal of this chapter is to teach you to do the following.**

1. Define all of the terms in the Chapter 13 Glossary.

### Section 13.1 Gases and Their Properties

2. For a typical gas, state the percentage of space inside a gas-filled container that is occupied by the gas particles themselves.
3. State the average distance that oxygen molecules, $O_2$, travel between collisions at normal room temperature and pressure.
4. Write the key assumptions that distinguish an ideal gas from a real gas.
5. Describe the process that gives rise to gas pressure.
6. State the accepted SI unit for gas pressure.
7. Convert between the names and abbreviations for the following pressure units: pascals (Pa), atmospheres (atm), millimeters of mercury (mmHg), and torr.
8. Convert a gas pressure described in pascals (Pa), atmospheres (atm), millimeters of mercury (mmHg), or torr to any of the other units.

9. Convert between the names and variables used to describe pressure ($P$), temperature ($T$), volume ($V$), and number of moles of gas ($n$).
10. For each of the following pairs of gas properties, (1) describe the relationship between the properties, (2) describe a simple system that could be used to demonstrate the relationship, and (3) explain the reason for the relationship.
    a. volume and pressure when moles of gas and temperature are constant
    b. pressure and temperature when volume and moles of gas are constant
    c. volume and temperature when pressure and moles of gas are constant
    d. moles of gas and pressure when volume and temperature are constant
    e. moles of gas and volume when pressure and temperature are constant
11. Explain why decreased volume for the gasoline–air mixture in the cylinders of a gasoline engine, increased number of moles of gas, and increased temperature lead to an increase in pressure in the cylinders.
12. Explain why air moves in and out of our lungs as we breathe.
13. Explain why balloons expand as they rise into the atmosphere.

### Section 13.2 Ideal Gas Calculations

14. Write at least one value for the universal gas constant, $R$, including units.
15. Given values for three of the following four variables, calculate the fourth: $P$, $V$, $n$, and $T$.
16. Given values for four of the following five variables, calculate the fifth: $P$, $V$, $g$, $M$, and $T$.
17. Given the pressure, temperature, and identity of a gas, calculate its density.
18. Given (directly or indirectly) values for seven of the following eight variables, calculate the eighth: $P_1$, $V_1$, $n_1$, $T_1$, $P_2$, $V_2$, $n_2$, and $T_2$.

### Section 13.3 Equation Stoichiometry and Ideal Gases

19. Convert between volume of gas and moles of gas in an equation stoichiometry problem, using either the molar volume at STP, the ideal gas equation, or $R$ as a conversion factor.

### Section 13.4 Dalton's Law of Partial Pressures

20. Explain why the total pressure of a mixture of ideal gases is equal to the sum of the partial pressures of all the gases taken separately.
21. Given all but one of the following properties for a mixture of gases, calculate the one not given: total pressure of the mixture of gases and partial pressure of each gas.
22. Given all but one of the following properties for a mixture of gases, calculate the one not given: total pressure of the mixture of gases, mass or moles of each gas, temperature, and volume of the container.

## Review Questions

1. Describe the particle nature of gases.
2. What is 265.2 °C on the Kelvin scale? Convert 565.7 K to °C.
3. About 55% of industrially produced sodium sulfate is used to make detergents. It is made from the reaction

$$4NaCl + 2SO_2 + 2H_2O + O_2 \rightarrow 2Na_2SO_4 + 4HCl$$

a. What is the maximum mass of sodium sulfate that can be produced in the reaction of 745 Mg of sodium chloride with excess $SO_2$, $H_2O$, and $O_2$?

b. What is the maximum mass of sodium sulfate that can be produced in the reaction of 745 Mg of sodium chloride with 150 Mg $H_2O$ and excess $SO_2$ and $O_2$?

c. If 868 Mg $Na_2SO_4$ is formed in the reaction of 745 Mg of sodium chloride with 150 Mg of $H_2O$ and excess $SO_2$ and $O_2$, what is the percent yield?

d. What volume of 2.0 M $Na_2SO_4$ can be formed from the reaction of 745 Mg of sodium chloride with excess $SO_2$, $H_2O$, and $O_2$?

## Key Ideas

**Complete the following statements by writing one of these words or phrases in each blank.**

0 °C
$10^{-7}$ m
0.1%
70%
99.9%
1 atm
101.325 kPa
273.15 K
760 mmHg
algebra
alone
attractions
attractive or repulsive
Avogadro's Law
collide
continuous
correct equation
degrees Celsius, °C
direction of motion
directly
empty space
force
force per unit area
greater force
increases
Kelvin
Kelvin temperature
liters, L
molar mass
molarity
moles, mol
moles of gas
necessary unit conversions
number of collisions
number of particles
pascal, Pa
point-masses
portion
pressure
rapid
real
repulsions
sum
temperature
temperature and volume
ten
velocity
volume

4. Under typical conditions, the average distance between gas particles is about ________________ times their diameter.
5. Because of the large distances between gas particles, the volume occupied by the particles themselves is very small compared to the volume of the ________________ around them.
6. For a gas at room temperature and pressure, the gas particles themselves occupy about ________________ of the total volume. The other ________________ of the total volume is empty space (whereas in liquids, about ________________ of the volume is occupied by particles).
7. Because of the large distances between gas particles, there are very few ________________ or ________________ between them.
8. The particles in a gas are in ________________ and ________________ motion.
9. As the temperature of a gas increases, the particles' velocity ________________.
10. The particles in a gas are constantly colliding with the walls of the container and with each other. Because of these collisions, the gas particles are constantly changing their ________________ and their ________________.
11. Oxygen, $O_2$, molecules at normal temperatures and pressures move an average of ________________ between collisions.
12. The particles of an ideal gas are assumed to be ________________, that is, particles that have a mass but occupy no volume.
13. There are no ________________ forces at all between the particles of an ideal gas.
14. The ideal gas model is used to predict changes in four related gas properties: ________________, ________________, ________________, and ________________.
15. Volumes of gases are usually described in ________________ or cubic meters, $m^3$, and numbers of particles are usually described in ________________.
16. Although gas temperatures are often measured with thermometers that report temperatures in ________________ scientists generally use ________________ temperatures for calculations.
17. Each time a gas particle collides with and ricochets off one of the walls of its container, it exerts a(n) ________________ against the wall. The sum of the forces of these ongoing collisions of gas particles against all the container's interior walls creates a continuous pressure upon those walls.
18. The accepted SI unit for gas pressure is the ________________.
19. ________________ gases deviate somewhat from predicted behavior of ideal gases.
20. The observation that the pressure of an ideal gas is inversely proportional to the volume it occupies if the ________________ and the temperature are constant is a statement of Boyle's Law.

21. When the volume of a chamber that contains a gas decreases but the moles of gas remain constant, there is an increase in the concentration (moles per liter) of the gas. This leads to an increase in the number of particles near any given area of the container walls at any time and to an increase in the number of collisions against the walls per unit area in a given time. More collisions mean an increase in the ________________, or pressure, of the gas.
22. The pressure of an ideal gas is directly proportional to the ________________ of the gas if the volume and moles of gas are constant. This relationship is sometimes called Gay-Lussac's Law.
23. Increased temperature means increased motion of the particles of a gas. If the particles are moving faster in the container, they will ________________ with the walls more often and with ________________ per collision. This leads to a greater overall force pushing on the walls and to a greater force per unit area or pressure.
24. For an ideal gas, volume and temperature described in kelvins are ________________ proportional if moles of gas and pressure are constant. This is a statement of Charles's Law.
25. The increase in the moles of gas in a constant-volume container leads to an increase in the ________________ with the walls per unit time. This leads to an increase in the force per unit area—that is, to an increase in gas pressure.
26. If the ________________ of an ideal gas are held constant, the moles of gas in a container and the gas pressure are directly proportional.
27. The relationship between moles of an ideal gas and volume is summarized by ________________, which states that the volume and moles of gas are directly proportional if the temperature and pressure are constant.
28. It is always a good idea to include the units in a solved equation as well as the numbers. If the units cancel to yield a reasonable unit for the unknown property, you can feel confident that you picked the ________________, that you did the ________________ correctly to solve for your unknown, and that you have made the ________________.
29. STP stands for "standard temperature and pressure." Standard temperature is ________________, or ________________, and standard pressure is ________________, or ________________, or ________________.
30. There are three different ways to convert between a measurable property and moles in equation stoichiometry problems. For pure liquids and solids, we can convert between mass and moles, using the ________________ as a conversion factor. For solutions, ________________ provides a conversion factor that enables us to convert between moles of solute and volume of solution.
31. The ________________ of the total pressure that one gas in a mixture of gases contributes is called the partial pressure of the gas.
32. Assuming ideal gas character, the partial pressure of any gas in a mixture is the pressure that the gas would exert if it were ________________ in the container.
33. Dalton's law of partial pressures states that the total pressure of a mixture of gases is equal to the ________________ of the partial pressures of gas all the gases.

## Chapter Problems

### Section 13.1 Gases and Their Properties

OBJECTIVE 2

34. For a gas under typical conditions, what approximate percentage of the volume is occupied by the gas particles themselves?

35. One of the car maintenance tasks you might do at a service station is to fill your tires with air. Why are the tires not really full when you get to the desired tire pressure? Why can you still add more air to the tire?

36. Why is it harder to walk through water than to walk through air?

OBJECTIVE 3

37. What is the average distance that molecules of oxygen gas, $O_2$, travel between collisions at room temperature and pressure?

OBJECTIVE 4

38. What are the key assumptions that distinguish an ideal gas from a real gas?

OBJECTIVE 5

39. Consider a child's helium-filled balloon. Describe the process that creates the gas pressure that keeps the balloon inflated.

40. A TV weather person predicts a storm for the next day and refers to the dropping barometric pressure to support the prediction. What causes the pressure in the air?

OBJECTIVE 6 OBJECTIVE 7

41. What is the accepted SI unit for gas pressure? What is its abbreviation?

OBJECTIVE 8

42. The pressure at the center of the earth is $4 \times 10^{11}$ pascals, Pa. What is this pressure in kilopascals, atmospheres, millimeters of mercury, and torr?

OBJECTIVE 8

43. The highest sustained pressure achieved in the laboratory is $1.5 \times 10^7$ kilopascals, kPa. What is this pressure in atmospheres, millimeters of mercury, and torr?

44. What does *inversely proportional* mean?

OBJECTIVE 10

45. For each of the following pairs of gas properties, (1) describe the relationship between the properties, (2) describe a simple system that could be used to demonstrate the relationship, and (3) explain the reason for the relationship.
    a. volume and pressure when moles of gas and temperature are constant
    b. pressure and temperature when volume and moles of gas are constant
    c. volume and temperature when pressure and moles of gas are constant
    d. moles of gas and pressure when volume and temperature are constant
    e. moles of gas and volume when pressure and temperature are constant

OBJECTIVE 11

46. Explain why decreased volume for the gasoline–air mixture in the cylinders of a gasoline engine, increased number of moles of gas, and increased temperature lead to an increase in pressure in the cylinders.

47. Ammonia, $NH_3$, is a gas that can be dissolved in water and used as a cleaner. When an ammonia solution is used to clean the wax off a floor, ammonia gas escapes from the solution, mixes easily with the gas particles in the air, and spreads throughout the room. Very quickly, everyone in the room can smell the ammonia gas. Explain why gaseous ammonia mixes so easily with air.

48. After gasoline vapor and air are mixed in the cylinders of a car, the piston in a cylinder moves up into the cylinder and compresses the gaseous mixture. Explain why these gases can be compressed.

OBJECTIVE 12

49. Explain why air moves in and out of our lungs as we breathe.

OBJECTIVE 13

50. Explain why balloons expand as they rise into the atmosphere.

51. With reference to the relationships between properties of gases, answer the following questions.
    a. In a common classroom demonstration, water is heated in a 1-gallon can, covered with a tight lid, and allowed to cool. As the can cools, it collapses. Why?
    b. Dents in ping-pong balls can often be removed by placing the balls in hot water. Why?
52. Use the relationships between properties of gases to answer the following questions.
    a. When a balloon is placed in the freezer, it shrinks. Why?
    b. Aerosol cans often explode in fires. Why?
    c. Why do your ears pop when you drive up a mountain?

### Section 13.2 Ideal Gas Calculations

53. Neon gas produced for use in luminous tubes comes packaged in 1.0-L containers at a pressure of 1.00 atm and a temperature of 19 °C. How many moles of Ne do these cylinders hold? OBJECTIVE 15
54. Fluorescent light bulbs contain a mixture of mercury vapor and a noble gas such as argon. How many moles of argon are in a 121-mL fluorescent bulb with a pressure of argon gas of 325 Pa at 40 °C? OBJECTIVE 15
55. A typical aerosol can is able to withstand 10–12 atm without exploding.
    a. If a 375-mL aerosol can contains 0.062 mole of gas, at what temperature would the gas pressure reach a pressure of 12 atm? OBJECTIVE 15
    b. Aerosol cans usually contain liquids as well as gas. If the 375-mL can described in part (a) contained liquid along with the 0.062 mole of gas, why would it explode at a lower temperature than if it contained only the gas? (*Hint*: What happens to a liquid when it is heated?)
56. Ethylene oxide is produced from the reaction of ethylene and oxygen at 270–290 °C and 8–20 atm. In order to prevent potentially dangerous pressure buildups, the container in which this reaction takes place has a safety valve set to release gas when the pressure reaches 25 atm. If a 15-$m^3$ reaction vessel contains $7.8 \times 10^3$ moles of gas, at what temperature will the pressure reach 25 atm? (There are $10^3$ L per cubic meter.) OBJECTIVE 15
57. Bromomethane, $CH_3Br$, commonly called methyl bromide, is used as a soil fumigant, but because it is a threat to the ozone layer, its use is being phased out. This colorless gas is toxic by ingestion, inhalation, or skin absorption, so the American Conference of Government Industrial Hygienists has assigned it a threshold limit value, or TLV—a maximum concentration to which a worker may be repeatedly exposed day after day without experiencing adverse effects. (TLV standards are meant to serve as guides in control of health hazards, not to provide definitive dividing lines between safe and dangerous concentrations.) Bromomethane reaches its TLV when 0.028 mole escapes into a room that has a volume of 180 $m^3$ and a temperature of 18 °C. What is the pressure in atmospheres of the gas under those conditions? (There are $10^3$ L per cubic meter.) OBJECTIVE 15
58. Sulfur hexafluoride, $SF_6$, is a very stable substance that is used as a gaseous insulator for electrical equipment. Although $SF_6$ is considered relatively safe, it reaches its TLV (defined in Problem 57) when 143 moles of it escapes into a room with a volume of 3500 $m^3$ and a temperature of 21 °C. What is its pressure in pascals for these conditions? OBJECTIVE 15

OBJECTIVE 15

59. Gaseous chlorine, $Cl_2$, which is used in water purification, is a dense, greenish-yellow substance with an irritating odor. It is formed from the electrolysis of molten sodium chloride. Chlorine reaches its TLV (defined in Problem 57) in a laboratory when 0.017 mole of $Cl_2$ is released, yielding a pressure of $7.5 \times 10^{-4}$ mmHg at a temperature of 19 °C. What is the volume, in liters, of the room?

OBJECTIVE 15

60. Hydrogen bromide gas, HBr, is a colorless gas used as a pharmaceutical intermediate. Hydrogen bromide reaches its TLV (defined in Problem 57) in a laboratory room when 0.33 mole of it is released, yielding a pressure of 0.30 Pa at a temperature of 22 °C. What is the volume, in cubic meters, of the room?

OBJECTIVE 16

61. Hydrogen sulfide, $H_2S$, which is used to precipitate sulfides of metals, is a colorless gas that smells like rotten eggs. It can be detected by the nose at about 0.6 $mg/m^3$. To reach this level in a 295-$m^3$ room, 0.177 g of $H_2S$ must escape into the room. What is the pressure in pascals of 0.177 g of $H_2S$ that escapes into a 295-$m^3$ laboratory room at 291 K?

OBJECTIVE 16

62. Nitrogen dioxide, $NO_2$, is a reddish-brown gas above 21.1 °C and a brown liquid below 21.1 °C. It is used to make nitric acid and as an oxidizer for rocket fuels. Its TLV (defined in Problem 57) is reached when 1.35 g of $NO_2$ escapes into a room that has a volume of 225 $m^3$. What is the pressure, in millimeters of mercury, of 1.35 g of $NO_2$ at 302 K in a 225-$m^3$ room?

OBJECTIVE 16

63. Hydrogen chloride, HCl, is used to make vinyl chloride, which is then used to make polyvinyl chloride (PVC) plastic. Hydrogen chloride can be purchased in liquefied form in cylinders that contain 8 lb of HCl. What volume in liters will 8.0 lb of HCl occupy if the compound becomes a gas at 1.0 atm and 85 °C?

OBJECTIVE 16

64. Sulfur dioxide, $SO_2$, forms in the combustion of the sulfur found in fossil fuels such as coal. In the air, $SO_2$ forms $SO_3$, which dissolves in water to form sulfuric acid. Thus, $SO_2$ is a major contributor to acid rain as well as a strong eye and lung irritant. The 1990 amendments to the Clean Air Act of 1967 called for a reduction in sulfur dioxide released from power plants to 10 million tons per year by January 1, 2000. This is about one-half the emission measured in 1990. The U.S. atmospheric standard for $SO_2$ is 0.35 $mg/m^3$, which would yield a pressure due to sulfur dioxide of 0.013 Pa at 17 °C. To what volume, in cubic meters, must $1.0 \times 10^7$ tons of $SO_2$ expand in order to yield a pressure of 0.013 Pa at 17 °C? (There are 2000 lb per ton.)

OBJECTIVE 16

65. The bulbs used for fluorescent lights have a mercury gas pressure of 1.07 Pa at 40 °C. How many milligrams of liquid mercury must evaporate at 40 °C to yield this pressure in a 1.39-L fluorescent bulb?

OBJECTIVE 16

66. Low-pressure sodium lamps have a sodium vapor pressure of 0.7 Pa at 260 °C. How many micrograms of Na must evaporate at 260 °C to yield a pressure of 0.7 Pa in a 27-mL bulb?

OBJECTIVE 16

67. Flashtubes are light bulbs that produce a high-intensity flash of light that is very short in duration. They are used in photography and for airport approach lighting. The xenon gas, Xe, in flashtubes has large atoms that can quickly carry the heat away from the filament to stop the radiation of light. At what temperature does 1.80 g of Xe in a 0.625-L flashtube have a pressure of 75.0 kPa?

68. As part of the production of vinyl chloride—which is used to make polyvinyl chloride (PVC) plastic—1,2-dichloroethane, $ClCH_2CH_2Cl$, is decomposed to form vinyl chloride and hydrogen chloride. If 515 kg of 1,2-dichloroethane gas is released into a 96-$m^3$ container, at what temperature will the pressure become 3.4 atm?

OBJECTIVE 16

69. When 0.690 g of an unknown gas is held in an otherwise-empty 285-mL container, the pressure is 756.8 mmHg at 19 °C. What is the molecular mass of the gas?

OBJECTIVE 16

70. A 1.02-g sample of an unknown gas in a 537-mL container generates a pressure of 101.7 kPa at 298 K. What is the molecular mass of the gas?

OBJECTIVE 16

71. Consider two helium tanks used to fill children's balloons. The tanks have equal volume and are at the same temperature, but tank A is new and therefore has a higher pressure of helium than tank B, which has been used to fill many balloons. Which tank holds gas with a greater density, tank A or tank B? Support your answer.

72. Two hot-air balloons have just been launched. They have equal volume and pressure, but balloon A has a more efficient heating system, so the gas in balloon A is at a higher temperature than the gas in balloon B. Which balloon holds gas with a greater density, balloon A or balloon B? Support your answer. Which balloon do you think will rise faster?

73. Butadiene, $CH_2CHCHCH_2$, a suspected carcinogen, is used to make plastics such as styrene-butadiene rubber (SBR). Because it is highly flammable and forms explosive peroxides when exposed to air, it is often shipped in insulated containers at about 2 °C. What is the density of butadiene when its pressure is 102 kPa and its temperature is 2 °C?

OBJECTIVE 17

74. Picture a gas in the apparatus shown in Figure 13.2. If the temperature and number of gas particles remain constant, how would you change the volume to increase the pressure of the gas?

75. In order to draw air into your lungs, your diaphragm and other muscles contract, increasing the lung volume. This lowers the air pressure of the lungs to below atmospheric pressure, and air flows in. When your diaphragm and other muscles relax, the volume of the lungs decreases, and air is forced out.

OBJECTIVE 18

a. If the total volume of your lungs at rest is 6.00 L and the initial pressure is 759 mmHg, what is the new pressure if the lung volume is increased to 6.02 L?

b. If the total volume of your lungs at rest is 6.00 L and the initial pressure is 759 mmHg, at what volume will the pressure be 763 mmHg?

OBJECTIVE 18

76. Dry air is 78% nitrogen gas, 21% oxygen gas, and 0.9% argon gas. These gases can be separated by distillation after the air is first converted to a liquid. In one of the steps in this process, filtered air is compressed enough to increase the pressure to about 5.2 atm.

a. To what volume must 175 L of air at 1.0 atm be compressed to yield a pressure of 5.2 atm?

b. The argon derived from the distillation of air can be used in fluorescent light bulbs. When 225 L of argon at 0.997 atm is distilled from the air and allowed to expand to a volume of 91.5 $m^3$, what is the new pressure of the gas in pascals?

OBJECTIVE 18

77. A scuba diver has 4.5 L of air in his lungs 66 ft below the ocean surface, where the pressure is 3.0 atm. What would the volume of this gas be at the surface, where the pressure is 1.0 atm? If the diver's lungs can hold no more than 7.0 L without rupturing, is he in trouble if he doesn't exhale as he rises to the surface?

78. Picture a gas in the apparatus shown in Figure 13.2. If the volume and temperature remain constant, how would you change the number of gas particles to increase the pressure of the gas?

79. Picture a gas in the apparatus shown in Figure 13.2. If the pressure and temperature remain constant, how would you change the number of gas particles to increase the volume of the gas?

OBJECTIVE 18

80. A balloon containing 0.62 mole of gas has a volume of 15 L. Assuming constant temperature and pressure, how many moles of gas does the balloon contain when enough gas leaks out to decrease the volume to 11 L?

81. Picture a gas in the apparatus shown in Figure 13.2. If the volume and number of gas particles remain constant, how would the temperature have to change to increase the pressure of the gas?

OBJECTIVE 18

82. Consider a weather balloon with a volume on the ground of 113 $m^3$ at 101 kPa and 14 °C.

a. If the balloon rises to a height where the temperature is −50 °C and the pressure is 2.35 kPa, what will its volume be? (In 1958, a balloon reached 101,500 ft or 30.85 km, the height at which the temperature is about −50 °C.)

b. When the balloon rises to 10 km, where the temperature is −40 °C, its volume increases to 129 $m^3$. If the pressures inside and outside the balloon are equal, what is the atmospheric pressure at this height?

c. If the balloon has a volume of 206 $m^3$ at 25 km, where the pressure is 45 kPa, what is the temperature at this height?

OBJECTIVE 18

83. Ethylene oxide, which is used as a rocket propellant, is formed from the reaction of ethylene gas with oxygen gas.

a. After 273 $m^3$ of ethylene oxide is formed at 748 kPa and 525 K, the gas is cooled at constant volume to 293 K. What is the new pressure?

b. After 273 $m^3$ of ethylene oxide at 748 kPa and 525 K is cooled to 293 K, it is allowed to expand to $1.10 \times 10^3$ $m^3$. What is the new pressure?

c. What volume of ethylene gas at 293 K and 102 kPa must be compressed to yield 295 $m^3$ of ethylene gas at 808 kPa and 585 K?

OBJECTIVE 18

84. The hydrogen gas used to manufacture ammonia can be made from small hydrocarbons such as methane in the so-called steam-reforming process, which is run at high temperature and pressure.

a. If 3.2 atm of methane gas at 21 °C is introduced into a container, to what temperature must the gas be heated to increase the pressure to 12 atm?

b. If $4.0 \times 10^3$ L of methane gas at 21 °C is heated and allowed to expand at a constant pressure, what will the volume become, in cubic meters, when the temperature reaches 815 °C?

c. What volume of methane gas at 21 °C and 1.1 atm must be compressed to yield $1.5 \times 10^4$ L of methane gas at 12 atm and 815 °C?

### Section 13.3 Equation Stoichiometry and Ideal Gases

85. Chlorine gas is used in the production of many other chemicals, including carbon tetrachloride, polyvinyl chloride plastic, and hydrochloric acid. It is produced from the electrolysis of molten sodium chloride. What minimum mass of sodium chloride, in megagrams, is necessary to make $2.7 \times 10^5$ L of chlorine gas at standard temperature and pressure, STP? OBJECTIVE 19

$$2NaCl(l) \xrightarrow{\text{Electrolysis}} 2Na(l) + Cl_2(g)$$

86. Sulfur is used in the production of vulcanized rubber, detergents, dyes, pharmaceuticals, insecticides, and many other substances. The final step in the Claus method for making sulfur is OBJECTIVE 19

$$SO_2(g) + 2H_2S(g) \rightarrow 3S(s) + 2H_2O(l)$$

a. What minimum volume of sulfur dioxide at STP would be necessary to produce 82.5 kg of sulfur?

b. What maximum mass of sulfur, in kilograms, could be produced by the reaction of $1.3 \times 10^4$ m$^3$ of $SO_2(g)$ with $2.5 \times 10^4$ m$^3$ of $H_2S(g)$, both at STP?

87. Air bags in cars inflate when sodium azide, $NaN_3$, decomposes to generate nitrogen gas. How many grams of $NaN_3$ must react to generate 112 L $N_2$ at 121 kPa and 305 K? OBJECTIVE 19

$$2NaN_3(s) \rightarrow 2Na(s) + 3N_2(g)$$

88. When natural gas is heated to 370 °C, the sulfur it contains is converted to hydrogen sulfide. The $H_2S$ gas can be removed by a reaction with sodium hydroxide that forms solid sodium sulfide and water. OBJECTIVE 19

$$H_2S + 2NaOH \rightarrow Na_2S + 2H_2O$$

What volume of $H_2S(g)$ at 370 °C and 1.1 atm can be removed by 2.7 Mg of NaOH?

89. The hydrogen chloride gas used to make polyvinyl chloride (PVC) plastic is made by reacting hydrogen gas with chlorine gas. What volume of $HCl(g)$ at 105 kPa and 296 K can be formed from 1150 m$^3$ of $H_2(g)$ at STP? OBJECTIVE 19

$$H_2(g) + Cl_2(g) \rightarrow 2HCl(g)$$

OBJECTIVE 19

90. The hydrogen gas needed to make ammonia, hydrogen chloride gas, and methanol can be obtained from small hydrocarbons such as methane or propane through the steam-reforming process. If $2.7 \times 10^7$ L of $C_3H_8(g)$ at 810 °C and 8.0 atm react in the first step of the process, shown below, what is the maximum volume in liters of CO(g) at STP that can form?

$$C_3H_8 + 3H_2O \rightarrow 3CO + 7H_2$$

$$3CO + 3H_2O \rightarrow 3CO_2 + 3H_2$$

OBJECTIVE 19

91. Ammonia and carbon dioxide combine to produce urea, $NH_2CONH_2$, and water through the following reaction. About 90% of the urea is used to make fertilizers, but some is also used to make animal feed, plastics, and pharmaceuticals. To get the optimum yield of urea, the ratio of $NH_3$ to $CO_2$ is carefully regulated at a 3:1 molar ratio of $NH_3$ to $CO_2$.

$$2NH_3(g) + CO_2(g) \rightarrow NH_2CONH_2(s) + H_2O(l)$$

a. What volume of $NH_3$ at STP must be combined with $1.4 \times 10^4$ L of $CO_2$ at STP to yield the 3:1 molar ratio?

b. The concentration of urea in a typical liquid fertilizer is about 1.0 M $NH_2CONH_2$. What minimum volume of $CO_2$ at 190 °C and 34.0 atm is needed to make $3.5 \times 10^4$ L of 1.0 M $NH_2CONH_2$?

OBJECTIVE 19

92. Alka-Seltzer® contains the base sodium hydrogen carbonate, $NaHCO_3$, which reacts with hydrochloric acid in the stomach to yield sodium chloride, carbon dioxide, and water. What volume of $CO_2$ gas at 104.5 kPa and 311 K will form when 10.0 mL of 1.0 M HCl is mixed with an excess of $NaHCO_3$?

OBJECTIVE 19

93. Some chlorofluorocarbons, CFCs, are formed from carbon tetrachloride, $CCl_4$, which can be made from the methane in natural gas, as shown below. What minimum volume, in cubic meters, of methane gas at STP must be added to react completely $1.9 \times 10^6$ L of $Cl_2(g)$ at STP?

$$CH_4(g) + 4Cl_2(g) \rightarrow CCl_4(l) + 4HCl(g)$$

OBJECTIVE 19

94. The carbon tetrachloride made in the previous problem can be used to make CFC-12, $CF_2Cl_2$, and CFC-11, $CFCl_3$:

$$3HF(g) + 2CCl_4(g) \rightarrow CF_2Cl_2(g) + CFCl_3(g) + 3HCl(g)$$

a. What is the minimum volume of HF(g), in cubic meters, at 1.2 atm and 19 °C that would be necessary to react with 235 kg of $CCl_4$ in this reaction?

b. What is the maximum volume of CFC-12 at 764 mmHg and 18 °C that can be formed from 724 $m^3$ of hydrogen fluoride gas at 1397 mmHg and 27 °C?

95. Ethylene oxide, $C_2H_4O$, is a cyclic compound; its two carbon atoms and its oxygen atom form a three-membered ring. The small ring is somewhat unstable, so ethylene oxide is a very reactive substance, useful as a rocket propellant, to sterilize medical plastic tubing, and to make polyesters and antifreeze. It is formed from the reaction of ethylene, $CH_2CH_2$, with oxygen at 270–290 °C and 8–20 atm in the presence of a silver catalyst.

OBJECTIVE 19

$$2CH_2CH_2(g) + O_2(g) \rightarrow 2\ CH_2{-}CH_2(g) \quad (\text{ring with } \ddot{O} \text{ bridging the two } CH_2)$$

a. What minimum mass, in megagrams, of liquefied ethylene, $CH_2CH_2$, is necessary to form $2.0 \times 10^2\ m^3$ of ethylene oxide, $C_2H_4O$, at 107 kPa and 298 K?

b. Ethylene can be purchased in cylinders containing 30 lb of liquefied ethylene. Assuming complete conversion of ethylene into ethylene oxide, determine how many of these cylinders would be necessary to make $2.0 \times 10^2\ m^3$ of ethylene oxide at 107 kPa and 298 K.

c. How many liters of ethylene oxide gas at 2.04 atm and 67 °C can be made from $1.4 \times 10^4$ L of $CH_2CH_2$ gas at 21 °C and 1.05 atm?

d. How many kilograms of liquefied ethylene oxide can be made from the reaction of $1.4 \times 10^4$ L of $CH_2CH_2$ gas at 1.05 atm and 21 °C and $8.9 \times 10^3$ L of oxygen gas at 0.998 atm and 19 °C?

96. Methanol, $CH_3OH$, a common solvent, is also used to make formaldehyde, acetic acid, and detergents. It is made from the methane in natural gas in two steps, the second of which is run at 200–300 °C and 50–100 atm using metal oxide catalysts:

OBJECTIVE 19

$$3CH_4(g) + 2H_2O(g) + CO_2(g) \rightarrow 4CO(g) + 8H_2(g)$$

$$CO(g) + 2H_2(g) \rightarrow CH_3OH(l)$$

a. In the second reaction, how many megagrams of liquid methanol can be produced from 187 $m^3$ of carbon monoxide, CO, at $2.0 \times 10^3$ kPa and 500 K?

b. In the second reaction, what minimum volume of $H_2$ gas at $4.0 \times 10^3$ kPa and 500 K is necessary to convert 187 $m^3$ of CO gas at $2.0 \times 10^3$ kPa and 500 K?

c. How many megagrams of liquid methanol can be made from the reaction of 187 $m^3$ of $CO(g)$ at $2.0 \times 10^3$ kPa and 500 K and 159 $m^3$ of $H_2(g)$ at $4.0 \times 10^3$ kPa and 500 K?

OBJECTIVE 19

**97.** Rutile ore is about 95% titanium(IV) oxide, $TiO_2$. The $TiO_2$ is purified in the two reactions that follow. In the first reaction, what minimum volume, in cubic meters, of $Cl_2$ at STP is necessary to convert all of the $TiO_2$ in $1.7 \times 10^4$ metric tons of rutile ore into $TiCl_4$?

$$3TiO_2 + 4C + 6Cl_2 \xrightarrow{900\ °C} 3TiCl_4 + 2CO + 2CO_2$$

$$TiCl_4 + O_2 \xrightarrow{1200\text{–}1400\ °C} TiO_2 + 2Cl_2$$

OBJECTIVE 19

98. Acetic acid, $CH_3CO_2H$, can be made from methanol and carbon dioxide in the following reaction, which uses a rhodium/iodine catalyst. What is the percent yield when 513 kg of liquid $CH_3CO_2H$ form from the reaction of 309 kg of $CH_3OH$ with 543 $m^3$ of CO(*g*) at 1 atm and 175 °C?

$$CH_3OH + CO \rightarrow CH_3CO_2H$$

### Section 13.4 Dalton's Law of Partial Pressures

OBJECTIVE 20

99. Explain why the total pressure of a mixture of ideal gases is equal to the sum of the partial pressures of all the individual gases.

OBJECTIVE 21 OBJECTIVE 22

**100.** A typical 100-watt incandescent light bulb contains argon gas and nitrogen gas.

a. If enough argon and nitrogen are added to this light bulb to yield a partial pressure of Ar of 98.5 kPa and a partial pressure of $N_2$ of 10.9 kPa, what is the total gas pressure in the bulb?

b. If a 125-mL light bulb contains $5.99 \times 10^{-4}$ mole of $N_2$ and $5.39 \times 10^{-3}$ mole of Ar, what is the pressure in kPa in the bulb at a typical operating temperature of 119 °C?

OBJECTIVE 21 OBJECTIVE 22

101. A typical fluorescent light bulb contains argon gas and mercury vapor.

a. If the total gas pressure in a fluorescent light bulb is 307.1 Pa and the partial pressure of the argon gas is 306.0 Pa, what is the partial pressure of the mercury vapor?

b. If a 935-mL fluorescent bulb contains 4108 μg of argon gas and 77 μg of mercury vapor, what is the total gas pressure in the fluorescent bulb at 313 K?

**102.** If a 135-mL luminous tube contains 1.2 mg of Ne and 1.2 mg of Ar, at what temperature will the total gas pressure be 13 mmHg?

103. The atmosphere of Venus contains carbon dioxide and nitrogen gases. At the planet's surface, the temperature is about 730 K, the total atmospheric pressure is 98 atm, and the partial pressure of carbon dioxide is 94 atm. If scientists wanted to collect 10.0 moles of gas from the surface of Venus, what volume of gas should they collect?

**104.** Natural gas is a mixture of gaseous hydrocarbons and other gases in small quantities. The percentage of each component varies depending on the source. Consider a 21.2-$m^3$ storage container that holds 11.8 kg of methane gas, $CH_4$; 2.3 kg of ethane gas, $C_2H_6$; 1.1 kg of propane gas, $C_3H_8(g)$; and an unknown amount of other gases.

a. If the total pressure in the container is 1.00 atm at 21 °C, how many moles of gases other than $CH_4$, $C_2H_6$, and $C_3H_6$ are in the container?

b. What percentage of the gas particles in the cylinder are methane molecules?

### Additional Problems

105. One balloon is filled with helium, and a second balloon is filled to the same volume with xenon. Assuming that the two balloons are at the same temperature and pressure and that the gases are ideal, which of the following statements is true?

a. The helium balloon contains more atoms than the xenon balloon.

b. The helium balloon contains fewer atoms than the xenon balloon.

c. The two balloons contain the same number of atoms.

**106.** Although the temperature decreases with increasing altitude in the troposphere (the lowest portion of the earth's atmosphere), it increases again in the upper portion of the stratosphere. Thus, at about 50 km, the temperature is back to about 20 °C. Compare the following properties of the air along the California coastline at 20 °C and air at the same temperature but an altitude of 50 km. Assume that the percent composition of the air in both places is essentially the same.

a. Do the particles in the air along the California coastline and the air at 50 km have the same average kinetic energy? If not, which has particles with the higher average kinetic energy? Explain your answer.

b. Do the particles in the air along the California coastline and the air at 50 km have the same velocity? If not, which has particles with the higher average velocity? Explain your answer.

c. Does the air along the California coastline and the air at 50 km have the same average distance between the particles? If not, which has the greater average distance between the particles? Explain your answer.

d. Do the particles in the air along the California coastline and the air at 50 km have the same average frequency of particle collisions? If not, which has the greater frequency of between particle collisions? Explain your answer.

e. Does the air along the California coastline and the air at 50 km have the same density? If not, which has the higher density? Explain your answer.

f. Does the air along the California coastline and the air at 50 km have the same gas pressure? If not, which has the higher gas pressure? Explain your answer.

107. Compare the following properties of air on a sunny 28 °C day at a Florida beach to the properties of air along the coast of Alaska, where the temperature is −4 °C. Assume that the barometric pressure is the same in both places and that the percent composition of the air is essentially the same.

   a. Do the air in Florida at 28 °C and the air at −4 °C in Alaska have the same density? If not, which has the higher density?

   b. Do the air in Florida at 28 °C and the air at −4 °C in Alaska have the same average distance between the particles? If not, which has the higher average distance between the particles?

108. With reference to the relationships between properties of gases, answer the following questions.

   a. The pressure in the tires of a car sitting in a garage is higher at noon than at midnight. Why?

   b. The pressure in a car's tires increases as it is driven from home to school. Why?

109. With reference to the relationships between properties of gases, answer the following questions.

   a. When the air is pumped out of a gallon can, the can collapses. Why?

   b. A child's helium balloon escapes and rises into the air where it expands and pops. Why?

110. When people inhale helium gas and talk, their voices sound strange, as if it had been recorded and played back at a faster rate. Why do you think that happens?

111. If the following reaction is run in a cylinder with a movable piston that allows the pressure of gas in the cylinder to remain constant, and if the temperature is constant, what will happen to the volume of gas as the reaction proceeds? Why?

$$N_2(g) + 3H_2(g) \rightarrow 2NH_3(g)$$

### Discussion Problems

112. Consider two identical boxes, one containing helium gas and one containing xenon gas, both at the same temperature. Each box has a small hole in the side that allows atoms to escape. Which gas will escape faster? Why?

113. Is it easier to drink from a straw at the top of Mount Everest or at the seashore in Monterey, California? What causes the liquid to move up through a straw?

114. Like any other object that has mass, gas particles are attracted to the earth by gravity. Why don't gas particles in a balloon settle to the bottom of the balloon?

115. When you look at a beam of sunlight coming through the window, you can see dancing particles of dust. Why are they moving? Why don't they settle to the floor?

116. Which of the following statements is/are true of a gas in a closed container?
    a. When the particles collide, the temperature rises in the container.
    b. When more gas is added to the container, the greater number of collisions between particles leads to an increase in temperature.
    c. If the gas is heated to a higher temperature, the number of collisions per second between particles will increase.
117. Baking powder contains sodium hydrogen carbonate, $NaHCO_3$, and a weak acid. Why does the reaction between these two components cause a donut to rise?
118. Use our model for gases to answer the following questions.
    a. How does a vacuum cleaner work?
    b. Why does a suction cup on a dart cause the dart to stick to a wall? Why does it help to wet the suction cup first?
    c. NASA is developing suction-cup shoes for use in space stations. Why will they work inside the space station but not outside the space station?
119. To ensure that tennis balls in Denver will have the same bounce as tennis balls along the California coast, manufacturers put different gas pressures in them. Would you expect the gas pressure to be greater in a tennis ball intended for Denver or in one intended for San Francisco? Why?
120. About 90% of the atoms in the universe are thought to be hydrogen atoms. Thus hydrogen accounts for about 75% of the mass of the universe. Earth's crust, waters, and atmosphere contain about 0.9% hydrogen. Why is there so much less hydrogen on the planet Earth than in the rest of the universe? The planet Jupiter is thought to consist of about 92% hydrogen. Why is there so much more hydrogen on Jupiter than on Earth?
121. When a scuba tank is filled from a storage tank, the scuba tank gets warmer and the temperature of the storage tank goes down. Why?
122. Why does a helium-filled balloon that escaped to the ceiling drop to the floor overnight?

*In this chapter, we will explore the changes between liquids and gases, helping us to visualize the changes taking place on this mist covered lake.*

# Liquids: Condensation, Evaporation, and Dynamic Equilibrium

CHAPTER 14



OVER THE PAST WEEKS, you have seen numerous examples of how chemistry can deepen your understanding of everyday phenomena. In this chapter, we revisit the topic of liquids and the changes they undergo, in order to explain some of the things you might experience on an unusually warm spring morning.

**7:47 A.M.** You get out of bed and take a quick shower. As you step out of the shower, you shiver with cold, even though the day is already warm.

**7:51 A.M.** You turn toward the bathroom mirror to comb your hair, but it's so steamed up you can hardly see yourself. What causes water to collect on the mirror's surface?

**7:54 A.M.** The water that dripped onto the floor has almost dried up now, but the water that collects in the bottom of your toothbrush cup never seems to disappear. Why?

**8:01 A.M.** Drops of cologne land on the counter. Why do they evaporate so much more quickly than water, and even *more* quickly in the heat from your blow dryer?

**8:03 A.M.** You notice a can of hair spray sitting on the window ledge in the hot sun. You've heard that if its temperature gets high enough, the can will explode. What causes that to happen?

**8:04 A.M.** Heating the water for your morning tea, you wonder what causes bubbles to form when the water boils—and why they *don't* form until the water becomes extremely hot. That reminds you—why did it take so long to boil the potatoes during your backpacking trip to the high Sierras last week?

**8:12 A.M.** Whew! That's a lot of wondering for the first few minutes of your day. Why not relax, drink your tea, and settle down to read this chapter? All of your questions are about to be answered.

*Why does water vapor condense on a bathroom mirror?*

## Review Skills

The presentation of information in this chapter assumes that you can already perform the tasks listed below. You can test your readiness to proceed by answering the Review Questions at the end of the chapter. This might also be a good time to read the Chapter Objectives, which precede the Review Questions.

- Describe the relationship between the temperature of a substance and the motion of its component particles. (Section 2.1)
- Compare the freedom of motion and the attractions between the component particles in solids, liquids, and gases. (Section 2.1)
- Given a formula for a compound, classify it as either a molecular compound or an ionic compound. (Section 3.2)
- Given the names or formulas for two elements, decide whether the bond that would form between them would be covalent or ionic. (Section 3.2)
- Given a name or chemical formula, tell whether it represents a binary ionic compound, an ionic compound with polyatomic ion(s), a binary covalent compound, a binary acid, or an oxyacid. (Section 5.3)
- Convert between names and chemical formulas for binary ionic compounds, ionic compounds with polyatomic ion(s), binary covalent compounds, binary acids, and oxyacids. (Section 5.3)
- Given the formula for a molecule or polyatomic ion, draw a reasonable Lewis structure for it. (Section 12.2)
- Given the Lewis structure or enough information to produce it, draw the geometric sketch of a molecule, including bond angles. (Section 12.4)

## 14.1 Changing from Gas to Liquid and from Liquid to Gas—An Introduction to Dynamic Equilibrium

Our discussion of liquids focuses on two opposing processes: **condensation,** in which liquids are formed from gases, and **evaporation,** in which liquids return to gases. An understanding of these processes will help us understand the dynamics of *all* reversible physical and chemical changes.

### The Process of Condensation

You have probably noticed that the steam from a hot shower deposits liquid droplets on all the surfaces of your bathroom, but have you ever thought about what's happening on the submicroscopic level of atoms and molecules when this occurs? How does water vapor differ from liquid water, and why does water change from one state to the other? (Note that we use the term **vapor** to describe the gaseous form of a substance that is liquid at normal temperatures and pressures. We also use it to describe gas that has recently come from a liquid.)

Picture yourself riding on a water molecule in a system in which water vapor is present at a relatively high temperature. The first image in Figure 14.1 shows how you might visualize this gas. The spheres in this figure represent the particles of any substance; they could be atoms or molecules. For our discussion, each of the spheres represents a water molecule.

OBJECTIVE 2

Initially, your molecule moves in a straight line, with no significant interactions with other particles. Then, all of a sudden, it collides violently with a slower-moving water molecule. Because of the collision, your particle changes direction and slows down, the other molecule speeds up, and both molecules move off along new, straight-line paths. The contact between the two particles was so brief that at no time did you detect any attraction or repulsion between them. They simply collided, bounced off each other, and moved apart.

Now picture the same system as we slowly decrease the vapor's temperature, reducing the average velocity of the molecules. The collisions between your molecule and others decrease in violence, so much so that sometimes two colliding molecules stick together for a while, held by a mutual attraction. (One of the goals of this chapter is to describe this

OBJECTIVE 2

attraction and other types of attractions between particles.) Initially, these attractions do not last very long. A faster-moving water molecule slams into your pair, knocking them apart, and all three particles continue on alone. As the temperature decreases further, however, pairs of molecules are less likely to be knocked apart. Instead, they begin to form trios and even larger clusters (Figure 14.1, center image).

The more slowly the particles move, the harder it becomes for them to escape their mutual attractions. The molecule clusters grow so large that they fall to the bottom of the container (or cling to its sides), where they

OBJECTIVE 2

combine to form liquid water (Figure 14.1, bottom image). This is the process of condensation.

In short, as the gaseous water molecules in the steam from a hot shower cool, they cluster—or condense—into droplets of liquid water, forming a fog that fills the room. These droplets then combine with other droplets and collect on every surface in the bathroom, including your bathroom mirror.

**Figure 14.1**
**The Change from Gas to Liquid**

At a high temperature, there are no significant attractions between the particles.

As the temperature is lowered, attractions between particles lead to the formation of very small clusters that remain in the gas phase.

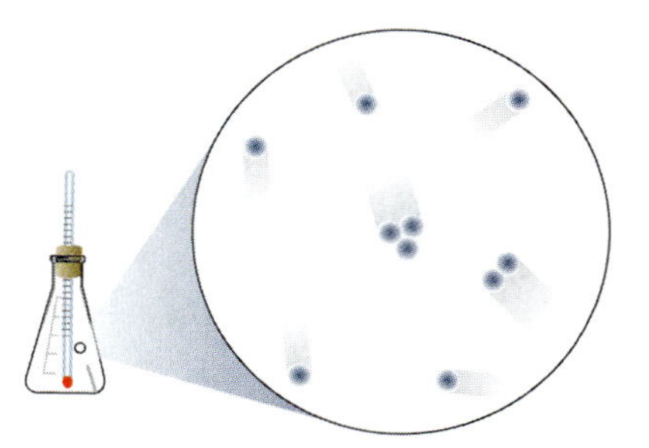

As the temperature is lowered further, the particles move slowly enough to form clusters so large that they drop to the bottom of the container and combine to form a liquid.

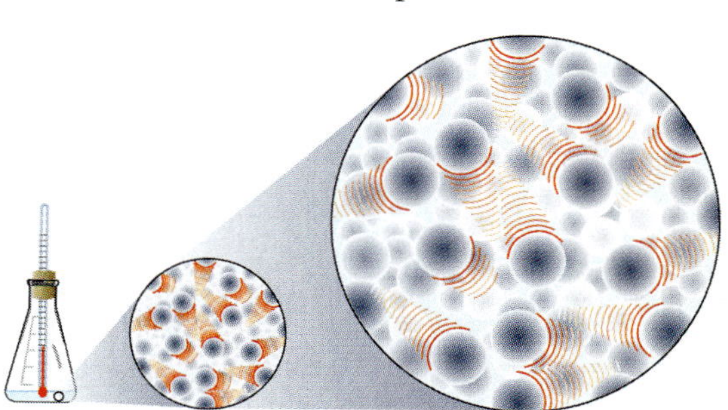

OBJECTIVE 2

## The Process of Evaporation

Why does rainwater on a flat street evaporate so much more rapidly than the same amount of water in a deep puddle? Why does nail polish remover evaporate so much more quickly than water, and why do all liquids evaporate more quickly when they are hot? Many factors affect the **rate of evaporation,** the amount of liquid changing to gas per second. We begin our exploration of them by reviewing some information first presented in Section 2.1.

To get an idea of the changes taking place when a liquid evaporates, picture yourself riding on a molecule in a liquid held inside a closed container. In a typical liquid, about 70% of the volume is occupied by particles, so there is not much empty space for you and your molecule to move into. Your movement is further hindered by the attractions between your molecule and the other molecules around it. Nevertheless, you bump and jostle to all parts of the container, colliding with other particles, changing velocity and direction, breaking old attractions, and making new ones.

Now picture your particle moving toward the surface of the liquid. If it continues along the same trajectory when it gets to the surface, its momentum (mass times velocity) will take you out of the liquid into the space above—until the attractions that hold the particles together in the liquid pull you back. If your particle is moving fast enough, though, its momentum can take it so far beyond the surface that the attractions will be broken. In that case, the particle escapes the liquid and joins the vapor above the liquid (see Figure 14.2).

OBJECTIVE 3

**Figure 14.2**
**Evaporation**

OBJECTIVE 3

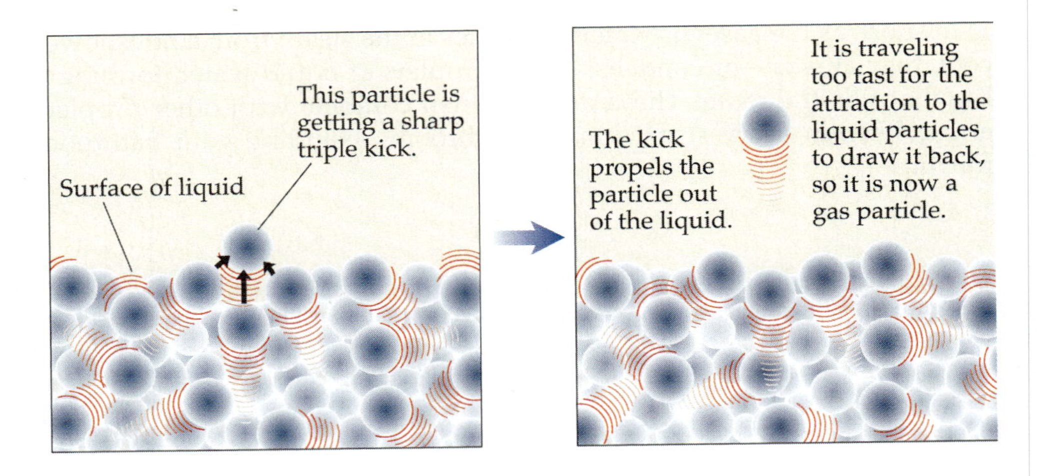
This particle is getting a sharp triple kick.
Surface of liquid
The kick propels the particle out of the liquid.
It is traveling too fast for the attraction to the liquid particles to draw it back, so it is now a gas particle.

OBJECTIVE 4

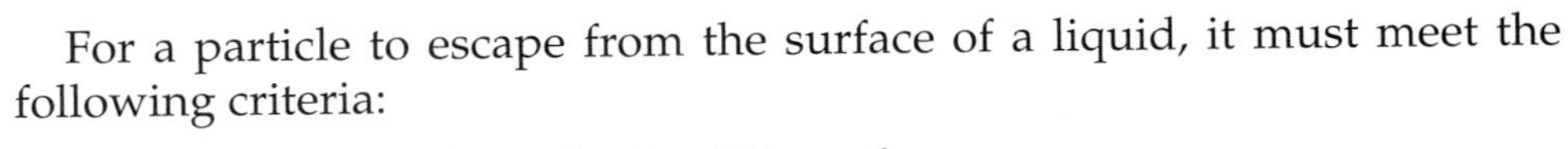

For a particle to escape from the surface of a liquid, it must meet the following criteria:

- The particle must be at the liquid's surface.
- Its direction of motion must take it beyond the liquid's surface.
- Its momentum must be great enough to overcome the backward pull of the other particles at the surface.

The fastest water molecules are escaping from the water on his skin. The average velocity, and thus the temperature, of the remaining water molecules are lower.

## How Evaporation Causes Cooling

Particles must have a relatively high velocity in order to break the attractions holding them in the liquid. Particles with relatively low velocity do not have enough momentum to escape. Thus, fast-moving particles escape, and slow-moving particles do not. After the more rapidly moving particles have escaped, the particles left in the liquid have a lower average velocity, and lower average velocity means lower temperature. In other words, evaporation lowers a liquid's temperature. We experience this change in temperature when we step out of a shower or a swimming pool.

## Rate of Evaporation

OBJECTIVE 5

The rate of evaporation—the number of particles moving from liquid to gas per second—depends on three factors:

- Surface area of the liquid
- Strength of attractions between the particles in the liquid
- Temperature of the liquid

OBJECTIVE 6

OBJECTIVE 7

Greater surface area means more particles at the surface of a liquid, which leads to a greater rate of evaporation. Picture two identical glasses of water. One glass is left on a table, and the second glass is emptied onto the floor. At the same temperature, the percentage of particles with the momentum needed to escape from the surface is the same for both samples of water, but the only particles that get an opportunity to escape are particles at or near the surface. Because there are a lot more particles at the surface of the water on the floor, the number of particles that escape from that water into the gas phase is much greater than from the water in the glass. This explains why rain evaporates so much more rapidly from a flat street than from a deep puddle.

Water evaporates much more quickly from a flat street than from a deep pothole.

Weaker attractions between particles also lead to a higher rate of evaporation. Consider the differences between liquid acetone and liquid water. Acetone, $CH_3COCH_3$, is a common laboratory solvent and is the main ingredient in many nail polish removers. If you have ever used it, you probably noticed that it evaporates much more rapidly than water. The attractions between acetone molecules are weaker than those between water molecules, so it is easier for particles to escape from the surface of liquid acetone than from that of liquid water. At the same temperature, more molecules will escape per second from the acetone than from the water (Figure 14.3).

OBJECTIVE 8

**Weaker attractions between particles**

↓

Lower momentum necessary for particles to escape the liquid

↓

At a constant temperature, a greater percentage of the particles have the momentum necessary to escape.

↓

**Higher rate of evaporation**

Figure 14.3 Attractions Influence Rate of Evaporation

OBJECTIVE 8

The rate of evaporation is also dependent on the liquid's temperature. Increased temperature increases the average velocity and momentum of the particles. As a result, a greater percentage of particles will have the minimum momentum necessary to escape, so the liquid will evaporate more quickly (Figure 14.4). This explains why rainwater on a city street evaporates much more rapidly after the sun comes out.

OBJECTIVE 9

Figure 14.4 Temperature Influences Rate of Evaporation

**Increased temperature**

↓

Increased velocity and momentum of the particles

↓

Increased percentage of particles having the minimum momentum to escape

↓

**Increased rate of evaporation**

## Dynamic Equilibrium Between Liquid and Vapor

In a closed container, evaporation and condensation take place at the same time. Our model helps us visualize this situation.

Picture a container that has been partly filled with liquid and then closed tightly so that nothing can escape. As soon as the liquid is poured in, particles begin to evaporate from it into the space above at a rate that is dependent on the surface area of the liquid, the strengths of attractions between the liquid particles, and the temperature. If these three factors remain constant, then the rate of evaporation will be constant.

Imagine yourself riding on one of the particles in the vapor phase above the liquid. Your rapidly moving particle collides with other particles, with the container walls, and with the surface of the liquid. When it collides with the surface of the liquid and its momentum carries it into the liquid, your particle returns to the liquid state. The number of gas particles that return to liquid per second is called the **rate of condensation.**

Now let's go back to the instant the liquid is poured into the container. If we assume that the container initially holds no vapor particles, then no condensation is occurring when the liquid is first added. As the liquid evaporates, however, particles of vapor gradually collect in the space above the liquid, and the condensation process slowly begins. As long as the rate of evaporation of the liquid is greater than the rate of condensation of the vapor, the concentration of vapor particles above the liquid will increase. However, as the concentration of vapor particles increases, the rate of collisions of vapor particles with the liquid increases, boosting the rate of condensation.

OBJECTIVE 10

If there is enough liquid in the container, not all of it will evaporate. Instead, the rising rate of condensation will eventually become equal to the rate of evaporation. At this point, for every particle that leaves the liquid, a particle somewhere else in the container returns to the liquid. Thus there is no net change in the amount of substance in the liquid state or in the amount of substance in the vapor state (Figures 14.5 and 14.6). There is no change in volume either (because our system is enclosed in a container), so there is no change in the concentration of vapor above the liquid and no change in the rate of collision with the surface of the liquid. Therefore, the rate of condensation stays constant.

OBJECTIVE 10

When two opposing rates of change are equal—such as the rate of evaporation and the rate of condensation in our closed system—we say the system has reached a **dynamic equilibrium.** Dynamic equilibrium is found in many of the systems that you will encounter in chemistry, so it is important to have a general understanding of the conditions necessary to create it. First, the system must exhibit two ongoing, opposing changes, from state A to state B and from state B to state A. In our example, state A is the liquid, state B is the vapor, and the two opposing changes are evaporation and condensation. For a dynamic equilibrium to exist, the rates of the two opposing changes must be equal, so that there are constant changes between state A and state B but no *net* change in the components of the system. In the dynamic equilibrium of our liquid–vapor system, the liquid is constantly changing to vapor and the vapor is constantly changing to liquid, but no net change in the amounts of liquid and vapor is observed (Figure 14.6).

OBJECTIVE 10

OBJECTIVE 11

An analogy might help with this visualization of dynamic equilibrium. It is common these days for children's indoor play areas to have a large bin filled with plastic balls. The bin is like a wading pool where the kids do not get wet. Picture two kids in a bin throwing balls at one another. Most of the balls that they throw hit the nets surrounding the bin, but a few escape through the openings and fall onto the floor (Figure 14.7a, page 592). The

OBJECTIVE 11

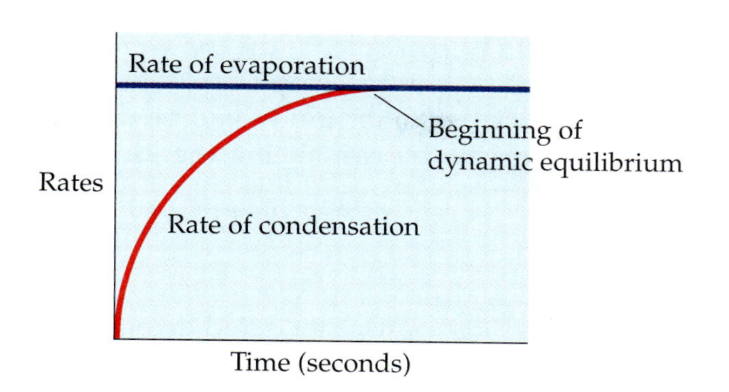


**Figure 14.5**
Rate of Evaporation, Rate of Condensation, and the Liquid–Vapor Equilibrium

OBJECTIVE 10

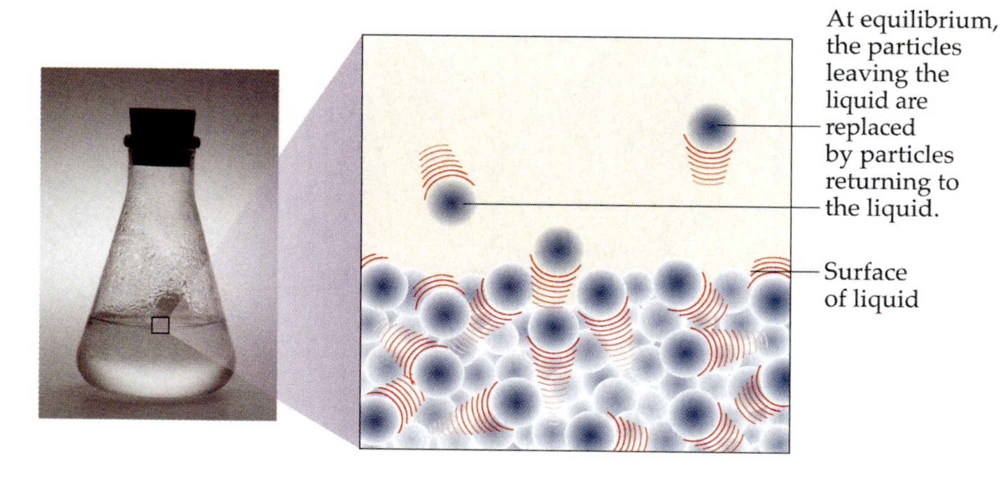

At equilibrium, the particles leaving the liquid are replaced by particles returning to the liquid.
Surface of liquid

**Figure 14.6**
Liquid–Vapor Equilibrium

OBJECTIVE 10

escape of the balls can be likened to that of particles evaporating from the surface of a liquid.

OBJECTIVE 11

The escaping balls attract the attention of an attendant, who rushes over to return them to the bin. At first, there are only a few balls on the floor, so the attendant has to move a relatively long way to get from one ball to another and throw them back in the bin. This makes the rate of return rather low, and the kids are able to throw the balls out faster than he can return them—a situation that results in a steady increase in the number of balls on the floor. Now there is less distance between the balls, and the attendant is able to return them to the bin at a faster and faster rate (Figure 14.7b). This situation is like the increasing rate of condensation that results from an increase in the concentration of gas above a liquid.

Eventually, the attendant is reaching the balls quickly enough to return them to the bin just as rapidly as the kids can throw them out. At this point, the process continues at a frantic pace but with no net change in the numbers of balls in the bin and on the floor. This is like the dynamic equilibrium reached between the rates of evaporation and condensation of liquid and vapor in a closed container. When the rates are equal, there is no net change in the amount of liquid or vapor in the system.

The dynamic equilibrium lasts until something happens to disrupt the system in a way that changes one or both of the rates. For example, the liquid–vapor system would be disrupted if the top were removed from the closed container, allowing vapor to escape and decreasing the rate of condensation. The plastic ball equilibrium would be disrupted if a parent came to the attendant's rescue by convincing the kids to stop throwing balls and go get ice cream.

**Figure 14.7**
**Plastic Ball Equilibrium**

(a) Kids throwing balls out faster than attendant can return them.

(b) Attendant is reaching and returning balls as fast as the kids throw them out.

## Equilibrium Vapor Pressure

OBJECTIVE 12

When a liquid is placed in a closed container and begins to evaporate, the vapor exerts a pressure against the container's walls (and against the surface of the liquid). When this system reaches a dynamic equilibrium between the rate of evaporation and the rate of condensation, the amount of vapor above the liquid remains constant. According to the ideal gas equation, the pressure of the vapor (or gas) is dependent on the vapor's concentration (moles divided by volume, or $n/V$) and its temperature:

$$PV = nRT \quad P = \frac{n}{V}RT = [\text{concentration}]RT$$

If the temperature of the system, as well as the concentration of the vapor, remains constant, then the pressure of the vapor stays constant as well.

OBJECTIVE 12

The space above a liquid in a closed container usually contains substances other than the vapor of the evaporating liquid. The pressure exerted by the vapor that has escaped the liquid is therefore the partial pressure of that substance in the total amount of gas inside the container. For example, if the container held both air and liquid at the time we closed it, any vapor that escapes the liquid will mix with the oxygen, nitrogen, and other gases in the air and contribute only a portion of the total pressure. The partial pressure of vapor above a liquid in a closed system with a dynamic equilibrium between the rate of evaporation and the rate of condensation is called the **equilibrium vapor pressure, $P_{vap}$.**

## SPECIAL TOPIC 14.1 Chemistry Gets the Bad Guys

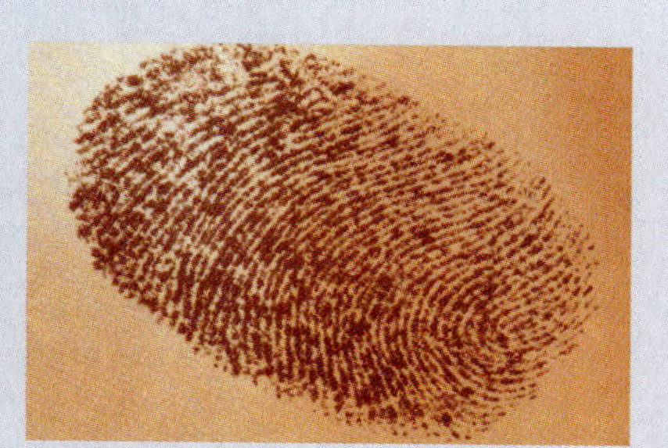

We've all seen it on TV and in the movies . . . the detective arrives at the crime scene and stalks into the room wearing a trench coat. The area is taped off, and the experts from the crime lab are already "dusting for prints." In the movies, this usually means spreading black powder that will stick to the fingerprints, but in the real world, forensic scientists have developed new and better ways to make fingerprints visible and help identify "the perp."

There are three types of fingerprints. If the people investigating a crime are lucky, the fingerprints will be either *plastic* (impressions left in soft material such as wax or paint) or *visible* (imprints made by blood, dirt, ink, or grease). Usually, however, the fingerprints are *latent* (invisible patterns that must be made visible before they can be of use). Latent fingerprints are composed of perspiration, oils, amino acids, proteins, inorganic salts (such as sodium and potassium chloride), and many other substances. In order to make latent fingerprints visible, the investigator covers the area with a substance that either sticks to them more than to the surfaces they're sitting on or, in some cases, reacts chemically with the fingerprint components.

A 1993 kidnapping case provided forensic chemists with a puzzling problem. An 8-year-old was kidnapped in Knoxville, Tennessee, but fortunately, she was able to escape and identify the car used in the kidnapping. When the car was checked for fingerprints, the investigators found many latent prints of the car's owner, but they could not find any fingerprints of the kidnapped girl. Knowing that this would raise doubts in the minds of the jury, they began to study the differences between the fingerprints of children and adults in hopes of finding an explanation. They found that an adult's fingerprints contain a relatively high concentration of molecules called esters that are produced by the oil glands on a grown person's face and then transferred to the fingers when they touch the face. These esters are not produced until after puberty, so they are not found in a child's fingerprints. The ester molecules are very large, with strong attractions that lead to low rates of evaporation. This discovery, which explained why the car owner's fingerprints could be found even after the child's prints had evaporated, helped convince the jury that the car's owner was guilty.

**Source:** Doris R. Kimbrough and Ronald DeLorenzo, "Solving the Mystery of Fading Fingerprints with London Dispersion Forces," *Journal of Chemical Education* 75, 10 (October 1998), pp. 1300–1301.

Different substances have different equilibrium vapor pressures at the same temperature. For example, the vapor pressure above liquid acetone in a closed container is higher than for water at the same temperature. Because the attractions between acetone molecules are weaker than the attractions between water molecules, it is easier for an acetone molecule to break them and move into the vapor phase. Therefore, the rate of evaporation from liquid acetone is greater than for water at the same temperature. When the dynamic equilibrium between evaporation and condensation for each liquid is reached, the rate of condensation for the acetone is higher than for water. The rate of condensation is higher because the concentration of vapor above the liquid is higher. The higher concentration of acetone particles creates a higher equilibrium vapor pressure. This logic is summarized in Figure 14.8, on page 594. The weaker the attractions between particles of a substance, the higher the equilibrium vapor pressure for that substance at a given temperature.

OBJECTIVE 13

**Figure 14.8**
Relative Equilibrium Vapor Pressures

OBJECTIVE 13

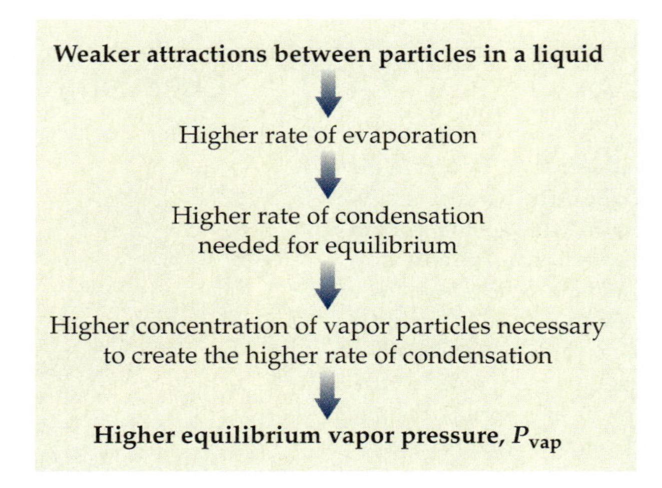

Weaker attractions between particles in a liquid
Higher rate of evaporation
Higher rate of condensation needed for equilibrium
Higher concentration of vapor particles necessary to create the higher rate of condensation
Higher equilibrium vapor pressure, $P_{vap}$


OBJECTIVE 14

The equilibrium vapor pressure of a liquid increases with increased temperature. At a higher temperature, the faster-moving particles can escape from the liquid more easily, so the rate of evaporation increases, leading to a higher rate of condensation at equilibrium. To reach this higher rate of condensation, the concentration of vapor above the liquid must rise to yield a higher rate of collisions with the surface of the liquid. A higher concentration of vapor leads to a higher vapor pressure (Figure 14.9). Figure 14.10 shows how the equilibrium vapor pressure varies with temperature for acetone and water.

**Figure 14.9**
Effect of Temperature on Equilibrium Vapor Pressure

OBJECTIVE 14

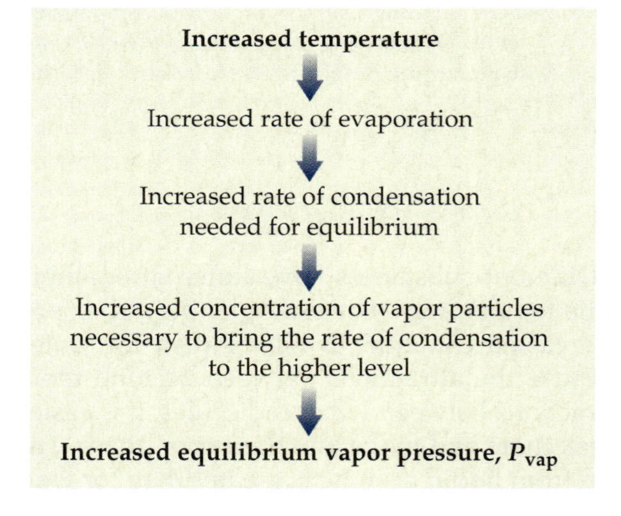

Increased temperature
Increased rate of evaporation
Increased rate of condensation needed for equilibrium
Increased concentration of vapor particles necessary to bring the rate of condensation to the higher level
Increased equilibrium vapor pressure, $P_{vap}$

Moscow firefighters, backed by a firefighting helicopter, battle a warehouse blaze that sent aerosol cans stored there soaring into the air and exploding.

A similar situation can arise in our children's play area. Seeing that the attendant is trying to return the balls to the bin, the kids might take on the challenge and begin to *try* to throw the balls out the opening. This increases the rate of escape and at first allows them to get ahead of the poor attendant, but as the concentration of the balls on the floor increases, his rate of return increases until the two rates are equal again.

We can now explain why high temperature can cause an aerosol can to explode. If you leave an aerosol can in the hot sun, the temperature of the substances in the can will go up, and the rates of evaporation of the liquids that it contains will increase. The concentration of vapor in the can will rise accordingly, and the vapor pressure will increase. If the pressure becomes too high, the can will explode. A typical aerosol can is built to withstand pressures up to about 10–12 times normal room pressure.

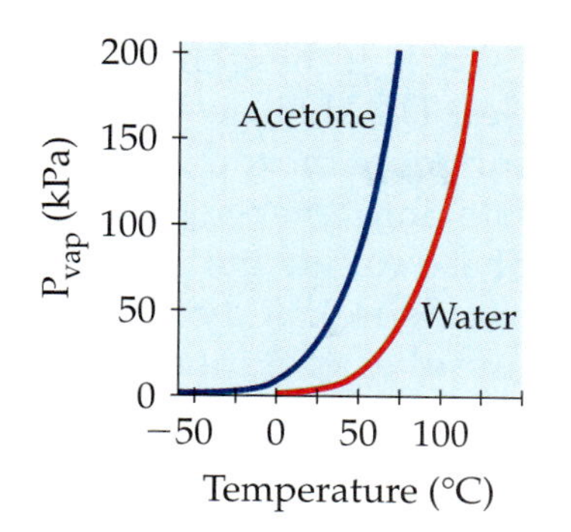


**Figure 14.10**
Equilibrium Vapor Pressure, $P_{vap}$, and Temperature

## 14.2 Boiling Liquids

Why do bubbles form in very hot water but not in water that is cold or even warm? Why does it take longer to boil your potatoes on a backpacking trip to the high Sierras? This section presents a model for visualizing the process of boiling and then uses it to answer questions like these.

### How Do Bubbles Form in Liquids?

OBJECTIVE 15

To understand why bubbles form in a boiling liquid, picture yourself riding on particle 1 in the liquid shown in Figure 14.11. Your particle is moving constantly, sometimes at very high velocity. When it collides with particles 2 and 3, it pushes them aside and moves in between them, leaving a small space in your wake. If at the same time particle 4 moves across this space and collides with particles 5 and 6, the space will grow in volume. Collisions like these are continuously causing spaces to form within the liquid. These spaces can be thought of as tiny bubbles (Figure 14.11).

OBJECTIVE 15

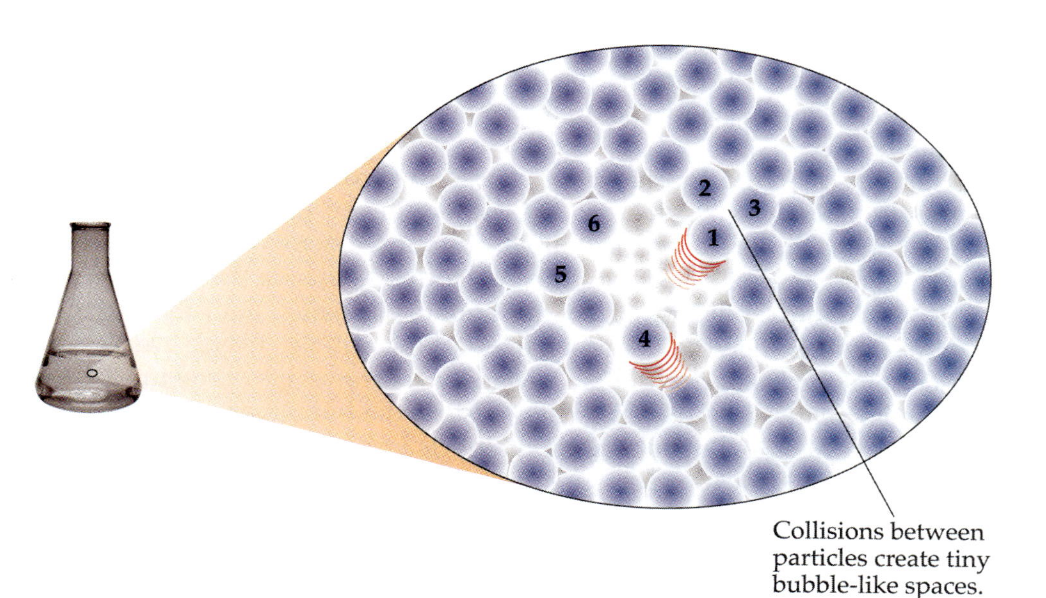


**Figure 14.11**
Spaces in Liquids

Each of these tiny bubbles is enclosed in a spherical shell of liquid particles. The shell acts as a surface that, except for its shape, is the same as the surface at the top of the liquid. Particles can escape from the surface (evaporate) into a vapor phase inside the bubble, and when particles in that vapor phase collide with the surface of the bubble, they return to the liquid state (condense). A dynamic equilibrium can form, in the bubble, between the rate of evaporation and the rate of condensation, just like the liquid–vapor equilibrium above the liquid in the closed container (Figure 14.12).

**Figure 14.12**
Bubble in a Liquid

OBJECTIVE 15

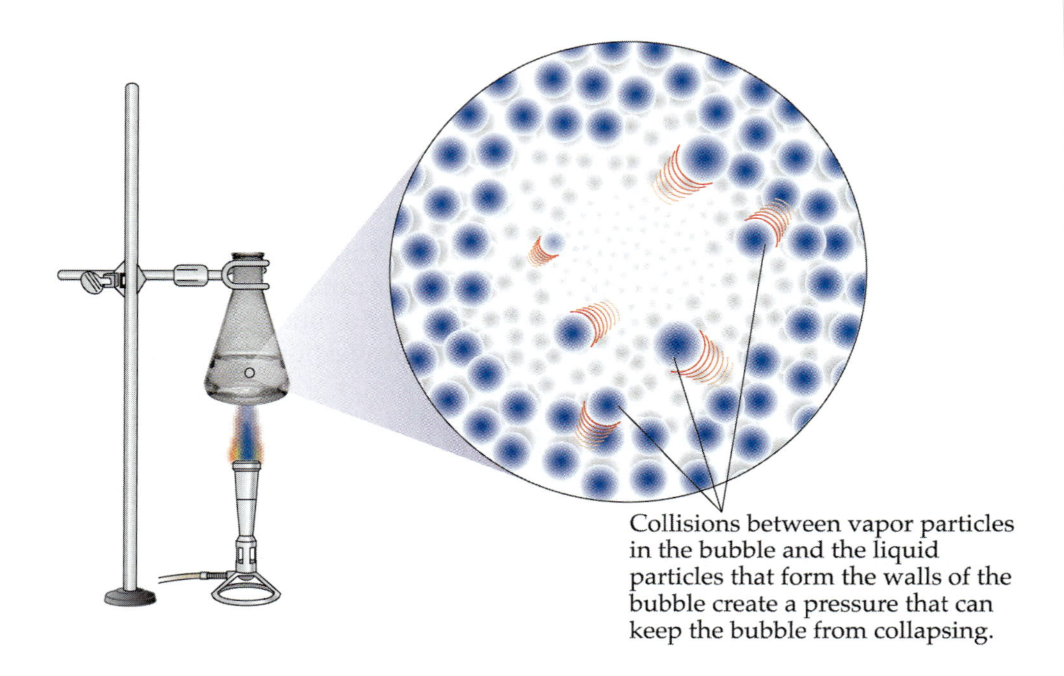

Collisions between vapor particles in the bubble and the liquid particles that form the walls of the bubble create a pressure that can keep the bubble from collapsing.

Each time a particle moves across a bubble's interior and collides with the bubble's surface, it exerts a tiny force pushing the wall of the bubble outward. All such collisions with the shell of the bubble combine to produce a gas pressure inside the bubble. This pressure is the same as the equilibrium vapor pressure for the vapor above the liquid in a closed container.

OBJECTIVE 16

There is an opposing pressure pushing inward on the bubble equal to the sum of the gas pressure on the upper surface of the liquid and the weight of the liquid above the bubble. You can view this external pressure as competing with the equilibrium vapor pressure inside the bubble. If the external pressure pushing in on the bubble is greater than the bubble's vapor pressure, then the liquid particles of the bubble surface are pushed closer together, and the bubble collapses. If the vapor pressure of the bubble is greater than the external pressure, the bubble grows. If the two pressures are equal, the bubble maintains its volume (Figure 14.13).

Therefore, if the vapor pressure of the bubble is greater than or equal to the external pressure acting on it, the bubble continues to exist. Because the vapor in bubbles is less dense than the liquid, the bubble will rise

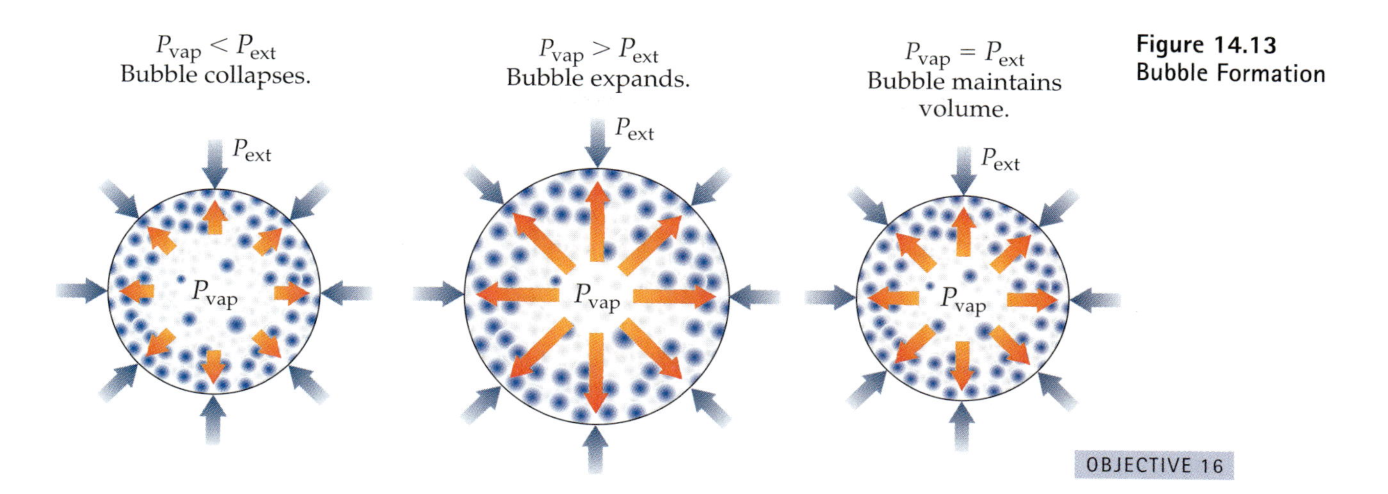


**Figure 14.13**
**Bubble Formation**

OBJECTIVE 16

through the liquid and escape from its upper surface. When this happens, the liquid is boiling. **Boiling** can be defined as the conversion of liquid into vapor anywhere in the liquid, rather than just at the top surface. The temperature at which a liquid boils is called the **boiling-point temperature**, or often just the *boiling point.*

Remember that an increase in temperature increases the equilibrium vapor pressure, which is also the pressure of bubbles as they form in the liquid. Below the boiling-point temperature, the vapor pressure is too low to counteract the greater external pressure, so our tiny bubbles that are always forming will quickly collapse. They do not last long enough to move to the surface, so the liquid does not boil. As the liquid is heated, the vapor pressure of the liquid increases until the temperature gets high enough to make the pressure within the bubble equal to the external pressure. At this temperature, the bubbles can maintain their volume, and the liquid boils. The boiling-point temperature can therefore be defined as the temperature at which the equilibrium vapor pressure of the liquid becomes equal to the external pressure acting on the liquid.

OBJECTIVE 16

## Response of Boiling-Point Temperature to External Pressure Changes

OBJECTIVE 17

If the external pressure acting on a liquid changes, then the internal vapor pressure needed to preserve a bubble changes, and therefore the boiling-point temperature changes. For example, if water is placed in a sealed container, and if the pressure above the water is decreased by pumping some of the gas out of the space above the water, the water can be made to boil at room temperature (Figure 14.14, on page 598). If the external gas pressure is lowered sufficiently, the particles have enough energy even at room temperature to escape into bubbles and maintain sufficient pressure inside them to

If the pressure above water is lowered enough, the water will boil at room temperature.

**Figure 14.14**
Effect of Changing External Pressure on Boiling-Point Temperatures

OBJECTIVE 17

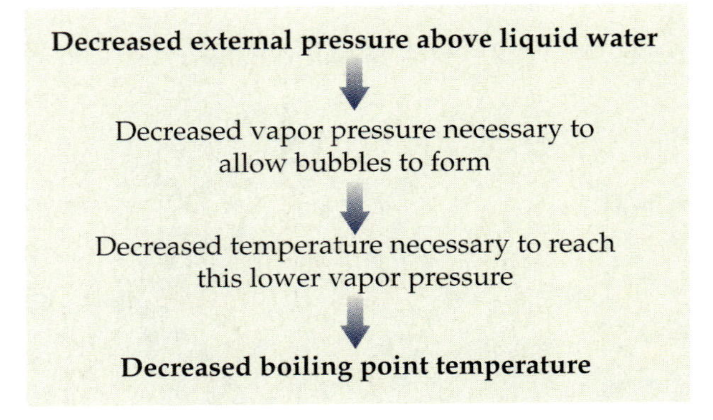


keep them from collapsing (remember, the external pressure pushing in on them is now lower).

**Decreased external pressure above liquid water**
Decreased vapor pressure necessary to allow bubbles to form
Decreased temperature necessary to reach this lower vapor pressure
**Decreased boiling point temperature**

OBJECTIVE 18

This helps us to explain why potatoes take longer to cook in boiling water in the mountains than at the seashore and why they cook much faster in a pressure cooker. When pure water begins to boil at a constant external pressure, its temperature levels off and stays constant at the boiling-point temperature as long as there is liquid in the container. (The heat being added to keep the liquid boiling works to increase the potential energy of the gas compared to that of the liquid, instead of increasing the kinetic energy of the liquid. If the average kinetic energy of the liquid is constant, the temperature remains constant.)

At 8840 meters above sea level, on the top of Mount Everest, the atmospheric pressure is about 34 kPa (0.34 atm). At this pressure, water boils at about 72 °C, and as we have just learned, the water remains at that temperature until all of it has boiled away. The potatoes will take much longer to cook at this temperature, which is 28 °C lower than the normal boiling temperature. If we could heat the water in a pressure cooker, however, the pressure above the water could increase to higher than the atmospheric pressure, causing the boiling point of the water to increase. For example, water boils at about 120 °C when the pressure above the water is 202 kPa (2 atm). Potatoes cook much faster at this temperature (Figure 14.15).

Because water boils at a lower temperature in the mountains, where the external pressure on the water is less, food in the boiling water cooks more slowly than it would at sea level.

Because the external pressure determines the temperature at which a liquid boils, we need to specify the external pressure associated with a reported boiling point. If the pressure is not specified, it is understood to be 1 atmosphere. The temperature at which the equilibrium vapor pressure of a liquid equals 1 atmosphere is called its **normal boiling-point temperature** or just its **normal boiling point.** If a liquid has an external pressure of one atmosphere acting on it, the normal boiling point is the temperature at which the liquid will boil.

For example, if you are told that the boiling point of hexane is 69 °C, you can assume that this is the temperature at which the equilibrium vapor pressure of the hexane reaches 1 atm. If the pressure acting on the liquid hexane is 1 atm, it will boil at 69 °C. If the pressure acting on the hexane is less than 1 atm, it will boil at a lower temperature. If the pressure acting on the hexane is greater than 1 atm, it will not boil until it is heated to a temperature above 69 °C.

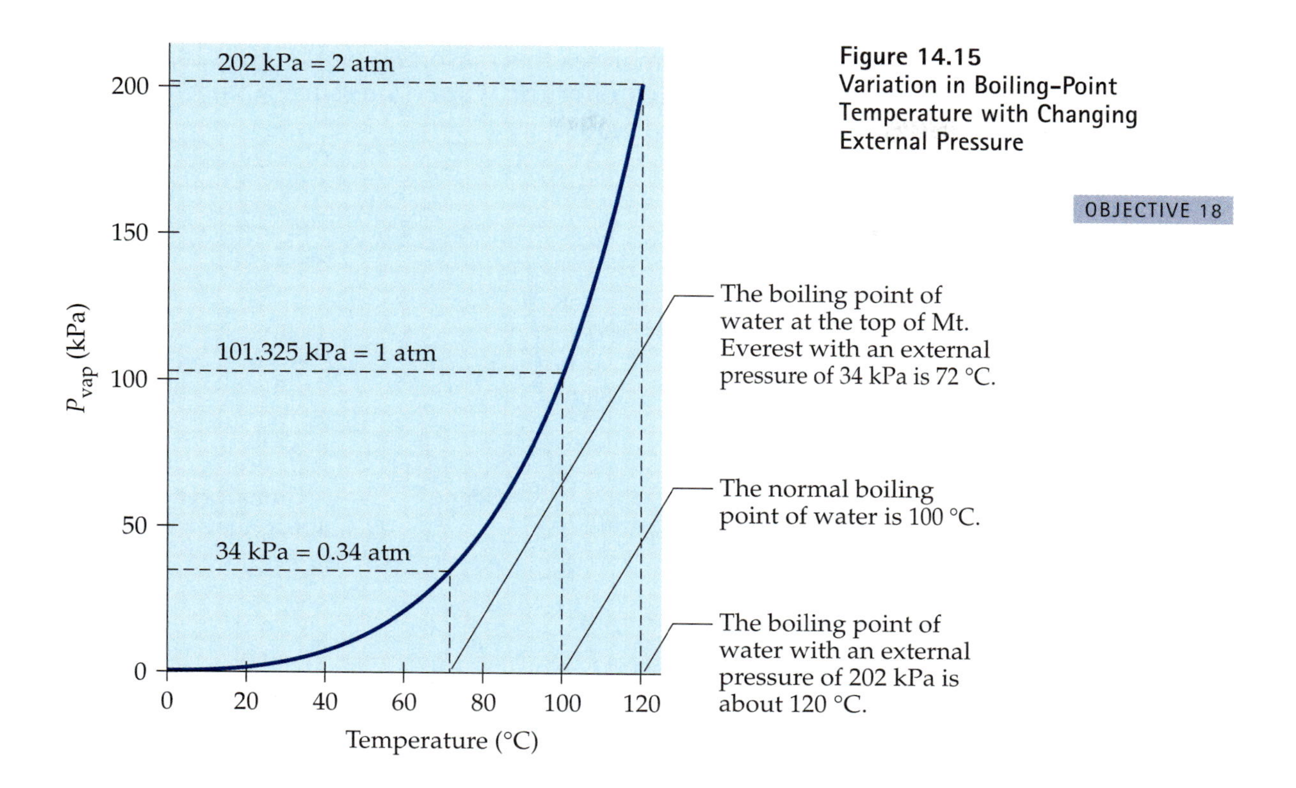


**Figure 14.15**
Variation in Boiling-Point Temperature with Changing External Pressure

OBJECTIVE 18

## Relative Boiling-Point Temperatures and Strengths of Attractions

OBJECTIVE 19

We can predict which of two substances has a higher boiling point by comparing the relative strengths of the attractions between the particles. We know that increased strength of attractions leads to decreased rate of evaporation, decreased rate of condensation at equilibrium, decreased concentration of vapor, and decreased vapor pressure at a given temperature. This leads to an increased temperature necessary to reach a vapor pressure of 1 atmosphere. Thus, stronger attractions holding the particles in the liquid lead to higher normal boiling points. For example, because the attractions between water molecules are stronger than the attractions between acetone molecules, the normal boiling point of water (100 °C) is higher than the normal boiling point of acetone

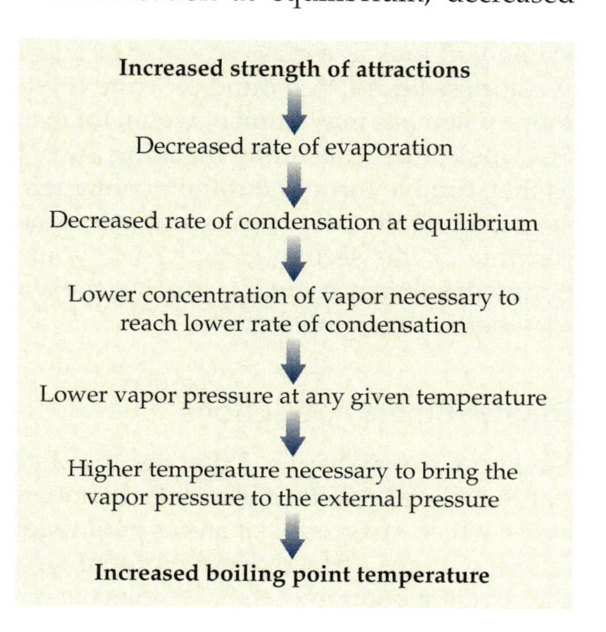


**Figure 14.16**
Strengths of Attractions and Boiling-Point Temperatures

OBJECTIVE 19

(56.5 °C). Figure 14.16 summarizes this reasoning, and Figure 14.17 shows graphs of the change in equilibrium vapor pressure with changes in temperature for acetone and water.

**Figure 14.17**
Boiling-Point Temperatures for Acetone and Water

OBJECTIVE 19

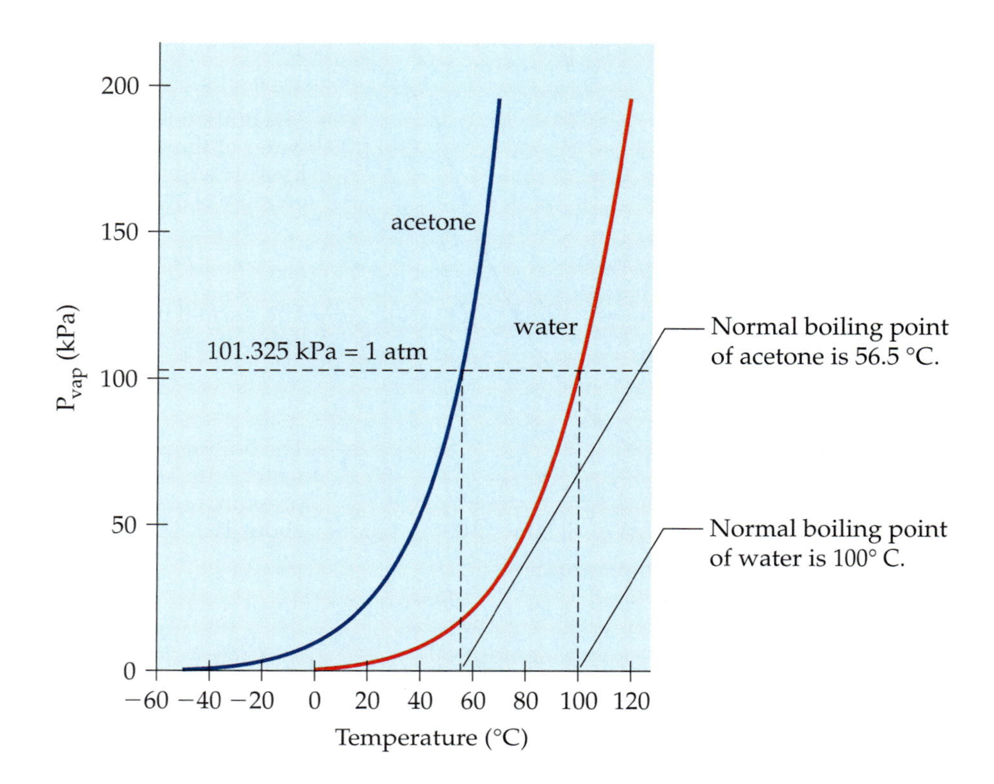


## 14.3 Particle-Particle Attractions

When you look at a glass of water or a teaspoon of salt, what do you see? A colorless liquid? A mound of white crystals? A chemist sees much, much more. Chemists may think of water, for example, as a vast crowd of jostling $H_2O$ molecules, constantly breaking away from and attracting one another as they tumble throughout their container. They may see a salt crystal as a system of alternating cations and anions held together by ionic bonds. The purpose of this section is to expand your ability to visualize the particle nature of substances by introducing the ways in which particles are bound together in the liquid form.

### Dipole-Dipole Attractions

We know from Chapter 3 that because hydrogen chloride consists of two nonmetallic elements, it is a molecular compound. It is composed of molecules, which are groups of atoms held together by covalent bonds. Because chlorine atoms attract electrons more strongly than hydrogen atoms do, the covalent bond in each HCl molecule is a polar covalent bond. The chlorine atom has a partial minus charge, and the hydrogen atom has a partial positive charge. A molecule that contains a partial positive charge at one end and a partial negative charge at the other is called a **molecular dipole,** or often just a **dipole.**

Under normal pressure, HCl gas condenses to a liquid at −84.9 °C. This is the temperature at which the attractions between the HCl molecules are strong enough to keep the molecules close together. The attraction between one HCl molecule and another is a **dipole–dipole attraction,** which is an attraction between the partial negative end of one molecule and the partial positive end of another molecule (Figure 14.18).

OBJECTIVE 20

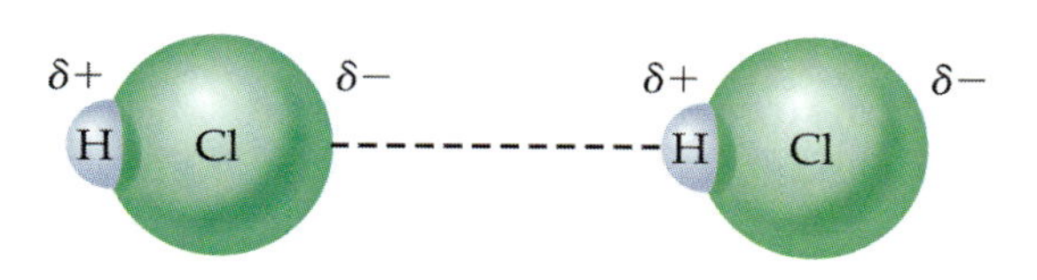


**Figure 14.18**
Dipole–Dipole Attraction Between HCl Molecules

OBJECTIVE 20

If you were riding on a molecule in a sample of liquid hydrogen chloride, you would see several other HCl molecules close by, and your particle would be attracted to each of them. You would move constantly from place to place, breaking old dipole–dipole attractions and making new ones (Figure 14.19).

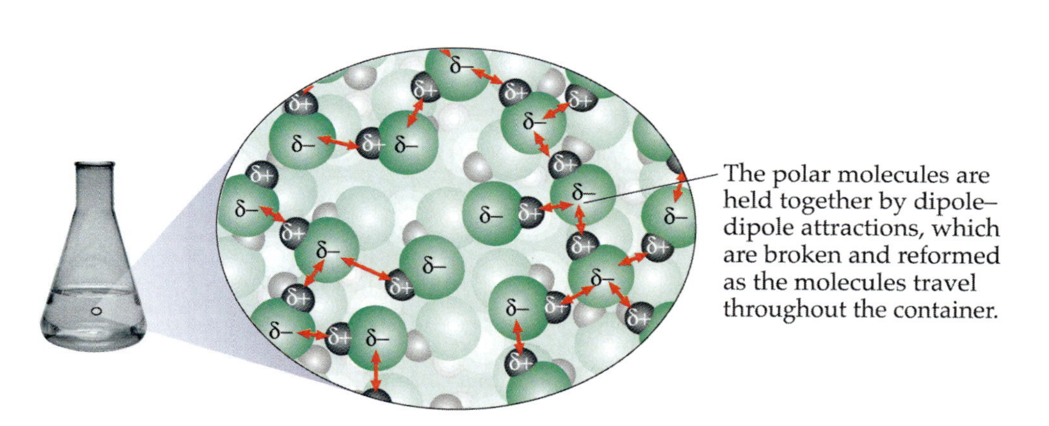

The polar molecules are held together by dipole–dipole attractions, which are broken and reformed as the molecules travel throughout the container.

**Figure 14.19**
Dipole–Dipole Attractions in a Liquid

OBJECTIVE 21

When a polar molecular substance, such as hydrogen chloride, is heated to convert it from liquid to gas, dipole–dipole attractions are broken. The molecules themselves remain unchanged. For example, when liquid HCl is boiled, the dipole–dipole attractions between HCl molecules are broken, but the covalent bonds between the hydrogen atoms and the chlorine atoms within the HCl molecules are unaffected.

## Predicting Bond Type

To predict whether molecules are polar or nonpolar, we need first to predict whether the bonds within them are polar or nonpolar. Remember that atoms form nonpolar covalent bonds, polar covalent bonds, or ionic bonds, depending on the relative electron-attracting ability of the atoms in the bond (Section 3.2). If the two atoms that form a bond have very different electron-attracting abilities, one or more electrons will be transferred between the atoms, forming ions and an ionic bond. At the other extreme, if there is no significant difference in the electron-attracting ability of the two atoms that form a chemical bond, the bond they form is a nonpolar covalent bond, with no significant transfer of electrons from one atom to another and therefore no significant charges on either of the atoms. Polar covalent bonds lie between these two extremes: The atoms differ just enough in their electron-attracting ability to result in a transfer that is significant but not complete. This leads to partial charges on the atoms and to a polar covalent bond.

**Electronegativity,** EN, is a term chemists use to describe the electron-attracting ability of an atom in a chemical bond. Figure 14.20 shows the electronegativities for many of the elements. The higher an element's electronegativity, the greater its ability to attract electrons from other elements. For example, because chlorine has a higher electronegativity (3.16) than hydrogen (2.20), we expect chlorine atoms to attract electrons more strongly than hydrogen atoms do. Comparing electronegativities enables us to predict whether a covalent bond is nonpolar covalent, polar covalent, or ionic.

**Figure 14.20**
A Table of Average Electronegativities

| 1 1A | 2 2A | 3 3B | 4 4B | 5 5B | 6 6B | 7 7B | 8 8B | 9 8B | 10 8B | 11 1B | 12 2B | 13 3A | 14 4A | 15 5A | 16 6A | 17 7A | 18 8A |
|---|---|---|---|---|---|---|---|---|---|---|---|---|---|---|---|---|---|
| | | | | | | | | | | H 2.20 | | | | | | | |
| Li 0.98 | Be 1.57 | | | | | | | | | | | B 2.04 | C 2.55 | N 3.04 | O 3.44 | F 3.98 | |
| Na 0.93 | Mg 1.31 | | | | | | | | | | | Al 1.61 | Si 1.9 | P 2.19 | S 2.58 | Cl 3.16 | |
| K 0.82 | Ca 1.0 | Sc 1.36 | Ti 1.54 | V 1.63 | Cr 1.66 | Mn 1.55 | Fe 1.83 | Co 1.88 | Ni 1.91 | Cu 1.9 | Zn 1.65 | Ga 1.81 | Ge 2.19 | As 2.18 | Se 2.55 | Br 2.96 | |
| Rb 0.82 | Sr 0.95 | Y 1.22 | Zr 1.33 | Nb 1.6 | Mo 2.16 | Tc 1.9 | Ru 2.2 | Rh 2.28 | Pd 2.2 | Ag 1.93 | Cd 1.69 | In 1.78 | Sn 1.96 | Sb 2.05 | Te 2.1 | I 2.66 | Xe 2.6 |
| Cs 0.79 | Ba 0.89 | Lu 1.3 | Hf 1.5 | Ta 2.36 | W 1.9 | Re 2.2 | Os 2.2 | Ir 2.28 | Pt 2.54 | Au 2 | Hg 1.8 | Tl 2.33 | Pb 2.02 | Bi 2.0 | Po 2.2 | | |
| Fr 0.89 | Ra 1.1 | | | | | | | | | | | | | | | | |

OBJECTIVE 22a

*If the difference in electronegativity (ΔEN) between the atoms in the bond is less than 0.4, we expect them to share electrons more or less equally, forming a nonpolar covalent bond.* For example, hydrogen has an electronegativity of 2.20, and carbon has an electronegativity of 2.55. The difference in electronegativity (ΔEN) is 0.35, which is less than 0.4. We expect the bond between carbon and hydrogen to be a nonpolar covalent bond.

OBJECTIVE 22a

*If the difference in electronegativity (ΔEN) between two atoms is between 0.4 and 1.7, we expect the bond between them to be a polar covalent bond.* For example, hydrogen has an electronegativity of 2.20, and chlorine has an electronegativity of 3.16. The difference in electronegativity (ΔEN) is 0.96, so the bond is a polar covalent bond, which is sometimes called a **bond dipole.**

OBJECTIVE 22a

*If the difference in electronegativity (ΔEN) between two atoms is greater than 1.7, we expect the bond between them to be ionic.* For example, sodium has an electronegativity of 0.93, and chlorine has an electronegativity of 3.16. The difference in electronegativity (ΔEN) is 2.23, so we predict their bond to be ionic.

There are exceptions to these guidelines. For example, the electronegativity difference is 1.78 in the H–F bond and 1.79 in the P–F bond. Our electronegativity guidelines lead us to predict that these will be ionic bonds, but experiments show them to be covalent bonds. Exceptions like these will not confuse you if you remember that the bond between two nonmetallic atoms is always a covalent bond (Figure 14.21).

Covalent bond
(nonmetal-nonmetal always covalent)

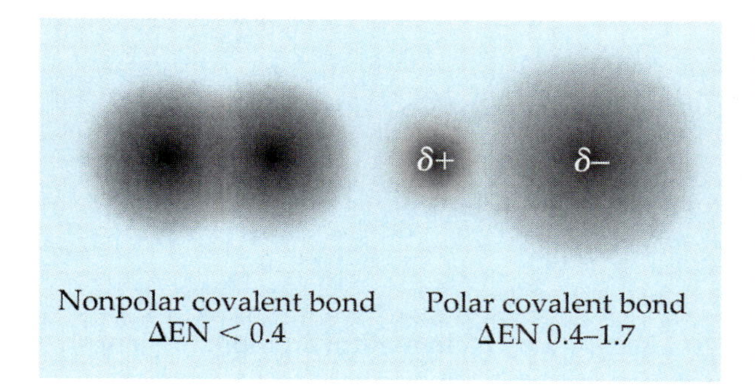


Ionic bond
(metal-nonmetal usually ionic)

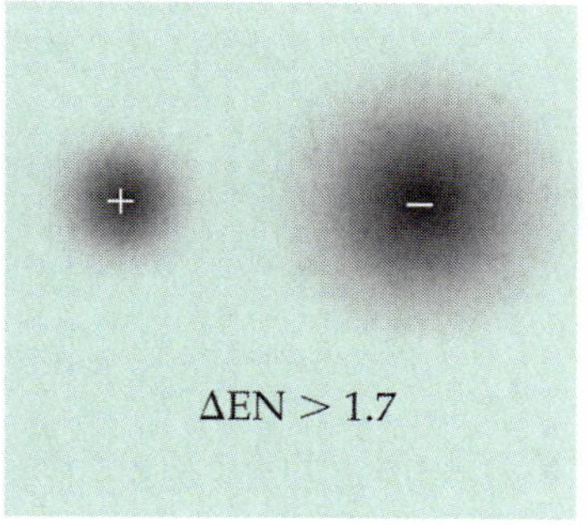


**Figure 14.21**
Identification of Bond Type

OBJECTIVE 22a

Electronegativities also give us a simple way of predicting which atom in a polar covalent bond has the partial negative charge and which has the partial positive charge. *The atom with the higher electronegativity has the partial negative charge. The atom with the lower electronegativity has the partial positive charge.* For example, the electronegativity of chlorine (3.16) is higher than that of hydrogen (2.20). In the H–Cl bond, we predict that the chlorine atom attracts electrons more strongly than the hydrogen atom and gets a partial negative charge. Because of its lower electronegativity, the hydrogen atom has a partial positive charge.

OBJECTIVE 22b

Similar logic enables us to predict which atom in an ionic bond forms the positive cation and which forms the negative anion. *The nonmetallic element has the higher electronegativity and forms the anion. The metallic element has the lower electronegativity and forms the cation.*

OBJECTIVE 22c

We can use another simple guideline to predict which of two bonds is more polar. *The bond with the greater difference in electronegativity (ΔEN) is likely to be the more polar bond.*

OBJECTIVE 22d

For example, ΔEN for the H–Cl bond is 0.96. The ΔEN for the H–F bond is 1.78. Because there is a greater difference between the electron-attracting abilities of hydrogen atoms and fluorine atoms than between those of hydrogen atoms and chlorine atoms, we expect larger partial charges in the H–F bond. That is, we expect the H–F bond to be more polar than the H–Cl bond.

The following sample study sheet summarizes the guidelines for using electronegativity to make predictions about bonds.

## Sample Study Sheet 14.1

### Electronegativity, Types of Chemical Bonds, and Bond Polarity

OBJECTIVE 22

**Tip-off** You wish to (1) classify a chemical bond as nonpolar covalent, polar covalent, or ionic, (2) identify which element in a polar covalent bond is partially negative and which is partially positive, (3) identify which element in an ionic bond forms the anion and which forms the cation, or (4) identify which of two bonds is more polar.

**General Steps**

- Use the following guidelines to identify a chemical bond as ionic, nonpolar covalent, or polar covalent.

  $\Delta EN < 0.4$ → nonpolar covalent

  $\Delta EN$ 0.4–1.7 → polar covalent

  $\Delta EN > 1.7$ → ionic

- Use the following guidelines to identify which element in a polar covalent bond is partially negative and which is partially positive.

  Higher electronegativity → partial negative charge

  Lower electronegativity → partial positive charge

- Use the following guidelines to identify which element in an ionic bond forms the anion and which forms the cation.

  Nonmetal, which has the higher electronegativity → anion

  Metal, which has the lower electronegativity → cation

- Use the following guideline to decide which of two bonds is more polar.

  The greater the $\Delta EN$, the more polar the bond.

**Example** See Examples 14.1 and 14.2.

## EXAMPLE 14.1 Using Electronegativities

OBJECTIVE 22

Classify the following bonds as nonpolar covalent, polar covalent, or ionic. If a bond is polar covalent, identify which atom has the partial negative charge (δ−) and which has the partial positive charge (δ+). If the bond is ionic, identify which element is negative and which is positive.

a. O bonded to H

b. C bonded to S

c. Li bonded to Cl

d. H bonded to F

***Solutions***

a. Both O and H represent nonmetallic elements, so the bond is covalent. Because oxygen's electronegativity is 3.44 and hydrogen's is 2.20, the difference in electronegativity (ΔEN) is 1.24, so this is a **polar covalent bond.** The higher electronegativity of the oxygen tells us that oxygen atoms attract electrons more strongly than hydrogen atoms. Therefore, the **oxygen atom** in the O–H bond is **δ−,** and the **hydrogen atom** in the O–H bond is **δ+**.

b. Both C and S represent nonmetallic elements, so the bond is covalent. The electronegativity of carbon is 2.55, and the electronegativity of sulfur is 2.58. The ΔEN is only 0.03, so the bond is a **nonpolar covalent bond.**

c. Lithium is a metal, and chlorine is a nonmetal. Bonds between metallic atoms and nonmetallic atoms are expected to be ionic. Because lithium's electronegativity is 0.98 and chlorine's is 3.16, the ΔEN is 2.18, so again we would predict the bond to be best described as an **ionic bond.** The nonmetallic **chlorine** has the **negative charge,** and the metallic **lithium** has the **positive charge.**

d. On the basis of hydrogen's electronegativity being 2.20 and fluorine's being 3.98, for a ΔEN of 1.78, we might expect these elements to form an ionic bond. However, because both of these elements are nonmetallic, the bond is instead a very **polar covalent bond.** The higher electronegativity of the fluorine tells us that fluorine atoms attract electrons more strongly than do hydrogen atoms. The **fluorine atom** in the H–F bond is **δ−,** and the **hydrogen atom** in the H–F bond is **δ+.**

## EXAMPLE 14.2 Comparing Bond Polarities

OBJECTIVE 22

Which bond would you expect to be more polar, H–S or H–Cl?

***Solution***

The ΔEN for the H–S bond is 0.38, and the ΔEN for the H–Cl bond is 0.96. We expect the bond with the greater difference, the **H–Cl bond,** to be **more polar.** The Cl in the H–Cl bond would have a larger **δ−** charge than the S in the H–S bond.

## Exercise 14.1 Using Electronegativities

OBJECTIVE 22

Classify the following bonds as nonpolar covalent, polar covalent, or ionic. If a bond is polar covalent, identify which atom has the partial negative charge and which has the partial positive charge. If a bond is ionic, identify which atom is negative and which is positive.

a. N bonded to H

b. N bonded to Cl

c. Ca bonded to O

d. P bonded to F

## Exercise 14.2 Comparing Bond Polarities

OBJECTIVE 22

Which bond would you expect to be more polar, P–H or P–F? Why?

### Predicting Molecular Polarity

Now that we can predict whether covalent bonds are polar or nonpolar, we can continue our discussion of polar molecules. Three questions will help you predict whether substances are composed of polar or nonpolar molecules. First,

*Is the substance molecular?*

If a substance is composed of all nonmetallic elements and does not contain the ammonium ion, you can assume that the answer to the first question is yes.

For reasons that we explore below, molecules are polar if they have an asymmetrical arrangement of polar covalent bonds, so another question is

*If the substance is molecular, do the molecules contain polar covalent bonds?*

OBJECTIVE 23

When there are no polar bonds in a molecule, there is no permanent charge difference between one part of the molecule and another, and the molecule is nonpolar. For example, the $Cl_2$ molecule has no polar bonds because the two atoms in each Cl–Cl bond attract electrons equally. It is therefore a nonpolar molecule. None of the bonds in hydrocarbon molecules, such as hexane, $C_6H_{14}$, are significantly polar, so hydrocarbons are nonpolar molecular substances. Lastly,

*If the molecules contain polar covalent bonds, are these bonds asymmetrically arranged?*

OBJECTIVE 23

A molecule can possess polar bonds and still be nonpolar. If the polar bonds are evenly (or symmetrically) distributed, the bond dipoles cancel and do not create a molecular dipole. For example, the two bonds in a molecule of $CO_2$ are significantly polar, but they are symmetrically arranged around the central carbon atom. No one side of the molecule has more negative or positive charge than another side, so the molecule is nonpolar:

$$\overset{\delta-}{:\ddot{O}}=\overset{\delta+}{C}=\overset{\delta-}{\ddot{O}:}$$

A water molecule is polar because (1) its O–H bonds are significantly polar, *and* (2) its bent geometry makes the distribution of those polar bonds asymmetrical. The side of the water molecule containing the more electronegative oxygen atom is partially negative, and the side of the molecule containing the less electronegative hydrogen atoms is partially positive.

OBJECTIVE 23

H δ+
|
:Ö—H δ+
δ−

You can see a more thorough description of the procedure used to predict whether molecules are polar or nonpolar at the Web address that follows. For now, it will be useful to remember that the following substances have polar molecules: water, $H_2O$; ammonia, $NH_3$; oxyacids (such as nitric acid, $HNO_3$, and acetic acid, $HC_2H_3O_2$); hydrogen halides (HCl, HF, HBr, and HI); methanol, $CH_3OH$; and ethanol, $C_2H_5OH$. You will also be expected to recognize the following as composed of nonpolar molecules: the elements composed of molecules ($H_2$, $N_2$, $O_2$, $F_2$, $Cl_2$, $Br_2$, $I_2$, $P_4$, $S_8$, and $Se_8$); carbon dioxide, $CO_2$; and hydrocarbons (such as pentane, $C_5H_{12}$, and hexane, $C_6H_{14}$).

OBJECTIVE 24

OBJECTIVE 25

The following Web site describes a procedure for determining whether molecules are polar:

www.chemplace.com/college/

## Hydrogen Bonds

Now let's return to our discussion of attractions between molecules (**intermolecular attractions**). Hydrogen fluoride, HF, is used to make uranium hexafluoride, $UF_6$, a chemical that plays an important role in the production of nuclear reactor fuel. Chemists have learned that gaseous hydrogen fluoride condenses to a liquid at 19.54 °C, a much higher temperature than the one at which hydrogen chloride condenses to a liquid (−84.9 °C). This difference suggests that the attractions between HF molecules are much stronger than the attractions between HCl molecules. (It is common to consider room temperature to be 20 °C, or 68 °F, so HF is a liquid at just slightly below room temperature.)

The main reason why the attractions are so much stronger between HF molecules than between HCl molecules is that the HF molecules have much higher partial positive and partial negative charges on the atoms: The difference in electronegativity between hydrogen and chlorine is 0.96, whereas the difference in electronegativity between hydrogen and fluorine is 1.78. As a result, the H–F bond is far more polar, with a very large partial positive charge on each hydrogen atom and a very large partial negative charge on each fluorine atom. These relatively large charges on HF molecules lead to such strong intermolecular attractions that chemists place them in a special category and call them hydrogen bonds. **Hydrogen bonds** are attractions that occur between a nitrogen, oxygen, or fluorine atom of one molecule and a hydrogen atom bonded to a nitrogen, oxygen, or fluorine atom in another molecule. Hydrogen fluoride is the only fluorine-containing compound that is capable of hydrogen bonding (Figure 14.22).

OBJECTIVE 26

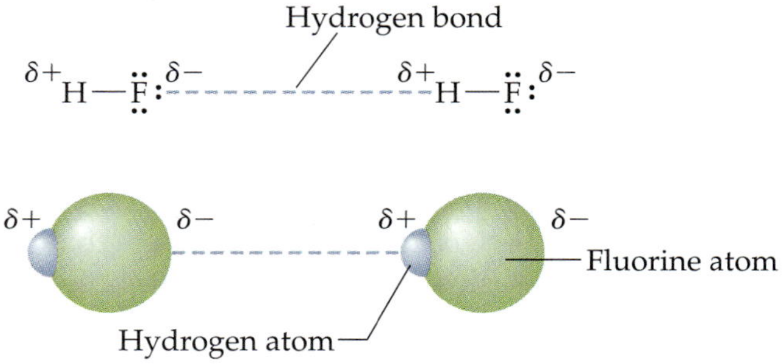

**Figure 14.22**
**Two Ways to Illustrate Hydrogen Bonding in HF**
In HF, the hydrogen bond is the attraction between the partial negative charge of a fluorine atom in one HF molecule and the partial positive charge of a hydrogen atom attached to a fluorine atom in another HF molecule.

OBJECTIVE 26

Hydrogen bond

Fluorine atom

Hydrogen atom

Hydrogen bonds are very important in the chemistry of life. The strands of DNA that carry our genetic information are held together by hydrogen bonds, and hydrogen bonds help to give the enzymes that regulate our body chemistry their characteristic shapes. Hydrogen bonding also explains the physical characteristics of water.

Life on earth, if it existed at all, would be far different if water were a gas at normal temperatures, like other, similar substances. For example, hydrogen sulfide, $H_2S$ (the substance that gives rotten eggs their distinctive smell) has a structure that is very similar to that of water, but it is a gas above −60.7 °C. The fact that water is a liquid suggests that the attractions between $H_2O$ molecules are stronger than the attractions in $H_2S$. Hydrogen sulfide molecules have relatively weak dipole–dipole attractions between them, but $H_2O$ molecules have strong hydrogen bonds between them because of the attraction between the partial negative charge on an oxygen atom of one water molecule and the partial positive charge on a hydrogen atom of another molecule (Figure 14.23).

As you discovered in Chapter 3, the three most common alcohols are methanol, $CH_3OH$, ethanol, $C_2H_5OH$, and 2-propanol, $C_3H_7OH$.

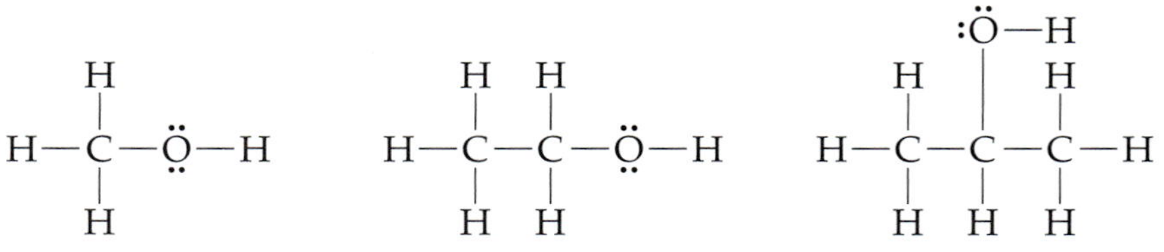

Methanol, $CH_3OH$ Ethanol, $C_2H_5OH$ 2-propanol, $C_3H_7OH$

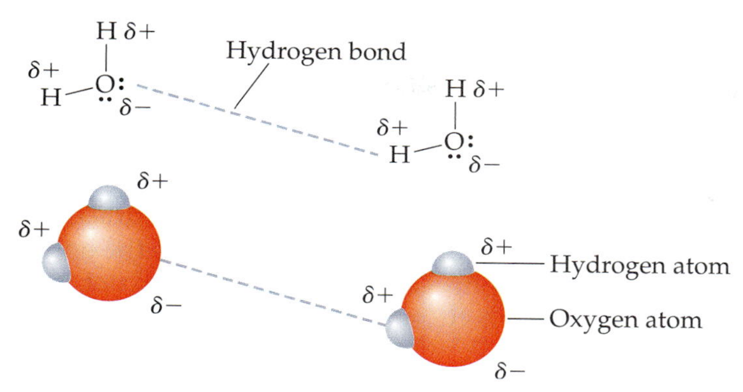


**Figure 14.23**
**Hydrogen Bonding Between Water Molecules**
In water, the hydrogen bond is the attraction between the partial negative charge of an oxygen atom in one molecule and the partial positive charge of a hydrogen atom attached to an oxygen atom in another molecule.

OBJECTIVE 26

Like water, alcohol molecules contain the O–H bond and therefore experience hydrogen bonding of their molecules in the liquid and solid form (Figure 14.24).

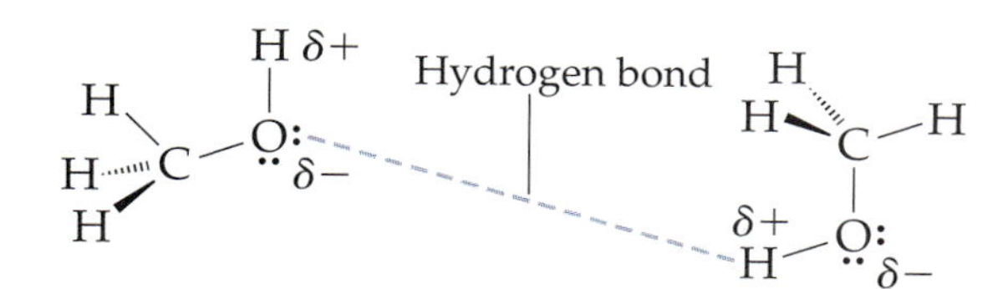


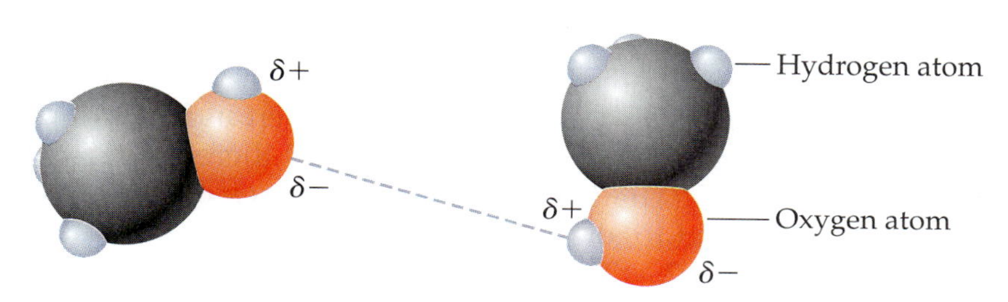


**Figure 14.24**
**Hydrogen Bonding in Methanol**
In methanol, the hydrogen bond is the attraction between the partial negative charge of an oxygen atom in one molecule and the partial positive charge of a hydrogen atom attached to an oxygen atom in another molecule.

OBJECTIVE 26

Ammonia, $NH_3$, a gas at room temperature, is used to make fertilizers, explosives, rocket fuel, and other nitrogen-containing products. Dissolved in water, it is used as a household cleaner. When $NH_3$ gas is cooled below −33.35 °C, the molecules slow down to the point where the hydrogen bonds they form during collisions keep them together long enough to form a liquid (Figure 14.25).

One use of nitrogen explosives made from ammonia

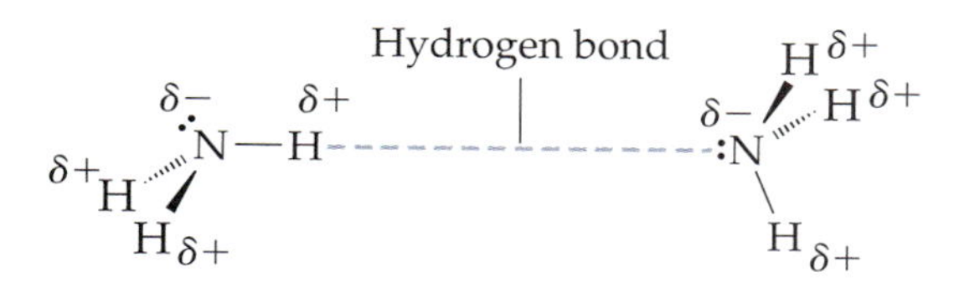


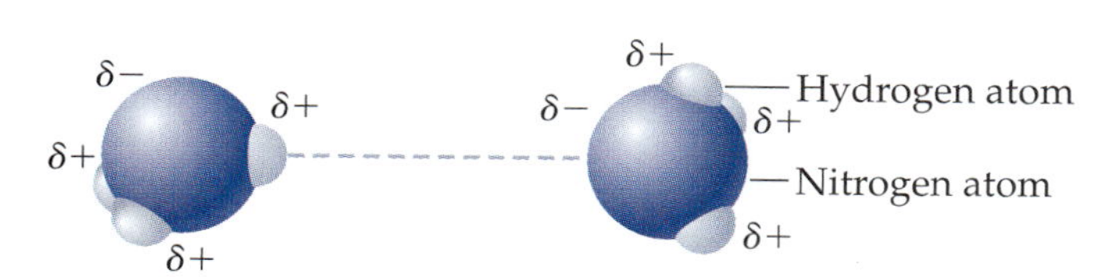


**Figure 14.25**
**Hydrogen Bonding in Ammonia**
In ammonia, the hydrogen bond is the attraction between the partial positive charge of a hydrogen atom attached to a nitrogen atom in one molecule and the partial negative charge of a nitrogen atom.

OBJECTIVE 26

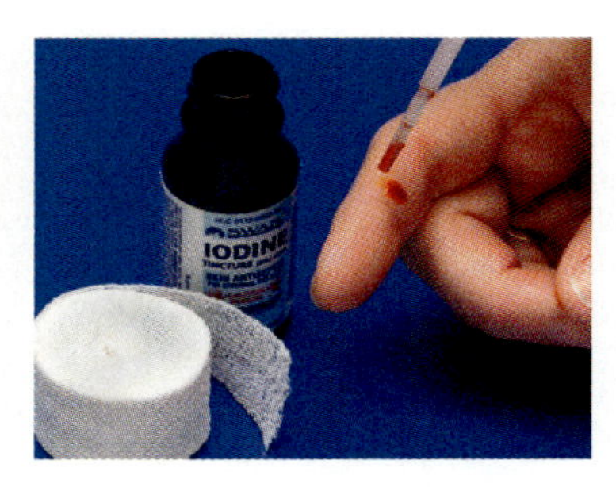


When solid iodine is dissolved in alcohol, the solution can be used as an antiseptic.

## London Forces

Are you old enough to remember *tincture of iodine*? Before the introduction of antibacterial ointments and sprays for the treatment of minor cuts and burns, tincture of iodine, a mixture of iodine and alcohol, was a common disinfectant for the skin. The pure iodine used to make this solution consists of diatomic molecules, $I_2$, whose mutual attractions are strong enough to form a solid at room temperature. Because the only bond within the $I_2$ molecule is the nonpolar I–I bond, we do not expect normal dipole–dipole attractions between iodine molecules. Why, then, do they attract each other strongly?

OBJECTIVE 27

Although we expect the I–I bond to be nonpolar, with a symmetrical distribution of the electrons, this arrangement is far from static. Even though the most probable distribution of charge in an isolated $I_2$ molecule is balanced, a sample of iodine that contains many billions of molecules will always include, by chance, some molecules whose electron clouds are shifted more toward one iodine atom than toward the other. The resulting dipoles are often called *instantaneous dipoles* because they may be short-lived (Figure 14.26). The constant collisions between molecules contribute to the formation of instantaneous dipoles. When $I_2$ molecules collide, the repulsion between their electron clouds will distort the clouds and shift them from their nonpolar state.

An instantaneous dipole can create a dipole in the molecule next to it. For example, the negative end of one instantaneous dipole will repel the negative electron cloud of a nonpolar molecule next to it, pushing the cloud toward the far side of the neighboring molecule. The new dipole is called an *induced dipole*. The induced dipole can then induce a dipole in another neighboring molecule. This continues until there are many polar molecules in the system. The resulting partial charges on instantaneous and induced dipoles lead to attractions between the opposite charges on the molecules (Figure 14.26). These attractions are called **London dispersion forces, London forces,** or **dispersion forces.**

OBJECTIVE 28

The larger the molecules of a substance, the stronger the London forces between them. A larger molecule has more electrons and a greater chance of having its electron cloud distorted from its nonpolar shape. Thus instantaneous dipoles are more likely to form in larger molecules. The electron clouds in larger molecules are also larger, so the average distance between the nuclei and the electrons is greater; as a result, the electrons are held less tightly and shift more easily to create a dipole.

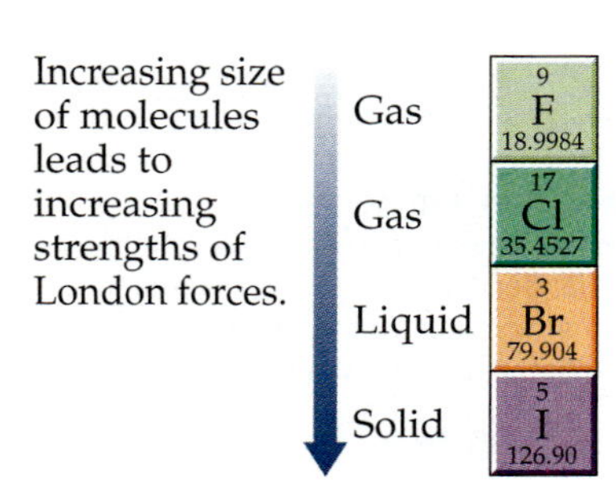


Increasing size of molecules leads to increasing strengths of London forces.

The increased strength of London forces with increased size of molecules explains a trend in the properties of halogens. Fluorine, $F_2$, and chlorine, $Cl_2$, which consist of relatively small molecules, are gases at room temperature and pressure. Bromine, $Br_2$, with larger molecules and therefore stronger London forces between them, is a liquid. Iodine, $I_2$, with still larger molecules, is a solid.

The effect of molecular size on the strengths of London forces can also be used to explain the different states observed among the byproducts of crude oil. Crude oil is a mixture of hydrocarbon molecules that can be separated into natural gas, gasoline, kerosene, lubricating oils, and

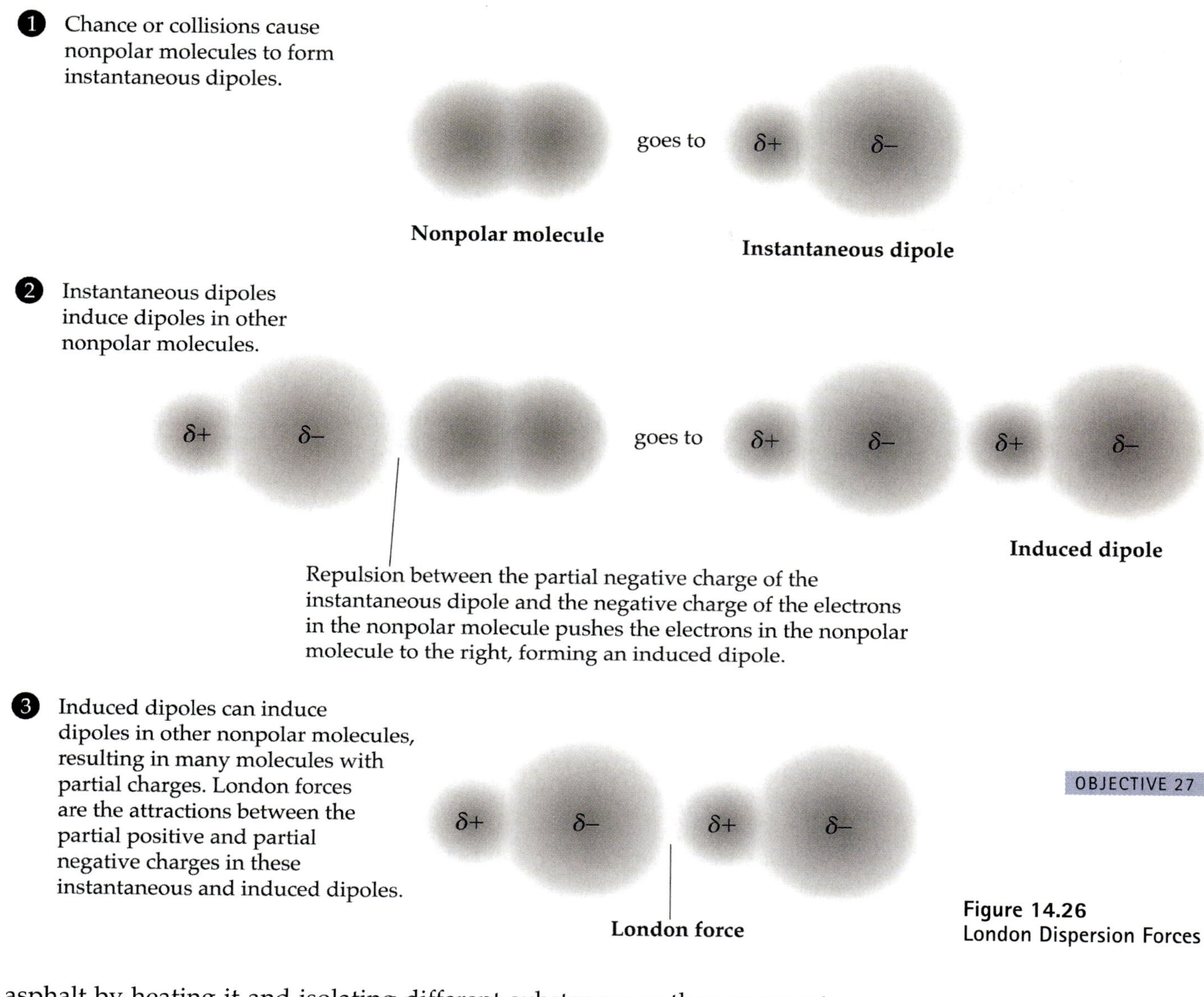

1 Chance or collisions cause nonpolar molecules to form instantaneous dipoles.
goes to
δ+
δ–
Nonpolar molecule
Instantaneous dipole
2 Instantaneous dipoles induce dipoles in other nonpolar molecules.
δ+
δ–
goes to
δ+
δ–
δ+
δ–
Induced dipole
Repulsion between the partial negative charge of the instantaneous dipole and the negative charge of the electrons in the nonpolar molecule pushes the electrons in the nonpolar molecule to the right, forming an induced dipole.
3 Induced dipoles can induce dipoles in other nonpolar molecules, resulting in many molecules with partial charges. London forces are the attractions between the partial positive and partial negative charges in these instantaneous and induced dipoles.
δ+
δ–
δ+
δ–
London force

OBJECTIVE 27

**Figure 14.26**
London Dispersion Forces

asphalt by heating it and isolating different substances as they evaporate from the mixture. The smallest hydrocarbons, with up to four carbon atoms, have the weakest London forces between molecules, so they evaporate from the crude oil mixture first and can be piped to our homes as natural gas. Hydrocarbons with 5 to 25 carbons have strong enough London forces to remain liquids at room temperature and pressure. After removal of the natural-gas molecules, further heating of the remaining mixture drives off molecules with 6 to 12 carbon atoms, which can be used as gasoline. Kerosene, a mixture of hydrocarbon molecules with 11 to 16 carbon atoms, evaporates next. Lubricating oils, which contain hydrocarbon molecules with 15 to 25 carbon atoms, are then driven off at an even higher temperature. Hydrocarbons with more than 25 carbons remain solid throughout this process. When the liquid components of crude oil have been separated, the solid asphalt that is left behind can be used to build roads.

OBJECTIVE 29

Hydrocarbons have many uses. Natural gas is used to heat our homes, cars require liquid gasoline and lubricating oils, and solid asphalt is used to make roads and driveways.

We know that polar molecules are attracted to each other by dipole-dipole attractions. Experiments have shown, though, that the actual strengths of the attractions between polar molecules are greater than we would predict from the polarity of the isolated molecules. The following Web site describes how London forces play a part in the attractions between polar molecules:

**www.chemplace.com/college/**

## Particle Interaction in Pure Elements

As we have noted, the $I_2$ molecules in solid iodine are held together by London forces, but what about the solid forms of other elements? Let's review what we discovered in Section 2.5 about the particles that form the structures of the pure elements and find out more about the attractions that hold them together.

OBJECTIVE 30

Most of the elements are metallic, and their pure solids are held together by metallic bonds. According to the sea-of-electrons model described in Section 2.5, the basic structure of a solid metal consists of cations that have lost one or more electrons. The lost electrons are free to move throughout the solid, so our image of the particles in a solid metal is one of island-like cations in a sea of electrons. **Metallic bonds** can be described as the attractions between the positive metal cations and the negative electrons that surround them.

OBJECTIVE 30

The nonmetallic elements exhibit more variety in the types of particles their solids are made of and the types of attractions between them. The particles that make up the solid structure of diamond are carbon atoms, and the attractions that hold them in the solid form are covalent bonds. The noble gases are composed of separate atoms. When a noble gas is cooled enough to form a liquid, London forces hold the particles together.

OBJECTIVE 30

Pure samples of all the other nonmetallic elements consist of molecules. Hydrogen, nitrogen, oxygen, and all of the halogens (group 17) are composed of diatomic molecules: $H_2$, $N_2$, $O_2$, $F_2$, $Cl_2$, $Br_2$, and $I_2$. Sulfur and selenium can take different forms, but the most common form for each is a molecule with eight atoms, $S_8$ and $Se_8$. Phosphorus also has different forms, but the most important forms consist of $P_4$ molecules. All of the molecules of these elements are nonpolar, and the attractions that hold their solids together are London forces.

## Summary of the Types of Particles and the Attractions Between Them

Table 14.1 summarizes the kinds of particles that make up different substances and the attractions between them. Figure 14.27 shows a scheme for predicting the types of attractions between particles.

Table 14.1
The Particles in Different Types of Substances
and the Attractions that Hold Them in the Solid or Liquid Form

OBJECTIVE 30

| Type of substance | Particles | Examples | Attraction in solid or liquid |
|---|---|---|---|
| **Elements** | | | |
| metal | cations in a sea of electrons | Au | metallic bond |
| noble gases | atoms | Xe | London forces |
| carbon (diamond) | carbon atoms | C(dia) | covalent bonds |
| other nonmetal elements | molecules | $H_2$, $N_2$, $O_2$, $F_2$, $Cl_2$, $Br_2$, $I_2$, $S_8$, $Se_8$, $P_4$ | London forces |
| **Ionic compounds** | | | |
| ionic | cations and anions | NaCl | ionic bond |
| **Molecular compounds** | | | |
| nonpolar molecular | molecules | $CO_2$ and hydrocarbons | London forces |
| polar molecules without H–F, O–H, or N–H bond | molecules | HF, HCl, HBr, and HI | dipole–dipole forces |
| molecules with H–F, O–H, or N–H bond | molecules | HF, $H_2O$, alcohols, $NH_3$ | hydrogen bonds |

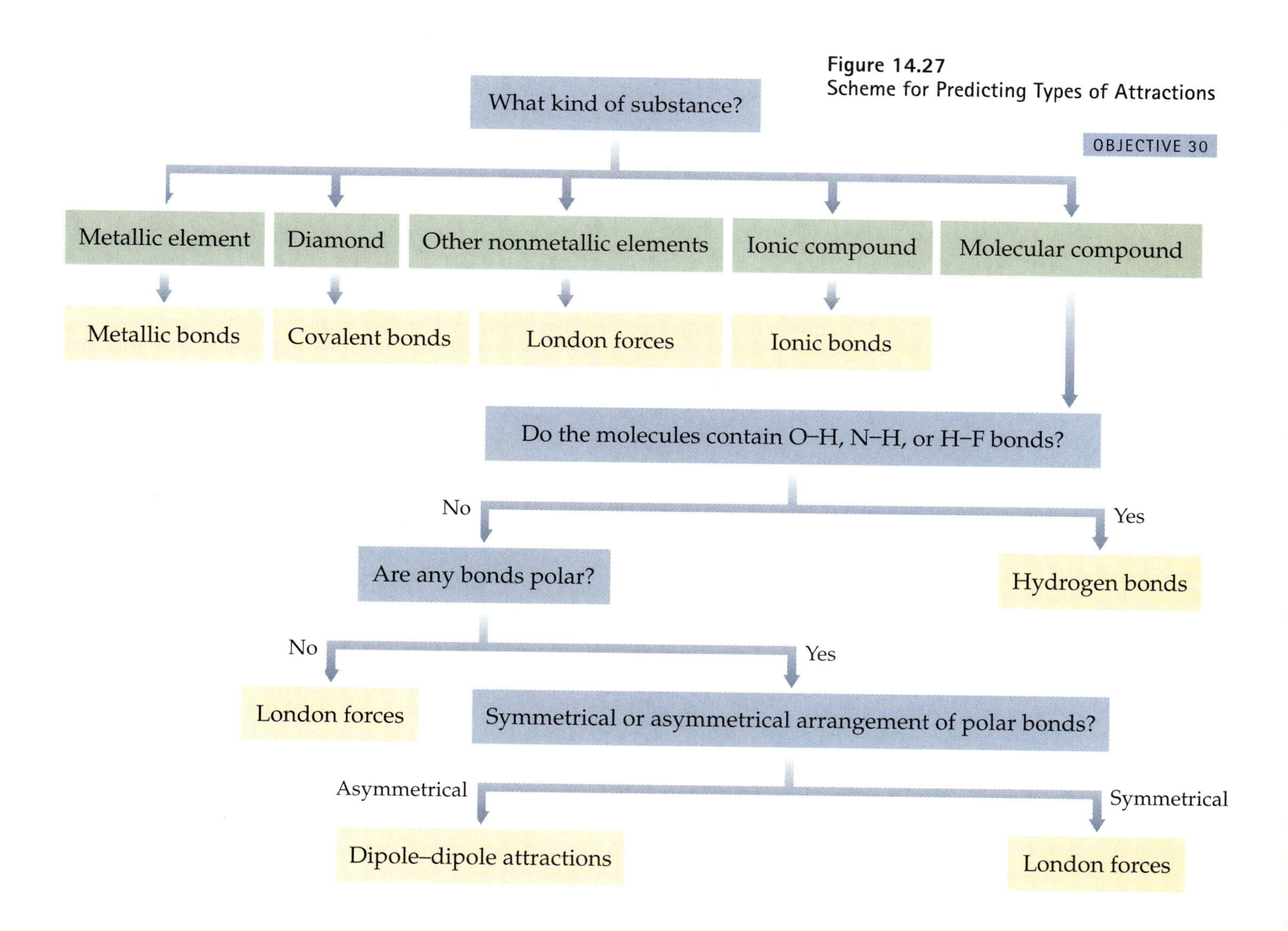


Figure 14.27
Scheme for Predicting Types of Attractions

OBJECTIVE 30

## EXAMPLE 14.3 Types of Particles and Types of Attractions

OBJECTIVE 30

For (a) $C_8H_{18}$, (b) aluminum, (c) $Br_2$, (d) LiBr, (e) ethanol, and (f) $CH_3NH_2$, specify (1) the type of particle that forms the substance's basic structure and (2) the name of the type of attraction that holds these particles in the solid and liquid form.

***Solution***

a. The compound $\mathbf{C_8H_{18}}$ is a hydrocarbon, so it is composed of **$\mathbf{C_8H_{18}}$ molecules** held together by **London forces.**

b. **Aluminum** is a metallic element, so it is composed of **Al cations in a sea of electrons.** The particles are held together by **metallic bonds.**

c. The formula $\mathbf{Br_2}$ describes bromine, a nonmetallic element. Except for carbon and the noble gases, the nonmetallic elements are composed of **molecules** held together by **London forces.**

d. Because **LiBr** consists of a metal and a nonmetal, we assume that it has ionic bonds and that it is an ionic compound with **cations and anions** held together by **ionic bonds.**

e. The formula for **ethanol** is $C_2H_5OH$. Ethanol, like all alcohols, has a structure formed by **molecules** held together by **hydrogen bonds.**

f. Because all of the symbols in the formula $\mathbf{CH_3NH_2}$ represent nonmetallic elements, we assume that $CH_3NH_2$ has all covalent bonds and therefore is a molecular substance composed of **$\mathbf{CH_3NH_2}$ molecules.** The $-NH_2$ in the formula suggests that the molecules have an N–H bond, which leads to **hydrogen bonding** between molecules.

## EXERCISE 14.3 Types of Particles and Types of Attractions

OBJECTIVE 30

For (a) iron, (b) iodine, (c) $CH_3OH$, (d) $NH_3$, (e) hydrogen chloride, (f) KF, and (g) carbon in the diamond form, specify (1) the type of particle that forms the substance's basic structure and (2) the name of the type of attraction that holds these particles in the solid and liquid form.

If you can predict the types of attractions between particles in two different substances, you can also predict the relative strengths of those attractions and then the relative boiling-point temperatures for the substances. The following Web site shows a general procedure for making these predictions:

www.chemplace.com/college/

## Chapter Glossary

**Condensation** The change from vapor to liquid.

**Evaporation** The change from liquid to vapor.

**Vapor** A gas derived from a substance that is liquid at normal temperatures and pressures. This term is also often used to describe gas that has recently come from a liquid.

**Rate of evaporation** The number of particles moving from liquid to gas per second.

**Rate of condensation** The number of particles moving from gas to liquid per second.

**Dynamic equilibrium** A system that has two equal and opposing rates of change, from state A to state B and from state B to state A. There are constant changes between state A and state B but no net change in the amount of the components in either state.

**Equilibrium vapor pressure, $P_{vap}$** The partial pressure of vapor above a liquid in a closed system with a dynamic equilibrium between the rate of evaporation and the rate of condensation.

**Boiling** The conversion of liquid to vapor anywhere in the liquid rather than just at the top surface.

**Boiling-point temperature** The temperature at which a liquid boils. It is also the temperature at which the equilibrium vapor pressure of the liquid becomes equal to the external pressure acting on the liquid.

**Normal boiling point** The temperature at which the equilibrium vapor pressure of the liquid equals 1 atmosphere.

**Molecular dipole** (or often just **dipole**) A molecule that contains an asymmetrical distribution of positive and negative charges.

**Dipole–dipole attraction** The intermolecular attraction between the partial negative end of one polar molecule and the partial positive end of another polar molecule.

**Electronegativity** A measure of the electron-attracting ability of an atom in a chemical bond.

**Bond dipole** A polar covalent bond, which has an atom with a partial positive charge and an atom with a partial negative charge.

**Intermolecular attraction** Attraction between molecules.

**Hydrogen bond** The intermolecular attraction between a nitrogen, oxygen, or fluorine atom of one molecule and a hydrogen atom bonded to a nitrogen, oxygen, or fluorine atom in another molecule.

**London dispersion forces, London forces, or dispersion forces** The attractions produced between molecules by instantaneous and induced dipoles.

**Metallic bond** The attraction between the positive metal cations that form the basic structure of a solid metal and the negative charge from the mobile sea of electrons that surrounds the cations.

You can test yourself on the glossary terms at the following Web site:

**www.chemplace.com/college/**

## Chapter Objectives

**The goal of this chapter is to teach you to do the following.**

1. Define all of the terms in the Chapter 14 Glossary.

**Section 14.1 Change from Gas to Liquid and from Liquid to Gas—An Introduction to Dynamic Equilibrium**

2. Describe the submicroscopic changes that occur when gas is converted from high-temperature gas to a lower-temperature gas to liquid.
3. Describe the process of evaporation from the surface of a liquid.
4. List three criteria that a particle must meet to escape from the surface of a liquid and move into the gas phase.
5. Explain why liquids cool as they evaporate.
6. List the three factors that affect the rate of evaporation of a liquid.
7. Explain why increased surface area of a liquid leads to a higher rate of evaporation.
8. Explain why weaker strength of attractions between particles in a liquid leads to a higher rate of evaporation.
9. Explain why increasing a liquid's temperature causes an increase in the rate of evaporation.
10. Describe the process by which a dynamic equilibrium is established between the rate of evaporation and the rate of condensation for a liquid in a closed container.
11. Write a general description of dynamic equilibrium.
12. Explain why the vapor pressure is constant above a liquid in a system that has a dynamic equilibrium between the rate of evaporation and the rate of condensation.
13. Explain why weaker attractions between particles in a liquid lead to higher equilibrium vapor pressures.
14. Explain why increasing a liquid's temperature leads to an increase in equilibrium vapor pressure.

**Section 14.2 Boiling Liquids**

15. Describe how a bubble can form in a liquid.
16. Explain why boiling does not take place in a liquid until the liquid reaches its boiling-point temperature.
17. Explain why a change in the external pressure acting on a liquid changes its boiling-point temperature.
18. Explain why potatoes take longer to cook in boiling water on the top of Mount Everest than at sea level and why they cook faster in water boiled in a pressure cooker.
19. Explain why liquids whose particles experience stronger mutual attractions have higher boiling points than liquids whose particles experience weaker mutual attractions.

**Section 14.3 Particle–Particle Attractions**

20. Describe dipole–dipole attractions between polar molecules.
21. Draw a sketch of polar molecules, such as HCl molecules, in the liquid form, showing the dipole–dipole attractions that hold the particles together.
22. Given a table of electronegativities, do the following:
    a. Classify bonds between elements as nonpolar covalent, polar covalent, or ionic.
    b. Identify which of two atoms in a polar covalent bond has a partial negative charge and which atom has a partial positive charge.
    c. Identify which of two atoms in an ionic bond has a negative charge and which atom has a positive charge.
    d. Given two bonds, determine which of the bonds would be expected to be more polar.
23. Explain why hydrocarbon molecules are nonpolar; why carbon dioxide, $CO_2$, molecules are nonpolar; and why water molecules are polar.
24. Identify the following substances as composed of polar molecules: water, $H_2O$; ammonia, $NH_3$; oxyacids (such as nitric acid, $HNO_3$, and acetic acid, $HC_2H_3O_2$); hydrogen halides (HF, HCl, HBr, and HI); methanol, $CH_3OH$; and ethanol, $C_2H_5OH$.
25. Identify the following substances as composed of nonpolar molecules: the elements composed of molecules ($H_2$, $N_2$, $O_2$, $F_2$, $Cl_2$, $Br_2$, $I_2$, $P_4$, $S_8$, and $Se_8$); carbon dioxide, $CO_2$; and hydrocarbons (such as pentane, $C_5H_{12}$, and hexane, $C_6H_{14}$).
26. Describe hydrogen bonds between molecules of HF, $H_2O$, $CH_3OH$, and $NH_3$.
27. Describe London dispersion forces between nonpolar molecules, including how the attractions form between molecules that initially have no dipole.
28. Explain why larger molecules have stronger London forces.
29. Describe the similarities and differences among natural gas, gasoline, kerosene, lubricating oils, and asphalt.
30. Given the name or chemical formula for a substance, (1) categorize the substance as a metallic element, carbon in the diamond form, another nonmetallic element, an ionic compound, a polar molecular compound with hydrogen bonds, a polar molecular compound without hydrogen bonds, or a nonpolar molecular compound; (2) identify the type of particle that forms its basic structure; and (3) identify the type of attraction that holds its particles in the solid and liquid form.

## Review Questions

1. For each of the following pairs of elements, decide whether a bond between them would be covalent or ionic.
   a. Ni and F
   b. S and O
2. Classify each of the following as either a molecular compound or an ionic compound.
   a. oxygen difluoride
   b. $Na_2O$
   c. calcium carbonate
   d. $C_3H_8$
3. Classify each of the following compounds as (1) a binary ionic compound, (2) an ionic compound with polyatomic ion(s), (3) a binary covalent compound, (4) a binary acid, (5) an alcohol, or (6) an oxyacid. Write the chemical formula that corresponds to each name.
   a. magnesium chloride
   b. hydrogen chloride
   c. sodium nitrate
   d. methane
   e. ammonia
   f. hydrochloric acid
   g. nitric acid
   h. ethanol
4. Classify each of the following compounds as (1) a binary ionic compound, (2) an ionic compound with polyatomic ion(s), (3) a binary covalent compound, (4) a binary acid, (5) an alcohol, or (6) an oxyacid. Write the name that corresponds to each chemical formula.
   a. HF
   b. $CH_3OH$
   c. LiBr
   d. $NH_4Cl$
   e. $C_2H_6$
   f. $BF_3$
   g. $H_2SO_4$
5. For each of the formulas listed below,
   - Draw a reasonable Lewis structure.
   - Write the name of the electron-group geometry around the central atom.
   - Draw the geometric sketch of the molecule, including bond angles.
   - Write the name of the molecular geometry around the central atom.

     a. $CCl_4$
     b. $SF_2$
     c. $PF_3$
     d. $BCl_3$

## Key Ideas

**Complete the following statements by writing one of these words or phrases in each blank.**

| | |
|---|---|
| 1 atmosphere | molecular compound |
| anywhere | momentum |
| attract | more |
| cancel | more rapidly |
| chance | negative |
| collisions | net |
| difference | number |
| dipole–dipole | nonpolar |
| direction of motion | opposing |
| equilibrium vapor | partial negative |
| external pressure | partial positive |
| greater than 1.7 | partial pressure |
| higher | percentage |
| higher electronegativity | polar |
| hydrogen | positive |
| increase | remain |
| increased temperature | repel |
| induced | significant |
| internal vapor pressure | strength of attractions |
| large | stronger |
| less than | surface |
| less tightly | surface area |
| lower | temperature |
| lower electronegativity | top surface |
| molecular | weaker |

6. Gases can be converted into liquids by a decrease in temperature. At a high temperature, there are no ________________ attractions between particles of a gas. As the temperature is lowered, attractions between particles lead to the formation of very small clusters that ________________ in the gas phase. As the temperature is lowered further, the particles move slowly enough to form clusters so ________________ that they drop to the bottom of the container and combine to form a liquid.

7. For a particle to escape from the surface of a liquid, it must be at the liquid's ________________, its ________________ must take it beyond the liquid's surface, and its ________________ must be great enough to overcome the backward pull of the other particles at the surface.

8. During the evaporation of a liquid, the ________________ moving particles escape, leaving the particles left in the liquid with a lower average velocity, and lower average velocity means ________________ temperature.

9. The rate of evaporation—the ________________ of particles moving from liquid to gas per second—depends on three factors: ________________ of the liquid, ________________ between the particles in the liquid, and ________________ of the liquid.
10. Greater surface area means ________________ particles at the surface of a liquid, which leads to a greater rate of evaporation.
11. Weaker attractions between particles lead to a(n) ________________ rate of evaporation.
12. Increased temperature increases the average velocity and momentum of the particles. As a result, a greater ________________ of particles will have the minimum momentum necessary to escape, so the liquid will evaporate more quickly.
13. As long as the rate of evaporation of the liquid is greater than the rate of condensation of the vapor, the concentration of vapor particles above the liquid will ________________. However, as the concentration of vapor particles increases, the rate of ________________ of vapor particles with the liquid increases, boosting the rate of condensation.
14. For a dynamic equilibrium to exist, the rates of the two ________________ changes must be equal, so that there are constant changes between state A and state B but no ________________ change in the components of the system.
15. The ________________ of vapor above a liquid in a closed system with a dynamic equilibrium between the rate of evaporation and the rate of condensation is called the equilibrium vapor pressure, $P_{vap}$.
16. The ________________ the attractions between particles of a substance, the higher the equilibrium vapor pressure for that substance at a given temperature.
17. Boiling can be defined as the conversion of liquid into vapor ________________ in the liquid rather than just at the ________________.
18. As a liquid is heated, the vapor pressure of the liquid increases until the temperature gets high enough to make the pressure within the bubbles that form equal to the ________________. At this temperature, the bubbles can maintain their volume, and the liquid boils.
19. The boiling-point temperature can be defined as the temperature at which the ________________ pressure of the liquid becomes equal to the external pressure acting on the liquid.
20. If the external pressure acting on a liquid changes, then the ________________ needed to preserve a bubble changes, and therefore the boiling-point temperature changes.
21. The temperature at which the equilibrium vapor pressure of a liquid equals ________________ is called its normal boiling-point temperature or just its normal boiling point.
22. We know that increased strength of attractions leads to decreased rate of evaporation, decreased rate of condensation at equilibrium, decreased concentration of vapor, and decreased vapor pressure at a given temperature. This leads to a(n) ________________ necessary to reach a vapor pressure of 1 atmosphere.

23. A molecule that contains a partial positive charge at one end and a partial negative charge at the other is called a(n) ____________ dipole, or often just a dipole.
24. A dipole–dipole attraction is an attraction between the ____________ end of one molecule and the ____________ end of another molecule.
25. When a polar molecular substance, such as hydrogen chloride, is heated to convert it from liquid to gas, ____________ attractions are broken.
26. The higher an element's electronegativity, the greater its ability to ____________ electrons from other elements.
27. If the difference in electronegativity (ΔEN) between the atoms in the bond is ____________ 0.4, we expect them to share electrons more or less equally, forming a nonpolar covalent bond.
28. If the difference in electronegativity (ΔEN) between two atoms is between 0.4 and 1.7, we expect the bond between them to be a(n) ____________ covalent bond.
29. If the difference in electronegativity (ΔEN) between two atoms is ____________, we expect the bond between them to be ionic.
30. The atom with the higher electronegativity has the partial ____________ charge. The atom with the lower electronegativity (ΔEN) has the partial ____________ charge.
31. The nonmetallic element in a bond between a metallic atom and a nonmetallic atom has the ____________ and forms the anion. The metallic element has the ____________ and forms the cation.
32. When comparing two covalent bonds, the bond with the greater ____________ in electronegativity is likely to be the more polar bond.
33. If a substance is composed of all nonmetallic elements and does not contain the ammonium ion, you can assume that it is a(n) ____________.
34. When there are no polar bonds in a molecule, there is no permanent charge difference between one part of the molecule and another, and the molecule is ____________.
35. If the polar bonds in a molecule are evenly (or symmetrically) distributed, the bond dipoles ____________ and do not create a molecular dipole.
36. ____________ bonds are attractions that occur between a nitrogen, oxygen, or fluorine atom of one molecule and a hydrogen atom bonded to a nitrogen, oxygen, or fluorine atom in another molecule.
37. An instantaneous dipole can create a dipole in the molecule next to it. For example, the negative end of one instantaneous dipole will ____________ the negative electron cloud of a nonpolar molecule next to it, pushing the cloud toward the far side of the neighboring molecule. The new dipole is called a(n) ____________ dipole.
38. The larger the molecules of a substance, the ____________ the London forces between them. A larger molecule has more electrons and a greater ____________ of having its electron cloud distorted from its nonpolar shape. Thus instantaneous dipoles are more likely to form in larger molecules. The electron clouds in larger molecules are also larger, so the average distance between the nuclei and the electrons is greater; as a result, the electrons are held ____________ and shift more easily to create a dipole.

## Chapter Problems

### Section 14.1 Change from Gas to Liquid and from Liquid to Gas—An Introduction to Dynamic Equilibrium

OBJECTIVE 2

39. A batch of corn whiskey is being made in a backwoods still. The ingredients are mixed and heated, and because the ethyl alcohol, $C_2H_5OH$, evaporates more rapidly than the other components, the vapor that forms above the mixture is enriched in $C_2H_5OH$. When the vapor is captured and condensed, the liquid it forms has a higher percentage of ethyl alcohol than the original mixture. Describe the submicroscopic changes that occur when pure, gaseous $C_2H_5OH$ is converted into a liquid, including a description of the transitions from a high-temperature gas to a lower-temperature gas to a liquid.

40. Why is dew more likely to form on a lawn at night than in the day? Describe the changes that take place as dew forms.

41. Acetone, $CH_3COCH_3$, is a laboratory solvent that is also commonly used as a nail polish remover.

OBJECTIVE 3

a. Describe the submicroscopic events that take place at the surface of liquid acetone when it evaporates.

OBJECTIVE 4

b. Do all of the acetone molecules moving away from the surface of the liquid escape? If not, why not? What three criteria must be met for a molecule to escape from the surface of the liquid and move into the gas phase?

OBJECTIVE 5

c. If you spill some nail polish remover on your hand, the spot will soon feel cold. Why?

OBJECTIVE 7

d. If you spill some acetone on a lab bench, it evaporates much more rapidly than the same amount of acetone in a test tube. Why?

OBJECTIVE 9

e. If you spill acetone on a hot plate in the laboratory, it evaporates much more quickly than the same amount of acetone spilled on the cooler lab bench. Why?

42. Consider two test tubes, each containing the same amount of liquid acetone. A student leaves one of the test tubes open overnight and covers the other one with a balloon so that gas cannot escape. When the student returns to the lab the next day, all of the acetone is gone from the open test tube, but most of it remains in the covered tube.

a. Explain why the acetone is gone from one test tube and not from the other.

b. Was the initial rate at which liquid changed to gas (the rate of evaporation) greater in one test tube than in the other? Explain your answer.

c. Consider the system after 30 minutes, with liquid remaining in both test tubes. Is condensation (vapor to liquid) taking place in both test tubes? Is the rate of condensation the same in both test tubes? Explain your answer.

OBJECTIVE 10

d. Describe the submicroscopic changes in the covered test tube that lead to a constant amount of liquid and vapor.

OBJECTIVE 12

e. The balloon expands slightly after it is placed over the test tube, suggesting an initial increase in pressure in the space above the liquid. Why? After this initial expansion, the balloon stays inflated by the same amount. Why doesn't the pressure inside the balloon change after the initial increase?

f. If the covered test tube is heated, the balloon expands. Part of this expansion is due to the increase in gas pressure that results from the rise in temperature of the gas, but the increase is greater than expected from this factor alone. What other factor accounts for the increase in pressure? Describe the submicroscopic changes that take place that lead to this other factor.

OBJECTIVE 14

Balloon is uninflated.

Slight inflation.

More inflation.

43. The attractions between ethanol molecules, $C_2H_5OH$, are stronger than the attractions between diethyl ether molecules, $CH_3CH_2OCH_2CH_3$.
    a. Which of these substances would you expect to have the higher rate of evaporation at room temperature? Why?
    b. Which of these substances would you expect to have the higher equilibrium vapor pressure? Why?

OBJECTIVE 13

44. Why do caregivers rub alcohol on the skin of feverish patients?
45. Picture a half-empty milk bottle in the refrigerator. The water in the milk will be constantly evaporating into the gas-filled space above the liquid, and the water molecules in this space will be constantly colliding with the liquid and returning to the liquid state. If the milk is tightly closed, a dynamic equilibrium forms between the rate of evaporation and the rate of condensation. If the bottle is removed from the refrigerator and left out in the room with its cap still on tightly, what happens to the rates of evaporation and condensation? An hour later when the milk has reached room temperature, will a dynamic equilibrium exist between evaporation and condensation?

### Section 14.2 Boiling Liquids

46. Describe the difference between boiling-point temperature and normal boiling-point temperature.
47. The normal boiling point of ethanol, $C_2H_5OH$, is 78.3 °C.
    a. Describe the submicroscopic events that occur when a bubble forms in liquid ethanol.

    OBJECTIVE 15

    b. Consider heating liquid ethanol in a system where the external pressure acting on the liquid is 1 atm. Explain why bubbles cannot form and escape from the liquid until the temperature reaches 78.3 °C.

    OBJECTIVE 16

    c. If the external pressure on the surface of the ethanol is increased to 2 atm, will its boiling-point temperature increase, decrease, or stay the same? Why?

    OBJECTIVE 17

OBJECTIVE 18

48. Explain why potatoes take longer to cook in boiling water on top of Mount Everest than at sea level and why they cook faster in water boiled in a pressure cooker.

49. At 86 m below sea level, Death Valley is the lowest point in the western hemisphere. The boiling point of water in Death Valley is slightly higher than water's normal boiling point. Explain why.

50. Butane is a gas at room temperature and pressure, but the butane found in some cigarette lighters is a liquid. How can this be?

OBJECTIVE 19

51. Explain why liquid substances with stronger interparticle attractions will have higher boiling points than liquid substances whose particles experience weaker interparticle attractions.

### Section 14.3 Particle–Particle Attractions

OBJECTIVE 20 OBJECTIVE 21

52. Bromine monofluoride, BrF, a substance that liquefies at −20 °C, consists of polar molecules that are very similar to HCl molecules. Each BrF molecule consists of a bromine atom with a polar covalent bond to a fluorine atom.
    a. Describe what you would see if you were small enough to ride on a BrF molecule in gaseous bromine monofluoride.
    b. Describe the process that takes place when bromine monofluoride gas is cooled enough to form a liquid.
    c. What type of attraction holds the molecules together in the liquid form?
    d. Draw a rough sketch of the structure of liquid bromine monofluoride.
    e. Describe what you would see if you were small enough to ride on a BrF molecule in liquid bromine monofluoride.

53. Do the electronegativities of the elements in a column on the periodic table increase or decrease from top to bottom? Do the electronegativities of the elements in a period on the periodic table increase or decrease from left to right? In general, the greater the separation of two elements on the periodic table, the greater the difference in electronegativity and the greater the expected polarity of the bond they form. Is this always true? If it is not always true, when is it true and when is it not necessarily true?

OBJECTIVE 22

54. Complete the following table by classifying each bond as nonpolar covalent, polar covalent, or ionic. If a bond is polar covalent, identify the atom that has the partial negative charge and the atom that has the partial positive charge. If a bond is ionic, identify the ion that has the negative charge and the ion that has the positive charge.

| Atoms | Is the bond polar covalent, nonpolar covalent, or ionic? | For polar covalent bonds, which atom is partial negative? For ionic bonds, which ion is negative? |
|---|---|---|
| C–N | | |
| C–H | | |
| H–Br | | |
| Li–F | | |
| C–Se | | |
| Se–S | | |
| F–S | | |
| O–P | | |
| O–K | | |
| F–H | | |

55. Complete the following table by classifying each bond as nonpolar covalent, polar covalent, or ionic. If a bond is polar covalent, identify the atom that has the partial negative charge and the atom that has the partial positive charge. If a bond is ionic, identify the ion that has the negative charge and the ion that has the positive charge.

OBJECTIVE 22

| Atoms | Is the bond polar covalent, nonpolar covalent, or ionic? | For polar covalent bonds, which atom is partial negative? For ionic bonds, which is the anion? |
|---|---|---|
| N–O | | |
| Al–Cl | | |
| Cl–N | | |
| H–I | | |
| Br–Cl | | |
| Cl–S | | |
| Se–I | | |
| N–Sr | | |
| O–F | | |
| F–P | | |

56. Identify the bond in each pair that you would expect to be more polar.

OBJECTIVE 22

a. C–O or C–H

b. P–H or H–Cl

57. Identify the bond in each pair that you would expect to be more polar.

OBJECTIVE 22

a. P–F or N–Cl

b. S–I or Se–F

58. Explain why water molecules are polar, why ethane, $C_2H_6$, molecules are nonpolar, and why carbon dioxide, $CO_2$, molecules are nonpolar.

OBJECTIVE 23

59. The C–F bond is more polar than the N–F bond, but $CF_4$ molecules are nonpolar and $NF_3$ molecules are polar. Explain why.

60. Ammonia has been used as a refrigerant. In the cooling cycle, gaseous ammonia is alternately compressed into a liquid and allowed to expand back to the gaseous state. What are the particles that form the basic structure of ammonia? What type of attraction holds these particles together? Draw a rough sketch of the structure of liquid ammonia.

OBJECTIVE 26

61. Methanol can be used as a home heating oil extender. What are the particles that form the basic structure of methanol? What type of attraction holds these particles together? Draw a rough sketch of the structure of liquid methanol.

OBJECTIVE 26

62. Bromine, $Br_2$, is used to make ethylene bromide, which is an antiknock additive in gasoline. The $Br_2$ molecules have nonpolar covalent bonds between the atoms, so we expect isolated $Br_2$ molecules to be nonpolar. Despite the nonpolar character of isolated $Br_2$ molecules, attractions form between bromine molecules that are strong enough to hold the particles in the liquid form at room temperature and pressure. What is the nature of these attractions? How do they arise? Describe what you would see if you were small enough to ride on a $Br_2$ molecule in liquid bromine.

OBJECTIVE 27

OBJECTIVE 28

63. Carbon tetrabromide, $CBr_4$, is used to make organic compounds, and carbon tetrachloride, $CCl_4$, is an industrial solvent the use of which is diminishing because of its toxicity. Both $CCl_4$ and $CBr_4$ molecules are nonpolar. Carbon tetrachloride is a liquid at room temperature, but carbon tetrabromide is a solid, which suggests that the attractions between $CCl_4$ molecules are weaker than those between $CBr_4$ molecules. Explain why.

64. Carbon disulfide, $CS_2$, which is used to make rayon, is composed of nonpolar molecules that are similar to carbon dioxide molecules, $CO_2$. Unlike carbon dioxide, carbon disulfide is liquid at room temperature. Why?

65. Milk bottles are often made of polyethylene plastic, which is composed of extremely large, nonpolar molecules. What type of attraction holds these molecules together in the solid form? In general, would you expect this type of attraction to be stronger or weaker than the hydrogen bonds that hold water molecules together? Because polyethylene is a solid at room temperature and water is a liquid, the attractions between polyethylene molecules must be stronger than those between water molecules. What makes the attractions between polyethylene molecules so strong?

66. Methanol, $CH_3OH$, is used to make formaldehyde, $CH_2O$, which is used in embalming fluids. The molecules of these substances have close to the same atoms and about the same molecular mass, so why is methanol a liquid at room temperature and formaldehyde a gas?

OBJECTIVE 28 OBJECTIVE 29

67. Describe the similarities and differences among natural gas, gasoline, and asphalt. Explain why the substances in natural gas are gaseous, why the substances in gasoline are liquid, and why the substances in asphalt are solid.

OBJECTIVE 30

68. Complete the following table by specifying (1) the name for the type of particle viewed as forming the structure of a solid, liquid, or gas of each substance and (2) the name of the type of attraction that holds these particles in the solid or liquid form.

| Substance | Particles to visualize | Type of attraction |
|---|---|---|
| silver | | |
| HCl | | |
| $C_2H_5OH$ | | |
| NaBr | | |
| carbon (diamond) | | |
| $C_5H_{12}$ | | |
| water | | |

69. Complete the following table by specifying (1) the name for the type of particle viewed as forming the structure of a solid, liquid, or gas of each substance and (2) the name of the type of attraction that holds these particles in the solid or liquid form.

OBJECTIVE 30

| Substance | Particles to visualize | Type of attraction |
|---|---|---|
| ammonia | | |
| KCl | | |
| $C_6H_{14}$ | | |
| Cu | | |
| iodine | | |
| hydrogen bromide | | |
| $C_3H_7OH$ | | |

70. Have you ever broken a mercury thermometer? If you have, you probably noticed that the mercury forms droplets on the surface on which it falls rather than spreading out and wetting it like water. Describe the difference between liquid mercury and liquid water that explains this different behavior. (*Hint:* Consider the attractions between particles.)

### Discussion Topic

71. Chlorofluorocarbons, CFCs, were used for years as aerosol propellants (Section 7.4). Although CFCs, such as CFC-12, are gases at normal pressure, under high pressures they can be converted into liquids. The liquid CFC in an aerosol can evaporates, and the resulting vapor pressure pushes substances out of the can. When the valve on the top of the can is pressed and substances escape, the pressure inside the can decreases at first, but it quickly returns to the pressure that existed before the valve was pushed. Describe what happens at the "submicroscopic" level as the pressure returns to its original value. Why does the pressure return to the same value as long as any liquid CFC remains in the can? If you were looking for a replacement for CFCs in aerosol cans, what criteria might you consider?

*The safety of this scuba diver depends on an understanding of the solubility of nitrogen and oxygen in the bloodstream.*

CHAPTER 15

# Solution Dynamics



FOR A CHEMICAL REACTION TO TAKE PLACE, the reacting particles must first collide—freely, frequently, and at high velocity. Particles in gases move freely and swiftly and collide continually, but gas phase reactions are difficult to contain and are sometimes dangerous. The particles in a liquid also move freely and quickly and have a high rate of collision, but very few pure substances are liquids at normal temperatures and pressures. Thus many important chemical reactions must take place in liquid solutions, where the reacting particles (dissolved gases, liquids, and solids) are free to move and collide in a medium that is very easy to contain. For example, "meals ready to eat" (MREs), which are used to feed soldiers in the field, include a packet that contains powdered magnesium metal and a solid acid. These substances are unable to react in the solid form. When water is added to dissolve the acid, hydronium ions are liberated, and they migrate throughout the mixture, colliding with magnesium metal, reacting with it in an extremely exothermic process, and thus releasing enough heat to cook the meal.

$$Mg(s) + 2H_3O^+(aq) \rightarrow Mg^{2+}(aq) + 2H_2O(l) + H_2(g) + \text{Energy}$$

In Chapters 4 and 5 we began to develop a mental image of how solutions form, but there are still a lot of questions to be addressed. Let's look at some situations you might encounter in a day spent doing chores around the house.

**8:43** After an early breakfast, you have time for some car maintenance. The fluid level in your cooling system is low, so you mix coolant and water and add it to the radiator. You wonder why the coolant and water mix so easily.

**9:21** Next the lawn needs attention, but before you can cut the grass, you need to gas up the mower. In the process, you spill a little gasoline on the garage floor and try unsuccessfully to clean it up with water. Why don't the gasoline and water mix?

**12:40** Washing the lunch dishes gives you plenty of time to wonder why the soap you are using helps clean grease away. You even have enough time to ask yourself why detergent works better than soap.

**12:52** Your neighbor Hope drops by for a visit. You brew coffee for her and pour some iced tea for yourself. You seem to be out of granular sugar, so you raid the supply of sugar cubes that you usually save for dinner parties. Why does the sugar cube take longer to dissolve than the same amount of granular sugar, and why does powdered sugar dissolve even faster? Why do they all dissolve faster if you stir them? And why does sugar dissolve so much faster in the hot coffee than in the iced tea?

**1:10** You convince Hope that it's time to get out of the kitchen and into the outside world, so it's off to Monterey Bay to do some scuba diving. On the way, you wonder

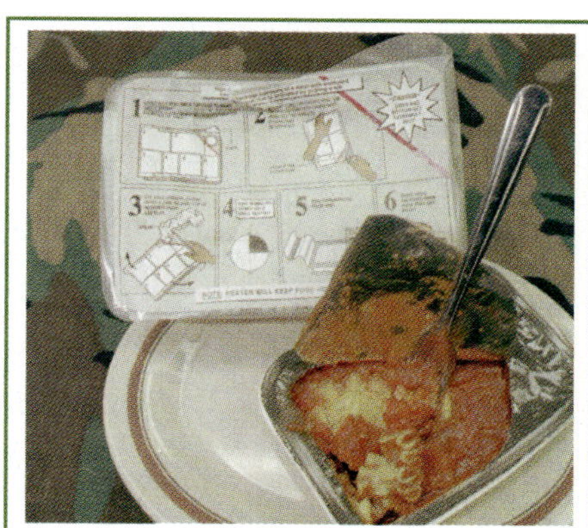

*Water dissolves the solids in this packet, allowing them to mix and react.*

about the reasons for some of the diving-safety rules you've been taught. Why can diving too deep make you feel drowsy and disoriented? Why is it important to ascend from deep water slowly?

The discussion of solution dynamics in this chapter will answer these questions and many more.

### Review Skills

The presentation of information in this chapter assumes that you can already perform the tasks listed below. You can test your readiness to proceed by answering the Review Questions at the end of the chapter. This might also be a good time to read the Chapter 15 Objectives, which precede the Review Questions.

- Describe the structure of liquid water. (Section 3.3)
- Given a description of a solution, identify the solute and solvent. (Section 4.2)
- Describe the process by which an ionic compound dissolves in water, referring to the nature of the particles in solution and the kind of attractions that form between them. (Section 4.2)
- Convert between names of compounds and their chemical formulas. (Section 5.3)
- Write a general description of dynamic equilibrium. (Section 14.1)
- Given the name or chemical formula for a compound, (1) categorize the substance as an ionic compound, a polar molecular compound with hydrogen bonds, a polar molecular compound without hydrogen bonds, or a nonpolar molecular compound; (2) identify the type of particle that forms its basic structure; and (3) identify the type of attraction that holds its particles in the solid and liquid form. (Section 14.3)

## 15.1 Entropy, Solutions, and Solubility

We start this stage of our exploration of the nature of solutions by looking at some of the reasons why certain substances mix to form solutions and why others do not. When you are finished reading this section, you will not only have the tools necessary for predicting the solubility of substances in liquids, but you will also learn a bit about the **Second Law of Thermodynamics,** which is the key scientific principle that enables us to explain and predict changes.

### Entropy and the Second Law of Thermodynamics

One way to state the Second Law of Thermodynamics is

**The entropy of the universe increases.**

**Entropy** is a measure of the *disorder* of a system; the more disordered the system is, the greater its entropy. This definition and the Second Law lead to the following general statement.

**A change from state A to state B tends to take place when state B has greater disorder than state A.**

To see why disordered systems are favored over ordered ones, let's take a look at a very simple system consisting of 4 particles, each of which can be found in any one of 9 positions. Figure 15.1 shows that there are 126 ways to arrange these 4 particles in the nine positions. We know that the particles in solids exhibit an *orderly* pattern and are as close together as possible, so in our simple system, let's assume that any arrangement that has the 4 particles clustered together is like a solid. Figure 15.1 shows that there are 4 ways to do this. In our system, we will consider any other, *less organized* arrangement as being like a gas. Figure 15.1 shows that there are 122 ways to position the particles in a gas-like arrangement. Thus over 96% of the possible arrangements lead to gas-like states. Therefore, if we assume that the particles can move freely between positions, they are more likely to be found in a gas-like state than in a solid-like state.

OBJECTIVE 2

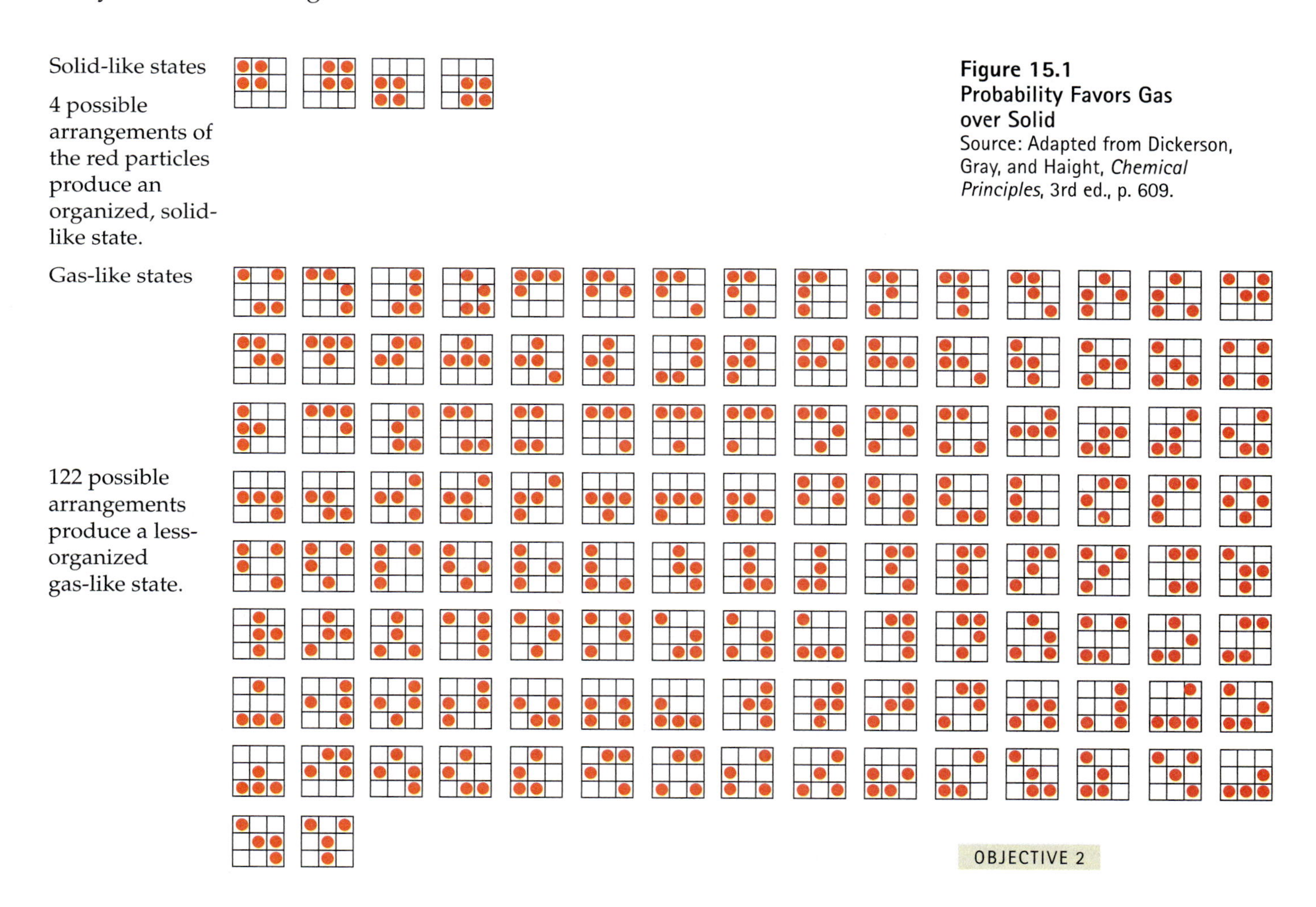

Solid-like states
4 possible arrangements of the red particles produce an organized, solid-like state.
Gas-like states
122 possible arrangements produce a less-organized gas-like state.

**Figure 15.1**
**Probability Favors Gas over Solid**
Source: Adapted from Dickerson, Gray, and Haight, *Chemical Principles*, 3rd ed., p. 609.

OBJECTIVE 2

If the four particles had 16 possible positions, there would be 1820 possible combinations. Nine of these would be in solid-like states, the other 1811 in gas-like states. Thus over 99.5% of the possible arrangements would represent gas-like states, as opposed to 96% for the system with 9 possible positions. This shows that an increase in the number of possible positions leads to an increase in the probability that the system will be in the more disordered, gas-like state. In real systems, which provide a huge number of possible positions for particles, there is an extremely high probability that substances will shift from the *more organized*, solid form, which represents fewer ways of arranging the particles, to the *more disorganized*, gas form, which represents more ways of arranging particles.

| Solid |
| --- |
| Fewer ways to arrange particles<br>Less probable<br>More ordered |
| **Gas** |
| More ways to arrange particles<br>More probable<br>Less ordered |

OBJECTIVE 3

Now we turn to another characteristic of matter that is related to entropy and the Second Law. **Particles of matter tend to become more dispersed (spread out).** The simple system shown in Figure 15.2 provides an example. It has two chambers that can be separated by a removable partition. Part (a) of the figure shows this system with gas on one side only. If the partition is removed, the motion of the gas particles causes them to move back and forth between the chambers. Because there are more possible arrangements for the gas particles when they are dispersed throughout both chambers, as in part (c), than when they are concentrated in one chamber, as in part (b), probability suggests that they will spread out to fill the total volume available to them.

**Figure 15.2**
Expansion of Gases

OBJECTIVE 3

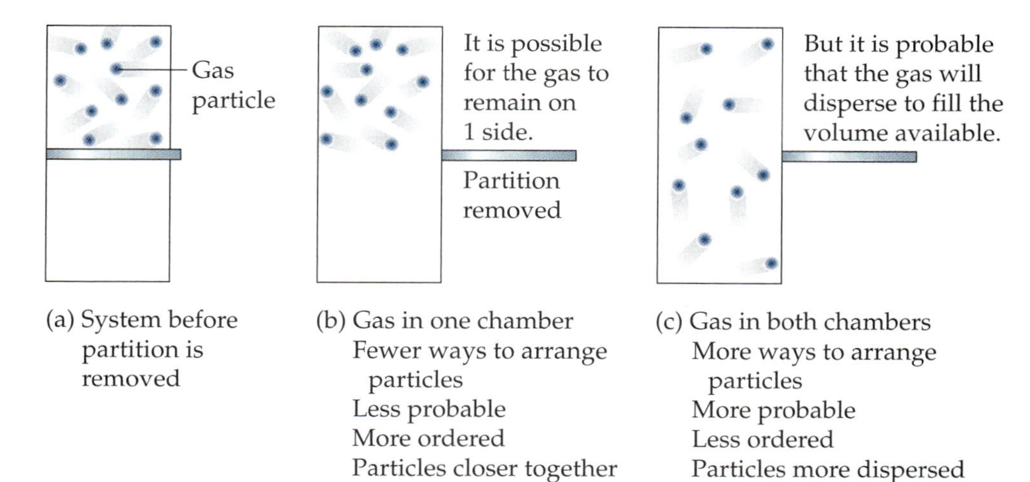


We can now restate the Second Law in either of the following ways:

> **A change from state A to state B tends to take place when state B has a greater number of possible arrangements.**

> **A change from state A to state B takes place when state B is more dispersed.**

## Why Do Solutions Form?

Although much of the explanation why certain substances mix and form solutions and why others do not is beyond the scope of this text, we can get an inkling of why solutions form by taking a look at the process by which ethanol, $C_2H_5OH$, dissolves in water. Ethanol is actually **miscible** in water, which means that the two liquids can be mixed in any proportion without any limit to their solubility. The Second Law accounts for this kind of change as well.

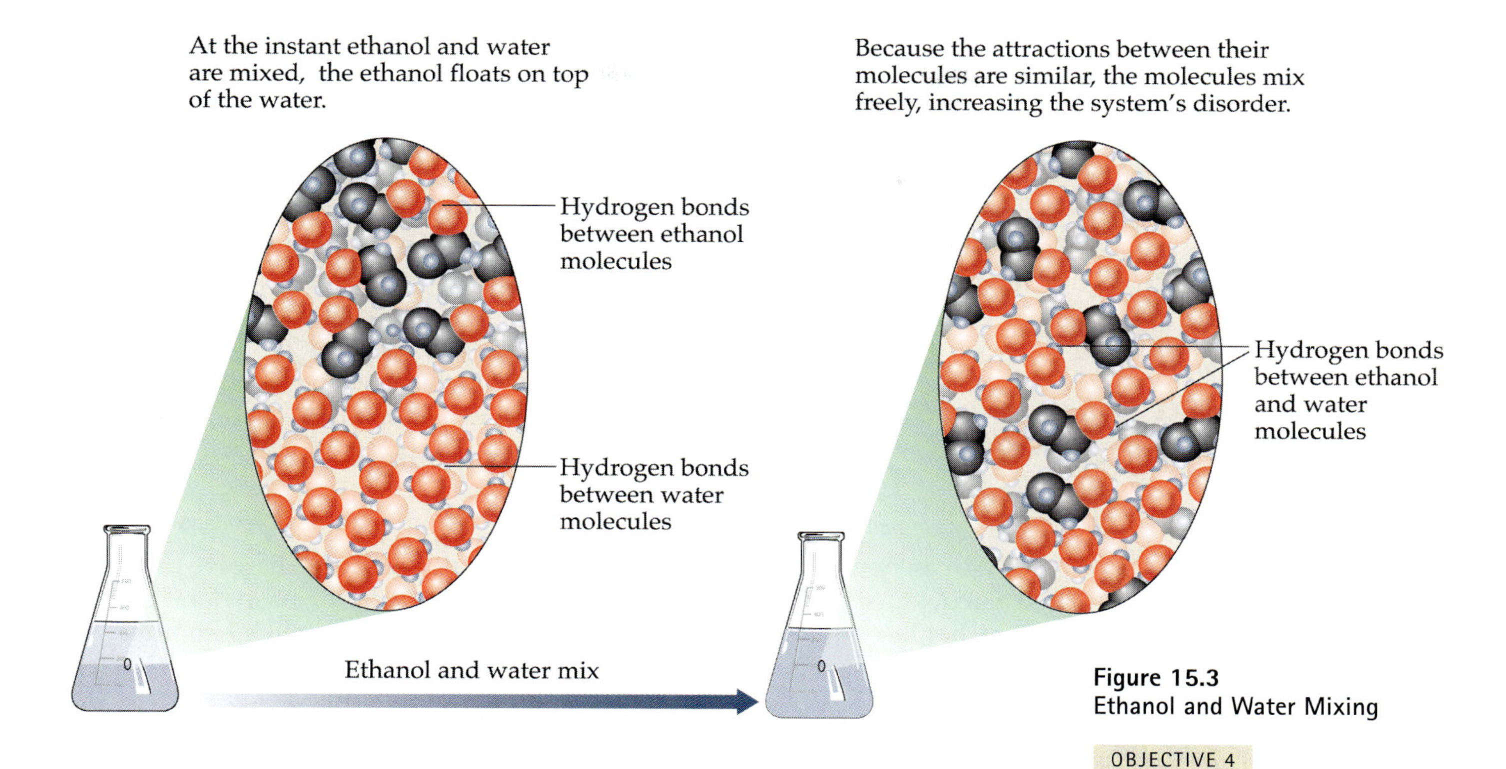


**Figure 15.3**
Ethanol and Water Mixing

OBJECTIVE 4

Picture a layer of ethanol being carefully added to the top of some water (Figure 15.3). Because the particles of a liquid are moving constantly, some of the ethanol particles at the boundary between the two liquids will immediately move into the water, and some of the water molecules will move into the ethanol. In this process, water–water and ethanol–ethanol attractions are broken, and ethanol–water attractions are formed. Because both the ethanol and the water are molecular substances with O–H bonds, the attractions broken between water molecules and between ethanol molecules are hydrogen bonds. The attractions that form between the ethanol and water molecules are also hydrogen bonds (Figure 15.4).

OBJECTIVE 4

Because the attractions between the particles are so similar, the freedom of movement of the ethanol molecules in the water solution is about the

**Figure 15.4**
The Attractions Broken and Formed When Ethanol Dissolves in Water

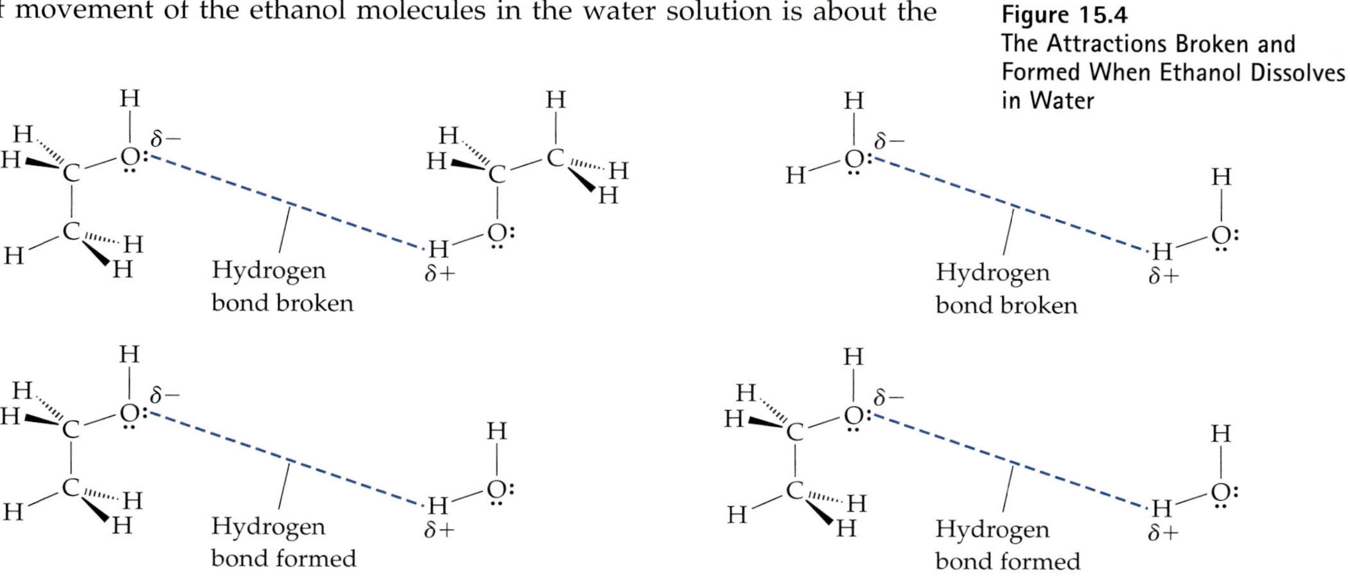

same as their freedom of movement in the pure ethanol. The same can be said for the water. Because of this freedom of movement, both liquids will spread out to fill the total volume of the combined liquids. In this way, they will shift to the most probable, most disordered, most dispersed, highest-entropy state available: the state of being completely mixed. There are many more possible arrangements for this system when the ethanol and water molecules are dispersed throughout a solution than when they are restricted to separate layers (Figure 15.3).

OBJECTIVE 4

At the following Internet site, you will find an animation that illustrates the solution of ethanol in water.
**www.chemplace.com/college/**

We can now explain why automobile radiator coolants dissolve in water. The coolants typically contain either ethylene glycol or propylene glycol, which, like ethanol and water, include hydrogen–bonding O–H bonds.

Why do coolant and water mix?

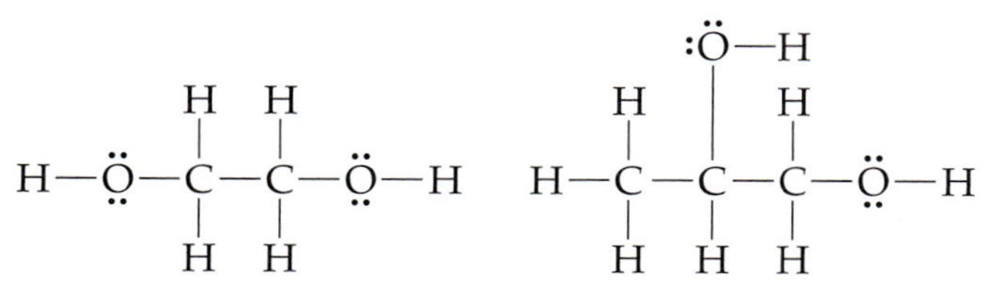

Ethylene glycol    Propylene glycol

These substances mix easily with water for the same reason that ethanol mixes easily with water: The attractions broken on mixing are hydrogen bonds, and the attractions formed are also hydrogen bonds. There is no reason why the particles of each liquid cannot move somewhat freely from one liquid to another, so they shift toward the most probable (most dispersed, most disordered, highest-entropy), mixed state.

For a more detailed description of entropy and solubility, see the following Web site:
**www.chemplace.com/college/**

## Predicting Solubility

The dividing line between what we call soluble and what we call insoluble is arbitrary, but the following are common criteria for describing substances as insoluble, soluble, or moderately soluble.

- If less than 1 gram of the substance will dissolve in 100 milliliters (or 100 g) of solvent, the substance is considered insoluble.

OBJECTIVE 5

- If more than 10 grams of substance will dissolve in 100 milliliters (or 100 g) of solvent, the substance is considered soluble.

- If between 1 and 10 grams of a substance will dissolve in 100 milliliters (or 100 g) of solvent, the substance is considered moderately soluble.

Although it is difficult to determine specific solubilities without either finding them by experiment or referring to a table of solubilities, we do have guidelines that enable us to predict *relative* solubilities. One such fundamental principle is

**Like dissolves like.**

For example, this guideline could be used to predict that ethanol, which is composed of *polar* molecules, would be soluble in water, which is also composed of *polar* molecules. Likewise, pentane ($C_5H_{12}$), which has *nonpolar* molecules, is miscible with hexane, which also has *nonpolar* molecules. We will use the *Like Dissolves Like* guideline to predict whether a substance is likely to be more soluble in water or in hexane. It can also be used to predict which of two substances is likely to be more soluble in water and which of two substances is likely to be more soluble in a nonpolar solvent, such as hexane:

OBJECTIVE 6 OBJECTIVE 7 OBJECTIVE 8 OBJECTIVE 9

- Polar substances are likely to dissolve in polar solvents. For example, ionic compounds, which are very polar, are often soluble in the polar solvent water.
- Nonpolar substances are likely to dissolve in nonpolar solvents. For example, nonpolar molecular substances are likely to dissolve in hexane, a common nonpolar solvent.

Two additional guidelines are derived from these:

OBJECTIVE 6 OBJECTIVE 7 OBJECTIVE 8 OBJECTIVE 9

- Nonpolar substances are not likely to dissolve to a significant degree in polar solvents. For example, nonpolar molecular substances, such as hydrocarbons, are likely to be insoluble in water.
- Polar substances are not likely to dissolve to a significant degree in non-polar solvents. For example, ionic compounds are insoluble in hexane.

It is more difficult to predict the solubility of polar molecular substances than to predict the solubility of ionic compounds and nonpolar molecular substances. Many polar molecular substances are soluble in both water and hexane. For example, ethanol is miscible with both water and hexane. The following generalization is helpful:

- Substances composed of small polar molecules, such as acetone and ethanol, are usually soluble in water. (They are also often soluble in hexane.)

Polar bond

$CH_3-C(=O)-CH_3$

Acetone

Polar bond

$CH_3-CH_2-O-H$

Ethanol

The guidelines we have discussed are summarized in Table 15.1

OBJECTIVE 6 OBJECTIVE 7

OBJECTIVE 8 OBJECTIVE 9

**Table 15.1**
**Summary of Solubility Guidelines**

| Type of substance | Soluble in water? | Soluble in hexane? |
|---|---|---|
| ionic compounds | often | no |
| molecular compounds with nonpolar molecules | no | yes |
| molecular compounds with small polar molecules | usually | often |

## EXAMPLE 15.1 Predicting Water Solubility

OBJECTIVE 6

Predict whether each of the following compounds is soluble in water.

a. nickel(II) chloride, $NiCl_2$ (used in nickel plating)

b. 2-propanol (isopropyl alcohol), $CH_3CH(OH)CH_3$ (in rubbing alcohol)

c. 1-pentene, $CH_2CHCH_2CH_2CH_3$ (blending agent for high-octane fuels)

d. The nonpolar molecular compound carbon disulfide, $CS_2$ (used to make rayon and cellophane)

e. The polar molecular compound propionic acid, $CH_3CH_2CO_2H$ (in the ionized form propionate, acts as a mold inhibitor in bread)

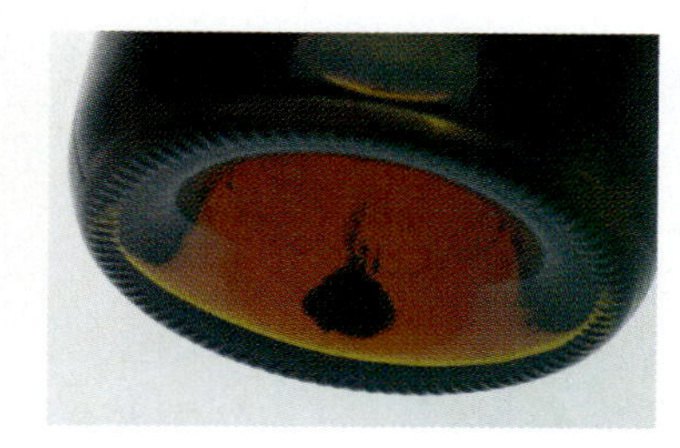

As the alcohol concentration of wine increases, the solvent's polarity decreases, and ionic compounds in the wine begin to precipitate.

***Solution***

a. Although there are many exceptions, we expect ionic compounds, such as nickel(II) chloride, $NiCl_2$, to be **soluble** in water.

b. Like all alcohols, 2-propanol, $CH_3CH(OH)CH_3$, is a polar molecular compound. Because this substance has relatively small polar molecules, we expect it to be **soluble** in water.

c. Like all hydrocarbons, 1-pentene, $CH_2CHCH_2CH_2CH_3$, is a nonpolar molecular compound and therefore is **insoluble** in water.

d. We expect carbon disulfide, $CS_2$, such as all nonpolar molecular compounds, to be **insoluble** in water.

$$:\ddot{S}=C=\ddot{S}:$$

e. We expect polar molecular compounds with relatively small molecules, such as propionic acid, $CH_3CH_2CO_2H$, to be **soluble** in water.

```
    H   H  :O:
    |   |   ||
H — C — C — C — Ö — H     — Polar group
    |   |
    H   H
```

## Exercise 15.1 Predicting Water Solubility

OBJECTIVE 6

Predict whether each of the following compounds is soluble in water. (See Section 14.3 if you need to review the list of compounds that you are expected to recognize as composed of either polar or nonpolar molecules.)

a. sodium fluoride, $NaF$ (used to fluoridate water)

b. acetic acid, $CH_3CO_2H$ (added to foods for tartness)

c. acetylene, $C_2H_2$ (used in torches designed for welding and cutting)

d. methanol, $CH_3OH$ (a common solvent)

## Example 15.2 Predicting Solubility in Hexane

OBJECTIVE 7

Predict whether each of the following compounds is soluble in hexane, $C_6H_{14}$.

a. ethane, $C_2H_6$ (in natural gas)

b. potassium sulfide (in depilatory preparations used for removing hair chemically)

***Solution***

a. Ethane, $C_2H_6$, like all hydrocarbons, is a nonpolar molecular compound, which are expected to be **soluble** in hexane.

b. Potassium sulfide, $K_2S$, is an ionic compound and therefore is expected to be **insoluble** in hexane.

## Exercise 15.2 Predicting Solubility in Hexane

OBJECTIVE 7

Predict whether each of the following compounds is soluble in hexane, $C_6H_{14}$.

a. sodium perchlorate, $NaClO_4$ (used to make explosives)

b. propylene, $CH_3CHCH_2$ (used to make polypropylene plastic for children's toys)

## Example 15.3 Predicting Relative Solubility in Water and Hexane

OBJECTIVE 8

Predict whether each of the following compounds is more soluble in water or in hexane.

a. pentane, $C_5H_{12}$ (used in low-temperature thermometers)

b. strontium nitrate, $Sr(NO_3)_2$ (used in railroad flares)

***Solution***

a. Pentane, $C_3H_8$, like all hydrocarbons, is a nonpolar molecular compound that is expected to be **more soluble in hexane** than in water.

b. Strontium nitrate is an ionic compound, which is expected to be **more soluble in water** than in hexane.

### Exercise 15.3 Predicting Relative Solubility in Water and Hexane

OBJECTIVE 8

Predict whether each of the following compounds is more soluble in water or in hexane.

a. sodium iodate, $NaIO_3$ (used as a disinfectant)

b. 2,2,4-trimethylpentane (sometimes called isooctane), $(CH_3)_3CCH_2CH(CH_3)_2$ (used as a standard in the octane rating of gasoline)

## Hydrophobic and Hydrophilic Substances

Organic compounds are often polar in one part of their structure and nonpolar in another part. The polar section, which is attracted to water, is called **hydrophilic** (literally, "water loving"), and the nonpolar part of the molecule, which is not expected to be attracted to water, is called **hydrophobic** ("water fearing"). If we need to predict the relative water solubility of two similar molecules, we can expect the one with the proportionally larger polar portion to have higher water solubility. We predict that the molecule with the proportionally larger nonpolar portion will be less soluble in water.

OBJECTIVE 9

As an example, we can compare the structures of epinephrine (adrenaline) and amphetamine. Epinephrine is a natural stimulant that is released in the body in times of stress. Amphetamine (sold under the trade name Benzedrine) is an artificial stimulant that causes many of the same effects as epinephrine. The three –OH groups and the N–H bond in the epinephrine structure cause a greater percentage of that molecule to be polar, so we predict that epinephrine will be more soluble in water than will amphetamine.

**Figure 15.5**
**Molecular Line Drawings**

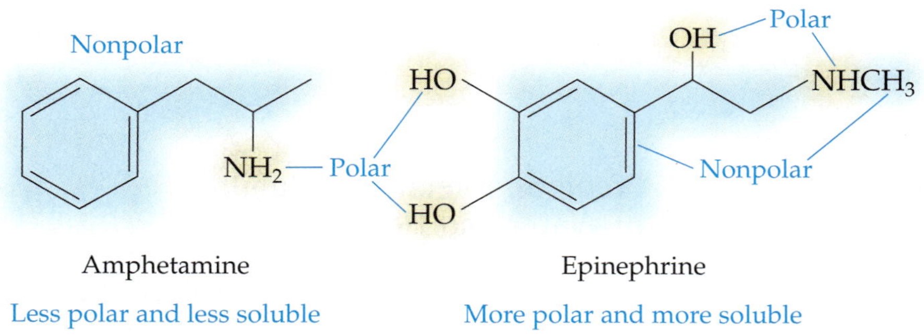

Amphetamine
Less polar and less soluble

Epinephrine
More polar and more soluble

The molecular representations in Figure 15.5 are known as *line drawings*. The corners, where two lines meet, represent carbon atoms, and the end of any line that does not have a symbol attached also represents a carbon atom. We assume that each carbon has enough hydrogen atoms attached to yield four bonds total. Compare the line drawings to the more detailed structures in Figure 15.6.

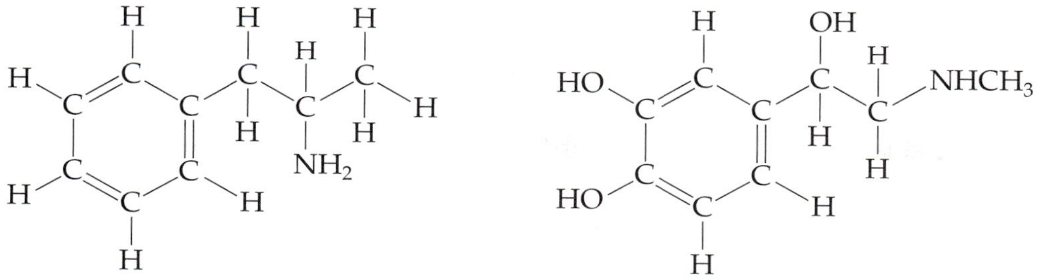

Amphetamine

Epinephrine

**Figure 15.6**
**Detailed Molecular Structures**

OBJECTIVE 10

The difference in polarity between epinephrine and amphetamine has an important physiological consequence. The cell membranes that separate the bloodstream from the brain cells have a nonpolar interior that tends to prevent polar substances in the blood from moving into the brain tissue. Epinephrine is too polar to move from the bloodstream into the brain, but amphetamine is not. The stimulant effects of amphetamine are in part due to its ability to pass through the blood–brain barrier.

OBJECTIVE 10

Methamphetamine is a stimulant that closely resembles amphetamine. With only one polar N–H bond, it has very low water solubility, but it can be induced to dissolve in water when converted to the much more polar, ionic form called methamphetamine hydrochloride. On the street, the solid form of this illicit ionic compound is known as *crystal meth* (Figure 15.7).

OBJECTIVE 11

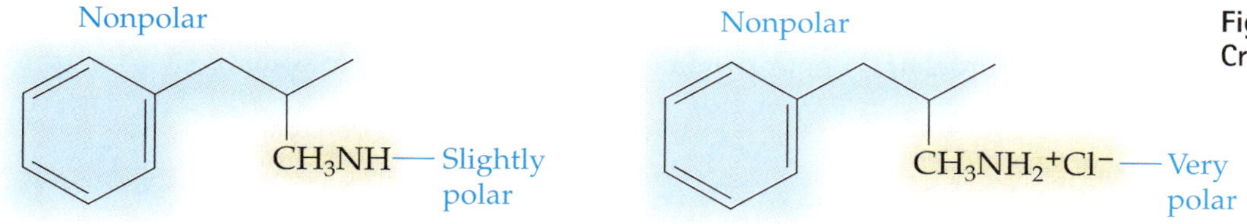

Nonpolar

$CH_3NH$ — Slightly polar

Methamphetamine

Nonpolar

$CH_3NH_2^+Cl^-$ — Very polar

Methamphetamine hydrochloride

**Figure 15.7**
**Crystal Meth**

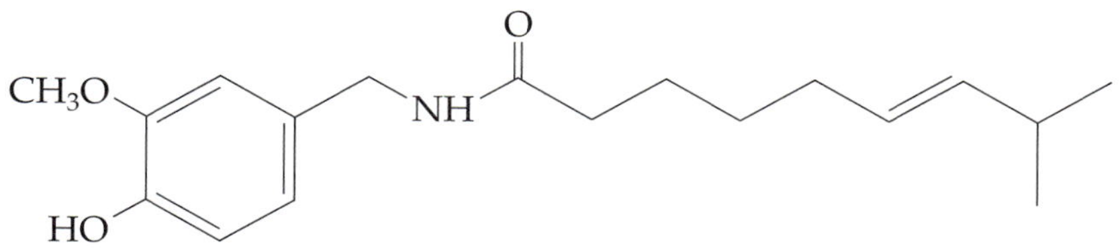

$CH_3O$ — HO — NH — O

Capsaicin

If you get a bite of a spicy dish that has too much chili pepper, do you grab a glass of water or a spoonful of sour cream to put out the flames? Capsaicin molecules, which are largely responsible for making green and red chili peppers hot, are mostly nonpolar. Therefore, the oil in the sour cream will more efficiently dissolve them, diluting them enough to minimize their effects. Because capsaicin is not water-soluble, the glass of water won't help much.

## Example 15.4 Predicting Relative Solubility in Water

OBJECTIVE 9

Butanoic acid is a foul-smelling substance that contributes to body odor; however, it reacts with various alcohols to form esters that have very pleasant odors and that are often used in flavorings and perfumes. Stearic acid is a natural fatty acid that can be derived from beef fat. Which of the following compounds would you expect to be more soluble in water?

```
     H   H   H  :O:
     |   |   |   ||
  H—C—C—C—C—O—H
     |   |   |      ..
     H   H   H
```

Butanoic acid

or

```
        :O: H  H  H  H  H  H  H  H  H  H  H  H  H  H  H  H  H
         ||  |  |  |  |  |  |  |  |  |  |  |  |  |  |  |  |  |
  H—O—C—C—C—C—C—C—C—C—C—C—C—C—C—C—C—C—C—C—H
    ..       |  |  |  |  |  |  |  |  |  |  |  |  |  |  |  |  |
             H  H  H  H  H  H  H  H  H  H  H  H  H  H  H  H  H
```

Stearic acid

*Solution*

**Butanoic acid** has a proportionally larger polar portion and is therefore **more soluble** in water than stearic acid is. In fact, butanoic acid is miscible with water (will mix with it in any proportion). The much higher proportion of the stearic acid structure that is nonpolar makes it almost insoluble in water.

## Exercise 15.4 Predicting Relative Solubility in Water

OBJECTIVE 9

The compound to the left below, 2-methyl-2-propanol, which is often called t-butyl alcohol, is an octane booster for unleaded gasoline. The other compound, menthol, when added to foods and medicines, affects the cold receptors on the tongue in such a way so as to produce a "cool" taste. Which of these two compounds would you expect to be more soluble in water?

```
            H
            |
         H—C—H
      H     |     H
      |     |     |
   H—C———C———C—H
      |     |     |
      H     |     H
          :O—H
           ..
```

2-methyl-2-propanol

OH

Menthol

## 15.2 Fats, Oils, Soaps, and Detergents

Solubility helps to explain why soap and detergent act to clean greasy dishes and oily clothes. As always, the properties of substances are a consequence of their chemical structures, so let's take a look at those structures.

Soap can be made from animal fats and vegetable oils, which in turn are composed of triacylglycerols, or triglycerides. (Although biochemists prefer the term *triacylglycerol*, we will use the more common term *triglyceride* for these compounds.) These are molecules that consist of long-chain hydrocarbon groups attached to a three-carbon backbone. In the general triglyceride structure shown in Figure 15.8, $R_1$, $R_2$, and $R_3$ represent the hydrocarbon groups (groups that contain only carbon and hydrogen atoms).

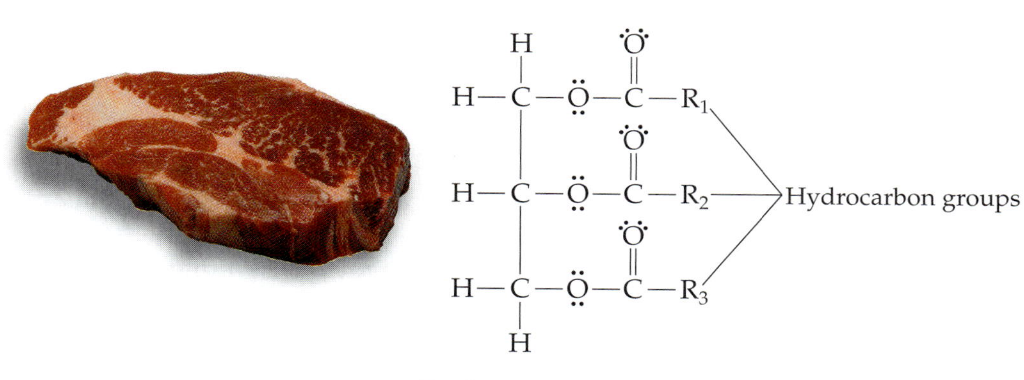


Triglyceride (triacylglycerol)

**Figure 15.8**
**Triglycerides**
The marbling in a steak is composed of triglycerides.

OBJECTIVE 12

An example of a triglyceride is tristearin, a typical fat molecule found in beef fat (Figure 15.9). Tristearin's structure can be depicted more efficiently in a line drawing (Figure 15.10, on page 642).

**Figure 15.9**
**Detailed Tristearin Structure**

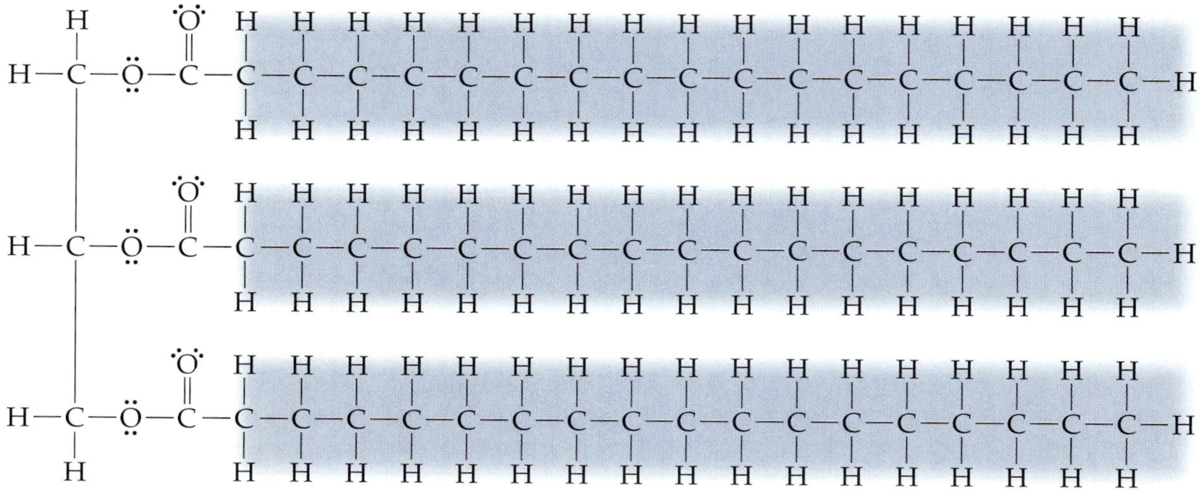

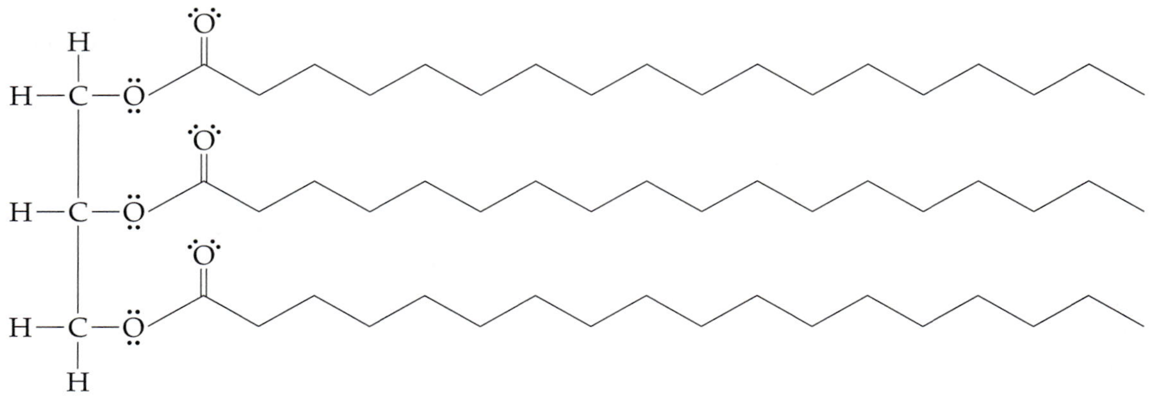

**Figure 15.10**
**Tristearin Line Drawing**

Most triglyceride molecules differ from tristearin in that three *different* hydrocarbon groups are attached to the carbon backbone. (In tristearin, those three hydrocarbon groups are all alike.) The hydrocarbon groups can differ in the length of the carbon chain and in the number of double bonds between carbon atoms. A higher number of carbon–carbon double bonds in its molecules causes a triglyceride to be liquid at room temperature. The triglycerides in vegetable oils, for example, have enough carbon–carbon double bonds to remain liquid (Figure 15.11). The triglycerides in animal fats have fewer carbon–carbon double bonds and tend to form a solid.

**Figure 15.11**
**Typical Liquid Triglyceride**
Liquid triglycerides are rich in carbon–carbon double bonds.

Triglyceride in fat or oil

**Figure 15.12**
Soap

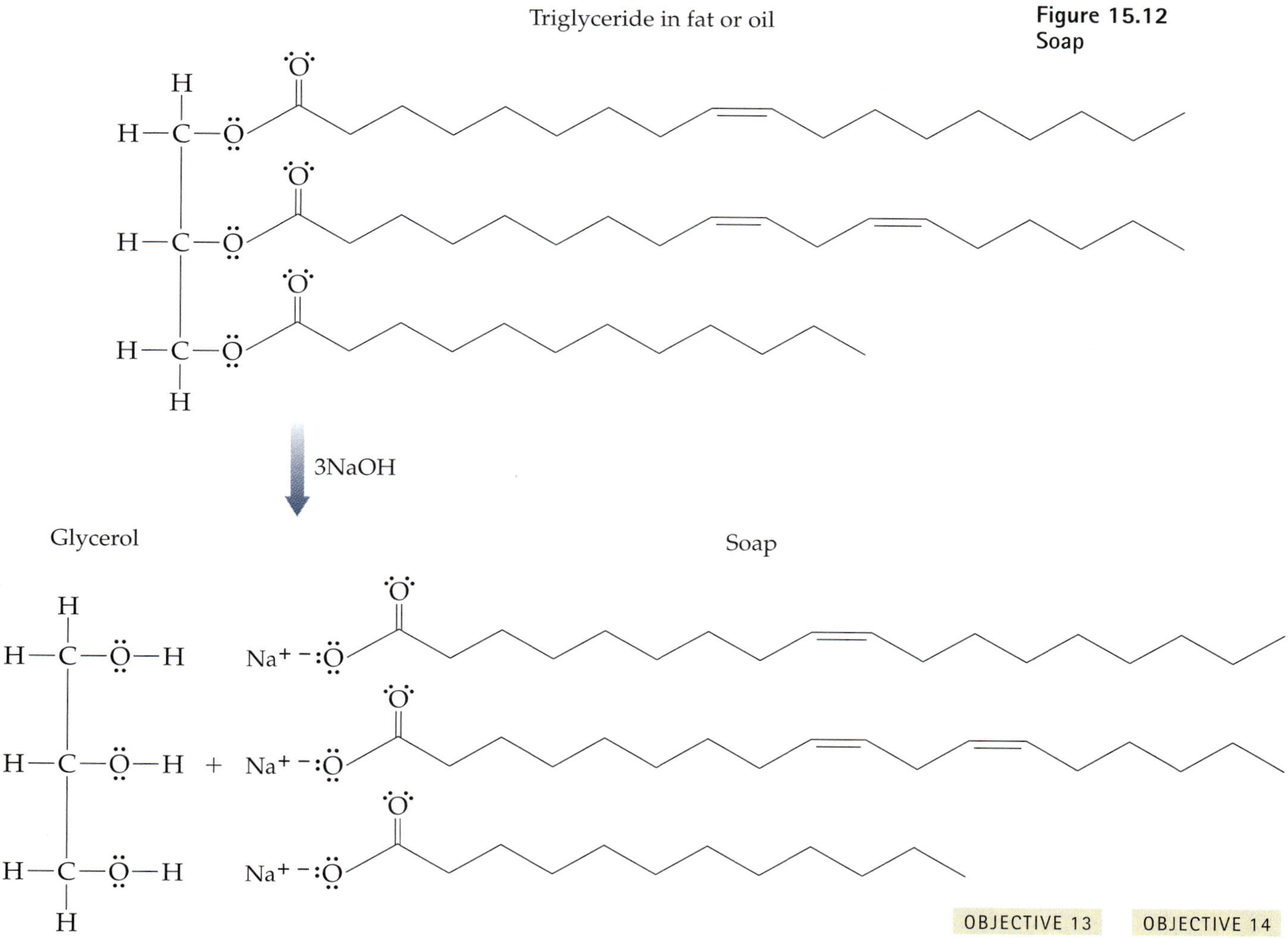

3NaOH

Glycerol

Soap

H—C—Ö—H

H—C—Ö—H + Na+ ⁻:Ö

H—C—Ö—H

OBJECTIVE 13 OBJECTIVE 14

A soap is an ionic compound composed of a cation (usually sodium or potassium) and an anion with the general form $RCO_2^-$, in which R represents a long-chain hydrocarbon group. Because the hydrocarbon chains in fats and oils differ, soaps are mixtures of various similar structures (Figure 15.12). They are produced in the reaction of triglyceride and sodium hydroxide, with glycerol as a byproduct. Old-fashioned lye soap was made by heating fats and oils (collected from cooking) with lye, which is sodium hydroxide. The hydrophobic, nonpolar hydrocarbon portion of each soap anion does not mix easily with water, but the hydrophilic, negatively charged end of each anion does form attractions to water molecules, and these attractions are strong enough to keep the anion in solution.

Now let's picture the events that take place at the molecular level in a tub of soapy water and greasy dishes. To clean a greasy dish, we first scrub its surface and agitate the water, but without soap (or detergent) in the water, the oil droplets would quickly regroup and return to the surface of the dish. If soap anions are in the solution, however, the agitation helps their nonpolar hydrocarbon ends to enter the nonpolar triglyceride droplets, while their anionic ends remain sticking out into the water. Soon every drop of fat or oil is surrounded by an outer coating of soap (Figure 15.13). When two droplets approach each other, the negative ends of the soap anions cause the droplets to repel each other, preventing the triglyceride droplets from recombining and redepositing on the object being cleaned. The droplets stay suspended in the solution and are washed down the drain with the dishwater (Figure 15.14).

OBJECTIVE 15

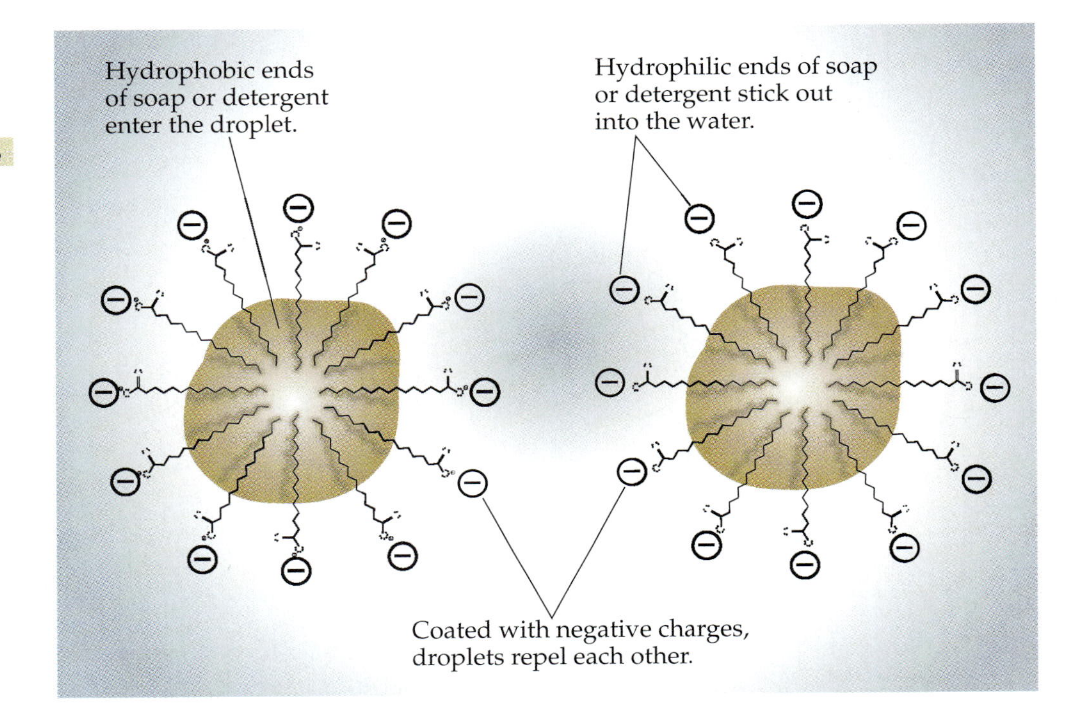
Hydrophobic ends
of soap or detergent
enter the droplet.
Hydrophilic ends of soap
or detergent stick out
into the water.
Coated with negative charges,
droplets repel each other.

**Figure 15.13**
**Oil Droplets and Soap or Detergents**

OBJECTIVE 15

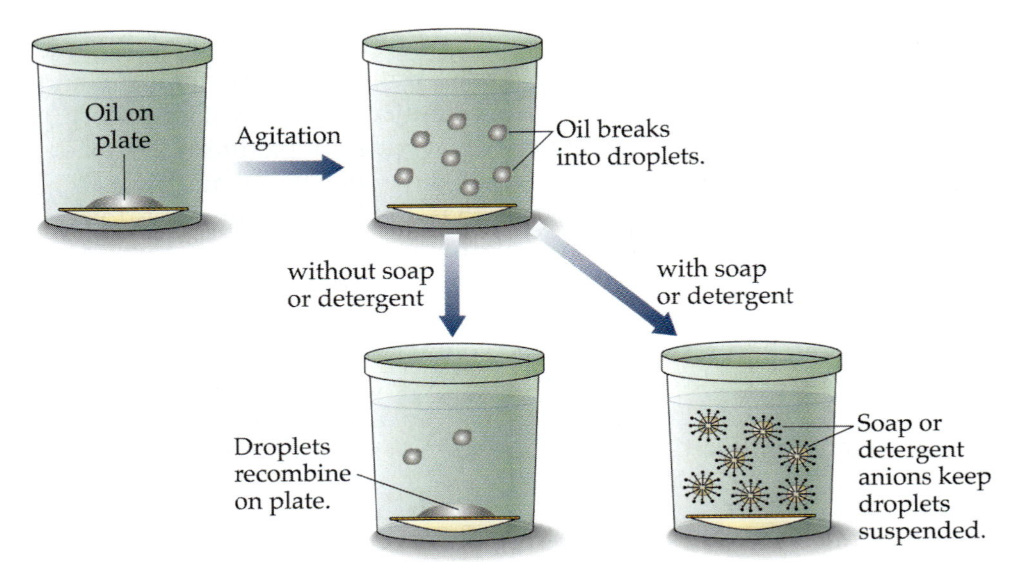
Oil on
plate
Agitation
Oil breaks
into droplets.
without soap
or detergent
with soap
or detergent
Droplets
recombine
on plate.
Soap or
detergent
anions keep
droplets
suspended.

**Figure 15.14**
**Cleaning Greasy Dishes**
Why do soaps and detergents help clean grease away?

OBJECTIVE 15

For soap to work, its anions must stay in solution. Unfortunately, they tend to precipitate from solution when the water is "hard." Hard water is water that contains dissolved calcium ions, $Ca^{2+}$, magnesium ions, $Mg^{2+}$, and often iron ions, $Fe^{2+}$ or $Fe^{3+}$. These ions bind strongly to soap anions, causing the soap to precipitate from hard-water solutions.

OBJECTIVE 16

Detergents have been developed to avoid this problem of soap in hard water. Their structures are similar to (although more varied than) that of soap, but they are less likely to form insoluble compounds with hard-water ions. Some detergents are ionic like soap, and some are molecular. For example, in sodium dodecyl sulfate (SDS), a typical ionic detergent, the $-CO_2^-$ portion of the conventional soap structure is replaced by an $-OSO_3^-$ group, which is less likely to link with hard-water cations and precipitate from the solution:

OBJECTIVE 17

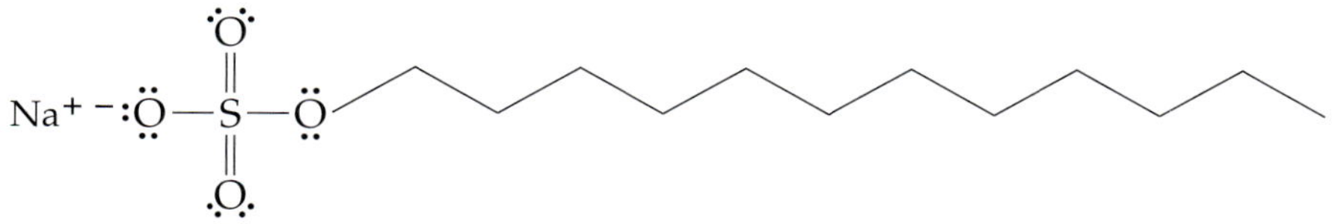

SDS, a typical ionic detergent

## 15.3 Saturated Solutions and Dynamic Equilibrium

Why is there a limit to the amount of salt that will dissolve in water? Why does it take longer for a sugar cube to dissolve than for the same amount of granular sugar, and why can you speed things up by stirring? Why does sugar dissolve so much more rapidly in hot coffee than in iced tea? To answer these questions, we need to know more about the rate at which solutes go into solution and the rate at which they return to the undissolved form.

### Two Opposing Rates of Change

When a solid is added to a liquid, a competition begins between two opposing rates of change. On the one hand, particles leave the surface of the solid and move into solution, and on the other hand, particles leave the solution and return to the solid form.

To examine these changes at the molecular or ionic level, let's picture what happens when table salt dissolves in water. (You might want to review Section 4.4, which describes this process in detail.) Imagine the NaCl solid sitting on the bottom of the container. The sodium and chloride ions at the solid's surface are constantly moving out into the water and then being pulled back by their attractions to the other ions still on the surface of the solid. Sometimes when an ion moves out into the water, a water molecule collides with it and pushes it farther out into the solution. Other water molecules move into the gap between the ion and the solid and shield the ion from being recaptured by the solid (Figure 4.4). The ion, now in solution, is kept stable by attractions between its charge and those of the polar water molecules that surround it. If the ion is a cation, it is

OBJECTIVE 18

surrounded by the negative oxygen atoms of the water molecules; if it is an anion, it is surrounded by their positive hydrogen ends (Figure 4.3).

OBJECTIVE 18

OBJECTIVE 19

Once the ions dissolve, they travel throughout the solution just like any other particle in a liquid. Eventually they collide with the surface of the remaining solid. When this happens, they are recaptured by the attractions that hold the particles of the solid together, and they are likely to return to the solid form. One of the most important factors that determine the rate at which solute particles return to solid form is the concentration of the solute in the solution. The more particles of solute per liter of solution, the more collisions between solute particles and the solid. A higher rate of collisions results in a greater rate of return of solute particles from the solution to the solid (Figures 15.15 and 15.16).

**Figure 15.15**
Solute Concentration and Rate of Return to Solid Form

OBJECTIVE 19

**Higher concentration of solute particles in solution**

↓

More collisions per second between solute and solid

↓

**Increased rate of return to the solid form**

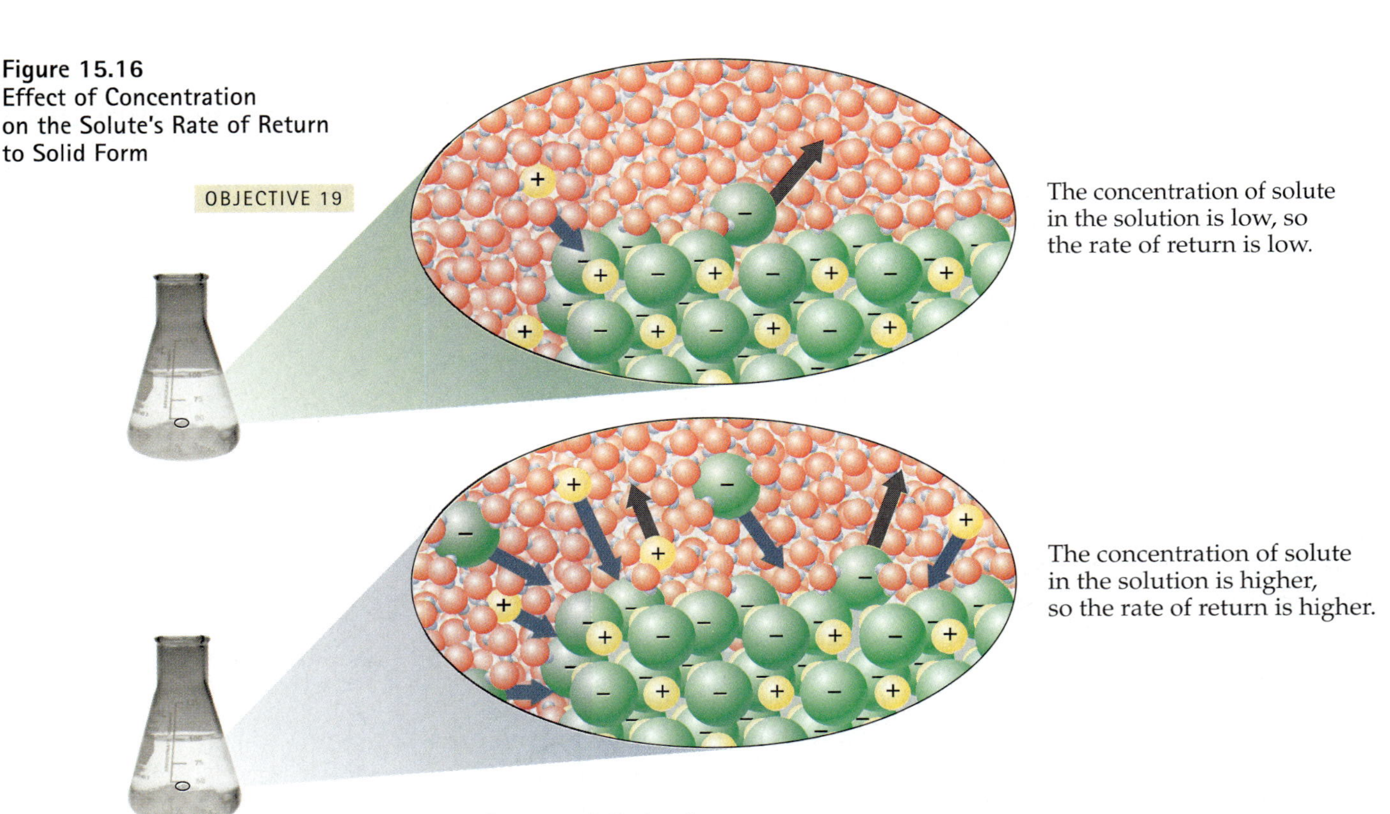

The concentration of solute in the solution is low, so the rate of return is low.
The concentration of solute in the solution is higher, so the rate of return is higher.


**Figure 15.16**
Effect of Concentration on the Solute's Rate of Return to Solid Form

OBJECTIVE 19

## Net Rate of Solution

When a solid is first added to a liquid, there are more particles leaving its surface and moving into solution than there are particles returning from the solution to the solid form. As long as the rate of solution is greater than the rate of return, there will be a net shift of particles into solution.

OBJECTIVE 20

The net rate of change in the number of particles in solution is equal to the number of particles that move from solid to solution per second minus the number of solute particles that return to solid form per second. This net rate of solution depends on three factors:

- Surface area of the solute
- Degree of agitation or stirring
- Temperature

OBJECTIVE 21

OBJECTIVE 22

To understand why surface area affects the rate of solution, picture a lump of sugar added to water. There are a certain number of particles at the surface of the sugar cube, and these are the only particles that have any possibility of escaping into the solution. The particles in the center of the cube cannot dissolve until the particles between them and the surface have escaped. If the sugar cube is split in half, however, and the two halves are separated, two new surfaces are exposed. Some particles that used to be in the center of the cube, and hence were unable to escape, are now at the surface and can dissolve. The cutting of the cube has therefore increased the rate of solution (Figure 15.17). When bakers want to dissolve sugar, they do not want to wait too long, so they add very finely powdered sugar. This sugar dissolves much more rapidly than granular sugar or sugar cubes, because it has a much greater total surface area.

Why does powdered sugar dissolve faster than granular sugar?

**Figure 15.17**
**Surface Area and Rate of Solution**

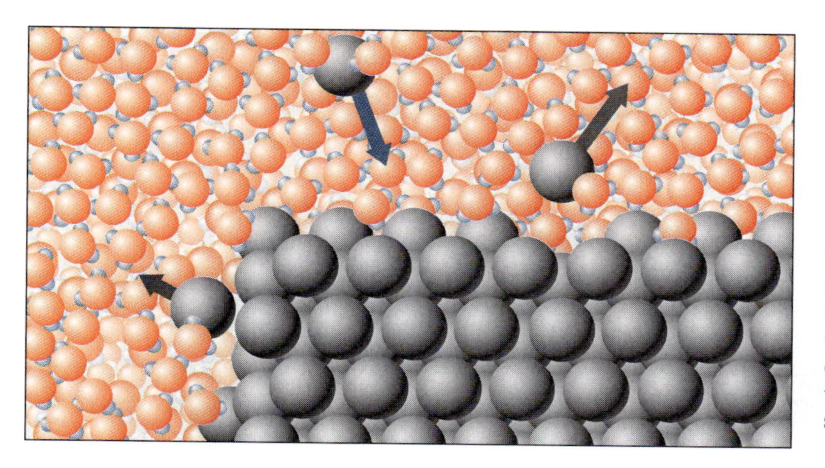

The rate of solution depends on the number of particles at the solid's surface.

OBJECTIVE 22

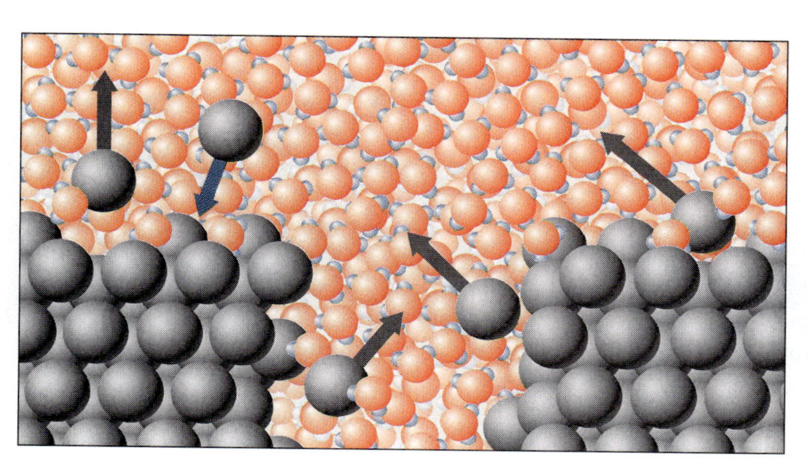

If the solid is fragmented, new surfaces are exposed, allowing more particles to escape into solution.

Near the solid, close to where particles are leaving the solid surface and entering solution, the concentrations of dissolved solute are higher than in more remote parts of the liquid. As we observed earlier, a higher concentration of solute leads to a higher rate of return to the solid form. This higher rate of return leads to a lower overall *net* rate of solution. However, if we stir or in some way agitate the solution, the solute particles near the solid are quickly carried away, and the relatively high local concentrations of solute are decreased. This diminishes the rate of return and increases the *net* rate of solution, which explains why stirring your coffee dissolves the sugar in the coffee more rapidly (Figures 15.18 and 15.19).

OBJECTIVE 23

**Figure 15.18**
Agitation and Rate of Solution

OBJECTIVE 23

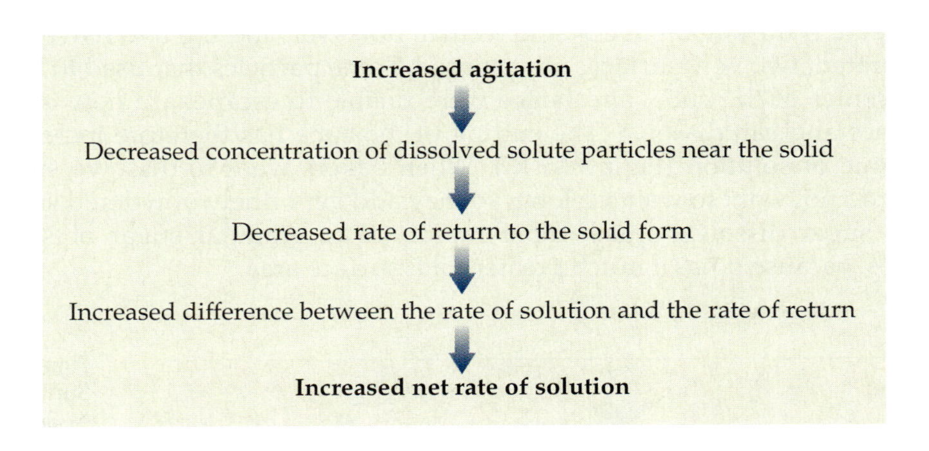

**Increased agitation**
Decreased concentration of dissolved solute particles near the solid
Decreased rate of return to the solid form
Increased difference between the rate of solution and the rate of return
**Increased net rate of solution**


OBJECTIVE 23

**Figure 15.19**
Agitation and Net Rate of Solution

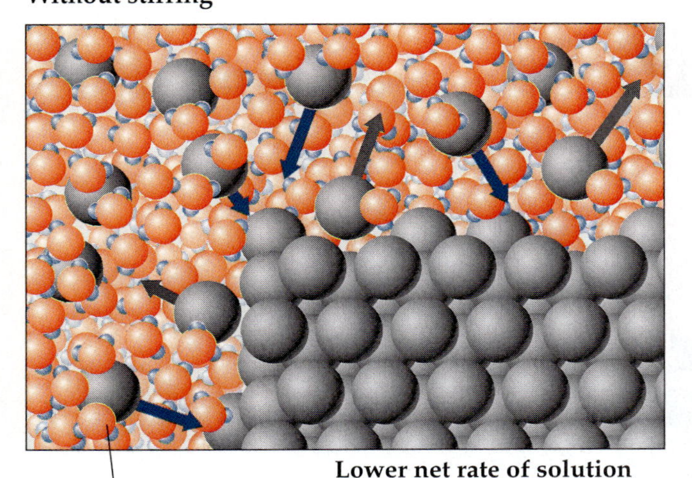


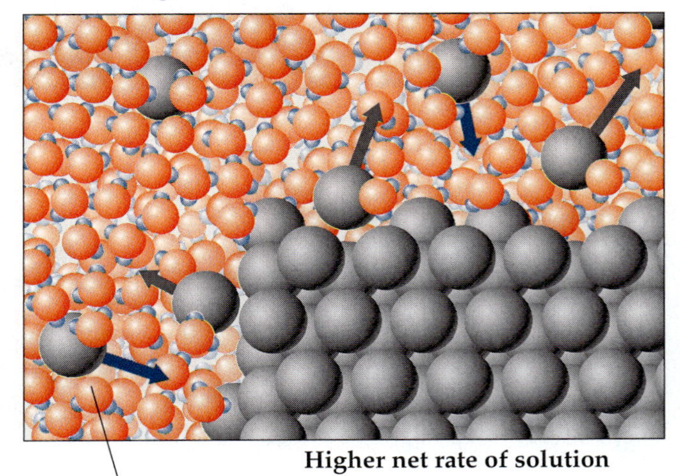

Finally, increasing the temperature of the system increases the average velocity of the solute and solvent particles. More rapidly moving particles in the solid escape more easily into the solution, increasing the rate of solution. Increased temperature also leads to a situation that is similar to agitation. The higher-velocity solute particles move away from the surface of the solid more quickly, minimizing the concentration of dissolved solute in the solid's vicinity (Figure 15.20). This decreases the rate of return of particles to the solid, increasing the net rate of solution, which explains why hot coffee will dissolve sugar more rapidly than cold iced tea.

OBJECTIVE 24

Why does sugar dissolve faster in hot water than in cold water?

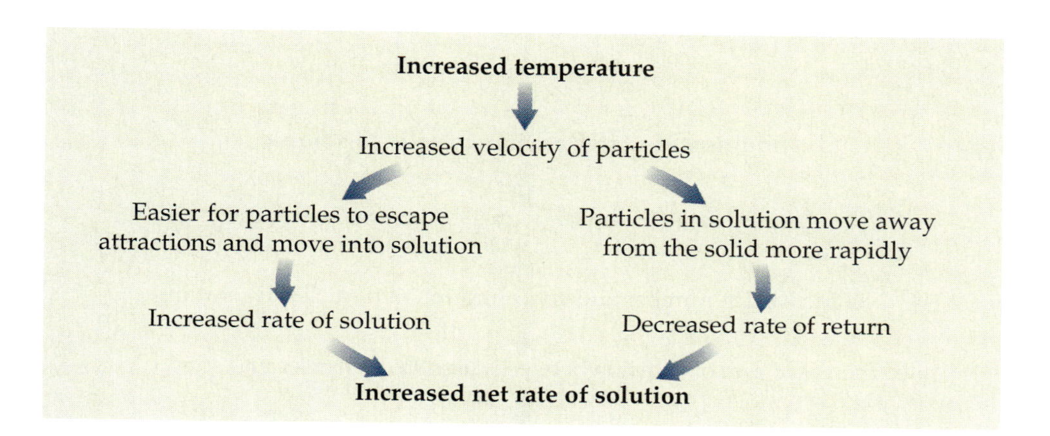


**Figure 15.20**
Relationship Between Temperature and Net Rate of Solution

OBJECTIVE 24

## Saturated Solutions

If you keep adding sugar to your coffee, eventually you reach a point where the maximum amount has dissolved, and adding more sugar merely leaves solid sugar on the bottom of the cup. At this point, you have reached the solubility limit of sugar in your coffee solution. Let's try to visualize what this means at the molecular level.

If the surface area, temperature, and degree of agitation remain essentially the same, the rate at which the particles move into the solution remains close to constant. Meanwhile, the rate at which the particles return to the solid will depend on the concentration of solute particles in solution. As long as the rate of solution is greater than the rate of return, there will be a net movement of particles into solution, but as the concentration of dissolved solute increases, the rate of return will increase as well.

OBJECTIVE 25

If there is enough solid in the solution, the rate of return will eventually become equal to the rate of solution, and the system will reach a dynamic equilibrium. (If too little solid has been added, it will all dissolve before equilibrium is reached.) At that point, for every particle that leaves the solid and moves into solution, a particle somewhere else in the system is returning from the dissolved form to the solid form. There is a constant movement of particles from the solid to the dissolved state and back again, but there is no *net* change in the amounts of solid and solute in the system. (This is like the plastic ball equilibrium described in Figure 14.7.) The solution has reached its solubility limit. When enough solute has been added to the solution to reach that limit, we call the solution **saturated.** (Figures 15.21 and 15.22). Table 15.2 lists water solubilities of some common substances. If a solution has less solute dissolved than is predicted by the solubility limit, it is **unsaturated.**

OBJECTIVE 26

A saturated solution need not be accompanied by excess solid in its container, although if it is, there will be a dynamic equilibrium between the rate of solution and the rate of return. A saturated solution without excess solid can be made by first adding solid to a solvent until no more will dissolve and then filtering out the excess solid.

**Figure 15.21**
Dynamic Equilibrium and Saturated Solutions

OBJECTIVE 25

**Addition of a large amount of solid to a liquid**

↓

Initially, rate of solution is greater than rate of return

↓

Net increase in number and concentration of particles in solution

↓

Increased rate of collision between dissolved particles and solid

↓

Increased rate of return . . .

↓

. . . Until rate of return equals rate of solution

↓

Constant changes from solid to dissolved solute and back, but no net changes in amounts of solid and dissolved solute

↓

**Saturated solution due to dynamic equilibrium**

Table 15.2
Water Solubilities of Common Substances at Room Temperature

| Substance | Approximate solubility at room temperature (g substance/100 mL $H_2O$) |
|---|---|
| ethanol, $C_2H_5OH$ (in alcoholic beverages) | miscible (infinite) |
| acetic acid, $HC_2H_3O_2$ (in vinegar) | miscible (infinite) |
| sucrose, $C_{12}H_{22}O_{11}$ (white table sugar) | 200 |
| hydrogen chloride, HCl | 69 |
| sodium chloride, NaCl (table salt) | 36 |
| carbon dioxide, $CO_2$ | 0.17 |
| methane, $CH_4$ | 0.0069 |
| oxygen, $O_2$ | 0.0045 |
| calcium carbonate, $CaCO_3$ (in limestone) | 0.0014 |

OBJECTIVE 26

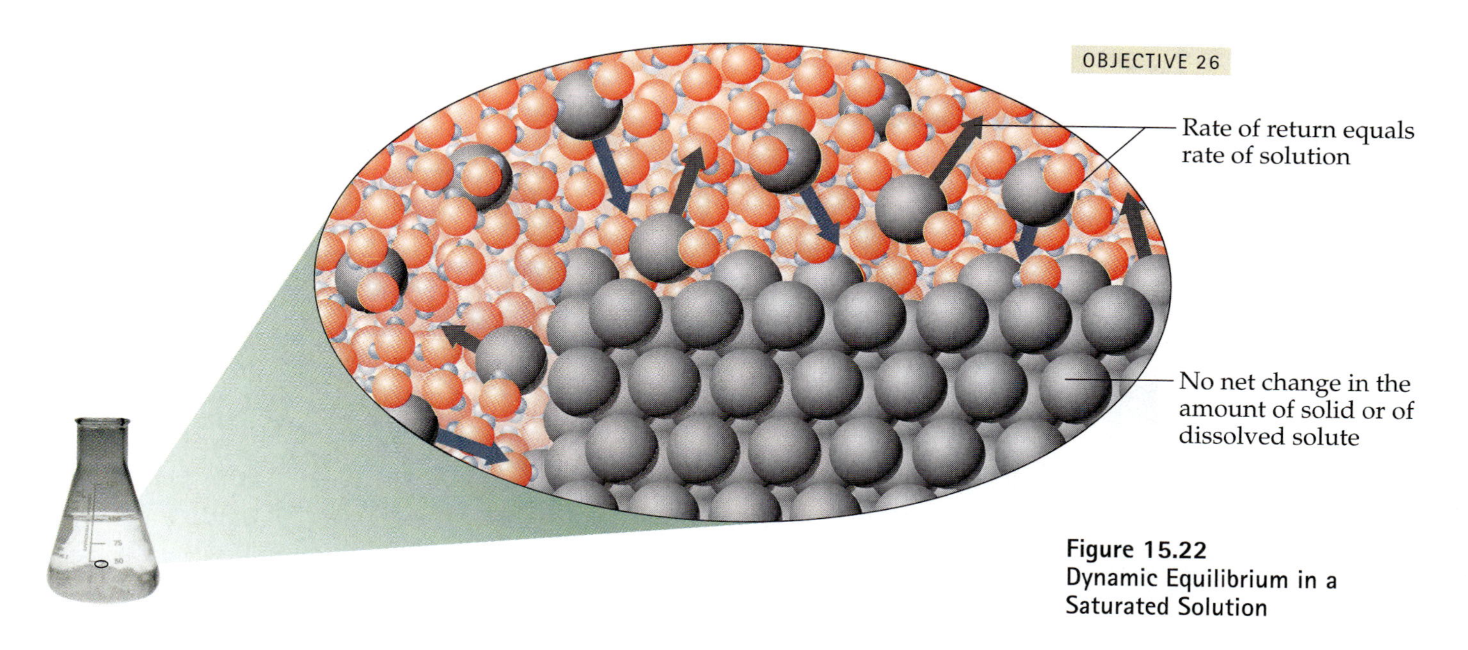
Rate of return equals rate of solution
No net change in the amount of solid or of dissolved solute

**Figure 15.22**
**Dynamic Equilibrium in a Saturated Solution**

Visit the following Web site to see how it is possible to make a supersaturated solution, which is a solution that contains more dissolved solid than predicted by the solubility limit.
www.chemplace.com/college/

## 15.4 Solutions of Gases in Liquids

Why do scuba divers get drowsy and disoriented when they dive too deeply? Why do divers get the bends when they rush to the surface too quickly? Why is moving to the surface too quickly like opening a can of soft drink? As we study the properties of solutions of gases in liquids, we will discover the answers to these questions.

OBJECTIVE 27

Imagine a system that consists of a gas above a liquid in a closed container (Figure 15.23, on the next page). The gas particles move rapidly, colliding with each other, with the walls of their container, and with the surface of the liquid. When such a particle collides with the liquid surface, its momentum takes it into the liquid at least a short way, and it forms attractions to the liquid particles. As this happens, the gas becomes a solute in the liquid.

OBJECTIVE 27

The dissolved gas particle moves throughout the liquid like any other particle in the liquid. It eventually makes its way back to the surface, and if it has enough energy to break the attractions that hold it in the liquid form, it can escape from the surface back into the gas phase above the liquid. Therefore, in a system with a gas above a liquid, there are two opposing rates of change: the rate of solution and the rate of escape.

**Figure 15.23**
**Dynamic Equilibrium for a Solution of a Gas in a Liquid**

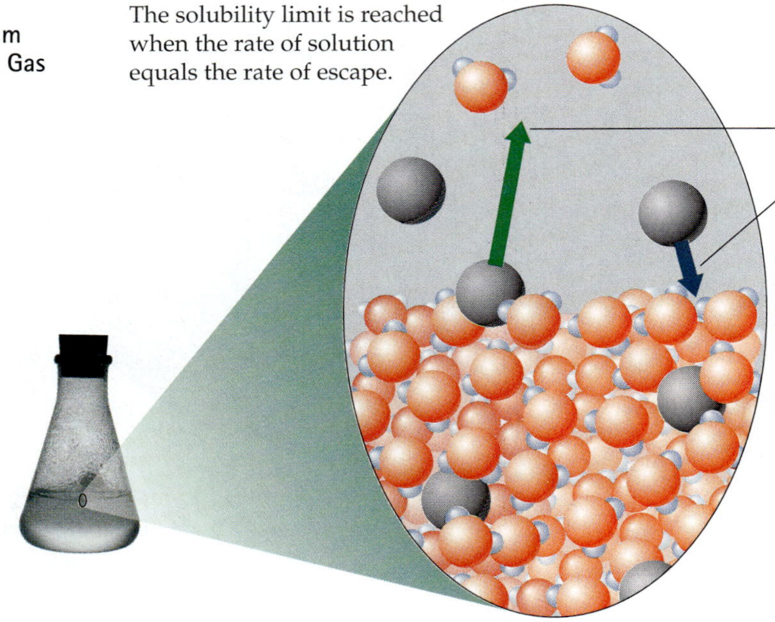

The solubility limit is reached when the rate of solution equals the rate of escape.

OBJECTIVE 27

For every gas particle that escapes from the liquid, another gas particle collides with the surface and goes into solution.

## Gas Solubility

The gas's rate of solution is dependent on the number of collisions with the surface of the liquid per second. The gas's rate of escape is dependent on the number of solute particles at the surface, on the strengths of attractions between the gas and liquid particles, and on the temperature of the system. The overall solubility of the gas in the liquid is dependent on the relative rates of solution and escape.

OBJECTIVE 27

Let's take a closer look at the changes that take place in this system. At the instant that a gas is placed over a pure liquid in a closed container, there are no particles of that gas in solution, so the rate of escape is zero. Immediately, however, the particles begin to collide with the surface of the liquid, so gas particles begin to move into solution right away. As long as the rate of solution is greater than the rate of escape, the concentration of dissolved gas particles increases. This increase in solute concentration leads to an increase in the number of solute particles at the surface of the liquid and to an increase in the rate of escape. Eventually, a dynamic equilibrium is reached where the rate of solution and the rate of escape are equal. At that point, there is no net change in the amount of gas above the liquid or in the concentration of dissolved gas in the solution. The solubility limit has been reached (Figures 15.23 and 15.24).

## Partial Pressure and Gas Solubility

If the partial pressure of a gas over a liquid is increased, the solubility of that gas is increased. To understand why, picture a system consisting of a gas above a liquid in a closed container. Assume that a dynamic equilibrium has formed between the rate of solution and the rate of escape. If

**Addition of a gas above a liquid in a closed container**

Initially, rate of solution is greater than rate of escape

Net shift of particles into solution

Increased concentration of dissolved gas

Increased rate of escape . . .

. . . Until rate of escape equals rate of solution

Constant changes between dissolved and undissolved gas, but no net change in amount of either

**Dynamic equilibrium (solubility limit)**

**Figure 15.24**
Gas Solubility

OBJECTIVE 27

more gas is pumped into the space above the liquid, the concentration of gas and therefore the partial pressure of the gas above the liquid increase. This leads to an increased rate of collision between the gas particles and the surface of the liquid—and to an increase in the rate of solution. At this stage, the rate of solution is greater than the rate of escape, and there is a net movement of particles into solution. Eventually, however, the concentration of solute in solution will be high enough to create a new dynamic equilibrium, with a higher rate of escape equal to the higher rate of solution. Figure 15.25 summarizes these events.

OBJECTIVE 28

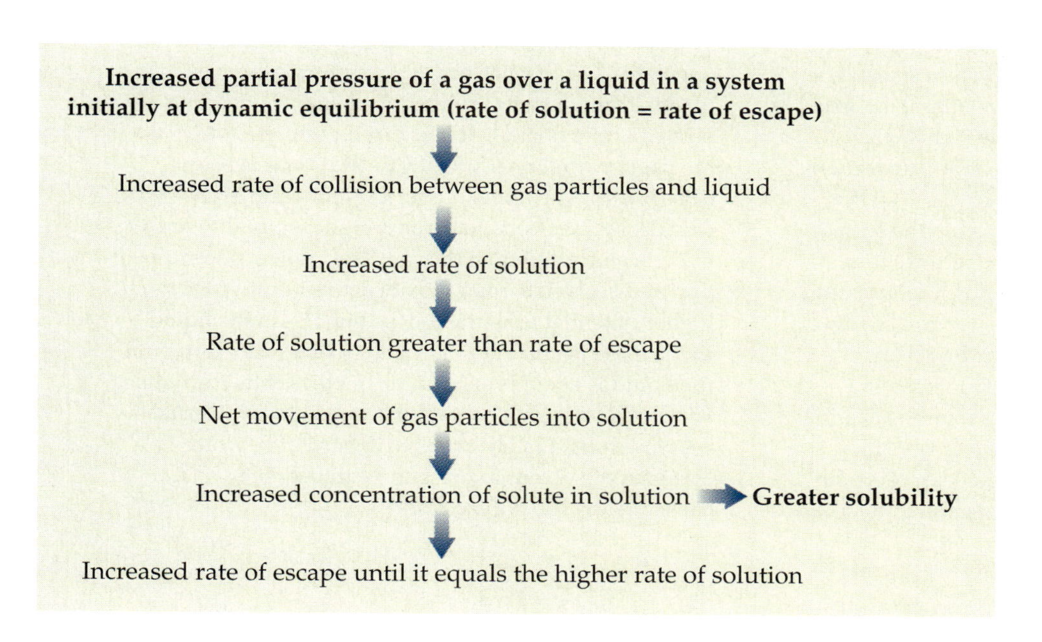


**Figure 15.25**
Partial Pressure and Gas Solubility

OBJECTIVE 28

## Gas Solubility and Breathing

When air is drawn into your lungs, it enters 300 million tiny air sacs called alveoli. Through the very thin, two-cell-width walls of the alveoli, an exchange of gases takes place between your blood and the inhaled air. The huge number of alveoli provide a very high surface area of about 60–80 $m^2$, which allows large amounts of gases to be exchanged very rapidly. Even so, this system never reaches equilibrium—but it still provides us with an example of the effect of partial pressures of gases on solubility.

## SPECIAL TOPIC 15.1 Gas Solubility, Scuba Diving, and Soft Drinks

When scuba divers descend into the ocean, the external pressure they experience is the sum of the atmospheric pressure on the top of the water and the pressure due to the weight of the water above them. Unless a diver is completely sealed in a diving capsule, the pressure of the air taken into the lungs is equal to this external pressure, which increases by 1 atmosphere, or 101 kPa, for each 10 meters that the diver descends.

As the total pressure of the air taken into the lungs increases, the partial pressure of the oxygen in the air increases, and the concentration of the oxygen in the blood goes up. If too much oxygen is taken into the blood, the excess oxygen can begin to oxidize enzymes and other chemicals and cause damage to the nervous system. In the extreme, it can lead to coma and death. For this reason, deep-sea divers commonly use mixtures of gases whose percentage of oxygen is lower than the amount in normal air.

There are other problems that scuba divers encounter as a result of the higher gas pressures of the air they breathe. The higher partial pressure of nitrogen that enters the lungs at greater depths leads to a higher concentration of nitrogen in the blood. This can cause a condition called nitrogen narcosis, or *rapture of the deep*. The accompanying disorientation, drowsiness, and other symptoms similar to those of alcohol intoxication can lead to serious consequences.

As long as the diver remains deep below the surface, bubbles of significant size do not form, because the external pressure is greater than their potential pressure, but when the diver begins to rise, the situation changes. If the diver rises slowly, the gases in the blood are expelled from the lungs rapidly enough to keep the potential pressure of the bubbles lower than the external pressure, and no bubbles form. If the diver rises too rapidly (perhaps disoriented by nitrogen narcosis), the gases do not leave the bloodstream fast enough to keep the potential pressure of the bubbles below the diminishing external pressure. Bubbles then form in the bloodstream, blocking the smooth flow of blood through the smaller vessels and causing severe muscle and joint pain, as well as more serious damage. This condition is called the bends.

Why do bubbles form when a soft drink is opened, and how is this similar to the bubbles that can form in a diver's blood if he goes to the surface too quickly?

One way of avoiding the bends is to substitute a less soluble gas, such as helium, for the nitrogen gas in the air tank. Less gas will enter the blood, leading to lower potential pressures of gas bubbles and less likelihood of experiencing the bends.

The formation of bubbles when a soft drink bottle is opened is very similar to the formation of bubbles in the bloodstream of someone with the bends. Soft drinks are bottled under about 4 atm of carbon dioxide, whereas the usual partial pressure of $CO_2$ in dry air is about 0.00035 atm. The much higher $CO_2$ pressure in the soft drink container leads to a much higher solubility of $CO_2$ in the water solution in the can. The higher concentration of dissolved carbon dioxide leads in turn to a much higher potential pressure for the bubbles in the liquid. In the closed container, the 4 atm of pressure exerted on the liquid is enough to keep the bubbles from forming. When the soft drink container is opened, the external pressure immediately drops to about 1 atm, which is less than the potential pressure of the bubbles, and consequently the bubbles form.

In dry air, about 21% of the particles are oxygen molecules. Because the overall atmospheric pressure changes with altitude, the partial pressure of oxygen in the air is different in different places. For example, in Mexico City, the site of the 1968 Olympics, the total atmospheric pressure is about 75 kPa, compared to 101 kPa at sea level. Thus the partial pressure of oxygen in Mexico City is about 16 kPa, compared to about 21 kPa at sea level. This lower partial pressure of oxygen gas leads to a lower rate of solution of $O_2$ into the blood and hence to lower concentrations of oxygen in the blood. Because of the need for a steady supply of oxygen to the muscles to maintain peak athletic performance, this was an important issue for Olympic athletes.

Lower partial pressures of oxygen at high altitudes mean that less oxygen enters the blood. Above, runners at the Olympics in Mexico. At right, alveoli in the lungs.

Visit the following Web site to see how changes in temperature affect solid and gas solubility.
www.chemplace.com/college/

OBJECTIVE 29

## SPECIAL TOPIC 15.2 Global Warming, Oceans, and $CO_2$ Torpedoes

Real-world situations are more complicated than the simple systems described in this chapter, but you can use what you have learned here to understand them better. For example, what you have learned about the solution of gases can be used to understand more about the solution of carbon dioxide in the ocean and its effect on global warming.

Greenhouse gases were defined in Special Topic 10.1: *Big Problems Require Bold Solutions—Global Warming and Limiting Reactants*. They are gases in the atmosphere that trap some of the infrared energy released by the earth as it cools. We have seen that carbon dioxide is a greenhouse gas and that scientists are searching for ways to reduce its concentration in the atmosphere.

As we saw in Special Topic 10.1, a large amount of carbon dioxide from the atmosphere eventually enters the ocean. The first step in this process can be understood in terms of the model used in this chapter to describe how gas dissolves in a liquid. The collisions between the $CO_2$ gas particles and the surface of the ocean lead to the movement of about $9.2 \times 10^{13}$ kg of $CO_2$ into the oceans every year. There is a competing rate of $CO_2$ escape from the ocean—about $9.0 \times 10^{13}$ kg $CO_2$ per year. Thus, the net rate of movement of $CO_2$ into the ocean is about $2 \times 10^{12}$ kg per year.

Several plans have been proposed for reducing the release of greenhouse gases into the atmosphere. One is to collect the carbon dioxide produced by fossil-fuel-burning power plants and place it directly into the ocean. This idea is controversial, however, because when carbon dioxide dissolves in water, it forms carbonic acid, $H_2CO_3$. Dissolving large quantities of $CO_2$ from power plants in ocean water could lead to significant changes in the acidity of the water and hence to major biological changes.

Another suggestion is to drop 100- to 1000-ton torpedo-shaped blocks of solid carbon dioxide (dry ice) into the ocean. It is believed that if these $CO_2$ torpedoes could be encased in an insulating layer of solid water (regular ice), they would survive the plunge to the bottom of the ocean and embed themselves about 10 m into the sediments on the ocean floor. The $CO_2$ would then combine with water to form structures that would trap the gas permanently, thus preventing it from changing the ocean's acidity.

## Chapter Glossary

**Entropy** A measure of the disorder in a system.

**Second Law of Thermodynamics** The scientific principle that the entropy of the universe increases. This principle helps us to predict that changes take place when they lead to an increase in the disorder and degree of dispersal in the universe.

**Miscible** Able to be mixed in any proportion without any limit to solubility.

**Hydrophilic ("water loving")** A polar molecule or ion (or a portion of a molecule or polyatomic ion) that is attracted to water.

**Hydrophobic ("water fearing")** A nonpolar molecule (or a portion of a molecule or polyatomic ion) that is not expected to mix with water.

**Saturated solution** A solution that has enough solute dissolved to reach the solubility limit.

**Unsaturated solution** A solution that has less solute dissolved than is predicted by the solubility limit.

You can test yourself on the glossary terms at the following Web site:

**www.chemplace.com/college/**

## Chapter Objectives

**The goal of this chapter is to teach you to do the following.**

1. Define all of the terms in the Chapter Glossary.

**Section 15.1 Entropy, Solutions, and Solubility**

2. Using the shift from solid to gas as an example, explain why particles change from more ordered states to more disordered states.
3. Using the expansion of gas as an example, explain why particles tend to become dispersed.
4. With reference to the Second Law of Thermodynamics, explain why water and compounds that are like water (that is, are composed of relatively small polar molecules with hydrogen bonds that link them) mix to form solutions.
5. Given the solubility of a substance in grams of substance that will dissolve in 100 g (or 100 mL) of solvent, classify the substance as soluble, insoluble, or moderately soluble in the solvent.
6. Given a description of a compound as ionic, polar molecular, or nonpolar molecular (or given enough information to determine this), predict whether the compound would be soluble or insoluble in water.
7. Given a description of a compound as ionic, polar molecular, or nonpolar molecular (or given enough information to determine this), predict whether the compound would be soluble or insoluble in hexane.

8. Given a description of a compound as ionic, polar molecular, or nonpolar molecular (or given enough information to determine this), predict whether the compound would be expected to be more soluble in water or in hexane.
9. Given descriptions of two compounds that differ in the percentages of their structures that are polar and nonpolar (hydrophilic and hydrophobic), predict their relative solubilities in water and in hexane.
10. Explain why amphetamine can pass through the blood–brain barrier more easily than epinephrine.
11. Explain why methamphetamine is often converted to methamphetamine hydrochloride.

**Section 15.2 Fats, Oils, Soaps, and Detergents**

12. Write or identify a description of the general structure of a triacylglycerol (triglyceride) molecule.
13. Given the chemical formula for a triacylglycerol (triglyceride), write the chemical formulas for the products of its reaction with sodium hydroxide.
14. Describe the structure of a typical soap, identifying the hydrophilic and hydrophobic portions of its structure.
15. Explain how soap and detergent enhance the cleaning process.
16. Describe hard water, and explain why soap does not work as well in hard water as it does in soft water.
17. Describe the difference in structure between a typical anionic detergent and a typical soap, and explain why detergent works better in hard water.

**Section 15.3 Saturated Solutions and Dynamic Equilibrium**

18. Describe the reversible change that takes place when an ionic compound, such as sodium chloride, is dissolved in water.
19. Explain why an increase in the concentration of solute particles in a solution containing undissolved solid leads to an increase in the rate of return of the solute to the solid form.
20. Describe the difference between a rate of solution and a *net* rate of solution.
21. List three factors that determine net rates of solution.
22. Explain why increased surface area of a solid solute leads to an increase in the rate of solution of the solid.
23. Explain why agitation or stirring leads to an increase in the rate of solution of a solid.
24. Explain why increased temperature typically leads to an increase in the rate of solution of a solid.
25. Explain why the addition of a relatively large amount of solid to a liquid leads to a dynamic equilibrium with no net change in the amount of undissolved solid and dissolved solute.
26. Describe the changes that take place in a system that contains undissolved solid in equilibrium with dissolved solute.

### Section 15.4 Solutions of Gases in Liquids

27. Describe the reversible change that takes place when a closed container holds a liquid with a space above it that contains a gas that dissolves in the liquid. Explain why this system comes to a dynamic equilibrium with no net change in the amount of gas above the liquid or the amount of gas dissolved in the liquid.
28. Explain why an increase in the partial pressure of a gaseous substance over a liquid leads to an increase in the solubility of the substance in the liquid.
29. With reference to the relationship between the partial pressure of substances and their solubility, explain why the concentration of oxygen in the blood decreases as you move to higher elevations.

## Review Questions

1. In the past, sodium bromide was used medically as a sedative. Describe the process by which this ionic compound dissolves in water, including the nature of the particles in solution and the attractions between the particles in the solution. Identify the solute and the solvent in this solution.
2. Draw a reasonable Lewis structure and a geometric sketch for each of the following molecules. Identify each compound as polar or nonpolar.
   a. $C_2H_6$
   b. $CH_3CH_2OH$
   c. $CH_3CO_2H$
3. For each of the following substances, write the name for the type of particle viewed as forming the structure of its solid, liquid, or gas, and write the name of the type of attraction that holds these particles in the solid or liquid form.
   a. heptane, $C_7H_{16}$
   b. formic acid, $HCO_2H$
   c. copper(II) sulfate, $CuSO_4$
   d. methanol, $CH_3OH$
   e. iodine, $I_2$
   f. carbon dioxide, $CO_2$
4. When liquid propane is pumped into an empty tank, some of the liquid evaporates, and after that, both liquid and gaseous propane are present in the tank. Explain why the system adjusts so that there is a constant amount of liquid and gas in the container. With reference to the constant changes that take place inside the container, explain why this system can be described as a dynamic equilibrium.

## Key Ideas

**Complete the following statements by writing one of these words or phrases in each blank.**

| | |
|---|---|
| 10 | liquid |
| A | long-chain |
| any proportion | moderately |
| attracted | net |
| B | nonpolar |
| cation | nonpolar solvents |
| degree of agitation or stirring | not |
| disorder | oils |
| disordered | polar solvents |
| dispersed | polar substances |
| fats | precipitate |
| hard | probability |
| hard-water | small |
| hydrophilic | solubility |
| hydrophobic | surface area of the solute |
| insoluble | temperature |
| larger | universe |

5. The Second Law of Thermodynamics is the key scientific principle that allows us to explain and predict changes. One way to state the Second Law of Thermodynamics is that the entropy of the ________________ increases.
6. Entropy is a measure of the ________________ of a system; the more ________________ the system is, the greater its entropy.
7. A change from state A to state B tends to take place when state ________________ has greater disorder than state ________________.
8. Particles of matter tend to become more ________________ (spread out).
9. Because there are more possible arrangements for gas particles when they are dispersed throughout a container than when they are concentrated in one corner of it, ________________ suggests that they will spread out to fill the total volume available to them.
10. Miscible liquids can be mixed in ________________ without any limit to their solubility.
11. If less than 1 gram of the substance will dissolve in 100 milliliters (or 100 g) of solvent, the substance is considered ________________.
12. If more than ________________ grams of substance will dissolve in 100 milliliters (or 100 g) of solvent, the substance is considered soluble.
13. If between 1 and 10 grams of a substance will dissolve in 100 milliliters (or 100 g) of solvent, the substance is considered ________________ soluble.
14. Polar substances are likely to dissolve in ________________.

15. Nonpolar substances are likely to dissolve in __________________.
16. Nonpolar substances are __________________ likely to dissolve to a significant degree in polar solvents.
17. __________________ are not likely to dissolve to a significant degree in nonpolar solvents.
18. Substances composed of __________________ polar molecules are usually soluble in water.
19. A polar section of a molecule, which is __________________ to water, is called hydrophilic (literally, "water loving"), and a(n) __________________ part of the molecule, which is not expected to be attracted to water, is called hydrophobic ("water fearing").
20. If we need to predict the relative water solubility of two similar molecules, we can expect the one with the proportionally __________________ polar portion to have higher water solubility.
21. Soap can be made from animal __________________ and vegetable __________________, which in turn are composed of triacylglycerols, or triglycerides.
22. A higher number of carbon–carbon double bonds in a triglyceride's molecules causes it to be __________________ at room temperature.
23. A soap is an ionic compound composed of a(n) __________________ (usually sodium or potassium) and an anion with the general form $RCO_2^-$, in which R represents a(n) __________________ hydrocarbon group.
24. The __________________, nonpolar hydrocarbon portion of each soap anion does not mix easily with water, but the __________________, negatively charged end of each anion does form attractions to water molecules, and these attractions are strong enough to keep the anion in solution.
25. __________________ water is water that contains dissolved calcium ions, $Ca^{2+}$, magnesium ions, $Mg^{2+}$, and often iron ions, $Fe^{2+}$ or $Fe^{3+}$. These ions bind strongly to soap anions, causing the soap to __________________ from hard-water solutions.
26. Detergent structures are similar to (although more varied than) that of soap, but they are less likely to form insoluble compounds with __________________ ions.
27. The net rate of solution depends on three factors: __________________, __________________, and __________________.
28. If we stir or in some way agitate the solution, the solute particles near the solid are quickly carried away, and the relatively high local concentrations of solute are decreased. This diminishes the rate of return and increases the __________________ rate of solution.
29. If the partial pressure of a gas over a liquid is increased, the __________________ of that gas is increased.

## Chapter Problems

### Section 15.1 Entropy, Solutions, and Solubility

OBJECTIVE 2

30. Dry ice is often used to keep ice cream cold at picnics. As the day goes on, the dry ice disappears as the solid sublimes (goes directly from solid to gas). With reference to the Second Law of Thermodynamics, explain why this change takes place.

OBJECTIVE 3

31. The apparatus shown below consists of two containers connected by an opening that is initially blocked. One side contains a gas, and the other side is empty. When the divider between the two containers is removed, the gas moves between the containers until it is evenly distributed. Explain, in terms of the Second Law of Thermodynamics, why this happens. Is the entropy of the system greater before or after? Why? Your explanation should compare the relative (1) disorder in these systems, (2) degree of dispersal of the particles, and (3) number of ways to arrange the particles in the systems.

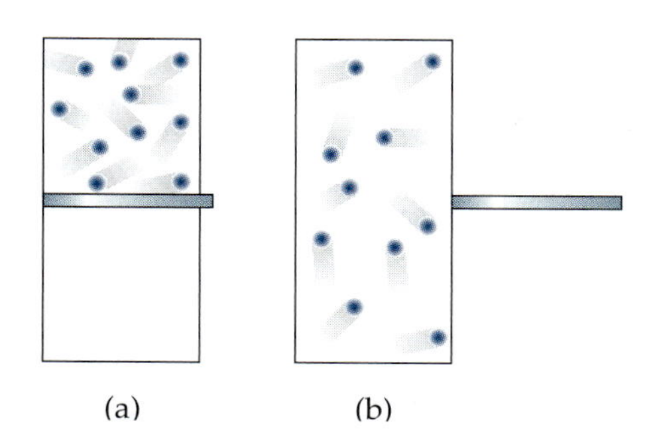


32. Acetone is commonly used to clean laboratory glassware when soap and water cannot do the job. Acetone works well because it dissolves certain substances that are not soluble in water. It is also a good choice because, being miscible with water, it can easily be rinsed from the cleaned glassware with water. Explain what it means when we say acetone is miscible with water.

OBJECTIVE 4

33. The primary components of vinegar are acetic acid and water, both of which are composed of polar molecules with hydrogen bonds that link them. These two liquids will mix in any proportion. With reference to the Second Law of Thermodynamics, explain why acetic acid and water are miscible.

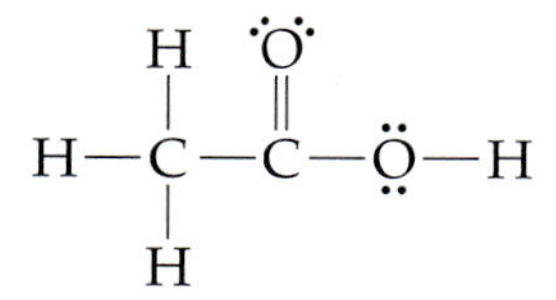


Acetic acid

34. The solubilities in water of several substances are listed below. Classify each of these substances as soluble, insoluble, or moderately soluble on the basis of the criteria presented in the chapter.

   a. camphor, $C_{10}H_{16}O$ (a topical anti-infective): 0.1 g per 100 mL of $H_2O$ at 25 °C

   b. sodium formate, $NaHCO_2$ (used in dyeing and printing fabrics): 97.2 g per 100 g of $H_2O$ at 20 °C

   c. cobalt(II) fluoride, $CoF_2$ (a catalyst in organic reactions): 1.5 g per 100 g of $H_2O$ at 25 °C

35. What is meant by *like* in the solubility guideline *like dissolves like?*
36. Would the following combinations be expected to be soluble or insoluble?
    a. polar solute and polar solvent
    b. nonpolar solute and polar solvent
    c. ionic solute and water
    d. molecular solute with small molecules and hexane
    e. hydrocarbon solute and water
37. Would the following combinations be expected to be soluble or insoluble?
    a. polar solute and nonpolar solvent
    b. nonpolar solute and nonpolar solvent
    c. ionic solute and hexane
    d. molecular solute with small molecules and water
    e. hydrocarbon solute and hexane
38. Write the chemical formula for the primary solute in each of the following solutions. Explain how the solubility guideline *like dissolves like* leads to the prediction that these substances would be soluble in water.
    a. vinegar
    b. household ammonia
39. Write the chemical formula for the primary solute in each of the following solutions. Explain how the solubility guideline *like dissolves like* leads to the prediction that these substances would be soluble in water.
    a. hydrochloric acid
    b. vodka

OBJECTIVE 6

40. Predict whether each of the following compounds is soluble in water.
    a. the polar molecular compound 1-propanol, $CH_3CH_2CH_2OH$ (a solvent for waxes and vegetable oils)
    b. cis-2-pentene, $CH_3CHCHCH_2CH_3$ (a polymerization inhibitor)
    c. the polar molecular compound formic acid, $HCO_2H$ (transmitted in ant bites)
    d. strontium chlorate, $Sr(ClO_3)_2$ (used in tracer bullets)

OBJECTIVE 6

41. Predict whether each of the following compounds is soluble in water.
    a. potassium hydrogen sulfate, $KHSO_4$ (used in wine making)
    b. the polar molecular compound propylene glycol, $CH_3CH(OH)CH_2OH$ (used in some antifreezes)
    c. benzene, $C_6H_6$ (used to produce many organic compounds)

OBJECTIVE 7

42. Predict whether each of the following compounds is soluble in hexane, $C_6H_{14}$.
    a. butane, $C_4H_{10}$ (fuel for cigarette lighters)
    b. potassium hydrogen oxalate, $KHC_2O_4$ (used to remove stains)

OBJECTIVE 7

43. Predict whether each of the following compounds is soluble in hexane, $C_6H_{14}$.
    a. lithium chromate, $Li_2CrO_4$ (a corrosion inhibitor in alcohol-based antifreezes and water-cooled reactors)
    b. ethylene, $CH_2CH_2$ (used in orchard sprays to accelerate fruit ripening)

44. Predict whether each of the following compounds would be more soluble in water or hexane. OBJECTIVE 8

   a. toluene, $C_6H_5CH_3$ (in aviation fuels)

   b. lithium perchlorate, $LiClO_4$ (in solid rocket fuels)

45. Predict whether each of the following compounds would be more soluble in water or hexane. OBJECTIVE 8

   a. magnesium nitrate, $Mg(NO_3)_2$ (in fireworks)

   b. heptane, $C_7H_{16}$ (a standard for gasoline octane rating)

46. Chlorine, $Cl_2$, helps to keep swimming pools clean, but it has to be constantly either added to the water or formed in the water from other substances. With reference to the solubility of chlorine in water, explain why this is necessary.

47. Would you expect acetone, $CH_3COCH_3$ (in some nail polish removers), or 2-hexanone, $CH_3COCH_2CH_2CH_2CH_3$ (a solvent), to be more soluble in water? Why? OBJECTIVE 9

48. Would you expect lactic acid, $CH_3CH(OH)CO_2H$ ( in sour milk), or myristic acid, $CH_3(CH_2)_{12}CO_2H$ (used to make cosmetics) to be more soluble in water? Why? OBJECTIVE 9

49. Would you expect ethane, $C_2H_6$ (in natural gas), or strontium perchlorate, $Sr(ClO_4)_2$ (in fireworks), to be more soluble in hexane, $C_6H_{14}$? Why? OBJECTIVE 9

50. Would you expect potassium phosphate, $K_3PO_4$ (used to make liquid soaps), or ethyl propionate, $CH_3CH_2CO_2CH_2CH_3$ (used, because of its pineapple smell, in fruit essences), to be more soluble in hexane, $C_6H_{14}$? Why? OBJECTIVE 9

51. Explain why amphetamine can pass through the blood–brain barrier more easily than epinephrine. OBJECTIVE 10

52. Explain why methamphetamine is often converted to methamphetamine hydrochloride. OBJECTIVE 11

### Section 15.2 Fats, Oils, Soaps, and Detergents

53. What are the products of the reaction between the following triglyceride and sodium hydroxide? OBJECTIVE 13

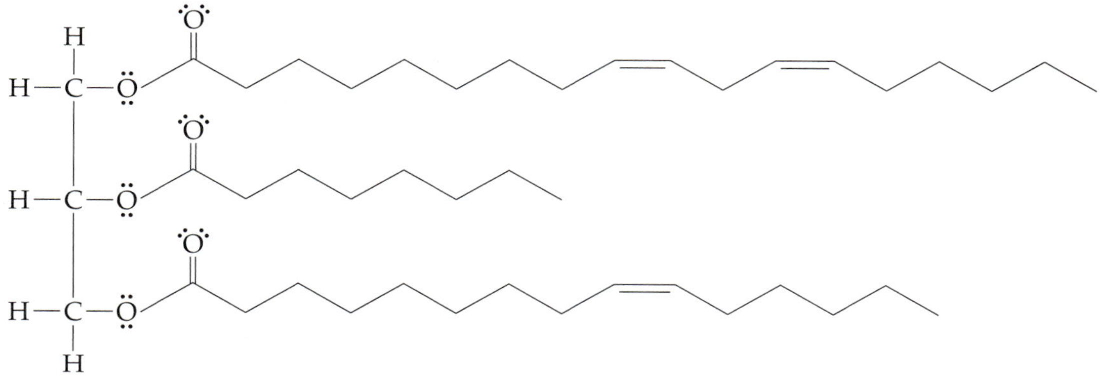

OBJECTIVE 13

54. What are the products of the reaction between the following triglyceride and sodium hydroxide?

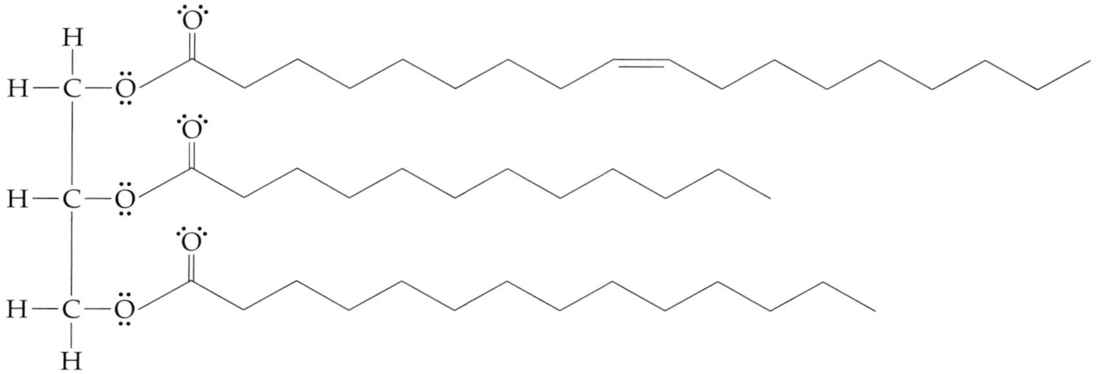

OBJECTIVE 14

55. What part of the following soap structure is hydrophilic and what portion is hydrophobic?

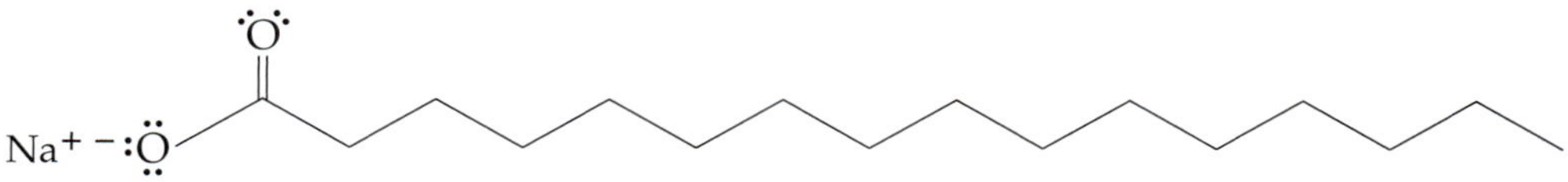

OBJECTIVE 15

56. You throw a backyard party that resembles your idea of a big Texas barbecue. The guests get more food than they can possibly eat, including a big juicy steak in the center of each plate. Referring to the interactions between particles and the corresponding changes that take place on the submicroscopic level, describe how soap or detergent can help you clean the greasy plates that you're left with when your guests go home.

OBJECTIVE 16 OBJECTIVE 17

57. If you live in an area that has very hard water, detergents are a better choice for cleaning agents than soap. Describe the difference in structure between a typical anionic detergent and a typical soap. Describe what makes water hard, and explain why detergents work better in hard water than soap does.

### Section 15.3 Saturated Solutions and Dynamic Equilibrium

58. List three ways to increase the rate at which salt dissolves in water.

59. Epsom salts, a common name for magnesium sulfate heptahydrate, $MgSO_4 \cdot 7H_2O$, can be used to help reduce swelling caused by injury. For example, if you want to reduce the swelling of a sprained ankle, you can soak the ankle in a saturated solution of magnesium sulfate prepared by adding an excess of Epsom salts to water and waiting until the greatest possible amount of solid dissolves.

OBJECTIVE 18

a. Describe the reversible change that takes place as the $MgSO_4$ dissolves. (You do not need to mention the water molecules that are attached to the magnesium sulfate.)

OBJECTIVE 19

b. Explain why an increase in the concentration of solute particles in this solution containing undissolved solid leads to an increase in the rate of return of the solute to the solid form.

OBJECTIVE 23

c. Why does the magnesium sulfate dissolve faster if the solution is stirred?

d. When the maximum amount of magnesium sulfate has dissolved, does the solid stop dissolving? Explain.

60. A person who has trouble swallowing tablets can dissolve aspirin in water and drink the solution. The fastest way to dissolve the aspirin is to grind it into a powder and then dissolve it in hot water.

    a. Why does powdered aspirin dissolve more rapidly than an aspirin tablet? OBJECTIVE 22

    b. Why does aspirin dissolve more rapidly in hot water than in cold water? OBJECTIVE 24

61. If you wanted to make sugar dissolve as quickly as possible, would you use

    a. room temperature water or hot water? Why?

    b. powdered sugar or granular sugar? Why?

62. Some of the minerals found in rocks dissolve in water as it flows over the rocks. Would these minerals dissolve more quickly

    a. at the bottom of a waterfall or in a still pond? Why?

    b. in a cold mountain stream or in the warmer water downstream? Why?

    c. over large rocks or over sand composed of the same material? Why?

63. What does it mean to say a solution is saturated? Describe the changes that take place at the particle level in a saturated solution of sodium chloride that contains an excess of solid NaCl. Is the NaCl still dissolving?

64. Solutions are called *dilute* when the concentration of the solute in solution is relatively low and *concentrated* when the concentration is relatively high. Can a solution be both dilute and saturated? Explain your answer.

65. Can a solution be both concentrated (with a relatively high concentration of solute) and unsaturated? Explain your answer.

### Section 15.4 Solutions of Gases in Liquids

66. The soft drinks sold at county fairs are often dispensed from large pressurized containers that contain carbon dioxide gas above the liquid at a partial pressure of about 4 atm, compared to carbon dioxide's normal partial pressure of 0.00035 atm in the air at sea level.

    a. Describe the reversible change that takes place inside one of these soft drink containers when the pressure of $CO_2$ above the liquid is first brought from 0.00035 atm to 4 atm. Explain why this system comes to a dynamic equilibrium in which there is no net change in the amount of gas above the liquid or in the amount of gas dissolved in the liquid. OBJECTIVE 27

    b. Explain why an increase in the partial pressure of $CO_2$ over a liquid leads to an increase in the solubility of the gas in the liquid. OBJECTIVE 28

67. Consider a soft drink bottle with a screw cap. When the cap is removed, the excess $CO_2$ in the space above the soft drink escapes into the room, leaving normal air above the liquid. Explain why the soft drink will lose its carbonation more quickly if the cap is left off than if the cap is immediately put back on tightly.

68. You visit a dude ranch located in a beautiful mountain valley at an altitude of 6000 feet. The day you arrive, you decide to take your usual evening run. On this evening, however, you don't run as far as usual, because you feel winded and tired much sooner than at home by the beach. Why does this happen? OBJECTIVE 29

### Discussion Question

69. The particles in the air of a typical classroom number about $10^{28}$. Is it possible for all of these particles to be found in one corner of the room? Would you change your answer if there were only five air particles in the room?

*An understanding of the process of chemical changes is important for the development of new chemicals, including the chemicals used to fertilize this field.*

CHAPTER 16

# The Process of Chemical Reactions



HAVE YOU EVER CONSIDERED BECOMING A CHEMICAL ENGINEER? The men and women in this profession develop industrial processes for the large-scale production of the chemicals we use to fertilize and protect our crops, to synthesize textiles, plastics, and other ubiquitous modern materials, to cure our diseases, and so much more. Or perhaps you have considered becoming a research chemist, who figures out new ways to make existing chemicals and ways to produce chemicals that have never existed before. Although these two careers require different sets of skills and aptitudes, they also have some concerns and traits in common. For example, both kinds of chemists need to understand the factors that affect the speed with which chemicals can be made, and both must know why chemical changes do not always proceed to 100% products. Armed with this knowledge, chemical engineers and research chemists can develop ways to make chemical products more efficiently, more safely, and more economically.

This chapter introduces a model for visualizing the changes that take place in a reaction mixture as a chemical reaction proceeds. The model describes the requirements that must be met before a reaction can occur and explains why certain factors speed the reaction up or slow it down. It will help us understand why some chemical reactions are significantly reversible and why such reactions reach a dynamic equilibrium with equal rates of change in both directions. It will also allow us to explore the factors that can push a chemical equilibrium forward to create more products or backward to create more reactants.

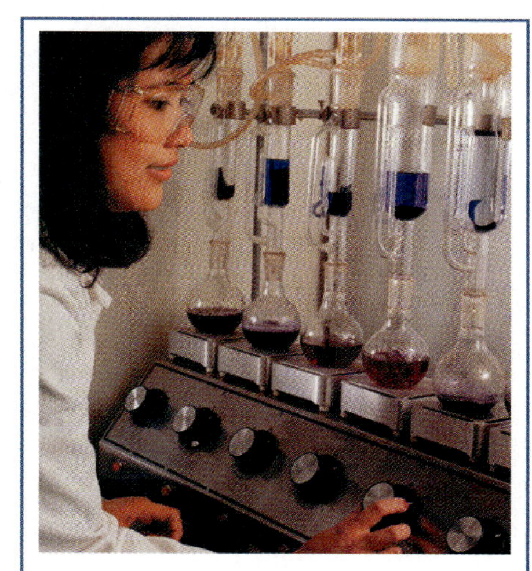

*Research chemists want to know how they can produce chemicals of the highest purity in the shortest time.*

## Review Skills

The presentation of information in this chapter assumes that you can already perform the tasks listed below. You can test your readiness to proceed by answering the Review Questions at the end of the chapter. This might also be a good time to read the Chapter Objectives, which precede the Review Questions.

- Describe the particle nature of gases. (Section 2.1)
- Write or recognize the definitions of the terms *energy*, *kinetic energy*, *potential energy*, *thermal energy*, *heat*, *radiant energy*, *exergonic*, *exothermic*, *endergonic*, *endothermic*, and *catalyst*. (Chapter 7 Glossary)
- Describe the relationship between stability and potential energy. (Section 7.1)
- Describe the relationship between average kinetic energy and temperature. (Section 7.1)
- Explain why energy is necessary to break chemical bonds. (Section 7.2)
- Explain why energy is released when chemical bonds are formed. (Section 7.2)
- Explain why some chemical reactions release heat to their surroundings. (Section 7.2)
- Explain why some chemical reactions absorb heat from their surroundings. (Section 7.2)
- Write a general description of dynamic equilibrium. (Section 14.1)

## 16.1 Collision Theory: A Model for the Reaction Process

Gasoline and oxygen coexist quietly until a spark from a spark plug propels them into violent reaction. Why? Why are ozone molecules in the stratosphere destroyed more rapidly in the presence of chlorine atoms released from CFC molecules? In order to understand these situations, we need to take a look at a model called *collision theory*, which is useful for visualizing the process of chemical change.

### The Basics of Collision Theory

We will demonstrate the basic assumptions of collision theory by using it to describe the reaction of an oxygen atom and an ozone molecule to form two oxygen molecules (Figure 16.1).

**Figure 16.1**
Collision Theory

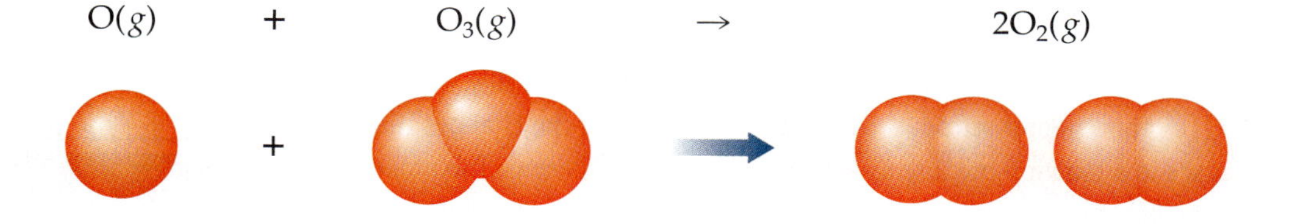


Try to picture oxygen atoms and ozone molecules moving in a random way in the stratosphere. Some of the particles are moving with a high velocity and some are moving more slowly. The particles are constantly colliding, changing their direction of motion, and speeding up or slowing down (Figure 16.2). Some of the collisions between oxygen atoms and ozone molecules lead to the production of two oxygen molecules, but some of the collisions do not. To understand why some collisions are productive and others are not, we need to take a closer look at the events that take place in the reaction process.

OBJECTIVE 2

STEP 1 **Reactants collide.** The process begins with a violent collision between an O atom and an $O_3$ molecule, which shakes them up and provides them with enough energy for the bond between two oxygen atoms in $O_3$ to *begin* to break. We saw in Section 7.2 that energy is required to break chemical bonds. At the same time as one O–O bond is breaking, another O–O bond *begins* to form between the original single oxygen atom and the oxygen atom breaking away from the $O_3$ molecule. We know from Chapter 7 that bond making releases energy, which in this case can supply some of the energy necessary for the bond breaking. Initially, the bond breaking predominates over the bond making, so the energy released in bond making is not enough to compensate for the energy necessary for bond breaking. The extra energy necessary for the reaction comes from the kinetic energy of the moving particles as they collide (Figure 16.3).

OBJECTIVE 2

OBJECTIVE 3

OBJECTIVE 4

OBJECTIVE 3

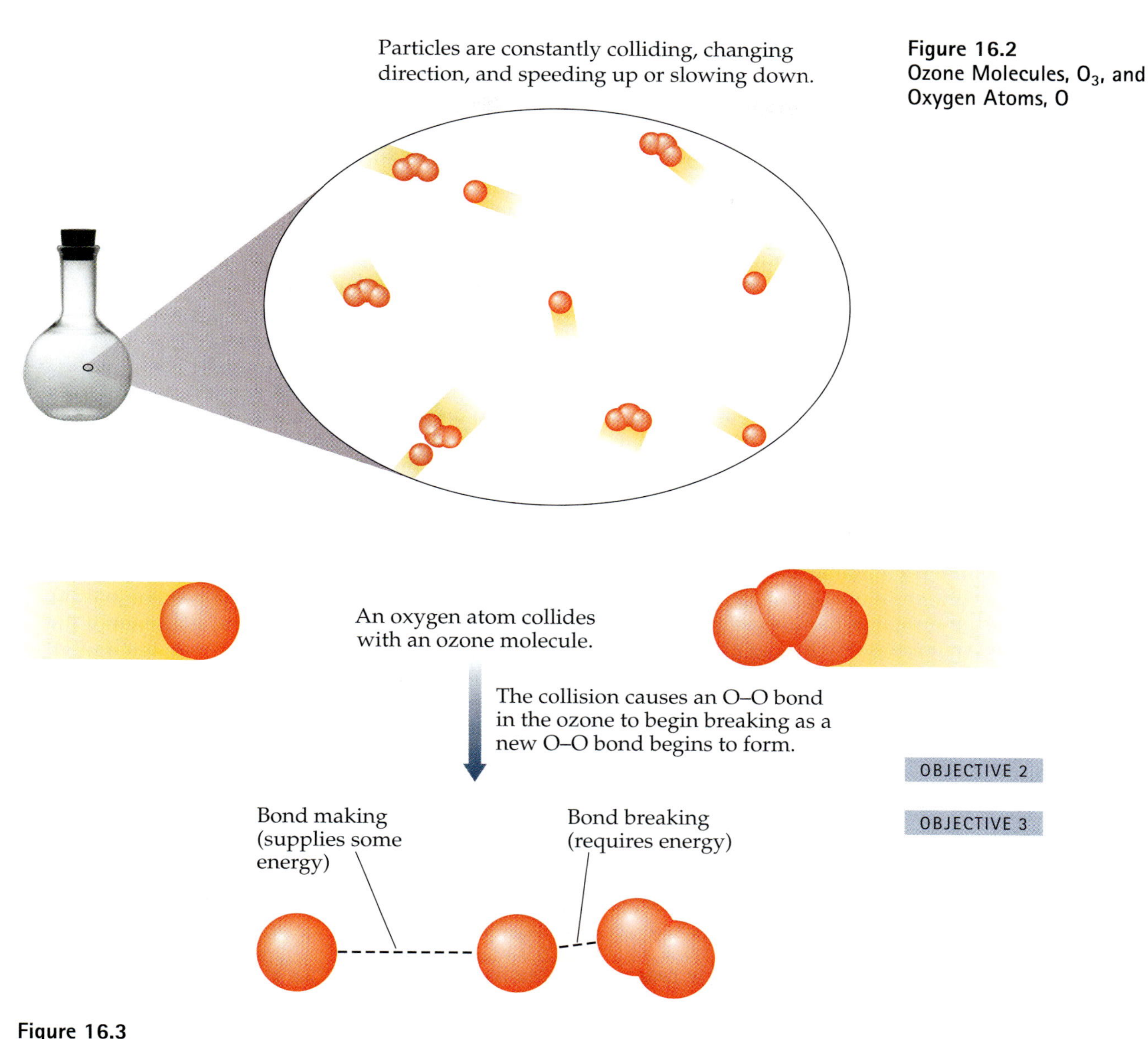
Particles are constantly colliding, changing direction, and speeding up or slowing down.
An oxygen atom collides with an ozone molecule.
The collision causes an O–O bond in the ozone to begin breaking as a new O–O bond begins to form.
Bond making (supplies some energy)
Bond breaking (requires energy)

**Figure 16.2**
**Ozone Molecules, $O_3$, and Oxygen Atoms, O**

OBJECTIVE 2

OBJECTIVE 3

**Figure 16.3**
**Initial Stage of a Reaction Between Two Particles**
Initially, the energy required for bond breaking is greater than the energy supplied by bond making. The extra energy necessary for the reaction comes from the kinetic energy of the moving particles as they collide.

OBJECTIVE 2

**STEP 2 Formation of activated complex.** As the oxygen atoms of the O–O bond in the $O_3$ molecule separate, the attraction between them decreases, and as the attraction decreases, less energy is needed for moving the atoms even farther apart. Meanwhile, the oxygen atoms that are forming the new bond move closer together, attracting each other more strongly and releasing more energy. At a certain stage in the progress of the reaction, bond breaking and bond making are of equal importance. In other words, the energy necessary for bond breaking is balanced by the energy supplied by bond making. At this turning point, the particles involved in the reaction are joined in a structure known as the *activated complex*, or *transition state*, which is the least stable (and therefore the highest-energy) intermediate in the most efficient pathway between reactants and products. In the activated complex for our reaction between an oxygen atom and an ozone molecule, the bond being broken and the bond forming have roughly equal strengths and lengths (Figure 16.4).

**Figure 16.4**
**The Activated Complex for the Reaction Between Oxygen Atoms and Ozone Molecules**

OBJECTIVE 2

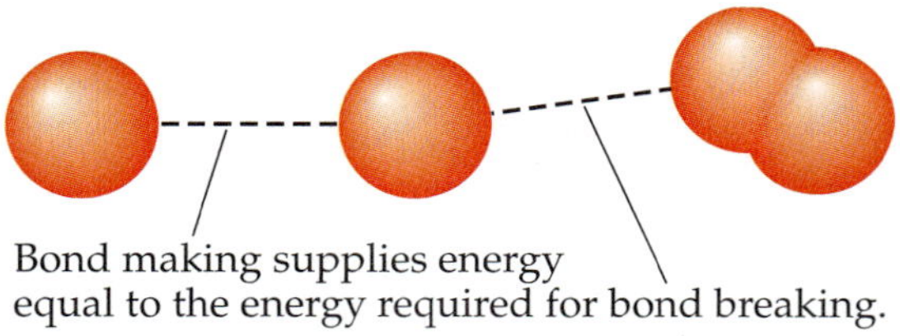

Bond making supplies energy equal to the energy required for bond breaking.

**STEP 3 Formation of product.** As the reaction continues beyond the activated complex, the energy released in bond making becomes greater than the energy necessary for bond breaking, and energy is released (Figure 16.5).

**Figure 16.5**
**Final Stage of a Reaction Between Two Particles**

OBJECTIVE 2

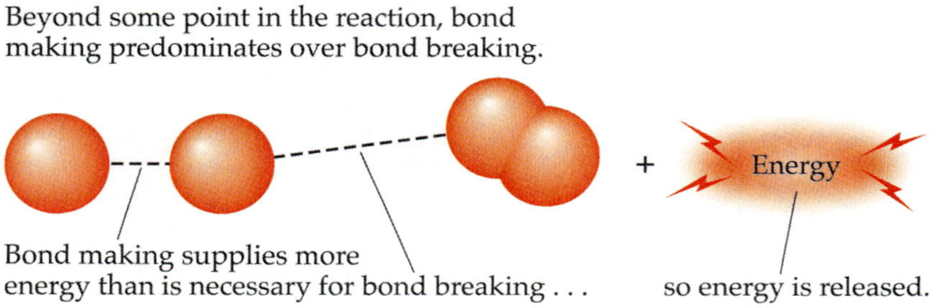

Beyond some point in the reaction, bond making predominates over bond breaking.

+ Energy

Bond making supplies more energy than is necessary for bond breaking . . .

so energy is released.

Thus, as the reaction begins, an input of energy is necessary to produce the activated complex; as the reaction proceeds, and the system shifts from the activated complex to products, energy is released. In a chemical reaction, the minimum energy necessary for reaching the activated complex and proceeding to products is called the **activation energy.** Only the collisions that provide a net kinetic energy equal to or greater than the activation energy can lead to products.

OBJECTIVE 6

Remember that at any instant in time, the particles in a gas, liquid, or solid have a wide range of velocities and thus a wide range of kinetic energies. If you were riding on a particle—in a gas, for example—you would be constantly colliding with other particles, speeding up or slowing down, and increasing or decreasing your kinetic energy. Sometimes you collide with a slow moving particle while moving slowly yourself. This collision is not too jarring. It has a low net kinetic energy. Sometimes you collide with a rapidly moving particle while moving rapidly yourself. This collision is much more violent and has a much higher net kinetic energy.

OBJECTIVE 6

If a collision does not provide enough kinetic energy to form an activated complex, then the reaction does not proceed to products. Instead, the bond that has begun to break pulls back together, the bond that has begun to form falls apart, and the particles accelerate apart unchanged. This is like the situation depicted in Figure 16.6, where a rolling ball rolls back down the same side of a hill it started up.

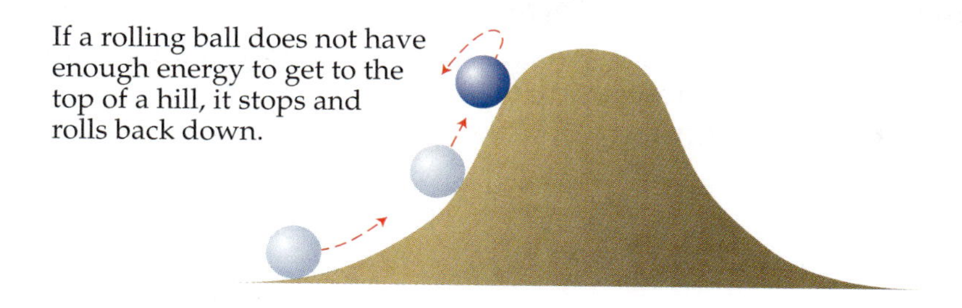

If a rolling ball does not have enough energy to get to the top of a hill, it stops and rolls back down.

**Figure 16.6**
Not Enough Kinetic Energy to Get Over the Hill

OBJECTIVE 6

The activation energy for the oxygen–ozone reaction is 17 kJ/mole $O_3$. If the collision between reactants yields a net kinetic energy equal to or greater than the activation energy, the reaction can proceed to products (Figure 16.7). This is like a ball rolling up a hill with enough kinetic energy to reach the top of the hill, from which it can roll down the other side (Figure 16.8).

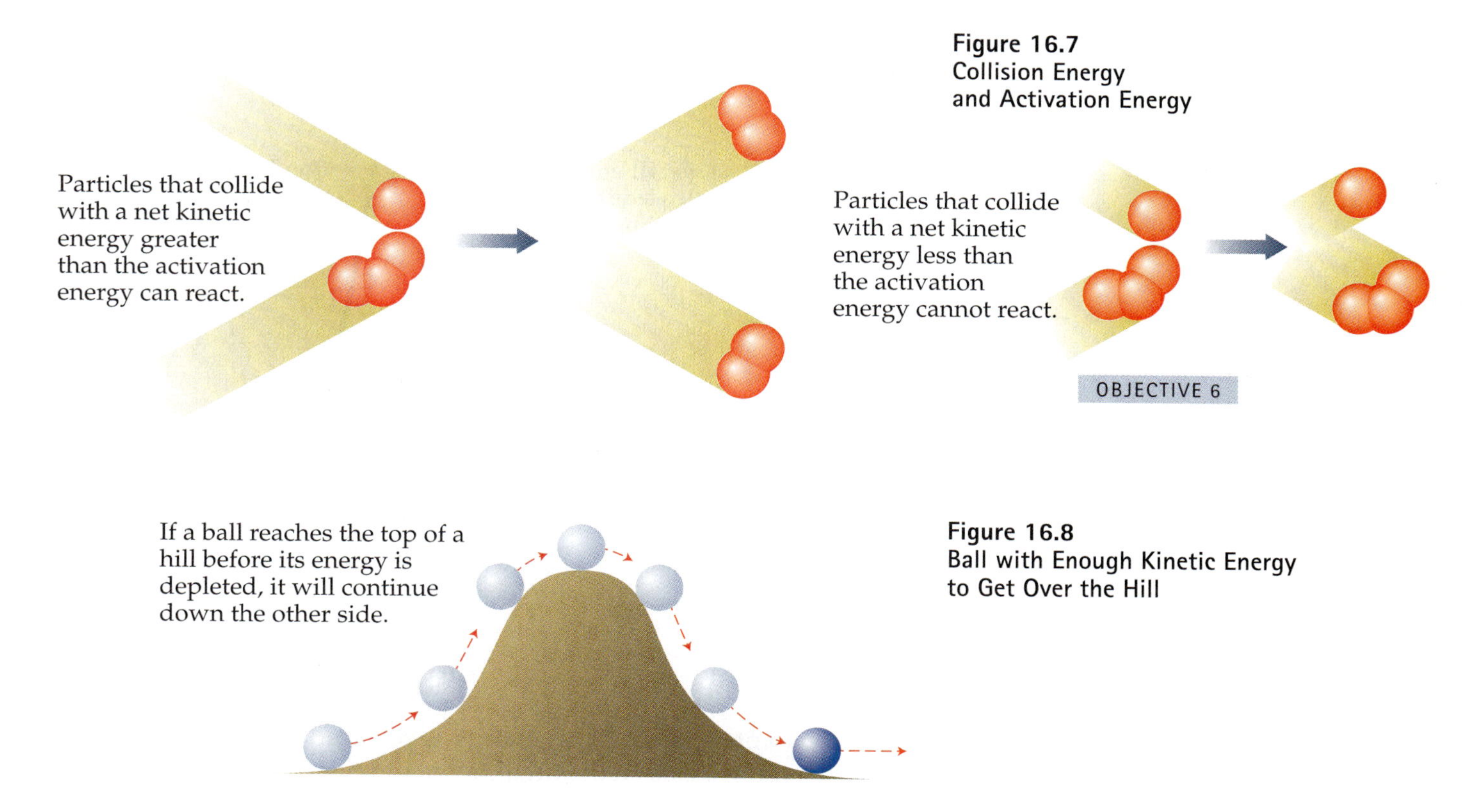


**Figure 16.7**
Collision Energy and Activation Energy

OBJECTIVE 6

**Figure 16.8**
Ball with Enough Kinetic Energy to Get Over the Hill

The bonds between oxygen atoms in $O_2$ molecules are stronger and more stable than the bonds between atoms in the ozone molecules, so more energy is released in the formation of the new bonds than is consumed in the breaking of the old bonds. This leads to an overall release of thermal energy, so the reaction is exothermic. The energies associated with chemical reactions are usually described in terms of kilojoules per

OBJECTIVE 5

mole, kJ/mole. If the energy is released, the value is described with a negative sign. Because 390 kJ of energy is evolved when 1 mole of ozone molecules reacts with oxygen atoms to form oxygen molecules, the energy of the reaction is −390 kJ/mole $O_3$.

Chemists often describe the progress of chemical reactions with energy diagrams such as Figure 16.9, which shows the energy changes associated with the $O/O_3$ reaction. It shows that the reactants must have enough kinetic energy to climb an energy hill before they can proceed on to products.

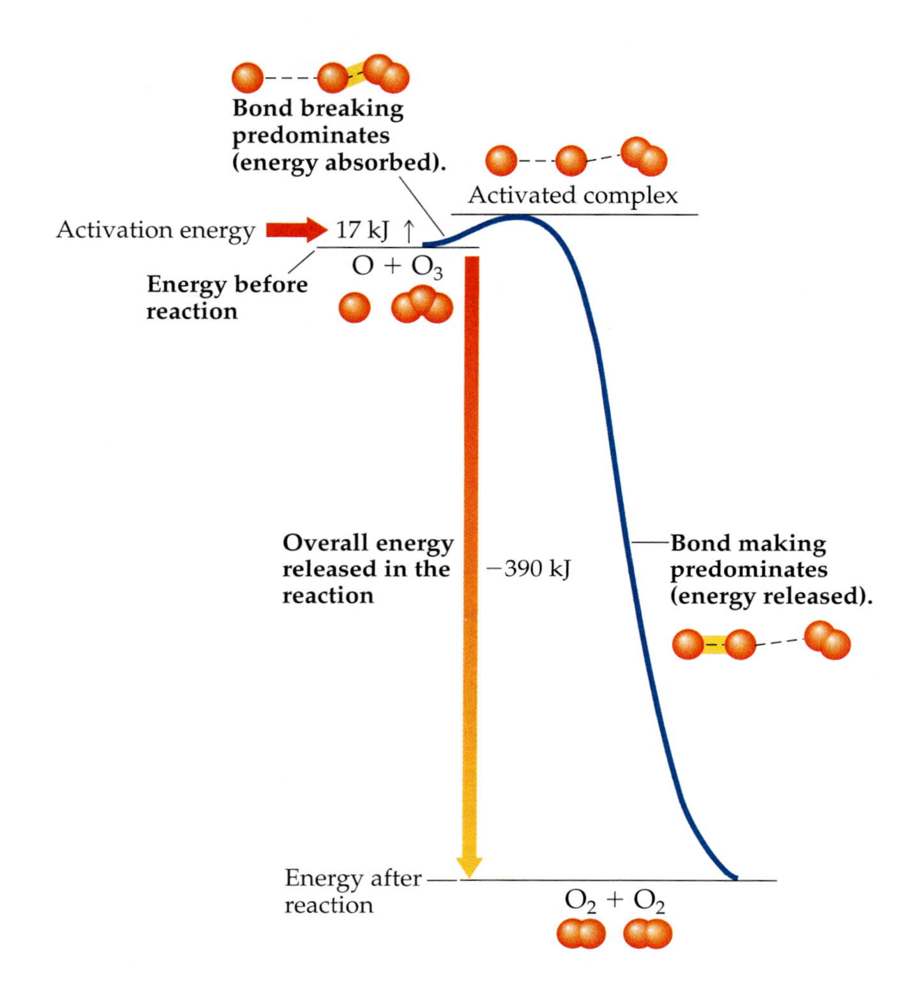


**Figure 16.9**
Energy Diagram for the Exergonic (Energy-Releasing) Reaction of an Oxygen Atom with an Ozone Molecule to Form Two Oxygen Molecules

OBJECTIVE 7

## Endergonic Reactions

We know from Chapter 7 that if less energy is released in the formation of new bonds than is consumed in the breaking of old bonds, a reaction will be endergonic, absorbing energy overall as it takes place. If this energy comes from the motion (kinetic energy) of the reactants, then the particles in the system will be moving more slowly after the reaction than before. The system will have less thermal energy, and the temperature will decrease. Because the system is at a lower temperature than the surroundings, it can absorb heat from the surroundings. Remember that when heat energy is absorbed in the process of a change, the change is called endothermic. The energies associated with endergonic (or endothermic) changes are described with positive values. For example, 178 kJ of energy is necessary to convert 1 mole of calcium carbonate, $CaCO_3$, to calcium oxide, CaO, and carbon dioxide, $CO_2$. Thus the energy of reaction is +178 kJ/mole.

OBJECTIVE 8

$$CaCO_3(s) + 178\text{ kJ} \rightarrow CaO(s) + CO_2(g)$$

Figure 16.10 shows the energy diagram for a typical endergonic reaction.

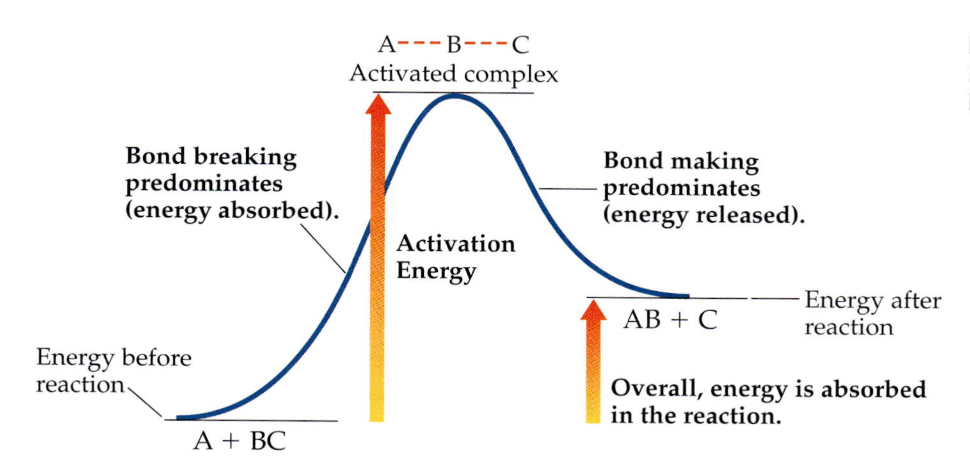


**Figure 16.10**
Energy Diagram for an Endergonic Reaction

OBJECTIVE 9

Even if a collision between two particles provides the activation energy, they still might not react. For the reaction to take place, the particles must collide in such a way as to allow the formation of the new bond or bonds to occur at the same time as the breaking of the old bond or bonds. For example, in the $O_3/O$ reaction, the new bond must form between the free oxygen atom and one of the *outer* atoms in the $O_3$ molecule. Therefore, the free oxygen atom must collide with one of the outer oxygen atoms rather than with the central atom. The orientation of the particles in the first collision shown in Figure 16.11 can lead to reaction if the collision provides the activation energy. The orientation shown in the second collision is not likely to proceed to products.

OBJECTIVE 10

**Figure 16.11**
Oxygen Atoms Collision Orientation

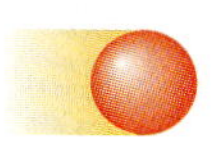

One favorable orientation

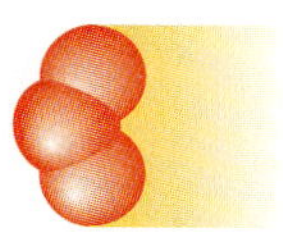

One unfavorable orientation

## Summary of Collision Theory

Collision theory stipulates that three requirements must be met before a reaction between two particles is likely to take place.

**1.** The reactant particles must collide.

The collision brings together the atoms that will form the new bonds, and the kinetic energy of the particles provides energy for the reaction to proceed.

OBJECTIVE 11

OBJECTIVE 11

2. The collision must provide at least the minimum energy necessary to produce the activated complex.

   It takes energy to initiate the reaction by converting the reactants into the activated complex. If the collision does not provide this energy, products cannot form.

OBJECTIVE 11

3. The orientation of the colliding particles must favor the formation of the activated complex, in which the new bond or bonds are able to form as the old bond or bonds break.

   Because the formation of the new bonds provides some of the energy necessary to break the old bonds, the making and breaking of bonds must occur more or less simultaneously. This is possible only when the particles collide in such a way that the bond-forming atoms are close to each other.

## 16.2 Rates of Chemical Reactions

At normal temperatures, the $CaCO_3$ in limestone can exist for centuries.

Heated to high temperature in an industrial plant, it is quickly converted into lime, CaO, and carbon dioxide.

As we saw in the last section, the calcium carbonate in limestone can be converted into calcium oxide (lime) and carbon dioxide. At normal temperatures, this reaction is so slow that most limestone formations remain unreacted for thousands of years. Why, then, does it take place rapidly at 1200 °C? Similarly, why does the combustion of gasoline take place more quickly when the fuel–air mixture in a cylinder of your car is compressed into a smaller volume by a moving piston? How does your car's catalytic converter speed the conversion of $NO(g)$ into $N_2(g)$ and $O_2(g)$? Now that we know more about the requirements for reaction, we can proceed to answer questions like these.

In this section, we focus on the factors that affect **rates of chemical reactions,** which for a gas can be described as the number of product molecules that form per liter per second.

$$\text{Rate of reaction} = \frac{\text{number of moles of product formed}}{\text{liter} \cdot \text{second}}$$

We can use the $O/O_3$ reaction described in the last section as an illustration of the factors affecting rate:

$$O(g) + O_3(g) \rightarrow 2O_2(g)$$

$$\text{Rate of reaction} = \frac{\text{mole } O_2 \text{ formed}}{\text{L} \cdot \text{s}}$$

In the last section, we found that for a reaction between two particles to form products, the particles must collide with enough energy to reach the activated complex and with an orientation that allows the new bond or bonds to form as the old bond or bonds break. Any factor that affects these conditions will also affect the rate of the reaction. Let's look again at the chemical reaction between oxygen atoms and ozone molecules.

## Temperature and Rates of Chemical Reactions

In general, increased temperature increases the rate of chemical reactions.[1] There are two reasons for this. One is that increased temperature means increased average velocity of the particles. As heat is transferred to a system containing $O_3$ and O, for example, the particles move faster. This leads to an increase in the number of particles colliding per second—and therefore an increase in the number of products formed per second.

OBJECTIVE 12

Increased temperature also means an increase in the average kinetic energy of the collisions between the particles in a system. This leads to an increase in the fraction of the collisions that have enough energy to reach the activated complex (the activation energy)—and thus to an increase in the rate of the reaction. For the reaction of $O_3$ and O, an increase in temperature from 25 °C (298 K) to 35 °C (308 K) increases the fraction of the particles with the activation energy from 0.0010 to 0.0013. Figure 16.12 summarizes the reasons why reaction rates are increased by increased temperature.

OBJECTIVE 12

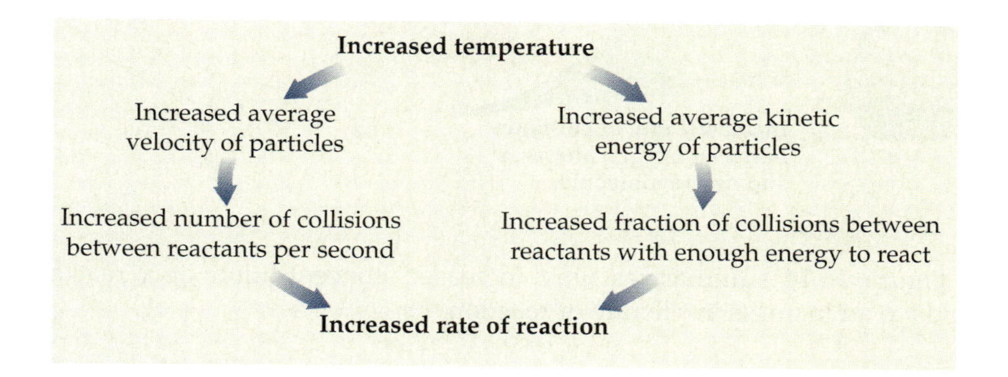


**Figure 16.12**
Temperature and Reaction Rate

OBJECTIVE 12

## Concentration and Rates of Chemical Reactions

Another way to increase the rate of a chemical reaction is to increase the concentration of the reactant particles. **Concentration** can be described as the number of particles per unit volume. For gases, it is usually described as moles of gas particles per liter of container. Gas concentration can be increased by either increasing the number of reactant particles in a container or decreasing the volume of the container. Once again, let us consider the $O_3$/O reaction. The concentration of oxygen atoms in the system described in Figure 16.2 can be increased either by adding more oxygen atoms, O, or by decreasing the volume of the container.

[1]The most important exception to this statement is the speed of reactions mediated by naturally occurring catalysts called enzymes. As a result of the change in shape of the enzyme, chemical reactions that depend on the presence of enzymes can be *slowed* by increased temperature.

OBJECTIVE 13

Increasing the concentration of one or both reactants will lead to shorter distances between reactant particles and to their occupying a larger percentage of the volume. With shorter distances between the particles and less empty space for them to move in without colliding, more collisions can be expected to occur per second. Figure 16.13 shows the container from Figure 16.2 after more oxygen atoms are added. At a constant temperature, increasing the concentration of oxygen atoms in the container will increase the rate of collisions between O and $O_3$—and will therefore increase the rate of the reaction. The same would be true if the concentration of $O_3$ were increased; increasing the concentration of $O_3$ would lead to an increase in the rate of the reaction.

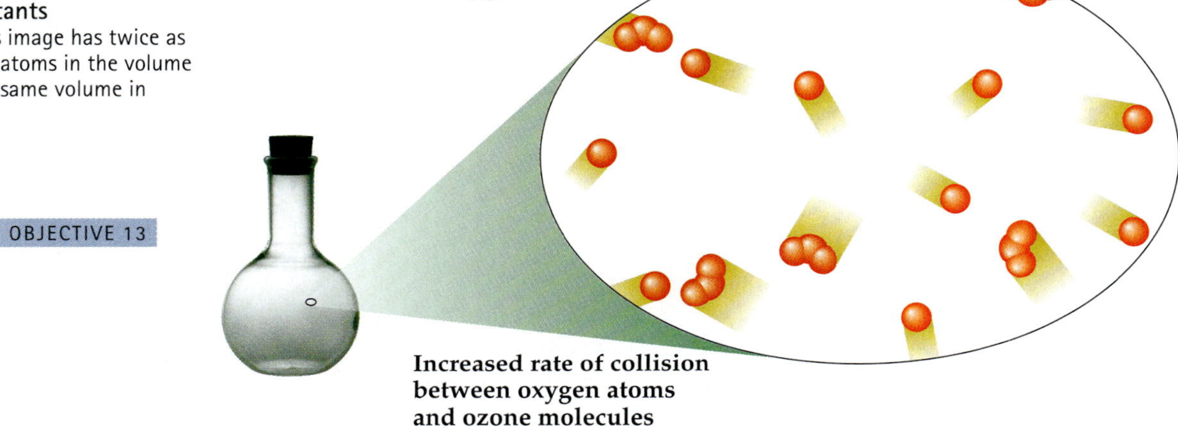
Increased concentration of oxygen atoms
Increased rate of collision between oxygen atoms and ozone molecules

**Figure 16.13**
**An Increase in One of the Reactants**
Note that this image has twice as many oxygen atoms in the volume shown as the same volume in Figure 16.2.

OBJECTIVE 13

Figure 16.14 summarizes why increased concentration of a reactant leads to an increase in the rate of reaction.

**Figure 16.14**
**Reactant Concentration and Reaction Rate**

OBJECTIVE 13

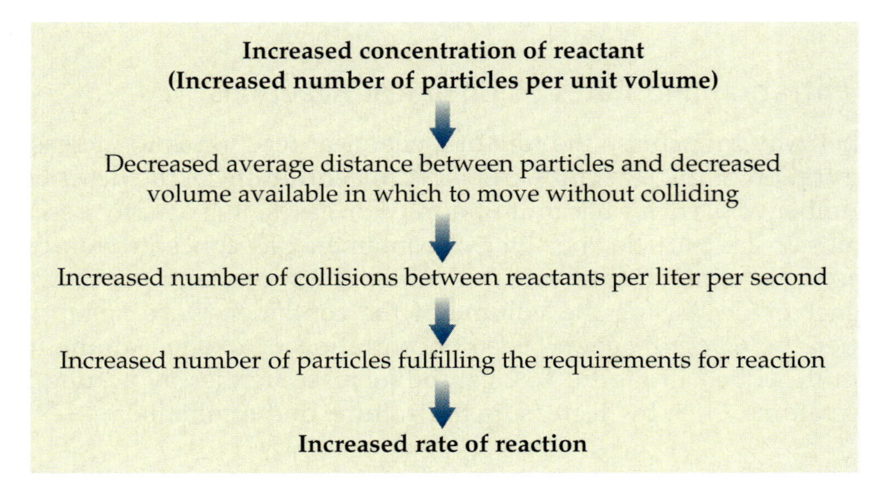
Increased concentration of reactant
(Increased number of particles per unit volume)
Decreased average distance between particles and decreased volume available in which to move without colliding
Increased number of collisions between reactants per liter per second
Increased number of particles fulfilling the requirements for reaction
Increased rate of reaction

We can use similar reasoning to explain why the compression of fuel–air mixtures in the cylinders of a car engine speeds the reaction between the gaseous fuel and oxygen from the air. When a piston moves into a cylinder, the decreased volume available to the reacting gases creates an increased concentration of each gas, which decreases the distance between the reacting molecules and decreases the volume available in which to move without colliding. This leads to more collisions between the reactants per second—and thus to a greater rate of reaction.

## Catalysts

A catalyst is a substance that accelerates a chemical reaction without being permanently altered itself. Catalysts are employed by chemical manufacturers to increase the rates of 60–70% of their reactions, and most of the important chemical reactions that take place in your body are regulated by catalysts called *enzymes*.

One of the ways in which catalysts accelerate chemical reactions is by providing an alternative pathway between reactants and products that has a lower activation energy. Once again, the reaction between ozone molecules and oxygen atoms in the stratosphere provides an example.

$$O_3(g) + O(g) \rightarrow 2O_2(g)$$

The activation energy for this reaction, as noted in Figure 16.9, is about 17 kJ/mole. At 25 °C (298 K), about 1 of every 1000 collisions (0.1%) between $O_3$ molecules and O atoms has a net kinetic energy large enough to form the activated complex and proceed to products. This does not mean that 1/1000 of the collisions lead to products, however. The collisions must still have the correct orientation in order for the bond making and bond breaking to proceed at the same time.

We saw in Chapter 7 that chlorine atoms can catalyze this reaction in two steps:

OBJECTIVE 14

$$O_3(g) + Cl(g) \rightarrow ClO(g) + O_2(g)$$

$$O(g) + ClO(g) \rightarrow Cl(g) + O_2(g)$$

The net reaction that summarizes these two intermediate reactions is

$$\text{Net reaction} \quad O_3(g) + O(g) \xrightarrow{Cl} 2O_2(g)$$

Thus chlorine atoms are a threat to the ozone layer just because they provide a second pathway for the conversion of $O_3$ and O to $O_2$, but there is another reason why the chorine reactions speed the destruction of ozone. The reaction between $O_3$ and Cl that forms ClO and $O_2$ has an activation energy of 2.1 kJ/mole. At 25 °C, about 3 of every 7 collisions (43%) have enough energy to form the activated complex. The reaction between O and ClO to form Cl and $O_2$ has an activation energy of only 0.4 kJ/mole. At 25 °C, about 85% of the collisions have at least this energy. Therefore, a much higher fraction of the collisions have the minimum energy necessary to react than in the direct reaction between $O_3$ and O, in which only 0.1% of the collisions have energy equal to or greater than the activation energy (Figure 16.15).

OBJECTIVE 14

**Figure 16.15**
Potential-Energy Diagram of the Catalyzed and Uncatalyzed Reactions That Convert $O_3$ and O into $O_2$

OBJECTIVE 14

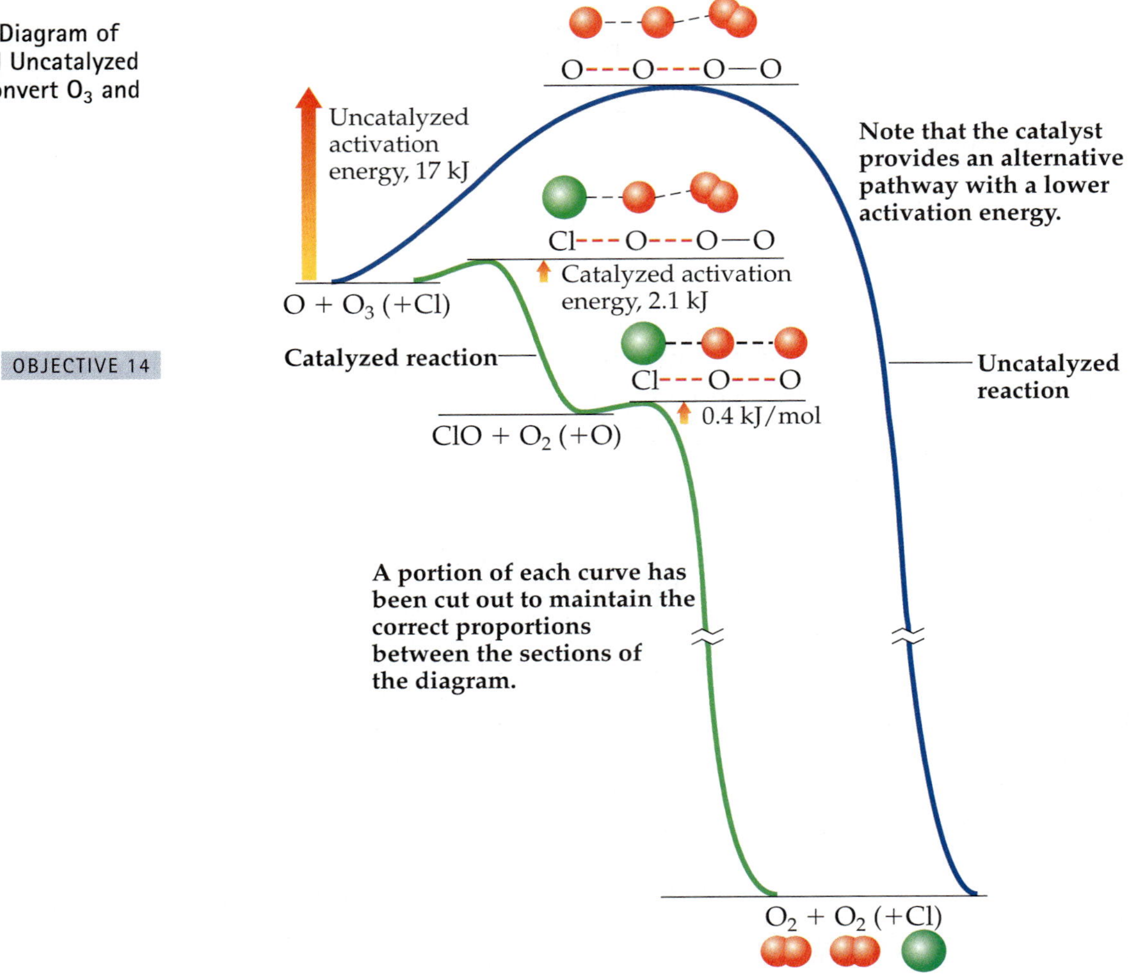
O---O---O—O
Uncatalyzed activation energy, 17 kJ
Note that the catalyst provides an alternative pathway with a lower activation energy.
Cl---O---O—O
Catalyzed activation energy, 2.1 kJ
O + O3 (+Cl)
Catalyzed reaction
Cl---O---O
0.4 kJ/mol
Uncatalyzed reaction
ClO + O2 (+O)
A portion of each curve has been cut out to maintain the correct proportions between the sections of the diagram.
O2 + O2 (+Cl)

Figure 16.16 summarizes why the rate of a catalyzed reaction would be expected to be greater than the rate of the same reaction without the catalyst.

**Figure 16.16**
Catalysts and Reaction Rates

OBJECTIVE 15

**The catalyst provides an alternative pathway with a lower activation energy.**

↓

A greater fraction of collisions have the activation energy.

↓

A greater fraction of collisions lead to products.

↓

**Increased rate of reaction**

## Homogeneous and Heterogeneous Catalysts

If the reactants and the catalyst are all in the same state (all gases or all in solution) the catalyst is a **homogeneous catalyst.** The chlorine atoms are homogeneous catalysts for the reaction that converts $O_3$ and O into $O_2$ because the chlorine atoms and the reactants are all gases.

If the catalyst is not in the same state as the reactants, the catalyst is called a **heterogeneous catalyst.** A heterogeneous catalyst is usually a solid, and the reactants are either gases or liquids. For example, in Special Topic 6.2: *Air Pollution and Catalytic Converters*, you read that the catalyst in a car's catalytic converter, which helps convert gaseous nitrogen monoxide, $NO(g)$, into nitrogen gas, $N_2(g)$, and oxygen gas, $O_2(g)$, is a solid mixture of transition metals and transition metal oxides.

OBJECTIVE 16

Heterogeneous catalysis is thought to proceed in the four steps shown in Figure 16.17, which uses the catalytic conversion of NO into $N_2$ and $O_2$ in automobiles as an example. The model suggests that as exhaust from a car's engine passes over the solid catalyst—a transition metal such as platinum, palladium, iridium, or rhodium—nitrogen monoxide molecules become attached to the catalyst's surface, bonds between nitrogen and oxygen atoms are broken, the atoms migrate across the surface, and new bonds form between nitrogen and oxygen atoms.

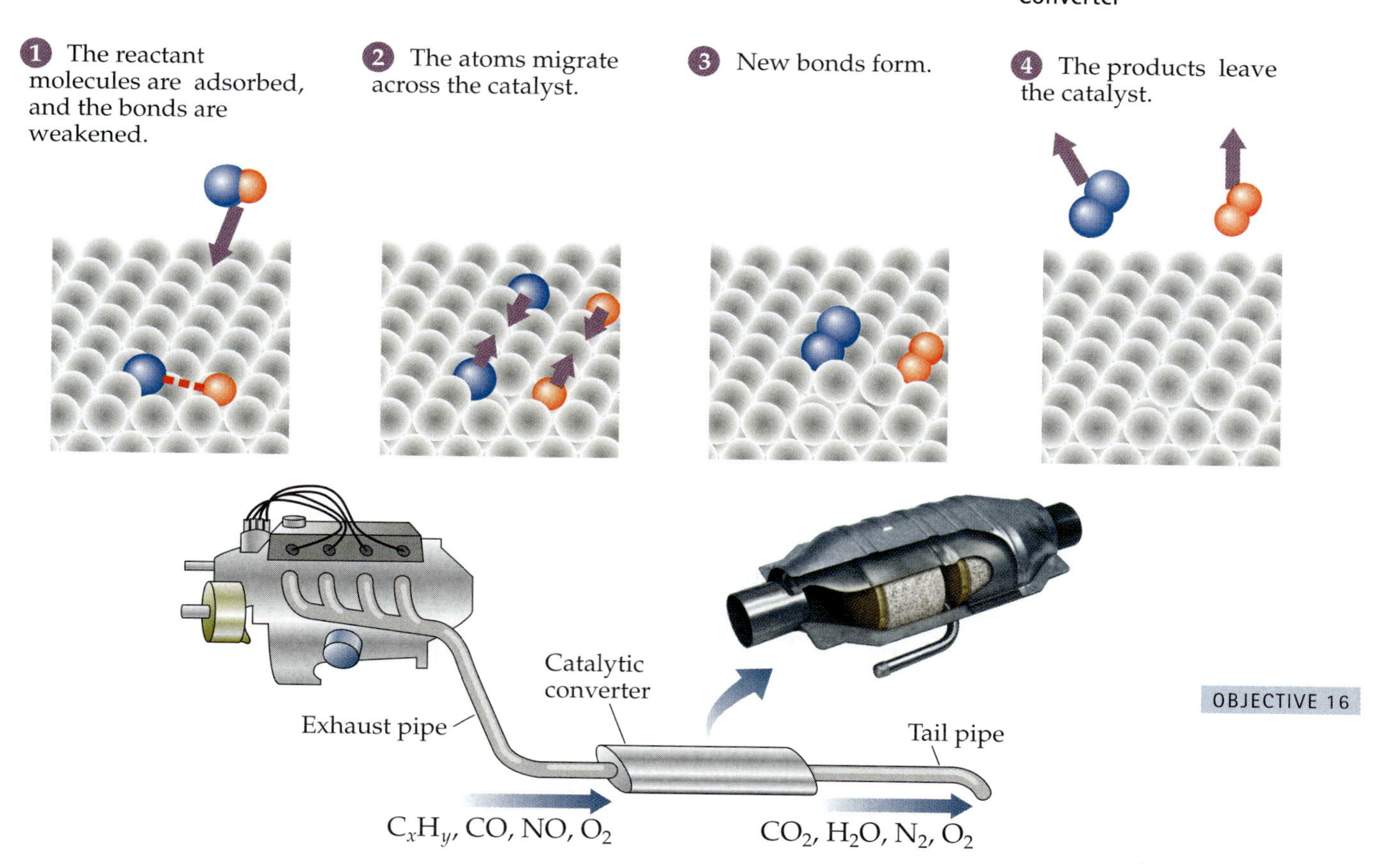
1 The reactant molecules are adsorbed, and the bonds are weakened.
2 The atoms migrate across the catalyst.
3 New bonds form.
4 The products leave the catalyst.
Catalytic converter
Exhaust pipe
Tail pipe
$C_xH_y$, CO, NO, $O_2$
$CO_2$, $H_2O$, $N_2$, $O_2$

**Figure 16.17**
A Proposed Mechanism for the Conversion of $NO(g)$ Into $N_2(g)$ and $O_2(g)$ Aided by a Transition Metal Solid in a Catalytic Converter

OBJECTIVE 16

## Special Topic 16.1 Green Chemistry—The Development of New and Better Catalysts

Because catalysts are used by the chemical industry to produce many essential products, the discovery of new and better catalysts is important for meeting the goals of Green Chemistry (Special Topic 1.1). As catalysts improve, less energy will be required for the production of industrial chemicals, and raw material and by-products will become more benign.

The Monsanto Company, as an example, received the 1996 Alternate Synthetic Pathways Award (a Presidential Green Chemistry Challenge Award) for improving the methods used in the production of their herbicide Roundup®. Formerly, the synthesis of this product required the use of ammonia, formaldehyde, hydrogen cyanide, and hydrochloric acid. All of these substances pose dangers to the workers using them and to the environment, but hydrogen cyanide in particular is extremely toxic. Another problem was that the overall process produced up to 1 kilogram of cyanide- and formaldehyde-containing waste for every 7 kilograms of product. When Monsanto developed a process that used a new metallic copper catalyst for making Roundup®, the company was able to eliminate cyanide and formaldehyde from the procedure and generate no waste at all.

Biocatalysts—catalysts produced by living organisms—are another important category of new catalysts being developed for industry. Enzymes are biocatalysts capable of forming very specific products very quickly. In some cases, their use in industrial reactions eliminates the need for hazardous reactants or the production of hazardous by-products. For example, the conventional synthesis of acrylamide (used to make dyes, adhesives, and permanent-press fabrics) employs sulfuric acid in the first step of a two-step process and ammonia in the second step. Both of these substances can be difficult or even dangerous to handle. The conventional synthesis forms acrylamide and the waste product ammonium sulfate. An alternative technique, using an enzyme called nitrile hydratase, eliminates the need for the sulfuric acid and ammonia and does not produce ammonium sulfate.

Old process

$$CH_2{=}CHCN + H_2O \xrightarrow[2.\ NH_3]{1.\ H_2SO_4} \underset{\text{Acrylamide}}{CH_2{=}CHCONH_2} + (NH_4)_2SO_4$$

New process

$$CH_2{=}CHCN + H_2O \xrightarrow{\text{Nitrile hydratase}} CH_2{=}CHCONH_2$$

## 16.3 Reversible Reactions and Chemical Equilibrium

Our next step in understanding chemical change is to apply what we have learned about rates of reactions to explore what happens when a reaction is reversible. Let's take a look at a real-world example that we can use to illustrate our main points.

Let's say you become so interested in chemistry that you decide to train to be an industrial chemist. Your first professional assignment is to design a procedure for making hydrogen gas that will then be used to make ammonia for fertilizers.

$$N_2(g) + 3H_2(g) \rightarrow 2NH_3(g)$$

After searching the chemical literature, you compile a list of all the reactions that lead to the formation of hydrogen gas. Then you shorten it by eliminating all but the reactions whose initial reactants are inexpensive, readily available, and relatively nontoxic. The following reaction between methane, $CH_4$ (the primary component of natural gas), and water vapor meets these criteria.

$$CH_4(g) + H_2O(g) \rightarrow CO(g) + 3H_2(g)$$

You decide to focus on this reaction and to research the conditions under which it is likely to produce high-purity hydrogen gas as quickly as possible. You know that rates of chemical reactions can be increased by increasing the concentration of reactants, by raising the temperature, and by adding a catalyst. Another search of the chemical literature reveals that the reaction between methane gas and water vapor runs most quickly (and yields a high concentration of product gases) when it takes place over a nickel catalyst at a total gas pressure of about 4000 kPa and a temperature between 760 °C and 980 °C. This process, which is called steam re-forming, is summarized by the equation that follows. The mixture of carbon monoxide and hydrogen gas formed in this process is used to make (synthesize) so many other chemicals that it is called *synthesis gas*.

$$CH_4(g) + H_2O(g) \xrightarrow[\text{Ni}]{\substack{\text{760–980 °C} \\ \text{4000 kPa}}} CO(g) + 3H_2(g)$$

The reaction between methane and water described above produces a significant amount of hydrogen gas, but industrial chemists are always searching for ways to increase the overall yield of their conversions. Therefore, your next step is to look for a way of using the carbon monoxide formed in the reaction to make even more hydrogen gas. You discover that carbon monoxide will react with water vapor to yield carbon dioxide and hydrogen gas (Figure 16.18, on the next page).

$$CO(g) + H_2O(g) \rightarrow CO_2(g) + H_2(g)$$

Your first thought is that this reaction should be run in the same reaction vessel as the first reaction, but further study shows that at the temperature of the first reaction (760–980 °C), the reaction between CO and $H_2O$ does not give a very high yield of hydrogen gas. At this temperature, the reaction is significantly reversible: As soon as $CO_2$ and $H_2$ form, they begin to react with each other to re-form CO and $H_2O$.

$$CO_2(g) + H_2(g) \rightarrow CO(g) + H_2O(g)$$

**Figure 16.18**
**Production and Use of Hydrogen Gas**

Chemical plants like the Air Products plant here in La Porte, Texas, make a mixture of hydrogen gas and carbon monoxide gas, called synthesis gas.

Hydrogen gas

Ammonia for fertilizers, explosives, plastics, and fibers

HCl for cleaning metals, acidifying oil wells, food processing, and the manufacture of many other chemicals

Reduction of metal oxides to form pure metals

Methanol, used to make formaldehyde, acetic acid, MTBE, and many other chemicals

The two competing changes are summarized in a single-reaction equation written with a double arrow (to show the reaction's reversible nature).

$$CO(g) + H_2O(g) \rightleftharpoons CO_2(g) + H_2(g)$$

If you were riding on one of the hydrogen atoms, you would find yourself sometimes on an $H_2$ molecule and sometimes on an $H_2O$ molecule, constantly going back and forth. Although the hydrogen yield of this reversible process is low at 760–980 °C, it is much higher at the lower temperature of 425 °C (for reasons explained later in the chapter). Therefore, your best overall plan for making $H_2(g)$ is to run the methane–water reaction in one vessel at 760–980 °C, transfer the products of this reaction to another container, add more water vapor, and run the reaction between carbon monoxide and water in the second container at 425 °C. The rate of the second reaction is further increased by allowing the reactants to combine over a chromium(III) oxide catalyst. (Note that because the reaction at 425 °C yields a high percentage of products, the equation that follows has a single arrow.)

$$CO(g) + H_2O(g) \xrightarrow[Cr_2O_3]{425\ °C} CO_2(g) + H_2(g)$$

## Reversible Reactions and Dynamic Equilibrium

Let's look more closely at the events taking place in the reaction between carbon monoxide gas and water vapor and at their effect on reaction rates. Because this reaction is significantly reversible at 870 °C, we write its equation at this temperature with a double arrow.

$$CO(g) + H_2O(g) \xrightleftharpoons{870\ ^\circ C} CO_2(g) + H_2(g)$$

Picture a system in which 1.0 mole of $CO(g)$ and 1.0 mole of $H_2O(g)$ are added to a 1.0-L container at 870 °C. As the reaction proceeds and the concentrations of CO and $H_2O$ diminish, the rate of the forward reaction decreases. At the start of the reaction, there is no carbon dioxide or hydrogen gas in the container, so the rate of the reverse reaction is initially zero. But as the concentrations of $CO_2$ and $H_2$ increase, the rate of the reverse reaction increases.

OBJECTIVE 17

OBJECTIVE 18

OBJECTIVE 19

As long as the rate of the forward reaction is greater than the rate of the reverse reaction, the concentrations of the reactants (CO and $H_2O$) will steadily decrease, and the concentrations of the products ($CO_2$ and $H_2$) will constantly increase. These changes lead to a decrease in the forward rate of the reaction and to an increase in the rate of the reverse reaction—a trend that continues until the two rates become equal (Figure 16.19, on the next page). At this point, our system has reached a dynamic equilibrium. Because all of the reactants and products are gaseous, this system is an example of a **homogeneous equilibrium,** an equilibrium system in which all of the components are in the same state.

OBJECTIVE 17

OBJECTIVE 20

In a dynamic equilibrium for a reversible chemical reaction, the forward and reverse reaction rates are equal, so although reactants and products are constantly changing back and forth, there is no *net* change in the amount of either. In our example, the CO and $H_2O$ are constantly reacting to form $CO_2$ and $H_2$, but $CO_2$ and $H_2$ are reacting to re-form CO and $H_2O$ at the same rate. Note that although the concentrations of reactants and products become constant, they do not become equal (Figure 16.20, next page).

OBJECTIVE 17

OBJECTIVE 21

As long as the system remains in a dynamic equilibrium, the concentrations of CO, $H_2O$, $CO_2$, and $H_2$ remain constant, as do the rates of the forward and reverse reactions (Figures 16.19 and 16.20). This condition continues unless something occurs to disrupt the equilibrium. (You will see a list of ways that equilibrium systems can be disrupted later in this section.)

An analogy helped us picture equilibrium systems in Chapter 14. Let's try a new analogy here. Suppose that you have left your job as an industrial chemist in order to start a ski rental shop near the Squaw Valley ski resort in California. To prepare for opening day, you purchase all the skis, boots, poles, and other accessories that you think will be necessary. When you open your doors for business, the skis begin to leave the store at a certain rate, $Rate_{leave}$, and by the end of the day, the skis are returned at a rate represented by $Rate_{return}$. This sets up a reversible process with two opposing rates of change.

$$\text{skis(in)} \xrightleftharpoons[\text{Rate}_{\text{return}}]{\text{Rate}_{\text{leave}}} \text{skis(out)}$$

**Figure 16.19**
The Rates of Reaction for the Reversible Reaction
$CO(g) + H_2O(g) \rightleftharpoons CO_2(g) + H_2(g)$

OBJECTIVE 17

OBJECTIVE 20

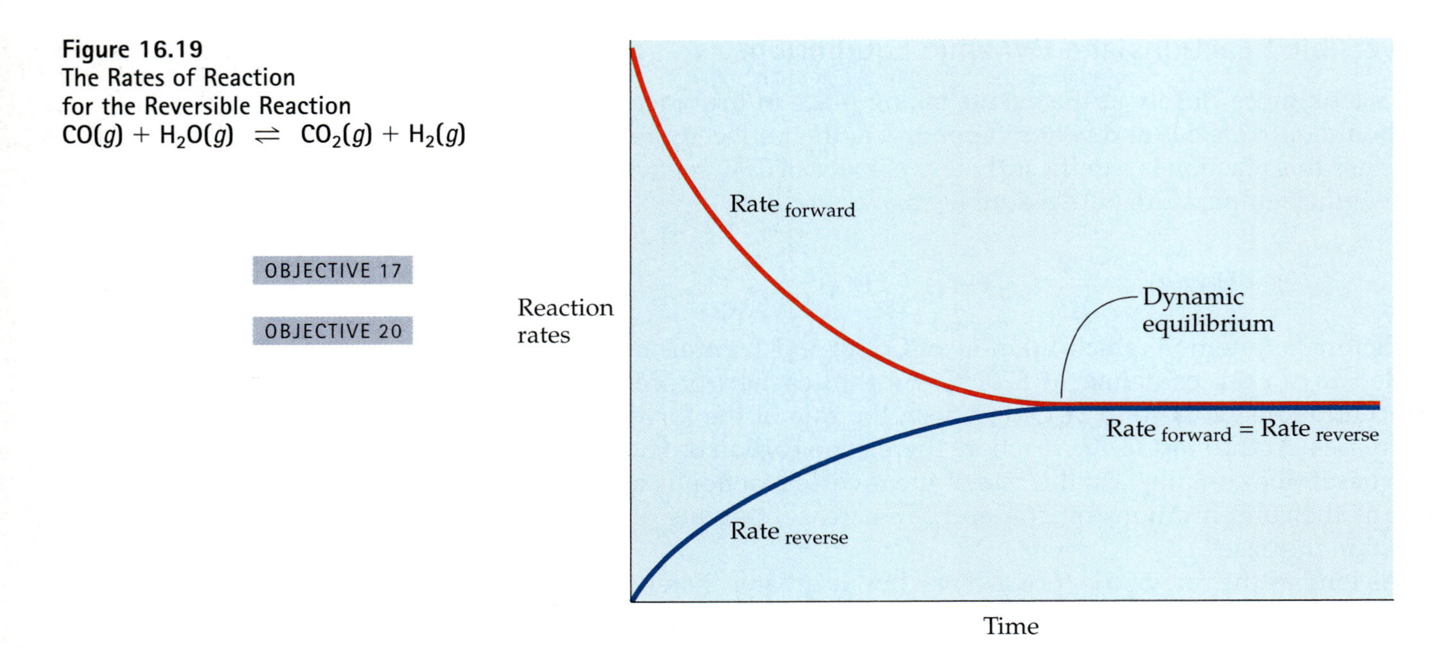


**Figure 16.20**
The Concentrations of CO, $H_2O$, $CO_2$, and $H_2$ for the Reversible Reaction
$CO(g) + H_2O(g) \rightleftharpoons CO_2(g) + H_2(g)$

OBJECTIVE 17

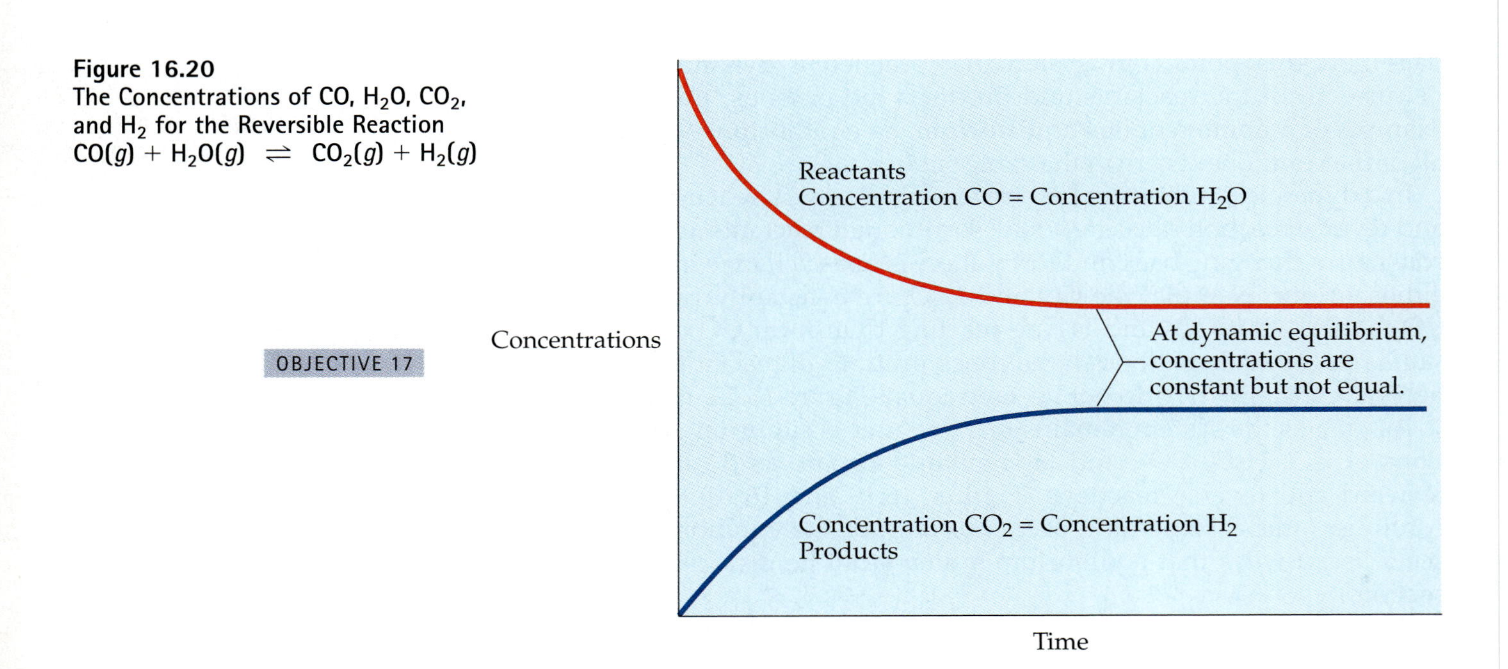


At the beginning of the first day of business, you have many different sizes and styles of ski boots and skis, so very few people are turned away without equipment. As the day progresses, you begin to run out of some sizes and styles that people want, so more people leave without renting anything. Thus the rate at which the skis leave the store is at its peak at the beginning of the first day and then diminishes steadily.

As long as the skis are going out of the store faster than they are coming in, the number of your skis on the ski slopes increases. This leads to a greater probability that some people will decide to stop skiing and return their equipment to the store. Thus the rate of return increases steadily. Eventually, the increasing $\text{Rate}_{\text{return}}$ becomes equal to the decreasing $\text{Rate}_{\text{leave}}$. At this point, although skis will be constantly going out and coming back in, there will be no net change in the number of skis in the store and the number of skis that your customers have outside the store. The system will have reached a dynamic equilibrium, and as long as the conditions for your business remain unchanged, this dynamic equilibrium will continue (Figure 16.21). (We will return to this analogy later to illustrate how changes in certain conditions can disrupt an equilibrium system.)

**Figure 16.21**
**Ski Shop Equilibrium**

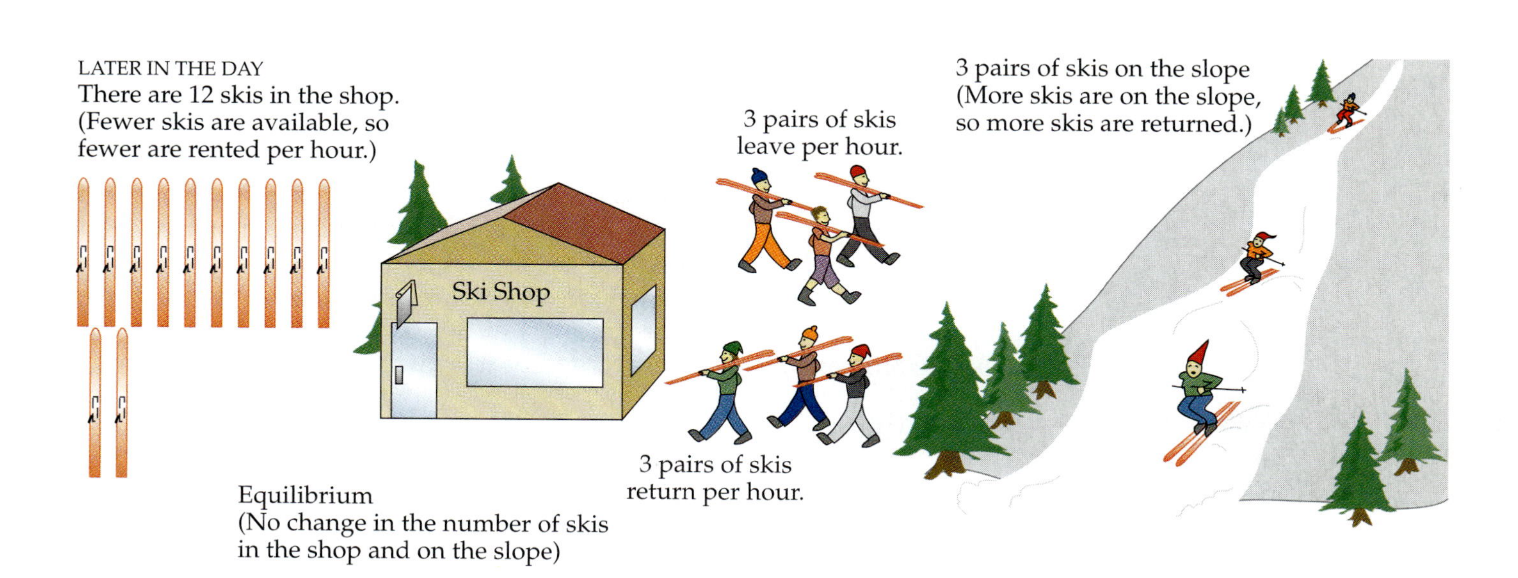
EARLY MORNING
Initially, there are 20 skis in the shop.
Ski Shop
5 pairs of skis leave per hour.
0 pairs of skis return per hour.
No skis on the slope
LATER IN THE DAY
There are 12 skis in the shop. (Fewer skis are available, so fewer are rented per hour.)
Ski Shop
3 pairs of skis leave per hour.
3 pairs of skis return per hour.
3 pairs of skis on the slope (More skis are on the slope, so more skis are returned.)
Equilibrium (No change in the number of skis in the shop and on the slope)

## Equilibrium Constants

All reversible chemical reactions spontaneously progress toward an equilibrium mixture of constant concentrations of reactants and products. At equilibrium, some reactions yield more products than reactants, and others yield more reactants than products. For example, sulfur dioxide, $SO_2$ (formed in forest fires), and nitrogen dioxide, $NO_2$ (formed in electrical storms), react in the atmosphere to yield sulfur trioxide, $SO_3$ (which reacts with water to form sulfuric acid, one of the components of acid rain) and nitrogen monoxide, NO. At normal temperatures, this reaction has a 99.92% yield, so it goes almost to completion.

$$SO_2(g) + NO_2(g) \rightleftharpoons SO_3(g) + NO(g)$$

At the other extreme, nitrogen gas and oxygen gas, which under the high-temperature conditions inside a car's engine react to form a significant amount of nitrogen monoxide gas, behave quite differently at room temperature. At 25 °C, the equilibrium mixture of $N_2$, $O_2$, and NO gases has about $1 \times 10^{-13}$% of the $N_2(g)$ and $O_2(g)$ converted into NO($g$).

$$N_2(g) + O_2(g) \rightleftharpoons 2NO(g)$$

The extent to which reversible reactions proceed toward products before reaching equilibrium can be described with an **equilibrium constant**, which is derived from the ratio of the concentrations of products to the concentrations of reactants at equilibrium. For homogeneous equilibria, the concentrations of all reactants and products can be described in moles per liter, and the concentration of each is raised to a power equal to its coefficient in a balanced equation for the reaction. The following shows the general form for the equilibrium constant expression:

OBJECTIVE 22

$$aA + bB + \cdots \rightarrow eE + fF + \cdots$$

OBJECTIVE 22

$$\text{Equilibrium constant} = K_C = \frac{[E]^e [F]^f \ldots}{[A]^a [B]^b \ldots}$$

In this general chemical equation, the lower-case letters represent coefficients, and the upper-case letters represent chemical formulas. The brackets around a formula represent the concentration of that substance in moles per liter. Because the key variables are *concentrations* of reactants and products, the equilibrium constant has the symbol $K_C$.

Gases in equilibrium constant expressions are often described with pressures instead of concentrations (mol/L). When an equilibrium constant is derived from gas pressures, we represent it with the symbol $K_P$.

$$aA + bB + \cdots \rightarrow eE + fF + \cdots$$

OBJECTIVE 23

$$\text{Equilibrium constant} = K_P = \frac{P_E{}^e P_F{}^f \ldots}{P_A{}^a P_B{}^b \ldots}$$

The values for $K_C$ and $K_P$ for the same reaction can be different, so it is important to specify whether you are using $K_P$ or $K_C$.

## EXAMPLE 16.1 Writing Equilibrium Constant Expressions

OBJECTIVE 22 OBJECTIVE 23

The first step in the process you designed for producing hydrogen was to react methane gas with water vapor to form carbon monoxide and hydrogen gases. Write the equilibrium constant expressions for $K_C$ and $K_P$ for the following equation for this reaction.

$$CH_4(g) + H_2O(g) \rightleftharpoons CO(g) + 3H_2(g)$$

***Solution***

$$CH_4(g) + H_2O(g) \rightleftharpoons CO(g) + 3H_2(g)$$

Coefficient of 3, so raise concentration or pressure to the third power.

$$K_C = \frac{[CO]\,[H_2]^3}{[CH_4]\,[H_2O]} \qquad K_P = \frac{P_{CO}\,P_{H_2}^3}{P_{CH_4}\,P_{H_2O}}$$

Note that the concentration of $H_2$ in the $K_C$ expression and the pressure of $H_2$ in the $K_P$ expression must be raised to the third power, because in the balanced equation, hydrogen has a coefficient of 3. Students commonly forget to do this.

## EXERCISE 16.1 Writing Equilibrium Constant Expressions

OBJECTIVE 22 OBJECTIVE 23

Sulfur dioxide, $SO_2$, one of the intermediates in the production of sulfuric acid, can be made from the reaction of hydrogen sulfide gas with oxygen gas. Write the equilibrium constant expressions for $K_C$ and $K_P$ for the following equation for this reaction.

$$2H_2S(g) + 3O_2(g) \rightleftharpoons 2SO_2(g) + 2H_2O(g)$$

## Determination of Equilibrium Constant Values

OBJECTIVE 24

One way to determine the value of an equilibrium constant is to run the reaction in the laboratory, wait for it to reach equilibrium, determine the concentrations or partial pressures of reactants and products, and plug them into the equilibrium constant expression (see Example 16.2). Note that equilibrium constants are usually described without units.

## EXAMPLE 16.2 Equilibrium Constant Calculation

OBJECTIVE 24

Methanol is a common substance used both as a solvent and as a reactant in the synthesis of many other substances. Methanol can be made from synthesis gas, a mixture of $CO(g)$ and $H_2(g)$. When a mixture of these gases is allowed to come to equilibrium over a copper oxide catalyst at 200 °C, the partial pressures of the gases are 0.96 atm for $CO(g)$, 1.92 atm for $H_2(g)$, and 0.11 atm for $CH_3OH(g)$. What is $K_P$ for this reaction at 200 °C?

$$CO(g) + 2H_2(g) \rightleftharpoons CH_3OH(g)$$

Methanol is used to make the preservative formaldehyde.

***Solution***

$$K_P = \frac{P_{CH_3OH}}{P_{CO}P_{H_2}^2} = \frac{0.11\text{ atm}}{0.96\text{ atm}\,(1.92\text{ atm})^2} = 0.031\ {}^{1}\!/\!_{\text{atm}^2},\text{ or } 0.031$$

Remember that the values of $K_P$ vary with temperature, so this value is valid only at 200 °C. (The reason for this variation with temperature is explained later in this section.)

## EXERCISE 16.2 Equilibrium Constant Calculation

OBJECTIVE 24

Ethanol, $C_2H_5OH$, can be made from the reaction of ethylene gas, $C_2H_4$, and water vapor. A mixture of $C_2H_4(g)$ and $H_2O(g)$ is allowed to come to equilibrium in a container at 110 °C, and the partial pressures of the gases are found to be 0.35 atm for $C_2H_4(g)$, 0.75 atm for $H_2O(g)$, and 0.11 atm for $C_2H_5OH(g)$. What is $K_P$ for this reaction at 110 °C?

$$C_2H_4(g) + H_2O(g) \rightleftharpoons C_2H_5OH(g)$$

Table 16.1 shows $K_P$ values for some gas phase homogeneous equilibria at 25 °C.

**Table 16.1**
Equilibrium Constants for Homogeneous Equilibria at 25 °C

| Reaction | $K_P$ expression | $K_P$ |
|---|---|---|
| $2SO_2(g) + O_2(g) \rightleftharpoons 2SO_3(g)$ | $K_P = \frac{P_{SO_3}^2}{P_{SO_2}^2 P_{O_2}}$ | $6.8 \times 10^{24}$ |
| $CO(g) + Cl_2(g) \rightleftharpoons COCl_2(g)$ | $K_P = \frac{P_{COCl_2}}{P_{CO_2} P_{Cl_2}}$ | $6.5 \times 10^{11}$ |
| $2NO(g) + O_2(g) \rightleftharpoons 2NO_2(g)$ | $K_P = \frac{P_{NO_2}^2}{P_{NO}^2 P_{O_2}}$ | $2.2 \times 10^{12}$ |
| $N_2(g) + 3H_2(g) \rightleftharpoons 2NH_3(g)$ | $K_P = \frac{P_{NH_3}^2}{P_{N_2} P_{H_2}^3}$ | $5.8 \times 10^{5}$ |
| $SO_2(g) + NO_2(g) \rightleftharpoons SO_3(g) + NO(g)$ | $K_P = \frac{P_{SO_3} P_{NO}}{P_{SO_2} P_{NO_2}}$ | $1.7 \times 10^{6}$ |
| $CO(g) + H_2O(g) \rightleftharpoons CO_2(g) + H_2(g)$ | $K_P = \frac{P_{CO_2} P_{H_2}}{P_{CO} P_{H_2O}}$ | $1.0 \times 10^{5}$ |
| $2NO_2(g) \rightleftharpoons N_2O_4(g)$ | $K_P = \frac{P_{N_2O_4}}{P_{NO_2}^2}$ | 6.7 |
| $SO_2(g) + Cl_2(g) \rightleftharpoons SO_2Cl_2(g)$ | $K_P = \frac{P_{SOCl_2}}{P_{SO_2} P_{Cl_2}}$ | $1.4 \times 10^{-18}$ |
| $C_2H_6(g) \rightleftharpoons C_2H_4(g) + H_2(g)$ | $K_P = \frac{P_{C_2H_4} P_{H_2}}{P_{C_2H_6}}$ | $1.2 \times 10^{-18}$ |
| $N_2(g) + O_2(g) \rightleftharpoons 2NO(g)$ | $K_P = \frac{P_{NO}^2}{P_{N_2} P_{O_2}}$ | $4.7 \times 10^{-31}$ |

## Equilibrium Constants and Extent of Reaction

Equilibrium constants indicate the extent to which reactions yield products as they progress toward equilibrium. The larger a value for $K$, the farther the reaction shifts toward products before the rates of the forward and reverse reactions become equal and the concentrations of reactants and products stop changing. The following generalizations provide guidelines for interpreting equilibrium constant values.

OBJECTIVE 25

| | |
|---|---|
| $K > 10^2$ | Products are favored at equilibrium. |
| $K < 10^{-2}$ | Reactants are favored at equilibrium. |
| $10^{-2} < K < 10^2$ | Neither reactants nor products are favored. |

### EXAMPLE 16.3 Predicting the Extent of Reaction

OBJECTIVE 25

Using the information in Table 16.1, predict whether each of the following reversible reactions favors reactants, products, or neither at 25 °C.

a. Carbon monoxide gas and chlorine gas combine to make phosgene gas, $COCl_2$, a poison gas used in World War I.

$$CO(g) + Cl_2(g) \rightleftharpoons COCl_2(g)$$

b. Ethylene, which is used to make polyethylene plastics, can be formed from the following reaction:

$$C_2H_6(g) \rightleftharpoons C_2H_4(g) + H_2(g)$$

*Solution*

a. **Favors products.** The equilibrium constant given in Table 16.1 is $6.5 \times 10^{11}$. Reactions such as this one, with extremely large equilibrium constants, can be considered to go to completion and are often described with single arrows.

$$CO(g) + Cl_2(g) \rightarrow COCl_2(g)$$

b. **Favors reactants.** The highly unfavorable equilibrium constant for this reaction ($1.2 \times 10^{-18}$) seems to suggest that this would be a poor way to make $C_2H_4(g)$, but in fact, the equilibrium constant gets much larger, and more favorable toward products, as the temperature rises. At 800–900 °C, ethylene gas is produced in a much larger proportion.

### EXERCISE 16.3 Predicting the Extent of Reaction

OBJECTIVE 25

Using the information in Table 16.1, predict whether each of the following reversible reactions favors reactants, products, or neither at 25 °C.

a. This reaction is partially responsible for the release of pollutants from automobiles.

$$2NO(g) + O_2(g) \rightleftharpoons 2NO_2(g)$$

b. The $NO_2(g)$ molecules formed in the reaction in part (a) can combine to form $N_2O_4$.

$$2NO_2(g) \rightleftharpoons N_2O_4(g)$$

The brown gas nitrogen dioxide, $NO_2$, is one cause of the haze in this smoggy Canadian city.

## Heterogeneous Equilibria

When the reactants and products for a reversible reaction are not all in the same phase (gas, liquid, solid, or aqueous), the equilibrium is called a **heterogeneous equilibrium.** For example, liquid phosphorus trichloride (used to make pesticides, gasoline additives, and other important substances) is made by passing chlorine gas over gently heated solid phosphorus.

$$P_4(s) + 6Cl_2(g) \rightleftharpoons 4PCl_3(l)$$

The equilibrium constant expression that follows is the one derived from the guidelines used for homogeneous equilibria.

$$K' = \frac{[PCl_3]^4}{[P_4][Cl_2]^6}$$

As we shall see, this is not the most common form of the equilibrium constant expression for the reaction. The more common form leaves out the pure solid and liquid.

$$K_C = \frac{1}{[Cl_2]^6}$$

OBJECTIVE 26

When all reactants are gases, the concentrations in the equilibrium expression are described in moles per liter. Pure liquids and solids are treated differently, however, because the concentrations of liquids and solids, unlike the concentrations of gases, remain constant even when the amount of substance or the volume of the container changes. To illustrate this, let's consider a reaction vessel with a constant volume in which a mixture of $P_4(s)$, $Cl_2(g)$, and $PCl_3(l)$ has come to equilibrium.

When more $Cl_2$ gas is added to the system, the new gas mixes easily with the gas already in the container, and because the moles of $Cl_2$ increase but the volume is constant, the concentration of $Cl_2$ increases. The concentration of $Cl_2$ could also be changed if the volume of the container could change. If the gas were compressed, the concentration would increase, and if the gas were allowed to expand, the concentration would decrease (Figure 16.22).

Note, however, that the concentrations of solids and liquids cannot be changed in these ways. If the number of moles of $P_4(s)$ in the container is doubled, its volume doubles too, leaving the concentration (mol/L) of the phosphorus constant. Increasing or decreasing the total volume of the container will not change the volume occupied by the solid phosphorus, so the concentration (mol/L) of the $P_4(s)$ also remains constant with changes in the volume of the container. Similarly, the concentration of liquid $PCl_3$ is also independent of the volume of the container and the amount of $PCl_3$ present (Figure 16.22). By convention, the constant concentrations of pure solids and liquids are understood to be incorporated into the equilibrium constant itself and are left out of the equilibrium constant expression.

$$K' = \frac{[PCl_3]^4}{[P_4][Cl_2]^6}$$

OBJECTIVE 22

OBJECTIVE 23

$$\frac{K'[P_4]}{[PCl_3]^4} = \frac{1}{[Cl_2]^6} = K_C \quad \text{or} \quad K_P = \frac{1}{P^6_{Cl_2}}$$

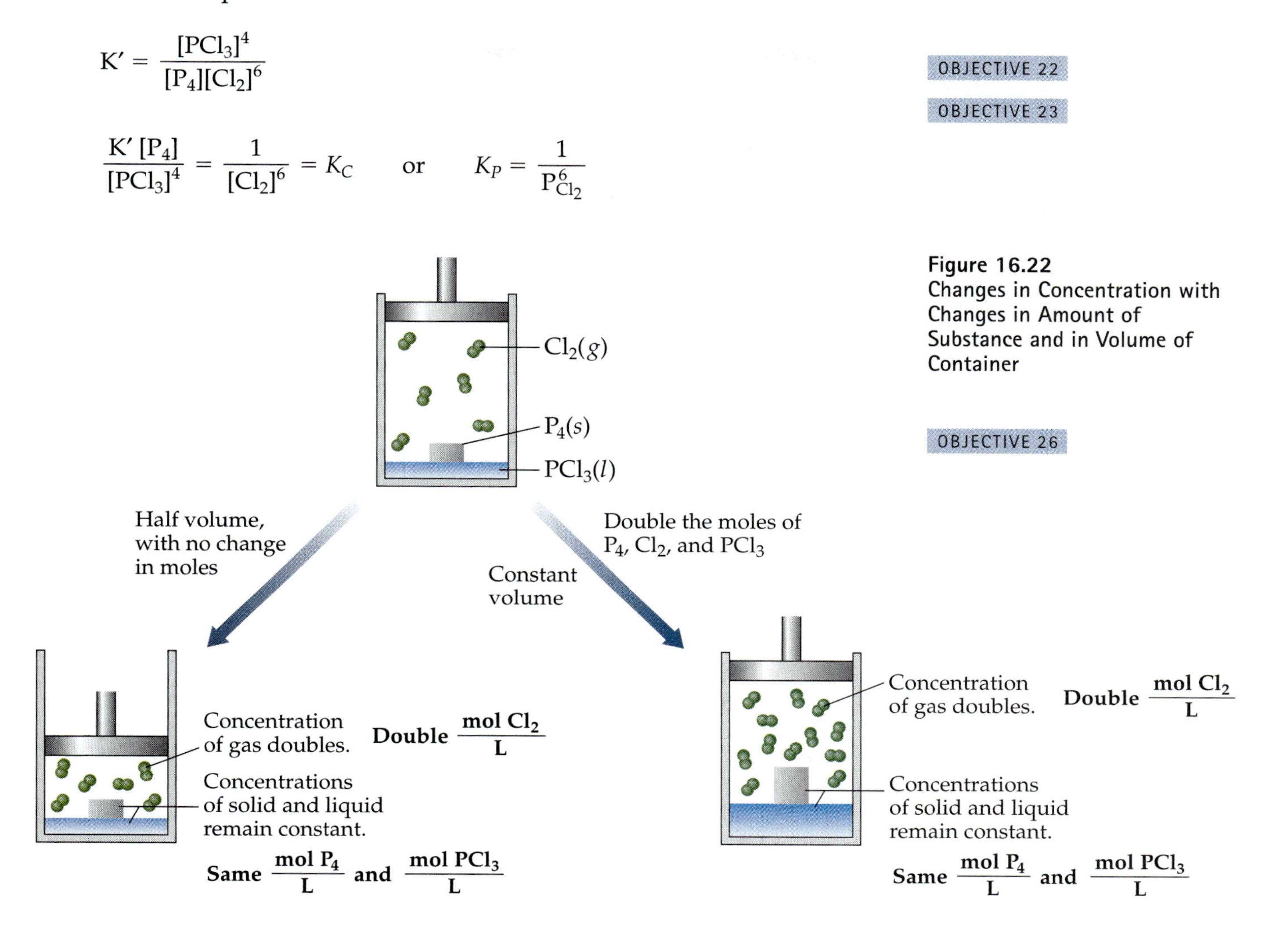


**Figure 16.22**
Changes in Concentration with Changes in Amount of Substance and in Volume of Container

OBJECTIVE 26

## EXAMPLE 16.4 Writing Equilibrium Constants for Heterogeneous Equilibria

OBJECTIVE 22

OBJECTIVE 23

One of the steps in the production of sulfuric acid is to make solid sulfur from the reaction of sulfur dioxide gas and hydrogen sulfide gas. Write $K_C$ and $K_P$ expressions for this reaction.

$$SO_2(g) + 2H_2S(g) \rightleftharpoons 3S(s) + H_2O(g)$$

***Solution***

$$SO_2(g) + 2H_2S(g) \rightleftharpoons 3S(s) + H_2O(g)$$

The solid does not appear in the $K_C$ and $K_P$ expressions.

$$K_C = \frac{[H_2O]}{[SO_2][H_2S]^2} \qquad K_P = \frac{P_{H_2O}}{P_{SO_2}P^2_{H_2S}}$$

## Exercise 16.4 Writing Equilibrium Constants for Heterogeneous Equilibria

OBJECTIVE 22

OBJECTIVE 23

The following equation describes one of the steps in the purification of titanium dioxide, which is used as a white pigment in paints. Liquid titanium(IV) chloride reacts with oxygen gas to form solid titanium oxide and chlorine gas. Write $K_C$ and $K_P$ expressions for this reaction.

$$TiCl_4(l) + O_2(g) \rightleftharpoons TiO_2(s) + 2Cl_2(g)$$

White paint contains titanium dioxide.

## Equilibrium Constants and Temperature

Changing temperature always causes a shift in equilibrium systems—sometimes toward more products and sometimes toward more reactants. For example, some industrial reactions that yield an equilibrium mixture with less than 1% products at room temperature yield more than 99% products at higher temperature. This shift in equilibrium systems is reflected in changes in equilibrium constants that accompany changes in temperature.

Before we tackle the question of why changing temperature affects equilibrium systems, let's quickly establish some key characteristics of energy changes in chemical reactions. Remember that chemical reactions are either endergonic (they absorb energy as the reaction forms weaker chemical bonds) or exergonic (they evolve energy as the reaction leads to stronger bonds). In a reversible reaction, if the forward reaction is exergonic, the reverse reaction is endergonic. If the forward reaction is endergonic, the reverse reaction is exergonic.

$$\text{reactants} \underset{\text{endergonic}}{\overset{\text{exergonic}}{\rightleftharpoons}} \text{products + energy}$$

or

$$\text{reactants + energy} \underset{\text{exergonic}}{\overset{\text{endergonic}}{\rightleftharpoons}} \text{products}$$

Increased temperature increases the rate of both the forward and the reverse reactions, but it increases the rate of the endergonic reaction more than it increases the rate of the exergonic reaction. Therefore, increasing the temperature of a chemical system at equilibrium disrupts the balance of the forward and reverse rates of reaction and shifts the system in the direction of the endergonic reaction.

OBJECTIVE 27

OBJECTIVE 28

- If the forward reaction is endergonic, increased temperature will shift the system toward more products, increasing the ratio of products to reactants and increasing the equilibrium constant.

- If the forward reaction is exergonic and the reverse reaction is endergonic, increased temperature will shift the system toward more reactants, decreasing the ratio of products to reactants and decreasing the equilibrium constant.

OBJECTIVE 28

For example, pure water ionizes in a reversible endergonic reaction that forms hydrogen ions, $H^+$ (which form hydronium ions, $H_3O^+$), and hydroxide ions, $OH^-$. The equilibrium constant for this reaction is called the water dissociation constant, $K_w$.

$$H_2O(l) + \text{energy} \rightleftharpoons H^+(aq) + OH^-(aq)$$

$$K_w = [H^+][OH^-] = 1.01 \times 10^{-14} \quad \text{at } 25\ °C$$

Because the forward reaction is endergonic, increased temperature drives the system to products, increasing the value for the water dissociation constant. Table 16.2 shows $K_w$ values for this reaction at different temperatures.

Table 16.2
Water Dissociation Constants at Various Temperatures

| Temperature | $K_w$ |
|---|---|
| 0 °C | $1.14 \times 10^{-15}$ |
| 10 °C | $2.92 \times 10^{-15}$ |
| 25 °C | $1.01 \times 10^{-14}$ |
| 30 °C | $1.47 \times 10^{-14}$ |
| 40 °C | $2.92 \times 10^{-14}$ |
| 50 °C | $5.47 \times 10^{-14}$ |
| 60 °C | $9.61 \times 10^{-14}$ |

The reaction between carbon monoxide gas and water vapor to form carbon dioxide gas and hydrogen gas is exothermic, which means that thermal energy is evolved as the reaction proceeds in the forward direction. Therefore, increased temperature favors the reverse reaction, shifting the system toward more reactants and decreasing the equilibrium constant for the reaction.

$$CO(g) + H_2O(g) \rightleftharpoons CO_2(g) + H_2(g) + \text{thermal energy}$$

The fact that higher temperature leads to a lower percentage of $CO(g)$ and $H_2O(g)$ converted to $CO_2(g)$ and $H_2(g)$ creates a dilemma for the industrial chemist designing the process for making hydrogen gas. To maximize the percentage yield, the reaction should be run at as low a temperature as

OBJECTIVE 29

possible, but at low temperature, the rates of the forward and reverse reactions are both very low, so it takes a long time for the system to come to equilibrium. Some chemical plants in the United States have solved this problem by setting up two chambers for the reaction, one at high temperature to convert some of the $CO(g)$ and $H_2O(g)$ to $CO_2(g)$ and $H_2(g)$ quickly, and a low-temperature section to complete the process.

At **www.chemplace.com/college/**
you can see how concentration or gas pressures of reactants and products at equilibrium can be calculated from initial concentrations of reactants and products and equilibrium constants.

And, now that you know about reversible reactions and how solutions are described in terms of molarity, you will be able to understand the origin of the pH scale for describing acidic and basic solutions.
You can find information about this at the Web site.

You can also see how weak acids are described with equilibrium constants.

## 16.4 Disruption of Equilibrium

Chemical systems at equilibrium are actually somewhat rare outside the laboratory. For example, the concentrations of the reactants and products involved in the reactions taking place in your body are constantly changing, leading to changes in the forward and reverse reaction rates. These constant fluctuations prevent the reactions from ever reaching equilibrium. In most industrial procedures, the products of reversible reactions are purposely removed as they form, preventing the reverse rate of the reaction from ever rising to match the rate of the forward reaction.

In general, equilibrium systems are easily disrupted. Every time you take a bottle of vinegar from a cool pantry and hold it in your warm hand, you increase the temperature of the solution and disrupt the equilibrium between acetic acid and its ions. To illustrate the conditions that lead to disruption of equilibria and to show why they do so, let's return to our example of making hydrogen gas from carbon monoxide and water vapor.

$$CO(g) + H_2O(g) \underset{\text{Rate}_{\text{reverse}}}{\overset{\text{Rate}_{\text{forward}}}{\rightleftharpoons}} CO_2(g) + H_2(g)$$

## The Effect of Changes in Concentrations on Equilibrium Systems

Any change in the concentration of a reactant or product in a chemical system at equilibrium will destroy the balance of the forward and reverse reaction rates and create a nonequilibrium situation, which will lead in some cases to a shift toward more reactants and in others to a shift toward more products. One common means of shifting an equilibrium system toward the formation of more products is to increase the concentration of one of the reactants. Imagine that our system containing $CO(g)$, $H_2O(g)$, $CO_2(g)$, and $H_2(g)$ is in equilibrium, so that the rate of the forward change is equal to the rate of the reverse change. If more water vapor is added to the system, collisions between CO molecules and $H_2O$ molecules will be more frequent, and $Rate_{forward}$ will increase. Initially, the rate of the reverse reaction, $Rate_{reverse}$, will not be affected, so the two rates will no longer be the same, and the system will no longer be at equilibrium.

OBJECTIVE 30

To reiterate, the addition of water vapor to our equilibrium system causes $Rate_{forward}$ to be greater than $Rate_{reverse}$, so the products will be made more rapidly than the reactants are re-formed. The concentration of the reactants $CO(g)$ and $H_2O(g)$ will decrease, and the concentration of the products $CO_2(g)$ and $H_2(g)$ will increase. Chemists describe this situation by saying that the *increase of one or more reactants shifts a system initially at equilibrium toward products* (Figure 16.23).

OBJECTIVE 30

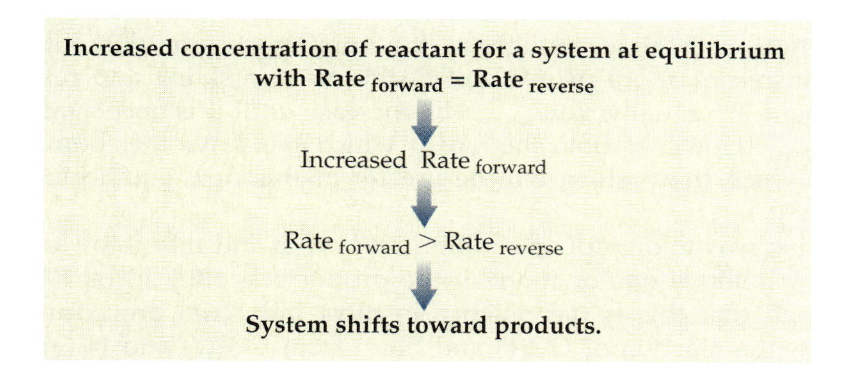
Increased concentration of reactant for a system at equilibrium with Rate forward = Rate reverse
Increased Rate forward
Rate forward > Rate reverse
System shifts toward products.

**Figure 16.23**
Effect of Increased Reactant Concentration on Equilibrium

OBJECTIVE 30

Eventually, however, as the concentrations of $CO_2(g)$ and $H_2(g)$ increase, $Rate_{reverse}$ increases, and as the concentrations of $CO(g)$ and $H_2O(g)$ decrease, $Rate_{forward}$ decreases. If there are no further disruptions of the system, it will arrive at a new equilibrium with $Rate_{forward}$ and $Rate_{reverse}$ once again equal. Note that at this new equilibrium, both $Rate_{forward}$ and $Rate_{reverse}$ are higher (Figure 16.24, on the next page).

Back at your ski shop, you might encounter a similar situation. When we last looked in on the business, we saw a dynamic equilibrium in which the rate at which skis left the shop, $Rate_{leave}$, was equal to the rate at which they were being returned, $Rate_{return}$. If you buy more skis and equipment, however, you will be more likely to have what your customers want, so you will turn fewer customers away without skis. Thus the rate at which the skis leave the shop, $Rate_{leave}$, will increase, making $Rate_{leave}$ greater

OBJECTIVE 30

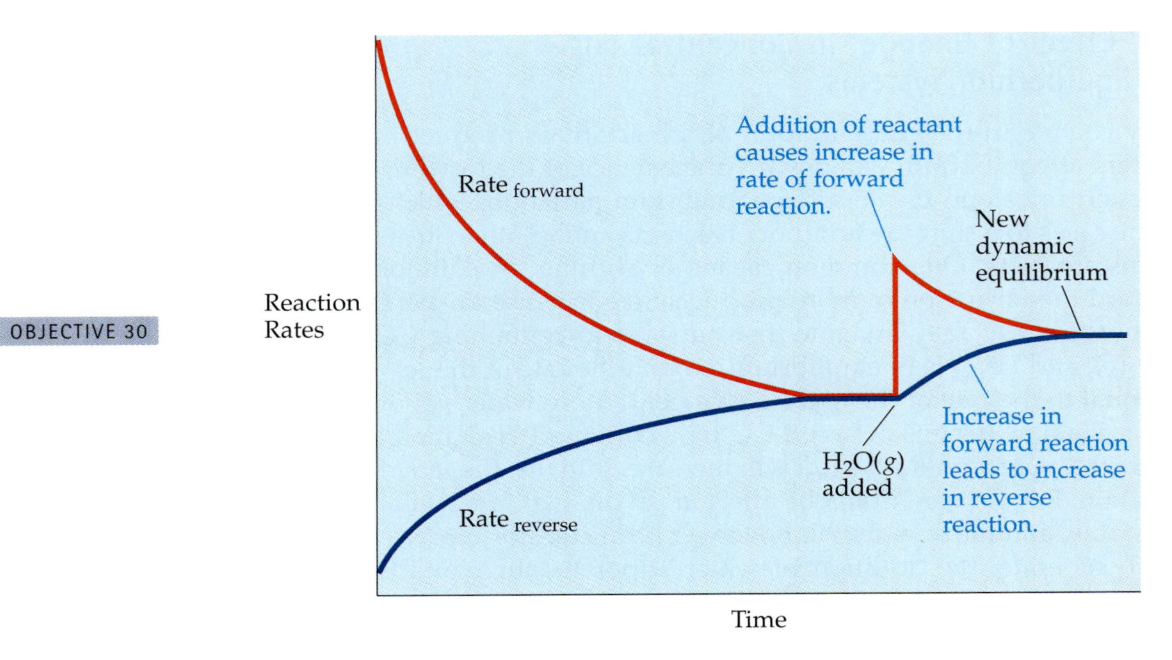
Reaction Rates
Rate forward
Addition of reactant causes increase in rate of forward reaction.
New dynamic equilibrium
$H_2O(g)$ added
Increase in forward reaction leads to increase in reverse reaction.
Rate reverse
Time

**Figure 16.24**
**Change in the Rates of Reaction When a Reactant Is Added to an Equilibrium System for the Reversible Reaction** $CO(g) + H_2O(g) \rightleftharpoons CO_2(g) + H_2(g)$

than $Rate_{return}$. In time, though, as the number of your skis out on the slopes increases, more people will decide to stop skiing and return the equipment. Eventually, $Rate_{return}$ will increase until it is once again equal to $Rate_{leave}$. However, both the rate at which skis leave the shop and the rate at which they return will be greater at this new equilibrium (Figure 16.25).

Another way to disrupt an equilibrium system and shift it toward products is to remove one or more of the products as they form. (We have mentioned that this is the practice in most industrial procedures.) For example, the reaction of $CO(g)$ and $H_2(g)$ from $CO_2(g)$ and $H_2(g)$ can be driven toward products by the removal of $CO_2(g)$ as it forms. This is done in the industrial process by reacting the $CO_2(g)$ with some of the excess water vapor and monoethanolamine, $HOCH_2CH_2NH_2$.

$$CO(g) + H_2O(g) \rightleftharpoons CO_2(g) + H_2(g)$$

$$CO_2(g) + H_2O(g) + HOCH_2CH_2NH_2(g) \rightarrow HOCH_2CH_2NH_3HCO_3(s)$$

The removal of $CO_2(g)$ decreases its concentration, decreasing the frequency of collisions between $CO_2(g)$ and $H_2(g)$ and therefore decreasing the rate of the reverse reaction, $Rate_{reverse}$. This change makes $Rate_{forward}$ greater than $Rate_{reverse}$, shifting the system toward the formation of more products (Figure 16.26, on page 698).

OBJECTIVE 31

As the concentration of products increases, $Rate_{reverse}$ increases. As the concentration of reactants decreases, $Rate_{forward}$ decreases. If no further changes are made to the system, it will return to a new dynamic equilibrium with $Rate_{forward}$ equaling $Rate_{reverse}$ (Figure 16.27, on page 698).

**Figure 16.25**
Ski Shop Equilibrium and Increased Stock

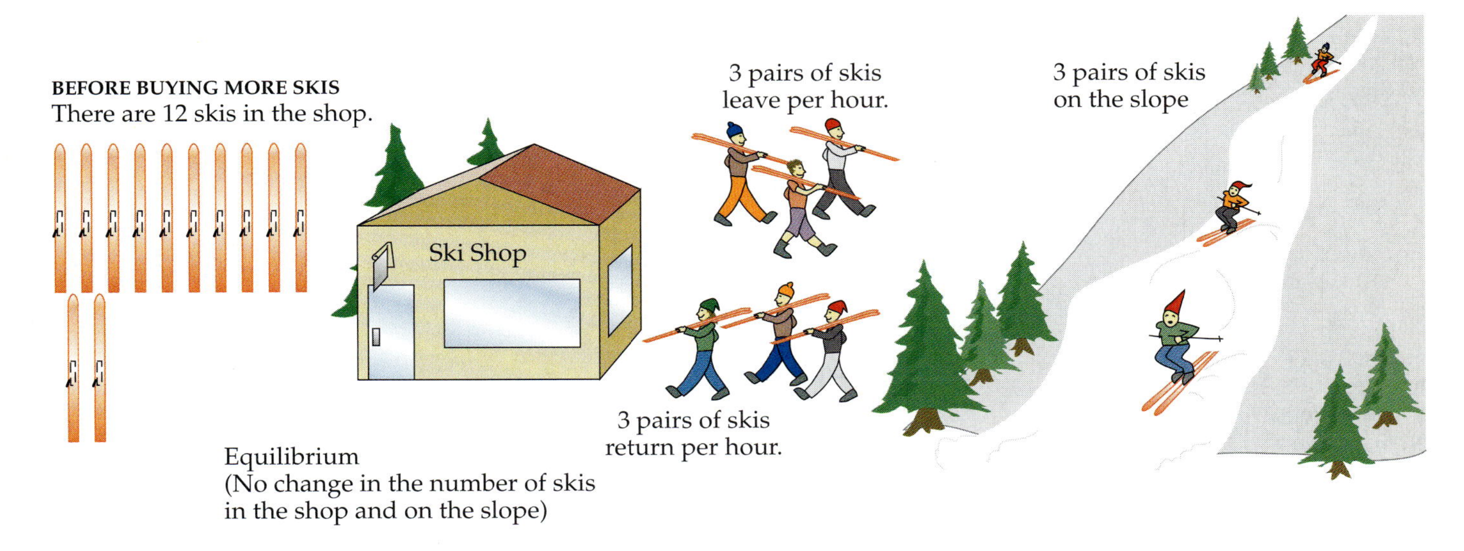
BEFORE BUYING MORE SKIS
There are 12 skis in the shop.
Ski Shop
3 pairs of skis leave per hour.
3 pairs of skis on the slope
3 pairs of skis return per hour.
Equilibrium
(No change in the number of skis in the shop and on the slope)

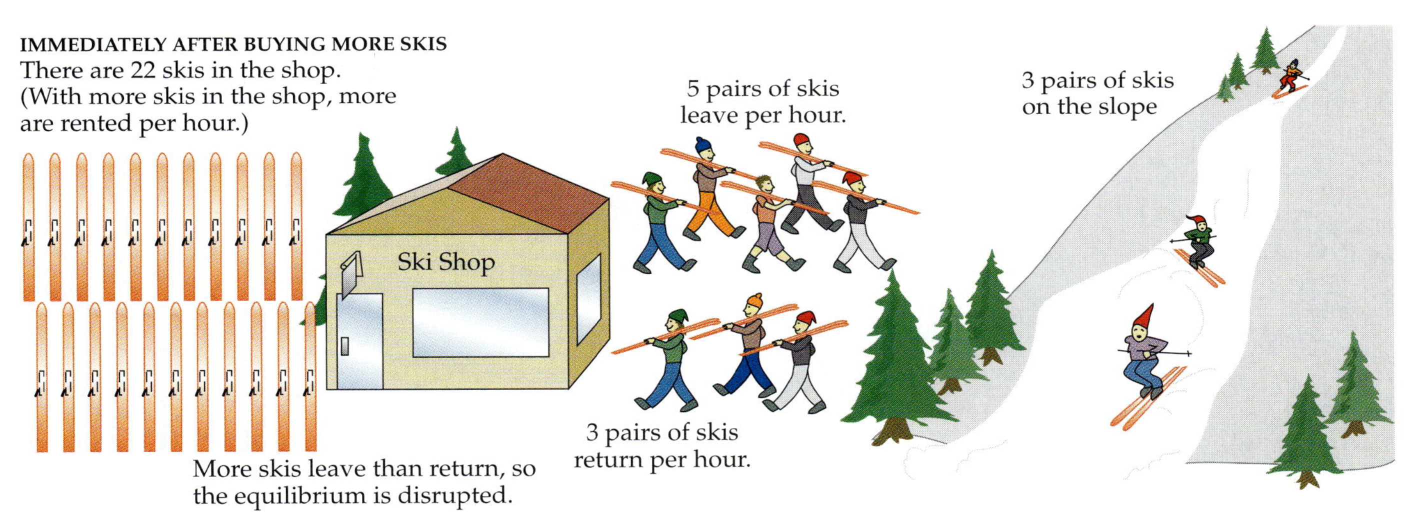
IMMEDIATELY AFTER BUYING MORE SKIS
There are 22 skis in the shop.
(With more skis in the shop, more are rented per hour.)
Ski Shop
5 pairs of skis leave per hour.
3 pairs of skis on the slope
3 pairs of skis return per hour.
More skis leave than return, so the equilibrium is disrupted.

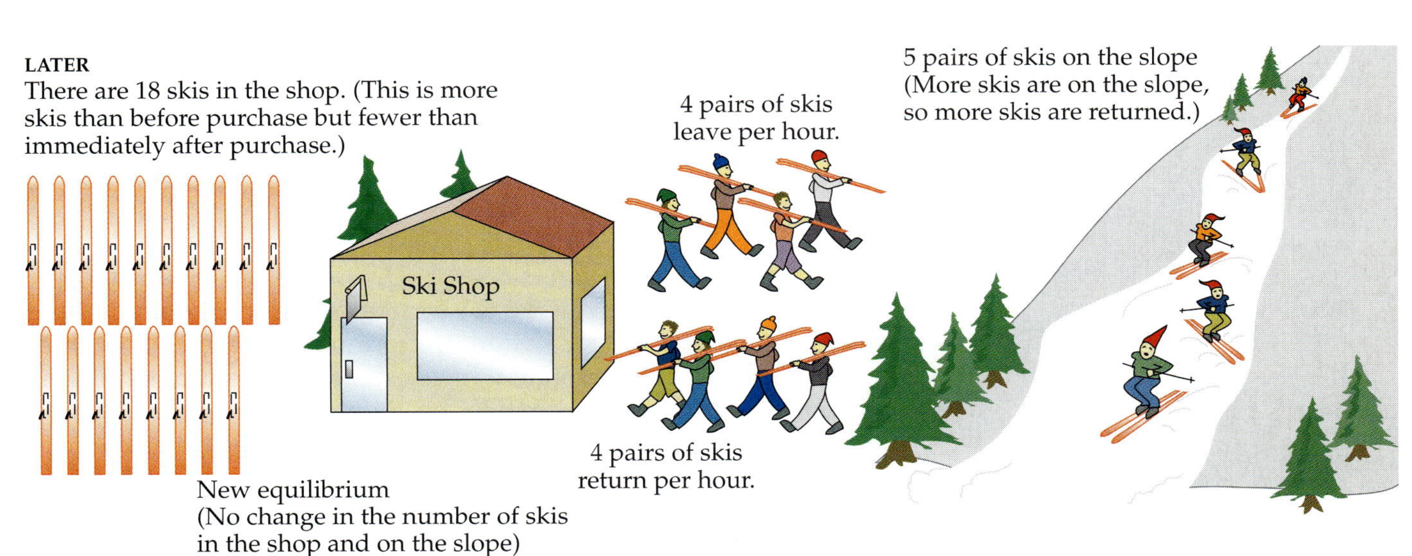
LATER
There are 18 skis in the shop. (This is more skis than before purchase but fewer than immediately after purchase.)
Ski Shop
4 pairs of skis leave per hour.
5 pairs of skis on the slope
(More skis are on the slope, so more skis are returned.)
4 pairs of skis return per hour.
New equilibrium
(No change in the number of skis in the shop and on the slope)

**Figure 16.26**
Effect of Decreased Concentration of Product on Equilibrium

OBJECTIVE 31

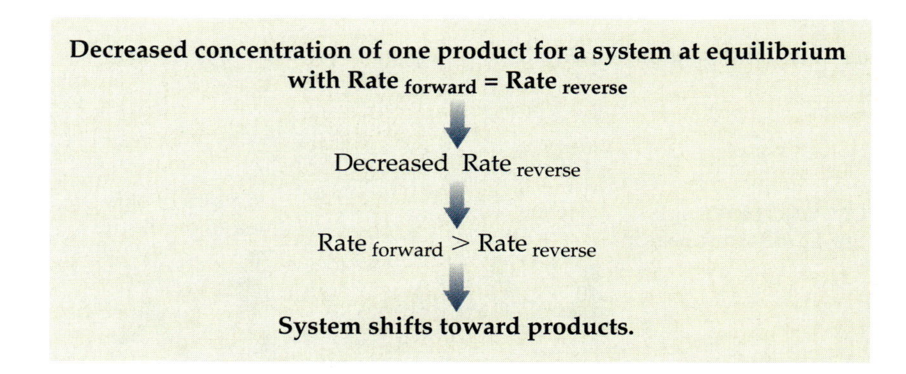

Decreased concentration of one product for a system at equilibrium with Rate forward = Rate reverse
Decreased Rate reverse
Rate forward > Rate reverse
System shifts toward products.


**Figure 16.27**
Change in the Rates of Reaction When a product Is Removed from an Equilibrium System
$CO(g) + H_2O(g) \rightleftharpoons CO_2(g) + H_2(g)$

OBJECTIVE 31

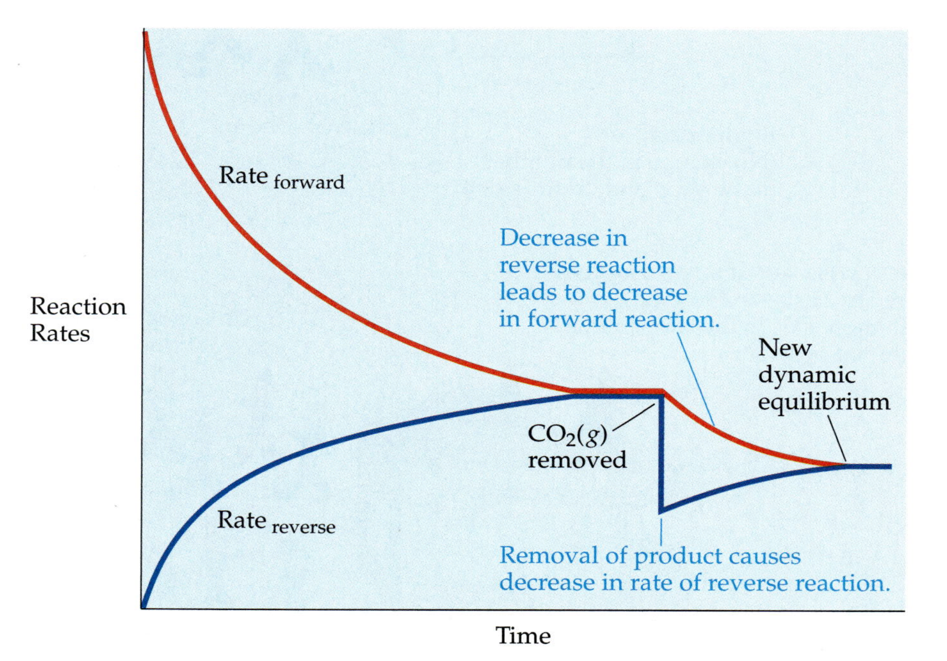


At the following Web site, you can read about how changing volume affects gas phase reactions at equilibrium.
**www.chemplace.com/college/**

## Le Châtelier's Principle

Whenever an equilibrium system is disrupted, it will move toward a new equilibrium by shifting either toward the formation of more products (and less reactants) or toward the formation of more reactants (and less products). We can predict the direction in which a system will shift by thinking about the effect of the proposed changes on $Rate_{forward}$ and $Rate_{reverse}$, but we can often arrive at the correct prediction more easily by applying a guideline called **Le Châtelier's principle.** This principle states that *if a system at equilibrium is altered in a way that disrupts the equilibrium, the system will shift in such a way as to counter the change.*

OBJECTIVE 32

OBJECTIVE 33

For example, we can use this principle to predict the direction in which the following equation shifts when water vapor is added to the system at equilibrium.

$$CO(g) + H_2O(g) \rightleftharpoons CO_2(g) + H_2(g)$$

Le Châtelier's principle tells us that the system will shift toward products in order to decrease the increased concentration of $H_2O(g)$, partially counteracting the change that disrupted the equilibrium system.

OBJECTIVE 32

OBJECTIVE 33

Likewise, we can use Le Châtelier's principle to predict the effect of temperature on the following equilibrium for which the forward reaction is exothermic.

$$CO(g) + H_2O(l) \rightleftharpoons CO_2(g) + H_2(g) + \text{thermal energy}$$

Le Châtelier's principle tells us that if we decrease the temperature for this system at equilibrium, the system will shift to partially counteract this change and increase the temperature. Thus we predict that decreased temperature will shift the system toward products, because the exothermic forward reaction releases thermal energy and leads to an increase in temperature.

OBJECTIVE 32

OBJECTIVE 33

Table 16.3 summarizes the effects of changes in the concentrations of reactants and products and changes in the temperature of equilibrium systems.

**Table 16.3**
Summary of the Shifts in Equilibrium Predicted by Le Châtelier's Principle

OBJECTIVE 32 OBJECTIVE 33

| Reaction | Cause of disruption | To counteract change | Direction of shift |
|---|---|---|---|
| all | add reactant(s) | decrease reactant(s) | to products |
| all | add product(s) | decrease product(s) | to reactants |
| all | remove reactant(s) | increase reactant(s) | to reactants |
| all | remove product(s) | increase products(s) | to products |
| endothermic forward reaction | increase temperature | decrease temperature | to products |
| endothermic forward reaction | decrease temperature | increase temperature | to reactants |
| exothermic forward reaction | increase temperature | decrease temperature | to reactants |
| exothermic forward reaction | decrease temperature | increase temperature | to products |

## The Effect of Catalysts on Equilibria

The presence of a catalyst has no effect on the equilibrium concentrations of reactants and products. Catalysts speed both the forward and reverse rates to the same degree. The system will reach equilibrium more quickly, but the ratio of the concentrations of products to reactants—and thus the value of the equilibrium constant—is unchanged.

OBJECTIVE 34

We can illustrate a situation of this kind using our ski rental analogy. You decide to "catalyze" your ski rental business by constructing a drive-up window to increase the convenience of renting and returning skis. Because checking out the skis is easier now, $\text{Rate}_{\text{leave}}$ increases. Because returning skis is also easier now, $\text{Rate}_{\text{return}}$ also increases. If both of these rates increase equally, there will be no net change in the number of skis in the shop or on the slopes at any time.

## Example 16.5 Predicting the Effect of Disruptions on Equilibrium

OBJECTIVE 32

OBJECTIVE 33

Ammonia gas, which is used in the manufacture of fertilizers and explosives, is made from the reaction of nitrogen gas and hydrogen gas. The forward reaction is exothermic. Predict whether each of the following changes in an equilibrium system of nitrogen, hydrogen, and ammonia will shift the system to more products, to more reactants, or to neither. Explain each answer in two ways: (1) by applying Le Châtelier's principle and (2) by describing the effect of the change on the forward and reverse reaction rates.

$$N_2(g) + 3H_2(g) \rightleftharpoons 2NH_3(g) + 92.2 \text{ kJ}$$

a. The concentration of $N_2$ is increased by the addition of more $N_2$.

b. The concentration of ammonia gas is decreased.

c. The temperature is increased from 25 °C to 500 °C.

d. An iron catalyst is added.

*Solution*

a. The concentration of $N_2$ is increased by the addition of more $N_2$.

   1. Using Le Châtelier's principle, we predict that the system will **shift to more products** to partially counteract the increase in $N_2$.
   2. The increase in the concentration of nitrogen speeds the forward reaction without initially affecting the rate of the reverse reaction. The equilibrium is disrupted, and the system **shifts toward more products** because the forward rate is greater than the reverse rate.

b. The concentration of ammonia is decreased.

   1. Using Le Châtelier's principle, we predict that the system will **shift toward more products** to partially counteract the decrease in the concentration of ammonia.
   2. The decrease in the concentration of ammonia decreases the rate of the reverse reaction. It does not initially affect the forward rate. The equilibrium is disrupted, and the system **shifts toward more products** because the forward rate is greater than the reverse rate.

c. The temperature is increased from 25 °C to 400 °C.

   1. Using Le Châtelier's principle, we predict that the system shifts in the endothermic direction to partially counteract the increase in temperature. Because the forward reaction is exothermic, the reverse reaction must be endothermic. As the system **shifts toward more reactants,** energy is absorbed, and the temperature decreases.
   2. The increased temperature increases the rates of both the forward and the reverse reactions, but it has a greater effect on the endothermic reaction. Thus the system **shifts toward more reactants** because the reverse rate becomes greater than the forward rate.

d. An iron catalyst is added.

   1. Le Châtelier's principle does not apply here.
   2. The catalyst speeds the forward and reverse rates equally. Thus there is **no shift in the equilibrium.** (The purpose of the catalyst is to bring the system to equilibrium more rapidly.)

## Exercise 16.5 Predicting the Effect of Disruptions on Equilibrium

OBJECTIVE 32

OBJECTIVE 33

Nitric acid can be made from the exothermic reaction of nitrogen dioxide gas and water vapor in the presence of a rhodium and platinum catalyst at 700–900 °C and 5–8 atm. Predict whether each of the following changes in the equilibrium system will shift the system to more products, to more reactants, or to neither. Explain each answer in two ways: (1) by applying Le Châtelier's principle and (2) by describing the effect of the change on the forward and reverse reaction rates.

$$3NO_2(g) + H_2O(g) \underset{\substack{750\text{–}920\ °C \\ 5\text{–}8\ \text{atm}}}{\overset{\text{Rh/Pt}}{\rightleftharpoons}} 2HNO_3(g) + NO(g) + 37.6\ \text{kJ}$$

a. The concentration of $H_2O$ is increased by the addition of more $H_2O$.

b. The concentration of $NO_2$ is decreased.

c. The concentration of $HNO_3(g)$ is decreased by removing the nitric acid as it forms.

d. The temperature is decreased from 1000 °C to 800 °C.

e. The Rh/Pt catalyst is added to the equilibrium system.

## Special Topic 16.2 The Big Question—How Did We Get Here?

How did the human species come to be? This huge, multifaceted question is being investigated from many different angles, not all of which are the province of science. Scientists have generally focused on four aspects of it:

- the creation of matter, including elements
- the formation of the molecules necessary for life
- the collection of these molecules into simple organisms
- the evolution of more complex organisms, including ourselves

Physicists are working to discover how matter was created, and biologists are attempting to explain how simple organisms formed and evolved into organisms that are more complex. Chemists get the middle ground, the problem of how biomolecules (molecules important to biological systems) originally formed from the elements. This alone is a vast topic, so we will limit our discussion here to a theory that attempts to explain how some simple precursors of important biomolecules might have formed on the earth billions of years ago.

To explain how living organisms grow and reproduce, we need to explain how certain chemicals, such as amino acids and proteins, are formed. Amino acids have the following general structure, where R represents one of several groups that make one amino acid different from another.

$$H_2N-\underset{\substack{| \\ R}}{CH}-\overset{\substack{O \\ \|}}{C}-OH$$

Amino acid

(continued)

## Special Topic 16.2 (Continued)

*(continued from previous page)*

**Could an important step in the creation of life have happened here?**

When two or more amino acids are linked together, the product is called a peptide (example shown below). Protein molecules, which are relatively large peptides, are important biomolecules that have many functions in our bodies, including acting as enzymes (naturally occurring catalysts).

Like any reasonable recipe, which includes both a list of ingredients and instructions for combining them under the proper conditions, a reasonable theory for the origin of biomolecules must include a plausible explanation of how the necessary reactants (ingredients) were available and of how the necessary conditions arose for the conversion of these reactants into a significant amount of product. As we have learned in this chapter, just combining the correct reactants does not necessarily lead to the formation of a significant amount of product. Some reactions are very slow, and even for those that are fast, the amount of product formed can be limited by the competition between the forward and reverse reactions.

### *Proposed Recipe for Making Proteins*

*Ingredients:* $CO$, $CO_2$, $NH_3$, $H_2S$

Boil water in the ocean's pressure cooker, add ingredients, and wait for proteins to form.

In 1977 the oceanographer Jack Corliss discovered hydrothermal vents on the deep ocean floor, complete with both the raw materials needed to make biomolecules and the necessary conditions for the reactions. Hydrothermal vents are springs of hot water (up to 350 °C) that form when seawater seeping into cracks in the ocean floor is heated by volcanic rock in the earth and then escapes from other openings in the sea bottom. The water that rushes out of these vents contains, among other substances, carbon monoxide, carbon dioxide, ammonia, and hydrogen sulfide. Corliss's suggestion that life could have originated near these vents was not immediately accepted by the scientific community because it was untested. Scientists are skeptical about even the most reasonable theories until experiments are done to confirm them.

In the late 1980s, Günter Wächtershäuser, a chemist at the University of Regensburg in Germany, described in more detail how amino acids and peptides could be made near hydrothermal vents. The general steps he proposed are seen on the next page.

$$H_2N-CH(CH_3)-C(=O)-N(H)-CH(CH_2OH)-C(=O)-N(H)-CH(H)-C(=O)-N(H)-CH(CH_2SH)-C(=O)-OH$$

Peptide

Special Topic 16.2 (Continued)

$$CO_2 + H_2O + H_2S + CO \rightarrow CH_3COSH$$

thioacetic acid

↓ combines with more CO

$$CH_3COCO_2H$$

pyruvic acid

↓ reacts with ammonia, $NH_3$

$$H_2NCH(CH_3)CO_2H$$

alanine, an amino acid

↓

$$H_2NCH(CH_3)CONHCH(CH_3)CO_2H$$

a dipeptide

The reaction sequence is extremely slow at normal temperatures and pressures, but Wächtershäuser suggested that hydrothermal vents not only could provide the raw materials necessary to start and maintain the reactions but also could provide the conditions necessary for relatively rapid rates of change. We know from Section 16.1 that the rates of chemical reactions can be increased by increasing the concentrations of reactants, by increasing the temperature, or by adding a catalyst. We also know from Section 16.4 that equilibrium systems can be pushed toward products by increasing the concentrations of reactants and increasing the temperature for reactions that are endothermic in the forward direction. Wächtershäuser suggested that the high temperature of the water coming out of the hydrothermal vents and the high concentrations of reactants caused by the high pressure on the ocean floor combine to speed the reactions and push the system toward products. Wächtershäuser also suggested that iron pyrite found around the vents could catalyze the reaction.

Experiments have now been done to duplicate the conditions found at hydrothermal vents, and they have confirmed that Wächtershäuser's theory is possible. In an article in the April 11, 1997, issue of *Science*, Wächtershäuser and Claudia Huber of the Technical University of Munich reported that they had produced thioacetic acid from the raw materials at 100 °C. Jay A. Brandes at the Carnegie Institution in Washington, DC, showed that this reaction takes place at the temperatures of the hydrothermal vents (350 °C) if the reactants are combined at high pressure.

One of the perceived weaknesses of Wächtershäuser's theory was the need for ammonia. Some scientists questioned whether it would be present in high enough concentrations for the conversion of pyruvic acid to amino acids, but Brandes has done experiments that show that ammonia is formed when a mixture of water and nitrogen oxides known to be found in the vents is heated to 500 °C and compressed to 500 atm. Experiments done by Brandes duplicating the conditions at hydrothermal vents also show a 40% conversion of pyruvic acid and ammonia to the amino acid alanine. Wächtershäuser and Huber have demonstrated the conversion of amino acids to peptides in the laboratory.

Do these experiments show that the first amino acids and proteins on earth were made in this way? No. There are other plausible explanations, and at this point, there is no general agreement on which explanation is best. In fact, as of this writing, Wächtershäuser's theory is a minority view, but it is gaining in popularity. The search for an explanation of the origin of biomolecules continues, each step leading us closer to a better understanding of the physical world and how it came to be.

## Chapter Glossary

**Collision theory** A model for the process of chemical change.

**Activation energy** The minimum energy necessary for reactants to form the activated complex and proceed to products.

**Rates of chemical reactions** The number of product molecules that form (perhaps described as moles of product formed) per liter of container per second.

**Concentration** The number of particles per unit volume. For gases, concentration is usually described in terms of moles of gas particles per liter of container. Substances in solution are described with molarity (moles of solute per liter of solution).

**Homogeneous catalyst** A catalyst that is in the same phase as the reactants (so that all substances are gases or all are in solution).

**Heterogeneous catalyst** A catalyst that is not in the same phase as the reactants—for example, a solid catalyst for a gas phase reaction.

**Homogeneous equilibrium** An equilibrium system in which all of the components are in the same phase (gas, liquid, solid, or aqueous).

**Equilibrium constant** A value that describes the extent to which reversible reactions proceed toward products before reaching equilibrium.

**Equilibrium constant expression** An expression showing the ratio of the concentrations (or gas pressures) of products to the concentrations (or gas pressures) of reactants for a reversible reaction at equilibrium. The concentration (or gas pressure) of each reactant and product is raised to a power equal to its coefficient in a balanced equation for the reaction. When the equilibrium constant expression includes concentrations, the symbol for the equilibrium constant is $K_C$. When it includes gas pressures, the symbol for the equilibrium constant is $K_P$.

**Heterogeneous equilibrium** An equilibrium in which the reactants and products are not all in the same phase (gas, liquid, solid, or aqueous).

**Water dissociation constant, $K_w$** The equilibrium constant for the reaction $H_2O(l) \rightleftharpoons H^+(aq) + OH^-(aq)$.

**Le Châtelier's principle** If a system at equilibrium is altered in a way that disrupts the equilibrium, the system will shift in such a way as to counter the change.

You can test yourself on the glossary terms at the following Web site:
**www.chemplace.com/college/**

## Chapter Objectives

**The goal of this chapter is to teach you to do the following.**

1. Define all of the terms in the Chapter Glossary.

### Section 16.1 Collision Theory: A Model for the Process of Chemical Reactions

2. Use collision theory to describe the general process that takes place as a system moves from reactants to products in a reaction between two reactants, such as the reaction between $O(g)$ and $O_3(g)$.

3. Explain why two particles must collide before a reaction between them can take place, referring to the role of the motion of the particles in providing energy for the reaction.
4. Explain why it is usually necessary for the new bonds in a chemical reaction's products to form at the same time as the old bonds in the reactants are broken.
5. Identify the sign (positive or negative) for the heat of reaction for an exothermic reaction.
6. Explain why a collision between reactant particles must have a certain minimum energy (activation energy) in order to proceed to products.
7. Draw an energy diagram for a typical exergonic reaction, showing the relative energies of the reactants, the activated complex, the products, and (using arrows) the activation energy and the overall energy of the reaction.
8. Identify the sign (positive or negative) for the overall energy of reaction for an endothermic reaction.
9. Draw an energy diagram for a typical endergonic reaction, showing the relative energies of the reactants, the activated complex, the products, and (using arrows) the activation energy and the overall energy of the reaction.
10. Explain why two reactant particles must collide with the correct orientation if a reaction between them is likely to take place.
11. List the three requirements that must be met before a reaction between two particles is likely to take place.

### Section 16.2 Rates of Chemical Reactions

12. Give two reasons why increased temperature usually increases the rate of chemical reactions.
13. Explain why increased concentration of a reactant leads to an increase in the rate of a chemical reaction.
14. Explain why chlorine atoms speed the conversion of ozone molecules, $O_3$, and oxygen atoms, O, into oxygen molecules, $O_2$. Include the change in the activation energy of the catalyzed reaction compared to the uncatalyzed reaction and the effect that this change has on the fraction of collisions that have the minimum energy necessary for reaction.
15. Explain why the rate of a catalyzed reaction is expected to be greater than the rate of the same reaction without the catalyst.
16. Describe the four steps thought to occur in heterogeneous catalysis.

### Section 16.3 Reversible Reactions and Chemical Equilibrium

17. Given an equation for a reversible chemical reaction, describe the changes that take place in the reaction vessel from the time when the reactants are first added to the container until a dynamic equilibrium is reached.
18. Explain why the rate of the forward reaction in a chemical change is at its peak when the reactants are first added to the container and why this rate diminishes with time.
19. Explain why the rate of the reverse reaction in a chemical change is zero when the reactants are first added to a container and why this rate increases with time.

20. With reference to the changing forward and reverse reaction rates, explain why any reversible reaction moves toward a dynamic equilibrium with equal forward and reverse reactions rates.
21. Describe the changes that take place in a chemical system at equilibrium, and explain why there are no changes in the concentrations of reactants and products in spite of these changes.
22. Given a balanced equation for a reversible chemical reaction, write its equilibrium constant, $K_C$, expression with the reactants and products described in terms of concentrations.
23. Given a balanced equation for a reversible chemical reaction that involves at least one gaseous substance, write its equilibrium constant, $K_P$, expression with the reactants and products described in terms of gas pressures.
24. Given a balanced equation for a reversible chemical change and gas pressures for reactants and products at equilibrium, calculate the reaction's $K_P$.
25. Given an equilibrium constant for a reversible chemical reaction, predict whether the reaction favors reactants, products, or neither.
26. Explain why pure solids and liquids are left out of equilibrium constant expressions for heterogeneous equilibria.
27. Explain why increased temperature drives a reversible chemical reaction in the endothermic direction.
28. Explain why increased temperature increases the equilibrium constant for a reversible chemical change that has an endothermic forward reaction and why increased temperature decreases the equilibrium constant for a change that has an exothermic forward reaction.
29. Explain why some exothermic chemical reactions are run at high temperature despite the fact that the ratio of products to reactants will be lower at equilibrium.

#### Section 16.4 Disruption of Equilibrium

30. Explain why an increase in the concentration of a reactant in a reversible reaction at equilibrium disrupts the equilibrium and shifts the system toward more products.
31. Explain why a decrease in the concentration of a product in a reversible reaction at equilibrium disrupts the equilibrium and shifts the system toward more products.
32. Given a complete, balanced equation for a reversible chemical reaction at equilibrium and being told whether it is endothermic or exothermic, predict whether each of the following changes would shift the system toward more products, more reactants, or neither: adding reactant, adding product, removing reactant, removing product, increasing temperature, decreasing temperature, or adding a catalyst.
33. Explain your answers for the previous objective in terms of Le Châtelier's principle and in terms of the effect of the changes on the forward and reverse reaction rates.
34. Explain why a catalyst has no effect on equilibrium concentrations for a reversible reaction.

## Review Questions

1. Describe what you visualize occurring inside a container of oxygen gas, $O_2$, at room temperature and pressure.
2. Write in each blank the word that best fits the definition.
   a. ____________________ is the capacity to do work.
   b. ____________________ is the capacity to do work due to the motion of an object.
   c. A(n) ____________________ change is a change that absorbs energy.
   d. A(n) ____________________ change is a change that releases energy.
   e. ____________________ is the energy associated with the random motion of particles.
   f. ____________________ is thermal energy that is transferred from a region of higher temperature to a region of lower temperature as a result of the collisions of particles.
   g. A(n) ____________________ change is a change that leads to heat energy being evolved from the system to the surroundings.
   h. A(n) ____________________ change is a change that leads the system to absorb heat energy from the surroundings.
   i. A(n) ____________________ is a substance that speeds a chemical reaction without being permanently altered itself.
3. When the temperature of the air changes from 62 °C at 4:00 A.M. to 84 °C at noon on a summer day, does the average kinetic energy of the particles in the air increase, decrease, or stay the same?
4. Explain why it takes energy to break an O–O bond in an $O_3$ molecule.
5. Explain why energy is released when two oxygen atoms come together to form an $O_2$ molecule.
6. Explain why some chemical reactions *release heat* to their surroundings.
7. Explain why some chemical reactions *absorb heat* from their surroundings.
8. What are the general characteristics of any dynamic equilibrium system?

## Key Ideas

**Complete the following statements by writing one of these words or phrases in each blank.**

| | |
|---|---|
| activated complex | kilojoules per mole, kJ/mol |
| alternative pathway | larger |
| balanced | lower activation energy |
| change | minimum |
| close to each other | moles per liter |
| collide | more quickly |
| collision | negative |
| coefficient | net |
| disrupts | new bonds |
| endergonic | no effect |
| endergonic reaction | old bonds |
| equal | orientation |
| equal to | phase |
| equilibrium constant | positive |
| exergonic | pressures |
| fraction | products |
| greater than | reactants |
| heterogeneous | released |
| homogeneous | shorter |
| increases | toward products |
| $K_P$ | transition state |

9. At a certain stage in the progress of a reaction, bond breaking and bond making are of equal importance. In other words, the energy necessary for bond breaking is ____________ by the energy supplied by bond making. At this turning point, the particles involved in the reaction are joined in a structure known as the activated complex, or ____________.

10. As a reaction continues beyond the activated complex, the energy released in bond making becomes greater than the energy necessary for bond breaking, and energy is ____________.

11. In a chemical reaction, the ____________ energy necessary for reaching the activated complex and proceeding to products is called the activation energy. Only the collisions that provide a net kinetic energy ____________ or ____________ the activation energy can lead to products.

12. The energies associated with chemical reactions are usually described in terms of ____________. If the energy is released, the value is described with a(n) ____________ sign.

13. The energies associated with endergonic (or endothermic) changes are described with ____________ values.

14. Collision theory stipulates that three requirements must be met before a reaction between two particles is likely to take place. The reactant particles must ________________. The ________________ must provide at least the minimum energy necessary to produce the ________________. The ________________ of the colliding particles must favor the formation of the activated complex, in which the new bond or bonds are able to form as the old bond or bonds break.
15. Because the formation of the ________________ provides some of the energy necessary to break the ________________, the making and breaking of bonds must occur more or less simultaneously. This is possible only when the particles collide in such a way that the bond-forming atoms are ________________.
16. In general, increased temperature ________________ the rate of chemical reactions.
17. Increased temperature means an increase in the average kinetic energy of the collisions between the particles in a system. This leads to an increase in the ________________ of the collisions that have enough energy to reach the activated complex (the activation energy).
18. Increasing the concentration of one or both reactants will lead to ________________ distances between reactant particles and to their occupying a larger percentage of the volume. Therefore, more collisions can be expected to occur per second.
19. One of the ways in which catalysts accelerate chemical reactions is by providing a(n) ________________ between reactants and products that has a(n) ________________.
20. If the reactants and the catalyst are all in the same state (all gases or all in solution) the catalyst is a(n) ________________ catalyst.
21. If the catalyst is not in the same state as the reactants, the catalyst is called a(n) ________________ catalyst.
22. In a dynamic equilibrium for a reversible chemical reaction, the forward and reverse reaction rates are ________________, so although reactants and products are constantly changing back and forth, there is no ________________ change in the amount of either.
23. The extent to which reversible reactions proceed toward products before reaching equilibrium can be described with a(n) ________________, which is derived from the ratio of the concentrations of products to the concentrations of reactants at equilibrium. For homogeneous equilibria, the concentrations of all reactants and products can be described in ________________, and the concentration of each is raised to a power equal to its ________________ in a balanced equation for the reaction.
24. Gases in equilibrium constant expressions are often described with ________________ instead of concentrations (mol/L). When an equilibrium constant is derived from gas pressures, we represent it with the symbol ________________.
25. The ________________ a value for *K*, the farther the reaction shifts toward products before the rates of the forward and reverse reactions become equal and the concentrations of reactants and products stop changing.
26. When the reactants and products for a reversible reaction are not all in the same ________________ (gas, liquid, solid, or aqueous), the equilibrium is called a heterogeneous equilibrium.

27. Changing temperature always causes a shift in equilibrium systems—sometimes toward more products and sometimes toward more ____________.

28. Increased temperature increases the rate of both the forward and the reverse reactions, but it increases the rate of the ____________ reaction more than it increases the rate of the ____________ reaction. Therefore, increasing the temperature of a chemical system at equilibrium will disrupt the balance of the forward and reverse rates of reaction and shift the system in the direction of the ____________.

29. If the forward reaction is endergonic, increased temperature will shift the system toward more ____________.

30. The increase of one or more reactants shifts a system initially at equilibrium ____________.

31. Le Châtelier's principle states that if a system at equilibrium is altered in a way that ____________ the equilibrium, the system will shift in such a way as to counter the ____________.

32. The presence of a catalyst has ____________ on the equilibrium concentrations of reactants and products. Catalysts speed both the forward and reverse rates to the same degree. The system will reach equilibrium ____________, but the ratio of the concentrations of products to reactants—and thus the value of the equilibrium constant—is unchanged.

## Chapter Problems

### Section 16.1 Collision Theory: A Model for the Reaction Process

33. Assume that the following reaction is a single-step reaction in which one of the O–O bonds in $O_3$ is broken and a new N–O bond is formed. The heat of reaction is –226 kJ/mol.

$$NO(g) + O_3(g) \rightarrow NO_2(g) + O_2(g) + 226\text{ kJ}$$

OBJECTIVE 2

a. With reference to collision theory, describe the general process that takes place as this reaction moves from reactants to products.

OBJECTIVE 11

b. List the three requirements that must be met before a reaction between $NO(g)$ and $O_3(g)$ is likely to take place.

OBJECTIVE 3

c. Explain why $NO(g)$ and $O_3(g)$ must collide before a reaction can take place.

OBJECTIVE 4

d. Explain why it is usually necessary for the new N–O bonds to form at the same time that the O–O bonds are broken.

e. Draw a rough sketch of the activated complex. (You do not need to show bond angles. Be sure to show the bond that is breaking and the bond that is being formed.)

OBJECTIVE 6

f. Explain why a collision between $NO(g)$ and $O_3(g)$ must have a certain minimum energy (activation energy) in order to proceed to products.

g. The activation energy for this reaction is 132 kJ/mol. Draw an energy diagram for this reaction, showing the relative energies of the reactants, the activated complex, the products, and (using arrows) the activation energy and heat of reaction.

OBJECTIVE 7

h. Is this reaction exothermic or endothermic?

OBJECTIVE 5 OBJECTIVE 8

i. Explain why $NO(g)$ and $O_3(g)$ molecules must collide with the correct orientation if a reaction between them is going to be likely to take place.

OBJECTIVE 10

34. Assume that the following reaction is a single-step reaction in which a C–Br bond is broken as the C–I bond is formed. The heat of reaction is +38 kJ/mol.

$$I^-(aq) + CH_3Br(aq) + 38\text{ kJ} \rightarrow CH_3I(aq) + Br^-(aq)$$

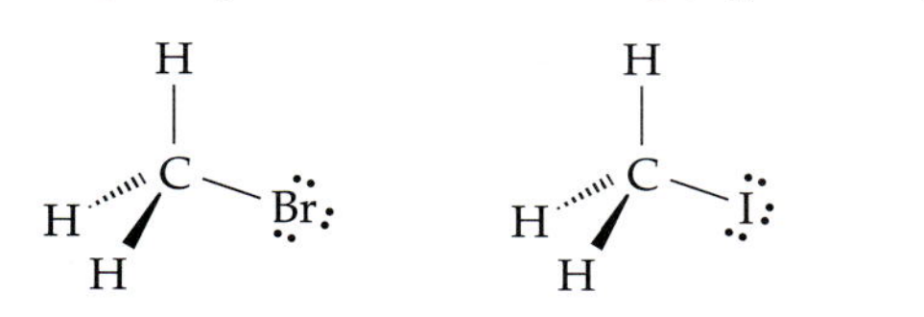

a. With reference to collision theory, describe the general process that takes place as this reaction moves from reactants to products.

OBJECTIVE 2

b. List the three requirements that must be met before a reaction between $I^-$ and $CH_3Br$ is likely to take place.

OBJECTIVE 11

c. Explain why an $I^-$ ion and a $CH_3Br$ molecule must collide before a reaction can take place.

OBJECTIVE 3

d. Explain why, in the process of this reaction, it is usually necessary for the new C–I bonds to form at the same time that the C–Br bonds are broken.

OBJECTIVE 4

e. Draw a rough sketch of the activated complex. (You do not need to show bond angles. Be sure to show the bond that is breaking and the bond that is being formed.)

f. Explain why a collision between an $I^-$ ion and a $CH_3Br$ molecule must have a certain minimum energy (activation energy) in order to proceed to products.

OBJECTIVE 6

g. The activation energy for this reaction is 76 kJ/mol. Draw an energy diagram for this reaction, showing the relative energies of the reactants, the activated complex, the products, and (using arrows) the activation energy and heat of reaction.

OBJECTIVE 7

h. Is this reaction exothermic or endothermic?

OBJECTIVE 5 OBJECTIVE 8

i. Explain why an $I^-$ ion and a $CH_3Br$ molecule must collide with the correct orientation if a reaction between them is to be likely to take place.

OBJECTIVE 10

### Section 16.2 Rates of Chemical Reactions

35. Consider the following general reaction for which gases A and B are mixed in a constant volume container.

$$A(g) + B(g) \rightarrow C(g) + D(g)$$

What happens to the rate of this reaction when

a. more gas A is added to the container?
b. the temperature is decreased?
c. a catalyst is added that lowers the activation energy?

36. Consider the following general reaction for which gases A and B are mixed in a constant volume container.

$$A(g) + B(g) \rightarrow C(g) + D(g)$$

What happens to the rate of this reaction when

a. gas B is removed from the container?
b. the temperature is increased?

37. The reactions listed below are run at the same temperature. The activation energy for the first reaction is 132 kJ/mol. The activation energy for the second reaction is 76 kJ/mol. In which of these reactions would a higher fraction of collisions between reactants have the minimum energy necessary to react (the activation energy)? Explain your answer.

$$NO(g) + O_3(g) \rightarrow NO_2(g) + O_2(g) \quad \text{Activation energy} = 132\ \text{kJ}$$

$$I^-(aq) + CH_3Br(aq) \rightarrow CH_3I(aq) + Br^-(aq) \quad \text{Activation energy} = 76\ \text{kJ}$$

38. Consider the following reaction.

$$NO(g) + O_3(g) \rightarrow NO_2(g) + O_2(g)$$

OBJECTIVE 12

a. Explain why increased temperature increases the rate of reaction.

OBJECTIVE 13

b. Will an increase in the concentration of $NO(g)$ increase, decrease, or not affect the rate of the reaction? Explain your answer.

39. Two reactions can be described by the energy diagrams below. What is the approximate activation energy for each reaction? Which reaction is exothermic and which is endothermic?

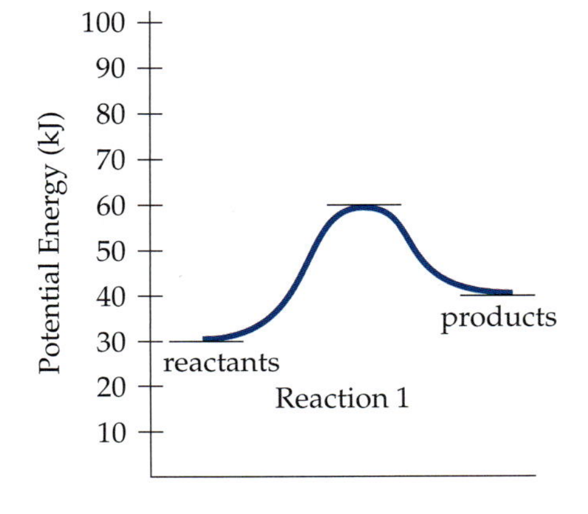


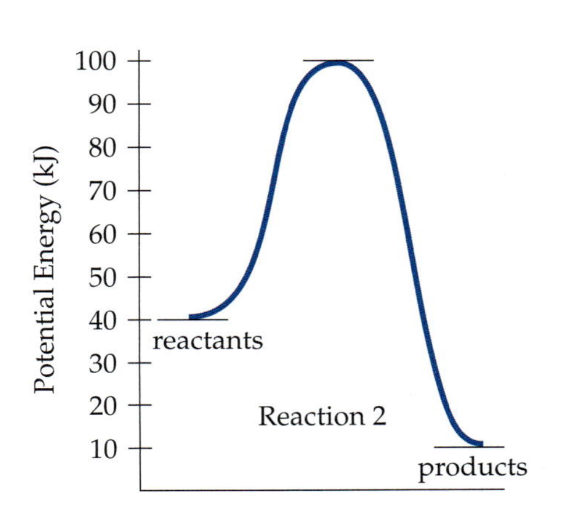

40. If both systems described by the energy diagrams in the previous problem are at the same temperature, if the concentrations of initial reactants are equivalent for each reaction, and if the orientation requirements for each reaction are about the same, which of these reactions would you expect to have the greatest forward reaction rate? Why?

41. Explain why chlorine atoms speed the conversion of ozone molecules, $O_3$, and oxygen atoms, O, into oxygen molecules, $O_2$. OBJECTIVE 14

42. Write a general explanation of why the rate of a catalyzed reaction would be expected to be greater than the rate of the same reaction without the catalyst. OBJECTIVE 15

43. Using the proposed mechanism for the conversion of $NO(g)$ into $N_2(g)$ and $O_2(g)$ as an example, write a description of the four steps thought to occur in heterogeneous catalysis. OBJECTIVE 16

**Section 16.3 Reversible Reactions and Chemical Equilibrium**

44. Reversible chemical reactions lead to dynamic equilibrium states. What is *dynamic* about these states? Why are they called equilibrium states?

45. Equilibrium systems have two opposing rates of change that are equal. For each of the following equilibrium systems that were mentioned in earlier chapters, describe what is changing in the two opposing rates.
    a. a solution of the weak acid acetic acid, $HC_2H_3O_2$ (Chapter 5)
    b. pure liquid in a closed container (Chapter 14)
    c. a closed bottle of carbonated water with 4 atm of $CO_2$ in the gas space above the liquid (Chapter 15)

46. Equilibrium systems have two opposing rates of change that are equal. For each of the following equilibrium systems that were mentioned in earlier chapters, describe what is changing in the two opposing rates.
    a. a solution of the weak base ammonia, $NH_3$ (Chapter 5)
    b. saturated solution of table salt, NaCl, with excess solid NaCl (Chapter 15)

47. Two gases, A and B, are added to an empty container. They react in the following reversible reaction.

$$A(g) + B(g) \rightleftharpoons C(g) + D(g)$$

    a. When is the forward reaction rate greatest: (1) when A and B are first mixed, (2) when the reaction reaches equilibrium, or (3) sometime between these two events?
    b. When is the reverse reaction rate greatest: (1) when A and B are first mixed, (2) when the reaction reaches equilibrium, or (3) sometime between these two events?

OBJECTIVE 17 OBJECTIVE 18 OBJECTIVE 19

48. Assume that in the following reversible reaction, both the forward and the reverse reactions take place in a single step.

$$NO(g) + O_3(g) \rightleftharpoons NO_2(g) + O_2(g)$$

Describe the changes that take place in the reaction vessel from the time when $NO(g)$ and $O_3(g)$ are first added to the container until a dynamic equilibrium is reached (including the changes in concentrations of reactants and products and the changes in the forward and reverse reaction rates).

49. Assume that in the following reversible reaction, both the forward and the reverse reactions take place in a single step.

$$I^-(aq) + CH_3Br(aq) \rightleftharpoons CH_3I(aq) + Br^-(aq)$$

OBJECTIVE 20

a. With reference to the changing forward and reverse reaction rates, explain why this reaction moves toward a dynamic equilibrium with equal forward and reverse reaction rates.

OBJECTIVE 21

b. Describe the changes that take place once the reaction reaches an equilibrium state. Are there changes in the concentrations of reactants and products at equilibrium? Explain your answer.

OBJECTIVE 22 OBJECTIVE 23

50. Write $K_C$ and $K_P$ expressions for each of the following equations.

a. $2CH_4(g) \rightleftharpoons C_2H_2(g) + 3H_2(g)$

b. $2N_2O(g) + O_2(g) \rightleftharpoons 4NO(g)$

c. $Sb_2S_3(s) + 3H_2(g) \rightleftharpoons 2Sb(s) + 3H_2S(g)$

OBJECTIVE 22 OBJECTIVE 23

51. Write $K_C$ and $K_P$ expressions for each of the following equations.

a. $4CuO(s) \rightleftharpoons 2Cu_2O(s) + O_2(g)$

b. $CH_4(g) + 4Cl_2(g) \rightleftharpoons CCl_4(l) + 4HCl(g)$

c. $2H_2S(g) + CH_4(g) \rightleftharpoons CS_2(g) + 4H_2(g)$

OBJECTIVE 24

52. A mixture of nitrogen dioxide and dinitrogen tetroxide is allowed to come to equilibrium at 30 °C, and the gases' partial pressures are found to be 1.69 atm $N_2O_4$ and 0.60 atm $NO_2$.

a. On the basis of these data, what is $K_P$ for the following equation?

$$NO_2(g) \rightleftharpoons \tfrac{1}{2}N_2O_4(g)$$

b. Based on this data, what is $K_P$ for the following equation?

$$2NO_2(g) \rightleftharpoons N_2O_4(g)$$

c. Table 16.1 lists the $K_P$ for the equation in part (b) as 6.7 at 25 °C. Explain why your answer to part (b) is not 6.7.

53. A mixture of nitrogen gas, hydrogen gas, and ammonia gas is allowed to come to equilibrium at 200 °C, and the gases' partial pressures are found to be 0.61 atm $N_2$, 1.1 atm $H_2$, and 0.52 atm $NH_3$.

OBJECTIVE 24

a. On the basis of these data, what is $K_P$ for the following equation?

$$N_2(g) + 3H_2(g) \rightleftharpoons 2NH_3(g)$$

b. On the basis of these data, what is $K_P$ for the following equation?

$$\tfrac{1}{2}N_2(g) + \tfrac{3}{2}H_2(g) \rightleftharpoons NH_3(g)$$

54. Predict whether each of the following reactions favors reactants, products, or neither at the temperature for which the equilibrium constant is given.

OBJECTIVE 25

a. $CH_3OH(g) + CO(g) \rightleftharpoons CH_3CO_2H(g)$ $K_P = 1.2 \times 10^{-22}$ at 25 °C

b. $CH_4(g) + 4Cl_2(g) \rightleftharpoons CCl_4(g) + 4HCl(g)$ $K_P = 3.3 \times 10^{68}$ at 25 °C

c. $CO(g) + Cl_2(g) \rightleftharpoons COCl_2(g)$ $K_P = 0.20$ at 600 °C

55. Predict whether each of the following reactions favors reactants, products, or neither at the temperature for which the equilibrium constant is given.

OBJECTIVE 25

a. $2COF_2(g) \rightleftharpoons CO_2(g) + CF_4(g)$ $K_P = 2$ at 1000 °C

b. $\tfrac{1}{8}S_8(s) + O_2(g) \rightleftharpoons SO_2(g)$ $K_P = 4.2 \times 10^{52}$ at 25 °C

c. $C_2H_6(g) \rightleftharpoons C_2H_4(g) + H_2(g)$ $K_P = 1.2 \times 10^{-18}$ at 25 °C

56. Write the $K_C$ expression for the following equation. Explain why the concentration of $CH_3OH$ is left out of the expression.

OBJECTIVE 22 OBJECTIVE 26

$$CO(g) + 2H_2(g) \rightleftharpoons CH_3OH(l)$$

57. Write the $K_C$ expression for the following equation. Explain why the concentration of $NH_4HS$ is left out of the expression.

OBJECTIVE 22 OBJECTIVE 26

$$NH_4HS(s) \rightleftharpoons NH_3(g) + H_2S(g)$$

58. Ethylene, $C_2H_4$, is one of the organic substances found in the air we breathe. It reacts with ozone in an endothermic reaction to form formaldehyde, $CH_2O$, which is one of the substances in smoggy air that cause eye irritation.

$$2C_2H_4(g) + 2O_3(g) + \text{energy} \rightleftharpoons 4CH_2O(g) + O_2(g)$$

a. Why does the forward reaction take place more rapidly in Los Angeles than in a wilderness area of Montana with the same air temperature?

b. For a variety of reasons, natural systems rarely reach equilibrium, but if this reaction was run in the laboratory, would increased temperature for the reaction at equilibrium shift the reaction to more reactants or more products?

59. When the temperature of an equilibrium system for the following reaction is increased, the reaction shifts toward more products. Is the forward reaction endothermic or exothermic?

$$2NOBr(g) \rightleftharpoons 2NO(g) + Br_2(l)$$

60. When the temperature of an equilibrium system for the following reaction is increased, the reaction shifts toward more reactants. Is the forward reaction endothermic or exothermic?

$$H_2(g) + Br_2(g) \rightleftharpoons 2HBr(g)$$

61. A lightning bolt creates a large quantity of nitrogen monoxide in its path by combining nitrogen gas and oxygen gas.

$$N_2(g) + O_2(g) + 180.5 \text{ kJ} \rightleftharpoons 2NO(g)$$

a. Explain why lightning increases the amount of $NO(g)$ produced by this reaction. (*Hint:* Lightning generates a lot of heat.)

b. Would increased temperature increase or decrease the equilibrium constant for this reaction?

62. Assume that you are picking up a few extra dollars to pay for textbooks by acting as a trainer's assistant for a heavyweight boxer. One of your jobs is to wave smelling salts under the nose of the fighter to clear his head between rounds. The smelling salts are ammonium carbonate, which decomposes in the following reaction to yield ammonia. The ammonia does the wakeup job. Suppose the fighter gets a particularly nasty punch to the head and needs an extra jolt to be brought back to his senses. How could you shift the following equilibrium to the right to yield more ammonia?

$$(NH_4)_2CO_3(s) + \text{energy} \rightleftharpoons 2NH_3(g) + CO_2(g) + H_2O(g)$$

OBJECTIVE 27 OBJECTIVE 28

63. Ethylene, $C_2H_4$, which is used to make polyethylene plastics, can be made from ethane, $C_2H_6$, one of the components of natural gas. The heat of reaction for the decomposition of ethane gas into ethylene gas and hydrogen gas is 136.94 kJ per mole of $C_2H_4$ formed, so the reaction is endothermic. The reaction is run at high temperature, in part because at 800–900 °C, the equilibrium constant for the reaction is much higher, indicating that a higher percentage of products forms at this temperature. Explain why increased temperature drives this reversible chemical reaction in the endergonic direction and why this leads to an increase in the equilibrium constant for the reaction.

$$C_2H_6(g) \rightleftharpoons C_2H_4(g) + H_2(g)$$

64. Formaldehyde, $CH_2O$, is one of the components of embalming fluids and has been used to make foam insulation and plywood. It can be made from methanol, $CH_3OH$ (often called wood alcohol). The heat of reaction for the combination of gaseous methanol and oxygen gas to form gaseous formaldehyde and water vapor is –199.32 kJ per mole of $CH_2O$ formed, so the reaction is exothermic.

$$2CH_3OH(g) + O_2(g) \rightleftharpoons CH_2O(g) + 2H_2O(g)$$

OBJECTIVE 27 OBJECTIVE 28

a. Increased temperature drives the reaction toward reactants and lowers the value for the equilibrium constant. Explain why this is true.

OBJECTIVE 29

b. This reaction is run by the chemical industry at 450–900 °C, even though the equilibrium ratio of product to reactant concentrations is lower than at room temperature. Explain why this exothermic chemical reaction is run at high temperature despite this fact.

65. Is the equilibrium constant for a particular reaction dependent on
    a. the initial concentration of reactants?
    b. the initial concentration of products?
    c. the temperature for the reaction?

### Section 16.4 Disruption of Equilibrium

66. Urea, $NH_2CONH_2$, is an important substance in the production of fertilizers. The equation shown below describes an industrial reaction that produces urea. The heat of reaction is –135.7 kJ per mole of urea formed. Predict whether each of the following changes in the equilibrium system will shift the system to more products, to more reactants, or to neither. Explain each answer in two ways: (1) by applying Le Châtelier's principle and (2) by describing the effect of the change on the forward and reverse reaction rates.

OBJECTIVE 30 OBJECTIVE 31 OBJECTIVE 32 OBJECTIVE 33

$$2NH_3(g) + CO_2(g) \rightleftharpoons NH_2CONH_2(s) + H_2O(g) + 135.7 \text{ kJ}$$

    a. The concentration of $NH_3$ is increased by the addition of more $NH_3$. (In the industrial production of urea, an excess of ammonia is added so that the ratio of $NH_3$ to $CO_2$ is 3:1.)
    b. The concentration of $H_2O(g)$ is decreased by removing water vapor.
    c. The temperature is increased from 25 °C to 190 °C. (In the industrial production of urea, ammonia and carbon dioxide are heated to 190 °C.)

67. Acetic acid, which is used to make many important compounds, is produced from methanol and carbon monoxide (which are themselves both derived from methane in natural gas) by a process called the Monsanto process. This endothermic reaction is run over a rhodium and iodine catalyst at 175 °C and 1 atm of pressure. Predict whether each of the following changes in the equilibrium system will shift the system to more products, to more reactants, or to neither. Explain each answer in two ways: (1) by applying Le Châtelier's principle and (2) by describing the effect of the change on the forward and reverse reaction rates.

OBJECTIVE 30 OBJECTIVE 31 OBJECTIVE 32 OBJECTIVE 33 OBJECTIVE 34

$$CH_3OH(g) + CO(g) + 207.9 \text{ kJ} \underset{\substack{175\ °C \\ 1\ \text{atm}}}{\overset{Rh/I_2}{\rightleftharpoons}} CH_3CO_2H(g)$$

    a. The concentration of CO is increased by the addition of more CO.
    b. The concentration of $CH_3OH$ is decreased.
    c. The concentration of $CH_3CO_2H(g)$ is decreased by removing the acetic acid as it forms.
    d. The temperature is decreased from 300 °C to 175 °C.
    e. The $Rh/I_2$ catalyst is added to the equilibrium system.

68. Hydriodic acid, which is used to make pharmaceuticals, is made from hydrogen iodide. The hydrogen iodide is made from hydrogen gas and iodine gas in the following exothermic reaction.

$$H_2(g) + I_2(g) \rightleftharpoons 2HI(g) + 9.4 \text{ kJ}$$

What changes could you make for this reaction at equilibrium to shift the reaction to the right and maximize the concentration of hydrogen iodide in the final product mixture?

69. Nitrosyl chloride is used to make synthetic detergents. It decomposes to form nitrogen monoxide and chlorine gas.

$$2NOCl(g) + \text{energy} \rightleftharpoons 2NO(g) + Cl_2(g)$$

Without directly changing the concentration of NOCl, what changes could you make for this reaction at equilibrium to shift the reaction to the left and minimize the decomposition of the NOCl?

70. Phosgene gas, $COCl_2$, which is a very toxic substance used to make pesticides and herbicides, is made by passing carbon monoxide gas and chlorine gas over solid carbon, which acts as a catalyst.

$$CO(g) + Cl_2(g) \rightleftharpoons COCl_2(g)$$

If the carbon monoxide concentration is increased by adding CO to an equilibrium system of this reaction, what effect, if any, does it have on the following? (Assume constant temperature.)

a. The concentration of $COCl_2$ after the system has shifted to come to a new equilibrium.

b. The concentration of $Cl_2$ after the system has shifted to come to a new equilibrium.

c. The equilibrium constant for the reaction.

71. The fertilizer ammonium carbamate decomposes in the following reaction.

$$NH_4CO_2NH_2(s) \rightleftharpoons 2NH_3(g) + CO_2(g)$$

If the ammonia concentration is increased by adding $NH_3$ to an equilibrium system of this reaction, what effect, if any, does it have on the following? (Assume constant temperature.)

a. The concentration of $CO_2$ after the system has shifted to come to a new equilibrium.

b. The equilibrium constant for the reaction.

### Discussion Problem

72. Illustrate the following characteristics of a reversible chemical change by developing an analogy based on a new college library.
    a. When reactants are first added to a container, the rate of the forward reaction is at a peak, and then it decreases steadily with time.
    b. If there are no products initially, the rate of the reverse reaction is zero at the instant that reactants are added to the container, but the reverse rate of change increases steadily as the reaction proceeds.
    c. At some point, the forward and reverse rates become equal, and if the system is not disrupted, the rates remain equal. This leads to a constant amount of reactants and products.
    d. The addition of one or more of the reactants to a system at equilibrium drives the system to produce more products.
    e. The removal of one or more of the products from a system at equilibrium also drives the system toward more products.

*Chemicals provide the energy necessary for all of our activities.*

# An Introduction to Organic Chemistry, Biochemistry, and Synthetic Polymers

CHAPTER 17



It's Friday night, and you don't feel like cooking so you head for your favorite eatery, the local 1950s-style diner. There you spend an hour talking and laughing with friends while downing a double hamburger, two orders of fries, and the thickest milkshake in town. After the food has disappeared, you're ready to dance the night away at a nearby club.

What's in the food that gives you the energy to talk, laugh, and dance? How do these substances get from your mouth to the rest of your body, and what happens to them once they get there? The branch of chemistry that answers these questions and many more is called **biochemistry,** the chemistry of biological systems. Because the scope of biochemistry is huge, we will attempt no more than a glimpse of it here by tracing some of the chemical and physical changes that food undergoes in your body. You will encounter the kinds of questions that biochemists ask and will see some of the answers that they provide. Because chemicals that are important to biological systems are often organic, or carbon-based, compounds, we start this chapter with an introduction to organic chemistry.

*How does the body manage to benefit from the nutrients in the foods we eat?*

It's not always apparent to the naked eye, but the structures of many plastics and synthetic fabrics are similar to the structures of biological substances. In fact, nylon was purposely developed to mimic the structural characteristics of protein. The last section in this chapter shows you how these substances are similar and how synthetic polymers are made and used.

## Review Skills

The presentation of information in this chapter assumes that you can already perform the tasks listed below. You can test your readiness to proceed by answering the Review Questions at the end of the chapter. This might also be a good time to read the Chapter Objectives, which precede the Review Questions.

- Give a general description of the information provided in a Lewis structure. (Section 3.3)
- Describe the information given by a space-filling model, a ball-and-stick model, and a geometric sketch. (Section 3.3)
- Given a Lewis structure or enough information to write one, draw a geometric sketch of the molecule, including bond angles (or approximate bond angles). (Section 12.4)

## 17.1 Organic Compounds

These men are surrounded by many different types of organic compounds, including alkanes and ether in their gas tank and carboxylic acid, aldehyde, ester, and arene on one man's spinach salad.

Two co-workers at a pharmaceutical company, John and Stuart, jump into John's car at noon to drive four blocks to get some lunch. The gasoline that fuels the car is composed of many different organic compounds, including some that belong to the category of organic compounds called *alkanes* and a fuel additive called methyl t-butyl *ether* (MTBE). When they get to the restaurant, Stuart orders a spinach and fruit salad. The spinach contains a *carboxylic acid* called oxalic acid, and the odor from the orange and pineapple slices is due in part to the *aldehyde* 3-methylbutanal and the *ester* ethyl butanoate. The salad dressing is preserved with BHT, which is an *arene*. John orders fish, but he sends it back. The smell of the *amine* called trimethylamine let him know that it was spoiled.

The number of natural and synthetic organic, or carbon-based, compounds runs into the millions. Fortunately, the task of studying them is not so daunting as their number would suggest, because organic compounds can be categorized according to structural similarities that lead to similarities in the compounds' important properties. For example, you discovered in Section 3.3 that alcohols are organic compounds possessing one or more –OH groups attached to a hydrocarbon group (a group that contains only carbon and hydrogen). Because of this structural similarity, all alcohols share certain chemical characteristics. Chemists are therefore able to describe the properties of alcohols in general, which is a great deal simpler than describing each substance individually.

After reading this section, you too will know how to recognize and describe alkanes, ethers, carboxylic acids, aldehydes, esters, arenes, amines, and other types of organic compounds.

### Formulas for Organic Compounds

Organic (carbon-based) compounds are often much more complex than inorganic compounds, so it is more difficult to deduce their structures from their chemical formulas. Moreover, many organic formulas represent two or more isomers, each with a Lewis structure of its own (Section 12.2). The formula $C_6H_{14}O$, for example, has numerous isomers, including

$$
\begin{array}{ccccccccccccccc}
 & & H & & H & & H & & H & & & & H & & H \\
 & & | & & | & & | & & | & & & & | & & | \\
H & - & C & - & C & - & C & - & C & - & \ddot{\underset{..}{O}} & - & C & - & C & - H \\
 & & | & & | & & | & & | & & & & | & & | \\
 & & H & & H & & H & & H & & & & H & & H
\end{array}
$$

Butyl ethyl ether

$$
\begin{array}{ccccccccccccccc}
 & & & & H & & H & & H & & H & & H & & H \\
 & & & & | & & | & & | & & | & & | & & | \\
H & - & \ddot{\underset{..}{O}} & - & C & - & C & - & C & - & C & - & C & - & C - H \\
 & & & & | & & | & & | & & | & & | & & | \\
 & & & & H & & H & & H & & H & & H & & H
\end{array}
$$

1-Hexanol

$$
\begin{array}{ccccccccccccc}
 & & & & & & :\ddot{O}-H & & & & & & \\
 & & H & & H & & | & & H & & H & & H \\
 & & | & & | & & | & & | & & | & & | \\
H & - & C & - & C & - & C & - & C & - & C & - & C - H \\
 & & | & & | & & | & & | & & | & & | \\
 & & H & & H & & H & & H & & H & & H
\end{array}
$$

3-Hexanol

Chemists have developed ways of writing organic formulas so as to describe their structures as well. For example, the formula for butyl ethyl ether can be written $CH_3CH_2CH_2CH_2OCH_2CH_3$, and the formula for 1-hexanol can be written $HOCH_2CH_2CH_2CH_2CH_2CH_3$ to show the order of the atoms in the structure.

```
   H  H  H  H       H  H                 H  H  H  H  H  H
   |  |  |  |  ..   |  |                 |  |  |  |  |  |
H—C—C—C—C—O—C—C—H          H—O—C—C—C—C—C—C—H
   |  |  |  |  ..   |  |          ..   |  |  |  |  |  |
   H  H  H  H       H  H                 H  H  H  H  H  H
```

$CH_3CH_2CH_2CH_2OCH_2CH_3$ $HOCH_2CH_2CH_2CH_2CH_2CH_3$

OBJECTIVE 2

Formulas like these that serve as a collapsed or condensed version of a Lewis structure are often called *condensed formulas* (even though they are longer than the molecular formulas). To simplify these formulas, the repeating $–CH_2–$ groups can be represented by $CH_2$ in parentheses followed by a subscript indicating the number of times it is repeated. In this convention, butyl ethyl ether becomes $CH_3(CH_2)_3OCH_2CH_3$, and 1-hexanol becomes $HOCH_2(CH_2)_4CH_3$.

The position of the –OH group in 3-hexanol can be shown with the condensed formula $CH_3CH_2CH(OH)CH_2CH_2CH_3$. The parentheses, which are often left out, indicate the location at which the –OH group comes off the chain of carbon atoms. According to this convention, the group in parentheses is attached to the carbon that precedes it in the condensed formula.

```
           ..
          :O—H
   H  H   |   H  H  H
   |  |   |   |  |  |
H—C—C—C—C—C—C—H
   |  |   |   |  |  |
   H  H   H   H  H  H
```

$CH_3CH_2CH(OH)CH_2CH_2CH_3$

Although Lewis structures are useful for describing the bonding within molecules, they can be time-consuming to draw, and they do not show the spatial relationships of the atoms well. For example, the Lewis structure of butyl ethyl ether seems to indicate that the bond angles around each carbon atom are either 90° or 180° and that the carbon atoms lie in a straight line. In contrast, the ball-and-stick and space-filling models in Figure 17.1 show that the angles are actually about 109° and that the carbons are in a zigzag arrangement. The highly simplified depiction known as a *line drawing,* which was introduced in Chapter 15, shows an organic structure's geometry better than a Lewis structure does and takes much less time to draw. Remember that in a line drawing, each corner represents a carbon, each line represents a bond (a double line is a double bond), and an end of a line without another symbol attached also represents a carbon. We assume that there are enough hydrogen atoms attached to each carbon to yield a total of four bonds.

OBJECTIVE 2

Study Figures 17.1, 17.2, and 17.3 and then practice converting Lewis structures into condensed formulas and line drawings, and vice versa.

**Figure 17.1**
Ways to Describe Butyl Ethyl Ether

OBJECTIVE 2

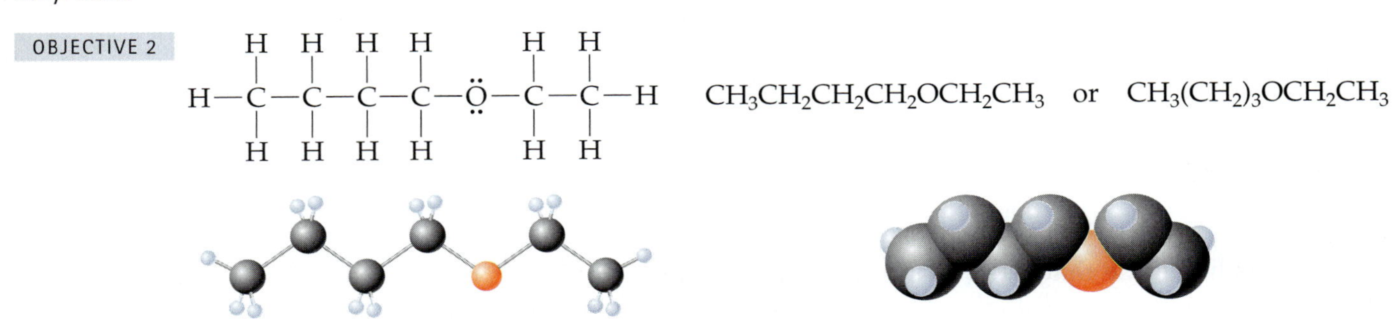


Carbon atoms with 2 hydrogen atoms attached

O

Carbon atoms with 3 hydrogen atoms attached

**Figure 17.2**
Ways to Describe 1-Hexanol

OBJECTIVE 2

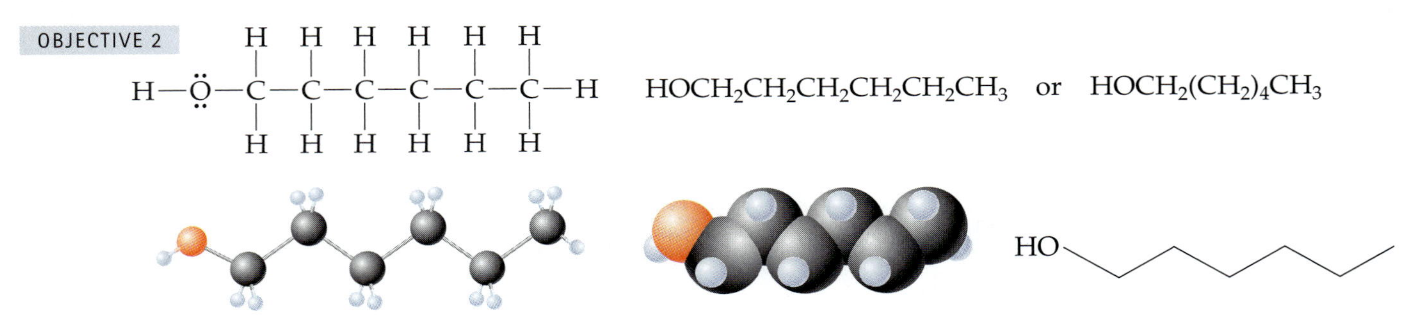


**Figure 17.3**
Ways to Describe 3-Hexanol

OBJECTIVE 2

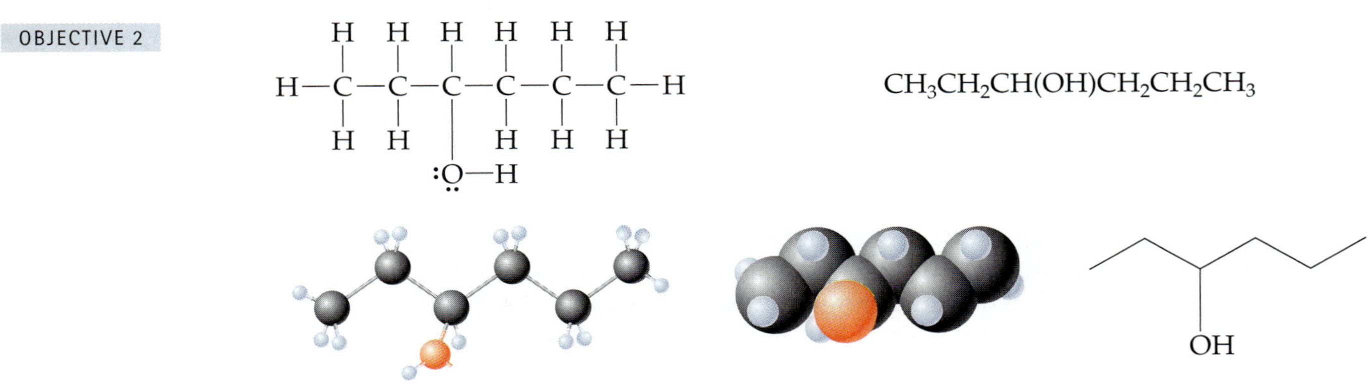

The remainder of this section lays a foundation for your future study of organic chemistry by offering brief descriptions of some of the most important families of organic compounds. Table 17.1 on page 734 provides a summary of these descriptions.

This is not the place to describe the process of naming organic compounds, which is much more complex than naming inorganic compounds, except to say that many of the better-known organic substances have both a systematic and a common name. In the examples that follow, the first name presented reflects the rules set up by the International Union of Pure and Applied Chemistry (IUPAC). Any alternative names are given in parentheses. Thereafter, we will refer to the compound by whichever name is more frequently used by chemists.

## Alkanes

Hydrocarbons (compounds composed of carbon and hydrogen) in which all of the carbon–carbon bonds are single bonds are called **alkanes.** An example is 2,2,4-trimethylpentane (or isooctane), depicted in Figure 17.4. To show that two methyl groups, $-CH_3$, come off the second carbon atom and that another comes off the fourth carbon atom, its formula can be described as $CH_3C(CH_3)_2CH_2CH(CH_3)CH_3$ or $(CH_3)_3CCH_2CH(CH_3)_2$.

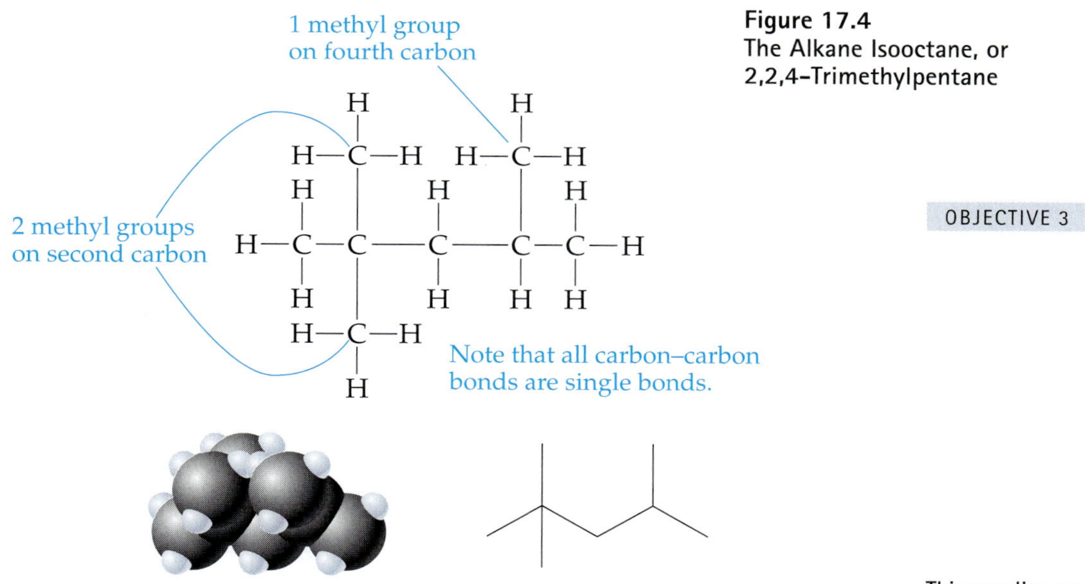


**Figure 17.4**
The Alkane Isooctane, or 2,2,4-Trimethylpentane

OBJECTIVE 3

Isooctane is used as a standard of comparison in the rating of gasoline. The "octane rating" you see at the gas pump is an average of a "research octane" value, R, determined under laboratory conditions and a "motor octane" value, M, based on actual road operation. Gasoline that has a research octane rating of 100 runs a test engine as efficiently as a fuel that is 100% isooctane. A gasoline that runs a test engine as efficiently (or, rather, as inefficiently) as 100% n-heptane, $CH_3(CH_2)_5CH_3$, has a zero research octane rating. A gasoline that has a research octane rating of 80 runs a test engine as efficiently as a mixture of 80% isooctane and 20% n-heptane.

This gasoline pump shows the octane rating for the gasoline it pumps.

## Alkenes

Hydrocarbons that have one or more carbon–carbon double bonds are called **alkenes.** The alkene 2-methylpropene (isobutene), $CH_2C(CH_3)CH_3$ or $CH_2C(CH_3)_2$, is used to make many other substances, including the gasoline additive MTBE and the antioxidant BHT (Figure 17.5).

All alkenes have very similar chemical and physical properties, which are determined primarily by the carbon–carbon double bond. When a small section of an organic molecule is largely responsible for the molecule's chemical and physical characteristics, that section is called a **functional group.**

**Figure 17.5**
The Alkene 2-Methylpropene (Isobutene)

OBJECTIVE 3

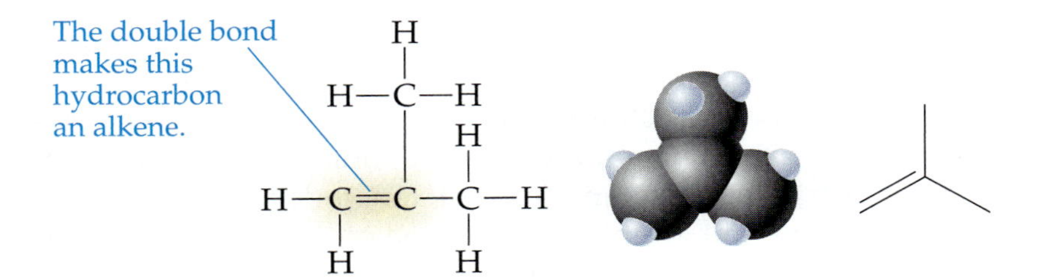

The double bond makes this hydrocarbon an alkene.

The energy necessary to weld metals with an oxyacetylene torch comes from the combustion of acetylene.

## Alkynes

Hydrocarbons that have one or more carbon–carbon triple bonds are called **alkynes.** The most common alkyne is ethyne (or acetylene), $C_2H_2$ (Figure 17.6). It is the gas used in oxyacetylene torches.

**Figure 17.6**
The Alkyne Acetylene, or Ethyne

OBJECTIVE 3

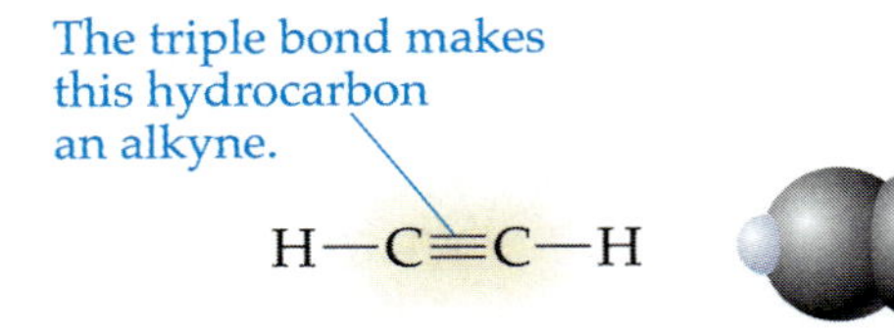

The triple bond makes this hydrocarbon an alkyne.
H—C≡C—H

## Arenes (Aromatics)

Benzene, $C_6H_6$, has six carbon atoms arranged in a ring.

or

Benzene

Compounds that contain the benzene ring are called **arenes** or **aromatics.** There are many important arenes, including butylated hydroxytoluene (BHT), which is an antioxidant commonly added to food containing fats and oils, and trinitrotoluene (TNT), the explosive (Figure 17.7).

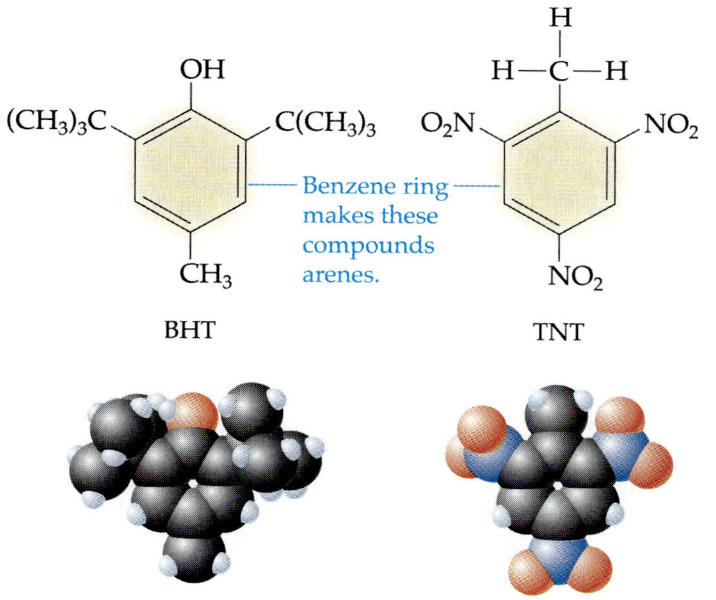

OH
$(CH_3)_3C$ $C(CH_3)_3$
$CH_3$
BHT

H
H—C—H
$O_2N$ $NO_2$
$NO_2$
TNT

Benzene ring makes these compounds arenes.

**Figure 17.7**
The Arenes Butylated Hydroxytoluene, BHT, and Trinitrotoluene, TNT

OBJECTIVE 3

## Alcohols

As you learned in Chapter 3, **alcohols** are compounds with one or more –OH groups attached to a hydrocarbon group—that is, to a group consisting of only carbon and hydrogen atoms. We have encountered methanol (methyl alcohol), $CH_3OH$, and ethanol (ethyl alcohol), $C_2H_5OH$, in earlier chapters; 2-propanol (isopropyl alcohol), $(CH_3)_2CH(OH)$, is a common rubbing alcohol, and 1,2-ethanediol (ethylene glycol), $HOCH_2CH_2OH$, is a common coolant and antifreeze.

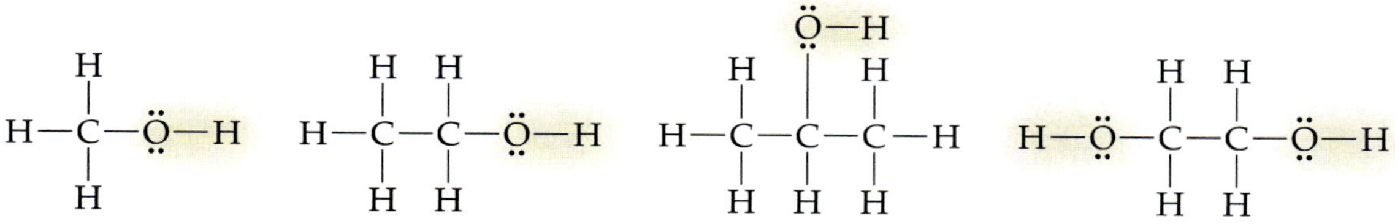

Methanol

Ethanol

2-Propanol

Ethylene glycol

The alcohol 1,2,3-propanetriol (which is more commonly called glycerol or glycerin), $HOCH_2CH(OH)\ CH_2OH$, is used as an emollient (smoother) and demulcent (softener) in cosmetics and as an antidrying agent in toothpaste and tobacco (Figure 17.8).

**Figure 17.8**
Glycerol, an Alcohol

OBJECTIVE 3

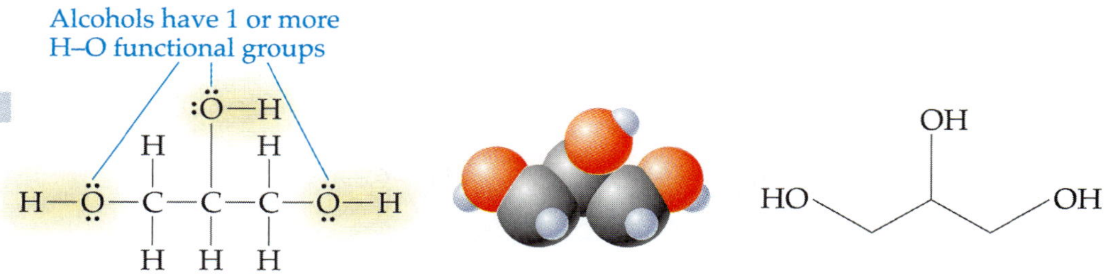

Alcohols have 1 or more H–O functional groups

## Carboxylic Acids

**Carboxylic acids** are organic compounds that have the general formula

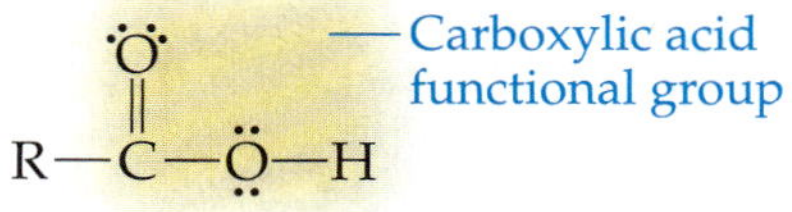

Carboxylic acid functional group

in which R represents either a hydrocarbon group (consisting entirely of carbon and hydrogen atoms) or a hydrogen atom. The carboxylic acid functional group can be written as –COOH or $–CO_2H$. Methanoic acid (formic acid), HCOOH or $HCO_2H$, is the substance that causes ant bites to sting and itch. Ethanoic acid (acetic acid), which is written as $CH_3COOH$, $CH_3CO_2H$, or $HC_2H_3O_2$, is the substance that gives vinegar its sour taste. Butanoic acid (butyric acid), $CH_3CH_2CH_2COOH$ or $CH_3CH_2CH_2CO_2H$, is the substance that gives rancid butter its awful smell. Oxalic acid, HOOCCOOH or $HO_2CCO_2H$, which has two carboxylic acid functional groups, is found in leafy green plants such as spinach. Stearic acid, $CH_3(CH_2)_{16}COOH$ or $CH_3(CH_2)_{16}CO_2H$, is a natural fatty acid found in beef fat (Figure 17.9).

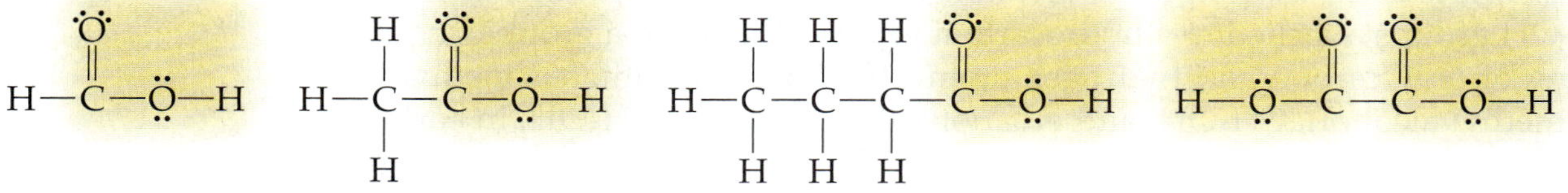

Formic acid

Acetic acid

Butanoic acid

Oxalic acid

The Lewis structure for stearic acid can be condensed to

OBJECTIVE 3

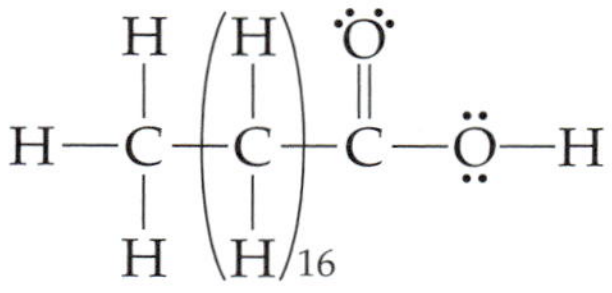

Stearic acid

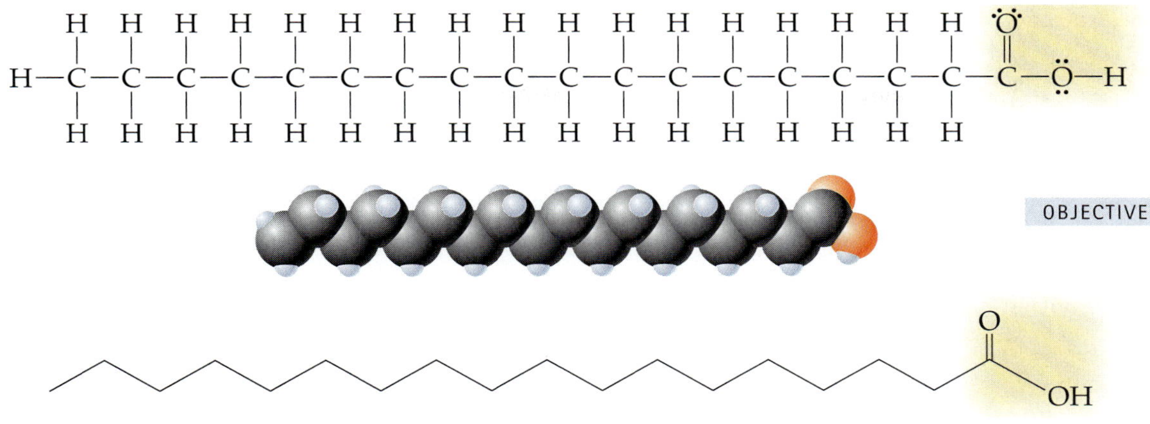

OBJECTIVE 3

**Figure 17.9**
Stearic Acid, a Carboxylic Acid

## Ethers

**Ethers** consist of two hydrocarbon groups surrounding an oxygen atom. One important ether is diethyl ether, $CH_3CH_2OCH_2CH_3$, which is used as an anesthetic. A group with a condensed formula of $CH_3CH_2–$ is called an ethyl group and is often described as $C_2H_5–$, so the formula of diethyl ether can also be $C_2H_5OC_2H_5$ or $(C_2H_5)_2O$ (Figure 17.10).

**Figure 17.10**
Diethyl Ether

OBJECTIVE 3

The ether tert-butyl methyl ether (methyl t-butyl ether or MTBE), $CH_3OC(CH_3)_3$, is added to gasoline to boost its octane rating.

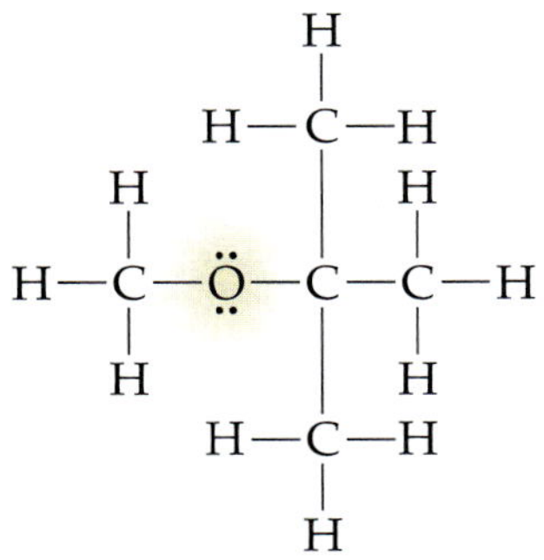

MTBE

## Aldehydes

Compounds called **aldehydes** have the general structure

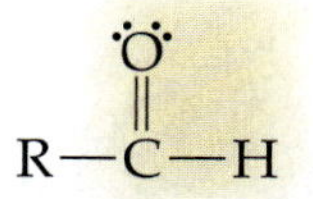

$$R-\overset{\overset{\large\ddot{O}}{\|}}{C}-H$$

Aldehyde

R can be a hydrogen atom or a hydrocarbon group. An aldehyde's functional group is usually represented by –CHO in condensed formulas. The simplest aldehyde is formaldehyde, HCHO, which has many uses, including the manufacture of polymeric resins.

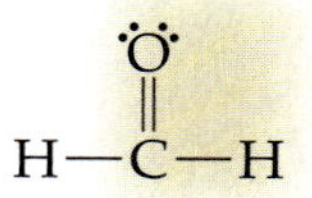

$$H-\overset{\overset{\large\ddot{O}}{\|}}{C}-H$$

Formaldehyde

Natural aldehydes contribute to the pleasant odors of food. For example, 3-methylbutanal (isovaleraldehyde), $(CH_3)_2CHCH_2CHO$, is found in oranges, lemons, and peppermint. In the line drawing for aldehydes, it is customary to show the hydrogen in the aldehyde functional group (Figure 17.11).

**Figure 17.11**
**Isovaleraldehyde, or 3-methylbutanal**

OBJECTIVE 3

## Ketones

**Ketones** have the general formula

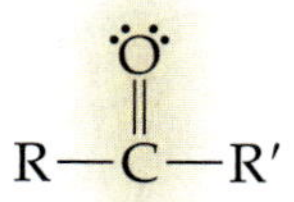

$$R-\overset{\overset{\large\ddot{O}}{\|}}{C}-R'$$

Ketone

The hydrocarbon groups represented by R and R′ can be identical or different. The two most common ketones are 2-propanone (acetone), $CH_3COCH_3$, and 2-butanone (methyl ethyl ketone or MEK), $CH_3COCH_2CH_3$. Both compounds are solvents frequently used in nail polish removers (Figure 17.12).

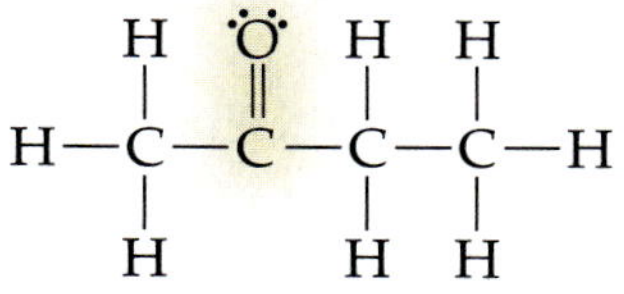

Methyl ethyl ketone (MEK)

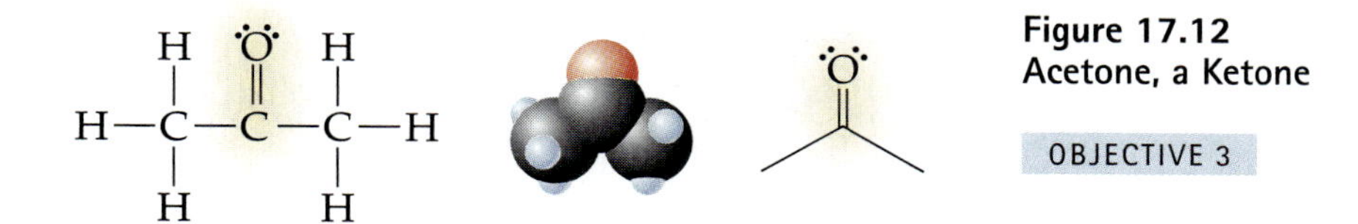

**Figure 17.12**
**Acetone, a Ketone**

OBJECTIVE 3

## Esters

**Esters** are pleasant-smelling substances whose general formula is

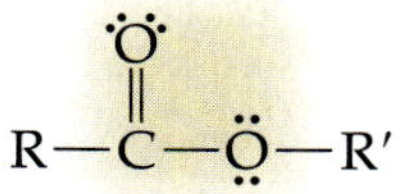

Ester

where R represents either a hydrocarbon group or a hydrogen atom, and R′ represents a hydrocarbon group. In condensed formulas, the ester functional group is indicated by either –COO– or $-CO_2-$. Ethyl butanoate (or ethyl butyrate), $CH_3CH_2CH_2COOCH_2CH_3$ or $CH_3CH_2CH_2CO_2CH_2CH_3$, is an ester that contributes to pineapples' characteristic odor (Figure 17.13).

**Figure 17.13**
**Ethyl Butanoate, an Ester**

## Amines

**Amines** have the general formula

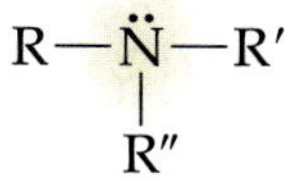

Amine

in which the R's represent hydrocarbon groups or hydrogen atoms (but at least one of the groups must be a hydrocarbon group). The amine 1-aminobutane (or n-butylamine), $CH_3CH_2CH_2CH_2NH_2$, is an intermediate that can be converted into pharmaceuticals, dyes, and insecticides. Amines can have more than one amine functional group. Amines often have distinctive and unpleasant odors. For example, 1,5-diaminopentane (cadaverine), $H_2N(CH_2)_5NH_2$, and 1,4-diaminobutane (putrescine), $H_2N(CH_2)_4NH_2$, form part of the odor of rotting flesh and in much smaller quantities, bad breath.

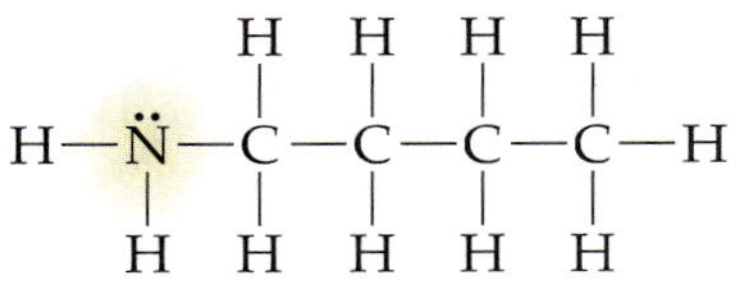

1-Aminobutane

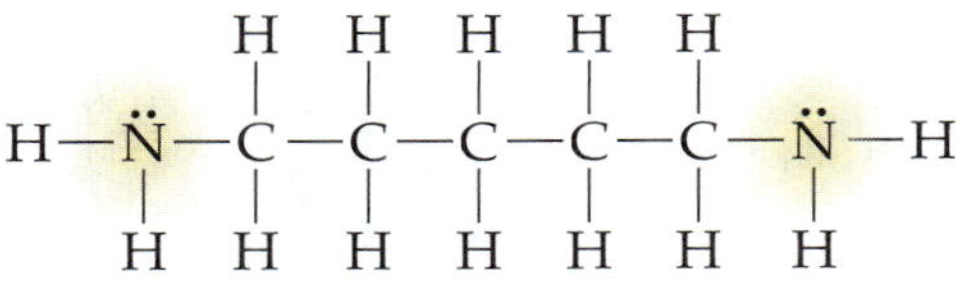

Cadaverine

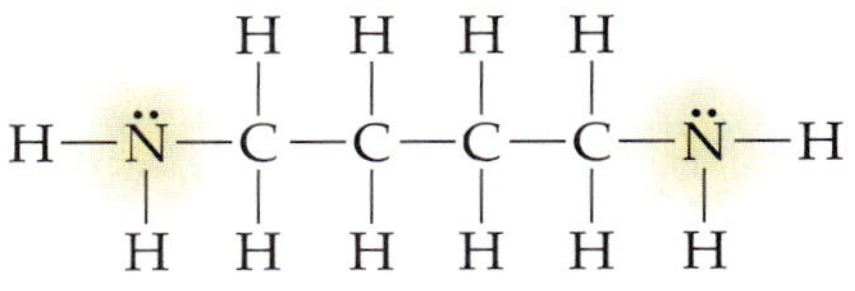

Putrescine

Trimethylamine, $(CH_3)_3N$, is partly responsible for the smell of spoiled fish (Figure 17.14).

**Figure 17.14**
**Trimethylamine, an Amine**

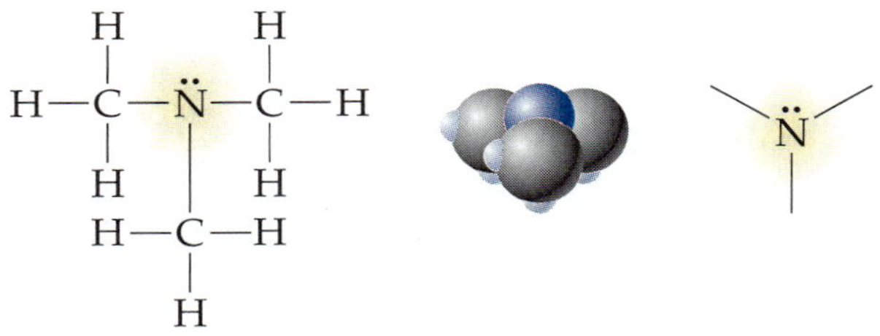

OBJECTIVE 3

## Amides

**Amides** have the general formula

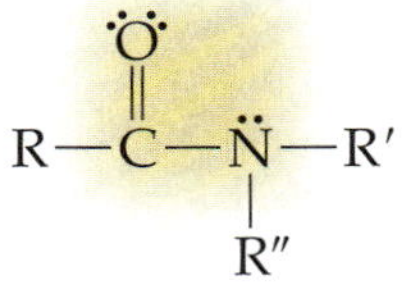

Amide

where the R's represent hydrocarbon groups or hydrogen atoms. In condensed formulas, the amide functional group is indicated by –CON–. The amide ethanamide (acetamide), $CH_3CONH_2$, has many uses, including the production of explosives (Figure 17.15).

Table 17.1 *(continued)*

OBJECTIVE 2 OBJECTIVE 3

| Type of Compound | General Structure | General Condensed Formula | General Structure |
|---|---|---|---|
| Ester | R—C(=Ö:)—Ö—R′<br>R ≠ H | RCOOR or $RCO_2R$ | Ethyl butanoate (ethyl butyrate), $CH_3CH_2CH_2COOCH_2CH_3$, smells like pineapples; used in artificial flavorings<br>H—C(H)(H)—C(H)(H)—C(H)(H)—C(=Ö:)—Ö—C(H)(H)—C(H)(H)—H |
| Amine | R—N̈—R′<br>│<br>R″<br>At least 1 R is a hydrocarbon group. | $R_3N$ | 1-Aminobutane (n-butyl amine), $CH_3CH_2CH_2CH_2NH_2$, intermediate in the production of pharmaceuticals, dyes, and insecticides<br>H—N̈(H)—C(H)(H)—C(H)(H)—C(H)(H)—C(H)(H)—H |
| Amide | R—C(=Ö:)—N̈(R″)—R′ | $RCONR_2$ | Ethanamide (acetamide), $CH_3CONH_2$, used to make explosives<br>H—C(H)(H)—C(=Ö:)—N̈(H)—H |

## EXERCISE 17.1 Organic Compounds

OBJECTIVE 3

Identify each of these structures as representing an alkane, alkene, alkyne, arene (aromatic), alcohol, carboxylic acid, ether, aldehyde, ketone, ester, amine, or amide.

```
       H   H   H   H   H   H   H
       |   |   |   |   |   |   |
a.  H—C—C—C—C—C—C—C—H
       |   |   |   |   |   |   |
       H   H   H   H   H   H   H

       H   H   H   H   H   H   H   H
       |   |   |   |   |   |   |   |    ..
b.  H—C—C—C—C—C—C—C—C—N—H
       |   |   |   |   |   |   |   |   |
       H   H   H   H   H   H   H   H   H

       H   H   H   H       H   H   H   H
       |   |   |   |   ..  |   |   |   |
c.  H—C—C—C—C—O—C—C—C—C—H
       |   |   |   |   ..  |   |   |   |
       H   H   H   H       H   H   H   H

       H   H  :O:      H   H   H   H
       |   |   ||  ..  |   |   |   |
d.  H—C—C—C—O—C—C—C—C—H
       |   |       ..  |   |   |   |
       H   H           H   H   H   H

       H   H   H  :O:  H   H
       |   |   |   ||  |   |
e.  H—C—C—C—C—C—C—H
       |   |   |       |   |
       H   H   H       H   H
```

```
f.      H  /H\   O
        |  | |   ||
    H — C — C — C — O — H
        |  | |
        H  \H/10
```

```
j.         H
           |
       H — C — H
      H    |        H   H
      |    |        |   |
  H — C — C = C — C — C — H
      |       |     |   |
      H       H     H   H
```

```
g.         H
           |
       H — C — H
      H    |     O
      |    |     ||
  H — C — C — C — N — H
      |    |          |
      H    H          H
```

```
k.    H   H         H
      |   |         |
  H — C — C — N — C — H
      |   |   |     |
      H   H   |     H
          H — C — H
              |
              H
```

```
h.         H
           |
       H — C — H
      H    |    H   H   O
      |    |    |   |   ||
  H — C — C — C — C — C — H
      |    |    |   |
      H    |    H   H
       H — C — H
           |
           H
```

```
l.                       H
                         |
                     H — C — H
      H                  |    H
      |                  |    |
  H — C — C ≡ C — C — C — H
      |                  |    |
      H                  |    H
                     H — C — H
                         |
                         H
```

```
i.         H
           |
       H — C — H
      H    |    H   H
      |    |    |   |
  H — C — C — C — C — H
      |    |    |   |
      H    |    H   H
           O — H
```

```
m.       H
         |
     H — C — H
         |
     (benzene ring)
         |
     H — C — H
         |
         H
```

## EXERCISE 17.2 Condensed Formulas

OBJECTIVE 2

Write condensed formulas to represent the Lewis structures in parts(a) through (l) of Exercise 17.1.

## EXERCISE 17.3 Line Drawings

OBJECTIVE 2

Make line drawings that represent the Lewis structures in parts (a) through (j) of Exercise 17.1.

## Special Topic 17.1 Rehabilitation of Old Drugs and Development of New Ones

Imagine that you are a research chemist who has been hired by a large pharmaceutical company to develop a new drug for treating AIDS. How are you going to do it? Modern approaches to drug development fall into four general categories.

***Old Drug, New Use*** One approach is to do a computer search of all drugs to try to find one that can be put to a new use. For example, imagine you want to develop a drug for combating the lesions seen in Kaposi's sarcoma, an AIDS-related condition. These lesions are caused by the abnormal proliferation of small blood vessels. A list of all of the drugs that are thought to inhibit the growth of blood vessels might include some that are effective in treating Kaposi's sarcoma.

One of the drugs on that list is thalidomide, originally developed as a sedative by a German pharmaceutical company in the 1950s. It was considered a safe alternative to other sedatives, which are lethal in large doses; but when it was also used to reduce nausea associated with pregnant women's "morning sickness," it caused birth defects in the babies they were carrying. Thalidomide never did receive approval in the United States, and it was removed from the European market in the 1960s. About 10,000 children were born with incompletely formed arms and legs as a result of thalidomide's effects.

Thalidomide is thought to inhibit the formation of limbs in the fetus by slowing the formation of blood vessels, but what can be disastrous for unborn children can be lifesaving for others. Today the drug is being used as a treatment for Kaposi's sarcoma, and it may also be helpful in treating AIDS-related weight loss and brain cancer.

***Old Drug, New Design*** Another approach to drug development is to take a chemical already known to have a certain desirable effect and alter it slightly in hopes of enhancing its potency. The chemists at the Celgene Corporation have taken this approach with thalidomide. They have developed a number of new drugs that are similar in structure to thalidomide but appear to be 400 to 500 times more potent.

***Rational Drug Design*** In a third, more direct, approach often called rational drug design, the researcher first tries to determine what chemicals in the body are leading to the trouble. Often these chemicals are enzymes, large molecules that contain an active site in their structure where other molecules must fit to cause a change in the body. Once the offending enzyme is identified, isolated, and purified, it is "photographed" by x-ray crystallography, which reveals the enzyme's three-dimensional structure, including the shape of the active site. The next step is to design a molecule that will fit into the active site and deactivate the enzyme. If the enzyme is important for the replication of viruses like the AIDS virus or a flu virus, the reproduction of the virus will be slowed.

***Combinatorial Chemistry*** The process of making a single new chemical, isolating it, and purifying it in quantities large enough for testing is time-consuming and expensive. If the chemical fails to work, all you can do is start again and hope for success with the next. Thus chemists are always looking for ways to make and test more new chemicals more rapidly. A new approach to the production of chemicals, called combinatorial chemistry, holds great promise for doing just that.

Instead of making one new chemical at a time, the strategy of combinatorial chemistry is to make and test thousands of similar chemicals at the same time. It therefore requires highly efficient techniques for isolating and identifying different compounds. One way of easily separating the various products from the solution in which they form is to run the reaction on the surface of tiny polymer beads that can be filtered from the reaction mixture after the reaction takes place. The beads must be tagged in some way so that the researcher can identify which ones contain which new substance. One of the more novel ways of doing this is to cause the reaction to take place inside a tiny capsule from which a microchip sends out an identifying signal.

After a library of new chemicals has been produced, the thousands of compounds need to be tested to see which have desirable properties. Unfortunately, the procedures for testing large numbers of chemicals are often less than precise. One approach is to test all of them in rapid succession for one characteristic that suggests a desired activity. A secondary library is then made with a range of structures similar to the structure of any substance that has that characteristic, and these new chemicals are also tested. In this way, the chemist can zero in on the chemicals that are most likely to have therapeutic properties. (Note again the connection between structure and properties.) The most likely candidates are then made in larger quantities, purified more carefully, and tested in more traditional ways. Combinatorial chemistry has shown promise for producing pharmaceuticals of many types, including anticancer drugs and drugs to combat AIDS.

## 17.2 Important Substances in Food

*Organic chemistry is the study of carbon compounds. Biochemistry is the study of carbon compounds that crawl.*

MIKE ADAMS
SCIENCE WRITER

Let's take a closer look at the fast-food dinner described in the introduction to this chapter. Our food is a mixture of many different kinds of substances, but the energy we need to run our bodies comes from three of them: carbohydrates (the source of 40%–50% of our energy), protein (11%–14%), and fat (the rest). Table 17.2 shows typical energy and mass values for a burger, a serving of fries, and a milkshake. In order to understand what happens to these substances when we eat them, you need to know a little bit more about their composition.

**Table 17.2**
Fast-Food Dinner (These energy and mass values are derived from the *USDA Nutrient Database for Standard Reference*).

| | Energy, calories | Energy, kJ | Total mass | Protein mass | Carbohydrate mass | Fat mass |
|---|---|---|---|---|---|---|
| Large double hamburger with condiment | 540.2 | 2260 | 226.0 g | 34.3 g | 40.3 g | 26.6 g |
| Fried potatoes | 663.0 | 2774 | 255.0 g | 7.2 g | 76.9 g | 38.1 g |
| Chocolate milkshake | 355.6 | 1488 | 300.0 g | 9.2 g | 63.5 g | 8.1 g |

### Carbohydrates

**Carbohydrate** is a general name for sugars, starches, and cellulose. It derives from an earlier belief that these substances were hydrates of carbon, because many of them have the general formula $(CH_2O)_n$. Today, chemists also refer to carbohydrates as **saccharides,** after the smaller units from which they are built. Sugars are monosaccharides and disaccharides. Starches and cellulose are polysaccharides. Carbohydrates serve many different functions in nature. For example, sugar and starch are important for energy storage and production in both plants and animals, and cellulose provides the support structure of woody plants.

OBJECTIVE 6

The most important **monosaccharides** are the sugars glucose, fructose, and galactose, isomers with the general formula $C_6H_{12}O_6$. Each of these sugars can exist in either of two ring forms or in an open-chain form (Figures 17.18 and 17.19). In solution, they are constantly shifting from one form to another. Note that glucose and galactose have aldehyde functional groups in the open-chain form and that fructose has a ketone functional group. Glucose and galactose differ only in the relative position of the –H and –OH groups on the fourth carbon from the top.

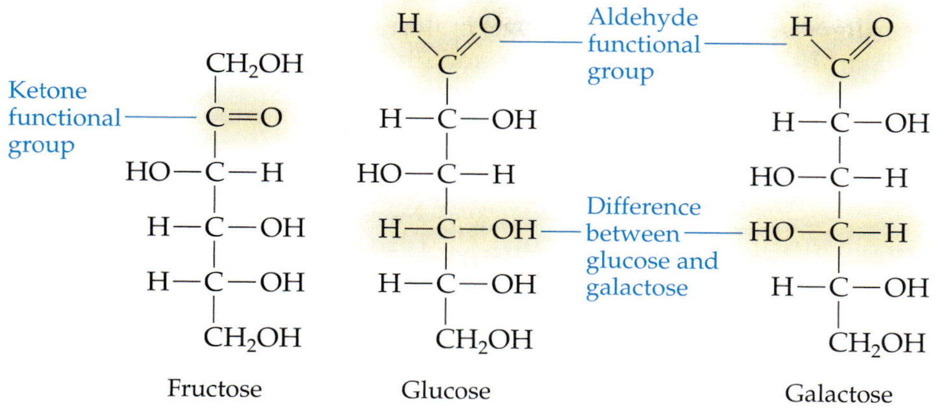

Ketone functional group

Aldehyde functional group

Difference between glucose and galactose

Fructose

Glucose

Galactose

**Figure 17.18**
Open-chain Form of Three Monosaccharides

OBJECTIVE 4

OBJECTIVE 5

OBJECTIVE 6

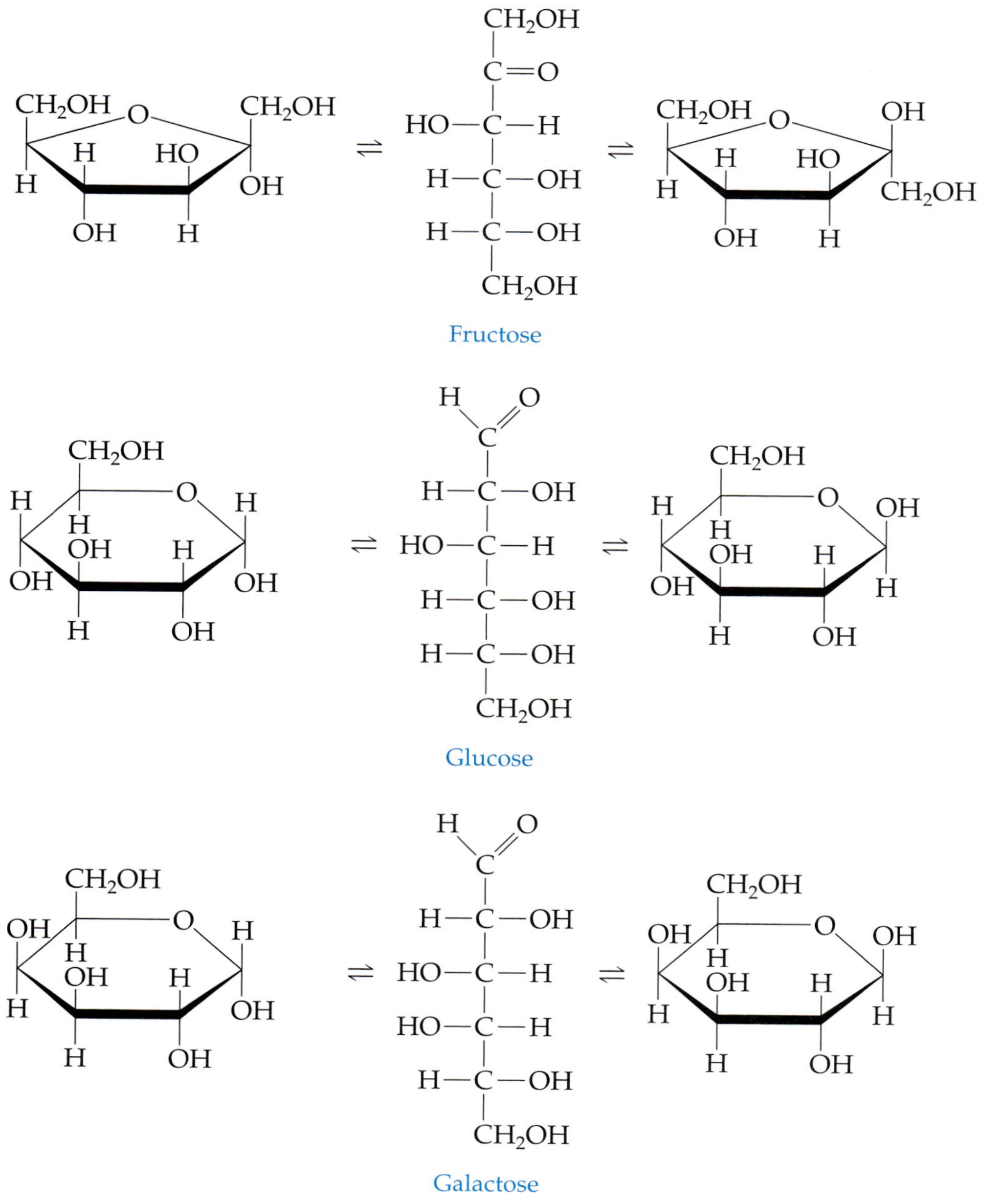

Fructose

Glucose

Galactose

**Figure 17.19**
Fructose, Glucose, and Galactose

OBJECTIVE 6

OBJECTIVE 5

OBJECTIVE 7

**Disaccharides** are composed of two monosaccharide units. Maltose, a disaccharide consisting of two glucose units, is formed in the brewing of beer from barley in a process called malting. Lactose, or milk sugar, is a disaccharide consisting of galactose and glucose; sucrose is a disaccharide that contains glucose and fructose (Figure 17.20).

OBJECTIVE 7

**Figure 17.20**
Three Disaccharides: Maltose, Lactose, and Sucrose

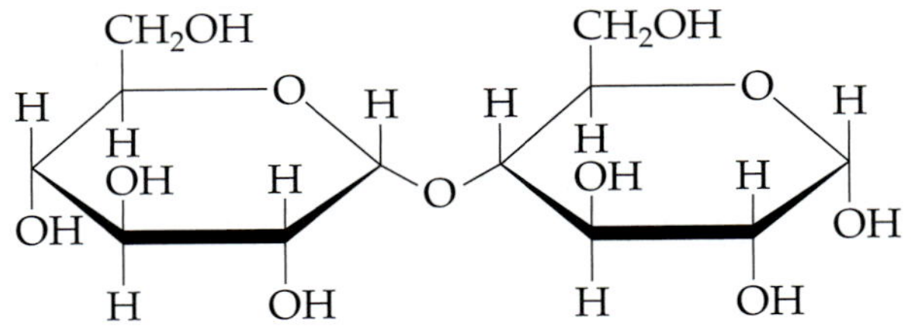

Maltose (glucose and glucose)

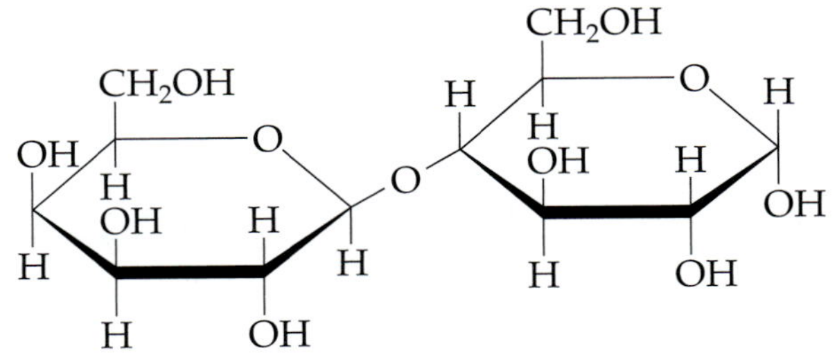

Lactose (galactose and glucose)

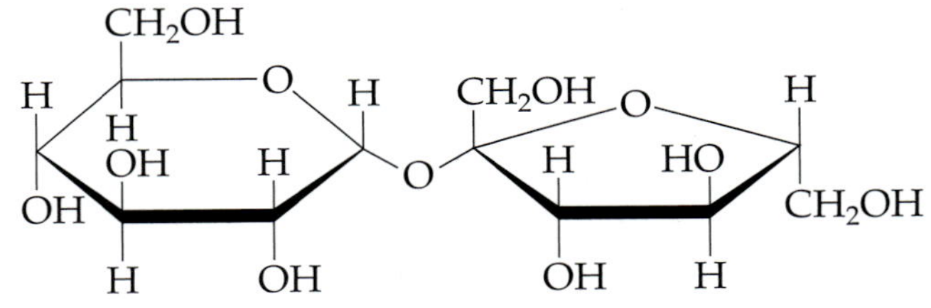

Sucrose (glucose and fructose)

OBJECTIVE 5

**Polysaccharides** consist of many saccharide units linked together to form long chains. The most common polysaccharides are starch, glycogen (sometimes called animal starch), and cellulose. All of these are composed of repeating glucose units, but they differ in the way the glucose units are attached.

OBJECTIVE 8

Almost every kind of plant cell has energy stored in the form of starch. Starch itself has two general forms, amylose and amylopectin. Amylose molecules are long, unbranched chains. Amylopectin molecules are long chains that branch (Figure 17.21). Glycogen is similar to amylopectin, but its branches are usually shorter and more numerous. Glycogen molecules are stored in liver and muscle cells of animals, where they can be converted into glucose molecules and used as a source of energy. All the polysaccharides are **polymers,** a general name for large molecules composed of repeating units called **monomers.**

**Figure 17.21**
**Polysaccharides**
Starch provides energy both for plants and for the animals that eat the plants. Cellulose is an indigestible polysaccharide that provides structure for plants and fiber in animal diets.

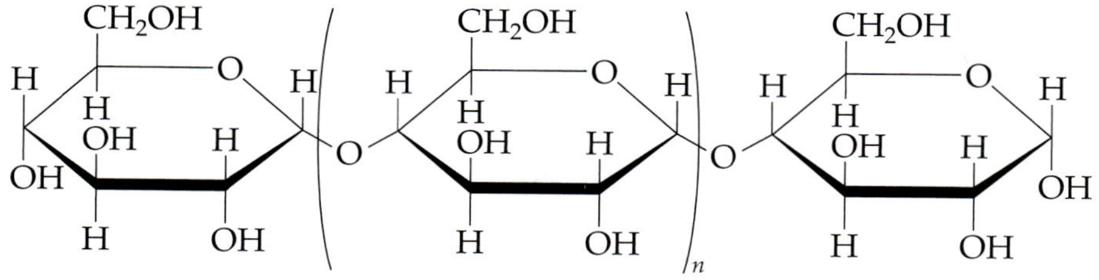

Amylose

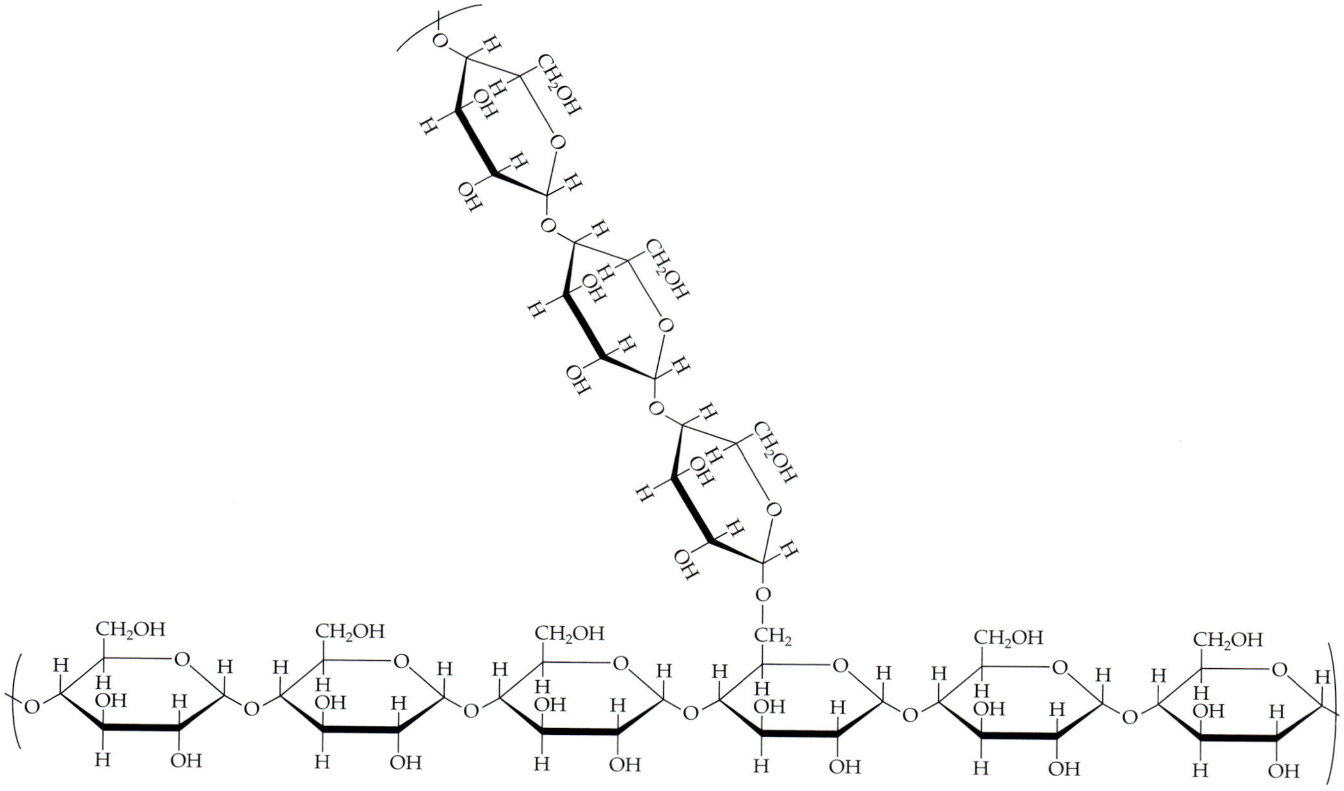

Amylopectin

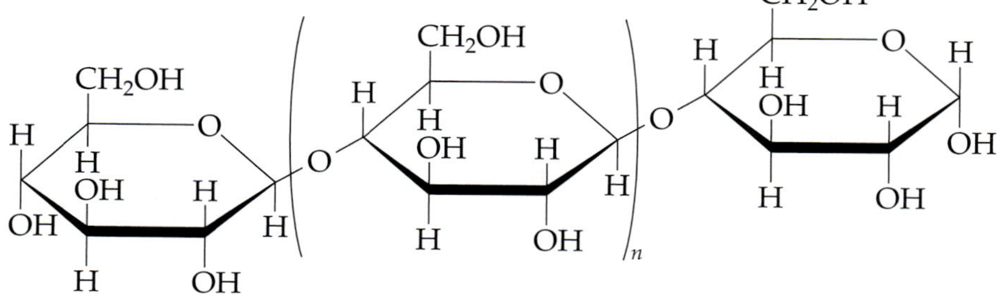

Cellulose

OBJECTIVE 8

Cellulose is the primary structural material in plants. Like starch, it is composed of large numbers of glucose molecules linked together; but in cellulose, the manner of linking produces very organized chains that can pack together closely, allowing strong attractions to form (Figure 17.21). The strong structures that result provide support and protection for plants. Our digestive enzymes[1] are able to break the linkages in starch to release energy-producing glucose, but they are unable to liberate glucose molecules from cellulose because they cannot break the linkages there. Cellulose passes through our digestive tract unchanged.

OBJECTIVE 8

## Amino Acids and Protein

Protein molecules are polymers composed of monomers called amino acids. Wonderfully varied in size and shape, they play a wide variety of roles in our bodies. For example, proteins provide the underlying structure of our cells, form antibodies that fight off invaders, regulate many necessary chemical changes, and help to transport molecules through the bloodstream.

All but one of the 20 kinds of amino acids found in proteins have the following general form:

OBJECTIVE 10

OBJECTIVE 4

Amine group

$$H_2N{-}\underset{R}{\overset{H}{C}}{-}\overset{\overset{\Large O}{\|}}{C}{-}OH \quad \text{or} \quad H_2N{-}\underset{R}{\overset{H}{C}}{-}CO_2H$$

Carboxylic acid group

The R represents a group called a side chain that distinguishes one amino acid from another.

One end of the amino acid has a carboxylic acid functional group that tends to lose an $H^+$ ion, and the other end has a basic amine group that attracts $H^+$ ions. Therefore, under physiological conditions (the conditions prevalent within our bodies), amino acids are likely to have the form

OBJECTIVE 11

$$H_3\overset{+}{N}{-}\underset{R}{\overset{H}{C}}{-}\overset{\overset{\Large O}{\|}}{C}{-}O^- \quad \text{or} \quad H_3\overset{+}{N}{-}\underset{R}{\overset{H}{C}}{-}CO_2^-$$

OBJECTIVE 10

The structures of the 20 amino acids that our bodies need are shown in Figure 17.22 on pages 743 and 744. Each amino acid is identified by either a three-letter or a one-letter abbreviation. Note that the amino acid proline has a slightly different form than the others.

[1]Remember that enzymes are naturally occurring catalysts—substances in plant and animal systems that speed chemical changes without being permanently altered themselves.

**Figure 17.22**
Amino Acid Structures

**Amino acids with hydrogen or hydrocarbon side chains**

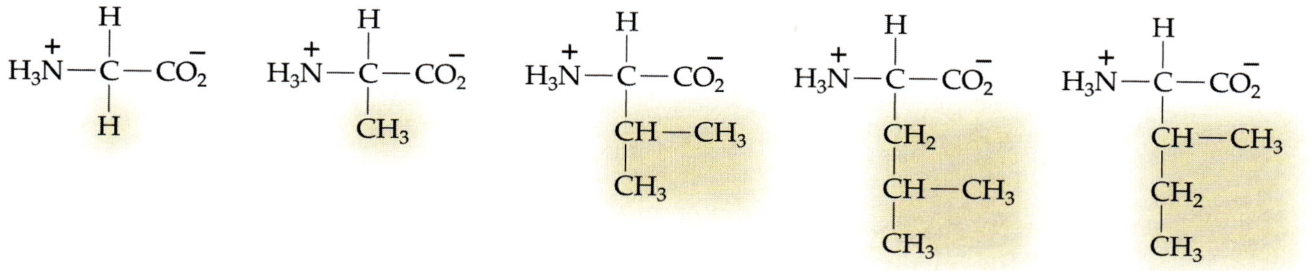

$H$
$H_3\overset{+}{N}-C-CO_2^-$
$H$

Glycine, Gly (G)

$H$
$H_3\overset{+}{N}-C-CO_2^-$
$CH_3$

Alanine, Ala (A)

$H$
$H_3\overset{+}{N}-C-CO_2^-$
$CH-CH_3$
$CH_3$

Valine, Val (V)

$H$
$H_3\overset{+}{N}-C-CO_2^-$
$CH_2$
$CH-CH_3$
$CH_3$

Leucine, Leu (L)

$H$
$H_3\overset{+}{N}-C-CO_2^-$
$CH-CH_3$
$CH_2$
$CH_3$

Isoleucine, Ile (I)

**Cyclic amino acid**

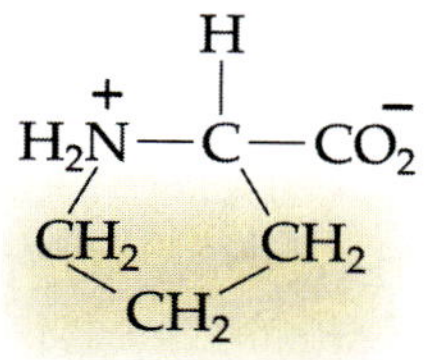

$H$
$H_2\overset{+}{N}-C-CO_2^-$
$CH_2$ $CH_2$
$CH_2$

Proline, Pro (P)

**Aromatic amino acids**

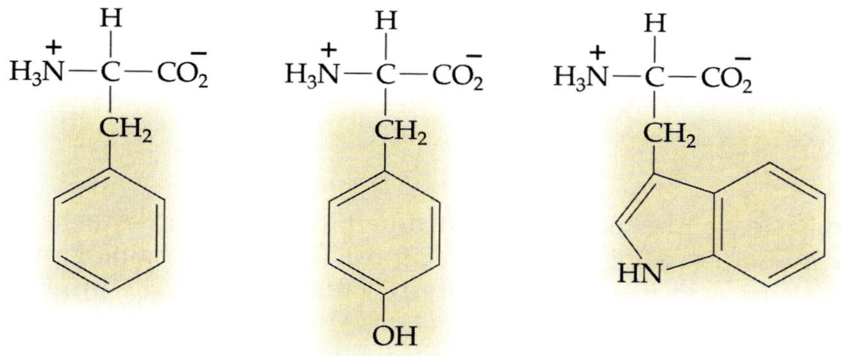

$H$
$H_3\overset{+}{N}-C-CO_2^-$
$CH_2$

Phenylalanine, Phe (F)

$H$
$H_3\overset{+}{N}-C-CO_2^-$
$CH_2$
$OH$

Tyrosine, Tyr (Y)

$H$
$H_3\overset{+}{N}-C-CO_2^-$
$CH_2$
HN

Tryptophan, Trp (W)

**Amino acids with hydroxl- or sulfur-containing side chains**

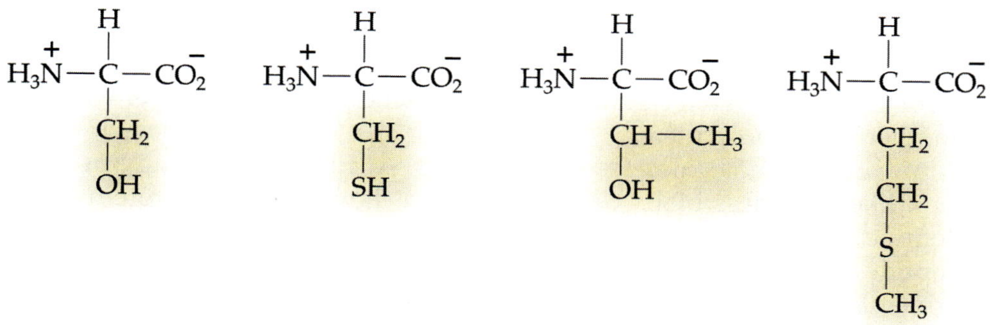

$H$
$H_3\overset{+}{N}-C-CO_2^-$
$CH_2$
$OH$

Serine, Ser (S)

$H$
$H_3\overset{+}{N}-C-CO_2^-$
$CH_2$
$SH$

Cysteine, Cys (C)

$H$
$H_3\overset{+}{N}-C-CO_2^-$
$CH-CH_3$
$OH$

Threonine, Thr (T)

$H$
$H_3\overset{+}{N}-C-CO_2^-$
$CH_2$
$CH_2$
$S$
$CH_3$

Methionine, Met (M)

**Basic amino acids**

Histidine, His (H) Lysine, Lys (K) Arginine, Arg (R)

**Acidic amino acids and their amides**

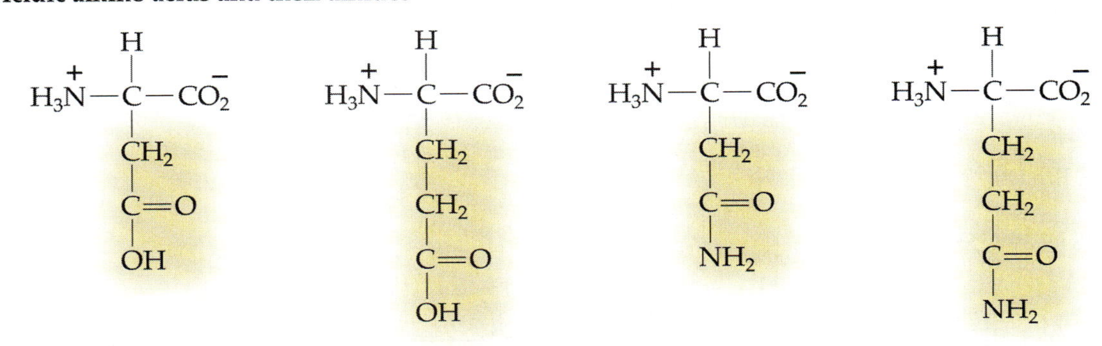

Aspartic acid, Asp (D) Glutamic acid, Glu (E) Asparagine, Asn (N) Glutamine, Gln (Q)

**Figure 17.22** *(continued)*

OBJECTIVE 12

OBJECTIVE 4

Amino acids are linked together by a **peptide bond,** created when the carboxylic acid group of one amino acid reacts with the amine group of another amino acid to form an amide functional group. The product is called a **peptide.** Although the language used to describe peptides is not consistent among scientists, small peptides are often called *oligopeptides*, and large peptides are called *polypeptides*. Figure 17.23 shows how alanine, serine, glycine, and cysteine can be linked to form a structure called a tetrapeptide (a peptide made from four amino acids). Because the reaction that links amino acids produces water as a by-product, it is an example of a **condensation reaction,** a chemical change in which a larger molecule is made from two smaller molecules accompanied by the release of water or another small molecule.

All **protein** molecules are polypeptides. At first glance, many of them look like shapeless blobs of atoms. In fact, however, each protein has a definite form that is determined by the order of the amino acids in the peptide chain and the interactions between them. To illustrate the general principles of protein structure, let's look at one of the most thoroughly studied of all proteins, a relatively small one called bovine pancreatic trypsin inhibitor (BPTI) (Figure 17.24).

Condensation reaction releases water.

$+\ 3H_2O$

Peptide bonds (amide functional groups)

OBJECTIVE 4 OBJECTIVE 12

**Figure 17.23**
**The Condensation Reaction That Forms the Tetrapeptide Ala-Ser-Gly-Cys**

Protein molecules are described in terms of their primary, secondary, and tertiary structures. The **primary structure** of a protein is the linear sequence of its amino acids. The primary structure for BPTI is

OBJECTIVE 13

Arg-Pro-Asp-Phe-Cys-Leu-Glu-Pro-Pro-Tyr-Thr-Gly-Pro-Cys-Lys-Ala-Arg-Ile-Ile-Arg-Tyr-Phe-Tyr-Asn-Ala-Lys-Ala-Gly-Leu-Cys-Gln-Thr-Phe-Val-Tyr-Gly-Gly-Cys-Arg-Ala-Lys-Arg-Asn-Asn-Phe-Lys-Ser-Ala-Glu-Asp-Cys-Leu-Arg-Thr-Cys-Gly-Gly-Ala

The arrangement of atoms that are close to each other in the polypeptide chain is called the **secondary structure** of the protein. Images of two such arrangements, an α-helix and a β-sheet, are shown in Figures 17.25 and 17.26.

OBJECTIVE 13

**Figure 17.24**
**A Ball-and-Stick Model of a Protein Called Bovine Pancreatic Trypsin Inhibitor (BPTI)**

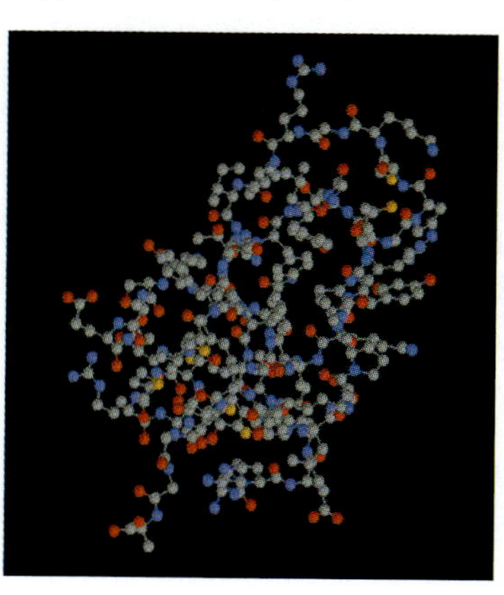

**Figure 17.25**
**α-Helix**

Space-filling model of a portion of the α-helical secondary structure of a protein molecule

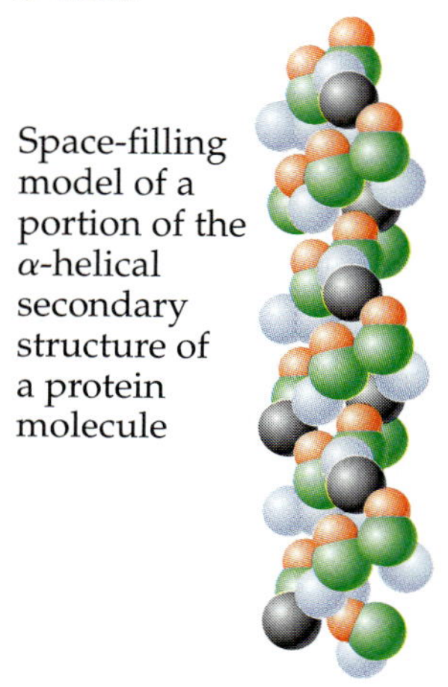

Ball-and-stick model of a portion of the α-helical secondary structure of a protein molecule

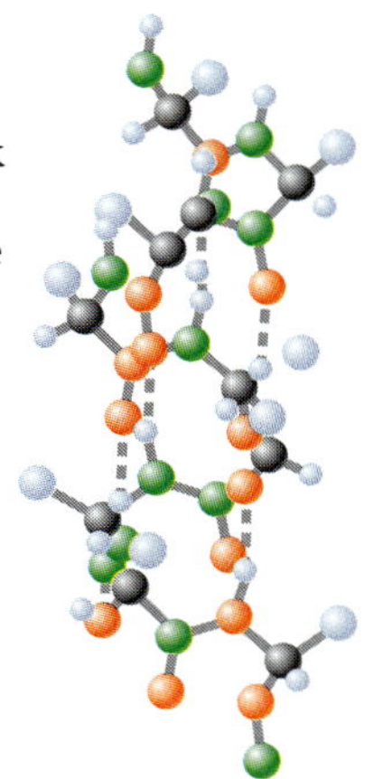

This ribbon model shows the general arrangement of atoms in a portion of the α-helical secondary structure of a protein molecule.

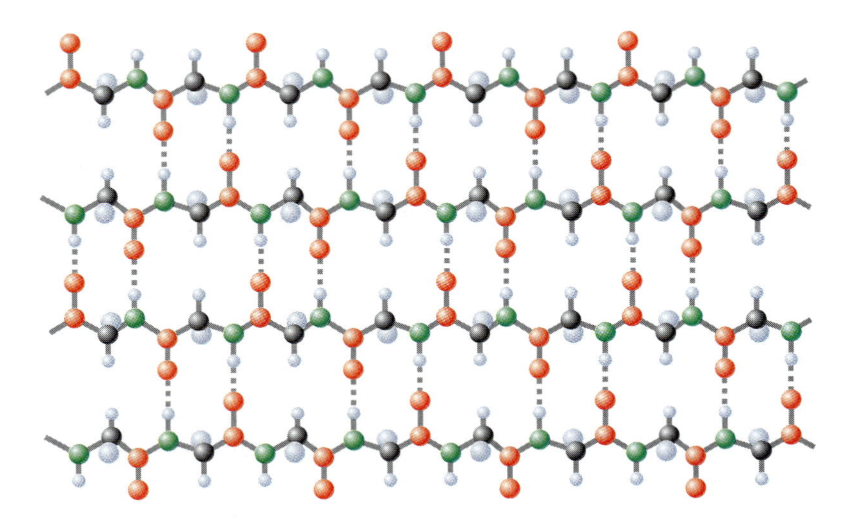

**Figure 17.26**
β-Sheet

OBJECTIVE 13

BPTI contains both α-helix and β-sheet secondary structures, separated by less regular arrangements of amino acids. Because of the complexity of protein molecules, simplified conventions are used in drawing them to clarify their secondary and tertiary structures. Figure 17.27 shows the ribbon convention, in which α-helices are depicted by coiled ribbons and β-sheets are represented by flat ribbons.

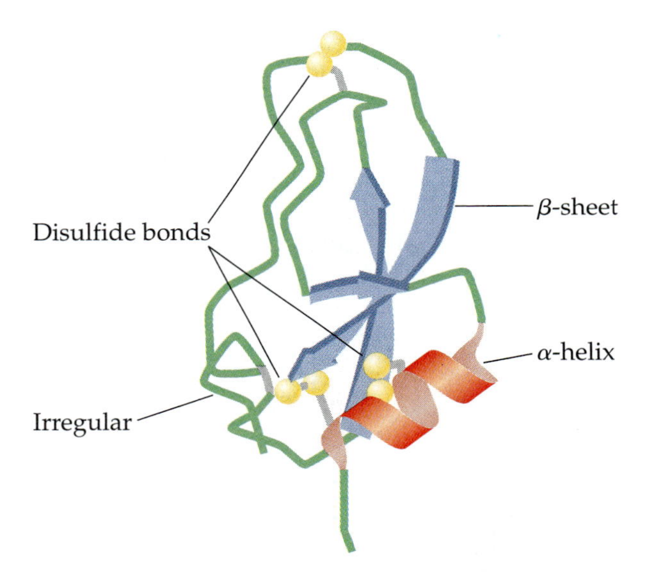


**Figure 17.27**
The Ribbon Structure of the Protein BPTI

OBJECTIVE 14

When the long chains of amino acids link to form protein structures, not only do they arrange themselves into secondary structures, but the whole chain also arranges itself into a very specific overall shape called the **tertiary structure** of the protein. The protein chain is held in its tertiary structure by interactions between the side chains of its amino acids. For example, covalent bonds that form between sulfur atoms in different parts of the chain help create and hold the BPTI molecule's specific shape (Figure 17.27). These bonds, which are called **disulfide bonds,** can form between two cysteine amino acids (Figure 17.28).

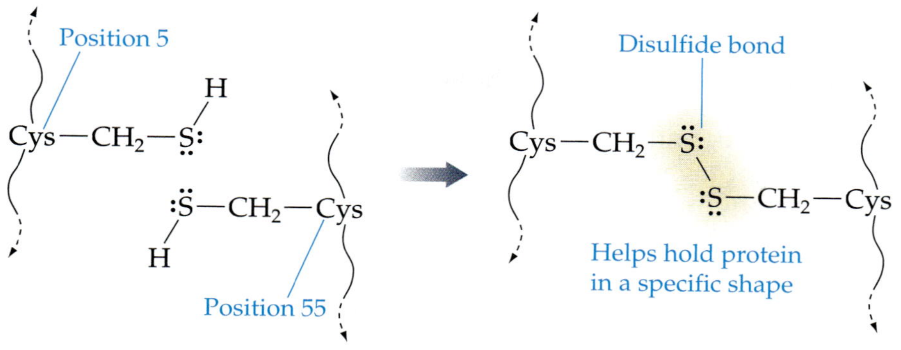

**Figure 17.28**
Disulfide Bonds Between Cysteine Amino Acid Side Chains in a Protein Molecule

OBJECTIVE 14

Hydrogen bonds can also help hold protein molecules in their specific tertiary shape. For example, the possibility of hydrogen bonding between their –OH groups will cause two serine amino acids in a protein chain to be attracted to each other (Figure 17.29).

OBJECTIVE 14

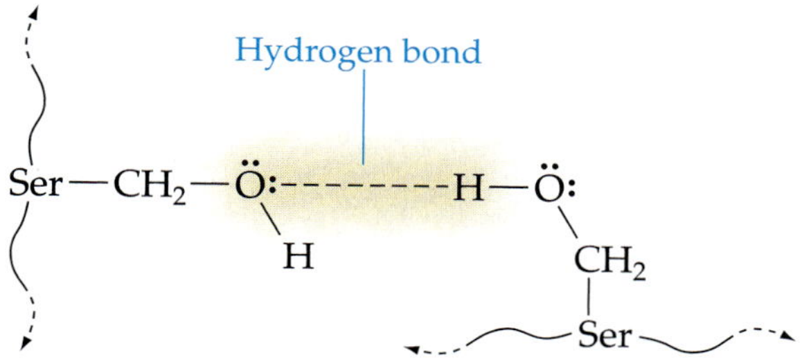

**Figure 17.29**
Hydrogen Bonding Between Two Serine Amino Acids in a Protein Molecule

OBJECTIVE 14

The tertiary structure is also determined by the creation of **salt bridges,** which consist of negatively charged side chains attracted to positively charged side chains (Figure 17.30).

OBJECTIVE 14

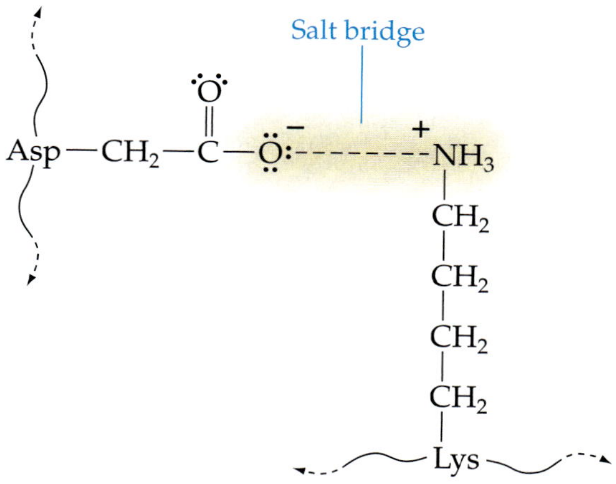

**Figure 17.30**
Salt Bridge Between an Aspartic Acid Side Chain at One Position in a Protein Molecule and a Lysine Amino Acid Side Chain in Another Position

OBJECTIVE 14

## Fat

OBJECTIVE 15

The fat stored in our bodies is our primary long-term energy source. A typical 70-kg human has fuel reserves of about 400,000 kJ in fat, 100,000 kJ in protein (mostly muscle protein), 2500 kJ in glycogen, and 170 kJ in total glucose. One of the reasons why it is more efficient to store energy as fat than as carbohydrate or protein is that fat produces 37 kJ/g, whereas carbohydrate and protein produce only 17 kJ/g.

OBJECTIVE 16

As we saw in Section 15.2, animal fats and vegetable oils are made up of **triglycerides,** which have many different structures but the same general design: long-chain hydrocarbon groups attached to a three-carbon backbone.

Hydrocarbon groups ($R_1$, $R_2$, $R_3$)

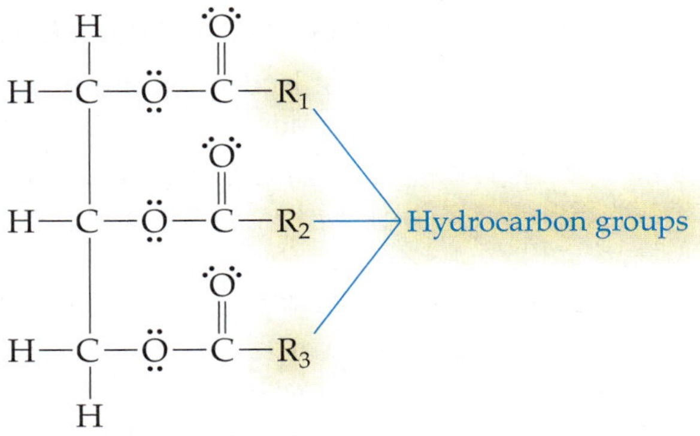

Triglyceride (triacylglycerol)

We saw that the hydrocarbon groups in triglycerides can differ in the length of the carbon chain and in the frequency of double bonds between their carbon atoms. The liquid triglycerides in vegetable oils have more carbon–carbon double bonds than the solid triglycerides in animal fats. The more carbon–carbon double bonds a triglyceride molecule has, the more likely it is to be liquid at room temperature.

A process called **hydrogenation** converts liquid triglycerides to solid triglycerides by adding hydrogen atoms to the double bonds and so converting them to single bonds. For example, the addition of hydrogen in the presence of a platinum catalyst changes corn oil into margarine.

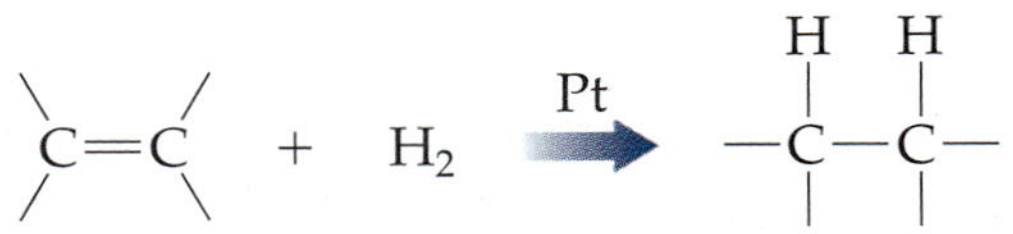

$$\text{C=C} + H_2 \xrightarrow{\text{Pt}} \text{—CH—CH—}$$

When enough hydrogen atoms are added to a triglyceride to convert all double bonds to single bonds, we call it a **saturated triglyceride** (or **fat**). It is saturated with hydrogen atoms. A triglyceride that still has one or more carbon–carbon double bonds is an **unsaturated triglyceride.** If enough hydrogen is added to an unsaturated triglyceride to convert some but not

OBJECTIVE 17

## Special Topic 17.2 Olestra

Should you eat less fat? Scientists doing medical research think you probably should; they recommend no more than 30% fat in our diets, whereas the average American diet is estimated to contain 34% fat. Perhaps you're convinced that you should cut down on fatty foods, but you can't imagine watching the Super Bowl without a big bag of chips at your side. The chemists at Procter & Gamble have been trying to resolve your dilemma by developing an edible substance with the rich taste and smooth texture of fat molecules but without the calories. Olestra seems to meet these criteria.

Fat digestion is an enzyme-mediated process that breaks fat molecules into glycerol and fatty acids, which are then able to enter the bloodstream. Olestra is a hexa-, hepta-, or octa-ester of fatty acids (derived from vegetable oil, such as soybean oil or cottonseed oil) and sucrose. Because the body contains no digestive enzymes that can convert Olestra's fat-like molecules into their smaller components of sucrose and fatty acids, and because Olestra is too large to enter the bloodstream undigested, the compound passes though our systems unchanged.

Olestra is stable at high temperature and can be used in fried or baked foods, such as potato chips. A 1-ounce serving of potato chips made with olestra has 0 g of fat and 70 Cal, comparing favorably with the 10 g of fat and 160 Cal of normal chips. But there are also drawbacks associated with Olestra. Studies have shown that it can cause gastrointestinal distress and prevent the absorption of the fat-soluble vitamins (A, D, E, and K) and carotenoids (members of a group of nutrients that includes beta-carotene). Because of these findings, foods with Olestra carry the following warning: "Olestra may cause abdominal cramping and loose stools. Olestra inhibits the absorption of some vitamins and other nutrients. Vitamins A, D, E, and K have been added." Adding vitamins to foods that contain Olestra keeps Olestra from blocking the absorption of the fat-soluble vitamins in our other foods. The decrease in absorption of beta-carotene is a problem only when foods with Olestra are eaten along with other foods that are rich in carotene, such as carrots.

Scientists will continue to study the pros and cons of Olestra-containing foods, but in the end, it will be up to you to decide whether the benefits of lower-fat, lower-calorie food products outweigh the potential problems associated with their consumption.

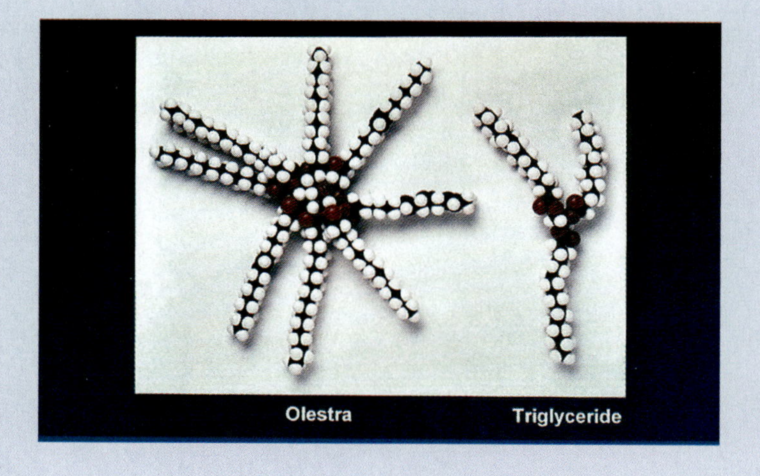

## SPECIAL TOPIC 17.3 Harmless Dietary Supplements or Dangerous Drugs?

*Despite the testimonials to muscle size and strength, there is no evidence that andro, creatine or any other substance enhances athletic performances over what could be attained by practice, training and proper nutrition.*

TODD B. NIPPOLDT, M.D.
MAYO CLINIC ENDOCRINOLOGIST

When serious athletes hear of natural substances that build muscles and provide energy, they are bound to wonder whether supplementing their diets with these substances could improve their athletic performance. Mark McGwire's incredible 70 home runs in the 1998 baseball season drew attention to two substances that he was taking at the time: the steroid androstenedione ("andro") and creatine, a compound found in the muscle tissue of vertebrates.

Should these substances be classified as dietary supplements or as drugs? This is a legal distinction with wide-ranging repercussions. Because both andro and creatine are classified as dietary supplements under the Dietary Supplement and Health Act of 1994, they can be sold over the counter to anyone, without first being subjected to the extensive scientific testing necessary for substances classified as drugs. Before you start seasoning your steaks with andro and creatine, however, several important questions should be answered.

*Are they safe?* Although small amounts of androstenedione and testosterone in the body are essential to good health, introducing larger than normal amounts into one's system has potentially serious side effects, including gland cancer, hair loss, impotence, and acne. Increased testosterone levels in women can lead to a deeper voice and facial hair.

*Are they effective?* It has been found that when a typical 100-mg dose of androstenedione is consumed, all but a small percentage is destroyed in the liver, and what is left boosts testosterone levels only temporarily. It is not clear whether this has a significant effect on muscle building. According to the National Strength and Conditioning Association (a professional society for athletic trainers, sports medicine physicians and researchers, professional coaches, and physical therapists), there is no reliable evidence that andro improves athletic performance.

Research suggests that taking creatine does lead to a small improvement in some physical tasks, but there is still doubt whether supplemental amounts have any significant value. Meat contains creatine. Assuming that adequate amounts of meat are eaten, one's liver will normally produce about 2 g of the substance per day. The creatine is stored in the muscles, but any excess is promptly removed by the kidneys.

*Are they legal?* As of this writing, andro is banned in the NFL, the Olympics, and the NCAA, but it is still permitted in baseball and basketball, which ban only illegal drugs. For this reason, Mark McGwire could take it, but shot putter Randy Barnes, the 1996 Olympic gold medallist and world record holder, was banned from Olympic competition for life for doing so. (Barnes claims that he was not told about the ban and is appealing the decision.)

Did andro and creatine make a significant difference in Mark McGwire's ability to hit home runs? Edward R. Laskowski, M.D., co-director of the Sports Medicine Center at the Mayo Clinic in Rochester, Minnesota, says, "Mark McGwire has all the tools within himself to do what he did. If you ask elite athletes in any sport what they did to get to the top, they often break it down to the basics—training, conditioning and practice."

all of the carbon–carbon double bonds to single bonds, we say it has been *partially hydrogenated*. Margarine is often described as being made with partially hydrogenated vegetable oils.

OBJECTIVE 18

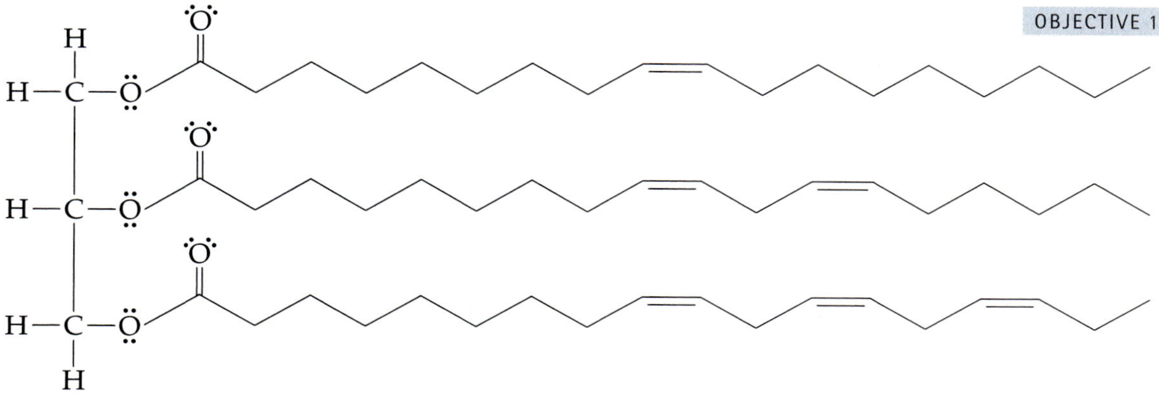

Unsaturated triglyceride
Liquid
Typical molecule in vegetable oil

$+ \ 5H_2(g)$

OBJECTIVE 19

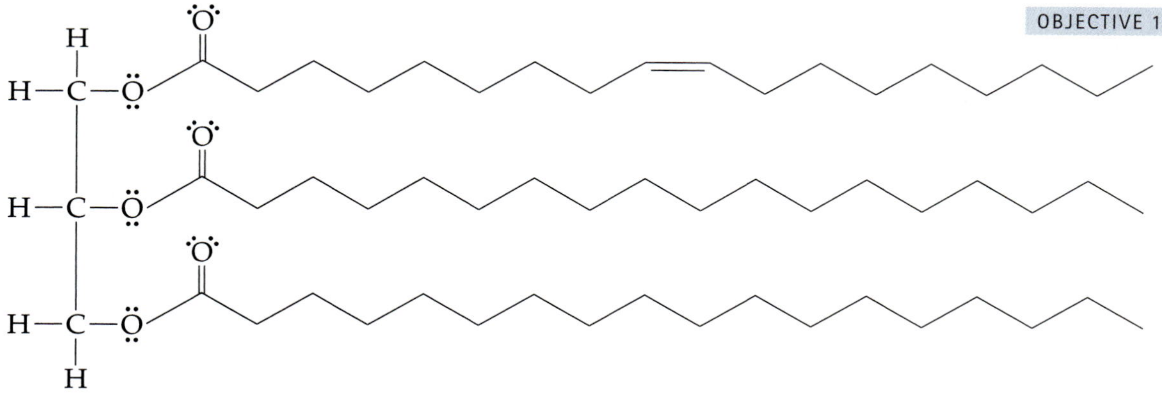

Partially hydrogenated triglyceride
Solid
Typical molecule in margarine

## Steroids

When they hear the word *steroid*, many people think first of the controversies over substances banned in the Olympic Games. In fact, **steroids** are important hormones produced in our bodies that help control inflammation, regulate our immune system, help maintain salt and water balance,

**Figure 17.31**
Formation of Testosterone from Progesterone

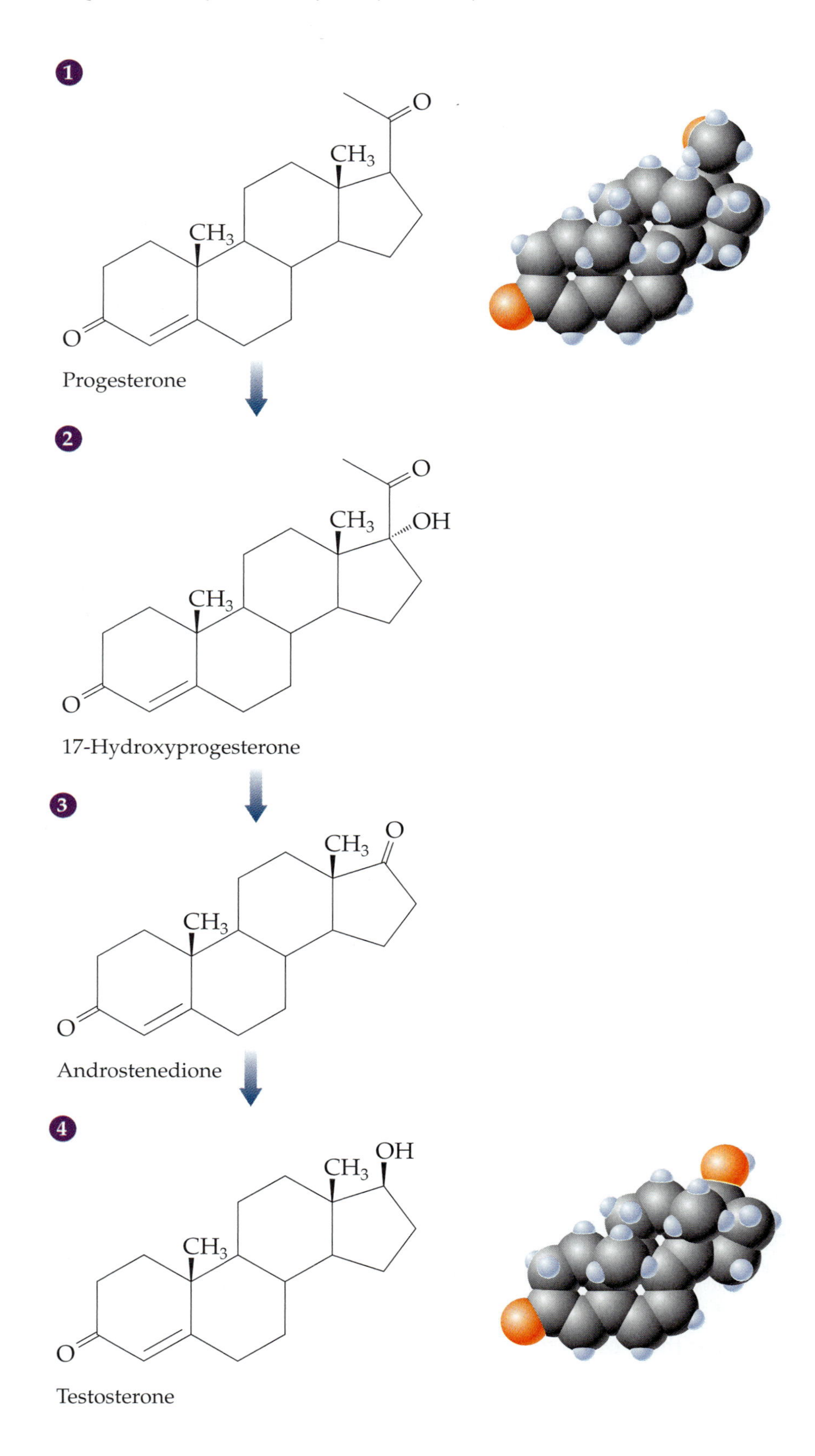

and control the development of sexual characteristics. Steroids are derivatives of the following four-ring structure. One important member of this group of biomolecules is cholesterol.

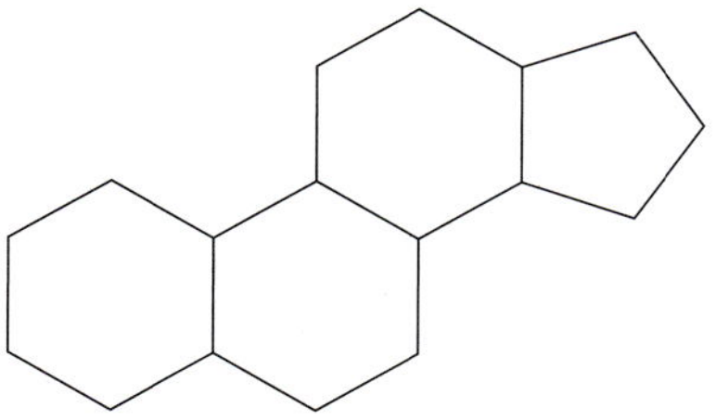

General four-ring structure of steroids

OBJECTIVE 4

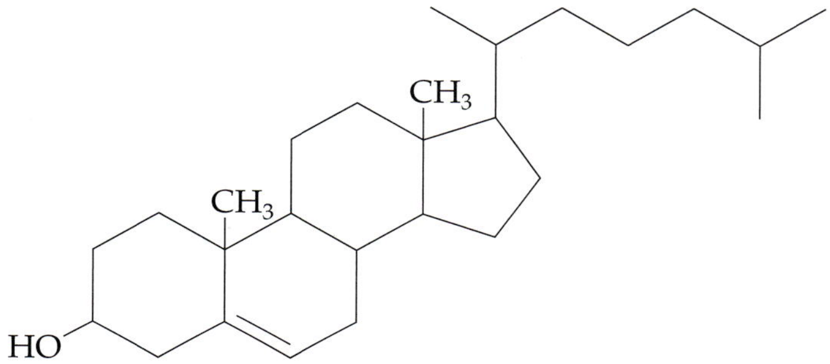

Cholesterol

Because cholesterol plays a role in the development of atherosclerosis, or hardening of the arteries, it too has gotten a bad reputation. The general public is largely unaware that as the starting material for the production of many important body chemicals, including hormones (compounds that help regulate chemical changes in the body), cholesterol is necessary for normal, healthy functioning of our bodies. For example, cholesterol is converted into the hormone progesterone, which is converted into other hormones, such as the male hormone testosterone (Figure 17.31, on on the facing page).

Estradiol, an important female hormone, is synthesized from testosterone. Estradiol and progesterone together regulate the monthly changes in the uterus and ovaries that are described collectively as the menstrual cycle (Figure 17.32).

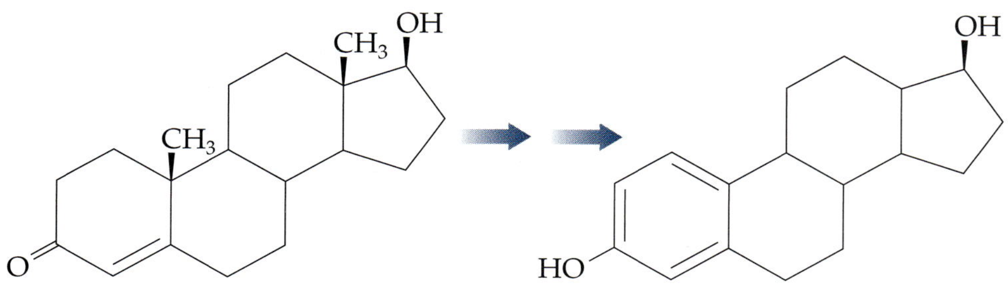

Testosterone

Estradiol

**Figure 17.32**
**Formation of the Female Sex Hormone Estradiol from Testosterone**

OBJECTIVE 3

## 17.3 Digestion

OBJECTIVE 20

Let's go back to that burger, fries, and milkshake and see what happens when the carbohydrate, protein, and fat molecules that they contain are digested. **Digestion** is the process of converting large molecules into small molecules capable of passing into the bloodstream to be carried throughout the body and used for many different purposes. Disaccharides are broken down into monosaccharides (glucose, galactose, and fructose), polysaccharides into glucose, protein into amino acids, and fat into glycerol and fatty acids (Table 17.3).

OBJECTIVE 20

Table 17.3
Products of Digestion

| Substance in food | Breakdown products |
|---|---|
| disaccharides (maltose, lactose, sucrose) | monosaccharides (glucose, galactose, and fructose) |
| polysaccharides (starch) | glucose |
| protein | amino acids |
| fat | glycerol and fatty acids |

The food that this woman eats will be digested to form the substances listed in Table 17.3.

Figure 17.33 shows a diagram of your digestive tract. When you eat that fast-food dinner, its digestion begins in a minor way in your mouth. The food next passes from your mouth through your esophagus to your stomach, where the first stages of protein digestion turn it into a pasty material called chyme. The chyme then travels to the small intestine, where most of the digestive process takes place. Because the purpose of this chapter is to give just an overview of the biochemistry involved, rather than a complete description, only protein digestion is described here.

**Figure 17.33**
The Human Digestive System

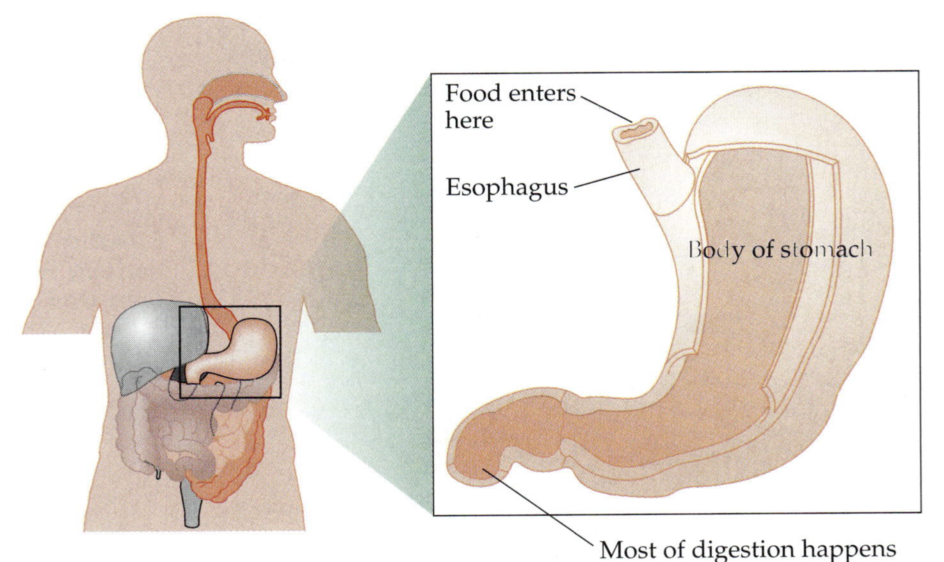
Food enters here
Esophagus
Body of stomach
Pasty material called chyme is formed in the stomach.
Most of digestion happens in the small intestine.

## Digestive Enzymes

The digestion process is regulated by **enzymes,** which not only increase the speed of chemical changes (by huge amounts) but also do so at the mild temperatures and, except in the stomach, close to neutral pH inside the body. Table 17.4 lists the sources of some of the more important digestive enzymes, what substance they digest, and the products that result from the digestion.

Table 17.4
Sources and Activities of Major Digestive Enzymes

| Organ | What it digests | Enzyme | Products |
|---|---|---|---|
| mouth | starch | amylase | maltose |
| stomach | protein | pepsin | shorter polypeptides |
| small intestine | starch | pancreatic amylase | maltose, maltriose, and short polysaccharides |
| | polypeptides | trypsin, chymotrypsin, carboxypeptidase | amino acids, dipeptides, and tripeptides |
| | triglyceride | pancreatic lipase | fatty acids and monoglycerides |
| | maltose | maltase | glucose |
| | sucrose | sucrase | glucose and fructose |
| | lactose | lactase | glucose and galactose |
| | polypeptides | aminopeptidase | amino acids, dipeptides, and tripeptides |

## Digestion of Protein

Certain undigested proteins can move from the digestive tract of babies into their blood (allowing newborns to get antibodies from their mother's first milk), but with rare exceptions, only amino acids (not proteins) move into an adult's bloodstream. For our cells to obtain the raw materials necessary for building the proteins of the human body, the proteins in our food must first be converted into amino acids.

The digestion of proteins begins in the stomach. The acidic conditions there weaken the links that maintain the protein molecules' tertiary structure. This process is called **denaturation,** because the loss of tertiary structure causes a corresponding loss of the protein's "natural" function. One of the reactions responsible for denaturation is shown in Figure 17.34: The $H^+$ ions in the stomach juices disrupt salt bridges within the protein molecules by binding to the negatively charged aspartic acid side chains.

OBJECTIVE 21

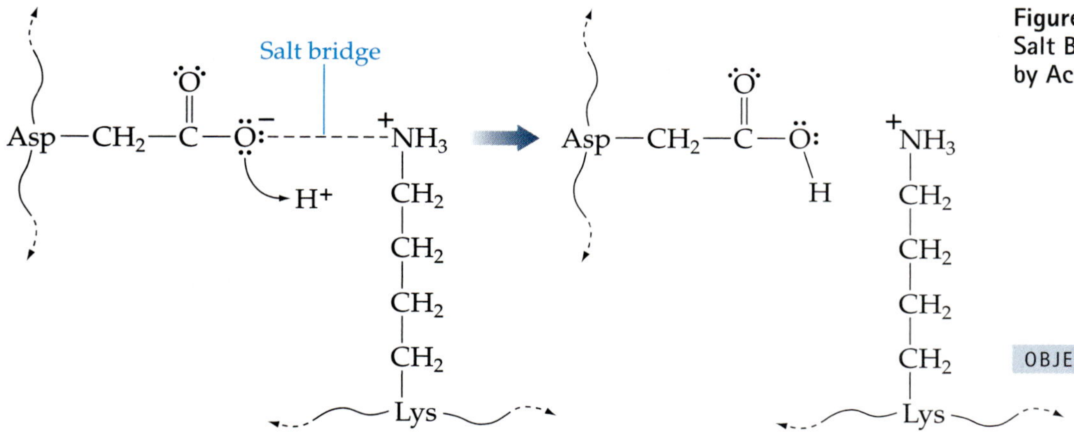

Figure 17.34
Salt Bridges Broken by Acidic Conditions

OBJECTIVE 21

Although an enzyme called pepsin begins to digest protein molecules while they are in the stomach, most of the digestion of protein takes place after the food leaves the stomach and moves into the small intestines. Here enzymes such as trypsin, chymotrypsin, elastase, carboxypeptidase, and aminopeptidase convert protein molecules into amino acids, dipeptides, and tripeptides. The dipeptides and tripeptides are converted into amino acids by other enzymes. Once the amino acids are free, they can move into the bloodstream and circulate throughout the body.

In all forms of digestion (whether of proteins, carbohydrates, or fats), larger molecules are broken down into smaller molecules by a reaction with water in which a water molecule is split in two, each part joining a different product molecule. This type of reaction is called **hydrolysis.** Remember that proteins are long chains of amino acids linked together by amide functional groups called peptide bonds. When protein molecules are digested, a series of hydrolysis reactions convert them into separate amino acids.

$$RCONR_2 + H_2O \rightleftharpoons RCO_2H + HNR_2$$

$$R{-}C({=}\ddot{O})-\ddot{N}(R){-}R \;+\; H{-}\ddot{O}{-}H \;\rightleftharpoons\; R{-}C({=}\ddot{O}){-}\ddot{O}{-}H \;+\; H{-}\ddot{N}(R){-}R$$

Amide Water Carboxylic acid Amine

In the laboratory, this reaction is very slow unless a strong acid catalyst is added to the mixture; yet in the small intestines, where the conditions are essentially neutral rather than acidic, most of the hydrolysis of proteins takes place rather quickly. The reason, as we have seen, is the presence of enzymes.

For an enzyme-mediated reaction to take place, the reacting molecule or molecules, which are called **substrates,** must fit into a specific section of the enzyme's structure called the **active site.** A frequently used analogy for the relationship of substrate to active site is the way a key must fit into a lock in order to do its job. Each active site has (1) a shape that fits a specific substrate or substrates only, (2) side chains that attract the enzyme's particular substrate(s), and (3) side chains specifically positioned to speed the reaction. Therefore, each enzyme will act on only a specific molecule or a specific type of molecule, and only in a specific way. For example, chymotrypsin's one enzymatic function is to accelerate the breaking of peptide bonds that link an amino acid that has a nonpolar side chain, such as phenylalanine, to another amino acid on the interior of polypeptide chains.

OBJECTIVE 22

Visit the following Web site to see a proposed mechanism for how chymotrypsin catalyzes the hydrolysis of certain peptide bonds in protein molecules.

**www.chemplace.com/college/**

## 17.4 Synthetic Polymers

Political events of the 1930s created an interesting crisis in fashion. Women wanted sheer stockings, but with the growing unrest in the world, manufacturers were having an increasingly difficult time obtaining the silk necessary to make them. Chemistry came to the rescue.

If you were a chemist trying to develop a substitute for silk, your first step would be to find out as much as you could about its chemical structure. Silk is a polyamide (polypeptide), a long-chain molecule (polymer) composed of amino acids linked together by amide functional groups (peptide bonds). Silk molecules are 44% glycine (the simplest of the amino acids, with a hydrogen for its distinguishing side chain) and 40% alanine (another very simple amino acid, with a $-CH_3$ side chain). Having acquired this information, you might decide to try synthesizing a simple polypeptide of your own.

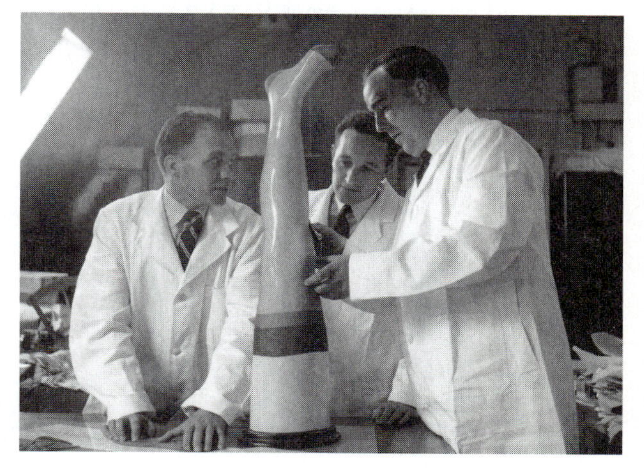

Nylon was designed by scientists to approximate the qualities of silk.

The next step in your project would be to plan a process for making the new polymer, perhaps using the process of protein formation in living organisms as a guide. We saw in Section 17.1 that polypeptides form in nature when the carboxylic acid group of one amino acid reacts with the amine group of another amino acid to form an amide functional group called a peptide bond. The reason why amino acids are able to form long chains in this way is that amino acids are difunctional: Each amino acid possesses both an amine functional group and a carboxylic acid functional group. After two amino acids are linked by a peptide bond, each of them still has either a carboxylic acid group or an amine group free to link to yet another amino acid (Figure 17.23).

### Nylon, a Synthetic Polypeptide

OBJECTIVE 23

W. H. Carothers, working for E.I. Du Pont de Nemours and Company, developed the first synthetic polyamide. He found a way to react adipic acid (a di-carboxylic acid) with hexamethylene diamine (which has two amine functional groups) to form long-chain polyamide molecules called Nylon 66. (The first "6" in the "66" indicates the number of carbon atoms in each portion of the polymer chain that are contributed by the diamine, and the second "6" shows the number of carbon atoms in each portion that are contributed by the di-carboxylic acid.) The reactants are linked together by condensation reactions in which an –OH group removed from a carboxylic functional group combines with a –H from an amine group to form water, and an amide linkage forms between the reacting molecules (Figure 17.35, on the next page). When small molecules, such as water, are released in the formation of a polymer, the polymer is called a **condensation** (or sometimes a **step-growth**) **polymer.**

**Figure 17.35**
**Nylon Formation**

OBJECTIVE 23

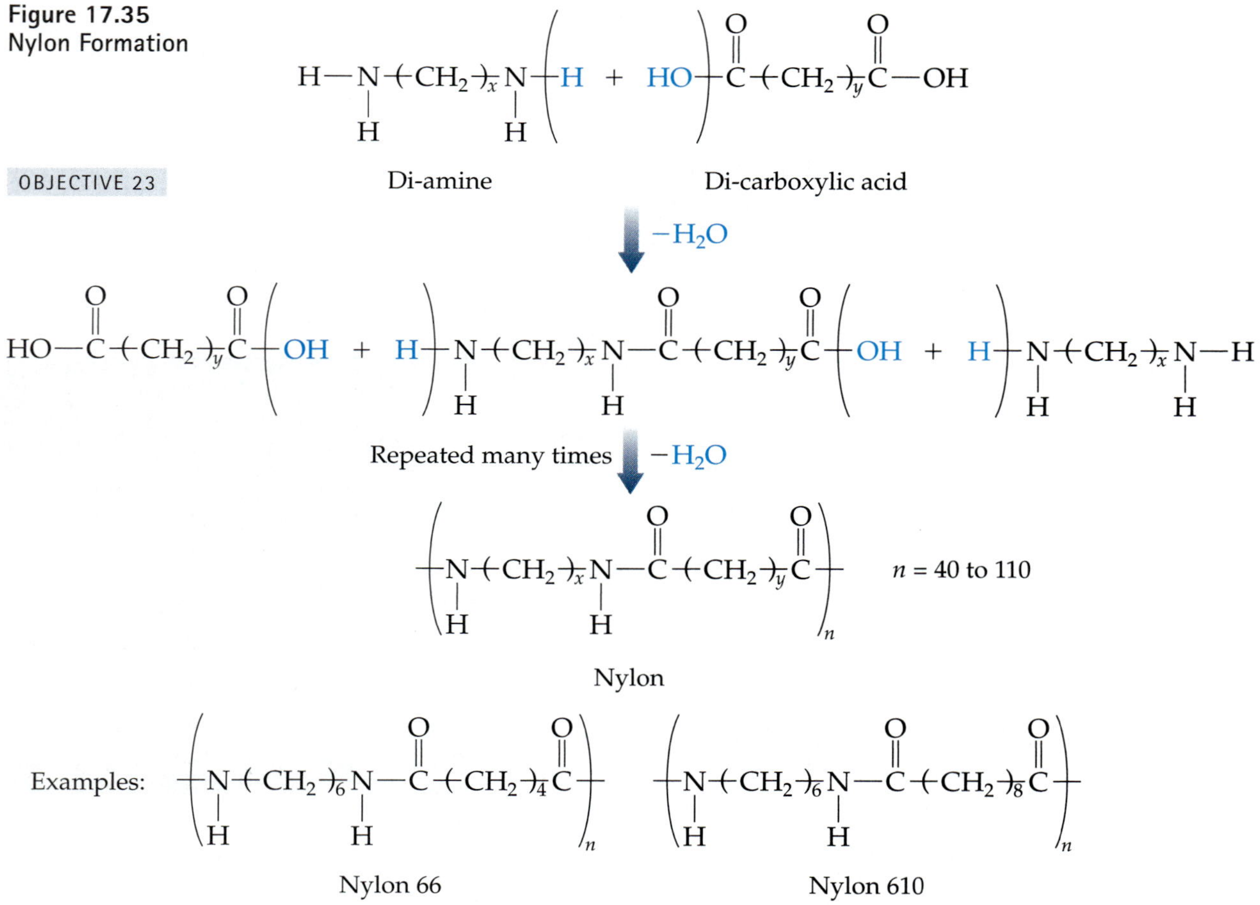

$$H{-}\underset{H}{\underset{|}{N}}{-}(CH_2)_x{-}\underset{H}{\underset{|}{N}}{-}H + HO{-}\overset{O}{\overset{\|}{C}}{-}(CH_2)_y{-}\overset{O}{\overset{\|}{C}}{-}OH$$

Di-amine    Di-carboxylic acid

$\downarrow -H_2O$

$$HO{-}\overset{O}{\overset{\|}{C}}{-}(CH_2)_y{-}\overset{O}{\overset{\|}{C}}{-}OH + H{-}\underset{H}{\underset{|}{N}}{-}(CH_2)_x{-}\underset{H}{\underset{|}{N}}{-}\overset{O}{\overset{\|}{C}}{-}(CH_2)_y{-}\overset{O}{\overset{\|}{C}}{-}OH + H{-}\underset{H}{\underset{|}{N}}{-}(CH_2)_x{-}\underset{H}{\underset{|}{N}}{-}H$$

Repeated many times $\downarrow -H_2O$

$$\left(\underset{H}{\underset{|}{N}}{-}(CH_2)_x{-}\underset{H}{\underset{|}{N}}{-}\overset{O}{\overset{\|}{C}}{-}(CH_2)_y{-}\overset{O}{\overset{\|}{C}}\right)_n \quad n = 40 \text{ to } 110$$

Nylon

Examples:

$$\left(\underset{H}{\underset{|}{N}}{-}(CH_2)_6{-}\underset{H}{\underset{|}{N}}{-}\overset{O}{\overset{\|}{C}}{-}(CH_2)_4{-}\overset{O}{\overset{\|}{C}}\right)_n$$

Nylon 66

$$\left(\underset{H}{\underset{|}{N}}{-}(CH_2)_6{-}\underset{H}{\underset{|}{N}}{-}\overset{O}{\overset{\|}{C}}{-}(CH_2)_8{-}\overset{O}{\overset{\|}{C}}\right)_n$$

Nylon 610

Chemists write chemical formulas for polymers by enclosing the repeating unit in parentheses followed by a subscript $n$ to indicate that the unit is repeated many times:

OBJECTIVE 27

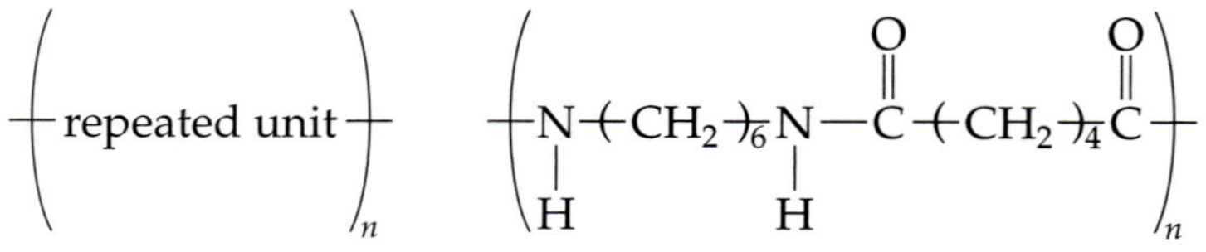

$$\left(\text{repeated unit}\right)_n$$

General polymer formula

$$\left(\underset{H}{\underset{|}{N}}{-}(CH_2)_6{-}\underset{H}{\underset{|}{N}}{-}\overset{O}{\overset{\|}{C}}{-}(CH_2)_4{-}\overset{O}{\overset{\|}{C}}\right)_n$$

Nylon 66

Nylon 66 was first made in 1935 and went into commercial production in 1940. Its fibers were strong, elastic, abrasion-resistant, lustrous, and easy to wash. With these qualities, nylon became more than just a good substitute for silk in stockings. Today nylon polymers are used in a multitude of products, including carpeting, upholstery fabrics, automobile tires, and turf for athletic fields.

One of the reasons for nylon's exceptional strength is the hydrogen bonding between amide functional groups. The higher the percentage of amide functional groups in nylon's polymer structure, the stronger the hydrogen bonding between the chains. Thus changing the number of carbon atoms in the diamine ($x$ in Figure 17.35) and in the di-carboxylic acid ($y$ in Figure 17.35) changes the nylon's properties. For example, Nylon 610, which has four more carbon atoms in its di-carboxylic acid molecules than are found in Nylon 66, is somewhat weaker than Nylon 66 and has a lower melting point. Nylon 610 is used for bristles in paintbrushes.

OBJECTIVE 24

Camping and backpacking equipment are often made of nylon.

## Polyesters

You've got a big day planned in the city: an afternoon at the ballpark watching Mark McGwire belt home runs, followed by dinner and disco dancing at a "retro" club called Saturday Night Fever. Mark's uniform and your own disco outfit are almost certainly made from polyester, which is a condensation polymer similar to nylon. Polyesters are made from the reaction of a diol (a compound with two alcohol functional groups) with a di-carboxylic acid. Figure 17.36 shows the steps in the formation of poly(ethylene terephthalate) from ethylene glycol and terephthalic acid.

OBJECTIVE 25

**Figure 17.36**
Polyester Formation

$$H-OCH_2CH_2O-H \;+\; HO-\overset{O}{\overset{\|}{C}}-C_6H_4-\overset{O}{\overset{\|}{C}}-OH$$

Ethylene glycol — Terephthalic acid

$\downarrow -H_2O$

$$HO-\overset{O}{\overset{\|}{C}}-C_6H_4-\overset{O}{\overset{\|}{C}}-OH \;+\; H-OCH_2CH_2O-\overset{O}{\overset{\|}{C}}-C_6H_4-\overset{O}{\overset{\|}{C}}-OH \;+\; H-OCH_2CH_2O-H$$

$\downarrow$ Repeated many times

OBJECTIVE 25 OBJECTIVE 27

$$\left(OCH_2CH_2O-\overset{O}{\overset{\|}{C}}-C_6H_4-\overset{O}{\overset{\|}{C}}\right)_n$$

$n$ = a large integer

Poly(ethylene terephthalate)

The transparency of which polyester is capable makes it a popular choice for photographic film and projection slides. Mylar, which is used to make long-lasting balloons, is a polyester, as is the polymer used for making eyeglass lenses. Polyesters have been used for fabrics that would once have been made from cotton, whose fundamental structure consists

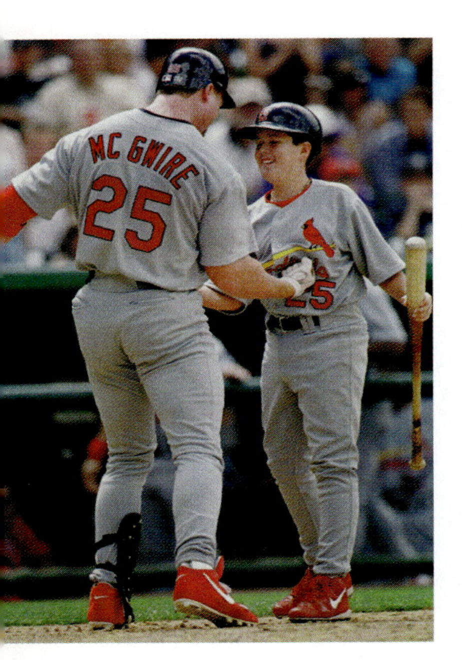


The Cardinals' Mark McGwire is greeted by his 11-year-old son, Matt, after hitting one of his many home runs. The uniforms worn by baseball players are made of polyester fibers.

of the polymer cellulose. Polyester fibers, such as the fibers of Dacron® and Fortrel®—made from poly(ethylene terephthalate)—are about three times as strong as cellulose fibers, so polyester fabrics or blends that include polyester last longer than fabrics made from pure cotton. The strength and elasticity of polyesters make them ideal for sports uniforms.

## Addition Polymers

Unlike condensation (step-growth) polymers, which release small molecules, such as water, as they form, the reactions that lead to **addition, or chain-growth, polymers** incorporate all of the reactants' atoms into the final product. Addition polymers are usually made from molecules that have the following general structure:

```
W       X
 \     /
  C = C
 /     \
Y       Z
```

Different W, X, Y, and Z groups distinguish one addition polymer from another.

Visit the following Web site to see one way in which addition polymers can be made.
**www.chemplace.com/college/**

If all of the atoms attached to the carbons of the monomer's double bond are hydrogen atoms, then the initial reactant is ethylene, and the polymer it forms is polyethylene.

OBJECTIVE 27

```
  H       H                          / H   H \
   \     /       polymerization     |  |   |  |
n   C = C      ----------------->  -|- C - C -|-      n = a very large integer
   /     \                          |  |   |  |
  H       H                          \ H   H / n

   Ethylene                           Polyethylene
```

OBJECTIVE 26

Polyethylene molecules can be made using different techniques. One process leads to branches that keep the molecules from fitting closely together. Other techniques have been developed to make polyethylene molecules with very few branches. These straight-chain molecules fit together more efficiently, yielding a high-density polyethylene, HDPE, that is more opaque, harder, and stronger than the low-density polyethylene, LDPE. HDPE is used for containers, such as milk bottles, and LDPE is used for filmier products, such as sandwich bags.

Table 17.5 shows other addition polymers that can be made using monomers with different groups attached to the carbons in the monomer's double bond.

**Table 17.5**
Addition (Chain-Growth) Polymers

OBJECTIVE 27

| Initial Reactant | Polymer | Examples of Uses |
|---|---|---|
| Ethylene | Polyethylene | packaging, beverage containers, food containers, toys, detergent bottles, plastic buckets, mixing bowls, oil bottles, plastic bags, drapes, squeeze bottles, wire, and cable insulation |
| Propylene | Polypropylene | clothing, home furnishings, indoor-outdoor carpeting, rope, automobile interior trim, battery cases, margarine and yogurt containers, grocery bags, caps for containers, carpet fiber, food wrap, plastic chairs, and luggage |
| Vinyl chloride | Poly(vinyl chloride) | "vinyl" seats in automobiles, "vinyl" siding for houses, rigid pipes, food wrap, vegetable oil bottles, blister packaging, rain coats, shower curtains, and flooring |
| Styrene | Polystyrene | foam insulation and packaging (Styrofoam®); plastic utensils; rigid, transparent salad containers; clothes hangers; foam cups; and plates |

## SPECIAL TOPIC 17.3 Recycling Synthetic Polymers

You finish off the last of the milk. What are you going to do with the empty bottle? If you toss it into the trash, it will almost certainly go into a landfill, taking up space and serving no useful purpose. But if you put it in the recycle bin, it's likely to be melted down to produce something new.

Between 50 and 60 billion pounds of synthetic polymers are manufactured each year in the United States—over 200 pounds per person. A large percentage of these polymers are tossed into our landfills after use. This represents a serious waste of precious raw materials (the petroleum products from which synthetic polymers are made) and exacerbates concern that the landfills are quickly filling up. These factors give the recycling of polymers a high priority.

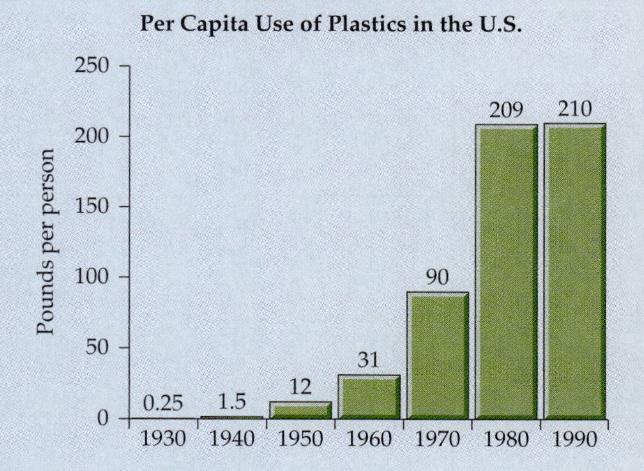


*(continued)*

## Special Topic 17.3 (Continued)

*(continued from previous page)*

Some synthetic polymers can be recycled and some cannot. So-called *thermoplastic polymers*, usually composed of linear or only slightly branched molecules, can be heated and formed and then reheated and re-formed. On the other hand, *thermosetting polymers*, which consist of molecules with extensive three-dimensional cross-linking, decompose when heated, so they cannot be reheated and re-formed. In general, thermosetting polymers cannot be recycled.

In 1988 the Plastic Bottling Institute suggested a system in which numbers embossed on objects made of polymers tell the recycling companies what type of polymer was used in the object's construction (Table 17.6). This numbering system facilitates the collection, sorting, and reprocessing of the polymers that can be recycled.

Uses of Thermosets

69% Building and construction
8% Consumer products
4% Transportation equipment
4% Adhesives and coatings
4% Electrical equipment
11% Other

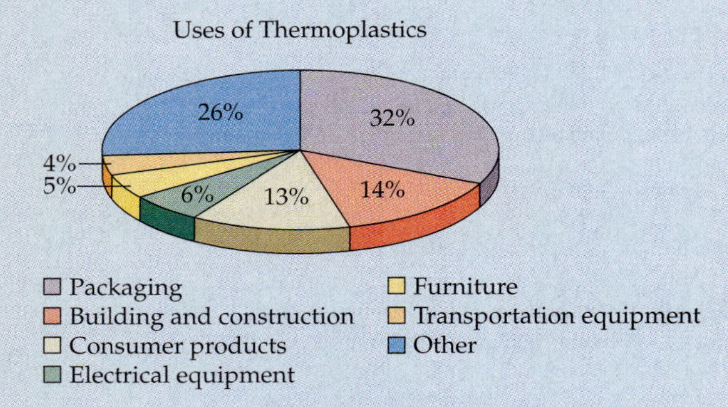


**Table 17.6**
Recyclable Thermoplastics

OBJECTIVE 28

| Symbol and abbreviation | Name of polymer | Examples of uses for virgin polymer | Examples of uses for recycled polymer |
|---|---|---|---|
| 1 PET | polyethylene terephthalate | beverage containers, boil-in food pouches, processed meat packages | detergent bottles, carpet fibers, fleece jackets |
| 2 HDPE | high-density polyethylene | milk bottles, detergent bottles, plastic buckets, mixing bowls, oil bottles, toys, plastic bags | compost bins, detergent bottles, crates, agricultural pipes, curbside recycling bins |
| 3 PVC | poly(vinyl chloride) | food wrap, vegetable oil bottles, blister packaging, plastic pipes | detergent bottles, tiles, plumbing pipe fittings |
| 4 LDPE | low-density polyethylene | shrink wrap, plastic sandwich bags, garment bags, squeeze bottles | films for industry and general packaging |
| 5 PP | polypropylene | margarine and yogurt containers, grocery bags, caps for containers, carpet fiber, food wrap, plastic chairs, luggage | compost bins, curbside recycling bins |
| 6 PS | polystyrene | plastic utensils, clothes hangers, foam cups and plates | coat hangers, office accessories, video/CD boxes |
| 7 OTHER | includes nylon | other | — |

## Chapter Glossary

**Biochemistry** The chemistry of biological systems.

**Alkanes** Hydrocarbons (compounds composed of carbon and hydrogen) in which all of the carbon–carbon bonds are single bonds.

**Alkenes** Hydrocarbons that have one or more carbon–carbon double bonds.

**Functional group** A small section of an organic molecule that to a large extent determines the chemical and physical characteristics of the molecule.

**Alkynes** Hydrocarbons that have one or more carbon–carbon triple bonds.

**Arenes or aromatics** Compounds that contain the benzene ring.

**Alcohols** Compounds with one or more –OH groups attached to a hydrocarbon group.

**Carboxylic acids** Compounds that have a hydrogen atom or a hydrocarbon group connected to a –COOH (or $-CO_2H$) group.

**Ethers** Compounds with two hydrocarbon groups surrounding an oxygen atom.

**Aldehydes** Compounds that have a hydrogen atom or a hydrocarbon group connected to a –CHO group.

**Ketones** Compounds that have the –CO– functional group surrounded by hydrocarbon groups.

**Esters** Compounds that have the general formula, $RCO_2R'$, where R can be a hydrogen atom or a hydrocarbon group and R′ is a hydrocarbon group.

**Amine** A compound with the general formula $R_3N$, where R represents a hydrogen atom or a hydrocarbon group and at least one R-group is a hydrocarbon group.

**Amides** Compounds with the general formula RCONR, where each R represents a hydrogen atom or a hydrocarbon group.

**Carbohydrates** Another name for saccharides. Many have the general formula $(CH_2O)_n$, which led early chemists to think they were hydrates of carbon.

**Saccharides** Sugar, starch, and cellulose.

**Monosaccharides** Sugar molecules with one saccharide unit.

**Disaccharides** Sugar molecules composed of two monosaccharide units.

**Polysaccharides** Molecules with many saccharide units.

**Polymer** A large molecule composed of repeating units.

**Monomer** The repeating unit in a polymer.

**Peptide bond** An amide functional group that forms when the carboxylic acid group on one amino acid reacts with the amine group of another amino acid.

**Peptide** A substance that contains two or more amino acids linked together by peptide bonds.

**Condensation reaction** A chemical reaction in which a larger molecule is made from connecting two smaller molecules accompanied by the release of water or another small molecule.

**Proteins** Natural polypeptides.

**Primary protein structure** The sequence of amino acids in a protein molecule.

**Secondary protein structure** The arrangement of atoms that are close to each other in a polypeptide chain. Examples of secondary structures include the α-helix and β-sheet.

**Tertiary protein structure** The overall arrangement of atoms in a protein molecule.

**Triglyceride** A compound with three hydrocarbon groups attached to a three-carbon backbone by ester functional groups.

**Hydrogenation** A process by which hydrogen is added to an unsaturated triglyceride to convert double bonds into single bonds. This can be done by combining the unsaturated triglyceride with hydrogen gas and a platinum catalyst.

**Saturated triglyceride** A triglyceride with single bonds between all of the carbon atoms.

**Unsaturated triglyceride** A triglyceride that has one or more carbon–carbon double bonds.

**Steroids** Compounds that contain the following four-ring structure:

**Digestion** The process of converting large molecules into small molecules that can move into the bloodstream to be carried throughout the body.

**Denature** To change the tertiary structure of a protein, causing it to lose its natural function.

**Hydrolysis** A chemical reaction in which larger molecules are broken down into smaller molecules by a reaction with water in which a water molecule is split in two, each part joining a different product molecule.

**Enzyme** A naturally occurring catalyst.

**Substrate** A molecule that an enzyme causes to react.

**Active site** A specific section of the protein structure of an enzyme in which the substrate fits and reacts.

**Condensation (or step-growth) polymer** A polymer formed in a reaction that releases small molecules, like water. This category includes nylon and polyester.

**Addition (or chain-growth) polymer** A polymer that contains all of the atoms of the original reactant in its structure. This category includes polyethylene, polypropylene, and poly(vinyl chloride).

You can test yourself on the glossary terms at the following Web site:

**www.chemplace.com/college/**

## Chapter Objectives

**The goal of this chapter is to teach you to do the following.**

1. Define all of the terms in the Chapter 17 Glossary.

### Section 17.1 Organic Compounds

2. Given a Lewis structure of an organic molecule, draw its condensed formula and line drawing.
3. Given a Lewis structure, a condensed formula, or a line drawing of an organic compound, identify it as representing an alkane, alkene, alkyne, arene (aromatic), alcohol, carboxylic acid, ether, aldehyde, ketone, ester, amine, or amide.

### Section 17.2 Important Substances in Food

4. Given a structure for a biomolecule, identify it as a carbohydrate, amino acid, peptide, triglyceride, or steroid.
5. Given a structure for a carbohydrate molecule, identify it as a monosaccharide, disaccharide, or polysaccharide.
6. Describe the general differences among glucose, galactose, and fructose.
7. Identify the saccharide units that form the disaccharides maltose, lactose, and sucrose.
8. Describe the similarities and differences among amylose, amylopectin, glycogen, and cellulose.
9. Explain why starch can be digested in our digestive tracts and why cellulose cannot.
10. Describe the general structure of amino acids.
11. Explain why amino acid molecules in our bodies usually have a positive end and a negative end.
12. Describe how amino acids are linked to form peptides.
13. Identify descriptions of the primary, secondary, and tertiary structures of proteins.
14. Describe how disulfide bonds, hydrogen bonds, and salt bridges help hold protein molecules together in specific tertiary structures.
15. Explain why it is more efficient to store energy in the body as fat than as carbohydrate or protein.
16. Write or identify a description of the general structure of a triglyceride molecule.
17. Given the chemical formulas for two triglycerides with different numbers of carbon–carbon double bonds, identify the one that is more likely to be a solid at room temperature and the one that is more likely to be a liquid.
18. Given the chemical formula for a triglyceride, identify it as saturated or unsaturated.
19. Given the chemical formula for an unsaturated triglyceride, draw the structure for the product of its complete hydrogenation.

### Section 17.3 Digestion

20. Identify the digestion products of disaccharides, polysaccharides, protein, and fat.
21. Describe how the digestion of protein molecules is facilitated by changes in the stomach.
22. Explain why each enzyme only acts on a specific molecule or a specific type of molecule.

### Section 17.4 Synthetic Polymers

23. Describe how Nylon 66 is made.
24. Explain why Nylon 66 is stronger than Nylon 610.
25. Describe how polyesters are made.
26. Describe the similarities and differences between the molecular structures of low-density polyethylene (LDPE) and high-density polyethylene (HDPE).
27. Given a structure for a polymer, identify it as representing nylon, polyester, polyethylene, poly(vinyl chloride), polypropylene, or polystyrene.
28. Given the recycling code for an object, identify the polymer used to make the object.

## Review Questions

1. Draw a Lewis structure, a geometric sketch, a ball-and-stick model, and a space-filling model for methane, $CH_4$.
2. Draw a Lewis structure, a geometric sketch, a ball-and-stick model, and a space-filling model for ammonia, $NH_3$.
3. Draw a Lewis structure, a geometric sketch, a ball-and-stick model, and a space-filling model for water, $H_2O$.
4. Draw a Lewis structure, a geometric sketch, a ball-and-stick model, and a space-filling model for methanol, $CH_3OH$.
5. The following Lewis structure represents a molecule of formaldehyde, $CH_2O$. Draw a geometric sketch, a ball-and-stick model, and a space-filling model for this molecule.

```
    :Ö:
     ||
  H—C—H
```

6. The following Lewis structure represents a molecule of hydrogen cyanide, HCN. Draw a geometric sketch, a ball-and-stick model, and a space-filling model for this molecule.

```
  H—C≡N:
```

7. The following Lewis structure represents a molecule of ethanamide, $CH_3CONH_2$. Draw a geometric sketch for this molecule.

```
     H  :Ö:
     |   ||   ..
  H—C—C—N—H
     |        |
     H        H
```

## Key Ideas

**Complete the following statements by writing one of these words or phrases in each blank.**

| | |
|---|---|
| 17 kJ/g | large |
| 37 kJ/g | linear sequence |
| acidic | liver |
| active site | long-term |
| addition | monomers |
| amide | monosaccharide |
| amino acids | monosaccharides |
| amylopectin | muscle cells |
| amylose | *n* |
| benzene ring | –OH |
| carbon-carbon | overall shape |
| cellulose | peptide |
| cholesterol | parentheses |
| close to each other | partially |
| condensation | polysaccharides |
| denaturation | protein |
| diol | proteins in our food |
| disaccharides | repeating units |
| double bonds | shape |
| energy | single |
| fatty acids | single bonds |
| fructose | small |
| galactose | small section |
| glucose | split in two |
| glucose units | starches |
| glycerol | step-growth |
| glycogen | substrates |
| hydrocarbon groups | sugars |
| hydrocarbons | water |
| hydrogenation | |

8. Hydrocarbons (compounds composed of carbon and hydrogen) in which all of the carbon–carbon bonds are ____________ bonds are called alkanes.

9. Hydrocarbons that have one or more ____________ double bonds are called alkenes.

10. When a(n) ________________ of an organic molecule is largely responsible for the molecule's chemical and physical characteristics, that section is called a functional group.
11. ________________ that have one or more carbon–carbon triple bonds are called alkynes.
12. Compounds that contain the ________________ are called arenes or aromatics.
13. Alcohols are compounds with one or more ________________ groups attached to a hydrocarbon group, that is, to a group consisting of only carbon and hydrogen atoms.
14. Ethers consist of two ________________ surrounding an oxygen atom.
15. Carbohydrate is a general name for ________________, ________________, and cellulose.
16. Sugars are monosaccharides and ________________. Starches and cellulose are ________________.
17. Disaccharides are composed of two ________________ units.
18. Maltose is a disaccharide consisting of two ________________ units.
19. Lactose, or milk sugar, is a disaccharide consisting of ________________ and glucose.
20. Sucrose is a disaccharide that contains glucose and ________________.
21. The most common polysaccharides are starch, ________________ (sometimes called animal starch), and cellulose. All of these are composed of repeating ________________, but they differ in the way the units are attached.
22. Almost every kind of plant cell has ________________ stored in the form of starch. Starch itself has two general forms, ________________ and ________________.
23. Glycogen molecules are stored in ________________ and ________________ of animals, where they can be converted into glucose molecules and be used as a source of energy.
24. All the polysaccharides are polymers, a general name for large molecules composed of ________________, called monomers.
25. Our digestive enzymes are able to break the linkages in starch to release energy-producing glucose, but they are unable to liberate glucose molecules from ________________ because they cannot break the linkages there.
26. Protein molecules are polymers composed of ________________ called ________________.
27. Amino acids are linked together by a(n) ________________ bond, created when the carboxylic acid group of one amino acid reacts with the amine group of another amino acid to form a(n) ________________ functional group.
28. A condensation reaction is a chemical change in which ________________ or other small molecules are released.

29. The primary structure of a protein is the ________________ of its amino acids.
30. The arrangement of atoms that are ________________ in the polypeptide chain is called the secondary structure of the protein.
31. The tertiary structure of a protein is its very specific ________________.
32. The fat stored in our bodies is our primary ________________ energy source.
33. One of the reasons why it is more efficient to store energy as fat than as carbohydrate or protein is that fat produces ________________, whereas carbohydrate and protein produce only ________________.
34. A process called ________________ converts liquid triglycerides to solid triglycerides by adding hydrogen atoms to the double bonds and so converting them to single bonds.
35. When enough hydrogen atoms are added to a triglyceride to convert all double bonds to ________________, we call it a saturated triglyceride (or fat). It is saturated with hydrogen atoms.
36. A triglyceride that still has one or more carbon–carbon ________________ is an unsaturated triglyceride.
37. If enough hydrogen is added to an unsaturated triglyceride to convert some but not all of the carbon–carbon double bonds to single bonds, we say it has been ________________ hydrogenated.
38. As the starting material for the production of many important body chemicals, including hormones (compounds that help regulate chemical changes in the body), the steroid ________________ is necessary for normal, healthy functioning of our bodies.
39. Digestion is the process of converting ________________ molecules into ________________ molecules capable of passing into the bloodstream to be carried throughout the body and used for many different purposes.
40. In digestion, disaccharides are broken down into ________________ (glucose, galactose, and fructose), polysaccharides into glucose, ________________ into amino acids, and fat into ________________ and ________________.
41. For our cells to obtain the raw materials necessary for building the proteins of the human body, the ________________ must first be converted into amino acids.
42. The digestion of proteins begins in the stomach. The ________________ conditions there weaken the links that maintain the protein molecules' tertiary structure. This process is called ________________, because the loss of tertiary structure causes a corresponding loss of the protein's "natural" function.
43. In all forms of digestion (whether of proteins, carbohydrates, or fats), larger molecules are broken down into smaller molecules by a reaction with water in which a water molecule is ________________, each part joining a different product molecule. This type of reaction is called hydrolysis.

44. For an enzyme-mediated reaction to take place, the reacting molecule or molecules, which are called ________________, must fit into a specific section of the enzyme's structure called the ________________. A frequently used analogy for the relationship of substrate to active site is the way a key must fit into a lock in order to do its job. Each active site has (1) a(n) ________________ that fits a specific substrate or substrates only, (2) side chains that attract the enzyme's particular substrate(s), and (3) side chains specifically positioned to speed the reaction.

45. The reactants that form nylon are linked together by ________________ reactions in which an –OH group removed from a carboxylic functional group combines with a –H from an amine group to form water, and an amide linkage forms between the reacting molecules.

46. When small molecules, such as water, are released in the formation of a polymer, the polymer is called a condensation (or sometimes ________________) polymer.

47. Chemists write chemical formulas for polymers by enclosing the repeating unit in ________________ followed by a subscript ________________ to indicate that the unit is repeated many times.

48. Polyesters are made from the reaction of a(n) ________________ (a compound with two alcohol functional groups) with a di-carboxylic acid.

49. The reactions that lead to ________________, or chain-growth, polymers incorporate all of the reactants' atoms into the final product.

## Chapter Problems

### Section 17.1 Organic Compounds

50. Classify each of the following as organic or inorganic (not organic) compounds.
    a. sodium chloride, NaCl, in table salt
    b. hexane, $C_6H_{14}$, in gasoline
    c. ethyl butanoate, $CH_3CH_2CH_2CO_2CH_2CH_3$, in a pineapple
    d. water, $H_2O$, in your body

51. Classify each of the following as organic or inorganic (not organic) compounds.
    a. an oil molecule, $C_{57}H_{102}O_6$, in corn oil
    b. silicon dioxide, $SiO_2$, in beach sand
    c. aluminum oxide, $Al_2O_3$, in a ruby
    d. sucrose, $C_{12}H_{22}O_{11}$, in a piece of hard candy

52. Identify each of these Lewis structures as representing an alkane, alkene, alkyne, arene (aromatic), alcohol, carboxylic acid, aldehyde, ketone, ether, ester, amine, or amide.

OBJECTIVE 3

a. H—C—C—C—C—C—C—C—H

b. H—C—C—C—C—C—C—H

c. H—C—(C)$_{12}$—C—O—H

d. H—C—C—C—C—N—H

e. H—C—C—O—C—C—H

f. H—C—C—C—H

g. H—C=C—C=C—H

h. H—C—C—C—O—C—H

i. H—C—C—C—C—C—C—C—H

j. H—C—C≡C—C—H

k. H—C—H

OBJECTIVE 3

53. Identify each of these Lewis structures as representing an alkane, alkene, alkyne, arene (aromatic), alcohol, carboxylic acid, aldehyde, ketone, ether, ester, amine, or amide.

```
                            H
                            |
                         H—C—H
       H  :Ö:        H   H  |   H
       |   ||        |   |  |   |
a.  H—C—C—Ö—C—C—C—C—H
       |             |   |  |   |
       H             H   H  H   H

       H   H   H   H   H
       |   |   |   |   |
b.  H—C—C—C—C—C—H
       |   |   |   |   |
       H   H   |   H   H
            H—C—H
               |
               H

          H
          |
       H—C—H
       H  |   H  :Ö:  H
       |  |   |   ||  |
c.  H—C—C—C—C—C—H
       |  |   |       |
       H  H   H       H

          H
          |
       H—C—H
       H  |   H   H        H
       |  |   |   |        |
d.  H—C—C—C—C—N̈—C—H
       |  |   |   |   |    |
       H  H   H   H   H    H

       H   H  :Ö:
       |   |   ||
e.  H—C—C—C—H
       |   |
       H   |
        H—C—H
           |
           H

          H
          |
       H—C—H
       H  |   H
       |  |   |
f.  H—C—C—C—H
       |  |   |
       H  |   H
         :Ö—H

          H
          |
       H—C—H
       H  |   H  :Ö:
       |  |   |   ||
g.  H—C—C—C—C—N̈—H
       |  |   |       |
       H  H   H       H

                                 H
                                 |
                              H—C—H
       H   H   H   H             |  H
       |   |   |   |             |  |
h.  H—C—C—C—C—Ö—C—C—H
       |   |   |   |             |  |
       H   H   H   H             H  H

       H   H   H   H
       |   |   |   |
i.  H—C—C═C—C—H
       |           |
       H           H

                 H
                 |
              H—C—H
                 |   H
                 |   |
j.  H—C≡C—C—C—H
                 |   |
                 H   H

       H     H
       |     |
k.  H—C——C—H
       |     |
       |     H
```

**54.** Write condensed chemical formulas to represent the Lewis structures in parts (a) through (j) of problem 52. For example, 2-propanol can be described as $CH_3CH(OH)CH_3$. OBJECTIVE 2

55. Write condensed chemical formulas to represent the Lewis structures in parts (a) through (j) of problem 53. For example, 2-propanol can be described as $CH_3CH(OH)CH_3$. OBJECTIVE 2

**56.** Write line drawings to represent the Lewis structures in parts (a) through (i) of problem 52. OBJECTIVE 2

57. Write line drawings to represent the Lewis structures in parts (a) through (i) of problem 53. OBJECTIVE 2

**58.** The chemical structure of the artificial sweetener aspartame is below. Identify all of the organic functional groups that it contains.

```
       H   O       H   O
       |   ||      |   ||
H2N—C—C—N—C—C—O—CH3
       |       |   |
      CH2      H  CH2
       |           |
       C=O      (benzene ring)
       |
      OH
```

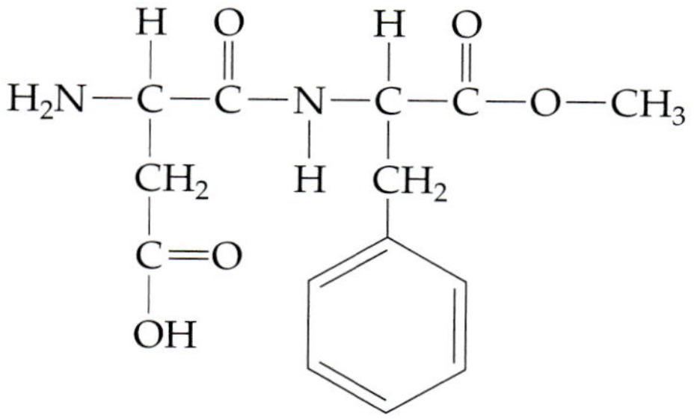

59. Mifepristone (often called RU-486) is a controversial *morning after* contraceptive pill. Identify all of the organic functional groups that it contains.

CH3   OH   C≡CH   O

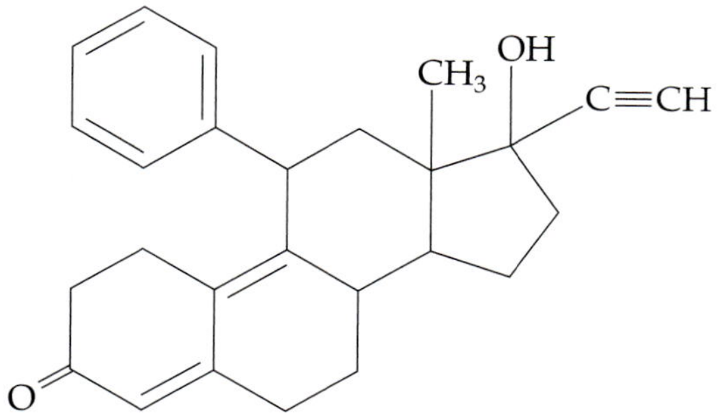

**60.** Draw geometric sketches, including bond angles, for each of the following organic molecules.

a. 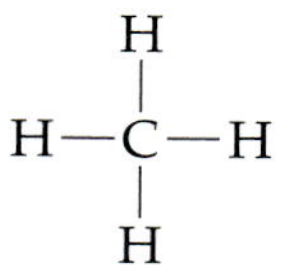
```
    H
    |
H—C—H
    |
    H
```

b. 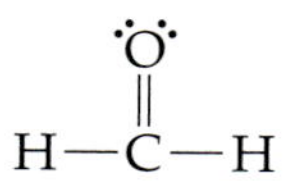
```
   :Ö:
    ||
H—C—H
```

c. H—C≡N:

61. Draw geometric sketches, including bond angles, for each of the following organic molecules.

```
         ••
        :O:
         ||
a.  :Cl—C—Cl:
     ••     ••
```

b. H—C≡C—H

```
        H
        |
c.  H—C—Br:
        |   ••
        H
```

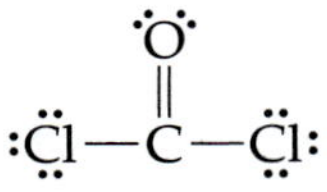

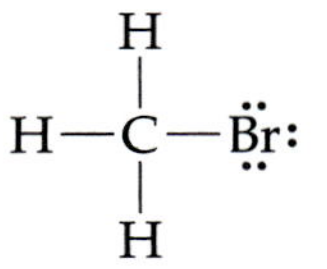

62. The four smallest alkanes have the following formulas: $CH_4$, $C_2H_6$, $C_3H_8$, and $C_4H_{10}$. Note the trend for the relationship between the number of carbon atoms and the number of hydrogen atoms. Based on this trend, what would the formula be for the alkane with 22 carbons?

63. Because the structure for a particular alkane can be drawn in different ways, two drawings of the same substance can look like isomers. Are each of the following pairs isomers or different representations of the same thing?

```
                                                 H
                                                 |
                                               H—C—H
       H  H  H  H  H                       H     |   H  H
       |  |  |  |  |                       |     |   |  |
a.  H—C—C—C—C—C—H       and        H—C—C—C—C—H
       |  |  |  |  |                       |     |   |  |
       H  H  H  H  H                       H     H   H  H

                                           H
                                           |
                                         H—C—H
       H  H  H  H  H                       |   H  H  H
       |  |  |  |  |                       |   |  |  |
b.  H—C—C—C—C—C—H       and        H—C—C—C—C—H
       |  |  |  |  |                       |   |  |  |
       H  H  H  H  H                       H   H  H  H
```

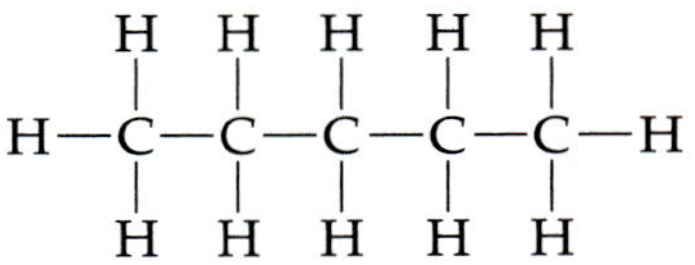

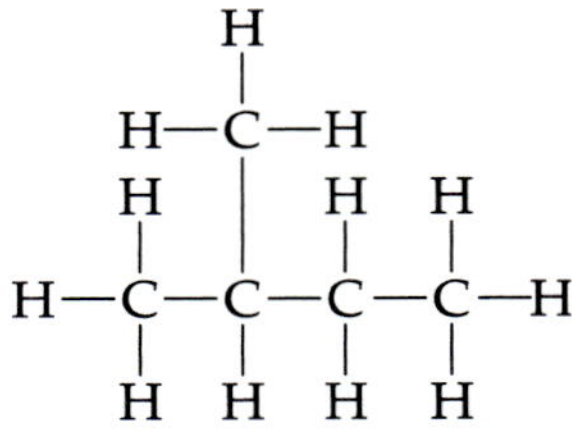

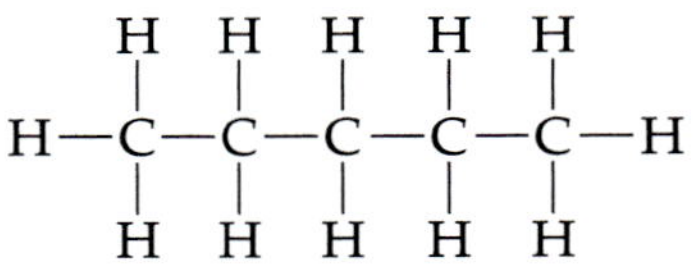

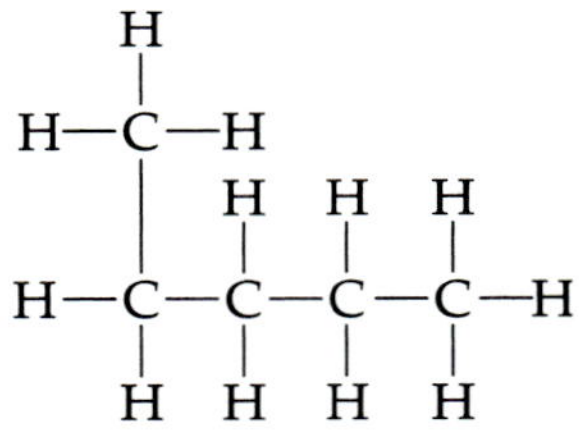

c.  and

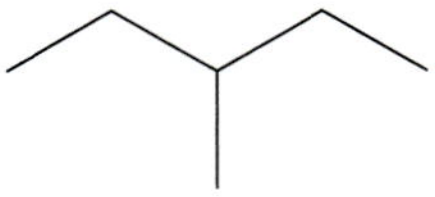

d. 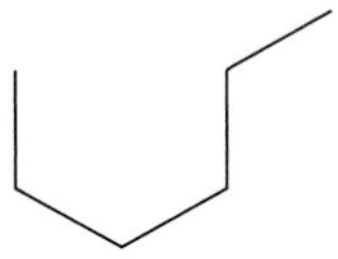 and

64. Are each of the following pairs isomers or different representations of the same thing?

a. and

b. and

c. 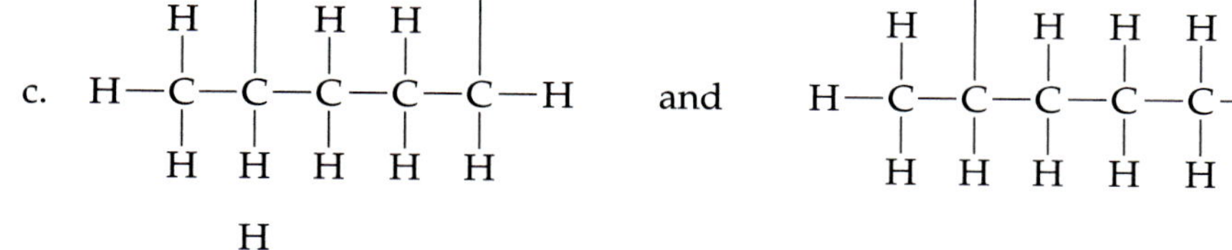 and

d. and

65. Draw line drawings for three isomers of $C_5H_{12}$.

66. Draw line drawings for three isomers of $C_4H_8$. (Draw each isomer with one double bond.)

67. Two of the three isomers of $C_3H_8O$ are alcohols and one is an ether. Draw condensed structures for these three isomers.

68. Draw the condensed structure for an isomer of $C_3H_6O_2$ that is a carboxylic acid, and draw another condensed structure for an isomer of $C_3H_6O_2$ that is an ester.

69. Draw a Lewis structure for an isomer of $C_2H_5NO$ that is an amide, and draw a second Lewis structure for a second isomer of $C_2H_5NO$ that has both an amine functional group and an aldehyde functional group.

70. Draw the Lewis structure for an isomer of $C_3H_6O$ that is a ketone, and draw another Lewis structure for an isomer of $C_3H_6O$ that is an aldehyde.

71. Ketones, aldehydes, carboxylic acids, esters, and amides all have a carbon–oxygen double bond (often called a carbonyl group). Explain how these classifications of organic compounds are different from each other.

### Section 17.2 Important Substances in Food

OBJECTIVE 4

72. Identify each of the following structures as representing a carbohydrate, amino acid, peptide, triglyceride, or steroid.

a. $H_3\overset{+}{N}-\underset{CH_3}{\overset{H}{C}}-CO_2^-$

b. $CH_2OH$, H, O, H, $CH_2OH$, O, H, H, OH, H, O, H, HO, OH, $CH_2OH$, H, OH, OH, H

c. H, H—C—Ö, :Ö:, H—C—Ö, :Ö:, H—C—Ö, :Ö:, H

d. O, $CH_3$, $CH_3$, O

e.

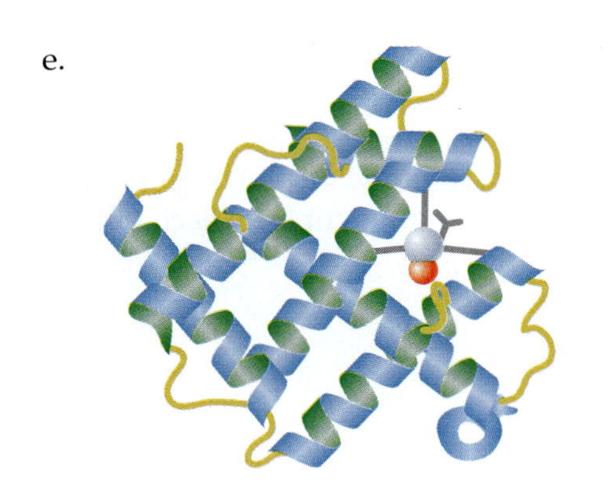

73. Identify each of the following structures as representing a carbohydrate, amino acid, peptide, triglyceride, or steroid.

OBJECTIVE 4

a.

H   Ö   H H H H H H H H H H H H H H H
H—C—Ö—C—C—C—C—C—C—C—C—C—C—C—C—C—C—C—C—H
H H H H H H H H H H H H H H H

Ö   H H H H H H H H H H H H H H H
H—C—Ö—C—C—C—C—C—C—C—C—C—C—C—C—C—C—C—C—H
H H H H H H H H H H H H H H H

Ö   H H H H H H H H H H H H H H H
H—C—Ö—C—C—C—C—C—C—C—C—C—C—C—C—C—C—C—C—H
H   H H H H H H H H H H H H H H H

b.

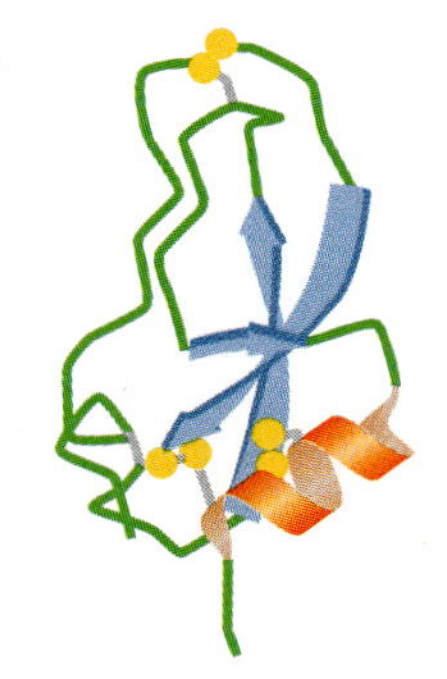

c.

H Ö   H Ö   H Ö   H Ö
H—N̈—C—C—N̈—C—C—N̈—C—C—N̈—C—C—Ö—H
H CH₃   H CH₂   H H   H CH₂
OH   SH

d.

H
$H_3\overset{+}{N}$—C—$CO_2^-$
$CH_2$
SH

e.

CH₂OH   CH₂OH   CH₂OH
H   O   H   O   H   O   H
H OH H   O   H OH H   O   H OH H   OH
OH   H   H   H OH   H   OH
H OH
( … )$_n$

OBJECTIVE 5

**74.** Identify each of the following structures as representing a monosaccharide, disaccharide, or polysaccharide.

a.

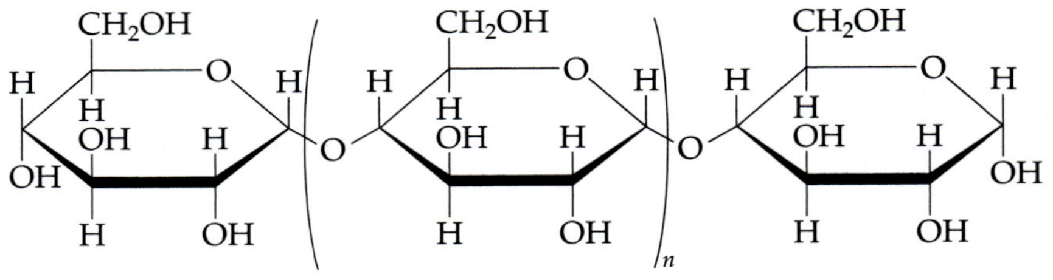

b.

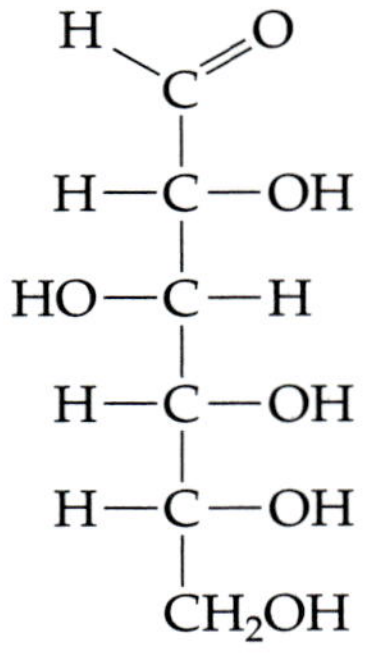

```
  H   O
   \ //
    C
    |
 H—C—OH
    |
HO—C—H
    |
 H—C—OH
    |
 H—C—OH
    |
   CH2OH
```

c.

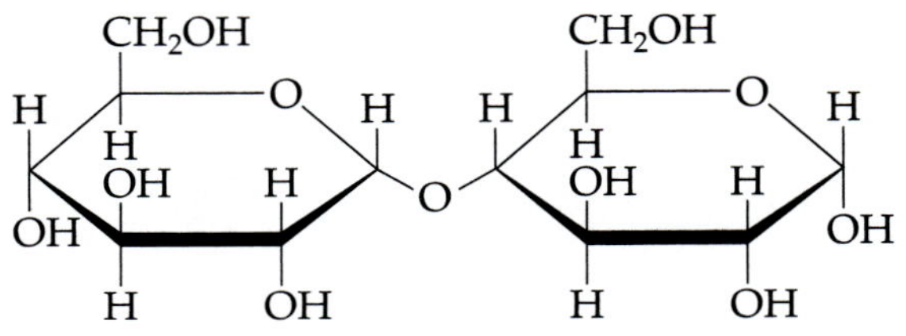

d.

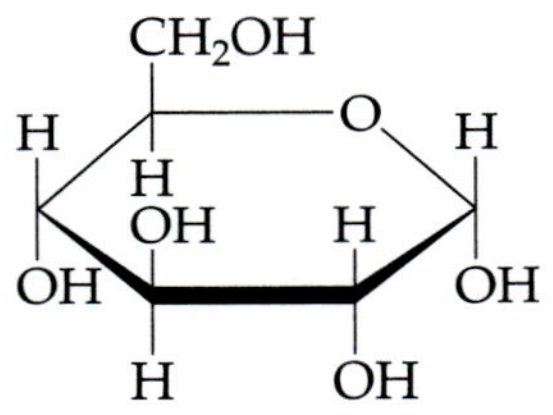

75. Identify each of the following structures as representing a monosaccharide, disaccharide, or polysaccharide.

OBJECTIVE 5

a.

b.

c.

$CH_2OH$
|
C=O
|
HO—C—H
|
H—C—OH
|
H—C—OH
|
$CH_2OH$

d.

**76.** Identify each of the following as a monosaccharide, disaccharide, or polysaccharide.

a. maltose

b. fructose

c. amylose

d. cellulose

77. Identify each of the following as a monosaccharide, disaccharide, or polysaccharide.

a. amylopectin

b. glucose

c. lactose

d. galactose

OBJECTIVE 6

**78.** How do glucose and galactose differ?

OBJECTIVE 6

79. How do glucose and fructose differ?

OBJECTIVE 7

**80.** What saccharide units form maltose, lactose, and sucrose?

OBJECTIVE 8

81. Describe the similarities and differences among amylose, amylopectin, and glycogen.

OBJECTIVE 8

**82.** Describe the similarities and differences between starches (such as amylose, amylopectin, and glycogen) and cellulose.

OBJECTIVE 9

83. Explain why the starch molecules found in a potato can be digested in our digestive tracts but the cellulose in the same potato cannot.

OBJECTIVE 11

**84.** Explain why glycine amino acid molecules in our bodies are usually found in the second form shown below rather than in the first.

$$H_2N{-}CH_2{-}CO_2H \qquad H_3\overset{+}{N}{-}CH_2{-}CO_2^-$$

**85.** Using Figure 17.22, draw the Lewis structure of the dipeptide that has alanine combined with serine. Circle the peptide bond in your structure.

86. Using Figure 17.22, draw the Lewis structure of the dipeptide that has cysteine combined with glycine. Circle the peptide bond in your structure.

OBJECTIVE 12

**87.** Show how the amino acids leucine, phenylalanine, and threonine can be linked together to form the tripeptide leu-phe-thr.

OBJECTIVE 12

88. Show how the amino acids tryptophan, aspartic acid, and asparagine can be linked together to form the tripeptide trp-asp-asn.

89. When the artificial sweetener aspartame is digested, it yields methanol as well as the amino acids aspartic acid and phenylalanine. Although methanol is toxic, the extremely low levels introduced into the body by eating aspartame are not considered dangerous, but for people who suffer from phenylketonuria (PKU), the phenylalanine can cause severe mental retardation. Babies are tested for this disorder at birth, and when it is detected, they are placed on diets that are low in phenylalanine. Using Figure 17.22, identify the portions of aspartame's structure that yield aspartic acid, phenylalanine, and methanol.

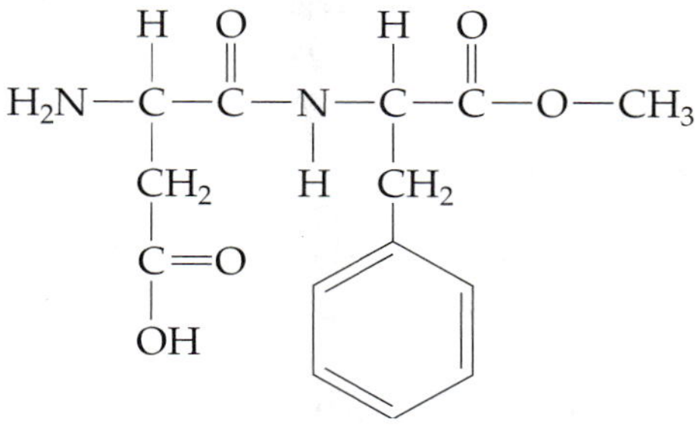

90. Describe the differences among the primary, secondary, and tertiary structures of proteins. OBJECTIVE 13

91. Describe how disulfide bonds, hydrogen bonds, and salt bridges help hold protein molecules together in specific tertiary structures. OBJECTIVE 14

92. Explain why it is more efficient to store energy in the body as fat than as carbohydrate or protein. OBJECTIVE 15

93. Identify each of the following triglycerides as saturated or unsaturated. Which is more likely to be a solid at room temperature, and which is more likely to be a liquid? OBJECTIVE 17 OBJECTIVE 18

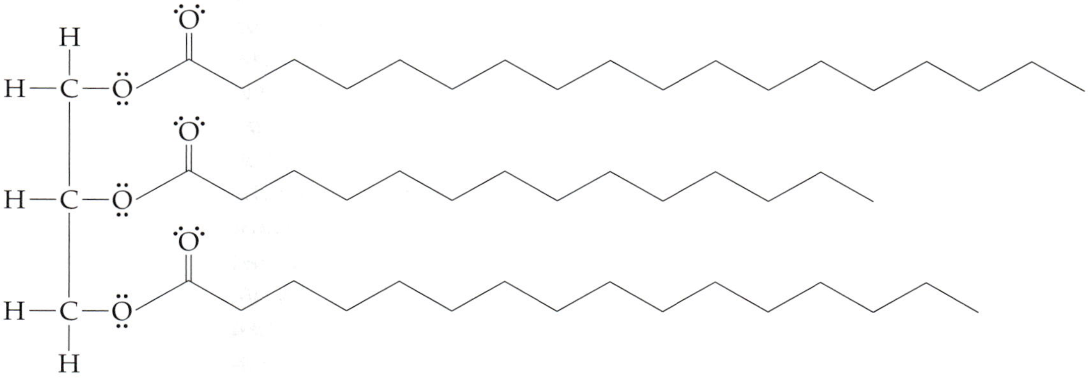

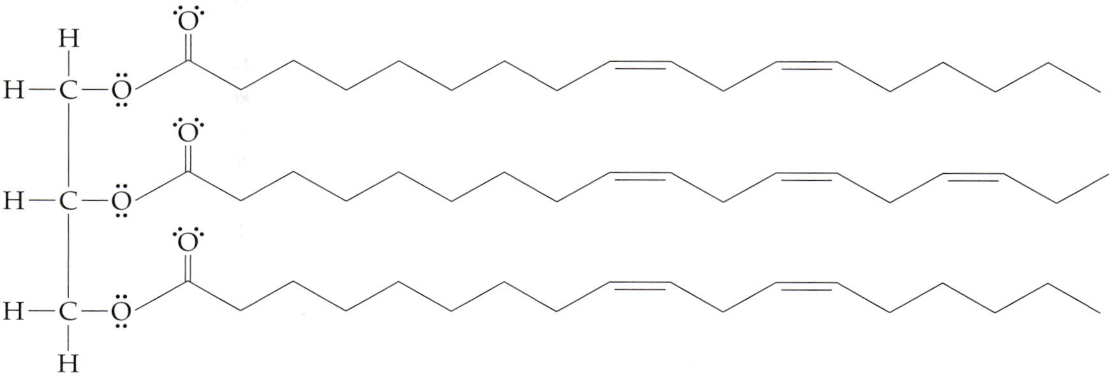

OBJECTIVE 19

94. Draw the structure of the triglyceride that would form from the complete hydrogenation of the following triglyceride.

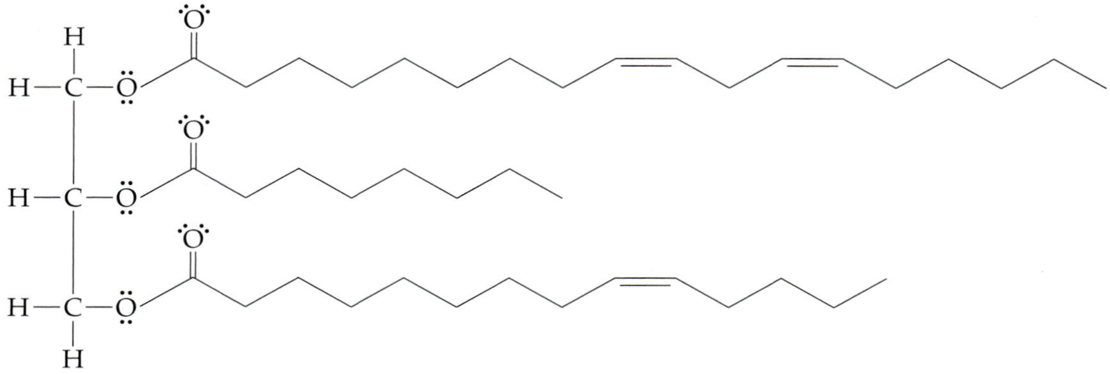

OBJECTIVE 19

95. Draw the structure of the triglyceride that would form from the complete hydrogenation of the following triglyceride.

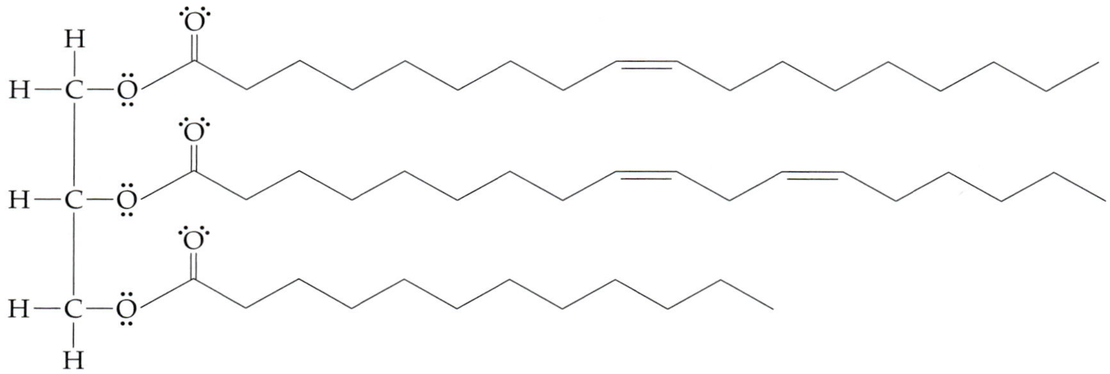

## Section 17.3 Digestion

OBJECTIVE 20

96. When you wash some fried potatoes down with a glass of milk, you deliver a lot of different nutritive substances to your digestive tract, including lactose (a disaccharide), protein, and fat from the milk and starch from the potatoes. What are the digestion products of disaccharides, polysaccharides, protein, and fat?

OBJECTIVE 21

97. Describe how the digestion of protein molecules is facilitated by conditions in the stomach.

OBJECTIVE 22

98. Explain why each enzyme acts only on a specific molecule or a specific type of molecule.

## Section 17.4 Synthetic Polymers

OBJECTIVE 23

99. Describe how Nylon 66 is made.

OBJECTIVE 24

100. Explain why Nylon 66 is stronger than Nylon 610.

OBJECTIVE 25

101. Describe how polyesters are made.

OBJECTIVE 26

102. Describe the similarities and differences between the molecular structures of low-density polyethylene (LDPE) and high-density polyethylene (HDPE).

103. Identify each of the following as representing nylon, polyester, polyethylene, poly(vinyl chloride), polypropylene, or polystyrene. (In each case, the *n* represents some large integer.)

OBJECTIVE 27

a. 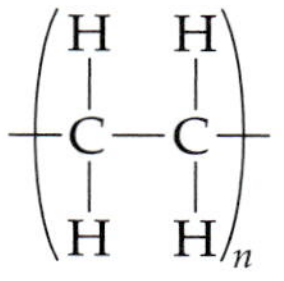 $\left( -\underset{\text{H}}{\overset{\text{H}}{\text{C}}}-\underset{\text{H}}{\overset{\text{H}}{\text{C}}}- \right)_n$

b. $\left( -OCH_2CH_2O-\overset{\text{O}}{\overset{\|}{\text{C}}}-C_6H_4-\overset{\text{O}}{\overset{\|}{\text{C}}}- \right)_n$

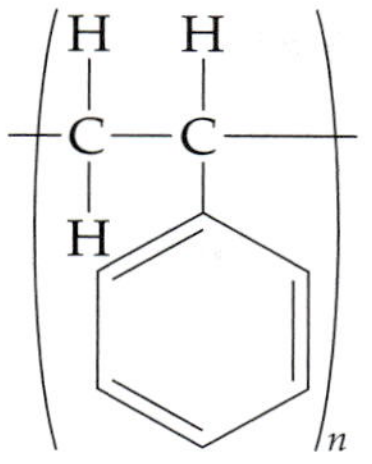

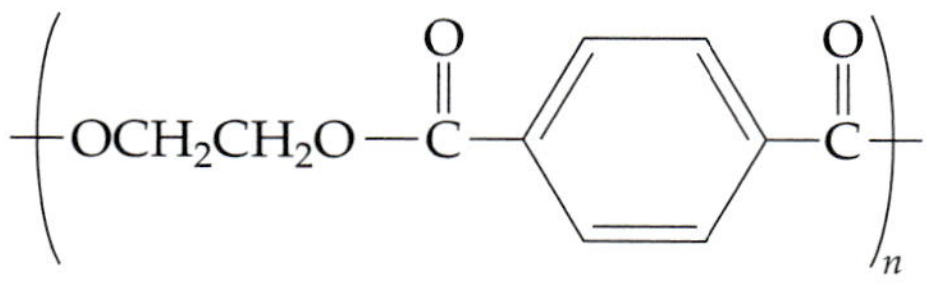

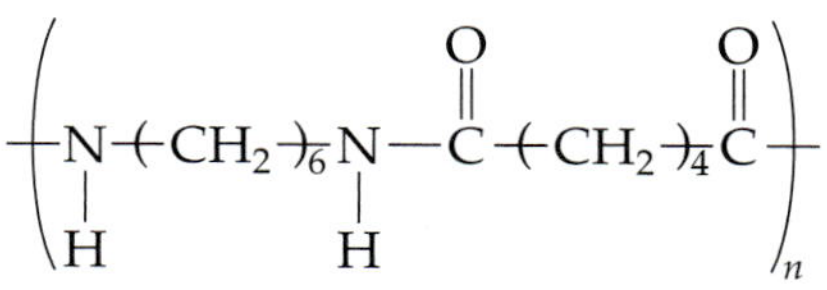

c. $\left( -\underset{\text{H}}{\overset{\text{H}}{\text{C}}}-\underset{\text{CH}_3}{\overset{\text{H}}{\text{C}}}- \right)_n$

d. $\left( -\underset{\text{H}}{\overset{\text{H}}{\text{C}}}-\underset{\text{C}_6\text{H}_5}{\overset{\text{H}}{\text{C}}}- \right)_n$

e. $\left( -\underset{\text{H}}{\text{N}}(CH_2)_6\underset{\text{H}}{\text{N}}-\overset{\text{O}}{\overset{\|}{\text{C}}}(CH_2)_4\overset{\text{O}}{\overset{\|}{\text{C}}}- \right)_n$

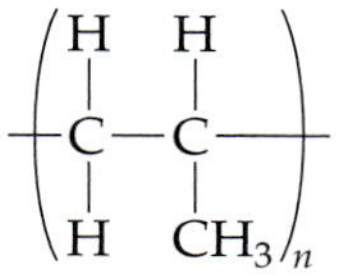

f. 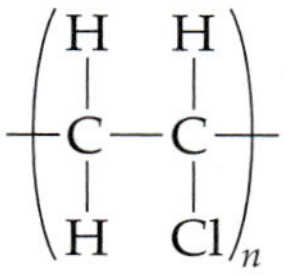 $\left( -\underset{\text{H}}{\overset{\text{H}}{\text{C}}}-\underset{\text{Cl}}{\overset{\text{H}}{\text{C}}}- \right)_n$

104. Both ethylene and polyethylene are composed of nonpolar molecules. Explain why ethylene is a gas at room temperature while polyethylene is a solid at the same temperature.

105. Find three plastic objects in your home that are labeled with a recycling code of 1. From what substance are these objects made? Are objects of this type recycled in your town?

OBJECTIVE 28

106. Find three objects in your home that are labeled with a recycling code of 2. From what substance are these objects made? Are objects of this type recycled in your town?

OBJECTIVE 28

107. Find one object representing each of the recycling codes 3, 4, 5, and 6. From what substance is each object made? Can these objects be recycled in your town?

OBJECTIVE 28

### Discussion Questions

108. Cyclopropane, $C_3H_6$, is a potent anesthetic that can be dangerous because it is very flammable. Develop a theory of why it is so reactive. *Hints:* Draw a Lewis structure for $C_3H_6$ that has all single bonds (note the "cyclo" portion of the name). Predict the bond angles between the carbon atoms in the structure on the basis of the number of electron groups around each carbon. Compare this angle to the bond angles between carbon atoms that cyclopropane must have on the basis of shape that you have drawn.

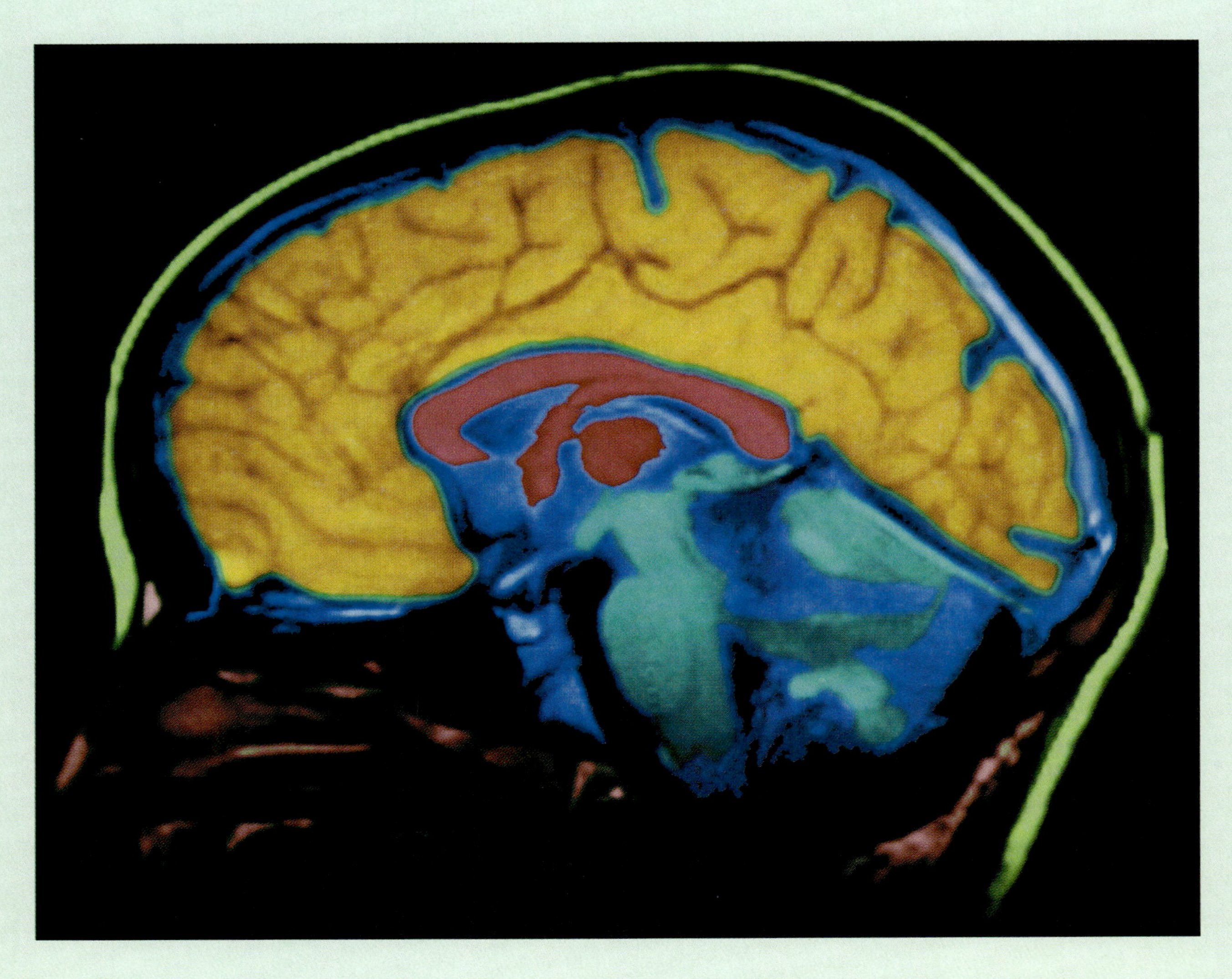

*Positron Emission Tomography (PET) scans can show which parts of the brain are being used when certain tasks are performed.*

CHAPTER 18

# Nuclear Chemistry



STAN IS GOING TO VISIT HIS SON FRED at the radiology department of a local research hospital, where Fred has been recording the brain activity of children with learning differences and comparing it to the brain activity of children who excel in normal school environments. To pursue this research, Fred uses imaging technology developed through the science of nuclear chemistry, the study of changes that occur within the nuclei of atoms. But even before getting into his car to go see Fred, Stan is already surrounded by substances that are undergoing nuclear reactions. In fact, nuclear reactions accompany Stan wherever he goes. He has strontium-90 in his bones and iodine-131 in his thyroid, and both substances are constantly undergoing nuclear reactions of a type known as beta emission. Stan is not unique in this respect. All of our bodies contain these substances and others like them.

Stan is surrounded by nuclear changes that take place outside his body, as well. The soil under his house contains a small amount of uranium-238, which undergoes nuclear reactions of a type called alpha emission. A series of changes in the nucleus of the uranium-238 leads to an even smaller amount of radon-222, which is a gas that Stan inhales in every breath he takes at home. Subsequently, radon-222 undergoes a nuclear reaction very similar to the reaction for uranium-238.

On Stan's way to the hospital, he passes a nuclear power plant that generates electricity for the homes and businesses in his city by means of yet another kind of nuclear reaction. When Stan gets to the hospital, Fred shows him the equipment he is using in his research. It is a positron emission tomography (PET) machine that enables Fred to generate images showing which parts of a child's brain are being used when the child does certain tasks. Positron emission is another type of nuclear change described in this chapter.

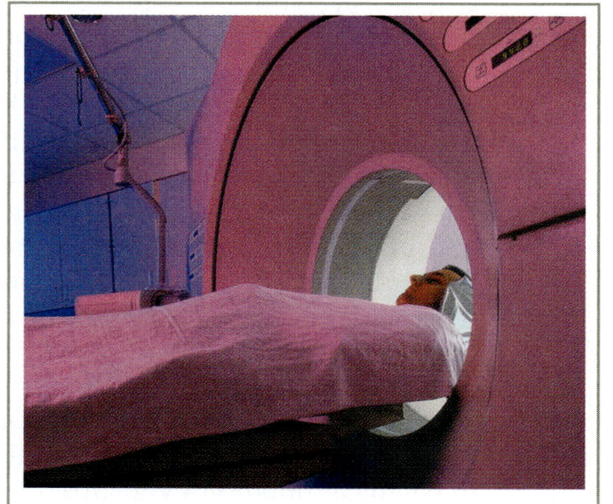

*The MRI scan being done here uses changes in the nuclei of atoms to create an image of the soft tissues of the body.*

There are good reasons why, in the preceding 17 chapters, our exploration of chemistry has focused largely on the behavior of electrons. Chemistry is *the study of the structure and behavior of matter*, and most of our understanding of such phenomena comes from studying the gain, loss, and sharing of electrons. At the same time, however, we have neglected the properties of the nuclei of atoms and the changes that some nuclei can undergo. In this chapter, we turn our attention toward the center of the atom to learn what *nuclear stability* means and to understand the various kinds of nuclear reactions.

## Review Skills

The presentation of information in this chapter assumes that you can already perform the tasks listed below. You can test your readiness to proceed by answering the Review Questions at the end of the chapter. This might also be a good time to read the Chapter Objectives, which precede the Review Questions.

- Describe the nuclear model of the atom, including the general location of the protons, neutrons, and electrons, the relative size of the nucleus compared to the size of the atom, and the modern description of the electron. (Section 2.4)
- Write the definitions for *isotope*, *atomic number*, and *mass number*. (Chapter 2 Glossary)
- Write the definitions for *energy*, *kinetic energy*, and *potential energy*. (Chapter 7 Glossary)
- Write or identify a description of the Law of Conservation of Energy. (Section 7.1)
- Describe the relationship between stability, capacity to do work, and potential energy. (Section 7.1)
- Write a brief description of radiant energy in terms of its particle and wave nature. (Section 7.1)
- Write or identify the relative energies and wavelengths of the following forms of radiant energy: gamma rays, X rays, ultraviolet (UV) rays, visible light, infrared (IR) rays, microwaves, and radio waves. (Section 7.1)
- Write the definitions for *excited state* and *ground state*. (Chapter 11 Glossary)

# 18.1 The Nucleus and Radioactivity

Our journey into the center of the atom begins with a brief review. You learned in Chapter 2 that the protons and neutrons in each atom are found in a tiny, central nucleus that measures about 1/100,000 the diameter of the atom itself. You also learned that the atoms of each element are not necessarily identical; they can differ with respect to the number of neutrons in their nuclei. When an element has two or more species of atoms, each with the same number of protons but a different number of neutrons, the different species are called isotopes. Different isotopes of the same element have the same atomic number, but they differ in mass number, which is the sum of the numbers of protons and neutrons in the nucleus. In the context of nuclear science, protons and neutrons are called **nucleons,** because they reside in the nucleus. The atom's mass number is often called the **nucleon number,** and a particular type of nucleus, characterized by a specific atomic number and nucleon number, is called a **nuclide.** Nuclides are represented in chemical notation by a subscript atomic number (Z) and a superscript nucleon number (A) on the left side of the element's symbol (X):

OBJECTIVE 2

Mass number (nucleon number)

$$^{A}_{Z}X$$

Atomic number    Element symbol

For example, the most abundant nuclide of uranium has 92 protons and 146 neutrons, so its atomic number is 92, its nucleon number is 238

(92 + 146), and its symbol is $^{238}_{92}U$. Often, the atomic number is left off the symbol. Nuclides can also be described with the name of the element followed by the nucleon number. Therefore, $^{238}_{92}U$ is commonly described as $^{238}U$ or uranium-238. Examples 18.1 and 18.2 provide practice in writing and interpreting nuclide symbols.

OBJECTIVE 3

OBJECTIVE 4

## EXAMPLE 18.1 Nuclide Symbols

OBJECTIVE 2 OBJECTIVE 3 OBJECTIVE 4

A nuclide that has 26 protons and 33 neutrons is used to study blood chemistry. Write its nuclide symbol in the form of $^{A}_{Z}X$. Write two other ways to represent this nuclide.

***Solution***

Because this nuclide has 26 protons, its atomic number, Z, is 26, which identifies the element as iron, Fe. This nuclide of iron has 59 total nucleons (26 protons + 33 neutrons), so its nucleon number, A, is 59.

$^{59}_{26}\mathbf{Fe}$ or $^{59}\mathbf{Fe}$ or **iron-59**

## EXERCISE 18.1 Nuclide Symbols

OBJECTIVE 2 OBJECTIVE 3 OBJECTIVE 4

One of the nuclides used in radiation therapy for the treatment of cancer has 39 protons and 51 neutrons. Write its nuclide symbol in the form of $^{A}_{Z}X$. Write two other ways to represent this nuclide.

## EXAMPLE 18.2 Nuclide Symbols

OBJECTIVE 2 OBJECTIVE 3 OBJECTIVE 4

Physicians can assess a patient's lung function with the help of krypton-81. What are this nuclide's atomic number and mass number? How many protons and how many neutrons are in the nucleus of each atom? Write two other ways to represent this nuclide.

***Solution***

The periodic table shows us that the atomic number for krypton is 36, so each krypton atom has 36 protons. The number following the element name in krypton-81 is this nuclide's mass number. The difference between the mass number (the sum of the numbers of protons and neutrons) and the atomic number (the number of protons) is equal to the number of neutrons, so krypton-81 has 45 neutrons (81 − 36).

**atomic number = 36; mass number = 81;**

**36 protons 45 neutrons;** $^{81}_{36}\mathbf{Kr}$ $^{81}\mathbf{Kr}$

## EXERCISE 18.2 Nuclide Symbols

OBJECTIVE 2 OBJECTIVE 3 OBJECTIVE 4

A nuclide with the symbol $^{201}Tl$ can be used to assess a patient's heart in a stress test. What are its atomic number and mass number? How many protons and how many neutrons are in the nucleus of each atom? Write two other ways to represent this nuclide.

## Nuclear Stability

OBJECTIVE 5

Two forces act upon the particles within the nucleus to produce the nuclear structure. One, called the **electrostatic force** (or electromagnetic force), is the force that causes opposite electrical charges to attract each other and like charges to repel each other. The positively charged protons in the nucleus of an atom have an electrostatic force pushing them apart. The other force within the nucleus, called the **strong force,** holds nucleons (protons and neutrons) together.

OBJECTIVE 5

If one proton were to encounter another, the electrostatic force pushing them apart would be greater than the strong force pulling them together, and the two protons would fly in separate directions. Therefore, nuclei that contain more than one proton and no neutrons do not exist. Neutrons can be described as the nuclear "glue" that enables protons to stay together in the nucleus. Because neutrons are uncharged, there are no electrostatic repulsions between them and other particles. At the same time, each neutron in the nucleus of an atom is attracted to other neutrons and to protons by the strong force. Therefore, adding neutrons to a nucleus increases the attractive forces holding the particles of the nucleus together without increasing the amount of repulsion between those particles. As a result, although a nucleus that consists of only two protons is unstable, a helium nucleus that consists of two protons and two neutrons is very stable. The increased stability is reflected in the significant amount of energy released in the reaction.

$$p + p + n + n \rightarrow {}^{4}_{2}He^{2+}$$

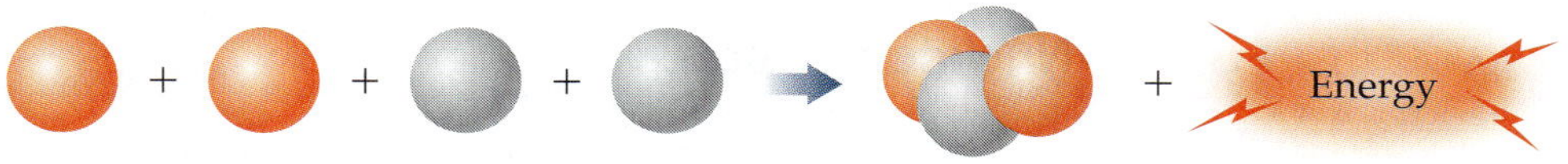

OBJECTIVE 5

For many of the lighter elements, the possession of an equal number of protons and neutrons leads to stable atoms. For example, carbon-12 atoms, ${}^{12}_{6}C$, with 6 protons and 6 neutrons, and oxygen-16 atoms, ${}^{16}_{8}O$, with 8 protons and 8 electrons, are both very stable. Larger atoms with more protons in their nuclei require a higher ratio of neutrons to protons to balance the increased electrostatic repulsion between protons. Table 18.1 shows the steady increase in the neutron-to-proton ratios of the most abundant isotopes of the elements in group 15 in the periodic table.

Table 18.1
Neutron-to-Proton Ratio for the Most Abundant Isotopes of the Group 15 Elements

| Element | Number of neutrons | Number of protons | Neutron-to-proton ratio |
|---|---|---|---|
| nitrogen, N | 7 | 7 | 1 to 1 |
| phosphorus, P | 16 | 15 | 1.07 to 1 |
| arsenic, As | 42 | 33 | 1.27 to 1 |
| antimony, Sb | 70 | 51 | 1.37 to 1 |
| bismuth, Bi | 126 | 83 | 1.52 to 1 |

There are 264 stable nuclides found in nature. The graph in Figure 18.1 shows the neutron-to-proton ratios of these stable nuclides. Collectively, these nuclides fall within what is known as the **band of stability.**

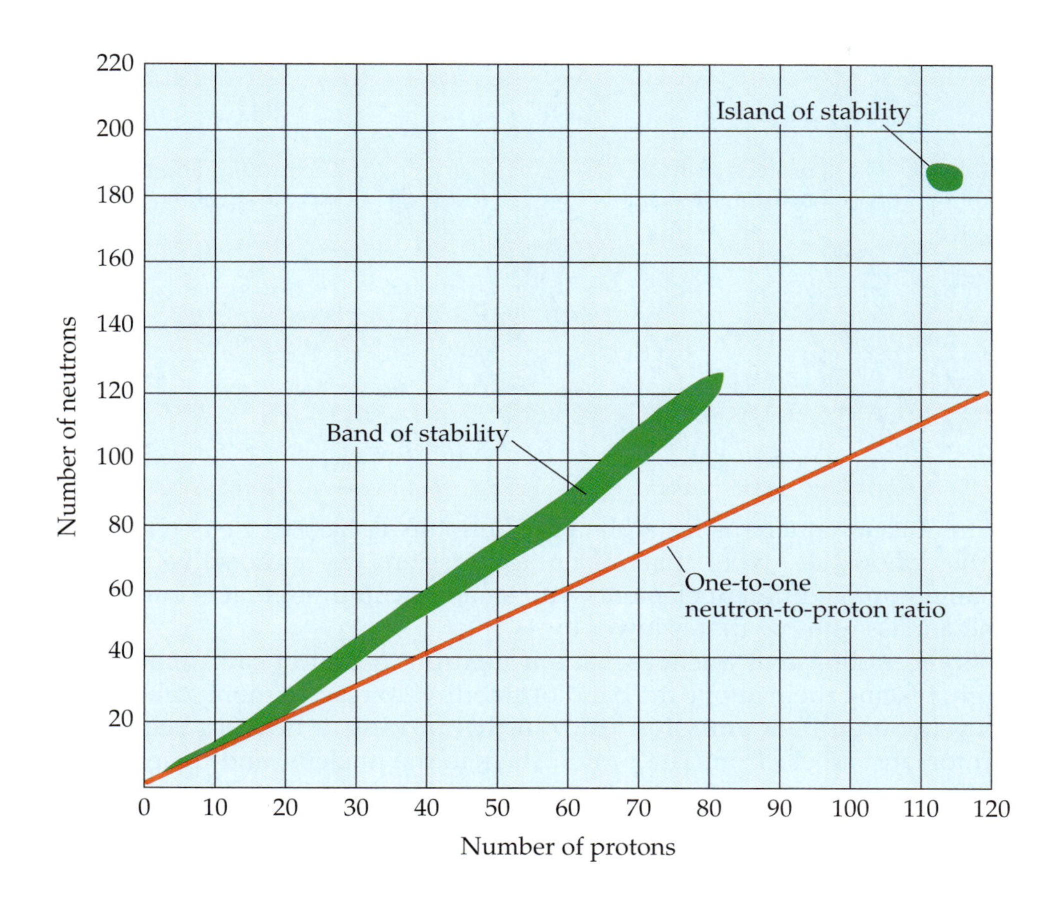


**Figure 18.1**
The Band of Stability

A nuclide containing numbers of protons and neutrons that place it outside this band of stability will be unstable until it undergoes one or more nuclear reactions that take it into the band of stability. We call these unstable atoms **radioactive nuclides,** and the changes they undergo to reach stability are called **radioactive decay.** Note that the band of stability stops at 83 protons. All of the known nuclides with more than 83 protons are radioactive, but scientists have postulated that there should be a small island of stability around the point representing 114 protons and 184 neutrons. The relative stability of the heaviest atoms that have so far been synthesized in the laboratory suggests that this is true. (See Special Topic 2.1: *Why Create New Elements?*)

## Types of Radioactive Emissions

OBJECTIVE 6

One of the ways in which nuclides with more than 83 protons change to reach the band of stability is to release 2 protons and 2 neutrons in the form of a helium nucleus, which in this context is called an **alpha particle.**

Natural uranium, which is found in many rock formations on earth, has three isotopes that all experience **alpha emission,** the release of alpha particles. The isotope composition of natural uranium is 99.27% uranium-238, 0.72% uranium-235, and a trace of uranium-234. The nuclear equation for the alpha emission of uranium-238, the most abundant isotope, is

$$^{238}_{92}\mathrm{U} \rightarrow {}^{234}_{90}\mathrm{Th} + {}^{4}_{2}\mathrm{He}$$

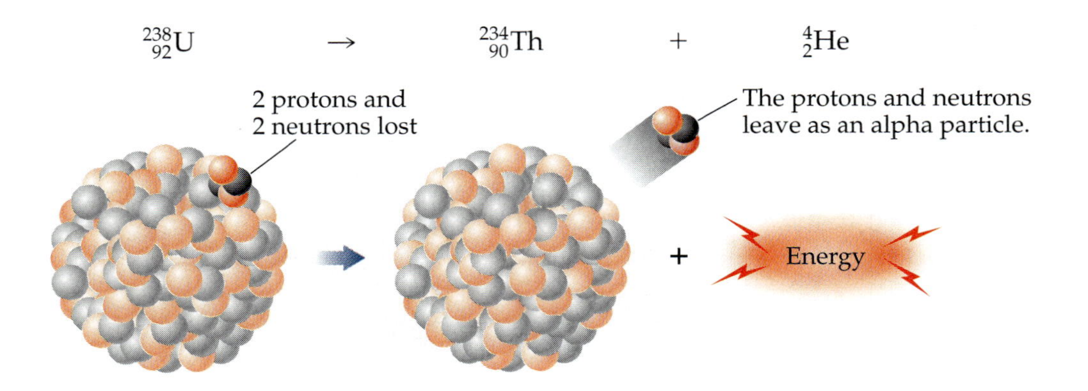
2 protons and
2 neutrons lost
The protons and neutrons
leave as an alpha particle.
+
Energy

OBJECTIVE 7

In nuclear equations for alpha emission, the alpha particle is written as either $\alpha$ or ${}^{4}_{2}\mathrm{He}$. Note that in alpha emission, the radioactive nuclide changes into a different element, with an atomic number that is lower by 2 and a mass number that is lower by 4.

OBJECTIVE 8

OBJECTIVE 6

Some radioactive nuclides have a neutron-to-proton ratio that is too high, placing them above the band of stability. To reach a more stable state, they undergo **beta emission ($\beta^-$).** In this process, a neutron becomes a proton and an electron. The proton stays in the nucleus, and the electron, which is called a **beta particle** in this context, is ejected from the atom.

$$\mathrm{n} \rightarrow \mathrm{p} + \mathrm{e}^-$$

OBJECTIVE 7

In nuclear equations for beta emission, the electron is written as $\beta$, $\beta^-$, or ${}^{0}_{-1}\mathrm{e}$. Iodine-131, which has several medical uses, including the measurement of iodine uptake by the thyroid, is a beta emitter:

$$^{131}_{53}\mathrm{I} \rightarrow {}^{131}_{54}\mathrm{Xe} + {}^{0}_{-1}\mathrm{e}$$

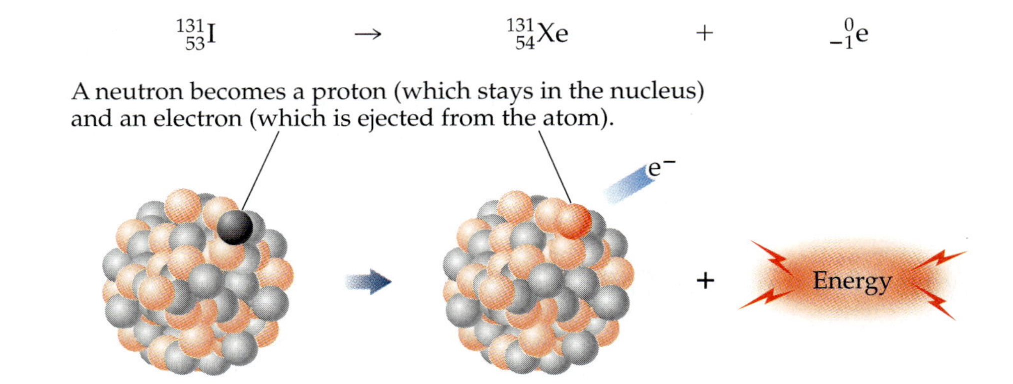
A neutron becomes a proton (which stays in the nucleus)
and an electron (which is ejected from the atom).
e−
+
Energy

Note that in beta emission, the radioactive nuclide changes into a different element, with an atomic number that is higher by 1 but the same mass number.

OBJECTIVE 8

If a radioactive nuclide has a neutron-to-proton ratio that is too low, placing it below the band of stability, it can move toward stability in one of two ways; positron emission or electron capture. **Positron emission ($\beta^+$)** is similar to beta emission, but in this case, a proton becomes a neutron and an antimatter electron, or antielectron.[1] The latter is also called a **positron** because, although it resembles an electron in most ways, it has a positive charge. The neutron stays in the nucleus, and the positron speeds out of the nucleus at high velocity.

OBJECTIVE 6

$$p \rightarrow n + e^+$$

OBJECTIVE 7

In nuclear equations for positron emission, the electron is written as $\beta^+$, ${}^{0}_{+1}e$, or ${}^{0}_{1}e$. Potassium-40, which is important in geologic dating, undergoes positron emission:

$${}^{40}_{19}K \rightarrow {}^{40}_{18}Ar + {}^{0}_{+1}e$$

A proton becomes a neutron (which stays in the nucleus) and a positron (which is ejected from the atom).

$e^+$ + Energy

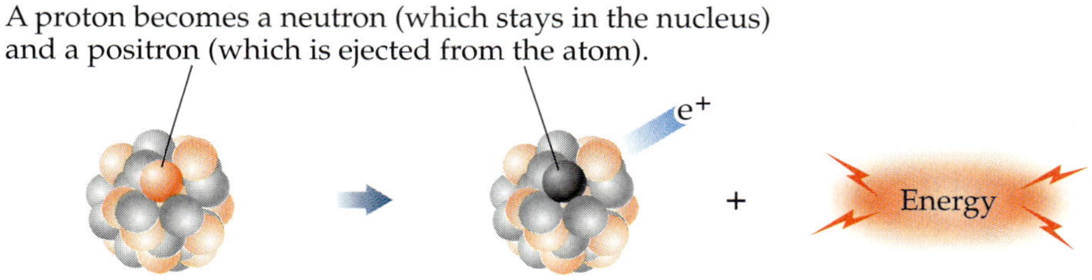

Note that in positron emission, the radioactive nuclide changes into a different element, with an atomic number that is lower by 1 but the same mass number.

OBJECTIVE 8

The second way in which an atom with an excessively low neutron-to-proton ratio can reach a more stable state is for a proton in its nucleus to capture one of the atom's electrons. In this process, called **electron capture,** the electron combines with the proton to form a neutron.

OBJECTIVE 6

$$e^- + p \rightarrow n$$

[1]Special Topic 11.1 describes antiparticles, such as antielectrons (positrons). Every particle has a twin antiparticle that formed along with it from very concentrated energy. When a particle meets an antimatter counterpart, they annihilate each other, leaving pure energy in their place. For example, when a positron collides with an electron, they both disappear, sending out two gamma ($\gamma$) photons in opposite directions.

Iodine-125, which is used to determine blood hormone levels, moves toward stability through electron capture.

$$^{0}_{-1}e + ^{125}_{53}I \rightarrow ^{125}_{52}Te$$

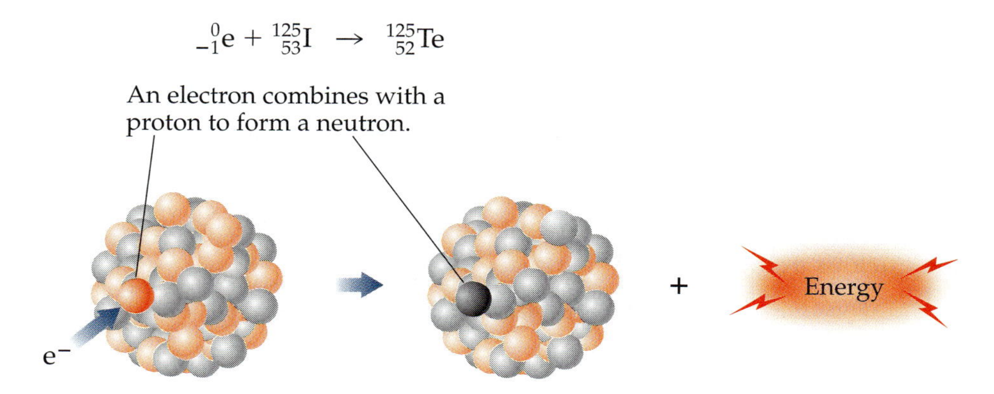
An electron combines with a proton to form a neutron.
e⁻
+
Energy

Like positron emission, electron capture causes the radioactive nuclide to change to a new element, with an atomic number that is lower by 1 but the same mass number.

Because radioactive decay leads to more stable products, it always releases energy. Some of this energy is released in the form of kinetic energy, adding to the motion of the product particles, but often some of it is given off in the form of radiant energy called gamma rays. **Gamma rays** can be viewed as streams of high-energy photons. For example, cobalt-60 is a beta emitter that also releases gamma radiation. The energy released in the beta emission leaves the product element, nickel-60, in an excited state. When the nickel-60 descends to its ground state, it gives off photons in the gamma-ray region of the radiant-energy spectrum. (See Section 7.1 for a review of the different forms of radiant energy.)

OBJECTIVE 9

$$^{60}_{27}Co \rightarrow ^{60}_{28}Ni^{*} + ^{0}_{-1}e \rightarrow ^{60}_{28}Ni + \gamma \text{ photon}$$

Excited state
e⁻ Beta emission
γ Gamma photon

## Nuclear Reactions and Nuclear Equations

OBJECTIVE 10

Now that we have seen some examples of nuclear reactions, let's look more closely at how they differ from the chemical reactions we have studied in the rest of this text.

- **Nuclear reactions** involve changes in the nucleus, whereas chemical reactions involve the loss, gain, and sharing of electrons.

- Different isotopes of the same element may undergo very different nuclear reactions, even though an element's isotopes all share the same chemical characteristics.
- Unlike chemical reactions, the rates of nuclear reactions are unaffected by temperature, pressure, and the presence of other atoms to which the radioactive atom may be bonded.
- Nuclear reactions, in general, give off much more energy than chemical reactions.

OBJECTIVE 11

The equations that describe nuclear reactions are different from those that describe chemical reactions because in nuclear equations, charge is disregarded. If you study the nuclear changes for alpha, beta, and positron emission that have already been described in this section, you will see that the products must be charged. For example, when an alpha particle is released from a uranium-238 nucleus, two positively charged protons are lost. Assuming that the uranium atom was uncharged initially, the thorium atom formed would have a −2 charge. Because the alpha particle is composed of two positively charged protons and two uncharged neutrons (and no electrons), it has a +2 overall charge.

$$^{238}_{92}\text{U} \rightarrow {}^{234}_{90}\text{Th}^{2-} + {}^{4}_{2}\text{He}^{2+}$$

The ions lose their charges quickly by exchanging electrons with other particles. Because we are usually not concerned about charges for nuclear reactions, and because these charges do not last very long, they are not usually mentioned in nuclear equations.

OBJECTIVE 12

Scientists may not be interested in the charges on the products of nuclear reactions, but they are very much interested in the changes that take place in the nuclei of the initial and final particles. Therefore, **nuclear equations** must clearly show the changes in the atomic numbers of the nuclides (the number of protons) and the changes in their mass numbers (the sum of the numbers of protons and neutrons). Note that in each of the following equations, the sum of the superscripts (mass numbers, A) for the reactants is equal to the sum of the superscripts for the products. Likewise, the sum of the subscripts (atomic numbers, Z) for the reactants is equal to the sum of the subscripts for the products. To show this to be true, beta particles are described as ${}^{0}_{-1}\text{e}$, and positrons are described as ${}^{0}_{+1}\text{e}$.

OBJECTIVE 12

*Alpha emission*

| | | | | | | | |
|---|---|---|---|---|---|---|---|
| mass number | 238 | | 234 | + | 4 | = | 238 |
| | $^{238}_{92}\textbf{U}$ | → | $^{234}_{90}\textbf{Th}$ | + | $^{4}_{2}\textbf{He}$ | | |
| atomic number | 92 | | 90 | + | 2 | = | 92 |

*Beta emission*

| | | | | | | | |
|---|---|---|---|---|---|---|---|
| mass number | 131 | | 131 | + | 0 | = | 131 |
| | $^{131}_{53}\textbf{I}$ | → | $^{131}_{54}\textbf{Xe}$ | + | $^{0}_{-1}\textbf{e}$ | | |
| atomic number | 53 | | 54 | + | (−1) | = | 53 |

*Positron emission*

| | | | | | | | |
|---|---|---|---|---|---|---|---|
| mass number | 40 | | 40 | + | 0 | = | 40 |
| | $^{40}_{19}K$ | → | $^{40}_{18}Ar$ | + | $^{0}_{+1}e$ | | |
| atomic number | 19 | | 18 | + | 1 | = | 19 |

*Electron capture*

| | | | | | | | |
|---|---|---|---|---|---|---|---|
| mass number | 0 | + | 125 | = | 125 | | 125 |
| | $^{0}_{-1}e$ | + | $^{125}_{53}I$ | | | → | $^{125}_{52}Te$ |
| atomic number | −1 | + | 53 | = | 52 | | 52 |

OBJECTIVE 12

The following general equations describe these nuclear changes:

*Alpha emission* $^{A}_{Z}X \rightarrow \, ^{A-4}_{Z-2}Y + \, ^{4}_{2}He$

*Beta emission* $^{A}_{Z}X \rightarrow \, _{Z+1}^{A}Y + \, _{-1}^{0}e$

*Positron emission* $^{A}_{Z}X \rightarrow \, _{Z-1}^{A}Y + \, _{+1}^{0}e$

*Electron capture* $_{-1}^{0}e + \, ^{A}_{Z}X \rightarrow \, _{Z-1}^{A}Y$

Table 18.2 summarizes these changes.

OBJECTIVE 12

**Table 18.2**
Nuclear Changes

| Type of change | Symbol | Change in protons (atomic number, Z) | Change in neutrons | Change in mass number, A |
|---|---|---|---|---|
| Alpha emission | $\alpha$ or $^{4}_{2}He$ | −2 | −2 | −4 |
| Beta emission | $\beta$, $\beta^-$, or $_{-1}^{0}e$ | +1 | −1 | 0 |
| Positron emission | $\beta^+$, $_{+1}^{0}e$, or $_{1}^{0}e$ | −1 | +1 | 0 |
| Electron capture | E. C. | −1 | +1 | 0 |
| Gamma emission | $\gamma$ or $^{0}_{0}\gamma$ | 0 | 0 | 0 |

Example 18.3 provides practice in writing nuclear equations for alpha emission, beta emission, positron emission, and electron capture.

## EXAMPLE 18.3 Nuclear Equations

OBJECTIVE 12

Write nuclear equations for (a) alpha emission by polonium-210, used in radiation therapy, (b) beta emission by gold-198, used to assess kidney activity, (c) positron emission by nitrogen-13, used in making brain, heart, and liver images, and (d) electron capture by gallium-67, used to do whole-body scans for tumors.

***Solution***

a. The symbol for polonium-210 is $^{210}_{84}Po$, and the symbol for an alpha particle is $^{4}_{2}He$. Therefore, the beginning of our equation is

$$^{210}_{84}Po \rightarrow \_\_\_\_\_\_ + \, ^{4}_{2}He$$

The first step in completing this equation is to determine the subscript for the missing formula by asking what number would make the sum of the subscripts on the right side of the arrow equal to the subscript on the left. That number gives us the atomic number of the missing nuclide. We then consult the periodic table to find out what element the missing nuclide represents. In this particular equation, the subscripts on the right must add up to 84, so the subscript for the missing nuclide must be 82. This is the atomic number of lead, so the symbol for the product nuclide is Pb. We next determine the superscript for the missing formula by asking what number would make the sum of the superscripts on the right side of the equation equal to the superscript on the left. The mass number for the product nuclide must be 206.

$$\mathbf{{}^{210}_{84}Po \rightarrow {}^{206}_{82}Pb + {}^{4}_{2}He}$$

b. The symbol for gold-198 is ${}^{198}_{79}\text{Au}$, and the symbol for a beta particle is ${}^{0}_{-1}\text{e}$. Therefore, the beginning of our equation is

$${}^{198}_{79}\text{Au} \rightarrow \_\_\_\_\_\_ + {}^{0}_{-1}\text{e}$$

To make the subscripts balance in our equation, the subscript for the missing nuclide must be 80, indicating that the symbol for the product nuclide should be Hg, for mercury. The mass number stays the same in beta emission, so we write 198.

$$\mathbf{{}^{198}_{79}Au \rightarrow {}^{198}_{80}Hg + {}^{0}_{-1}e}$$

c. The symbol for nitrogen-13 is ${}^{13}_{7}\text{N}$, and the symbol for a positron is ${}^{0}_{+1}\text{e}$. Therefore, the beginning of our equation is

$${}^{13}_{7}\text{N} \rightarrow \_\_\_\_\_\_ + {}^{0}_{+1}\text{e}$$

To make the subscripts balance, the subscript for the missing nuclide must be 6, so the symbol for the product nuclide is C, for carbon. The mass number stays the same in positron emission, so we write 13.

$$\mathbf{{}^{13}_{7}N \rightarrow {}^{13}_{6}C + {}^{0}_{+1}e}$$

d. The symbol for gallium-67 is ${}^{67}_{31}\text{Ga}$, and the symbol for an electron is ${}^{0}_{-1}\text{e}$. Therefore, the beginning of our equation is

$${}^{67}_{31}\text{Ga} + {}^{0}_{-1}\text{e} \rightarrow \_\_\_\_\_\_$$

To balance the subscripts, the atomic number for our missing nuclide must be 30, so the symbol for the product nuclide is Zn, for zinc. The mass number stays the same in electron capture, so we write 67.

$$\mathbf{{}^{67}_{31}Ga + {}^{0}_{-1}e \rightarrow {}^{67}_{30}Zn}$$

## EXERCISE 18.3 Nuclear Equations

OBJECTIVE 12

Write nuclear equations for (a) alpha emission by plutonium-239, one of the substances formed in nuclear power plants; (b) beta emission by sodium-24, used to detect blood clots; (c) positron emission by oxygen-15, used to assess the efficiency of the lungs; and (d) electron capture by copper-64, used to diagnose lung disease.

Example 18.4 shows how you can complete a nuclear equation when one of the symbols for a particle is missing.

## EXAMPLE 18.4 Nuclear Equations

OBJECTIVE 13

Glenn Seaborg and his team of scientists at the Lawrence Laboratory at the University of California, Berkeley, have created a number of new elements, some of which—berkelium, californium, lawrencium—have been named in honor of their work. Complete the following nuclear equations that describe the processes used to create these elements.

a. ${}^{244}_{96}\text{Cm} + {}^{4}_{2}\text{He} \rightarrow$ _______ $+ {}^{1}_{1}\text{H} + 2{}^{1}_{0}\text{n}$

b. ${}^{238}_{92}\text{U} +$ _______ $\rightarrow {}^{246}_{98}\text{Cf} + 4{}^{1}_{0}\text{n}$

c. _______ $+ {}^{10}_{5}\text{B} \rightarrow {}^{257}_{103}\text{Lr} + 5{}^{1}_{0}\text{n}$

***Solution***

First, determine the subscript for the missing formula by asking what number would make the sum of the subscripts on the left side of the arrow equal the sum of the subscripts on the right. That number is the atomic number of the missing nuclide and leads us to the element symbol for that nuclide. Next, determine the superscript for the missing formula by asking what number would make the sum of the superscripts on the left side of the arrow equal to the sum of the superscripts on the right.

a. ${}^{244}_{96}\text{Cm} + {}^{4}_{2}\text{He} \rightarrow \mathbf{{}^{245}_{97}Bk} + {}^{1}_{1}\text{H} + 2{}^{1}_{0}\text{n}$

b. ${}^{238}_{92}\text{U} + \mathbf{{}^{12}_{6}C} \rightarrow {}^{246}_{98}\text{Cf} + 4{}^{1}_{0}\text{n}$

c. $\mathbf{{}^{252}_{98}Cf} + {}^{10}_{5}\text{B} \rightarrow {}^{257}_{103}\text{Lr} + 5{}^{1}_{0}\text{n}$

## EXERCISE 18.4 Nuclear Equations

OBJECTIVE 13

Complete the following nuclear equations.

a. ${}^{14}_{7}\text{N} + {}^{4}_{2}\text{He} \rightarrow$ _______ $+ {}^{1}_{1}\text{H}$

b. ${}^{238}_{92}\text{U} +$ _______ $\rightarrow {}^{247}_{99}\text{Es} + 5{}^{1}_{0}\text{n}$

c. _______ $+ {}^{2}_{1}\text{H} \rightarrow {}^{239}_{93}\text{Np} + {}^{1}_{0}\text{n}$

## Rates of Radioactive Decay

Because the different radioactive nuclides have different stabilities, the rates at which they decay differ as well. These rates are described in terms of a nuclide's **half-life,** the time it takes for one-half of a sample to disappear. For example, radioactive carbon-14, which decays to form nitrogen-14 by emitting a beta particle, has a half-life of 5730 years. After 5730 years, one-half of a sample remains, and one-half has become nitrogen-14. After 11,460 years (2 half-lives), half of that remainder will have decayed to form nitrogen-14, bringing the sample down to one-fourth of its original amount. After 17,190 years (3 half-lives), half of what remained after 11,460 years will have decayed to form nitrogen-14, so one-eighth of the original sample will remain. This continues, with one-half of the sample decaying each half-life.

Imagine having a pie and being told that you are allowed to eat only one-half of whatever amount is on the plate each day. The first day you eat a lot of pie: one-half of it. The next day you eat half of what is there, but that's only one-fourth of a pie ($\frac{1}{2} \times \frac{1}{2}$). The next day you eat one-eighth of the original pie ($\frac{1}{2} \times \frac{1}{4}$, or $\frac{1}{2} \times \frac{1}{2} \times \frac{1}{2}$), and the next day one-sixteenth ($\frac{1}{2} \times \frac{1}{8}$ or $\frac{1}{2} \times \frac{1}{2} \times \frac{1}{2} \times \frac{1}{2}$). On the fifth day (after 5 half-lives), the piece you eat is only $\frac{1}{32}$ of the original pie ($\frac{1}{2} \times \frac{1}{16}$ or $\frac{1}{2} \times \frac{1}{2} \times \frac{1}{2} \times \frac{1}{2} \times \frac{1}{2}$). The process continues until there is not enough pie to bother eating any. The situation with radioactive nuclides is similar. One-half of their amount disappears each half-life until there's no significant amount left. The length of time necessary for a radioactive sample to dwindle to insignificance depends on its half-life and on the amount that was present to begin with (Figure 18.2).

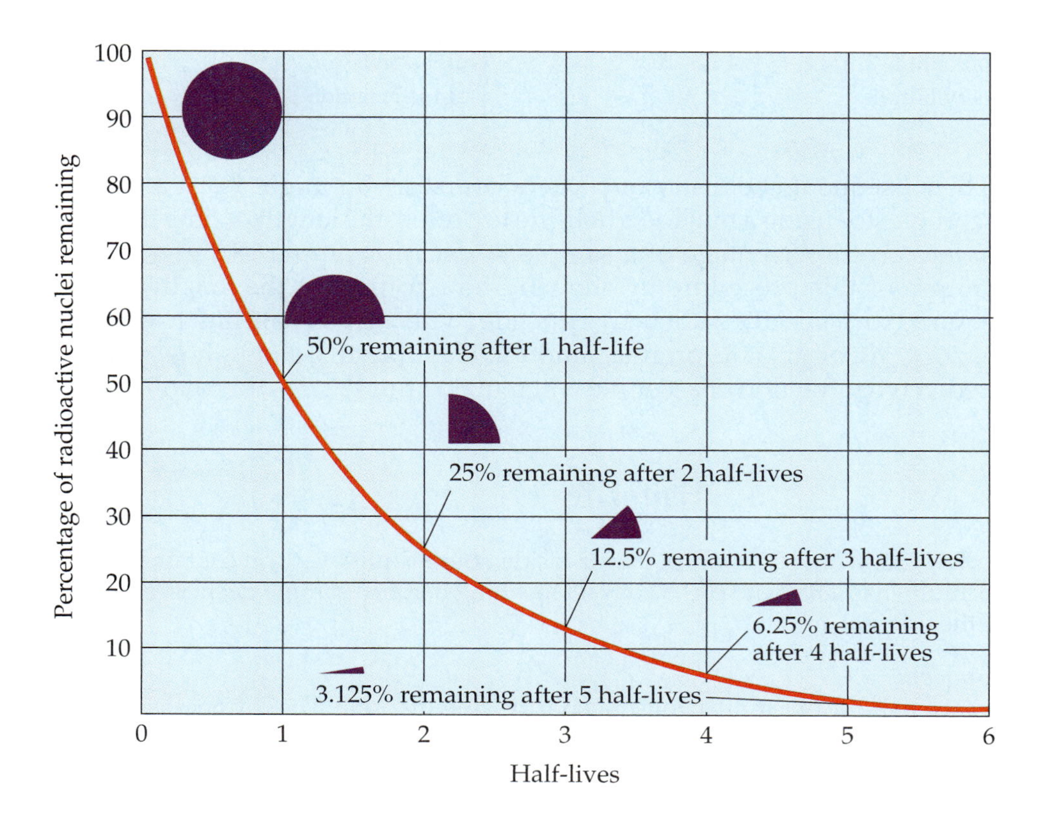


**Figure 18.2**
**Half-Life and Radioactive Decay**

OBJECTIVE 14

Table 18.3 shows some common radioactive isotopes and their half-lives.

**Table 18.3**
Half-Lives of Common Radioactive Isotopes

| Nuclide | Half-life | Type of change |
|---|---|---|
| rubidium-87 | $5.7 \times 10^{10}$ years | beta emission |
| thorium-232 | $1.39 \times 10^{10}$ years | alpha emission |
| uranium-238 | $4.51 \times 10^{9}$ years | alpha emission |
| uranium-235 | $7.13 \times 10^{9}$ years | alpha emission |
| plutonium-239 | $2.44 \times 10^{4}$ years | alpha emission |
| carbon-14 | 5730 years | beta emission |
| radium-226 | 1622 years | alpha emission |
| cesium-133 | 30 years | beta emission |
| strontium-90 | 29 years | beta emission |
| hydrogen-3 | 12.26 years | beta emission |
| cobalt-60 | 5.26 years | beta emission |
| iron-59 | 45 days | beta emission |
| phosphorus-32 | 14.3 days | beta emission |
| barium-131 | 11.6 days | electron capture and positron emission |
| iodine-131 | 8.06 days | beta emission |
| radon-222 | 3.82 days | alpha emission |
| gold-198 | 2.70 days | beta emission |
| krypton-79 | 34.5 hours | electron capture and positron emission |
| carbon-11 | 20.4 min | positron emission |
| fluorine-17 | 66 s | positron emission |
| polonium-213 | $4.2 \times 10^{-6}$ s | alpha emission |
| beryllium-8 | $1 \times 10^{-16}$ s | alpha emission |

In subsequent chemistry or physics courses, you might learn a general technique for using a nuclide's half-life to predict the length of time required for *any* given percentage of a sample to decay. Example 18.5 gives you a glimpse of this procedure by showing how to predict the length of time required for a specific radioactive nuclide (with a given half-life) to decay to ½, ¼, ⅛, 1/16, or 1/32 of its original amount. Example 18.6 shows how you can predict what fraction of a sample will remain after 1, 2, 3, 4, or 5 half-lives.

## EXAMPLE 18.5 Half-Life

OBJECTIVE 14

Radon-222, which is found in the air inside houses built over soil containing uranium, has a half-life of 3.82 days. How long before a sample decreases to 1/32 of the original amount?

***Solution***
In each half-life of a radioactive nuclide, the amount diminishes by one-half. The fraction 1/32 is ½ × ½ × ½ × ½ × ½, so 5 half-lives are needed to reduce the sample to that extent. For radon-222, then, 5 half-lives are **19.1 days** (5 × 3.82 days).

### Exercise 18.5 Half-Life

OBJECTIVE 14

One of the radioactive nuclides formed in nuclear power plants is hydrogen-3, called tritium, which has a half-life of 12.26 years. How long before a sample decreases to ⅛ of its original amount?

### Example 18.6 Half-Life

OBJECTIVE 15

One of the problems associated with the storage of radioactive wastes from nuclear power plants is that some of the nuclides remain radioactive for a very long time. An example is plutonium-239, which has a half-life of $2.44 \times 10^4$ years. What fraction of plutonium-239 is left after $9.76 \times 10^4$ years?

***Solution***
The length of time divided by the half-life yields the number of half-lives:

$$\frac{9.76 \times 10^4 \text{ years}}{2.44 \times 10^4 \text{ year}}\text{s} = 4 \text{ half-lives}$$

In each half-life of a radioactive nuclide, the amount diminishes by one-half, so the fraction remaining would be **1/16** (½ × ½ × ½ × ½).

### Exercise 18.6 Half-Life

OBJECTIVE 15

Uranium-238 is one of the radioactive nuclides that is sometimes found in soil. It has a half-life of $4.51 \times 10^9$ years. What fraction of a sample is left after $9.02 \times 10^9$ years?

## Radioactive Decay Series

OBJECTIVE 16

Many of the naturally occurring radioactive nuclides have relatively short half-lives. Radon-222, which according to U.S. Environmental Protection Agency estimates causes between 5000 and 20,000 deaths per year from lung cancer, has a half-life of only 3.82 days. With such a short half-life, why are this and other short-lived nuclides still around? The answer is that although they disappear relatively quickly once they form, these nuclides are constantly being replenished because they are products of other radioactive decays.

Three relatively abundant and long-lived radioactive nuclides are responsible for producing many of the other natural radioactive isotopes on earth. One of them is uranium-238, with a half-life of 4.51 billion years,

OBJECTIVE 16

which changes to lead-206 in a series of 8 alpha decays and 6 beta decays (Figure 18.3). Chemists call such a sequence a **nuclear decay series.** Because this sequence of decays is happening constantly in soil and rocks containing uranium, all of the radioactive intermediates between uranium-238 and lead-206 are constantly being formed and are therefore still found in nature.

OBJECTIVE 16

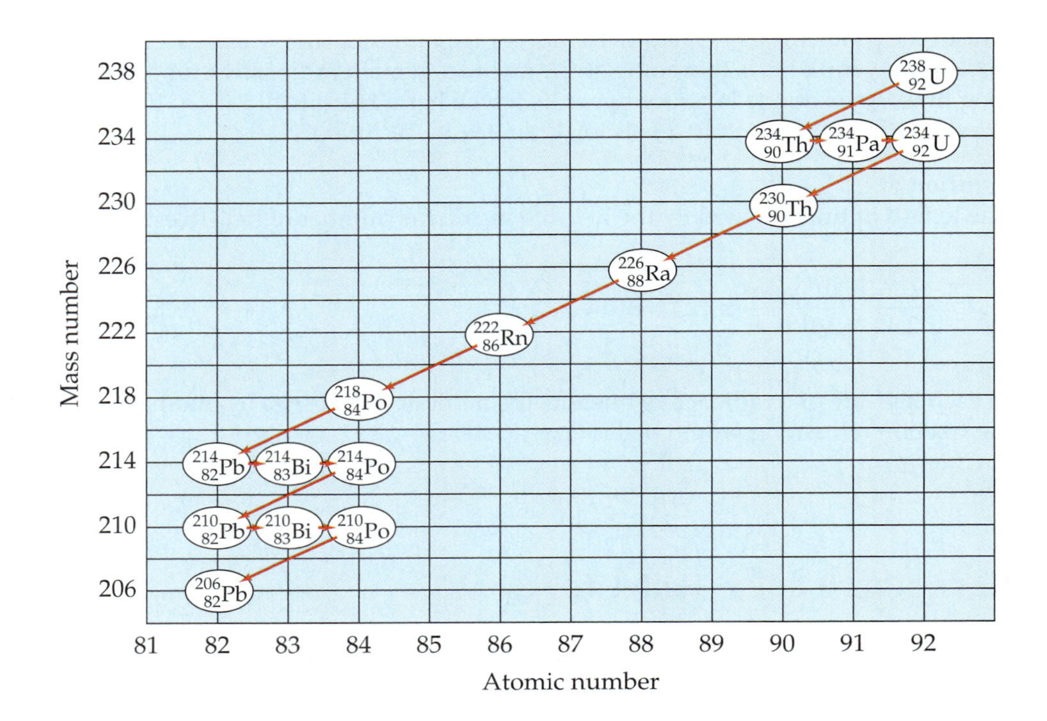


**Figure 18.3**
Uranium-238 Decay Series

OBJECTIVE 17

OBJECTIVE 18

You can see in Figure 18.3 that one of the products of the uranium-238 decay series is radium-226. This nuclide, with a half-life of 1622 years, is thought to be the second leading cause of lung cancer (after smoking). The next step in this same decay series forms radon-222. Radon-222 is also thought to cause cancer, but it does not do so directly. Radon-222 is a gas and enters our lungs through the air. Then, because of its fairly short half-life, a significant amount of it decays to form polonium-218 while it is still in our lungs. Polonium and all of the radioactive nuclides that follow it in the decay series are solids that remain in the lining of the lungs, emitting alpha particles, beta particles, and gamma rays. Houses built above earth that contains uranium can harbor significant concentrations of radon, so commercial test kits have been developed to detect it. Radon is a bigger problem in colder climates and at colder times of the year because it accumulates inside houses that are sealed up tight to trap warm air.

Radon test kit.

The other two important decay series are the one in which uranium-235 (with a half-life of $7.13 \times 10^8$ years) decays in 11 steps to lead-207 and the one in which thorium-232 (which has a half-life of $1.39 \times 10^{10}$ years) decays in ten steps to lead-208.

## The Effect of Radiation on the Body

Alpha particles, beta particles, and gamma photons are often called **ionizing radiation,** because as they travel through a substance, they strip electrons from its atoms, leaving a trail of ions in their wake. Let's explore why this happens and take a look at the effects that ionizing radiation has on our bodies.

OBJECTIVE 19

Picture an alpha particle moving through living tissue at up to 10% of the speed of light. Remember that alpha particles are helium nuclei, so each has a +2 charge. As such a particle moves past, say, an uncharged water molecule (a large percentage of our body is water), it attracts the molecule's electrons. One of these electrons might be pulled toward the passing alpha particle enough to escape from the water molecule, but it might not be able to catch up to the fast-moving alpha particle. Instead, the electron is quickly incorporated into another atom or molecule, forming an anion, while the water molecule that lost the electron becomes positively charged. The alpha particle continues on its way, creating many ions before slowing down enough for electrons to catch up with it and neutralize its charge. When a neutral water molecule, which has all of its electrons paired, loses one electron, the cation that is formed has an unpaired electron. Particles with unpaired electrons are called **free radicals.**

OBJECTIVE 20

Free radical

$$H_2O \xrightarrow{\alpha\ \text{particle}} H_2O^{\bullet +} + e^-$$

OBJECTIVE 19

In beta radiation, the repulsion between the beta particles' negative charges and the electrons on the atoms and molecules of our tissue causes electrons to be pushed off the uncharged particles, creating ions like the ones created by alpha particles. Gamma photons produce their damage by exciting electrons enough to actually remove them from atoms.

The water cations, $H_2O^{\bullet +}$, and free electrons produced by ionizing radiation react with uncharged water molecules to form other ions and free radicals.

OBJECTIVE 20

$$H_2O^{\bullet +} + H_2O \rightarrow H_3O^+ + {\bullet}OH$$

$$H_2O + e^- \rightarrow H{\bullet} + OH^-$$

OBJECTIVE 21

These very reactive ions and free radicals then react with important substances in the body, leading to immediate tissue damage and also to delayed problems, such as cancer. The cells that reproduce most rapidly are the ones most vulnerable to harm, because they are the sites of greatest chemical activity. This is why nuclear emissions have a greater effect on children, who have larger numbers of rapidly reproducing cells, than on adults. The degree of damage is, of course, related to the length of exposure, but it also depends on the kind of radiation and on whether the source is inside or outside the body.

Some radioactive nuclides are especially damaging because they tend to concentrate in particular parts of the body. For example, because both strontium and calcium are alkaline earth metals in group 2 in the periodic table, they combine with other elements in similar ways. Therefore, if radioactive strontium-90 is ingested, it concentrates in the bones in substances that would normally contain calcium. This can lead to bone cancer or leukemia. For similar reasons, radioactive cesium-137 can enter the cells of the body in place of its fellow alkali metal potassium, leading to tissue damage. Nonradioactive iodine and radioactive iodine-131 are both absorbed by thyroid glands. Because iodine-131 is one of the radioactive nuclides produced in nuclear power plants, the Chernobyl accident released large quantities of it. To reduce the likelihood of thyroid damage, people were directed to take large quantities of salt containing nonradioactive iodine-127. This flooding of the thyroid glands with the nondamaging form of iodine made it less likely that the iodine-131 would be absorbed.

OBJECTIVE 22

Because alpha particles are relatively large and slow-moving compared to other emissions from radioactive atoms, it is harder for them to slip between the atoms in the matter through which they pass. Alpha particles are blocked by 0.02 mm to 0.04 mm of water or about 0.05 mm of human tissue. Therefore, alpha particles that strike the outside of the body enter no farther than the top layer of skin. Because beta particles are much smaller and can move up to 90% the speed of light, they are about 100 times as penetrating as alpha particles. Thus beta particles are stopped by 2 mm to 4 mm of water or by 5 mm to 10 mm of human tissue. Beta particles from a source outside the body may penetrate to the lower layers of skin, but they will be stopped before they reach the vital organs. Gamma photons are much more penetrating, so gamma radiation from outside the body can damage internal organs.

OBJECTIVE 23

Although alpha and beta radiation are less damaging to us than gamma rays when emitted from external sources, both forms of radiation can do significant damage when emitted from within the body (by a source that has been eaten or inhaled). Because they lose all of their energy over a very short distance, alpha or beta particles can do more damage to localized areas in the body than the same number of gamma photons would.

OBJECTIVE 24

## 18.2 Uses of Radioactive Substances

As you saw in the last section, radioactive substances can be damaging to our bodies, but scientists have figured out ways to use some of their properties for our benefit. For example, radioactive nuclides are employed to diagnose lung and liver disease, to treat thyroid problems and cancer, and to determine the ages of archaeological finds. Let's examine some of these beneficial uses.

## Medical Uses

Cobalt-60 emits ionizing radiation in the form of beta particles and gamma photons. You saw in the last section that gamma photons, which penetrate the body and damage the tissues, do more damage to rapidly reproducing cells than to others. This characteristic, coupled with the fact that cancer cells reproduce very rapidly, underlies the strategy of using radiation to treat cancer. Typically, a focused beam of gamma photons from cobalt-60 is directed at a cancerous tumor. The ions and free radicals that the gamma photons produce inside the tumor damage its cells and cause the tumor to shrink.

OBJECTIVE 25

Like many other radioactive nuclides used in medicine, cobalt-60 is made by bombarding atoms of another element (in this case, iron) with neutrons. The iron contains a small percentage of iron-58, which forms cobalt-60 via the following steps:

$$^{58}_{26}\text{Fe} + ^{1}_{0}\text{n} \rightarrow ^{59}_{26}\text{Fe}$$

$$^{59}_{26}\text{Fe} \rightarrow ^{59}_{27}\text{Co} + ^{0}_{-1}\text{e}$$

$$^{59}_{27}\text{Co} + ^{1}_{0}\text{n} \rightarrow ^{60}_{27}\text{Co}$$

One of the most significant advances in medicine in recent years has been the development of computer-imaging techniques. One such technique that relies on the characteristics of atomic nuclei is called magnetic resonance imaging (MRI). The patient is placed in a strong magnetic field and exposed to radio wave radiation, which is absorbed and re-emitted in different ways by different tissues in the body. Computer analysis of the re-emitted radio waves yields an image of the tissues of the body.

OBJECTIVE 26

A simplified description of how MRI works begins with the fact that protons act like tiny magnets. When patients are put in the strong magnetic field, the proton magnets in the hydrogen atoms in their bodies line up either with or against the field (these orientations are called parallel and antiparallel, respectively). The antiparallel arrangement is slightly higher in energy than the parallel position, and the energy difference is equivalent to the energy of photons in the radio wave range of the radiant-energy spectrum. As a result, when radio waves are directed at the patient, they are absorbed by protons in the parallel orientation and excite the protons into the antiparallel orientation. As the protons return to the more stable parallel orientation, they re-emit energy, which can be detected by scanners placed around the patient's body.

OBJECTIVE 26

$$\text{proton parallel to field} + \text{absorbed energy} \rightarrow \text{proton antiparallel to field}$$

$$\text{proton antiparallel to field} \rightarrow \text{proton parallel to field} + \text{emitted energy}$$

Because the soft tissues of the body contain a lot of water (with a lot of hydrogen atoms) and bones do not, the MRI process is especially useful for creating images of the soft tissues of the body. Hydrogen atoms absorb and re-emit radio wave photons in different ways, depending on their

OBJECTIVE 26

molecular environment, so the computer analysis of the data leads to detailed images of the soft tissues.

Special Topic 11.1: *Why Does Matter Exist, and Why Should We Care About Answering This Question?* describes positron emission tomography (PET), which is the newest computer-imaging diagnostic technique. In this technique, a solution containing a positron-emitting substance is introduced into the body. The positrons collide with electrons, and the two species annihilate each other, creating two gamma photons that move apart in opposite directions.

OBJECTIVE 27

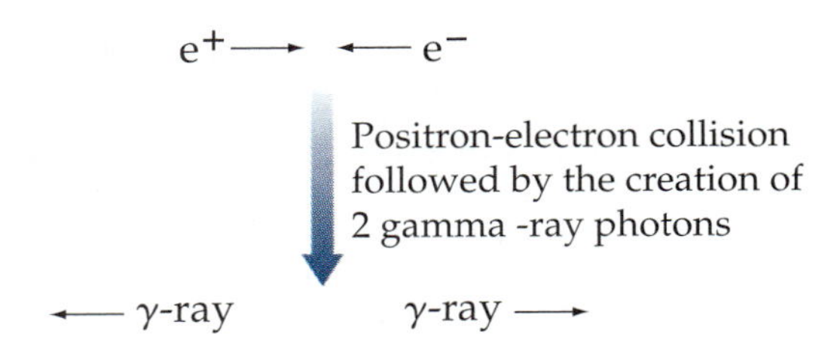


These photons can be detected and the data computer-analyzed to yield images showing where in the body the radioactive substances collected. Different nuclides are used to study different parts of the body. For example, fluorine-18-containing substances collect in the bones, so that nuclide is used in bone scanning. Glucose molecules are used throughout the body but are especially concentrated in the brain, so glucose constructed with carbon-11 can be used to study brain function. PET scans reveal which parts of the brain use glucose most actively during different activities. This ability to study dynamic processes in the body, such as brain activity and blood flow, makes the PET scan a valued research and diagnostic tool.

OBJECTIVE 27

OBJECTIVE 28

The nuclides used for PET scans have half-lives that range from about 2 minutes to about 110 minutes, so one of the challenges in refining this technology has been to find ways to incorporate the positron emitters quickly into the substances, such as glucose, that will be introduced into the body.

## Carbon-14 Dating

*[If not for radiocarbon dating,] we would still be floundering in a sea of imprecisions sometimes bred of inspired guesswork but more often of imaginative speculations.*
DESMOND CLARK, ANTHROPOLOGIST

As anthropologists such as Desmond Clark attempt to piece together the strands of human history, they often call on other scientists for help. One important contribution that nuclear science has made in this area is the ability to determine the age of ancient artifacts. There are several techniques for doing this, but the most common process for dating objects of up to about 50,000 years of age is called **radiocarbon dating.**

Natural carbon is composed of three isotopes: It is 98.89% carbon-12, 1.11% carbon-13, and 0.00000000010% carbon-14. The last of these, carbon-14, is most important in radiocarbon dating. Carbon-14 atoms are constantly being produced in our upper atmosphere through neutron bombardment of nitrogen atoms.

OBJECTIVE 29

$$^{14}_{7}\text{N} + ^{1}_{0}\text{n} \rightarrow ^{14}_{6}\text{C} + ^{1}_{1}\text{H}$$

OBJECTIVE 29

Once formed, the carbon-14 is quickly oxidized to produce carbon dioxide, $CO_2$, which is then converted into many different substances in plants.

When animals eat the plants, the carbon-14 becomes part of the animals too. For these reasons, carbon-14 is found in all living things. This radioactive isotope is a beta emitter with a half-life of 5730 years (±40 years), so as soon as it becomes part of a plant or animal, it begins to disappear.

$$^{14}_{6}\text{C} \rightarrow\ ^{14}_{7}\text{N} +\ ^{0}_{-1}\text{e}$$

As long as a plant or animal remains alive, its intake of carbon-14 balances the isotope's continuous decay, so that the ratio of $^{14}C$ to $^{12}C$ in living tissues remains constant at about 1 in 1,000,000,000,000. When the plant or animal dies, it stops taking in fresh carbon, but the carbon-14 it contains continues to decay. Thus the ratio of $^{14}C$ to $^{12}C$ drops steadily. Therefore, to date an artifact, scientists analyze a portion of it to determine the $^{14}C/^{12}C$ ratio, which is then used to calculate its age.

OBJECTIVE 29

Initially, researchers expected that once the $^{14}C/^{12}C$ ratio was known, calculating an object's age would be simple. For example, if the $^{14}C/^{12}C$ ratio had dropped to one-half of the ratio found in the air today, the object would have been described as about 5730 years old. A $^{14}C/^{12}C$ ratio of one-fourth of the ratio found in the air today would date it as 11,460 years old (2 half-lives). However, this works only if we can assume that the $^{14}C/^{12}C$ ratio in the air was the same when the object died as it is now, and scientists have discovered that this is not strictly true. For example, it is now believed that the large quantities of carbon periodically released into the atmosphere by volcanoes have in many cases been isolated from the air for so long that they have much lower than average levels of carbon-14. The levels of cosmic radiation also fluctuate, and lower levels of neutron bombardment produce lower levels of carbon-14.

OBJECTIVE 29

Information derived from Bristlecone pines in California has helped scientists to date artifacts.

By checking the $^{14}C/^{12}C$ ratio of the wood in tree rings (which are formed once a year), scientists discovered that the ratio has varied by about ±5% over the last 1500 years. Further study of very old trees, such as the bristlecone pines shown here, has enabled researchers to develop calibration curves for radiocarbon dating that go back about 10,000 years. These calibration curves are now used to get more precise dates for objects.

Radiocarbon dating has been used to date charcoal from ancient fires, fragments of bone, shells from the ocean, hair, insect remains, paper, cloth, and many other carbon-containing substances. Two of the most famous artifacts that have been dated by this technique are the Shroud of Turin and the Dead Sea Scrolls (see the photographs on the next page). Because these objects are very precious, they were dated only after techniques had been developed for testing very small amounts of substance (as little as 100 mg of material). The Dead Sea Scrolls are a collection of about 600 Hebrew and Aramaic manuscripts discovered in caves near Khirbat Qumrân in Jordan, near the Dead Sea. Carbon-14 dating confirmed the age that had already been predicted from other clues, such as the kinds of handwriting on the scrolls and the materials used to make them. These scrolls, which contain hymn books, biblical commentaries, and parts of every book in the Old Testament except Esther, are thought to have been written between 200 B.C. and A.D. 68; this makes them about 1000 years older than any other surviving biblical transcripts.

The Dead Sea Scrolls have been dated by the radiocarbon dating technique.

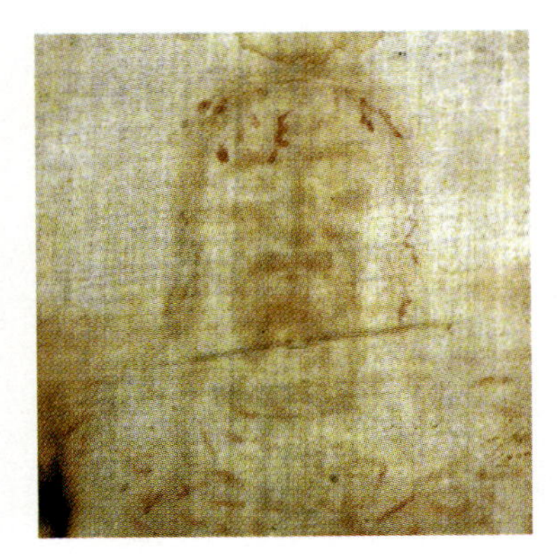

Above, a regular photo of the Shroud of Turin. Below, a photonegative of one portion of the Shroud of Turin

The Shroud of Turin has been dated by the carbon-14 process.

The dating of the Shroud of Turin in 1988 has been more controversial. The shroud is a linen cloth with a faint image of an adult who seems to have been crucified. Some people believe it to be the burial shroud of Jesus of Nazareth, but the results of radiocarbon dating suggest that the shroud is only about 600 years old. If these results are correct, they rule out the possibility of the shroud's having been used to bury Jesus.

The development of more sensitive ways to measure levels of radioactive substances has enabled scientists to take advantage of the decay of nuclides other than carbon-14. For example, chlorine-36 can be used to date ground water, marine sediments can be dated by measuring levels of beryllium-11 and aluminum-26, and krypton-81 has been used to estimate the age of glacial ice.

## Other Uses for Radioactive Nuclides

OBJECTIVE 30

There are many other uses for radioactive nuclides. For example, some smoke detectors contain the alpha emitter americium-241. The alpha particles it releases ionize the air in the detector's interior, allowing an electric current to pass through. When particles of smoke enter the detector, they block the alpha particles, thus decreasing the number of ions in the air, reducing the electric current, and triggering the alarm.

OBJECTIVE 31

Iridium-192 is used to check for faulty connections in pipes. Film is wrapped around the outside of a welded junction, and then a radioactive substance is run through the pipe. If there is a crack in the connection, radiation leaks out and exposes the film.

OBJECTIVE 32

One of the more controversial uses of radioactive nuclides is in food irradiation. Gamma-ray beams, X rays, and electron beams have been directed at food for a variety of purposes. Radiation inhibits the sprouting of potatoes and onions, retards the growth of mold on strawberries, and kills bacteria in poultry and fish. Cobalt-60 and cesium-137 have been used for these purposes. The controversy lies in whether the radiation causes changes in the food that could have adverse health consequences. Although the food is not any more radioactive after the treatment than before, the radiation does create ions and free radicals in the food. Most of these recombine to form harmless substances, such as water, but some form other, more worrisome chemicals, such as $H_2O_2$. Cooking and pasteurization also form substances in food that would not be there otherwise, so the formation of new substances, in and of itself, is not necessarily unusual or bad. Researchers have been attempting to identify the new substances in each situation and consider what problems, if any, they might create. Because people in certain parts of the world lose up to 50% of their food to spoilage, questions about the safety of using radiation to preserve food are of major importance. At this point, the scientific consensus is that the benefits of irradiating certain foods outweigh the potential dangers.

OBJECTIVE 33

Unstable nuclides have also been employed as **radioactive tracers** in scientific research. For example, scientists have used carbon-14 to study many aspects of photosynthesis. Because the radiation emitted from carbon-14 atoms can be detected outside of the system into which the carbon-14-containing molecules have been placed, the location of changes involving carbon can be traced. Likewise, phosphorus-32 atoms can be used to trace phosphorus-containing chemicals as they move from the soil into plants. Carbon-14, hydrogen-3, and sulfur-35 have been used to trace the biochemical changes that take place in our bodies. Table 18.4 lists many other uses for radioactive nuclides.

**Table 18.4**
Uses for Radioactive Nuclides

| Nuclide | Nuclear change | Application |
|---|---|---|
| argon-41 | beta emission | measure flow of gases from smokestacks |
| barium-131 | electron capture | detect bone tumors |
| carbon-11 | positron emission | PET brain scan |
| carbon-14 | beta emission | archaeological dating |
| cesium-133 | beta emission | radiation therapy |
| cobalt-60 | gamma emission | cancer therapy |
| copper-64 | beta emission, positron emission, electron capture | lung and liver disease diagnosis |
| chromium-51 | electron capture | determine blood volume and red blood cell lifetime |
| | | diagnose gastrointestinal disorders |
| fluorine-18 | beta emission, positron emission, electron capture | bone scanning; study of cerebral sugar metabolism |
| gallium-67 | electron capture | diagnosis of lymphoma and Hodgkin's disease; whole-body scan for tumors |
| gold-198 | beta emission | assess kidney activity |
| hydrogen-3 | beta emission | biochemical tracer; measurement of the water content of the of the body |
| indium-111 | gamma emission | label blood platelets |
| iodine-125 | electron capture | determination of blood hormone levels |
| iodine-131 | beta emission | measure thyroid uptake of iodine |
| iron-59 | beta emission | assessment of blood iron metabolism and diagnosis of anemia |
| krypton-79 | positron emission and electron capture | assessment of cardiovascular function |
| nitrogen-13 | positron emission | brain, heart, and liver imaging |
| oxygen-15 | positron emission | lung function test |
| phosphorus-32 | beta emission | leukemia therapy, detection of eye tumors, radiation therapy, and detection of breast carcinoma |
| polonium-210 | alpha emission | radiation therapy |
| potassium-40 | beta emission | geologic dating |
| radium-226 | alpha emission | radiation therapy |
| selenium-75 | beta emission and electron capture | measure size and shape of pancreas |
| sodium-24 | beta emission | blood studies and detection of blood clots |
| technetium-99 | gamma emission | bone scans and detection of blood clots |
| xenon-133 | beta emission | measurement of lung capacity |

## 18.3 Nuclear Energy

In Section 18.1, we saw that energy is released when nucleons (protons and neutrons) combine to form nuclei.

$$p + p + n + n \rightarrow {}^{4}_{2}He^{2+}$$

+ + + + Energy

Let's look more closely at this energy.

OBJECTIVE 34

The amount of energy released when a nucleus is formed is a reflection of the strength with which the nucleons are bound. This amount is therefore called the atom's **binding energy.** The alpha particle whose creation is illustrated above has a binding energy of $4.54 \times 10^{-12}$ J, a small amount that scientists often prefer to describe in terms of the more convenient **electron volt,** which is equivalent to $1.6 \times 10^{-19}$ joule. The binding energy of the alpha particle is 28.4 MeV (million electron volts). Small as this amount may be, it is still significantly larger than the energies associated with electrons. It takes about 10,000 times as much energy to remove a proton or a neutron from the nucleus of a hydrogen-2 atom as to remove its single electron.

OBJECTIVE 35

One way of comparing the relative stabilities of different nuclei is to look at their binding energies in terms of the binding energy per nucleon—that is, the nuclide's binding energy divided by the number of nucleons. A higher binding energy per nucleon means more stable and more tightly bound nucleons in the nucleus. The binding energy per nucleon for the helium nucleus is 7.10 MeV (28.4 MeV total binding energy divided by 4 nucleons). The chart in Figure 18.4 shows that the binding energy per nucleon varies for different nuclei, which suggests that the stabilities of nuclei vary. The different half-lives found in Table 18.3 reflect this varying stability. The binding energy (and therefore the stability) starts low for hydrogen-2, rises (with some exceptions) with atomic mass for mass numbers up to around 56, and then drops again for larger atoms.

OBJECTIVE 37

There are several important conclusions that we can draw from Figure 18.4. First, it shows that certain nuclides are more stable than we might expect them to be if the trend for the change in binding energy per nucleon were smooth. The stabilities of ${}^{4}_{2}He$, ${}^{12}_{6}C$, ${}^{16}_{8}O$, and ${}^{20}_{10}Ne$ are all high. This is explained by the fact that each of them has an even number of protons and an even number of neutrons. In short, paired nucleons (like paired electrons) are more stable than unpaired ones. Thus, of the 264 stable isotopes in nature, 160 of them have an even number of protons and an even number of neutrons, 50 have an even number of protons and an odd number of neutrons, 50 have an odd number of protons and an even number of neutrons, and only 4 (${}^{2}_{1}H$, ${}^{6}_{3}Li$, ${}^{10}_{5}B$, and ${}^{14}_{7}N$) have an odd number of protons and an odd number of neutrons.

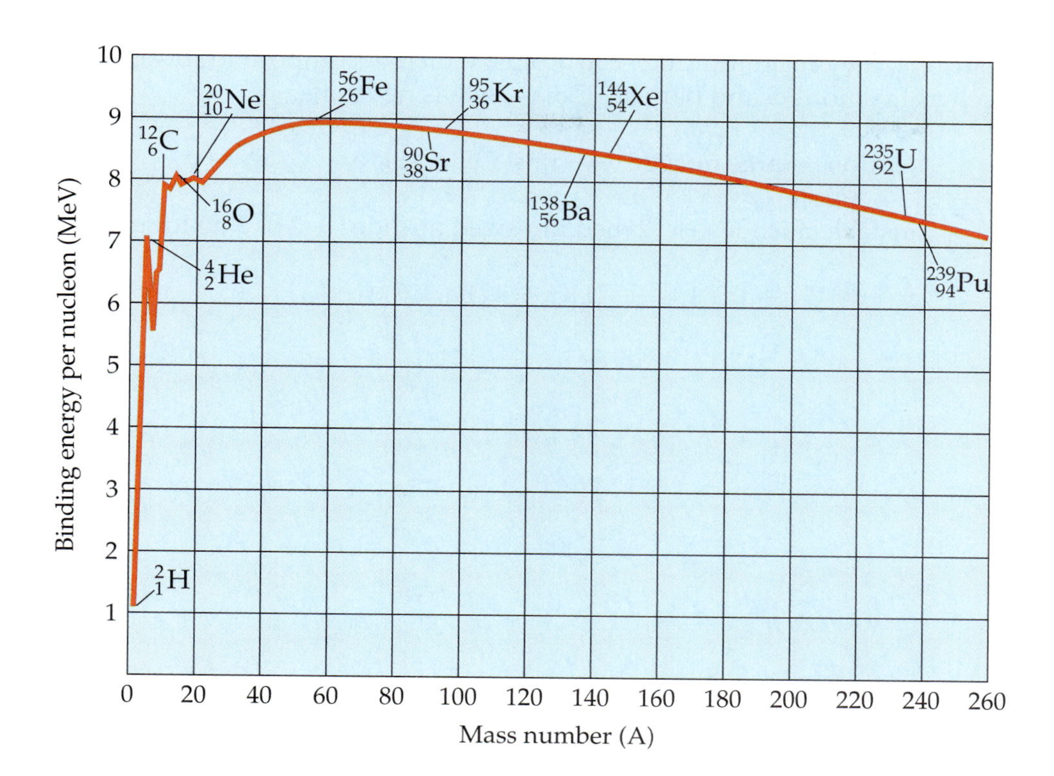


**Figure 18.4**
Binding Energy per Nucleon Versus Mass Number

OBJECTIVE 36

In Chapter 11, we found that there was something special about having the same number of electrons as the noble gases (2, 10, 18, 36, 54, and 86). There also appears to be something stable about having 2, 8, 20, 28, 50, 82, or 126 protons or neutrons. The stability of nuclides with double magic numbers (the aforementioned numbers of protons and neutrons are often called "magic numbers") is very high. For example, $^{4}_{2}$He, $^{16}_{8}$O, $^{40}_{20}$Ca, and $^{208}_{82}$Pb are especially stable.

OBJECTIVE 38

Note that initially, as we read left to right along the curve in Figure 18.4, the binding energy per nucleon generally increases. This means that more energy is released per nucleon to form a nucleus approaching the size of the iron-56 nucleus than to form a nucleus that is smaller. Thus, as a rule, when atoms are much smaller than iron-56, energy is released when they combine to form larger atoms. As we will see, this process of combining smaller atoms to make larger ones (called **fusion**) is the process that fuels the sun. Likewise, the chart shows that when atoms are larger than iron-56, splitting them to form more stable, smaller atoms should also release energy. This process, called **fission,** is the process that fuels nuclear reactors used to make electricity.

OBJECTIVE 39

## Nuclear Fission and Electric Power Plants

In a typical nuclear fission process, a neutron collides with a large atom, such as uranium-235, and forms a much less stable nuclide that spontaneously decomposes into two medium-sized atoms and two or three neutrons. For example, when uranium-235 atoms are bombarded with

OBJECTIVE 39

neutrons, they form uranium-236 atoms, which decompose to form atoms such as krypton-95 and barium-138 as well as neutrons.

Neutron + large nuclide $\rightarrow$ unstable nuclide

Unstable nuclide $\rightarrow$ 2 medium-sized nuclides + 2 or 3 neutrons

$$^{1}_{0}n + ^{235}_{92}U \rightarrow ^{236}_{92}U \rightarrow ^{95}_{36}Kr + ^{138}_{56}Ba + 3\,^{1}_{0}n$$

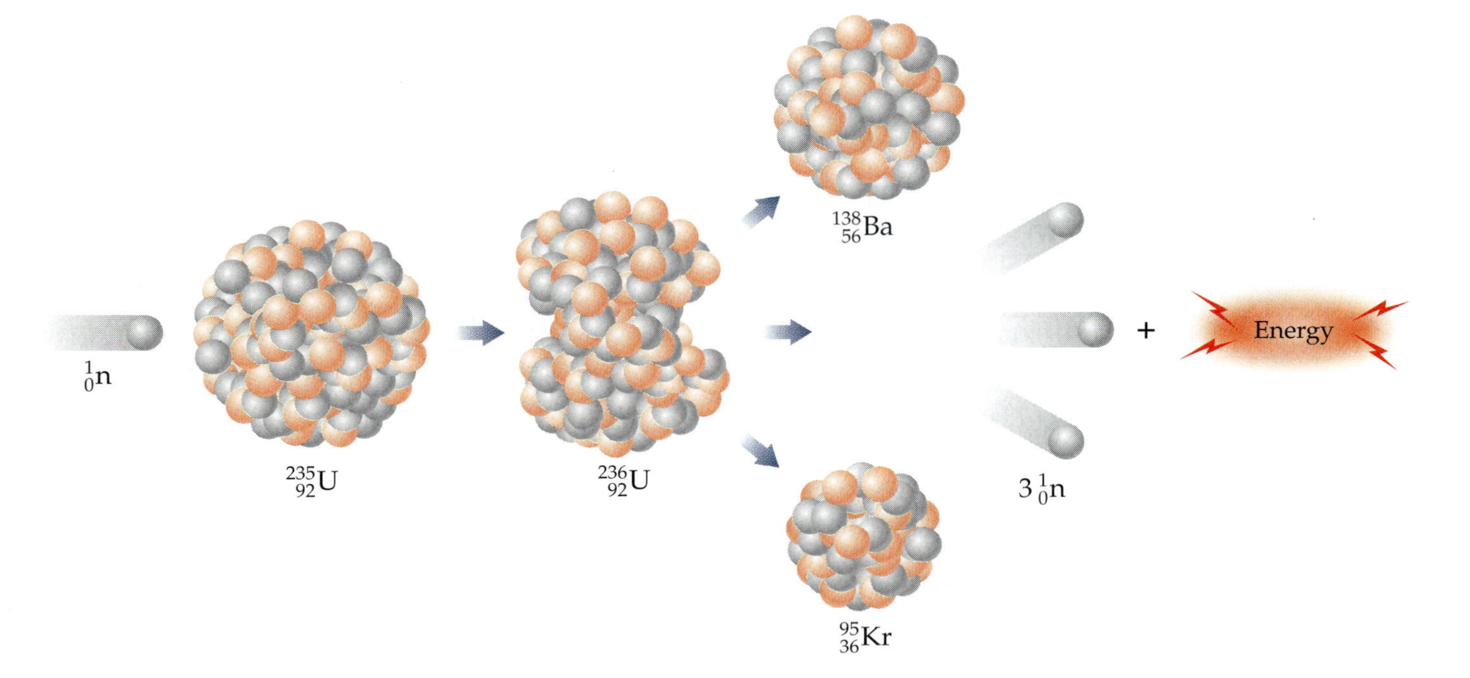


OBJECTIVE 39

The nuclides produced in the reaction pictured above are only 2 of many possible fission products of uranium-235. More than 200 different nuclides form, representing 35 different elements. Another possible reaction is

$$^{1}_{0}n + ^{235}_{92}U \rightarrow ^{236}_{92}U \rightarrow ^{144}_{54}Xe + ^{90}_{38}Sr + 2\,^{1}_{0}n$$

Nuclear reactions like these are used to power electricity generating plants.

The nuclear reactor in a nuclear power plant is really just a big furnace whose job is to generate heat and thus convert liquid water to steam in order to turn a steam turbine generator that produces electricity. The electricity-generating portion of a nuclear power plant is typically no different from the electricity-generating portion of a plant that generates heat from burning fossil fuels. Therefore, instead of describing that part of the power plant's setup, we will focus exclusively on how the fission process generates heat.

OBJECTIVE 40

Figure 18.4 shows that energy is released when larger nuclides with lower binding energy per nucleon are converted into medium-sized nuclides with a higher binding energy per nucleon. The reason why the

OBJECTIVE 39

fission of uranium-235 can generate a lot of energy in a short period of time is that under the right circumstances, it can initiate a **chain reaction,** a process in which one of the products of a reaction initiates another identical reaction. In the fission of uranium-235, one or more of the neutrons formed in the reaction can collide with another uranium-235 atom and cause it to undergo fission too (Figure 18.5).

OBJECTIVE 39

**Figure 18.5**
**Uranium-235 Chain Reaction**

OBJECTIVE 39

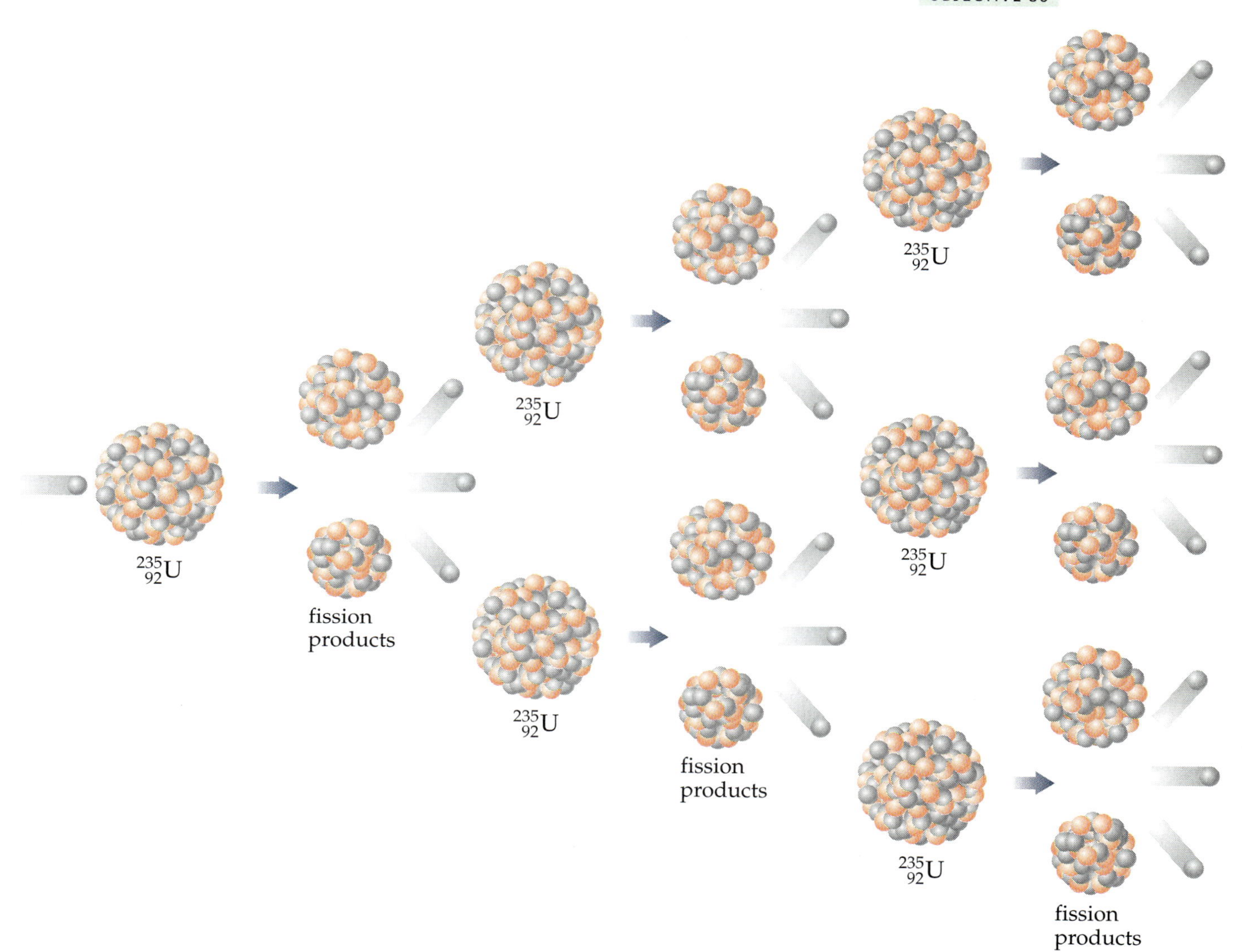


To sustain a chain reaction for the fission of uranium-235, an average of at least one of the neutrons generated in each reaction must go on to cause another reaction. If this does not occur, the series of reactions slows down and eventually stops. When natural uranium is bombarded by neutrons, the chain reaction cannot be sustained. Too many neutrons are absorbed by other entities without leading to fission. To see why, we need to look again at the natural composition of uranium, described earlier in the chapter as 99.27% uranium-238, 0.72% uranium-235, and a trace of uranium-234. The only one of these isotopes that undergoes fission is uranium-235.

OBJECTIVE 41

Uranium-238 does absorb neutrons, but the uranium-239 that then forms reacts by beta emission rather than nuclear fission.

$$^{238}_{92}\text{U} + ^{1}_{0}\text{n} \rightarrow ^{239}_{92}\text{U}$$

$$^{239}_{92}\text{U} \rightarrow ^{239}_{93}\text{Np} + ^{0}_{-1}\text{e}$$

$$^{239}_{93}\text{Np} \rightarrow ^{239}_{94}\text{Pu} + ^{0}_{-1}\text{e}$$

When natural uranium is bombarded by neutrons, the uranium-238 absorbs so many of the neutrons released in the fission reactions of uranium-235 that a chain reaction cannot be sustained. One part of the solution to this problem for nuclear power plants is to create a uranium mixture that is enriched in uranium-235 (to about 3%). A typical 1000-megawatt power plant will have from 90,000 to 100,000 kilograms of this enriched fuel packed in 100 to 200 zirconium rods about 4 meters long (Figure 18.6).

OBJECTIVE 41

OBJECTIVE 40

There is another way to increase the likelihood that neutrons will be absorbed by uranium-235 instead of uranium-238. Both uranium-235 and uranium-238 absorb fast neutrons, but if the neutrons are slowed down, they are much more likely to be absorbed by uranium-235 atoms. Therefore, in a nuclear reactor, the fuel rods are surrounded by a substance called a **moderator** that slows the neutrons as they pass through it. Several substances have been used as moderators, but normal water is most commonly employed (Figure 18.6).

OBJECTIVE 42

OBJECTIVE 40

Another problem associated with the absorption of neurons by uranium-238 is that it leads to the creation of plutonium-239 (look again at the series of equations presented above). This is cause for concern because plutonium-239 undergoes nuclear fission more readily than uranium-235 and is thus a possible nuclear weapon fuel. Moreover, plutonium has a relatively long half-life, so the radioactive wastes from nuclear reactors (which contain many other unstable nuclides besides plutonium) must be carefully isolated from the environment for a very long time.

OBJECTIVE 43

An efficient nuclear reactor needs to sustain the chain reaction but should not allow the fission reactions to take place too rapidly. For this reason, nuclear power plants have **control rods** containing substances such as cadmium or boron, which are efficient neutron absorbers. At the first sign of trouble, these control rods are inserted between the fuel rods, absorbing the neutrons that would have passed from one fuel rod to another and preventing them from causing more fission reactions. The deployment of the control rods halts the chain reaction and stops the production of heat.

OBJECTIVE 44

The control rods serve another purpose in the normal operation of the power plant. When fresh fuels rods are introduced, the control rods are partially inserted to absorb some of the neutrons released. As the uranium-235 reacts and its percentage of the total mixture in the fuel rods decreases, the control rods are progressively withdrawn. In this way, a constant rate of fission can be maintained, even as the percentage of the fissionable uranium-235 diminishes (Figure 18.6).

Nuclear power is a major source of energy for the generation of electricity worldwide. Nuclear power plants are found in over 30 countries and generate about 17% of the world's electricity. France gets about 76% of its

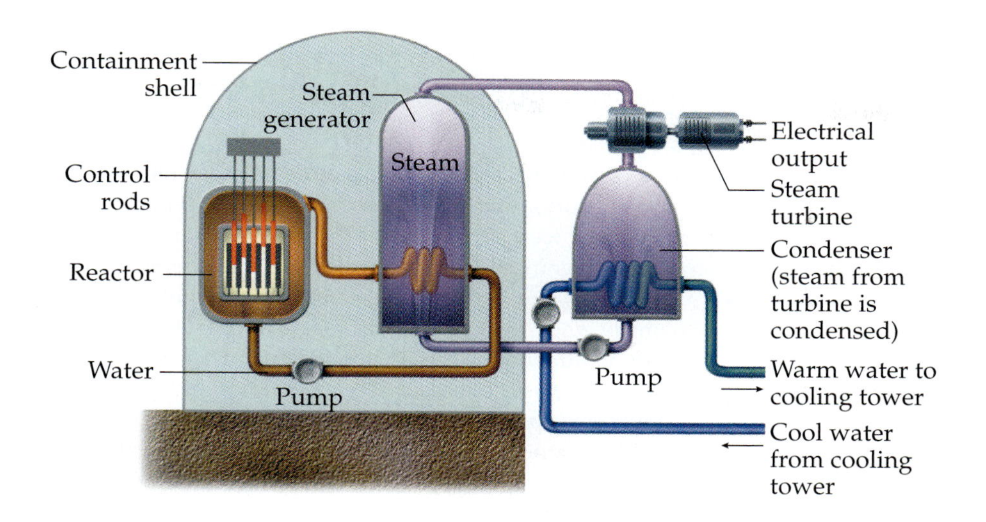


**Figure 18.6**
**Generating Electricity from Nuclear Power**

electricity from nuclear power, Japan about 33%, and the United States about 22%. Special Topic 18.1: *A New Treatment for Brain Cancer* on page 814 describes another use for a fission reaction.

## Nuclear Fusion and the Sun

OBJECTIVE 45

Whereas nuclear *fission* reactions yield energy by splitting large nuclei to form medium-sized ones, nuclear *fusion* reactions release energy by combining small nuclei into larger and more stable species. For example, the sun releases energy at a rate of about $3.8 \times 10^{26}$ J/s from the fusion of hydrogen nuclei to form helium. This is equivalent to burning $3 \times 10^{18}$ gallons of gasoline per second. The change takes place in three steps:

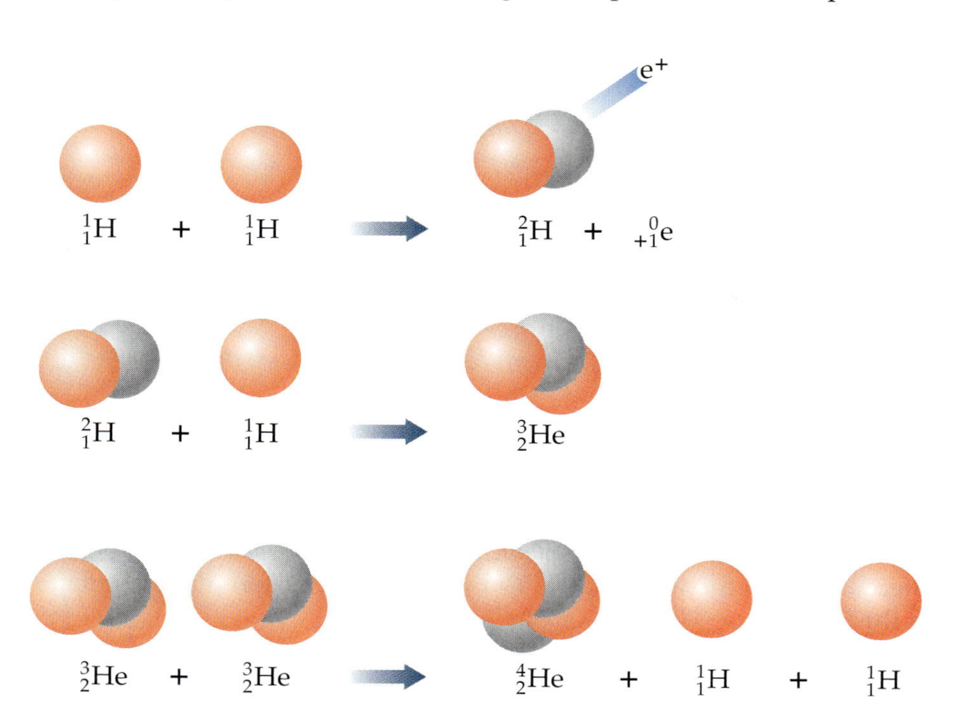

## SPECIAL TOPIC 18.1 A New Treatment for Brain Cancer

A promising new experimental treatment for brain tumors makes use of the fission reaction of boron-10, an isotope that represents 19.9% of natural boron. The patient is given a boron-containing compound that is selectively absorbed by the tumor cells, after which a low-energy neutron beam from a nuclear reactor or linear accelerator is directed toward the tumor. Most of the neutrons pass through normal cells without affecting them, but when they strike a boron-10 atom in a tumor cell, the atom absorbs the neutron to form an unstable boron-11 atom, which then splits in two to form a helium-4 atom and a lithium-7 atom. These products rush away from each other at a very high velocity, doing serious damage to the tumor cell as they go.

$$^{1}_{0}n + ^{10}_{5}B \rightarrow ^{11}_{5}B \rightarrow ^{4}_{2}He + ^{7}_{3}Li + \text{energy}$$

The diameter of a tumor cell is about 10 μm, and the helium and lithium atoms lose the energy derived from their creation in about 5 μm to 8 μm. Therefore, the product atoms do most of their damage to the cell in which they were produced. If the normal cells do not absorb a significant amount of the boron-containing compound, they do not incur a significant amount of damage. The chemist's primary role in the development of this treatment is to design boron-containing compounds that deliver more boron to tumor cells and as little as possible to normal ones.

This treatment is still experimental, but the clinical trials that have been done in several places in the world seem promising for the treatment of glioma, a form of brain cancer that has low survival rates. Tests are also being done to see whether this treatment could be used for other cancers, such as oral cancer and thyroid cancer, and for other medical problems, such as rheumatoid arthritis.

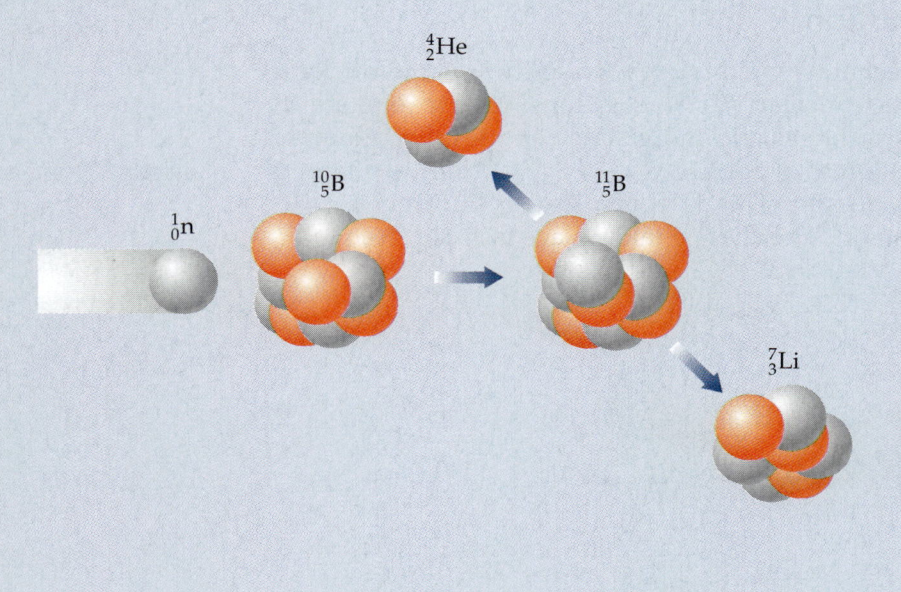


It seems fitting that the end of this text should take us back to the creation of the elements, which is the very beginning of chemistry. Special Topic 18.2: *The Origin of the Elements* describes how scientists believe the chemical elements were made. You will see that fusion played a major role in this process.

## SPECIAL TOPIC 18.2 The Origin of the Elements

Scientists believe that nuclear fusion played an important role in the creation of the chemical elements. To get a rough idea of what they think occurred, we have to begin at the beginning of the universe.

The standard model for the origin of the universe suggests that it all started about 15 billion years ago with a tremendous explosion called the Big Bang. About 1 second after the Big Bang, the universe is thought to have consisted of an expanding sea of light and elementary particles, including electrons, protons, and neutrons. The model describes three ways in which different elements formed.

The first of these processes took place in the first minutes following the Big Bang. The expanding cloud cooled enough for hydrogen atoms to form and for a series of nuclear reactions to occur. In each of these reactions, very-high-velocity particles collided and fused to produce a new and larger nucleus:

$$^{1}_{1}H + ^{1}_{0}n \rightarrow ^{2}_{1}H + \text{energy}$$

$$^{2}_{1}H + ^{1}_{0}n \rightarrow ^{3}_{1}H + \text{energy}$$

$$^{3}_{1}H + ^{1}_{1}H \rightarrow ^{4}_{2}He + \text{energy}$$

$$^{2}_{1}H + ^{1}_{1}H \rightarrow ^{3}_{2}He + \text{energy}$$

$$^{3}_{2}He + ^{1}_{0}n \rightarrow ^{4}_{2}He + \text{energy}$$

$$^{3}_{2}He + ^{4}_{2}He \rightarrow ^{7}_{4}Be + \text{energy}$$

$$^{7}_{4}Be + ^{0}_{-1}e \rightarrow ^{7}_{3}Li + \text{energy}$$

After this first phase of element synthesis, the universe consisted mostly of hydrogen and helium, with extremely small amounts of lithium and beryllium. Even today, about 90% of the atoms and 75% of the mass of today's universe is hydrogen, and most of the rest is helium.

By chance, the material of the expanding universe was unevenly distributed. In the regions of higher concentrations of matter, gravity pulled the particles even closer together, and stars began to form. As the cosmic dust was compressed, the material in the core of the forming stars began to heat up, providing the energy for hydrogen atoms to fuse and form helium atoms. In the extremely high temperatures of a star's core, colliding atoms attain enough speed to overcome the repulsion between their positive nuclei and approach each other closely enough for the nuclear attraction to fuse the nuclei together. This is the second way in which elements formed: by fusion reactions occurring in the cores of stars. Because the heat released in the fusion reactions would normally cause the gases to expand, it counteracts the gravitational force that would otherwise cause the star to collapse. When the two forces balance, the star finds a stable size.

Eventually, much of the hydrogen in the core of a young star is converted into helium, so that the star chiefly consists of a helium core and a hydrogen outer shell. The ultimate fate of a star depends on its size. If the star is relatively small, it will simply burn out when the hydrogen in the core is depleted, and it will become a white dwarf star. The situation is different for a very large star. When there is no longer enough hydrogen to convert to helium and supply the heat that keeps the star from imploding, the star begins to collapse, becoming what astronomers call a "red giant." The increased velocity of the particles as they rush toward the center of the star means an increase in the temperature of the gases. This change supplies the heat necessary for a new set of fusion reactions to occur that form carbon and oxygen. If the star is large enough, and if its fusion reactions produce enough energy, other elements (up to the size of iron) are also formed.

A star whose mass is over 10 times the mass of our sun can reach a stage where the reactions in its core cannot overcome gravity. At that point, the star collapses very rapidly, heating to tremendous temperatures as a result, and then explodes as a supernova. During this explosion—which can take place in a matter of seconds—elements larger than iron are thought to be formed very rapidly. The explosion also launches these atoms far out into space.

In the last of the three ways in which elements are thought to have been formed, high-energy hydrogen or helium atoms participate in fusion reactions with gas and dust in the space between stars.

Because the heavier elements are formed in localized areas of the universe, the distribution of these elements is uneven. The elemental composition of the earth, for example, is very different from that of most of the rest of the universe. Table 18.5 lists the percent abundance of elements in the earth's crust, waters, and atmosphere.

*(continued)*

SPECIAL TOPIC 18.2 (CONTINUED)

*(continued from previous page)*
Note that 11 elements make up over 99% of the planet's mass. Some elements that play major roles in our culture and technology—such as copper, tin, zinc, and gold—are actually very rare.

**Table 18.5**
Percent Abundance of the Elements in the Earth's Crust, Waters, and Atmosphere

| Element | Percent abundance | Element | Percent abundance |
|---|---|---|---|
| oxygen | 49.20 | chlorine | 0.19 |
| silicon | 25.70 | phosphorus | 0.11 |
| aluminum | 7.50 | manganese | 0.09 |
| iron | 4.71 | carbon | 0.08 |
| calcium | 3.39 | sulfur | 0.06 |
| sodium | 2.63 | barium | 0.04 |
| potassium | 2.40 | nitrogen | 0.03 |
| magnesium | 1.93 | fluorine | 0.03 |
| hydrogen | 0.87 | all others | 0.49 |
| titanium | 0.58 | | |

## Chapter Glossary

**Nuclear chemistry** The study of the properties and behavior of atomic nuclei.

**Nucleons** The particles that reside in the nucleus of atoms (protons and neutrons).

**Nucleon number** The sum of the numbers of protons and neutrons (nucleons) in the nucleus of an atom. It is also called the mass number.

**Nuclide** A particular type of nucleus that is characterized by a specific atomic number (Z) and nucleon number (A).

**Electrostatic force (or electromagnetic force)** The force between electrically charged particles.

**Strong force** The force that draws nucleons (protons and neutrons) together.

**Band of stability** On a graph of the numbers of neutrons versus protons in the nuclei of atoms, the portion that represents stable nuclides.

**Radioactive nuclide** An unstable nuclide whose numbers of protons and neutrons place it outside the band of stability.

**Radioactive decay** One of several processes that transform a radioactive nuclide into a more stable product or products.

**Alpha particle** The emission from radioactive nuclides that is composed of two protons and two neutrons in the form of a helium nucleus.

**Alpha emission** The release of an alpha particle by atoms that have too many protons to be stable.

**Beta emission ($\beta^-$)** The conversion of a neutron into a proton, which stays in the nucleus, and an electron (called a beta particle in this context), which is ejected from the atom.

**Beta particle** A high-velocity electron released from radioactive nuclides that have too many neutrons.

**Positron emission ($\beta^+$)** In radioactive nuclides that have too few neutrons, the conversion of a proton into a neutron, which stays in the nucleus, and a positron, which is ejected from the nucleus.

**Positron** A high-velocity antielectron released from radioactive nuclides that have too few neutrons.

**Electron capture** In radioactive nuclides that have too few neutrons, the combination of an electron with a proton to form a neutron, which stays in the nucleus.

**Gamma ray** A stream of high-energy photons.

**Nuclear reaction** A process that results in a change in an atomic nucleus (as opposed to a chemical reaction, which involves the loss, gain, or sharing of electrons).

**Nuclear equation** The shorthand notation that describes nuclear reactions. It shows changes in the participating nuclides' atomic numbers (the number of protons) and mass numbers (the sum of the numbers of protons and neutrons)

**Half-life** The time it takes for one-half of a sample to disappear by radioactive decay.

**Nuclear decay series** A series of radioactive decays that lead from a large unstable nuclide, such as uranium-238, to a stable nuclide, such as lead-206.

**Ionizing radiation** Alpha particles, beta particles, and gamma photons, which are all able to strip electrons from atoms as they move through matter, leaving ions in their wake.

**Free radicals** Particles with unpaired electrons.

**Radiocarbon (or carbon-14) dating** The process of determining the age of an artifact that contains material from formerly living plants or animals by comparing the ratio of carbon-14 to carbon-12 in the object to the ratio thought to have characterized living organisms when the object was originally produced.

**Radioactive tracer** A radioactive nuclide that is incorporated into substances that can then be tracked through detection of the nuclide's emissions.

**Binding energy** The amount of energy released when a nucleus is formed.

**Electron volt** An energy unit equivalent to $1.6 \times 10^{-19}$ joule. It is often used to describe the energy associated with nuclear changes.

**Fusion** Nuclear reaction that yields energy by combining smaller atoms to make larger, more stable ones.

**Fission** Nuclear reaction that yields energy by splitting larger atoms to form smaller, more stable atoms.

**Chain reaction** A process in which one of the products of a reaction initiates another identical reaction.

**Moderator** A substance in a nuclear reactor that slows neutrons as they pass through it.

**Control rods** Rods containing substances such as cadmium or boron (which are efficient neutron absorbers), used to regulate the rate of nuclear fission in a power plant and to stop the fission process if necessary.

You can test yourself on the glossary terms at the following Web site:
**www.chemplace.com/college/**

## Chapter Objectives

**The goal of this chapter is to teach you to do the following.**

1. Define all of the terms in the Chapter 18 Glossary.

### Section 18.1 The Nucleus and Radioactivity

2. Given a symbol for a nuclide, identify its atomic number and mass number (nucleon number).
3. Given a symbol for a nuclide, identify the numbers of protons and neutrons that its nucleus contains, or, given the numbers of protons and neutrons that a nuclide's nucleus contains, write its symbol.
4. Given one of the following ways to describe nuclides, write the other two:

   (element name) − (mass number),

   $^{\text{mass number}}_{\text{atomic number}}$(element symbol),

   $^{\text{mass number}}$(element symbol)

   For example, one of the nuclides of uranium can be described as uranium-238, $^{238}_{92}U$, or $^{238}U$.
5. Describe the two opposing forces between particles in the nucleus, and, with reference to these forces, explain why the optimum ratio of neutrons to protons (for the stability of the nuclide) increases with increasing atomic number.
6. Write descriptions of alpha emission, beta emission, positron emission, and electron capture.
7. Write the symbols used for alpha particles, beta particles, positrons, and gamma photons.
8. Given the symbols for three nuclides of the same element—one that is stable and non-radioactive, one that is radioactive and has a lower mass number than the stable nuclide, and one that is radioactive and has a higher mass number than the stable nuclide—predict which of the radioactive nuclides would be more likely to undergo beta emission and which would be more likely to undergo positron emission (or electron capture).
9. Explain why gamma rays often accompany alpha emission, beta emission, positron emission, and electron capture.

10. Describe the differences between nuclear reactions and chemical reactions.
11. Describe the difference between nuclear equations and chemical equations.
12. Given a description of a radioactive nuclide and whether it undergoes alpha emission, beta emission, positron emission, or electron capture, write a nuclear equation for the reaction.
13. Given an incomplete nuclear equation, write the symbol for the missing component.
14. Given the half-life for a radioactive nuclide, predict how long before a sample decreases to $\frac{1}{2}$, to $\frac{1}{4}$, to $\frac{1}{8}$, to $\frac{1}{16}$, and to $\frac{1}{32}$ of its original amount.
15. Indicate what fraction of the initial amount of a radioactive nuclide will remain after 1, 2, 3, 4, and 5 half-lives.
16. Explain why short-lived radioactive nuclides are found in nature.
17. Describe the source of the radium-226 and radon-222 found in nature, and describe the problems that these nuclides cause.
18. Explain why radon-222 is considered an indirect rather than a direct cause of cancer.
19. Explain why alpha particles, beta particles, and gamma photons are all considered ionizing radiation.
20. Describe how alpha particles, beta particles, and gamma photons interact with water to form ions and free radicals, and explain why these products can damage the body.
21. Describe the types of tissues that are most sensitive to damage from radioactive emission, and explain why radiation treatments do more damage to cancer cells than to regular cells and why children are more affected by radiation than adults are.
22. Explain how strontium-90 atoms can cause bone cancer or leukemia, how cesium-137 atoms can cause tissue damage, and how iodine-131 can damage thyroid glands.
23. Describe the relative penetrating ability of alpha particles, beta particles, and gamma photons, and use this description to explain why gamma photons emitted outside the body can damage internal organs but alpha and beta emitters must be ingested to do damage.
24. Explain why alpha particles from a source inside the body do more damage to tissues than the same number of gamma photons.

### Section 18.2 Uses of Radioactive Substances

25. Describe how cobalt-60 is used to treat cancer.
26. Explain how MRI produces images of the soft tissues of the body.
27. Explain how PET can show dynamic processes in the body, such as brain activity and blood flow.
28. Explain how fluorine-18 is used to study bones and how carbon-11 is used to study brain activity.
29. Describe how carbon-14 (radiocarbon) dating of artifacts is done.
30. Explain how smoke detectors that contain americium-241 work.
31. Describe how iridium-192 can be used to find leaks in welded pipe joints.
32. Describe the pros and cons of food irradiation.
33. Explain how scientists use radioactive tracers.

### Section 18.3 Nuclear Energy

34. Explain how the binding energy of a nucleus reflects its stability.
35. Explain how the binding energy per nucleon for a nuclide can be used to compare its stability to that of other nuclides.
36. Describe the general trend in the variation in binding energy per nucleon for the natural nuclides, and use it to explain how energy is released in both nuclear fusion and nuclear fission.
37. Explain why ${}^{4}_{2}He$, ${}^{12}_{6}C$, ${}^{16}_{8}O$, and ${}^{20}_{10}Ne$ are especially stable.
38. Explain why ${}^{40}_{20}Ca$, and ${}^{208}_{82}Ne$ are especially stable.
39. Describe the fission reaction of uranium-235, and explain how it can lead to a chain reaction.
40. Describe how heat is generated in a nuclear power plant.
41. Explain why uranium must be enriched in uranium-235 before it can be used as fuel in a typical nuclear reactor.
42. Describe the role of the moderator in a nuclear reactor.
43. Describe the problems associated with the production of plutonium-239 in nuclear reactors.
44. Describe the role of the control rods in a nuclear reactor.
45. Describe how energy is generated in the sun.

## Review Questions

1. Describe the nuclear model of the atom, including the general location of the protons, neutrons, and electrons, the relative size of the nucleus compared to the size of the atom, and the modern description of the electron.
2. With reference to both their particle nature and their wave nature, describe the similarities and differences between gamma radiation and radio waves. Which has higher energy?

*Complete the following statements by writing in each blank the word or phrase that best completes the thought.*

3. Atoms that have the same number of protons but different numbers of neutrons are called ________________. They have the same atomic number but different mass numbers.
4. The ________________ for an atom is equal to the number of protons in the atom's nucleus. It establishes the element's identity.
5. The ________________ for an atom is equal to the sum of the numbers of protons and neutrons in the atom's nucleus.
6. ________________ is the capacity to do work.
7. ________________ is the capacity to do work that results from the motion of an object.
8. The ________________ states that energy can be neither created nor destroyed, but it can be transferred from one system to another and changed from one form to another.
9. ________________ is a retrievable, stored form of energy that an object possesses by virtue of its position or state.
10. The more stable a system is, the ________________ (higher or lower) its potential energy.

11. When a system shifts from a less stable to a more stable state, energy is ________________ (absorbed or released).
12. The ________________ of an atom is the condition in which its electrons are in the orbitals that give it the lowest possible potential energy.
13. The ________________ of an atom is the condition in which one or more of its electrons are in orbitals that do not represent the lowest possible potential energy.

## Key Ideas

**Complete the following statements by writing one of these words or phrases in each blank.**

| | |
|---|---|
| 35 | neutrons |
| 83 protons | nitrogen |
| 100 | nucleus |
| 200 | one-half |
| 10,000 | one lower |
| A | photons |
| alpha | protons |
| attractions | pull |
| binding energy per nucleon | pushed |
| $^{14}C/^{12}C$ ratio | rapidly reproducing |
| charge | ratio |
| different | released |
| double magic numbers | releases |
| electrostatic | repulsion |
| energy | skin |
| excite | smaller than |
| furnace | strong |
| glue | too high |
| identical | too low |
| internal organs | unaffected |
| ionizing | Z |
| larger than | |

14. Because ________________ and ________________ reside in the nucleus of atoms, they are called nucleons.
15. The symbols used for nucleons have the atomic number (________________) as a subscript in front of the element symbol and the nucleon number (________________) as a superscript above the atomic number.
16. There are two forces among the particles within the nucleus. The first, called the ________________ force, is the force between electrically charged particles. The second force, called the ________________ force, holds nucleons (protons and neutrons) together.

17. You can think of neutrons as the nuclear ________________ that allows protons to stay together in the nucleus. Adding neutrons to a nucleus leads to more ________________ holding the particles of the nucleus together without causing increased ________________ between those particles.

18. Larger atoms with more protons in their nuclei require a greater ________________ of neutrons to protons to balance the increased repulsion between protons.

19. If a nucleus contains more than ________________, the nucleus cannot be made completely stable no matter how many neutrons are added.

20. One of the ways in which heavy nuclides change to move back into the band of stability is to release two protons and two neutrons in the form of a helium nucleus, called a(n) ________________ particle.

21. When a radioactive nuclide has a neutron-to-proton ratio that is ________________, it undergoes beta emission ($\beta^-$). In this process, a neutron becomes a proton and an electron. The proton stays in the nucleus, and the electron, which is called a beta particle in this context, is ejected from the atom.

22. When a radioactive nuclide has a neutron-to-proton ratio that is ________________, it can move toward stability in one of two ways: positron emission or electron capture. In positron emission ($\beta^+$), a proton becomes a neutron and a positron. The neutron stays in the nucleus, and the positron speeds out of the nucleus at high velocity.

23. In electron capture, an electron combines with the proton to form a neutron. Like positron emission, electron capture causes the radioactive nuclide to change to a new element with an atomic number that is ________________ but with the same mass number.

24. Because radioactive decay leads to more stable products, it always ________________ energy, some in the form of kinetic energy of the moving product particles, some in the form of gamma rays. Gamma rays can be viewed as a stream of high-energy ________________.

25. Nuclear reactions involve changes in the ________________, whereas chemical reactions involve the loss, gain, and sharing of electrons.

26. Different isotopes of the same element, which share the same chemical characteristics, often undergo very ________________ nuclear reactions.

27. Unlike chemical reactions, the rates of nuclear reactions are ________________ by temperature, pressure, and the other atoms to which the radioactive atom is bonded.

28. Nuclear reactions, in general, give off a lot more ________________ than chemical reactions.

29. The equations that describe nuclear reactions are different than those that describe chemical reactions because in nuclear equations, ________________ is disregarded.

30. Rates of radioactive decay are described in terms of half-life, the time it takes for ________________ of a sample to disappear.

31. Alpha particles, beta particles, and gamma photons are often called ________________ radiation, because they are all able to strip electrons from atoms as they move through matter, leaving ions in their wake.

32. As alpha particles, which move at up to 10% of the speed of light, move through the tissues of our bodies, they ____________ electrons away from the tissue's atoms.
33. The repulsion between negatively charged beta particles and the electrons on atoms and molecules of our tissues leads to electrons being ____________ off the uncharged particles.
34. Gamma photons are ionizing radiation, because they can ____________ electrons enough to actually remove them from atoms.
35. Alpha particles that strike the outside of the body are stopped by the top layer of ____________.
36. Because beta particles are smaller than alpha particles, and because they can move at up to 90% of the speed of light, they are about ____________ times as penetrating as alpha particles.
37. Gamma photons are much more penetrating than alpha and beta particles, so gamma radiation from outside the body can damage ____________.
38. Gamma photons that penetrate the body do more damage to ____________ cells than to other cells.
39. Carbon-14 atoms are constantly being produced in our upper atmosphere through neutron bombardment of ____________ atoms.
40. To date an artifact, a portion of it is analyzed to determine the ____________, which can be used to determine its age.
41. Because the amount of energy ____________ when a nucleus is formed is a reflection of the strength with which nucleons are bound, it is called the atom's binding energy.
42. It takes about ____________ times as much energy to remove a proton or a neutron from the nucleus of a hydrogen-2 atom as to remove that atom's single electron.
43. A higher ____________ reflects more stable and more tightly bound nucleons in the nucleus.
44. There appears to be something stable about having 2, 8, 20, 28, 50, 82, or 126 protons or neutrons. The nuclides with ____________ have very high stability.
45. For atoms ____________ iron-56, energy is released when smaller atoms combine to form larger ones.
46. For atoms ____________ iron-56, splitting larger atoms to form smaller, more stable atoms releases energy.
47. The fission reactions of uranium-235 yield more than ____________ different nuclides of ____________ different elements.
48. The nuclear reactor in a nuclear power plant is really just a big ____________ that generates heat to convert liquid water to steam that turns a steam turbine generator to produce electricity.
49. A chain reaction is a process in which one of the products of the reaction initiates another ____________ reaction.

## Chapter Problems

### Section 18.1 The Nucleus and Radioactivity

OBJECTIVE 2 OBJECTIVE 4

50. A radioactive nuclide that has an atomic number of 88 and a mass (nucleon) number of 226 is used in radiation therapy. Write its nuclide symbol in the form of ${}^{A}_{Z}X$. Write two other ways to represent this nuclide.

OBJECTIVE 2 OBJECTIVE 4

51. A radioactive nuclide that has an atomic number of 54 and a mass (nucleon) number of 133 is used to determine lung capacity. Write its nuclide symbol in the form of ${}^{A}_{Z}X$. Write two other ways to represent this nuclide.

OBJECTIVE 3 OBJECTIVE 4

52. A radioactive nuclide that has 6 protons and 5 neutrons is used to generate positron emission tomography (PET) brain scans. Write its nuclide symbol in the form of ${}^{A}_{Z}X$. Write two other ways to represent this nuclide.

OBJECTIVE 3 OBJECTIVE 4

53. A radioactive nuclide that has 29 protons and 35 neutrons is used to diagnose liver disease. Write its nuclide symbol in the form of ${}^{A}_{Z}X$. Write two other ways to represent this nuclide.

OBJECTIVE 2 OBJECTIVE 4

54. A radioactive nuclide with the symbol ${}^{40}_{19}K$ is used for geologic dating. What are its atomic number and mass (nucleon) number? Write two other ways to represent this nuclide.

OBJECTIVE 2 OBJECTIVE 4

55. A radioactive nuclide with the symbol ${}^{198}_{79}Au$ is used in the measurement of kidney activity. What are its atomic number and mass (nucleon) number? Write two other ways to represent this nuclide.

OBJECTIVE 3 OBJECTIVE 4

56. A radioactive nuclide with the symbol ${}^{111}_{49}In$ is used to label blood platelets. How many protons and how many neutrons does each atom have? Write two other ways to represent this nuclide.

OBJECTIVE 3 OBJECTIVE 4

57. A radioactive nuclide with the symbol ${}^{18}_{9}F$ is used in bone scans. How many protons and how many neutrons does each atom have? Write two other ways to represent this nuclide.

OBJECTIVE 2 OBJECTIVE 3 OBJECTIVE 4

58. Barium-131 is used to detect bone tumors. What are its atomic number and mass number? How many protons and how many neutrons are in the nucleus of each atom? Write two other ways to represent this nuclide.

OBJECTIVE 2 OBJECTIVE 3 OBJECTIVE 4

59. Polonium-210 is used in radiation therapy. What are its atomic number and mass number? How many protons and how many neutrons are in the nucleus of each atom? Write two other ways to represent this nuclide.

OBJECTIVE 2 OBJECTIVE 3 OBJECTIVE 4

60. The radioactive nuclide with the symbol ${}^{75}Se$ is used to measure the shape of the pancreas. What are its atomic number and mass number? How many protons and how many neutrons are in the nucleus of each atom? Write two other ways to represent this nuclide.

OBJECTIVE 2 OBJECTIVE 3 OBJECTIVE 4

61. The radioactive nuclide with the symbol ${}^{125}I$ is used to measure blood hormone levels. What are its atomic number and mass number? How many protons and how many neutrons are in the nucleus of each atom? Write two other ways to represent this nuclide.

OBJECTIVE 5

62. Describe the two opposing forces between particles in the nucleus, and, with reference to these forces, explain why the ratio of neutrons to protons required for a stable nuclide increases as the number of protons in a nucleus increases.

63. Explain why the ${}^{2}_{2}He$ nuclide does not exist.

64. Write a general description of the changes that take place in alpha emission. Write the two symbols used for an alpha particle. Write the general equation for alpha emission, using X for the reactant element symbol, Y for the product element symbol, Z for atomic number, and A for mass number. OBJECTIVE 6 OBJECTIVE 7

65. Write a general description of the changes that take place in beta emission. Write the three symbols used for a beta particle. Write the general equation for beta emission, using X for the reactant element symbol, Y for the product element symbol, Z for atomic number, and A for mass number. OBJECTIVE 6 OBJECTIVE 7

66. Write a general description of the changes that take place in positron emission. Write the three symbols used for a positron. Write the general equation for positron emission, using X for the reactant element symbol, Y for the product element symbol, Z for atomic number, and A for mass number. OBJECTIVE 6 OBJECTIVE 7

67. Write a general description of the changes that take place in electron capture. Write the general equation for electron capture, using X for the reactant element symbol, Y for the product element symbol, Z for atomic number, and A for mass number. OBJECTIVE 6 OBJECTIVE 7

68. Consider three isotopes of bismuth: ${}^{202}_{83}\text{Bi}$, ${}^{209}_{83}\text{Bi}$, and ${}^{215}_{83}\text{Bi}$. Bismuth-209 is stable. One of the other nuclides undergoes beta emission, and the remaining nuclide undergoes electron capture. Identify the isotope that makes each of these changes, and explain your choices. OBJECTIVE 8

69. Consider three isotopes of nitrogen: ${}^{13}_{7}\text{N}$, ${}^{14}_{7}\text{N}$, and ${}^{16}_{7}\text{N}$. Nitrogen-14 is stable. One of the other nuclides undergoes beta emission, and the remaining nuclide undergoes positron emission. Identify the isotope that makes each of these changes, and explain your choices. OBJECTIVE 8

70. Explain why gamma rays often accompany alpha emission, beta emission, positron emission, and electron capture. OBJECTIVE 9

71. What nuclear process or processes lead to each of the results listed below? The possibilities are alpha emission, beta emission, positron emission, electron capture, and gamma emission. OBJECTIVE 6
    a. Atomic number increases by 1.
    b. Mass number decreases by 4
    c. No change occurs in atomic number or mass number.
    d. The number of protons decreases by 1.
    e. The number of neutrons decreases by 1.
    f. The number of protons decreases by 2.

72. What nuclear process or processes lead to each of the results listed below? The possibilities are alpha emission, beta emission, positron emission, electron capture, and gamma emission. OBJECTIVE 6
    a. The number of neutrons increases by 1.
    b. Atomic number decreases by 2.
    c. The number of neutrons decreases by 2.
    d. Atomic number decreases by 1.
    e. The number of protons increases by 1.
    f. No change occurs in the number of protons and neutrons.

OBJECTIVE 10

**73.** Describe the differences between nuclear reactions and chemical reactions.

74. Explain why $^{38}_{17}Cl$ and $^{38}_{17}Cl^-$ are very different chemically but undergo identical nuclear reactions.

OBJECTIVE 11

75. Describe the difference between nuclear equations and chemical equations.

OBJECTIVE 12

**76.** Marie Curie won the Nobel Prize in physics in 1903 for her study of radioactive nuclides, including polonium-218 (which was named after her native country, Poland). Polonium-218 undergoes alpha emission. Write the nuclear equation for this change.

OBJECTIVE 12

77. Americium-243 is an alpha emitter used in smoke detectors. Write the nuclear equation for its alpha emission.

OBJECTIVE 12

**78.** Cobalt-60, which is the most common nuclide used in radiation therapy for cancer, undergoes beta emission. Write the nuclear equation for this reaction.

OBJECTIVE 12

79. Radioactive iron-59, which is used to assess blood iron changes, shifts toward stability by emitting beta particles. Write the nuclear equation for this reaction.

OBJECTIVE 12

**80.** Carbon-11 is used in PET brain scans because it emits positrons. Write the nuclear equation for the positron emission of carbon-11.

OBJECTIVE 12

81. Oxygen-13 atoms undergo positron emission, so they can be used to generate PET scans. Write the nuclear equation for this reaction.

OBJECTIVE 12

**82.** Mercury-197 was used in the past for brain scans. Its decay can be detected, because this nuclide undergoes electron capture, which forms an excited atom that then releases a gamma photon that escapes the body and strikes a detector. Write the nuclear equation for the electron capture by mercury-197.

OBJECTIVE 12

83. Your cardiovascular system can be assessed by using krypton-79, which shifts to a more stable nuclide by electron capture. Write an equation that describes this change.

OBJECTIVE 13

**84.** Complete the following nuclear equations.

a. $^{90}_{38}Sr \rightarrow ^{90}_{39}Y + \_\_\_\_$

b. $^{17}_{9}F \rightarrow ^{17}_{8}O + \_\_\_\_$

c. $^{222}_{86}Rn \rightarrow ^{218}_{84}Po + \_\_\_\_$

d. $^{18}_{9}F + \_\_\_\_ \rightarrow ^{18}_{8}O$

e. $^{235}_{92}U \rightarrow \_\_\_\_ + ^{4}_{2}He$

f. $^{7}_{4}Be + ^{0}_{-1}e \rightarrow \_\_\_\_$

g. $^{52}_{26}Fe \rightarrow \_\_\_\_ + ^{0}_{+1}e$

h. $^{3}_{1}H \rightarrow \_\_\_\_ + ^{0}_{-1}e$

i. $\_\_\_\_ \rightarrow ^{14}_{7}N + ^{0}_{-1}e$

j. $\_\_\_\_ \rightarrow ^{118}_{53}I + ^{0}_{+1}e$

k. $\_\_\_\_ + ^{0}_{-1}e \rightarrow ^{204}_{83}Bi$

l. $\_\_\_\_ \rightarrow ^{234}_{90}Th + ^{4}_{2}He$

85. Complete the following nuclear equations. OBJECTIVE 13

a. ______ $\rightarrow$ ${}^{18}_{8}O + {}^{0}_{+1}e$

b. ${}^{26}_{13}Al \rightarrow {}^{26}_{14}Si +$ ______

c. ______ $+ {}^{0}_{-1}e \rightarrow {}^{58}_{26}Fe$

d. ${}^{242}_{96}Cm \rightarrow {}^{238}_{94}Pu +$ ______

e. ______ $\rightarrow {}^{210}_{88}Ra + {}^{4}_{2}He$

f. ${}^{210}_{84}Po \rightarrow$ ______ $+ {}^{4}_{2}He$

g. ${}^{30}_{15}P \rightarrow$ ______ $+ {}^{0}_{+1}e$

h. ${}^{104}_{43}Tc \rightarrow {}^{104}_{44}Ru +$ ______

i. ${}^{44}_{22}Ti + {}^{0}_{-1}e \rightarrow$ ______

j. ${}^{137}_{55}Cs \rightarrow$ ______ $+ {}^{0}_{-1}e$

k. ______ $\rightarrow {}^{32}_{16}S + {}^{0}_{-1}e$

l. ${}^{64}_{29}Cu +$ ______ $\rightarrow {}^{64}_{28}Ni$

86. Silver-117 atoms undergo three beta emissions before they reach a stable nuclide. What is the final product?

87. Molybdenum-105 atoms undergo four beta emissions before they reach a stable nuclide. What is the final product?

88. Tellurium-116 atoms undergo two electron captures before they reach a stable nuclide. What is the final product?

89. Cesium-127 atoms undergo two electron captures before they reach a stable nuclide. What is the final product?

90. Samarium-142 atoms undergo two positron emissions before they reach a stable nuclide. What is the final product?

91. Indium-107 atoms undergo a positron emission and an electron capture before they reach a stable nuclide. What is the final product?

92. Bismuth-211 atoms undergo an alpha emission and beta emission before they reach a stable nuclide. What is the final product?

93. Polonium-214 atoms undergo an alpha emission, two beta emissions, and another alpha emission before they reach a stable nuclide. What is the final product?

94. Complete the following nuclear equations, which describe the changes that led to the formation of previously undiscovered nuclides. OBJECTIVE 13

a. ${}^{246}_{96}Cm + {}^{12}_{6}C \rightarrow$ ______ $+ 6{}^{1}_{0}n$

b. ______ $+ {}^{16}_{8}O \rightarrow {}^{263}_{106}Sg + 2{}^{1}_{0}n$

c. ${}^{240}_{95}Am + {}^{4}_{2}He \rightarrow {}^{243}_{97}Bk +$ ______

d. ${}^{252}_{98}Cf +$ ______ $\rightarrow {}^{257}_{103}Lr + 5{}^{1}_{0}n$

OBJECTIVE 13

95. Complete the following nuclear equations, which describe the changes that led to the formation of previously undiscovered nuclides.

a. $^{238}_{92}U + ^{12}_{6}C \rightarrow$ ________ $+ 6\,^{1}_{0}n$

b. ________ $+ ^{16}_{8}O \rightarrow ^{252}_{102}No + 5\,^{1}_{0}n$

c. $^{253}_{99}Es + ^{4}_{2}He \rightarrow ^{256}_{101}Md +$ ________

d. $^{246}_{96}Cm +$ ________ $\rightarrow ^{254}_{102}No + 5\,^{1}_{0}n$

OBJECTIVE 13

**96.** In February 1981, the first atoms of the element bohrium-262, $^{262}_{107}Bh$, were made from the bombardment of bismuth-209 atoms by chromium-54 atoms. Write a nuclear equation for this reaction. (One or more neutrons may be released in this type of nuclear reaction.)

OBJECTIVE 13

97. In December 1994, the nuclide that has been temporarily named unununium-272, $^{272}_{111}Uuu$, was made from the bombardment of bismuth-209 atoms with nickel-64 atoms. Write a nuclear equation for this reaction. (One or more neutrons may be released in this type of nuclear reaction.)

OBJECTIVE 14

**98.** Cesium-133, which is used in radiation therapy, has a half-life of 30 years. How long will it be before a sample decreases to ¼ of what was originally there?

OBJECTIVE 14

99. Gold-198, which is used to assess kidney function, has a half-life of 2.70 days. How long will it be before a sample decreases to 1/16 of what was originally there?

OBJECTIVE 15

**100.** Phosphorus-32, which is used for leukemia therapy, has a half-life of 14.3 days. What fraction of a sample is left in 42.9 days?

OBJECTIVE 15

101. Cobalt-60, which is used in radiation therapy, has a half-life of 5.26 years. What fraction of a sample is left in 26.3 years?

OBJECTIVE 16

**102.** Explain why short-lived radioactive nuclides are found in nature.

OBJECTIVE 17

103. Describe the source of the radium-226 and radon-222 found in nature. What problems do these nuclides cause?

OBJECTIVE 18

104. Explain why radon-222 is considered an indirect rather than a direct cause of cancer.

**105.** The first six steps of the decay series for uranium-235 consist of the changes alpha emission, beta emission, alpha emission, beta emission, alpha emission, and alpha emission. Write the products formed after each of these six steps.

106. The last five steps of the decay series for uranium-235 consist of alpha emission from $^{219}_{86}Rn$, followed by alpha emission, beta emission, beta emission, and alpha emission. Write the products formed after each of these steps.

**107.** In the first five steps of the decay series for thorium-232, the products are $^{228}_{88}Ra$, $^{228}_{89}Ac$, $^{228}_{90}Th$, $^{224}_{88}Ra$, and $^{220}_{86}Rn$. Identify each of these steps as alpha emissions or beta emissions.

108. The last five steps of the decay series for thorium-232 start with $^{220}_{86}Rn$, and the products are $^{216}_{84}Po$, $^{212}_{82}Pb$, $^{212}_{83}Bi$, $^{212}_{84}Po$, and $^{208}_{82}Pb$. Identify each of these steps as alpha emissions or beta emissions.

OBJECTIVE 19

**109.** Explain why alpha particles are considered ionizing radiation.

OBJECTIVE 19

110. Explain why beta particles are considered ionizing radiation.

OBJECTIVE 19

**111.** Explain why gamma photons are considered ionizing radiation.

OBJECTIVE 20

112. Describe how alpha particles, beta particles, and gamma photons interact with water to form ions and free radicals, and explain why these products can damage the body.

113. What types of tissues are most sensitive to emission from radioactive nuclides? Why do radiation treatments do more damage to cancer cells than to regular cells? Why are children more affected by radiation than adults are? OBJECTIVE 21

114. Explain how strontium-90 atoms can cause bone cancer or leukemia, how cesium-137 atoms can cause tissue damage, and how iodine-131 can damage thyroid glands. OBJECTIVE 22

115. Why do you think radium-226 concentrates in our bones?

116. Describe the relative penetrating ability of alpha particles, beta particles, and gamma photons, and use this description to explain why gamma photons emitted outside the body can damage internal organs but alpha and beta emitters must be inside the body to do damage. OBJECTIVE 23

117. Why are alpha particles more damaging to tissues when the source is ingested than the same number of gamma photons would be? OBJECTIVE 24

### Section 18.2 Uses of Radioactive Substances

118. Describe how cobalt-60 is used to treat cancer. OBJECTIVE 25

119. Explain how MRI produces images of the soft tissues of the body. OBJECTIVE 26

120. Explain how PET can show dynamic processes in the body, such as brain activity and blood flow. OBJECTIVE 27

121. Explain how fluorine-18 is used to study bones and how carbon-11 is used to study brain activity. OBJECTIVE 28

122. Describe how carbon-14 (radiocarbon) dating of artifacts is done. OBJECTIVE 29

123. Explain how smoke detectors containing americium-241 work. OBJECTIVE 30

124. Describe how iridium-192 can be used to find leaks in welded pipe joints. OBJECTIVE 31

125. Outline the pros and cons of food irradiation. OBJECTIVE 32

126. Explain how scientists use radioactive tracers. OBJECTIVE 33

### Section 18.3 Nuclear Energy

127. Explain how the binding energy of a nucleus reflects its stability. OBJECTIVE 34

128. Explain how the binding energy per nucleon can be used to compare the stability of nuclides. OBJECTIVE 35

129. Describe the general trend in binding energy per nucleon for the natural nuclides, and use it to explain how energy is released in both nuclear fusion and nuclear fission. OBJECTIVE 36

130. Explain why ${}^{4}_{2}He$, ${}^{12}_{6}C$, ${}^{16}_{8}O$, and ${}^{20}_{10}Ne$ are especially stable. OBJECTIVE 37

131. Explain why ${}^{40}_{20}Ca$, and ${}^{208}_{82}Pb$ are especially stable. OBJECTIVE 38

132. Give two reasons why ${}^{16}_{8}O$ is more stable than ${}^{15}_{8}O$.

133. Describe the fission reaction of uranium-235, and explain how it can lead to a chain reaction. OBJECTIVE 39

134. Describe how heat is generated in a nuclear power plant. OBJECTIVE 40

135. Explain why uranium must be enriched in uranium-235 before it can be used as fuel in a typical nuclear reactor. OBJECTIVE 41

136. Describe the role of the moderator in a nuclear reactor. OBJECTIVE 42

OBJECTIVE 43

137. Describe the problems associated with the production of plutonium-239 in nuclear reactors.

138. Nuclear wastes must be isolated from the environment for a very long time, because they contain relatively long-lived radioactive nuclides, such as technetium-99 with a half-life of over $2.1 \times 10^5$ years. One proposed solution is to bombard the waste with neutrons to convert the long-lived nuclides into nuclides that decay more quickly. When technetium-99 absorbs a neutron, it forms technetium-100, which has a half-life of 16 seconds and forms stable ruthenium-100 by emitting a beta particle. Write the nuclear equations for these two changes.

OBJECTIVE 44

139. Describe the role of the control rods in a nuclear reactor.

OBJECTIVE 45

140. Describe how energy is generated in the sun.

### Additional Problems

141. A radioactive nuclide that has an atomic number of 53 and a mass (nucleon) number of 131 is used to measure thyroid function. Write its nuclide symbol in the form of ${}^{A}_{Z}X$. Write two other ways to represent this nuclide.

142. A radioactive nuclide that has an atomic number of 6 and a mass (nucleon) number of 14 is used to determine the age of artifacts. Write its nuclide symbol in the form of ${}^{A}_{Z}X$. Write two other ways to represent this nuclide.

143. A radioactive nuclide that has 11 protons and 13 neutrons is used to detect blood clots. Write its nuclide symbol in the form of ${}^{A}_{Z}X$. Write two other ways to represent this nuclide.

144. A radioactive nuclide that has 43 protons and 56 neutrons is used in bone scans. Write its nuclide symbol in the form of ${}^{A}_{Z}X$. Write two other ways to represent this nuclide.

145. A radioactive nuclide with the symbol ${}^{133}_{55}Cs$ is used in radiation therapy. What are its atomic number and mass (nucleon) number? Write two other ways to represent this nuclide.

146. A radioactive nuclide with the symbol ${}^{79}_{36}Kr$ is used to assess cardiovascular function. What are its atomic number and mass (nucleon) number? Write two other ways to represent this nuclide.

147. A radioactive nuclide with the symbol ${}^{51}_{24}Cr$ is used to determine blood volume. How many protons and neutrons does each atom have? Write two other ways to represent this nuclide.

148. A radioactive nuclide with the symbol ${}^{15}_{8}O$ is used to test lung function. How many protons and neutrons does each atom have? Write two other ways to represent this nuclide.

149. Gallium-67 is used to diagnose lymphoma. What are its atomic number and mass number? How many protons and how many neutrons are in the nucleus of each atom? Write two other ways to represent this nuclide.

150. Nitrogen-13 is used in heart imaging. What are its atomic number and mass number? How many protons and how many neutrons are in the nucleus of each atom? Write two other ways to represent this nuclide.

151. The radioactive nuclide with the symbol $^{32}P$ is used to detect eye tumors. What are its atomic number and mass number? How many protons and how many neutrons are in the nucleus of each atom? Write two other ways to represent this nuclide.

152. The radioactive nuclide with the symbol $^{41}Ar$ is used to measure the flow of gases from smokestacks. What are its atomic number and mass number? How many protons and how many neutrons are in the nucleus of each atom? Write two other ways to represent this nuclide.

153. Consider three isotopes of neon: $^{18}_{10}Ne$, $^{20}_{10}Ne$, and $^{24}_{10}Ne$. Neon-20, which is the most abundant isotope of neon, is stable. One of the other nuclides undergoes beta emission, and the remaining nuclide undergoes positron emission. Identify the isotope that makes each of these changes, and explain your choices.

154. Consider three isotopes of chromium: $^{48}_{24}Cr$, $^{52}_{24}Cr$, and $^{56}_{24}Cr$. Chromium-52, the most abundant, is stable. One of the other nuclides undergoes beta emission, and the remaining nuclide undergoes electron capture. Identify the isotope that makes each of these changes, and explain your choices.

155. Write the nuclear equation for the alpha emission of bismuth-189.

156. Write the nuclear equation for the alpha emission of radium-226, which is used in radiation therapy.

157. Phosphorus-32, which is used to detect breast cancer, undergoes beta emission. Write the nuclear equation for this reaction.

158. Radioactive xenon-133 is used to measure lung capacity. It shifts toward stability by emitting beta particles. Write the nuclear equation for this reaction.

159. Write the nuclear equation for the positron emission of potassium-40.

160. Liver disease can be diagnosed with the help of radioactive copper-64, which is a positron emitter. Write the nuclear equation for this reaction.

161. Radioactive selenium-75, which is used to determine the shape of the pancreas, shifts to a more stable nuclide via electron capture. Write the nuclear equation for this change.

162. Intestinal fat absorption can be measured using iodine-125, which undergoes electron capture. Write an equation that describes this change.

163. Germanium-78 atoms undergo two beta emissions before they reach a stable nuclide. What is the final product?

164. Iron-61 atoms undergo two beta emissions before they reach a stable nuclide. What is the final product?

165. Iron-52 atoms undergo one positron emission and one electron capture before they reach a stable nuclide. What is the final product?

166. Titanium-43 atoms undergo two positron emissions before they reach a stable nuclide. What is the final product?

167. Arsenic-69 atoms undergo one positron emission and one electron capture before they reach a stable nuclide. What is the final product?

168. Radon-217 atoms undergo two alpha emissions and a beta emission before they reach a stable nuclide. What is the final product?

169. Astatine-216 atoms undergo an alpha emission, a beta emission, and another alpha emission before they reach a stable nuclide. What is the final product?

170. Complete the following nuclear equations.

    a. ${}^{249}_{98}\text{Cf} + {}^{15}_{7}\text{N} \rightarrow$ ______ $+ 5\,{}^{1}_{0}\text{n}$

    b. ______ $+ {}^{10}_{5}\text{B} \rightarrow {}^{257}_{103}\text{Lr} + 2\,{}^{1}_{0}\text{n}$

    c. ${}^{121}_{51}\text{Sb} + {}^{1}_{1}\text{H} \rightarrow {}^{121}_{52}\text{Te} +$ ______

171. Complete the following nuclear equations.

    a. ${}^{10}_{5}\text{B} + {}^{1}_{0}\text{n} \rightarrow$ ______ $+ {}^{1}_{1}\text{H}$

    b. ______ $+ {}^{4}_{2}\text{He} \rightarrow {}^{124}_{53}\text{I} + {}^{1}_{0}\text{n}$

    c. ${}^{239}_{94}\text{Pu} + {}^{4}_{2}\text{He} \rightarrow$ ______ $+ {}^{1}_{1}\text{H} + 2\,{}^{1}_{0}\text{n}$

172. Nitrogen-containing explosives carried by potential terrorists can be detected at airports by bombarding suspicious luggage with low-energy neutrons. The nitrogen-14 atoms absorb the neutrons, forming nitrogen-15 atoms. The nitrogen-15 atoms emit gamma photons of a characteristic wavelength that can be detected outside the luggage. Write a nuclear equation for the reaction that forms nitrogen-15 from nitrogen-14.

173. In September 1982, the element meitnerium-266, ${}^{266}_{109}\text{Mt}$, was made from the bombardment of bismuth-209 atoms with iron-58 atoms. Write a nuclear equation for this reaction. (One or more neutrons may be released in this type of nuclear reaction.)

174. In March 1984, the nuclide hassium-265, ${}^{265}_{108}\text{Hs}$, was made from the bombardment of lead-208 atoms with iron-58 atoms. Write a nuclear equation for this reaction. (One or more neutrons may be released in this type of nuclear reaction.)

175. In November 1994, the nuclide that has been temporarily named ununnilium-269, ${}^{269}_{110}\text{Uun}$, was made from the bombardment of lead-208 atoms with nickel-62 atoms. Write a nuclear equation for this reaction. (One or more neutrons may be released in this type of nuclear reaction.)

176. Krypton-79, which is used to assess cardiovascular function, has a half-life of 34.5 hours. How long before a sample decreases to $1/8$ of what was originally there?

177. Strontium-90 has a half-life of 29 years. How long before a sample decreases to $1/32$ of what was originally there?

178. Iron-59, which is used to diagnose anemia, has a half-life of 45 days. What fraction of it is left in 90 days?

179. Fluorine-17 has a half-life of 66 seconds. What fraction of it is left in 264 seconds?

### Discussion Questions

180. Suggest a way in which radioactive sulfur-35 could be used to show that the following reversible change takes place in a saturated solution of silver sulfide with excess solid on the bottom.

$$Ag_2S(s) \rightarrow 2Ag^+(aq) + S^{2-}(aq)$$

181. Vitamin $B_{12}$ is water-soluble vitamin that can be derived from oysters, salmon, liver, and kidney. At the core of the complex structure of vitamin $B_{12}$ is a cobalt ion. Suggest a way in which radioactive cobalt-57 could be used to determine which tissues of the body adsorb the most vitamin $B_{12}$.

# Appendix A

## Measurement and Units

Table A.1
Common Units and Their Abbreviations
*[Note that the abbreviation for inch (in.) is the only abbreviation that ends in a period.]*

| Type of measurement | Unit | Abbreviation |
|---|---|---|
| English mass | ton | ton |
| | pound | lb |
| | ounce | oz |
| English length | mile | mi or mile |
| | yard | yd |
| | foot | ft |
| | inch | in. |
| English volume | gallon | gal |
| | quart | qt |
| | pint | pt |
| | fluid ounce | fl oz |
| | cubic foot | cu ft |
| | cubic inch | cu in. |
| Time | year | yr or year |
| | day | d or day |
| | hour | h or hr |
| | minute | min |
| | second | s or sec |
| Temperature | Degree Celsius | °C |
| | Degree Fahrenheit | °F |
| | kelvin | K |
| Energy | joule | J |
| | calorie | cal |
| | dietary calorie | Cal |

**Table A.2**
The Relationship Between English and Metric Length Units
*Both four and eight digits are given for each relationship. Four digits will be used for nearly all of the English–metric conversions done in this text.*

| Metric length | English length |
|---|---|
| kilometer (km) | 0.6214 mi<br>0.62137119 mi |
| | 3281 ft<br>3280.8399 ft |
| meter (m) | 3.281 ft<br>3.2808399 ft |
| | 39.37 in.<br>39.370079 in. |
| | 1.094 yd<br>1.0936133 yd |
| centimeter (cm) | 0.3937 in.<br>0.39370079 in. |
| | 0.03281 ft<br>0.032808399 ft |
| millimeter (mm) | $3.937 \times 10^{-2}$ in.<br>$3.93780079 \times 10^{-2}$ in. |
| micrometer (μm) | $3.937 \times 10^{-5}$ in.<br>$3.9370079 \times 10^{-5}$ in. |
| nanometer (nm) | $3.937 \times 10^{-8}$ in.<br>$3.9370079 \times 10^{-8}$ in. |

| English length | Metric length |
|---|---|
| mile (mi) | 1.609 km<br>1.609344 km (exact) |
| | 1609 m<br>1609.344 m (exact) |
| yard (yd) | 0.9144 m (exact)<br>91.44 cm (exact) |
| foot (ft) | 30.48 cm (exact)<br>0.3048 m (exact) |
| inch (in.) | 2.54 cm (exact)<br>0.0254 m (exact) |

**Table A.3**
The Relationship Between English Mass Units and Metric Mass Units
*Both four and eight digits are given for each relationship. Four digits will be used for nearly all of the English–metric conversions done in this text.*

| Metric mass | English mass |
| --- | --- |
| metric ton (t) | 1.102 ton (short)<br>1.1023113 ton (short) |
| | 2205 lb<br>2204.6226 lb |
| kilogram (kg) | 2.205 lb<br>2.2046226 lb |
| | 0.001102 ton<br>0.0011023113 ton |
| gram (g) | 0.002205 lb<br>0.0022046226 lb |
| | 0.03527 oz<br>0.035273962 oz |
| milligram (mg) | $3.527 \times 10^{-5}$ oz<br>$3.5273962 \times 10^{-5}$ oz |
| microgram (μg) | $3.527 \times 10^{-8}$ oz<br>$3.5273962 \times 10^{-8}$ oz |

| English mass | Metric mass |
| --- | --- |
| ton (ton) | 0.9072 t<br>0.90718474 t |
| | 907.2 kg<br>907.18474 kg |
| pound (lb) | 453.6 g<br>453.59237 g |
| | 0.4536 kg<br>0.45359237 kg |
| ounce (oz) | 28.35 g<br>28.349523 g |
| | $2.835 \times 10^{4}$ mg<br>$2.8349523 \times 10^{4}$ mg |

**Table A.4**
The Relationship Between English and Metric Volume Units
*Both four and seven digits are given for each relationship. Four digits will be used for nearly all of the English–Metric conversions.*

| Metric volume | English volume |
|---|---|
| liter (L) | 1.057 qt |
| | 1.056718 qt |
| | 0.2642 gal |
| | 0.2641795 gal |
| | 33.81 fl oz |
| | 33.81498 fl oz |
| | 61.03 cu in. |
| | 61.02545 cu in. |
| millimeter (mL) | 0.03381 fl oz |
| | 0.03381497 fl oz |
| | 0.06103 cu in. |
| | 0.06102545 cu in. |
| microliter (μL) | $3.381 \times 10^{-5}$ fl oz |
| | $3.381497 \times 10^{-5}$ fl oz |
| cubic centimeter ($cm^3$) | 0.03381 fl oz |
| | 0.03381497 fl oz |
| | 0.06103 cu in. |
| | 0.06102545 cu in. |

| English volume | Metric volume |
|---|---|
| gallon[a] (gal) | 3.785 L |
| | 3.785306 L |
| | 3785 mL |
| | 3785.306 mL |
| quart[a] (qt) | 0.9463 L |
| | 0.9463264 L |
| | 946.3 mL |
| | 946.3264 mL |
| fluid ounce[a] (fl oz) | 0.02957 L |
| | 0.02957270 L |
| | 29.57 mL |
| | 29.57270 mL |
| cubic inch (cu in.) | 16.39 $cm^3$ |
| | 16.387064 $cm^3$ (exact) |
| | 16.39 mL |
| | 16.387064 mL (exact) |
| | 0.1634 L |
| | 0.16387064 L (exact) |

[a] There are three kinds of gallons, quarts, and fluid ounces: British, U.S. dry, and U.S. liquid. The values given here are for U.S. liquid gallons, quarts, and fluid ounces.

**Table A.5**
English–English Conversion Factors

| Type of measurement | Conversion factors |
| --- | --- |
| length | $\left(\frac{12\text{ in.}}{1\text{ ft}}\right)$ $\left(\frac{3\text{ ft}}{1\text{ yd}}\right)$ $\left(\frac{5280\text{ ft}}{1\text{ mi}}\right)$ |
| mass | $\left(\frac{16\text{ oz}}{1\text{ lb}}\right)$ $\left(\frac{2000\text{ lb}}{1\text{ ton}}\right)$ |
| volume | $\left(\frac{4\text{ qt}}{1\text{ gal}}\right)$ $\left(\frac{32\text{ fl oz}}{1\text{ qt}}\right)$ $\left(\frac{2\text{ pt}}{1\text{ qt}}\right)$ |

# Appendix B
## Scientific Notation

Numbers expressed in scientific notation have the following form:

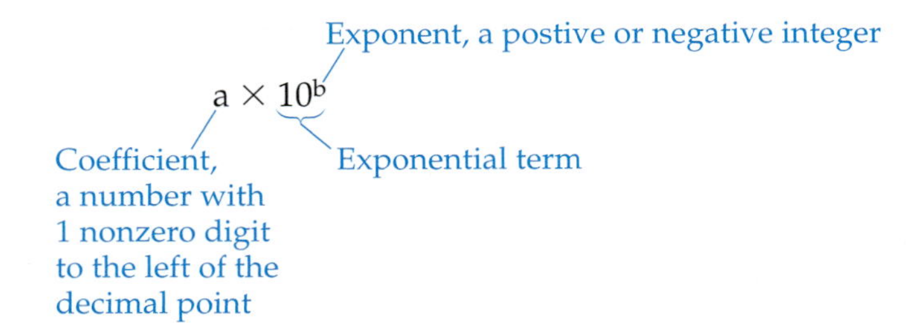


For example, there are about $5.5 \times 10^{21}$ carbon atoms in a 0.55-carat diamond. In the number $5.5 \times 10^{21}$, 5.5 is the coefficient, $10^{21}$ is the exponential term, and 21 is the exponent. The coefficient, which should have just one nonzero digit to the left of the decimal point, reflects the number's uncertainty. Scientists agree that unless otherwise stated, numbers are plus or minus one in the last position reported so a scientist reading $5.5 \times 10^{21}$ carbon atoms in a 0.55-carat diamond assumes that there are from $5.4 \times 10^{21}$ to $5.6 \times 10^{21}$ carbon atoms in the stone.

The exponential term shows the size of the number. Positive exponents are used for large numbers, and negative exponents are used for small numbers. For example, the moon orbits the sun at $2.2 \times 10^4$, or 22,000, mi/hr.

$$2.2 \times 10^4 = 2.2 \times 10 \times 10 \times 10 \times 10 = 22{,}000$$

A red blood cell has a diameter of about $5.6 \times 10^{-4}$, or 0.00056, inch.

$$5.6 \times 10^{-4} = 5.6 \times \frac{1}{10^4} = \frac{5.6}{10 \times 10 \times 10 \times 10} = 0.00056$$

Use the following steps to convert from a decimal number to scientific notation.

- Shift the decimal point until there is one nonzero number to the left of the decimal point, counting the number of positions that the decimal point moves.
- Write the resulting coefficient times an exponential term in which the exponent is positive if the decimal point was moved to the left and negative if the decimal point was moved to the right. The number in the

exponent is equal to the number of positions that the decimal point was shifted.

For example, when 22,000 is converted into scientific notation, the decimal point is shifted four positions to the left, so the exponential term has an exponent of 4.

$$22{,}000 = 2.2 \times 10^4$$

When 0.00056 is converted into scientific notation, the decimal point is shifted four positions to the right, so the exponential term has an exponent of −4.

$$0.00056 = 5.6 \times 10^{-4}$$

To convert from scientific notation to a decimal number, shift the decimal point in the coefficient to the right if the exponent is positive and to the left if it is negative. The number in the exponent tells you the number of positions to shift the decimal point.

For example, when $2.2 \times 10^4$ is converted to a decimal number, the decimal point is shifted four positions to the right because the exponent is 4.

$$2.2 \times 10^4 \quad \text{goes to} \quad 22{,}000$$

When $5.6 \times 10^{-4}$ is converted to a decimal number, the decimal point is shifted four positions to the left because the exponent is −4.

$$5.6 \times 10^{-4} \quad \text{goes to} \quad 0.00056$$

There are two reasons for using scientific notation. The first is for convenience. It takes a lot less time and space to report the mass of an electron as $9.1096 \times 10^{-28}$ g than to report it as 0.00000000000000000000000000091096 g. The second reason is to report the uncertainty of a value more clearly. For example, a typical peanut butter sandwich provides our bodies about $1.4 \times 10^3$ kJ of energy. Because of the variation in the type and amount of peanut butter added to the sandwich, there's some variation in the energy provided. The value $1.4 \times 10^3$ kJ suggests that the energy from a typical peanut butter sandwich could range from $1.3 \times 10^3$ kJ to $1.5 \times 10^3$ kJ. If the value were reported as 1400 kJ, its uncertainty would not be so clear. It could be $1400 \pm 1$, $1400 \pm 10$, or $1400 \pm 100$.

When multiplying exponential terms, add exponents.

$$10^3 \cdot 10^6 = 10^{3+6} = 10^9$$

$$10^3 \cdot 10^{-6} = 10^{3+(-6)} = 10^{-3}$$

$$3.2 \times 10^{-4} \cdot 1.5 \times 10^9 = 3.2 \cdot 1.5 \times 10^{-4+9} = 4.8 \times 10^5$$

When dividing exponential terms, subtract exponents.

$$\frac{10^{12}}{10^{3}} = 10^{12-3} = 10^{9}$$

$$\frac{10^{6}}{10^{-3}} = 10^{6-(-3)} = 10^{9}$$

$$\frac{9.0 \times 10^{11}}{1.5 \times 10^{-6}} = \frac{9.0}{1.5} \times 10^{11-(-6)} = 6.0 \times 10^{17}$$

$$\frac{10^{2} \cdot 10^{-3}}{10^{6}} = 10^{2+(-3)-6} = 10^{-7}$$

$$\frac{1.5 \times 10^{4} \cdot 4.0 \times 10^{5}}{2.0 \times 10^{12} \cdot 10^{3}} = \frac{1.5 \cdot 4.0}{2.0} \times 10^{4+5-12-3} = 3.0 \times 10^{-6}$$

When raising exponential terms to a power, multiply exponents.

$$(10^{4})^{3} = 10^{4 \cdot 3} = 10^{12}$$

$$(3 \times 10^{5})^{2} = (3)^{2} \times (10^{5})^{2} = 9 \times 10^{10}$$

# Appendix C
## Using a Scientific Calculator

The procedures for doing calculations using scientific calculators are not always self-evident. This appendix will take you step-by-step through an example problem in Chapter 1 and some of the example problems in Chapter 8, showing how two common kinds of scientific calculators would be used to solve them. If the procedures shown here do not apply to your particular calculator, you may have to consult its instruction manual.

Section 1.6 shows that the following dimensional analysis setup can be used to convert 947 mm into the equivalent length in centimeters.

$$? \text{ cm} = 947 \cancel{\text{mm}} \left(\frac{1 \cancel{\text{m}}}{10^3 \cancel{\text{mm}}}\right)\left(\frac{10^2 \text{ cm}}{1 \cancel{\text{m}}}\right)$$

The general approach to calculating an answer to a problem presented in this form is to input the first number in the mathematical expression (the number in the first value to the right of the equals sign) and then multiply or divide by the remaining numbers. *When a number is on the top of a ratio, you multiply by it, and when it is on the bottom, you divide by it.* Therefore, we input 947, divide by $10^3$, and then multiply by $10^2$. What buttons must be pushed to get this done depends on the calculator. The steps presented below work with many inexpensive scientific calculators. The letters and symbols in boxes represent buttons on the calculator.

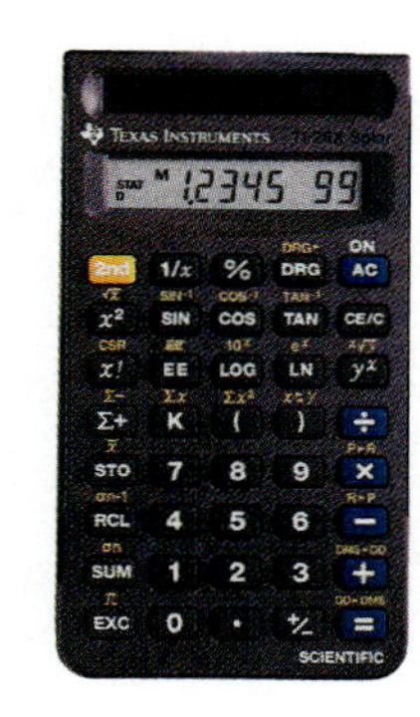

947 [÷] 1 [EE] 3 [×] 1 [EE] 2 [=]

Most calculators have either a [EE] or a [EXP] button that allows you to input scientific notation. To use one of these buttons, enter the coefficient first, press the [EE] or [EXP] button (which takes the place of the "× *10*" in scientific notation), and then enter the exponent. For example, to input $1.4 \times 10^4$, you would push 1.4 [EE] (or [EXP]) 4. For a number like $10^3$, which has no explicit coefficient, many calculators require you to push 1 for the implied coefficient. In other words, $10^3$ is the same as $1 \times 10^3$, so we push 1 [EE] (or [EXP]) 3.

With a more expensive graphing calculator, a different procedure is used. The [2nd] button on such a calculator gives you access to the functions described above the buttons. Thus you would push

947 [÷] [2nd] [EE] 3 [×] [2nd] [EE] 2 [ENTER]

$$? \text{ cm} = 947 \text{ mm}\left(\frac{1 \text{ m}}{10^3 \text{ mm}}\right)\left(\frac{10^2 \text{ cm}}{1 \text{ m}}\right) = \mathbf{94.7\ cm}$$

The calculation above can also be done with conversion factors that have negative exponents.

$$? \text{ cm} = 947 \text{ mm}\left(\frac{10^{-3} \text{ m}}{1 \text{ mm}}\right)\left(\frac{1 \text{ cm}}{10^{-2} \text{ m}}\right) = \mathbf{94.7\ cm}$$

On a typical inexpensive calculator, the answer to this calculation can be determined by using the following sequences:

947 [×] 1 [EE] 3 [+/−] [÷] 1 [EE] 2 [+/−] [=]

The [+/−] button converts positive to negative and negative to positive. With a more expensive, graphing calculator, you could use the following sequence:

947 [×] [2nd] [EE] [(−)] 3 [÷] [2nd] [EE] [(−)] 2 [ENTER]

The [(−)] button makes the next number that is pushed negative.

Example 8.2 provides us with another example:

$$? \text{ km} = 365 \text{ nm}\left(\frac{1 \text{ m}}{10^9 \text{ nm}}\right)\left(\frac{1 \text{ km}}{10^3 \text{ m}}\right)$$

We could use the following sequences to arrive at the answer using one type of calculator.

365 [÷] 1 [EE] 9 [÷] 1 [EE] 3 [=]

When we push the sequence of buttons listed above as the procedure for solving Example 8.2, the displays of many common calculators show *3.65 −10*, which means $3.65 \times 10^{-10}$. The *–10* on the right side of the display indicates the exponent for the scientific notation.

$$? \text{ km} = 365 \text{ nm}\left(\frac{1 \text{ m}}{10^9 \text{ nm}}\right)\left(\frac{1 \text{ km}}{10^3 \text{ m}}\right) = \mathbf{3.65 \times 10^{-10}\ km}$$

The sequence below could also be used for this problem.

365 [÷] [2nd] [EE] 9 [÷] [2nd] [EE] 3 [=]

On a typical graphing calculator, the display shows *3.65E-10*, where the *E* represents × *10*.

To do the calculation laid out in Example 8.3,

$$? \text{ lb} = 1.67 \times 10^{-18} \cancel{\mu g} \left( \frac{1 \cancel{g}}{10^6 \cancel{\mu g}} \right) \left( \frac{1 \text{ lb}}{453.6 \cancel{g}} \right)$$

you might use the following procedure (or some variation that fits your calculator):

1.67 [EE] 18 [+/−] [÷] 1 [EE] 6 [÷] 453.6 [=]

The answer displayed on the calculator would be *3.6817 –27*, which you would report as $3.68 \times 10^{-27}$ to reflect the correct number of significant figures (Section 1A.2 or Section 8.2).

Another possible procedure is to push

1.67 [2nd] [EE] [(−)] 18 [÷] [2nd] [EE] 6 [÷] 453.6 [ENTER]

The display shows *3.68165785E-27*, which you would round down to three significant figures: $3.68 \times 10^{-27}$.

$$? \text{ lb} = 1.67 \times 10^{-18} \cancel{\mu g} \left( \frac{1 \cancel{g}}{10^6 \cancel{\mu g}} \right) \left( \frac{1 \text{ lb}}{453.6 \cancel{g}} \right) = \mathbf{3.68 \times 10^{-27} \text{ lb}}$$

Example 8.5

$$? \text{ min} = 6.0 \cancel{ft} \left( \frac{12 \cancel{in.}}{1 \cancel{ft}} \right) \left( \frac{2.54 \cancel{cm}}{1 \cancel{in.}} \right) \left( \frac{1 \cancel{m}}{10^2 \cancel{cm}} \right) \left( \frac{1 \cancel{s}}{0.01 \cancel{m}} \right) \left( \frac{1 \text{ min}}{60 \cancel{s}} \right)$$

$$= \mathbf{3 \text{ min}}$$

can be done using one of the following sequences:

6 [×] 12 [×] 2.54 [÷] 1 [EE] 2 [÷] .01 [÷] 60 [=]

or

6 [×] 12 [×] 2.54 [÷] [2nd] [EE] 2 [÷] .01 [÷] 60 [ENTER]

The calculators report the answer as *3.048*, but we round it to 3 to reflect the correct number of significant figures.

## Chapter 8 Review Questions

**1.** Length, meter (m); mass, gram (g); volume, liter (L); energy, joule (J); gas pressure, pascal (Pa)
**2.** milliliter, volume, mL; microgram, mass, μg; kilojoule, energy, kJ; kelvin, temperature, K
**3. (a)** $10^{-6}$ m = 1 μm **(b)** $10^6$ g = 1 Mg **(c)** $10^{-3}$ L = 1 mL **(d)** $10^{-9}$ m = 1 nm **(e)** 1 $cm^3$ = 1 mL **(f)** $10^3$ L = $m^3$ **(g)** $10^3$ kg = 1 t **(h)** 1 Mg = 1 t
**4.** We assume that each reported number is ±1 in the last decimal position reported. Therefore, we assume the mass of the graduated cylinder between 1124.1 g and 1124.3 g, the volume of the methanol is between 1.19 L and 1.21 L, and the total mass is between 2073.8 g and 2074.0 g.

## Chapter 8 Key Ideas

**5.** unit conversion, correct **7.** unwanted, desired **9.** cancel **11.** one **13.** inexact, fewest **15.** counting **17.** decimal places, fewest decimal places **19.** grams per milliliter, grams per cubic centimeter **21.** identity, known **23.** part, whole **25.** definitions

## Chapter 8 Problems

**26. (a)** $6.7294 \times 10^4$ **(b)** $4.38763102 \times 10^8$ **(c)** $7.3 \times 10^{-5}$ **(d)** $4.35 \times 10^{-8}$
**28. (a)** 4,097 **(b)** 15,541.2 **(c)** 0.0000234 **(d)** 0.000000012
**30. (a)** 2,877 **(b)** 316 **(c)** 5.5 **(d)** 2.827
**32. (a)** $10^{12}$ **(b)** $10^9$ **(c)** $10^7$ **(d)** $10^5$ **(e)** $10^{29}$ **(f)** $10^3$
**34. (a)** $7.6 \times 10^{15}$ **(b)** $1.02 \times 10^{16}$ **(c)** $3.3 \times 10^5$ **(d)** $1.87 \times 10^{-3}$ or 0.00187 **(e)** $8.203 \times 10^{13}$ **(f)** $2.111 \times 10^4$

**36. (a)** $\left(\frac{10^3\text{ m}}{1\text{ km}}\right)$ **(b)** $\left(\frac{10^2\text{ cm}}{1\text{ m}}\right)$ **(c)** $\left(\frac{10^3\text{ mm}}{1\text{ m}}\right)$ **(d)** $\left(\frac{1\text{ cm}^3}{1\text{ mL}}\right)$ **(e)** $\left(\frac{2.54\text{ cm}}{1\text{ in.}}\right)$ **(f)** $\left(\frac{453.6\text{ g}}{1\text{ lb}}\right)$

**38. (a)** $\left(\frac{10^3\text{ g}}{1\text{ kg}}\right)$ **(b)** $\left(\frac{10^3\text{ mg}}{1\text{ g}}\right)$ **(c)** $\left(\frac{1.094\text{ yd}}{1\text{ m}}\right)$ **(d)** $\left(\frac{2.205\text{ lb}}{1\text{ kg}}\right)$
**40.** $9.1093897 \times 10^{-28}$ g
**42.** 3.2 μm
**44.** 0.04951 lb or 0.049507 lb (if use more significant figures in the gram-to-pound conversion factor)
**46.** $5.908 \times 10^{-26}$ oz or $5.908138 \times 10^{-26}$ oz (if use more significant figures in the gram-to-pound conversion factor)
**48.** 2.2 μm
**50. (a)** not exact—2 significant figures **(b)** exact **(c)** not exact—2 significant figures **(d)** exact **(e)** exact **(f)** exact **(g)** not exact—4 significant figures **(h)** not exact—6 significant figures **(i)** not exact—4 significant figures **(j)** not exact—4 significant figures
**52. (a)** 5 **(b)** 3 **(c)** 3 **(d)** 5 **(e)** 4
**54. (a)** 3 **(b)** 4 **(c)** 4
**56. (a)** 34.6 **(b)** 193 **(c)** 24.0 **(d)** 0.00388 **(e)** 0.0230 **(f)** $2.85 \times 10^3$ **(g)** $7.84 \times 10^4$
**58. (a)** 103 **(b)** 6.2 **(c)** $5.71 \times 10^{11}$
**60. (a)** 100.14 **(b)** 1
**62.** 0.1200 g/mL **64.** 0.1850 kg **66.** 0.8799 qt **68.** $1.9 \times 10^{19}$ kg $Na^+$ **70.** 94.3 L to brain **72.** 0.18 mg to energy **74.** 33 km/hr **76.** 0.275 s **78.** 3.5 hr **80.** $3.326 \times 10^5$ kJ **82.** 0.444 kg C **84.** 2.7 lb beans per day **86.** $1.6 \times 10^8$ kg HCl **88.** 52 days **90.** 9812 L air **92.** 44.0 minutes **94.** 28 qt **96.** $7.6 \times 10^5$ beats **98.** $3.6 \times 10^9$ molecules **100.** 6.9 days **102.** 95 ft **103.** 88 °F, 304 K **105.** 108 °C, 381 K **107.** $2.750 \times 10^3$ °C, 4982 °F **109.** $5.5 \times 10^3$ °C, $5.8 \times 10^3$ K

## Chapter 9 Exercises

**9.1. (a)** 196.9665 (from periodic table)
**(b)** 196.9665 g (There are $6.022 \times 10^{23}$ atoms per mole of atoms, and one mole of an element has a mass in grams equal to its atomic mass.)

**(c)** $\left(\frac{196.9665\text{ g Au}}{1\text{ mol Au}}\right)$

**(d)** $? \text{ g Au} = 0.20443 \text{ mol Au}\left(\frac{196.9665\text{ g Au}}{1\text{ mol Au}}\right) = \mathbf{40.266\text{ g Au}}$

**(e)** $? \text{ mg Au} = 7.046 \times 10^{-3} \text{ mol Au}\left(\frac{196.9665\text{ g Au}}{1\text{ mol Au}}\right)\left(\frac{10^3\text{ mg}}{1\text{ g}}\right)$

$= \mathbf{1388\text{ mg Au}}$

**(f)** $? \text{ mol Au} = 1.00 \text{ troy oz Au}\left(\frac{31.10\text{ g}}{1\text{ troy oz}}\right)\left(\frac{1\text{ mol Au}}{196.9665\text{ g Au}}\right)$

$= \mathbf{0.158\text{ mol Au}}$

**9.2. (a)** 2(12.011) + 6(1.00794) + 1(15.9994) = 46.069
**(b)** 46.069 g (One mole of a molecular compound has a mass in grams equal to its molecular mass.)

**(c)** $\left(\frac{46.069\text{ g } C_2H_5OH}{1\text{ mol } C_2H_5OH}\right)$

**(d)** $? \text{ mol } C_2H_5OH = 16\text{ g } C_2H_5OH\left(\frac{1\text{ mol } C_2H_5OH}{46.069\text{ g } C_2H_5OH}\right)$

$= \mathbf{0.35\text{ mol } C_2H_5OH}$

**(e)** ? 1 mL $C_2H_5OH$

$= 1.0\text{ mol } C_2H_5OH\left(\frac{46.069\text{ g } C_2H_5OH}{1\text{ mol } C_2H_5OH}\right)\left(\frac{1\text{ mL } C_2H_5OH}{0.7893\text{ g } C_2H_5OH}\right)$

$= \mathbf{58\text{ mL } C_2H_5OH}$

**9.3. (a)** Formula Mass
= 1(22.9898) + 1(1.00794) + 1(12.011) + 3(15.9994) = 84.007
**(b)** 84.007 g (One mole of an ionic compound has a mass in grams equal to its formula mass.)

**(c)** $\left(\frac{84.007\text{ g } NaHCO_3}{1\text{ mol } NaHCO_3}\right)$

**(d)** ? mol $NaHCO_3$

$= 0.4\text{ g } NaHCO_3\left(\frac{1\text{ mol } NaHCO_3}{84.007\text{ g } NaHCO_3}\right) = \mathbf{5 \times 10^{23}\text{ mol } NaHCO_3}$

**9.4. (a)** $\frac{6\text{ mol H}}{1\text{ mol } C_2H_5OH}$ **(b)** $\frac{3\text{ mol O}}{1\text{ mol } NaHCO_3}$

**(c)** There is 1 mole of $HCO_3^-$ per 1 mole of $NaHCO_3$.

**9.5.** ? g $S_2Cl_2$

$= 123.8\text{ g S}\left(\frac{1\text{ mol S}}{32.066\text{ g S}}\right)\left(\frac{1\text{ mol } S_2Cl_2}{2\text{ mol S}}\right)\left(\frac{135.037\text{ g } S_2Cl_2}{1\text{ mole } S_2Cl_2}\right)$

$= \mathbf{260.7\text{ g } S_2Cl_2}$

**9.6.** $? \text{ kg V} = 2.3\text{ kg } V_2O_5\left(\frac{10^3\text{ g}}{1\text{ kg}}\right)\left(\frac{1\text{ mol } V_2O_5}{181.880\text{ g } V_2O_5}\right)$

$\left(\frac{2\text{ mol V}}{1\text{ mol } V_2O_5}\right)\left(\frac{50.9415\text{ g V}}{1\text{ mol V}}\right)\left(\frac{1\text{ kg}}{10^3\text{ g}}\right) = \mathbf{1.3\text{ kg V}}$

**9.7.** $? \text{ mol Bi} = 32.516\text{ g Bi}\left(\frac{1\text{ mol Bi}}{208.9804\text{ g Bi}}\right)$

$= 0.15559\text{ mol Bi} \div 0.15559 = 1\text{ mol Bi} \times 2 = 2\text{ mol Bi}$

$? \text{ mol S} = 7.484\text{ g S}\left(\frac{1\text{ mol S}}{32.066\text{ gS}}\right)$

$= 0.2334\text{ mol S} \div 0.15559 \cong 1\frac{1}{2}\text{ mol S} \times 2 = 3\text{ mol S}$
Our empirical formula is $\mathbf{Bi_2S_3}$ or bismuth(III) sulfide.

**9.8.** $? \text{ mol K} = 35.172\text{ g K}\left(\frac{1\text{ mol K}}{39.0983\text{ K}}\right) = 0.89958\text{ mol K} \div 0.89958$

$= 1\text{ mol K} \times 2 = 2\text{ mol K}$

# Selected Answers

## CHAPTER 1 EXERCISES

**1.1. (a)** 1 megagram = $10^6$ gram **(b)** 1 milliliter = $10^{-3}$ liter
**1.2. (a)** 71 mL to 73 mL **(b)** 8.22 m to 8.24 m **(c)** $4.54 \times 10^{-5}$ g to $4.56 \times 10^{-5}$ g
**1.3.** 2.30 g because the reported values differ by about ± 0.01.

## CHAPTER 1 KEY IDEAS

**1.** observation, data, hypothesis, research or experimentation, research, published, applications, hypothesizing and testing
**3.** meter, m **5.** second, time **7.** base units, derive **9.** mass, distance
**11.** Celsius, Kelvin **13.** certain, estimated

## CHAPTER 1 PROBLEMS

**15.** megagram, mass, Mg; milliliter, volume, mL; nanometer, length, nm; kelvin, temperature, K
**17. (a)** $10^3$ **(b)** $10^9$ **(c)** $10^{-3}$ **(d)** $10^{-9}$
**19. (e)** 10,000,000 **(f)** 1,000,000,000,000 **(g)** 0.0000001 **(h)** 0.000000000001
**21. (e)** $10^3$ m = 1 km **(f)** $10^{-3}$ L = 1 mL **(g)** $10^6$ g = 1 Mg **(h)** 1 $cm^3$ = 1 mL **(i)** $10^3$ kg = 1 t
**23. (a)** meter **(b)** centimeter **(c)** millimeter **(d)** kilometer
**25.** inch is larger
**27. (a)** milliliter **(b)** cubic meter **(c)** liter
**29.** fluid ounce is larger
**31.** Mass is usually defined as a measure of the amount of matter in an object. The weight of an object, on the Earth, is a measure of the force of gravitational attraction between the object and the Earth. The more mass an object has, the greater the gravitational attraction between it and another object. The farther an object gets from the earth, the less that attraction is, and the lower its weight. Unlike the weight of an object, the mass of an object is independent of location. Mass is described with mass units, like grams and kilograms. Weight can be described with force units, like newtons.
**32.** ounce is larger **36.** degree Celsius is larger **38.** 10 °F
**41. (a)** 30.5 m means 30.5 ± 0.1 m or 30.4 m to 30.6 m.
**(b)** 612 g means 612 ± 1 g or 611 g to 613 g.
**(c)** 1.98 m means 1.98 ± 0.01 m or 1.97 m to 1.99 m.
**(d)** $9.1096 \times 10^{-28}$ g means $(9.1096 \pm 0.0001) \times 10^{-28}$ g or $9.1095 \times 10^{-28}$ g to $9.1097 \times 10^{-28}$ g.
**(e)** $1.5 \times 10^{18}$ $m^3$ means $(1.5 \pm 0.1) \times 10^{18}$ $m^3$ or $1.4 \times 10^{18}$ $m^3$ to $1.6 \times 10^{18}$ $m^3$.
**43. (a)** It's difficult to estimate the hundredth position accurately. For the object on the left, we might report 7.67 cm, 7.68 cm, or 7.69. The end of the right object seems to be right on the 9 cm mark, so we report 9.00 cm. **(b)** 7.7 cm and 9.0 cm
**45.** Our uncertainty is in the tenth position, so we report 10.4 s.
**47. (a)** 27.2410 g **(b)** Our convention calls for only reporting one uncertain digit in our value. Because we are uncertain about the thousandth position, we might report 27.241 g (or perhaps even 27.24 g).

## INTER-CHAPTER 1A EXERCISES

**1A.1.** $\left(\frac{1\text{ mm}}{10^{-3}\text{ m}}\right)\left(\frac{10^3\text{ m}}{1\text{m}}\right)\left(\frac{1\text{ mm}}{10^3\text{ mm}}\right)\left(\frac{10^3\text{ mm}}{1\text{ m}}\right)$

**1A.2.** $? \text{ kg} = 612 \text{ g}\left(\frac{1\text{ kg}}{10^3\text{ g}}\right) = \mathbf{0.612\ kg}$

**1A.3** $? \ \mu\text{L} = 0.010 \text{ mL}\left(\frac{1\text{ L}}{10^3\text{ mL}}\right)\left(\frac{10^6\ \mu\text{L}}{1\text{ L}}\right) = \mathbf{10\ \mu L}$

**1A.4.** ? people =

$$10.7 \times 10^{11}\text{ Mg carbohydrates}\left(\frac{10^6\text{ g}}{1\text{ Mg}}\right)\left(\frac{1\text{ kg}}{10^3\text{ g}}\right)\left(\frac{1\text{ person}}{2.8 \times 10^4\text{ kg}}\right)$$

**STEP 1:** The numbers $1.7 \times 10^{11}$ and $2.8 \times 10^4$ are not defined, and they cannot be counted so they are not exact. We can assume that they come from a mixture of measurements and calculations. Both $10^6$ and $10^3$ come from definitions of metric prefixes, so they are exact.
**STEP 2:** Nonzero digits are significant and the exponential term of a number expressed in scientific notation is considered exact, so $1.7 \times 10^{11}$ and $2.8 \times 10^4$ both have two significant figures.
**STEP 3:** When multiplying and dividing, we round our answer off to the same number of significant figures as the value that contains the fewest significant figures, so we report two significant figures in this answer. The calculator reports $6.0714 \times 10^9$, which we round off to **$6.1 \times 10^4$ people.** We need scientific notation to unambiguously report two significant figures.
**1A.5.** 295.7 K − 273.15 = **22.6 °C**
**STEP 1:** The number 295.7 comes from a measurement, so it is not exact. The number 273.15 comes from a definition, so it is exact.
**STEP 2:** The number 295.7 is precise to the tenths position.
**STEP 3:** We round our answer to the tenth position.

**1A.6.** $? \text{ g} = 834.6 \text{ mL}\left(\frac{0.0960\text{ g}}{1\text{ mL}}\right)\left(\frac{1\text{ kg}}{10^3\text{ g}}\right) = \mathbf{0.801\ kg}$

**1A.7.** $\frac{?\text{ g}}{\text{mL}} = \frac{(83.836 - 47.356)\text{ g}}{45.6\text{ mL}} = \frac{36.480\text{ g}}{45.6\text{ mL}} = \mathbf{0.800\ g/mL}$

## INTER-CHAPTER 1A KEY IDEAS

**1.** unit-conversion, numerical problems **3.** relationship between two units **5.** base unit **7.** degree of uncertainty, greater **9.** exact
**11.** never **13.** decimal places, fewest decimal places

## INTER-CHAPTER 1A PROBLEMS

**14. (a)** $4.239 \times 10^3$ **(b)** $5.723845 \times 10^6$ **(c)** $4.15 \times 10^{-4}$ **(d)** $1.623 \times 10^{-9}$
**16. (a)** 276,423 **(b)** 634.2 **(c)** 0.00000088 **(d)** 0.0005667
**18. (a)** 192.5 **(b)** 498 **(c)** 0.02 **(d)** 3.12
**20. (a)** $10^{15}$ **(b)** $10^4$ **(c)** $10^6$ **(d)** $10^{-4}$ **(e)** $10^{11}$ **(f)** $10^{-1}$
**22. (a)** $8.4 \times 10^{11}$ **(b)** $3.13 \times 10^{15}$ **(c)** $2.5 \times 10^6$ **(d)** 63 **(e)** $3.279 \times 10^{23}$ **(f)** $2.08 \times 10^{11}$

**24. (a)** $\left(\frac{10^3\text{ g}}{1\text{ kg}}\right)$ **(b)** $\left(\frac{453.6\text{ g}}{1\text{ lb}}\right)$ **(c)** $\left(\frac{10^6\ \mu\text{m}}{1\text{ m}}\right)$ **(d)** $\left(\frac{10^{12}\text{ pm}}{1\text{ m}}\right)$

**(e)** $\left(\frac{60\text{ min}}{1\text{ h}}\right)$ **(f)** $\left(\frac{2.54\text{ cm}}{1\text{ in.}}\right)$ **(g)** $\left(\frac{12\text{ in.}}{1\text{ ft}}\right)$ **(h)** $\left(\frac{3.785\text{ L}}{1\text{ gal}}\right)$

**26. (a)** 0.772 kg **(b)** 45 cm **(c)** 1255 L **(d)** 0.084 m
**28. (a)** 0.4565 Mg **(b)** 1549 μm
**30.** $10^4$ Gg **32.** 4.940 MJ
**34. (a)** 5 **(b)** 3 **(c)** 4 **(d)** 5 **(e)** 6
**36. (a)** Not exact—4 **(b)** exact **(c)** Not exact—4 **(d)** exact **(e)** exact **(f)** exact **(g)** Not exact—4 **(h)** exact **(i)** 34.100 g (Not exact—4), 11.0 mL (Not exact—3), 3.10 g/mL (Not exact—3)
**38. (a)** 2 **(b)** 1000.05 **40.** 0.0023 m, 0.23 cm **42.** 4.2 g
**44.** $6.80 \times 10^{-7}$ m, $6.80 \times 10^5$ pm **46.** $1.989 \times 10^{33}$ g, $1.989 \times 10^{24}$ Gg, $4.385 \times 10^{30}$ lb **48.** 2.14 m, 84.3 in., 7.02 ft **50.** $2.4 \times 10^{13}$ g, $2.4 \times 10^{10}$ kg, $5.3 \times 10^{10}$ lb, $2.6 \times 10^7$ ton **52.** 106.0 g, 0.1060 kg, 0.2338 lb **54.** $2.9 \times 10^6$ m, $1.8 \times 10^3$ mi **56.** 17.32 kg, 4.433 kg
**58.** $2.8 \times 10^7$ L **60.** 2.6 g/mL **62.** 4.5 min **64.** 6.8 mi **66.** $1 \times 10^{14}$ $

**68.** $1.38 \times 10^3$ yr **70.** $3 \times 10^9$ Gg H **72.** $3.2 \times 10^7$ vehicles with air bags **74.** 1.1 Mg car **76.** 758 mmHg, $6.2 \times 10^2$ mmHg, $1.8 \times 10^2$ mmHg **78.** $5.78 \times 10^3$ K **80.** 1.4 cm **82.** 750 cm or $7.50 \times 10^2$ cm, 7.50 m **84.** $2 \times 10^{-10}$ m, $2 \times 10^{-4}$ μm
**86.** $3.4 \times 10^2$ cm, $1.3 \times 10^2$ in., 11 ft **88.** $9.6 \times 10^2$ cm, $3.8 \times 10^2$ in., 31 ft **90.** $5.900 \times 10^9$ km, $3.667 \times 10^9$ mi **92.** $9.142 \times 10^7$ mi
**94.** $5.8 \times 10^6$ Mg **96.** $2.0 \times 10^5$ flight operations
**98.** $3.8 \times 10^3$ vibrations
**100.** $1.7 \times 10^{18}$ km, 18 orbits

## Chapter 2 Exercises

**2.1.** *aluminum,* Al, 13, 3A or IIIA, metal, representative (or main-group) element, 3, solid; *silicon,* Si, 14, 4A or IVA, metalloid, representative (or main-group) element, 3, solid; *nickel,* Ni, 10, 8B or VIIIB, metal, transition metal, 4, solid; *sulfur,* S, 16, 6A or VIA, nonmetal, representative (or main-group) element, 3, solid; *fluorine,* F, 17, 7A or VIIA, nonmetal, representative (or main-group) element, 2, gas; *potassium,* K, 1, 1A or IA, metal, representative (or main-group) element, 4, solid; *mercury,* Hg, 12, 2B or IIB, metal, transition metal, 6, liquid; *uranium,* U, (No group number), metal, inner transition metal, 7, solid; *manganese,* Mn, 7, 7B or VIIB, metal, transition metal, 4, solid; *calcium,* Ca, 2, 2A or IIA, metal, representative (or main-group) element, 4, solid; *bromine,* Br, 17, nonmetal, representative (or main-group) element, 4, liquid; *silver,* Ag, 1B, metal, transition metal, 5, solid; *carbon,* C, 14, nonmetal, representative (or main-group) element, 2, solid
**2.2.** **(a)** noble gases **(b)** halogens **(c)** alkaline earth metals **(d)** alkali metals
**2.3.** **(a)** +2 cation **(b)** −1 anion

## Chapter 2 Review Questions

**1.** Matter is anything that occupies space and has mass.
**2.** The distance between the floor and a typical doorknob is about one meter. A penny weighs about three grams.

## Chapter 2 Key Ideas

**3.** simplified but useful **5.** motion **7.** attract **9.** empty space, expands **11.** escape **13.** straight-line path **15.** simpler
**17.** vertical column **19.** solid, liquid, gas **21.** sun **23.** $10^{-15}$
**25.** cloud **27.** gains **29.** chemical **31.** single atom

## Chapter 2 Problems

**33.** **(a)** Strong attractions between the particles keep each particle at the same average distance from other particles and in the same general position with respect to its neighbors.
**(b)** The velocity of the particles increases, causing more violent collisions between them. This causes them to move apart, so the solid expands. See Figure 2.1.
**(c)** The particles break out of their positions in the solid and move more freely throughout the liquid, constantly breaking old attractions and making new ones. Although the particles are still close together in the liquid, they are more disorganized, and there is more empty space between them.
**35.** **(a)** The attractions between liquid particles are not strong enough to keep the particles in position like the solid. The movement of particles allows the liquid to take the shape of its container. The attractions are strong enough to keep the particles at the same average distance, leading to constant volume.
**(b)** The velocity of the particles increases, so they will move throughout the liquid more rapidly. The particles will collide with more force. This causes them to move apart, so the liquid expands slightly.
**(c)** Particles that are at the surface of the liquid and that are moving away from the surface fast enough to break the attractions that pull them back will escape to the gaseous form. The gas particles will disperse throughout the neighborhood as they mix with the particles in the air. See Figures 2.3 and 2.4.
**37.** The air particles are moving faster.
**39.** There is plenty of empty space between particles in a gas.
**40.** **(a)** Cl **(b)** Zn **(c)** P **(d)** U
**43.** **(a)** carbon **(b)** copper **(c)** neon **(d)** potassium
**46.** *sodium,* Na, 1 or 1A or IA, metal, Representative (or main-group) element, 3; *tin,* Sn, 14 or 4A or IVA, metal, Representative (or main-group) element, 5; *helium,* He, 18 or 8A or VIIIA, nonmetal, Representative (or main-group) element, 1; *nickel,* Ni, 10 or 8B or VIIIB, metal, Transition metal, 4; *silver,* Ag, 11 or 1B or IB, metal, Transition metal, 5; *aluminum,* Al, 13 or 3A or IIIA, metal, Representative (or main-group) element, 3; *silicon,* Si, 14 or 4A or IVA, metalloid, Representative (or main-group) element, 3; *sulfur,* S, 16, nonmetal, Representative (or main-group) element, 3; *mercury,* Hg, 2B, metal, Transition metal, 6
**48.** **(a)** halogens **(b)** noble gases **(c)** alkali metals **(d)** alkaline earth metals
**50.** **(a)** gas **(b)** liquid **(c)** solid **(d)** gas **(e)** solid **(f)** solid
**52.** lithium and potassium
**54.** **(a)** chlorine, Cl **(b)** potassium, K **(c)** silicon, Si
**56.** Because manganese is a metal, we expect it to be malleable.
**58.** Protons and neutrons are in a tiny core of the atom called the nucleus, which has a diameter about 1/100,000 the diameter of the atom. The position and motion of the electrons are uncertain, but they generate a negative charge that is felt in the space that surrounds the nucleus.
**60.** **(a)** +1 cation **(b)** −2 anion
**63.** **(a)** 8 **(b)** 12 **(c)** 92 **(d)** 3 **(e)** 82 **(f)** 25
**66.** **(a)** cobalt, Co **(b)** tin, Sn **(c)** calcium, Ca **(d)** fluorine, F
**68.** See Figure 2.13. The cloud around the two hydrogen nuclei represents the negative charge cloud generated by the two electrons in the covalent bond that holds the atoms together in the $H_2$ molecule.
**70.** **(a)** Neon is composed of separate neon atoms. Its structure is very similar to the structure of He shown in Figure 2.12.
**(b)** Bromine is composed of $Br_2$ molecules. See Figure 2.16.
**(c)** Nitrogen is composed of $N_2$ molecules. Its structure is very similar to the structure of $H_2$ shown in Figure 2.15.
**72.** Each atom in a metallic solid has released one or more electrons, allowing the electrons to move freely throughout the solid. When the atoms lose these electrons, they become cations, which form the organized structure we associate with solids. The released electrons flow between the stationary cations like water flows between islands in the ocean. See Figure 2.18.

## Chapter 3 Exercises

**3.1.** **(a)** Element **(b)** Compound **(c)** Mixture
**3.2.** **(a)** all nonmetal atoms—molecular **(b)** metal-nonmetal—ionic

**3.3.**

```
            ..                          ..    ..
           :I:                         :Cl:  :Cl:
            |                            |     |
(a)  :Ï̤—N̤—Ï̤:         (b)  :C̤̈l—C—C—C̤̈l:
                                         |     |
                                        :C̤l:  :C̤l:

(c)  H—Ö̤—Ö̤—H          (d)  H—C═C—H
                                 |  |
                                 H  H
```

**3.4.** **(a)** diphosphorus pentoxide **(b)** phosphorus trichloride **(c)** carbon monoxide **(d)** dihydrogen monosulfide or hydrogen sulfide **(e)** ammonia
**3.5.** **(a)** $S_2F_{10}$ **(b)** $NF_3$ **(c)** $C_3H_8$ **(d)** HCl
**3.6.** **(a)** magnesium ion **(b)** fluoride ion **(c)** tin(II) ion
**3.7.** **(a)** $Br^-$ **(b)** $Al^{3+}$ **(c)** $Au^+$
**3.8.** **(a)** lithium chloride **(b)** chromium(III) sulfate **(c)** ammonium hydrogen carbonate
**3.9.** **(a)** $Al_2O_3$ **(b)** $CoF_3$ **(c)** $FeSO_4$ **(d)** $(NH_4)_2HPO_4$ **(e)** $KHCO_3$

## Chapter 3 Review Questions

**1.** ion **2.** cation **3.** anion **4.** covalent **5.** molecule **6.** diatomic
**7.** See Figures 2.1, 2.2, and 2.4.
**8.** Protons and neutrons are in a tiny core of the atom called the nucleus, which has a diameter about 1/100,000 the diameter of the atom. The position and motion of the electrons are uncertain, but they generate a negative charge that is felt in the space that surrounds the nucleus.
**9.** The hydrogen atoms are held together by a covalent bond formed due to the sharing of two electrons. See Figure 2.13.

**10.** lithium, Li, 1 or 1A, metal; carbon, C, 14 or 4A, nonmetal; chlorine, Cl, 17 or 7A, nonmetal; oxygen, O, 16 or 6A, nonmetal; copper, Cu, 11 or 1B, metal; calcium, Ca, 2 or 2A, metal; scandium, Sc, 3 or 3B, metal
**11. (a)** halogens **(b)** noble gases **(c)** alkali metals **(d)** alkaline earth metals

## Chapter 3 Key Ideas

**12.** whole-number ratio **14.** symbols, subscripts **16.** variable **18.** cations, anions **20.** negative, positive, partial negative, partial positive **22.** usually ionic **24.** octet **26.** four **28.** arranged in space **30.** electron-charge clouds **32.** same average distance apart, constantly breaking, forming new attractions **34.** nonmetallic **36.** noble gas **38.** larger, smaller **40.** $H^+$

## Chapter 3 Problems

**42. (a)** mixture—variable composition
**(b)** element and pure substance—The symbol K is on the periodic table of the elements. All elements are pure substances.
**(c)** The formula shows the constant composition, so ascorbic acid is a pure substance. The three element symbols in the formula indicate a compound.
**44. (a)** $N_2O_3$ **(b)** $SF_4$ **(c)** $AlCl_3$ **(d)** $Li_2CO_3$

**46.**

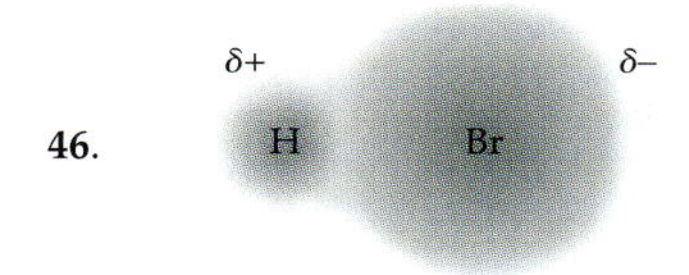


**48.** The metallic potassium atoms lose one electron and form +1 cations, and the nonmetallic fluorine atoms gain one electron and form −1 anions.

$$\underset{19p/19e^-}{K} \rightarrow \underset{19p/18e^-}{K^+} + e^-$$

$$\underset{9p/9e^-}{F} + e^- \rightarrow \underset{9p/10e^-}{F^-}$$

The ionic bonds are the attractions between $K^+$ cations and $F^-$ anions.
**50.** See Figure 3.6.
**52. (a)** covalent . . . nonmetal-nonmetal **(b)** ionic . . . metal-nonmetal
**54. (a)** all nonmetallic atoms—molecular **(b)** metal-nonmetal—ionic
**56. (a)** 7 **(b)** 4
**58.** Each of the following answers is based on the assumption that nonmetallic atoms tend to form covalent bonds in order to get an octet (8) of electrons around each atom, like the very stable noble gases (other than helium). Covalent bonds (represented by lines in Lewis structures) and lone pairs each contribute two electrons to the octet.
**(a)** oxygen

:Ö· If oxygen atoms form two covalent bonds, they will have an octet of electrons around them. Water is an example:

H—Ö—H

**(b)** fluorine

:F̤̈· If fluorine atoms form one covalent bond, they will have an octet of electrons around them. Hydrogen fluoride, HF, is an example:

H—F̤̈:

**(c)** carbon

·Ċ· If carbon atoms form four covalent bonds, they will have an octet of electrons around them. Methane, $CH_4$, is an example:

```
    H
    |
H — C — H
    |
    H
```

**(d)** phosphorus

·P̤̈· If phosphorus atoms form three covalent bonds, they will have an octet of electrons around them. Phosphorus trichloride, $PCl_3$, is an example:

```
:Cl — P — Cl:
      |
     :Cl:
```

**60.** The molecule contains a carbon atom, two chlorine atoms, and two fluorine atoms. There are two covalent C–Cl bonds and two covalent C–F bonds. The Cl and F atoms have three lone pairs each.
**62. (a)** H—one bond, zero lone pairs
**(b)** iodine—one bond, three lone pairs
**(c)** sulfur—two bonds, two lone pairs
**(d)** N—three bonds, one lone pair

**64.**

```
                          :Br:
                           |
(a) :F—O—F:    (b)  H — C — Br:    (c) :I — P — I:
                           |                 |
                          :Br:              :I:
```

**66.**

```
(a) H—C≡N:    (b) :Cl — C = C — Cl:
                        |   |
                      :Cl: :Cl:
```

**68. (a)** methanol and methyl alcohol **(b)** ethanol and ethyl alcohol
**(c)** 2-propanol and isopropyl alcohol

**70.** H—C—H (with H above and below C)

Lewis structure | Space-filling model | Ball-and-stick model | Geometric sketch (109.5°)

The Lewis structure shows the four covalent bonds between the carbon atoms and the hydrogen atoms. The space-filling model provides the most accurate representation of the electron-charge clouds for the atoms in $CH_4$. The ball-and-stick model emphasizes the molecule's correct molecular shape and shows the covalent bonds more clearly. Each ball represents an atom, and each stick represents a covalent bond between two atoms. The geometric sketch shows the three-dimensional tetrahedral structures with a two-dimensional drawing. Picture the hydrogen atoms connected to the central carbon atom with solid lines as being in the same plane as the carbon atom. The hydrogen atom connected to the central carbon with a solid wedge comes out of the plane toward you. The hydrogen atom connected to the carbon atom by a dashed wedge is located back behind the plane of the page.

**72.** H—Ö—H

Lewis structure | Space-filling model | Ball-and-stick model | Geometric sketch (105°)

The Lewis structure shows the two O–H covalent bonds and the two lone pairs on the oxygen atom. The space-filling model provides the most accurate representation of the electron-charge clouds for the atoms and the bonding electrons. The ball-and-stick model emphasizes the molecule's correct molecular shape and shows the covalent bonds more clearly. The geometric sketch shows the structure with a two-dimensional drawing.
**74.** Water is composed of $H_2O$ molecules. Pairs of these molecules are attracted to each other by the attraction between partially positive hydrogen atom of one molecule and the partially negative oxygen atom of the other molecule. See Figure 3.13. Each water molecule is

moving constantly, breaking the attractions to some molecules, and making new attractions to other molecules. Figure 3.14 shows the model we will use to visualize liquid water.

**76.** $BrF_5$—bromine pentafluoride

**78.** $SCl_2$—sulfur dichloride

**80. (a)** diiodine pentoxide **(b)** bromine trifluoride **(c)** iodine monobromide **(d)** methane **(e)** hydrogen bromide or hydrogen monobromide

**82. (a)** $C_3H_8$ **(b)** ClF **(c)** $P_4S_7$ **(d)** $CBr_4$ **(e)** HF

**84.** Because metallic atoms hold some of their electrons relatively loosely, they tend to lose electrons and form cations. Because nonmetallic atoms attract electrons more strongly than metallic atoms, they tend to gain electrons and form anions. Thus, when a metallic atom and a nonmetallic atom combine, the nonmetallic atom often pulls one or more electrons far enough away from the metallic atom to form ions and an ionic bond.

**85. (a)** 4 protons and 2 electrons **(b)** 16 protons and 18 electrons

**87. (a)** calcium ion **(b)** lithium ion **(c)** chromium (II) ion **(d)** fluoride **(e)** silver ion or silver(I) ion **(f)** scandium ion **(g)** phosphide **(h)** lead(II) ion

**89. (a)** $Mg^{2+}$ **(b)** $Na^+$ **(c)** $S^{2-}$ **(d)** $Fe^{3+}$ **(e)** $Sc^{3+}$ **(f)** $N^{3-}$ **(g)** $Mn^{3+}$ **(h)** $Zn^{2+}$

**91.** The metallic silver atoms form cations, and the nonmetallic bromine atoms form anions. The anions and cations alternate in the ionic solid with each cation surrounded by six anions and each anion surrounded by six cations. See Figure 3.18.

**93. (a)** ammonium **(b)** acetate **(c)** hydrogen sulfate

**96. (a)** sodium oxide **(b)** nickel(III) oxide **(c)** lead(II) nitrate **(d)** barium hydroxide **(e)** potassium hydrogen carbonate

**98. (a)** $K_2S$ **(b)** $Zn_3P_2$ **(c)** $NiCl_2$ **(d)** $Mg(H_2PO_4)_2$ **(e)** $LiHCO_3$

## Chapter 4 Exercises

**4.1. (a)** $P_4(s) + 6Cl_2(g) \rightarrow 4PCl_3(l)$
**(b)** $3PbO(s) + 2NH_3(g) \rightarrow 3Pb(s) + N_2(g) + 3H_2O(l)$
**(c)** $P_4O_{10}(s) + 6H_2O(l) \rightarrow 4H_3PO_4(aq)$
**(d)** $3Mn(s) + 2CrCl_3(aq) \rightarrow 3MnCl_2(aq) + 2Cr(s)$
**(e)** $C_2H_2(g) + \frac{5}{2}O_2(g) \rightarrow 2CO_2(g) + H_2O(l)$
or $2C_2H_2(g) + 5O_2(g) \rightarrow 4CO_2(g) + 2H_2O(l)$
**(f)** $3Co(NO_3)_2(aq) + 2Na_3PO_4(aq) \rightarrow Co_3(PO_4)_2(s) + 6NaNO_3(aq)$
**(g)** $2CH_3NH_2(g) + \frac{9}{2}O_2(g) \rightarrow 2CO_2(g) + 5H_2O(l) + N_2(g)$
or $4CH_3NH_2(g) + 9O_2(g) \rightarrow 4CO_2(g) + 10H_2O(l) + 2N_2(g)$
**(h)** $2FeS(s) + \frac{9}{2}O_2(g) + 2H_2O(l) \rightarrow Fe_2O_3(s) + 2H_2SO_4(aq)$
or $4FeS(s) + 9O_2(g) + 4H_2O(l) \rightarrow 2Fe_2O_3(s) + 4H_2SO_4(aq)$

**4.2. (a)** soluble **(b)** insoluble **(c)** soluble **(d)** insoluble **(e)** insoluble

**4.3. (a)** $3CaCl_2(aq) + 2Na_3PO_4(aq) \rightarrow Ca_3(PO_4)_2(s) + 6NaCl(aq)$
**(b)** $3KOH(aq) + Fe(NO_3)_3(aq) \rightarrow 3KNO_3(aq) + Fe(OH)_3(s)$
**(c)** $NaC_2H_3O_2(aq) + CaSO_4(aq)$ No Reaction
**(d)** $K_2SO_4(aq) + Pb(NO_3)_2(aq) \rightarrow 2KNO_3(aq) + PbSO_4(s)$

## Chapter 4 Review Questions

**1.** $H_2$, $N_2$, $O_2$, $F_2$, $Cl_2$, $Br_2$, $I_2$

**2. (a)** ionic **(b)** covalent **(c)** ionic **(d)** covalent

**3.** Water is composed of $H_2O$ molecules that are attracted to each other due to the attraction between partially positive hydrogen atoms and the partially negative oxygen atoms of other molecules. See Figure 3.13. Each water molecule is moving constantly, breaking the attractions to some molecules, and making new attractions to other molecules. See Figure 3.14.

**4. (a)** $NH_3$ **(b)** $CH_4$ **(c)** $C_3H_8$ **(d)** $H_2O$

**5. (a)** $NO_2$ **(b)** $CBr_4$ **(c)** $Br_2O$ **(d)** NO

**6. (a)** LiF **(b)** $Pb(OH)_2$ **(c)** $K_2O$ **(d)** $Na_2CO_3$ **(e)** $CrCl_3$ **(f)** $Na_2HPO_4$

## Chapter 4 Key Ideas

**7.** converted into **9.** shorthand description **11.** continuous, delta, Δ **13.** subscripts **15.** same proportions **17.** solute, solvent **19.** minor, major **21.** organized, repeating **23.** left out

## Chapter 4 Problems

**25.** For each two particles of solid copper(I) hydrogen carbonate that are heated strongly, one particle of solid copper(I) carbonate, one molecule of liquid water, and one molecule of gaseous carbon dioxide are formed.

**27. (a)** $N_2(g) + 3H_2(g) \rightarrow 2NH_3(g)$
**(b)** $4Cl_2(g) + 2CH_4(g) + O_2(g) \rightarrow 8HCl(g) + 2CO(g)$
**(c)** $B_2O_3(s) + 6NaOH(aq) \rightarrow 2Na_3BO_3(aq) + 3H_2O(l)$
**(d)** $2Al(s) + 2H_3PO_4(aq) \rightarrow 2AlPO_4(s) + 3H_2(g)$
**(e)** $CO(g) + \frac{1}{2}O_2(g) \rightarrow CO_2(g)$
or $2CO(g) + O_2(g) \rightarrow 2CO_2(g)$
**(f)** $C_6H_{14}(l) + \frac{19}{2}O_2(g) \rightarrow 6CO_2(g) + 7H_2O(l)$
or $2C_6H_{14}(l) + 19O_2(g) \rightarrow 12CO_2(g) + 14H_2O(l)$
**(g)** $Sb_2S_3(s) + \frac{9}{2}O_2(g) \rightarrow Sb_2O_3(s) + 3SO_2(g)$
or $2Sb_2S_3(s) + 9O_2(g) \rightarrow 2Sb_2O_3(s) + 6SO_2(g)$
**(h)** $2Al(s) + 3CuSO_4(aq) \rightarrow Al_2(SO_4)_3(aq) + 3Cu(s)$
**(i)** $3P_2H_4(l) \rightarrow 4PH_3(g) + \frac{1}{2}P_4(s)$
or $6P_2H_4(l) \rightarrow 8PH_3(g) + P_4(s)$

**29.** $CS_2 + 3Cl_2 \rightarrow S_2Cl_2 + CCl_4$
$4CS_2 + 8S_2Cl_2 \rightarrow 3S_8 + 4CCl_4$
$S_8 + 4C \rightarrow 4CS_2$

**31.** $2HF + CHCl_3 \rightarrow CHClF_2 + 2HCl$

**33.** When solid lithium iodide is added to water, all of the ions at the surface of the solid can be viewed as vibrating back and forth between moving out into the water and returning to the solid surface. Sometimes when an ion vibrates out into the water, a water molecule collides with it, helping to break the ionic bond, and pushing it out into the solution. Water molecules move into the gap between the ion in solution and the solid and shield the ion from the attraction to the solid. The ions are kept stable and held in solution by attractions between them and the polar water molecules. The negatively charged oxygen ends of water molecules surround the lithium ions, and the positively charged hydrogen ends of water molecules surround the iodide ions. (See Figures 4.4 and 4.5 with $Li^+$ in the place of $Na^+$ and $I^-$ in the place of $Cl^-$.)

**37.** In a solution of a solid in a liquid, the solid is generally considered the solute, and the liquid is the solvent. Therefore, camphor is the solute in this solution, and ethanol is the solvent.

**40.** At the instant that the solution of silver nitrate is added to the aqueous sodium bromide, there are four different ions in solution surrounded by water molecules, $Ag^+$, $NO_3^-$, $Na^+$, and $Br^-$. The oxygen ends of the water molecules surround the silver and sodium cations, and the hydrogen ends of water molecules surround the nitrate and bromide anions. When silver ions and bromide ions collide, they stay together long enough for other silver ions and bromide ions to collide with them, forming clusters of ions that precipitate from the solution. The sodium and nitrate ions are unchanged in the reaction. They were separate and surrounded by water molecules at the beginning of the reaction, and they are still separate and surrounded by water molecules at the end of the reaction. See Figures 4-7 to 4-9 with silver ions in the place of calcium ions and bromide ions in the place of carbonate ions.

**42. (a)** soluble **(b)** soluble **(c)** insoluble **(d)** insoluble

**44. (a)** insoluble **(b)** soluble **(c)** soluble **(d)** insoluble

**46. (a)** $Co(NO_3)_2(aq) + Na_2CO_3(aq) \rightarrow CoCO_3(s) + 2NaNO_3(aq)$
**(b)** $2KI(aq) + Pb(C_2H_3O_2)_2(aq) \rightarrow 2KC_2H_3O_2(aq) + PbI_2(s)$
**(c)** $CuSO_4(aq) + LiNO_3(aq)$ No reaction
**(d)** $3Ni(NO_3)_2(aq) + 2Na_3PO_4(aq) \rightarrow Ni_3(PO_4)_2(s) + 6NaNO_3(aq)$
**(e)** $K_2SO_4(aq) + Ba(NO_3)_2(aq) \rightarrow 2KNO_3(aq) + BaSO_4(s)$

**48.** $Al^{3+}(aq) + PO_4^{3-}(aq) \rightarrow AlPO_4(s)$

**50.** $Cd(C_2H_3O_2)_2(aq) + 2NaOH(aq) \rightarrow Cd(OH)_2(s) + 2NaC_2H_3O_2(aq)$

**52. (a)** $SiCl_4 + 2H_2O \rightarrow SiO_2 + 4HCl$
**(b)** $2H_3BO_3 \rightarrow B_2O_3 + 3H_2O$
**(c)** $I_2 + 3Cl_2 \rightarrow 2ICl_3$
**(d)** $2Al_2O_3 + 3C \rightarrow 4Al + 3CO_2$

**54. (a)** $4NH_3 + Cl_2 \rightarrow N_2H_4 + 2NH_4Cl$
**(b)** $Cu + 2AgNO_3 \rightarrow Cu(NO_3)_2 + 2Ag$
**(c)** $Sb_2S_3 + 6HNO_3 \rightarrow 2Sb(NO_3)_3 + 3H_2S$
**(d)** $Al_2O_3 + 3Cl_2 + 3C \rightarrow 2AlCl_3 + 3CO$

**56.** $2Ca_3(PO_4)_2 + 6SiO_2 + 10C \rightarrow P_4 + 10CO + 6CaSiO_3$
$P_4 + 5O_2 + 6H_2O \rightarrow 4H_3PO_4$

**58.** **(a)** soluble **(b)** soluble **(c)** insoluble **(d)** insoluble
**60.** **(a)** $NaCl(aq) + Al(NO_3)_3(aq)$ No reaction
**(b)** $Ni(NO_3)_2(aq) + 2NaOH(aq) \rightarrow Ni(OH)_2(s) + 2NaNO_3(aq)$
**(c)** $3MnCl_2(aq) + 2Na_3PO_4(aq) \rightarrow Mn_3(PO_4)_2(s) + 6NaCl(aq)$
**(d)** $Zn(C_2H_3O_2)_2(aq) + Na_2CO_3(aq) \rightarrow ZnCO_3(s) + 2NaC_2H_3O_2(aq)$
**62.** $H_2 + Cl_2 \rightarrow 2HCl$
**64.** $Al_2O_3 + 3H_2SO_4 \rightarrow Al_2(SO_4)_3 + 3H_2O$
**66.** $CaF_2 + H_2SO_4 \rightarrow 2HF + CaSO_4$
**68.** $Na_2CO_3 + Ca(OH)_2 \rightarrow 2NaOH + CaCO_3$
**71.** $CaCO_3 + 2NaCl \rightarrow Na_2CO_3 + CaCl_2$
**73.** $NH_3 + 2O_2 \rightarrow HNO_3 + H_2O$
**75.** $C_3H_8(g) + 6H_2O(g) \rightarrow 3CO_2(g) + 10H_2(g)$
**77.** **(a)** $C + \frac{1}{2}O_2 \rightarrow CO$ or $2C + O_2 \rightarrow 2CO$
**(b)** $Fe_2O_3 + 3CO \rightarrow 2Fe + 3CO_2$
**(c)** $2CO \rightarrow C + CO_2$
**79.** Add a water solution of an ionic compound that forms an insoluble substance with aluminum ions and a soluble substance with sodium ions. For example, a potassium carbonate solution would precipitate the aluminum ions as $Al_2(CO_3)_3$.
**81.** $FeCl_3(aq) + 3AgNO_3(aq) \rightarrow Fe(NO_3)_3(aq) + 3AgCl(s)$
**83.** $2C_5H_{11}NSO_2(s) + {}^{31}\!/_{2}O_2(g)$
$\rightarrow 10CO_2(g) + 11H_2O(l) + 2SO_2(g) + N_2(g)$
or $4C_5H_{11}NSO_2(s) + 31O_2(g)$
$\rightarrow 20CO_2(g) + 22H_2O(l) + 4SO_2(g) + 2N_2(g)$

## Chapter 5 Exercises

**5.1.** **(a)** $HF(aq)$ **(b)** $H_3PO_4$
**5.2.** **(a)** hydriodic acid **(b)** acetic acid
**5.3.** **(a)** aluminum fluoride **(b)** phosphorus trifluoride **(c)** phosphoric acid **(d)** calcium sulfate **(e)** calcium hydrogen sulfate **(f)** copper(II) chloride **(g)** ammonium fluoride **(h)** hydrochloric acid **(i)** ammonium phosphate
**5.4.** **(a)** $NH_4NO_3$ **(b)** $HC_2H_3O_2$ **(c)** $NaHSO_4$ **(d)** KBr **(e)** $MgHPO_4$ **(f)** $HF(aq)$ **(g)** $P_2O_4$ **(h)** $Al_2(CO_3)_3$ **(i)** $H_2SO_4$
**5.5.** **(a)** strong acid **(b)** strong base **(c)** weak base **(d)** weak acid
**5.6.** **(a)** $HCl(aq) + NaOH(aq) \rightarrow H_2O(l) + NaCl(aq)$
**(b)** $HF(aq) + LiOH(aq) \rightarrow H_2O(l) + LiF(aq)$
**(c)** $H_3PO_4(aq) + 3LiOH(aq) \rightarrow 3H_2O(l) + Li_3PO_4(aq)$
**(d)** $Fe(OH)_3(s) + 3HNO_3(aq) \rightarrow Fe(NO_3)_3(aq) + 3H_2O(l)$
**5.7.** $Na_2CO_3(aq) + 2HBr(aq) \rightarrow 2NaBr(aq) + H_2O(l) + CO_2(g)$
**5.8.** **(a)** $HNO_2$ **(b)** $H_2CO_3$ **(c)** $H_3O^+$ **(d)** $HPO_4^{2-}$
**5.9.** **(a)** $HC_2O_4^-$ **(b)** $BrO_4^-$ **(c)** $NH_2^-$ **(d)** $HPO_4^{2-}$
**5.10.** **(a)** $HNO_2$ (B/L acid), NaBrO (B/L base)
**(b)** $HNO_2$ (B/L acid), $H_2AsO_4^-$ (B/L base)
**(c)** $H_2AsO_4^-$ (B/L acid), $OH^-$ (B/L base)

## Chapter 5 Review Questions

**1.** **(a)** aqueous = Water solutions are called aqueous solutions.
**(b)** spectator ion = Ions that are important for delivering other ions into solution to react, but do not actively participate in the reaction themselves are called spectator ions.
**(c)** double-displacement reaction = A chemical reaction that has the following form is called a double-displacement reaction.

$$AB + CD \rightarrow AD + CB$$

**(d)** net ionic equation = A net ionic equation is a chemical equation for which the spectator ions have been eliminated leaving only the substances actively involved in the reaction.
**2.** **(a)** carbonate **(b)** hydrogen carbonate
**3.** **(a)** $H_2PO_4^-$ **(b)** $C_2H_3O_2^-$
**4.** **(a)** Ionic **(b)** Not ionic **(c)** Ionic **(d)** Ionic **(e)** Not ionic
**5.** **(a)** potassium bromide **(b)** copper(II) nitrate **(c)** ammonium hydrogen phosphate
**6.** **(a)** $Ni(OH)_2$ **(b)** $NH_4Cl$ **(c)** $Ca(HCO_3)_2$
**7.** **(a)** insoluble **(b)** insoluble **(c)** soluble **(d)** soluble
**8.** When solid sodium hydroxide, NaOH, is added to water, all of the sodium ions, $Na^+$, and hydroxide ions, $OH^-$, at the surface of the solid can be viewed as shifting back and forth between moving out into the water and returning to the solid surface. Sometimes when an ion moves out into the water, a water molecule collides with it, helping to break the ionic bond, and pushing it out into the solution. Water molecules move into the gap between the ion in solution and the solid and shield the ion from the attraction to the solid. The ions are kept stable and held in solution by attractions between them and the polar water molecules. The negatively charged oxygen ends of water molecules surround the sodium ions, and the positively charged hydrogen ends of water molecules surround the hydroxide ions. (See Figures 4.4 and 4.5 with $OH^-$ in the place of $Cl^-$.)
**9.** $3ZnCl_2(aq) + 2Na_3PO_4(aq) \rightarrow Zn_3(PO_4)_2(s) + 6NaCl(aq)$

## Chapter 5 Key Ideas

**10.** sour **12.** acidic **14.** $H_aX_bO_c$ **16.** nearly one, significantly less than one **18.** weak **20.** hydro, :ic, acid **22.** hydroxide ions, $OH^-$ **24.** hydroxides **26.** fewer **28.** greater than 7, higher **30.** neutralize **32.** double-displacement **34.** donor, acceptor, transfer **36.** removed **38.** Brønsted-Lowry, Arrhenius

## Chapter 5 Problems

**39.** When $HNO_3$ molecules dissolve in water, each $HNO_3$ molecule donates a proton, $H^+$, to water forming hydronium ion, $H_3O^+$, and nitrate ion, $NO_3^-$. This reaction goes to completion, and the solution of the $HNO_3$ contains essentially no uncharged acid molecules. Once the nitrate ion and the hydronium ion are formed, the negatively charged oxygen atoms of the water molecules surround the hydronium ion and the positively charged hydrogen atoms of the water molecules surround the nitrate ion. Figure 5.12 shows you how you can picture this solution.
**41.** Each sulfuric acid molecule loses its first hydrogen ion completely.

$$H_2SO_4(aq) + H_2O(l) \rightarrow H_3O^+(aq) + HSO_4^-(aq)$$

The second hydrogen ion is not lost completely.

$$HSO_4^-(aq) + H_2O(l) \rightleftharpoons H_3O^+(aq) + SO_4^{2-}(aq)$$

In a typical solution of sulfuric acid, for each 100 sulfuric acid molecules added to water, the solution contains about 101 hydronium ions, $H_3O^+$, 99 hydrogen sulfate ions, $HSO_4^-$, and 1 sulfate ion, $SO_4^{2-}$.
**43.** A weak acid is a substance that is incompletely ionized in water due to a reversible reaction with water that forms hydronium ion, $H_3O^+$. A strong acid is a substance that is completely ionized in water due to a completion reaction with water that forms hydronium ions, $H_3O^+$.
**45.** **(a)** weak **(b)** strong **(c)** weak
**47.** **(a)** weak **(b)** weak **(c)** weak
**49.** **(a)** $HNO_2(aq) + H_2O(l) \rightleftharpoons H_3O^+(aq) + NO_2^-(aq)$
**(b)** $HBr(aq) + H_2O(l) \rightarrow H_3O^+(aq) + Br^-(aq)$
**51.** **(a)** $HNO_3$, nitric acid **(b)** $H_2CO_3$, carbonic acid **(c)** $H_3PO_4$, phosphoric acid
**53.** **(a)** oxyacid, $H_3PO_4$ **(b)** ionic compound with polyatomic ion, $NH_4Br$ **(c)** binary covalent compound, $P_2I_4$
**(d)** ionic compound with polyatomic ion, $LiHSO_4$
**(e)** binary acid, $HCl(aq)$ **(f)** binary ionic compound, $Mg_3N_2$
**(g)** oxyacid, $HC_2H_3O_2$ **(h)** ionic compound with polyatomic ion, $PbHPO_4$
**55.** **(a)** binary acid, hydrobromic acid **(b)** binary covalent compound, chlorine trifluoride **(c)** binary ionic compound, calcium bromide **(d)** ionic compound with polyatomic ion, iron(III) sulfate **(e)** oxyacid, carbonic acid **(f)** ionic compound with polyatomic ion, ammonium sulfate **(g)** ionic compound with polyatomic ion, potassium hydrogen sulfate
**58.** **(a)** weak acid **(b)** strong base **(c)** weak acid **(d)** weak base **(e)** weak base **(f)** weak acid **(g)** strong acid **(h)** weak acid
**60.** **(a)** pH $< 7$ so acidic **(b)** pH $> 7$ so basic
**(c)** pH about 7 so essentially neutral (or more specifically, very slightly acidic)
**62.** The lower the pH is, the more acidic the solution. Carbonated water is more acidic than milk.
**64.** **(a)** acid **(b)** acid **(c)** base

**66.** Because hydrochloric acid, HCl(*aq*), is an acid, it reacts with water to form hydronium ions, $H_3O^+$, and chloride ions, $Cl^-$. Because it is a strong acid, the reaction is a completion reaction, leaving only $H_3O^+$ and $Cl^-$ in solution with no HCl remaining.

$$HCl(aq) + H_2O(l) \rightarrow H_3O^+(aq) + Cl^-(aq)$$

$$\text{or } HCl(aq) \rightarrow H^+(aq) + Cl^-(aq)$$

Because NaOH is a water-soluble ionic compound, it separates into sodium ions, $Na^+$, and hydroxide ions, $OH^-$, when it dissolves in water.

Thus, at the instant that the two solutions are mixed, the solution contains water molecules, hydronium ions, $H_3O^+$, chloride ions, $Cl^-$, sodium ions, $Na^+$, and hydroxide ions, $OH^-$.

When the hydronium ions collide with the hydroxide ions, they react to form water. If an equivalent amount of acid and base are added together, the $H_3O^+$ and the $OH^-$ will be completely reacted.

$$H_3O^+(aq) + OH^-(aq) \rightarrow 2H_2O(l)$$

$$\text{or } H^+(aq) + OH^-(aq) \rightarrow H_2O(l)$$

The sodium ions and chloride ions remain in solution with the water molecules.

**68.** A solution with an insoluble ionic compound, like $Ni(OH)_2$, at the bottom has a constant escape of ions from the solid into the solution balanced by constant return of ions to the solid due to collisions of ions with the surface of the solid. Thus, even though $Ni(OH)_2$ has very low solubility in water, there are always a few $Ni^{2+}$ and $OH^-$ ions in solution.

If a nitric acid solution is added to water with solid $Ni(OH)_2$ at the bottom, a neutralization reaction takes place. Because the nitric acid is a strong acid, it is ionized in solution, so the nitric acid solution contains hydronium ions, $H_3O^+$, and nitrate ions, $NO_3^-$. The hydronium ions will react with the basic hydroxide ions in solution to form water molecules.

Because the hydronium ions remove the hydroxide anions from solution, the return of ions to the solid is stopped. The nickel(II) cations cannot return to the solid unless they are accompanied by anions to balance their charge. The escape of ions from the surface of the solid continues. When hydroxide ions escape, they react with the hydronium ions and do not return to the solid. Thus, there is a steady movement of ions into solution, and the solid that contains the basic anion dissolves.

The complete equation for this reaction is below.

$$Ni(OH)_2(s) + 2HNO_3(aq) \rightarrow Ni(NO_3)_2(aq) + 2H_2O(l)$$

**70.** Because hydrochloric acid, HCl(*aq*), is an acid, it reacts with water to form hydronium ions, $H_3O^+$, and chloride ions, $Cl^-$. Because it is a strong acid, the reaction is a completion reaction, leaving only $H_3O^+$ and $Cl^-$ in a HCl(*aq*) solution with no HCl remaining.

$$HCl(aq) + H_2O(l) \rightarrow H_3O^+(aq) + Cl^-(aq)$$

$$\text{or } HCl(aq) \rightarrow H^+(aq) + Cl^-(aq)$$

Because $K_2CO_3$ is a water-soluble ionic compound, it separates into potassium ions, $K^+$, and carbonate ions, $CO_3^{2-}$, when it dissolves in water. The carbonate ions are weakly basic, so they react with water in a reversible reaction to form hydrogen carbonate, $HCO_3^-$ and hydroxide, $OH^-$.

$$CO_3^{2-}(aq) + H_2O(l) \rightleftharpoons HCO_3^-(aq) + OH^-(aq)$$

Thus, at the instant that the two solutions are mixed, the solution contains water molecules, hydronium ions, $H_3O^+$, chloride ions, $Cl^-$, potassium ions, $K^+$, carbonate ions, $CO_3^{2-}$, hydrogen carbonate ions, $HCO_3^-$, and hydroxide ions, $OH^-$.

The hydronium ions react with hydroxide ions, carbonate ions, and hydrogen carbonate ions. When the hydronium ions collide with the hydroxide ions, they react to form water. When the hydronium ions collide with the carbonate ions or hydrogen carbonate ions, they react to form carbonic acid, $H_2CO_3$. The carbonate with its minus two charge requires two $H^+$ ions to yield a neutral compound, and the hydrogen carbonate requires one $H^+$ to neutralize its minus one charge.

$$2H_3O^+(aq) + CO_3^{2-}(aq) \rightarrow H_2CO_3(aq) + 2H_2O(l)$$

$$\text{or } 2H^+(aq) + CO_3^{2-}(aq) \rightarrow H_2CO_3(aq)$$

$$H_3O^+(aq) + HCO_3^-(aq) \rightarrow H_2CO_3(aq) + H_2O(l)$$

$$\text{or } H^+(aq) + HCO_3^-(aq) \rightarrow H_2CO_3(aq)$$

The carbonic acid is unstable in water and decomposes to form carbon dioxide gas and water.

$$H_2CO_3(aq) \rightarrow CO_2(g) + H_2O(l)$$

If an equivalent amount of acid and base are added together, the $H_3O^+$, $OH^-$, $CO_3^{2-}$, and $HCO_3^-$, will be completely reacted.

The potassium ions and chloride ions remain in solution with the water molecules.

**72. (a)** $HCl(aq) + LiOH(aq) \rightarrow H_2O(l) + LiCl(aq)$
**(b)** $H_2SO_4(aq) + 2NaOH(aq) \rightarrow 2H_2O(l) + Na_2SO_4(aq)$
**(c)** $KOH(aq) + HF(aq) \rightarrow KF(aq) + H_2O(l)$
**(d)** $Cd(OH)_2(s) + 2HCl(aq) \rightarrow CdCl_2(aq) + 2H_2O(l)$
**74.** $2HI(aq) + CaCO_3(s) \rightarrow H_2O(l) + CO_2(g) + CaI_2(aq)$
**76.** $2Fe(OH)_3(s) + 3H_2SO_4(aq) \rightarrow Fe_2(SO_4)_3(aq) + 6H_2O(l)$
$Fe_2(SO_4)_3(aq) + 6NaOH(aq) \rightarrow 2Fe(OH)_3(s) + 3Na_2SO_4(aq)$
**78. (a)** $HCl(aq) + NaOH(aq) \rightarrow H_2O(l) + NaCl(aq)$
**(b)** $H_2SO_4(aq) + 2LiOH(aq) \rightarrow 2H_2O(l) + Li_2SO_4(aq)$
**(c)** $2HCl(aq) + K_2CO_3(aq) \rightarrow H_2O(l) + CO_2(g) + 2KCl(aq)$
**81. (a)** $HIO_3$ **(b)** $H_2SO_3$ **(c)** $HPO_3^{2-}$ **(d)** $H_2$
**83. (a)** $ClO_4^-$ **(b)** $SO_3^{2-}$ **(c)** $H_2O$ **(d)** $H_2PO_2^-$
**85.** The same substance can donate an $H^+$ in one reaction (and act as a Brønsted–Lowry acid) and accept an $H^+$ in another reaction (and act as a Brønsted–Lowry base). For example, consider the following net ionic equations for the reaction of dihydrogen phosphate ion.

$$\underset{\text{B/L base}}{H_2PO_4^-(aq)} + \underset{\text{B/L acid}}{HCl(aq)} \rightarrow H_3PO_4(aq) + Cl^-(aq)$$

$$\underset{\text{B/L acid}}{H_2PO_4^-(aq)} + \underset{\text{B/L base}}{2OH^-(aq)} \rightarrow PO_4^{3-}(aq) + 2H_2O(l)$$

**87. (a)** $HC_2H_3O_2$ (B/L acid), NaCN (B/L base)
**(b)** HF (B/L acid), $H_2PO_3^-$ (B/L base)
**(c)** $H_2PO_3^-$ (B/L acid), $OH^-$ (B/L base)
**(d)** $H_3PO_3$ (B/L acid), NaOH (B/L base)
**89.** Conjugate base—$CH_3CH_2CH_2CO_2^-$
B/L Acid—$CH_3CH_2CH_2CO_2H$
B/L Base—$H_2O$
**91.** $HCl(aq) + HS^-(aq) \rightarrow Cl^-(aq) + H_2S(aq)$
$HS^-(aq) + OH^-(aq) \rightarrow S^{2-}(aq) + H_2O(l)$
**93. (a)** $HBr(aq) + NaOH(aq) \rightarrow H_2O(l) + NaBr(aq)$
**(b)** $H_2SO_3(aq) + 2LiOH(aq) \rightarrow 2H_2O(l) + Li_2SO_3(aq)$
**(c)** $KHCO_3(aq) + HF(aq) \rightarrow KF(aq) + H_2O(l) + CO_2(g)$
**(c)** $Al(OH)_3(s) + 3HNO_3(aq) \rightarrow Al(NO_3)_3(aq) + 3H_2O(l)$
**95. (a)** acidic **(b)** basic **(c)** very slightly basic (essentially neutral)
**97.** acidic
**99.** $CaCO_3(s) + 2H^+(aq) \rightarrow Ca^{2+}(aq) + CO_2(g) + H_2O(l)$
**100.** $H_2SO_4(aq) + Mg(OH)_2(s) \rightarrow 2H_2O(l) + MgSO_4(aq)$
**102.** $CH_3CH_2CO_2H(aq) + NaOH(aq)$
$\rightarrow NaCH_3CH_2CO_2(aq) + H_2O(l)$
**104.** $HO_2CCH_2CH(OH)CO_2H(aq) + 2NaOH(aq)$
$\rightarrow Na_2O_2CCH_2CH(OH)CO_2(aq) + 2H_2O(l)$
**106.** B/L Acid—$NaHSO_4(aq)$
B/L Base—$NaHS(aq)$
Conjugate acid-base pairs: $HS^-/H_2S$ and $HSO_4^-/SO_4^{2-}$
or $NaHS/H_2S$ and $NaHSO_4^-/Na_2SO_4$

## Chapter 6 Exercises

**6.1.** They are all redox reactions.

$$2\overset{0}{C}(s) + \overset{0}{O_2}(g) \rightarrow 2\overset{+2\,-2}{CO}(g)$$

C(*s*) is oxidized and is the reducing agent. $O_2$ is reduced and is the oxidizing agent.

$$\overset{+3\,-2}{Fe_2O_3}(s) + \overset{+2\,-2}{CO}(g) \rightarrow 2\overset{0}{Fe}(l) + 3\overset{+4\,-2}{CO_2}(g)$$

Each carbon atom in CO(*g*) is oxidized and CO(*g*) is the reducing agent. Each Fe atom in $Fe_2O_3$ is reduced and $Fe_2O_3$ is the oxidizing agent.

$$2\overset{+2\,-2}{CO}(g) \rightarrow \overset{0}{C}(\text{in iron}) + \overset{+4\,-2}{CO_2}(g)$$

Carbon atoms in CO are both oxidized and reduced, and CO is both the oxidizing agent and the reducing agent.
**6.2. (a)** $2C_4H_{10}(g) + 13O_2(g) \rightarrow 8CO_2(g) + 10H_2O(l)$
**(b)** $2C_3H_7OH(l) + 9O_2(g) \rightarrow 6CO_2(g) + 8H_2O(l)$
**(c)** $2C_4H_9SH(l) + 15O_2(g) \rightarrow 8CO_2(g) + 10H_2O(l) + 2SO_2(g)$
**6.3. (a)** decomposition **(b)** combustion **(c)** single-displacement **(d)** combination

## Chapter 6 Review Questions

**1. (a)** $Fe^{3+}$ and $Br^-$ **(b)** $Co^{2+}$ and $PO_4^{3-}$ **(c)** $Ag^+$ and $Cl^-$ **(d)** $NH_4^+$ and $SO_4^{2-}$
**2. (a)** molecular **(b)** ionic with polyatomic ion **(c)** binary ionic **(d)** molecular **(e)** molecular **(f)** molecular **(g)** ionic with polyatomic ion **(h)** molecular
**3. (a)** $C_8H_{18}(l) + {}^{25}\!/_2O_2(g) \rightarrow 8CO_2(g) + 9H_2O(l)$
or $2C_8H_{18}(l) + 25O_2(g) \rightarrow 16CO_2(g) + 18H_2O(l)$
**(b)** $4P_4(s) + 5S_8(s) \rightarrow 8P_2S_5(s)$

## Chapter 6 Key Ideas

**4.** loses **6.** rarely, oxidized, reduced, oxidation, reduction **8.** half-reactions **10.** oxidizing agent **12.** flow of electrons **14.** two or more **16.** heat, light **18.** water **20.** pure element **21.** voltaic cells, electrical energy, chemical energy **23.** electrical conductors **25.** reduction, positive electrode **27.** force, electrolysis

## Chapter 6 Problems

**30. (a)** Complete—Ionic bonds formed **(b)** Incomplete—Polar covalent bonds formed
**32. (a)** Incomplete—Polar covalent bonds formed
**(b)** Incomplete—Polar covalent bonds formed
**34.** $2Al \rightarrow 2Al^{3+} + 6e^-$ oxidation—loss of electrons
$3Br_2 + 6e^- \rightarrow 6Br^-$ reduction—gain of electrons
**36. (a)** S is zero. **(b)** S is −2. **(c)** Each Na is +1, and the S is −2. **(d)** The Fe is +2, and the S is −2.
**38. (a)** Each Sc is +3, and each O is −2. **(b)** The Rb is +1, and the H is −1. **(c)** N is zero. **(d)** H is +1. N is −3.
**40. (a)** H is +1. F is −1. C is +2. **(b)** H is +1. O is −1. **(c)** H is +1. O is −2. S is +6.
**42. (a)** H is +1. O is −2. P is +5. **(b)** Ni is +2. O is −2. S is +6. **(c)** O is −2. N is +3. **(d)** Mn is +2. O is −2. P is +1.

**45.** $\overset{0}{Cl_2}(g) + \overset{0}{H_2}(g) \rightarrow 2\overset{+1\,-1}{HCl}(g)$

Yes, it's redox.
H atoms in $H_2$ are oxidized, Cl atoms in $Cl_2$ are reduced, $Cl_2$ is the oxidizing agent, and $H_2$ is the reducing agent.

**47.** $\overset{0}{Mg}(s) + 2\overset{+1\,-2}{H_2O}(l) \rightarrow \overset{+2\,-2\,+1}{Mg(OH)_2}(aq) + \overset{0}{H_2}(g) + \text{heat}$

Yes, it's redox.
Mg atoms in Mg(*s*) are oxidized, H atoms in $H_2O$ are reduced, $H_2O$ is the oxidizing agent, and Mg is the reducing agent.

$$2\overset{0}{Mg}(s) + \overset{+4\,-2}{CO_2}(g) \rightarrow 2\overset{+2\,-2}{MgO}(s) + \overset{0}{C}(s) + \text{heat}$$

Yes, it's redox.
Mg atoms in Mg(*s*) are oxidized, C atoms in $CO_2$ are reduced, $CO_2$ is the oxidizing agent, and Mg is the reducing agent.

**49.** $2\overset{-2\,+1\,-2\,+1}{CH_3OH} + \overset{0}{O_2} \rightarrow 2\overset{0\,+1\,-2}{CH_2O} + 2\overset{+1\,-2}{H_2O}$

Yes, it's redox.
C atoms in $CH_3OH$ are oxidized, O atoms in $O_2$ are reduced, $O_2$ is the oxidizing agent, and $CH_3OH$ is the reducing agent.

**51.** $\overset{0}{Co}(s) + 2\overset{+1\,+5\,-2}{AgNO_3}(aq) \rightarrow \overset{+2\,+5\,-2}{Co(NO_3)_2}(aq) + 2\overset{0}{Ag}(s)$

Co in Co(*s*) is oxidized, and Co(*s*) is the reducing agent. Ag in $AgNO_3$ is reduced, and $AgNO_3$ is the oxidizing agent.

$$\overset{+5\,-2}{V_2O_5}(s) + 5\overset{0}{Ca}(l) \xrightarrow{\Delta} 2\overset{0}{V}(l) + 5\overset{+2\,-2}{CaO}(s)$$

V in $V_2O_5$ is reduced, and $V_2O_5$ is the oxidizing agent. Ca in Ca(*l*) is oxidized, and Ca(*l*) is the reducing agent.

$$\overset{+2\,+4\,-2}{CaCO_3}(aq) + \overset{+4\,-2}{SiO_2}(s) \rightarrow \overset{+2\,+4\,-2}{CaSiO_3}(s) + \overset{+4\,-2}{CO_2}(g) \quad \text{Not redox}$$

$$2\overset{+1\,-1}{NaH}(s) \xrightarrow{\Delta} 2\overset{0}{Na}(s) + \overset{0}{H_2}(g)$$

Na in NaH is reduced, and NaH (or $Na^+$ in NaH) is the oxidizing agent. H in NaH is oxidized, and NaH (or $H^-$ in NaH) is also the reducing agent.

$$5\overset{+3\,-2}{As_4O_6}(s) + 8\overset{+1\,+7\,-2}{KMnO_4}(aq) + 18\overset{+1\,-2}{H_2O}(l) + 52\overset{+1\,-1}{KCl}(aq) \rightarrow 20\overset{+1\,+5\,-2}{K_3AsO_4}(aq) + 8\overset{+2\,-1}{MnCl_2}(aq) + 36\overset{+1\,-1}{HCl}(aq)$$

As in $As_4O_6$ is oxidized, and $As_4O_6$ is the reducing agent. Mn in $KMnO_4$ is reduced, and $KMnO_4$ is the oxidizing agent.

**53.** ${}^1\!/_8\overset{0}{S_8}(s) + \overset{0}{O_2}(g) \rightarrow \overset{+4\,-2}{SO_2}(g)$

Yes, it's redox.
S atoms in $S_8$ are oxidized, O atoms in $O_2$ are reduced, $O_2$ is the oxidizing agent, and $S_8$ is the reducing agent.

$$\overset{+4\,-2}{SO_2} + {}^1\!/_2\overset{0}{O_2} \rightarrow \overset{+6\,-2}{SO_3}$$

Yes, it's redox.
S atoms in $SO_2$ are oxidized, O atoms in $O_2$ are reduced, $O_2$ is the oxidizing agent, and $SO_2$ is the reducing agent.

$$\overset{+6\,-2}{SO_3} + \overset{+1\,-2}{H_2O} \rightarrow \overset{+1\,+6\,-2}{H_2SO_4} \quad \text{Not redox}$$

**55. (a)** decomposition **(b)** single displacement **(c)** combustion **(d)** single displacement **(e)** combination and combustion
**57. (a)** combination and combustion **(b)** combustion **(c)** single displacement **(d)** combustion **(e)** decomposition
**59. (a)** $C_3H_8(g) + 5O_2(g) \rightarrow 3CO_2(g) + 4H_2O(l)$
**(b)** $C_4H_9OH(l) + 6O_2(g) \rightarrow 4CO_2(g) + 5H_2O(l)$
**(c)** $CH_3COSH(l) + {}^7\!/_2O_2(g) \rightarrow 2CO_2(g) + 2H_2O(l) + SO_2(g)$
or $2CH_3COSH(l) + 7O_2(g) \rightarrow 4CO_2(g) + 4H_2O(l) + 2SO_2(g)$
**61. (a)** Because nickel(II) nitrate is a water soluble ionic compound, the $Ni(NO_3)_2$ solution contains $Ni^{2+}$ ions surrounded by the negatively charged oxygen ends of water molecules and separate $NO_3^-$ ions surrounded by the positively charged hydrogen ends of water molecules. These ions move throughout the solution colliding with each other, with water molecules, and with the walls of their container. When solid magnesium is added to the solution, nickel ions begin to collide with the surface of the magnesium. When each $Ni^{2+}$ ion collides with an uncharged magnesium atom, two electrons are transferred from the magnesium atom to the nickel(II) ion. Magnesium ions go into solution, and uncharged nickel solid forms on the surface of the magnesium. See Figure 6.4. Picture $Ni^{2+}$ in the place of $Cu^{2+}$ and magnesium metal in the place of zinc metal. Because the magnesium atoms lose electrons and change their oxidation number from 0 to +2, they are oxidized and act as the reducing agent. The $Ni^{2+}$ ions gain electrons and decrease their oxidation number from +2 to 0, so they are reduced and act as the oxidizing agent. The half reaction equations and the net ionic equation for this reaction are below.

oxidation: $Mg(s) \rightarrow Mg^{2+}(aq) + 2e^-$
reduction: $Ni^{2+}(aq) + 2e^- \rightarrow Ni(s)$
Net Ionic Equation: $Mg(s) + Ni^{2+}(aq) \rightarrow Mg^{2+}(aq) + Ni(s)$

**62.** **(a)** $\mathrm{Mn}(s) + \mathrm{Pb(NO_3)_2}(aq) \rightarrow \mathrm{Pb}(s) + \mathrm{Mn(NO_3)_2}\,(aq)$
See Figure 6.5, substituting Mn for Zn, $Mn^{2+}$ for $Zn^{2+}$ and $NO_3^-$ for $SO_4^{2-}$ in the left half-cell and substituting Pb for Cu, $Pb^{2+}$ for $Cu^{2+}$ and $NO_3^-$ for $SO_4^{2-}$ in the right half-cell. The voltaic cell that utilizes the redox reaction between manganese metal and lead(II) ions is composed of two half-cells. The first half-cell consists of a strip of manganese metal in a solution of manganese(II) nitrate. The second half-cell consists of a strip of lead metal in a solution of lead(II) nitrate. In the $Mn/Mn^{2+}$ half-cell, manganese atoms lose two electrons and are converted to manganese ions. The electrons pass through the wire to the $Pb/Pb^{2+}$ half-cell where $Pb^{2+}$ ions gain the two electrons to form uncharged lead atoms. Mn is oxidized to $Mn^{2+}$ at the manganese electrode, so this electrode is the anode. $Pb^{2+}$ ions are reduced to uncharged lead atoms at the lead strip, so metallic lead is the cathode.

**63.** $\overset{0}{\mathrm{Zn}}(s) + 2\overset{+4}{\mathrm{Mn}}\overset{-2}{\mathrm{O}}_2(s) + 2\overset{-3}{\mathrm{N}}\overset{+1}{\mathrm{H}}_4{}^+(aq)$

$\rightarrow \overset{+2}{\mathrm{Zn}}{}^{2+}(aq) + \overset{+3}{\mathrm{Mn}}_2\overset{-2}{\mathrm{O}}_3(s) + 2\overset{-3}{\mathrm{N}}\overset{+1}{\mathrm{H}}_3(aq) + \overset{+1}{\mathrm{H}}_2\overset{-2}{\mathrm{O}}(l)$

Zn is oxidized and is the reducing agent. Mn in $MnO_2$ is reduced, so $MnO_2$ is the oxidizing agent.

**65.** $\overset{0}{\mathrm{Pb}}(s) + \overset{+4}{\mathrm{Pb}}\overset{-2}{\mathrm{O}}_2(s) + 2\overset{+1}{\mathrm{H}}\overset{+6}{\mathrm{S}}\overset{-2}{\mathrm{O}}_4{}^-(aq) + 2\overset{+1}{\mathrm{H}}_3\overset{-2}{\mathrm{O}}{}^+(aq) \rightarrow 2\overset{+2}{\mathrm{Pb}}\overset{+6}{\mathrm{S}}\overset{-2}{\mathrm{O}}_4(s) + 4\overset{+1}{\mathrm{H}}_2\overset{-2}{\mathrm{O}}(l)$

Pb is oxidized and is the reducing agent. Pb in $PbO_2$ is reduced, so $PbO_2$ is the oxidizing agent.

**67.** $2\overset{+1}{\mathrm{Na}}\overset{+1}{\mathrm{H}}\overset{+4}{\mathrm{C}}\overset{-2}{\mathrm{O}}_3(s) \xrightarrow{\Delta} \overset{+4}{\mathrm{C}}\overset{-2}{\mathrm{O}}_2(g) + \overset{+1}{\mathrm{Na}}_2\overset{+4}{\mathrm{C}}\overset{-2}{\mathrm{O}}_3(s) + \overset{+1}{\mathrm{H}}_2\overset{-2}{\mathrm{O}}(g)$

Because none of the atoms change their oxidation number, this is not redox.

**69.** $\overset{+2}{\mathrm{Hg}}\overset{-2}{\mathrm{O}}(s) + \overset{0}{\mathrm{Zn}}(s) \rightarrow \overset{+2}{\mathrm{Zn}}\overset{-2}{\mathrm{O}}(s) + \overset{0}{\mathrm{Hg}}(l)$

Yes, it's redox. Zn atoms in Zn(s) are oxidized, Hg atoms in HgO are reduced, HgO is the oxidizing agent, and Zn is the reducing agent.

**71.** $\overset{+2}{\mathrm{Mn}}(\overset{-2}{\mathrm{O}}\overset{+1}{\mathrm{H}})_2(s) + \overset{+1}{\mathrm{H}}_3\overset{+5}{\mathrm{P}}\overset{-2}{\mathrm{O}}_4(aq) \rightarrow \overset{+2}{\mathrm{Mn}}_3(\overset{+5}{\mathrm{P}}\overset{-2}{\mathrm{O}}_4)_2(s) + 3\overset{+1}{\mathrm{H}}_2\overset{-2}{\mathrm{O}}(l)$ Not redox

**73.** $\overset{0}{\mathrm{Xe}} + \overset{0}{\mathrm{F}}_2 \rightarrow \overset{+6}{\mathrm{Xe}}\overset{-1}{\mathrm{F}}_6$

Yes, it's redox. Xe atoms are oxidized, F atoms in $F_2$ are reduced, $F_2$ is the oxidizing agent, and Xe is the reducing agent.

$\overset{+6}{\mathrm{Xe}}\overset{-1}{\mathrm{F}}_6 + \overset{+1}{\mathrm{H}}_2\overset{-2}{\mathrm{O}} \rightarrow \overset{+6}{\mathrm{Xe}}\overset{-2}{\mathrm{O}}\overset{-1}{\mathrm{F}}_4 + 2\overset{+1}{\mathrm{H}}\overset{-1}{\mathrm{F}}$

None of the atoms change their oxidation number, so it's not redox.

$\overset{+6}{\mathrm{Xe}}\overset{-1}{\mathrm{F}}_6 + \overset{-2}{\mathrm{O}}\overset{+5}{\mathrm{P}}\overset{-1}{\mathrm{F}}_3 \rightarrow \overset{+6}{\mathrm{Xe}}\overset{-2}{\mathrm{O}}\overset{-1}{\mathrm{F}}_4 + \overset{+5}{\mathrm{P}}\overset{-1}{\mathrm{F}}_5$

None of the atoms change their oxidation number, so it's not redox.

**75.** $\overset{+1}{\mathrm{Na}}\overset{+5}{\mathrm{Br}}\overset{-2}{\mathrm{O}}_3 + \overset{+2}{\mathrm{Xe}}\overset{-1}{\mathrm{F}}_2 + \overset{+1}{\mathrm{H}}_2\overset{-2}{\mathrm{O}} \rightarrow \overset{+1}{\mathrm{Na}}\overset{+7}{\mathrm{Br}}\overset{-2}{\mathrm{O}}_4 + 2\overset{+1}{\mathrm{H}}\overset{-1}{\mathrm{F}} + \overset{0}{\mathrm{Xe}}$

Yes, it's redox. Br atoms in $NaBrO_3$ are oxidized, Xe atoms in $XeF_2$ are reduced, $XeF_2$ is the oxidizing agent, and $NaBrO_3$ is the reducing agent.

$\overset{+1}{\mathrm{Na}}\overset{+5}{\mathrm{Br}}\overset{-2}{\mathrm{O}}_3 + \overset{0}{\mathrm{F}}_2 + 2\overset{+1}{\mathrm{Na}}\overset{-2}{\mathrm{O}}\overset{+1}{\mathrm{H}} \rightarrow \overset{+1}{\mathrm{Na}}\overset{+7}{\mathrm{Br}}\overset{-2}{\mathrm{O}}_4 + 2\overset{+1}{\mathrm{Na}}\overset{-1}{\mathrm{F}} + \overset{+1}{\mathrm{H}}_2\overset{-2}{\mathrm{O}}$

Yes, it's redox. Br atoms in $NaBrO_3$ are oxidized, F atoms in $F_2$ are reduced, $F_2$ is the oxidizing agent, and $NaBrO_3$ is the reducing agent.

**77.** $\overset{+1}{\mathrm{I}}\overset{-1}{\mathrm{F}}(g) + 2\overset{0}{\mathrm{F}}_2(g) \rightarrow \overset{+5}{\mathrm{I}}\overset{-1}{\mathrm{F}}_5$

Yes, it's redox. I atoms in IF are oxidized, F atoms in $F_2$ are reduced, $F_2$ is the oxidizing agent, and IF is the reducing agent.

$\overset{+1}{\mathrm{Br}}\overset{-1}{\mathrm{F}}(g) + 2\overset{0}{\mathrm{F}}_2(g) \rightarrow \overset{+5}{\mathrm{Br}}\overset{-1}{\mathrm{F}}_5(g)$

Yes, it's redox. Br atoms in BrF are oxidized, F atoms in $F_2$ are reduced, $F_2$ is the oxidizing agent, and BrF is the reducing agent.

$\overset{0}{\mathrm{Cl}}_2(g) + 3\overset{0}{\mathrm{F}}_2(g) \rightarrow 2\overset{+3}{\mathrm{Cl}}\overset{-1}{\mathrm{F}}_3(g)$

Yes, it's redox. Cl atoms in $Cl_2$ are oxidized, F atoms in $F_2$ are reduced, $F_2$ is the oxidizing agent, and $Cl_2$ is the reducing agent.

**79.** $4\overset{+1}{\mathrm{Na}}\overset{-1}{\mathrm{Cl}} + 2\overset{+4}{\mathrm{S}}\overset{-2}{\mathrm{O}}_2 + 2\overset{+1}{\mathrm{H}}_2\overset{-2}{\mathrm{O}} + \overset{0}{\mathrm{O}}_2 \rightarrow 2\overset{+1}{\mathrm{Na}}_2\overset{+6}{\mathrm{S}}\overset{-2}{\mathrm{O}}_4 + 4\overset{+1}{\mathrm{H}}\overset{-1}{\mathrm{Cl}}$

Yes, it's redox. S atoms in $SO_2$ are oxidized, O atoms in $O_2$ are reduced, $O_2$ is the oxidizing agent, and $SO_2$ is the reducing agent.

**81.** $3\overset{+1}{\mathrm{H}}_2\overset{-2}{\mathrm{S}} + 2\overset{0}{\mathrm{O}}_2 \rightarrow 3\overset{+4}{\mathrm{S}}\overset{-2}{\mathrm{O}}_2 + 3\overset{+1}{\mathrm{H}}_2\overset{-2}{\mathrm{O}}$

Yes, it's redox. S atoms in $H_2S$ are oxidized, O atoms in $O_2$ are reduced, $O_2$ is the oxidizing agent, and $H_2S$ is the reducing agent.

$\overset{+4}{\mathrm{S}}\overset{-2}{\mathrm{O}}_2 + 2\overset{+1}{\mathrm{H}}_2\overset{-2}{\mathrm{S}} \rightarrow 3\overset{0}{\mathrm{S}} + 3\overset{+1}{\mathrm{H}}_2\overset{-2}{\mathrm{O}}$

Yes, it's redox. S atoms in $H_2S$ are oxidized, S atoms in $SO_2$ are reduced, $SO_2$ is the oxidizing agent, and $H_2S$ is the reducing agent.

**83.** $\overset{+2}{\mathrm{Pb}}\overset{-1}{\mathrm{Br}}_2 \xrightarrow{\text{sunlight}} \overset{0}{\mathrm{Pb}} + \overset{0}{\mathrm{Br}}_2$

Yes, it's redox. Br atoms in $PbBr_2$ are oxidized, and Pb atoms in $PbBr_2$ are reduced.

**85.** $\overset{+2}{\mathrm{Ca}}\overset{+4}{\mathrm{C}}\overset{-2}{\mathrm{O}}_3 \rightarrow \overset{+2}{\mathrm{Ca}}\overset{-2}{\mathrm{O}} + \overset{+4}{\mathrm{C}}\overset{-2}{\mathrm{O}}_2$

$\overset{+2}{\mathrm{Ca}}\overset{-2}{\mathrm{O}} + \overset{+1}{\mathrm{H}}_2\overset{-2}{\mathrm{O}} \rightarrow \overset{+2}{\mathrm{Ca}}(\overset{-2}{\mathrm{O}}\overset{+1}{\mathrm{H}})_2$

$\overset{+4}{\mathrm{S}}\overset{-2}{\mathrm{O}}_2 + \overset{+1}{\mathrm{H}}_2\overset{-2}{\mathrm{O}} \rightarrow \overset{+1}{\mathrm{H}}_2\overset{+4}{\mathrm{S}}\overset{-2}{\mathrm{O}}_3$

$\overset{+2}{\mathrm{Ca}}(\overset{-2}{\mathrm{O}}\overset{+1}{\mathrm{H}})_2 + \overset{+1}{\mathrm{H}}_2\overset{+4}{\mathrm{S}}\overset{-2}{\mathrm{O}}_3 \rightarrow \overset{+2}{\mathrm{Ca}}\overset{+4}{\mathrm{S}}\overset{-2}{\mathrm{O}}_3 + 2\overset{+1}{\mathrm{H}}_2\overset{-2}{\mathrm{O}}$

Because none of the atoms change their oxidation number, none of these reactions are redox reactions.

**87.** $10\overset{0}{\mathrm{Al}}(s) + 6\overset{-3}{\mathrm{N}}\overset{+1}{\mathrm{H}}_4\overset{+7}{\mathrm{Cl}}\overset{-2}{\mathrm{O}}_4(s) \rightarrow 5\overset{+3}{\mathrm{Al}}_2\overset{-2}{\mathrm{O}}_3(s) + 6\overset{+1}{\mathrm{H}}\overset{-1}{\mathrm{Cl}}(g) + 3\overset{0}{\mathrm{N}}_2(g) + 9\overset{+1}{\mathrm{H}}_2\overset{-1}{\mathrm{O}}(g)$

Yes, it's redox. Al atoms in Al and N atoms in $NH_4ClO_4$ are oxidized, and Cl atoms in $NH_4ClO_4$ are reduced.

**89.** **(a)**

$\overset{+1}{\mathrm{K}}_2\overset{+6}{\mathrm{Cr}}_2\overset{-2}{\mathrm{O}}_7(aq) + 14\overset{+1}{\mathrm{H}}\overset{-1}{\mathrm{Cl}}(aq) \rightarrow 2\overset{+1}{\mathrm{K}}\overset{-1}{\mathrm{Cl}}(aq) + 2\overset{+3}{\mathrm{Cr}}\overset{-1}{\mathrm{Cl}}_3(aq) + 7\overset{+1}{\mathrm{H}}_2\overset{-2}{\mathrm{O}}(l) + 3\overset{0}{\mathrm{Cl}}_2(g)$

Yes, it's redox. Cl in HCl is oxidized, Cr in $K_2Cr_2O_7$ is reduced, HCl is the reducing agent, and $K_2Cr_2O_7$ is the oxidizing agent.

**(b)** $\overset{0}{\mathrm{Ca}}(s) + 2\overset{+1}{\mathrm{H}}_2\overset{-2}{\mathrm{O}}(l) \rightarrow \overset{+2}{\mathrm{Ca}}(\overset{-2}{\mathrm{O}}\overset{+1}{\mathrm{H}})_2(s) + \overset{0}{\mathrm{H}}_2(g)$

Yes, it's redox. Ca is oxidized, H in $H_2O$ is reduced, Ca is the reducing agent, and $H_2O$ is the oxidizing agent.

**91.** **(a)** $\overset{0}{\mathrm{Ca}}(s) + \overset{0}{\mathrm{F}}_2(g) \rightarrow \overset{+2}{\mathrm{Ca}}\overset{-1}{\mathrm{F}}_2(s)$

Yes, it's redox. Ca is oxidized and is the reducing agent. F in $F_2$ is reduced, and $F_2$ is the oxidizing agent.

**(b)** $2\overset{0}{\mathrm{Al}}(s) + 3\overset{+1}{\mathrm{H}}_2\overset{-2}{\mathrm{O}}(g) \xrightarrow{\Delta} \overset{+3}{\mathrm{Al}}_2\overset{-2}{\mathrm{O}}_3(s) + 3\overset{0}{\mathrm{H}}_2(g)$

Yes, it's redox. Al is oxidized and is the reducing agent. H in $H_2O$ is reduced, so $H_2O$ is the oxidizing agent.

**93.** **(a)** combination **(b)** single displacement **(c)** decomposition

**95.** $CO_2(g) + 2H_2(g) \rightarrow C(s) + 2H_2O(g)$
**97.** $TiCl_4 + 2Mg \rightarrow Ti + 2MgCl_2$
**99.** $2K(s) + 2H_2O(l) \rightarrow 2KOH(aq) + H_2(g)$
**101.** $Ca(s) + Br_2(l) \rightarrow CaBr_2(s)$
**103.** $2HCl(aq) + Mg(OH)_2(s) \rightarrow MgCl_2(aq) + 2H_2O(l)$

## Chapter 7 Exercise

**7.1. (a)** HO and $NO_2$ have higher potential energy than $HNO_3$. Separated atoms are less stable and higher potential energy than atoms in a chemical bond, so energy is required to break a chemical bond. Thus, energy is required to separate the nitrogen and oxygen atoms being held together by mutual attraction in a chemical bond. The energy supplied goes to an increased potential energy of the separate HO and $NO_2$ compared to $HNO_3$. If the bond is reformed, the potential energy is converted into a form of energy that could be used to do work.
**(b)** A nitrogen dioxide molecule with a velocity of 439 m/s has greater kinetic energy than the same molecule with a velocity of 399 m/s. Any object in motion can collide with another object and move it, so any object in motion has the capacity to do work. This capacity to do work resulting from the motion of an object is called kinetic energy, KE. The particle with the higher velocity will move another object (like another molecule) farther, so it can do more work. It must therefore have more energy.
**(c)** The more massive nitrogen dioxide molecule has greater kinetic energy than the less massive nitrogen monoxide molecule with the same velocity. The moving particle with the higher mass can move another object (like another molecule) farther, so it can do more work. It must therefore have more energy.
**(d)** Gaseous nitrogen has higher potential energy than liquid nitrogen. When nitrogen goes from liquid to gas, the attractions that link the $N_2$ molecules together are broken. The energy that the nitrogen liquid must absorb to break these attractions goes to an increased potential energy of the nitrogen gas. If the nitrogen returns to the liquid form, attractions are reformed, and potential energy is converted into a form of energy that could be used to do work.
**(e)** Separate BrO and $NO_2$ molecules have a higher potential energy than the $BrONO_2$ molecule that they form. Atoms in a chemical bond are more stable and have lower potential energy than separated atoms, so energy is released when chemical bonds form. When BrO and $NO_2$ are converted into $BrONO_2$, a new bond is formed, and some of the potential energy of the BrO and $NO_2$ is released. The energy could be used to do some work. For example, if some of the potential energy is converted into increased kinetic energy of a molecule like $N_2$, the faster moving molecule could bump into something and move it and therefore do work.

$$BrO(g) + NO_2(g) \rightarrow BrONO_2(g)$$

**(f)** An alpha particle and two separate electrons have higher potential energy than an uncharged helium atom. The attraction between the alpha particle and the electrons will pull them together, and as they move together, they could bump into something, move it, and do work.

## Chapter 7 Review Questions

**1. (a)** The particles in the liquid are close together, but there is generally more empty space between them than in the solid. The particles of a liquid fill about 70% of the total volume. Because the particles in a liquid are moving faster than in a solid, and because there is more empty space between them, they break the attractions to the particles around them and constantly move into new positions to form new attractions. This leads to a less organized arrangement of particles compared to that of the solid. The following image shows the structure of a typical liquid. See Figure 2.2.
**(b)** The particles of a gas are much farther apart than in the solid or liquid. For a typical gas, the average distance between particles is about ten times the diameter of each particle. This leads to the gas particles themselves taking up only about 0.1% of the total volume. The other 99.9% of the total volume is empty space. According to our model, each particle in a gas moves freely in a straight-line path until it collides with another gas particle or with a liquid or solid. The particles are usually moving fast enough to break any attraction that might form between them, so after two particles collide, they bounce off each other and continue on their way alone. The following image shows the structure of a typical gas. See Figure 2.4.
**(c)** Particles that are at the surface of the liquid and that are moving away from the surface fast enough to break the attractions that pull them back will escape to the gaseous form. See Figure 2.3.
**(d)** Increased temperature leads to an increase in the average velocity of the particles in the liquid. This makes it easier for the particles to break the attractions between them and move from one position to another, including away from the surface into the gaseous form.
**2. (a)** According to our model, the particles of a solid can be pictured as spheres that are packed as closely together as possible. The spheres for NaCl are alternating $Na^+$ cations and $Cl^-$ anions. Strong attractions hold these particles in the same general position, but the particles are still constantly moving (Figure 2.1). Each particle is constantly changing its direction and speeding up and slowing down. Despite the constant changes in direction and velocity, at a constant temperature, the strong attractions between particles keep them the same average distance apart and in the same general orientation to each other.
**(b)** When a solid is heated, the average velocity of the particles increases. The more violent collisions between the faster moving particles usually cause each particle to push its neighbors farther away. Therefore, increased temperature usually leads to an expansion of solids. See Figure 2.1.
**(c)** The sodium ions and chloride ions break out of their positions in the solid and move more freely throughout the liquid, constantly breaking old attractions and making new ones. Although the particles are still close together in the liquid, they are more disorganized, and there is more empty space between them.

## Chapter 7 Key Ideas

**3.** energy, resistance **5.** mass **7.** created, destroyed, transferred, changed **9.** change **11.** increases **13.** decrease **15.** 4.184 **17.** thermal **19.** collisions **21.** massless **23.** peaks, cycle **25.** released **27.** absorbed **29.** oxygen, $O_2$, ozone, $O_3$ **31.** cars, sun **33.** shorter, longer **35.** 10, 50 **37.** UV-B, sunburn, premature aging, skin cancer **39.** 242 nm, 240 nm to 320 nm **41.** stable

## Chapter 7 Problems

**43. (a)** An ozone molecule, $O_3$, with a velocity of 410 m/s has greater kinetic energy than the same molecule with a velocity of 393 m/s. Any object in motion can collide with another object and move it, so any object in motion has the capacity to do work. This capacity to do work resulting from the motion of an object is called kinetic energy, KE. The particle with the higher velocity will move another object (like another molecule) farther, so it can do more work. It must therefore have more energy.
**(b)** An ozone molecule, $O_3$, has greater kinetic energy than an $O_2$ molecule with the same velocity. The moving particle with the higher mass can move another object (like another molecule) farther, so it can do more work. It must therefore have more energy.
**(c)** The attraction between the separated electron and a proton will pull them together, and as they move together, they could bump into something, move it, and do work. Therefore, a proton and an electron farther apart have higher potential energy than a proton and an electron close together.
**(d)** Separated atoms are less stable than atoms in a chemical bond, so the potential energy of OH and Cl is higher than HOCl. Energy is required to separate the oxygen atom and the chlorine atom being held together by mutual attraction in a chemical bond. The energy supplied goes to an increased potential energy of the separate OH and Cl compared to HOCl. If the bond is reformed, the potential energy is converted into a form of energy that could be used to do work.
**(e)** Separated atoms are less stable than atoms in a chemical bond, so the potential energy of two separate Cl atoms is greater than one $Cl_2$ molecule. When two Cl atoms are converted into a $Cl_2$ molecule, a new bond is formed, and some of the potential energy of the Cl atoms is released. The energy could be used to do some work.

**(f)** Gaseous water has higher potential energy than liquid water. When water evaporates, the hydrogen bonds that link the water molecules together are broken. The energy that the water must absorb to break these attractions goes to an increased potential energy of the water vapor. If the water returns to the liquid form, hydrogen bonds are reformed, and the potential energy is converted into a form of energy that could be used to do work.
**46. (a)** If you nudge the brick off the top of the building, its potential energy will be converted into kinetic energy as it falls. If it hits the roof of a parked car, it will move the metal of the roof down, making a dent. When an object is moved, work is done.
**(b)** You can shoot the rubber band across the room at a paper airplane. When you release the rubber band, its potential energy is converted into kinetic energy, which is used to do the work of moving the airplane.
**(c)** It is possible to run a car on pure alcohol. When the alcohol is burned, its potential energy is converted into energy that does the work of moving the car.
**48. (a)** Some of the kinetic energy of the moving hand is transferred to the string to set it moving and to the tips of the bow as it bends. Some of this kinetic energy is converted into potential energy of the stretched string and bow.
**(b)** Some of the potential energy of the stretched string and bow are converted into kinetic energy of the moving arrow.
**51. (a)** Because a chemical bond is broken in this reaction, energy would need to be absorbed to supply the energy necessary to move to the higher potential energy products. In the stratosphere, this energy comes from the radiant energy of a photon.
**(b)** Because a chemical bond is made in this reaction, energy would be released as the system moves to the lower potential energy product. This potential energy could be converted into kinetic energy.
**53.** A speeding bullet has a certain kinetic energy that is related to its overall mass and its velocity. This is its external kinetic energy. The bullet is also composed of silver atoms that, like all particles, are moving in a random way. The particles within the bullet are constantly moving, colliding with their neighbors, changing their direction of motion, and changing their velocities. The kinetic energy associated with this internal motion is the internal kinetic energy. The internal motion is independent of the overall motion of the bullet.
**55.** A typical thermometer used in the chemical laboratory consists of a long cylindrical glass container with a bulb at the bottom that contains a reservoir of mercury and a thin tube running up the inside of the thermometer that the mercury can rise into as it expands.

When the thermometer is first placed in the hot water, the particles in the water have a greater average kinetic energy than the particles of the glass and mercury in the thermometer. When the more energetic water molecules collide with the particles in the glass, the particles of water slow down, and the particles in the glass speed up. The average kinetic energy of the water molecules decreases and the average kinetic energy of the particles of the glass increases. Thermal energy has been transferred from the water to the glass. The glass particles then collide with the less energetic mercury atoms, speeding them up and slowing down themselves. The average kinetic energy of the mercury increases, as thermal energy is transferred from the glass to the mercury. Thus, some of the kinetic energy of the water is transferred to the glass, which then transfers some of this energy to the mercury. This will continue until the particles of water, glass, and mercury all have the same average kinetic energy.

Try to picture the atoms of mercury in the liquid mercury. Now that they are moving faster, they collide with the particles around them with greater force. This pushes the particles farther apart and causes the liquid mercury to expand. Because the mercury has a larger volume, it moves further up the thin column in the center of the thermometer.
**56.** In the particle view, radiant energy is a stream of tiny, massless packets of energy called photons. Different forms of radiant energy differ with respect to the energy of each of their photons. The energies of the photons of visible light are lower than for ultraviolet radiation.

In the wave view, as radiant energy moves away from the source, it has an effect on the space around it that can be described as a wave consisting of an oscillating electric field perpendicular to an oscillating magnetic field (Figure 7.11). Different forms of radiant energy differ with respect to the wavelengths and frequencies of these oscillating waves. The waves associated with visible light have longer wavelength than the waves associated with ultraviolet radiation.
**58. (a)** radio waves $<$ microwaves $<$ infrared radiation $<$ visible light $<$ ultraviolet radiation $<$ X rays $<$ gamma rays
**(b)** gamma rays $<$ X rays $<$ ultraviolet radiation $<$ visible light $<$ infrared radiation $<$ microwaves $<$ radio waves
**60.** The bonds in the products must be less stable and therefore higher potential energy than the bonds in the reactants. Energy is absorbed in the reaction to supply the energy necessary to increase the potential energy of the products compared to reactants.

$$\text{More stable bonds} + \text{energy} \rightarrow \text{Less stable bonds}$$

$$\text{lower PE} + \text{energy} \rightarrow \text{higher PE}$$

**62. (a)** The bonds in the products must be more stable and therefore lower potential energy than the bonds in the reactants. The potential energy difference between reactants and products is released.
**(b)** Some of the potential energy of the reactants is converted into kinetic energy of the products, making the average kinetic energy of the products higher than the average kinetic energy of the reactants.
**(c)** The particles of the higher-temperature products collide with the particles of the lower-temperature container with greater average force than the particles of the container. Therefore, collisions between the particles of the products and the container speed up the particles of the container, increasing its internal kinetic energy or thermal energy, while slowing the particles of the products, decreasing their thermal energy. In this way, thermal energy is transferred from the products to the container. Likewise, the container, which is now at a higher temperature, transfers thermal energy to the lower-temperature surroundings. Heat has been transferred from the products to the container to the surroundings.
**65. (a)** exothermic **(b)** endothermic **(c)** exothermic
**67.** See Figure 7.13.
**68.** Ozone is a very powerful oxidizing agent. This can be useful. For example, ozone mixed with oxygen can be used to sanitize hot tubs, and it is used in industry to bleach waxes, oils, and textiles. Conversely, when the levels in the air get too high, the highly reactive nature of ozone becomes a problem. For example, $O_3$ is a very strong respiratory irritant that can lead to shortness of breath, chest pain when inhaling, wheezing, and coughing. It also damages rubber and plastics, leading to premature deterioration of products made with these materials. According to the Agricultural Research Service of North Carolina State University, ozone damages plants more than all other pollutants combined.
**70.** Shorter wavelengths of radiant energy are associated with higher energy photons. Radiant energy of wavelengths less than 400 nm has enough energy to break N–O bonds in $NO_2$ molecules, but radiant energy with wavelengths longer than 400 nm does not supply enough energy to separate the atoms.
**72.** The shorter wavelength UV-B radiation (from about 290 nm to 320 nm) has higher energy than the UV-A radiation. Radiation in this portion of the spectrum has high enough energy so that excessive exposure can cause sunburn, premature skin aging, and skin cancer.
**74.** $O_2$ molecules absorb UV radiation with wavelengths less than 242 nm, and $O_3$ molecules absorb radiant energy with wavelengths from 240 nm to 320 nm. See Figure 7.16.
**76.** Gases are removed from the lower atmosphere in two general ways. They either dissolve in the clouds and are rained out, or they react chemically to be converted into other substances. Neither of these mechanisms are important for CFCs. Chlorofluorocarbons are insoluble in water, and they are so stable that they can exist in the lower atmosphere for years. During this time, the CFC molecules wander around in the atmosphere, moving wherever the air currents take them. They can eventually make their way up into the stratosphere.
**79. (a)** When the natural gas burns, some of its potential energy is used to do the work of moving the bus.
**(b)** If you put an object, such as a small ball of paper, on one end of the compressed spring and then release the spring, the potential

energy stored in the compressed spring will be converted into kinetic energy of the moving spring as it stretches out. The kinetic energy of the moving spring will do the work of moving the object.
**(c)** When a squirrel chews off the pinecone, allowing it to fall to the ground, the potential energy that it has because it is higher than the ground will be converted into kinetic energy as it falls. The falling pinecone can collide with another pinecone and do the work of knocking the second pinecone off its branch.
**81. (a)** The kinetic energy of the mover's hands is transferred to the kinetic energy of the moving ropes, which is transferred to the kinetic energy of the moving piano. Some of this kinetic energy is converted into potential energy as the piano rises.
**(b)** Some of the potential energy of the piano is converted into kinetic energy as it falls.
**83. (a)** Some of the potential energy of the gasoline molecules is converted into kinetic energy of moving product particles. This is reflected in an increase in the temperature of the gaseous products.
**(b)** Some of the kinetic energy of the moving gaseous product particles is transferred into kinetic energy of the moving piston.
**(c)** Some of the kinetic energy of the moving piston is transferred into kinetic energy of the moving crankshaft and ultimately into kinetic energy of the moving wheels.
**85. (a)** exothermic **(b)** endothermic

## Chapter 8 Exercises

**8.1. (a)** $\frac{10^3\text{ J}}{1\text{ kJ}}$ and $\frac{1\text{ kJ}}{10^3\text{ J}}$ **(b)** $\frac{10^2\text{ cm}}{1\text{ m}}$ and $\frac{1\text{ m}}{10^2\text{ cm}}$

**(c)** $\frac{10^9\text{ L}}{1\text{ GL}}$ and $\frac{1\text{ GL}}{10^9\text{ L}}$ **(d)** $\frac{10^6\ \mu\text{g}}{1\text{ g}}$ and $\frac{1\text{ g}}{10^6\ \mu\text{g}}$

**(e)** $\frac{10^6\text{ g}}{1\text{ Mg}}$ and $\frac{1\text{ Mg}}{10^6\text{ g}}$

**8.2.** $? \text{ Mg} = 4.352\ \mu\text{g}\left(\frac{1\text{ g}}{10^6\mu\text{g}}\right)\left(\frac{1\text{ Mg}}{10^6\text{g}}\right) = \mathbf{4.352 \times 10^{-12}\ Mg}$

**8.3.** $? \text{ gal} = 1.5 \times 10^{18}\text{ kL}\left(\frac{10^3\text{ L}}{1\text{ kL}}\right)\left(\frac{1\text{ gal}}{3.785\text{ L}}\right) = \mathbf{4.0 \times 10^{20}\ gal}$

**8.4.** $? \text{ oz} = 10.5\text{ g}\left(\frac{1\text{ lb}}{453.6\text{ g}}\right)\left(\frac{16\text{ oz}}{1\text{ lb}}\right) = \mathbf{0.370\ oz}$

The 10.5 comes from a measurement and has three significant figures. The 453.6 is calculated and rounded off. It has four significant figures. The 16 comes from a definition and is exact. We report three significant figures in our answer.

**8.5.** $? \text{ hr} = 25.0\text{ mi}\left(\frac{5280\text{ ft}}{1\text{ mi}}\right)\left(\frac{12\text{ in.}}{1\text{ ft}}\right)\left(\frac{2.54\text{ cm}}{1\text{ in.}}\right)\left(\frac{1\text{ m}}{10^2\text{ cm}}\right)\left(\frac{1\text{ km}}{10^3\text{ m}}\right)$

$\left(\frac{1\text{ s}}{11.0\text{ km}}\right)\left(\frac{1\text{ min}}{60\text{ s}}\right)\left(\frac{1\text{ hr}}{60\text{ min}}\right) = \mathbf{1.02 \times 10^{-3}\ hr}$

The 25.0 comes from a measurement and has three significant figures. The 11.0 comes from a blend of measurement and calculations. It also has three significant figures. All of the other numbers come from definitions and are therefore exact. We report three significant figures in our answer.

**8.6. (a)** 684 − 595.325 = **89** **(b)** 92.771 + 9.3 = **102.1**

**8.7. (a)** $? \text{ kg} = 15.6\text{ gal}\left(\frac{3.785\text{ L}}{1\text{ gal}}\right)\left(\frac{10^3\text{ mL}}{1\text{ L}}\right)\left(\frac{0.70\text{ g gas}}{1\text{ mL gas}}\right)\left(\frac{1\text{ kg}}{10^3\text{ g}}\right)$

or $? \text{ kg} = 15.6\text{ gal}\left(\frac{3.785\text{ L}}{1\text{ gal}}\right)\left(\frac{0.70\text{ kg gas}}{1\text{ L gas}}\right) = \mathbf{41\ kg\ gasoline}$

**(b)** $? \text{ L} = 242.6\text{ met. tons}\left(\frac{10^3\text{ kg}}{1\text{ met. ton}}\right)\left(\frac{10^3\text{ g}}{1\text{ kg}}\right)\left(\frac{1\text{ mL Fe}}{7.86\text{ g Fe}}\right)\left(\frac{1\text{ L}}{10^3\text{ mL}}\right)$

or $? \text{ L} = 242.6\text{ met. tons}\left(\frac{10^3\text{ kg}}{1\text{ met. ton}}\right)\left(\frac{1\text{ L Fe}}{7.86\text{ kg Fe}}\right) = \mathbf{3.09 \times 10^4\ L\ Fe}$

**8.8. (a)** $\frac{?\text{ g}}{\text{mL}} = \frac{(57.452 - 48.737)\text{ g}}{13.2\text{ mL}} = \mathbf{0.660\ g/mL}$

**(b)** $\frac{?\text{ g}}{1\text{ mL}} = \frac{1.2 \times 10^4\text{ kg}}{2.4 \times 10^4\text{ L}}\left(\frac{10^3\text{ g}}{1\text{ kg}}\right)\left(\frac{1\text{ L}}{10^3\text{ mL}}\right) = \mathbf{0.50\ g/mL}$

**8.9. (a)**
$? \text{ lb } HCO_3^- = 1.8 \times 10^{21}\text{ kg ocean}\left(\frac{0.014\text{ kg } HCO_3^-}{100\text{ kg ocean}}\right)\left(\frac{2.205\text{ lb}}{1\text{ kg}}\right)$
$= \mathbf{5.6 \times 10^{17}\ lb\ HCO_3^-}$

**(b)** $? \text{ L blood to muscles} = 5.2\text{ L blood total}\left(\frac{78\text{ L blood to muscles}}{100\text{ L blood total}}\right)$
$= \mathbf{4.1\ L\ blood\ to\ muscles}$

**8.10. (a)** $? \text{ nm} = 2 \times 10^{-15}\text{ m}\left(\frac{10^9\text{ nm}}{1\text{ m}}\right) = \mathbf{2 \times 10^{-6}\ nm}$

**(b)** $? \text{ lb} = 9.1093897 \times 10^{-31}\text{ kg}\left(\frac{10^3\text{ g}}{1\text{ kg}}\right)\left(\frac{10^9\text{ ng}}{1\text{ g}}\right)$
$= \mathbf{9.1093897 \times 10^{-19}\ ng}$

**(c)** $? \text{ kg} = 4.070 \times 10^6\text{ lb}\left(\frac{453.6\text{ g}}{1\text{ lb}}\right)\left(\frac{1\text{ kg}}{10^3\text{ g}}\right) = \mathbf{1.846 \times 10^6\ kg}$

**(d)** $\frac{?\text{ g}}{\text{mL}} = \frac{88.978\text{ g}}{2.9659\text{ L}}\left(\frac{1\text{ L}}{10^3\text{ mL}}\right) = \mathbf{0.030000\ g/mL}$

**(e)** $? \text{ kg} = 2.5\text{ L}\left(\frac{10^3\text{ mL}}{1\text{ L}}\right)\left(\frac{1.03\text{ g}}{1\text{ mL}}\right)\left(\frac{1\text{ kg}}{10^3\text{ g}}\right)$

or $? \text{ kg} = 2.5\text{ L}\left(\frac{1.03\text{ kg}}{1\text{ L}}\right) = \mathbf{2.6\ kg}$

**(f)** $? \text{ s} = 6.0\text{ ft}\left(\frac{12\text{ in.}}{1\text{ ft}}\right)\left(\frac{2.54\text{ cm}}{1\text{ in.}}\right)\left(\frac{1\text{ m}}{10^2\text{ cm}}\right)\left(\frac{1\text{ s}}{18\text{ m}}\right) = \mathbf{0.10\ s}$

**(g)** $\frac{?\text{ km}}{\text{hr}}$

$= \frac{22\text{ in.}}{6.2 \times 10^{-9}\text{ s}}\left(\frac{2.54\text{ cm}}{1\text{ in.}}\right)\left(\frac{1\text{ m}}{10^2\text{ cm}}\right)\left(\frac{1\text{ km}}{10^3\text{ m}}\right)\left(\frac{60\text{ s}}{1\text{ min}}\right)\left(\frac{60\text{ min}}{1\text{ hr}}\right)$
$= \mathbf{3.2 \times 10^8\ km/hr}$

**(h)** $? \text{ ton } Ca^{2+}$
$= 1.8 \times 10^{21}\text{ kg ocean}\left(\frac{0.041\text{ kg } Ca^{2+}}{100\text{ kg ocean}}\right)\left(\frac{2.205\text{ lb}}{1\text{ kg}}\right)\left(\frac{1\text{ ton}}{2000\text{ lb}}\right)$
$= \mathbf{8.1 \times 10^{14}\ ton\ Ca^{2+}}$

**(i)** $? \text{ L to brain} = 1.0\text{ hr}\left(\frac{60\text{ min}}{1\text{ hr}}\right)\left(\frac{5.0\text{ L total}}{1\text{ min}}\right)\left(\frac{15\text{ L to brain}}{100\text{ L total}}\right) = \mathbf{45\ L}$

**8.11. (a)** $°\text{F} = 2.5\ °\text{C}\left(\frac{1.8\ °\text{F}}{1\ °\text{C}}\right) + 32\ °\text{F} = \mathbf{36.5\ °F}$

$\text{K} = 2.5\ °\text{C} + 273.15 = \mathbf{275.7\ K}$

**(b)** $°\text{C} = (5.4\ °\text{F} - 32\ °\text{F})\frac{1\ °\text{C}}{1.8\ °\text{F}} = \mathbf{-14.8\ °C}$

$\text{K} = -14.8\ °\text{C} + 273.15 = \mathbf{258.4\ K}$

**(c)** $°\text{C} = 2.15 \times 10^3\text{ K} - 273.15 = \mathbf{1.88 \times 10^3\ °C}$

$°\text{F} = 1.88 \times 10^3\ °\text{C}\left(\frac{1.8\ °\text{F}}{1\ °\text{C}}\right) + 32\ °\text{F} = \mathbf{3.42 \times 10^3\ °F}$

or $\mathbf{3.41 \times 10^3\ °F}$ if the unrounded answer to the first calculation is used in the second calculation.

## CHAPTER 8 REVIEW QUESTIONS

**1.** Length, meter (m); mass, gram (g); volume, liter (L); energy, joule (J); gas pressure, pascal (Pa)
**2.** milliliter, volume, mL; microgram, mass, μg; kilojoule, energy, kJ; kelvin, temperature, K
**3. (a)** $10^{-6}$ m = 1 μm **(b)** $10^6$ g = 1 Mg **(c)** $10^{-3}$ L = 1 mL **(d)** $10^{-9}$ m = 1 nm **(e)** 1 $cm^3$ = 1 mL **(f)** $10^3$ L = $m^3$ **(g)** $10^3$ kg = 1 t **(h)** 1 Mg = 1 t
**4.** We assume that each reported number is ±1 in the last decimal position reported. Therefore, we assume the mass of the graduated cylinder between 1124.1 g and 1124.3 g, the volume of the methanol is between 1.19 L and 1.21 L, and the total mass is between 2073.8 g and 2074.0 g.

## CHAPTER 8 KEY IDEAS

**5.** unit conversion, correct **7.** unwanted, desired **9.** cancel **11.** one **13.** inexact, fewest **15.** counting **17.** decimal places, fewest decimal places **19.** grams per milliliter, grams per cubic centimeter **21.** identity, known **23.** part, whole **25.** definitions

## CHAPTER 8 PROBLEMS

**26. (a)** $6.7294 \times 10^4$ **(b)** $4.38763102 \times 10^8$ **(c)** $7.3 \times 10^{-5}$ **(d)** $4.35 \times 10^{-8}$
**28. (a)** 4,097 **(b)** 15,541.2 **(c)** 0.0000234 **(d)** 0.000000012
**30. (a)** 2,877 **(b)** 316 **(c)** 5.5 **(d)** 2.827
**32. (a)** $10^{12}$ **(b)** $10^9$ **(c)** $10^7$ **(d)** $10^5$ **(e)** $10^{29}$ **(f)** $10^3$
**34. (a)** $7.6 \times 10^{15}$ **(b)** $1.02 \times 10^{16}$ **(c)** $3.3 \times 10^5$ **(d)** $1.87 \times 10^{-3}$ or 0.00187 **(e)** $8.203 \times 10^{13}$ **(f)** $2.111 \times 10^4$
**36. (a)** $\left(\frac{10^3 \text{ m}}{1 \text{ km}}\right)$ **(b)** $\left(\frac{10^2 \text{ cm}}{1 \text{ m}}\right)$ **(c)** $\left(\frac{10^3 \text{ mm}}{1 \text{ m}}\right)$ **(d)** $\left(\frac{1 \text{ cm}^3}{1 \text{ mL}}\right)$ **(e)** $\left(\frac{2.54 \text{ cm}}{1 \text{ in.}}\right)$ **(f)** $\left(\frac{453.6 \text{ g}}{1 \text{ lb}}\right)$
**38. (a)** $\left(\frac{10^3 \text{ g}}{1 \text{ kg}}\right)$ **(b)** $\left(\frac{10^3 \text{ mg}}{1 \text{ g}}\right)$ **(c)** $\left(\frac{1.094 \text{ yd}}{1 \text{ m}}\right)$ **(d)** $\left(\frac{2.205 \text{ lb}}{1 \text{ kg}}\right)$
**40.** $9.1093897 \times 10^{-28}$ g
**42.** 3.2 μm
**44.** 0.04951 lb or 0.049507 lb (if use more significant figures in the gram-to-pound conversion factor)
**46.** $5.908 \times 10^{-26}$ oz or $5.908138 \times 10^{-26}$ oz (if use more significant figures in the gram-to-pound conversion factor)
**48.** 2.2 μm
**50. (a)** not exact—2 significant figures **(b)** exact **(c)** not exact—2 significant figures **(d)** exact **(e)** exact **(f)** exact **(g)** not exact—4 significant figures **(h)** not exact—6 significant figures **(i)** not exact—4 significant figures **(j)** not exact—4 significant figures
**52. (a)** 5 **(b)** 3 **(c)** 3 **(d)** 5 **(e)** 4
**54. (a)** 3 **(b)** 4 **(c)** 4
**56. (a)** 34.6 **(b)** 193 **(c)** 24.0 **(d)** 0.00388 **(e)** 0.0230 **(f)** $2.85 \times 10^3$ **(g)** $7.84 \times 10^4$
**58. (a)** 103 **(b)** 6.2 **(c)** $5.71 \times 10^{11}$
**60. (a)** 100.14 **(b)** 1
**62.** 0.1200 g/mL **64.** 0.1850 kg **66.** 0.8799 qt **68.** $1.9 \times 10^{19}$ kg $Na^+$ **70.** 94.3 L to brain **72.** 0.18 mg to energy **74.** 33 km/hr **76.** 0.275 s **78.** 3.5 hr **80.** $3.326 \times 10^5$ kJ **82.** 0.444 kg C **84.** 2.7 lb beans per day **86.** $1.6 \times 10^8$ kg HCl **88.** 52 days **90.** 9812 L air **92.** 44.0 minutes **94.** 28 qt **96.** $7.6 \times 10^5$ beats **98.** $3.6 \times 10^9$ molecules **100.** 6.9 days **102.** 95 ft **103.** 88 °F, 304 K **105.** 108 °C, 381 K **107.** $2.750 \times 10^3$ °C, 4982 °F **109.** $5.5 \times 10^3$ °C, $5.8 \times 10^3$ K

## CHAPTER 9 EXERCISES

**9.1. (a)** 196.9665 (from periodic table)
**(b)** 196.9665 g (There are $6.022 \times 10^{23}$ atoms per mole of atoms, and one mole of an element has a mass in grams equal to its atomic mass.)
**(c)** $\left(\frac{196.9665 \text{ g Au}}{1 \text{ mol Au}}\right)$

**(d)** $? \text{ g Au} = 0.20443 \text{ mol Au}\left(\frac{196.9665 \text{ g Au}}{1 \text{ mol Au}}\right) = \mathbf{40.266 \text{ g Au}}$

**(e)** $? \text{ mg Au} = 7.046 \times 10^{-3} \text{ mol Au}\left(\frac{196.9665 \text{ g Au}}{1 \text{ mol Au}}\right)\left(\frac{10^3 \text{ mg}}{1 \text{ g}}\right)$
$= \mathbf{1388 \text{ mg Au}}$

**(f)** $? \text{ mol Au} = 1.00 \text{ troy oz Au}\left(\frac{31.10 \text{ g}}{1 \text{ troy oz}}\right)\left(\frac{1 \text{ mol Au}}{196.9665 \text{ g Au}}\right)$
$= \mathbf{0.158 \text{ mol Au}}$

**9.2. (a)** 2(12.011) + 6(1.00794) + 1(15.9994) = 46.069
**(b)** 46.069 g (One mole of a molecular compound has a mass in grams equal to its molecular mass.)
**(c)** $\left(\frac{46.069 \text{ g } C_2H_5OH}{1 \text{ mol } C_2H_5OH}\right)$

**(d)** $? \text{ mol } C_2H_5OH = 16 \text{ g } C_2H_5OH\left(\frac{1 \text{ mol } C_2H_5OH}{46.069 \text{ g } C_2H_5OH}\right)$
$= \mathbf{0.35 \text{ mol } C_2H_5OH}$

**(e)** $? \text{ 1 mL } C_2H_5OH$
$= 1.0 \text{ mol } C_2H_5OH\left(\frac{46.069 \text{ g } C_2H_5OH}{1 \text{ mol } C_2H_5OH}\right)\left(\frac{1 \text{ mL } C_2H_5OH}{0.7893 \text{ g } C_2H_5OH}\right)$
$= \mathbf{58 \text{ mL } C_2H_5OH}$

**9.3. (a)** Formula Mass
= 1(22.9898) + 1(1.00794) + 1(12.011) + 3(15.9994) = 84.007
**(b)** 84.007 g (One mole of an ionic compound has a mass in grams equal to its formula mass.)
**(c)** $\left(\frac{84.007 \text{ g } NaHCO_3}{1 \text{ mol } NaHCO_3}\right)$

**(d)** $? \text{ mol } NaHCO_3$
$= 0.4 \text{ g } NaHCO_3\left(\frac{1 \text{ mol } NaHCO_3}{84.007 \text{ g } NaHCO_3}\right) = \mathbf{5\ 3\ 10^{23} \text{ mol } NaHCO_3}$

**9.4. (a)** $\frac{6 \text{ mol H}}{1 \text{ mol } C_2H_5OH}$ **(b)** $\frac{3 \text{ mol O}}{1 \text{ mol } NaHCO_3}$
**(c)** There is 1 mole of $HCO_3^-$ per 1 mole of $NaHCO_3$.

**9.5.** $? \text{ g } S_2Cl_2$
$= 123.8 \text{ g S}\left(\frac{1 \text{ mol S}}{32.066 \text{ g S}}\right)\left(\frac{1 \text{ mol } S_2Cl_2}{2 \text{ mol S}}\right)\left(\frac{135.037 \text{ g } S_2Cl_2}{1 \text{ mole } S_2Cl_2}\right)$
$= \mathbf{260.7 \text{ g } S_2Cl_2}$

**9.6.** $? \text{ kg V} = 2.3 \text{ kg } V_2O_5\left(\frac{10^3 \text{ g}}{1 \text{ kg}}\right)\left(\frac{1 \text{ mol } V_2O_5}{181.880 \text{ g } V_2O_5}\right)$
$\left(\frac{2 \text{ mol V}}{1 \text{ mol } V_2O_5}\right)\left(\frac{50.9415 \text{ g V}}{1 \text{ mol V}}\right)\left(\frac{1 \text{ kg}}{10^3 \text{ g}}\right) = \mathbf{1.3 \text{ kg V}}$

**9.7.** $? \text{ mol Bi} = 32.516 \text{ g Bi}\left(\frac{1 \text{ mol Bi}}{208.9804 \text{ g Bi}}\right)$
= 0.15559 mol Bi ÷ 0.15559 = 1 mol Bi × 2 = 2 mol Bi
$? \text{ mol S} = 7.484 \text{ g S}\left(\frac{1 \text{ mol S}}{32.066 \text{ gS}}\right)$
= 0.2334 mol S ÷ 0.15559 ≅ 1½ mol S × 2 = 3 mol S
Our empirical formula is $\mathbf{Bi_2S_3}$ or bismuth(III) sulfide.

**9.8.** $? \text{ mol K} = 35.172 \text{ g K}\left(\frac{1 \text{ mol K}}{39.0983 \text{ K}}\right) = 0.89958 \text{ mol K} \div 0.89958$
= 1 mol K × 2 = 2 mol K

$$? \text{ mol S} = 28.846 \text{ g S}\left(\frac{1 \text{ mol S}}{32.066 \text{ g S}}\right) = 0.89958 \text{ mol S} \div 0.89958$$

$$= 1 \text{ mol S} \times 2 = 2 \text{ mol S}$$

$$? \text{ mol O} = 35.982 \text{ g O}\left(\frac{1 \text{ mol O}}{15.9994 \text{ g O}}\right) = 2.2490 \text{ mol O} \div 0.89958$$

$$\cong 2\tfrac{1}{2} \text{ mol O} \times 2 = 5 \text{ mol O}$$

Empirical Formula $\mathbf{K_2S_2O_5}$

**9.9.** $$? \text{ mol C} = 39.94 \text{ g C}\left(\frac{1 \text{ mol C}}{12.011 \text{ g C}}\right) = 3.325 \text{ mol C} \div 1.11$$

$$= 3 \text{ mol C} \times 2 = 6 \text{ mol C}$$

$$? \text{ mol H} = 1.12 \text{ g H}\left(\frac{1 \text{ mol H}}{1.00794 \text{ g H}}\right) = 1.11 \text{ mol H} \div 1.11$$

$$= 1 \text{ mol H} \times 2 = 2 \text{ mol H}$$

$$? \text{ mol Cl} = 58.94 \text{ g Cl}\left(\frac{1 \text{ mol Cl}}{35.4527 \text{ g Cl}}\right) = 1.662 \text{ mol Cl} \div 1.11$$

$$\cong 1\tfrac{1}{2} \text{ cl} \times 2 = 3 \text{ mol Cl}$$

Empirical Formula $\mathbf{C_6H_2Cl_3}$

$$n = \frac{\text{molecular mass}}{\text{empirical formula mass}} = \frac{360.88}{180.440} \cong 2$$

Molecular Formula = $(C_6H_2Cl_3)_2$ or $\mathbf{C_{12}H_4Cl_6}$

## Chapter 9 Review Questions

**1. (a)** $\left(\frac{10^3 \text{ g}}{1 \text{ kg}}\right)$ **(b)** $\left(\frac{10^3 \text{ mg}}{1 \text{ g}}\right)$ **(c)** $\left(\frac{10^3 \text{ kg}}{1 \text{ metric ton}}\right)$ **(d)** $\left(\frac{10^6\ \mu\text{g}}{1 \text{ g}}\right)$
**2.** $3.45 \times 10^7$ g **3.** 0.184570 kg **4.** 4.5000 Mg **5.** $8.71 \times 10^8$ g
**6.** 695 kg $Al_2O_3$

## Chapter 9 Key Ideas

**7.** impossible **9.** weighted, naturally **11.** atomic mass **13.** formula unit, kinds, numbers **15.** formula units **17.** simplest
**19.** whole-number

## Chapter 9 Problems

**21. (a)** 22.9898 u **(b)** 15.9994 u
**23. (a)** 32.066 g **(b)** 18.9984 g
**25. (a)** 65.39 g/mol **(b)** 26.9815 g/mol
**27. (a)** $\left(\frac{55.845 \text{ g Fe}}{1 \text{ mol Fe}}\right)$ **(b)** $\left(\frac{83.80 \text{ g Kr}}{1 \text{ mol Kr}}\right)$
**29.** $6.3 \times 10^{-7}$ mol Se **31.** 8.9 mg Fe
**33. (a)** 65.9964, $\left(\frac{65.9964 \text{ g } H_3PO_2}{1 \text{ mol } H_3PO_2}\right)$
**(b)** 93.128, $\left(\frac{93.128 \text{ g } C_6H_5NH_2}{1 \text{ mol } C_6H_5NH_2}\right)$
**35. (a)** $7.398 \times 10^{-3}$ mol $C_8H_9NO$ **(b)** $2.03 \times 10^3$ g $C_8H_9NO$
**38. (a)** molecular compound—molecules **(b)** ionic compound—formula units **(c)** ionic compound—formula units **(d)** molecular compound—molecules
**40. (a)** 448.69, $\left(\frac{448.69 \text{ g } BiBr_3}{1 \text{ mol } BiBr_3}\right)$ **(b)** 342.154, $\left(\frac{342.154 \text{ g } Al_2(SO_4)_3}{1 \text{ mol } Al_2(SO_4)_3}\right)$
**42. (a)** $5.00 \times 10^{-3}$ mol $CaCO_3$ **(b)** 10.01 kg $CaCO_3$
**44.** $1.5 \times 10^{-3}$ mol $Al_2O_3$ **46.** $\frac{2 \text{ mol N}}{1 \text{ mol } N_2O_5}$ **48.** $\frac{2 \text{ mol Cr}}{1 \text{ mol } Cr_2O_3}$
**50.** There are 2 moles of ammonium ions in one mole of $(NH_4)_2C_2O_4$.
**52. (a)** $\left(\frac{1 \text{ mol Ca}}{1 \text{ mol } CaCO_3}\right)$ **(b)** 0.162 g Ca **(c)** 16.2% of daily value Ca
**54.** 24 μg $NaVO_3$ **56.** 101 kg Ti **59.** $Ag_3N$, silver nitride
**61.** $Ca_2P_2O_7$ **63.** $Ni_3As_2O_8$ **65.** $CaI_2O_6$ **67.** $C_2N_2H_8$ **69.** $C_6H_6Cl_6$
**71.** $C_3H_6N_6$ **73.** 2.88 g, $1.1 \times 10^4$ screws, 74 gross screws
**75.** 4.515 mol $Al_2O_3$ **77.** $1.1 \times 10^2$ mol Be
**79.** 12 carats
**81. (a)** $5.3 \times 10^{-5}$ mol $C_6H_5OH$ **(b)** 0.08720 kg $C_6H_5OH$
**83.** 50.60 kg Be **85.** $MoSi_2$ **87.** $KHC_2O_4$ **89.** $Na_2S_2O_5$
**91.** $MnH_4P_2O_8$ or $Mn(H_2PO_4)_2$, manganese(II) dihydrogen phosphate
**93.** $C_{13}H_{10}O_4N_2$ **95.** $7.30 \times 10^3$ kg Zn
**97.** 47 kg As **99. (a)** $1.3 \times 10^2$ t Mg **(b)** $2.9 \times 10^2$ t ore
**101.** 6.8 t ore **103.** $2.3 \times 10^2$ g baking powder **105.** 11 t zircon sand

## Chapter 10 Exercises

**10.1. (a)** $\left(\frac{8 \text{ mol } C_2H_4Cl_2}{6 \text{ mol } Cl_2}\right)$ or $\left(\frac{8 \text{ mol } C_2H_4Cl_2}{7 \text{ mol } O_2}\right)$ or

$\left(\frac{8 \text{ mol } C_2H_4Cl_2}{4 \text{ mol } C_2HCl_3}\right)$ or $\left(\frac{8 \text{ mol } C_2H_4Cl_2}{4 \text{ mol } C_2Cl_4}\right)$ or $\left(\frac{8 \text{ mol } C_2H_4Cl_2}{14 \text{ mol } H_2O}\right)$ or

$\left(\frac{6 \text{ mol } Cl_2}{7 \text{ mol } O_2}\right)$ or $\left(\frac{6 \text{ mol } Cl_2}{4 \text{ mol } C_2HCl_3}\right)$ or $\left(\frac{6 \text{ mol } Cl_2}{4 \text{ mol } C_2Cl_4}\right)$ or

$\left(\frac{6 \text{ mol } Cl_2}{14 \text{ mol } H_2O}\right)$ or $\left(\frac{7 \text{ mol } O_2}{4 \text{ mol } C_2HCl_3}\right)$ or $\left(\frac{7 \text{ mol } O_2}{4 \text{ mol } C_2Cl_4}\right)$ or

$\left(\frac{7 \text{ mol } O_2}{14 \text{ mol } H_2O}\right)$ or $\left(\frac{4 \text{ mol } C_2HCl_3}{4 \text{ mol } C_2Cl_4}\right)$ or $\left(\frac{4 \text{ mol } C_2HCl_3}{14 \text{ mol } H_2O}\right)$ or

$\left(\frac{4 \text{ mol } C_2Cl_4}{14 \text{ mol } H_2O}\right)$

**(b)** ? g $H_2O$

$$= 362.47 \text{ g } C_2Cl_4\left(\frac{1 \text{ mol } C_2Cl_4}{165.833 \text{ g } C_2Cl_4}\right)\left(\frac{14 \text{ mol } H_2O}{4 \text{ mol } C_2Cl_4}\right)\left(\frac{18.0153 \text{ g } H_2O}{1 \text{ mol } H_2O}\right)$$

$$\text{or } ? \text{ g } H_2O = 362.47 \text{ g } C_2Cl_4\left(\frac{14 \times 18.0153 \text{ g } H_2O}{4 \times 165.833\ C_2Cl_4}\right) = \mathbf{137.82 \text{ g } H_2O}$$

**(c)** $$? \text{ kg } C_2Cl_4 = 23.75 \text{ kg } C_2H_4Cl_2\left(\frac{10^3\text{g}}{1 \text{ kg}}\right)\left(\frac{1 \text{ mol } C_2H_4Cl_2}{98.959 \text{ g } C_2H_4Cl_2}\right)$$

$$\left(\frac{4 \text{ mol } C_2Cl_4}{8 \text{ mol } C_2H_4Cl_2}\right)\left(\frac{165.833 \text{ g } C_2Cl_4}{1 \text{ mol } C_2Cl_4}\right)\left(\frac{1 \text{ kg}}{10^3 \text{ g}}\right) \text{ or}$$

$$? \text{ kg } C_2Cl_4 = 23.75 \text{ kg } C_2H_4Cl_2\left(\frac{4 \times 165.833 \text{ kg } C_2Cl_4}{8 \times 98.959 \text{ kg } C_2H_4Cl_2}\right)$$

$$= \mathbf{19.90 \text{ kg } C_2Cl_4}$$

**10.2. (a)** $$? \text{ Mg } UO_2F_2 = 24.543 \text{ Mg } UF_6\left(\frac{10^6 \text{ g}}{1 \text{ Mg}}\right)\left(\frac{1 \text{ mol } UF_6}{352.019 \text{ g } UF_6}\right)$$

$$\left(\frac{1 \text{ mol } UO_2F_2}{1 \text{ mol } UF_6}\right)\left(\frac{308.0245 \text{ g } UO_2F_2}{1 \text{ mol } UO_2F_2}\right)\left(\frac{1 \text{ Mg}}{10^6 \text{ g}}\right) \text{ or}$$

$$? \text{ Mg } UO_2F_2 = 24.543 \text{ Mg } UF_6\left(\frac{1 \times 308.0245 \text{ Mg } UO_2F_2}{1 \times 352.019 \text{ Mg } UF_6}\right)$$

$$= \mathbf{21.476 \text{ Mg } UO_2F_2}$$

$$? \text{ Mg } UO_2F_2 = 8.0 \text{ Mg } H_2O\left(\frac{10^6 \text{ g}}{1 \text{ Mg}}\right)\left(\frac{1 \text{ mol } H_2O}{18.0153 \text{ g } H_2O}\right)$$

$$\left(\frac{1 \text{ mol } UO_2F_2}{2 \text{ mol } H_2O}\right)\left(\frac{308.0245 \text{ g } UO_2F_2}{1 \text{ Mol } UO_2F_2}\right)\left(\frac{1 \text{ Mg}}{10^6 \text{ g}}\right) \text{ or}$$

$$? \text{ Mg } UO_2F_2 = 8.0 \text{ Mg } H_2O\left(\frac{1 \times 308.0245 \text{ Mg } UO_2F_2}{2 \times 18.0153 \text{ Mg } H_2O}\right)$$

$$= 68 \text{ Mg } UO_2F_2$$

**(b)** Water is much less toxic and less expensive than the radioactive and rare uranium compound. Water in the form of either liquid or steam is also very easy to separate from the solid product mixture.

**10.3.** $? \text{ kg } Na_2CrO_4 = 1.0 \text{ kg } FeCr_2O_4 \left(\frac{10^3 \text{ g}}{1 \text{ kg}}\right)\left(\frac{1 \text{ mol } FeCr_2O_4}{223.835 \text{ g } FeCr_2O_4}\right)$

$\left(\frac{8 \text{ mol } Na_2CrO_4}{4 \text{ mol } FeCr_2O_4}\right)\left(\frac{161.9733 \text{ g } Na_2CrO_4}{1 \text{ mol } Na_2CrO_4}\right)\left(\frac{1 \text{ kg}}{10^3 \text{ g}}\right)$ or

$$? \text{ kg } Na_2CrO_4 = 1.0 \text{ kg } FeCr_2O_4 \left(\frac{8 \times 161.9733 \text{ kg } Na_2CrO_4}{4 \times 223.835 \text{ kg } FeCr_2O_4}\right)$$

$$= 1.4 \text{ kg } Na_2CrO_4$$

$$\text{Percent Yield} = \frac{\text{actual yield}}{\text{theoretical yield}} \times 100 = \frac{1.2 \text{ kg } Na_2CrO_4}{1.4 \text{ kg } Na_2CrO_4} \times 100$$

$$= \mathbf{86\% \text{ yield}}$$

**10.4.** $\text{Molarity} = \frac{? \text{ mol } AgClO_4}{1 \text{ L } AgClO_4 \text{ soln}}$

$$= \frac{29.993 \text{ g } AgClO_4}{50.0 \text{ mL } AgClO_4 \text{ soln}}\left(\frac{1 \text{ mol } AgClO_4}{207.3185 \text{ g } AgClO_4}\right)\left(\frac{10^3 \text{ mL}}{1 \text{ L}}\right)$$

$$= \frac{2.893 \text{ mol } AgClO_4}{1 \text{ L } AgClO_4 \text{ soln}}$$

$$= \mathbf{2.893 \text{ M } AgClO_4}$$

**10.5.** $2HNO_3(aq) + Na_2CO_3(aq) \rightarrow H_2O(l) + CO_2(g) + 2NaNO_3(aq)$

$? \text{ mL } HNO_3 \text{ soln} = 75.0 \text{ mL } Na_2CO_3$

$$\left(\frac{0.250 \text{ mol } Na_2CO_3}{10^3 \text{ mL } Na_2CO_3}\right)\left(\frac{2 \text{ mol } HNO_3}{1 \text{ mol } Na_2CO_3}\right)\left(\frac{10^3 \text{ mL } HNO_3 \text{ soln}}{6.00 \text{ mol } HNO_3}\right)$$

$$= \mathbf{6.25 \text{ mL } HNO_3 \text{ soln}}$$

**10.6.** $2AgNO_3(aq) + MgCl_2(aq) \rightarrow 2AgCl(s) + Mg(NO_3)_2(aq)$

$? \text{ g AgCl} = 25.00 \text{ mL } MgCl_2 \left(\frac{0.050 \text{ mol } MgCl_2}{10^3 \text{ mL } MgCl_2}\right)$

$$\left(\frac{2 \text{ mol AgCl}}{1 \text{ mol } MgCl_2}\right)\left(\frac{143.3209 \text{ g AgCl}}{1 \text{ mol AgCl}}\right) = \mathbf{0.36 \text{ g AgCl}}$$

## Chapter 10 Review Questions

**1. (a)** $4HF + SiO_2 \rightarrow SiF_4 + 2H_2O$
**(b)** $4NH_3 + 5O_2 \rightarrow 4NO + 6H_2O$
**(c)** $3Ni(C_2H_3O_2)_2 + 2Na_3PO_4 \rightarrow Ni_3(PO_4)_2 + 6NaC_2H_3O_2$
**(d)** $H_3PO_4 + 3KOH \rightarrow 3H_2O + K_3PO_4$
**2. (a)** $Ca(NO_3)_2(aq) + Na_2CO_3(aq) \rightarrow CaCO_3(s) + 2NaNO_3(aq)$
**(b)** $3HNO_3(aq) + Al(OH)_3(s) \rightarrow 3H_2O(l) + Al(NO_3)_3(aq)$
**3.** 0.83888 mol $H_3PO_3$ **4.** 0.1202 kg $Na_2SO_4$

## Chapter 10 Key Ideas

**5.** stoichiometry **7.** stoichiometric **9.** optimize **11.** easy to separate **13.** two or more amounts, amount of product **15.** reversible **17.** slow **19.** volumes **21.** molar mass, mass, volume of solution

## Chapter 10 Problems

**22. (a)** $2HF(l) \xrightarrow{\text{electric current}} H_2(g) + F_2(g)$

**(b)**

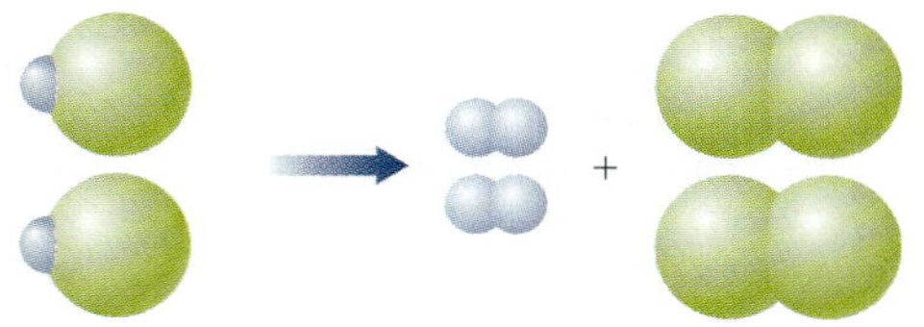

**(c)** $\left(\frac{1 \text{ mol } F_2}{2 \text{ mol HF}}\right)$ or $\left(\frac{2 \text{ mol HF}}{1 \text{ mol } F_2}\right)$
**(d)** 0.5 mol $F_2$
**(e)** 6.904 mol HF
**24. (a)** $3Mg + N_2 \rightarrow Mg_3N_2$
**(b)** $\left(\frac{3 \text{ mol Mg}}{1 \text{ mol } Mg_3N_2}\right)$ or $\left(\frac{1 \text{ mol } Mg_3N_2}{3 \text{ mol Mg}}\right)$
**(c)** 0.33 mol $Mg_3N_2$
**(d)** $\left(\frac{1 \text{ mol } N_2}{1 \text{ mol } mg_3N_2}\right)$ or $\left(\frac{1 \text{ mol } Mg_3N_2}{1 \text{ mol } N_2}\right)$
**(e)** 3.452 mol $N_2$
**26. (a)** $\left(\frac{2 \text{ mol HF}}{1 \text{ mol } XeF_2}\right)$ or $\left(\frac{1 \text{ mol } XeF_2}{2 \text{ mol HF}}\right)$
**(b)** 8.0 mol $XeF_2$ **(c)** 2 mol $NaBrO_4$ **(d)** 2 mole of $NaBrO_4$
**(e)** $\left(\frac{2 \text{ mol HF}}{1 \text{ mol } NaBrO_4}\right)$ or $\left(\frac{1 \text{ mol } NaBrO_4}{2 \text{ mol HF}}\right)$
**(f)** 11.64 mol HF
**28. (a)** $BrF(g) + 2F_2(g) \rightarrow BrF_5(l)$
**(b)** $\left(\frac{1 \text{ mol } BrF_5}{2 \text{ mol } F_2}\right)$ or $\left(\frac{2 \text{ mol } F_2}{1 \text{ mol } BrF_5}\right)$
**(c)** 3 mol $BrF_5$ **(d)** 6 mol $BrF_5$
**(e)** $\left(\frac{1 \text{ mol } BrF_5}{1 \text{ mol BrF}}\right)$ or $\left(\frac{1 \text{ mol BrF}}{1 \text{ mol } BrF_5}\right)$
**(f)** 0.78 mol BrF
**31. (a)** $\left(\frac{9 \text{ mol Fe}}{4 \text{ mol } C_6H_5NO_2}\right)$ or $\left(\frac{4 \text{ mol } C_6H_5NO_2}{9 \text{ mol Fe}}\right)$
**(b)** 827.2 g Fe
**(c)** $\left(\frac{4 \text{ mol } C_6H_5NH_2}{4 \text{ mol } C_6H_5NO_2}\right)$ or $\left(\frac{1 \text{ mol } C_6H_5NH_2}{1 \text{ mol } C_6H_5NO_2}\right)$ or $\left(\frac{4 \text{ mol } C_6H_5NO_2}{4 \text{ mol } C_6H_5NH_2}\right)$

or $\left(\frac{1 \text{ mol } C_6H_5NO_2}{1 \text{ mol } C_6H_5NH_2}\right)$
**(d)** 613.1 g $C_6H_5NH_2$
**(e)** $\left(\frac{3 \text{ mol } Fe_3O_4}{4 \text{ mol } C_6H_5NH_2}\right)$ or $\left(\frac{4 \text{ mol } C_6H_5NH_2}{3 \text{ mol } Fe_3O_4}\right)$
**(f)** 1143 g $Fe_3O_4$ or 1.143 kg $Fe_3O_4$ **(g)** 78.00% yield
**33. (a)** $2Na_2CrO_4 + H_2SO_4 \rightarrow Na_2Cr_2O_7 + Na_2SO_4 + H_2O$
**(b)** 104.8 kg $Na_2CrO_4$ **(c)** 45.94 kg $Na_2SO_4$
**35. (a)** 1.98 kg $Na_2Cr_2O_7$ **(b)** 1.08 Mg $Na_2SO_4$
**38.** 71.65 g $Cr_2O_3$
**40. (a)** 12.239 kg $B_4C$
**(b)** There are two reasons why we are not surprised that the carbon is in excess. We would expect carbon to be less expensive than the less common boric acid, and the excess carbon can be separated easily from the solid $B_4C$ by converting it to gaseous carbon dioxide or carbon monoxide.
**42. (a)** 613.1 g $C_6H_5NH_2$
**(b)** Both iron and water would be less expensive than nitrobenzene. They would also be expected to be less toxic than nitrobenzene.
**44. (a)** $CaO + 3C \rightarrow CaC_2 + CO$
**(b)** The carbon is probably best to have in excess. We would expect carbon to be less expensive than the calcium oxide, and the excess carbon can be separated easily from the solid $CaC_2$ by converting it to gaseous carbon dioxide or carbon monoxide. Thus, the CaO would be limiting.
**(c)** We would add 752.8 g CaO and well over 483.7 g C.
**46.** (1) Many chemical reactions are significantly reversible. Because there is a constant conversion of reactants to products and products to reactants, the reaction never proceeds completely to products. (2) It is common, especially in reactions involving organic compounds, to have side reactions. These reactions form products other than the

desired product. (3) Sometimes a reaction is so slow that it has not reached the maximum yield by the time that the product is isolated. (4) Even if 100% of the limiting reactant proceeds to products, the product still usually needs to be separated from the other components in the product mixture. (The other components include excess reactants, products of side reactions, and other impurities.) This separation generally involves some loss of product.
**48.** Although the maximum (or theoretical) yield of a reaction is determined by the limiting reactant rather than reactants in excess, reactants that are in excess can affect the actual yield of an experiment. Sometimes the actual yield is less than the theoretical yield because the reaction is reversible. Adding a large excess of one of the reactants ensures that the limiting reactant reacts as completely as possible (by speeding up the forward rate in the reversible reaction and driving the reaction toward a greater actual yield of products).
**50.** 0.4378 M $Al_2(SO_4)_3$
**52. (a)** 48.5 mL $H_2SO_4$ soln **(b)** 824 g $UO_2SO_4$
**54. (a)** $Na_2SO_3(aq) + FeCl_2(aq) \rightarrow 2NaCl(aq) + FeSO_3(s)$
**(b)** 3.428 g $FeSO_3$

**56. (a)** $\left(\frac{1 \text{ mol } HNO_3}{1 \text{ mol } KOH}\right)$ **(b)** 41.8 mL $HNO_3$ soln

**58. (a)** $\left(\frac{1 \text{ mol } H_2SO_4}{2 \text{ mol } NaOH}\right)$ **(b)** 9.9 mL $H_2SO_4$ soln

**60. (a)** $\left(\frac{2 \text{ mol } HCl}{1 \text{ mol } Co(OH)_2}\right)$ **(b)** 8.87 L HCl soln

**62. (a)** $\left(\frac{3 \text{ mol } HNO_3}{1 \text{ mol } Cr(OH)_3}\right)$ **(b)** 7.534 L $HNO_3$ soln

**64. (a)** $\left(\frac{3 \text{ mol } H_2}{1 \text{ mol } HCP}\right)$ or $\left(\frac{1 \text{ mol } HCP}{3 \text{ mol } H_2}\right)$

**(b)** 3 mol HCP

**(c)** $\left(\frac{3 \text{ mol } H_2}{1 \text{ mol } CH_4}\right)$ or $\left(\frac{1 \text{ mol } CH_4}{3 \text{ mol } H_2}\right)$

**(d)** 5.6502 mol $H_2$
**66. (a)** $5IF \rightarrow 2I_2 + IF_5$ **(b)** 6.0 mol $I_2$ **(c)** 1.588 mol $IF_5$
**70.** $1.38 \times 10^3$ g $I_2$ or 1.38 kg $I_2$
**72. (a)** $2KCl + SO_2 + \frac{1}{2}O_2 + H_2O \rightarrow K_2SO_4 + 2HCl$ or $4KCl + 2SO_2 + O_2 + 2H_2O \rightarrow 2K_2SO_4 + 4HCl$
**(b)** $3.23 \times 10^5$ kg $K_2SO_4$
**(c)** 91.0% yield
**74. (a)** 13.06 Mg C **(b)** 13.47 Mg P **(c)** 36.58 Mg CaO **(d)** 82.63% yield
**76.** 70.68% yield
**78. (a)** Since the $Na_2CrO_4$ forms the least product (70.860 g $Na_2Cr_2O_7$), it is the limiting reactant. **(b)** Both water and carbon dioxide are very inexpensive and nontoxic. Since $CO_2$ is a gas and since water can be easily converted to steam, they are also very easily separated from solid products. Adding an excess of these substances drives the reversible reaction toward products and yields a more complete conversion of $Na_2CrO_4$ to $Na_2Cr_2O_7$.
**80.** 0.359 M $SnBr_2$
**82. (a)** $3Na_2CO_3(aq) + 2Cr(NO_3)_3(aq) \rightarrow Cr_2(CO_3)_3(s) + 6NaNO_3(aq)$
**(b)** 0.142 g $Cr_2(CO_3)_3$

**86. (a)** $\left(\frac{2 \text{ mol } HCl}{1 \text{ mol } ZnCO_3}\right)$ **(b)** 17.9 mL HCl soln

**88.** 136 mL $H_2SO_4$ solution
**90. (a)** $3H_2SO_4 + Al_2O_3 \rightarrow Al_2(SO_4)_3 + 3H_2O$
**(b)** $4.8 \times 10^3$ kg $Al_2(SO_4)_3$
**92.** $2.4 \times 10^4$ kg $Na_2CO_3$
**94.** $4 \times 10^2$ g $NH_2CONHCONH_2$ **96. (a)** $2.8 \times 10^5$ lb $TiCl_4$
**(b)** Carbon is inexpensive, nontoxic, and easy to convert to gaseous CO or $CO_2$, which are easy to separate from solid and liquid products. Although chlorine gas is a more dangerous substance, it is inexpensive and easy to separate from the product mixture. Because the ultimate goal is to convert the titanium in $TiO_2$ into $TiCl_4$, the $TiO_2$ is the more important reactant, so it is limiting.
**98.** 12.44 kg $CaHPO_4$

## Chapter 11 Exercises

**11.1.** $1s^2 2s^2 2p^6 3s^2 3p^6 4s^2 3d^{10} 4p^6 5s^2 4d^{10} 5p^3$

5s ⇅ 5p ↑ ↑ ↑ 4d ⇅ ⇅ ⇅ ⇅ ⇅
4s ⇅ 4p ⇅ ⇅ ⇅ 3d ⇅ ⇅ ⇅ ⇅ ⇅
3s ⇅ 3p ⇅ ⇅ ⇅
2s ⇅ 2p ⇅ ⇅ ⇅
1s ⇅

**11.2. (a)** $[Kr]\ 5s^1$
**(b)** $[Ar]\ 4s^2 3d^8$
**(c)** $[Xe]\ 6s^2 4f^{14} 5d^{10} 6p^3$

## Chapter 11 Review Questions

**1.** Protons and neutrons are in a tiny core of the atom called the nucleus, which has a diameter about 1/100,000 the diameter of the atom. The position and motion of the electrons are uncertain, but they generate a negative charge that is felt in the space that surrounds the nucleus.
**2.** Increased stability of the components of a system leads to decreased potential energy, and decreased stability of the components of a system leads to increased potential energy.

## Chapter 11 Key Ideas

**3.** strange **5.** impossible, exactly **7.** probably, definitely **9.** negative charge **11.** intensity of movement, intensity of the negative charge **13.** possible **15.** other elements **17.** high percentage, volume **19.** cloud **21.** strength, decreased **23.** spontaneously returns **25.** sublevel **27.** *n* **29.** seventh **31.** spinning **33.** 2, spins **35.** *s* orbitals **37.** *d*

## Chapter 11 Problems

**39.** The possible waveforms are limited by the fact that the string is tied down and cannot move at the ends. In theory, there are an infinite number of possible waveforms that allow the string to remain stationary at the ends.
**41.** The negative-charge distribution of an electron in a 1*s* orbital of a hydrogen atom looks like the image that follows. The cloud is pictured as surrounding the nucleus and represents the variation in the intensity of the negative charge at different positions outside the nucleus. The negative charge is most intense at the nucleus and diminishes with increasing distance from the nucleus. The variation in charge intensity for this waveform is the same in all directions, so the waveform is a sphere. Theoretically, the charge intensity decreases toward zero as the distance from the nucleus approaches infinity. The 1*s* orbital can be described as a sphere that contains a high percentage (for example 90% or 99%) of the charge of the 1*s* electron. See Figure 11.3.

According to the particle interpretation of the wave character of the electron, a 1*s* orbital is a surface within which we have a high probability of finding the electron. In the particle view, the electron cloud can be compared to a multiple-exposure photograph of the electron (once again, we must resort to an analogy to describe electron behavior). If we were able to take a series of sharply focused photos of an electron over a period of time without advancing the film, our final picture would look like the image that follows. We would find a high density of dots near the nucleus (because most of the times when the shutter snaps, the electron would be near the nucleus) and a decrease in density with increasing distance from the nucleus (because some of the times the shutter snaps, the electron would be farther away from the nucleus). This arrangement of dots would bear out the wave equation's prediction of the probability of finding the electron at any given distance from the nucleus. See Figure 11.5.
**43.** The 2*s* orbital for an electron in a hydrogen atom is spherical like the 1*s* orbital, but it is a larger sphere. For an electron in the 2*s* orbital,

the charge is most intense at the nucleus, it diminishes in intensity to a minimum with increasing distance from the nucleus, it increases again to a maximum, and finally it diminishes again. The section of the 2s orbital where the charge intensity goes to zero is called a node. Figure 11.7 shows cutaway, quarter section views of the 1s and 2s orbitals.

**45.** The 3*p* orbital is larger than the 2*p* orbital. Because the average distance between the positively charged nucleus and the negative charge of an electron in a 2*p* orbital would be less than for an electron in a 3*p* orbital, the attraction between a 2*p*-electron and the nucleus would be stronger. This makes an electron in a 2*p* orbital more stable and lower potential energy than an electron in a 3*p* orbital.

**47.** The three 2*p* orbitals are identical in shape and size, but each is 90° from the other two. Because they can be viewed as being on the x, y and z axes of a three-dimensional coordinate system, they are often called the $2p_x$, $2p_y$, and $2p_z$ orbitals. One electron with a 2*p* waveform has its negative charge distributed in two lobes on opposite sides of the nucleus. We will call this a dumbbell shape. Figures 11.8 and 11.9 show two ways to visualize these orbitals, and Figure 11.10 shows the three 2*p* orbitals together.

**50.** 3

**52.** There are 9 orbitals in the third principal energy level: 1 in the 3*s* sublevel, 3 in the 3*p* sublevel, and 5 in the 3*d* sublevel.

**54.** **(a)** exist **(b)** exist **(c)** not exist **(d)** exist

**57.** 2 The maximum number of electrons in *any* orbital is 2.

**59.** The maximum number of electrons in any *p* sublevel is 6. The maximum number of electrons in any *d* sublevel is 10.

**61.** The third principal energy level can hold up to 18 electrons: 2 in the 3*s*, 6 in the 3*p*, and 10 in the 3*d*.

**63.** **(a)** 2*s* **(b)** 3*s* **(c)** 4*s* **(d)** 6*s*

**65.** **(a)** $1s^2 2s^2 2p^2$

2s ⇅ 2p ↑ ↑ _

1s ⇅

**(b)** $1s^2 2s^2 2p^6 3s^2 3p^3$

3s ⇅ 3p ↑ ↑ ↑

2s ⇅ 2p ⇅ ⇅ ⇅

1s ⇅

**(c)** $1s^2 2s^2 2p^6 3s^2 3p^6 4s^2 3d^3$

4s ⇅ 3d ↑ ↑ ↑ _ _

3s ⇅ 3p ⇅ ⇅ ⇅

2s ⇅ 2p ⇅ ⇅ ⇅

1s ⇅

**(d)** $1s^2 2s^2 2p^6 3s^2 3p^6 4s^2 3d^{10} 4p^6 5s^2 4d^{10} 5p^5$

5s ⇅ 5p ⇅ ⇅ ↑ 4d ⇅ ⇅ ⇅ ⇅ ⇅

4s ⇅ 4p ⇅ ⇅ ⇅ 3d ⇅ ⇅ ⇅ ⇅ ⇅

3s ⇅ 3p ⇅ ⇅ ⇅

2s ⇅ 2p ⇅ ⇅ ⇅

1s ⇅

**(e)** $1s^2 2s^2 2p^6 3s^2 3p^6 4s^2 3d^{10} 4p^6 5s^2 4d^{10} 5p^6 6s^2 4f^{14} 5d^{10}$

6s ⇅ 5d ⇅ ⇅ ⇅ ⇅ ⇅ 4f ⇅ ⇅ ⇅ ⇅ ⇅ ⇅ ⇅

5s ⇅ 5p ⇅ ⇅ ⇅ 4d ⇅ ⇅ ⇅ ⇅ ⇅

4s ⇅ 4p ⇅ ⇅ ⇅ 3d ⇅ ⇅ ⇅ ⇅ ⇅

3s ⇅ 3p ⇅ ⇅ ⇅

2s ⇅ 2p ⇅ ⇅ ⇅

1s ⇅

**67.** **(a)** Be **(b)** Na **(c)** Br **(d)** Pb

**69.** **(a)** excited **(b)** ground **(c)** excited **(d)** ground

**71.** **(a)** [He] $2s^2 2p^5$ **(b)** [Ne] $3s^2 3p^2$ **(c)** [Ar] $4s^2 3d^7$ **(d)** [Kr] $5s^2 4d^{10} 5p^1$ **(e)** [Xe] $6s^2 4f^{14} 5d^{10} 6p^4$

**73.** **(a)** 7*s* **(b)** 3*d* **(c)** 5*f* **(d)** 6*p*

**75.** **(a)** Al **(b)** Ca **(c)** Mn

**77.** The pair "a" and "c" represent the alkaline earth metals magnesium and strontium.

**79.** **(a)** 30 **(b)** 2 **(c)** 128

**82.** [Uuo] $8s^2 5g^1$

## Chapter 12 Exercises

**12.1** (a) $CCl_4$: :C̈l—C(—C̈l:)(:C̈l:)(:C̈l:) (b) :C̈l—Ö—C̈l: (c) :F̈—C(=Ö)—F̈: (d) :C̈l—C(:C̈l:)(:C̈l:)—C(:C̈l:)(:C̈l:)—C̈l:

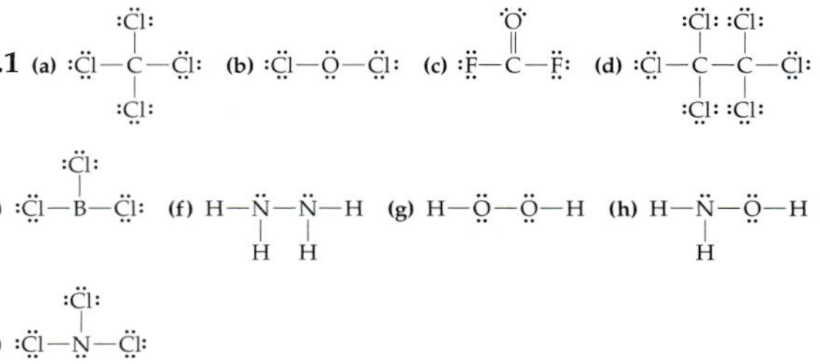

(e) :C̈l—B(—:C̈l:)—C̈l: (f) H—N̈(H)—N̈(H)—H (g) H—Ö—Ö—H (h) H—N̈(H)—Ö—H

(i) :C̈l—N̈(—:C̈l:)—C̈l:

**12.2** [:Ö—C(=Ö)—Ö:]²⁻ ↔ [Ö=C(—:Ö:)—Ö:]²⁻ ↔ [:Ö—C(—:Ö:)=Ö]²⁻ [O⋯C(⋯O)⋯O]²⁻

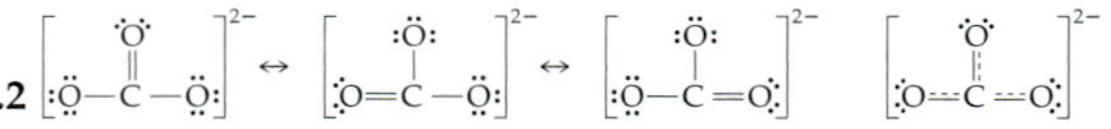

**12.3** **(a)** Electron Group Geometry—**tetrahedral**

:B̈r: 109.5° C B̈r B̈r B̈r

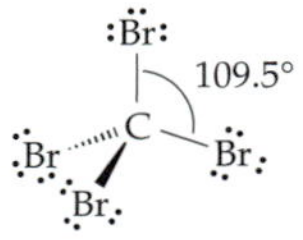

Molecular Geometry—**tetrahedral**

**(b)** Electron Group Geometry—**tetrahedral**

Äs F̈ F̈ F̈ ≈109.5°

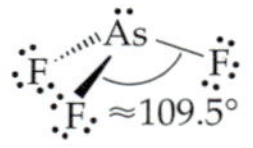

Molecular Geometry—**trigonal pyramid**

**(c)** Electron Group Geometry—**linear**

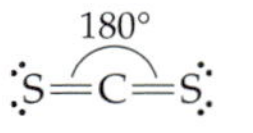

180° S̈=C=S̈

Molecular Geometry—**linear**

**(d)** Electron Group Geometry—**tetrahedral**

:F̈: ≈109.5° :Ö F̈:

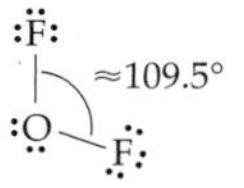

Molecular Geometry—**bent**

**(e)** Electron Group Geometry—**trigonal planar**

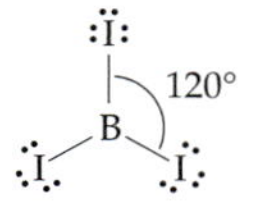

Molecular Geometry—**trigonal planar**

**(f)** Electron Group Geometry—**trigonal planar**

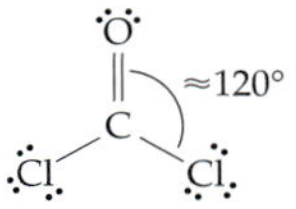

Molecular Geometry—**trigonal planar**

**(g)** Electron Group Geometry for left carbon—**tetrahedral**
Electron Group Geometry for right carbon—**linear**

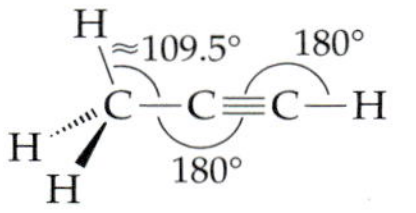

Molecular Geometry for left carbon—**tetrahedral**
Molecular Geometry for right carbon—**linear**

**(h)** Electron Group Geometry—**tetrahedral**

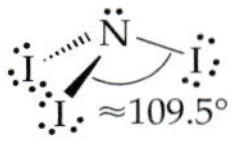

Molecular Geometry—**trigonal pyramid**

## Chapter 12 Review Questions

**1.** 7A

**2.** 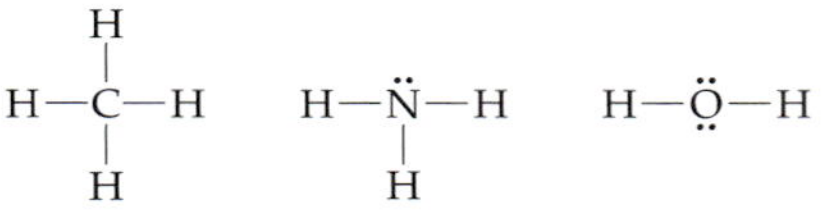 H—C—H (with H above and below C)  H—N̈—H (with H below N)  H—Ö—H

**3.** Orbital can be defined as the volume that contains a high percentage of the electron charge generated by an electron in an atom. It can also be defined as the volume within which an electron has a high probability of being found.

**4. (a)** $1s^2 2s^2 2p^4$

2s ⇅  2p ⇅ ↑ ↑

1s ⇅

**(b)** $1s^2 2s^2 2p^6 3s^2 3p^3$

3s ⇅  3p ↑ ↑ ↑

2s ⇅  2p ⇅ ⇅ ⇅

1s ⇅

## Chapter 12 Key Ideas

**5.** true, useful **7.** change **9.** *s* and *p* **11.** lone pairs **13.** four **15.** most common, polyatomic ions **17.** cations **19.** never **21.** fewest **23.** oxygen **25.** two **27.** structural **29.** as far apart **31.** electron group, molecular

## Chapter 12 Problems

**32.** Our models come with advantages and disadvantages. They help us to visualize, explain, and predict chemical changes, but we need to remind ourselves now and then that they are only models, and as models, they have their limitations. For example, because a model is a simplified version of what we think is true, the processes it depicts are sometimes described using the phrase "*as if*." When you read, "It is *as if* an electron were promoted from one orbital to another," the phrase is a reminder that we do not necessarily think this is what really happens. We find it *useful* to talk about the process *as if* this is the way it happens.

One characteristic of models is that they change with time. Because our models are a simplification of what we think is real, we are not surprised when they sometimes fail to explain experimental observation. When this happens, the model is altered to fit the new observations.

**34. (a)** 5 valence electrons—$2s^2\,2p^3$ **(b)** 6 valence electrons—$3s^2\,3p^4$ **(c)** 7 valence electrons—$5s^2\,5p^5$ **(d)** 8 valence electrons—$3s^2\,3p^6$

**36. (a)** 5 ·N̈· **(b)** 6 :S̈· **(c)** 7 :Ï· **(d)** 8 :Är:

**38. (a)** Group 15 or 5A **(b)** Group 18 or 8A **(c)** Group 13 or 3A

**40.** 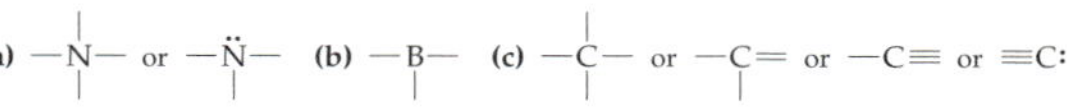 (a) —N— or —N̈— (b) —B— (c) —C— or —C= or —C≡ or ≡C:

**42.** The answers to each of these problems is based on the following assumptions of the valence bond model.

- Only the highest energy electrons participate in bonding.
- Covalent bonds usually form to pair unpaired electrons.

**(a)** Fluorine is in group 7A, so it has seven valence electrons per atom. The orbital diagram for the valence electrons of fluorine is below.

2p ⇅ ⇅ ↑

2s ⇅  :F̈·

The one unpaired electron leads to one bond, and the three pairs of electrons give fluorine atoms three lone pairs.

**(b)** Carbon is in group 4A, so it has four valence electrons per atom. It is *as if* one electron is promoted from the 2*s* orbital to the 2*p* orbital.

2p ↑ ↑ _  →  2p ↑ ↑ ↑

2s ⇅  :Ċ·  2s ↑  ·Ċ·

The four unpaired electrons lead to four covalent bonds. Because there are no pairs of electrons, carbon atoms have no lone pairs when they form four bonds.

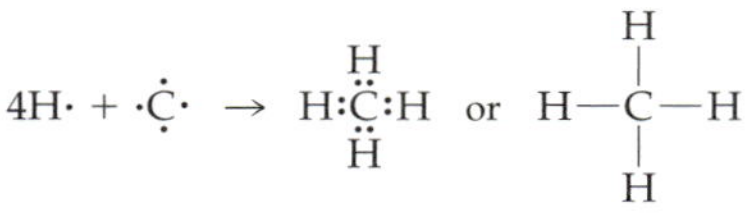

4H· + ·Ċ· → H:C̈:H (H above and below) or H—C—H (H above and below)

**(c)** Nitrogen is in group 5A, so it has five valence electrons per atom. The orbital diagram for the valence electrons of nitrogen is below.

2p ↑ ↑ ↑

2s ⇅  ·N̈·

The three unpaired electrons lead to three bonds, and the one pair of electrons gives nitrogen atoms one lone pair.

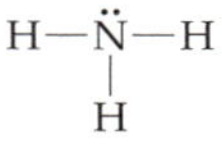

H—N̈—H (H below N)

**(d)** Sulfur is in group 6A, so it has six valence electrons per atom. The orbital diagram for the valence electrons of sulfur is below.

3p ⇅ ↑ ↑

3s ⇅  :S̈·

The two unpaired electrons lead to two bonds, and the two pairs of electrons give sulfur atoms two lone pairs.

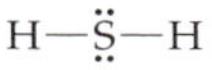

H—S̈—H

**(e)** Oxygen is in group 6A, so it has six valence electrons per atom. If it gains one electron, it will have a total of seven.

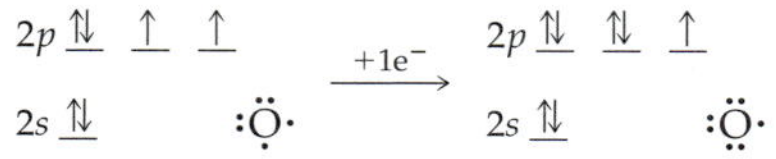

$2p$ ⇅ ↑ ↑ $\xrightarrow{+1e^-}$ $2p$ ⇅ ⇅ ↑

$2s$ ⇅ :Ö· $2s$ ⇅ :Ö·

The one unpaired electron leads to one bond, and the three pairs of electrons give oxygen atoms with an extra electron three lone pairs.

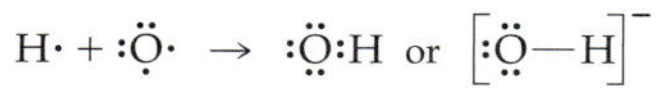

H· + :Ö· → :Ö:H or [:Ö—H]$^-$

**44. (a)** $CH_4$ **(b)** $H_2S$ **(c)** $BF_3$

**46.** Single bonds, Single bonds, Lone pairs, Double bond

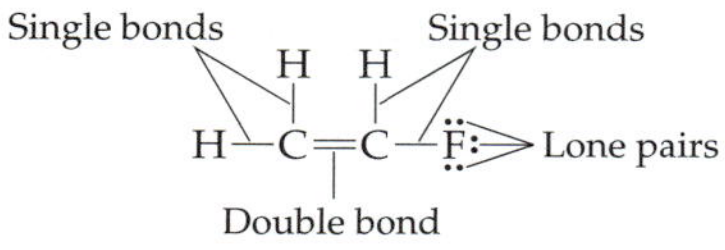

**48. (a)** Carbon—The element with the fewest atoms in the formula is often in the center. The atom that is capable of making the most bonds is often in the center. Carbon atoms usually form 4 bonds, and bromine atoms usually form 1 bond.
**(b)** Sulfur—The element with the fewest atoms in the formula is often in the center. Oxygen atoms are rarely in the center. Oxygen atoms rarely bond to other oxygen atoms.
**(c)** Sulfur—The element with the fewest atoms in the formula is often in the center. The atom that is capable of making the most bonds is often in the center. Sulfur atoms usually form 2 bonds, and hydrogen atoms form 1 bond. Hydrogen atoms are never in the center.
**(d)** Nitrogen—The atom that is capable of making the most bonds is often in the center. Nitrogen atoms usually form 3 bonds, oxygen atoms usually form 2 bonds, and fluorine atoms form 1 bond. Fluorine atoms are never in the center.
**50. (a)** $1 + 5 + 3(6) = 24$ valence electrons
**(b)** $2(4) + 3(1) + 7 = 18$ valence electrons

**52. (a)** (b) :F̤—Ö—Ö—F̤: (c) H—C≡C—F̤:

**(d)** H—N—C̤l: **(e)** H—P—H **(f)** :F̤—S̤—S̤—F̤:

**(g)** H—Ö—N̈=Ö **(h)** :F̤—N—N—F̤: **(i)** H—C=C—C—H

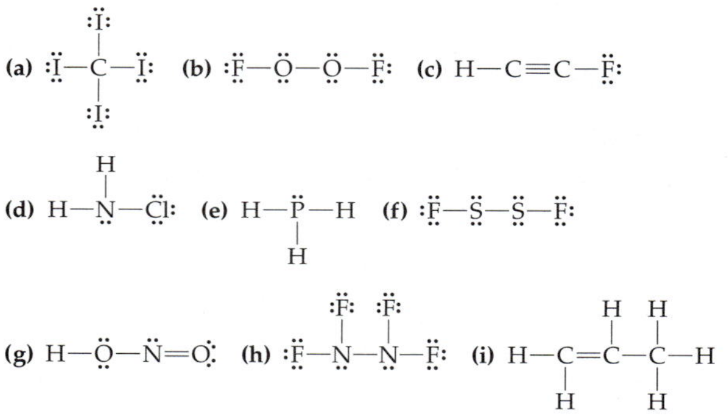

**54.** Ö=Ö—Ö: ↔ :Ö—Ö=Ö    Ö--Ö--Ö

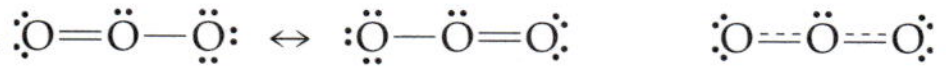

**56.** Atoms are arranged in molecules to keep the electron groups around the central atom as far apart as possible. An electron group is either (1) a single bond, (2) a multiple bond (double or triple), or (3) a lone pair. The Lewis structure for $CO_2$ shows that the carbon atom has 2 electron groups around it. The best way to get two things as far apart as possible is in a linear arrangement.

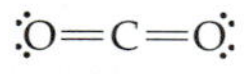

Ö=C=Ö

The Lewis structure for $H_2O$ shows that the oxygen atom has 4 electron groups around it. The best way to get four things as far apart as possible is in a tetrahedral arrangement.

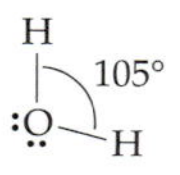

105°

180°

**58.** Y—X—Y

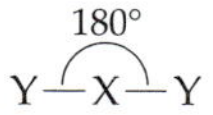

**61. (a)** :F̤—C—F̤: tetrahedral 109.5° tetrahedral

**(b)** Ö=N̈—C̤l: trigonal planar ≈120° bent

**(c)** :B̤r—B—B̤r: trigonal planar ≈120° trigonal planar

**(d)** :B̤r—Äs—B̤r: tetrahedral ≈109.5° trigonal pyramid

**(e)** :B̤r—C—B̤r: trigonal planar ≈120° trigonal planar

**(f)** Ö=C=S̈ linear 180° linear

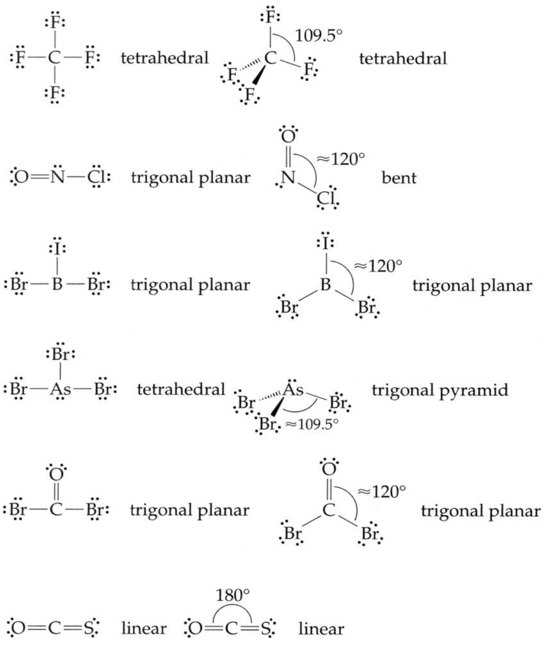

## CHAPTER 13 EXERCISES

**13.1. (a)** $PV = nRT$

$$n = \frac{PV}{RT} = \frac{117\text{ kPa }(175\text{ mL})}{8.3145\,\frac{\text{L}\cdot\text{kPa}}{\text{K}\cdot\text{mol}}\,294.8\text{ K}}\left(\frac{1\text{ L}}{10^3\text{ mL}}\right) = \mathbf{8.35 \times 10^{-3}\ mol\ Kr}$$

**(b)** $V = ?$ $g = 1.196$ g $T = 97\ °C + 273.15 = 370$ K $P = 1.70$ atm

$$PV = \frac{g}{M}RT$$

$$V = \frac{gRT}{PM} = \frac{1.196\text{ g Kr}\left(\frac{0.082058\text{ L}\cdot\text{atm}}{\text{K}\cdot\text{mole}}\right)370\text{ K}}{1.70\text{ atm}\left(83.80\,\frac{\text{g}}{\text{mol}}\right)} = \mathbf{0.255\ L\ Kr}$$

**(c)** $\frac{g}{V} = ?$ $P = 762$ mmHg $T = 18.2\ °C + 273.15 = 291.4$ K

$$PV = \frac{g}{M}RT$$

$$\frac{g}{V} = \frac{PM}{RT} = \frac{762\text{ mmHg}\left(83.80\,\frac{\text{g}}{\text{mol}}\right)}{0.082058\,\frac{\text{L}\cdot\text{atm}}{\text{K}\cdot\text{mole}}\,(291.4\text{ K})}\left(\frac{1\text{ atm}}{760\text{ mmHg}}\right) = \mathbf{3.51\ g/L}$$

**13.2.** $P_1 = 102$ kPa $T_1 = 18\ °C + 273.15 = 291$ K
$V_1 = 1.6 \times 10^4$ L $P_2 = ?$
$T_2 = -8.6\ °C + 273.15 = 264.6$ K $V_2 = 4.7 \times 10^4$ L

$$\frac{P_1V_1}{n_1T_1} = \frac{P_2V_2}{n_2T_2} \text{ to } \frac{P_1V_1}{T_1} = \frac{P_2V_2}{T_2} \quad P_2 = P_1\left(\frac{T_2}{T_1}\right)\left(\frac{V_1}{V_2}\right)$$

$$= 102\text{ kPa}\left(\frac{264.6\text{ K}}{291\text{ K}}\right)\left(\frac{1.6 \times 10^4\text{ L}}{4.7 \times 10^4\text{ L}}\right) = \mathbf{32\ kPa}$$

**13.3. (a)** $? \text{ L } O_2 = 125 \text{ Mg C} \left(\frac{10^6 \text{ g}}{1 \text{ Mg}}\right)\left(\frac{1 \text{ mol C}}{12.011 \text{ g C}}\right)\left(\frac{1 \text{ mol } O_2}{2 \text{ mol C}}\right)$

$\left(\frac{22.414 \text{ L } O_2}{1 \text{ mol } O_2}\right)_{STP} = \mathbf{1.17 \times 10^8 \text{ L } O_2 \text{ or } 1.17 \times 10^5 \text{ m}^3 \text{ } O_2}$

**(b)** $? \text{ L CO} = 8.74 \times 10^5 \text{ L } O_2 \left(\frac{\text{K} \cdot \text{mol}}{0.082058 \text{ L} \cdot \text{atm}}\right)\left(\frac{0.994 \text{ atm}}{300 \text{ K}}\right)$

$\left(\frac{2 \text{ mol CO}}{1 \text{ mol } O_2}\right)\left(\frac{0.082058 \text{ L} \cdot \text{atm}}{\text{K} \cdot \text{mol}}\right)\left(\frac{308 \text{ K}}{1.05 \text{ atm}}\right) = \mathbf{1.70 \times 10^6 \text{ L CO}}$

**13.4. (a)** $? \text{ L } Cl_2 = 3525 \text{ L NaOH soln} \left(\frac{12.5 \text{ mol NaOH}}{1 \text{ L NaOH soln}}\right)$

$\left(\frac{1 \text{ mol } Cl_2}{2 \text{ mol NaOH}}\right)\left(\frac{8.3145 \text{ L} \cdot \text{kPa}}{\text{K} \cdot \text{mol}}\right)\left(\frac{291.0 \text{ K}}{101.4 \text{ kPa}}\right) = \mathbf{5.26 \times 10^5 \text{ L } Cl_2}$

**13.5. (a)** $P_{total} = P_{Ne} + P_{Ar}$
$P_{Ar} = P_{total} - P_{Ne} = 1.30 \text{ kPa} - 0.27 \text{ kPa} = \mathbf{1.03 \text{ kPa}}$

**(b)** $? \text{ mol Ar} = 6.3 \text{ mg Ar} \left(\frac{1 \text{ g}}{10^3 \text{ mg}}\right)\left(\frac{1 \text{ mol Ar}}{39.948 \text{ g Ar}}\right) = 1.6 \times 10^{-4} \text{ mol Ar}$

$? \text{ mol Ne} = 1.2 \text{ mg Ne} \left(\frac{1 \text{ g}}{10^3 \text{ mg}}\right)\left(\frac{1 \text{ mol Ne}}{20.1797 \text{ g Ne}}\right) = 5.9 \times 10^{-5} \text{ mol Ne}$

$P_{total} = \left(\sum n\right)\frac{RT}{V} = (1.6 \times 10^{-4} \text{ mol} + 5.9 \times 10^{-5} \text{ mol})$

$\frac{\frac{0.082058 \text{ L} \cdot \text{atm}}{\text{K} \cdot \text{mol}}(291 \text{ K})}{375 \text{ mL}}\left(\frac{10^3 \text{ mL}}{1 \text{ L}}\right)\left(\frac{760 \text{ mmHg}}{1 \text{ atm}}\right) = \mathbf{11 \text{ mmHg}}$

## Chapter 13 Review Questions

**1.** The Kinetic Molecular Theory provides a simple model of the nature of matter. It has the following components:

- All matter is composed of tiny particles.
- These particles are in constant motion. The amount of motion is proportional to temperature. Increased temperature means increased motion.
- Solids, gases, and liquids differ in the degree of motion of their particles and the extent to which the particles interact.

Because the particles of a gas are much farther apart than for the solid or liquid, the particles do not have significant attractions between them. The particles in a gas move freely in straight-line paths until they collide with another particle or the walls of the container. If you were riding on a particle in the gas state, your ride would be generally boring with regular interruptions caused by violent collisions. Between these collisions, you would not even know there were other particles in the container. Because the particles are moving with different velocities and in different directions, the collisions lead to constant changes in the direction and velocity of the motion of each particle. The rapid, random movement of the gas particles allows gases to adjust to the shape and volume of their container.
**2.** 538.4 K, 292.6 °C
**3. (a)** 905 Mg **(b)** 905 Mg **(c)** 95.9% yield **(d)** $3.2 \times 10^6$ L $Na_2SO_4$ solution

## Chapter 13 Key Ideas

**4.** ten **6.** 0.1%, 99.9%, 70% **8.** rapid, continuous **10.** direction of motion, velocity **12.** point-masses **14.** pressure, volume, number of particles, temperature **16.** degrees Celsius, °C, Kelvin **18.** pascal, Pa **20.** moles of gas **22.** Kelvin temperature **24.** directly **26.** temperature and volume **28.** correct equation, algebra, necessary unit conversions **30.** molar mass, molarity **32.** alone

## Chapter 13 Problems

**34.** 0.1%
**36.** When we walk through air, we push the air particles out of the way as we move. Because the particles in a liquid occupy about 70% of the space that contains the liquid (as opposed to 0.1% of the space occupied by gas particles), there are a lot more particles to push out of the way as you move through water.
**38.** The particles are assumed to be point-masses, that is, particles that have a mass but occupy no volume. There are no attractive or repulsive forces between the particles in an ideal gas.
**40.** The particles in the air ($N_2$, $O_2$, Xe, $CO_2$, and others) are constantly moving and constantly colliding with everything surrounded by the air. Each of these collisions exerts a tiny force against the object with which they collide. The total force of these collisions per unit area is the atmospheric pressure.
**42.** $4 \times 10^8$ kPa, $4 \times 10^6$ atm, $3 \times 10^9$ mmHg, $3 \times 10^9$ torr
**44.** X and Y are inversely proportional if a decease in X leads to a proportional increase in Y or an increase in X leads to a proportional decrease in Y. For example, volume of gas and its pressure are inversely proportional if the temperature and moles of gas are constant. If the volume is decreased to one-half its original value, the pressure of the gas will double. If the volume is doubled, the pressure decreases to one-half its original value. The following expression summarizes this inverse relationship:

$P \propto \frac{1}{V}$ if n and T are constant

**47.** In gases, there is plenty of empty space between the particles and essentially no attractions between them, so there is nothing to stop gases, like ammonia and the gases in air, from mixing readily and thoroughly.
**49.** When the muscles of your diaphragm contract and your chest expands, the volume of your lungs increases. This leads to a decrease in the number of particles per unit volume inside the lungs, which leaves fewer particles near any given area of the inner surface of the lungs. There are then fewer collisions per second per unit area of lungs and a decrease in force per unit area or gas pressure. During quiet, normal breathing, this increase in volume decreases the pressure in the lungs to about 0.4 kilopascals lower than the atmospheric pressure. The larger volume causes air to move into the lungs faster than it moves out, bringing in fresh oxygen. When the muscles relax, the lungs return to their original volume, and the decrease in volume causes the pressure in the lungs to increase to about 0.4 kilopascals above atmospheric pressure. Air now goes out of the lungs faster than it comes in.
**51. (a)** As the can cools, the water vapor in the can condenses to liquid, leaving fewer moles of gas. Decreased moles of gas and decreased temperature both lead to decreased gas pressure in the can. Because the external pressure pushing on the outside of the can is then greater than the internal pressure pushing outward, the can collapses.
**(b)** The increased temperature causes the internal pressure of the ping-pong ball to increase. This leads to the pressure pushing out on the shell of the ball to be greater than the external pressure pushing in on the shell. If the difference in pressure is enough, the dents are pushed out.
**53.** 0.042 moles Ne
**55. (a)** $8.8 \times 10^2$ K or $6.1 \times 10^2$ °C
**(b)** As the temperature is increased, the liquid would evaporate more rapidly, increasing the amount of gas in the container and increasing the pressure.
**57.** $3.7 \times 10^{-6}$ atm **59.** $4.1 \times 10^5$ L **61.** 0.0426 Pa **63.** $2.9 \times 10^3$ L **65.** 0.115 mg **67.** 411 K or 138 °C **69.** 58.3 g/mol
**71.** Tank A—The following shows that density (g/V) is proportional to pressure when temperature is constant. Higher pressure means higher density at a constant temperature.

$PV = \frac{g}{M}RT \qquad \frac{g}{V} = P\left(\frac{M}{RT}\right)$

**73.** 2.41 g/L **75. (a)** 756 mmHg **(b)** 5.97 L
**77.** 13.5 L The diver better exhale as he goes up.
**78.** Increased number of gas particles leads to increased pressure when temperature and volume are constant.
**80.** 0.45 mol **82. (a)** $3.77 \times 10^3$ m$^3$ **(b)** 71.8 kPa **(c)** 233 K
**85.** 1.4 Mg NaCl **87.** 232 g $NaN_3$ **89.** $2.41 \times 10^3$ m$^3$ HCl

**91. (a)** $4.2 \times 10^4$ L $NH_3$ **(b)** $3.9 \times 10^4$ L $CO_2$ **93.** 475 $m^3$ $CH_4$
**95. (a)** 0.24 Mg $CH_2CH_2$ **(b)** 18 cylinders **(c)** $8.3 \times 10^3$ L $C_2H_4O$
**(d)** 27 kg $C_2H_4O$ **97.** $9.1 \times 10^6$ $m^3$ $Cl_2$ **100. (a)** 109.4 kPa **(b)** 156 kPa
**102.** 316 K or 43 °C **104. (a)** 42 mol other **(b)** 83.7%
**106. (a)** Yes, the average kinetic energy is dependent on the temperature of the particles, so the average kinetic energy is the same when the temperature is the same.
**(b)** Yes. If the composition of the air is the same, the average mass of the particles is the same. Because the average kinetic energy of the particles is the same for the same temperature, the average velocity must also be the same.

$$KE_{average} = \frac{1}{2} m\mu^2_{average}$$

**(c)** No, because there are fewer particles per unit volume at 50 km, the average distance between the particles is greater than at sea level.
**(d)** No, because the average velocity of the particles is the same and the average distance between the particles at 50 km is longer, the average distance between the collisions at 50 km is greater than at sea level. Thus the frequency of collisions is greater at sea level.
**(e)** No, because there are fewer particles per unit volume, the density of the air at 50 km is much less than at sea level. Thus the sea level air has the higher density.
**(f)** No. Because the particles in the air in the two locations would have the same average mass and the same average velocity, they would collide with the walls of a container with the same force per collision. Because there are more gas particles per unit volume at sea level than at 50 km, there would be more collisions with the walls of a container that holds a sample of the air along the California coastline. This would lead to a greater force pushing on the walls of the container and a greater gas pressure for the gas at sea level.
**108. (a)** The higher temperature at noon causes the pressure to increase.
**(b)** The interaction between the moving tires and the stationary road causes the particles in the tires to increase their velocity, so the temperature of the tires and ultimately of the gas in the tires goes up. The increase in the temperature of the gas in the tire increases the pressure of the gas.
**110.** Helium is less dense than air, so it is easier for our vocal cords to vibrate. Because of this, they vibrate faster, making our voices sound strange.

## Chapter 14 Exercises

**14.1. (a)** polar covalent; N is partial negative and H is partial positive. **(b)** nonpolar covalent **(c)** ionic; O is negative, and Ca is positive. **(d)** polar covalent; F is partial negative and P is partial positive.
**14.2.** P–F bond; greater difference in electronegativity
**14.3.** iron, Fe cations in a sea of electrons, metallic bonds
iodine, $I_2$ molecules, London forces
$CH_3OH$, $CH_3OH$ molecules, hydrogen bonds
$NH_3$, $NH_3$ molecules, hydrogen bonds
hydrogen chloride, HCl molecules, dipole–dipole attractions
KF, Cations and anions, ionic bond
C (diamond), atoms, covalent bonds

## Chapter 14 Review Questions

**1. (a)** ionic bonds **(b)** covalent bonds
**2. (a)** $OF_2$ all nonmetallic elements (no ammonium), so molecular **(b)** metal-nonmetal, so ionic **(c)** $CaCO_3$ metal-polyatomic ion, so ionic **(d)** all nonmetallic elements (no ammonium), so molecular
**3. (a)** ionic, $MgCl_2$ **(b)** binary covalent, HCl **(c)** ionic with polyatomic ion, $NaNO_3$ **(d)** binary covalent (hydrocarbon), $CH_4$ **(e)** binary covalent, $NH_3$ **(f)** binary acid, HCl(*aq*) **(g)** oxyacid, $HNO_3$ **(h)** alcohol, $C_2H_5OH$
**4. (a)** binary covalent, hydrogen fluoride **(b)** alcohol, methanol **(c)** ionic, lithium bromide **(d)** ionic with polyatomic ion, ammonium chloride **(e)** binary covalent (hydrocarbon), ethane **(f)** binary covalent, boron trifluoride **(g)** oxyacid, sulfuric acid

**5. (a)** :Cl—C(—Cl:)(—Cl:)—Cl: tetrahedral; C with four Cl, 109.5° tetrahedral
**(b)** :F—S—F: tetrahedral; S with two F, ≈109.5° bent
**(c)** :F—P(—F:)—F: tetrahedral; P with three F, ≈109.5° trigonal pyramid
**(d)** :Cl—B(—Cl:)—Cl: trigonal planar; B with three Cl, 120° trigonal planar

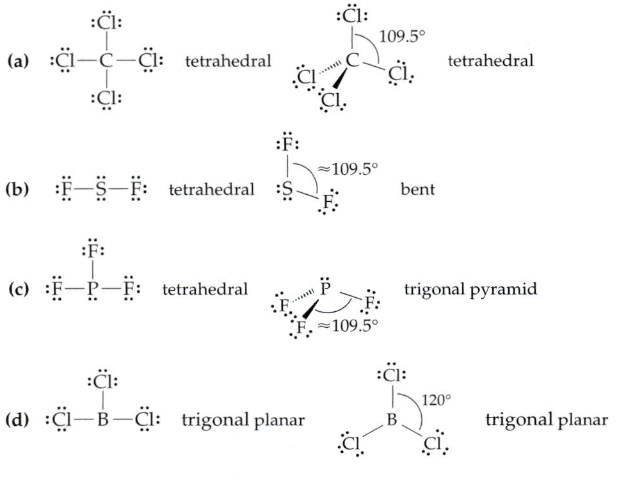

## Chapter 14 Key Ideas

**6.** significant, remain, large **8.** more rapidly, lower **10.** more **12.** percentage **14.** opposing, net **16.** weaker **18.** external pressure **20.** internal vapor pressure **22.** increased temperature **24.** partial positive, partial negative **26.** attract **28.** polar **30.** negative, positive **32.** difference **34.** nonpolar **36.** Hydrogen **38.** stronger, chance, less tightly

## Chapter 14 Problems

**40.** At the lower temperature during the night, the average velocity of the water molecules in the air is lower, making it more likely that they will stay together when they collide. They stay together long enough for other water molecules to collide with them forming clusters large enough for gravity to pull them down to the grass where they combine with other clusters to form the dew.
**42. (a)** In the closed test tube, the vapor particles that have escaped from the liquid are trapped in the space above the liquid. The concentration of acetone vapor rises quickly to the concentration that makes the rate of condensation equal to the rate of evaporation, so there is no net change in the amount of liquid or vapor in the test tube. In the open test tube, the acetone vapor escapes into the room. The concentration of vapor never gets high enough to balance the rate of evaporation, so all of the liquid finally disappears.
**(b)** No, the rate of evaporation is dependent on the strengths of attractions between particles in the liquid, the liquid's surface area, and temperature. All of these factors are the same for the two systems, so the initial rate of evaporation is the same for each.
**(c)** There will be some vapor above both liquids, so some vapor molecules will collide with the surface of the liquid and return to the liquid state. Thus, there will be condensation in both test tubes. The concentration of acetone vapor above the liquid in the closed container will be higher, so the rate of collision between the vapor particles and the liquid surface will be higher. Thus, the rate of condensation in the closed container will be higher.
**(d)** The liquid immediately begins to evaporate with a rate of evaporation that is dependent on the surface area of the liquid, the strengths of attractions between the liquid particles, and temperature. If these three factors remain constant, the rate of evaporation will be constant. If we assume that the container initially holds no vapor particles, there is no condensation of vapor when the liquid is first added. As the liquid evaporates, the number of vapor particles above the liquid increases, and the condensation process begins. As long as the rate of evaporation of the liquid is greater than the rate of condensation of the vapor, the concentration of vapor particles above the liquid will increase. As the concentration of vapor particles increases, the rate of collisions of vapor particles with the liquid increases, increasing the rate of condensation. If there is enough liquid in the container to avoid all of it evaporating, the rising rate of condensation eventually becomes equal to the rate of evaporation. At

this point, for every particle that leaves the liquid, a particle somewhere else in the container returns to the liquid. Thus, there is no net change in the amount of substance in the liquid form or the amount of substance in the vapor form.
**(e)** While the rate of evaporation is greater than the rate of condensation, there is a steady increase in the amount of vapor above the liquid. This increases the total pressure of gas above the liquid, so the balloon expands. When the rates of evaporation and condensation become equal, the amount of vapor and the total gas pressure remain constant, so the balloon maintains the same degree of inflation.
**(f)** The rate of evaporation is dependent on the temperature of the liquid. Increased temperature increases the average velocity and momentum of the particles in the liquid. This increases the percentage of particles that have the minimum velocity necessary to escape and increases the rate of evaporation. More particles escape per second, and the partial pressure due to the vapor above the liquid increases.
**43. (a)** The weaker attractions between diethyl ether molecules are easier to break, allowing a higher percentage of particles to escape from the surface of liquid diethyl ether than from liquid ethanol. If the surface area and temperature is the same for both liquids, more particles will escape per second from the diethyl ether than from the ethanol.
**(b)** Because the attractions between diethyl ether molecules are weaker than those between ethanol molecules, it is easier for a diethyl ether molecule to break them and move in to the vapor phase. Therefore, the rate of evaporation from liquid diethyl ether is greater than for ethanol at the same temperature. When the dynamic equilibrium between evaporation and condensation for the liquids is reached and the two rates become equal, the rate of condensation for the diethyl ether is higher than for ethanol. Because the rate of condensation is determined by the concentration of vapor above the liquid, the concentration of diethyl ether vapor at equilibrium is higher than for the ethanol. The higher concentration of diethyl ether particles leads to a higher equilibrium vapor pressure.
**45.** As the temperature of the liquid milk increases, its rate of evaporation increases. This will disrupt the equilibrium, making the rate of evaporation greater than the rate of condensation. This leads to an increase in the concentration of water molecules in the gas-space above the liquid, which increases the rate of condensation until it increases enough to once again become equal to the rate of evaporation. At this new dynamic equilibrium, the rates of evaporation and condensation will both be higher.
**47. (a)** Ethanol molecules are moving constantly, sometimes at very high velocity. When they collide with other particles, they push them out of their positions, leaving small spaces in the liquid. Other particles move across the spaces and collide with other particles, and the spaces grow in volume. These spaces can be viewed as tiny bubbles. The surface of each of these tiny bubbles is composed of a spherical shell of liquid particles. Except for shape, this surface is the same as the surface at the top of the liquid. Particles can escape from the surface (evaporate) into the vapor phase in the bubble, and when particles in the vapor phase collide again with the surface of the bubble, they return to the liquid state (condense). A dynamic equilibrium between the rate of evaporation and the rate of condensation is set up in the bubble just like the liquid-vapor equilibrium above the liquid in the closed container.
**(b)** Each time a particle moves across a bubble and collides with the surface of the bubble, it exerts a tiny force pushing the wall of the bubble out. All of the collisions with the shell of the bubble combine to yield a gas pressure inside the bubble. This pressure is the same as the equilibrium vapor pressure for the vapor above the liquid in a closed container. If the 1 atm of external pressure pushing on the bubble is greater that the vapor pressure of the bubble, the liquid particles are pushed closer together, and the bubble collapses. If the vapor pressure of the bubble is greater than the external pressure, the bubble will grow. If the two pressures are equal, the bubble maintains its volume. The vapor pressure of the bubbles in ethanol does not reach 1 atm until the temperature rises to 78.3 °C.
**(c)** If the external pressure acting on the bubbles in ethanol rises to 2 atm, the vapor pressure inside the bubbles must rise to 2 atm also to allow boiling. This requires an increase in the temperature, so the boiling point increases.
**49.** The pressure of the earth's atmosphere decreases with increasing distance from the center of the earth. Therefore, the average atmospheric pressure in Death Valley is greater than at sea level where it is 1 atmosphere. This greater external pressure acting on liquid water, increases the vapor pressure necessary for the water to maintain bubbles and boil. This leads to a higher temperature necessary to reach the higher vapor pressure. Because the boiling point temperature of a liquid is the temperature at which the vapor pressure of the liquid reaches the external pressure acting on it, the boiling point temperature is higher in Death Valley.
**51.** Increased strength of attractions leads to decreased rate of evaporation, decreased rate of condensation at equilibrium, decreased concentration of vapor, and decreased vapor pressure at a given temperature. This leads to an increased temperature necessary to reach a vapor pressure of one atmosphere.
**54.** C–N, Polar Covalent, N
C–H, Nonpolar Covalent
H–Br, Polar Covalent, Br
Li–F, Ionic, F
C–Se, Nonpolar Covalent
Se–S, Nonpolar Covalent
F–S, Polar Covalent, F
O–P, Polar Covalent, O
O–K, Ionic, O
F–H, Polar Covalent, F
**56. (a)** C–O **(b)** H–Cl
**58.** Water molecules have an asymmetrical distribution of polar bonds, so they are polar.

$H^{\delta+}$
|
$\delta-$:Ö—$H^{\delta+}$

All of the bonds in ethane molecules are nonpolar, so the molecules are nonpolar.

H H
| |
H—C—C—H
| |
H H

Carbon dioxide molecules have a symmetrical distribution of polar bonds, so they are nonpolar.

$\delta-$ $\delta+$ $\delta-$
:O=C=O:

**60.** Ammonia is composed of $NH_3$ molecules that are attracted by hydrogen bonds between the partially positive hydrogen atoms and the partially negative nitrogen atoms of other molecules. See Figure 14.25. The liquid would look much like the image for liquid water shown in Figure 3.14, except with $NH_3$ molecules in the place of $H_2O$ molecules.
**62.** The attractions are London forces. Because the Br–Br bond is nonpolar, the expected distribution of the electrons in the $Br_2$ molecule is a symmetrical arrangement around the two bromine nuclei, but this arrangement is far from static. Even though the most probable distribution of charge in an isolated $Br_2$ molecule is balanced, in a sample of bromine that contains many billions of molecules, there is a chance that a few of these molecules will have their electron clouds shifted more toward one bromine atom than the other. The resulting dipoles are often called *instantaneous dipoles* because they may be short-lived. Remember also that in all states of matter, there are constant collisions between molecules. When $Br_2$ molecules collide, the repulsion between their electron clouds will distort the clouds and shift them from their nonpolar state. The dipoles that form are also called instantaneous dipoles. An instantaneous dipole can create a dipole in the molecule next to it. For example, the negative end of one instantaneous dipole will repel the negative electron cloud of a nonpolar molecule next to it, pushing the cloud to the far side of the neighboring molecule. The new dipole is called an *induced dipole*. The

induced dipole can then induce a dipole in the molecule next to it. This continues until there are many polar molecules in the system. The resulting partial charges on these polar molecules lead to attractions between the opposite charges on the molecules. See Figure 14.26.

**64.** Both of these substances have nonpolar molecules held together by London forces. Because the $CS_2$ molecules are larger, they have stronger London forces that raise carbon disulfide's boiling point to above room temperature.

**66.** Because of the O–H bond in methanol, the attractions between $CH_3OH$ molecules are hydrogen bonds. The hydrogen atoms in $CH_2O$ molecules are bonded to the carbon atom, not the oxygen atom, so there is no hydrogen bonding for formaldehyde. The C–O bond in each formaldehyde molecule is polar, and when there is only one polar bond in a molecule, the molecule is polar. Therefore, $CH_2O$ molecules are held together by dipole–dipole attractions. For molecules of about the same size, hydrogen bonds are stronger than dipole–dipole attractions. The stronger hydrogen bonds between $CH_3OH$ molecules raise its boiling point above room temperature, making it a liquid.

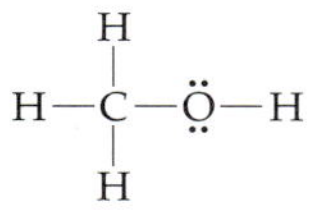

Methanol 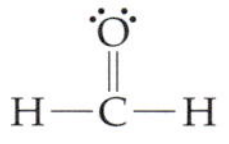 Formaldehyde

**68.** Silver, Ag cations in a sea of electrons, Metallic Bonds
HCl, HCl molecules, Dipole–Dipole Attractions
$C_2H_5OH$, $C_2H_5OH$ molecules, Hydrogen Bonds
NaBr, Cations and anions, Ionic Bonds
Carbon (diamond), Carbon atoms, Covalent Bonds
$C_5H_{12}$, $C_5H_{12}$ molecules, London Forces
water, $H_2O$ molecules, Hydrogen Bonds

**70.** As a liquid spreads out on a surface, some of the attractions between liquid particles are broken. Because the metallic bonds between mercury atoms are much stronger than the hydrogen bonds between water molecules, they keep mercury from spreading out like water.

## Chapter 15 Exercises

**15.1.** **(a)** soluble **(b)** soluble **(c)** insoluble **(d)** soluble
**15.2.** **(a)** Ionic compound so insoluble **(b)** Nonpolar molecular compound so soluble
**15.3.** **(a)** Ionic compound so more soluble in water
**(b)** Nonpolar molecular compound so more soluble in hexane
**15.4.** The 2-methyl-2-propanol has a greater percentage of its structure that is polar, so we expect it to be more soluble in water.

## Chapter 15 Review Questions

**1.** When solid sodium bromide is added to water, all of the ions at the surface of the solid can be viewed as shifting back and forth between moving out into the water and returning to the solid surface. Sometimes when an ion moves out into the water, a water molecule collides with it, helping to break the ionic bond, and pushing it out into the solution. Water molecules move into the gap between the ion in solution and the solid and shield the ion from the attraction to the solid. The ions are kept stable and held in solution by attractions between them and the polar water molecules. The negatively charged oxygen ends of water molecules surround the sodium ions, and the positively charged hydrogen ends of water molecules surround the bromide ions. (See Figures 4.4 and 4.5 with $Br^-$ in the place of $Cl^-$.) The sodium bromide is the solute, and the water is the solvent.

**2.** **(a)**

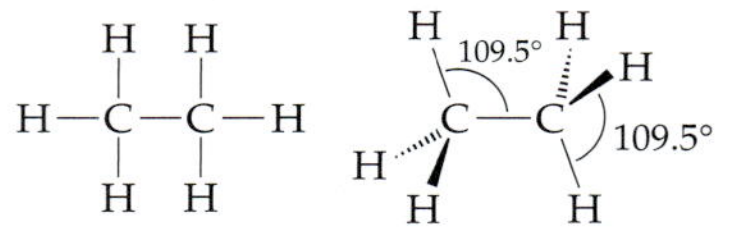

Nonpolar (no polar bonds)

**(b)**

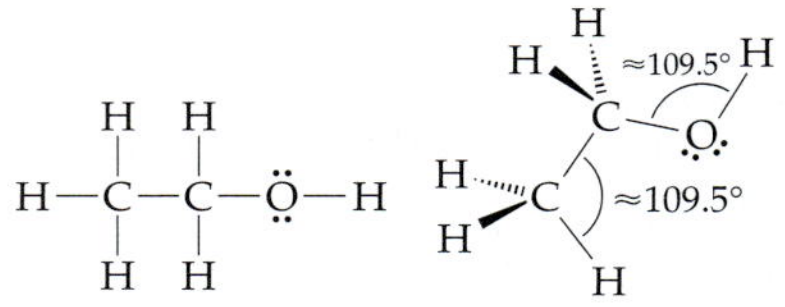

Polar (asymmetrical distribution of polar bonds)

**(c)**

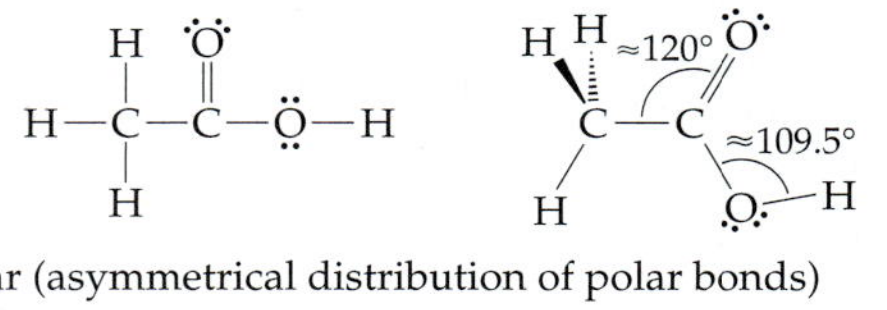

Polar (asymmetrical distribution of polar bonds)

**3.** **(a)** nonpolar molecules / London forces
**(b)** polar molecules / hydrogen bonds
**(c)** cations and anions / ionic bonds
**(d)** polar molecules / hydrogen bonds
**(e)** nonpolar molecules / London forces
**(f)** nonpolar molecules / London forces

**4.** As soon as the liquid propane is added to the tank, the liquid begins to evaporate. In the closed container, the vapor particles that have escaped from the liquid are trapped. The concentration of propane vapor rises quickly to the concentration that makes the rate of condensation equal to the rate of evaporation, so there is no net change in the amount of liquid or vapor in the tank. There are constant changes (from liquid to vapor and vapor to liquid), but because the rates of these two changes are equal (the rate of evaporation equals the rate of condensation), there is no net change in the system (the amount of liquid and vapor remains constant). Thus, the system is a dynamic equilibrium.

## Chapter 15 Key Ideas

**5.** universe **7.** B, A **9.** probability **11.** insoluble **13.** moderately **15.** nonpolar solvents **17.** Polar substances **19.** attracted, nonpolar **21.** fats, oils **23.** cation, long-chain **25.** Hard, precipitate **27.** surface area of the solute, degree of agitation or stirring, temperature **29.** solubility

## Chapter 15 Problems

**31.** The Second Law of Thermodynamics states that the entropy of the universe increases. Entropy is a measure of disorder and degree of dispersal. This leads to the general statement that changes tend to take place when they lead to an increase in the disorder or degree of dispersal of the system. The particles in (b) are more dispersed and more disordered than in (a), so (b) is a higher entropy state than (a). Therefore, we expect that the system will shift from (a) to (b). The reason for this shift is that there are more ways to arrange our system in the more dispersed form—(b)—than in the more concentrated form—(a). Because the particles can move freely in the container, they will shift to the more probable, more dispersed, and more disordered form.

| Gas in one chamber | → | Gas in both chambers |
|---|---|---|
| Fewer ways to arrange particles | | More ways to arrange particles |
| Less probable | | More probable |
| More ordered | | Less ordered |
| Particles closer together | | Particles more dispersed |

**33.** Picture a layer of acetic acid that is carefully added to water (Figure 15.3). Because the particles of a liquid are moving constantly, some of the acetic acid particles at the boundary between the two liquids will immediately move into the water, and some of the water molecules will move into the acetic acid. In this process, water-water and acetic acid-acetic acid attractions are broken and acetic acid-water attractions are formed. Both acetic acid and water are molecular substances with O–H bonds, so the attractions broken between water molecules and the attractions broken between acetic acid molecules are hydrogen bonds. The attractions that form between the acetic acid and water molecules are also hydrogen bonds. We expect the hydrogen bonds that form between water molecules and acetic acid molecules to be similar in strength to the hydrogen bonds that are broken. Because the attractions between the particles are so similar, the freedom of movement of the acetic acid molecules in the water solution is about the same as their freedom of movement in the pure acetic acid. The same can be said for the water. Because of this freedom of movement, both liquids will spread out to fill the total volume of the combined liquids. In this way, they will shift to the most probable, most disordered, most dispersed, highest entropy state available, the state of being completely mixed. There are many more possible arrangements for this system when the acetic acid and water molecules are dispersed throughout a solution than when they are restricted to separate layers. (Figure 15.3).
**36.** **(a)** soluble **(b)** insoluble **(c)** often soluble **(d)** often soluble **(e)** insoluble
**38.** **(a)** Acetic acid, $HC_2H_3O_2$, is like water because both are composed of small, polar molecules with hydrogen bonding between them. **(b)** Ammonia, $NH_3$, is like water because both are composed of small, polar molecules with hydrogen bonding between them.
**40.** **(a)** soluble **(b)** insoluble **(c)** soluble **(d)** soluble
**42.** **(a)** soluble **(b)** insoluble **44.** **(a)** hexane **(b)** water
**47.** acetone **49.** ethane
**51.** The three −OH groups and the N−H bond in the epinephrine structure give it a greater percentage of its structure that is polar, so we predict that epinephrine would be more soluble in water than amphetamine. The cell membranes that separate the blood stream from the brain cells have a nonpolar interior that tends to block polar substances from moving into the brain. Epinephrine is too polar to move from the blood stream into the brain, but the stimulant effects of the less polar amphetamine are in part due to its ability to pass through the blood-brain barrier. See Figure 15.5.

**53.**

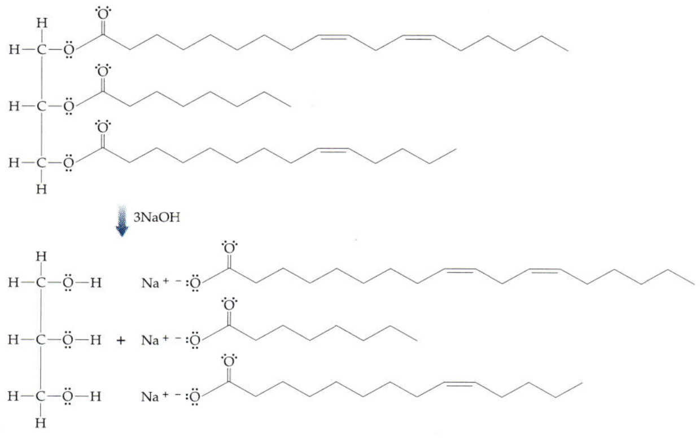

3NaOH

**55.** 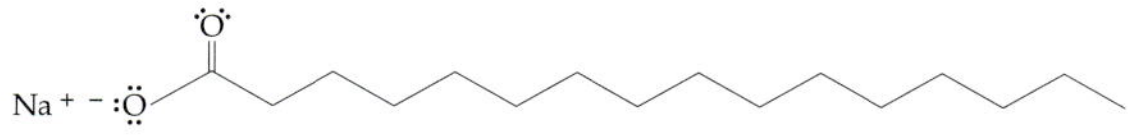 hydrophilic hydrophobic

The very polar, ionic end on the left is hydrophilic (attracted to water), and the nonpolar, hydrocarbon portion of the structure is hydrophobic.
**57.** For soap to work, its anions must stay in solution. Unfortunately, they tend to precipitate from solution when the water is "hard." Hard water is water that contains dissolved calcium ions, $Ca^{2+}$, magnesium ions, $Mg^{2+}$, and often iron ions, $Fe^{2+}$ or $Fe^{3+}$. These ions bind strongly to soap anions, causing the soap to precipitate from hard water solutions. Detergents have been developed to avoid this problem of soap in hard water. Their structures are similar to (although more varied than) soap but less likely to form insoluble compounds with hard water ions. Some detergents are ionic like soap, and some are molecular. For example, in sodium dodecyl sulfate (SDS), a typical ionic detergent, the $-CO_2^-$ portion of the conventional soap structure is replaced by an $-OSO_3^-$ group, which is less likely to link with hard water cations and precipitate from the solution:

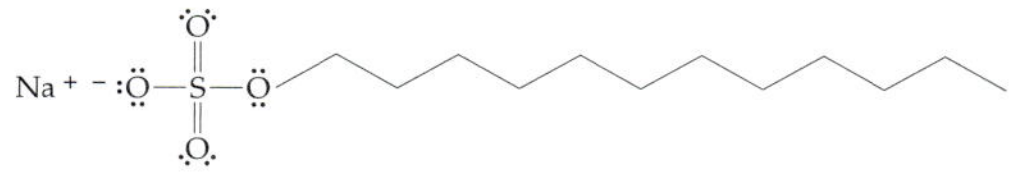

SDS, a typical ionic detergent

**59.** **(a)** Picture the $MgSO_4$ solid sitting on the bottom of the container. The magnesium and sulfate ions at the surface of the solid are constantly moving out into the water and being pulled back by the attractions to the other ions still on the surface of the solid. Sometimes when an ion moves out into the water, a water molecule collides with it and pushes it farther out into the solution. Other water molecules move into the gap between the ion in solution and the solid and shield the ion from the attraction to the solid. The ion is kept stable and held in solution by attractions between the polar water molecules and the charged ions. The negative oxygen atoms of the water molecules surround the cations, and the positive hydrogen ends of water molecules surround the anions. Once the ions are in solution, they move throughout the solution like any particle in a liquid. Eventually they collide with the surface of the solid. When this happens, they come back under the influence of the attractions that hold the particles in the solid, and they are likely to return to the solid form.
**(b)** The more particles of solute there are per liter of solution, the more collisions there will be between solute particles and the solid. More collisions lead to a greater rate of return of solute particles from the solution to the solid (Figures 15.15 and 15.16).
**(c)** As particles leave the solid and go into solution, localized high concentrations of dissolved solute form around the surface of the solid. Remember that the higher concentration of solute leads to a higher rate of return to the solid form. A higher rate of return leads to a lower overall *net* rate of solution. If you stir or in some way agitate the solution, the solute particles near the solid will be moved more quickly away from the solid, and the localized high concentrations of solute will be avoided. This will diminish the rate of return and increase the net rate of solution. This is why stirring the mixture of water and Epsom salts dissolves the magnesium sulfate more rapidly (Figures 15.19 and 15.20).
**(d)** The solution becomes saturated when the rate of solution and the rate of return become equal. Solid continues to dissolve, but particles return to the solid from solution at the same rate. Even though the specific particles in solution are constantly changing, there is no net change in the total amount of solid or the amount of ions in solution.
**61.** **(a)** Higher temperature increases the rate at which particles escape from a solid, increasing the rate of solution. The higher temperature also helps the escaped particles to move away from the solid more quickly, minimizing the rate of return. Together these two factors make the net rate of solution in the hot water higher.
**(b)** Only particles at the surface of the solid sugar have a possibility of escaping into the solution. The particles in the center each sugar crystal have to wait until the particles between them and the surface dissolve to have any chance of escaping. Powdered sugar has much smaller crystals, so a much higher percentage of the sugar particles are at the surface. This increases the rate of solution (Figures 15.17 and 15.18).
**63.** A saturated solution has enough solute dissolved to reach the solubility limit. In a saturated solution of NaCl, $Na^+$ and $Cl^-$ ions are constantly escaping from the surface of the solid and moving into solution, but other $Na^+$ and $Cl^-$ ions in solution are colliding with the solid and returning to the solid at a rate equal to the rate of solution. Because the rate of solution and the rate of return to the solid are equal, there is no net shift to more or less salt dissolved.

**65.** Yes, a solution can be both concentrated and unsaturated. If the solubility of a substance is high, the concentration of the solute in solution can be high even in a solution where the rate of return to the undissolved solute is still below the rate of solution.
**67.** At the same temperature, the rate at which carbon dioxide molecules escape from the soft drink is the same whether the container is open or closed. When a soft drink is open to the air, the carbon dioxide molecules that escape can move farther from the surface of the liquid and are therefore less likely to collide with the surface of the liquid and return to the solution. This means a lower rate of return in the drink open to the air. Therefore, the difference between the rate of escape of the $CO_2$ form the soft drink and the rate of return will be greater in the bottle open to the air, so there is a greater net rate of escape of the gas from the soft drink.

## Chapter 16 Exercises

**16.1.** $K_C = \dfrac{[SO_2]^2[H_2O]^2}{[H_2S]^2[O_2]^3}$ $\quad K_P = \dfrac{P_{SO_2}{}^2 P_{H_2O}{}^2}{P_{H_2S}{}^2 P_{O_2}{}^3}$

**16.2.** $K_P = \dfrac{P_{C_2H_5OH}}{P_{C_2H_4} P_{H_2O}} = \dfrac{0.11 \text{ atm}}{0.35 \text{ atm } (0.75 \text{ atm})} = \mathbf{0.42\ 1/atm\ or\ 0.42}$

**16.3. (a)** According to Table 16.1, the $K_P$ for this reaction is $2.2 \times 10^{12}$, so it favors products.
**(b)** According to Table 16.1, the $K_P$ for this reaction is 6.7. Neither reactants nor products are favored.

**16.4.** $K_C = \dfrac{[Cl_2]^2}{[O_2]}$ $\quad K_P = \dfrac{P_{Cl_2}{}^2}{P_{O_2}}$

**16.5. (a)** (1) Using Le Châtelier's Principle, we predict that the system will shift to more products to partially counteract the increase in $H_2O$. (2) The increase in the concentration of water vapor speeds the forward reaction without initially affecting the rate of the reverse reaction. The equilibrium is disrupted, and the system shifts to more products because the forward rate is greater than the reverse rate.
**(b)** (1) Using Le Châtelier's Principle, we predict that the system will shift to more reactants to partially counteract the decrease in $NO_2$. (2) The decrease in the concentration of $NO_2(g)$ slows the forward reaction without initially affecting the rate of the reverse reaction. The equilibrium is disrupted, and the system shifts toward more reactants because the reverse rate is greater than the forward rate.
**(c)** (1) Using Le Châtelier's Principle, we predict that the system will shift to more products to partially counteract the decrease in $HNO_3$. (2) The decrease in the concentration of $HNO_3(g)$ slows the reverse reaction without initially affecting the rate of the forward reaction. The equilibrium is disrupted, and the system shifts toward more products because the forward rate is greater than the reverse rate.
**(d)** (1) Using Le Châtelier's Principle, we predict that the system shifts in the exothermic direction to partially counteract the decrease in temperature. As the system shifts toward more products, energy is released, and the temperature increases. (2) The decreased temperature decreases the rates of both the forward and reverse reactions, but it has a greater effect on the endothermic reaction. Because the forward reaction is exothermic, the reverse reaction must be endothermic. Therefore, the reverse reaction is slowed more than the forward reaction. The system shifts toward more products because the forward rate becomes greater than the reverse rate.
**(e)** (1) Le Châtelier's Principle does not apply here. (2) The catalyst speeds both the forward and the reverse rates equally. Thus, there is no shift in the equilibrium. The purpose of the catalyst is to bring the system to equilibrium faster.

## Chapter 16 Review Questions

**1.** The gas is composed of $O_2$ molecules that are moving constantly in the container. For a typical gas, the average distance between particles is about ten times the diameter of each particle. This leads to the gas particles themselves taking up only about 0.1% of the total volume. The other 99.9% of the total volume is empty space.
According to our model, each $O_2$ molecule moves freely in a straight-line path until it collides with another $O_2$ molecule or one of the walls of the container. The particles are moving fast enough to break any attraction that might form between them, so after two particles collide, they bounce off each other and continue on alone. Due to collisions, each particle is constantly speeding up and slowing down, but its average velocity stays constant as long as the temperature stays constant.
**2. (a)** Energy **(b)** Kinetic energy **(c)** endergonic **(d)** exergonic **(e)** Thermal **(f)** Heat **(g)** exothermic **(h)** endothermic **(i)** catalyst
**3.** Increased temperature means increased average kinetic energy.
**4.** Any time a change leads to decreased forces of attraction, it leads to increased potential energy. The Law of Conservation of Energy states that energy cannot be created or destroyed, so energy must be added to the system. It always takes energy to break attractions between particles.
**5.** Any time a change leads to increased forces of attraction, it leads to decreased potential energy. The Law of Conservation of Energy states that energy cannot be created or destroyed, so energy is released from the system. Energy is always released when new attractions between particles are formed.
**6.** If the bonds in the products are stronger and lower potential energy than in the reactants, energy will be released from the system. If the energy released is due to the conversion of potential energy to kinetic energy, the temperature of the products will be higher than the original reactants. The higher temperature products are able to transfer heat to the surroundings, and the temperature of the surroundings increases.
**7.** If the bonds in the products are weaker and higher potential energy than in the reactants, energy must be absorbed. If the energy absorbed is due to the conversion of kinetic energy to potential energy, the temperature of the products will be lower than the original reactants. The lower temperature products are able to absorb heat from the surroundings, and the temperature of the surroundings decreases.
**8.** The system must have two opposing changes, from state A to state B and from state B to state A. For a dynamic equilibrium to exist, the rates of the two opposing changes must be equal, so there are constant changes between state A and state B but no net change in the components of the system.

## Chapter 16 Key Ideas

**9.** balanced, transition state **11.** minimum, equal to, greater than **13.** positive **15.** new bonds, old bonds, close to each other **17.** fraction **19.** alternative pathway, lower activation energy **21.** heterogeneous **23.** equilibrium constant, moles per liter, coefficient **25.** larger **27.** reactants **29.** products **31.** disrupts, change

## Chapter 16 Problems

**33. (a)** NO and $O_3$ molecules are constantly moving in the container, sometimes with a high velocity and sometimes more slowly. The particles are constantly colliding, changing their direction of motion, and speeding up or slowing down. If the molecules collide in a way that puts the nitrogen atom in NO near one of the outer oxygen atoms in $O_3$, one of the O–O bonds in the $O_3$ molecule begins to break, and a new bond between one of the oxygen atoms in the ozone molecule and the nitrogen atom in NO begins to form. If the collision yields enough energy to reach the activated complex, it proceeds on to products. If the molecules do not have the correct orientation, or if they do not have enough energy, they separate without a reaction taking place.
**(b)** NO and $O_3$ molecules must collide, they must collide with the correct orientation to form an N–O bond at the same time that an O–O bond is broken, and they must have the minimum energy necessary to reach the activated complex (the activation energy).
**(c)** The collision brings the atoms that will form the new bonds close, and the net kinetic energy in the collision provides the energy necessary to reach the activated complex and proceed to products.
**(d)** It takes a significant amount of energy to break O–O bonds, and collisions between particles are not likely to provide enough. As N–O bonds form, they release energy, so the formation of the new bonds can provide energy to supplement the energy provided by the colli-

sions. The sum of the energy of collision and the energy released in bond formation is more likely to provide enough energy for the reaction.

**(e)** N—O- - - - -O- - - - -O—O

Bond making   Bond breaking

**(f)** In the initial stage of the reaction, the energy released in bond making is less than the energy absorbed by bond breaking. Therefore, energy must be available from the colliding particles to allow the reaction to proceed. At some point in the change, the energy released in bond formation becomes equal to the energy absorbed in bond breaking. If the colliding particles have enough energy to reach this point (in other words, if they have the activation energy), the reaction proceeds to products.

**(g)**

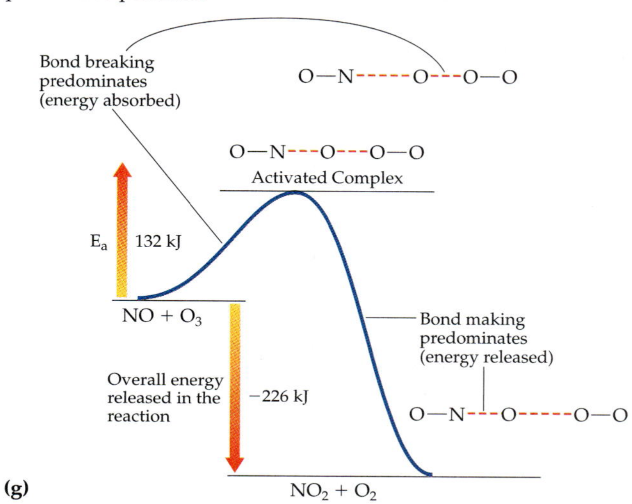


**(h)** The negative sign for the heat of reaction shows that energy is released overall, so the reaction is exothermic.
**(i)** For a reaction to be likely, new bonds must be made at the same time as other bonds are broken. Therefore, the nitrogen atom in NO must collide with one of the outer oxygen atoms in $O_3$.
**35. (a)** Increased concentration of reactant A leads to increased rate of collision between A and B and therefore to leads to increased rate of reaction.
**(b)** Decreased temperature leads to decreased average kinetic energy of collisions between A and B. This leads to a decrease in the percentage of collisions with the minimum energy necessary for the reaction and therefore leads to decreased rate of reaction.
**(c)** With a lower activation energy, there is a greater percentage of collisions with the minimum energy necessary for the reaction and therefore an increased rate of reaction.
**37.** At a particular temperature, the lower the activation energy is, the higher the percentage of collisions with at least that energy or more will be. Thus, the second reaction would have the higher fraction of collisions with the activation energy.
**39.** The approximate activation energy for reaction 1 is 30 kJ and for reaction 2 is 60 kJ. Reaction 1 is endothermic, and reaction 2 is exothermic.
**41.** In part, chlorine atoms are a threat to the ozone layer just because they provide another pathway for the conversion of $O_3$ and O to $O_2$, but there is another reason. The reaction between $O_3$ and Cl that forms ClO and $O_2$ has an activation energy of 2.1 kJ/mole. At 25 °C, about three of every seven collisions (or 43%) have enough energy to reach the activated complex. The reaction between O and ClO to form Cl and $O_2$ has an activation energy of only 0.4 kJ/mole. At 25 °C, about 85% of the collisions have at least this energy. The uncatalyzed reaction has an activation energy of about 17 kJ/mole. At 25 °C (298 K), about one of every one thousand collisions (or 0.1%) between $O_3$ molecules and O atoms has a net kinetic energy large enough to form the activated complex and proceed to products. Thus, a much higher fraction of the collisions have the minimum energy necessary to react for the catalyzed reaction than for the direct reaction between $O_3$ and O. Thus, a much greater fraction of the collisions has the minimum energy necessary for the reaction to proceed for the catalyzed reaction than for the uncatalyzed reaction. Figures 16.15 and 16.16 illustrate this.
**43.** Step #1: The reactants (NO molecules) collide with the surface of the catalyst where they bind to the catalyst. This step is called adsorption. The bonds within the reactant molecules are weakened or even broken as the reactants are adsorbed. (N–O bonds are broken.) Step #2: The adsorbed particles (separate N and O atoms) move over the surface of the catalyst. Step #3: The adsorbed particles combine to form products ($N_2$ and $O_2$). Step #4: The products ($N_2$ and $O_2$) leave the catalyst. See Figure 16.17.
**45. (a)** Acetic acid molecules react with water to form hydronium ions and acetate ions, and at the same time, hydronium ions react with acetate ions to return to acetic acid molecules and water.

$$HC_2H_3O_2(aq) + H_2O(l) \rightleftharpoons H_3O^+(aq) + C_2H_3O_2^-(aq)$$

**(b)** Liquids evaporate to form vapor at a rate that is balanced by the return of vapor to liquid.
**(c)** Carbon dioxide escapes from the solution at a rate that is balanced by the return of $CO_2$ to the solution.
**47. (a)** The forward reaction rate is at its peak when A and B are first mixed. Because A and B concentrations are diminishing as they form C and D, the rate of the forward reaction declines steadily until equilibrium is reached.
**(b)** The reverse reaction rate is at its peak when the reaction reaches equilibrium. Because C and D concentrations are increasing as they form from A and B, the rate of the reverse reaction increases steadily until equilibrium is reached.
**49. (a)** When $I^-$ ions and $CH_3Br$ molecules are added to a container, they begin to collide and react. As the reaction proceeds, the concentrations of $I^-$ and $CH_3Br$ diminish, so the rate of the forward reaction decreases. Initially, there are no $CH_3I$ molecules or $Br^-$ ions in the container, so the rate of the reverse reaction is initially zero. As the concentrations of $CH_3I$ and $Br^-$ increase, the rate of the reverse reaction increases. As long as the rate of the forward reaction is greater than the rate of the reverse reaction, the concentrations of the reactants ($I^-$ and $CH_3Br$) will steadily decrease, and the concentrations of products ($CH_3I$ and $Br^-$) will constantly increase. This leads to a decrease in the forward rate of the reaction and an increase in the rate of the reverse reaction. This continues until the two rates become equal. At this point, our system has reached a dynamic equilibrium.
**(b)** In a dynamic equilibrium for reversible chemical reactions, the forward and reverse reaction rates are equal, so although there are constant changes between reactants and products, there is no net change in the amounts of each. $I^-$ and $CH_3Br$ are constantly reacting to form $CH_3I$ and $Br^-$, but $CH_3I$ and $Br^-$ are reacting to reform $CH_3Br$ and $I^-$ at the same rate. Thus, there is no net change in the amounts of $I^-$, $CH_3Br$, $CH_3I$, or $Br^-$.

**50. (a)** $K_C = \dfrac{[C_2H_2][H_2]^3}{[CH_4]^2}$ $\quad K_P = \dfrac{P_{C_2H_2}P^3_{H_2}}{P^2_{CH_4}}$

**(b)** $K_C = \dfrac{[NO^4]}{[NO_2]^2[O_2]}$ $\quad K_P = \dfrac{P^4_{NO}}{P^2_{N_2O}P_{O_2}}$

**(c)** $K_C = \dfrac{[H_2S]^3}{[H_2]^3}$ $\quad K_P = \dfrac{P^3_{H_2S}}{P^3_{H_2}}$

**52. (a)** $K_P = \dfrac{P^{1/2}_{N_2O_4}}{P_{NO_2}} = \dfrac{(1.69)^{1/2}}{0.60} = 2.2$

**(b)** $K_P = \dfrac{P_{N_2O_4}}{P^2_{NO_2}} = \dfrac{1.69}{(0.60)^2} = 4.7$

**(c)** Changing temperature leads to a change in the value for an equilibrium constant. (Because $K_P$ for this reaction decreases with increasing temperature, the reaction must be exothermic.)
**54. (a)** $K_P < 10^{-2}$ so reactants favored **(b)** $K_P > 10^2$ so products favored **(c)** $10^{-2} < K_P < 10^2$ so neither favored

**56.** $K_C = \dfrac{1}{[CO][H_2]^2}$

If the number of moles of $CH_3OH(l)$ in the container is doubled, its volume doubles too, leaving the concentration (mol/L) of the methanol constant. Increasing or decreasing the total volume of the container will not change the volume occupied by the liquid methanol, so the concentration (mol/L) of the $CH_3OH(l)$ also remains constant with changes in the volume of the container. The constant concentration of methanol can be incorporated into the equilibrium constant itself and left out of the equilibrium constant expression.

$$K' = \frac{[CH_3OH]}{[CO][H_2]^2} \quad \frac{K'}{[CH_3OH]} = \frac{1}{[CO][H_2]^2} = K_C$$

**58. (a)** Los Angeles has a much higher ozone concentration than in the Montana wilderness.
**(b)** Toward more products (Increased temperature favors the endothermic direction of reversible reactions.)
**60.** Increased temperature favors the endothermic direction of reversible reactions, so this reaction is endothermic in the reverse direction and exothermic in the forward direction.
**62.** Increased temperature will drive this endothermic reaction toward products, so warming the smelling salt container in your hands will increase the amount of ammonia released.
**64. (a)** Increased temperature increases the rate of both the forward and the reverse reactions, but it increases the rate of the endergonic reaction more than it increases the rate of the exergonic reaction. Therefore, changing the temperature of a chemical system at equilibrium will disrupt the balance of the forward and reverse rates of reaction and shift the system in the direction of the endergonic reaction. Because this reaction is exothermic in the forward direction, it must be endothermic in the reverse direction. Increased temperature shifts the system toward more reactants, decreasing the ratio of products to reactants and, therefore, decreasing the equilibrium constant.
**(b)** To maximize the percentage yield at equilibrium, the reaction should be run at as low a temperature as possible, but at low temperature, the rates of the forward and reverse reactions are both very low, so it takes a long time for the system to come to equilibrium. In this case, it is best to run the reaction at high temperature to get to equilibrium quickly. (The unreacted methanol can be recycled back into the original reaction vessel after the formaldehyde has been removed from the product mixture. )
**66. (a)** Using Le Châtelier's Principle, we predict that the system will shift to more products to partially counteract the increase in $NH_3$. The increase in the concentration of ammonia speeds the forward reaction without initially affecting the rate of the reverse reaction. The equilibrium is disrupted, and the system shifts to more products because the forward rate is greater than the reverse rate.
**(b)** Using Le Châtelier's Principle, we predict that the system will shift to more products to partially counteract the decrease in $H_2O$. The decrease in the concentration of $H_2O(g)$ slows the reverse reaction without initially affecting the rate of the forward reaction. The equilibrium is disrupted, and the system shifts toward more products because the forward rate is greater than the reverse rate.
**(c)** Using Le Châtelier's Principle, we predict that the system shifts in the endothermic direction to partially counteract the increase in temperature. Because the forward reaction is exothermic, the reverse reaction must be endothermic. As the system shifts toward more reactants, energy is absorbed, and the temperature decreases. The increased temperature increases the rates of both the forward and reverse reactions, but it has a greater effect on the endothermic reaction. Thus, the system shifts toward more reactants because the reverse rate becomes greater than the forward rate.
**68.** The addition of either $H_2$ or $I_2$ (or both) would increase the concentrations of reactants, increasing the rate of collision between them, increasing the forward rate, and shifting the system toward more product. Lower temperature favors the exothermic direction of the reaction, so lower temperature would shift this reaction to a higher percentage of products at equilibrium.

**70. (a)** The system will shift toward products, which leads to increased $COCl_2$.
**(b)** The system will shift toward products, which leads to decreased $Cl_2$.
**(c)** Equilibrium constants are unaffected by reactant and product concentrations, so the equilibrium constant remains the same.

## CHAPTER 17 EXERCISES

**17.1. (a)** alkane **(b)** amine **(c)** ether **(d)** ester **(e)** ketone **(f)** carboxylic acid **(g)** amide **(h)** aldehyde **(i)** alcohol **(j)** alkene **(k)** amine **(l)** alkyne **(m)** arene
**17.2. (a)** $CH_3CH_2CH_2CH_2CH_2CH_2CH_3$ or $CH_3(CH_2)_5CH_3$
**(b)** $CH_3CH_2CH_2CH_2CH_2CH_2CH_2CH_2NH_2$ or $CH_3(CH_2)_7NH_2$
**(c)** $CH_3CH_2CH_2CH_2OCH_2CH_2CH_2CH_3$
**(d)** $CH_3CH_2CO_2CH_2CH_2CH_2CH_3$ or $CH_3CH_2COOCH_2CH_2CH_2CH_3$
**(e)** $CH_3CH_2CH_2COCH_2CH_3$
**(f)** $CH_3CH_2CH_2CH_2CH_2CH_2CH_2CH_2CH_2CH_2CH_2CO_2H$ or $CH_3(CH_2)_{10}CO_2H$
or $CH_3CH_2CH_2CH_2CH_2CH_2CH_2CH_2CH_2CH_2CH_2COOH$ or $CH_3(CH_2)_{10}COOH$
**(g)** $(CH_3)_2CHCONH_2$
**(h)** $(CH_3)_3CCH_2CH_2CHO$
**(i)** $(CH_3)_2C(OH)CH_2CH_3$
**(j)** $(CH_3)_2CCHCH_2CH_3$
**(k)** $CH_3CH_2N(CH_3)_2$
**(l)** $CH_3CCC(CH_3)_3$

**17.3.**
**(a)**
**(b)** $NH_2$
**(c)** O
**(d)** O, O
**(e)** O
**(f)** O, OH
**(g)** O, $NH_2$
**(h)** O, H
**(i)** OH
**(j)**

## Chapter 17 Review Questions

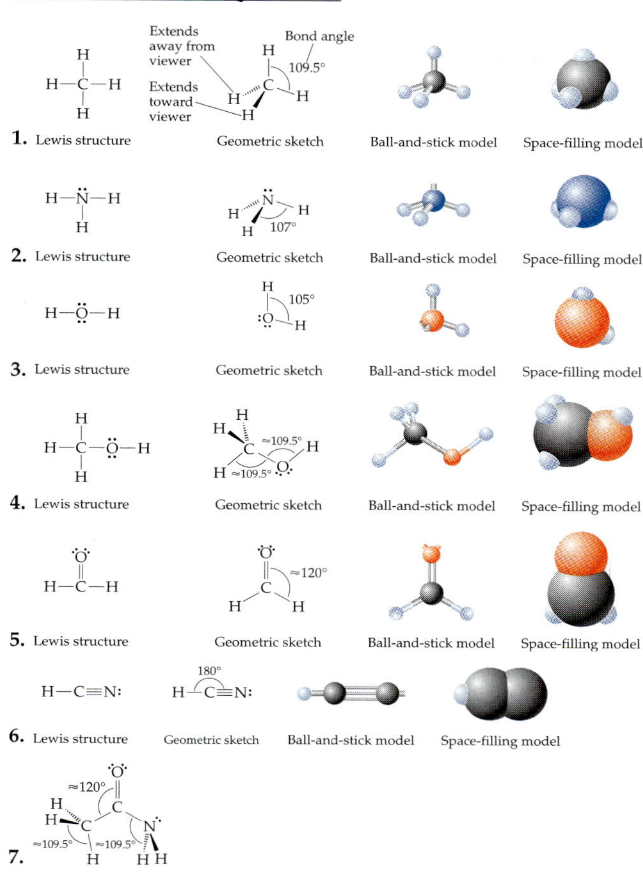


## Chapter 17 Key Ideas

**8.** single **10.** small section **12.** benzene ring **14.** hydrocarbon groups **16.** disaccharides, polysaccharides **18.** glucose **20.** fructose **22.** energy, amylose, amylopectin **24.** repeating units **26.** monomers, amino acids **28.** water **30.** close to each other **32.** long-term **34.** hydrogenation **36.** double bonds **38.** cholesterol **40.** monosaccharides, protein, glycerol, fatty acids **42.** acidic, denaturation **44.** substrates, active site, shape **46.** step-growth **48.** diol

## Chapter 17 Problems

**50. (a)** inorganic **(b)** organic **(c)** organic **(d)** inorganic
**52. (a)** ketone **(b)** alkane **(c)** carboxylic acid **(d)** amide **(e)** ether **(f)** aldehyde **(g)** alkene **(h)** ester **(i)** alcohol **(j)** alkyne **(k)** arene
**54. (a)** $CH_3(CH_2)_4COCH_3$ or $CH_3CH_2CH_2CH_2CH_2COCH_3$
**(b)** $CH_3CH_2CH(CH_3)CH_2CH(CH_3)_2$ or $CH_3CH_2CH(CH_3)CH_2CH(CH_3)CH_3$
**(c)** $CH_3(CH_2)_{12}COOH$ or $CH_3(CH_2)_{12}CO_2H$
**(d)** $CH_3CH_2CH_2CONH_2$
**(e)** $CH_3CH_2OCH(CH_3)_2$ or $CH_3CH_2OCH(CH_3)CH_3$
**(f)** $(CH_3)_2CHCHO$ or $CH_3CH(CH_3)CHO$
**(g)** $CH_2C(CH_3)CHCH_2$
**(h)** $CH_3CH_2COOCH_3$ or $CH_3CH_2CO_2CH_3$
**(i)** $CH_3CH_2CH_2CH(OH)CH(CH_2OH)CH_2CH_3$
**(j)** $CH_3CCCH_3$

**56.**
(a)
(b)
(c) OH
(d) $NH_2$
(e)
(f) H
(g)
(h)
(i) OH HO

**58.**
amine amide ester
$H_2N$—C—C—N—C—C—O—$CH_3$
carboxylic acid
$CH_2$ H $CH_2$
C=O
OH
benzene ring

**60. (a)** 109.5°
**(b)** ≈120°
**(c)** H—C≡N: 180°

**63. (a)** isomers **(b)** same **(c)** isomers **(d)** same

**65.**

**67.** $CH_3CH_2CH_2OH$ $CH_3CH(OH)CH_3$ $CH_3OCH_2CH_3$

**69.** 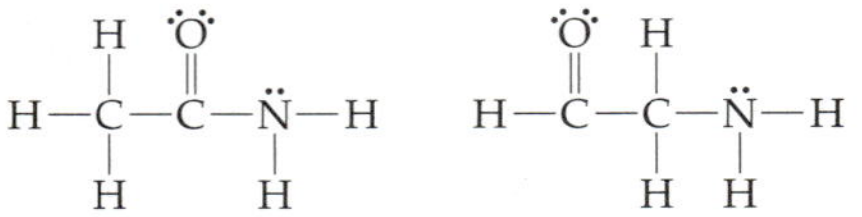 H—C(H)(H)—C(=O)—N(H)—H   H—C(=O)—C(H)(H)—N(H)—H

**72.** **(a)** amino acid **(b)** carbohydrate **(c)** triglyceride **(d)** steroid **(e)** peptide

**74.** **(a)** polysaccharide **(b)** monosaccharide **(c)** disaccharide **(d)** monosaccharide

**76.** **(a)** disaccharide **(b)** monosaccharide **(c)** polysaccharide **(d)** polysaccharide

**78.** Glucose and galactose differ in the relative positions of an –H and an –OH on one of their carbon atoms. In the standard notation for the open-chain form, glucose and galactose differ only in the relative position of the –H and –OH groups on the fourth carbon from the top. In the standard notation for the ring structures, the –OH group is down on the number 4 carbon of glucose and up on the number 4 carbon of galactose. See Figures 17.18 and 17.19.

**80.** Maltose—2 glucose units; Lactose—glucose and galactose; Sucrose—glucose and fructose

**82.** Starch and cellulose molecules are composed of many glucose molecules linked together, but cellulose has different linkages between the molecules than starch. See Figure 17.21.

**84.** One end of the amino acid has a carboxylic acid group that tends to lose an $H^+$ ion, and the other end has a basic amine group that attracts $H^+$ ions. Therefore, in the conditions found in our bodies, amino acids are likely to be in the second form.

**85.** 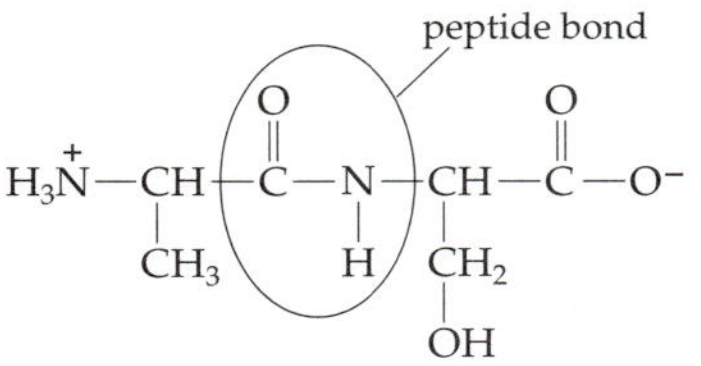 $H_3N^+$—CH($CH_3$)—C(=O)—N(H)—CH($CH_2$OH)—C(=O)—$O^-$ (peptide bond)

**87.** 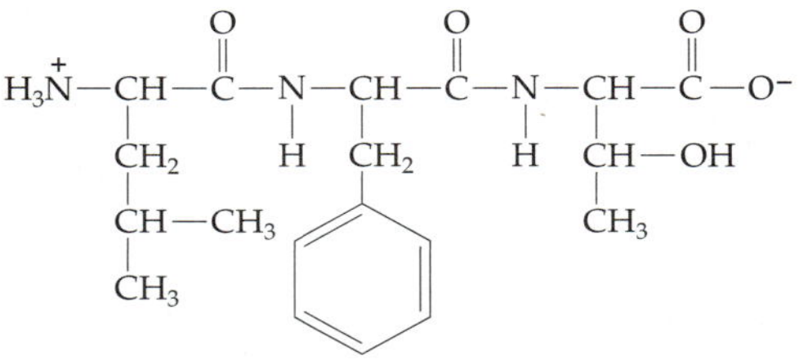 $H_3N^+$—CH($CH_2$CH($CH_3$)$CH_3$)—C(=O)—N(H)—CH($CH_2$$C_6H_5$)—C(=O)—N(H)—CH(CH(OH)$CH_3$)—C(=O)—$O^-$

**89.** $H_2N$—C(H)($CH_2$C(=O)OH)—C(=O)—N(H)—C(H)($CH_2$$C_6H_5$)—C(=O)—O—$CH_3$

aspartic acid  phenylalanine  methanol

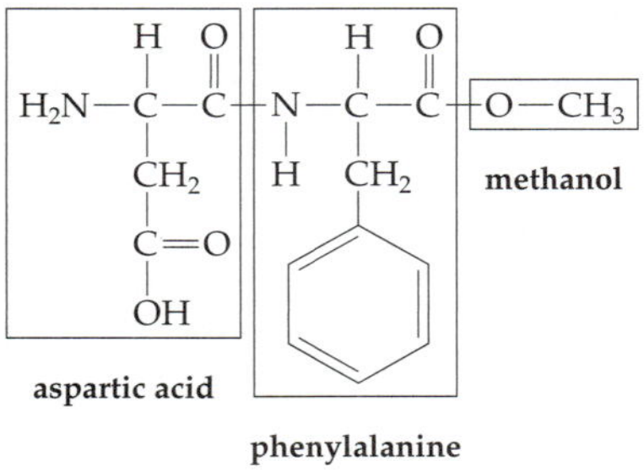

**91.** Each of these interactions draw specific amino acids in a protein chain close together, leading to a specific shape of the protein molecule. Disulfide bonds are covalent bonds between sulfur atoms from two cysteine amino acids (Figure 17.28). Hydrogen bonding forms between –OH groups in two amino acids, like serine or threonine, in a protein chain (Figure 17.29). Salt bridges are attractions between negatively charged side chains and positively charged side chains. For example, the carboxylic acid group of an aspartic acid side chain can lose its $H^+$, leaving the side chain with a negative charge. The basic side chain of a lysine amino acid can gain an $H^+$ and a positive charge. When these two charges form, the negatively charged aspartic acid is attracted to the positively charged lysine by a salt bridge (Figure 17.30).

**94.** 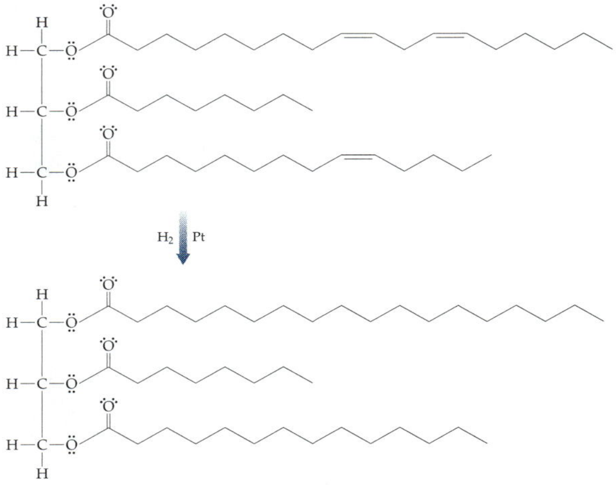 $H_2$ / Pt

**96.** Disaccharides—monosaccharides (glucose and galactose from lactose)
Polysaccharides—glucose
Protein—amino acids
Fat—glycerol and fatty acids

**98.** Before an enzyme reaction takes place, the molecule or molecules that are going to react (called substrates) must fit into a specific section of the protein structure called the active site. Because the active site has a shape that fits specific substrates, because it has side chains that attract particular substrates, and because it has side chains in distinct positions that speed the reaction, each enzyme will only act on a specific molecule or a specific type of molecule.

**100.** One of the reasons for the exceptional strength of nylon is the hydrogen bonding between amide functional groups. A higher percentage of amide functional groups in nylon molecules' structures leads to stronger hydrogen bonds between them. Thus, changing the number of carbon atoms in the diamine and in the di-carboxylic acid changes the properties of nylon. Nylon 610, which has four more carbon atoms in the di-carboxylic acid molecules that form it than for Nylon 66, is somewhat weaker than Nylon 66 and has a lower melting point.

**102.** Polyethylene molecules can be made using different techniques. One process leads to branches that keep the molecules from fitting closely together. Other techniques have been developed to make polyethylene molecules with very few branches. These straight-chain molecules fit together more efficiently, yielding a high-density polyethylene, HDPE, that is more opaque, harder, and stronger than the low-density polyethylene, LDPE.

**104.** Nonpolar molecules are attracted to each other by London forces, and increased size of molecules lead to stronger London forces. Polyethylene molecules are much larger than the ethylene molecules that are used to make polyethylene, so polyethylene molecules have much stronger attractions between them, making them solids at room temperature.

## Chapter 18 Exercises

**18.1.** Because this nuclide has 39 protons, its atomic number, Z, is 39. This identifies the element as yttrium. This nuclide of yttrium has 90 total nucleons (39 protons + 51 neutrons), so its nucleon number, A, is 90.

$^{90}_{39}Y$   $^{90}Y$   yttrium-90

**18.2.** The periodic table shows us that the atomic number for thallium is 81, so each thallium atom has 81 protons. The superscript in the symbol $^{201}Tl$ is this nuclide's mass number. The difference between the mass number (the sum of the numbers of protons and neutrons) and the atomic number (the number of protons) is equal to the

number of neutrons, so this nuclide has 120 neutrons (201 − 81).
Atomic number = 81   mass number = 201   81 protons
120 neutrons

$^{201}_{81}Tl$   thallium-201

**18.3. (a)** $^{239}_{94}Pu \rightarrow ^{235}_{92}U + ^{4}_{2}He$

**(b)** $^{24}_{11}Na \rightarrow ^{24}_{12}Mg + ^{0}_{-1}e$

**(c)** $^{15}_{8}O \rightarrow ^{15}_{7}N + ^{0}_{+1}e$

**(d)** $^{64}_{29}Cu + ^{0}_{-1}e \rightarrow ^{64}_{28}Ni$

**18.4. (a)** $^{14}_{7}N + ^{4}_{2}He \rightarrow ^{17}_{8}O + ^{1}_{1}H$

**(b)** $^{238}_{92}U + ^{14}_{7}N \rightarrow ^{247}_{99}Es + 5^{1}_{0}n$

**(c)** $^{238}_{92}U + ^{2}_{1}H \rightarrow ^{239}_{93}Np + ^{1}_{0}n$

**18.5.** In each half-life of a radioactive nuclide, the amount diminishes by one-half. The fraction ⅛ is ½ × ½ × ½, so it takes three half-lives to diminish to ⅛ remaining. Therefore, it will take 36.78 years for tritium to decrease to ⅛ of what was originally there.
**18.6.** The length of time divided by the half-life yields the number of half-lives.

$$\frac{9.02 \times 10^9 \text{ years}}{4.51 \times 10^9 \text{ years}} = 2 \text{ half-lives}$$

Therefore, the fraction remaining would be ¼ (½ × ½).

## Chapter 18 Review Questions

**1.** Protons and neutrons are in a tiny core of the atom called the nucleus, which has a diameter about 1/100,000 the diameter of the atom. The position and motion of the electrons are uncertain, but they generate a negative charge that is felt in the space that surrounds the nucleus.
**2.** In the particle view, radiant energy is a stream of tiny, massless packets of energy called photons. Different forms of radiant energy differ with respect to the energy of each of their photons. The energies of the photons of radio waves are much lower than for gamma radiation. In the wave view, as radiant energy moves away from the source, it has an effect on the space around it that can be described as a wave consisting of an oscillating electric field perpendicular to an oscillating magnetic field. Different forms of radiant energy differ with respect to the wavelengths and frequencies of these oscillating waves. The waves associated with radio waves have a much longer wavelength than the waves associated with gamma radiation.
**3.** isotopes **4.** atomic number **5.** mass number **6.** Energy **7.** Kinetic energy **8.** Law of Conservation of Energy **9.** Potential energy **10.** lower **11.** released **12.** ground state **13.** excited state

## Chapter 18 Key Ideas

**14.** protons, neutrons **16.** electrostatic, strong **18.** ratio **20.** alpha **22.** too low **24.** releases, photons **26.** different **28.** energy **30.** one-half **32.** pull **34.** excite **36.** 100 **38.** rapidly reproducing **40.** $^{14}C/^{12}C$ ratio **42.** 10,000 **44.** double magic numbers **46.** larger than **48.** furnace

## Chapter 18 Problems

**50.** $^{226}_{88}Ra$   $^{226}Ra$   radium-226

**52.** $^{11}_{6}C$   $^{11}C$   carbon-11

**54.** atomic number = 19   mass number = 40
$^{40}K$   potassium-40
**56.** 49 protons   62 neutrons   $^{111}In$   indium-111
**58.** atomic number = 56   mass number = 131   56 protons
75 neutrons   $^{131}_{56}Ba$   $^{131}Ba$
**60.** atomic number = 34   mass number = 75   34 protons
42 neutrons   $^{75}_{34}Se$   selenium-75

**62.** The first force among the particles in the nucleus is called the electrostatic force (or electromagnetic force). It is the force between electrically charged particles. Opposite charges attract each other, and like charges repel each other, so the positively charged protons in the nucleus of an atom have an electrostatic force pushing them apart. The second force, called the strong force, holds nucleons (protons and neutrons) together. You can think of neutrons as the nuclear glue that allows protons to stay together in the nucleus. Because neutrons are uncharged, there are no electrostatic repulsions among them and other particles, but each neutron in the nucleus of an atom is attracted to other neutrons and to protons by the strong force. Therefore, adding neutrons to a nucleus leads to more attractions holding the particles of the nucleus together without causing increased repulsion between those particles. Larger atoms with more protons in their nuclei require a greater ratio of neutrons to protons to balance the increased electrostatic repulsion between protons.
**64.** One of the ways that heavy nuclides change to move back into the band of stability is to release two protons and two neutrons in the form of a helium nuclei, called an alpha particle. In nuclear equations for alpha emission, the alpha particle is described as either $\alpha$ or $^{4}_{2}He$. In alpha emission, the radioactive nuclide changes into a different element that has an atomic number that is two lower and a mass number that is four lower.

$$^{A}_{Z}X \rightarrow ^{A-4}_{Z-2}Y + ^{4}_{2}He$$

**66.** In positron emission ($\beta^+$), a proton becomes a neutron and an anti-electron. The neutron stays in the nucleus, and the positron speeds out of the nucleus at high velocity.

$$p \rightarrow n + e^+$$

In nuclear equations for positron emission, the electron is described as either $\beta^+$, $^{0}_{+1}e$, or $^{0}_{1}e$. In positron emission, the radioactive nuclide changes into a different element that has an atomic number that is one lower but that has the same mass number.

$$^{A}_{Z}X \rightarrow ~_{Z-1}^{A}Y + ^{0}_{+1}e$$

**68.** Bismuth-202, which has a lower neutron to proton ratio than the stable bismuth-209, undergoes electron capture, which increases the neutron to proton ratio. Bismuth-215, which has a higher neutron to proton ratio than the stable bismuth-209, undergoes beta emission, which decreases the neutron to proton ratio.
**71. (a)** beta emission **(b)** alpha emission **(c)** gamma emission **(d)** positron emission or electron capture **(e)** beta emission **(f)** alpha emission
**73.** (1) Nuclear reactions involve changes in the nucleus, as opposed to chemical reactions that involve the loss, gain, and sharing of electrons. (2) Different isotopes of the same element often undergo very different nuclear reactions, whereas they all share the same chemical characteristics. (3) Unlike chemical reactions, the rates of nuclear reactions are unaffected by temperature, pressure, and the other atoms to which the radioactive atom is bonded. (4) Nuclear reactions, in general, give off a lot more energy than chemical reactions.
**74.** These two particles differ only in the number of their electrons. Because chemical reactions involve the loss, gain, and sharing of electrons, the number of electrons for an atom is very important for chemical changes. The attractions between charges are also very important for chemical changes, so the different charges of these particles change how they act chemically. Nuclear reactions are determined by the stability of nuclei, which is related to the number of protons and neutrons in the nuclei. They are unaffected by the number of electrons and unaffected by the overall charge of the particles. Therefore, these particles, which both have 17 protons and 21 neutrons in their nuclei, have the same nuclear stability and undergo the same nuclear changes.

**76.** $^{218}_{84}Po \rightarrow ^{214}_{82}Pb + ^{4}_{2}He$

**78.** $^{60}_{27}Co \rightarrow ^{60}_{28}Ni + ^{0}_{-1}e$

**80.** $^{11}_{6}C \rightarrow ^{11}_{5}B + ^{0}_{+1}e$

**82.** ${}^{197}_{80}\text{Hg} + {}^{0}_{-1}\text{e} \rightarrow {}^{197}_{79}\text{Au}$

**84. (a)** ${}^{90}_{38}\text{Sr} \rightarrow {}^{90}_{39}\text{Y} + {}^{0}_{-1}\text{e}$

**(b)** ${}^{17}_{9}\text{F} \rightarrow {}^{17}_{8}\text{O} + {}^{0}_{+1}\text{e}$

**(c)** ${}^{222}_{86}\text{Rn} \rightarrow {}^{218}_{84}\text{Po} + {}^{4}_{2}\text{He}$

**(d)** ${}^{18}_{9}\text{F} + {}^{0}_{-1}\text{e} \rightarrow {}^{18}_{8}\text{O}$

**(e)** ${}^{235}_{92}\text{U} \rightarrow {}^{231}_{90}\text{Th} + {}^{4}_{2}\text{He}$

**(f)** ${}^{7}_{4}\text{Be} + {}^{0}_{-1}\text{e} \rightarrow {}^{7}_{3}\text{Li}$

**(g)** ${}^{52}_{26}\text{Fe} \rightarrow {}^{52}_{25}\text{Mn} + {}^{0}_{+1}\text{e}$

**(h)** ${}^{3}_{1}\text{H} \rightarrow {}^{3}_{2}\text{He} + {}^{0}_{-1}\text{e}$

**(i)** ${}^{14}_{6}\text{C} \rightarrow {}^{14}_{7}\text{N} + {}^{0}_{-1}\text{e}$

**(j)** ${}^{118}_{54}\text{Xe} \rightarrow {}^{118}_{53}\text{I} + {}^{0}_{+1}\text{e}$

**(k)** ${}^{204}_{84}\text{Po} + {}^{0}_{-1}\text{e} \rightarrow {}^{204}_{83}\text{Bi}$

**(l)** ${}^{238}_{92}\text{U} \rightarrow {}^{234}_{90}\text{Th} + {}^{4}_{2}\text{He}$

**86.** ${}^{117}_{47}\text{Ag} \rightarrow {}^{117}_{48}\text{Cd} \rightarrow {}^{117}_{49}\text{In} \rightarrow {}^{117}_{50}\text{Sn}$

**88.** ${}^{116}_{52}\text{Te} \rightarrow {}^{116}_{51}\text{Sb} \rightarrow {}^{116}_{50}\text{Sn}$

**90.** ${}^{142}_{62}\text{Sm} \rightarrow {}^{142}_{61}\text{Pm} \rightarrow {}^{142}_{60}\text{Nd}$

**92.** ${}^{211}_{83}\text{Bi} \rightarrow {}^{207}_{81}\text{Tl} \rightarrow {}^{207}_{82}\text{Pb}$

**94. (a)** ${}^{246}_{96}\text{Cm} + {}^{12}_{6}\text{C} \rightarrow {}^{252}_{102}\text{No} + 6\,{}^{1}_{0}\text{n}$

**(b)** ${}^{249}_{98}\text{Cf} + {}^{16}_{8}\text{O} \rightarrow {}^{263}_{106}\text{Sg} + 2\,{}^{1}_{0}\text{n}$

**(c)** ${}^{240}_{95}\text{Am} + {}^{4}_{2}\text{He} \rightarrow {}^{243}_{97}\text{Bk} + {}^{1}_{0}\text{n}$

**(d)** ${}^{252}_{98}\text{Cf} + {}^{10}_{5}\text{B} \rightarrow {}^{257}_{103}\text{Lr} + 5\,{}^{1}_{0}\text{n}$

**96.** ${}^{209}_{83}\text{Bi} + {}^{54}_{24}\text{Cr} \rightarrow {}^{262}_{107}\text{Bh} + {}^{1}_{0}\text{n}$

**98.** It takes 2 half-lives for a radioactive nuclide to decay to ¼ of its original amount (½ × ½). Therefore, it will take 60 years for cesium-133 to decrease to ¼ of what was originally there.
**100.** The 42.9 days is 3 half-lives (42.9/14.3), so the fraction remaining would be ⅛ (½ × ½ × ½).
**102.** Although short-lived radioactive nuclides disappear relatively quickly once they form, they are constantly being replenished because they are products of other radioactive decays. There are three long-lived radioactive nuclides (uranium-235, uranium-238, and thorium-232) that are responsible for many of the natural radioactive isotopes.

**105.** ${}^{235}_{92}\text{U} \rightarrow {}^{231}_{90}\text{Th} \rightarrow {}^{231}_{91}\text{Pa} \rightarrow {}^{227}_{89}\text{Ac} \rightarrow {}^{227}_{90}\text{Th} \rightarrow {}^{223}_{88}\text{Ra} \rightarrow {}^{219}_{86}\text{Rn}$

**107.** alpha, beta, beta, alpha, alpha
**109.** As alpha particles, which move at up to 10% the speed of light, move through the tissues of our bodies, they drag electrons away from the tissue's atoms. Remember that alpha particles are helium nuclei, so they each have a +2 charge. Thus, as the alpha particle moves past an atom or molecule, it attracts the particle's electrons. One of the electrons might be pulled toward the passing alpha particle enough to escape, but it might not be able to catch up to the fast moving alpha particle. The electron lags behind the alpha particle and is quickly incorporated into another atom or molecule, forming an anion, and the particle that lost the electron becomes positively charged. The alpha particle continues on its way creating many ions before it is slowed enough for electrons to catch up to it and neutralize its charge.
**111.** Gamma photons can excite electrons enough to actually remove them from atoms.
**113.** The greatest effect is on tissues with rapidly reproducing cells where there are more frequent chemical changes. This is why nuclear emissions have a greater effect on cancerous tumors with their rapidly reproducing cells and on children, who have more rapidly reproducing cells than adults.
**115.** Because both radium and calcium are alkaline earth metals in group 2 on the periodic table, they combine with other elements in similar ways. Therefore, if radioactive radium-226 is ingested, it concentrates in the bones in substances that would normally contain calcium.
**117.** Alpha and beta particles lose all of their energy over a very short distance, so they can do more damage to localized areas in the body than the same number of gamma photons would.
**118.** Cobalt-60 emits ionizing radiation in the form of beta particles and gamma photons. Gamma photons penetrate the body and do more damage to rapidly reproducing cells that others. Typically, a focused beam of gamma photons from cobalt-60 is directed at a cancerous tumor. The gamma photons enter the tumor and create ions and free radicals that damage the tumor cells to shrink the tumor.
**120.** To get a PET scan of a patient, a solution that contains a positron-emitting substance is introduced into the body. The positrons that the radioactive atoms emit collide with electrons, and they annihilate each other, creating two gamma photons that move out in opposite directions.

$e^+ \longrightarrow \quad \longleftarrow e^-$

positron–electron collision
followed by the creation of
2 gamma ray photons

$\longleftarrow$ γ-ray    γ-ray $\longrightarrow$

These photons can be detected, and the data can be computer-analyzed to yield images that show where in the body the radioactive substances collected. Depending on the nuclide used and the substance into which it is incorporated, the radioactive substance will move to a specific part of the body.
**122.** Carbon-14 atoms are constantly being produced in our upper atmosphere through neutron bombardment of nitrogen atoms.

$${}^{14}_{7}\text{N} + {}^{1}_{0}\text{n} \rightarrow {}^{14}_{6}\text{C} + {}^{1}_{1}\text{H}$$

The carbon-14 formed is quickly oxidized to form carbon dioxide, $CO_2$, which is converted into many different substances in plants. When animals eat the plants, the carbon-14 becomes part of the animal too. For these reasons, carbon-14 is found in all living things. The carbon-14 is a beta-emitter with a half-life of 5730 years (±40 years), so as soon as it becomes part of a plant or animal, it begins to disappear.

$${}^{14}_{6}\text{C} \rightarrow {}^{14}_{7}\text{N} + {}^{0}_{-1}\text{e}$$

As long as a plant or animal is alive, the intake of carbon-14 balances the decay so that the ratio of ${}^{14}C$ to ${}^{12}C$ remains constant at about 1 in 1,000,000,000,000. When the plant or animal dies, it stops taking in fresh carbon, but the carbon-14 it contains continues to decay. Thus the ratio of ${}^{14}C$ to ${}^{12}C$ drops steadily. Therefore, to date an artifact, a portion of it is analyzed to determine the ${}^{14}C/{}^{12}C$ ratio, which can be used to calculate its age.

Initially, it was thought that determination of the age of something using this technique would be simple. For example, if the ${}^{14}C/{}^{12}C$ ratio had dropped to one-half of the ratio found in the air today, it would be considered to be about 5730 years old. A ${}^{14}C/{}^{12}C$ ratio of one-fourth of the ratio found in the air today would date it as

11,460 years old (two half-lives). This only works if we can assume that the $^{14}C/^{12}C$ ratio in the air was the same when the object died as it is now, and scientists have discovered that this is not strictly true. Study of very old trees, such as the bristlecone pines in California, have allowed researchers to develop calibration curves that adjust the results of radiocarbon dating experiments for the variation in the $^{14}C/^{12}C$ ratio that go back about 10,000 years. These calibration curves are now used to get more precise dates for objects.

**124.** The radioactive Iridium-192 is introduced to the pipe, and the connection is wrapped on the outside with film. If there is a crack in the connection, radiation leaks out and exposes the film.

**126.** Unstable nuclides have been used as radioactive tracers that help researchers discover a wide range of things. For example, incorporating carbon-14 into molecules helped scientists to study many of the aspects of photosynthesis. Because the radiation emitted from the carbon-14 atoms can be detected outside of the system into which the molecules are placed, the changes that involve carbon can be traced. Phosphorus-32 atoms can be used to trace phosphorus-containing chemicals as they move from the soil into plants under various conditions. Carbon-14, hydrogen-3, and sulfur-35 have been used to trace the biochemical changes that take place in our bodies.

**128.** The greater the strengths of the attractions between nucleons, the more stable the nucleus and the greater the difference in potential energy between the separate nucleons and the nucleus. This is reflected in a greater binding energy per nucleon.

**130.** This can be explained by the fact that they have an even number of protons and an even number of the neutrons. Paired nucleons (like paired electrons) are more stable than unpaired ones.

**132.** Paired nucleons are more stable than unpaired ones, and oxygen-16 with 8 protons and 8 neutrons would have its nucleons paired. Oxygen-15 with an odd number of neutrons (7) would be less stable. Oxygen-16 also has double magic numbers, making it especially stable. Finally, oxygen-15 has too few neutrons to be stable.

**134.** Heat is generated by the chain reaction of uranium-235, which is initiated by the bombardment of uranium fuel that is about 3% uranium-235. The products are significantly more stable than the initial reactants, so the system shifts from higher potential energy to lower potential energy, releasing energy as increased kinetic energy of the product particles. Higher kinetic energy means higher temperature.

**136.** Both uranium-235 and uranium-238 absorb fast neutrons, but if the neutrons are slowed down, they are much more likely to be absorbed by uranium-235 atoms than uranium-238 atoms. Therefore, in a nuclear reactor, the fuel rods are surrounded by a substance called a moderator, which slows the neutrons as they pass through it. Several substances have been used as moderators, but normal water is most common.

**138.** ${}^{99}_{43}Tc + {}^{0}_{1}n \rightarrow {}^{100}_{43}Tc$

${}^{100}_{43}Tc \rightarrow {}^{100}_{44}Ru + {}^{0}_{-1}e$

**141.** ${}^{131}_{53}I$ $\quad {}^{131}I$ $\quad$ iodine-131

**143.** ${}^{24}_{11}Na$ $\quad {}^{24}Na$ $\quad$ sodium-24

**145.** atomic number = 55 mass number = 133 ${}^{133}Cs$ cesium-133

**147.** 24 protons 27 neutrons ${}^{51}Cr$ chromium-51

**149.** atomic number = 31 mass number = 67 31 protons 36 neutrons ${}^{67}_{31}Ga$ ${}^{67}Ga$

**151.** atomic number = 15 mass number = 32 15 protons 17 neutrons ${}^{32}_{15}P$ phosphorus-32

**153.** Neon-18, which has a lower neutron to proton ratio than the stable neon-20, undergoes positron emission, which increases the neutron to proton ratio. Neon-24, which has a higher neutron to proton ratio than the stable neon-20, undergoes beta emission, which decreases the neutron to proton ratio.

**155.** ${}^{189}_{83}Bi \rightarrow {}^{185}_{81}Tl + {}^{4}_{2}He$

**157.** ${}^{32}_{15}P \rightarrow {}^{32}_{16}S + {}^{0}_{-1}e$

**159.** ${}^{40}_{19}K \rightarrow {}^{40}_{18}Ar + {}^{0}_{+1}e$

**161.** ${}^{75}_{34}Se + {}^{0}_{-1}e \rightarrow {}^{75}_{33}As$

**163.** ${}^{78}_{32}Ge \rightarrow {}^{78}_{33}As \rightarrow {}^{78}_{34}Se$

**165.** ${}^{52}_{26}Fe \rightarrow {}^{52}_{25}Mn \rightarrow {}^{52}_{24}Cr$

**167.** ${}^{69}_{33}As \rightarrow {}^{69}_{32}Ge \rightarrow {}^{69}_{31}Ga$

**170. (a)** ${}^{249}_{98}Cf + {}^{15}_{7}N \rightarrow {}^{259}_{105}Db + 5\,{}^{1}_{0}n$

**(b)** ${}^{249}_{98}Cf + {}^{10}_{5}B \rightarrow {}^{257}_{103}Lr + 2\,{}^{1}_{0}n$

**(c)** ${}^{121}_{51}Sb + {}^{1}_{1}H \rightarrow {}^{121}_{52}Te + {}^{1}_{0}n$

**172.** ${}^{14}_{7}N + {}^{1}_{0}n \rightarrow {}^{15}_{7}N$

**174.** ${}^{208}_{82}Pb + {}^{58}_{26}Fe \rightarrow {}^{265}_{108}Hs + {}^{1}_{0}n$

**176.** It takes 3 half-lives for a radioactive nuclide to decay to ⅛ of its original amount (½ × ½ × ½). Therefore, it will take 103.5 hours (104 hours to three significant figures) for krypton-79 to decrease to ⅛ of what was originally there.

**178.** The 90 days is 2 half-lives (90/45), so the fraction remaining would be ¼ (½ × ½).

# Credits

**Chapter 1**
**Opener,** Courtesy of NASA; **page 3,** © Image Port/Index Stock; **page 5,** Courtesy of Rohm & Haas; **page 10,** Michael Newman/PhotoEdit; **page 11,** Courtesy of BIPM; **page 13,** © Terry Why/Index Stock Imagery/PictureQuest; **page 14,** © David Parker/SPL/ Photo Researchers, Inc.; **page 17,** AP/Wide World Photos.

**Chapter 2**
**Opener,** © Bob Daemmrich/Stock Boston Inc./PictureQuest; **page 59,** © 2001 Sidney Harris; **page 60,** © Peter Beck/corbisstockmarket.com; **page 60,** Barrington Brown/Photo Researchers, Inc.; **page 62,** © CORBIS; **page 64,** © 2001 PhotoDisc; **page 65,** © Eric Simmons/Stock Boston, Inc./PictureQuest; **page 67,** © CORBIS; **page 67,** © Mitchell Gerber/CORBIS; **page 70,** © Earl & Nazima Kowall/CORBIS; **page 72,** Courtesy of Andre P. Labonte, Department of Physics, Purdue University; **page 78,** Courtesy of GSI; **page 80,** © CORBIS.

**Chapter 3**
**Opener,** © 2001 PhotoDisc; **page 95,** © Phil Schermeister/CORBIS; **page 101,** © Enzo & Paolo Ragazzini/CORBIS; **page 108,** © Mark E. Gibson/CORBIS; **page 120,** © CORBIS.

**Chapter 4**
**Opener,** © David Keaton/ corbisstockmarket.com; **page 151,** © 2001 PhotoDisc; **page 152,** Courtesy of IBM Corporation; **page 168,** © 2001 PhotoDisc; **page 171,** © D.G. Arnold/CMSP.

**Chapter 5**
**Opener,** © Brian Sytnyk/Masterfile; **page 188,** © Robert Holmes/CORBIS; **page 195,** © Carl & Ann Purcell/Hearst Castle/CA Park Service/CORBIS; **page 195,** © Carl & Ann Purcell/Hearst Castle/CA Park Service/CORBIS; **page 201,** © Anna Clopet/CORBIS; **page 204,** Michael Newman/PhotoEdit; **page 208,** AP/Wide World Photos; **page 208,** © 2001 PhotoDisc; **page 214,** © CORBIS; **page 230,** © Dave Bartruff/CORBIS.

**Chapter 6**
**Opener,** © 2001 PhotoDisc; **page 235,** © Rob Lewine/ corbisstockmarket.com; **page 236,** © Ben Martin/TimePix; **page 240,** Paul Fusco/Magnum/PictureQuest; **page 241,** © Charles E. Rotkin/CORBIS; **page 248,** © Uwe Walz/CORBIS; **page 249,** © CORBIS; **page 251,** Courtesy of Car Sound Exhaust, Inc.; **page 255,** © Bettmann/CORBIS; **page 259,** Courtesy of Electric Fuel Corporation.

**Chapter 7**
**Opener,** © Bob Witkowski/CORBIS; page 279, © Michal Keller/ Corbisstockmarket.com; **page 289,** Courtesy of NASA; **page 296,** © Ted Spiegel/CORBIS; **page 297,** Courtesy of US EPA; **page 298,** Courtesy of NASA; **page 302,** © Roger Ressmeyer/CORBIS.

**Chapter 8**
**Opener,** © CORBIS; **page 319,** © 2001 PhotoDisc; **page 324,** © REUTERS/Dan Chung/Hulton Archive; **page 325,** © CORBIS; **page 334,** © CORBIS; **page 335,** Courtesy of NASA; **page 340,** © AFP/CORBIS; **page 348,** © Stockbyte/Picturequest; **page 354,** Courtesy of NASA; **page 355,** © 2001 PhotoDisc; **page 357,** © CORBIS.

**Chapter 9**
**Opener,** © Peggy & Ronald Barnett/corbisstockmarket.com; **page 361,** © 2001 PhotoDisc; **page 361,** © CORBIS; **page 361,** © Stephen Frisch/PictureQuest; **page 363,** © CORBIS; **page 367,** SuperStock; **page 382,** Robert Mayer/South Florida Sun Sentinel/Corbis Sygma; **page 385,** © Bettmann/CORBIS.

**Chapter 10**
**Opener,** © L. Lefkowitz/FPG; **page 403,** © CORBIS; **page 418,** © W. & D. Mcintyre/Photo Researchers; **page 422,** © D.P. Wilson/Science Source/Photo Researchers.

**Chapter 11**
**Opener,** IBM Corporation, Research Division, Almaden Research Center; **page 453,** © Dennis Kunkel/Phototake; **page 459,** © Denis Anthony Valentine/corbisstockmarket.com; **page 461,** © CORBIS; **page 479,** © Roger Ressmeyer/CORBIS.

**Chapter 12**
**Opener,** © Roger Ressmeyer/CORBIS; **page 491,** © Chuck Savage/ corbisstockmarket.com; **page 493,** © Cathy Crawford/CORBIS.

**Chapter 13**
**Opener,** © Robert Holmes/CORBIS; **page 538,** Richard Megna/Fundamental Photographs; **page 542,** © CORBIS; **page 551,** © Geoffrey Clements/CORBIS; **page 555,** Courtesy of Mississippi Chemical Corporation; **page 564,** © Roger Ball/ corbisstockmarket.com.

**Chapter 14**
**Opener,** © Frank Lane Picture Agency/CORBIS; **page 585,** © 2001 Photo24/Brand X Pictures/PictureQuest; **page 588,** © Photex/CORBIS; **page 589,** © Daryl Benson/Masterfile; **page 593,** © CORBIS; **page 594,** AP/Wide World Photos; **page 598,** © Paul A. Souders/CORBIS; **page 609,** © Bruce Forster/Stone; **page 612,** © Image Port/Index Stock.

**Chapter 15**
**Opener,** © CORBIS; **page 629,** AP/Wide World Photos; **page 655,** © Lester V. Bergman/CORBIS; **page 655,** © Bettmann/CORBIS.

**Chapter 16**
**Opener,** © Tom Bean/corbisstockmarket.com; **page 667,** © CORBIS; **page 674,** © Scott T. Smith/CORBIS; **page 674,** SuperStock; **page 682,** Photo courtesy of Air Products and Chemicals, Inc.; **page 688,** © Greig Cranna/ Stock, Boston Inc./PictureQuest; **page 690,** © Didier Dorval/Masterfile; **page 702,** © Ralph White/CORBIS.

**Chapter 17**
**Opener,** © Ronnie Kaufman/corbisstockmarket.com; **page 722,** © Michael Mahovlich/Masterfile; © Bernard Boutrit/Woodfin Camp/PictureQuest; **page 726,** © Craig Aumess/CORBIS; **page 745,** Courtesy of Stephen Bond, UCSD; **page 749,** © The Procter & Gamble Company. Used by permission.; **page 754,** © Holly Harris/Stone; **page 757,** © Hulton Archive; **page 759,** © Randy M. Ury/corbisstock-market.com; **page 760,** AP/Wide World Photos.

**Chapter 18**
**Opener,** © CORBIS; page 785, © CORBIS; **page 800,** © Dennis MacDonald/PhotoEdit; **page 805,** © Ecoscene/CORBIS; **page 806,** © Bettmann/CORBIS; **page 806,** AP/Wide World Photos; **page 806,** AP/Wide World Photos; Appendix, Reprinted with permission from Texas Instruments.

# Glossary/Index

Abbreviated electron configurations, of multi-electron atoms, 475–480
**Absolute zero** Zero kelvins (0 K), the lowest possible temperature, equivalent to +273.15 °C. It is the point beyond which motion can no longer be decreased. 18–19
**Accuracy** How closely a measured value approaches the true value of a property.
in reporting measurement values, 20
Acetaldehyde, Lewis structure of, 510–511
Acetamide, molecular structure of, 733
Acetate, 419
Acetate ion, 128. *See also* Acetic acid
solubility of compounds with, 168
Acetic acid, 188, 217, 419
freezing point of, 191
glacial, 191
molecular structure of, 728
as organic acid, 190–191
taste of, 204
water solubility of, 637
as weak acid, 192, 193, 194
Acetone
boiling point of, 599–600
evaporation of, 589
molecular structure of, 635, 731
solubility of, 635
Acetylene
molecular structure of, 496, 517–518
properties of, 726
water solubility of, 637
Acid–base reactions, 207–216
tooth decay and, 212
Acidic paper, preserving books with, 215
**Acidic solution** A solution with a significant concentration of hydronium ions, $H_3O^+$. 205–207
defined, 188
Acid rain, 195, 206
Acids, 187, 188–199
Arrhenius, 188–196
binary, 189, 196, 196–197
bleach and, 216
Brønsted–Lowry, 188, 216–220
defined, 188, 216
diprotic, 191, 193, 211
identifying, 203, 205
in meals ready to eat, 629
monoprotic, 191, 193, 213
names for, 196–199
organic, 190–191, 196
polyprotic, 191, 211
in saliva, 212
strong, 191–196
taste of, 204
triprotic, 191
weak, 191–196
Acrylamide, 680
Actinium (Ac), as element, 71
Activated complex, in chemical reactions, 669–670
**Activation energy** The minimum energy necessary for reactants to form the activated complex and proceed to products.
in chemical reactions, 670–672
of oxygen–ozone reaction, 677, 678
**Active site** A specific section of the protein structure of an enzyme in which the substrate fits and reacts.
Active sites, in enzymes, 756
**Actual yield** The amount of product that is actually obtained in a chemical reaction.
from chemical reactions, 418–420
Adams, Mike, 738
Addition, rounding off for, 43, 332–334
**Addition (or chain-growth) polymer** A polymer that contains all of the atoms of the original reactant in its structure. This category includes polyethylene, polypropylene, and poly(vinyl chloride). 760–761
Adipic acid
empirical formula for, 384–385
in manufacture of nylon, 386
molecular formula for, 384–385
Adults
effects of ionizing radiation on, 801–802
fingerprints of, 593
Aerosol cans, 595
Aging, oxidizing agents and, 240
Agitation, rate of solution and, 647, 648
Agricultural Research Service, 296
Air
breathing of, 654–655
density of, 334
gases in, 560, 561
in lungs, 541–542
as mixture, 560, 561
Air pollution
catalytic converters and, 251
ozone and, 296–300
volatile organic compounds and, 564
Alanine (Ala, A)
molecular structure of, 743
in silk, 757
**Alcohols** Compounds with one or more –OH groups attached to a hydrocarbon group. 110, 722, 733
phosphorus tribromide and, 420
properties of, 727–728, 734
tincture of iodine and, 610
**Aldehydes** Compounds that have a hydrogen atom or a hydrocarbon group connected to a –CHO group. 722, 733
properties of, 730, 734
Aldol, molecular structure of, 733
Algebra, 319
Alkali metals. *See also* Cesium (Cs); Lithium (Li); Potassium (K); Sodium (Na)
electron configurations of, 475–476
ion charges of, 123, 125
in periodic table, 69
Alkaline batteries, chemistry of, 259
Alkaline earth metals. *See also* Barium (Ba); Beryllium (Be); Calcium (Ca); Magnesium (Mg); Radium (Ra); Strontium (Sr)
ion charges of, 123, 125
in periodic table, 69
**Alkanes** Hydrocarbons (compounds composed of carbon and hydrogen) in which all of the carbon–carbon bonds are single bonds.
properties of, 725, 734. *See also* Hydrocarbons